T0189852

Lecture Notes in Computer Science 12350

More information about this series at http://www.springer.com/series/7412

Andrea Vedaldi · Horst Bischof ·
Thomas Brox · Jan-Michael Frahm (Eds.)

Computer Vision – ECCV 2020

16th European Conference
Glasgow, UK, August 23–28, 2020
Proceedings, Part V

 Springer

Editors
Andrea Vedaldi ⓘ
University of Oxford
Oxford, UK

Horst Bischof ⓘ
Graz University of Technology
Graz, Austria

Thomas Brox ⓘ
University of Freiburg
Freiburg im Breisgau, Germany

Jan-Michael Frahm
University of North Carolina at Chapel Hill
Chapel Hill, NC, USA

ISSN 0302-9743   ISSN 1611-3349 (electronic)
Lecture Notes in Computer Science
ISBN 978-3-030-58557-0   ISBN 978-3-030-58558-7 (eBook)
https://doi.org/10.1007/978-3-030-58558-7

LNCS Sublibrary: SL6 – Image Processing, Computer Vision, Pattern Recognition, and Graphics

This Springer imprint is published by the registered company Springer Nature Switzerland AG
The registered company address is: Gewerbestrasse 11, 6330 Cham, Switzerland

Foreword

Hosting the European Conference on Computer Vision (ECCV 2020) was certainly an exciting journey. From the 2016 plan to hold it at the Edinburgh International Conference Centre (hosting 1,800 delegates) to the 2018 plan to hold it at Glasgow's Scottish Exhibition Centre (up to 6,000 delegates), we finally ended with moving online because of the COVID-19 outbreak. While possibly having fewer delegates than expected because of the online format, ECCV 2020 still had over 3,100 registered participants.

Although online, the conference delivered most of the activities expected at a face-to-face conference: peer-reviewed papers, industrial exhibitors, demonstrations, and messaging between delegates. In addition to the main technical sessions, the conference included a strong program of satellite events with 16 tutorials and 44 workshops.

Furthermore, the online conference format enabled new conference features. Every paper had an associated teaser video and a longer full presentation video. Along with the papers and slides from the videos, all these materials were available the week before the conference. This allowed delegates to become familiar with the paper content and be ready for the live interaction with the authors during the conference week. The live event consisted of brief presentations by the oral and spotlight authors and industrial sponsors. Question and answer sessions for all papers were timed to occur twice so delegates from around the world had convenient access to the authors.

As with ECCV 2018, authors' draft versions of the papers appeared online with open access, now on both the Computer Vision Foundation (CVF) and the European Computer Vision Association (ECVA) websites. An archival publication arrangement was put in place with the cooperation of Springer. SpringerLink hosts the final version of the papers with further improvements, such as activating reference links and supplementary materials. These two approaches benefit all potential readers: a version available freely for all researchers, and an authoritative and citable version with additional benefits for SpringerLink subscribers. We thank Alfred Hofmann and Aliaksandr Birukou from Springer for helping to negotiate this agreement, which we expect will continue for future versions of ECCV.

August 2020

Vittorio Ferrari
Bob Fisher
Cordelia Schmid
Emanuele Trucco

Preface

Welcome to the proceedings of the European Conference on Computer Vision (ECCV 2020). This is a unique edition of ECCV in many ways. Due to the COVID-19 pandemic, this is the first time the conference was held online, in a virtual format. This was also the first time the conference relied exclusively on the Open Review platform to manage the review process. Despite these challenges ECCV is thriving. The conference received 5,150 valid paper submissions, of which 1,360 were accepted for publication (27%) and, of those, 160 were presented as spotlights (3%) and 104 as orals (2%). This amounts to more than twice the number of submissions to ECCV 2018 (2,439). Furthermore, CVPR, the largest conference on computer vision, received 5,850 submissions this year, meaning that ECCV is now 87% the size of CVPR in terms of submissions. By comparison, in 2018 the size of ECCV was only 73% of CVPR.

The review model was similar to previous editions of ECCV; in particular, it was double blind in the sense that the authors did not know the name of the reviewers and vice versa. Furthermore, each conference submission was held confidentially, and was only publicly revealed if and once accepted for publication. Each paper received at least three reviews, totalling more than 15,000 reviews. Handling the review process at this scale was a significant challenge. In order to ensure that each submission received as fair and high-quality reviews as possible, we recruited 2,830 reviewers (a 130% increase with reference to 2018) and 207 area chairs (a 60% increase). The area chairs were selected based on their technical expertise and reputation, largely among people that served as area chair in previous top computer vision and machine learning conferences (ECCV, ICCV, CVPR, NeurIPS, etc.). Reviewers were similarly invited from previous conferences. We also encouraged experienced area chairs to suggest additional chairs and reviewers in the initial phase of recruiting.

Despite doubling the number of submissions, the reviewer load was slightly reduced from 2018, from a maximum of 8 papers down to 7 (with some reviewers offering to handle 6 papers plus an emergency review). The area chair load increased slightly, from 18 papers on average to 22 papers on average.

Conflicts of interest between authors, area chairs, and reviewers were handled largely automatically by the Open Review platform via their curated list of user profiles. Many authors submitting to ECCV already had a profile in Open Review. We set a paper registration deadline one week before the paper submission deadline in order to encourage all missing authors to register and create their Open Review profiles well on time (in practice, we allowed authors to create/change papers arbitrarily until the submission deadline). Except for minor issues with users creating duplicate profiles, this allowed us to easily and quickly identify institutional conflicts, and avoid them, while matching papers to area chairs and reviewers.

Papers were matched to area chairs based on: an affinity score computed by the Open Review platform, which is based on paper titles and abstracts, and an affinity score computed by the Toronto Paper Matching System (TPMS), which is based on the paper's full text, the area chair bids for individual papers, load balancing, and conflict avoidance. Open Review provides the program chairs a convenient web interface to experiment with different configurations of the matching algorithm. The chosen configuration resulted in about 50% of the assigned papers to be highly ranked by the area chair bids, and 50% to be ranked in the middle, with very few low bids assigned.

Assignments to reviewers were similar, with two differences. First, there was a maximum of 7 papers assigned to each reviewer. Second, area chairs recommended up to seven reviewers per paper, providing another highly-weighed term to the affinity scores used for matching.

The assignment of papers to area chairs was smooth. However, it was more difficult to find suitable reviewers for all papers. Having a ratio of 5.6 papers per reviewer with a maximum load of 7 (due to emergency reviewer commitment), which did not allow for much wiggle room in order to also satisfy conflict and expertise constraints. We received some complaints from reviewers who did not feel qualified to review specific papers and we reassigned them wherever possible. However, the large scale of the conference, the many constraints, and the fact that a large fraction of such complaints arrived very late in the review process made this process very difficult and not all complaints could be addressed.

Reviewers had six weeks to complete their assignments. Possibly due to COVID-19 or the fact that the NeurIPS deadline was moved closer to the review deadline, a record 30% of the reviews were still missing after the deadline. By comparison, ECCV 2018 experienced only 10% missing reviews at this stage of the process. In the subsequent week, area chairs chased the missing reviews intensely, found replacement reviewers in their own team, and managed to reach 10% missing reviews. Eventually, we could provide almost all reviews (more than 99.9%) with a delay of only a couple of days on the initial schedule by a significant use of emergency reviews. If this trend is confirmed, it might be a major challenge to run a smooth review process in future editions of ECCV. The community must reconsider prioritization of the time spent on paper writing (the number of submissions increased a lot despite COVID-19) and time spent on paper reviewing (the number of reviews delivered in time decreased a lot presumably due to COVID-19 or NeurIPS deadline). With this imbalance the peer-review system that ensures the quality of our top conferences may break soon.

Reviewers submitted their reviews independently. In the reviews, they had the opportunity to ask questions to the authors to be addressed in the rebuttal. However, reviewers were told not to request any significant new experiment. Using the Open Review interface, authors could provide an answer to each individual review, but were also allowed to cross-reference reviews and responses in their answers. Rather than PDF files, we allowed the use of formatted text for the rebuttal. The rebuttal and initial reviews were then made visible to all reviewers and the primary area chair for a given paper. The area chair encouraged and moderated the reviewer discussion. During the discussions, reviewers were invited to reach a consensus and possibly adjust their ratings as a result of the discussion and of the evidence in the rebuttal.

After the discussion period ended, most reviewers entered a final rating and recommendation, although in many cases this did not differ from their initial recommendation. Based on the updated reviews and discussion, the primary area chair then made a preliminary decision to accept or reject the paper and wrote a justification for it (meta-review). Except for cases where the outcome of this process was absolutely clear (as indicated by the three reviewers and primary area chairs all recommending clear rejection), the decision was then examined and potentially challenged by a secondary area chair. This led to further discussion and overturning a small number of preliminary decisions. Needless to say, there was no in-person area chair meeting, which would have been impossible due to COVID-19.

Area chairs were invited to observe the consensus of the reviewers whenever possible and use extreme caution in overturning a clear consensus to accept or reject a paper. If an area chair still decided to do so, she/he was asked to clearly justify it in the meta-review and to explicitly obtain the agreement of the secondary area chair. In practice, very few papers were rejected after being confidently accepted by the reviewers.

This was the first time Open Review was used as the main platform to run ECCV. In 2018, the program chairs used CMT3 for the user-facing interface and Open Review internally, for matching and conflict resolution. Since it is clearly preferable to only use a single platform, this year we switched to using Open Review in full. The experience was largely positive. The platform is highly-configurable, scalable, and open source. Being written in Python, it is easy to write scripts to extract data programmatically. The paper matching and conflict resolution algorithms and interfaces are top-notch, also due to the excellent author profiles in the platform. Naturally, there were a few kinks along the way due to the fact that the ECCV Open Review configuration was created from scratch for this event and it differs in substantial ways from many other Open Review conferences. However, the Open Review development and support team did a fantastic job in helping us to get the configuration right and to address issues in a timely manner as they unavoidably occurred. We cannot thank them enough for the tremendous effort they put into this project.

Finally, we would like to thank everyone involved in making ECCV 2020 possible in these very strange and difficult times. This starts with our authors, followed by the area chairs and reviewers, who ran the review process at an unprecedented scale. The whole Open Review team (and in particular Melisa Bok, Mohit Unyal, Carlos Mondragon Chapa, and Celeste Martinez Gomez) worked incredibly hard for the entire duration of the process. We would also like to thank René Vidal for contributing to the adoption of Open Review. Our thanks also go to Laurent Charling for TPMS and to the program chairs of ICML, ICLR, and NeurIPS for cross checking double submissions. We thank the website chair, Giovanni Farinella, and the CPI team (in particular Ashley Cook, Miriam Verdon, Nicola McGrane, and Sharon Kerr) for promptly adding material to the website as needed in the various phases of the process. Finally, we thank the publication chairs, Albert Ali Salah, Hamdi Dibeklioglu, Metehan Doyran, Henry Howard-Jenkins, Victor Prisacariu, Siyu Tang, and Gul Varol, who managed to compile these substantial proceedings in an exceedingly compressed schedule. We express our thanks to the ECVA team, in particular Kristina Scherbaum for allowing open access of the proceedings. We thank Alfred Hofmann from Springer who again serve as the publisher. Finally, we thank the other chairs of ECCV 2020, including in particular the general chairs for very useful feedback with the handling of the program.

August 2020   Andrea Vedaldi
                                                                Horst Bischof
                                                                Thomas Brox
                                                            Jan-Michael Frahm

Organization

General Chairs

Vittorio Ferrari	Google Research, Switzerland
Bob Fisher	University of Edinburgh, UK
Cordelia Schmid	Google and Inria, France
Emanuele Trucco	University of Dundee, UK

Program Chairs

Andrea Vedaldi	University of Oxford, UK
Horst Bischof	Graz University of Technology, Austria
Thomas Brox	University of Freiburg, Germany
Jan-Michael Frahm	University of North Carolina, USA

Industrial Liaison Chairs

Jim Ashe	University of Edinburgh, UK
Helmut Grabner	Zurich University of Applied Sciences, Switzerland
Diane Larlus	NAVER LABS Europe, France
Cristian Novotny	University of Edinburgh, UK

Local Arrangement Chairs

Yvan Petillot	Heriot-Watt University, UK
Paul Siebert	University of Glasgow, UK

Academic Demonstration Chair

Thomas Mensink	Google Research and University of Amsterdam, The Netherlands

Poster Chair

Stephen Mckenna	University of Dundee, UK

Technology Chair

Gerardo Aragon Camarasa	University of Glasgow, UK

Tutorial Chairs

Carlo Colombo	University of Florence, Italy
Sotirios Tsaftaris	University of Edinburgh, UK

Publication Chairs

Albert Ali Salah	Utrecht University, The Netherlands
Hamdi Dibeklioglu	Bilkent University, Turkey
Metehan Doyran	Utrecht University, The Netherlands
Henry Howard-Jenkins	University of Oxford, UK
Victor Adrian Prisacariu	University of Oxford, UK
Siyu Tang	ETH Zurich, Switzerland
Gul Varol	University of Oxford, UK

Website Chair

Giovanni Maria Farinella	University of Catania, Italy

Workshops Chairs

Adrien Bartoli	University of Clermont Auvergne, France
Andrea Fusiello	University of Udine, Italy

Area Chairs

Lourdes Agapito	University College London, UK
Zeynep Akata	University of Tübingen, Germany
Karteek Alahari	Inria, France
Antonis Argyros	University of Crete, Greece
Hossein Azizpour	KTH Royal Institute of Technology, Sweden
Joao P. Barreto	Universidade de Coimbra, Portugal
Alexander C. Berg	University of North Carolina at Chapel Hill, USA
Matthew B. Blaschko	KU Leuven, Belgium
Lubomir D. Bourdev	WaveOne, Inc., USA
Edmond Boyer	Inria, France
Yuri Boykov	University of Waterloo, Canada
Gabriel Brostow	University College London, UK
Michael S. Brown	National University of Singapore, Singapore
Jianfei Cai	Monash University, Australia
Barbara Caputo	Politecnico di Torino, Italy
Ayan Chakrabarti	Washington University, St. Louis, USA
Tat-Jen Cham	Nanyang Technological University, Singapore
Manmohan Chandraker	University of California, San Diego, USA
Rama Chellappa	Johns Hopkins University, USA
Liang-Chieh Chen	Google, USA

Yung-Yu Chuang	National Taiwan University, Taiwan
Ondrej Chum	Czech Technical University in Prague, Czech Republic
Brian Clipp	Kitware, USA
John Collomosse	University of Surrey and Adobe Research, UK
Jason J. Corso	University of Michigan, USA
David J. Crandall	Indiana University, USA
Daniel Cremers	University of California, Los Angeles, USA
Fabio Cuzzolin	Oxford Brookes University, UK
Jifeng Dai	SenseTime, SAR China
Kostas Daniilidis	University of Pennsylvania, USA
Andrew Davison	Imperial College London, UK
Alessio Del Bue	Fondazione Istituto Italiano di Tecnologia, Italy
Jia Deng	Princeton University, USA
Alexey Dosovitskiy	Google, Germany
Matthijs Douze	Facebook, France
Enrique Dunn	Stevens Institute of Technology, USA
Irfan Essa	Georgia Institute of Technology and Google, USA
Giovanni Maria Farinella	University of Catania, Italy
Ryan Farrell	Brigham Young University, USA
Paolo Favaro	University of Bern, Switzerland
Rogerio Feris	International Business Machines, USA
Cornelia Fermuller	University of Maryland, College Park, USA
David J. Fleet	Vector Institute, Canada
Friedrich Fraundorfer	DLR, Austria
Mario Fritz	CISPA Helmholtz Center for Information Security, Germany
Pascal Fua	EPFL (Swiss Federal Institute of Technology Lausanne), Switzerland
Yasutaka Furukawa	Simon Fraser University, Canada
Li Fuxin	Oregon State University, USA
Efstratios Gavves	University of Amsterdam, The Netherlands
Peter Vincent Gehler	Amazon, USA
Theo Gevers	University of Amsterdam, The Netherlands
Ross Girshick	Facebook AI Research, USA
Boqing Gong	Google, USA
Stephen Gould	Australian National University, Australia
Jinwei Gu	SenseTime Research, USA
Abhinav Gupta	Facebook, USA
Bohyung Han	Seoul National University, South Korea
Bharath Hariharan	Cornell University, USA
Tal Hassner	Facebook AI Research, USA
Xuming He	Australian National University, Australia
Joao F. Henriques	University of Oxford, UK
Adrian Hilton	University of Surrey, UK
Minh Hoai	Stony Brooks, State University of New York, USA
Derek Hoiem	University of Illinois Urbana-Champaign, USA

Timothy Hospedales	University of Edinburgh and Samsung, UK
Gang Hua	Wormpex AI Research, USA
Slobodan Ilic	Siemens AG, Germany
Hiroshi Ishikawa	Waseda University, Japan
Jiaya Jia	The Chinese University of Hong Kong, SAR China
Hailin Jin	Adobe Research, USA
Justin Johnson	University of Michigan, USA
Frederic Jurie	University of Caen Normandie, France
Fredrik Kahl	Chalmers University, Sweden
Sing Bing Kang	Zillow, USA
Gunhee Kim	Seoul National University, South Korea
Junmo Kim	Korea Advanced Institute of Science and Technology, South Korea
Tae-Kyun Kim	Imperial College London, UK
Ron Kimmel	Technion-Israel Institute of Technology, Israel
Alexander Kirillov	Facebook AI Research, USA
Kris Kitani	Carnegie Mellon University, USA
Iasonas Kokkinos	Ariel AI, UK
Vladlen Koltun	Intel Labs, USA
Nikos Komodakis	Ecole des Ponts ParisTech, France
Piotr Koniusz	Australian National University, Australia
M. Pawan Kumar	University of Oxford, UK
Kyros Kutulakos	University of Toronto, Canada
Christoph Lampert	IST Austria, Austria
Ivan Laptev	Inria, France
Diane Larlus	NAVER LABS Europe, France
Laura Leal-Taixe	Technical University Munich, Germany
Honglak Lee	Google and University of Michigan, USA
Joon-Young Lee	Adobe Research, USA
Kyoung Mu Lee	Seoul National University, South Korea
Seungyong Lee	POSTECH, South Korea
Yong Jae Lee	University of California, Davis, USA
Bastian Leibe	RWTH Aachen University, Germany
Victor Lempitsky	Samsung, Russia
Ales Leonardis	University of Birmingham, UK
Marius Leordeanu	Institute of Mathematics of the Romanian Academy, Romania
Vincent Lepetit	ENPC ParisTech, France
Hongdong Li	The Australian National University, Australia
Xi Li	Zhejiang University, China
Yin Li	University of Wisconsin-Madison, USA
Zicheng Liao	Zhejiang University, China
Jongwoo Lim	Hanyang University, South Korea
Stephen Lin	Microsoft Research Asia, China
Yen-Yu Lin	National Chiao Tung University, Taiwan, China
Zhe Lin	Adobe Research, USA

Haibin Ling	Stony Brooks, State University of New York, USA
Jiaying Liu	Peking University, China
Ming-Yu Liu	NVIDIA, USA
Si Liu	Beihang University, China
Xiaoming Liu	Michigan State University, USA
Huchuan Lu	Dalian University of Technology, China
Simon Lucey	Carnegie Mellon University, USA
Jiebo Luo	University of Rochester, USA
Julien Mairal	Inria, France
Michael Maire	University of Chicago, USA
Subhransu Maji	University of Massachusetts, Amherst, USA
Yasushi Makihara	Osaka University, Japan
Jiri Matas	Czech Technical University in Prague, Czech Republic
Yasuyuki Matsushita	Osaka University, Japan
Philippos Mordohai	Stevens Institute of Technology, USA
Vittorio Murino	University of Verona, Italy
Naila Murray	NAVER LABS Europe, France
Hajime Nagahara	Osaka University, Japan
P. J. Narayanan	International Institute of Information Technology (IIIT), Hyderabad, India
Nassir Navab	Technical University of Munich, Germany
Natalia Neverova	Facebook AI Research, France
Matthias Niessner	Technical University of Munich, Germany
Jean-Marc Odobez	Idiap Research Institute and Swiss Federal Institute of Technology Lausanne, Switzerland
Francesca Odone	Universita di Genova, Italy
Takeshi Oishi	The University of Tokyo, Tokyo Institute of Technology, Japan
Vicente Ordonez	University of Virginia, USA
Manohar Paluri	Facebook AI Research, USA
Maja Pantic	Imperial College London, UK
In Kyu Park	Inha University, South Korea
Ioannis Patras	Queen Mary University of London, UK
Patrick Perez	Valeo, France
Bryan A. Plummer	Boston University, USA
Thomas Pock	Graz University of Technology, Austria
Marc Pollefeys	ETH Zurich and Microsoft MR & AI Zurich Lab, Switzerland
Jean Ponce	Inria, France
Gerard Pons-Moll	MPII, Saarland Informatics Campus, Germany
Jordi Pont-Tuset	Google, Switzerland
James Matthew Rehg	Georgia Institute of Technology, USA
Ian Reid	University of Adelaide, Australia
Olaf Ronneberger	DeepMind London, UK
Stefan Roth	TU Darmstadt, Germany
Bryan Russell	Adobe Research, USA

Mathieu Salzmann	EPFL, Switzerland
Dimitris Samaras	Stony Brook University, USA
Imari Sato	National Institute of Informatics (NII), Japan
Yoichi Sato	The University of Tokyo, Japan
Torsten Sattler	Czech Technical University in Prague, Czech Republic
Daniel Scharstein	Middlebury College, USA
Bernt Schiele	MPII, Saarland Informatics Campus, Germany
Julia A. Schnabel	King's College London, UK
Nicu Sebe	University of Trento, Italy
Greg Shakhnarovich	Toyota Technological Institute at Chicago, USA
Humphrey Shi	University of Oregon, USA
Jianbo Shi	University of Pennsylvania, USA
Jianping Shi	SenseTime, China
Leonid Sigal	University of British Columbia, Canada
Cees Snoek	University of Amsterdam, The Netherlands
Richard Souvenir	Temple University, USA
Hao Su	University of California, San Diego, USA
Akihiro Sugimoto	National Institute of Informatics (NII), Japan
Jian Sun	Megvii Technology, China
Jian Sun	Xi'an Jiaotong University, China
Chris Sweeney	Facebook Reality Labs, USA
Yu-wing Tai	Kuaishou Technology, China
Chi-Keung Tang	The Hong Kong University of Science and Technology, SAR China
Radu Timofte	ETH Zurich, Switzerland
Sinisa Todorovic	Oregon State University, USA
Giorgos Tolias	Czech Technical University in Prague, Czech Republic
Carlo Tomasi	Duke University, USA
Tatiana Tommasi	Politecnico di Torino, Italy
Lorenzo Torresani	Facebook AI Research and Dartmouth College, USA
Alexander Toshev	Google, USA
Zhuowen Tu	University of California, San Diego, USA
Tinne Tuytelaars	KU Leuven, Belgium
Jasper Uijlings	Google, Switzerland
Nuno Vasconcelos	University of California, San Diego, USA
Olga Veksler	University of Waterloo, Canada
Rene Vidal	Johns Hopkins University, USA
Gang Wang	Alibaba Group, China
Jingdong Wang	Microsoft Research Asia, China
Yizhou Wang	Peking University, China
Lior Wolf	Facebook AI Research and Tel Aviv University, Israel
Jianxin Wu	Nanjing University, China
Tao Xiang	University of Surrey, UK
Saining Xie	Facebook AI Research, USA
Ming-Hsuan Yang	University of California at Merced and Google, USA
Ruigang Yang	University of Kentucky, USA

Kwang Moo Yi	University of Victoria, Canada	
Zhaozheng Yin	Stony Brook, State University of New York, USA	
Chang D. Yoo	Korea Advanced Institute of Science and Technology, South Korea	
Shaodi You	University of Amsterdam, The Netherlands	
Jingyi Yu	ShanghaiTech University, China	
Stella Yu	University of California, Berkeley, and ICSI, USA	
Stefanos Zafeiriou	Imperial College London, UK	
Hongbin Zha	Peking University, China	
Tianzhu Zhang	University of Science and Technology of China, China	
Liang Zheng	Australian National University, Australia	
Todd E. Zickler	Harvard University, USA	
Andrew Zisserman	University of Oxford, UK	

Technical Program Committee

Sathyanarayanan N. Aakur	Samuel Albanie	Pablo Arbelaez
Wael Abd Almgaeed	Shadi Albarqouni	Shervin Ardeshir
Abdelrahman Abdelhamed	Cenek Albl	Sercan O. Arik
Abdullah Abuolaim	Hassan Abu Alhaija	Anil Armagan
Supreeth Achar	Daniel Aliaga	Anurag Arnab
Hanno Ackermann	Mohammad S. Aliakbarian	Chetan Arora
Ehsan Adeli	Rahaf Aljundi	Federica Arrigoni
Triantafyllos Afouras	Thiemo Alldieck	Mathieu Aubry
Sameer Agarwal	Jon Almazan	Shai Avidan
Aishwarya Agrawal	Jose M. Alvarez	Angelica I. Aviles-Rivero
Harsh Agrawal	Senjian An	Yannis Avrithis
Pulkit Agrawal	Saket Anand	Ismail Ben Ayed
Antonio Agudo	Codruta Ancuti	Shekoofeh Azizi
Eirikur Agustsson	Cosmin Ancuti	Ioan Andrei Bârsan
Karim Ahmed	Peter Anderson	Artem Babenko
Byeongjoo Ahn	Juan Andrade-Cetto	Deepak Babu Sam
Unaiza Ahsan	Alexander Andreopoulos	Seung-Hwan Baek
Thalaiyasingam Ajanthan	Misha Andriluka	Seungryul Baek
Kenan E. Ak	Dragomir Anguelov	Andrew D. Bagdanov
Emre Akbas	Rushil Anirudh	Shai Bagon
Naveed Akhtar	Michel Antunes	Yuval Bahat
Derya Akkaynak	Oisin Mac Aodha	Junjie Bai
Yagiz Aksoy	Srikar Appalaraju	Song Bai
Ziad Al-Halah	Relja Arandjelovic	Xiang Bai
Xavier Alameda-Pineda	Nikita Araslanov	Yalong Bai
Jean-Baptiste Alayrac	Andre Araujo	Yancheng Bai
	Helder Araujo	Peter Bajcsy
		Slawomir Bak

Mahsa Baktashmotlagh
Kavita Bala
Yogesh Balaji
Guha Balakrishnan
V. N. Balasubramanian
Federico Baldassarre
Vassileios Balntas
Shurjo Banerjee
Aayush Bansal
Ankan Bansal
Jianmin Bao
Linchao Bao
Wenbo Bao
Yingze Bao
Akash Bapat
Md Jawadul Hasan Bappy
Fabien Baradel
Lorenzo Baraldi
Daniel Barath
Adrian Barbu
Kobus Barnard
Nick Barnes
Francisco Barranco
Jonathan T. Barron
Arslan Basharat
Chaim Baskin
Anil S. Baslamisli
Jorge Batista
Kayhan Batmanghelich
Konstantinos Batsos
David Bau
Luis Baumela
Christoph Baur
Eduardo
    Bayro-Corrochano
Paul Beardsley
Jan Bednavr'ik
Oscar Beijbom
Philippe Bekaert
Esube Bekele
Vasileios Belagiannis
Ohad Ben-Shahar
Abhijit Bendale
Róger Bermúdez-Chacón
Maxim Berman
Jesus Bermudez-cameo

Florian Bernard
Stefano Berretti
Marcelo Bertalmio
Gedas Bertasius
Cigdem Beyan
Lucas Beyer
Vijayakumar Bhagavatula
Arjun Nitin Bhagoji
Apratim Bhattacharyya
Binod Bhattarai
Sai Bi
Jia-Wang Bian
Simone Bianco
Adel Bibi
Tolga Birdal
Tom Bishop
Soma Biswas
Mårten Björkman
Volker Blanz
Vishnu Boddeti
Navaneeth Bodla
Simion-Vlad Bogolin
Xavier Boix
Piotr Bojanowski
Timo Bolkart
Guido Borghi
Larbi Boubchir
Guillaume Bourmaud
Adrien Bousseau
Thierry Bouwmans
Richard Bowden
Hakan Boyraz
Mathieu Brédif
Samarth Brahmbhatt
Steve Branson
Nikolas Brasch
Biagio Brattoli
Ernesto Brau
Toby P. Breckon
Francois Bremond
Jesus Briales
Sofia Broomé
Marcus A. Brubaker
Luc Brun
Silvia Bucci
Shyamal Buch

Pradeep Buddharaju
Uta Buechler
Mai Bui
Tu Bui
Adrian Bulat
Giedrius T. Burachas
Elena Burceanu
Xavier P. Burgos-Artizzu
Kaylee Burns
Andrei Bursuc
Benjamin Busam
Wonmin Byeon
Zoya Bylinskii
Sergi Caelles
Jianrui Cai
Minjie Cai
Yujun Cai
Zhaowei Cai
Zhipeng Cai
Juan C. Caicedo
Simone Calderara
Necati Cihan Camgoz
Dylan Campbell
Octavia Camps
Jiale Cao
Kaidi Cao
Liangliang Cao
Xiangyong Cao
Xiaochun Cao
Yang Cao
Yu Cao
Yue Cao
Zhangjie Cao
Luca Carlone
Mathilde Caron
Dan Casas
Thomas J. Cashman
Umberto Castellani
Lluis Castrejon
Jacopo Cavazza
Fabio Cermelli
Hakan Cevikalp
Menglei Chai
Ishani Chakraborty
Rudrasis Chakraborty
Antoni B. Chan

Kwok-Ping Chan
Siddhartha Chandra
Sharat Chandran
Arjun Chandrasekaran
Angel X. Chang
Che-Han Chang
Hong Chang
Hyun Sung Chang
Hyung Jin Chang
Jianlong Chang
Ju Yong Chang
Ming-Ching Chang
Simyung Chang
Xiaojun Chang
Yu-Wei Chao
Devendra S. Chaplot
Arslan Chaudhry
Rizwan A. Chaudhry
Can Chen
Chang Chen
Chao Chen
Chen Chen
Chu-Song Chen
Dapeng Chen
Dong Chen
Dongdong Chen
Guanying Chen
Hongge Chen
Hsin-yi Chen
Huaijin Chen
Hwann-Tzong Chen
Jianbo Chen
Jianhui Chen
Jiansheng Chen
Jiaxin Chen
Jie Chen
Jun-Cheng Chen
Kan Chen
Kevin Chen
Lin Chen
Long Chen
Min-Hung Chen
Qifeng Chen
Shi Chen
Shixing Chen
Tianshui Chen

Weifeng Chen
Weikai Chen
Xi Chen
Xiaohan Chen
Xiaozhi Chen
Xilin Chen
Xingyu Chen
Xinlei Chen
Xinyun Chen
Yi-Ting Chen
Yilun Chen
Ying-Cong Chen
Yinpeng Chen
Yiran Chen
Yu Chen
Yu-Sheng Chen
Yuhua Chen
Yun-Chun Chen
Yunpeng Chen
Yuntao Chen
Zhuoyuan Chen
Zitian Chen
Anchieh Cheng
Bowen Cheng
Erkang Cheng
Gong Cheng
Guangliang Cheng
Jingchun Cheng
Jun Cheng
Li Cheng
Ming-Ming Cheng
Yu Cheng
Ziang Cheng
Anoop Cherian
Dmitry Chetverikov
Ngai-man Cheung
William Cheung
Ajad Chhatkuli
Naoki Chiba
Benjamin Chidester
Han-pang Chiu
Mang Tik Chiu
Wei-Chen Chiu
Donghyeon Cho
Hojin Cho
Minsu Cho

Nam Ik Cho
Tim Cho
Tae Eun Choe
Chiho Choi
Edward Choi
Inchang Choi
Jinsoo Choi
Jonghyun Choi
Jongwon Choi
Yukyung Choi
Hisham Cholakkal
Eunji Chong
Jaegul Choo
Christopher Choy
Hang Chu
Peng Chu
Wen-Sheng Chu
Albert Chung
Joon Son Chung
Hai Ci
Safa Cicek
Ramazan G. Cinbis
Arridhana Ciptadi
Javier Civera
James J. Clark
Ronald Clark
Felipe Codevilla
Michael Cogswell
Andrea Cohen
Maxwell D. Collins
Carlo Colombo
Yang Cong
Adria R. Continente
Marcella Cornia
John Richard Corring
Darren Cosker
Dragos Costea
Garrison W. Cottrell
Florent Couzinie-Devy
Marco Cristani
Ioana Croitoru
James L. Crowley
Jiequan Cui
Zhaopeng Cui
Ross Cutler
Antonio D'Innocente

Rozenn Dahyot
Bo Dai
Dengxin Dai
Hang Dai
Longquan Dai
Shuyang Dai
Xiyang Dai
Yuchao Dai
Adrian V. Dalca
Dima Damen
Bharath B. Damodaran
Kristin Dana
Martin Danelljan
Zheng Dang
Zachary Alan Daniels
Donald G. Dansereau
Abhishek Das
Samyak Datta
Achal Dave
Titas De
Rodrigo de Bem
Teo de Campos
Raoul de Charette
Shalini De Mello
Joseph DeGol
Herve Delingette
Haowen Deng
Jiankang Deng
Weijian Deng
Zhiwei Deng
Joachim Denzler
Konstantinos G. Derpanis
Aditya Deshpande
Frederic Devernay
Somdip Dey
Arturo Deza
Abhinav Dhall
Helisa Dhamo
Vikas Dhiman
Fillipe Dias Moreira
   de Souza
Ali Diba
Ferran Diego
Guiguang Ding
Henghui Ding
Jian Ding

Mingyu Ding
Xinghao Ding
Zhengming Ding
Robert DiPietro
Cosimo Distante
Ajay Divakaran
Mandar Dixit
Abdelaziz Djelouah
Thanh-Toan Do
Jose Dolz
Bo Dong
Chao Dong
Jiangxin Dong
Weiming Dong
Weisheng Dong
Xingping Dong
Xuanyi Dong
Yinpeng Dong
Gianfranco Doretto
Hazel Doughty
Hassen Drira
Bertram Drost
Dawei Du
Ye Duan
Yueqi Duan
Abhimanyu Dubey
Anastasia Dubrovina
Stefan Duffner
Chi Nhan Duong
Thibaut Durand
Zoran Duric
Iulia Duta
Debidatta Dwibedi
Benjamin Eckart
Marc Eder
Marzieh Edraki
Alexei A. Efros
Kiana Ehsani
Hazm Kemal Ekenel
James H. Elder
Mohamed Elgharib
Shireen Elhabian
Ehsan Elhamifar
Mohamed Elhoseiny
Ian Endres
N. Benjamin Erichson

Jan Ernst
Sergio Escalera
Francisco Escolano
Victor Escorcia
Carlos Esteves
Francisco J. Estrada
Bin Fan
Chenyou Fan
Deng-Ping Fan
Haoqi Fan
Hehe Fan
Heng Fan
Kai Fan
Lijie Fan
Linxi Fan
Quanfu Fan
Shaojing Fan
Xiaochuan Fan
Xin Fan
Yuchen Fan
Sean Fanello
Hao-Shu Fang
Haoyang Fang
Kuan Fang
Yi Fang
Yuming Fang
Azade Farshad
Alireza Fathi
Raanan Fattal
Joao Fayad
Xiaohan Fei
Christoph Feichtenhofer
Michael Felsberg
Chen Feng
Jiashi Feng
Junyi Feng
Mengyang Feng
Qianli Feng
Zhenhua Feng
Michele Fenzi
Andras Ferencz
Martin Fergie
Basura Fernando
Ethan Fetaya
Michael Firman
John W. Fisher

Matthew Fisher
Boris Flach
Corneliu Florea
Wolfgang Foerstner
David Fofi
Gian Luca Foresti
Per-Erik Forssen
David Fouhey
Katerina Fragkiadaki
Victor Fragoso
Jean-Sébastien Franco
Ohad Fried
Iuri Frosio
Cheng-Yang Fu
Huazhu Fu
Jianlong Fu
Jingjing Fu
Xueyang Fu
Yanwei Fu
Ying Fu
Yun Fu
Olac Fuentes
Kent Fujiwara
Takuya Funatomi
Christopher Funk
Thomas Funkhouser
Antonino Furnari
Ryo Furukawa
Erik Gärtner
Raghudeep Gadde
Matheus Gadelha
Vandit Gajjar
Trevor Gale
Juergen Gall
Mathias Gallardo
Guillermo Gallego
Orazio Gallo
Chuang Gan
Zhe Gan
Madan Ravi Ganesh
Aditya Ganeshan
Siddha Ganju
Bin-Bin Gao
Changxin Gao
Feng Gao
Hongchang Gao

Jin Gao
Jiyang Gao
Junbin Gao
Katelyn Gao
Lin Gao
Mingfei Gao
Ruiqi Gao
Ruohan Gao
Shenghua Gao
Yuan Gao
Yue Gao
Noa Garcia
Alberto Garcia-Garcia
Guillermo
    Garcia-Hernando
Jacob R. Gardner
Animesh Garg
Kshitiz Garg
Rahul Garg
Ravi Garg
Philip N. Garner
Kirill Gavrilyuk
Paul Gay
Shiming Ge
Weifeng Ge
Baris Gecer
Xin Geng
Kyle Genova
Stamatios Georgoulis
Bernard Ghanem
Michael Gharbi
Kamran Ghasedi
Golnaz Ghiasi
Arnab Ghosh
Partha Ghosh
Silvio Giancola
Andrew Gilbert
Rohit Girdhar
Xavier Giro-i-Nieto
Thomas Gittings
Ioannis Gkioulekas
Clement Godard
Vaibhava Goel
Bastian Goldluecke
Lluis Gomez
Nuno Gonçalves

Dong Gong
Ke Gong
Mingming Gong
Abel Gonzalez-Garcia
Ariel Gordon
Daniel Gordon
Paulo Gotardo
Venu Madhav Govindu
Ankit Goyal
Priya Goyal
Raghav Goyal
Benjamin Graham
Douglas Gray
Brent A. Griffin
Etienne Grossmann
David Gu
Jiayuan Gu
Jiuxiang Gu
Lin Gu
Qiao Gu
Shuhang Gu
Jose J. Guerrero
Paul Guerrero
Jie Gui
Jean-Yves Guillemaut
Riza Alp Guler
Erhan Gundogdu
Fatma Guney
Guodong Guo
Kaiwen Guo
Qi Guo
Sheng Guo
Shi Guo
Tiantong Guo
Xiaojie Guo
Yijie Guo
Yiluan Guo
Yuanfang Guo
Yulan Guo
Agrim Gupta
Ankush Gupta
Mohit Gupta
Saurabh Gupta
Tanmay Gupta
Danna Gurari
Abner Guzman-Rivera

JunYoung Gwak
Michael Gygli
Jung-Woo Ha
Simon Hadfield
Isma Hadji
Bjoern Haefner
Taeyoung Hahn
Levente Hajder
Peter Hall
Emanuela Haller
Stefan Haller
Bumsub Ham
Abdullah Hamdi
Dongyoon Han
Hu Han
Jungong Han
Junwei Han
Kai Han
Tian Han
Xiaoguang Han
Xintong Han
Yahong Han
Ankur Handa
Zekun Hao
Albert Haque
Tatsuya Harada
Mehrtash Harandi
Adam W. Harley
Mahmudul Hasan
Atsushi Hashimoto
Ali Hatamizadeh
Munawar Hayat
Dongliang He
Jingrui He
Junfeng He
Kaiming He
Kun He
Lei He
Pan He
Ran He
Shengfeng He
Tong He
Weipeng He
Xuming He
Yang He
Yihui He

Zhihai He
Chinmay Hegde
Janne Heikkila
Mattias P. Heinrich
Stéphane Herbin
Alexander Hermans
Luis Herranz
John R. Hershey
Aaron Hertzmann
Roei Herzig
Anders Heyden
Steven Hickson
Otmar Hilliges
Tomas Hodan
Judy Hoffman
Michael Hofmann
Yannick Hold-Geoffroy
Namdar Homayounfar
Sina Honari
Richang Hong
Seunghoon Hong
Xiaopeng Hong
Yi Hong
Hidekata Hontani
Anthony Hoogs
Yedid Hoshen
Mir Rayat Imtiaz Hossain
Junhui Hou
Le Hou
Lu Hou
Tingbo Hou
Wei-Lin Hsiao
Cheng-Chun Hsu
Gee-Sern Jison Hsu
Kuang-jui Hsu
Changbo Hu
Di Hu
Guosheng Hu
Han Hu
Hao Hu
Hexiang Hu
Hou-Ning Hu
Jie Hu
Junlin Hu
Nan Hu
Ping Hu

Ronghang Hu
Xiaowei Hu
Yinlin Hu
Yuan-Ting Hu
Zhe Hu
Binh-Son Hua
Yang Hua
Bingyao Huang
Di Huang
Dong Huang
Fay Huang
Haibin Huang
Haozhi Huang
Heng Huang
Huaibo Huang
Jia-Bin Huang
Jing Huang
Jingwei Huang
Kaizhu Huang
Lei Huang
Qiangui Huang
Qiaoying Huang
Qingqiu Huang
Qixing Huang
Shaoli Huang
Sheng Huang
Siyuan Huang
Weilin Huang
Wenbing Huang
Xiangru Huang
Xun Huang
Yan Huang
Yifei Huang
Yue Huang
Zhiwu Huang
Zilong Huang
Minyoung Huh
Zhuo Hui
Matthias B. Hullin
Martin Humenberger
Wei-Chih Hung
Zhouyuan Huo
Junhwa Hur
Noureldien Hussein
Jyh-Jing Hwang
Seong Jae Hwang

Sung Ju Hwang
Ichiro Ide
Ivo Ihrke
Daiki Ikami
Satoshi Ikehata
Nazli Ikizler-Cinbis
Sunghoon Im
Yani Ioannou
Radu Tudor Ionescu
Umar Iqbal
Go Irie
Ahmet Iscen
Md Amirul Islam
Vamsi Ithapu
Nathan Jacobs
Arpit Jain
Himalaya Jain
Suyog Jain
Stuart James
Won-Dong Jang
Yunseok Jang
Ronnachai Jaroensri
Dinesh Jayaraman
Sadeep Jayasumana
Suren Jayasuriya
Herve Jegou
Simon Jenni
Hae-Gon Jeon
Yunho Jeon
Koteswar R. Jerripothula
Hueihan Jhuang
I-hong Jhuo
Dinghuang Ji
Hui Ji
Jingwei Ji
Pan Ji
Yanli Ji
Baoxiong Jia
Kui Jia
Xu Jia
Chiyu Max Jiang
Haiyong Jiang
Hao Jiang
Huaizu Jiang
Huajie Jiang
Ke Jiang

Lai Jiang
Li Jiang
Lu Jiang
Ming Jiang
Peng Jiang
Shuqiang Jiang
Wei Jiang
Xudong Jiang
Zhuolin Jiang
Jianbo Jiao
Zequn Jie
Dakai Jin
Kyong Hwan Jin
Lianwen Jin
SouYoung Jin
Xiaojie Jin
Xin Jin
Nebojsa Jojic
Alexis Joly
Michael Jeffrey Jones
Hanbyul Joo
Jungseock Joo
Kyungdon Joo
Ajjen Joshi
Shantanu H. Joshi
Da-Cheng Juan
Marco Körner
Kevin Köser
Asim Kadav
Christine Kaeser-Chen
Kushal Kafle
Dagmar Kainmueller
Ioannis A. Kakadiaris
Zdenek Kalal
Nima Kalantari
Yannis Kalantidis
Mahdi M. Kalayeh
Anmol Kalia
Sinan Kalkan
Vicky Kalogeiton
Ashwin Kalyan
Joni-kristian Kamarainen
Gerda Kamberova
Chandra Kambhamettu
Martin Kampel
Meina Kan

Christopher Kanan
Kenichi Kanatani
Angjoo Kanazawa
Atsushi Kanehira
Takuhiro Kaneko
Asako Kanezaki
Bingyi Kang
Di Kang
Sunghun Kang
Zhao Kang
Vadim Kantorov
Abhishek Kar
Amlan Kar
Theofanis Karaletsos
Leonid Karlinsky
Kevin Karsch
Angelos Katharopoulos
Isinsu Katircioglu
Hiroharu Kato
Zoltan Kato
Dotan Kaufman
Jan Kautz
Rei Kawakami
Qiuhong Ke
Wadim Kehl
Petr Kellnhofer
Aniruddha Kembhavi
Cem Keskin
Margret Keuper
Daniel Keysers
Ashkan Khakzar
Fahad Khan
Naeemullah Khan
Salman Khan
Siddhesh Khandelwal
Rawal Khirodkar
Anna Khoreva
Tejas Khot
Parmeshwar Khurd
Hadi Kiapour
Joe Kileel
Chanho Kim
Dahun Kim
Edward Kim
Eunwoo Kim
Han-ul Kim

Hansung Kim
Heewon Kim
Hyo Jin Kim
Hyunwoo J. Kim
Jinkyu Kim
Jiwon Kim
Jongmin Kim
Junsik Kim
Junyeong Kim
Min H. Kim
Namil Kim
Pyojin Kim
Seon Joo Kim
Seong Tae Kim
Seungryong Kim
Sungwoong Kim
Tae Hyun Kim
Vladimir Kim
Won Hwa Kim
Yonghyun Kim
Benjamin Kimia
Akisato Kimura
Pieter-Jan Kindermans
Zsolt Kira
Itaru Kitahara
Hedvig Kjellstrom
Jan Knopp
Takumi Kobayashi
Erich Kobler
Parker Koch
Reinhard Koch
Elyor Kodirov
Amir Kolaman
Nicholas Kolkin
Dimitrios Kollias
Stefanos Kollias
Soheil Kolouri
Adams Wai-Kin Kong
Naejin Kong
Shu Kong
Tao Kong
Yu Kong
Yoshinori Konishi
Daniil Kononenko
Theodora Kontogianni
Simon Korman

Adam Kortylewski
Jana Kosecka
Jean Kossaifi
Satwik Kottur
Rigas Kouskouridas
Adriana Kovashka
Rama Kovvuri
Adarsh Kowdle
Jedrzej Kozerawski
Mateusz Kozinski
Philipp Kraehenbuehl
Gregory Kramida
Josip Krapac
Dmitry Kravchenko
Ranjay Krishna
Pavel Krsek
Alexander Krull
Jakob Kruse
Hiroyuki Kubo
Hilde Kuehne
Jason Kuen
Andreas Kuhn
Arjan Kuijper
Zuzana Kukelova
Ajay Kumar
Amit Kumar
Avinash Kumar
Suryansh Kumar
Vijay Kumar
Kaustav Kundu
Weicheng Kuo
Nojun Kwak
Suha Kwak
Junseok Kwon
Nikolaos Kyriazis
Zorah Lähner
Ankit Laddha
Florent Lafarge
Jean Lahoud
Kevin Lai
Shang-Hong Lai
Wei-Sheng Lai
Yu-Kun Lai
Iro Laina
Antony Lam
John Wheatley Lambert

Xiangyuan lan
Xu Lan
Charis Lanaras
Georg Langs
Oswald Lanz
Dong Lao
Yizhen Lao
Agata Lapedriza
Gustav Larsson
Viktor Larsson
Katrin Lasinger
Christoph Lassner
Longin Jan Latecki
Stéphane Lathuilière
Rynson Lau
Hei Law
Justin Lazarow
Svetlana Lazebnik
Hieu Le
Huu Le
Ngan Hoang Le
Trung-Nghia Le
Vuong Le
Colin Lea
Erik Learned-Miller
Chen-Yu Lee
Gim Hee Lee
Hsin-Ying Lee
Hyungtae Lee
Jae-Han Lee
Jimmy Addison Lee
Joonseok Lee
Kibok Lee
Kuang-Huei Lee
Kwonjoon Lee
Minsik Lee
Sang-chul Lee
Seungkyu Lee
Soochan Lee
Stefan Lee
Taehee Lee
Andreas Lehrmann
Jie Lei
Peng Lei
Matthew Joseph Leotta
Wee Kheng Leow

Gil Levi
Evgeny Levinkov
Aviad Levis
Jose Lezama
Ang Li
Bin Li
Bing Li
Boyi Li
Changsheng Li
Chao Li
Chen Li
Cheng Li
Chenglong Li
Chi Li
Chun-Guang Li
Chun-Liang Li
Chunyuan Li
Dong Li
Guanbin Li
Hao Li
Haoxiang Li
Hongsheng Li
Hongyang Li
Houqiang Li
Huibin Li
Jia Li
Jianan Li
Jianguo Li
Junnan Li
Junxuan Li
Kai Li
Ke Li
Kejie Li
Kunpeng Li
Lerenhan Li
Li Erran Li
Mengtian Li
Mu Li
Peihua Li
Peiyi Li
Ping Li
Qi Li
Qing Li
Ruiyu Li
Ruoteng Li
Shaozi Li

Sheng Li
Shiwei Li
Shuang Li
Siyang Li
Stan Z. Li
Tianye Li
Wei Li
Weixin Li
Wen Li
Wenbo Li
Xiaomeng Li
Xin Li
Xiu Li
Xuelong Li
Xueting Li
Yan Li
Yandong Li
Yanghao Li
Yehao Li
Yi Li
Yijun Li
Yikang LI
Yining Li
Yongjie Li
Yu Li
Yu-Jhe Li
Yunpeng Li
Yunsheng Li
Yunzhu Li
Zhe Li
Zhen Li
Zhengqi Li
Zhenyang Li
Zhuwen Li
Dongze Lian
Xiaochen Lian
Zhouhui Lian
Chen Liang
Jie Liang
Ming Liang
Paul Pu Liang
Pengpeng Liang
Shu Liang
Wei Liang
Jing Liao
Minghui Liao

Renjie Liao
Shengcai Liao
Shuai Liao
Yiyi Liao
Ser-Nam Lim
Chen-Hsuan Lin
Chung-Ching Lin
Dahua Lin
Ji Lin
Kevin Lin
Tianwei Lin
Tsung-Yi Lin
Tsung-Yu Lin
Wei-An Lin
Weiyao Lin
Yen-Chen Lin
Yuewei Lin
David B. Lindell
Drew Linsley
Krzysztof Lis
Roee Litman
Jim Little
An-An Liu
Bo Liu
Buyu Liu
Chao Liu
Chen Liu
Cheng-lin Liu
Chenxi Liu
Dong Liu
Feng Liu
Guilin Liu
Haomiao Liu
Heshan Liu
Hong Liu
Ji Liu
Jingen Liu
Jun Liu
Lanlan Liu
Li Liu
Liu Liu
Mengyuan Liu
Miaomiao Liu
Nian Liu
Ping Liu
Risheng Liu

Sheng Liu
Shu Liu
Shuaicheng Liu
Sifei Liu
Siqi Liu
Siying Liu
Songtao Liu
Ting Liu
Tongliang Liu
Tyng-Luh Liu
Wanquan Liu
Wei Liu
Weiyang Liu
Weizhe Liu
Wenyu Liu
Wu Liu
Xialei Liu
Xianglong Liu
Xiaodong Liu
Xiaofeng Liu
Xihui Liu
Xingyu Liu
Xinwang Liu
Xuanqing Liu
Xuebo Liu
Yang Liu
Yaojie Liu
Yebin Liu
Yen-Cheng Liu
Yiming Liu
Yu Liu
Yu-Shen Liu
Yufan Liu
Yun Liu
Zheng Liu
Zhijian Liu
Zhuang Liu
Zichuan Liu
Ziwei Liu
Zongyi Liu
Stephan Liwicki
Liliana Lo Presti
Chengjiang Long
Fuchen Long
Mingsheng Long
Xiang Long

Yang Long
Charles T. Loop
Antonio Lopez
Roberto J. Lopez-Sastre
Javier Lorenzo-Navarro
Manolis Lourakis
Boyu Lu
Canyi Lu
Feng Lu
Guoyu Lu
Hongtao Lu
Jiajun Lu
Jiasen Lu
Jiwen Lu
Kaiyue Lu
Le Lu
Shao-Ping Lu
Shijian Lu
Xiankai Lu
Xin Lu
Yao Lu
Yiping Lu
Yongxi Lu
Yongyi Lu
Zhiwu Lu
Fujun Luan
Benjamin E. Lundell
Hao Luo
Jian-Hao Luo
Ruotian Luo
Weixin Luo
Wenhan Luo
Wenjie Luo
Yan Luo
Zelun Luo
Zixin Luo
Khoa Luu
Zhaoyang Lv
Pengyuan Lyu
Thomas Möllenhoff
Matthias Müller
Bingpeng Ma
Chih-Yao Ma
Chongyang Ma
Huimin Ma
Jiayi Ma

K. T. Ma
Ke Ma
Lin Ma
Liqian Ma
Shugao Ma
Wei-Chiu Ma
Xiaojian Ma
Xingjun Ma
Zhanyu Ma
Zheng Ma
Radek Jakob Mackowiak
Ludovic Magerand
Shweta Mahajan
Siddharth Mahendran
Long Mai
Ameesh Makadia
Oscar Mendez Maldonado
Mateusz Malinowski
Yury Malkov
Arun Mallya
Dipu Manandhar
Massimiliano Mancini
Fabian Manhardt
Kevis-kokitsi Maninis
Varun Manjunatha
Junhua Mao
Xudong Mao
Alina Marcu
Edgar Margffoy-Tuay
Dmitrii Marin
Manuel J. Marin-Jimenez
Kenneth Marino
Niki Martinel
Julieta Martinez
Jonathan Masci
Tomohiro Mashita
Iacopo Masi
David Masip
Daniela Massiceti
Stefan Mathe
Yusuke Matsui
Tetsu Matsukawa
Iain A. Matthews
Kevin James Matzen
Bruce Allen Maxwell
Stephen Maybank

Helmut Mayer
Amir Mazaheri
David McAllester
Steven McDonagh
Stephen J. Mckenna
Roey Mechrez
Prakhar Mehrotra
Christopher Mei
Xue Mei
Paulo R. S. Mendonca
Lili Meng
Zibo Meng
Thomas Mensink
Bjoern Menze
Michele Merler
Kourosh Meshgi
Pascal Mettes
Christopher Metzler
Liang Mi
Qiguang Miao
Xin Miao
Tomer Michaeli
Frank Michel
Antoine Miech
Krystian Mikolajczyk
Peyman Milanfar
Ben Mildenhall
Gregor Miller
Fausto Milletari
Dongbo Min
Kyle Min
Pedro Miraldo
Dmytro Mishkin
Anand Mishra
Ashish Mishra
Ishan Misra
Niluthpol C. Mithun
Kaushik Mitra
Niloy Mitra
Anton Mitrokhin
Ikuhisa Mitsugami
Anurag Mittal
Kaichun Mo
Zhipeng Mo
Davide Modolo
Michael Moeller

Pritish Mohapatra
Pavlo Molchanov
Davide Moltisanti
Pascal Monasse
Mathew Monfort
Aron Monszpart
Sean Moran
Vlad I. Morariu
Francesc Moreno-Noguer
Pietro Morerio
Stylianos Moschoglou
Yael Moses
Roozbeh Mottaghi
Pierre Moulon
Arsalan Mousavian
Yadong Mu
Yasuhiro Mukaigawa
Lopamudra Mukherjee
Yusuke Mukuta
Ravi Teja Mullapudi
Mario Enrique Munich
Zachary Murez
Ana C. Murillo
J. Krishna Murthy
Damien Muselet
Armin Mustafa
Siva Karthik Mustikovela
Carlo Dal Mutto
Moin Nabi
Varun K. Nagaraja
Tushar Nagarajan
Arsha Nagrani
Seungjun Nah
Nikhil Naik
Yoshikatsu Nakajima
Yuta Nakashima
Atsushi Nakazawa
Seonghyeon Nam
Vinay P. Namboodiri
Medhini Narasimhan
Srinivasa Narasimhan
Sanath Narayan
Erickson Rangel
    Nascimento
Jacinto Nascimento
Tayyab Naseer

Lakshmanan Nataraj
Neda Nategh
Nelson Isao Nauata
Fernando Navarro
Shah Nawaz
Lukas Neumann
Ram Nevatia
Alejandro Newell
Shawn Newsam
Joe Yue-Hei Ng
Trung Thanh Ngo
Duc Thanh Nguyen
Lam M. Nguyen
Phuc Xuan Nguyen
Thuong Nguyen Canh
Mihalis Nicolaou
Andrei Liviu Nicolicioiu
Xuecheng Nie
Michael Niemeyer
Simon Niklaus
Christophoros Nikou
David Nilsson
Jifeng Ning
Yuval Nirkin
Li Niu
Yuzhen Niu
Zhenxing Niu
Shohei Nobuhara
Nicoletta Noceti
Hyeonwoo Noh
Junhyug Noh
Mehdi Noroozi
Sotiris Nousias
Valsamis Ntouskos
Matthew O'Toole
Peter Ochs
Ferda Ofli
Seong Joon Oh
Seoung Wug Oh
Iason Oikonomidis
Utkarsh Ojha
Takahiro Okabe
Takayuki Okatani
Fumio Okura
Aude Oliva
Kyle Olszewski

Björn Ommer
Mohamed Omran
Elisabeta Oneata
Michael Opitz
Jose Oramas
Tribhuvanesh Orekondy
Shaul Oron
Sergio Orts-Escolano
Ivan Oseledets
Aljosa Osep
Magnus Oskarsson
Anton Osokin
Martin R. Oswald
Wanli Ouyang
Andrew Owens
Mete Ozay
Mustafa Ozuysal
Eduardo Pérez-Pellitero
Gautam Pai
Dipan Kumar Pal
P. H. Pamplona Savarese
Jinshan Pan
Junting Pan
Xingang Pan
Yingwei Pan
Yannis Panagakis
Rameswar Panda
Guan Pang
Jiahao Pang
Jiangmiao Pang
Tianyu Pang
Sharath Pankanti
Nicolas Papadakis
Dim Papadopoulos
George Papandreou
Toufiq Parag
Shaifali Parashar
Sarah Parisot
Eunhyeok Park
Hyun Soo Park
Jaesik Park
Min-Gyu Park
Taesung Park
Alvaro Parra
C. Alejandro Parraga
Despoina Paschalidou

Nikolaos Passalis
Vishal Patel
Viorica Patraucean
Badri Narayana Patro
Danda Pani Paudel
Sujoy Paul
Georgios Pavlakos
Ioannis Pavlidis
Vladimir Pavlovic
Nick Pears
Kim Steenstrup Pedersen
Selen Pehlivan
Shmuel Peleg
Chao Peng
Houwen Peng
Wen-Hsiao Peng
Xi Peng
Xiaojiang Peng
Xingchao Peng
Yuxin Peng
Federico Perazzi
Juan Camilo Perez
Vishwanath Peri
Federico Pernici
Luca Del Pero
Florent Perronnin
Stavros Petridis
Henning Petzka
Patrick Peursum
Michael Pfeiffer
Hanspeter Pfister
Roman Pflugfelder
Minh Tri Pham
Yongri Piao
David Picard
Tomasz Pieciak
A. J. Piergiovanni
Andrea Pilzer
Pedro O. Pinheiro
Silvia Laura Pintea
Lerrel Pinto
Axel Pinz
Robinson Piramuthu
Fiora Pirri
Leonid Pishchulin
Francesco Pittaluga

Daniel Pizarro
Tobias Plötz
Mirco Planamente
Matteo Poggi
Moacir A. Ponti
Parita Pooj
Fatih Porikli
Horst Possegger
Omid Poursaeed
Ameya Prabhu
Viraj Uday Prabhu
Dilip Prasad
Brian L. Price
True Price
Maria Priisalu
Veronique Prinet
Victor Adrian Prisacariu
Jan Prokaj
Sergey Prokudin
Nicolas Pugeault
Xavier Puig
Albert Pumarola
Pulak Purkait
Senthil Purushwalkam
Charles R. Qi
Hang Qi
Haozhi Qi
Lu Qi
Mengshi Qi
Siyuan Qi
Xiaojuan Qi
Yuankai Qi
Shengju Qian
Xuelin Qian
Siyuan Qiao
Yu Qiao
Jie Qin
Qiang Qiu
Weichao Qiu
Zhaofan Qiu
Kha Gia Quach
Yuhui Quan
Yvain Queau
Julian Quiroga
Faisal Qureshi
Mahdi Rad

Filip Radenovic
Petia Radeva
Venkatesh
  B. Radhakrishnan
Ilija Radosavovic
Noha Radwan
Rahul Raguram
Tanzila Rahman
Amit Raj
Ajit Rajwade
Kandan Ramakrishnan
Santhosh
  K. Ramakrishnan
Srikumar Ramalingam
Ravi Ramamoorthi
Vasili Ramanishka
Ramprasaath R. Selvaraju
Francois Rameau
Visvanathan Ramesh
Santu Rana
Rene Ranftl
Anand Rangarajan
Anurag Ranjan
Viresh Ranjan
Yongming Rao
Carolina Raposo
Vivek Rathod
Sathya N. Ravi
Avinash Ravichandran
Tammy Riklin Raviv
Daniel Rebain
Sylvestre-Alvise Rebuffi
N. Dinesh Reddy
Timo Rehfeld
Paolo Remagnino
Konstantinos Rematas
Edoardo Remelli
Dongwei Ren
Haibing Ren
Jian Ren
Jimmy Ren
Mengye Ren
Weihong Ren
Wenqi Ren
Zhile Ren
Zhongzheng Ren

Zhou Ren
Vijay Rengarajan
Md A. Reza
Farzaneh Rezaeianaran
Hamed R. Tavakoli
Nicholas Rhinehart
Helge Rhodin
Elisa Ricci
Alexander Richard
Eitan Richardson
Elad Richardson
Christian Richardt
Stephan Richter
Gernot Riegler
Daniel Ritchie
Tobias Ritschel
Samuel Rivera
Yong Man Ro
Richard Roberts
Joseph Robinson
Ignacio Rocco
Mrigank Rochan
Emanuele Rodolà
Mikel D. Rodriguez
Giorgio Roffo
Grégory Rogez
Gemma Roig
Javier Romero
Xuejian Rong
Yu Rong
Amir Rosenfeld
Bodo Rosenhahn
Guy Rosman
Arun Ross
Paolo Rota
Peter M. Roth
Anastasios Roussos
Anirban Roy
Sebastien Roy
Aruni RoyChowdhury
Artem Rozantsev
Ognjen Rudovic
Daniel Rueckert
Adria Ruiz
Javier Ruiz-del-solar
Christian Rupprecht

Chris Russell
Dan Ruta
Jongbin Ryu
Ömer Sümer
Alexandre Sablayrolles
Faraz Saeedan
Ryusuke Sagawa
Christos Sagonas
Tonmoy Saikia
Hideo Saito
Kuniaki Saito
Shunsuke Saito
Shunta Saito
Ken Sakurada
Joaquin Salas
Fatemeh Sadat Saleh
Mahdi Saleh
Pouya Samangouei
Leo Sampaio
  Ferraz Ribeiro
Artsiom Olegovich
  Sanakoyeu
Enrique Sanchez
Patsorn Sangkloy
Anush Sankaran
Aswin Sankaranarayanan
Swami Sankaranarayanan
Rodrigo Santa Cruz
Amartya Sanyal
Archana Sapkota
Nikolaos Sarafianos
Jun Sato
Shin'ichi Satoh
Hosnieh Sattar
Arman Savran
Manolis Savva
Alexander Sax
Hanno Scharr
Simone Schaub-Meyer
Konrad Schindler
Dmitrij Schlesinger
Uwe Schmidt
Dirk Schnieders
Björn Schuller
Samuel Schulter
Idan Schwartz

William Robson Schwartz
Alex Schwing
Sinisa Segvic
Lorenzo Seidenari
Pradeep Sen
Ozan Sener
Soumyadip Sengupta
Arda Senocak
Mojtaba Seyedhosseini
Shishir Shah
Shital Shah
Sohil Atul Shah
Tamar Rott Shaham
Huasong Shan
Qi Shan
Shiguang Shan
Jing Shao
Roman Shapovalov
Gaurav Sharma
Vivek Sharma
Viktoriia Sharmanska
Dongyu She
Sumit Shekhar
Evan Shelhamer
Chengyao Shen
Chunhua Shen
Falong Shen
Jie Shen
Li Shen
Liyue Shen
Shuhan Shen
Tianwei Shen
Wei Shen
William B. Shen
Yantao Shen
Ying Shen
Yiru Shen
Yujun Shen
Yuming Shen
Zhiqiang Shen
Ziyi Shen
Lu Sheng
Yu Sheng
Rakshith Shetty
Baoguang Shi
Guangming Shi

Hailin Shi
Miaojing Shi
Yemin Shi
Zhenmei Shi
Zhiyuan Shi
Kevin Jonathan Shih
Shiliang Shiliang
Hyunjung Shim
Atsushi Shimada
Nobutaka Shimada
Daeyun Shin
Young Min Shin
Koichi Shinoda
Konstantin Shmelkov
Michael Zheng Shou
Abhinav Shrivastava
Tianmin Shu
Zhixin Shu
Hong-Han Shuai
Pushkar Shukla
Christian Siagian
Mennatullah M. Siam
Kaleem Siddiqi
Karan Sikka
Jae-Young Sim
Christian Simon
Martin Simonovsky
Dheeraj Singaraju
Bharat Singh
Gurkirt Singh
Krishna Kumar Singh
Maneesh Kumar Singh
Richa Singh
Saurabh Singh
Suriya Singh
Vikas Singh
Sudipta N. Sinha
Vincent Sitzmann
Josef Sivic
Gregory Slabaugh
Miroslava Slavcheva
Ron Slossberg
Brandon Smith
Kevin Smith
Vladimir Smutny
Noah Snavely

Roger
   D. Soberanis-Mukul
Kihyuk Sohn
Francesco Solera
Eric Sommerlade
Sanghyun Son
Byung Cheol Song
Chunfeng Song
Dongjin Song
Jiaming Song
Jie Song
Jifei Song
Jingkuan Song
Mingli Song
Shiyu Song
Shuran Song
Xiao Song
Yafei Song
Yale Song
Yang Song
Yi-Zhe Song
Yibing Song
Humberto Sossa
Cesar de Souza
Adrian Spurr
Srinath Sridhar
Suraj Srinivas
Pratul P. Srinivasan
Anuj Srivastava
Tania Stathaki
Christopher Stauffer
Simon Stent
Rainer Stiefelhagen
Pierre Stock
Julian Straub
Jonathan C. Stroud
Joerg Stueckler
Jan Stuehmer
David Stutz
Chi Su
Hang Su
Jong-Chyi Su
Shuochen Su
Yu-Chuan Su
Ramanathan Subramanian
Yusuke Sugano

Masanori Suganuma

Yumin Suh

Mohammed Suhail

Yao Sui

Heung-Il Suk

Josephine Sullivan

Baochen Sun

Chen Sun

Chong Sun

Deqing Sun

Jin Sun

Liang Sun

Lin Sun

Qianru Sun

Shao-Hua Sun

Shuyang Sun

Weiwei Sun

Wenxiu Sun

Xiaoshuai Sun

Xiaoxiao Sun

Xingyuan Sun

Yifan Sun

Zhun Sun

Sabine Susstrunk

David Suter

Supasorn Suwajanakorn

Tomas Svoboda

Eran Swears

Paul Swoboda

Attila Szabo

Richard Szeliski

Duy-Nguyen Ta

Andrea Tagliasacchi

Yuichi Taguchi

Ying Tai

Keita Takahashi

Kouske Takahashi

Jun Takamatsu

Hugues Talbot

Toru Tamaki

Chaowei Tan

Fuwen Tan

Mingkui Tan

Mingxing Tan

Qingyang Tan

Robby T. Tan

Xiaoyang Tan

Kenichiro Tanaka

Masayuki Tanaka

Chang Tang

Chengzhou Tang

Danhang Tang

Ming Tang

Peng Tang

Qingming Tang

Wei Tang

Xu Tang

Yansong Tang

Youbao Tang

Yuxing Tang

Zhiqiang Tang

Tatsunori Taniai

Junli Tao

Xin Tao

Makarand Tapaswi

Jean-Philippe Tarel

Lyne Tchapmi

Zachary Teed

Bugra Tekin

Damien Teney

Ayush Tewari

Christian Theobalt

Christopher Thomas

Diego Thomas

Jim Thomas

Rajat Mani Thomas

Xinmei Tian

Yapeng Tian

Yingli Tian

Yonglong Tian

Zhi Tian

Zhuotao Tian

Kinh Tieu

Joseph Tighe

Massimo Tistarelli

Matthew Toews

Carl Toft

Pavel Tokmakov

Federico Tombari

Chetan Tonde

Yan Tong

Alessio Tonioni

Andrea Torsello

Fabio Tosi

Du Tran

Luan Tran

Ngoc-Trung Tran

Quan Hung Tran

Truyen Tran

Rudolph Triebel

Martin Trimmel

Shashank Tripathi

Subarna Tripathi

Leonardo Trujillo

Eduard Trulls

Tomasz Trzcinski

Sam Tsai

Yi-Hsuan Tsai

Hung-Yu Tseng

Stavros Tsogkas

Aggeliki Tsoli

Devis Tuia

Shubham Tulsiani

Sergey Tulyakov

Frederick Tung

Tony Tung

Daniyar Turmukhambetov

Ambrish Tyagi

Radim Tylecek

Christos Tzelepis

Georgios Tzimiropoulos

Dimitrios Tzionas

Seiichi Uchida

Norimichi Ukita

Dmitry Ulyanov

Martin Urschler

Yoshitaka Ushiku

Ben Usman

Alexander Vakhitov

Julien P. C. Valentin

Jack Valmadre

Ernest Valveny

Joost van de Weijer

Jan van Gemert

Koen Van Leemput

Gul Varol

Sebastiano Vascon

M. Alex O. Vasilescu

Subeesh Vasu
Mayank Vatsa
David Vazquez
Javier Vazquez-Corral
Ashok Veeraraghavan
Erik Velasco-Salido
Raviteja Vemulapalli
Jonathan Ventura
Manisha Verma
Roberto Vezzani
Ruben Villegas
Minh Vo
MinhDuc Vo
Nam Vo
Michele Volpi
Riccardo Volpi
Carl Vondrick
Konstantinos Vougioukas
Tuan-Hung Vu
Sven Wachsmuth
Neal Wadhwa
Catherine Wah
Jacob C. Walker
Thomas S. A. Wallis
Chengde Wan
Jun Wan
Liang Wan
Renjie Wan
Baoyuan Wang
Boyu Wang
Cheng Wang
Chu Wang
Chuan Wang
Chunyu Wang
Dequan Wang
Di Wang
Dilin Wang
Dong Wang
Fang Wang
Guanzhi Wang
Guoyin Wang
Hanzi Wang
Hao Wang
He Wang
Heng Wang
Hongcheng Wang

Hongxing Wang
Hua Wang
Jian Wang
Jingbo Wang
Jinglu Wang
Jingya Wang
Jinjun Wang
Jinqiao Wang
Jue Wang
Ke Wang
Keze Wang
Le Wang
Lei Wang
Lezi Wang
Li Wang
Liang Wang
Lijun Wang
Limin Wang
Linwei Wang
Lizhi Wang
Mengjiao Wang
Mingzhe Wang
Minsi Wang
Naiyan Wang
Nannan Wang
Ning Wang
Oliver Wang
Pei Wang
Peng Wang
Pichao Wang
Qi Wang
Qian Wang
Qiaosong Wang
Qifei Wang
Qilong Wang
Qing Wang
Qingzhong Wang
Quan Wang
Rui Wang
Ruiping Wang
Ruixing Wang
Shangfei Wang
Shenlong Wang
Shiyao Wang
Shuhui Wang
Song Wang

Tao Wang
Tianlu Wang
Tiantian Wang
Ting-chun Wang
Tingwu Wang
Wei Wang
Weiyue Wang
Wenguan Wang
Wenlin Wang
Wenqi Wang
Xiang Wang
Xiaobo Wang
Xiaofang Wang
Xiaoling Wang
Xiaolong Wang
Xiaosong Wang
Xiaoyu Wang
Xin Eric Wang
Xinchao Wang
Xinggang Wang
Xintao Wang
Yali Wang
Yan Wang
Yang Wang
Yangang Wang
Yaxing Wang
Yi Wang
Yida Wang
Yilin Wang
Yiming Wang
Yisen Wang
Yongtao Wang
Yu-Xiong Wang
Yue Wang
Yujiang Wang
Yunbo Wang
Yunhe Wang
Zengmao Wang
Zhangyang Wang
Zhaowen Wang
Zhe Wang
Zhecan Wang
Zheng Wang
Zhixiang Wang
Zilei Wang
Jianqiao Wangni

Anne S. Wannenwetsch
Jan Dirk Wegner
Scott Wehrwein
Donglai Wei
Kaixuan Wei
Longhui Wei
Pengxu Wei
Ping Wei
Qi Wei
Shih-En Wei
Xing Wei
Yunchao Wei
Zijun Wei
Jerod Weinman
Michael Weinmann
Philippe Weinzaepfel
Yair Weiss
Bihan Wen
Longyin Wen
Wei Wen
Junwu Weng
Tsui-Wei Weng
Xinshuo Weng
Eric Wengrowski
Tomas Werner
Gordon Wetzstein
Tobias Weyand
Patrick Wieschollek
Maggie Wigness
Erik Wijmans
Richard Wildes
Olivia Wiles
Chris Williams
Williem Williem
Kyle Wilson
Calden Wloka
Nicolai Wojke
Christian Wolf
Yongkang Wong
Sanghyun Woo
Scott Workman
Baoyuan Wu
Bichen Wu
Chao-Yuan Wu
Huikai Wu
Jiajun Wu

Jialin Wu
Jiaxiang Wu
Jiqing Wu
Jonathan Wu
Lifang Wu
Qi Wu
Qiang Wu
Ruizheng Wu
Shangzhe Wu
Shun-Cheng Wu
Tianfu Wu
Wayne Wu
Wenxuan Wu
Xiao Wu
Xiaohe Wu
Xinxiao Wu
Yang Wu
Yi Wu
Yiming Wu
Ying Nian Wu
Yue Wu
Zheng Wu
Zhenyu Wu
Zhirong Wu
Zuxuan Wu
Stefanie Wuhrer
Jonas Wulff
Changqun Xia
Fangting Xia
Fei Xia
Gui-Song Xia
Lu Xia
Xide Xia
Yin Xia
Yingce Xia
Yongqin Xian
Lei Xiang
Shiming Xiang
Bin Xiao
Fanyi Xiao
Guobao Xiao
Huaxin Xiao
Taihong Xiao
Tete Xiao
Tong Xiao
Wang Xiao

Yang Xiao
Cihang Xie
Guosen Xie
Jianwen Xie
Lingxi Xie
Sirui Xie
Weidi Xie
Wenxuan Xie
Xiaohua Xie
Fuyong Xing
Jun Xing
Junliang Xing
Bo Xiong
Peixi Xiong
Yu Xiong
Yuanjun Xiong
Zhiwei Xiong
Chang Xu
Chenliang Xu
Dan Xu
Danfei Xu
Hang Xu
Hongteng Xu
Huijuan Xu
Jingwei Xu
Jun Xu
Kai Xu
Mengmeng Xu
Mingze Xu
Qianqian Xu
Ran Xu
Weijian Xu
Xiangyu Xu
Xiaogang Xu
Xing Xu
Xun Xu
Yanyu Xu
Yichao Xu
Yong Xu
Yongchao Xu
Yuanlu Xu
Zenglin Xu
Zheng Xu
Chuhui Xue
Jia Xue
Nan Xue

Tianfan Xue
Xiangyang Xue
Abhay Yadav
Yasushi Yagi
I. Zeki Yalniz
Kota Yamaguchi
Toshihiko Yamasaki
Takayoshi Yamashita
Junchi Yan
Ke Yan
Qingan Yan
Sijie Yan
Xinchen Yan
Yan Yan
Yichao Yan
Zhicheng Yan
Keiji Yanai
Bin Yang
Ceyuan Yang
Dawei Yang
Dong Yang
Fan Yang
Guandao Yang
Guorun Yang
Haichuan Yang
Hao Yang
Jianwei Yang
Jiaolong Yang
Jie Yang
Jing Yang
Kaiyu Yang
Linjie Yang
Meng Yang
Michael Ying Yang
Nan Yang
Shuai Yang
Shuo Yang
Tianyu Yang
Tien-Ju Yang
Tsun-Yi Yang
Wei Yang
Wenhan Yang
Xiao Yang
Xiaodong Yang
Xin Yang
Yan Yang

Yanchao Yang
Yee Hong Yang
Yezhou Yang
Zhenheng Yang
Anbang Yao
Angela Yao
Cong Yao
Jian Yao
Li Yao
Ting Yao
Yao Yao
Zhewei Yao
Chengxi Ye
Jianbo Ye
Keren Ye
Linwei Ye
Mang Ye
Mao Ye
Qi Ye
Qixiang Ye
Mei-Chen Yeh
Raymond Yeh
Yu-Ying Yeh
Sai-Kit Yeung
Serena Yeung
Kwang Moo Yi
Li Yi
Renjiao Yi
Alper Yilmaz
Junho Yim
Lijun Yin
Weidong Yin
Xi Yin
Zhichao Yin
Tatsuya Yokota
Ryo Yonetani
Donggeun Yoo
Jae Shin Yoon
Ju Hong Yoon
Sung-eui Yoon
Laurent Younes
Changqian Yu
Fisher Yu
Gang Yu
Jiahui Yu
Kaicheng Yu

Ke Yu
Lequan Yu
Ning Yu
Qian Yu
Ronald Yu
Ruichi Yu
Shoou-I Yu
Tao Yu
Tianshu Yu
Xiang Yu
Xin Yu
Xiyu Yu
Youngjae Yu
Yu Yu
Zhiding Yu
Chunfeng Yuan
Ganzhao Yuan
Jinwei Yuan
Lu Yuan
Quan Yuan
Shanxin Yuan
Tongtong Yuan
Wenjia Yuan
Ye Yuan
Yuan Yuan
Yuhui Yuan
Huanjing Yue
Xiangyu Yue
Ersin Yumer
Sergey Zagoruyko
Egor Zakharov
Amir Zamir
Andrei Zanfir
Mihai Zanfir
Pablo Zegers
Bernhard Zeisl
John S. Zelek
Niclas Zeller
Huayi Zeng
Jiabei Zeng
Wenjun Zeng
Yu Zeng
Xiaohua Zhai
Fangneng Zhan
Huangying Zhan
Kun Zhan

Xiaohang Zhan
Baochang Zhang
Bowen Zhang
Cecilia Zhang
Changqing Zhang
Chao Zhang
Chengquan Zhang
Chi Zhang
Chongyang Zhang
Dingwen Zhang
Dong Zhang
Feihu Zhang
Hang Zhang
Hanwang Zhang
Hao Zhang
He Zhang
Hongguang Zhang
Hua Zhang
Ji Zhang
Jianguo Zhang
Jianming Zhang
Jiawei Zhang
Jie Zhang
Jing Zhang
Juyong Zhang
Kai Zhang
Kaipeng Zhang
Ke Zhang
Le Zhang
Lei Zhang
Li Zhang
Lihe Zhang
Linguang Zhang
Lu Zhang
Mi Zhang
Mingda Zhang
Peng Zhang
Pingping Zhang
Qian Zhang
Qilin Zhang
Quanshi Zhang
Richard Zhang
Rui Zhang
Runze Zhang
Shengping Zhang
Shifeng Zhang

Shuai Zhang
Songyang Zhang
Tao Zhang
Ting Zhang
Tong Zhang
Wayne Zhang
Wei Zhang
Weizhong Zhang
Wenwei Zhang
Xiangyu Zhang
Xiaolin Zhang
Xiaopeng Zhang
Xiaoqin Zhang
Xiuming Zhang
Ya Zhang
Yang Zhang
Yimin Zhang
Yinda Zhang
Ying Zhang
Yongfei Zhang
Yu Zhang
Yulun Zhang
Yunhua Zhang
Yuting Zhang
Zhanpeng Zhang
Zhao Zhang
Zhaoxiang Zhang
Zhen Zhang
Zheng Zhang
Zhifei Zhang
Zhijin Zhang
Zhishuai Zhang
Ziming Zhang
Bo Zhao
Chen Zhao
Fang Zhao
Haiyu Zhao
Han Zhao
Hang Zhao
Hengshuang Zhao
Jian Zhao
Kai Zhao
Liang Zhao
Long Zhao
Qian Zhao
Qibin Zhao

Qijun Zhao
Rui Zhao
Shenglin Zhao
Sicheng Zhao
Tianyi Zhao
Wenda Zhao
Xiangyun Zhao
Xin Zhao
Yang Zhao
Yue Zhao
Zhichen Zhao
Zijing Zhao
Xiantong Zhen
Chuanxia Zheng
Feng Zheng
Haiyong Zheng
Jia Zheng
Kang Zheng
Shuai Kyle Zheng
Wei-Shi Zheng
Yinqiang Zheng
Zerong Zheng
Zhedong Zheng
Zilong Zheng
Bineng Zhong
Fangwei Zhong
Guangyu Zhong
Yiran Zhong
Yujie Zhong
Zhun Zhong
Chunluan Zhou
Huiyu Zhou
Jiahuan Zhou
Jun Zhou
Lei Zhou
Luowei Zhou
Luping Zhou
Mo Zhou
Ning Zhou
Pan Zhou
Peng Zhou
Qianyi Zhou
S. Kevin Zhou
Sanping Zhou
Wengang Zhou
Xingyi Zhou

Yanzhao Zhou
Yi Zhou
Yin Zhou
Yipin Zhou
Yuyin Zhou
Zihan Zhou
Alex Zihao Zhu
Chenchen Zhu
Feng Zhu
Guangming Zhu
Ji Zhu
Jun-Yan Zhu
Lei Zhu
Linchao Zhu
Rui Zhu
Shizhan Zhu
Tyler Lixuan Zhu

Wei Zhu
Xiangyu Zhu
Xinge Zhu
Xizhou Zhu
Yanjun Zhu
Yi Zhu
Yixin Zhu
Yizhe Zhu
Yousong Zhu
Zhe Zhu
Zhen Zhu
Zheng Zhu
Zhenyao Zhu
Zhihui Zhu
Zhuotun Zhu
Bingbing Zhuang
Wei Zhuo

Christian Zimmermann
Karel Zimmermann
Larry Zitnick
Mohammadreza
    Zolfaghari
Maria Zontak
Daniel Zoran
Changqing Zou
Chuhang Zou
Danping Zou
Qi Zou
Yang Zou
Yuliang Zou
Georgios Zoumpourlis
Wangmeng Zuo
Xinxin Zuo

Additional Reviewers

Victoria Fernandez
    Abrevaya
Maya Aghaei
Allam Allam
Christine
    Allen-Blanchette
Nicolas Aziere
Assia Benbihi
Neha Bhargava
Bharat Lal Bhatnagar
Joanna Bitton
Judy Borowski
Amine Bourki
Romain Brégier
Tali Brayer
Sebastian Bujwid
Andrea Burns
Yun-Hao Cao
Yuning Chai
Xiaojun Chang
Bo Chen
Shuo Chen
Zhixiang Chen
Junsuk Choe
Hung-Kuo Chu

Jonathan P. Crall
Kenan Dai
Lucas Deecke
Karan Desai
Prithviraj Dhar
Jing Dong
Wei Dong
Turan Kaan Elgin
Francis Engelmann
Erik Englesson
Fartash Faghri
Zicong Fan
Yang Fu
Risheek Garrepalli
Yifan Ge
Marco Godi
Helmut Grabner
Shuxuan Guo
Jianfeng He
Zhezhi He
Samitha Herath
Chih-Hui Ho
Yicong Hong
Vincent Tao Hu
Julio Hurtado

Jaedong Hwang
Andrey Ignatov
Muhammad
    Abdullah Jamal
Saumya Jetley
Meiguang Jin
Jeff Johnson
Minsoo Kang
Saeed Khorram
Mohammad Rami Koujan
Nilesh Kulkarni
Sudhakar Kumawat
Abdelhak Lemkhenter
Alexander Levine
Jiachen Li
Jing Li
Jun Li
Yi Li
Liang Liao
Ruochen Liao
Tzu-Heng Lin
Phillip Lippe
Bao-di Liu
Bo Liu
Fangchen Liu

Hanxiao Liu
Hongyu Liu
Huidong Liu
Miao Liu
Xinxin Liu
Yongfei Liu
Yu-Lun Liu
Amir Livne
Tiange Luo
Wei Ma
Xiaoxuan Ma
Ioannis Marras
Georg Martius
Effrosyni Mavroudi
Tim Meinhardt
Givi Meishvili
Meng Meng
Zihang Meng
Zhongqi Miao
Gyeongsik Moon
Khoi Nguyen
Yung-Kyun Noh
Antonio Norelli
Jaeyoo Park
Alexander Pashevich
Mandela Patrick
Mary Phuong
Bingqiao Qian
Yu Qiao
Zhen Qiao
Sai Saketh Rambhatla
Aniket Roy
Amelie Royer
Parikshit Vishwas
    Sakurikar
Mark Sandler
Mert Bülent Sarıyıldız
Tanner Schmidt
Anshul B. Shah

Ketul Shah
Rajvi Shah
Hengcan Shi
Xiangxi Shi
Yujiao Shi
William A. P. Smith
Guoxian Song
Robin Strudel
Abby Stylianou
Xinwei Sun
Reuben Tan
Qingyi Tao
Kedar S. Tatwawadi
Anh Tuan Tran
Son Dinh Tran
Eleni Triantafillou
Aristeidis Tsitiridis
Md Zasim Uddin
Andrea Vedaldi
Evangelos Ververas
Vidit Vidit
Paul Voigtlaender
Bo Wan
Huanyu Wang
Huiyu Wang
Junqiu Wang
Pengxiao Wang
Tai Wang
Xinyao Wang
Tomoki Watanabe
Mark Weber
Xi Wei
Botong Wu
James Wu
Jiamin Wu
Rujie Wu
Yu Wu
Rongchang Xie
Wei Xiong

Yunyang Xiong
An Xu
Chi Xu
Yinghao Xu
Fei Xue
Tingyun Yan
Zike Yan
Chao Yang
Heran Yang
Ren Yang
Wenfei Yang
Xu Yang
Rajeev Yasarla
Shaokai Ye
Yufei Ye
Kun Yi
Haichao Yu
Hanchao Yu
Ruixuan Yu
Liangzhe Yuan
Chen-Lin Zhang
Fandong Zhang
Tianyi Zhang
Yang Zhang
Yiyi Zhang
Yongshun Zhang
Yu Zhang
Zhiwei Zhang
Jiaojiao Zhao
Yipu Zhao
Xingjian Zhen
Haizhong Zheng
Tiancheng Zhi
Chengju Zhou
Hao Zhou
Hao Zhu
Alexander Zimin

Contents – Part V

BLSM: A Bone-Level Skinned Model of the Human Mesh

Haoyang Wang[1,2(✉)], Riza Alp Güler[1,2], Iasonas Kokkinos[1],
George Papandreou[1], and Stefanos Zafeiriou[1,2]

[1] Ariel AI, London, UK
{hwang,alpguler,iasonas,gpapan,szafeiriou}@arielai.com
[2] Imperial College London, London, UK
{haoyang.wang15,r.guler,s.zafeiriou}@imperial.ac.uk

Abstract. We introduce BLSM, a bone-level skinned model of the human body mesh where bone scales are set prior to template synthesis, rather than the common, inverse practice. BLSM first sets bone lengths and joint angles to specify the skeleton, then specifies identity-specific surface variation, and finally bundles them together through linear blend skinning. We design these steps by constraining the joint angles to respect the kinematic constraints of the human body and by using accurate mesh convolution-based networks to capture identity-specific surface variation.

We provide quantitative results on the problem of reconstructing a collection of 3D human scans, and show that we obtain improvements in reconstruction accuracy when comparing to a SMPL-type baseline. Our decoupled bone and shape representation also allows for out-of-box integration with standard graphics packages like Unity, facilitating full-body AR effects and image-driven character animation. Additional results and demos are available from the project webpage: http://arielai.com/blsm.

Keywords: 3D human body modelling · Graph convolutional networks

1 Introduction

Mesh-level representations of the human body form a bridge between computer graphics and computer vision, facilitating a broad array of applications in motion capture, monocular 3D reconstruction, human synthesis, character animation, and augmented reality. The articulated human body deformations can be captured by rigged modelling where a skeleton animates a template shape; this is used in all graphics packages for human modelling and animation, and also in state-of-the-art statistical models such as SMPL or SCAPE [6,23].

Our work aims at increasing the accuracy of data-driven rigged mesh representations. Our major contribution consists in revisiting the template synthesis

Electronic supplementary material The online version of this chapter (https://doi.org/10.1007/978-3-030-58558-7_1) contains supplementary material, which is available to authorized users.

© Springer Nature Switzerland AG 2020
A. Vedaldi et al. (Eds.): ECCV 2020, LNCS 12350, pp. 1–17, 2020.
https://doi.org/10.1007/978-3-030-58558-7_1

process prior to rigging. Current models, such as SMPL, first synthesize the template mesh in a canonical pose through an expansion on a linear basis. The skeleton joints are then estimated post-hoc by regressing from the synthesized mesh to the joints. Our approach instead disentangles bone length variability from acquired body traits dependent e.g. on exercise or dietary habits.

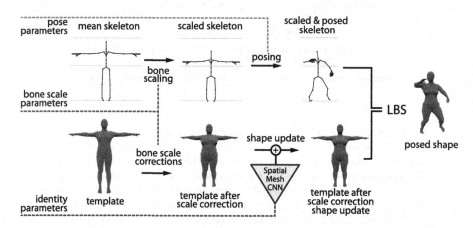

Fig. 1. Overview of our Bone-Level Skinned Model (BLSM): The top row shows skeleton synthesis: starting from a canonical, bind pose, we first scale the bone lengths and then apply an articulated transformation. The bottom row shows shape control: the canonical mesh template is affected by the bone scaling transform through Bone-Scaling Blend Shapes, and then further updated to capture identity-specific shape variation. The skeleton drives the deformation of the resulting template through Linear Blend Skinning, yielding the posed shape.

Based on this, we first model bone length-driven mesh variability in isolation, and then combine it with identity-specific updates to represent the full distribution of bodies. As we show experimentally, this disentangled representation results in more compact models, allowing us to obtain highly-accurate reconstructions with a low parameter count.

Beyond this intuitive motivation, decoupling bone lengths from identity-specific variation is important when either is fixed; e.g. when re-targeting an outfit to a person we can scale the rigged outfit's lengths to match those of the person, while preserving the bone length-independent part of the outfit shape. In particular, we model the mesh synthesis as the sequential specification of identity-specific bone length, pose-specific joint angles, and identity-specific surface variation, bundled together through linear blend skinning.

We further control and strengthen the individual components of this process: Firstly, we constrain joint angles to respect the kinematic constraints of human body, reducing body motion to 43 pose atoms, amounting to joint rotations around a single axis. Alternative techniques require either restricting the form of the regressor [14] or penalizing wrong estimates through adversarial training [18].

Secondly, we introduce accurate mesh convolution-based networks to capture identity-specific surface variation. We show that these largely outperform their linear basis counterparts, demonstrating for the first time the merit of mesh convolutions in rigged full-body modelling (earlier works [21] were applied to the setup of the face mesh).

We provide quantitative results on the problem of reconstructing a collection of 3D human scans, and show that we obtain systematic gains in average vertex reconstruction accuracy when comparing to a SMPL-type baseline. We note that this is true even though we do not use the pose-corrective blendhapes of [23]; these can be easily integrated, but we leave this for future work.

Beyond quantitative evaluation, we also show that our decoupled bone and shape representation facilitates accurate character animation in-the-wild. Our model formulation allows for out-of-box integration with standard graphics packages like Unity, leading to full-body augmented reality experiences.

2 Related Work

3D Human Body Modelling. Linear Blend Skinning (LBS) is widely used to model 3D human bodies due to its ability to represent articulated motions. Some early works have focused on synthesizing realistic 3D humans by modifying the LBS formulation. PSD [20] defines deformations as a function of articulated pose. [3] use the PSD approach learned from 3D scans of real human bodies. Other authors have focused on learning parametric model of human body shapes independently from the pose [2,25,30]. Following these works, [6,11,13,15,16] model both body shape and pose changes with triangle deformations. These work has been extended to also model dynamic soft-tissue motion [26].

Closely related to [4], SMPL [23] propose an LBS-based statistical model of the human mesh, working directly on a vertex coordinate space: T-posed shapes are first generated from a PCA-based basis, and then posed after updating joint locations. More recent works have focused on improving the representational power of the model by combining part models, e.g. for face and hands [17,29], without however modifying the body model. One further contribution of [23] consists in handling artefacts caused by LBS around the joints when posing the template through the use of pose-corrective blend shapes [23]. Our formulation can be easily extended to incorporate these, but in this work we focus on our main contribution which is modelling of the shape at the bone level.

Graph Convolutions for 3D Human Bodies. Different approaches have been proposed to extend convolutional neural networks to non-euclidean data such as graphs and manifolds [9,10,12,22,27,32]. Among these, [9,10,22,32] have attempted to model and reconstruct 3D human bodies using convolutional operators defined on meshes. While these methods achieve good performance on shape reconstruction and learning correspondences, their generalisation is not comparable to LBS based methods. Furthermore, the process of synthesising new articulated bodies using mesh convolutional networks is not easy to control since the latent vector typically encodes both shape and pose information.

3   Bone-Level Skinned Model

We start with a high-level overview, before presenting in detail the components of our approach. As shown in Fig. 1, when seen as a system, our model takes as input bone scales, joint angles, and shape coefficients and returns an array of 3-D vertex locations. In particular, BLSM operates along two streams, whose results are combined in the last stage. The upper stream, detailed in Subsect. 3.1, determines the internal skeleton by first setting the bone scales through bone scaling coefficients c_b, delivering a bind pose. This is in turn converted to a new pose by specifying joint angles θ, yielding the final skeleton $T(c_b, \theta)$.

The bottom stream, detailed in Subsect. 3.2, models the person-specific template synthesis process: Starting from a mesh corresponding to an average body type, $\bar{V}$, we first absorb the impact of bone scaling by adding a shape correction term, V_b. This is in turn augmented by an identity-specific shape update V_s, modelled by a mesh-convolutional network. The person template is obtained as

$$V = \bar{V} + V_b + V_s. \tag{1}$$

Finally, we bundle the results of these two streams using Linear Blend Skinning, as described in Sect. 3.3, delivering the posed template $\hat{V}$:

$$\hat{V} = LBS(V, T(c_b, \theta)). \tag{2}$$

3.1   Skeleton Modeling

Kinematic Model. Our starting point for human mesh modelling is the skeleton. As is common in graphics, the skeleton is determined by a tree-structured graph that ties together human bones through joint connections.

Starting with a single bone, its 'bind pose' is expressed by a template rotation matrix R^t and translation vector O^t that indicate the displacement and rotation between the coordinate systems at the two bone joints. We model the transformation with respect to the bind pose through a rotation matrix R and a scaling factor s, bundled together in a 4×4 matrix T:

$$T = \underbrace{\begin{bmatrix} sI & 0 \\ 0 & 1 \end{bmatrix} \begin{bmatrix} R & 0 \\ 0 & 1 \end{bmatrix}}_{\text{deformation}} \underbrace{\begin{bmatrix} R^t & O^t \\ 0 & 1 \end{bmatrix}}_{\text{resting bone}}. \tag{3}$$

We note that common models for character modelling use $s = 1$ and only allow for limb rotation. Any change in object scale, or bone length is modelled by modifying the displacement at the bind pose, O. This is done only implicitly, by regressing the bind pose joints from a 3D synthesized shape. By contrast, our approach gives us a handle on the scale of a limb through the parameter s, making the synthesis of the human skeleton explicitly controllable.

The full skeleton is constructed recursively, propagating from the root node to the leaf nodes along a kinematic chain. Every bone transformation encodes a displacement, rotation, and scaling between two adjacent bones, i and j, where i is the parent and j is the child node. To simplify notation, we will describe the modelling along a single kinematic chain, meaning $j = i+1$, and denote the local transformation of a bone by $\mathbf{T}^i$. The global transformation $\mathbf{T}_j$ from the local coordinates of bone j to world coordinates is given by: $\mathbf{T}_j = \prod_{i \le j} \mathbf{T}^i$, where we compose the transformations for every bone on the path from the root to the j-th node. This product accumulates the effects of consecutive transformations: for instance a change in the scale of a bone will incur the same scaling for all of its descendants. These descendants can in turn have their own scale parameters, which are combined with those of their ancestors. The 3D position of each bone j can be read from the last column of $\mathbf{T}_j$, while the upper-left 3×3 part of $\mathbf{T}_j$ provides the scaling and orientation of its coordinate system.

Parametric Bone Scaling. We model human proportions by explicitly scaling each bone. For this we perform PCA on bone lengths, as detailed in Sect. 4.2 and use the resulting principal components to express individual bone scales as:

$$\mathbf{b} = \bar{\mathbf{b}} + \mathbf{c}_b \mathbf{P}_b \qquad (4)$$

where $\mathbf{c}_b$ are the bone scaling coefficients, $\mathbf{P}_b$ is the bone-scaling matrix, and $\bar{\mathbf{b}}$ is the mean bone scale.

From Eq. 4 we obtain individual bone scales. However, the bone scales s that appear in Eq. 3 are meant to be used through the kinematic chain recursion, meaning that the product of parent scales delivers the actual bone scale, $\mathbf{b}_j = \prod_{i \le j} s_i$; this can be used to transform the predictions of Eq. 4 into a form that can be used in Eq. 3:

$$s_i = \begin{cases} \mathbf{b}_i/\mathbf{b}_{i-1}, \, i > 0 \\ 1 \qquad i = 0 \end{cases} \qquad (5)$$

Kinematically Feasible Posing. We refine our modeling of joint angles to account for the kinematic constraints of the human body. For instance the knee has one degree of freedom, the wrist has two, and the neck has three. For each joint we set the invalid degrees of freedom to be identically equal to zero, and constrain the remaining angles to be in a plausble range (e.g. $\pm45°$ for an elbow). In Fig. 2 we show sample meshes synthesized by posing a template along one valid degree of freedom.

For this, for each such degree of freedom we use an unconstrained variable $x \in R$ and map it to a valid Euler angle $\theta \in [\theta_{min}, \theta_{max}]$ by using a hyperbolic tangent unit:

$$\theta = \frac{\theta_{max} - \theta_{min}}{2} \tanh(x) + \frac{\theta_{min} + \theta_{max}}{2} \qquad (6)$$

This allows us to perform unconstrained optimization when fitting our model to data, while delivering kinematically feasible poses. The resulting per-joint Euler angles are converted into a rotation matrix, delivering the matrix R in Eq. 3.

Fig. 2. 7 out of the 47 degrees of freedom corresponding to kinematically feasible joint rotations for our skeleton.

Using Eq. 6 alleviates the need for restricting the regressor form [14] or adversarial training [18], while at the same time providing us with a compact, interpretable dictionary of 47 body motions. We provide samples of all such motions in the supplemental material.

3.2   Template Synthesis

Having detailed skeleton posing, we now turn to template synthesis. We start by modeling the effect of bone length on body shape, and then turn to modelling identity-specific variability.

Bone-Dependent Shape Variations. Bone length can be used to account for a substantial part of body shape variability. For example, longer bones correlate with a male body-shape, while limb proportions can correlate with ectomorph, endomorph and mesomorph body-type variability. We represent the bone-length dependent deformation of the template surface through a linear update:

$$\mathbf{V}_b = \mathbf{c}_b \mathbf{P}_{bc} \tag{7}$$

where $\mathbf{P}_{bc}$ is the matrix of bone-corrective blendshapes (Fig. 3).

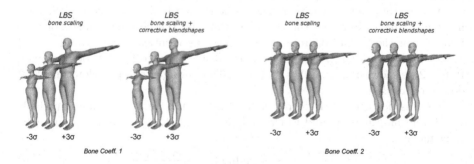

Fig. 3. Impact of bone length variation on the template. Plain linear blend skinning results in artifacts. The linear, bone-corrective blendshapes eliminate these artifacts, and capture correlations of bone lengths with gender and body type.

Graph Convolutional Shape Modelling. Having accounted for the bone length-dependent part of shape variability, we turn to the remainder of the person-specific variability. The simplest approach is to use a linear update:

$$\mathbf{V}_s = \mathbf{c}_s \mathbf{P}_s \tag{8}$$

where $\mathbf{c}_s$ are the shape coefficients, and $\mathbf{P}_s$ is the matrix of shape components; we refer to this baseline as the linear model. By contrast, we propose a more powerful, mesh-convolutional update. For this we use multi-layer mesh convolution decoder that precisely models the nonlinear manifold of plausible shapes in its output space.

We represent the triangular mesh as a graph $(\mathbf{V}, \mathcal{E})$ with vertices $\mathbf{V}$ and edges $\mathcal{E}$ and denote the convolution operator on a mesh as:

$$(f \star g)_x = \sum_{l=1}^{L} g_l f(x_l) \tag{9}$$

where g_l denotes the filter weights, and $f(x_l)$ denotes the feature of the l-th neighbour of vertex x. The neighbours are ordered consistently across all meshes, allowing us to construct a one-to-one mapping between the neighbouring features and the filter weights. Here we adapt the setting of [10], where the ordering is defined by a spiral starting from the vertex x, followed by the d-ring of the vertex, i.e. for a vertex x, x_l is defined by the ordered sequence:

$$S(x) = \{x, R_1^1(x), R_2^1(x), ..., R_{\|R^h\|}^h\}, \tag{10}$$

where h is the patch radius and $R_j^d(x)$ is the j-th element in the d-ring.

We use a convolutional mesh decoder to model the normalised deformations from the bone-updated shape. The network consists of blocks of convolution-upsampling layers similar to [27]. We pre-compute the decimated version of the template shape with quadratic edge collapse decimation to obtain the upsampling matrix. Given the latent vector $\mathbf{z}$, shape variation is represented as

$$\mathbf{V}_s = \mathcal{D}(\mathbf{z}) \tag{11}$$

where $\mathcal{D}$ is the learned mesh convolutional decoder.

3.3 Linear Blend Skinning

Having detailed the skeleton and template synthesis processes, we now turn to posing the synthesized template based on the skeleton. We use Linear Blend Skinning (LBS), where the deformation of a template mesh $\mathbf{V}$ is determined by the transformations of the skeleton. We consider that the bind pose of the skeleton is described by the matrices $\hat{\mathbf{T}}_j$, where the 3D mesh vertices take their canonical values $\mathbf{v}_i \in \mathbf{V}$, while the target pose is described by $\mathbf{T}_j$.

According to LBS, each vertex is influenced by every bone j according to a weight w_{ij}; the positions of the vertices $\hat{\mathbf{v}}_i$ at the target pose are given by:

$$\hat{\mathbf{v}}_k = \sum_j w_{ij} \mathbf{T}_j \hat{\mathbf{T}}_j^{-1} \mathbf{v}_k. \tag{12}$$

In the special case where $\mathbf{T} = \hat{\mathbf{T}}$, we recover the template shape, while in the general case, Eq. 12 can be understood as first charting every point $\mathbf{v}_k$ with respect to the bind bone (by multiplying it with $\hat{\mathbf{T}}_j^{-1}$), and then transporting to the target bone (by multiplying with $\mathbf{T}_j$).

4   Model Training

Having specified BLSM, we now turn to learning its parameters from data. For this we use the CAESAR dataset [28] to train the shape model, which contains high resolution 3D scans of 4400 subjects wearing tight clothing. This minimal complexity due to extraneous factors has made CAESAR appropriate for the estimation of statistical body models, such as SMPL. For training skinning weights we use D-FAUST [8] dataset. Our training process consists in minimising the reconstruction error of CAESAR and D-FAUST through BLSM.

Since BLSM is implemented as a multi-layer network in pytorch, one could try to directly minimize the reconstruction loss with respect to the model parameters using any standard solver. Unfortunately however, this is a nonlinear optimisation problem with multiple local minima; we therefore use a carefully engineered pipeline that solves successively demanding optimization problems, as detailed below, and use automatic differentiation to efficiently compute any derivatives required during optimization.

4.1   Unconstrained Landmark-Based Alignment

Each CAESAR scan $\mathbf{S}^n$ is associated with 73 anatomical landmarks, $\mathbf{L}^n$ that have been localised in 3D. We start by fitting our template to these landmarks by gradient descent on the joint angles $\boldsymbol{\theta}^n$ and bone scales $\mathbf{s}^n$, so as to minimize the 3D distances between the landmark positions and the respective template vertices. More specifically, the following optimization problem is solved:

$$\boldsymbol{\theta}^n, \mathbf{s}^n = \operatorname{argmin}_{\boldsymbol{\theta}, \mathbf{s}} \| \mathbf{A} \, LBS(\mathbf{V}_T, T(\mathbf{s}, \boldsymbol{\theta})) - \mathbf{L}^n \|^2 \tag{13}$$

where $\mathbf{A}$ selects the subset of landmarks from the template.

This delivers an initial fitting which we further refine by registering our BLSM-based prediction $\hat{\mathbf{S}}^n = LBS(\mathbf{V}_T, T(\mathbf{s}^n, \boldsymbol{\theta}^n))$ to each scan $\mathbf{S}^n$ with Non-Rigid ICP (NICP) [5]. This alignment stage does not use yet a statistical model to constrain the parameter estimates, and as such can be error-prone; the following steps recover shapes that are more regularized, but the present result acts like a proxy to the scan that is in correspondence with the template vertices.

4.2   Bone Basis and Bone-Corrective Blendshapes

We start learning our model by estimating a linear basis for bone scales. For each shape $\hat{\mathbf{S}}^n$ we estimate the lengths of the bones obtained during the optimization process described in the previous section.

We perform PCA on the full set of CAESAR subjects and observe that linear bases capture 97% of bone length variability on the first three eigenvectors. We convert the PCA-based mean vector and basis results from bone lengths into the mean bone scaling factor $\bar{\mathbf{b}}$ and bone scaling basis $\mathbf{P}_b$ used in Eq. 4 by dividing them by the mean length of the respective bone along each dimension.

Having set the bone scaling basis, we use it as a regularizer to re-estimate the pose $\boldsymbol{\theta}^n$ and bone scale coefficients $\mathbf{c}_b^n$ used to match our template $\mathbf{V}_T$ to each registration $\hat{\mathbf{S}}^n$ by solving the following optimisation problem:

$$\boldsymbol{\theta}^n, \mathbf{c}_b^n = \mathrm{argmin}_{\boldsymbol{\theta}, \mathbf{c}_b} \| LBS(\mathbf{V}_T, T(\mathbf{c}_b, \boldsymbol{\theta})) - \hat{\mathbf{S}}^n \|^2 \tag{14}$$

Finally we optimize over the bone-corrective basis $\mathbf{P}_b$ and mean shape $\bar{\mathbf{V}}$:

$$\mathbf{P}_{bc}^*, \bar{\mathbf{V}}^* = \mathrm{argmin}_{\mathbf{P}_b, \bar{\mathbf{V}}} \sum_{n=1}^{N} \| LBS(\bar{\mathbf{V}} + \mathbf{c}_b^n \mathbf{P}_{bc}, T(\mathbf{c}_b^n, \boldsymbol{\theta}^n)) - \hat{\mathbf{S}}^n \|^2 \tag{15}$$

Given that $\mathbf{V}_T$ and $\hat{\mathbf{S}}^n$ are in one-to-one correspondence, we no longer need ICP to optimize Eq. 14 and Eq. 15, allowing us instead to exploit automatic differentiation and GPU computation for gradient descent-based optimization.

4.3   Shape Blendshapes

Once bone-corrected blendshapes have been used to improve the fit of our model to the registered shape $\hat{\mathbf{S}}^n$, the residual in the reconstruction is attributed only to identity-specific shape variability. We model these residuals as vertex displacements $\mathbf{V_D}^n$ and estimate them for each registration $\hat{\mathbf{S}}^n$ by setting:

$$LBS(\bar{\mathbf{V}} + \mathbf{V}_b^n + \mathbf{V_D}^n, T(\mathbf{c}_b^n, \boldsymbol{\theta}^n)) = \hat{\mathbf{S}}^n \tag{16}$$

to ensure that the residual is defined in the T-pose coordinate system.

For the linear alternative described in Subsect. 3.2 the shape basis $\mathbf{P}_s$ is computed by performing PCA analysis of $\{\mathbf{V_D}^n\}$. To train the graph convolutional system described in Subsect. 3.2, we learn the parameters of the spiral mesh convolutional decoder $\mathcal{D}$ and the latent vectors $\mathbf{z}^n$ that minimize the following loss:

$$\mathrm{argmin}_{\mathcal{D}, \mathbf{z}} \sum_{n=1}^{N} \| \mathbf{V_D}^n - \mathcal{D}(\mathbf{z}^n) \| \tag{17}$$

4.4 Blending Weights

So far the blending weights of our LBS formulation are manually initialised, which can be further improved from the data. For this purpose we use the D-FAUST dataset [8], which contains registrations of a variety of identity and poses. For each registration $\mathbf{S}^n$ in the dataset, we first estimate the parameters of our model, namely c_b^n, c_s^n, $\boldsymbol{\theta}^n$, as well as the residual $\hat{\mathbf{V}}_D^n$ which is the error on the T-pose coordinate system after taking into account the shape blendshapes. Then we optimize instead the blending weights to minimize the following error:

$$\text{argmin}_{\mathbf{W}} \sum_{n=1}^{N} \|\mathbf{S}_n - LBS_{\mathbf{W}}(\bar{\mathbf{V}} + \mathbf{V}_b^n + \mathbf{V}_s^n + \hat{\mathbf{V}}_{\mathbf{D}}^n, T(c_b^n, \boldsymbol{\theta}^n))\|^2 \qquad (18)$$

where we use the mapping:

$$\mathbf{W} = \frac{f(\mathbf{W}')}{\sum_j f(\mathbf{W}')_{ij}} \quad \text{with} \quad f(\mathbf{X}) = \sqrt{\mathbf{X}^2 + \varepsilon} \qquad (19)$$

to optimize freely $\mathbf{W}'$ while ensuring the output weights $\mathbf{W}$ satisfy the LBS blending weights constraints: $\sum_j \mathbf{W}_{ij} = \mathbf{1}$, and $mathbf{W}_{ij} \geq 0$.

5 Evaluation

5.1 Implementation Details

Baseline Implementation. The publicly available SMPL model [23] has 10 shape bases, a mesh topology that is different to that of our model, and pose-corrective blendshapes, making any direct comparison to our model inconclusive.

In order to have directly comparable results across multiple shape coefficient dimensionalities we train a SMPL-like model (referred to as SMPL-reimpl) using the mesh topology, kinematic structure, and blending weight implementation of our model, and SMPL's PCA-based modeling of shape variability in the T-pose. We further remove any pose-corrective blendshape functionality, allowing us to directly assess the impact of our disentangled, bone-driven modeling of mesh variability against a baseline that does not use it.

In order to train SMPL-reimpl, we first define manually the joint regressor required by [23] by taking the mean of the ring of vertex that lies around a certain joint; we then train the blending weights and joint regressor on the D-FAUST dataset, as described in [23]. The shape blendshapes are then trained with the CAESAR dataset using the same method described in [23].

We further note that our evaluations focus on the gender-neutral versions of both SMPL-reimpl and BLSM - and may therefore be skewed in favour of BLSM's ability to easily capture large-scale, gender-dependent bone variations. This is however relevant to the performance of most downstream, CNN-driven human mesh reconstruction systems that do not know in advance the subject's gender [14,18,19], and have therefore adopted the neutral model. More recent work [24] has shown improvements based on exploiting gender attributes in tandem with the gender-specific SMPL models. We leave a more thorough ablation of the interplay between gender and reconstruction accuracy in future work.

Mesh Convolutional Networks. For graph convolutional shape modelling, we train networks with 4 convolutional layers, with $(48, 32, 32, 16)$ filters for each layer, respectively. Convolutional layers are followed by batch normalisation and upsampling layers with factors $(2, 2, 2, 4)$ respectively. For the convolutional layers, we use ELU as the activation function. Finally, the output layer is a convolutional layer with filter size 3 and linear activation, which outputs the normalised vertex displacements. We train our network with an Adam optimiser, with a learning rate of 1e-3 and weight decay of 5e-5 for the network parameters, and learning rate of 0.1 and weight decay 1e-7 for the latent vectors. The learning rates are multiplied by a factor of 0.99 after each epoch.

5.2 Quantitative Evaluation

We evaluate the representation power of our proposed BLSM model on the CAESAR dataset and compare its generalisation ability against the SMPL-type baseline on D-FAUST dataset and our in-house testset. D-FAUST contains 10 subjects, each doing 14 different motions. We further expand the testset with our in-house dataset. Captured with a custom-built multi-camera active stereo system (3dMD LLC, Atlanta, GA), our in-house testset consists of 4D sequences at 30 FPS of 20 individuals spanning different body types and poses. Each instance contains around 50K vertices. These scans are registered to our template as described in Subsect. 4.1, while using NICP with temporal consistency constraints.

The models that we compare are aligned to the registered meshes by minimising the L2 distance between each vertex. We use an Adam optimiser with learning rate 0.1 to optimise parameters for all models, and reduce the learning rate by a factor of 0.9 on plateau. To avoid local minima, we use a multi-stage optimization approach as in [7]. We first fit the vertices on the torso (defined by the blending weights of the torso bones on our template) by optimising over the shape coefficients and the joint angles of the torso bones. Then for second and third stage, upper-limbs and lower-limbs are added respectively. In the last stage, all the vertices are used to fine-tune the fitted parameters. In the following, we report the mean absolute vertex errors (MABS) of gender neutral models.

CAESAR. In Fig. 4 we plot the fitting errors on the CAESAR dataset as a function of the number of shape coefficients, namely shape blendshapes for SMPL-reimpl, bone blendshapes and shape blendshapes for BLSM-linear and latent space dimension for BLSM-spiral.

We observe that our BLSM-linear model attains lower reconstruction error compared to the SMPL-reimpl baseline. The sharpest decrease happens for the first three coefficients, corresponding to bone-level variability modelling. Starting from the fourth coefficient the error decreases more slowly for the linear model, but the BLSM-spiral variant further reduces errors. These results suggest that our BLSM method captures more of the shape variation with fewer coefficients compared to the SMPL-reimpl baseline.

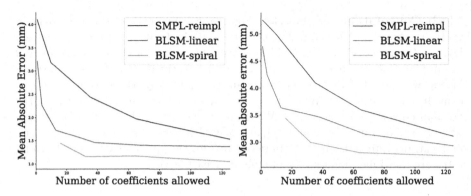

Fig. 4. Mean absolute vertex error on the CAESAR dataset (left) and our in-house testset (right) against number of shape coefficients.

D-FAUST. For D-FAUST, we select one male and one female subject (50009 and 50021) for evaluation, and the rest for training blending weights for both models and joint regressors for SMPL-reimpl. We evaluate first shape generalisation error by fitting the models to all sequences of the test subjects (Fig. 5 left). We observe that our BLSM-linear model obtain lower generalisation error compared to SMPL-reimpl baseline, and the result is improved further with BLSM-spiral.

We also evaluate the pose generalisation error of our models (Fig. 5 right). The errors are obtained by first fitting the models to one random frame of each subject, then fit the pose parameter to rest of the frames while keeping the shape coefficients fixed. This metric suggests how well a fitted shape generalise to new poses. We observe that both of our linear and spiral models generalises better than our SMPL-reimpl baseline. We argue that by introducing bone scales to the model, the fitted poses are well regularized, thus during training it is more straight forward to decouple the shape and pose variations in the dataset, while avoiding the need to learn subject specific shapes and joints as in SMPL.

In-House Testset. In Fig. 4, we also report the average MABS across all sequences in our in-house testset as a function of the number of shape coefficients used. We observe that our proposed models are able to generalise better than the SMPL-reimpl model on our testset. Our proposed models are compact and able to represent variations in our testset with a smaller number of shape coefficients than SMPL-reimpl.

In Fig. 6, we show the mean per-vertex error heatmaps on all sequences and on some example registrations in our testset. Compared to SMPL-reimpl, our proposed models are able to fit closely across the full body, while the SMPL-reimpl model produces larger error on some of the vertices. The result suggests that our proposed model generalise better on surface details than the SMPL-reimpl baseline model.

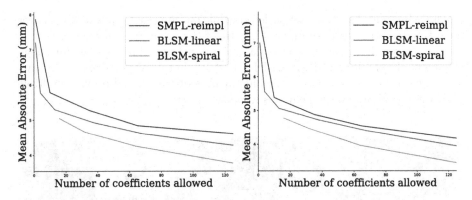

Fig. 5. Shape generalisation error (left) and pose generalisation error (right) on D-FAUST dataset against number of shape coefficients.

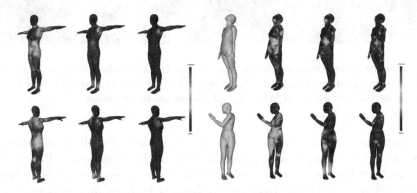

Fig. 6. Mean absolute vertex error and example of reconstructions on the testset. Left to right: SMPL-reimpl, BLSM-Linear, BLSM-Spiral. For linear models we show result with 125 coefficients allowed. For BLSM-spiral the latent size is 128.

5.3 Qualitative Evaluation

In Fig. 8 we show samples from our linear model by varying the bone bases, as well as identity-specific shape coefficients from -3σ to $+3\sigma$. We observe that our model captures a variety of body shapes and the method successfully decouples bone length-dependent variations and identities specific shape variations.

This decoupling allows us to perform simple and accurate character animation driven by persons in unconstrained environments as shown in Fig. 7. In an offline stage, we rig several characters from [1] to our model's skeleton. Given an image of a person, we first fit our model to it using a method similar to [18]. We then apply the estimated bone transformations (scales and rotations) to the rigged characters. This allows accurate image-driven character animation within any standard graphics package like Unity. Alternative methods require either solving a deformation transfer problem [31,33], fixed shape assumptions, or approximations to a constant skeleton, while our approach can exactly recover the estimated skeleton position as it is part of the mesh construction.

Please note that many recent works that predict model parameters for image alignment are applicable to our model [14,18,19]; in this work we focus on showing the merit of our model once the alignment is obtained, and provide multiple CNN-driven results on the project's webpage.

Fig. 7. Image-driven character animation: we rig two characters from [1] using our model's bone structure. This allows us to transform any person into these characters, while preserving the pose and body type of the person in the image.

We also assess the representational power of our mesh convolutional networks by examining the samples from each dimension of the latent space (Fig. 8). We observe that while capturing large deformations such as gender and body type, the network also captures details such as different body fat distributions.

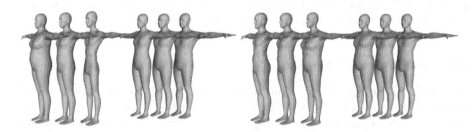

Fig. 8. Linear (left) vs. graph convolutional (right) modeling of shape variation.

6   Conclusion

In this paper we propose BLSM, a bone-level skinned model of the 3D human body mesh where bone modelling and identity-specific variations are decoupled. We introduce a data-driven approach for learning skeleton, skeleton-conditioned shape variations and identity-specific variations. Our formulation facilitates the use of mesh convolutional networks to capture identity specific variations, while explicitly modeling the range of articulated motion through built-in constraints.

We provide quantitative results showing that our model outperforms existing SMPL-like baseline on the 3D reconstruction problem. Qualitatively, we also show that by virtue of being bone-level our formulation allows us to perform accurate character retargeting in-the-wild.

Acknowledgements. The work of S. Zafeiriou and R.A. Guler was funded in part by EPSRC Project EP/S010203/1 DEFORM.

References

1. Mixamo (2019). https://www.mixamo.com
2. Allen, B., Curless, B., Curless, B., Popović, Z.: The space of human body shapes: reconstruction and parameterization from range scans. ACM Trans. Graph. (TOG) **22**, 587–594 (2003). ACM
3. Allen, B., Curless, B., Popović, Z.: Articulated body deformation from range scan data. ACM Trans. Graph. (TOG) **21**, 612–619 (2002). ACM
4. Allen, B., Curless, B., Popović, Z., Hertzmann, A.: Learning a correlated model of identity and pose-dependent body shape variation for real-time synthesis. In: Proceedings of the 2006 ACM SIGGRAPH/Eurographics Symposium on Computer Animation, pp. 147–156. Eurographics Association (2006)
5. Amberg, B., Romdhani, S., Vetter, T.: Optimal step nonrigid ICP algorithms for surface registration. In: 2007 IEEE Conference on Computer Vision and Pattern Recognition, pp. 1–8. IEEE (2007)
6. Anguelov, D., Srinivasan, P., Koller, D., Thrun, S., Rodgers, J., Davis, J.: Scape: shape completion and animation of people. ACM Trans. Graph. (TOG) **24**, 408–416 (2005)
7. Bogo, F., Kanazawa, A., Lassner, C., Gehler, P., Romero, J., Black, M.J.: Keep it SMPL: automatic estimation of 3D human pose and shape from a single image. In: Leibe, B., Matas, J., Sebe, N., Welling, M. (eds.) ECCV 2016. LNCS, vol. 9909, pp. 561–578. Springer, Cham (2016). https://doi.org/10.1007/978-3-319-46454-1_34
8. Bogo, F., Romero, J., Pons-Moll, G., Black, M.J.: Dynamic faust: registering human bodies in motion. In: Proceedings of the IEEE Conference on Computer Vision and Pattern Recognition, pp. 6233–6242 (2017)
9. Boscaini, D., Masci, J., Rodolà, E., Bronstein, M.: Learning shape correspondence with anisotropic convolutional neural networks. In: Advances in Neural Information Processing Systems, pp. 3189–3197 (2016)
10. Bouritsas, G., Bokhnyak, S., Ploumpis, S., Bronstein, M., Zafeiriou, S.: Neural 3D morphable models: spiral convolutional networks for 3D shape representation learning and generation. In: Proceedings of the IEEE International Conference on Computer Vision, pp. 7213–7222 (2019)
11. Chen, Y., Liu, Z., Zhang, Z.: Tensor-based human body modeling. In: Proceedings of the IEEE Conference on Computer Vision and Pattern Recognition, pp. 105–112 (2013)
12. Defferrard, M., Bresson, X., Vandergheynst, P.: Convolutional neural networks on graphs with fast localized spectral filtering. In: Advances in Neural Information Processing Systems, pp. 3844–3852 (2016)

13. Guan, P., Reiss, L., Hirshberg, D.A., Weiss, A., Black, M.J.: Drape: dressing any person
14. Guler, R.A., Kokkinos, I.: Holopose: holistic 3D human reconstruction in-the-wild. In: Proceedings of the IEEE Conference on Computer Vision and Pattern Recognition, pp. 10884–10894 (2019)
15. Hasler, N., Stoll, C., Sunkel, M., Rosenhahn, B., Seidel, H.P.: A statistical model of human pose and body shape. Comput. Graph. Forum **28**, 337–346 (2009)
16. Hirshberg, D.A., Loper, M., Rachlin, E., Black, M.J.: Coregistration: simultaneous alignment and modeling of articulated 3D shape. In: Fitzgibbon, A., Lazebnik, S., Perona, P., Sato, Y., Schmid, C. (eds.) ECCV 2012. LNCS, vol. 7577, pp. 242–255. Springer, Heidelberg (2012). https://doi.org/10.1007/978-3-642-33783-3_18
17. Joo, H., Simon, T., Sheikh, Y.: Total capture: a 3D deformation model for tracking faces, hands, and bodies. In: Proceedings of the IEEE Conference on Computer Vision and Pattern Recognition, pp. 8320–8329 (2018)
18. Kanazawa, A., Black, M.J., Jacobs, D.W., Malik, J.: End-to-end recovery of human shape and pose. In: Computer Vision and Pattern Recognition (CVPR) (2018)
19. Kolotouros, N., Pavlakos, G., Black, M.J., Daniilidis, K.: Learning to reconstruct 3D human pose and shape via model-fitting in the loop. In: Proceedings of the IEEE International Conference on Computer Vision, pp. 2252–2261 (2019)
20. Lewis, J.P., Cordner, M., Fong, N.: Pose space deformation: a unified approach to shape interpolation and skeleton-driven deformation. In: Proceedings of the 27th Annual Conference on Computer Graphics and Interactive Techniques, pp. 165–172. ACM Press/Addison-Wesley Publishing Co. (2000)
21. Li, T., Bolkart, T., Black, M.J., Li, H., Romero, J.: Learning a model of facial shape and expression from 4D scans. ACM Trans. Graph. (TOG) **36**(6), 194 (2017)
22. Lim, I., Dielen, A., Campen, M., Kobbelt, L.: A simple approach to intrinsic correspondence learning on unstructured 3D meshes. In: Leal-Taixé, L., Roth, S. (eds.) ECCV 2018. LNCS, vol. 11131, pp. 349–362. Springer, Cham (2019). https://doi.org/10.1007/978-3-030-11015-4_26
23. Loper, M., Mahmood, N., Romero, J., Pons-Moll, G., Black, M.J.: SMPL: a skinned multi-person linear model. ACM Trans. Graph. (TOG) **34**(6), 248 (2015)
24. Pavlakos, G., et al.: Expressive body capture: 3D hands, face, and body from a single image. In: Proceedings of the IEEE Conference on Computer Vision and Pattern Recognition (CVPR) (2019)
25. Pishchulin, L., Wuhrer, S., Helten, T., Theobalt, C., Schiele, B.: Building statistical shape spaces for 3D human modeling. Pattern Recogn. **67**, 276–286 (2017)
26. Pons-Moll, G., Romero, J., Mahmood, N., Black, M.J.: Dyna: a model of dynamic human shape in motion. ACM Trans. Graph. (TOG) **34**(4), 120 (2015)
27. Ranjan, A., Bolkart, T., Sanyal, S., Black, M.J.: Generating 3D faces using convolutional mesh autoencoders. In: Ferrari, V., Hebert, M., Sminchisescu, C., Weiss, Y. (eds.) ECCV 2018. LNCS, vol. 11207, pp. 725–741. Springer, Cham (2018). https://doi.org/10.1007/978-3-030-01219-9_43
28. Robinette, K.M., Blackwell, S., Daanen, H., Boehmer, M., Fleming, S.: Civilian American and European Surface Anthropometry Resource (CAESAR), final report, vol. 1. Summary. Technical report, SYTRONICS INC DAYTON OH (2002)
29. Romero, J., Tzionas, D., Black, M.J.: Embodied hands: modeling and capturing hands and bodies together. ACM Trans. Graph. (TOG) **36**(6), 245 (2017)
30. Seo, H., Cordier, F., Magnenat-Thalmann, N.: Synthesizing animatable body models with parameterized shape modifications. In: Proceedings of the 2003 ACM SIGGRAPH/Eurographics Symposium on Computer Animation, pp. 120–125. Eurographics Association (2003)

31. Sumner, R.W., Popović, J.: Deformation transfer for triangle meshes. ACM Trans. Graph. (TOG) **23**(3), 399–405 (2004)
32. Verma, N., Boyer, E., Verbeek, J.: Feastnet: feature-steered graph convolutions for 3D shape analysis. In: Proceedings of the IEEE Conference on Computer Vision and Pattern Recognition, pp. 2598–2606 (2018)
33. Wang, J., et al.: Neural pose transfer by spatially adaptive instance normalization. In: Proceedings of the IEEE/CVF Conference on Computer Vision and Pattern Recognition, pp. 5831–5839 (2020)

Associative Alignment for Few-Shot Image Classification

Arman Afrasiyabi[1]([✉]), Jean-François Lalonde[1], and Christian Gagné[1,2]

[1] Université Laval, Quebec City, Canada
arman.afrasiyabi.1@ulaval.ca, {jflalonde,christian.gagne}@gel.ulaval.ca
[2] Canada CIFAR AI Chair, Mila, Montreal, Canada
https://lvsn.github.io/associative-alignment/

Abstract. Few-shot image classification aims at training a model from only a few examples for each of the "novel" classes. This paper proposes the idea of associative alignment for leveraging part of the base data by aligning the novel training instances to the closely related ones in the base training set. This expands the size of the effective novel training set by adding extra "related base" instances to the few novel ones, thereby allowing a constructive fine-tuning. We propose two associative alignment strategies: 1) a metric-learning loss for minimizing the distance between related base samples and the centroid of novel instances in the feature space, and 2) a conditional adversarial alignment loss based on the Wasserstein distance. Experiments on four standard datasets and three backbones demonstrate that combining our centroid-based alignment loss results in absolute accuracy improvements of 4.4%, 1.2%, and 6.2% in 5-shot learning over the state of the art for object recognition, fine-grained classification, and cross-domain adaptation, respectively.

Keywords: Associative alignment · Few-shot image classification

1 Introduction

Despite recent progress, generalizing on new concepts with little supervision is still a challenge in computer vision. In the context of image classification, few-shot learning aims to obtain a model that can learn to recognize novel image classes when very few training examples are available.

Meta-learning [9,36,42,47] is a possible approach to achieve this, by extracting common knowledge from a large amount of labeled data (the "base" classes) to train a model that can then learn to classify images from "novel" concepts with only a few examples. This is achieved by repeatedly sampling small subsets from the large pool of base images, effectively simulating the few-shot scenario. Standard transfer learning has also been explored as an alternative method [3,14,34]. The idea is to pre-train a network on the base samples and then fine-tune the

---

Electronic supplementary material The online version of this chapter (https://doi.org/10.1007/978-3-030-58558-7_2) contains supplementary material, which is available to authorized users.

A. Vedaldi et al. (Eds.): ECCV 2020, LNCS 12350, pp. 18–35, 2020.
https://doi.org/10.1007/978-3-030-58558-7_2

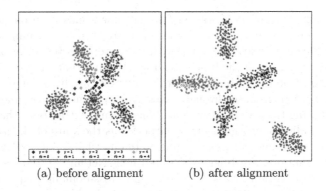

(a) before alignment   (b) after alignment

Fig. 1. The use of many related bases (circles) in addition to few novel classes samples (diamonds) allows better discriminative models: (a) using directly related bases may not properly capture the novel classes; while (b) aligning both related base and novel training instances (in the feature space) provides more relevant training data for classification. Plots are generated with t-SNE [30] applied to the ResNet-18 feature embedding before (a) and after (b) the application of the centroid alignment. Points are color-coded by class.

classification layer on the novel examples. Interestingly, Chen *et al.* [3] demonstrated that doing so performs on par with more sophisticated meta-learning strategies. It is, however, necessary to freeze the feature encoder part of the network when fine-tuning on the novel classes since the network otherwise overfits the novel examples. We hypothesize that this hinders performance and that gains could be made if the entire network is adapted to the novel categories.

In this paper, we propose an approach that simultaneously prevents overfitting without restricting the learning capabilities of the network for few-shot image classification. Our approach relies on the standard transfer learning strategy [3] as a starting point, but subsequently exploits base categories that are most similar (in the feature space) to the few novel samples to effectively provide additional training examples. We dub these similar categories the "related base" classes. Of course, the related base classes represent different concepts than the novel classes, so fine-tuning directly on them could confuse the network (see Fig. 1-(a)). The key idea of this paper is to *align*, in feature space, the novel examples with the related base samples (Fig. 1-(b)).

To this end, we present two possible solutions for associative alignment: by 1) centroid alignment, inspired by ProtoNet [42], benefits from explicitly shrinking the intra-class variations and is more stable to train, but makes the assumption that the class distribution is well-approximated by a single mode. Adversarial alignment, inspired by WGAN [1], does not make that assumption, but its train complexity is greater due to the critic network. We demonstrate, through extensive experiments, that our centroid-based alignment procedure achieves state-of-the-art performance in few-shot classification on several standard benchmarks. Similar results are obtained by our adversarial alignment, which shows the effectiveness of our associative alignment approach.

We present the following contributions. First, we propose two approaches for aligning novel to related base classes in the feature space, allowing for effective training of entire networks for few-shot image classification. Second, we introduce a strong baseline that combines standard transfer learning [3] with an additive angular margin loss [6], along with early stopping to regularize the network while pre-training on the base categories. We find that this simple baseline actually improves on the state of the art, in the best case by 3% in overall accuracy. Third, we demonstrate through extensive experiments—on four standard datasets and using three well-known backbone feature extractors—that our proposed centroid alignment significantly outperforms the state of the art in three types of scenarios: generic object recognition (gain of 1.7%, 4.4% 2.1% in overall accuracy for 5-shot on *mini*-ImageNet, tieredImageNet and FC100 respectively), fine-grained classification (1.2% on CUB), and cross-domain adaptation (6.2% from *mini*-ImageNet to CUB) using the ResNet-18 backbone.

2   Related Work

The main few-shot learning approaches can be broadly categorized into meta-learning and standard transfer learning. In addition, data augmentation and regularization techniques (typically in meta-learning) have also been used for few-shot learning. We briefly review relevant works in each category below. Note that several different computer vision problems such as object counting [58], video classification [59], motion prediction [16], and object detection [52] have been framed as few-shot learning. Here, we mainly focus on works from the image classification literature.

Meta-Learning. This family of approaches frames few-shot learning in the form of episodic training [7,9,36,39,42,46,52,54]. An episode is defined by pretending to be in a few-shot regime while training on the base categories, which are available in large quantities. Initialization- and metric-based approaches are two variations on the episodic training scheme relevant for this work. Initialization-based methods [9,10,22] learn an initial model able to adapt to few novel samples with a small number of gradient steps. In contrast, our approach performs a larger number of updates, but requires that the alignment be maintained between the novel samples and their related base examples. Metric-based approaches [2,12,21,25,27,33,42,44,45,47,53,57] learn a metric with the intent of reducing the intra-class variations while training on base categories. For example, ProtoNet [42] were proposed to learn a feature space where instances of a given class are located close to the corresponding prototype (centroid), allowing accurate distance-based classification. Our centroid alignment strategy borrows from such distance-based criteria but uses it to match the distributions in the feature space instead of building a classifier.

Standard Transfer Learning. The strategy behind this method is to pre-train a network on the base classes and subsequently fine-tune it on the novel examples [3,14,34]. Despite its simplicity, Chen *et al.* [3] recently demonstrated that such an approach could result in similar generalization performance compared to meta-learning when deep backbones are employed as feature extractors. However, they have also shown that the weights of the pre-trained feature extractor must remain frozen while fine-tuning due to the propensity for overfitting. Although the training procedure we are proposing is similar to standard fine-tuning in base categories, our approach allows the training of the entire network, thereby increasing the learned model capacity while improving classification accuracy.

Regularization Trick. Wang *et al.* [51] proposed regression networks for regularization purposes by refining the parameters of the fine-tuning model to be close to the pre-trained model. More recently, Lee *et al.* [24] exploited the implicit differentiation of a linear classifier with hinge loss and $\mathcal{L}_2$ regularization to the CNN-based feature learner. Dvornik *et al.* [8] uses an ensemble of networks to decrease the classifiers variance.

Data Augmentation. Another family of techniques relies on additional data for training in a few-shot regime, most of the time following a meta-learning training procedure [4,5,11,15,17,31,40,49,55,56]. Several ways of doing so have been proposed, including Feature Hallucination (FH) [17], which learns mappings between examples with an auxiliary generator that then hallucinates extra training examples (in the feature space). Subsequently, Wang *et al.* [49] proposed to use a GAN for the same purpose, and thus address the poor generalization of the FH framework. Unfortunately, it has been shown that this approach suffers from mode collapse [11]. Instead of generating artificial data for augmentation, others have proposed methods to take advantage of additional unlabeled data [13,26,37,50]. Liu *et al.* [29] propose to propagate labels from few labeled data to many unlabeled data, akin to our detection of related bases. We also rely on more data for training, but in contrast to these approaches, our method does not need any new data, nor does it require to generate any. Instead, we exploit the data that is *already available* in the base domain and align the novel domain to the relevant base samples through fine-tuning.

Previous work has also exploited base training data, most related to ours are the works of [4] and [28]. Chen *et al.* [4] propose to use an embedding and deformation sub-networks to leverage additional training samples, whereas we rely on a single feature extractor network which is much simpler to implement and train. Unlike random base example sampling [4] for interpolating novel example deformations in the image space, we propose to borrow the internal distribution structure of the detected related classes in feature space. Besides, our alignment strategies introduce extra criteria to keep the focus of the learner on the novel classes, which prevents the novel classes from becoming outliers. Focused on object detection, Lim *et al.* [28] proposes a model to search similar object categories using a sparse grouped Lasso framework. Unlike [28], we propose and evaluate two associative alignments in the context of few-shot image classification.

From the alignment perspective, our work is related to Jiang *et al.* [20] which stays in the context of zero-shot learning, and proposes a coupled dictionary matching in visual-semantic structures to find matching concepts. In contrast, we propose associative base-novel class alignments along with two strategies for enforcing the unification of the related concepts.

3   Preliminaries

Let us assume that we have a large base dataset $\mathcal{X}^b = \{(\mathbf{x}_i^b, y_i^b)\}_{i=1}^{N^b}$, where $\mathbf{x}_i^b \in \mathbb{R}^d$ is the i-th data instance of the set and $y_i^b \in \mathcal{Y}^b$ is the corresponding class label. We are also given a small amount of novel class data $\mathcal{X}^n = \{(\mathbf{x}_i^n, y_i^n)\}_{i=1}^{N^n}$, with labels $y_i^n \in \mathcal{Y}^n$ from a set of distinct classes $\mathcal{Y}^n$. Few-shot classification aims to train a classifier with only a few examples from each of the novel classes (e.g., 5 or even just 1). In this work, we used the standard transfer learning strategy of Chen et al. [3], which is organized into the following two stages.

Pre-training Stage. The learning model is a neural network composed of a feature extractor $f(\cdot|\theta)$, parameterized by θ, followed by a linear classifier $c(\mathbf{x}|\mathbf{W}) \equiv \mathbf{W}^\top f(\mathbf{x}|\theta)$, described by matrix $\mathbf{W}$, ending with a scoring function such as softmax to produce the output. The network is trained from scratch on examples from the base categories $\mathcal{X}^b$.

Fine-tuning Stage. In order to adapt the network to the novel classes, the network is subsequently fine-tuned on the few examples from $\mathcal{X}^n$. Since overfitting is likely to occur if all the network weights are updated, the feature extractor weights θ are frozen, with only the classifier weights $\mathbf{W}$ being updated in this stage.

4   Associative Alignment

Freezing the feature extractor weights θ indeed reduces overfitting, but also limits the learning capacity of the model. In this paper, we strive for the best of both worlds and present an approach which controls overfitting while maintaining the original learning capacity of the model. We borrow the internal distribution structure of a subset of *related* base categories, $\mathcal{X}^{rb} \subset \mathcal{X}^b$. To account for the discrepancy between the novel and related base classes, we propose to *align* the novel categories to the related base categories in feature space. Such a mapping allows for a bigger pool of training data while making instances of these two sets more coherent. Note that, as opposed to [4], we do not modify the related base instances in any way: we simply wish to align novel examples to the distributions of their related class instances.

In this section, we first describe how the related base classes are determined. Then, we present our main contribution: the "centroid associative alignment" method, which exploits the related base instances to improve classification performance on novel classes. We conclude by presenting an alternative associative alignment strategy, which relies on an adversarial framework.

4.1   Detecting the Related Bases

We develop a simple, yet effective procedure to select a set of base categories related to a novel category. Our method associates B base categories to each novel class. After training $c(f(\cdot|\theta)|\mathbf{W})$ on $\mathcal{X}^b$, we first fine-tune $c(\cdot|\mathbf{W})$ on $\mathcal{X}^n$

Fig. 2. Results of related base algorithm in a 5-way 5-shot scenario. Each column represents a different novel class. The top row shows the 5 novel instances, while the bottom row shows 60 randomly selected related base instances with $B = 10$.

while keeping θ fixed. Then, we define $\mathbf{M} \in \mathbb{R}^{K^b \times K^n}$ as a base-novel similarity matrix, where K^b and K^n are respectively the number of classes in $\mathcal{X}^b$ and $\mathcal{X}^n$. An element $m_{i,j}$ of the matrix $\mathbf{M}$ corresponds to the ratio of examples associated to the i-th base class that are classified as the j-th novel class:

$$m_{i,j} = \frac{1}{|\mathcal{X}_i^b|} \sum_{(\mathbf{x}_l^b, \cdot) \in \mathcal{X}_i^b} \mathbb{I}\left[j = \underset{k=1}{\overset{K^n}{\arg\max}} \left(c_k(f(\mathbf{x}_l^b | \theta) \,|\, \mathbf{W}) \right) \right],\tag{1}$$

where $c_k(f(\mathbf{x}|\theta)|\mathbf{W})$ is the classifier output $c(\cdot|\mathbf{W})$ for class k. Then, the B base classes with the highest score for a given novel class are kept as the related base for that class. Figure 2 illustrates example results obtained with this method in a 5-shot, 5-way scenario.

4.2 Centroid Associative Alignment

Let us assume the set of instances $\mathcal{X}_i^n$ belonging to the i-th novel class $i \in \mathcal{Y}^n$, $\mathcal{X}_i^n = \{(\mathbf{x}_j^n, y_j^n) \in \mathcal{X}^n \,|\, y_j^n = i\}$, and the set of related base examples $\mathcal{X}_i^{rb}$ belonging to the same novel class i according to the $g(\cdot|\mathbf{M})$ mapping function, $\mathcal{X}_i^{rb} = \{(\mathbf{x}_j^b, y_j^b) \in \mathcal{X}^{rb} \,|\, g(y_j|\mathbf{M}) = i\}$. The function $g(y_j|\mathbf{M}) : \mathcal{Y}^b \to \mathcal{Y}^n$ maps base class labels to the novel ones according to the similarity matrix $\mathbf{M}$. We wish to find an alignment transformation for matching probability densities $p(f(\mathbf{x}_{i,k}^n \,|\, \theta))$ and $p(f(\mathbf{x}_{i,l}^{rb} \,|\, \theta))$. Here, $\mathbf{x}_{i,k}^n$ is the k-th element from class i in the novel set, and $\mathbf{x}_{i,l}^{rb}$ is the l-th element from class i in the related base set. This approach has the added benefit of allowing the fine-tuning of all of the model parameters θ and $\mathbf{W}$ with a reduced level of overfitting.

We propose a metric-based centroid distribution alignment strategy. The idea is to enforce intra-class compactness during the alignment process. Specifically, we explicitly push the training examples from the i-th novel class $\mathcal{X}_i^n$ towards the centroid of their related examples $\mathcal{X}_i^{rb}$ in feature space. The centroid $\boldsymbol{\mu}_i$ of $\mathcal{X}_i^{rb}$ is computed by

$$\boldsymbol{\mu}_i = \frac{1}{|\mathcal{X}_i^{rb}|} \sum_{(\mathbf{x}_j, \cdot) \in \mathcal{X}_i^{rb}} f(\mathbf{x}_j | \theta),\tag{2}$$

Algorithm 1:
Centroid alignment.

Input: pre-trained model $c(f(\cdot|\theta)|\mathbf{W})$,
novel class $\mathcal{X}^n$, related base set $\mathcal{X}^{rb}$.
Output: fine-tuned $c(f(\cdot|\theta)|\mathbf{W})$.
while *not done* **do**

$\widetilde{\mathcal{X}}^n \leftarrow$ sample a batch from $\mathcal{X}^n$
$\widetilde{\mathcal{X}}^{rb} \leftarrow$ sample a batch from $\mathcal{X}^{rb}$

evaluate $\mathcal{L}_{\mathrm{ca}}(\widetilde{\mathcal{X}}^n, \widetilde{\mathcal{X}}^{rb})$, (Eq. 3)
$\theta \leftarrow \theta - \eta_{\mathrm{ca}} \nabla_\theta \mathcal{L}_{\mathrm{ca}}(\widetilde{\mathcal{X}}^n, \widetilde{\mathcal{X}}^{rb})$

evaluate $\mathcal{L}_{\mathrm{clf}}(\widetilde{\mathcal{X}}^{rb})$, (Eq. 7)
$\mathbf{W} \leftarrow \mathbf{W} - \eta_{\mathrm{clf}} \nabla_{\mathbf{W}} \mathcal{L}_{\mathrm{clf}}(\widetilde{\mathcal{X}}^{rb})$
evaluate $\mathcal{L}_{\mathrm{clf}}(\widetilde{\mathcal{X}}^n)$, (Eq. 7)
$\mathbf{W} \leftarrow \mathbf{W} - \eta_{\mathrm{clf}} \nabla_{\mathbf{W}} \mathcal{L}_{\mathrm{clf}}(\widetilde{\mathcal{X}}^n)$
$\theta \leftarrow \theta - \eta_{\mathrm{clf}} \nabla_\theta \mathcal{L}_{\mathrm{clf}}(\widetilde{\mathcal{X}}^n)$

end

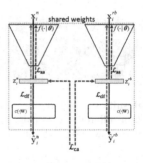

Fig. 3. Schematic overview of our centroid alignment. The feature learner $f(\cdot|\theta)$ takes an example from novel category $\mathbf{x}^n$ and an example related base $\mathbf{x}_i^{rb}$. A Euclidean centroid based alignment loss $\mathcal{L}_{\mathrm{ca}}$ (red arrow) aligns the encoded $\mathbf{x}_i^n$ and $\mathbf{x}_i^{rb}$. Blue arrows represent classification loss $\mathcal{L}_{\mathrm{clf}}$. (Color figure online)

where N^n and N^{rb} are the number of examples in $\mathcal{X}^n$ and $\mathcal{X}^{rb}$, respectively. This allows the definition of the centroid alignment loss as

$$\mathcal{L}_{\mathrm{ca}}(\mathcal{X}^n) = -\frac{1}{N^n N^{rb}} \sum_{i=1}^{K^n} \sum_{(\mathbf{x}_j, \cdot) \in \mathcal{X}_i^n} \log \frac{\exp[-\|f(\mathbf{x}_j|\theta) - \boldsymbol{\mu}_i\|_2^2]}{\sum_{k=1}^{K^n} \exp[-\|f(\mathbf{x}_j|\theta) - \boldsymbol{\mu}_k\|_2^2]}. \quad (3)$$

Our alignment strategy bears similarities to [42] which also uses Eq. 3 in a meta-learning framework. In our case, we use that same equation to match distributions. Figure 3 illustrates our proposed centroid alignment, and Algorithm 1 presents the overall procedure. First, we update the parameters of the feature extraction network $f(\cdot|\theta)$ using Eq. 3. Second, the entire network is updated using a classification loss $\mathcal{L}_{\mathrm{clf}}$ (defined in Sect. 5).

4.3 Adversarial Associative Alignment

As an alternative associative alignment strategy, and inspired by WGAN [1], we experiment with training the encoder $f(\cdot|\theta)$ to perform adversarial alignment using a conditioned critic network $h(\cdot|\phi)$ based on Wasserstein-1 distance between two probability densities p_x and p_y:

$$D(p_x, p_y) = \sup_{\|h\|_L \le 1} \mathbb{E}_{x \sim p_x}[h(x)] - \mathbb{E}_{x \sim p_y}[h(x)], \quad (4)$$

where sup is the supremum, and h is a 1-Lipschitz function. Similarly to Arjovsky et al. [1], we use a parameterized critic network $h(\cdot|\phi)$ conditioned by the concatenation of the feature embedding of either $\mathbf{x}_i^n$ or $\mathbf{x}_j^{rb}$, along with the corresponding label y_i^n encoded as a one-hot vector. Conditioning $h(\cdot|\phi)$ helps the

Algorithm 2:
Adversarial alignment

Input: pre-trained model $c(f(\cdot|\theta)|\mathbf{W})$, novel class $\mathcal{X}^n$, related base set $\mathcal{X}^{rb}$.
Output: fine-tuned $c(f(\cdot|\theta)|\mathbf{W})$.
while *not done* **do**

  $\widetilde{\mathcal{X}}^n \leftarrow$ sample a batch from $\mathcal{X}^n$
  $\widetilde{\mathcal{X}}^{rb} \leftarrow$ sample a batch from $\mathcal{X}^{rb}$

  for $i = 0, \ldots, n_{\text{critic}}$ **do**
    evaluate $\mathcal{L}_h(\widetilde{\mathcal{X}}^n, \widetilde{\mathcal{X}}^{rb})$, (Eq. 5)
    ▷ update critic:
    $\phi \leftarrow \phi + \eta_h \nabla_\phi \mathcal{L}_h(\widetilde{\mathcal{X}}^n, \widetilde{\mathcal{X}}^{rb})$
    $\phi \leftarrow \text{clip}(\phi, -0.01, 0.01)$
  end

  evaluate $\mathcal{L}_{\text{aa}}(\widetilde{\mathcal{X}}^n)$, (Eq. 6)
  $\theta \leftarrow \theta - \eta_{\text{aa}} \nabla_\theta \mathcal{L}_{\text{aa}}(\widetilde{\mathcal{X}}^n)$
  evaluate $\mathcal{L}_{\text{clf}}(\widetilde{\mathcal{X}}^{rb})$, (Eq. 7)
  $\mathbf{W} \leftarrow \mathbf{W} - \eta_{\text{clf}} \nabla_\mathbf{W} \mathcal{L}_{\text{clf}}(\widetilde{\mathcal{X}}^{rb})$
  evaluate $\mathcal{L}_{\text{clf}}(\widetilde{\mathcal{X}}^n)$, (Eq. 7)
  $\mathbf{W} \leftarrow \mathbf{W} - \eta_{\text{clf}} \nabla_\mathbf{W} \mathcal{L}_{\text{clf}}(\widetilde{\mathcal{X}}^n)$
  $\theta \leftarrow \theta - \eta_{\text{clf}} \nabla_\theta \mathcal{L}_{\text{clf}}(\widetilde{\mathcal{X}}^n)$
end

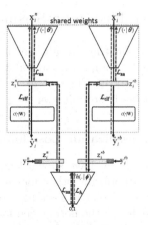

Fig. 4. Overview of our adversarial alignment. The feature learner $f(\cdot|\theta)$ takes an image $\mathbf{x}_i^n$ from the i-th novel class and an example $\mathbf{x}_i^{rb}$ of the related base. The critic $h(\cdot|\phi)$ takes the feature vectors and the one-hot class label vector. Green, red and blue arrows present the critic $\mathcal{L}_h$, adversarial $\mathcal{L}_{\text{aa}}$ and classification $\mathcal{L}_{\text{clf}}$ losses respectively. (Color figure online)

critic in matching novel categories and their corresponding related base categories. The critic $h(\cdot|\phi)$ is trained with loss

$$\mathcal{L}_h(\mathcal{X}^n, \mathcal{X}^{rb}) = \frac{1}{N^{rb}} \sum_{(\mathbf{x}_i^{rb}, y_i^{rb}) \in \mathcal{X}^{rb}} h\left([f(\mathbf{x}_i^{rb}|\theta)\ y_i^{rb}]\,|\,\phi\right)$$
$$- \frac{1}{N^n} \sum_{(\mathbf{x}_i^n, y_i^n) \in \mathcal{X}^n} h\left([f(\mathbf{x}_i^n|\theta)\ y_i^n]\,|\,\phi\right), \quad (5)$$

where, $[\cdot]$ is the concatenation operator. Then, the encoder parameters θ are updated using

$$\mathcal{L}_{\text{aa}}(\mathcal{X}^n) = \frac{1}{K^n} \sum_{(\mathbf{x}_i^n, y_i^n) \in \mathcal{X}^n} h\left([f(\mathbf{x}_i^n|\theta)\ y_i^n]|\phi\right). \quad (6)$$

Algorithm 2 summarizes our adversarial alignment method. First, we perform the parameter update of critic $h(\cdot|\phi)$ using Eq. 5. Similar to WGAN [1], we perform n_{critic} iterations to optimize h, before updating $f(\cdot|\theta)$ using Eq. 6. Finally, the entire network is updated by a classification loss $\mathcal{L}_{\text{clf}}$ (defined in Sect. 5).

5   Establishing a Strong Baseline

Before evaluating our alignment strategies in Sect. 6, we first establish a strong baseline for comparison by following the recent literature. In particular, we build on the work of Chen *et al.* [3] but incorporate a different loss function and episodic early stopping on the pre-training stage.

5.1   Classification Loss Functions

Deng *et al.* [6] have shown that an additive angular margin ("arcmax" hereafter) outperforms other metric learning algorithms for face recognition. The arcmax has a metric learning property since it enforces a geodesic distance margin penalty on the normalized hypersphere, which we think can be beneficial for few-shot classification by helping keep class clusters compact and well-separated.

Let $\mathbf{z}$ be the representation of $\mathbf{x}$ in feature space. As per [6], we transform the logit as $\mathbf{w}_j^\top \mathbf{z} = \|\mathbf{w}_j\| \|\mathbf{z}\| \cos \varphi_j$, where φ_j is the angle between $\mathbf{z}$ and $\mathbf{w}_j$, the j-th column in the weight matrix $\mathbf{W}$. Each weight $\|\mathbf{w}_j\| = 1$ by l_2 normalization. Arcmax adds an angular margin m to the distributed examples on a hypersphere:

$$\mathcal{L}_{\text{clf}} = -\frac{1}{N} \sum_{i=1}^{N} \log \frac{\exp(s \cos(\varphi_{y_i} + m))}{\exp(s \cos(\varphi_{y_i} + m)) + \sum_{\forall j \neq y_i} \exp(s \cos \varphi_j)}, \quad (7)$$

where s is the radius of the hypersphere on which $\mathbf{z}$ is distributed, N the number of examples, and m and s are hyperparameters (see Sect. 6.1). The overall goal of the margin is to enforce inter-class discrepancy and intra-class compactness.

5.2   Episodic Early Stopping

A fixed number of epochs in the pre-training stage has been commonly used (e.g., [3,9,42,47]), but this might hamper performance in the fine-tuning stage. Using validation error, we observe the necessity of early-stopping in pre-training phase (see supp. mat. for a validation error plot). We thus make the use of episodic early stopping using validation set at pre-training time, specifically by stopping the training when the mean accuracy over a window of recent epochs starts to decrease. The best model in the window is selected as the final result.

6   Experimental Validation

In the following, we are conducting an experimental evaluation and comparison of the proposed associative alignment strategies for few-shot learning. First, we introduce the datasets used and evaluate the strong baseline from Sect. 5.

6.1   Datasets and Implementation Details

Datasets. We present experiments on four benchmarks: *mini*-ImageNet [47], tieredImageNet [37], and FC100 [33] for generic object recognition; and CUB-200-2011 (CUB) [48] for fine-grained image classification. *mini*-ImageNet is a subset of the ImageNet ILSVRC-12 dataset [38] containing 100 categories and 600 examples per class. We used the same splits as Ravi and Larochelle [36], where 64, 16, and 20 classes are used for the base, validation, and novel classes, respectively. As a larger benchmark, the tieredImageNet [37] is also a subset of ImageNet ILSVRC-12 dataset [38], this time with 351 base, 97 validation, and 160 novel classes respectively. Derived from CIFAR-100 [23], the FC100 dataset [33] contains 100 classes grouped into 20 superclasses to minimize class overlap. Base, validation and novel splits contain 60, 20, 20 classes belonging to 12, 5, and 5 superclasses, respectively. The CUB dataset [48] contains 11,788 images from 200 bird categories. We used the same splits as Hilliard *et al.* [19] using 100, 50, and 50 classes for the base, validation, and novel classes, respectively.

Network Architectures. We experiment with three backbones for the feature learner $f(\cdot|\theta)$: 1) a 4-layer convolutional network ("Conv4") with input image resolution of 84×84, similar to [9,36,42]; 2) a ResNet-18 [18] with input size of 224×224; and 3) a 28-layers Wide Residual Network ("WRN-28-10") [41] with input size of 80×80 in 3 steps of dimension reduction. We use a single hidden layer MLP of 1024 dimensions as the critic network $h(\cdot|\phi)$ (c.f. Sect. 4.3).

Implementation Details. Recall from Sect. 3 that training consists of two stages: 1) pre-training using base categories $\mathcal{X}^b$; and 2) fine-tuning on novel categories $\mathcal{X}^n$. For pre-training, we use the early stopping algorithm from Sect. 5.2 with a window size of 50. Standard data augmentation approaches (i.e., color jitter, random crops, and left-right flips as in [3]) have been employed, and the Adam algorithm with a learning rate of 10^{-3} and batch size of 64 is used for both pre-training and fine-tuning. The arcmax loss (Eq. 7) is configured with $s = 20$ and $m = 0.1$ which are set by cross validation. In the fine-tuning stage, episodes are defined by randomly selecting $N = 5$ classes from the novel categories $\mathcal{X}^n$. k examples for each category are subsequently sampled ($k = 1$ and $k = 5$ in our experiments). As in Chen *et al.* [3], no standard data augmentation was used in this stage. We used episodic cross-validation to find s and m with a fixed encoder. More specifically, (s, m) were found to be $(5, 0.1)$ for the Conv4 and $(5, 0.01)$ for the WRN-28-10 and ResNet-18 backbones. The learning rate for Adam was set to 10^{-3} and 10^{-5} for the centroid and adversarial alignments respectively. Similarly to [1], 5 iterations (inner loop of Algorithm 2) were used to train the critic $h(\cdot|\phi)$. We fix the number of related base categories as $B = 10$ (see supp. mat. for an ablation study on B). For this reason, we used a relatively large number of categories (50 classes out of the 64 available in *mini*-ImageNet).

Table 1. Preliminary evaluation using *mini*-ImageNet and CUB, presenting 5-way classification accuracy using the Conv4 backbone, with ± indicating the 95% confidence intervals over 600 episodes. The best result is boldfaced, while the best result *prior to this work* is highlighted in blue. Throughout this paper, "–" indicates when a paper does not report results in the corresponding scenario.

	Method	*mini*-ImageNet		CUB	
		1-shot	5-shot	1-shot	5-shot
Meta learning	Meta-LSTM [36]	43.44 ± 0.77	55.31 ± 0.71	–	–
	MatchingNet[c] [47]	43.56 ± 0.84	55.31 ± 0.73	60.52 ± 0.88	75.29 ± 0.75
	ProtoNet[c] [42]	49.42 ± 0.78	68.20 ± 0.66	51.31 ± 0.91	70.77 ± 0.69
	MAML[c] [10]	48.07 ± 1.75	63.15 ± 0.91	55.92 ± 0.95	72.09 ± 0.76
	RelationNet[c] [44]	50.44 ± 0.82	65.32 ± 0.70	62.45 ± 0.98	76.11 ± 0.69
Tr. learning	Softmax[a]	46.40 ± 0.72	64.37 ± 0.59	47.12 ± 0.74	64.16 ± 0.71
	Softmax[a,b]	46.99 ± 0.73	65.33 ± 0.60	45.68 ± 0.86	66.94 ± 0.84
	Cosmax[a]	50.92 ± 0.76	67.29 ± 0.59	60.53 ± 0.83	79.34 ± 0.61
	Cosmax[a,b]	52.04 ± 0.82	68.47 ± 0.60	60.66 ± 1.04	79.79 ± 0.75
	Our baseline (Sect. 5)	51.90 ± 0.79	69.07 ± 0.62	60.85 ± 1.07	79.74 ± 0.64
Align.	Adversarial	52.13 ± 0.99	70.78 ± 0.60	**63.30** ± 0.94	**81.35** ± 0.67
	Centroid	**53.14** ± 1.06	**71.45** ± 0.72	62.71 ± 0.88	80.48 ± 0.81

[a]Our implementation
[b]With early stopping
[c]Implementation from [3] for CUB.

6.2 *mini*-ImageNet and CUB with a Shallow Conv4 Backbone

We first evaluate the new baseline presented in Sect. 5 and our associative alignment strategies using the Conv4 backbone on the *mini*-ImageNet (see supp. mat. for evaluations in higher number of ways) and CUB datasets, with corresponding results presented in Table 1. We note that arcmax with early stopping improves on using cosmax and softmax with and without early stopping for both the 1- and 5-shot scenarios, on both the *mini*-ImageNet and CUB datasets. We followed the same dataset split configuration, network architecture, and implementation details given in [3] for our testing. Our centroid associative alignment outperforms the state of the art in all the experiments, with gains of 1.24% and 2.38% in 1- and 5-shot over our baseline on *mini*-ImageNet. For CUB, the adversarial alignment provides an additional gain of 0.6% and 0.87% over the centroid one.

6.3 *mini*-ImageNet and tieredimageNet with Deep Backbones

We now evaluate our proposed associative alignment on both the *mini*-ImageNet and tieredimageNet datasets using two deep backbones: ResNet-18 and WRN-28-10. Table 2 compares our proposed alignment methods with several approaches.

Table 2. *mini*-ImageNet and tieredImageNet results using ResNet-18 and WRN-28-10 backbones. ± denotes the 95% confidence intervals over 600 episodes.

	Method	*mini*-ImageNet		tieredImageNet	
		1-shot	5-shot	1-shot	5-shot
ResNet-18	TADAM [33]	58.50 ± 0.30	76.70 ± 0.30	–	–
	ProtoNet[a] [42]	54.16 ± 0.82	73.68 ± 0.65	61.23 ± 0.77	80.00 ± 0.55
	SNAIL [32]	55.71± 0.99	68.88 ± 0.92	–	–
	IDeMe-Net [4]	59.14 ± 0.86	74.63 ± 0.74	–	–
	Activation to Param. [35]	59.60 ± 0.41	73.74 ± 0.19	–	–
	MTL [43]	61.20 ± 1.80	75.50 ± 0.80	–	–
	TapNet [54]	61.65 ± 0.15	76.36 ± 0.10	63.08 ± 0.15	80.26 ± 0.12
	VariationalFSL [57]	61.23 ± 0.26	77.69 ± 0.17	–	–
	MetaOptNet[b] [24]	**62.64** ± 0.61	78.63 ± 0.46	65.99 ± 0.72	81.56 ± 0.53
	our baseline (Sect. 5)	58.07 ± 0.82	76.62 ± 0.58	65.08 ± 0.19	83.67 ± 0.51
	adversarial alignment	58.84 ± 0.77	77.92 ± 0.82	66.44 ± 0.61	85.12 ± 0.53
	centroid alignment	59.88 ± 0.67	**80.35** ± 0.73	**69.29** ± 0.56	**85.97** ± 0.49
WRN-28-10	LEO [39]	61.76 ± 0.08	77.59 ± 0.12	66.33 ± 0.09	81.44 ± 0.12
	wDAE [15]	61.07 ± 0.15	76.75 ± 0.11	68.18 ± 0.16	83.09 ± 0.12
	CC+rot [13]	62.93 ± 0.45	79.87 ± 0.33	70.53 ± 0.51	84.98 ± 0.36
	Robust-dist++ [39]	63.28 ± 0.62	81.17 ± 0.43	–	–
	Transductive-ft [7]	65.73 ± 0.68	78.40 ± 0.52	73.34 ± 0.71	85.50 ± 0.50
	our baseline (Sect. 5)	63.28 ±0.71	78.31 ±0.57	68.47 ±0.86	84.11 ±0.65
	adversarial alignment	64.79 ±0.93	82.02 ±0.88	73.87 ±0.76	84.95 ±0.59
	centroid alignment	**65.92** ± 0.60	**82.85** ± 0.55	**74.40** ± 0.68	**86.61** ±0.59

[a] Results are from [3] for *mini*-ImageNet and from [24] for tieredImageNet
[b] ResNet-12.

mini-ImageNet. Our centroid associative alignment strategy achieves the best 1- and 5-shot classification tasks on both the ResNet-18 and WRN-28-10 backbones, with notable absolute accuracy improvements of 2.72% and 1.68% over MetaOptNet [24] and Robust-dist++ [8] respectively. The single case where a previous method achieves superior results is that of MetaOptNet, which outperforms our method by 2.76% in 1-shot. For the WRN-28-10 backbone, we achieve similar results to Transductive-ft [7] for 1-shot, but outperform their method by 4.45% in 5-shot. Note that unlike IDeMe-Net [4], SNAIL [32] and TADAM [33], which make use of extra modules, our method achieves significant improvements over these methods without any changes to the backbone.

tieredImageNet. Table 2 also shows that our centroid associative alignment outperforms the compared methods on tieredImageNet in both 1- and 5-shot scenarios. Notably, our centroid alignment results in a gain of 3.3% and 4.41% over MetaOptNet [24] using the ResNet-18. Likewise, our centroid alignment gains 1.06% and 1.11% over the best of the compared methods using WRN-28-10.

Table 3. Results on the FC100 and CUB dataset using ResNet-18 backbones. ± denotes the 95% confidence intervals over 600 episodes. The best result is boldfaced, while the best result *prior to this work* is highlighted in blue.

Method	FC100		CUB	
	1-shot	5-shot	1-shot	5-shot
Robust-20 [8]	–	–	58.67 ± 0.65	75.62 ± 0.48
GNN-LFT [45]	–	–	51.51 ± 0.80	73.11 ± 0.68
RelationNet[a] [44]	–	–	67.59 ± 1.02	82.75 ± 0.58
ProtoNet[a] [42]	40.5 ± 0.6	55.3 ± 0.6	71.88 ± 0.91	87.42 ± 0.48
TADAM [33]	40.1 ± 0.4	56.1 ± 0.4	–	–
MetaOptNet[a] [24]	41.1 ± 0.6	55.5 ± 0.6	–	–
MTL [43]	45.1 ± 1.8	57.6 ± 0.9	–	–
Transductive-ft [7]	43.2 ± 0.6	57.6 ± 0.6	–	–
our baseline (Sect. 5)	40.84 ± 0.71	57.02 ± 0.63	71.71 ± 0.86	85.74 ± 0.49
adversarial	43.44 ± 0.71	58.69 ± 0.56	70.80 ± 1.12	88.04 ± 0.54
centroid	**45.83** ± 0.48	**59.74** ± 0.56	**74.22** ± 1.09	**88.65** ± 0.55

[a]Implementation from [3] for CUB, and from [24] for FC100.

6.4   FC100 and CUB with a ResNet-18 Backbone

We present additional results on the FC100 and CUB datasets with a ResNet-18 backbone in Table 3. In FC100, our centroid alignment gains 0.73% and 2.14% over MTL [43] in 1- and 5-shot respectively. We also observe improvements in CUB with our associative alignment approaches, with the centroid alignment outperforming ProtoNet [42] by 2.3% in 1-shot and 1.2% in 5-shot. We outperform Robust-20 [8], an ensemble of 20 networks, by 4.03% and 4.15% on CUB.

6.5   Cross-Domain Evaluation

We also evaluate our alignment strategies in cross-domain image classification. Here, following [3], the base categories are drawn from *mini*-ImageNet, but the novel categories are from CUB. As shown in Table 4, we gain 1.3% and 5.4% over the baseline in the 1- and 5-shot, respectively, with our proposed centroid alignment. Adversarial alignment falls below the baseline in 1-shot by -1.2%, but gains 5.9% in 5-shot. Overall, our centroid alignment method shows absolute accuracy improvements over the state of the art (i.e., cosmax [3]) of 3.8% and 6.0% in 1- and 5- shot respectively. We also outperform Robust-20 [8], an ensemble of 20 networks, by 4.65% for 5-shot on *mini*-ImageNet to CUB cross-domain. One could argue that the three bird categories (i.e., house finch, robin, and toucan) in *mini*-ImageNet bias the cross-domain evaluation. Re-training the approach by excluding these classes resulted in a similar performance as shown in Table 4.

Table 4. Cross-domain results from *mini*-ImageNet to CUB in 1-shot, 5-shot, 10-shot scenarios using a ResNet-18 backbone.

Method	1-shot	5-shot	10-shot
ProtoNet[c] [49]	–	62.02 ± 0.70	–
MAML[c] [10]	–	51.34 ± 0.72	–
RelationNet[c] [44]	–	57.71 ± 0.73	–
Diverse 20 [8]	–	66.17 ± 0.73	–
cosmax[b] [3]	43.06 ± 1.01	64.38 ± 0.86	67.56 ± 0.77
our baseline (Sect. 5)	45.60 ± 0.94	64.93 ± 0.95	68.95 ± 0.78
adversarial	44.37 ± 0.94	70.80 ± 0.83	79.63 ± 0.71
adversarial[a]	44.65 ± 0.88	71.48 ± 0.96	78.52 ± 0.70
centroid	46.85 ± 0.75	70.37 ± 1.02	**79.98 ± 0.80**
centroid[a]	**47.25 ± 0.76**	**72.37 ± 0.89**	79.46 ± 0.72

[a]Without birds (house finch, robin, toucan) in base classes
[b]Our implementation, with early stopping
[c]Implementation from [3].

7   Discussion

This paper presents the idea of associative alignment for few-shot image classification, which allows for higher generalization performance by enabling the training of the entire network, still while avoiding overfitting. To do so, we design a procedure to detect related base categories for each novel class. Then, we proposed a centroid-based alignment strategy to keep the intra-class alignment while performing updates for the classification task. We also explored an adversarial alignment strategy as an alternative. Our experiments demonstrate that our approach, specifically the centroid-based alignment, outperforms previous works in almost all scenarios. The current limitations of our work provide interesting future research directions. First, the alignment approach (Sect. 4) might include irrelevant examples from the base categories, so using categorical semantic information could help filter out bad samples. An analysis showed that ~12% of the samples become out-of-distribution (OOD) using a centroid nearest neighbour criteria on *mini*ImageNet in 5-way 1- and 5-shot using ResNet-18. Classification results were not affected significantly by discarding OOD examples at each iteration. Second, the multi-modality of certain base categories look inevitable and might degrade the generalization performance compared to the single-mode case assumed by our centroid alignment strategy. Investigating the use of a mixture family might therefore improve generalization performance. Finally, our algorithms compute the related base once and subsequently keep them fixed during an episode, not taking into account the changes applied to the latent space during the episodic training. Therefore, a more sophisticated dynamic sampling mechanism could be helpful in the finetuning stage.

Acknowledgement. This project was supported by funding from NSERC-Canada, Mitacs, Prompt-Québec, and E Machine Learning. We thank Ihsen Hedhli, Saed Moradi, Marc-André Gardner, and Annette Schwerdtfeger for proofreading of the manuscript.

References

1. Arjovsky, M., Chintala, S., Bottou, L.: Wasserstein GAN. arXiv preprint arXiv:1701.07875 (2017)
2. Bertinetto, L., Henriques, J.F., Torr, P., Vedaldi, A.: Meta-learning with differentiable closed-form solvers. In: The International Conference on Learning Representations (2019)
3. Chen, W.Y., Liu, Y.C., Kira, Z., Wang, Y.C.F., Huang, J.B.: A closer look at few-shot classification. arXiv preprint arXiv:1904.04232 (2019)
4. Chen, Z., Fu, Y., Wang, Y.X., Ma, L., Liu, W., Hebert, M.: Image deformation meta-networks for one-shot learning. In: The Conference on Computer Vision and Pattern Recognition (2019)
5. Chu, W.H., Li, Y.J., Chang, J.C., Wang, Y.C.F.: Spot and learn: a maximum-entropy patch sampler for few-shot image classification. In: The Conference on Computer Vision and Pattern Recognition (2019)
6. Deng, J., Guo, J., Xue, N., Zafeiriou, S.: Arcface: additive angular margin loss for deep face recognition. In: The Conference on Computer Vision and Pattern Recognition (2019)
7. Dhillon, G.S., Chaudhari, P., Ravichandran, A., Soatto, S.: A baseline for few-shot image classification. arXiv preprint arXiv:1909.02729 (2019)
8. Dvornik, N., Schmid, C., Mairal, J.: Diversity with cooperation: ensemble methods for few-shot classification. In: The International Conference on Computer Vision (2019)
9. Finn, C., Abbeel, P., Levine, S.: Model-agnostic meta-learning for fast adaptation of deep networks. In: The International Conference on Machine Learning (2017)
10. Finn, C., Xu, K., Levine, S.: Probabilistic model-agnostic meta-learning. In: Advances in Neural Information Processing Systems (2018)
11. Gao, H., Shou, Z., Zareian, A., Zhang, H., Chang, S.F.: Low-shot learning via covariance-preserving adversarial augmentation networks. In: Advances in Neural Information Processing Systems (2018)
12. Garcia, V., Bruna, J.: Few-shot learning with graph neural networks. arXiv preprint arXiv:1711.04043 (2017)
13. Gidaris, S., Bursuc, A., Komodakis, N., Pérez, P., Cord, M.: Boosting few-shot visual learning with self-supervision. In: The International Conference on Computer Vision (2019)
14. Gidaris, S., Komodakis, N.: Dynamic few-shot visual learning without forgetting. In: The Conference on Computer Vision and Pattern Recognition (2018)
15. Gidaris, S., Komodakis, N.: Generating classification weights with GNN denoising autoencoders for few-shot learning. arXiv preprint arXiv:1905.01102 (2019)
16. Gui, L.-Y., Wang, Y.-X., Ramanan, D., Moura, J.M.F.: Few-shot human motion prediction via meta-learning. In: Ferrari, V., Hebert, M., Sminchisescu, C., Weiss, Y. (eds.) ECCV 2018. LNCS, vol. 11212, pp. 441–459. Springer, Cham (2018). https://doi.org/10.1007/978-3-030-01237-3_27

17. Hariharan, B., Girshick, R.: Low-shot visual recognition by shrinking and hallucinating features. In: The International Conference on Computer Vision (2017)

18. He, K., Zhang, X., Ren, S., Sun, J.: Deep residual learning for image recognition. In: The Conference on Computer Vision and Pattern Recognition (2016)

19. Hilliard, N., Phillips, L., Howland, S., Yankov, A., Corley, C.D., Hodas, N.O.: Few-shot learning with metric-agnostic conditional embeddings. arXiv preprint arXiv:1802.04376 (2018)

20. Jiang, H., Wang, R., Shan, S., Chen, X.: Learning class prototypes via structure alignment for zero-shot recognition. In: Ferrari, V., Hebert, M., Sminchisescu, C., Weiss, Y. (eds.) ECCV 2018. LNCS, vol. 11214, pp. 121–138. Springer, Cham (2018). https://doi.org/10.1007/978-3-030-01249-6_8

21. Kim, J., Oh, T.H., Lee, S., Pan, F., Kweon, I.S.: Variational prototyping-encoder: one-shot learning with prototypical images. In: The Conference on Computer Vision and Pattern Recognition (2019)

22. Kim, T., Yoon, J., Dia, O., Kim, S., Bengio, Y., Ahn, S.: Bayesian model-agnostic meta-learning. arXiv preprint arXiv:1806.03836 (2018)

23. Krizhevsky, A., Nair, V., Hinton, G.: CIFAR-10 and CIFAR-100 datasets (2009). https://www.cs.toronto.edu/kriz/cifar.html

24. Lee, K., Maji, S., Ravichandran, A., Soatto, S.: Meta-learning with differentiable convex optimization. In: The Conference on Computer Vision and Pattern Recognition (2019)

25. Li, W., Wang, L., Xu, J., Huo, J., Gao, Y., Luo, J.: Revisiting local descriptor based image-to-class measure for few-shot learning. In: The Conference on Computer Vision and Pattern Recognition (2019)

26. Li, X., et al.: Learning to self-train for semi-supervised few-shot classification. In: Advances in Neural Information Processing Systems (2019)

27. Lifchitz, Y., Avrithis, Y., Picard, S., Bursuc, A.: Dense classification and implanting for few-shot learning. In: The Conference on Computer Vision and Pattern Recognition (2019)

28. Lim, J.J., Salakhutdinov, R.R., Torralba, A.: Transfer learning by borrowing examples for multiclass object detection. In: Advances in Neural Information Processing Systems (2011)

29. Liu, B., Wu, Z., Hu, H., Lin, S.: Deep metric transfer for label propagation with limited annotated data. In: The IEEE International Conference on Computer Vision (ICCV) Workshops, October 2019

30. van der Maaten, L., Hinton, G.: Visualizing data using t-SNE. Journal of Machine Learning Research **9**, 2579–2605 (2008)

31. Mehrotra, A., Dukkipati, A.: Generative adversarial residual pairwise networks for one shot learning. arXiv preprint arXiv:1703.08033 (2017)

32. Mishra, N., Rohaninejad, M., Chen, X., Abbeel, P.: A simple neural attentive meta-learner. arXiv preprint arXiv:1707.03141 (2017)

33. Oreshkin, B., López, P.R., Lacoste, A.: Tadam: task dependent adaptive metric for improved few-shot learning. In: Advances in Neural Information Processing Systems (2018)

34. Qi, H., Brown, M., Lowe, D.G.: Low-shot learning with imprinted weights. In: The Conference on Computer Vision and Pattern Recognition (2018)

35. Qiao, S., Liu, C., Shen, W., Yuille, A.L.: Few-shot image recognition by predicting parameters from activations. In: The Conference on Computer Vision and Pattern Recognition (2018)

36. Ravi, S., Larochelle, H.: Optimization as a model for few-shot learning (2016)

37. Ren, M., et al.: Meta-learning for semi-supervised few-shot classification. arXiv preprint arXiv:1803.00676 (2018)
38. Russakovsky, O., et al.: Imagenet large scale visual recognition challenge. Int. J. Comput. Vis. (2015)
39. Rusu, A.A., et al.: Meta-learning with latent embedding optimization. arXiv preprint arXiv:1807.05960 (2018)
40. Schwartz, E., et al.: Delta-encoder: an effective sample synthesis method for few-shot object recognition. In: Advances in Neural Information Processing Systems (2018)
41. Sergey, Z., Nikos, K.: Wide residual networks. In: British Machine Vision Conference (2016)
42. Snell, J., Swersky, K., Zemel, R.: Prototypical networks for few-shot learning. In: Advances in Neural Information Processing Systems (2017)
43. Sun, Q., Liu, Y., Chua, T.S., Schiele, B.: Meta-transfer learning for few-shot learning. In: The Conference on Computer Vision and Pattern Recognition (2019)
44. Sung, F., Yang, Y., Zhang, L., Xiang, T., Torr, P.H., Hospedales, T.M.: Learning to compare: relation network for few-shot learning. In: The Conference on Computer Vision and Pattern Recognition (2018)
45. Tseng, H.Y., Lee, H.Y., Huang, J.B., Yang, M.H.: Cross-domain few-shot classification via learned feature-wise transformation. arXiv preprint arXiv:2001.08735 (2020)
46. Vilalta, R., Drissi, Y.: A perspective view and survey of meta-learning. Artif. Intell. Rev. (2002)
47. Vinyals, O., Blundell, C., Lillicrap, T., Wierstra, D., et al.: Matching networks for one shot learning. In: Advances in Neural Information Processing Systems (2016)
48. Wah, C., Branson, S., Welinder, P., Perona, P., Belongie, S.: The caltech-UCSD birds-200-2011 dataset (2011)
49. Wang, Y.X., Girshick, R., Hebert, M., Hariharan, B.: Low-shot learning from imaginary data. In: The Conference on Computer Vision and Pattern Recognition (2018)
50. Wang, Y.X., Hebert, M.: Learning from small sample sets by combining unsupervised meta-training with CNNs. In: Advances in Neural Information Processing Systems (2016)
51. Wang, Y.-X., Hebert, M.: Learning to learn: model regression networks for easy small sample learning. In: Leibe, B., Matas, J., Sebe, N., Welling, M. (eds.) ECCV 2016. LNCS, vol. 9910, pp. 616–634. Springer, Cham (2016). https://doi.org/10.1007/978-3-319-46466-4_37
52. Wang, Y.X., Ramanan, D., Hebert, M.: Meta-learning to detect rare objects. In: The International Conference on Computer Vision (2019)
53. Wertheimer, D., Hariharan, B.: Few-shot learning with localization in realistic settings. In: The Conference on Computer Vision and Pattern Recognition (2019)
54. Yoon, S.W., Seo, J., Moon, J.: Tapnet: Neural network augmented with task-adaptive projection for few-shot learning. arXiv preprint arXiv:1905.06549 (2019)
55. Zhang, H., Zhang, J., Koniusz, P.: Few-shot learning via saliency-guided hallucination of samples. In: The Conference on Computer Vision and Pattern Recognition (2019)
56. Zhang, H., Cisse, M., Dauphin, Y.N., Lopez-Paz, D.: mixup: beyond empirical risk minimization. arXiv preprint arXiv:1710.09412 (2017)
57. Zhang, J., Zhao, C., Ni, B., Xu, M., Yang, X.: Variational few-shot learning. In: The International Conference on Computer Vision (2019)

58. Zhao, F., Zhao, J., Yan, S., Feng, J.: Dynamic conditional networks for few-shot learning. In: Ferrari, V., Hebert, M., Sminchisescu, C., Weiss, Y. (eds.) ECCV 2018. LNCS, vol. 11219, pp. 20–36. Springer, Cham (2018). https://doi.org/10.1007/978-3-030-01267-0_2

59. Zhu, L., Yang, Y.: Compound memory networks for few-shot video classification. In: The European Conference on Computer Vision (2018)

Cyclic Functional Mapping: Self-supervised Correspondence Between Non-isometric Deformable Shapes

Dvir Ginzburg[✉] and Dan Raviv

Tel Aviv University, Tel Aviv, Israel
dvirginzburg@mail.tau.ac.il, darav@tauex.tau.ac.il

Abstract. We present the first spatial-spectral joint consistency network for self-supervised dense correspondence mapping between non-isometric shapes. The task of alignment in non-Euclidean domains is one of the most fundamental and crucial problems in computer vision. As 3D scanners can generate highly complex and dense models, the mission of finding dense mappings between those models is vital. The novelty of our solution is based on a cyclic mapping between metric spaces, where the distance between a pair of points should remain invariant after the full cycle. As the same learnable rules that generate the point-wise descriptors apply in both directions, the network learns invariant structures without any labels while coping with non-isometric deformations. We show here state-of-the-art-results by a large margin for a variety of tasks compared to known self-supervised and supervised methods.

Keywords: Dense shape correspondence · Self-supervision · One-shot learning · Spectral decomposition · 3D alignment

Reference   Network output

Fig. 1. TOSCA dataset results - similar colors represents correspondence mapping - we show excellent generalization after training for a single epoch on the TOSCA dataset with a pre-trained model on FAUST (Subsect. 6.5)

ⓒ Springer Nature Switzerland AG 2020
A. Vedaldi et al. (Eds.): ECCV 2020, LNCS 12350, pp. 36–52, 2020.
https://doi.org/10.1007/978-3-030-58558-7_3

1   Introduction

Alignment of non-rigid shapes is a fundamental problem in computer vision and plays an important role in multiple applications such as pose transfer, cross-shape texture mapping, 3D body scanning, and simultaneous localization and mapping (SLAM). The task of finding dense correspondence is especially challenging for non-rigid shapes, as the number of variables needed to define the mapping is vast, and local deformations might occur. To this end, we have seen a variety of papers focusing on defining unique key-points. These features capture the local uniqueness of the models using curvature [31], normals [48], or heat [45], for example, and further exploited for finding a dense mapping [5,12].

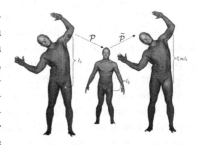

Fig. 2. Self-supervised dense correspondence using a cycle mapping architecture. By minimizing the geodesic distortion only on the source shape, we can learn complex deformations between structures.

A different approach used for alignment is based on pair-wise distortions, where angles [6,24] or distances [9,15,34,35], between pairs of points are minimized. Formulating this as a linear [52] or quadratic [2,9,47] optimization scheme showed a significant enhancement but with a painful time complexity even for small models. An interesting approach that addressed local deformations while still using Riemannian geometry introduced new metrics. Specifically, scale invariant [4,37], equi-affine invariant [32] and affine-invariant [36] metric showed superior results for alignment tasks. To confront the challenges in alignment between stretchable shapes we recognize non-metric methods based on conformal mapping [24], experimenting with alternative metrics such as scale [4] or affine invariant metrics [33], or attempts to embed the shapes into a plane or a cone [9,11], for example. A significant milestone named functional maps [30] has shown that such a mapping can be performed on the spectral domain, by aligning functions overlaid on top of the shapes.

Recently, a substantial improvement in dense alignment emerged using data-driven solutions, where axiomatic shape models and deformations were replaced by learnable counterparts. Among those methods a highly successful research direction was based on learning local features overlaid on the vertices of the shapes [25], where ResNet [19] like architecture is used to update SHOT descriptors [48].

The main challenge new data-driven geometric alignment algorithms need to face is the lack of data to train on or labeled data used for supervised learning. In many cases, the labeled data is expensive to generate or even infeasible to acquire, as seen, for example, in medical imaging.

A recent approach [18] showed that self-supervised learning could be applied for non-rigid alignment between isometric shapes by preserving the pair-wise distance measured on the source and on the target. While showing good results, an isometric limitation is a strong constraint that is misused in many scenarios.

On a different note, self-supervised learning was recently addressed in images, where a cyclic mapping between pictures, known as cyclic-GAN, was introduced [28,51,53]. The authors showed that given unpaired collection of images from different domains, a cyclic-loss that measures the distortion achieves robust state-of-the-art results in unsupervised learning for domain transfer. To this end, some papers have even shown promising insights when linking cyclic-constraints and alignment of the spectra [20,21,40,42]. Here we show that using the structure's metric one can align the spectrum of non-isometric shapes.

In this work, we claim that one can learn dense correspondence in a self-supervised manner in between non-isometric structures. We present a new learnable cyclic mechanism, where the same model is used both for forward and backward mapping learning to compensate for deformed parts. We measure the pair-wise distance distortion of the cyclic mapping on randomly chosen pair of points only from the source manifold. We show here state-of-the-art-results by a large margin for a variety of tasks compared to self-supervised and supervised methods, in isometric and non-isometric setups.

2   Contribution

We present an unsupervised learning scheme for dense 3D correspondence between shapes, based on learnable self similarities between metric spaces. The proposed approach has multiple advantages over other known methods. First, there is no need to define a model for the shapes or the deformations; Second, it has no need for labeled data with dense correspondence mappings. Third, the proposed method can handle isometric or non-isometric deformations and partial matching. The cyclic mapping approach allows our system to learn the geometric representation of manifolds by feeding it pairs of matching shapes, even without any labels, by measuring a geometric criterion (pair-wise distance) only on the source.

Our main contribution is based on the understanding that a cyclic mapping between metric spaces which follows the same rules, forces the network to learn invariant parts. We built the cyclic mapping using the functional maps framework [30], optimizing for a soft correspondence between shapes on the spectral domains by updating a local descriptor per point. The proposed approach can be adapted to any dimension, and here we provide state-of-the-art results on surfaces. We show results that are comparable to supervised learning [17,23,25] methods in the rare case we possess dense correspondence labels, and outperforms self-supervised learning approaches [17,18] when the shapes are isometric. Once the deformations are not isometric, our method stands out, and outperforms other methods by a large margin.

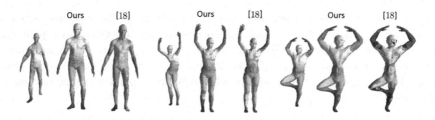

Fig. 3. Alignment between non-isometric shapes, where similar parts appear in similar colors. The shapes were locally scaled and stretched while changing their pose. Our approach learns the correct matching while [18] fails under local stretching.

3   Background

Our cyclic mapping is built on top of functional maps architecture. To explain the foundations of this approach, we must elaborate on distance matrices, functional maps and how to weave deep learning into functional maps. Finally, we discuss an isometric unsupervised approach for the alignment task and its limitations, which motivated this work.

3.1   Riemannian 2-Manifolds

We model 3D shapes as a Riemannian 2-manifold $(\mathcal{X}, g)$, where $\mathcal{X}$ is a real smooth manifold, equipped with an inner product g_p on the tangent space $T_p\mathcal{X}$ at each point p that varies smoothly from point to point in the sense that if $\mathcal{U}$ and $\mathcal{V}$ are differentiable vector fields on $\mathcal{X}$, then $p \to g_p(\mathcal{U}|_p, \mathcal{V}|_p)$ is a smooth function.

We equip the manifolds with a distance function $d\mathcal{X} : \mathcal{X} \times \mathcal{X} \to \Re$ induced by the standard volume form $d\mathcal{X}$. We state the distance matrix $\mathcal{D}_\mathcal{X}$, as a square symmetric matrix, represents the manifold's distance function $d\mathcal{X}$ such that

$$\mathcal{D}_{\mathcal{X}ij} = d\mathcal{X}(X_i, X_j). \tag{1}$$

3.2   Functional Maps

Functional maps [30] stands for matching real-valued functions in between manifolds instead of performing a straight forward point matching. Using a spectral basis, one can extract a compact representation for a match on the spectral domain. The clear advantage here is that many natural constraints on the map become linear constraints on the functional map. Given two manifolds $\mathcal{X}$ and $\mathcal{Y}$, and functional spaces on top $F(\mathcal{X})$ and $F(\mathcal{Y})$, we can define a functional map using orthogonal bases ϕ and ψ

$$\begin{aligned} Tf = T\sum_{i\geq 1} \langle f, \phi_i \rangle_\mathcal{X} \phi_i &= \sum_{i\geq 1} \langle f, \phi_i \rangle_\mathcal{X} T\phi_i \\ &= \sum_{i,j\geq 1} \langle f, \phi_i \rangle_\mathcal{X} \underbrace{\langle T\phi_i, \psi_j \rangle_\mathcal{Y}}_{c_{ji}} \psi_j, \end{aligned} \tag{2}$$

where $C \in \mathbb{R}^{k \times k}$ represents the mapping in between the domains given k matched functions, and every pair of corresponding functions on-top of the manifolds impose a linear constraint on the mapping. The coefficient matrix C is deeply dependent on the choice of the bases ϕ, ψ, and as shown in prior works [18,30,38] a good choice for such bases is the Laplacian eigenfunctions of the shapes.

3.3 Deep Functional Maps

Deep functional maps were first introduced in [25], where the mapping C between shapes was refined by learning new local features per point. The authors showed that using a ResNet [19] like architecture on-top of SHOT [48] descriptors, they can revise the local features in such a way that the global mapping is more accurate. The mapping is presented as a soft correspondence matrix P where P_{ji} is the probability $\mathcal{X}_i$ corresponds to $\mathcal{Y}_j$. The loss of the network is based on the geodesic distortion between the corresponding mapping and the ground truth, reading

$$\mathcal{L}_{sup}(\mathcal{X}, \mathcal{Y}) = \frac{1}{|\mathcal{X}|} \left\| \left(P \circ (\mathcal{D}_\mathcal{Y} \Pi^*) \right) \right\|_F^2, \tag{3}$$

where $|\mathcal{X}|$ is the number of vertices of shape $\mathcal{X}$, and if $|\mathcal{X}| = n$, and $|\mathcal{Y}| = m$, then $\Pi^* \in \Re^{m \times n}$ is the ground-truth mapping between $\mathcal{X}$ and $\mathcal{Y}$, $\mathcal{D}_\mathcal{Y} \in \Re^{m \times m}$ is the geodesic distance matrix of $\mathcal{Y}$, $\circ$ denote a point-wise multiplication, and $\|_F$ is the Frobenius norm. For each target vertex, the loss penalizes the distance between the actual corresponding vertex and the assumed one, multiplied by the amount of certainty the network has in that correspondence. Hence the loss is zero if

$$P_{ji} = 1 \Leftrightarrow \Pi^*(\mathcal{X}_i) = \mathcal{Y}_j, \tag{4}$$

as $\mathcal{D}(y, y) = 0 \; \forall y \in \mathcal{Y}$.

3.4 Self-Supervised Deep Functional Maps

The main drawback of deep functional maps is the need for ground truth labels. Obtaining alignment maps for domains such as ours is a strong requirement, and is infeasible in many cases either due to the cost to generate those datasets, or even impractical to collect. In a recent paper [18], the authors showed that for isometric deformations, we can replace the ground truth requirement with a different geometric criterion based on pair-wise distances. In practice, they married together the Gromov-Hausdorff framework with the deep functional maps architecture.

The Gromov-Hausdorff distance which measures the distance between metric spaces, reads

$$d_{GH}(\mathcal{X}, \mathcal{Y}) = \frac{1}{2} \inf_{\pi} (dis(\pi)), \tag{5}$$

where the infimum is taken over all correspondence distortions of a given mapping $\pi \colon \mathcal{X} \to \mathcal{Y}$. This distortion can be translated to a pair-wise distance [15,22] notation, which was used by [18] as a geometric criterion in the cost function of a deep functional map setup. Unfortunately, the pair-wise distance constraint is an extreme demand, forcing the models to be isometric, and can not be fulfilled in many practical scenarios.

4   Cyclic Self-Supervised Deep Functional Maps

The main contribution of this paper is the transition from the pair-wise distance comparison between source and target manifolds to a method that only examines the metric in the source manifold. Every pair of distances are mapped to the target and re-mapped back to the source. We use the same model for the forward and backward mapping to avoid a mode collapse, and we measure the distortion once a cyclic map has been completed, forcing the model to learn how to compensate for the deformations.

4.1   Correspondence Distortion

A mapping $\pi \colon X \to Y$ between two manifolds generates a pair-wise distortion

$$dis_{\pi}(\mathcal{X}, \mathcal{Y}) = \sum_{x_1, x_2 \in \mathcal{X}} \rho(d_{\mathcal{X}}(x_1, x_2), d_{\mathcal{Y}}(\pi(x_1), \pi(x_2)), \tag{6}$$

where ρ is usually an L_p norm metric, and $p = 2$ is a useful choice of the parameter.

As isometric mapping preserves pair-wise distances, minimizing the distances between those pairs provides a good metric-dependent correspondence. Specifically,

$$\pi^{iso}(\mathcal{X}, \mathcal{Y}) = \operatorname*{argmin}_{\pi \colon \mathcal{X} \to \mathcal{Y}} dis_{\pi}(\mathcal{X}, \mathcal{Y}). \tag{7}$$

Solving (7) takes the form of a quadratic assignment problem. The main drawback of this criterion, as the name suggests, is the isometric assumption. While it is a powerful tool for isometric mappings, natural phenomena do not follow that convention as stretching exists in the data. To overcome those limitations, we present here the cyclic distortion criterion.

4.2   Cyclic Distortion

We define a **cyclic distortion**[1] π^{cyc} as a composition of two mappings $\pi_\rightarrow$: $\mathcal{X} \rightarrow \mathcal{Y}$ and $\pi_\leftarrow : \mathcal{Y} \rightarrow \mathcal{X}$, which leads to a cyclic distortion

$$dis^{cyc}_{(\pi_\rightarrow, \pi_\leftarrow)}(\mathcal{X}, \mathcal{Y}) =$$
$$\sum_{x_1, x_2 \in \mathcal{X}} \rho(d_\mathcal{X}(x_1, x_2), d_\mathcal{X}(\tilde{x}_1, \tilde{x}_2)), \tag{8}$$

where $\tilde{x}_1 = \pi_\leftarrow(\pi_\rightarrow(x_1))$ and $\tilde{x}_2 = \pi_\leftarrow(\pi_\rightarrow(x_2))$.

$\pi_\rightarrow$ and $\pi_\leftarrow$ are being optimized using the same sub-network, implemented as shared weights in the learning process. Every forward mapping $\pi_\rightarrow$ induce a backward mapping $\pi_\leftarrow$ and vice-versa. We call this coupled pair $\pi = (\pi_\rightarrow, \pi_\leftarrow)$ a *conjugate mapping*, and denote the space of all conjugate mappings by $\mathcal{S}$. We define the cyclic mapping as

$$\pi^{cyc}(\mathcal{X}, \mathcal{Y}) = \underset{\pi:(\mathcal{X}\rightarrow\mathcal{Y}, \mathcal{Y}\rightarrow\mathcal{X})\in\mathcal{S}}{\mathrm{argmin}} \; dis^{cyc}_\pi(\mathcal{X}, \mathcal{Y}). \tag{9}$$

4.3   Deep Cyclic Mapping

Following the functional map convention, given C, Φ, Ψ the *soft correspondence matrix* mapping between $\mathcal{X}$ to $\mathcal{Y}$ reads [25]

$$P = \left|\Phi C \Psi^T\right|_{\mathcal{F}_c}, \tag{10}$$

where each entry P_{ji} is the probability point j in $\mathcal{X}$ corresponds to point i in $\mathcal{Y}$. We further use $|\cdot|_{\mathcal{F}_c}$ notation for column normalization, to emphasize the statistical interpretation of P.

Let P represents the forward mapping $\pi_\rightarrow$ soft correspondence and $\tilde{P}$ the backward mapping $\pi_\leftarrow$. The cyclic distortion is defined by

$$\mathcal{L}_{cyclic}(\mathcal{X}, \mathcal{Y}) = \frac{1}{|\mathcal{X}|^2} \left\| \left(D_\mathcal{X} - (\tilde{P}P)D_\mathcal{X}(\tilde{P}P)^T\right) \right\|^2_{\mathcal{F}}, \tag{11}$$

where $|\mathcal{X}|$ is the number of samples point pairs on $\mathcal{X}$.

Note that if we assumed the shapes were isometric, then we would have expected $D_\mathcal{Y}$ to be similar or even identical to $PD_\mathcal{X}P^T$, which yields once plugged into (11) the isometric constraint

$$\mathcal{L}_{isometric}(\mathcal{X}, \mathcal{Y}) = \frac{1}{|\mathcal{X}|^2} \left\| \left(D_\mathcal{X} - \tilde{P}D_\mathcal{Y}\tilde{P}^T\right) \right\|^2_{\mathcal{F}}. \tag{12}$$

The cyclic distortion (11) is self-supervised, as no labels are required, and only use the pair-wise distances on the source manifold $\mathcal{X}$. The conjugate mappings are based on the functional maps architecture and use the geometry of both spaces, source and target. Since we constrain the mapping on the source's geometry, the mapping copes with stretching, and thus learning invariant representations.

---

[1] An illustration of the distortion process is shown in Fig. 2.

5   Implementation

5.1   Hardware

The network was developed in TensorFlow [1], running on a GeForce GTX 2080 Ti GPU. The SHOT descriptor [48] was implemented in MATLAB, while the Laplace Beltrami Operator (LBO) [41] and geodesic distances were calculated in Python.

5.2   Pre-processing

We apply a sub-sampling process for shapes with more than 10,000 points using qSlim [16] algorithm. SHOT descriptor was computed on the sub-sampled shapes, generating a descriptor of length $s = 350$ per vertex. Finally, the LBO eigenfunctions corresponding to the most significant 70 eigenvalues (lowest by magnitude) were computed for each shape. The distance matrices were computed using the Fast Marching algorithm [43]. In order to initialize the conjugate mapping, we found that a hard constraint on P and $\tilde{P}$ coupling provides good results. Specifically we minimized in the first epoch the cost function $||P\tilde{P} - I||_{\mathcal{F}}^2$ before applying the soft cyclic criterion (11). Equation 11 consists of four probability matrices multiplication which leads to low values in the cost function and its derivatives. Hence, we found an advantage to start the optimization process with a simpler cost function in the first epoch. We examined several configurations of loss functions, including pre-training on non-cyclic loss, yet, we did not observe any improvement over direct optimization of Eq. (9).

5.3   Network Architecture

The architecture is motivated by [18,25] and shown in Fig. 4. The input to the first layer is the raw 3D triangular mesh representations of the two figures given by a list of vertices and faces. We apply a multi-layer SHOT [48] descriptor by evaluating the SHOT on $m \sim 5$ global scaled versions of the input. The figures vary from 0.2 to 2 times the size of the original figures, followed by a 1×1 convolution layers with $2\,m$ filters, to a 1×1 convolution layer with one filter, generating an output of $n \times s$ descriptor to the network. Besides, the relevant eigenfunctions and pair-wise distance matrix of the source shape are provided as parameters to the network.

The next stage is the ResNet [19] layers with the shared weights applied to both figures. Subsequently, the non-linear descriptors are multiplied by the $n \times k$ LBO eigenfunctions. We calculate the forward and backward mappings C and $\tilde{C}$ using the same network and evaluate the corresponding forward and backward mappings P and $\tilde{P}$, which are fed into the soft cyclic loss (11).

The fact we do not use geodesic distance metric on the target shape $D_{\mathcal{Y}}$ contributed to faster inference time, and lighter training batches, which accelerated the learning with comparison to [18,25].

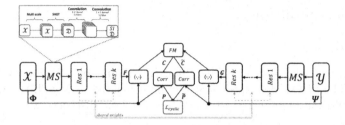

Fig. 4. Cyclic functional mapper in between two manifolds $\mathcal{X}$ and $\mathcal{Y}$ (left and right sides). The multi-scaled descriptors (top left, marked MS) based on shot [48] are passed to a ResNet like network, resulting in two corresponding coefficient matrices F and G. By projecting the refined descriptors onto the spectral space, two mappings, C and $\tilde{C}$, are computed. The two soft correspondence matrices P and $\tilde{P}$ are further used as part of the network cyclic loss $\mathcal{L}_{cyclic}$ as shown in Eq. (11).

6   Experiments

In this section, we present multiple experiments in different settings; synthetic and real layouts, transfer learning tasks, non-isometric transformations, partial matching and one-shot learning. We show benchmarks, as well as comparisons to state-of-the-art solutions, both for axiomatic and learned algorithms.

6.1   Mesh Error Evaluation

The measure of error for the correspondence mapping between two shapes will be according to the Princeton benchmark [22], that is, given a mapping $\pi_\rightarrow(\mathcal{X}, \mathcal{Y})$ and the ground truth $\pi^*_\rightarrow(\mathcal{X}, \mathcal{Y})$ the error of the correspondence matrix is the sum of geodesic distances between the mappings for each point in the source figure, divided by the area of the target figure.

$$\epsilon(\pi_\rightarrow) = \sum_{x\in\mathcal{X}} \frac{\mathcal{D}_\mathcal{Y}(\pi_\rightarrow(x), \pi^*_\rightarrow(x))}{\sqrt{area(\mathcal{Y})}}, \tag{13}$$

where the approximation of $area(\bullet)$ for a triangular mesh is the sum of its triangles area.

6.2   Synthetic FAUST

We compared our alignment on FAUST dataset [7] versus supervised [25] and unsupervised [18] methods. We followed the experiment as described in [25] and used the synthetic human shapes dataset, where the first 80 shapes (8 subjects with 10 different poses each) are devoted to training, and 20 shapes made of 2 different unseen subjects are used for testing.

For a fair comparison between methods, we did not run the PMF cleanup filter [50] as this procedure is extremely slow and takes about 15 min for one shape built of ∼7 k vertices on an i9 desktop.

We do not perform any triangular mesh preprocessing on the dataset, that is, we learn on the full resolution of 6890 vertices.

Each mini-batch is of size 4 (i.e. 4 pairs of figures), using $k = 120$ eigenfunctions, and 10 bins in SHOT with a radius set to be 5% of the geodesic distance from each vertex. We report superior results for inter-subject and intra-subject tasks in Table 1, while converging faster (see Fig. 6).

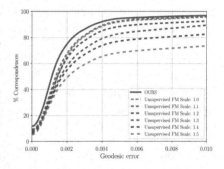

Fig. 5. Scale-Invariant Attributes - Network performance under global-scaling transformations

As shown in Fig. 5 global scaling is transparent to our cyclic consistent loss. We provide several evaluations on the Synthetic FAUST dataset where the target shape was global scaled by different factors during training. While we converge to the same result in each experiment, unsupervised methods as [18] that assume isometry fail to learn in such frameworks.

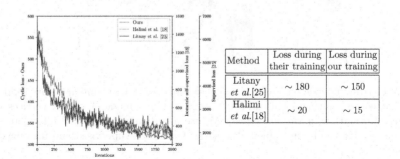

Method	Loss during their training	Loss during our training
Litany et al.[25]	∼ 180	∼ 150
Halimi et al.[18]	∼ 20	∼ 15

Fig. 6. We visualize our cyclic loss, the isometric constrained unsupervised loss [18], and the supervised loss [25] during the training of our cyclic loss on the synthetic FAUST dataset. We show that minimization of the cyclic loss on isometric structures provides lower loss values on previous architectures, without any geometric assumption.

6.3 Real Scans

We tested our method on real 3D scans of humans from the FAUST [7] dataset. While the synthetic samples had ∼6 k vertices, each figure in this set has ∼150 k vertices, creating the amount of plausible cyclic mappings extremely high. The dataset consists of multiple subjects in a variety of poses, where none of the poses (e.g., women with her hands up) in the test set were present in the training set.

The samples were acquired using a full-body 3D stereo capture system, resulting in missing vertices, and open-end meshes.

The dataset is split into two test cases as before, the intra and inter subjects (60 and 40 pairs respectively), and ground-truth correspondences is not given. Hence, the geodesic error evaluation is provided as an online service. As suggested in [25], after evaluating the soft correspondence mappings, we input our map to the PMF algorithm [50] for a smoother bijective correspondence refined map. We report

Table 1. Average error on the FAUST dataset measured as distance between mapped points and the ground truth. We compared between our approach and other supervised and unsupervised methods.

Method	Scans [cm]		Synthetic [cm]	
	Inter	Intra	Inter	Intra
Ours	**4.068**	**2.12**	**2.327**	**2.112**
Litany *et al.* [25]	4.826	2.436	2.452	2.125
Halimi *et al.* [18]	4.883	2.51	3.632	2.213
Groueix[a] *et al.* [17]	4.833	2.789	—	—
Li *et al.* [23]	4.079	2.161	—	—
Chen *et al.* [13]	8.304	4.86	—	—

[a]Unsupervised training.

state of the art results on both inter and intra class mappings in comparison to all the unsupervised techniques. We provide visualization in Fig. 7 and qualitative results in Table 1.

Fig. 7. Correspondence on FAUST real scans dataset, where similar colors represent the same correspondence. This dataset contains shapes made of ∼100 k vertices with missing information in various poses. We use a post-matching PMF filter [50], and show qualitative results in Table 1. We outperform both supervised and unsupervised methods.

6.4 Non-Isometric Deformations

An even bigger advantage of the proposed method is its ability to cope with local stretching. Due to the cyclic mapping approach, we learn local matching features directly between the models and are not relying on a base shape in the latent space or assume isometric consistency. We experimented with models generated in Autodesk Maya that were locally stretched and bent. We remark that none of the stretched samples was present during training, which was done on the "standard" FAUST real scans dataset. We show visual results in Fig. 3. The proposed approach successfully handles large non-isometric deformations.

6.5 TOSCA

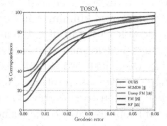
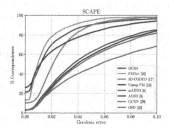

Fig. 8. Geodesic error on TOSCA dataset. We report superior results against other supervised and unsupervised learnable methods. Note that the compared methods did not run a post-processing optimization-based filter, or received partial matching as input.

Fig. 9. Geodesic error on SCAPE dataset. Our network was trained on FAUST dataset and used to predict the mapping on SCAPE. We provide superior results on all unspervised and almost all supervised methods showing good generalization properties.

The TOSCA dataset [10] consists of 80 objects from different domains as animals and humans in different poses. Although the animals are remarkably different in terms of LBO decomposition, as well as geometric characteristics, our model achieves excellent performance in terms of a geodesic error on the dataset after training for a single epoch on it, using the pre-trained model from the real scans FAUST dataset (Fig. 1).

In Fig. 8, we show a comparison between our and other supervised and unsupervised approaches and visualize a few samples in Fig. 1. Compared methods results were taken from [18]. Our network was trained for a single epoch on the dataset, with a pre-trained model of the real scans FAUST data and yet, shows great performance. We report state of the art results, compared to axiomatic, supervised, and unsupervised methods. Also note that while other methods mention training on each class separately, we achieve state-of-the-art results while training jointly.

6.6 SCAPE

To further emphasize the generalization capabilities of our network, we present our results on the SCAPE dataset [46], which is an artificial human shape dataset, digitally generated, with completely different properties from the FAUST dataset in any aspect (geometric entities, scale, ratio, for example). Nevertheless, our network that was trained on the real scan FAUST dataset performs remarkably well. See Fig. 9. Compared methods results were taken from [18].

6.7 One-Shot Single Pair Learning

Following the experiment shown in [18], we demonstrate that we can map in between two shapes seen for the first time without training on a large dataset. Compared to optimization approaches we witness improved running time due to optimized hardware and software dedicated to deep learning in recent years. In Fig. 10 we show such a mapping in between highly deformed shapes, and we found it intriguing that a learning method based on just two samples can converge to a feasible solution even without strong geometric assumptions. Note that in that case methods based on isometric criterion fail to converge due to the large non-isometric deformation. Running multiple epochs on two shapes not only converged to a pleasing mapping, but we found it to work well without any priors on the shapes or a need to engineer the initialization process. That is true for inter-class as well as highly deformed shapes (See Fig. 10) In this experiment we used our multi-SHOT pre-trained weights before we ran our cyclic mapper.

Fig. 10. Left - Highly deformed single pair, inter-class experiment. Right - Single pair (one shot) learning on deformable non-isometric shapes. Supervised methods as [25] are irrelevant, where isometric self-learning approach fails [18].

6.8 Partial Shapes Correspondence

The partial shapes correspondence task is inherently more complicated than the full figure alignment problems. While in most experiments shown above, the number of vertices in both shapes differed by less than 5%, in the partial shapes task, we consider mappings between objects that differ by a large margin of up to 75% in their vertex count. To this end, numerous bijective solutions, such as [44,49,50] show degraded performance on the partial challenge, resulting in targeted algorithms [26,38] for the mission. With that in mind, we show our results on the SHREC 2016 [14] partial shapes dataset (Fig. 12 and Fig. 11). We remark that our formulation does not require the map to be bijective. As $P \in \mathbb{R}^{n_1 \times n_2}$, $\tilde{P} \in \mathbb{R}^{n_2 \times n_1}$ where n_1, n_2 are the number of vertices in the shapes no bijection is enforced, rather than the composition is close to the identity. Thus, the distortion minimization function finds the best mapping it can even for partial to full correspondence.

Fig. 11. Partial to Full correspondence - Although the mapping is ill-posed by nature in the missing segments, the cyclic mapper acheives near-perfect results within the valid areas of the source shape.

We use the same architecture as described earlier, given hyperparameters and trained weights from the TOSCA (Subsect. 6.5) experiment, showing our network's generalization capabilities. As before, we have trained the network on this dataset only for a single epoch.

Fig. 12. Partial shapes correspondence on SHREC 16 [14] dataset after removing substantial parts (up to 75%). In every pair, we mapped the left shape into the right one, where similar mapped points share the color. Our method is robust to missing information.

7   Limitations

Our method uses functional maps architecture, which requires us to pre-compute sets of bases functions. To that end, this process can not be done in real-time in the current setup, and there might be an inconsistency in bases functions between shapes due to noise or large non-isometric deformations. While this method works well for isometric or stretchable domains, once the deformations are significantly large, we found that the current system does not converge to a reasonable geodesic error in terms of a pleasant visual solution, which makes it challenging to use in cross-domain alignments. We believe that the proposed approach can be used as part of semantic-correspondence to overcome those limitations. Furthermore, experimenting with mappings between partial shapes showed us (see Fig. 11), that missing parts can suffers from ambiguity and can appear as a non-smooth mapping.

8   Summary

We presented here a cyclic architecture for dense correspondence between shapes. In the heart of our network is the bidirectional mapper, which jointly learns the mapping from the source to the target and back via a siamese-like architecture. We introduced a novel loss function which measures the distortion only on the source, while still using both geometries, allowing us to cope with non-isometric deformations. This approach is self-supervised, can cope with local stretching as well as non-rigid isometric deformations. While our concept specifically addresses the wrong assumption of isometry for inter-class subjects, we see superior results even for intra-class datasets. Our method outperforms other unsupervised and supervised approaches on tested examples, and we report state-of-the-art results in several scenarios, including real 3D scans and partial matching task.

Acknowledgment. D.R. is partially funded by the Zimin Institute for Engineering Solutions Advancing BetterLives, the Israeli consortiums for soft robotics and autonomous driving, and the Shlomo Shmeltzer Institute for Smart Transportation.

References

1. Abadi, M., et al.: TensorFlow: Large-Scale Machine Learning on Heterogeneous Systems (2015). Software available from http://tensorflow.org/
2. Aflalo, Y., Bronstein, A., Kimmel, R.: On convex relaxation of graph isomorphism. Proc. Natl. Acad. Sci. **112**(10), 2942–2947 (2015)
3. Aflalo, Y., Dubrovina, A., Kimmel, R.: Spectral generalized multi-dimensional scaling. Int. J. Comput. Vision **118**(3), 380–392 (2016)
4. Aflalo, Y., Kimmel, R., Raviv, D.: Scale invariant geometry for nonrigid shapes. SIAM J. Imaging Sci. **6**(3), 1579–1597 (2013)
5. Aubry, M., Schlickewei, U., Cremers, D.: The wave kernel signature: a quantum mechanical approach to shape analysis. In: 2011 IEEE International Conference on Computer Vision Workshops (ICCV Workshops), pp. 1626–1633. IEEE (2011)
6. Ben-Chen, M., Gotsman, C., Bunin, G.: Conformal flattening by curvature prescription and metric scaling. Comput. Graph. Forum **27**, 449–458 (2008). Wiley Online Library
7. Bogo, F., Romero, J., Loper, M., Black, M.J.: FAUST: dataset and evaluation for 3D mesh registration. In: Proceedings of the IEEE Conference on Computer Vision and Pattern Recognition (CVPR). IEEE, Piscataway, June 2014
8. Boscaini, D., Masci, J., Rodolà, E., Bronstein, M.M., Cremers, D.: Anisotropic diffusion descriptors. Comput. Graph. Forum **35**, 431–441 (2016)
9. Bronstein, A.M., Bronstein, M.M., Kimmel, R.: Generalized multidimensional scaling: a framework for isometry-invariant partial surface matching. Proc. Natl. Acad. Sci. **103**(5), 1168–1172 (2006)
10. Bronstein, A.M., Bronstein, M.M., Kimmel, R.: Numerical Geometry of Non-Rigid Shapes. Springer, New York (2008)
11. Bronstein, M.M., Bronstein, A.M., Kimmel, R., Yavneh, I.: Multigrid multidimensional scaling. Numer. Linear Algebra Appl. **13**(2–3), 149–171 (2006)
12. Bronstein, M.M., Kokkinos, I.: Scale-invariant heat kernel signatures for non-rigid shape recognition. In: 2010 IEEE Computer Society Conference on Computer Vision and Pattern Recognition, pp. 1704–1711. IEEE (2010)
13. Chen, Q., Koltun, V.: Robust nonrigid registration by convex optimization. In: Proceedings of the IEEE International Conference on Computer Vision, pp. 2039–2047 (2015)
14. Cosmo, L., Rodolà, E., Bronstein, M.M., Torsello, A., Cremers, D., Sahillioglu, Y.: SHREC'16: Partial matching of deformable shapes
15. Elad, A., Kimmel, R.: On bending invariant signatures for surfaces. IEEE Trans. Pattern Anal. Mach. Intell. **25**(10), 1285–1295 (2003)
16. Garland, M., Heckbert, P.S.: Surface simplification using quadric error metrics. In: Proceedings of the 24th Annual Conference on Computer Graphics and Interactive Techniques, pp. 209–216. ACM Press/Addison-Wesley Publishing Co. (1997)
17. Groueix, T., Fisher, M., Kim, V.G., Russell, B.C., Aubry, M.: 3D-CODED : 3D Correspondences by Deep Deformation. CoRR abs/1806.05228 (2018), http://arxiv.org/abs/1806.05228

18. Halimi, O., Litany, O., Rodola, E., Bronstein, A.M., Kimmel, R.: Unsupervised learning of dense shape correspondence. In: Proceedings of the IEEE Conference on Computer Vision and Pattern Recognition, pp. 4370–4379 (2019)
19. He, K., Zhang, X., Ren, S., Sun, J.: Deep residual learning for image recognition. CoRR abs/1512.03385 (2015). http://arxiv.org/abs/1512.03385
20. Huang, Q.X., Guibas, L.: Consistent shape maps via semidefinite programming. Comput. Graph. Forum **32**, 177–186 (2013). Wiley Online Library
21. Huang, Q., Wang, F., Guibas, L.: Functional map networks for analyzing and exploring large shape collections. ACM Trans. Graph. (TOG) **33**(4), 1–11 (2014)
22. Kim, V.G., Lipman, Y., Funkhouser, T.: Blended intrinsic maps. ACM Trans. Graphics (TOG) **30**, 79 (2011). ACM
23. Li, C.L., Simon, T., Saragih, J., Póczos, B., Sheikh, Y.: LBS Autoencoder: self-supervised fitting of articulated meshes to point clouds. In: Proceedings of the IEEE Conference on Computer Vision and Pattern Recognition, pp. 11967–11976 (2019)
24. Lipman, Y., Daubechies, I.: Conformal Wasserstein distances: comparing surfaces in polynomial time. Adv. Math. **227**(3), 1047–1077 (2011)
25. Litany, O., Remez, T., Rodolà, E., Bronstein, A.M., Bronstein, M.M.: Deep functional maps: structured prediction for dense shape correspondence. CoRR abs/1704.08686 (2017). http://arxiv.org/abs/1704.08686
26. Litany, O., Rodolà, E., Bronstein, A.M., Bronstein, M.M.: Fully spectral partial shape matching. Comput. Graph. Forum **36**, 247–258 (2017). Wiley Online Library
27. Litman, R., Bronstein, A.M.: Learning spectral descriptors for deformable shape correspondence. IEEE Trans. Pattern Anal. Mach. Intell. **36**(1), 171–180 (2013)
28. Liu, M.Y., Breuel, T., Kautz, J.: Unsupervised image-to-image translation networks. In: Advances in Neural Information Processing Systems, pp. 700–708 (2017)
29. Masci, J., Boscaini, D., Bronstein, M., Vandergheynst, P.: Geodesic convolutional neural networks on Riemannian manifolds. In: Proceedings of the IEEE International Conference on Computer Vision Workshops, pp. 37–45 (2015)
30. Ovsjanikov, M., Ben-Chen, M., Solomon, J., Butscher, A., Guibas, L.: Functional maps: a flexible representation of maps between shapes. ACM Trans. Graph. (TOG) **31**(4), 30 (2012)
31. Pottmann, H., Wallner, J., Huang, Q.X., Yang, Y.L.: Integral invariants for robust geometry processing. Comput. Aided Geometr. Des. **26**(1), 37–60 (2009)
32. Raviv, D., Bronstein, A.M., Bronstein, M.M., Waisman, D., Sochen, N., Kimmel, R.: Equi-affine invariant geometry for shape analysis. J. Math. Imaging Vis. **50**(1–2), 144–163 (2014)
33. Raviv, D., Bronstein, M.M., Bronstein, A.M., Kimmel, R., Sochen, N.: Affine-invariant diffusion geometry for the analysis of deformable 3D shapes. In: CVPR 2011, pp. 2361–2367. IEEE (2011)
34. Raviv, D., Dubrovina, A., Kimmel, R.: Hierarchical matching of non-rigid shapes. In: Bruckstein, A.M., ter Haar Romeny, B.M., Bronstein, A.M., Bronstein, M.M. (eds.) SSVM 2011. LNCS, vol. 6667, pp. 604–615. Springer, Heidelberg (2012). https://doi.org/10.1007/978-3-642-24785-9_51
35. Raviv, D., Dubrovina, A., Kimmel, R.: Hierarchical framework for shape correspondence. Numer. Math. Theory Methods Appl. **6**(1), 245–261 (2013)
36. Raviv, D., Kimmel, R.: Affine invariant geometry for non-rigid shapes. Int. J. Comput. Vision **111**(1), 1–11 (2015)
37. Raviv, D., Raskar, R.: Scale invariant metrics of volumetric datasets. SIAM J. Imaging Sci. **8**(1), 403–425 (2015)

38. Rodolà, E., Cosmo, L., Bronstein, M.M., Torsello, A., Cremers, D.: Partial functional correspondence. Comput. Graph. Forum **36**, 222–236 (2017). Wiley Online Library
39. Rodolà, E., Rota Bulo, S., Windheuser, T., Vestner, M., Cremers, D.: Dense non-rigid shape correspondence using random forests. In: Proceedings of the IEEE Conference on Computer Vision and Pattern Recognition, pp. 4177–4184 (2014)
40. Roufosse, J.M., Sharma, A., Ovsjanikov, M.: Unsupervised deep learning for structured shape matching. In: Proceedings of the IEEE International Conference on Computer Vision, pp. 1617–1627 (2019)
41. Rustamov, R.M.: Laplace-Beltrami eigenfunctions for deformation invariant shape representation. In: Proceedings of the Fifth Eurographics Symposium on Geometry Processing, SGP 2007, pp. 225–233. Eurographics Association, Aire-la-Ville (2007). http://dl.acm.org/citation.cfm?id=1281991.1282022
42. Rustamov, R.M., Ovsjanikov, M., Azencot, O., Ben-Chen, M., Chazal, F., Guibas, L.: Map-based exploration of intrinsic shape differences and variability. ACM Trans. Graph. (TOG) **32**(4), 1–12 (2013)
43. Sethian, J.A.: A fast marching level set method for monotonically advancing fronts. Proc. Natl. Acad. Sci. **93**(4), 1591–1595 (1996)
44. Starck, J., Hilton, A.: Spherical matching for temporal correspondence of non-rigid surfaces. In: Tenth IEEE International Conference on Computer Vision (ICCV 2005), Volume 1, vol. 2, pp. 1387–1394. IEEE (2005)
45. Sun, J., Ovsjanikov, M., Guibas, L.: A concise and provably informative multi-scale signature based on heat diffusion. Comput. Graph. Forum **28**, 1383–1392 (2009). Wiley Online Library
46. Szeliski, R., et al.: SCAPE: shape completion and animation of people, vol. 24 (2005)
47. Tevs, A., Berner, A., Wand, M., Ihrke, I., Seidel, H.P.: Intrinsic shape matching by planned landmark sampling. Comput. Graph. Forum **30**, 543–552 (2011). Wiley Online Library
48. Tombari, F., Salti, S., Di Stefano, L.: Unique signatures of histograms for local surface description. In: Daniilidis, K., Maragos, P., Paragios, N. (eds.) ECCV 2010. LNCS, vol. 6313, pp. 356–369. Springer, Heidelberg (2010). https://doi.org/10.1007/978-3-642-15558-1_26
49. Vestner, M., et al.: Efficient deformable shape correspondence via kernel matching. In: 2017 International Conference on 3D Vision (3DV), pp. 517–526. IEEE (2017)
50. Vestner, M., Litman, R., Rodolà, E., Bronstein, A., Cremers, D.: Product manifold filter: non-rigid shape correspondence via kernel density estimation in the product space. In: Proceedings of the IEEE Conference on Computer Vision and Pattern Recognition, pp. 3327–3336 (2017)
51. Yi, Z., Zhang, H., Tan, P., Gong, M.: Dualgan: unsupervised dual learning for image-to-image translation. In: Proceedings of the IEEE International Conference on Computer Vision, pp. 2849–2857 (2017)
52. Zaharescu, A., Boyer, E., Varanasi, K., Horaud, R.: Surface feature detection and description with applications to mesh matching. In: 2009 IEEE Conference on Computer Vision and Pattern Recognition, pp. 373–380. IEEE (2009)
53. Zhu, J.Y., Park, T., Isola, P., Efros, A.A.: Unpaired image-to-image translation using cycle-consistent adversarial networks. In: Proceedings of the IEEE International Conference on Computer Vision, pp. 2223–2232 (2017)

View-Invariant Probabilistic Embedding
for Human Pose

Jennifer J. Sun[1]([✉]), Jiaping Zhao[2], Liang-Chieh Chen[2], Florian Schroff[2],
Hartwig Adam[2], and Ting Liu[2]

[1] California Institute of Technology, Pasadena, USA
jjsun@caltech.edu
[2] Google Research, Los Angeles, USA
{jiapingz,lcchen,fschroff,hadam,liuti}@google.com

Abstract. Depictions of similar human body configurations can vary
with changing viewpoints. Using only 2D information, we would like to
enable vision algorithms to recognize similarity in human body poses
across multiple views. This ability is useful for analyzing body movements and human behaviors in images and videos. In this paper, we propose an approach for learning a compact view-invariant embedding space
from 2D joint keypoints alone, without explicitly predicting 3D poses.
Since 2D poses are projected from 3D space, they have an inherent ambiguity, which is difficult to represent through a deterministic mapping.
Hence, we use probabilistic embeddings to model this input uncertainty.
Experimental results show that our embedding model achieves higher
accuracy when retrieving similar poses across different camera views, in
comparison with 2D-to-3D pose lifting models. We also demonstrate the
effectiveness of applying our embeddings to view-invariant action recognition and video alignment. Our code is available at https://github.com/
google-research/google-research/tree/master/poem.

Keywords: Human pose embedding · Probabilistic embedding ·
View-invariant pose retrieval

1   Introduction

When we represent three dimensional (3D) human bodies in two dimensions
(2D), the same human pose can appear different across camera views. There
can be significant visual variations from a change in viewpoint due to changing
relative depth of body parts and self-occlusions. Despite these variations, humans
have the ability to recognize similar 3D human body poses in images and videos.
This ability is useful for computer vision tasks where changing viewpoints should
not change the labels of the task. We explore how we can embed 2D visual

J.J. Sun—This work was done during the author's internship at Google.

Electronic supplementary material The online version of this chapter (https://
doi.org/10.1007/978-3-030-58558-7_4) contains supplementary material, which is available to authorized users.

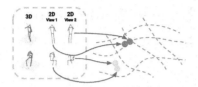
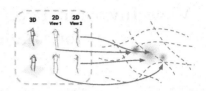
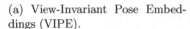

(a) View-Invariant Pose Embed-   (b) Probabilistic View-Invariant
dings (VIPE).   Pose Embeddings (Pr-VIPE).

Fig. 1. We embed 2D poses such that our embeddings are (a) view-invariant (2D projections of similar 3D poses are embedded close) and (b) probabilistic (embeddings are distributions that cover different 3D poses projecting to the same input 2D pose).

information of human poses to be consistent across camera views. We show that these embeddings are useful for tasks such as view-invariant pose retrieval, action recognition, and video alignment.

Inspired by 2D-to-3D lifting models [32], we learn view invariant embeddings directly from 2D pose keypoints. As illustrated in Fig. 1, we explore whether view invariance of human bodies can be achieved from 2D poses alone, without predicting 3D pose. Typically, embedding models are trained from images using deep metric learning techniques [8,14,35]. However, images with similar human poses can appear different because of changing viewpoints, subjects, backgrounds, clothing, etc. As a result, it can be difficult to understand errors in the embedding space from a specific factor of variation. Furthermore, multi-view image datasets for human poses are difficult to capture in the wild with 3D groundtruth annotations. In contrast, our method leverages existing 2D keypoint detectors: using 2D keypoints as inputs allows the embedding model to focus on learning view invariance. Our 2D keypoint embeddings can be trained using datasets in lab environments, while having the model generalize to in-the-wild data. Additionally, we can easily augment training data by synthesizing multi-view 2D poses from 3D poses through perspective projection.

Another aspect we address is input uncertainty. The input to our embedding model is 2D human pose, which has an inherent ambiguity. Many valid 3D poses can project to the same or very similar 2D pose [1]. This input uncertainty is difficult to represent using deterministic mappings to the embedding space (point embeddings) [24,37]. Our embedding space consists of probabilistic embeddings based on multivariate Gaussians, as shown in Fig. 1b. We show that the learned variance from our method correlates with input 2D ambiguities. We call our approach Pr-VIPE for **Pr**obabilistic **V**iew-**I**nvariant **P**ose **E**mbeddings. The non-probabilistic, point embedding formulation will be referred to as VIPE.

We show that our embedding is applicable to subsequent vision tasks such as pose retrieval [21,35], video alignment [11], and action recognition [18,60]. One direct application is pose-based image retrieval. Our embedding enables users to search images by fine-grained pose, such as jumping with hands up, riding bike with one hand waving, and many other actions that are potentially difficult to pre-define. The importance of this application is further highlighted by works such as [21,35]. Compared with using 3D keypoints with alignment for retrieval, our embedding enables efficient similarity comparisons in Euclidean space.

Contributions. Our main contribution is the method for learning an embedding space where 2D pose embedding distances correspond to their similarities in absolute 3D pose space. We also develop a probabilistic formulation that captures 2D pose ambiguity. We use cross-view pose retrieval to evaluate the view-invariant property: given a monocular pose image, we retrieve the same pose from different views without using camera parameters. Our results suggest 2D poses are sufficient to achieve view invariance without image context, and we do not have to predict 3D pose coordinates to achieve this. We also demonstrate the use of our embeddings for action recognition and video alignment.

2   Related Work

Metric Learning. We are working to understand similarity in human poses across views. Most works that aim to capture similarity between inputs generally apply techniques from metric learning. Objectives such as contrastive loss (based on pair matching) [4,12,37] and triplet loss (based on tuple ranking) [13,50,56,57] are often used to push together/pull apart similar/dissimilar examples in embedding space. The number of possible training tuples increases exponentially with respect to the number of samples in the tuple, and not all combinations are equally informative. To find informative training tuples, various mining strategies are proposed [13,38,50,58]. In particular, semi-hard triplet mining has been widely used [42,50,58]. This mining method finds negative examples that are fairly hard as to be informative but not too hard for the model. The hardness of a negative sample is based on its embedding distance to the anchor. Commonly, this distance is the Euclidean distance [13,50,56,57], but any differentiable distance function could be applied [13]. [16,19] show that alternative distance metrics also work for image and object retrieval.

In our work, we learn a mapping from Euclidean embedding distance to a probabilistic similarity score. This probabilistic similarity captures closeness in 3D pose space from 2D poses. Our work is inspired by the mapping used in soft contrastive loss [37] for learning from an occluded N-digit MNIST dataset.

Most of the papers discussed above involve deterministically mapping inputs to point embeddings. There are works that also map inputs to probabilistic embeddings. Probabilistic embeddings have been used to model specificity of word embeddings [55], uncertainty in graph representations [3], and input uncertainty due to occlusion [37]. We will apply probabilistic embeddings to address inherent ambiguities in 2D pose due to 3D-to-2D projection.

Human Pose Estimation. 3D human poses in a global coordinate frame are view-invariant, since images across views are mapped to the same 3D pose. However, as mentioned by [32], it is difficult to infer the 3D pose in an arbitrary global frame since any changes to the frame does not change the input data. Many approaches work with poses in the camera coordinate system [6,7,32,43, 46,48,52,53,62], where the pose description changes based on viewpoint. While our work focuses on images with a single person, there are other works focusing on describing poses of multiple people [47].

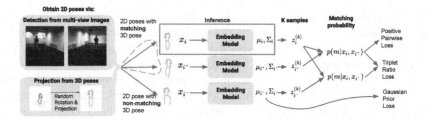

Fig. 2. Overview of Pr-VIPE model training and inference. Our model takes keypoint input from a single 2D pose (detected from images and/or projected from 3D poses) and predicts embedding distributions. Three losses are applied during training.

Our approach is similar in setup to existing 3D lifting pose estimators [6,9,32,43,46] in terms of using 2D pose keypoints as input. The difference is that lifting models are trained to regress to 3D pose keypoints, while our model is trained using metric learning and outputs an embedding distribution. Some recent works also use multi-view datasets to predict 3D poses in the global coordinate frame [20,26,44,49,54]. Our work differs from these methods with our goal (view-invariant embeddings), task (cross-view pose retrieval), and approach (metric learning). Another work on pose retrieval [35] embeds images with similar 2D poses in the same view close together. Our method focuses on learning view invariance, and we also differ from [35] in method (probabilistic embeddings).

View Invariance and Object Retrieval. When we capture a 3D scene in 2D as images or videos, changing the viewpoint often does not change other properties of the scene. The ability to recognize visual similarities across viewpoints is helpful for a variety of vision tasks, such as motion analysis [22,23], tracking [39], vehicle and human re-identification [8,61], object classification and retrieval [14,15,27], and action recognition [28,29,45,59].

Some of these works focus on metric learning for object retrieval. Their learned embedding spaces place different views of the same object class close together. Our work on human pose retrieval differs in a few ways. Our labels are continuous 3D poses, whereas in object recognition tasks, each embedding is associated with a discrete class label. Furthermore, we embed 2D poses, while these works embed images. Our approach allows us to investigate the impact of input 2D uncertainty with probabilistic embeddings and explore confidence measures to cross-view pose retrieval. We hope that our work provides a novel perspective on view invariance for human poses.

3   Our Approach

The training and inference framework of Pr-VIPE is illustrated in Fig. 2. Our goal is to embed 2D poses such that distances in the embedding space correspond to similarities of their corresponding absolute 3D poses in Euclidean space. We achieve this view invariance property through our triplet ratio loss (Sect. 3.2), which pushes together/pull apart 2D poses corresponding to similar/dissimilar 3D poses. The positive pairwise loss (Sect. 3.3) is applied to increase the matching probability of similar poses. Finally, the Gaussian prior loss (Sect. 3.4) helps regularize embedding magnitude and variance.

3.1   Matching Definition

The 3D pose space is continuous, and two 3D poses can be trivially different without being identical. We define two 3D poses to be matching if they are visually similar regardless of viewpoint. Given two sets of 3D keypoints $(\boldsymbol{y}_i, \boldsymbol{y}_j)$, we define a matching indicator function

$$
m_{ij} = \begin{cases} 1, & \text{if NP-MPJPE}(\boldsymbol{y}_i, \boldsymbol{y}_j) \leqslant \kappa \\ 0, & \text{otherwise,} \end{cases} \tag{1}
$$

where κ controls visual similarity between matching poses. Here, we use mean per joint position error (MPJPE) [17] between the two sets of 3D pose keypoints as a proxy to quantify their visual similarity. Before computing MPJPE, we normalize the 3D poses and apply Procrustes alignment between them. The reason is that we want our model to be view-invariant and to disregard rotation, translation, or scale differences between 3D poses. We refer to this normalized, Procrustes aligned MPJPE as **NP-MPJPE**.

3.2   Triplet Ratio Loss

The triplet ratio loss aims to embed 2D poses based on the matching indicator function (1). Let n be the dimension of the input 2D pose keypoints $\boldsymbol{x}$, and d be the dimension of the output embedding. We would like to learn a mapping $f : \mathbb{R}^n \to \mathbb{R}^d$, such that $D(\boldsymbol{z}_i, \boldsymbol{z}_j) < D(\boldsymbol{z}_i, \boldsymbol{z}_{j'}), \forall m_{ij} > m_{ij'}$, where $\boldsymbol{z} = f(\boldsymbol{x})$, and $D(\boldsymbol{z}_i, \boldsymbol{z}_j)$ is an embedding space distance measure.

For a pair of input 2D poses $(\boldsymbol{x}_i, \boldsymbol{x}_j)$, we define $p(m|\boldsymbol{x}_i, \boldsymbol{x}_j)$ to be the probability that their corresponding 3D poses $(\boldsymbol{y}_i, \boldsymbol{y}_j)$ match, that is, they are visually similar. While it is difficult to define this probability directly, we propose to assign its values by estimating $p(m|\boldsymbol{z}_i, \boldsymbol{z}_j)$ via metric learning. We know that if two 3D poses are identical, then $p(m|\boldsymbol{x}_i, \boldsymbol{x}_j) = 1$, and if two 3D poses are sufficiently different, $p(m|\boldsymbol{x}_i, \boldsymbol{x}_j)$ should be small. For any given input triplet $(\boldsymbol{x}_i, \boldsymbol{x}_{i+}, \boldsymbol{x}_{i-})$ with $m_{i,i+} > m_{i,i-}$, we want

$$
\frac{p(m|\boldsymbol{z}_i, \boldsymbol{z}_{i+})}{p(m|\boldsymbol{z}_i, \boldsymbol{z}_{i-})} \geqslant \beta, \tag{2}
$$

where $\beta > 1$ represents the ratio of the matching probability of a similar 3D pose pair to that of a dissimilar pair. Applying negative logarithm to both sides, we have

$$
(-\log p(m|\boldsymbol{z}_i, \boldsymbol{z}_{i+})) - (-\log p(m|\boldsymbol{z}_i, \boldsymbol{z}_{i-})) \leqslant -\log \beta. \tag{3}
$$

Notice that the model can be trained to satisfy this with the triplet loss framework [50]. Given batch size N, we define triplet ratio loss $\mathcal{L}_{\text{ratio}}$ as

$$
\mathcal{L}_{\text{ratio}} = \sum_{i=1}^{N} \max(0, D_m(\boldsymbol{z}_i, \boldsymbol{z}_{i+}) - D_m(\boldsymbol{z}_i, \boldsymbol{z}_{i-}) + \alpha)), \tag{4}
$$

with distance kernel $D_m(z_i, z_j) = -\log p(m|z_i, z_j)$ and margin $\alpha = \log \beta$. To form a triplet (x_i, x_{i+}, x_{i-}), we set the anchor x_i and positive x_{i+} to be projected from the same 3D pose and perform online semi-hard negative mining [50] to find x_{i-}.

It remains for us to compute matching probability using our embeddings. To compute $p(m|z_i, z_j)$, we use the formulation proposed by [37]:

$$p(m|z_i, z_j) = \sigma(-a||z_i - z_j||_2 + b), \tag{5}$$

where σ is a sigmoid function, and the trainable scalar parameters $a > 0$ and $b \in \mathbb{R}$ calibrate embedding distances to probabilistic similarity.

3.3 Positive Pairwise Loss

The positive pairs in our triplets have identical 3D poses. We would like them to have high matching probabilities, which can be encouraged by adding the positive pairwise loss

$$\mathcal{L}_{\text{positive}} = \sum_{i=1}^{N} -\log p(m|z_i, z_{i+}). \tag{6}$$

The combination of $\mathcal{L}_{\text{ratio}}$ and $\mathcal{L}_{\text{positive}}$ can be applied to training point embedding models, which we refer to as VIPE in this paper.

3.4 Probabilistic Embeddings

In this section, we discuss the extension of VIPE to the probabilistic formulation Pr-VIPE. The inputs to our model, 2D pose keypoints, are inherently ambiguous, and there are many valid 3D poses projecting to similar 2D poses [1]. This input uncertainty can be difficult to model using point embeddings [24, 37]. We investigate representing this uncertainty using distributions in the embedding space by mapping 2D poses to probabilistic embeddings: $x \rightarrow p(z|x)$. Similar to [37], we extend the input matching probability (5) to using probabilistic embeddings as $p(m|x_i, x_j) = \int p(m|z_i, z_j)p(z_i|x_i)p(z_j|x_j)dz_idz_j$, which can be approximated using Monte-Carlo sampling with K samples drawn from each distribution as

$$p(m|x_i, x_j) \approx \frac{1}{K^2} \sum_{k_1=1}^{K} \sum_{k_2=1}^{K} p(m|z_i^{(k_1)}, z_j^{(k_2)}). \tag{7}$$

We model $p(z|x)$ as a d-dimensional Gaussian with a diagonal covariance matrix. The model outputs mean $\mu(x) \in \mathbb{R}^d$ and covariance $\Sigma(x) \in \mathbb{R}^d$ with shared base network and different output layers. We use the reparameterization trick [25] during sampling.

In order to prevent variance from collapsing to zero and to regularize embedding mean magnitudes, we place a unit Gaussian prior on our embeddings with KL divergence by adding the Gaussian prior loss

$$\mathcal{L}_{\text{prior}} = \sum_{i=1}^{N} D_{\text{KL}}(\mathcal{N}(\mu(x_i), \Sigma(x_i)) \, \| \, \mathcal{N}(\mathbf{0}, \mathbf{I})). \tag{8}$$

Inference. At inference time, our model takes a single 2D pose (either from detection or projection) and outputs the mean and the variance of the embedding Gaussian distribution.

3.5 Camera Augmentation

Our triplets can be made of detected and/or projected 2D keypoints as shown in Fig. 2. When we train only with detected 2D keypoints, we are constrained to the camera views in training images. To reduce overfitting to these camera views, we perform camera augmentation by generating triplets using detected keypoints alongside projected 2D keypoints at random views.

To form triplets using multi-view image pairs, we use detected 2D keypoints from different views as anchor-positive pairs. To use projected 2D keypoints, we perform two random rotations to a normalized input 3D pose to generate two 2D poses from different views for anchor/positive. Camera augmentation is then performed by using a mixture of detected and projected 2D keypoints. We find that training using camera augmentation can help our models learn to generalize better to unseen views (Sect. 4.2.2).

3.6 Implementation Details

We normalize 3D poses similar to [7], and we perform instance normalization to 2D poses. The backbone network architecture for our model is based on [32]. We use $d = 16$ as a good trade-off between embedding size and accuracy. To weigh different losses, we use $w_{ratio} = 1$, $w_{positive} = 0.005$, and $w_{prior} = 0.001$. We choose $\beta = 2$ for the triplet ratio loss margin and $K = 20$ for the number of samples. The matching NP-MPJPE threshold is $\kappa = 0.1$ for all training and evaluation. Our approach does not rely on a particular 2D keypoint detector, and we use PersonLab [40] for our experiments. For random rotation in camera augmentation, we uniformly sample azimuth angle between $\pm 180°$, elevation between $\pm 30°$, and roll between $\pm 30°$. Our implementation is in TensorFlow, and all the models are trained with CPUs. More details and ablation studies on hyperparameters are provided in the supplementary materials.

4 Experiments

We demonstrate the performance of our model through pose retrieval across different camera views (Sect. 4.2). We further show our embeddings can be directly applied to downstream tasks, such as action recognition (Sect. 4.3.1) and video alignment (Sect. 4.3.2), without any additional training.

4.1 Datasets

For all the experiments in this paper, we only train on a subset of the Human3.6M [17] dataset. For pose retrieval experiments, we validate on the

Human3.6M hold-out set and test on another dataset (MPI-INF-3DHP [33]), which is unseen during training and free from parameter tuning. We also present qualitative results on MPII Human Pose [2], for which 3D groundtruth is not available. Additionally, we directly use our embeddings for action recognition and sequence alignment on Penn Action [60].

Human3.6M (H3.6M). H3.6M is a large human pose dataset recorded from 4 chest level cameras with 3D pose groundtruth. We follow the standard protocol [32]: train on Subject 1, 5, 6, 7, and 8, and hold out Subject 9 and 11 for validation. For evaluation, we remove near-duplicate 3D poses within 0.02 NP-MPJPE, resulting in a total of 10910 evaluation frames per camera. This process is camera-consistent, meaning if a frame is selected under one camera, it is selected under all cameras, so that the perfect retrieval result is possible.

MPI-INF-3DHP (3DHP). 3DHP is a more recent human pose dataset that contains 14 diverse camera views and scenarios, covering more pose variations than H3.6M [33]. We use 11 cameras from this dataset and exclude the 3 cameras with overhead views. Similar to H3.6M, we remove near-duplicate 3D poses, resulting in 6824 frames per camera. We use all 8 subjects from the train split of 3DHP. **This dataset is only used for testing.**

MPII Human Pose (2DHP). This dataset is commonly used in 2D pose estimation, containing 25K images from YouTube videos. Since groundtruth 3D poses are not available, we show qualitative results on this dataset.

Penn Action. This dataset contains 2326 trimmed videos for 15 pose-based actions from different views. We follow the standard protocol [36] for our action classification and video alignment experiments.

4.2   View-Invariant Pose Retrieval

Given multi-view human pose datasets, we query using detected 2D keypoints from one camera view and find the nearest neighbors in the embedding space from a different camera view. We iterate through all camera pairs in the dataset as query and index. Results averaged across all cameras pairs are reported.

4.2.1   Evaluation Procedure

We report Hit@k with $k = 1$, 10, and 20 on pose retrievals, which is the percentage of top-k retrieved poses that have at least one accurate retrieval. A retrieval is considered accurate if the 3D groundtruth from the retrieved pose satisfies the matching function (1) with $\kappa = 0.1$.

Baseline Approaches. We compare Pr-VIPE with 2D-to-3D lifting models [32] and $L2$-VIPE. $L2$-VIPE outputs $L2$-normalized point embeddings, and is trained with the squared $L2$ distance kernel, similar to [50].

For fair comparison, we use the same backbone network architecture for all the models. Notably, this architecture [32] has been tuned for lifting tasks on H3.6M. Since the estimated 3D poses in camera coordinates are not view-invariant, we apply normalization and Procrustes alignment to align the estimated 3D poses

Table 1. Comparison of cross-view pose retrieval results Hit@k (%) on H3.6M and 3DHP with chest-level cameras and all cameras. $*$ indicates that normalization and Procrustes alignment are performed on query-index pairs.

Dataset	H3.6M			3DHP (Chest)			3DHP (All)		
k	1	10	20	1	10	20	1	10	20
2D keypoints*	28.7	47.1	50.9	5.20	14.0	17.2	9.80	21.6	25.5
3D lifting*	69.0	89.7	92.7	24.9	54.4	62.4	24.6	53.2	61.3
$L2$-VIPE	73.5	94.2	96.6	23.8	56.7	66.5	18.7	46.3	55.7
$L2$-VIPE (w/aug.)	70.4	91.8	94.5	24.9	55.4	63.6	23.7	53.0	61.4
Pr-VIPE	**76.2**	**95.6**	**97.7**	25.4	59.3	69.3	19.9	49.1	58.8
Pr-VIPE (w/aug.)	73.7	93.9	96.3	**28.3**	**62.3**	**71.4**	**26.4**	**58.6**	**67.9**

between index and query for retrieval. In comparison, our embeddings do not require any alignment or other post-processing during retrieval.

For Pr-VIPE, we retrieve poses using nearest neighbors in the embedding space with respect to the sampled matching probability (7), which we refer to as retrival confidence. We present the results on the VIPE models with and without camera augmentation. We applied similar camera augmentation to the lifting model, but did not see improvement in performance. We also show the results of pose retrieval using aligned 2D keypoints only. The poor performance of using input 2D keypoints for retrieval from different views confirms the fact that models must learn view invariance from inputs for this task.

We also compare with the image-based EpipolarPose model [26]. Please refer to the supplementary materials for the experiment details and results.

4.2.2 Quantitative Results

From Table 1, we see that Pr-VIPE (with augmentation) outperforms all the baselines for H3.6M and 3DHP. The H3.6M results shown are on the hold-out set, and 3DHP is unseen during training, with more diverse poses and views. When we use all the cameras from 3DHP, we evaluate the generalization ability of models to new poses and new views. When we evaluate using only the 5 chest-level cameras from 3DHP, where the views are more similar to the training set in H3.6M, we mainly evaluate for generalization to new poses. When we evaluate using only the 5 chest-level cameras from 3DHP, the views are more similar to H3.6M, and generalization to new poses becomes more important. Our model is robust to the choice of β and the number of samples K (analysis in supplementary materials).

Table 1 shows that Pr-VIPE without camera augmentation is able to perform better than the baselines for H3.6M and 3DHP (chest-level cameras). This shows that Pr-VIPE is able to generalize as well as other baseline methods to new poses. However, for 3DHP (all cameras), the performance for Pr-VIPE without augmentation is worse compared with chest-level cameras. This observation

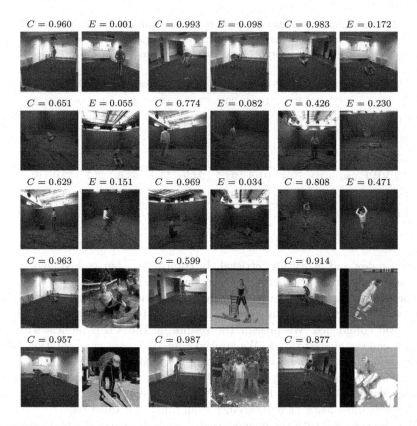

Fig. 3. Visualization of pose retrieval results. The first row is from H3.6M; the second and the third row are from 3DHP; the last two rows are using queries from H3.6M to retrieve from 2DHP. On each row, we show the query pose on the left for each image pair and the top-1 retrieval using the Pr-VIPE model (w/aug.) on the right. We display retrieval confidences ("C") and top-1 NP-MPJPEs ("E", if 3D pose groundtruth is available).

indicates that when trained on chest-level cameras only, Pr-VIPE does not generalize as well to new views. The same results can be observed for $L2$-VIPE between chest-level and all cameras. In contrast, the 3D lifting models are able to generalize better to new views with the help of additional Procrustes alignment, which requires expensive SVD computation for every index-query pair.

We further apply camera augmentation to training the Pr-VIPE and the $L2$-VIPE model. Note that this step does not require camera parameters or additional groundtruth. The results in Table 1 on Pr-VIPE show that the augmentation improves performance for 3DHP (all cameras) by 6% to 9%. This step also increases chest-level camera accuracy slightly. For $L2$-VIPE, we can observe a similar increase on all views. Camera augmentation reduces accuracy on H3.6M for both models. This is likely because augmentation reduces overfitting to the training camera views. By performing camera augmentation, Pr-VIPE is able to generalize better to new poses and new views.

4.2.3   Qualitative Results

Figure 3 shows qualitative retrieval results using Pr-VIPE. As shown in the first row, the retrieval confidence of the model is generally high for H3.6M. This indicates that the retrieved poses are close to their queries in the embedding space. Errors in 2D keypoint detection can lead to retrieval errors as shown by the rightmost pair. In the second and third rows, the retrieval confidence is lower for 3DHP. This is likely because there are new poses and views unseen during training, which has the nearest neighbor slightly further away in the embedding space. We see that the model can generalize to new views as the images are taken at different camera elevations from H3.6M. Interestingly, the rightmost pair on row 2 shows that the model can retrieve poses with large differences in roll angle, which is not present in the training set. The rightmost pair on row 3 shows an example of a large NP-MPJPE error due to mis-detection of the left leg in the index pose.

We show qualitative results using queries from the H3.6M hold-out set to retrieve from 2DHP in the last two rows of Fig. 3. The results on these in-the-wild images indicate that as long as the 2D keypoint detector works reliably, our model is able to retrieve poses across views and subjects. More qualitative results are provided in the supplementary materials.

4.3   Downstream Tasks

We show that our pose embedding can be directly applied to pose-based down-stream tasks using simple algorithms. We compare the performance of Pr-VIPE (**only trained on H3.6M, with no additional training**) on the Penn Action dataset against other approaches specifically trained for each task on the target dataset. In all the following experiments in this section, we compute our Pr-VIPE embeddings on single video frames and use the negative logarithm of the matching probability (7) as the distance between two frames. Then we apply temporal averaging within an atrous kernel of size 7 and rate 3 around the two center frames and use this averaged distance as the frame matching distance. Given the matching distance, we use standard dynamic time warping (DTW) algorithm to align two action sequences by minimizing the sum of frame matching distances. We further use the averaged frame matching distance from the alignment as the distance between two video sequences.

4.3.1   Action Recognition

We evaluate our embeddings for action recognition using nearest neighbor search with the sequence distance described above. Provided person bounding boxes in each frame, we estimate 2D pose keypoints using [41]. On Penn Action, we use the standard train/test split [36]. Using all the testing videos as queries, we conduct two experiments: (1) we use all training videos as index to evaluate overall performance and compare with state-of-the-art methods, and (2) we use training videos only under one view as index and evaluate the effectiveness of our embeddings in terms of view-invariance. For this second experiment, actions

Table 2. Comparison of action recognition results on Penn Action.

Methods	Input			Accuracy (%)
	RGB	Flow	Pose	
Nie et al. [36]	✓		✓	85.5
Iqbal et al. [18]			✓	79.0
Cao et al. [5]		✓	✓	95.3
	✓	✓		98.1
Du et al. [10]	✓	✓	✓	97.4
Liu et al. [30]	✓		✓	91.4
Luvizon et al. [31]	✓		✓	98.7
Ours			✓	97.5
Ours (1-view index)			✓	92.1

Table 3. Comparison of video alignment results on Penn Action.

Methods	Kendall's Tau
SaL [34]	0.6336
TCN [51]	0.7353
TCC [11]	0.7328
TCC + SaL [11]	0.7286
TCC + TCN [11]	0.7672
Ours	0.7476
Ours (same-view only)	0.7521
Ours (different-view only)	0.7607

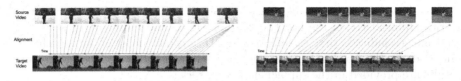

Fig. 4. Video alignment results using Pr-VIPE. The orange dots correspond to the visualized frames, and the blue line segments illustrate the frame alignment (Color figure online).

with zero or only one sample under the index view are ignored, and accuracy is averaged over different views.

From Table 2 we can see that without any training on the target domain or using image context information, our embeddings can achieve highly competitive results on pose-based action classification, outperforming the existing best baseline that only uses pose input and even some other methods that rely on image context or optical flow. As shown in the last row in Table 2, our embeddings can be used to classify actions from different views using index samples from only one single view with relatively high accuracy, which further demonstrates the advantages of our view-invariant embeddings.

4.3.2   Video Alignment

Our embeddings can be used to align human action videos from different views using DTW algorithm as described earlier in Sect. 4.3. We measure the alignment quality of our embeddings quantitatively using Kendall's Tau [11], which reflects how well an embedding model can be applied to align unseen sequences if we use nearest neighbor in the embedding space to match frames for video pairs. A value of 1 corresponds to perfect alignment. We also test the view-invariant properties of our embeddings by evaluating Kendall's Tau on aligning videos pairs from the same view, and aligning pairs with different views.

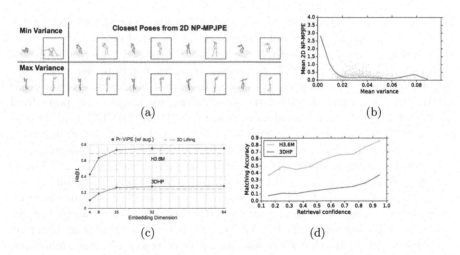

(a)   (b)

(c)   (d)

Fig. 5. Ablation study: (a) Top retrievals by 2D NP-MPJPE from the H3.6M hold-out subset for queries with largest and smallest variance. 2D poses are shown in the boxes. (b) Relationship between embedding variance and 2D NP-MPJPE to top-10 nearest 2D pose neighbors from the H3.6M hold-out subset. The orange curve represents the best fitting 5th degree polynomial. (c) Comparison of Hit@1 with different embedding dimensions. The 3D lifting baseline predicts 39 dimensions. (d) Relationship between retrieval confidence and matching accuracy.

In Table 3, we compare our results with other video embedding baselines that are trained for the alignment task on Penn Action, from which we observe that Pr-VIPE performs better than all the method that use a single type of loss. While Pr-VIPE is slightly worse than the combined TCC+TCN loss, our embeddings are able to achieve this without being explicitly trained for this task or taking advantage of image context. In the last two rows of Table 3, we show the results from evaluating video pairs only from the same or different views. We can see that our embedding achieves consistently high performance regardless of whether the aligned video pair is from the same or different views, which demonstrate its view-invariant property. In Fig. 4, we show action video synchronization results from different views using Pr-VIPE. We provide more synchronized videos for all actions in the supplementary materials.

4.4 Ablation Study

Point vs. Probabilistic Embeddings. We compare VIPE point embedding formulation with Pr-VIPE. When trained on detected keypoints, the Hit@1 for VIPE and Pr-VIPE are 75.4% and 76.2% on H3.6M, and 19.7% and 20.0% on 3DHP, respectively. When we add camera augmentation, the Hit@1 for VIPE and Pr-VIPE are 73.8% and 73.7% on H3.6M, and 26.1% and 26.5% on 3DHP, respectively. Despite the similar retrieval accuracies, Pr-VIPE is generally more accurate and, more importantly, has additional desirable properties in that the variance can model 2D input ambiguity as to be discussed next.

A 2D pose is ambiguous if there are similar 2D poses that can be projected from very different poses in 3D. To measure this, we compute the average 2D NP-MPJPE between a 2D pose and its top-10 nearest neighbors in terms of 2D NP-MPJPE. To ensure the 3D poses are different, we sample 1200 poses from H3.6M hold-out set with a minimum gap of 0.1 3D NP-MPJPE. If a 2D pose has small 2D NP-MPJPE to its neighbors, it means there are many similar 2D poses corresponding to different 3D poses and so the 2D pose is ambiguous.

Figure 5a shows that the 2D pose with the largest variance is ambiguous as it has similar 2D poses in H3.6M with different 3D poses. In contrast, we see that the closest 2D poses corresponding to the smallest variance pose on the first row of Fig. 5a are clearly different. Figure 5b further shows that as the average variance increases, the 2D NP-MPJPE between similar poses generally decreases, which means that 2D poses with larger variances are more ambiguous.

Embedding Dimensions. Figure 5c demonstrates the effect of embedding dimensions on H3.6M and 3DHP. The lifting model lifts 13 2D keypoints to 3D, and therefore has a constant output dimension of 39. We see that Pr-VIPE (with augmentation) is able to achieve a higher accuracy than lifting at 16 dimensions. Additionally, we can increase the number of embedding dimensions to 32, which increases accuracy of Pr-VIPE from 73.7% to 75.5%.

Retrieval Confidence. In order to validate the retrieval confidence values, we randomly sample 100 queries along with their top-5 retrievals (using Pr-VIPE retrieval confidence) from each query-index camera pair. This procedure forms 6000 query-retrieval sample pairs for H3.6M (4 views, 12 camera pairs) and 55000 for 3DHP (11 views, 110 camera pairs), which we bin by their retrieval confidences. Figure 5d shows the matching accuracy for each confidence bin. We can see that the accuracy positively correlates with the confidence values, which suggest our retrieval confidence is a valid indicator to model performance.

What if 2D Keypoint Detectors were Perfect? We repeat our pose retrieval experiments using groundtruth 2D keypoints to simulate a perfect 2D keypoint detector on H3.6M and 3DHP. All experiments use the 4 views from H3.6M for training following the standard protocol. For the baseline lifting model in camera frame, we achieve 89.9% Hit@1 on H3.6M, 48.2% on 3DHP (all), and 48.8% on 3DHP (chest). For Pr-VIPE, we achieve 97.5% Hit@1 on H3.6M, 44.3% on 3DHP (all), and 66.4% on 3DHP (chest). These results follow the same trend as using detected keypoints inputs in Table 1. Comparing the results with using detected keypoints, the large improvement in performance using groundtruth keypoints suggests that a considerable fraction of error in our model is due to imperfect 2D keypoint detections. Please refer to the supplementary materials for more ablation studies and embedding space visualization.

5 Conclusion

We introduce Pr-VIPE, an approach to learning probabilistic view-invariant embeddings from 2D pose keypoints. By working with 2D keypoints, we can use camera augmentation to improve model generalization to unseen views. We also demonstrate that our probabilistic embedding learns to capture input ambiguity. Pr-VIPE has a simple architecture and can be potentially applied to object and hand poses. For cross-view pose retrieval, 3D pose estimation models require expensive rigid alignment between query-index pair, while our embeddings can be applied to compare similarities in simple Euclidean space. In addition, we demonstrated the effectiveness of our embeddings on downstream tasks for action recognition and video alignment. Our embedding focuses on a single person, and for future work, we will investigate extending it to multiple people and robust models that can handle missing keypoints from input.

Acknowledgment. We thank Yuxiao Wang, Debidatta Dwibedi, and Liangzhe Yuan from Google Research, Long Zhao from Rutgers University, and Xiao Zhang from University of Chicago for helpful discussions. We appreciate the support of Pietro Perona, Yisong Yue, and the Computational Vision Lab at Caltech for making this collaboration possible. The author Jennifer J. Sun is supported by NSERC (funding number PGSD3-532647-2019) and Caltech.

References

1. Akhter, I., Black, M.J.: Pose-conditioned joint angle limits for 3D human pose reconstruction. In: CVPR (2015)
2. Andriluka, M., Pishchulin, L., Gehler, P., Schiele, B.: 2D human pose estimation: new benchmark and state of the art analysis. In: CVPR (2014)
3. Bojchevski, A., Günnemann, S.: Deep Gaussian embedding of graphs: Unsupervised inductive learning via ranking. In: ICLR (2018)
4. Bromley, J., Guyon, I., LeCun, Y., Säckinger, E., Shah, R.: Signature verification using a "siamese" time delay neural network. In: NeurIPS (1994)
5. Cao, C., Zhang, Y., Zhang, C., Lu, H.: Body joint guided 3-D deep convolutional descriptors for action recognition. IEEE Trans. Cybern. **48**(3), 1095–1108 (2017)
6. Chen, C.H., Ramanan, D.: 3D human pose estimation = 2D pose estimation + matching. In: CVPR (2017)
7. Chen, C.H., Tyagi, A., Agrawal, A., Drover, D., Stojanov, S., Rehg, J.M.: Unsupervised 3D pose estimation with geometric self-supervision. In: CVPR (2019)
8. Chu, R., Sun, Y., Li, Y., Liu, Z., Zhang, C., Wei, Y.: Vehicle re-identification with viewpoint-aware metric learning. In: ICCV (2019)
9. Drover, D., M. V, R., Chen, C.-H., Agrawal, A., Tyagi, A., Huynh, C.P.: Can 3D pose be learned from 2D projections alone? In: Leal-Taixé, L., Roth, S. (eds.) ECCV 2018. LNCS, vol. 11132, pp. 78–94. Springer, Cham (2019). https://doi.org/10.1007/978-3-030-11018-5_7
10. Du, W., Wang, Y., Qiao, Y.: RPAN: an end-to-end recurrent pose-attention network for action recognition in videos. In: ICCV (2017)
11. Dwibedi, D., Aytar, Y., Tompson, J., Sermanet, P., Zisserman, A.: Temporal cycle-consistency learning. In: CVPR (2019)

12. Hadsell, R., Chopra, S., LeCun, Y.: Dimensionality reduction by learning an invariant mapping. In: CVPR (2006)
13. Hermans, A., Beyer, L., Leibe, B.: In defense of the triplet loss for person re-identification. arXiv:1703.07737 (2017)
14. Ho, C.H., Morgado, P., Persekian, A., Vasconcelos, N.: PIEs: pose invariant embeddings. In: CVPR, pp. 12377–12386 (2019)
15. Hu, W., Zhu, S.C.: Learning a probabilistic model mixing 3D and 2D primitives for view invariant object recognition. In: CVPR (2010)
16. Huang, C., Loy, C.C., Tang, X.: Local similarity-aware deep feature embedding. In: NeurIPS (2016)
17. Ionescu, C., Papava, D., Olaru, V., Sminchisescu, C.: Human3.6M: large scale datasets and predictive methods for 3D human sensing in natural environments. IEEE TPAMI **36**, 1325–1339 (2013)
18. Iqbal, U., Garbade, M., Gall, J.: Pose for action-action for pose. In: FG (2017)
19. Iscen, A., Tolias, G., Avrithis, Y., Chum, O.: Mining on manifolds: metric learning without labels. In: CVPR (2018)
20. Iskakov, K., Burkov, E., Lempitsky, V., Malkov, Y.: Learnable triangulation of human pose. In: ICCV (2019)
21. Jammalamadaka, N., Zisserman, A., Eichner, M., Ferrari, V., Jawahar, C.: Video retrieval by mimicking poses. In: ACM ICMR (2012)
22. Ji, X., Liu, H.: Advances in view-invariant human motion analysis: a review. IEEE Trans. Syst. Man Cybern. Part C (Appl. Rev.) **40**(1), 13–24 (2009)
23. Ji, X., Liu, H., Li, Y., Brown, D.: Visual-based view-invariant human motion analysis: a review. In: Lovrek, I., Howlett, R.J., Jain, L.C. (eds.) KES 2008. LNCS (LNAI), vol. 5177, pp. 741–748. Springer, Heidelberg (2008). https://doi.org/10.1007/978-3-540-85563-7_93
24. Kendall, A., Gal, Y.: What uncertainties do we need in Bayesian deep learning for computer vision? In: NeurIPS (2017)
25. Kingma, D.P., Welling, M.: Auto-encoding variational Bayes. In: ICLR (2014)
26. Kocabas, M., Karagoz, S., Akbas, E.: Self-supervised learning of 3D human pose using multi-view geometry. In: CVPR (2019)
27. LeCun, Y., Huang, F.J., Bottou, L., et al.: Learning methods for generic object recognition with invariance to pose and lighting. In: CVPR (2004)
28. Li, J., Wong, Y., Zhao, Q., Kankanhalli, M.: Unsupervised learning of view-invariant action representations. In: NeurIPS (2018)
29. Liu, J., Akhtar, N., Ajmal, M.: Viewpoint invariant action recognition using RGB-D videos. IEEE Access **6**, 70061–70071 (2018)
30. Liu, M., Yuan, J.: Recognizing human actions as the evolution of pose estimation maps. In: CVPR (2018)
31. Luvizon, D.C., Tabia, H., Picard, D.: Multi-task deep learning for real-time 3D human pose estimation and action recognition. arXiv:1912.08077 (2019)
32. Martinez, J., Hossain, R., Romero, J., Little, J.J.: A simple yet effective baseline for 3D human pose estimation. In: ICCV (2017)
33. Mehta, D., et al.: Monocular 3D human pose estimation in the wild using improved CNN supervision. In: 3DV (2017)
34. Misra, I., Zitnick, C.L., Hebert, M.: Shuffle and learn: unsupervised learning using temporal order verification. In: Leibe, B., Matas, J., Sebe, N., Welling, M. (eds.) ECCV 2016. LNCS, vol. 9905, pp. 527–544. Springer, Cham (2016). https://doi.org/10.1007/978-3-319-46448-0_32
35. Mori, G., et al.: Pose embeddings: A deep architecture for learning to match human poses. arXiv:1507.00302 (2015)

36. Nie, B.X., Xiong, C., Zhu, S.C.: Joint action recognition and pose estimation from video. In: CVPR (2015)
37. Oh, S.J., Murphy, K., Pan, J., Roth, J., Schroff, F., Gallagher, A.: Modeling uncertainty with hedged instance embedding. In: ICLR (2019)
38. Oh Song, H., Xiang, Y., Jegelka, S., Savarese, S.: Deep metric learning via lifted structured feature embedding. In: CVPR (2016)
39. Ong, E.J., Micilotta, A.S., Bowden, R., Hilton, A.: Viewpoint invariant exemplar-based 3D human tracking. CVIU **104**, 178–189 (2006)
40. Papandreou, G., Zhu, T., Chen, L.-C., Gidaris, S., Tompson, J., Murphy, K.: PersonLab: person pose estimation and instance segmentation with a bottom-up, part-based, geometric embedding model. In: Ferrari, V., Hebert, M., Sminchisescu, C., Weiss, Y. (eds.) Computer Vision – ECCV 2018. LNCS, vol. 11218, pp. 282–299. Springer, Cham (2018). https://doi.org/10.1007/978-3-030-01264-9_17
41. Papandreou, G., et al.: Towards accurate multi-person pose estimation in the wild. In: CVPR (2017)
42. Parkhi, O.M., Vedaldi, A., Zisserman, A., et al.: Deep face recognition. In: BMVC (2015)
43. Pavllo, D., Feichtenhofer, C., Grangier, D., Auli, M.: 3D human pose estimation in video with temporal convolutions and semi-supervised training. In: CVPR (2019)
44. Qiu, H., Wang, C., Wang, J., Wang, N., Zeng, W.: Cross View Fusion for 3D Human Pose Estimation. In: ICCV (2019)
45. Rao, C., Shah, M.: View-invariance in action recognition. In: CVPR (2001)
46. Hossain, M.R.I., Little, J.J.: Exploiting temporal information for 3D human pose estimation. In: Ferrari, V., Hebert, M., Sminchisescu, C., Weiss, Y. (eds.) ECCV 2018. LNCS, vol. 11214, pp. 69–86. Springer, Cham (2018). https://doi.org/10.1007/978-3-030-01249-6_5
47. Rhodin, H., Constantin, V., Katircioglu, I., Salzmann, M., Fua, P.: Neural scene decomposition for multi-person motion capture. In: CVPR (2019)
48. Rhodin, H., Salzmann, M., Fua, P.: Unsupervised geometry-aware representation for 3D human pose estimation. In: Ferrari, V., Hebert, M., Sminchisescu, C., Weiss, Y. (eds.) ECCV 2018. LNCS, vol. 11214, pp. 765–782. Springer, Cham (2018). https://doi.org/10.1007/978-3-030-01249-6_46
49. Rhodin, H., et al.: Learning monocular 3D human pose estimation from multi-view images. In: CVPR (2018)
50. Schroff, F., Kalenichenko, D., Philbin, J.: FaceNet: a unified embedding for face recognition and clustering. In: CVPR (2015)
51. Sermanet, P., et al.: Time-contrastive networks: self-supervised learning from video. In: ICRA (2018)
52. Sun, X., Xiao, B., Wei, F., Liang, S., Wei, Y.: Integral human pose regression. In: Ferrari, V., Hebert, M., Sminchisescu, C., Weiss, Y. (eds.) ECCV 2018. LNCS, vol. 11210, pp. 536–553. Springer, Cham (2018). https://doi.org/10.1007/978-3-030-01231-1_33
53. Tekin, B., Márquez-Neila, P., Salzmann, M., Fua, P.: Learning to fuse 2D and 3D image cues for monocular body pose estimation. In: ICCV (2017)
54. Tome, D., Toso, M., Agapito, L., Russell, C.: Rethinking pose in 3D: multi-stage refinement and recovery for markerless motion capture. In: 3DV (2018)
55. Vilnis, L., McCallum, A.: Word representations via Gaussian embedding. In: ICLR (2015)
56. Wang, J., et al.: Learning fine-grained image similarity with deep ranking. In: CVPR (2014)

57. Wohlhart, P., Lepetit, V.: Learning descriptors for object recognition and 3D pose estimation. In: CVPR (2015)
58. Wu, C.Y., Manmatha, R., Smola, A.J., Krahenbuhl, P.: Sampling matters in deep embedding learning. In: ICCV (2017)
59. Xia, L., Chen, C.C., Aggarwal, J.K.: View invariant human action recognition using histograms of 3D joints. In: CVPRW (2012)
60. Zhang, W., Zhu, M., Derpanis, K.G.: From actemes to action: a strongly-supervised representation for detailed action understanding. In: ICCV (2013)
61. Zheng, L., Huang, Y., Lu, H., Yang, Y.: Pose invariant embedding for deep person re-identification. IEEE TIP **28**, 4500–4509 (2019)
62. Zhou, X., Huang, Q., Sun, X., Xue, X., Wei, Y.: Towards 3D human pose estimation in the wild: a weakly-supervised approach. In: ICCV (2017)

Contact and Human Dynamics
from Monocular Video

Davis Rempe[1,2]([envelope]), Leonidas J. Guibas[1], Aaron Hertzmann[2], Bryan Russell[2], Ruben Villegas[2], and Jimei Yang[2]

[1] Stanford University, Stanford, USA
drempe@stanford.edu
[2] Adobe Research, San Jose, USA
https://geometry.stanford.edu/projects/human-dynamics-eccv-2020/

Abstract. Existing deep models predict 2D and 3D kinematic poses from video that are approximately accurate, but contain visible errors that violate physical constraints, such as feet penetrating the ground and bodies leaning at extreme angles. In this paper, we present a physics-based method for inferring 3D human motion from video sequences that takes initial 2D and 3D pose estimates as input. We first estimate ground contact timings with a novel prediction network which is trained without hand-labeled data. A physics-based trajectory optimization then solves for a physically-plausible motion, based on the inputs. We show this process produces motions that are significantly more realistic than those from purely kinematic methods, substantially improving quantitative measures of both kinematic and dynamic plausibility. We demonstrate our method on character animation and pose estimation tasks on dynamic motions of dancing and sports with complex contact patterns.

1 Introduction

Recent methods for human pose estimation from monocular video [1,17,30,43] estimate accurate overall body pose with small absolute differences from the true poses in body-frame 3D coordinates. However, the recovered motions in world-frame are visually and physically implausible in many ways, including feet that float slightly or penetrate the ground, implausible forward or backward body lean, and motion errors like jittery, vibrating poses. These errors would prevent many subsequent uses of the motions. For example, inference of actions, intentions, and emotion often depends on subtleties of pose, contact and acceleration, as does computer animation; human perception is highly sensitive to physical inaccuracies [14,34]. Adding more training data would not solve these problems, because existing methods do not account for physical plausibility.

Physics-based trajectory optimization presents an appealing solution to these issues, particularly for dynamic motions like walking or dancing. Physics imposes

---

Electronic supplementary material The online version of this chapter (https://doi.org/10.1007/978-3-030-58558-7_5) contains supplementary material, which is available to authorized users.

© Springer Nature Switzerland AG 2020
A. Vedaldi et al. (Eds.): ECCV 2020, LNCS 12350, pp. 71–87, 2020.
https://doi.org/10.1007/978-3-030-58558-7_5

Input Video   MTC (side view)   Ours (side view)

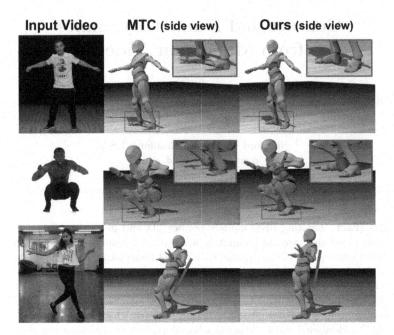

Fig. 1. Our contact prediction and physics-based optimization corrects numerous physically implausible artifacts common in 3D human motion estimations from, e.g., Monocular Total Capture (MTC) [43] such as foot floating (top row), foot penetrations (middle), and unnatural leaning (bottom).

important constraints that are hard to express in pose space but easy in terms of dynamics. For example, feet in static contact do not move, the body moves smoothly overall relative to contacts, and joint torques are not large. However, full-body dynamics is notoriously difficult to optimize [36], in part because contact is discontinuous, and the number of possible contact events grows exponentially in time. As a result, combined optimization of contact and dynamics is enormously sensitive to local minima.

This paper introduces a new strategy for extracting dynamically valid full-body motions from monocular video (Fig. 1), combining learned pose estimation with physical reasoning through trajectory optimization. As input, we use the results of kinematic pose estimation techniques [4,43], which produce accurate overall poses but inaccurate contacts and dynamics. Our method leverages a reduced-dimensional body model with centroidal dynamics and contact constraints [7,42] to produce a physically-valid motion that closely matches these inputs. We first infer foot contacts from 2D poses in the input video which are then used in a physics-based trajectory optimization to estimate 6D center-of-mass motion, feet positions, and contact forces. We show that a contact prediction network can be accurately trained on synthetic data. This allows us to separate initial contact estimation from motion optimization, making the optimization more tractable. As a result, our method is able to handle highly dynamic motions without sacrificing physical accuracy.

We focus on single-person dynamic motions from dance, walking, and sports. Our approach substantially improves the realism of inferred motions over state-of-the-art methods, and estimates numerous physical properties that could be useful for further inference of scene properties and action recognition. We primarily demonstrate our method on character animation by retargeting captured motion from video to a virtual character. We evaluate our approach using numerous kinematics and dynamics metrics designed to measure the physical plausibility of the estimated motion. The proposed method takes an important step to incorporating physical constraints into human motion estimation from video, and shows the potential to reconstruct realistic, dynamic sequences.

2 Related Work

We build on several threads of work in computer vision, computer animation, and robotics, each with a long history [9]. Recent vision results are detailed here.

Recent progress in pose estimation can accurately detect 2D human keypoints [4,12,27] and infer 3D pose [1,17,30] from a single image. Several recent methods extract 3D human motions from monocular videos by exploring various forms of temporal cues [18,26,43,44]. While these methods focus on explaining human motion in pixel space, they do not account for physical plausibility. Several recent works interpret interactions between people and their environment in order to make inferences about each [6,11,45]; each of these works uses only static kinematic constraints. Zou et al. [46] infer contact constraints to optimize 3D motion from video. We show how dynamics can improve inference of human-scene interactions, leading to more physically plausible motion capture.

Some works have proposed physics constraints to address the issues of kinematic tracking. Brubaker et al. [3] propose a physics-based tracker based on a reduced-dimensional walking model. Wei and Chai [41] track body motion from video, assuming keyframe and contact constraints are provided. Similar to our own work, Brubaker and Fleet [2] perform trajectory optimization for full-body motion. To jointly optimize contact and dynamics, they use a continuous approximation to contact. However, soft contact models introduce new difficulties, including inaccurate transitions and sensitivity to stiffness parameters, while still suffering from local minima issues. Moreover, their reduced-dimensional model includes only center-of-mass positional motion, which does not handle rotational motion well. In contrast, we obtain accurate contact initialization in a preprocessing step to simplify optimization, and we model rotational inertia.

Li et al. [23] estimate dynamic properties from videos. We share the same overall pipeline of estimating pose and contacts, followed by trajectory optimization. Whereas they focus on the dynamics of human-object interactions, we focus on videos where the human motion itself is much more dynamic, with complex variation in pose and foot contact; we do not consider human-object interaction. They use a simpler data term, and perform trajectory optimization in full-body dynamics unlike our reduced representation. Their classifier training requires hand-labeled data, unlike our automatic dataset creation method.

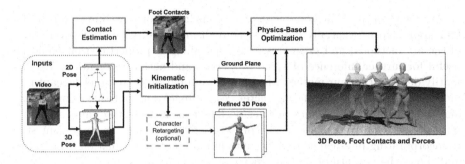

Fig. 2. Method overview. Given an input video, our method starts with initial estimates from existing 2D and 3D pose methods [4,43]. The lower-body 2D joints are used to infer foot contacts (orange box). Our optimization framework contains two parts (blue boxes). Inferred contacts and initial poses are used in a kinematic optimization that refines the 3D full-body motion and fits the ground. These are given to a reduced-dimensional physics-based trajectory optimization that applies dynamics.

Prior methods learn character animation controllers from video. Vondrak et al. [38] train a state-machine controller using image silhouette features. Peng et al. [32] train a controller to perform skills by following kinematically-estimated poses from input video sequences. They demonstrate impressive results on a variety of skills. They do not attempt accurate reconstruction of motion or contact, nor do they evaluate for these tasks, rather they focus on control learning.

Our optimization is related to physics-based methods in computer animation, e.g., [8,16,20,24,25,33,40]. Two unique features of our optimization are the use of low-dimensional dynamics optimization that includes 6D center-of-mass motion and contact constraints, thereby capturing important rotational and footstep quantities without requiring full-body optimization, and the use of a classifier to determine contacts before optimization.

3   Physics-Based Motion Estimation

This section describes our approach, which is summarized in Fig. 2. The core of our method is a physics-based trajectory optimization that enforces dynamics on the input motion (Sect. 3.1). Foot contact timings are estimated in a preprocess (Sect. 3.2), along with other inputs to the optimization (Sect. 3.3). Similar to previous work [23,43], in order to recover full-body motion we assume there is no camera motion and that the full body is visible.

3.1   Physics-Based Trajectory Optimization

The core of our framework is an optimization which enforces dynamics on an initial motion estimate given as input (see Sect. 3.3). The goal is to improve the plausibility of the motion by applying physical reasoning through the objective and constraints. We aim to avoid common perceptual errors, e.g., jittery, unnatural motion with feet skating and ground penetration, by generating a smooth trajectory with physically-valid momentum and static feet during contact.

The optimization is performed on a reduced-dimensional body model that captures overall motion, rotation, and contacts, but avoids the difficulty of optimizing all joints. Modeling rotation is necessary for important effects like arm swing and counter-oscillations [13,20,25], and the reduced-dimensional *centroidal* dynamics model can produce plausible trajectories for humanoid robots [5,7,28]. Our method is based on a recent robot motion planning algorithm from Winkler et al. [42] that leverages a simplified version of centroidal dynamics, which treats the robot as a rigid body with a fixed mass and moment of inertia. Their method finds a feasible trajectory by optimizing the position and rotation of the center-of-mass (COM) along with feet positions, contact forces, and contact durations as described in detail below. We modify this algorithm to suit our computer vision task: we use a temporally varying inertia tensor which allows for changes in mass distribution (swinging arms) and enables estimating the dynamic motions of interest, we add energy terms to match the input kinematic motion and foot contacts, and we add new kinematics constraints for our humanoid skeleton.

Inputs. The method takes initial estimates of: COM position $\bar{\mathbf{r}}(t) \in \mathbb{R}^3$ and orientation $\bar{\boldsymbol{\theta}}(t) \in \mathbb{R}^3$ trajectories, body-frame inertia tensor trajectory $\mathbf{I}_b(t) \in \mathbb{R}^{3\times3}$, and trajectories of the foot joint positions $\bar{\mathbf{p}}_{1:4}(t) \in \mathbb{R}^3$. There are four foot joints: left toe base, left heel, right toe base, and right heel, indexed as $i \in \{1,2,3,4\}$. These inputs are at discrete timesteps, but we write them here as functions for clarity. The 3D ground plane height h_{floor} and upward normal is provided. Additionally, for each foot joint at each time, a binary label is provided indicating whether the foot is in contact with the ground. These labels determine initial estimates of contact durations for each foot joint $\bar{T}_{i,1}, \bar{T}_{i,2}, \ldots, \bar{T}_{i,n_i}$ as described below. The distance from toe to heel ℓ_{foot} and maximum distance from toe to hip ℓ_{leg} are also provided. All quantities are computed from video input as described in Sects. 3.2 and 3.3, and are used to both initialize the optimization variables and as targets in the objective function.

Optimization Variables. The optimization variables are the COM position and Euler angle orientation $\mathbf{r}(t), \boldsymbol{\theta}(t) \in \mathbb{R}^3$, foot joint positions $\mathbf{p}_i(t) \in \mathbb{R}^3$ and contact forces $\mathbf{f}_i(t) \in \mathbb{R}^3$. These variables are continuous functions of time, represented by piece-wise cubic polynomials with continuity constraints. We also optimize contact timings. The contacts for each foot joint are independently parameterized by a sequence of phases that alternate between contact and flight. The optimizer cannot change the type of each phase (contact or flight), but it can modify their durations $T_{i,1}, T_{i,2}, \ldots, T_{i,n_i} \in \mathbb{R}$ where n_i is the number of total contact phases for the ith foot joint.

Objective. Our complete formulation is shown in Fig. 3. E_{data} and E_{dur} seek to keep the motion and contacts as close as possible to the initial inputs, which are

$$\min \quad \sum_{t=0}^{T} \Big(E_{data}(t) + E_{vel}(t) + E_{acc}(t)\Big) + E_{dur}$$

$$\text{s.t.} \quad m\ddot{\mathbf{r}}(t) = \sum_{i=1}^{4} \mathbf{f}_i(t) + m\mathbf{g} \qquad \text{(dynamics)}$$

$$\mathbf{I}_w(t)\dot{\boldsymbol{\omega}}(t) + \boldsymbol{\omega}(t) \times \mathbf{I}_w(t)\boldsymbol{\omega}(t) = \sum_{i=1}^{4} \mathbf{f}_i(t) \times (\mathbf{r}(t) - \mathbf{p}_i(t))$$

$$\dot{\mathbf{r}}(0) = \dot{\bar{\mathbf{r}}}(0), \dot{\mathbf{r}}(T) = \dot{\bar{\mathbf{r}}}(T) \qquad \text{(velocity boundaries)}$$

$$||\mathbf{p}_1(t) - \mathbf{p}_2(t)|| = ||\mathbf{p}_3(t) - \mathbf{p}_4(t)|| = \ell_{foot} \qquad \text{(foot kinematics)}$$

for every foot joint i :

$$||\mathbf{p}_i(t) - \mathbf{p}_{hip,i}(t)|| \leq \ell_{leg} \qquad \text{(leg kinematics)}$$

$$\sum_{j=1}^{n_i} T_{i,j} = T \qquad \text{(contact durations)}$$

for foot joint i in contact at time t :

$$\dot{\mathbf{p}}_i(t) = 0 \qquad \text{(no slip)}$$

$$p_i^z(t) = h_{floor}(\mathbf{p}_i^{xy}) \qquad \text{(on floor)}$$

$$0 \leq \mathbf{f}_i(t)^T \hat{\mathbf{n}} \leq f_{max} \qquad \text{(pushing/max force)}$$

$$|\mathbf{f}_i(t)^T \hat{\mathbf{t}}_{1,2}| < \mu \mathbf{f}_i(t)^T \hat{\mathbf{n}} \qquad \text{(friction pyramid)}$$

for foot joint i in flight at time t :

$$p_i^z(t) \geq h_{floor}(\mathbf{p}_i^{xy}) \qquad \text{(above floor)}$$

$$\mathbf{f}_i(t) = 0 \qquad \text{(no force in air)}$$

Fig. 3. Physics-based trajectory optimization formulation. Please see text for details.

derived from video, at discrete steps over the entire duration T:

$$E_{data}(t) = w_r ||\mathbf{r}(t) - \bar{\mathbf{r}}(t)||^2 + w_\theta ||\boldsymbol{\theta}(t) - \bar{\boldsymbol{\theta}}(t)||^2$$
$$+ w_p \sum_{i=1}^{4} ||\mathbf{p}_i(t) - \bar{\mathbf{p}}_i(t)||^2 \qquad (1)$$

$$E_{dur} = w_d \sum_{i=1}^{4} \sum_{j=1}^{n_i} (T_{i,j} - \bar{T}_{i,j})^2 \qquad (2)$$

We weigh these terms with $w_d = 0.1$, $w_r = 0.4$, $w_\theta = 1.7$, $w_p = 0.3$.

The remaining objective terms are regularizers that prefer small velocities and accelerations resulting in a smoother optimal trajectory:

$$E_{vel}(t) = \gamma_r ||\dot{\mathbf{r}}(t)||^2 + \gamma_\theta ||\dot{\boldsymbol{\theta}}(t)||^2 + \gamma_p \sum_{i=1}^{4} ||\dot{\mathbf{p}}_i(t)||^2 \qquad (3)$$

$$E_{acc}(t) = \beta_r ||\ddot{\mathbf{r}}(t)||^2 + \beta_\theta ||\ddot{\boldsymbol{\theta}}(t)||^2 + \beta_p \sum_{i=1}^{4} ||\ddot{\mathbf{p}}_i(t)||^2 \qquad (4)$$

with $\gamma_r = \gamma_\theta = 10^{-3}$, $\gamma_p = 0.1$ and $\beta_r = \beta_\theta = \beta_p = 10^{-4}$.

Constraints. The first set of constraints strictly enforce valid rigid body mechanics, including linear and angular momentum. This enforces important properties of motion, for example, during flight the COM must follow a parabolic arc according to Newton's Second Law. During contact, the body motion acceleration is limited by the possible contact forces e.g., one cannot walk at a 45° lean.

At each timestep, we use the world-frame inertia tensor $\mathbf{I}_w(t)$ computed from the input $\mathbf{I}_b(t)$ and the current orientation $\boldsymbol{\theta}(t)$. This assumes that the final output poses will not be dramatically different from those of the input: a reasonable assumption since our optimization does not operate on upper-body joints and changes in feet positioning are typically small (though perceptually important). We found that using a constant inertia tensor (as in Winkler et al. [42]) made convergence difficult to achieve. The gravity vector is $\mathbf{g} = -9.8\hat{\mathbf{n}}$, where $\hat{\mathbf{n}}$ is the ground normal. The angular velocity $\boldsymbol{\omega}$ is a function of the rotations $\boldsymbol{\theta}$ [42].

The contact forces are constrained to ensure that they push away from the floor but are not greater than $f_{max} = 1000\,\mathrm{N}$ in the normal direction. With 4 ft joints, this allows $4000\,\mathrm{N}$ of normal contact force: about the magnitude that a $100\,\mathrm{kg}$ ($220\,\mathrm{lb}$) person would produce for extremely dynamic dancing motion [19]. We assume no feet slipping during contact, so forces must also remain in a friction pyramid defined by friction coefficient $\mu = 0.5$ and floor plane tangents $\hat{\mathbf{t}}_1, \hat{\mathbf{t}}_2$. Lastly, forces should be zero at any foot joint not in contact.

Foot contact is enforced through constraints. When a foot joint is in contact, it should be stationary (no-slip) and at floor height h_{floor}. When not in contact, feet should always be on or above the ground. This avoids feet skating and penetration with the ground.

In order to make the optimized motion valid for a humanoid skeleton, the toe and heel of each foot should maintain a constant distance of ℓ_{foot}. Finally, no foot joint should be farther from its corresponding hip than the length of the leg ℓ_{leg}. The hip position $\mathbf{p}_{hip,i}(t)$ is computed from the COM orientation at that time based on the hip offset in the skeleton detailed in Sect. 3.3.

Optimization Algorithm. We optimize with IPOPT [39], a nonlinear interior point optimizer, using analytical derivatives. We perform the optimization in stages: we first use fixed contact phases and no dynamics constraints to fit the polynomial representation for COM and feet position variables as close as possible to the input motion. Next, we add in dynamics constraints to find a physically valid motion, and finally we allow contact phase durations to be optimized to further refine the motion if possible.

Following the optimization, we compute a full-body motion from the physically-valid COM and foot joint positions using Inverse Kinematics (IK) on a desired skeleton $\mathbf{S}_{tgt}$ (see supplement for details).

3.2   Learning to Estimate Contacts

Before performing our physics-based optimization, we need to infer when the subject's feet are in contact with the ground, given an input video. These contacts are a target for the physics optimization objective and their accuracy is crucial to its success. To do so, we train a network that, for each video frame, classifies whether the toe and heel of each foot are in contact with the ground.

The main challenge is to construct a suitable dataset and feature representation. There is currently no publicly-available dataset of videos with labeled foot contacts and a wide variety of dynamic motions. Manually labeling a large, varied dataset would be difficult and costly. Instead, we generate synthetic data using motion capture (mocap) sequences. We automatically label contacts in the mocap and then use 2D joint position features from OpenPose [4] as input to our model, rather than image features from the raw rendered video frames. This allows us to train on synthetic data but then apply the model to real inputs.

Dataset. To construct our dataset, we obtained 65 mocap sequences for the 13 most human-like characters from www.mixamo.com, ranging from dynamic dancing motions to idling. Our set contains a diverse range of mocap sequences, retargeted to a variety of animated characters. At each time of each motion sequence, four possible contacts are automatically labeled by a heuristic: a toe or heel joint is considered to be in contact when (i) it has moved less than 2 cm from the previous time, and (ii) it is within 5 cm from the known ground plane. Although more sophisticated labeling [15,21] could be used, we found this approach sufficiently accurate to learn a model for the videos we evaluated on.

We render these motions (see Fig. 5(c)) on their rigged characters with motion blur, randomized camera viewpoint, lighting, and floor texture. For each sequence, we render two views, resulting in over 100k frames of video with labeled contacts and 2D and 3D poses. Finally, we run a 2D pose estimation algorithm, OpenPose [4], to obtain the 2D skeleton which our model uses as input.

Model and Training. The classification problem is to map from 2D pose in each frame to the four contact labels for the feet joints. As we demonstrate in Sect. 4.1, simple heuristics based on 2D velocity do not accurately label contacts due to the ambiguities of 3D projection and noise.

For a given time t, our labeling neural network takes as input the 2D poses over a temporal window of duration w centered on the target frame at t. The 2D joint positions over the window are normalized to place the root position of the target frame at $(0,0)$, resulting in relative position and velocity. We set $w = 9$ video frames and use the 13 lower-body joint positions as shown in Fig. 4. Additionally, the OpenPose confidence c for each joint position is included as input. Hence, the input to the network is a vector of (x, y, c) values of dimension $3 * 13 * 9 = 351$. The model outputs four contact labels (left/right toe, left/right heel) for a window of 5 frames centered around the target. At test time, we use majority voting at overlapping predictions to smooth labels across time.

We use a five-layer multilayer perceptron (MLP) (sizes 1024, 512, 128, 32, 20) with ReLU non-linearities [29]. We train the network entirely on our synthetic dataset split 80/10/10 for train/validation/test based on motions per character, i.e., no motion will be in both train and test on the same character, but a training motion may appear in the test set retargeted to a different character. Although 3D motions may be similar in train and test, the resulting 2D motions (the network input) will be very different after projecting to differing camera viewpoints. The network is trained using a standard binary cross-entropy loss.

3.3   Kinematic Initialization

Along with contact labels, our physics-based optimization requires as input a ground plane and initial trajectories for the COM, feet, and inertia tensor. In order to obtain these, we compute an initial 3D full-body motion from video. Since this stage uses standard elements, e.g., [10], we summarize the algorithm here, and provide full details in the supplement.

First, Monocular Total Capture [43] (MTC) is applied to the input video to obtain an initial noisy 3D pose estimate for each frame. Although MTC accounts for motion through a texture-based refinement step, the output still contains a number of artifacts (Fig. 1) that make it unsuitable for direct use in our physics optimization. Instead, we initialize a skeleton $\mathbf{S}_{src}$ containing 28 body joints from the MTC input poses, and then use a kinematic optimization to solve for an optimal root translation and joint angles over time, along with parameters of the ground plane. The objective for this optimization contains terms to smooth the motion, ensure feet are stationary and on the ground when in contact, and to stay close to both the 2D OpenPose and 3D MTC pose inputs.

We first optimize so that the feet are stationary, but not at a consistent height. Next, we use a robust regression to find the ground plane which best fits the foot joint contact positions. Finally, we continue the optimization to ensure all feet are on this ground plane when in contact.

The full-body output motion of the kinematic optimization is used to extract inputs for the physics optimization. Using a predefined body mass (73 kg for all experiments) and distribution [22], we compute the COM and inertia tensor trajectories. We use the orientation about the root joint as the COM orientation, and the feet joint positions are used directly.

4   Results

Here we present extensive qualitative and quantitative evaluations of our contact estimation and motion optimization.

4.1   Contact Estimation

We evaluate our learned contact estimation method and compare to baselines on the synthetic test set (78 videos) and 9 real videos with manually-labeled foot contacts. The real videos contain dynamic dancing motions and include 700 labeled frames in total. In Table 1, we report classification accuracy for our method and numerous baselines.

We compare to using a velocity heuristic on foot joints, as described in Sect. 3.2, for both the 2D OpenPose and 3D MTC estimations. We also compare to using different subsets of joint positions. Our MLP using all lower-body joints is substantially more accurate on both synthetic and real videos than all baselines. Using upper-body joints down to the knees yields surprisingly good results.

Table 1. Classification accuracy of estimating foot contacts from video. Left: comparison to various baselines, Right: ablations using subsets of joints as input features.

Baseline method	Synthetic accuracy	Real accuracy	MLP input joints	Synthetic accuracy	Real accuracy
Random	0.507	0.480	Upper down to hips	0.919	0.692
Always contact	0.677	0.647	Upper down to knees	0.935	0.865
2D velocity	0.853	0.867	Lower up to ankles	0.933	0.923
3D velocity	0.818	0.875	**Lower up to hips**	**0.941**	**0.935**

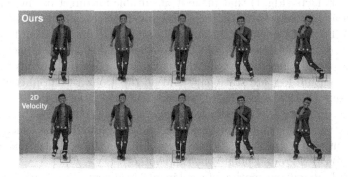

Fig. 4. Foot contact estimation on a video using our learned model compared to a 2D velocity heuristic. All visualized joints are used as input to the network which outputs four contact labels (left toes, left heel, right toes, right heel). Red joints are labeled as contacting. Key differences are shown with orange boxes (Color figure online).

In order to test the benefit of contact estimation, we compared our full optimization pipeline on the synthetic test set using network-predicted contacts versus contacts predicted using a velocity heuristic on the 3D joints from MTC input. Optimization using network-predicted contacts converged for 94.9% of the test set videos, compared to 69.2% for the velocity heuristic. This illustrates how contact prediction is crucial to the success of motion optimization.

Qualitative results of our contact estimation method are shown in Fig. 4. Our method is compared to the 2D velocity baseline which has difficulty for planted feet when detections are noisy, and often labels contacts for joints that are stationary but off the ground (e.g. heels).

4.2 Qualitative Motion Evaluation

Our method provides key qualitative improvements over prior kinematic approaches. We urge the reader to **view the supplementary video** in order to fully appreciate the generated motions. For qualitative evaluation, we demonstrate animation from video by retargeting captured motion to a computer-animated character. Given a target skeleton $\mathbf{S}_{tgt}$ for a character, we insert an

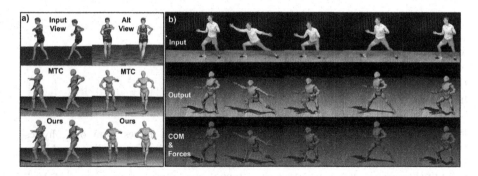

Fig. 5. Qualitative results on synthetic and real data. a) results on a synthetic test video with a ground truth alternate view. Two nearby frames are shown for the input video and the alternate view. We fix penetration, floating and leaning prevalent in our method's input from MTC. b) dynamic exercise video (top) and the output full-body motion (middle) and optimized COM trajectory and contact forces (bottom).

IK retargeting step following the kinematic optimization as shown in Fig. 2 (see supplement for details), allowing us to perform the usual physics-based optimization on this new skeleton. We use the same IK procedure to compare to MTC results directly targeted to the character.

Figure 1 shows that our proposed method fixes artifacts such as foot floating (top row), foot penetrations (middle), and unnatural leaning (bottom). Figure 5(a) shows frames comparing the MTC input to our final result on a synthetic video for which we have a ground truth alternate view. For this example only, we use the true ground plane as input to our method for a fair comparison (see Sect. 4.3). From the input view, our method fixes feet floating and penetration. From the first frame of the alternate view, we see that the MTC pose is in fact extremely unstable, leaning backward while balancing on its heels; our method has placed the contacting feet in a stable position to support the pose, better matching the true motion.

Figure 5(b) shows additional qualitative results on a real video. We faithfully reconstruct dynamic motion with complex contact patterns in a physically accurate way. The bottom row shows the outputs of the physics-based optimization stage of our method at multiple frames: the COM trajectory and contact forces at the heel and toe of each foot.

4.3   Quantitative Motion Evaluation

Quantitative evaluation of high-quality motion estimation presents a significant challenge. Recent pose estimation work evaluates average positional errors of joints in the local body frame up to various global alignment methods [31]. However, those pose errors can be misleading: a motion can be pose-wise close to ground truth on average, but produce extremely implausible dynamics, including vibrating positions and extreme body lean. These errors can be perceptually objectionable when remapping the motion onto an animated character, and prevent the use of inferred dynamics for downstream vision tasks.

Therefore, we propose to use a set of metrics inspired by the biomechanics literature [2,13,16], namely, to evaluate *plausibility* of physical quantities based on known properties of human motion.

We use two baselines: MTC, which is the state-of-the-art for pose estimation, and our kinematic-only initialization (Sect. 3.3), which transforms the MTC input to align with the estimated contacts from Sect. 3.2. We run each method on the synthetic test set

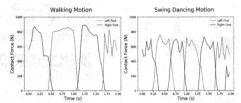

Fig. 6. Contact forces from our physics-based optimization for a walking and dancing motion. The net contact forces around 1000 N are 140% of the assumed body weight (73 kg), a reasonable estimate compared to prior force plate data [2].

of 78 videos. For these quantitative evaluations only, we use the ground truth floor plane as input to our method to ensure a fair comparison. Note that our method does not *need* the ground truth floor, but using it ensures a proper evaluation of our primary contributions rather than that of the floor fitting procedure, which is highly dependent on the quality of MTC input (see supplement for quantitative results using the estimated floor).

Dynamics Metrics. To evaluate dynamic plausibility, we estimate net ground reaction forces (GRF), defined as $\mathbf{f}_{GRF}(t) = \sum_i \mathbf{f}_i(t)$. For our full pipeline, we use the physics-based optimized GRFs which we compare to implied forces from the kinematic-only initialization and MTC input. In order to infer the GRFs implied by the kinematic optimization and MTC, we estimate the COM trajectory of the motion using the same mass and distribution as for our physics-based optimization (73 kg). We then approximate the acceleration at each time step and solve for the implied GRFs for all time steps (both in contact and flight).

We assess plausibility using GRFs measured in force plate studies, e.g., [2,13,35]. For walking, GRFs typically reach 80% of body weight; for a dance jump, GRFs can reach up to about 400% of body weight [19]. Since we do not know body weights of our subjects, we use a conservative range of 50kg–80kg for evaluation. Figure 6 shows the optimized GRFs produced by our method for a walking and swing dancing motion. The peak GRFs produced by our method match the data: for the walking motion, 115–184% of body weight, and 127–204% for dancing. In contrast, the kinematic-only GRFs are 319–510% (walking) and 765–1223% (dancing); these are implausibly high, a consequence of noisy and unrealistic joint accelerations.

We also measure GRF plausibility across the whole test set (Table 2(left)). GRF values are measured as a percentage of the GRF exerted by an idle 73 kg person. On average, our estimate is within 1% of the idle force, while the kinematic motion implies GRFs as if the person were 24.4% heavier. Similarly, the peak force of the kinematic motion is equivalent to the subject carrying an extra 830 kg of weight, compared to only 174 kg after physics optimization. The Max GRF for MTC is even less plausible, as the COM motion is jittery before smoothing during kinematic and dynamics optimization. *Ballistic GRF* measures the

Table 2. Physical plausibility evaluation on synthetic test set. *Mean/Max GRF* are contact forces as a proportion of body weight; see text for discussion of plausible values. *Ballistic GRF* are unexplained forces during flight; smaller values are better. Foot position metrics measure the percentage of frames containing typical foot contact errors per joint; smaller values are better.

Method	Dynamics (contact forces)			Kinematics (foot positions)		
	Mean GRF	Max GRF	Ballistic GRF	Floating	Penetration	Skate
MTC [43]	143.0%	9055.3%	115.6%	58.7%	21.1%	16.8%
Kinematics (ours)	124.4%	1237.5%	255.2%	**2.3%**	2.8%	**1.6%**
Physics (ours)	**99.0%**	**338.6%**	**0.0%**	8.2%	**0.3%**	3.6%

median GRF on the COM when no feet joints should be in contact according to ground truth labels. The GRF should be exactly 0%, meaning there are no contact forces and only gravity acts on the COM; the kinematic method obtains results of 255%, as if the subject were wearing a powerful jet pack.

Kinematics Metrics. We consider three kinematic measures of plausibility (Table 2(right)). These metrics evaluate accuracy of foot contact measurements. Specifically, given ground truth labels of foot contact we compute instances of foot *Floating*, *Penetration*, and *Skate* for heel and toe joints. *Floating* is the fraction of foot joints more than 3 cm off the ground when they should be in contact. *Penetration* is the fraction penetrating the ground more than 3 cm at any time. *Skate* is the fraction moving more than 2 cm when in contact.

After our kinematics initialization, the scores on these metrics are best (lower is better for all metrics) and degrade slightly after adding physics. This is due to the IK step which produces full-body motion following the physics-based optimization. Both the kinematic and physics optimization results substantially outperform MTC, which is rarely at a consistent foot height.

Positional Metrics. For completeness, we evaluate the 3D pose output of our method on variations of standard positional metrics. Results are shown in Table 3. In addition to our synthetic test set, we evaluate on all walking sequences from the training split of HumanEva-I [37] using the known ground plane as input. We measure the mean **global** per-joint position error (mm) for ankle and toe joints (*Feet* in Table 3) and over all joints (*Body*). We also report the error after aligning the root joint of only the first frame of each sequence to the ground truth skeleton (*Body-Align 1*), essentially removing any spurious constant offset from the predicted trajectory. Note that this differs from the common practice of aligning the roots at every frame, since this would negate the effect of our trajectory optimization and thus does not provide an informative performance measure. The errors between all methods are comparable, showing at most a difference of 5 cm which is very small considering global joint position. Though the goal of our method is to improve physical plausibility, it does not negatively affect the pose on these standard measures.

Table 3. Pose evaluation on synthetic and HumanEva-I walking datasets. We measure mean global per-joint 3D position error (no alignment) for feet and full-body joints. For full-body joints, we also report errors after root alignment on only the first frame of each sequence. We remain competitive while providing key physical improvements.

Method	Synthetic data			HumanEva-I walking		
	Feet	Body	Body-Align 1	Feet	Body	Body-Align 1
MTC [43]	581.095	**560.090**	**277.215**	511.59	532.286	**402.749**
Kinematics (ours)	573.097	562.356	281.044	**496.671**	525.332	407.869
Physics (ours)	**571.804**	573.803	323.232	508.744	**499.771**	421.931

5 Discussion

Contributions. The method described in this paper estimates physically-valid motions from initial kinematic pose estimates. As we show, this produces motions that are visually and physically much more plausible than the state-of-the-art methods. We show results on retargeting to characters, but it could also be used for further vision tasks that would benefit from dynamical properties of motion.

Estimating accurate human motion entails numerous challenges, and we have focused on one crucial sub-problem. There are several other important unknowns in this space, such as motion for partially-occluded individuals, and ground plane position. Each of these problems and the limitations discussed below are an enormous challenge in their own right and are therefore reserved for future work. However, we believe that the ideas in this work could contribute to solving these problems and open multiple avenues for future exploration.

Limitations. We make a number of assumptions to keep the problem manageable, all of which can be relaxed in future work: we assume that feet are unoccluded, there is a single ground plane, the subject is not interacting with other objects, and we do not handle contact from other body parts like knees or hands. These assumptions are permissible for the character animation from video mocap application, but should be considered in a general motion estimation approach. Our optimization is expensive. For a 2 s (60 frame) video clip, the physical optimization usually takes from 30 min to 1 h. This runtime is due primarily to the adapted implementation from prior work [42] being ill-suited for the increased size and complexity of human motion optimization. We expect a specialized solver and optimized implementation to speed up execution.

Acknowledgments. This work was in part supported by NSF grant IIS-1763268, grants from the Samsung GRO program and the Stanford SAIL Toyota Research Center, and a gift from Adobe Corporation. We thank the following YouTube channels for permitting us to use their videos: Dance FreaX (Fig. 4), Dancercise Studio (Fig. 1 and 2), Fencer's Edge (Fig. 5), and MihranTV (Fig. 1).

References

1. Bogo, F., Kanazawa, A., Lassner, C., Gehler, P., Romero, J., Black, M.J.: Keep It SMPL: automatic estimation of 3D human pose and shape from a single image. In: Leibe, B., Matas, J., Sebe, N., Welling, M. (eds.) ECCV 2016. LNCS, vol. 9909, pp. 561–578. Springer, Cham (2016). https://doi.org/10.1007/978-3-319-46454-1_34
2. Brubaker, M.A., Sigal, L., Fleet, D.J.: Estimating contact dynamics. In: The IEEE International Conference on Computer Vision (ICCV), pp. 2389–2396. IEEE (2009)
3. Brubaker, M.A., Fleet, D.J., Hertzmann, A.: Physics-based person tracking using the anthropomorphic walker. Int. J. Comput. Vis. **87**(1), 140–155 (2010). https://doi.org/10.1007/s11263-009-0274-5
4. Cao, Z., Hidalgo, G., Simon, T., Wei, S.E., Sheikh, Y.: OpenPose: realtime multi-person 2D pose estimation using part affinity fields. IEEE Trans. Pattern Anal. Mach. Intell., 1 (2019). IEEE
5. Carpentier, J., Mansard, N.: Multicontact locomotion of legged robots. IEEE Trans. Rob. **34**(6), 1441–1460 (2018)
6. Chen, Y., Huang, S., Yuan, T., Qi, S., Zhu, Y., Zhu, S.C.: Holistic++ scene understanding: single-view 3D holistic scene parsing and human pose estimation with human-object interaction and physical commonsense. In: The IEEE International Conference on Computer Vision (ICCV). pp. 8648–8657. IEEE (2019)
7. Dai, H., Valenzuela, A., Tedrake, R.: Whole-body motion planning with centroidal dynamics and full kinematics. In: IEEE-RAS International Conference on Humanoid Robots, pp. 295–302. IEEE (2014)
8. Fang, A.C., Pollard, N.S.: Efficient synthesis of physically valid human motion. ACM Trans. Graph. **22**(3), 417–426 (2003)
9. Forsyth, D.A., Arikan, O., Ikemoto, L., O'Brien, J., Ramanan, D.: Computational studies of human motion: part 1, tracking and motion synthesis. Found. Trends. Comput. Graph. Vis. **1**(2–3), 77–254 (2006)
10. Gleicher, M.: Retargetting motion to new characters. In: Proceedings of the 25th Annual Conference on Computer Graphics and Interactive Techniques, SIGGRAPH 1998, pp. 33–42. ACM, New York, NY, USA (1998)
11. Hassan, M., Choutas, V., Tzionas, D., Black, M.J.: Resolving 3D human pose ambiguities with 3D scene constraints. In: The IEEE International Conference on Computer Vision (ICCV), pp. 2282–2292. IEEE (October 2019)
12. He, K., Gkioxari, G., Dollar, P., Girshick, R.: Mask R-CNN. In: The IEEE International Conference on Computer Vision (ICCV), pp. 2961–2969. IEEE (October 2017)
13. Herr, H., Popovic, M.: Angular momentum in human walking. J. Exp. Biol. **211**(4), 467–481 (2008)
14. Hoyet, L., McDonnell, R., O'Sullivan, C.: Push it real: perceiving causality in virtual interactions. ACM Trans. Graph. **31**(4), 90:1–90:9 (2012)
15. Ikemoto, L., Arikan, O., Forsyth, D.: Knowing when to put your foot down. In: Proceedings of the 2006 Symposium on Interactive 3D Graphics and Games, I3D 2006, pp. 49–53. ACM, New York (2006)
16. Jiang, Y., Van Wouwe, T., De Groote, F., Liu, C.K.: Synthesis of biologically realistic human motion using joint torque actuation. ACM Trans. Graph. **38**(4), 72:1–72:12 (2019)
17. Kanazawa, A., Black, M.J., Jacobs, D.W., Malik, J.: End-to-end recovery of human shape and pose. In: The IEEE Conference on Computer Vision and Pattern Recognition (CVPR), pp. 7122–7131. IEEE (June 2018)

18. Kanazawa, A., Zhang, J.Y., Felsen, P., Malik, J.: Learning 3D human dynamics from video. In: The IEEE Conference on Computer Vision and Pattern Recognition (CVPR), pp. 5614–5623. IEEE (June 2019)

19. Kulig, K., Fietzer, A.L., Popovich Jr., J.M.: Ground reaction forces and knee mechanics in the weight acceptance phase of a dance leap take-off and landing. J. Sports Sci. **29**(2), 125–131 (2011)

20. de Lasa, M., Mordatch, I., Hertzmann, A.: Feature-based locomotion controllers. In: ACM SIGGRAPH 2010 Papers, SIGGRAPH 2010, pp. 131:1–131:10. ACM, New York (2010)

21. Le Callennec, B., Boulic, R.: Robust kinematic constraint detection for motion data. In: ACM SIGGRAPH/Eurographics Symposium on Computer Animation, SCA 2006, pp. 281–290. Eurographics Association, Aire-la-Ville (2006)

22. de Leva, P.: Adjustments to Zatsiorsky-Seluyanov's segment inertia parameters. J. Biomech. **29**(9), 1223–1230 (1996)

23. Li, Z., Sedlar, J., Carpentier, J., Laptev, I., Mansard, N., Sivic, J.: Estimating 3D motion and forces of person-object interactions from monocular video. In: The IEEE Conference on Computer Vision and Pattern Recognition (CVPR), pp. 8640–8649. IEEE (June 2019)

24. Liu, C.K., Hertzmann, A., Popović, Z.: Learning physics-based motion style with nonlinear inverse optimization. ACM Trans. Graph. **24**(3), 1071–1081 (2005)

25. Macchietto, A., Zordan, V., Shelton, C.R.: Momentum control for balance. In: ACM SIGGRAPH 2009 Papers, SIGGRAPH 2009, pp. 80:1–80:8. ACM, New York (2009)

26. Mehta, D., et al.: VNect: real-time 3D human pose estimation with a single RGB camera. ACM Trans. Graph. **36**(4), 44:1–44:14 (2017)

27. Newell, A., Yang, K., Deng, J.: Stacked hourglass networks for human pose estimation. In: Leibe, B., Matas, J., Sebe, N., Welling, M. (eds.) ECCV 2016. LNCS, vol. 9912, pp. 483–499. Springer, Cham (2016). https://doi.org/10.1007/978-3-319-46484-8_29

28. Orin, D.E., Goswami, A., Lee, S.H.: Centroidal dynamics of a humanoid robot. Auton. Robots **35**(2–3), 161–176 (2013). https://doi.org/10.1007/s10514-013-9341-4

29. Paszke, A., et al.: Pytorch: an imperative style, high-performance deep learning library. In: Wallach, H., et al. (eds.) Advances in Neural Information Processing Systems, vol. 32. pp. 8026–8037. Curran Associates, Inc. (2019)

30. Pavlakos, G., et al.: Expressive body capture: 3D hands, face, and body from a single image. In: The IEEE Conference on Computer Vision and Pattern Recognition (CVPR), pp. 10975–10985. IEEE (June 2019)

31. Pavllo, D., Feichtenhofer, C., Grangier, D., Auli, M.: 3D human pose estimation in video with temporal convolutions and semi-supervised training. In: The IEEE Conference on Computer Vision and Pattern Recognition (CVPR), pp. 7753–7762. IEEE (June 2019)

32. Peng, X.B., Kanazawa, A., Malik, J., Abbeel, P., Levine, S.: SFV: reinforcement learning of physical skills from videos. ACM Trans. Graph. **37**(6), 178:1–178:14 (2018)

33. Popović, Z., Witkin, A.: Physically based motion transformation. In: Proceedings of the 26th Annual Conference on Computer Graphics and Interactive Techniques, SIGGRAPH 1999, pp. 11–20. ACM Press/Addison-Wesley Publishing Co., New York (1999)

34. Reitsma, P.S.A., Pollard, N.S.: Perceptual metrics for character animation: sensitivity to errors in ballistic motion. ACM Trans. Graph. **22**(3), 537–542 (2003)

35. Robertson, D.G.E., Caldwell, G.E., Hamill, J., Kamen, G., Whittlesey, S.N.: Research Methods in Biomechanics. Human Kinetics, Champaign (2004)
36. Safonova, A., Hodgins, J.K., Pollard, N.S.: Synthesizing physically realistic human motion in low-dimensional, behavior-specific spaces. In: ACM SIGGRAPH 2004 Papers, SIGGRAPH 2004, pp. 514–521. ACM, New York (2004)
37. Sigal, L., Balan, A., Black, M.: HumanEva: synchronized video and motion capture dataset and baseline algorithm for evaluation of articulated human motion. Int. J. Comput. Vis. **87**, 4–27 (2010)
38. Vondrak, M., Sigal, L., Hodgins, J., Jenkins, O.: Video-based 3D motion capture through biped control. ACM Trans. Graph. **31**(4), 27:1–27:12 (2012)
39. Wächter, A., Biegler, L.T.: On the implementation of an interior-point filter line-search algorithm for large-scale nonlinear programming. Math. Program. **106**(1), 25–57 (2006)
40. Wang, J.M., Hamner, S.R., Delp, S.L., Koltun, V.: Optimizing locomotion controllers using biologically-based actuators and objectives. ACM Trans. Graph. **31**(4), 25 (2012)
41. Wei, X., Chai, J.: VideoMocap: modeling physically realistic human motion from monocular video sequences. In: ACM SIGGRAPH 2010 Papers, SIGGRAPH 2010, pp. 42:1–42:10. ACM, New York (2010)
42. Winkler, A.W., Bellicoso, D.C., Hutter, M., Buchli, J.: Gait and trajectory optimization for legged systems through phase-based end-effector parameterization. IEEE Robot. Autom. Lett. (RA-L) **3**, 1560–1567 (2018)
43. Xiang, D., Joo, H., Sheikh, Y.: Monocular total capture: posing face, body, and hands in the wild. In: The IEEE Conference on Computer Vision and Pattern Recognition (CVPR), pp. 10965–10974. IEEE (June 2019)
44. Xu, W., et al.: MonoPerfCap: human performance capture from monocular video. ACM Trans. Graph. **37**(2), 27:1–27:15 (2018)
45. Zanfir, A., Marinoiu, E., Sminchisescu, C.: Monocular 3D pose and shape estimation of multiple people in natural scenes-the importance of multiple scene constraints. In: The IEEE Conference on Computer Vision and Pattern Recognition (CVPR), pp. 2148–2157. IEEE (June 2018)
46. Zou, Y., Yang, J., Ceylan, D., Zhang, J., Perazzi, F., Huang, J.B.: Reducing footskate in human motion reconstruction with ground contact constraints. In: The IEEE Winter Conference on Applications of Computer Vision (WACV), pp. 459–468. IEEE (March 2020)

PointPWC-Net: Cost Volume on Point Clouds for (Self-)Supervised Scene Flow Estimation

Wenxuan Wu[1(✉)], Zhi Yuan Wang[2(✉)], Zhuwen Li[2(✉)], Wei Liu[2(✉)],
and Li Fuxin[1(✉)]

[1] CORIS Institute, Oregon State University, Corvallis, USA
{wuwen,lif}@oregonstate.edu
[2] Nuro, Inc., Mountain View, USA
{wangd,zli,w}@nuro.ai

Abstract. We propose a novel end-to-end deep scene flow model, called PointPWC-Net, that directly processes 3D point cloud scenes with large motions in a coarse-to-fine fashion. Flow computed at the coarse level is upsampled and warped to a finer level, enabling the algorithm to accommodate for large motion without a prohibitive search space. We introduce novel cost volume, upsampling, and warping layers to efficiently handle 3D point cloud data. Unlike traditional cost volumes that require exhaustively computing all the cost values on a high-dimensional grid, our point-based formulation discretizes the cost volume onto input 3D points, and a PointConv operation efficiently computes convolutions on the cost volume. Experiment results on FlyingThings3D and KITTI outperform the state-of-the-art by a large margin. We further explore novel self-supervised losses to train our model and achieve comparable results to state-of-the-art trained with supervised loss. Without any fine-tuning, our method also shows great generalization ability on the KITTI Scene Flow 2015 dataset, outperforming all previous methods. The code is released at https://github.com/DylanWusee/PointPWC.

Keywords: Cost volume · Self-supervision · Coarse-to-fine · Scene flow

1 Introduction

Scene flow is the 3D displacement vector between each surface point in two consecutive frames. As a fundamental tool for low-level understanding of the world, scene flow can be used in many 3D applications including autonomous driving. Traditionally, scene flow was estimated directly from RGB data [44,45,72,74]. But recently, due to the increasing application of 3D sensors such as LiDAR, there is interest on directly estimating scene flow from 3D point clouds.

Fueled by recent advances in 3D deep networks that learn effective feature representations directly from point cloud data, recent work adopt ideas from

Electronic supplementary material The online version of this chapter (https://doi.org/10.1007/978-3-030-58558-7_6) contains supplementary material, which is available to authorized users.

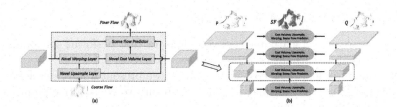

Fig. 1. (a) illustrates how the pyramid features are used by the novel cost volume, warping, and upsampling layers in one level. (b) shows the overview structure of PointPWC-Net. At each level, PointPWC-Net first warps features from the first point cloud using the upsampled scene flow. Then, the cost volume is computed using features from the warped first point cloud and the second point cloud. Finally, the scene flow predictor predicts finer flow at the current level using features from the first point cloud, the cost volume, and the upsampled flow. (Best viewed in color) (Color figure online)

2D deep optical flow networks to 3D to estimate scene flow from point clouds. FlowNet3D [36] operates directly on points with PointNet++ [54], and proposes a *flow embedding* which is computed in one layer to capture the correlation between two point clouds, and then propagates it through finer layers to estimate the scene flow. HPLFlowNet [20] computes the correlation jointly from multiple scales utilizing the upsampling operation in bilateral convolutional layers.

An important piece in deep optical flow estimation networks is the cost volume [31,65,81], a 3D tensor that contains matching information between neighboring pixel pairs from consecutive frames. In this paper, we propose a novel learnable point-based cost volume where we discretize the cost volume to input point pairs, avoiding the creation of a dense 4D tensor if we naively extend from the image to point cloud. Then we apply the efficient PointConv layer [80] on this irregularly discretized cost volume. We experimentally show that it outperforms previous approaches for associating point cloud correspondences, as well as the cost volume used in 2D optical flow. We also propose efficient upsampling and warping layers to implement a coarse-to-fine flow estimation framework.

As in optical flow, it is difficult and expensive to acquire accurate scene flow labels for point clouds. Hence, beyond supervised scene flow estimation, we also explore self-supervised scene flow which does not require human annotations. We propose new self-supervised loss terms: Chamfer distance [14], smoothness constraint and Laplacian regularization. These loss terms enable us to achieve state-of-the-art performance without any supervision.

We conduct extensive experiments on FlyingThings3D [44] and KITTI Scene Flow 2015 [46,47] datasets with both supervised loss and the proposed self-supervised losses. Experiments show that the proposed PointPWC-Net outperforms all previous methods by a large margin. The self-supervised version is comparable with some of the previous supervised methods on FlyingThings3D, such as SPLATFlowNet [63]. On KITTI where supervision is not available, our self-supervised version achieves better performance than the supervised version trained on FlyingThings3D, far surpassing state-of-the-art. We also ablate each critical component of PointPWC-Net to understand their contributions.

The key contributions of our work are:

- We propose a novel learnable cost volume layer that performs convolution on the cost volume without creating a dense 4D tensor.
- With the novel learnable cost volume layer, we present a novel model, called PointPWC-Net, that estimates scene flow from two consecutive point clouds in a coarse-to-fine fashion.
- We introduce self-supervised losses that can train PointPWC-Net without any ground truth label. To our knowledge, we are among the first to propose such an idea in 3D point cloud deep scene flow estimation.
- We achieve state-of-the-art performance on FlyingThing3D and KITTI Scene Flow 2015, far surpassing previous state-of-the-art.

2   Related Work

Deep Learning on Point Clouds. Deep learning methods on 3D point clouds have gained more attention in the past several years. Some latest work [19,25,34,53,54,57,63,68,73] directly take raw point clouds as input. [53,54,57] use a shared multi-layer perceptron (MLP) and max pooling layer to obtain features of point clouds. Other work [22,30,61,78–80] propose to learn continuous convolutional filter weights as a nonlinear function from 3D point coordinates, approximated with MLP. [22,80] use a density estimation to compensate the non-uniform sampling, and [80] significantly improves the memory efficiency by a change of summation trick, allowing these networks to scale up and achieving comparable capabilities with 2D convolution.

Optical Flow Estimation. Optical flow estimation is a core computer vision problem and has many applications. Traditionally, top performing methods often adopt the energy minimization approach [24] and a coarse-to-fine, warping-based method [4,7,8]. Since FlowNet [13], there were many recent work using a deep network to learn optical flow. [28] stacks several FlowNets into a larger one. [56] develops a compact spatial pyramid network. [65] integrates the widely used traditional pyramid, warping, and cost volume technique into CNNs for optical flow, and outperform all the previous methods with high efficiency. We utilized a basic structure similar to theirs but proposed novel cost volume, warping and upsampling layers appropriate for point clouds.

Scene Flow Estimation. 3D scene flow is first introduced by [72]. Many works [26,45,75] estimate scene flow using RGB data. [26] introduces a variational method to estimate scene flow from stereo sequences. [45] proposes an object-level scene flow estimation approach and introduces a dataset for 3D scene flow. [75] presents a piecewise rigid scene model for 3D scene flow estimation.

Recently, there are some works [12,70,71] that estimate scene flow directly from point clouds using classical techniques. [12] introduces a method that formulates the scene flow estimation problem as an energy minimization problem with assumptions on local geometric constancy and regularization for motion smoothness. [71] proposes a real-time four-steps method of constructing occupancy grids, filtering the background, solving an energy minimization problem, and refining with a filtering framework. [70] further improves the method in [71] by using an encoding network to learn features from an occupancy grid.

In some most recent work [20,36,78], researchers attempt to estimate scene flow from point clouds using deep learning in a end-to-end fashion. [78] uses PCNN to operate on LiDAR data to estimate LiDAR motion. [36] introduces FlowNet3D based on PointNet++ [54]. FlowNet3D uses a flow embedding layer to encode the motion of point clouds. However, it requires encoding a large neighborhood in order to capture large motions. [20] presents HPLFlowNet to estimate the scene flow using Bilateral Convolutional Layers (BCL), which projects the point cloud onto a permutohedral lattice. [3] estimates scene flow with a network that jointly predicts 3D bounding boxes and rigid motions of objects or background in the scene. Different from [3], we do not require the rigid motion assumption and segmentation level supervision to estimate scene flow.

Self-supervised Scene Flow. There are several recent works [27,32,35,82,87] which jointly estimate multiple tasks, i.e. depth, optical flow, ego-motion and camera pose without supervision. They take 2D images as input, which have ambiguity when used in scene flow estimation. In this paper, we investigate self-supervised learning of scene flow from 3D point clouds with our PointPWC-Net. Concurrently, Mittal et al. [49] introduced Nearest Neighbor (NN) Loss and Cycle Consistency Loss to self-supervised scene flow estimation from point clouds. However, they does not take the local structure properties of 3D point clouds into consideration. In our work, we propose to use smoothness and Laplacian coordinates to preserve local structure for scene flow.

Traditional Point Cloud Registration. Point cloud registration has been extensively studied well before deep learning [21,66]. Most of the work [10,18,23, 43,48,60,69,85] only works when most of the motion in the scene is globally rigid. Many methods are based on the iterative closest point (ICP) [5] and its variants [52]. Several works [1,6,29,50] deal with non-rigid point cloud registration. Coherent Point Drift (CPD) [50] introduces a probabilistic method for both rigid and non-rigid point set registration. However, the computation overhead makes it hard to apply on real world data in real-time. Many algorithms are proposed to extend the CPD method [2,11,15–17,33,37–42,51,55,59,67,76,83,84,86]. Some algorithms require additional information for point set registration. The work [11,59] takes the color information along with the spatial location into account. [1] requires meshes for non-rigid registration. In [55], the regression and clustering for point set registration in a Bayesian framework are presented. All the aforementioned work require optimization at inference time, which has significantly higher computation cost than our method which run in a fraction of a second during inference.

3 Approach

To compute optical flow with high accuracy, one of the most important components is the cost volume. In 2D images, the cost volume can be computed by aggregating the cost in a square neighborhood on a grid. However, computing cost volume across two point clouds is difficult since 3D point clouds are unordered with a nonuniform sampling density. In this section, we introduce a novel learnable cost volume layer, and use it to construct a deep network with the help of other auxiliary layers that outputs high quality scene flow.

3.1 The Cost Volume Layer

As one of the key components of optical flow estimation, most state-of-the-art algorithms, both traditional [58,64] and modern deep learning based ones [9,65,81], use the cost volume to estimate optical flow. However, computing cost volumes on point clouds is still an open problem. There are several works [20,36] that compute some kind of flow embedding or correlation between point clouds. [36] proposes a flow embedding layer to aggregate feature similarities and spatial relationships to encode point motions. However, the motion information between points can be lost due to the max pooling operation in the flow embedding layering. [20] introduces a CorrBCL layer to compute the correlation between two point clouds, which requires to transfer two point clouds onto the same permutohedral lattice.

To address these issues, we present a novel learnable cost volume layer directly on the features of two point clouds. Suppose $f_i \in \mathbb{R}^c$ is the feature for point $p_i \in P$ and $g_j \in \mathbb{R}^c$ the feature for point $q_j \in Q$, the matching cost between p_i and q_j can be defined as:

$$Cost(p_i, q_j) = h(f_i, g_j, q_j, p_i) \tag{1}$$
$$= MLP(concat(f_i, g_j, q_j - p_i)) \tag{2}$$

where *concat* stands for concatenation. In our network, the feature f_i and g_j are either the raw coordinates of the point clouds, or the convolution output from previous layers. The intuition is that, as a universal approximator, MLP should be able to learn the potentially nonlinear relationship between the two points. Due to the flexibility of the point cloud, we also add a direction vector $(q_j - p_i)$ to the computation besides the point features f_i and g_j.

Once we have the matching costs, they can be aggregated as a cost volume for predicting the movement between two point clouds. In 2D images, aggregating the cost is simply by applying some convolutional layers as in PWC-Net [65]. However, traditional convolutional layers can not be applied directly on point clouds due to their unorderness. [36] uses max-pooing to aggregate features in the second point cloud. [20] uses CorrBCL to aggregate features on a permutohedral lattice. However, their methods only aggregate costs in a point-to-point manner, which is sensitive to outliers. To obtain robust and stable cost volumes, in this

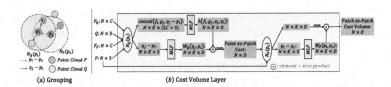

(a) Grouping   (b) Cost Volume Layer

Fig. 2. (a) **Grouping.** For a point p_c, we form its K-NN neighborhoods in each point cloud as $N_P(p_c)$ and $N_Q(p_c)$ for cost volume aggregation. We first aggregate the cost from the patch $N_Q(p_c)$ in point cloud Q. Then, we aggregate the cost from patch $N_P(p_c)$ in the point cloud P. (b) **Cost Volume Layer.** The features of neighboring points in $N_Q(p_c)$ are concatenated with the direction vector ($q_i - p_c$) to learn a point-to-patch cost between p_c and Q with PointConv. Then the point-to-patch costs in $N_P(p_c)$ are further aggregated with PointConv to construct a patch-to-patch cost volume

work, we propose to aggregate costs in a patch-to-patch manner similar to the cost volumes on 2D images [31,65].

For a point p_c in P, we first find a neighborhood $N_P(p_c)$ around p_c in P. For each point $p_i \in N_P(p_c)$, we find a neighborhood $N_Q(p_i)$ around p_i in Q. The cost volume for p_c is defined as:

$$CV(p_c) = \sum_{p_i \in N_P(p_c)} W_P(p_i, p_c) \sum_{q_j \in N_Q(p_i)} W_Q(q_j, p_i)\, cost(q_j, p_i) \quad (3)$$

$$W_P(p_i, p_c) = MLP(p_i - p_c) \quad (4)$$

$$W_Q(q_j, p_i) = MLP(q_j - p_i) \quad (5)$$

where $W_P(p_i, p_c)$ and $W_Q(q_j, p_i)$ are the convolutional weights *w.r.t* the direction vectors that are used to aggregate the costs from the patches in P and Q. It is learned as a continuous function of the directional vectors ($q_i - p_c$) $\in \mathbb{R}^3$ and ($q_j - p_i$) $\in \mathbb{R}^3$, respectively with an MLP, as in [80] and PCNN [78]. The output of the cost volume layer is a tensor with shape (n_1, D), where n_1 is the number of points in P, and D is the dimension of the cost volume, which encodes all the motion information for each point. The patch-to-patch idea used in the cost volume is illustrated in Fig. 2.

There are two major differences between this cost volume for scene flow of 3D point clouds and conventional 2D cost volumes for stereo and optical flow. The first one is that we introduce a learnable function $cost(\cdot)$ that can dynamically learn the cost or correlation within the point cloud structures. Ablation studies in Sect. 5.3 show that this novel learnable design achieve better results than traditional cost volume [65] in scene flow estimation. The second one is that this cost volume is discretized irregularly on the two input point clouds and their costs are aggregated with point-based convolution. Previously, in order to compute the cost volume for optical flow in a $d \times d$ area on a $W \times H$ 2D image, all the values in a $d^2 \times W \times H$ tensor needs to be populated, which is already slow to compute in 2D, but would be prohibitively costly in the 3D space. With (volumetric) 3D convolution, one needs to search a d^3 area to obtain a cost volume in 3D space. Our cost volume discretizes on input points and avoids this costly operation, while essentially creating the same capabilities to perform convolutions on the cost volume. With the proposed cost volume layer, we only need to find two neighborhoods $N_P(p_c)$ and $N_Q(p_i)$ of size K, which is much cheaper and does not depend on the number of points in a point cloud. In our experiments, we fix $|N_P(p_c)| = |N_Q(p_i)| = 16$. If a larger neighborhood is needed, we could subsample the neighborhood which would bring it back to the same speed. This subsampling operation is only applicable to the sparse point cloud convolution and not possible for conventional volumetric convolutions. We anticipate this novel cost volume layer to be widely useful beyond scene flow estimation. Table 2 shows that it is better than [36]'s MLP+Maxpool strategy.

3.2   PointPWC-Net

Given the proposed learnable cost volume layer, we construct a deep network for scene flow estimation. As demonstrated in 2D optical flow estimation, one of the most effective methods for dense estimation is the coarse-to-fine structure. In this section, we introduce some novel auxiliary layers for point clouds that construct a coarse-to-fine network for scene flow estimation along with the proposed learnable cost volume layer. The network is called *"PointPWC-Net"* following [65].

As shown in Fig. 1, PointPWC-Net predicts dense scene flow in a coarse-to-fine fashion. The input to PointPWC-Net is two consecutive point clouds, $P = \{p_i \in \mathbb{R}^3\}_{i=1}^{n_1}$ with n_1 points, and $Q = \{q_j \in \mathbb{R}^3\}_{j=1}^{n_2}$ with n_2 points. We first construct a feature pyramid for each point cloud. Afterwards, we build a cost volume using features from both point clouds at each layer. Then, we use the feature from P, the cost volume, and the upsampled flow to estimate the finer scene flow. We take the predicted scene flow as the coarse flow, upsample it to a finer flow, and warp points from P onto Q. Note that both the upsampling and the warping layers are efficient with no learnable parameters.

Feature Pyramid from Point Cloud. To estimate scene flow with high accuracy, we need to extract strong features from the input point clouds. We generate an L-level pyramid of feature representations, with the top level being the input point clouds, i.e., $l_0 = P/Q$. For each level l, we use furthest point sampling [54] to downsample the points by a factor of 4 from previous level $l-1$, and use Point-Conv [80] to perform convolution on the features from level $l-1$. As a result, we can generate a feature pyramid with L levels for each input point cloud. After this, we enlarge the receptive field at level l of the pyramid by upsampling the feature in level $l+1$ and concatenate it to the feature at level l.

Upsampling Layer. The upsampling layer can propagate the scene flow estimated from a coarse layer to a finer layer. We use a distance based interpolation to upsample the coarse flow. Let P^l be the point cloud at level l, SF^l be the estimated scene flow at level l, and p^{l-1} be the point cloud at level $l-1$. For each point p_i^{l-1} in the finer level point cloud P^{l-1}, we can find its K nearest neighbors $N(p_i^{l-1})$ in its coarser level point cloud P^l. The interpolated scene flow of finer level SF^{l-1} is computed using inverse distance weighted interpolation:

$$SF^{l-1}(p_i) = \frac{\sum_{j=1}^{k} w(p_i^{l-1}, p_j^l) SF^l(p_j^l)}{\sum_{j=1}^{k} w(p_i^{l-1}, p_j^l)} \tag{6}$$

where $w(p_i^{l-1}, p_j^l) = 1/d(p_i^{l-1}, p_j^l)$, $p_i^{l-1} \in P^{l-1}$, and $p_j^l \in N(p_i^{l-1})$. $d(p_i^{l-1}, p_j^l)$ is a distance metric. We use Euclidean distance in this work.

Warping Layer. Warping would "apply" the computed flow so that only the residual flow needs to be estimated afterwards, hence the search radius can be smaller when constructing the cost volume. In our network, we first up-sample the scene flow from the previous coarser level and then warp it before computing the cost volume. Denote the upsampled scene flow as $SF = \{sf_i \in \mathbb{R}^3\}_{i=1}^{n_1}$, and the warped point cloud as $P_w = \{p_{w,i} \in \mathbb{R}^3\}_{i=1}^{n_1}$. The warping layer is simply an element-wise addition between the upsampled and computed scene flow $P_w = \{p_{w,i} = p_i + sf_i | p_i \in P, sf_i \in SF\}_{i=1}^{n_1}$. A similar warping operation is used for visualization to compare the estimated flow with the ground truth in [20,36], but not used in coarse-to-fine estimation. [20] uses an offset strategy to reduce search radius which is specific to the permutohedral lattice.

Scene Flow Predictor. In order to obtain a flow estimate at each level, a convolutional scene flow predictor is built as multiple layers of PointConv and MLP. The inputs of the flow predictor are the cost volume, the feature of the first point cloud, the up-sampled flow from previous layer and the up-sampled feature of the second last layer from previous level's scene flow predictor, which we call the predictor feature. The output is the scene flow $SF = \{sf_i \in \mathbb{R}^3\}_{i=1}^{n_1}$ of the first point cloud P. The first several PointConv layers are used to merge the feature locally, and the following MLP is used to estimate the scene flow on each point. We keep the flow predictor structure at different levels the same, but the parameters are not shared.

4   Training Loss Functions

In this section, we introduce two loss functions to train PointPWC-Net for scene flow estimation. One is the standard multi-scale supervised training loss, which has been explored in deep optical flow estimation [65] in 2D images. We use this supervised loss to train the model for fair comparison with previous scene flow estimation work, including FlowNet3D [36] and HPLFlowNet [20]. Due to that acquiring densely labeled 3D scene flow dataset is extremely hard, we also propose a novel self-supervised loss to train our PointPWC-Net without any supervision.

4.1   Supervised Loss

We adopt the multi-scale loss function in FlowNet [13] and PWC-Net [65] as a supervised learning loss to demonstrate the effectiveness of the network structure and the design choice. Let SF_{GT}^l be the ground truth flow at the l-th level. The multi-scale training loss $\ell(\Theta) = \sum_{l=l_0}^L \alpha_l \sum_{p \in P} \left\| SF_\Theta^l(p) - SF_{GT}^l(p) \right\|_2$ is used where $\|\cdot\|_2$ computes the L_2-norm, α_l is the weight for each pyramid level l, and Θ is the set of all the learnable parameters in our PointPWC-Net, including the feature extractor, cost volume layer and scene flow predictor at different pyramid levels. Note that the flow loss is not squared as in [65] for robustness.

4.2   Self-supervised Loss

Obtaining the ground truth scene flow for 3D point clouds is difficult and there are not many publicly available datasets for scene flow learning from point clouds. In this section, we propose a self-supervised learning objective function to learn the scene flow in 3D point clouds without supervision. Our loss function contains three parts: *Chamfer distance*, *Smoothness constraint*, and *Laplacian regularization* [62,77]. To the best of our knowledge, we are the first to study self-supervised deep learning of scene flow estimation from 3D point clouds, concurrent with [49].

Chamfer Loss. The goal of Chamfer loss is to estimate scene flow by moving the first point cloud as close as the second one. Let SF_Θ^l be the scene flow predicted at level l. Let P_w^l be the point cloud warped from the first point cloud P^l according to SF_Θ^l in level l, Q^l be the second point cloud at level l. Let p_w^l and q^l be points in P_w^l and Q^l. The Chamfer loss ℓ_C^l can be written as:

$$P_w^l = P^l + SF_\Theta^l \tag{7}$$

$$\ell_C^l(P_w^l, Q^l) = \sum_{p_w^l \in P_w^l} \min_{q^l \in Q^l} \left\| p_w^l - q^l \right\|_2^2 + \sum_{q^l \in Q^l} \min_{p_w^l \in P_w^l} \left\| p_w^l - q^l \right\|_2^2$$

Smoothness Constraint. In order to enforce local spatial smoothness, we add a smoothness constraint ℓ_S^l, which assumes that the predicted scene flow $SF_\Theta^l(p_j^l)$ in a local region $N(p_i^l)$ of p_i^l should be similar to the scene flow at p_i^l:

$$\ell_S^l(SF^l) = \sum_{p_i^l \in P^l} \frac{1}{|N(p_i^l)|} \sum_{p_j^l \in N(p_i^l)} \left\| SF^l(p_j^l) - SF^l(p_i^l) \right\|_2^2 \tag{8}$$

where $|N(p_i^l)|$ is the number of points in the local region $N(p_i^l)$.

Laplacian Regularization. The Laplacian coordinate vector approximates the local shape characteristics of the surface [62]. The Laplacian coordinate vector $\delta^l(p_i^l)$ is computed as:

$$\delta^l(p_i^l) = \frac{1}{|N(p_i^l)|} \sum_{p_j^l \in N(p_i^l)} (p_j^l - p_i^l) \tag{9}$$

For scene flow, the warped point cloud P_w^l should have the same Laplacian coordinate vector with the second point cloud Q^l at the same position. Hence, we firstly compute the Laplacian coordinates $\delta^l(p_i^l)$ for each point in second

Table 1. Evaluation Results on the FlyingThings3D and KITTI Datasets. *Self* means self-supervised, *Full* means fully-supervised. All approaches are (at least) trained on FlyingThings3D. On KITTI, *Self* and *Full* refer to the respective models trained on FlyingThings3D that is directly evaluated on KITTI, while *Self+Self* means the model is firstly trained on FlyingThings3D with self-supervision, then fine-tuned on KITTI with self-supervision as well. *Full+Self* means the model is trained with full supervision on FlyingThings3D, then fine-tuned on KITTI with self-supervision. ICP [5], FGR [85], and CPD [50] are traditional method that does not require training. Our model outperforms all baselines by a large margin on all metrics

Dataset	Method	Sup.	EPE3D (m)↓	Acc3DS↑	Acc3DR↑	Outliers3D↓	EPE2D (px)↓	Acc2D↑
Flyingthings3D	ICP (rigid) [5]	*Self*	0.4062	0.1614	0.3038	0.8796	23.2280	0.2913
	FGR (rigid) [85]	*Self*	0.4016	0.1291	0.3461	0.8755	28.5165	0.3037
	CPD (non-rigid) [50]	*Self*	0.4887	0.0538	0.1694	0.9063	26.2015	0.0966
	PointPWC-Net	*Self*	**0.1213**	**0.3239**	**0.6742**	**0.6878**	**6.5493**	**0.4756**
	FlowNet3D [36]	*Full*	0.1136	0.4125	0.7706	0.6016	5.9740	0.5692
	SPLATFlowNet [63]	*Full*	0.1205	0.4197	0.7180	0.6187	6.9759	0.5512
	original BCL [20]	*Full*	0.1111	0.4279	0.7551	0.6054	6.3027	0.5669
	HPLFlowNet [20]	*Full*	0.0804	0.6144	0.8555	0.4287	4.6723	0.6764
	PointPWC-Net	*Full*	**0.0588**	**0.7379**	**0.9276**	**0.3424**	**3.2390**	**0.7994**
KITTI	ICP (rigid) [5]	*Self*	0.5181	0.0669	0.1667	0.8712	27.6752	0.1056
	FGR (rigid) [85]	*Self*	0.4835	0.1331	0.2851	0.7761	18.7464	0.2876
	CPD (non-rigid) [50]	*Self*	0.4144	0.2058	0.4001	0.7146	27.0583	0.1980
	PointPWC-Net (w/o ft)	*Self*	*0.2549*	*0.2379*	*0.4957*	*0.6863*	*8.9439*	*0.3299*
	PointPWC-Net (w/ft)	*Self+Self*	**0.0461**	**0.7951**	**0.9538**	**0.2275**	**2.0417**	**0.8645**
	FlowNet3D [36]	*Full*	0.1767	0.3738	0.6677	0.5271	7.2141	0.5093
	SPLATFlowNet [63]	*Full*	0.1988	0.2174	0.5391	0.6575	8.2306	0.4189
	original BCL [20]	*Full*	0.1729	0.2516	0.6011	0.6215	7.3476	0.4411
	HPLFlowNet [20]	*Full*	0.1169	0.4783	0.7776	0.4103	4.8055	0.5938
	PointPWC-Net (w/o ft)	*Full*	*0.0694*	*0.7281*	*0.8884*	*0.2648*	*3.0062*	*0.7673*
	PointPWC-Net (w/ft)	*Full+Self*	**0.0430**	**0.8175**	**0.9680**	**0.2072**	**1.9022**	**0.8669**

point cloud Q^l. Then, we interpolate the Laplacian coordinate of Q^l to obtain the Laplacian coordinate on each point p_w^l. We use an inverse distance-based interpolation method similar to Eq. (6) to interpolate the Laplacian coordinate δ^l. Let $\delta^l(p_w^l)$ be the Laplacian coordinate of point p_w^l at level l, $\delta^l(q_{inter}^l)$ be the interpolated Laplacian coordinate from Q^l at the same position as p_w^l.

The Laplacian regularization ℓ_L^l is defined as:

$$\ell_L^l(\delta^l(p_w^l), \delta^l(q_{inter}^l)) = \sum_{p_w^l \in P_w^l} \left\| \delta^l(p_w^l) - \delta^l(q_{inter}^l) \right\|_2^2 \tag{10}$$

The overall loss is a weighted sum of all losses across all pyramid levels as:

$$\ell(\Theta) = \sum_{l=l_0}^{L} \alpha_l(\beta_1 \ell_C^l + \beta_2 \ell_S^l + \beta_3 \ell_L^l) \tag{11}$$

where α_l is the factor for pyramid level l, $\beta_1, \beta_2, \beta_3$ are the scale factors for each loss respectively. With the self-supervised loss, our model is able to learn the scene flow from 3D point cloud pairs without any ground truth supervision.

5   Experiments

In this section, we train and evaluate our PointPWC-Net on the FlyingThings3D dataset [44] with the supervised loss and the self-supervised loss, respectively. Then, we evaluate the generalization ability of our model by first applying the model on the real-world KITTI Scene Flow dataset [46,47] *without any fine-tuning*. Then, with the proposed self-supervised losses, we further fine-tune our pre-trained model on the KITTI dataset to study the best performance we could obtain without supervision. Besides, we also compare the runtime of our model with previous work. Finally, we conduct ablation studies to analyze the contribution of each part of the model and the loss function.

Implementation Details. We build a 4-level feature pyramid from the input point cloud. The weights α are set to be $\alpha_0 = 0.02$, $\alpha_1 = 0.04$, $\alpha_2 = 0.08$, and $\alpha_3 = 0.16$, with weight decay 0.0001. The scale factor β in self-supervised learning are set to be $\beta_1 = 1.0$, $\beta_2 = 1.0$, and $\beta_3 = 0.3$. We train our model starting from a learning rate of 0.001 and reducing by half every 80 epochs. All the hyperparameters are set using the validation set of FlyingThings3D with 8,192 points in each input point cloud.

Evaluation Metrics. For fair comparison, we adopt the evaluation metrics that are used in [20]. Let SF_Θ denote the predicted scene flow, and SF_{GT} be the ground truth scene flow. The evaluate metrics are computed as follows:

- *EPE3D (m)*: $\|SF_\Theta - SF_{GT}\|_2$ averaged over each point in meters.
- *Acc3DS*: the percentage of points with *EPE3D* < 0.05 m or relative error $< 5\%$.
- *Acc3DR*: the percentage of points with *EPE3D* < 0.1 m or relative error $< 10\%$.
- *Outliers3D*: the percentage of points with *EPE3D* > 0.3 m or relative error $> 10\%$.
- *EPE2D (px)*: 2D end point error obtained by projecting point clouds back to the image plane.
- *Acc2D*: the percentage of points whose *EPE2D* $< 3px$ or relative error $< 5\%$.

5.1   Supervised Learning

First we conduct experiments with supervised loss. To our knowledge, there is no publicly available large-scale real-world dataset that has scene flow ground truth from point clouds (The input to the KITTI scene flow benchmark is 2D), thus we train our PointPWC-Net on the synthetic Flyingthings3D dataset, following [20]. Then, the pre-trained model is directly evaluated on KITTI Scene Flow 2015 dataset without any fine-tuning.

Train and Evaluate on FlyingThings3D. The FlyingThings3D training dataset includes 19,640 pairs of point clouds, and the evaluation dataset includes 3,824 pairs of point clouds. Our model takes $n = 8,192$ points in each point cloud. We first train the model with $\frac{1}{4}$ of the training set (4,910 pairs), and then fine-tune it on the whole training set, to speed up training.

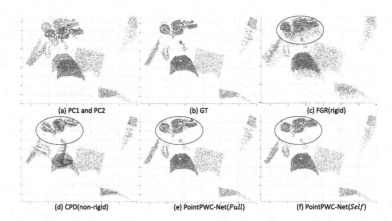

Fig. 3. Results on the FlyingThings3D Dataset. In (a), 2 point clouds PC1 and PC2 are presented in Magenta and Green, respectively. In (b–f), PC1 is warped to PC2 based on the (computed) scene flow. (b) shows the ground truth; (c) Results from FGR (rigid) [85]; (d) Results from CPD (non-rigid) [50]; (e) Results from PointPWC-Net (*Full*); (f) Results from PointPWC-Net (*Self*). Red ellipses indicate locations with significant non-rigid motion. Enlarge images for better view. (Best viewed in color) (Color figure online)

Table 1 shows the quantitative evaluation results on the Flyingthings3D dataset. Our method outperforms all the methods on all metrics by a large margin. Comparing to SPLATFlowNet, original BCL, and HPLFlowNet, our method avoids the preprocessing step of building a permutohedral lattice from the input. Besides, our method outperforms HPLFlowNet on *EPE3D* by *26.9%*. And, we are the only method with *EPE2D* under 4px, which improves over HPLFlowNet by *30.7%*. See Fig. 3(e) for example results.

Evaluate on KITTI w/o Fine-Tune. To study the generalization ability of our PointPWC-Net, we directly take the model trained using FlyingThings3D and evaluate it on KITTI Scene Flow 2015 [46,47] *without any fine-tuning*. KITTI Scene Flow 2015 consists of 200 training scenes and 200 test scenes. To evaluate our PointPWC-Net, we use ground truth labels and trace raw point clouds associated with the frames, following [20,36]. Since no point clouds and ground truth are provided on test set, we evaluate on all 142 scenes in the training set with available point clouds. We remove ground points with height < 0.3 m following [20] for fair comparison with previous methods.

From Table 1, our PointPWC-Net outperforms all the state-of-the-art methods, which demonstrates the generalization ability of our model. For *EPE3D*, our model is the only one below 10 cm, which improves over HPLFlowNet by *40.6%*. For *Acc3DS*, our method outperforms both FlowNet3D and HPLFlowNet by *35.4%* and *25.0%* respectively. See Fig. 4(e) for example results.

Table 2. Model Design. A learnable cost volume preforms much better than inner product cost volume used in PWC-Net [65]. Using our cost volume instead of the MLP+Maxpool used in FlowNet3D's flow embedding layer improves performance by 20.6%. Compared to no warping, the warping layer improves the performance by 40.2%

Component	Status	EPE3D (m)↓
Cost volume	PWC-Net [65]	0.0821
	MLP+Maxpool (learnable) [36]	0.0741
	Ours (learnable)	**0.0588**
Warping layer	w/o	0.0984
	w	**0.0588**

5.2   Self-supervised Learning

Acquiring or annotating dense scene flow from real-world 3D point clouds is very expensive, so it would be interesting to evaluate the performance of our self-supervised approach. We train our model using the same procedure as in supervised learning, i.e. first train the model with one quarter of the training dataset, then fine-tune with the whole training set. Table 1 gives the quantitative results on PointPWC-Net with self-supervised learning. We compare our method with ICP (rigid) [5], FGR (rigid) [85] and CPD (non-rigid) [50]. Because traditional point registration methods are not trained with ground truth, we can view them as self/un-supervised methods.

Train and Evaluate on FlyingThings3D. We can see that our PointPWC-Net outperforms traditional methods on all the metrics with a large margin. See Fig. 3(f) for example results.

Evaluate on KITTI w/o Fine-Tuning. Even only trained on FlyingThings3D without ground truth labels, our method can obtain *0.2549* m on *EPE3D* on KITTI, which improves over CPD (non-rigid) by *38.5%*, FGR (rigid) by *47.3%*, and ICP (rigid) by *50.8%*.

Fine-Tune on KITTI. With proposed self-supervised loss, we are able to fine-tune the FlyingThings3D trained models on KITTI without using any ground truth. In Table 1, the row *PointPWC-Net (w/ft) Full+Self* and *PointPWC-Net (w/ft) Self+Self* show the results. *Full+Self* means the model is trained with supervision on FlyingThings3D, then fine-tuned on KITTI without supervision. *Self+Self* means the model is firstly trained on FlyingThings3D, then fine-tuned on KITTI both using self-supervised loss. With KITTI fine-tuning, our PointPWC-Net can achieve *EPE3D* < 5 cm. Especially, our *PointPWC-Net (w/ft) Self+Self*, which is fully trained without any ground truth information, achieves similar performance on KITTI as the one that utilized FlyingThings3D ground truth. See Fig. 4(f) for example results.

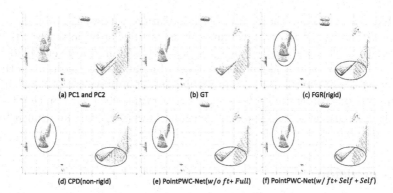

(a) PC1 and PC2   (b) GT   (c) FGR(rigid)

(d) CPD(non-rigid)   (e) PointPWC-Net(w/o ft+ Full)   (f) PointPWC-Net(w/ ft+ Self + Self)

Fig. 4. Results on the KITTI Scene Flow 2015 Dataset. In (a), 2 point clouds PC1 and PC2 are presented in Magenta and Green, respectively. In (b–f), PC1 is warped to PC2 based on the (computed) scene flow. (b) shows the ground truth; (c) Results from FGR (rigid) [85]; (d) Results from CPD (non-rigid) [50]; (e) Results from PointPWC-Net (*w/o ft+Full*) that is trained with supervision on FlyingThings3D, and directly evaluate on KITTI without any fine-tuning; (f) Results from PointPWC-Net (*w/ft+Self + Self*) which is trained on FlyingThings3D and fine-tuned on KITTI using the proposed self-supervised loss. Red ellipses indicate locations with significant non-rigid motion. Enlarge images for better view. (Best viewed in color) (Color figure online)

5.3   Ablation Study

We further conduct ablation studies on model design choices and the self-supervised loss function. On model design, we evaluate the different choices of cost volume layer and removing the warping layer. On the loss function, we investigate removing the smoothness constraint and Laplacian regularization in the self-supervised learning loss. All models in the ablation studies are trained using FlyingThings3D, and tested on the FlyingThings3D evaluation dataset.

Table 3. Loss Functions. The Chamfer loss is not enough to estimate a good scene flow. With the smoothness constraint, the scene flow result improves by 38.2%. Laplacian regularization also improves slightly

Chamfer	Smoothness	Laplacian	EPE3D (m)↓
✓	-	-	0.2112
✓	✓	-	0.1304
✓	✓	✓	**0.1213**

Table 4. Runtime. Average runtime (ms) on Flyingthings3D. The runtime for FlowNet3D and HPLFlowNet is reported from [20] on a single Titan V. The runtime for our PointPWC-Net is reported on a single 1080Ti

Method	Runtime (ms)↓
FlowNet3D [36]	130.8
HPLFlowNet [20]	98.4
PointPWC-Net	117.4

Tables 2 and 3 show the results of the ablation studies. In Table 2 we can see that our design of the cost volume obtains significantly better results than the inner product-based cost volume in PWC-Net [65] and FlowNet3D [36], and the warping layer is crucial for performance. In Table 3, we see that both the smoothness constraint and Laplacian regularization improve the performance in self-supervised learning. In Table 4, we report the runtime of our PointPWC-Net, which is comparable with other deep learning based methods and much faster than traditional ones.

6 Conclusion

To better estimate scene flow directly from 3D point clouds, we proposed a novel learnable cost volume layer along with some auxiliary layers to build a coarse-to-fine deep network, called PointPWC-Net. Because of the fact that real-world ground truth scene flow is hard to acquire, we introduce a loss function that train the PointPWC-Net without supervision. Experiments on the FlyingThings3D and KITTI datasets demonstrates the effectiveness of our PointPWC-Net and the self-supervised loss function, obtaining state-of-the-art results that outperform prior work by a large margin.

Acknowledgement. Wenxuan Wu and Li Fuxin were partially supported by the National Science Foundation (NSF) under Project #1751402, USDA National Institute of Food and Agriculture (USDA-NIFA) under Award 2019-67019-29462, as well as by the Defense Advanced Research Projects Agency (DARPA) under Contract No. N66001-17-12-4030 and N66001-19-2-4035. Any opinions, findings and conclusions or recommendations expressed in this material are those of the author(s) and do not necessarily reflect the views of the funding agencies.

References

1. Amberg, B., Romdhani, S., Vetter, T.: Optimal step nonrigid ICP algorithms for surface registration. In: 2007 IEEE Conference on Computer Vision and Pattern Recognition, pp. 1–8. IEEE (2007)
2. Bai, L., Yang, X., Gao, H.: Nonrigid point set registration by preserving local connectivity. IEEE Trans. Cybern. **48**(3), 826–835 (2017)
3. Behl, A., Paschalidou, D., Donné, S., Geiger, A.: PointFlowNet: learning representations for rigid motion estimation from point clouds. In: Proceedings of the IEEE Conference on Computer Vision and Pattern Recognition, pp. 7962–7971 (2019)
4. Bergen, J.R., Anandan, P., Hanna, K.J., Hingorani, R.: Hierarchical model-based motion estimation. In: Sandini, G. (ed.) ECCV 1992. LNCS, vol. 588, pp. 237–252. Springer, Heidelberg (1992). https://doi.org/10.1007/3-540-55426-2_27
5. Besl, P.J., McKay, N.D.: Method for registration of 3-D shapes. In: Sensor Fusion IV: Control Paradigms and Data Structures, vol. 1611, pp. 586–606. International Society for Optics and Photonics (1992)
6. Brown, B.J., Rusinkiewicz, S.: Global non-rigid alignment of 3-D scans. In: ACM SIGGRAPH 2007 papers, p. 21-es (2007)

7. Brox, T., Bruhn, A., Papenberg, N., Weickert, J.: High accuracy optical flow estimation based on a theory for warping. In: Pajdla, T., Matas, J. (eds.) ECCV 2004. LNCS, vol. 3024, pp. 25–36. Springer, Heidelberg (2004). https://doi.org/10.1007/978-3-540-24673-2_3

8. Bruhn, A., Weickert, J., Schnörr, C.: Lucas/Kanade meets Horn/Schunck: combining local and global optic flow methods. Int. J. Comput. Vis. **61**(3), 211–231 (2005). https://doi.org/10.1023/B:VISI.0000045324.43199.43

9. Chabra, R., Straub, J., Sweeney, C., Newcombe, R., Fuchs, H.: StereoDRNet: dilated residual StereoNet. In: Proceedings of the IEEE Conference on Computer Vision and Pattern Recognition, pp. 11786–11795 (2019)

10. Choi, S., Zhou, Q.Y., Koltun, V.: Robust reconstruction of indoor scenes. In: Proceedings of the IEEE Conference on Computer Vision and Pattern Recognition, pp. 5556–5565 (2015)

11. Danelljan, M., Meneghetti, G., Shahbaz Khan, F., Felsberg, M.: A probabilistic framework for color-based point set registration. In: Proceedings of the IEEE Conference on Computer Vision and Pattern Recognition, pp. 1818–1826 (2016)

12. Dewan, A., Caselitz, T., Tipaldi, G.D., Burgard, W.: Rigid scene flow for 3D LiDAR scans. In: 2016 IEEE/RSJ International Conference on Intelligent Robots and Systems (IROS), pp. 1765–1770. IEEE (2016)

13. Dosovitskiy, A., et al.: FlowNet: learning optical flow with convolutional networks. In: Proceedings of the IEEE International Conference on Computer Vision, pp. 2758–2766 (2015)

14. Fan, H., Su, H., Guibas, L.J.: A point set generation network for 3D object reconstruction from a single image. In: Proceedings of the IEEE Conference on Computer Vision and Pattern Recognition, pp. 605–613 (2017)

15. Fu, M., Zhou, W.: Non-rigid point set registration via mixture of asymmetric Gaussians with integrated local structures. In: 2016 IEEE International Conference on Robotics and Biomimetics (ROBIO), pp. 999–1004. IEEE (2016)

16. Ge, S., Fan, G.: Non-rigid articulated point set registration with local structure preservation. In: Proceedings of the IEEE Conference on Computer Vision and Pattern Recognition Workshops, pp. 126–133 (2015)

17. Ge, S., Fan, G., Ding, M.: Non-rigid point set registration with global-local topology preservation. In: Proceedings of the IEEE Conference on Computer Vision and Pattern Recognition Workshops, pp. 245–251 (2014)

18. Gelfand, N., Mitra, N.J., Guibas, L.J., Pottmann, H.: Robust global registration. In: Symposium on Geometry Processing, Vienna, Austria, vol. 2, p. 5 (2005)

19. Groh, F., Wieschollek, P., Lensch, H.P.A.: Flex-convolution. In: Jawahar, C.V., Li, H., Mori, G., Schindler, K. (eds.) ACCV 2018. LNCS, vol. 11361, pp. 105–122. Springer, Cham (2019). https://doi.org/10.1007/978-3-030-20887-5_7

20. Gu, X., Wang, Y., Wu, C., Lee, Y.J., Wang, P.: HPLFlowNet: hierarchical permutohedral lattice FlowNet for scene flow estimation on large-scale point clouds. In: Proceedings of the IEEE Conference on Computer Vision and Pattern Recognition, pp. 3254–3263 (2019)

21. Guo, Y., Bennamoun, M., Sohel, F., Lu, M., Wan, J.: 3D object recognition in cluttered scenes with local surface features: a survey. IEEE Trans. Pattern Anal. Mach. Intell. **36**(11), 2270–2287 (2014)

22. Hermosilla, P., Ritschel, T., Vázquez, P.P., Vinacua, À., Ropinski, T.: Monte Carlo convolution for learning on non-uniformly sampled point clouds. In: SIGGRAPH Asia 2018 Technical Papers, p. 235. ACM (2018)

23. Holz, D., Ichim, A.E., Tombari, F., Rusu, R.B., Behnke, S.: Registration with the point cloud library: a modular framework for aligning in 3-D. IEEE Robot. Autom. Mag. **22**(4), 110–124 (2015)
24. Horn, B.K., Schunck, B.G.: Determining optical flow. Artif. Intell. **17**(1–3), 185–203 (1981)
25. Hua, B.S., Tran, M.K., Yeung, S.K.: Pointwise convolutional neural networks. In: Proceedings of the IEEE Conference on Computer Vision and Pattern Recognition, pp. 984–993 (2018)
26. Huguet, F., Devernay, F.: A variational method for scene flow estimation from stereo sequences. In: 2007 IEEE 11th International Conference on Computer Vision, pp. 1–7. IEEE (2007)
27. Hur, J., Roth, S.: Self-supervised monocular scene flow estimation. In: Proceedings of the IEEE/CVF Conference on Computer Vision and Pattern Recognition, pp. 7396–7405 (2020)
28. Ilg, E., Mayer, N., Saikia, T., Keuper, M., Dosovitskiy, A., Brox, T.: FlowNet 2.0: evolution of optical flow estimation with deep networks. In: Proceedings of the IEEE Conference on Computer Vision and Pattern Recognition, pp. 2462–2470 (2017)
29. Jain, V., Zhang, H., van Kaick, O.: Non-rigid spectral correspondence of triangle meshes. Int. J. Shape Model. **13**(1), 101–124 (2007)
30. Jia, X., De Brabandere, B., Tuytelaars, T., Gool, L.V.: Dynamic filter networks. In: Advances in Neural Information Processing Systems, pp. 667–675 (2016)
31. Kendall, A., et al.: End-to-end learning of geometry and context for deep stereo regression. In: Proceedings of the IEEE International Conference on Computer Vision, pp. 66–75 (2017)
32. Lee, M., Fowlkes, C.C.: CeMNet: self-supervised learning for accurate continuous ego-motion estimation. In: Proceedings of the IEEE Conference on Computer Vision and Pattern Recognition Workshops. pp. 354–363 (2019)
33. Lei, H., Jiang, G., Quan, L.: Fast descriptors and correspondence propagation for robust global point cloud registration. IEEE Trans. Image Process. **26**(8), 3614–3623 (2017)
34. Li, Y., Bu, R., Sun, M., Wu, W., Di, X., Chen, B.: PointCNN: convolution on x-transformed points. In: Advances in Neural Information Processing Systems, pp. 820–830 (2018)
35. Liu, L., Zhai, G., Ye, W., Liu, Y.: Unsupervised learning of scene flow estimation fusing with local rigidity. In: Proceedings of the 28th International Joint Conference on Artificial Intelligence, pp. 876–882. AAAI Press (2019)
36. Liu, X., Qi, C.R., Guibas, L.J.: FlowNet3D: learning scene flow in 3D point clouds. In: Proceedings of the IEEE Conference on Computer Vision and Pattern Recognition, pp. 529–537 (2019)
37. Lu, M., Zhao, J., Guo, Y., Ma, Y.: Accelerated coherent point drift for automatic three-dimensional point cloud registration. IEEE Geosci. Remote Sens. Lett. **13**(2), 162–166 (2015)
38. Ma, J., Chen, J., Ming, D., Tian, J.: A mixture model for robust point matching under multi-layer motion. PLOS One **9**(3), e92282 (2014)
39. Ma, J., Jiang, J., Zhou, H., Zhao, J., Guo, X.: Guided locality preserving feature matching for remote sensing image registration. IEEE Trans. Geosci. Remote Sens. **56**(8), 4435–4447 (2018)
40. Ma, J., Zhao, J., Jiang, J., Zhou, H.: Non-rigid point set registration with robust transformation estimation under manifold regularization. In: 31st AAAI Conference on Artificial Intelligence (2017)

41. Ma, J., Zhao, J., Yuille, A.L.: Non-rigid point set registration by preserving global and local structures. IEEE Trans. Image Process. **25**(1), 53–64 (2015)
42. Ma, J., Zhou, H., Zhao, J., Gao, Y., Jiang, J., Tian, J.: Robust feature matching for remote sensing image registration via locally linear transforming. IEEE Trans. Geosci. Remote Sens. **53**(12), 6469–6481 (2015)
43. Makadia, A., Patterson, A., Daniilidis, K.: Fully automatic registration of 3D point clouds. In: 2006 IEEE Computer Society Conference on Computer Vision and Pattern Recognition, CVPR 2006, vol. 1, pp. 1297–1304. IEEE (2006)
44. Mayer, N., et al.: A large dataset to train convolutional networks for disparity, optical flow, and scene flow estimation. In: Proceedings of the IEEE Conference on Computer Vision and Pattern Recognition, pp. 4040–4048 (2016)
45. Menze, M., Geiger, A.: Object scene flow for autonomous vehicles. In: Proceedings of the IEEE Conference on Computer Vision and Pattern Recognition, pp. 3061–3070 (2015)
46. Menze, M., Heipke, C., Geiger, A.: Joint 3D estimation of vehicles and scene flow. In: ISPRS Workshop on Image Sequence Analysis (ISA) (2015)
47. Menze, M., Heipke, C., Geiger, A.: Object scene flow. ISPRS J. Photogram. Remote Sens. (JPRS) **140**, 60–76 (2018)
48. Mian, A.S., Bennamoun, M., Owens, R.: Three-dimensional model-based object recognition and segmentation in cluttered scenes. IEEE Trans. Pattern Anal. Mach. Intell. **28**(10), 1584–1601 (2006)
49. Mittal, H., Okorn, B., Held, D.: Just go with the flow: self-supervised scene flow estimation. In: Proceedings of the IEEE/CVF Conference on Computer Vision and Pattern Recognition, pp. 11177–11185 (2020)
50. Myronenko, A., Song, X.: Point set registration: coherent point drift. IEEE Trans. Pattern Anal. Mach. Intell. **32**(12), 2262–2275 (2010)
51. Nguyen, T.M., Wu, Q.J.: Multiple kernel point set registration. IEEE Trans. Med. Imaging **35**(6), 1381–1394 (2015)
52. Pomerleau, F., Colas, F., Siegwart, R., Magnenat, S.: Comparing ICP variants on real-world data sets. Auton. Robot. **34**(3), 133–148 (2013)
53. Qi, C.R., Su, H., Mo, K., Guibas, L.J.: PointNet: deep learning on point sets for 3D classification and segmentation. In: Proceedings of the IEEE Conference on Computer Vision and Pattern Recognition, pp. 652–660 (2017)
54. Qi, C.R., Yi, L., Su, H., Guibas, L.J.: Pointnet++: deep hierarchical feature learning on point sets in a metric space. In: Advances in Neural Information Processing Systems, pp. 5099–5108 (2017)
55. Qu, H.B., Wang, J.Q., Li, B., Yu, M.: Probabilistic model for robust affine and non-rigid point set matching. IEEE Trans. Pattern Anal. Mach. Intell. **39**(2), 371–384 (2016)
56. Ranjan, A., Black, M.J.: Optical flow estimation using a spatial pyramid network. In: Proceedings of the IEEE Conference on Computer Vision and Pattern Recognition, pp. 4161–4170 (2017)
57. Ravanbakhsh, S., Schneider, J., Poczos, B.: Deep learning with sets and point clouds. arXiv preprint arXiv:1611.04500 (2016)
58. Revaud, J., Weinzaepfel, P., Harchaoui, Z., Schmid, C.: EpicFlow: edge-preserving interpolation of correspondences for optical flow. In: Proceedings of the IEEE Conference on Computer Vision and Pattern Recognition, pp. 1164–1172 (2015)
59. Saval-Calvo, M., Azorin-Lopez, J., Fuster-Guillo, A., Villena-Martinez, V., Fisher, R.B.: 3D non-rigid registration using color: color coherent point drift. Comput. Vis. Image Underst. **169**, 119–135 (2018)

60. Shin, J., Triebel, R., Siegwart, R.: Unsupervised discovery of repetitive objects. In: 2010 IEEE International Conference on Robotics and Automation, pp. 5041–5046. IEEE (2010)
61. Simonovsky, M., Komodakis, N.: Dynamic edge-conditioned filters in convolutional neural networks on graphs. In: Proceedings of the IEEE Conference on Computer Vision and Pattern Recognition, pp. 3693–3702 (2017)
62. Sorkine, O.: Laplacian mesh processing. In: Eurographics (STARs), pp. 53–70 (2005)
63. Su, H., et al.: SPLATNet: sparse lattice networks for point cloud processing. In: Proceedings of the IEEE Conference on Computer Vision and Pattern Recognition, pp. 2530–2539 (2018)
64. Sun, D., Roth, S., Black, M.J.: A quantitative analysis of current practices in optical flow estimation and the principles behind them. Int. J. Comput. Vis. **106**(2), 115–137 (2014)
65. Sun, D., Yang, X., Liu, M.Y., Kautz, J.: PWC-Net: CNNs for optical flow using pyramid, warping, and cost volume. In: Proceedings of the IEEE Conference on Computer Vision and Pattern Recognition, pp. 8934–8943 (2018)
66. Tam, G.K., et al.: Registration of 3D point clouds and meshes: a survey from rigid to nonrigid. IEEE Trans. Visual Comput. Graph. **19**(7), 1199–1217 (2012)
67. Tao, W., Sun, K.: Asymmetrical Gauss mixture models for point sets matching. In: Proceedings of the IEEE Conference on Computer Vision and Pattern Recognition, pp. 1598–1605 (2014)
68. Tatarchenko, M., Park, J., Koltun, V., Zhou, Q.Y.: Tangent convolutions for dense prediction in 3D. In: Proceedings of the IEEE Conference on Computer Vision and Pattern Recognition, pp. 3887–3896 (2018)
69. Theiler, P.W., Wegner, J.D., Schindler, K.: Globally consistent registration of terrestrial laser scans via graph optimization. ISPRS J. Photogram. Remote Sens. **109**, 126–138 (2015)
70. Ushani, A.K., Eustice, R.M.: Feature learning for scene flow estimation from LiDAR. In: Conference on Robot Learning, pp. 283–292 (2018)
71. Ushani, A.K., Wolcott, R.W., Walls, J.M., Eustice, R.M.: A learning approach for real-time temporal scene flow estimation from LIDAR data. In: 2017 IEEE International Conference on Robotics and Automation (ICRA), pp. 5666–5673. IEEE (2017)
72. Vedula, S., Baker, S., Rander, P., Collins, R., Kanade, T.: Three-dimensional scene flow. In: Proceedings of the 7th IEEE International Conference on Computer Vision, vol. 2, pp. 722–729. IEEE (1999)
73. Verma, N., Boyer, E., Verbeek, J.: FeaStNet: feature-steered graph convolutions for 3D shape analysis. In: Proceedings of the IEEE Conference on Computer Vision and Pattern Recognition, pp. 2598–2606 (2018)
74. Vogel, C., Schindler, K., Roth, S.: Piecewise rigid scene flow. In: Proceedings of the IEEE International Conference on Computer Vision, pp. 1377–1384 (2013)
75. Vogel, C., Schindler, K., Roth, S.: 3D scene flow estimation with a piecewise rigid scene model. Int. J. Comput. Vis. **115**(1), 1–28 (2015)
76. Wang, G., Chen, Y.: Fuzzy correspondences guided Gaussian mixture model for point set registration. Knowl. Based Syst. **136**, 200–209 (2017)
77. Wang, N., Zhang, Y., Li, Z., Fu, Y., Liu, W., Jiang, Y.-G.: Pixel2Mesh: generating 3D mesh models from single RGB images. In: Ferrari, V., Hebert, M., Sminchisescu, C., Weiss, Y. (eds.) ECCV 2018. LNCS, vol. 11215, pp. 55–71. Springer, Cham (2018). https://doi.org/10.1007/978-3-030-01252-6_4

78. Wang, S., Suo, S., Ma, W.C., Pokrovsky, A., Urtasun, R.: Deep parametric continuous convolutional neural networks. In: Proceedings of the IEEE Conference on Computer Vision and Pattern Recognition, pp. 2589–2597 (2018)
79. Wang, Y., Sun, Y., Liu, Z., Sarma, S.E., Bronstein, M.M., Solomon, J.M.: Dynamic graph CNN for learning on point clouds. ACM Trans. Graph. (TOG) **38**(5), 146 (2019)
80. Wu, W., Qi, Z., Fuxin, L.: PointConv: deep convolutional networks on 3D point clouds. In: Proceedings of the IEEE Conference on Computer Vision and Pattern Recognition, pp. 9621–9630 (2019)
81. Xu, J., Ranftl, R., Koltun, V.: Accurate optical flow via direct cost volume processing. In: Proceedings of the IEEE Conference on Computer Vision and Pattern Recognition, pp. 1289–1297 (2017)
82. Yin, Z., Shi, J.: GeoNet: unsupervised learning of dense depth, optical flow and camera pose. In: Proceedings of the IEEE Conference on Computer Vision and Pattern Recognition, pp. 1983–1992 (2018)
83. Yu, D.: Fast rotation-free feature-based image registration using improved N-SIFT and GMM-based parallel optimization. IEEE Trans. Biomed. Eng. **63**(8), 1653–1664 (2015)
84. Zhang, S., Yang, K., Yang, Y., Luo, Y., Wei, Z.: Non-rigid point set registration using dual-feature finite mixture model and global-local structural preservation. Pattern Recogn. **80**, 183–195 (2018)
85. Zhou, Q.-Y., Park, J., Koltun, V.: Fast global registration. In: Leibe, B., Matas, J., Sebe, N., Welling, M. (eds.) ECCV 2016. LNCS, vol. 9906, pp. 766–782. Springer, Cham (2016). https://doi.org/10.1007/978-3-319-46475-6_47
86. Zhou, Z., Zheng, J., Dai, Y., Zhou, Z., Chen, S.: Robust non-rigid point set registration using student's-t mixture model. PLOS One **9**(3), e91381 (2014)
87. Zou, Y., Luo, Z., Huang, J.-B.: DF-Net: unsupervised joint learning of depth and flow using cross-task consistency. In: Ferrari, V., Hebert, M., Sminchisescu, C., Weiss, Y. (eds.) ECCV 2018. LNCS, vol. 11209, pp. 38–55. Springer, Cham (2018). https://doi.org/10.1007/978-3-030-01228-1_3

Points2Surf Learning Implicit Surfaces from Point Clouds

Philipp Erler[1]([✉])(ID), Paul Guerrero[2](ID), Stefan Ohrhallinger[1,3](ID),
Niloy J. Mitra[2,4](ID), and Michael Wimmer[1](ID)

[1] TU Wien, Favoritenstrasse 9-11/E193-02, 1040 Vienna, AT, Austria
`{perler,ohrhallinger,wimmer}@cg.tuwien.ac.at`
[2] Adobe, 1 Old Street Yard, London EC1Y 8AF, UK
`{guerrero,nimitra}@adobe.com`
[3] VRVis, Donau-City-Straße 11, 1220 Vienna, AT, Austria
`ohrhallinger@vrvis.at`
[4] University College London, 66 Gower St, London WC1E 6BT, UK
`n.mitra@cs.ucl.ac.uk`

Abstract. A key step in any scanning-based asset creation workflow is to convert unordered point clouds to a surface. Classical methods (e.g., Poisson reconstruction) start to degrade in the presence of noisy and partial scans. Hence, deep learning based methods have recently been proposed to produce complete surfaces, even from partial scans. However, such data-driven methods struggle to generalize to new shapes with large geometric and topological variations. We present Points2Surf, a novel *patch-based* learning framework that produces accurate surfaces directly from raw scans without normals. Learning a prior over a combination of detailed local patches and coarse global information improves generalization performance and reconstruction accuracy. O5ur extensive comparison on both synthetic and real data demonstrates a clear advantage of our method over state-of-the-art alternatives on previously unseen classes (on average, Points2Surf brings down reconstruction error by 30% over SPR and by 270%+ over deep learning based SotA methods) at the cost of longer computation times and a slight increase in small-scale topological noise in some cases. Our source code, pre-trained model, and dataset are available at: https://github.com/ErlerPhilipp/points2surf.

Keywords: Surface reconstruction · Implicit surfaces · Point clouds · Patch-based · Local and global · Deep learning · Generalization

1 Introduction

Converting unstructured point clouds to surfaces is a key step of most scanning-based asset creation workflows, including games and AR/VR applications. While scanning technologies have become more easily accessible (e.g., depth cameras on smart phones, portable scanners), algorithms for producing a surface mesh

Electronic supplementary material The online version of this chapter (https://doi.org/10.1007/978-3-030-58558-7_7) contains supplementary material, which is available to authorized users.

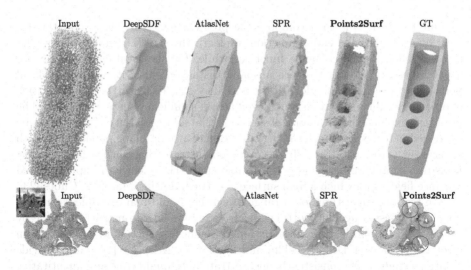

Fig. 1. We present POINTS2SURF, a method to reconstruct an accurate implicit surface from a noisy point cloud. Unlike current data-driven surface reconstruction methods like DeepSDF and AtlasNet, it is patch-based, improves detail reconstruction, and unlike Screened Poisson Reconstruction (SPR), a learned prior of low-level patch shapes improves reconstruction accuracy. Note the quality of reconstructions, both geometric and topological, against the original surfaces. The ability of POINTS2SURF to generalize to new shapes makes it the first learning-based approach with significant generalization ability under both geometric and topological variations.

remain limited. A good surfacing algorithm should be able to handle raw point clouds with noisy and varying sampling density, work with different surface topologies, and generalize across a large range of scanned shapes.

Screened Poisson Reconstruction (SPR) [19] is the most commonly used method to convert an unstructured point cloud, along with its per-point normals, to a surface mesh. While the method is general, in absence of any data-priors, SPR typically produces smooth surfaces, can incorrectly close off holes and tunnels in noisy or non-uniformly sampled scans (see Fig. 1), and further degenerates when per-point normal estimates are erroneous.

Hence, several data-driven alternatives [8,13,21,30] have recently been proposed. These methods specialize to particular object categories (e.g., cars, planes, chairs), and typically regress a global latent vector from any input scan. The networks can then decode a final shape (e.g., a collection of primitives [13] or a mesh [30]) from the estimated global latent vector. While such data-specific approaches handle noisy and partial scans, the methods do *not* generalize to new surfaces with varying shape and topology (see Fig. 1).

As a solution, we present POINTS2SURF, a method that learns to produce implicit surfaces directly from raw point clouds. During test time, our method can reliably handle raw scans to reproduce fine-scale data features even from noisy and non-uniformly sampled point sets, works for objects with varying topological attributes, and generalizes to new objects (see Fig. 1).

Our key insight is to decompose the problem into learning a global and a local function. For the global function, we learn the sign (i.e., inside or outside) of an implicit signed distance function, while, for the local function, we use a patch-based approach to learn absolute distance fields with respect to local point cloud patches. The global task is coarse (i.e., to learn the inside/outside of objects) and hence can be generalized across significant shape variations. The local task exploits the geometric observation that a large variety of shapes can be expressed in terms of a much smaller collection of atomic shape patches [2], which generalizes across arbitrary shapes. We demonstrate that such a factorization leads to a simple, robust, and generalizable approach to learn an implicit signed distance field, from which a final surface is extracted using Marching Cubes [23].

We test our algorithms on a range of synthetic and real examples, compare on unseen classes against both classical (reduction in reconstruction error by 30% over SPR) and learning based strong baselines (reduction in reconstruction error by 470% over DeepSDF [30] and 270% over AtlasNet [13]), and provide ablations studies. We consistently demonstrate both qualitative and quantitative improvement over all the methods that can be applied directly on raw scans.

2   Related Work

Several methods have been proposed to reconstruct surfaces from point clouds. We divide these into methods that aggregate information from a large dataset into a data-driven prior, and methods that do not use a data-driven prior.

Non-data-driven Surface Reconstruction. Berger et al. [4] present an in-depth survey that is focused primarily on non-data-driven methods. Here we focus on approaches that are most relevant to our method. Scale space meshing [10] applies iterative mean curvature motion to smooth the points for meshing. It preserves multi-resolution features well. Ohrhallinger et al. propose a combinatorial method [27] which compares favorably with previous methods such as Wrap [11], TightCocone [9] and Shrink [5] especially for sparse sampling and thin structures. However, these methods are not designed to process noisy point clouds. Another line of work deforms initial meshes [22,33] or parametric patches [36] to fit a noisy point cloud. These approaches however, cannot change the topology and connectivity of the original meshes or patches, usually resulting in a different connectivity or topology than the ground truth. The most widely-used approaches to reconstruct surfaces with arbitrary topology from noisy point clouds fit implicit functions to the point cloud and generate a surface as a level set of the function. Early work by Hoppe et al. introduced this approach [16], and since then several methods have focused on different representations of the implicit functions, like Fourier coefficients [17], wavelets [24], radial-basis functions [29] or multi-scale approaches [26,28]. Alliez et al. [1] use a PCA of 3D Voronoi cells to estimate gradients and fit an implicit function by solving an eigenvalue problem. This approach tends to over-smooth geometric detail. Poisson reconstruction [18,19] is the current gold standard for non-data-driven surface reconstruction from point clouds. None of the above methods make use of a prior that distills information about typical surface shapes from a large dataset. Hence, while they are very general, they fail to handle partial and/or noisy input. We provide extensive comparisons to Screened Poisson Reconstruction (SPR) [19] in Sect. 4.

Data-Driven Surface Reconstruction. Recently, several methods have been proposed to learn a prior of typical surface shapes from a large dataset. Early work was done by Sauerer et al. [21], where a decision tree is trained to predict the absolute distance part of an SDF, but ground truth normals are still required to obtain the sign (inside/outside) of the SDF. More recent data-driven methods represent surfaces with a single latent feature vector in a learned feature space. An encoder can be trained to obtain the feature vector from a point cloud. The feature representation acts as a strong prior, since only shapes that are representable in the feature space are reconstructed. Early methods use voxel-based representations of the surfaces, with spatial data-structures like octrees offsetting the cost of a full volumetric grid [34,35]. Scan2Mesh [8] reconstructs a coarse mesh, including vertices and triangles, from a scan with impressive robustness to missing parts. However, the result is typically very coarse and not watertight or manifold, and does not apply to arbitrary new shapes. AtlasNet [13] uses multiple parametric surfaces as representation that jointly form a surface, achieving impressive accuracy and cross-category generalization. More recently, several approaches learn implicit function representations of surfaces [6,25,30]. These methods are trained to learn a functional that maps a latent encoding of a surface to an implicit function that can be used to extract a continuous surface. The implicit representation is more suitable for surfaces with complex topology and tends to produce aesthetically pleasing smooth surfaces.

The single latent feature vector that the methods above use to represent a surface acts as a strong prior, allowing these methods to reconstruct surfaces even in the presence of strong noise or missing parts; but it also limits the generality of these methods. The feature space typically captures only shapes that are similar to the shapes in the training set, and the variety of shapes that can be captured by the feature space is limited by the fixed capacity of the latent feature vector. Instead, we propose to decompose the SDF that is used to reconstruct the surface into a coarse global sign and a detailed local absolute distance. Separate feature vectors are used to represent the global and local parts, allowing us to represent detailed local geometry, without losing coarse global information about the shape.

3   Method

Our goal is to reconstruct a watertight surface S from a 3D point cloud $P = \{p_1, \ldots, p_N\}$ that was sampled from the surface S through a noisy sampling process, like a 3D scanner. We represent a surface as the zero-set of a Signed Distance Function (SDF) f_S:

$$S = L_0(f_S) = \{x \in \mathbb{R}^3 \mid f_S(x) = 0\}. \tag{1}$$

Recent work [6,25,30] has shown that such an implicit representation of the surface is particularly suitable for neural networks, which can be trained as functionals that take as input a latent representation of the point cloud and output an approximation of the SDF:

$$f_S(x) \approx \tilde{f}_P(x) = s_\theta(x|z), \text{ with } z = e_\phi(P), \tag{2}$$

where z is a latent description of surface S that can be encoded from the input point cloud with an encoder e, and s is implemented by a neural network that is conditioned on the latent vector z. The networks s and e are parameterized by θ and ϕ, respectively. This representation of the surface is continuous, usually produces watertight meshes, and can naturally encode arbitrary topology. Different from non-data-driven methods like SPR [19], the trained network obtains a strong prior from the dataset it was trained on, that allows robust reconstruction of surfaces even from unreliable input, such as noisy and sparsely sampled point clouds. However, encoding the entire surface with a single latent vector imposes limitations on the accuracy and generality of the reconstruction, due to the limited capacity of the latent representation.

In this work, we factorize the SDF into the absolute distance f^d and the sign of the distance f^s, and take a closer look at the information needed to approximate each factor. To estimate the absolute distance $\tilde{f}^d(x)$ at a query point x, we only need a *local* neighborhood of the query point:

$$\tilde{f}_P^d(x) = s_\theta^d(x|z_x^d), \text{ with } z_x^d = e_\phi^d(\mathbf{p}_x^d), \tag{3}$$

where $\mathbf{p}_x^d \subset P$ is a local neighborhood of the point cloud centered around x. Estimating the distance from an encoding of a local neighborhood gives us more accuracy than estimating it from an encoding of the entire shape, since the local encoding z_x^d can more accurately represent the local neighborhood around x than the global encoding z. Note that in a point cloud without noise and sufficiently dense sampling, the single closest point $p^* \subset P$ to the query x would be enough to obtain a good approximation of the absolute distance. But since we work with noisy and sparsely sampled point clouds, using a larger local neighborhood increases robustness.

In order to estimate the sign $\tilde{f}^s(x)$ at the query point x, we need *global* information about the entire shape, since the interior/exterior of a watertight surface cannot be estimated reliably from a local patch. Instead, we take a global sub-sample of the point cloud P as input:

$$\tilde{f}_P^s(x) = \text{sgn}(\tilde{g}_P^s(x)) = \text{sgn}(s_\theta^s(x|z_x^s)), \text{ with } z_x^s = e_\psi^s(\mathbf{p}_x^s), \tag{4}$$

where $\mathbf{p}_x^s \subset P$ is a global subsample of the point cloud, ψ are the parameters of the encoder, and $\tilde{g}_P^s(x)$ are logits of the probability that x has a positive sign. Working with logits avoids discontinuities near the surface, where the sign changes. Since it is more important to have accurate information closer to the query point, we sample $\mathbf{p}_x^s$ with a density gradient that is highest near the query point and falls off with distance from the query point.

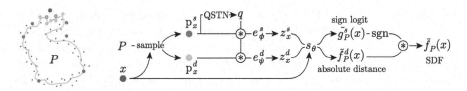

Fig. 2. Points2Surf Architecture. Given a query point x (red) and a point cloud P (gray), we sample a local patch (yellow) and a coarse global subsample (purple) of the point cloud. These are encoded into two feature vectors that are fed to a decoder, which outputs a logit of the sign probability and the absolute distance of the SDF at the query point x. (Color figure online)

We found that sharing information between the two latent descriptions z_x^s and z_x^d benefits both the absolute distance and the sign of the SDF, giving us the formulation we use in Points2Surf:

$$\left(\tilde{f}_P^d(x), \tilde{g}_P^s(x)\right) = s_\theta(x|z_x^d, z_x^s), \text{ with } z_x^d = e_\phi^d(\mathbf{p}_x^d) \text{ and } z_x^s = e_\psi^s(\mathbf{p}_x^s). \quad (5)$$

We describe the architecture of our encoders and decoder, the training setup, and our patch sampling strategy in more detail in Sect. 3.1.

To reconstruct the surface S, we apply Marching Cubes [23] to a sample grid of the estimated SDF $\tilde{f}^d(x) * \tilde{f}^s(x)$. In Sect. 3.2, we describe a strategy to improve performance by only evaluating a subset of the grid samples.

3.1   Architecture and Training

Figure 2 shows an overview of our architecture. Our approach estimates the absolute distance $\tilde{f}_P^d(x)$ and the sign logits $\tilde{g}_P^s(x)$ at a query point based on two inputs: the query point x and the point cloud P.

Pointset Sampling. The local patch $\mathbf{p}_x^d$ and the global sub-sample $\mathbf{p}_x^s$ are both chosen from the point cloud P based on the query point x. The set $\mathbf{p}_x^d$ is made of the n_d nearest neighbors of the query point (we choose $n_d = 300$ but also experiment with other values). Unlike a fixed radius, the nearest neighbors are suitable for query points with arbitrary distance from the point cloud. The global sub-sample $\mathbf{p}_x^s$ is sampled from P with a density gradient that decreases with distance from x:

$$\rho(p_i) = \frac{v(p_i)}{\sum_{p_j \in P} v(p_j)}, \text{ with } v(p_i) = \left[1 - 1.5\frac{\|p_i - x\|_2}{\max_{p_j \in P} \|p_j - x\|_2}\right]_{0.05}^1, \quad (6)$$

where ρ is the sample probability for a point $p_i \in P$, v is the gradient that decreases with distance from x, and the square brackets denote clamping. The minimum value for the clamping ensures that some far points are taken and the sub-sample can represent a closed object. We sample n_s points from P according to this probability (we choose $n_s = 1000$ in our experiments).

Pointset Normalization. Both $\mathbf{p}_x^d$ and $\mathbf{p}_x^s$ are normalized by centering them at the query point, and scaling them to unit radius. After running the network, the estimated distance is scaled back to the original size before comparing to the ground truth. Due to the centering, the query point is always at the origin of the normalized coordinate frame and does not need to be passed explicitly to the network. To normalize the orientation of both point subsets, we use a data-driven approach: a Quaternion Spatial Transformer Network (QSTN) [15] takes as input the global subset $\mathbf{p}_x^s$ and estimates a rotation represented as quaternion q that is used to rotate both point subsets. We take the global subset as input, since the global information can help the network with finding a more consistent rotation. The QSTN is trained end-to-end with the full architecture, without direct supervision for the rotation.

Encoder and Decoder Architecture. The local encoder e_ϕ^d, and the global encoder e_ψ^s are both implemented as PointNets [31], sharing the same architecture, but not the parameters. Following the PointNet architecture, a feature representation for each point is computed using 5 MLP layers, with a spatial transformer in feature space after the third layer. Each layer except the last one uses batch normalization and ReLU. The point feature representations are then aggregated into point set feature representations $z_x^d = e_\phi^d(\mathbf{p}_x^d)$ and $z_x^s = e_\psi^s(\mathbf{p}_x^s)$ using a channel-wise maximum. The decoder s_θ is implemented as 4-layer MLP that takes as input the concatenated feature vectors z_x^d and z_x^s and outputs both the absolute distance $\tilde{f}^d(x)$ and sign logits $\tilde{g}^s(x)$.

Losses and Training. We train our networks end-to-end to regress the absolute distance of the query point x from the watertight ground-truth surface S and classify the sign as positive (outside S) or negative (inside S). We assume that ground-truth surfaces are available during training for supervision. We use an L_2-based regression for the absolute distance:

$$\mathcal{L}^d(x, P, S) = \|\tanh(|\tilde{f}_P^d(x)|) - \tanh(|d(x, S)|)\|_2^2, \qquad (7)$$

where $d(x, S)$ is the distance of x to the ground-truth surface S. The tanh function gives more weight to smaller absolute distances, which are more important for an accurate surface reconstruction. For the sign classification, we use the binary cross entropy H as loss:

$$\mathcal{L}^s(x, P, S) = H\Big(\sigma(\tilde{g}_P^s(x)), \ [f_S(x) > 0]\Big), \qquad (8)$$

where σ is the logistic function to convert the sign logits to probabilities, and $[f_S(x) > 0]$ is 1 if x is outside the surface S and 0 otherwise. In our optimization, we minimize these two losses for all shapes and query points in the training set:

$$\sum_{(P,S)\in\mathcal{S}} \sum_{x\in\mathcal{X}_S} \mathcal{L}^d(x, P, S) + \mathcal{L}^s(x, P, S), \qquad (9)$$

where $\mathcal{S}$ is the set of surfaces S and corresponding point clouds P in the training set and $\mathcal{X}_S$ the set of query points for shape S. Estimating the sign as a classification task instead of regressing the signed distance allows the network to express confidence through the magnitude of the sign logits, improving performance.

3.2 Surface Reconstruction

At inference time, we want to reconstruct a surface $\tilde{S}$ from an estimated SDF $\tilde{f}(x) = \tilde{f}^d(x) * \tilde{f}^s(x)$. A straight-forward approach is to apply Marching Cubes [23] to a volumetric grid of SDF samples. Obtaining a high-resolution result, however, would require evaluating a prohibitive number of samples for each shape. We observe that in order to reconstruct a surface, a Truncated Signed Distance Field (TSDF) is sufficient, where the SDF is truncated to the interval $[-\epsilon, \epsilon]$ (we set ϵ to three times the grid spacing in all our experiments). We only need to evaluate the SDF for samples that are inside this interval, while samples outside the interval merely need the correct sign. We leave samples with a distance larger than ϵ to the nearest point in P blank, and in a second step, we propagate the signed distance values from non-blank to blank samples, to obtain the correct sign in the truncated regions of the TSDF. We iteratively apply a box filter of size 3^3 at the blank samples until convergence. In each step, we update initially unknown samples only if the filter response is greater than a user-defined confidence threshold (we use 13 in our experiments). After each step, the samples are set to -1 if the filter response was negative or to $+1$ for a positive response.

4 Results

We compare our method to SPR as the gold standard for non-data-driven surface reconstruction and to two state-of-the-art data-driven surface reconstruction methods. We provide both qualitative and quantitative comparisons on several datasets in Sect. 4.2, and perform an ablation study in Sect. 4.3.

4.1 Datasets

We train and evaluate on the ABC dataset [20] which contains a large number and variety of high-quality CAD meshes. We pick 4950 clean watertight meshes for training and 100 meshes for the validation and test sets. Note that each mesh produces a large number of diverse patches as training samples. Operating on local patches also allows us to generalize better, which we demonstrate on two additional test-only datasets: a dataset of 22 diverse meshes which are well-known in geometry processing, such as the Utah teapot and the Stanford Bunny, which we call the FAMOUS dataset, and 3 REAL scans of complex objects used in several denoising papers [32,37]. Examples from each dataset are shown in Fig. 3. The ABC dataset contains predominantly CAD models of mechanical parts, while the FAMOUS dataset contains more organic shapes, such as characters and animals. Since we train on the ABC dataset, the FAMOUS dataset serves to test the generalizability of our method versus baselines.

ABC   Famous

Fig. 3. Dataset examples. Examples of the ABC dataset and its three variants are shown on the left, examples of the famous dataset and its five variants on the right.

Pointcloud Sampling. As a pre-processing step, we center all meshes at the origin and scale them uniformly to fit within the unit cube. To obtain point clouds P from the meshes S in the datasets, we simulate scanning them with a time-of-flight sensor from random viewpoints using BlenSor [14]. BlenSor realistically simulates various types of scanner noise and artifacts such as backfolding, ray reflections, and per-ray noise. We scan each mesh in the FAMOUS dataset with 10 scans and each mesh in the ABC dataset with a random number of scans, between 5 and 30. For each scan, we place the scanner at a random location on a sphere centered at the origin, with the radius chosen randomly in $U[3L, 5L]$, where L is the largest side of the mesh bounding box. The scanner is oriented to point at a location with small random offset from the origin, between $U[-0.1L, 0.1L]$ along each axis, and rotated randomly around the view direction. Each scan has a resolution of 176×144, resulting in roughly 25k points, minus some points missing due to simulated scanning artifacts. The point clouds of multiple scans of a mesh are merged to obtain the final point cloud.

Dataset Variants. We create multiple versions of both the ABC and FAMOUS datasets, with varying amount of per-ray noise. This Gaussian noise added to the depth values simulates inaccuracies in the depth measurements. For both datasets, we create a noise-free version of the point clouds, called ABC *no-noise* and FAMOUS *no-noise*. Also, we make versions with strong noise (standard deviation is $0.05L$) called ABC *max-noise* and FAMOUS *max-noise*. Since we need varying noise strength for the training, we create a version of ABC where the standard deviation is randomly chosen in $U[0, 0.05L]$ (ABC *var-noise*), and a version with a constant noise strength of $0.01L$ for the test set (FAMOUS *med-noise*). Additionally we create sparser (5 scans) and denser (30 scans) point clouds in comparison to the 10 scans of the other variants of FAMOUS. Both variants have a medium noise strength of $0.01L$. The training set uses the ABC *var-noise* version, all other variants are used for evaluation only.

Query Points. The training set also contains a set $\mathcal{X}_S$ of query points for each (point cloud, mesh) pair $(P, S) \in \mathcal{S}$. Query points closer to the surface are more important for the surface reconstruction and more difficult to estimate. Hence, we randomly sample a set of 1000 points on the surface and offset them in the normal direction by a uniform random amount in $U[-0.02L, 0.02L]$. An additional 1000 query points are sampled randomly in the unit cube that contains the surface, for a total of 2000 query points per mesh. During training, we randomly sample a subset of 1000 query points per mesh in each epoch.

Table 1. Comparison of reconstruction errors. We show the Chamfer distance between reconstructed and ground-truth surfaces averaged over all shapes in a dataset. Both the absolute value of the error multiplied by 100 (abs.), and the error relative to Point2Surf (rel.) are shown to facilitate the comparison. Our method consistently performs better than the baselines, due to its strong and generalizable prior.

	DeepSDF		AtlasNet		SPR		Points2Surf	
	abs.	rel.	abs.	rel.	abs.	rel.	abs.	rel.
ABC no-noise	8.41	4.68	4.69	2.61	2.49	1.39	**1.80**	**1.00**
ABC var-noise	12.51	5.86	4.04	1.89	3.29	1.54	**2.14**	**1.00**
ABC max-noise	11.34	4.11	4.47	1.62	3.89	1.41	**2.76**	**1.00**
Famous no-noise	10.08	7.14	4.69	3.33	1.67	1.18	**1.41**	**1.00**
Famous med-noise	9.89	6.57	4.54	3.01	1.80	1.20	**1.51**	**1.00**
Famous max-noise	13.17	5.23	4.14	1.64	3.41	1.35	**2.52**	**1.00**
Famous sparse	10.41	5.41	4.91	2.55	2.17	1.12	**1.93**	**1.00**
Famous dense	9.49	7.15	4.35	3.28	1.60	1.21	**1.33**	**1.00**
Average	10.66	5.77	4.48	2.49	2.54	1.30	**1.92**	**1.00**

4.2   Comparison to Baselines

We compare our method to recent data-driven surface reconstruction methods, AtlasNet [13] and DeepSDF [30], and to SPR [19], which is still the gold standard in non-data-driven surface reconstruction from point clouds. Both AtlasNet and DeepSDF represent a full surface as a single latent vector that is decoded into either a set of parametric surface patches in AtlasNet, or an SDF in DeepSDF. In contrast, SPR solves for an SDF that has a given sparse set of point normals as gradients, and takes on values of 0 at the point locations. We use the default values and training protocols given by the authors for all baselines (more details in the Supplementary) and re-train the two data-driven methods on our training set. We provide SPR with point normals as input, which we estimate from the input point cloud using the recent PCPNet [15].

Error Metric. To measure the reconstruction error of each method, we sample both the reconstructed surface and the ground-truth surface with 10k points and compute the Chamfer distance [3,12] between the two point sets:

$$d_{\mathrm{ch}}(A, B) := \frac{1}{|A|} \sum_{p_i \in A} \min_{p_j \in B} \|p_i - p_j\|_2^2 + \frac{1}{|B|} \sum_{p_j \in B} \min_{p_i \in A} \|p_j - p_i\|_2^2, \qquad (10)$$

where A and B are point sets sampled on the two surfaces. The Chamfer distance penalizes both false negatives (missing parts) and false positives (excess parts).

Quantitative and Qualitative Comparison. A quantitative comparison of the reconstruction quality is shown in Table 1, and Fig. 4 compares a few reconstructions qualitatively. All methods were trained on the training set of the

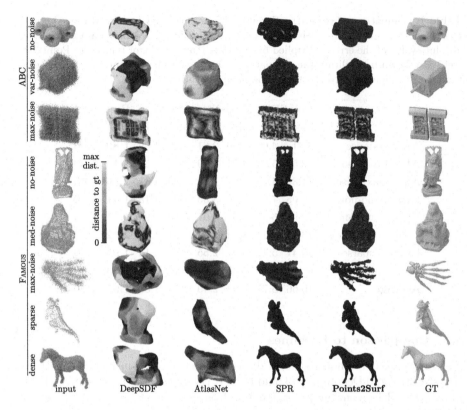

Fig. 4. Qualitative comparison of surface reconstructions. We evaluate one example from each dataset variant with each method. Colors show the distance of the reconstructed surface to the ground-truth surface.

ABC *var-noise* dataset, which contains predominantly mechanical parts, while the more organic shapes in the FAMOUS dataset test how well each method can generalize to novel types of shapes.

Both DeepSDF and AtlasNet use a global shape prior, which is well suited for a dataset with high geometric consistency among the shapes, like cars in ShapeNet, but struggles with the significantly larger geometric and topological diversity in our datasets, reflected in a higher reconstruction error than SPR or POINTS2SURF. This is also clearly visible in Fig. 4, where the surfaces reconstructed by DeepSDF and AtlasNet appear over-smoothed and inaccurate.

In SPR, the full shape space does not need to be encoded into a prior with limited capacity, resulting in a better accuracy. But this lack of a strong prior also prevents SPR from robustly reconstructing typical surface features, such as holes or planar patches (see Figs. 1 and 4).

POINTS2SURF learns a prior of local surface details, instead of a prior for global surface shapes. This local prior helps recover surface details like holes and planar patches more robustly, improving our accuracy over SPR. Since there

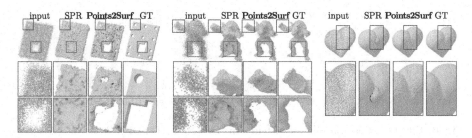

Fig. 5. Comparison of reconstruction details. Our learned prior improves the reconstruction robustness for geometric detail compared to SPR.

is less variety and more consistency in local surface details compared to global surface shapes, our method generalizes better and achieves a higher accuracy than the data-driven methods that use a prior over the global surface shape.

Generalization. A comparison of our generalization performance against Atlas-Net and DeepSDF shows an advantage for our method. In Table 1, we can see that the error for DeepSDF and AtlasNet increases more when going from the ABC dataset to the FAMOUS dataset than the error for our method. This suggests that our method generalizes better from the CAD models in the ABC dataset set to the more organic shapes in the FAMOUS dataset.

Topological Quality. Figure 5 shows examples of geometric detail that benefits from our prior. The first example shows that small features such as holes can be recovered from a very weak geometric signal in a noisy point cloud. Concavities, such as the space between the legs of the Armadillo, and fine shape details like the Armadillo's hand are also recovered more accurately in the presence of strong noise. In the heart example, the concavity makes it hard to estimate the correct normal direction based on only a local neighborhood, which causes SPR to produce artifacts. In contrast, the global information we use in our patches helps us estimate the correct sign, even if the local neighborhood is misleading.

Effect of Noise. Examples of reconstructions from point clouds with increasing amounts of noise are shown in Fig. 6. Our learned prior for local patches and our coarse global surface information makes it easier to find small holes and large concavities. In the medium noise setting, we can recover the small holes and the large concavity of the surface. With maximum noise, it is very difficult to detect the small holes, but we can still recover the concavity.

Real-World Data. The real-world point clouds in Fig. 1 bottom and Fig. 7 bottom both originate from a multi-view dataset [37] and were obtained with a plane-sweep algorithm [7] from multiple photographs of an object. We additionally remove outliers using the recent PointCleanNet [32]. Figure 7 top was obtained by the authors through an SfM approach. DeepSDF and AtlasNet do not generalize well to unseen shape categories. SPR performs significantly better but

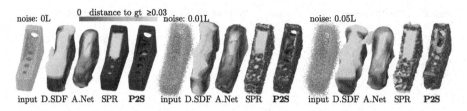

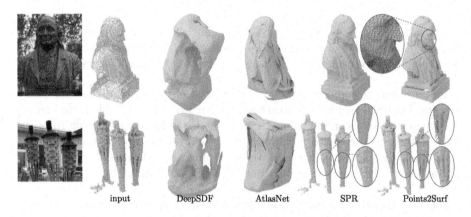

Fig. 6. Effect of noise on our reconstruction. DeepSDF (D.SDF), AtlasNet (A.Net), SPR and Point2Surf (P2S) are applied to increasingly noisy point clouds. Our patch-based data-driven approach is more accurate than DeepSDF and AtlasNet, and can more robustly recover small holes and concavities than SPR.

Fig. 7. Reconstruction of real-world point clouds. Snapshots of the real-world objects are shown on the left. DeepSDF and AtlasNet do not generalize well, resulting in inaccurate reconstructions, while the smoothness prior of SPR results in loss of detail near concavities and holes. Our data-driven local prior better preserves these details.

its smoothness prior tends to over-smooth shapes and close holes. Points2Surf better preserves holes and details, at the cost of a slight increase in topological noise. Our technique also generalizes to unseen point-cloud acquisition methods.

4.3   Ablation Study

We evaluate several design choices in (e_{vanilla}) using an ablation study, as shown in Table 2. We evaluate the number of nearest neighbors $k = 300$ that form the local patch by decreasing and increasing k by a factor of 4 (k_{small} and k_{large}), effectively halving and doubling the size of the local patch. A large k performs significantly worse because we lose local detail with a larger patch size. A small k still works reasonably well, but gives a lower performance, especially with strong noise. We also test a fixed radius for the local patch, with three different sizes ($r_{\text{small}} := 0.05L$, $r_{\text{med}} := 0.1L$ and $r_{\text{large}} := 0.2L$). A fixed patch size is less suit-able than nearest neighbors when computing the distance at query points that

Table 2. Ablation Study. We compare POINTS2SURF (e_{vanilla}) to several variants and show the Chamfer distance relative to POINTS2SURF. Please see the text for details.

	r_{small}	r_{med}	r_{large}	k_{small}	k_{large}	e_{shared}	$e_{\text{no\_QSTN}}$	e_{uniform}	e_{vanilla}
ABC var-noise	1.12	1.07	1.05	1.08	1.87	1.02	0.94	0.91	**1.00**
FAMOUS no-noise	1.09	1.08	1.17	1.05	8.10	1.06	0.97	0.96	**1.00**
FAMOUS med-noise	1.06	1.05	1.10	1.04	7.89	1.05	0.97	0.97	**1.00**
FAMOUS max-noise	1.07	1.19	1.13	1.09	1.79	1.01	1.05	0.89	**1.00**
Average	1.08	1.11	1.11	1.07	4.14	1.03	0.99	0.92	**1.00**

are far away from the surface, giving a lower performance than the standard nearest neighbor setting. The next variant is using a single shared encoder (e_{shared}) for both the global sub-sample $\mathbf{p}_x^s$ and the local patch $\mathbf{p}_x^d$, by concatenating the two before encoding them. The performance of e_{shared} is competitive, but shows that using two separate encoders increases performance. Omitting the QSTN ($e_{\text{no\_QSTN}}$) speeds-up the training by roughly 10% and yields slightly better results. The reason is probably that our outputs are rotation-invariant in contrast to the normals of PCPNet. Using a uniform global sub-sample in e_{uniform} increases the quality over the distance-dependent sub-sampling in e_{vanilla}. This uniform sub-sample preserves more information about the far side of the object, which benefits the inside/outside classification. Due to resource constraints, we trained all models in Table 2 for 50 epochs only. For applications where speed, memory and simplicity is important, we recommend using a shared encoder without the QSTN and with uniform sub-sampling.

5   Conclusion

We have presented POINTS2SURF as a method for surface reconstruction from raw point clouds. Our method reliably captures both geometric and topological details, and generalizes to unseen shapes more robustly than current methods.

The distance-dependent global sub-sample may cause inconsistencies between the outputs of neighboring patches, which results in bumpy surfaces.

One interesting future direction would be to perform multi-scale reconstructions, where coarser levels provide consistency for finer levels, reducing the bumpiness. This should also decrease the computation times significantly. Finally, it would be interesting to develop a differentiable version of Marching Cubes to jointly train SDF estimation and surface extraction.

Acknowledgement. This work has been supported by the FWF projects no. P24600, P27972 and P32418 and the ERC Starting Grant SmartGeometry (StG-2013-335373).

References

1. Alliez, P., Cohen-Steiner, D., Tong, Y., Desbrun, M.: Voronoi-based variational reconstruction of unoriented point sets. In: Symposium on Geometry Processing, vol. 7, pp. 39–48 (2007)
2. Badki, A., Gallo, O., Kautz, J., Sen, P.: Meshlet priors for 3D mesh reconstruction. arXiv (2020)
3. Barrow, H.G., Tenenbaum, J.M., Bolles, R.C., Wolf, H.C.: Parametric correspondence and chamfer matching: two new techniques for image matching. In: Proceedings of the 5th International Joint Conference on Artificial Intelligence, IJCAI 1977, vol. 2, pp. 659–663. Morgan Kaufmann Publishers Inc., San Francisco (1977). http://dl.acm.org/citation.cfm?id=1622943.1622971
4. Berger, M., et al.: A survey of surface reconstruction from point clouds. In: Computer Graphics Forum, vol. 36, pp. 301–329. Wiley Online Library (2017)
5. Chaine, R.: A geometric convection approach of 3-D reconstruction. In: Proceedings of the 2003 Eurographics/ACM SIGGRAPH Symposium on Geometry Processing, pp. 218–229. Eurographics Association (2003)
6. Chen, Z., Zhang, H.: Learning implicit fields for generative shape modeling. In: Proceedings of the CVPR (2019)
7. Collins, R.T.: A space-sweep approach to true multi-image matching. In: Proceedings of the CVPR IEEE Computer Society Conference on Computer Vision and Pattern Recognition, pp. 358–363. IEEE (1996)
8. Dai, A., Nießner, M.: Scan2Mesh: from unstructured range scans to 3D meshes. In: Proceedings of the Computer Vision and Pattern Recognition (CVPR). IEEE (2019)
9. Dey, T.K., Goswami, S.: Tight cocone: a water-tight surface reconstructor. In: Proceedings of the 8th ACM Symposium on Solid Modeling and Applications, pp. 127–134. ACM (2003)
10. Digne, J., Morel, J.M., Souzani, C.M., Lartigue, C.: Scale space meshing of raw data point sets. In: Computer Graphics Forum, vol. 30, pp. 1630–1642. Wiley Online Library (2011)
11. Edelsbrunner, H.: Surface reconstruction by wrapping finite sets in space. In: Aronov, B., Basu, S., Pach, J., Sharir, M. (eds.) Discrete and Computational Geometry. Algorithms and Combinatorics, vol. 25, pp. 379–404. Springer, Heidelberg (2003). https://doi.org/10.1007/978-3-642-55566-4_17
12. Fan, H., Su, H., Guibas, L.J.: A point set generation network for 3D object reconstruction from a single image. In: Proceedings of the IEEE Conference on Computer Vision and Pattern Recognition, pp. 605–613 (2017)
13. Groueix, T., Fisher, M., Kim, V.G., Russell, B., Aubry, M.: AtlasNet: a Papier-Mâché approach to learning 3D surface generation. In: Proceedings IEEE Conference on Computer Vision and Pattern Recognition (CVPR) (2018)
14. Gschwandtner, M., Kwitt, R., Uhl, A., Pree, W.: BlenSor: blender sensor simulation toolbox. In: Bebis, G., et al. (eds.) ISVC 2011. LNCS, vol. 6939, pp. 199–208. Springer, Heidelberg (2011). https://doi.org/10.1007/978-3-642-24031-7_20
15. Guerrero, P., Kleiman, Y., Ovsjanikov, M., Mitra, N.J.: PCPNet: learning local shape properties from raw point clouds. Comput. Graph. Forum $37(2)$, 75–85 (2018). https://doi.org/10.1111/cgf.13343
16. Hoppe, H., DeRose, T., Duchamp, T., McDonald, J., Stuetzle, W.: Surface reconstruction from unorganized points. ACM SIGGRAPH Comput. Graph. 26, 71–78 (1992)

17. Kazhdan, M.: Reconstruction of solid models from oriented point sets. In: Proceedings of the 3rd Eurographics Symposium on Geometry Processing, p. 73. Eurographics Association (2005)
18. Kazhdan, M., Bolitho, M., Hoppe, H.: Poisson surface reconstruction. In: Proceedings of the Eurographics Symposium on Geometry Processing (2006)
19. Kazhdan, M., Hoppe, H.: Screened Poisson surface reconstruction. ACM Trans. Graph. (ToG) **32**(3), 29 (2013)
20. Koch, S., et al.: ABC: a big CAD model dataset for geometric deep learning. In: The IEEE Conference on Computer Vision and Pattern Recognition (CVPR) (June 2019)
21. Ladicky, L., Saurer, O., Jeong, S., Maninchedda, F., Pollefeys, M.: From point clouds to mesh using regression. In: Proceedings of the IEEE International Conference on Computer Vision, pp. 3893–3902 (2017)
22. Li, G., Liu, L., Zheng, H., Mitra, N.J.: Analysis, reconstruction and manipulation using arterial snakes. In: ACM SIGGRAPH Asia 2010 Papers, SIGGRAPH ASIA 2010. Association for Computing Machinery, New York (2010). https://doi.org/10.1145/1866158.1866178
23. Lorensen, W.E., Cline, H.E.: Marching cubes: a high resolution 3D surface construction algorithm. ACM SIGGRAPH Comput. Graph. **21**, 163–169 (1987)
24. Manson, J., Petrova, G., Schaefer, S.: Streaming surface reconstruction using wavelets. Comput. Graph. Forum **27**, 1411–1420 (2008). Wiley Online Library
25. Mescheder, L., Oechsle, M., Niemeyer, M., Nowozin, S., Geiger, A.: Occupancy networks: Learning 3D reconstruction in function space. In: Proceedings of CVPR (2019)
26. Nagai, Y., Ohtake, Y., Suzuki, H.: Smoothing of partition of unity implicit surfaces for noise robust surface reconstruction. Comput. Graph. Forum **28**, 1339–1348 (2009). Wiley Online Library
27. Ohrhallinger, S., Mudur, S., Wimmer, M.: Minimizing edge length to connect sparsely sampled unstructured point sets. Comput. Graph. **37**(6), 645–658 (2013)
28. Ohtake, Y., Belyaev, A., Seidel, H.P.: A multi-scale approach to 3D scattered data interpolation with compactly supported basis functions. In: 2003 Shape Modeling International, pp. 153–161. IEEE (2003)
29. Ohtake, Y., Belyaev, A., Seidel, H.P.: 3D scattered data interpolation and approximation with multilevel compactly supported RBFs. Graph. Models **67**(3), 150–165 (2005)
30. Park, J.J., Florence, P., Straub, J., Newcombe, R., Lovegrove, S.: DeepSDF: learning continuous signed distance functions for shape representation. arXiv preprint arXiv:1901.05103 (2019)
31. Qi, C.R., Su, H., Mo, K., Guibas, L.J.: PointNet: Deep learning on point sets for 3d classification and segmentation. arXiv preprint arXiv:1612.00593 (2016)
32. Rakotosaona, M.J., La Barbera, V., Guerrero, P., Mitra, N.J., Ovsjanikov, M.: PointCleanNet: learning to denoise and remove outliers from dense point clouds. Comput. Graph. Forum **39**, 185–203 (2019)
33. Sharf, A., Lewiner, T., Shamir, A., Kobbelt, L., Cohen-Or, D.: Competing fronts for coarse-to-fine surface reconstruction. Comput. Graph. Forum **25**, 389–398 (2006). Wiley Online Library
34. Tatarchenko, M., Dosovitskiy, A., Brox, T.: Octree generating networks: efficient convolutional architectures for high-resolution 3D outputs. In: Proceedings of the IEEE International Conference on Computer Vision, pp. 2088–2096 (2017)
35. Wang, P.S., Sun, C.Y., Liu, Y., Tong, X.: Adaptive O-CNN: a patch-based deep representation of 3D shapes. ACM Trans. Graph. (TOG) **37**(6), 1–11 (2018)

36. Williams, F., Schneider, T., Silva, C.T., Zorin, D., Bruna, J., Panozzo, D.: Deep geometric prior for surface reconstruction. In: Proceedings of the IEEE Conference on Computer Vision and Pattern Recognition, pp. 10130–10139 (2019)
37. Wolff, K., et al.: Point cloud noise and outlier removal for image-based 3D reconstruction. In: 2016 4th International Conference on 3D Vision (3DV), pp. 118–127 (October 2016). https://doi.org/10.1109/3DV.2016.20

Few-Shot Scene-Adaptive Anomaly Detection

Yiwei Lu[1]([✉])[iD], Frank Yu[1][iD], Mahesh Kumar Krishna Reddy[1][iD],
and Yang Wang[1,2][iD]

[1] University of Manitoba, Winnipeg, Canada
{luy2,kumark,ywang}@cs.umanitoba.ca
[2] Huawei Technologies Canada, Markham, Canada

Abstract. We address the problem of anomaly detection in videos. The goal is to identify unusual behaviours automatically by learning exclusively from normal videos. Most existing approaches are usually data-hungry and have limited generalization abilities. They usually need to be trained on a large number of videos from a target scene to achieve good results in that scene. In this paper, we propose a novel few-shot scene-adaptive anomaly detection problem to address the limitations of previous approaches. Our goal is to learn to detect anomalies in a previously unseen scene with only a few frames. A reliable solution for this new problem will have huge potential in real-world applications since it is expensive to collect a massive amount of data for each target scene. We propose a meta-learning based approach for solving this new problem; extensive experimental results demonstrate the effectiveness of our proposed method. All codes are released in https://github.com/yiweilu3/Few-shot-Scene-adaptive-Anomaly-Detection.

Keywords: Anomaly detection · Few-shot learning · Meta-learning

1 Introduction

We consider the problem of anomaly detection in surveillance videos. Given a video, the goal is to identify frames where abnormal events happen. This is a very challenging problem since the definition of "anomaly" is ambiguous – any event that does not conform to "normal" behaviours can be considered as an anomaly. As a result, we cannot solve this problem via a standard classification framework since it is impossible to collect training data that cover all possible abnormal events. Existing literature usually addresses this problem by training a model using only normal data to learn a generic distribution for normal behaviours. During testing, the model classifies anomaly using the distance between the given sample and the learned distribution.

A lot of prior work (e.g. [1,2,6,8,20,31,32]) in anomaly detection use frame reconstruction. These approaches learn a model to reconstruct the normal training data and use the reconstruction error to identify anomalies. Alternatively,

Electronic supplementary material The online version of this chapter (https://doi.org/10.1007/978-3-030-58558-7_8) contains supplementary material, which is available to authorized users.

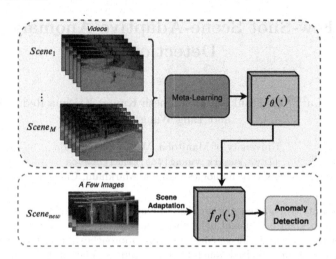

Fig. 1. An overview of our proposed problem setting. During training (1st row), we have access to videos collected from M different camera scenes. From such training data, we use a meta-learning method to obtain a model f_θ with parameters θ. Given a target scene (2nd row), we have access to a small number of frames from this target scene. Our goal is to produce a new model $f_{\theta'}$ where the model parameters θ' are specifically adapted to this scene. Then we can use $f_{\theta'}(\cdot)$ to perform anomaly detection on the remaining videos from this target scene.

[14,16,17,22,25] use future frame prediction for anomaly detection. These methods learn a model that takes a sequence of consecutive frames as the input and predicts the next frame. The difference between the predicted frame and the actual frame at the next time step is used to indicate the probability of an anomaly.

However, existing anomaly detection approaches share common limitations. They implicitly assume that the model (frame reconstruction, or future frame prediction) learned from the training videos can be directly used in unseen test videos. This is a reasonable assumption only if training and testing videos are from the same scene (e.g. captured by the same camera). In the experiment section, we will demonstrate that if we learn an anomaly detection model from videos captured from one scene and directly test the model in a completely different scene, the performance will drop. Of course, one possible way of alleviating this problem is to train the anomaly detection model using videos collected from diverse scenes. Then the learned model will likely generalize to videos from new scenes. However, this approach is also not ideal. In order to learn a model that can generalize well to diverse scenes, the model requires a large capacity. In many real-world applications, the anomaly detection system is often deployed on edge devices with limited computing powers. As a result, even if we can train a huge model that generalizes well to different scenes, we may not be able to deploy this model.

Our work is motivated by the following key observation. In real-world anomaly detection applications, we usually only need to consider one particular scene for testing since the surveillance cameras are normally installed at fixed locations. As long as a model works well in this particular scene, it does not matter at all whether the same model works on images from other scenes. In other words, we would like to have a model specifically adapted to the scene where the model is deployed. In this paper, we propose a novel problem called the *few-shot scene-adaptive anomaly detection* illustrated in Fig. 1. During training, we assume that we have access to videos collected from multiple scenes. During testing, the model is given a few frames in a video from a new target scene. Note that the learning algorithm does not see any images from the target scene during training. Our goal is to produce an anomaly detection model specifically adapted to this target scene using these few frames. We believe this new problem setting is closer to real-world applications. If we have a reliable solution to this problem, we only need a few frames from a target camera to produce an anomaly detection model that is specifically adapted to this camera. In this paper, we propose a meta-learning based approach to this problem. During training, we learn a model that can quickly adapt to a new scene by using only a few frames from it. This is accomplished by learning from a set of tasks, where each task mimics the few-shot scene-adaptive anomaly detection scenario using videos from an available scene.

This paper makes several contributions. First, we introduce a new problem called few-shot scene-adaptive anomaly detection, which is closer to the real-world deployment of anomaly detection systems. Second, we propose a novel meta-learning based approach for solving this problem. We demonstrate that our proposed approach significantly outperforms alternative methods on several benchmark datasets.

2 Related Work

Anomaly Detection in Videos: Recent research in anomaly detection for surveillance videos can be categorized as either reconstruction-based or prediction-based methods. Reconstruction-based methods train a deep learning model to reconstruct the frames in a video and use the reconstruction error to differentiate the normal and abnormal events. Examples of reconstruction models include convolutional auto-encoders [2,6,8,20,31], latent autoregressive models [1], deep adversarial training [32], etc. Prediction-based detection methods define anomalies as anything that does not conform to the prediction of a deep learning model. Sequential models like Convolutional LSTM (ConvLSTM) [40] have been widely used for future frame prediction and utilized to the task of anomaly detection [17,22]. Popular generative networks like generative adversarial networks (GANs) [7] and variational autoencoders (VAEs) [10] are also applied in prediction-based anomaly detection. Liu et al. [14] propose a conditional GAN based model with a low level optical flow [4] feature. Lu et al. [16] incorporate a sequential model in generative networks (VAEs) and propose a convolutional VRNN model. Moreover, [6] apply optical flow prediction constraint on a reconstruction based model.

Few-Shot and Meta Learning: To mimic the fast and flexible learning ability of humans, few-shot learning aims at adapting quickly to a new task with only a few training samples [13]. In particular, meta learning (also known as *learning to learn*) has been shown to be an effective solution to the few-shot learning problem. The research in meta-learning can be categorized into three common approaches: metric-based [11,36,38], model-based [24,33] and optimization-based approaches [5,29]. Metric-based approaches typically apply Siamese [11], matching [38], relation [36] or prototypical networks [34] for learning a metric or distance function over data points. Model-based approaches are devised for fast learning from the model architecture perspective [24,33], where rapid parameter updating during training steps is usually achieved by the architecture itself. Lastly, optimization-based approaches modify the optimization algorithm for quick adaptation [5,29]. These methods can quickly adapt to a new task through the meta-update scheme among multiple tasks during parameter optimization. However, most of the approaches above are designed for simple tasks like image classification. In our proposed work, we follow a similar optimization-based meta-learning approach proposed in [5] and apply it to the much more challenging task of anomaly detection. To the best of our knowledge, we are the first to cast anomaly detection as meta-learning from multiple scenes.

3   Problem Setup

We first briefly summarize the standard anomaly detection framework. Then we describe our problem setup of *few-shot scene-adaptive anomaly detection*.

Anomaly Detection: The anomaly detection framework can be roughly categorized into reconstruction-based or prediction-based methods. For reconstruction-based methods, given a image I, the model $f_\theta(\cdot)$ generates a reconstructed image $\hat{I}$. For prediction-based methods, given t consecutive frames $I_1, I_2, ..., I_t$ in a video, the goal is to learn a model $f_\theta(x_{1:t})$ with parameters θ that takes these t frames as its input and predicts the next frame at time $t + 1$. We use $\hat{I}_{t+1}$ to denote the predicted frame at time $t + 1$. The anomaly detection is determined by the difference between the predicted/reconstructed frame and the actual frame. If this difference is larger than a threshold, this frame is considered an anomaly.

During training, the goal is to learn the future frame prediction/reconstruction model $f_\theta(\cdot)$ from a collection of normal videos. Note that the training data only contain normal videos since it is usually difficult to collect training data with abnormal events for real-world applications.

Few-Shot Scene-Adaptive Anomaly Detection: The standard anomaly detection framework described above have some limitations that make it difficult to apply it in real-world scenarios. It implicitly assumes that the model $f_\theta(\cdot)$ (either reconstruction-based or prediction-based) learned from the training videos can generalize well on test videos. In practical applications, it is unrealistic to collect training videos from the target scene where the system will be deployed. In most cases, training and test videos will come from different scenes. The anomaly detection model $f_\theta(\cdot)$ can easily overfit to the particular training scene and will not generalize to a different scene during testing. We will empirically demonstrate this in the experiment section.

In this paper, we introduce a new problem setup that is closer to real-world applications. This setup is motivated by two crucial observations. First of all, in most anomaly detection applications, the test images come from a particular scene captured by the same camera. In this case, we only need the learned model to perform well on this particular scene. Second, although it is unrealistic to collect a large number of videos from the target scene, it is reasonable to assume that we will have access to a small number of images from the target scene. For example, when a surveillance camera is installed, there is often a calibration process. We can easily collect a few images from the target environment during this calibration process.

Motivated by these observations, we propose a problem setup called *few-shot scene-adaptive anomaly detection*. During training, we have access to videos collected from different scenes. During testing, the videos will come from a target scene that never appears during training. Our model will learn to adapt to this target scene from only a few initial frames. The adapted model is expected to work well in the target scene.

4 Our Approach: MAML for Scene-Adaptive Anomaly Detection

We propose to learn few-shot scene-adaptive anomaly detection models using a meta-learning framework, in particular, the MAML algorithm [5] for meta-learning. Figure 2 shows an overview of the proposed approach. The meta-learning framework consists of a meta-training phase and a meta-testing phase. During meta-training, we have access to videos collected from multiple scenes. The goal of meta-training is learning to quickly adapt to a new scene based on a few frames from it. During this phase, the model is trained from a large number of few-shot scene-adaptive anomaly detection tasks constructed using the videos available in meta-training, where each task corresponds to a particular scene. In each task, our method learns to adapt a pre-trained future frame prediction model using a few frames from the corresponding scene. The learning procedure (meta-learner) is designed in a way such that the adapted model will work well on other frames from the same scene. Through this meta-training process, the model will learn to effectively perform few-shot adaptation for a new scene. During meta-testing, given a few frames from a new target scene, the meta-learner is used to adapt a pre-trained model to this scene. Afterwards, the adapted model is expected to work well on other frames from this target scene.

Our proposed meta-learning framework can be used in conjunction with any anomaly detection model as the backbone architecture. We first introduce the meta-learning approach for scene-adaptive anomaly detection in a general way that is independent of the particular choice of the backbone architecture,

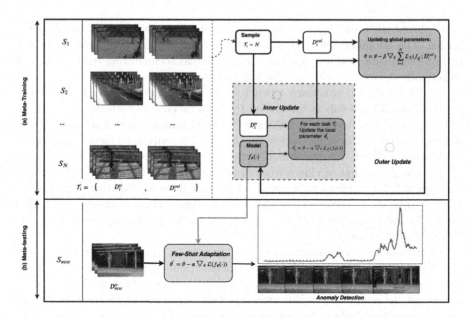

Fig. 2. An overview of our proposed approach. Our approach involves two phases: (a) meta-training and (b) meta-testing. In each iteration of the meta-training (a), we first sample a batch of N scenes $S_1, S_2, ..., S_N$. We then construct a task $\mathcal{T}_i = \{D_i^{tr}, D_i^{val}\}$ for each scene S_i with a training set D_i^{tr} and a validation set D_i^{val}. D_i^{tr} is used for *inner update* through gradient descent to obtain the updated parameters θ_i' for each task. Then D_i^{val} is used to measure the performance of θ_i'. An *outer update* procedure is used to update the model parameters θ by taking into account of all the sampled tasks. In meta-testing (b), given a new scene S_{new}, we use only a few frames to get the adapted parameters θ' for this specific scene. The adapted model is used for anomaly detection in other frames from this scene.

we then describe the details of the proposed backbone architectures used in this paper.

Our goal of few-shot scene-adaptive anomaly detection is to learn a model that can quickly adapt to a new scene using only a few examples from this scene. To accomplish this, the model is trained during a meta-training phase using a set of tasks where it learns to quickly adapt to a new task using only a few samples from the task. The key to applying meta-learning for our application is how to construct these tasks for the meta-training. Intuitively, we should construct these tasks so that they mimic the situation during testing.

Tasks in Meta-Learning: We construct the tasks for meta-training as follows. (1) Let us consider a future frame prediction model $f_\theta(I_{1:t}) \rightarrow \hat{I}_{t+1}$ that maps t observed frames $I_1, I_2, ..., I_t$ to the predicted frame $\hat{I}_{t+1}$ at $t+1$. We have access to M scenes during meta-training, denoted as $S_1, S_2, ..., S_M$. For a given scene S_i, we can construct a corresponding task $\mathcal{T}_i = (\mathcal{D}_i^{tr}, \mathcal{D}_i^{val})$, where $\mathcal{D}_i^{tr}$ and $\mathcal{D}_i^{val}$ are the training and the validation sets in the task $\mathcal{T}_i$. We first split videos from S_i into many overlapping consecutive segments of length $t+1$. Let us consider a segment $(I_1, I_2, ..., I_t, I_{t+1})$. We then consider the first t frames as the input x and the last frame as the output y, i.e. $x = (I_1, I_2, ..., I_t)$ and $y = I_{t+1}$. This will form an input/output pair (x, y). The future frame prediction model can be equivalently written as $f_\theta : x \to y$. In the training set $\mathcal{D}_i^{tr}$, we randomly sample K input/output pairs from $\mathcal{T}_i$ to learn future frame prediction model, i.e. $\mathcal{D}^{tr} = \{(x_1, y_1), (x_2, y_2), ..., (x_K, y_K)\}$. Note that to match the testing scheme, we make sure that all the samples in $\mathcal{D}^{tr}$ come from the same video. We also randomly sample K input/output pairs (excluding those in $\mathcal{D}_i^{tr}$) to form the test data $\mathcal{D}_i^{val}$.

(2) Similarly, for reconstruction-based models, we construct task $\mathcal{T}_i = (\mathcal{D}_i^{tr}, \mathcal{D}_i^{val})$ using individual frames. Since the groundtruth label for each image is itself, we randomly sample K images from one video as $\mathcal{D}_i^{tr}$ and sample K images from the same video as $\mathcal{D}_i^{val}$.

Meta-Training: Let us consider a pre-trained anomaly detection model $f_\theta :$ $x \to y$ with parameters θ. Following MAML [5], we adapt to a task $\mathcal{T}_i$ by defining a loss function on the training set $\mathcal{D}_i^{tr}$ of this task and use one gradient update to change the parameters from θ to θ_i':

$$\theta_i' = \theta - \alpha \bigtriangledown_\theta \mathcal{L}_{\mathcal{T}_i}(f_\theta; \mathcal{D}_i^{tr}), \text{ where} \tag{1a}$$

$$\mathcal{L}_{\mathcal{T}_i}(f_\theta; \mathcal{D}_i^{tr}) = \sum_{(x_j, y_j) \in \mathcal{D}_i^{tr}} L(f_\theta(x_j), y_j) \tag{1b}$$

where α is the step size. Here $L(f_\theta(x_j), y_j)$ measures the difference between the predicted frame $f_\theta(x_j)$ and the actual future frame y_j. We define $L(\cdot)$ by combine the least absolute deviation (L_1 loss) [28], multi-scale structural similarity measurement (L_{ssm} loss) [39] and gradient difference (L_{gdl} loss) [21]:

$$L(f_\theta(x_j), y_j) = \lambda_1 L_1(f_\theta(x_j), y_j) + \lambda_2 L_{ssm}(f_\theta(x_j), y_j) + \lambda_3 L_{gdl}(f_\theta(x_j), y_j), \tag{2}$$

where $\lambda_1, \lambda_2, \lambda_3$ are coefficients that weight between different terms of the loss function.

The updated parameters θ' are specifically adapted to the task $\mathcal{T}_i$. Intuitively we would like θ' to perform on the validation set $\mathcal{D}_i^{val}$ of this task. We measure the performance of θ' on $\mathcal{D}_i^{val}$ as:

$$\mathcal{L}_{\mathcal{T}_i}(f_{\theta'}; \mathcal{D}_i^{val}) = \sum_{(x_j, y_j) \in \mathcal{D}_i^{val}} L(f_{\theta'}(x_j), y_j) \tag{3}$$

The goal of meta-training is to learn the initial model parameters θ, so that the scene-adapted parameters θ' obtained via Eq. 1 will minimize the loss in Eq. 3 across all tasks. Formally, the objective of meta-learning is defined as:

$$\min_\theta \sum_{i=1}^{M} \mathcal{L}_{\mathcal{T}_i}(f_{\theta'}; \mathcal{D}_i^{val}) \tag{4}$$

Algorithm 1: Meta-training for few-shot scene-adaptive anomaly detection

---

Input: Hyper-parameters α, β
Initialize θ with a pre-trained model $f_\theta(\cdot)$;
while *not done* **do**
>   Sample a batch of scenes $\{S_i\}_{i=1}^N$;
>   **for** *each S_i* **do**
>   >   Construct $\mathcal{T}_i = (\mathcal{D}_i^{tr}, \mathcal{D}_i^{val})$ from S_i;
>   >   Evaluate $\nabla_\theta \mathcal{L}_{\mathcal{T}_i}(f_\theta; \mathcal{D}_i^{tr})$ in Eq. 1;
>   >   Compute scene-adaptative parameters $\theta_i' = \theta - \alpha \nabla_\theta \mathcal{L}_{\mathcal{T}_i}(f_\theta; \mathcal{D}_i^{tr})$;
>
>   **end**
>   Update $\theta \leftarrow \theta - \beta \sum_{i=1}^N \nabla_\theta \mathcal{L}_{\mathcal{T}_i}(f_{\theta_i'}; \mathcal{D}_i^{val})$ using each $\mathcal{D}_i^{val}$ and $\mathcal{L}_{\mathcal{T}_i}$ in Eq.3;

end

The loss in Eq. 4 involves summing over all tasks during meta-training. In practice, we sample a mini-batch of tasks in each iteration. Algorithm 1 summarizes the entire learning algorithm.

Meta-Testing: After meta-training, we obtain the learned model parameters θ. During meta-testing, we are given a new target scene S_{new}. We simply use Eq. 1 to obtain the adapted parameters θ' based on K examples in S_{new}. Then we apply θ' on the remaining frames in the S_{new} to measure the performance. We use the first several frames of one video in S_{new} for adaptation and use the remaining frames for testing. This is similar to real-world settings where it is only possible to obtain the first several frames for a new camera.

Backbone Architecture: Our scene-adaptive anomaly detection framework is general. In theory, we can use any anomaly detection network as the backbone architecture. In this paper, we propose a future frame prediction based backbone architecture similar to [14]. Following [14], we build our model based on conditional GAN. One limitation of [14] is that it requires additional low-level feature (i.e. optical flows) and is not trained end-to-end. To capture spatial-temporal information of the videos, we propose to combine generative models and sequential modelling. Specifically, we build a model using ConvLSTM and adversarial training. This model consists of a generator and a discriminator. To build the generator, we apply a U-Net [30] to predict the future frame and pass the prediction to a ConvLSTM module [40] to retain the information of the previous steps. The generator and discriminator are adversarially trained. We call our model r-GAN. Since the backbone architecture is not the main focus of the paper, we skip the details and refers readers to the supplementary material for the detailed architecture of this backbone. In the experiment section, we will demonstrate that our backbone architecture outperforms [14] even though we do not use optical flows.

We have also experiment with other variants of the backbone architecture. For example, we have tried using the ConvLSTM module in the latent space of an autoencoder. We call this variant *r-GAN**. Another variant is to use a variational autoencoder instead of GAN. We call this variant *r-VAE*. Readers are referred to the supplementary material for the details of these different variants. In the experiment, we will show that r-GAN achieves the best performance among all these different variants. So we use r-GAN as the backbone architecture in the meta learning framework.

5   Experiments

In this section, we first introduce our datasets and experimental setup in Sect. 5.1. We then describe some baseline approaches used for comparison in Sect. 5.2. Lastly, we show our experimental results and the ablation study results in Sect. 5.3.

5.1   Datasets and Setup

Fig. 3. Example frames from the datasets used for meta-training. The first row shows examples of different scenes from the Shanghai Tech dataset. The second row shows examples of different scenes from the UCF crime dataset.

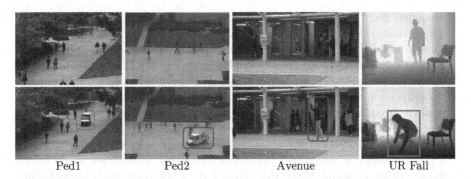

| Ped1 | Ped2 | Avenue | UR Fall |

Fig. 4. Example frames from datasets used in meta-testing. The first row shows examples of normal frames for four datasets, and the second row shows the abnormal frames. Note that training videos only contain normal frames. Videos with abnormal frames are only used for testing.

Table 1. Comparison of anomaly detection performance among our backbone architecture (r-GAN), its variants, and existing state-of-the-art in the standard setup (i.e. without scene adaptation). We report AUC (%) of different methods on UCSD Ped1 (Ped1), UCSD Ped2 (Ped2), CUHK Avenue (CUHK) and Shanghai Tech (ST) datasets. **We use the same train/test split as prior work on each dataset (i.e. without adaptation).** Our proposed backbone architecture outperforms the existing state-of-the-art on almost all datasets.

Category	Method	Ped1	Ped2	CUHK	ST
Feature	MPCCA [9]	59.0	69.3	-	-
	Del et al.[3]	-	-	78.3	-
Reconstruction	Conv-AE [8]	75.0	85.0	80.0	60.9
	Unmasking [37]	68.4	82.2	80.6	-
	LSA [1]	-	95.4	-	72.5
	ConvLSTM-AE [17]	75.5	88.1	77.0	-
	MemAE [6]	-	94.1	83.3	71.2
Prediction	Stacked RNN [18]	-	92.2	81.7	68.0
	FFP [14]	83.1	95.4	84.9	72.8
	MPED-RNN [23]	-	-	-	73.4
	Conv-VRNN [16]	**86.3**	96.1	85.8	-
	Nguyen et al. [25]	-	**96.2**	**86.9**	-
Our backbones	**r-VAE**	82.4	89.2	81.8	72.7
	r-GAN*	83.7	95.9	85.3	73.7
	r-GAN	**86.3**	**96.2**	85.8	**77.9**

Datasets: This paper addresses a new problem. In particular, the problem setup requires training videos from multiple scenes and test videos from different scenes. There are no existing datasets that we can directly use for this problem setup. Instead, we repurpose several available datasets.

- Shanghai Tech [18]: This dataset contains 437 videos collected from 13 scenes. The training videos only contain normal events, while the test videos may contain anomalies. In the standard split in [18], both training and test sets contain videos from these 13 scenes. This split does not fit our problem setup where test scenes should be distinct from those in training. In our experiment, we propose a new train/test split more suitable for our problem. We also perform cross-dataset testing where we use the original Shanghai Tech dataset during meta-training and other datasets for meta-testing.
- UCF crime [35]: This dataset contains normal and crime videos collected from a large number of real-world surveillance cameras where each video comes from a different scene. Since this dataset does not come with ground-truth frame-level annotations, we cannot use it for testing since we do not have the ground-truth to calculate the evaluation metrics. Therefore, we only use the 950 normal videos from this dataset for meta-training, then test the model on

Table 2. Comparison of anomaly detection in terms of AUC (%) of different methods on the UR fall detection dataset. This dataset contains depth images. We simply treat those as RGB images. **We use the same train/test split as prior work on this dataset (i.e. without adaptation).** Our proposed backbone architecture is state-of-the-art among all the methods.

Method	AUC (%)
DAE [20]	75.0
CAE [20]	76.0
CLSTMAE [27]	82.0
DSTCAE [26]	89.0
r-VAE	90.3
r-GAN*	89.6
r-GAN	**90.6**

Table 3. Comparison of K-shot scene-adaptive anomaly detection on the Shanghai Tech dataset. We use 6 scenes for training and the remaining 7 scenes for testing. We report results in terms of AUC (%) for $K = 1, 5, 10$. The proposed approach outperforms two baselines.

Methods	$K = 1$	$K = 5$	$K = 10$
Pre-trained	70.11	70.11	70.11
Fine-tuned	71.61	70.47	71.59
Ours	**74.51**	**75.28**	**77.36**

other datasets. This dataset is much more challenging than Shanghai Tech when being used for meta-training, since the scenes are diverse and very dissimilar to our test sets. Our insight is that if our model can adapt to a target dataset by meta-training on UCF crime, our model can be trained with similar surveillance videos.

– UCSD Pedestrian 1 [19], UCSD Pedestrian 2 (Ped 2) [19], and CUHK Avenue [15]: Each of these datasets contains videos from only one scene but different times. They contain 36, 12 and 21 test videos, respectively, including a total number of 99 abnormal events such as moving bicycles, vehicles, people throwing things, wandering and running. We use the model trained from Shanghai Tech or UCF crime datasets and test on these datasets.
– UR fall [12]: This dataset contains 70 depth videos collected with a Microsoft Kinect camera in a nursing home. Each frame is represented as a 1-channel grayscale image capturing the depth information. In our case, we convert each frame to an RGB image by duplicating the grayscale value among 3 color channels for every pixel. This dataset is originally collected for research in fall detection. We follow previous work in [26] which considers a person falling as the anomaly. Again, we use this dataset for testing. Since this dataset is drastically different from other anomaly detection datasets, good performance on this dataset will be very strong evidence of the generalization power of our approach.

Figures 3 and 4 show some example frames from the datasets we used in meta-training and meta-testing.

Evaluation Metrics: Following prior work [14,17,19], we evaluate the performance using the area under the ROC curve (AUC). The ROC curve is obtained by varying the threshold for the anomaly score for each frame-wise prediction.

Implementation Details: We implement our model in PyTorch. We use a fixed learning rate of 0.0001 for pre-training. We fix the hyperparameters α and β in meta-learning at 0.0001. During meta-training, we select the batch size of task/scenes in each epoch to be 5 on ShanghaiTech, and 10 on UCF crime.

5.2   Baselines

To the best of our knowledge, this is the first work on the scene-adaptive anomaly detection problem. Therefore, there is no prior work that we can directly compare with. Nevertheless, we define the following baselines for comparison.

Pre-trained: This baseline learns the model from videos available during training, then directly applies the model in testing without any adaptation.

Fine-Tuned: This baseline first learns a pre-trained model. Then it adapts to the target scene using the standard fine-tuning technique on the few frames from the target scene.

5.3   Experimental Results

Sanity Check on Backbone Architecture: We first perform an experiment as a sanity check to show that our proposed backbone architecture is comparable to the state-of-the-art. Note that this sanity check uses the standard training/test setup (training set and testing set are provided by the original datasets), and our model can be directly compared with other existing methods. Table 1 shows the comparisons among our proposed architecture (r-GAN), its variants (r-GAN* and r-VAE), and other methods when using the standard anomaly detection training/test setup on several anomaly detection datasets. Table 2 shows the comparison on the fall detection dataset. We can see that our backbone architecture r-GAN outperforms its variants and the existing state-of-the-art methods on almost all the datasets. As a result, we use r-GAN as our backbone architecture to test our few-shot scene-adaptive anomaly detection algorithm in this paper.

Results on Shanghai Tech: In this experiment, we use Shanghai Tech for both training and testing. In the train/test split used in [14], both training and test sets contain videos from the same set of 13 scenes. This split does not fit our problem. Instead, we propose a split where the training set contains videos of 6 scenes from the original training set, and the test set contains videos of the remaining 7 scenes from the original test set. This will allow us to demonstrate

Table 4. Comparison of K-shot ($K = 1, 5, 10$) scene-adaptive anomaly detection under the cross-dataset testing setting. We report results in terms of AUC (%) using the Shanghai Tech dataset and UCF crime dataset for meta-training. We compare our results with two baseline methods. Our results demonstrate the effectiveness of our method on few-shot scene-adaptive anomaly detection.

Target	Methods	1-shot (K = 1)	5-shot (K = 5)	10-shot (K = 10)
Shanghai Tech				
UCSD Ped 1	Pre-trained	73.1	73.1	73.1
	Fine-tuned	76.99	77.85	78.23
	Ours	**80.6**	**81.42**	**82.38**
UCSD Ped 2	Pre-trained	81.95	81.95	81.95
	Fine-tuned	85.64	89.66	91.11
	Ours	**91.19**	**91.8**	**92.8**
CUHK Avenue	Pre-trained	71.43	71.43	71.43
	Fine-tuned	75.43	76.52	77.77
	Ours	**76.58**	**77.1**	**78.79**
UR Fall	Pre-trained	64.08	64.08	64.08
	Fine-tuned	64.48	64.75	62.89
	Ours	**75.51**	**78.7**	**83.24**
UCF crime				
UCSD Ped 1	Pre-trained	66.87	66.87	66.87
	Fine-tuned	71.7	74.52	74.68
	Ours	**78.44**	**81.43**	**81.62**
UCSD Ped 2	Pre-trained	62.53	62.53	62.53
	Fine-tuned	65.58	72.63	78.32
	Ours	**83.08**	**86.41**	**90.21**
CUHK Avenue	Pre-trained	64.32	64.32	64.32
	Fine-tuned	66.7	67.12	70.61
	Ours	**72.62**	**74.68**	**79.02**
UR Fall	Pre-trained	50.87	50.87	50.87
	Fine-tuned	57.02	58.08	62.82
	Ours	**74.59**	**79.08**	**81.85**

the generalization ability of the proposed meta-learning approach. Table 3 shows the average AUC score over our test split of this dataset (7 scenes). Our model outperforms the two baselines.

Cross-Dataset Testing: To demonstrate the generalization power of our approach, we also perform cross-dataset testing. In this experiment, we use either Shanghai Tech (the original training set) or UCF crime for meta-training, then

Table 5. Ablation study for using different number of sampled tasks ($N = 1$ or $N = 5$) during each epoch of meta-training. The results show that even the performance of training with one task is better than fine-tuning. However, a larger number of tasks is able to train an improved model.

Target	Methods	K = 1	K = 5	K = 10
Ped1	Fine-tuned	76.99	77.85	78.23
	Ours ($N = 1$)	79.94	80.44	78.88
	Ours ($N = 5$)	**80.6**	**81.42**	**82.38**
Ped2	Fine-tuned	85.64	89.66	91.11
	Ours ($N = 1$)	90.73	91.5	91.11
	Ours ($N = 5$)	**91.19**	**91.8**	**92.8**
CUHK	Fine-tuned	75.43	76.52	77.77
	Ours ($N = 1$)	76.05	76.53	77.31
	Ours ($N = 5$)	**76.58**	**77.1**	**78.79**

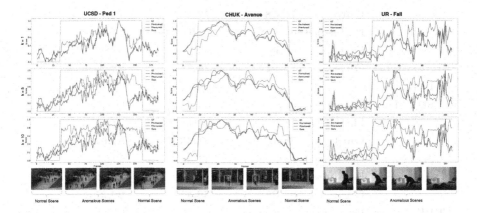

Fig. 5. Qualitative results on three benchmark datasets using a pre-trained model on the Shanghai Tech dataset. Different columns represent results on different datasets. Each row shows few-shot scene-adaptive anomaly detection results with different numbers of training samples K. The red bounding boxes showing the abnormal event localization are for visualization purposes. They are not the outputs of our model which only predicts an anomaly score at the frame level. (Color figure online)

use the other datasets (UCSD Ped1, UCSD Ped2, CUHK Avenue and UR Fall) for meta-testing. We present our cross-dataset testing results in Table 4. Compared with Table 3, the improvement of our approach over the baselines in Table 4 is even more significant (e.g. more than 20% in some cases). It is particularly exciting that our model can successfully adapt to the UR Fall dataset, considering this dataset contains depth images and scenes that are drastically different from those used during meta-training.

Ablation Study: In this study, we show the effect of the batch size (i.e. the number of sampled scenes) during the meta-training process. For this study, we train r-GAN on the Shanghai Tech dataset and test on Ped 1, Ped 2 and CUHK. We experiment with sampling either one ($N = 1$) or five ($N = 5$) tasks in each epoch during meta-training. Table 5 shows the comparison. Overall, using our approach with $N = 1$ performs better than simple fine-tuning, but not as good as $N = 5$. One explanation is that by having access to multiple scenes in one epoch, the model is less likely to overfit to any specific scene.

Qualitative Results: Figure 5 shows qualitative examples of detected anomalies. We visualize the anomaly scores on the frames in a video. We compare our method with the baselines in one graph for different values of K and different datasets.

6 Conclusion

We have introduced a new problem called *few-shot scene-adaptive anomaly detection*. Given a few frames captured from a new scene, our goal is to produce an anomaly detection model specifically adapted to this scene. We believe this new problem setup is closer to the real-world deployment of anomaly detection systems. We have developed a meta-learning based approach to this problem. During meta-training, we have access to videos from multiple scenes. We use these videos to construct a collection of tasks, where each task is a few-shot scene-adaptive anomaly detection task. Our model learns to effectively adapt to a new task with only a few frames from the corresponding scene. Experimental results show that our proposed approach significantly outperforms other alternative methods.

Acknowledgement. This work was supported by the NSERC and UMGF funding. We thank NVIDIA for donating some of the GPUs used in this work.

References

1. Abati, D., Porrello, A., Calderara, S., Cucchiara, R.: Latent space autoregression for novelty detection. In: CVPR (2019)
2. Chalapathy, R., Menon, A.K., Chawla, S.: Robust, deep and inductive anomaly detection. In: Ceci, M., Hollmén, J., Todorovski, L., Vens, C., Džeroski, S. (eds.) ECML PKDD 2017. LNCS (LNAI), vol. 10534, pp. 36–51. Springer, Cham (2017). https://doi.org/10.1007/978-3-319-71249-9_3
3. Del Giorno, A., Bagnell, J.A., Hebert, M.: A discriminative framework for anomaly detection in large videos. In: Leibe, B., Matas, J., Sebe, N., Welling, M. (eds.) ECCV 2016. LNCS, vol. 9909, pp. 334–349. Springer, Cham (2016). https://doi.org/10.1007/978-3-319-46454-1_21
4. Dosovitskiy, A., et al.: FlowNet: learning optical flow with convolutional networks. In: ICCV (2015)

5. Finn, C., Abbeel, P., Levine, S.: Model-agnostic meta-learning for fast adaptation of deep networks. In: ICML (2017)
6. Gong, D., et al.: Memorizing normality to detect anomaly: memory-augmented deep autoencoder for unsupervised anomaly detection. In: ICCV (2019)
7. Goodfellow, I., et al.: Generative adversarial nets. In: NeurIPS (2014)
8. Hasan, M., Choi, J., Neumann, J., Roy-Chowdhury, A.K., Davis, L.S.: Learning temporal regularity in video sequences. In: CVPR (2016)
9. Kim, J., Grauman, K.: Observe locally, infer globally: a space-time MRF for detecting abnormal activities with incremental updates. In: CVPR (2009)
10. Kingma, D.P., Welling, M.: Auto-encoding variational Bayes. arXiv preprint arXiv:1312.6114 (2013)
11. Koch, G., Zemel, R., Salakhutdinov, R.: Siamese neural networks for one-shot image recognition. In: ICML Deep Learning Workshop (2015)
12. Kwolek, B., Kepski, M.: Human fall detection on embedded platform using depth maps and wireless accelerometer. Comput. Methods Programs Biomed. **117**, 489–501 (2014)
13. Lake, B.M., Salakhutdinov, R., Tenenbaum, J.B.: Human-level concept learning through probabilistic program induction. Science **350**, 1332–1338 (2015)
14. Liu, W., Luo, W., Lian, D., Gao, S.: Future frame prediction for anomaly detection-a new baseline. In: CVPR (2018)
15. Lu, C., Shi, J., Jia, J.: Abnormal event detection at 150 FPS in Matlab. In: ICCV (2013)
16. Lu, Y., Reddy, M.K.K., Nabavi, S.S., Wang, Y.: Future frame prediction using convolutional VRNN for anomaly detection. In: AVSS (2019)
17. Luo, W., Liu, W., Gao, S.: Remembering history with convolutional LSTM for anomaly detection. In: ICME (2017)
18. Luo, W., Liu, W., Gao, S.: A revisit of sparse coding based anomaly detection in stacked RNN framework. In: ICCV (2017)
19. Mahadevan, V., Li, W., Bhalodia, V., Vasconcelos, N.: Anomaly detection in crowded scenes. In: CVPR (2010)
20. Masci, J., Meier, U., Cireşan, D., Schmidhuber, J.: Stacked convolutional autoencoders for hierarchical feature extraction. In: Honkela, T., Duch, W., Girolami, M., Kaski, S. (eds.) ICANN 2011. LNCS, vol. 6791, pp. 52–59. Springer, Heidelberg (2011). https://doi.org/10.1007/978-3-642-21735-7_7
21. Mathieu, M., Couprie, C., LeCun, Y.: Deep multi-scale video prediction beyond mean square error. In: ICLR (2016)
22. Medel, J.R., Savakis, A.: Anomaly detection in video using predictive convolutional long short-term memory networks. arXiv preprint arXiv:1612.00390 (2016)
23. Morais, R., Le, V., Tran, T., Saha, B., Mansour, M., Venkatesh, S.: Learning regularity in skeleton trajectories for anomaly detection in videos. In: CVPR (2019)
24. Munkhdalai, T., Yu, H.: Meta networks. In: ICML (2017)
25. Nguyen, T.N., Meunier, J.: Anomaly detection in video sequence with appearance-motion correspondence. In: ICCV (2019)
26. Nogas, J., Khan, S.S., Mihailidis, A.: DeepFall-non-invasive fall detection with deep spatio-temporal convolutional autoencoders. arXiv preprint arXiv:1809.00977 (2018)
27. Nogas, J., Khan, S.S., Mihailidis, A.: Fall detection from thermal camera using convolutional LSTM autoencoder. Technical report (2019)
28. Pollard, D.: Asymptotics for least absolute deviation regression estimators. Econom. Theory **7**, 186–199 (1991)

29. Ravi, S., Larochelle, H.: Optimization as a model for few-shot learning (2016)
30. Ronneberger, O., Fischer, P., Brox, T.: U-Net: convolutional networks for biomedical image segmentation. In: Navab, N., Hornegger, J., Wells, W.M., Frangi, A.F. (eds.) MICCAI 2015. LNCS, vol. 9351, pp. 234–241. Springer, Cham (2015). https://doi.org/10.1007/978-3-319-24574-4_28
31. Sabokrou, M., Fathy, M., Hoseini, M.: Video anomaly detection and localization based on the sparsity and reconstruction error of auto-encoder. Electron. Lett. **52**, 1222–1224 (2016)
32. Sabokrou, M., Khalooei, M., Fathy, M., Adeli, E.: Adversarially learned one-class classifier for novelty detection. In: CVPR (2018)
33. Santoro, A., Bartunov, S., Botvinick, M., Wierstra, D., Lillicrap, T.: Meta-learning with memory-augmented neural networks. In: ICML (2016)
34. Snell, J., Swersky, K., Zemel, R.: Prototypical networks for few-shot learning. In: NeurIPS (2017)
35. Sultani, W., Chen, C., Shah, M.: Real-world anomaly detection in surveillance videos. In: CVPR (2018)
36. Sung, F., Yang, Y., Zhang, L., Xiang, T., Torr, P.H., Hospedales, T.M.: Learning to compare: relation network for few-shot learning. In: CVPR (2018)
37. Tudor Ionescu, R., Smeureanu, S., Alexe, B., Popescu, M.: Unmasking the abnormal events in video. In: ICCV (2017)
38. Vinyals, O., Blundell, C., Lillicrap, T., Wierstra, D., et al.: Matching networks for one shot learning. In: NeurIPS (2016)
39. Wang, Z., Simoncelli, E.P., Bovik, A.C.: Multiscale structural similarity for image quality assessment. In: The Thrity-Seventh Asilomar Conference on Signals, Systems & Computers (2003)
40. Xingjian, S., Chen, Z., Wang, H., Yeung, D.Y., Wong, W.K., Woo, W.: Convolutional LSTM network: a machine learning approach for precipitation nowcasting. In: NeurIPS (2015)

Personalized Face Modeling for Improved Face Reconstruction and Motion Retargeting

Bindita Chaudhuri[1]([✉]), Noranart Vesdapunt[2], Linda Shapiro[1], and Baoyuan Wang[2]

[1] University of Washington, Seattle, USA
{bindita,shapiro}@cs.washington.edu
[2] Microsoft Cloud and AI, Redmond, USA
{noves,baoyuanw}@microsoft.com

Abstract. Traditional methods for image-based 3D face reconstruction and facial motion retargeting fit a 3D morphable model (3DMM) to the face, which has limited modeling capacity and fail to generalize well to in-the-wild data. Use of deformation transfer or multilinear tensor as a personalized 3DMM for blendshape interpolation does not address the fact that facial expressions result in different local and global skin deformations in different persons. Moreover, existing methods learn a single albedo per user which is not enough to capture the expression-specific skin reflectance variations. We propose an end-to-end framework that jointly learns a personalized face model per user and per-frame facial motion parameters from a large corpus of in-the-wild videos of user expressions. Specifically, we learn user-specific expression blendshapes and dynamic (expression-specific) albedo maps by predicting personalized corrections on top of a 3DMM prior. We introduce novel training constraints to ensure that the corrected blendshapes retain their semantic meanings and the reconstructed geometry is disentangled from the albedo. Experimental results show that our personalization accurately captures fine-grained facial dynamics in a wide range of conditions and efficiently decouples the learned face model from facial motion, resulting in more accurate face reconstruction and facial motion retargeting compared to state-of-the-art methods.

Keywords: 3D face reconstruction · Face modeling · Face tracking · Facial motion retargeting

1 Introduction

With the ubiquity of mobile phones, AR/VR headsets and video games, communication through facial gestures has become very popular, leading to extensive

B. Chaudhuri—This work was done when the author visited Microsoft.

Electronic supplementary material The online version of this chapter (https://doi.org/10.1007/978-3-030-58558-7_9) contains supplementary material, which is available to authorized users.

A. Vedaldi et al. (Eds.): ECCV 2020, LNCS 12350, pp. 142–160, 2020.
https://doi.org/10.1007/978-3-030-58558-7_9

research in problems like 2D face alignment, 3D face reconstruction and facial motion retargeting. A major component of these problems is to estimate the 3D face, i.e., face geometry, appearance, expression, head pose and scene lighting, from 2D images or videos. 3D face reconstruction from monocular images is ill-posed by nature, so a typical solution is to leverage a parametric 3D morphable model (3DMM) trained on a limited number of 3D face scans as prior knowledge [2,11,14,24,28,35,38,47,51]. However, the low dimensional space limits their modeling capacity as shown in [21,45,50] and scalability using more 3D scans is expensive. Similarly, the texture model of a generic 3DMM is learned in a controlled environment and does not generalize well to in-the-wild images. Tran et al. [49,50] overcomes these limitations by learning a non-linear 3DMM from a large corpus of in-the-wild images. Nevertheless, these reconstruction-based approaches do not easily support facial motion retargeting.

In order to perform tracking for retargeting, blendshape interpolation technique is usually adopted where the users' blendshapes are obtained by deformation transfer [43], but this alone cannot reconstruct expressions realistically as shown in [14,26]. Another popular technique is to use a multilinear tensor-based 3DMM [4,5,51], where the expression is coupled with the identity implying that same identities should share the same expression blendshapes. However, we argue that facial expressions are characterized by different skin deformations on different persons due to difference in face shape, muscle movements, age and other factors. This kind of user-specific local skin deformations cannot be accurately represented by a linear combination of predefined blendshapes. For example, smiling and raising eyebrows create different cheek folds and forehead wrinkle patterns respectively on different persons, which cannot be represented by simple blendshape interpolation and require correcting the corresponding blendshapes. Some optimization-based approaches [14,20,26,36] have shown that modeling user-specific blendshapes indeed results in a significant improvement in the quality of face reconstruction and tracking. However, these approaches are computationally slow and require additional preprocessing (e.g. landmark detection) during test time, which significantly limits real-time applications with in-the-wild data on the edge devices. The work [8] trains a deep neural network instead to perform retargeting in real-time on typical mobile phones, but its use of predefined 3DMM limits its face modeling accuracy. Tewari et al. [44] leverage in-the-wild videos to learn face identity and appearance models from scratch, but they still use expression blendshapes generated by deformation transfer.

Moreover, existing methods learn a single albedo map for a user. The authors in [17] have shown that skin reflectance changes with skin deformations, but it is not feasible to generate a separate albedo map for every expression during retargeting. Hence it is necessary to learn the static reflectance separately, and associate the expression-specific dynamic reflectance with the blendshapes so that the final reflectance can be obtained by interpolation similar to blendshape interpolation, as in [33]. Learning dynamic albedo maps in addition to static albedo map also helps to capture the fine-grained facial expression details like folds and wrinkles [34], thereby resulting in reconstruction of higher fidelity.

To address these issues, we introduce a novel end-to-end framework that leverages a large corpus of in-the-wild user videos to jointly learn personalized face modeling and face tracking parameters. Specifically, we design a modeling network which learns geometry and reflectance corrections on top of a 3DMM prior to generate user-specific expression blendshapes and dynamic (expression-specific) albedo maps. In order to ensure proper disentangling of the geometry from the albedo, we introduce the face parsing loss inspired by [57]. Note that [57] uses parsing loss in a fitting based framework whereas we use it in a learning based framework. We also ensure that the corrected blendshapes retain their semantic meanings by restricting the corrections to local regions using attention maps and by enforcing a blendshape gradient loss. We design a separate tracking network which predicts the expression blendshape coefficients, head pose and scene lighting parameters. The decoupling between the modeling and tracking networks enables our framework to perform reconstruction as well as retargeting (by tracking one user and transferring the facial motion to another user's model). Our main contributions are:

1. We propose a deep learning framework to learn user-specific expression blend-shapes and dynamic albedo maps that accurately capture the complex user-specific expression dynamics and high-frequency details like folds and wrinkles, thereby resulting in photorealistic 3D face reconstruction.
2. We bring two novel constraints into the end-to-end training: face parsing loss to reduce the ambiguity between geometry and reflectance and blendshape gradient loss to retain the semantic meanings of the corrected blendshapes.
3. Our framework jointly learns user-specific face model and user-independent facial motion in disentangled form, thereby supporting motion retargeting.

2   Related Work

Face Modeling: Methods like [19,25,30,32,41,48,53] leverage user images captured with varying parameters (e.g. multiple viewpoints, expressions etc.) at least during training with the aim of user-specific 3D face reconstruction (not necessarily retargeting). Monocular video-based optimization techniques for 3D face reconstruction [13,14] leverage the multi-frame consistency to learn the facial details. For single image based reconstruction, traditional methods [59] regress the parameters of a 3DMM and then learn corrective displacement [18,19,22] or normal maps [37,40] to capture the missing details. Recently, several deep learning based approaches have attempted to overcome the limited representation power of 3DMM. Tran et al. [49,50] proposed to train a deep neural network as a non-linear 3DMM. Tewari et al. [45] proposed to learn shape and reflectance correctives on top of the linear 3DMM. In [44], Tewari et al. learn new identity and appearance models from videos. However, these methods use expression blendshapes obtained by deformation transfer [43] from a generic 3DMM to their own face model and do not optimize the blendshapes based on the user's identity. In addition, these methods predict a single static albedo map to represent the face texture, which fail to capture adequate facial details.

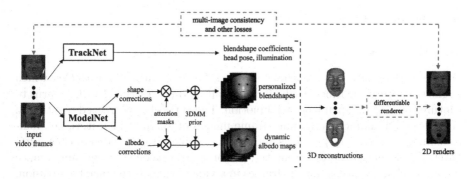

Fig. 1. Our end-to-end framework. Our framework takes frames from in-the-wild video(s) of a user as input and generates per-frame tracking parameters via the *TrackNet* and personalized face model via the *ModelNet*. The networks are trained together in an end-to-end manner (marked in red) by projecting the reconstructed 3D outputs into 2D using a differentiable renderer and computing multi-image consistency losses and other regularization losses. (Color figure online)

Personalization: Optimization based methods like [7,20,26] have demonstrated the need to optimize the expression blendshapes based on user-specific facial dynamics. These methods alternately update the blendshapes and the corresponding coefficients to accurately fit some example poses in the form of 3D scans or 2D images. For facial appearance, existing methods either use a generic texture model with linear or learned bases or use a GAN [15] to generate a static texture map. But different expressions result in different texture variations, and Nagano et al. [33] and Olszewski et al. [34] addressed this issue by using a GAN to predict the expression-specific texture maps given the texture map in neutral pose. However, the texture variations with expression also vary from person to person. Hence, hallucinating an expression-specific texture map for a person by learning the expression dynamics of other persons is not ideal. Besides, these methods requires fitted geometry as a preprocessing step, thereby limiting the accuracy of the method by the accuracy of the geometry fitting mechanism.

Face Tracking and Retargeting: Traditional face tracking and retargeting methods [3,27,52] generally optimize the face model parameters with occasional correction of the expression blendshapes using depth scans. Recent deep learning based tracking frameworks like [8,23,47,53] either use a generic face model and fix the model during tracking, or alternate between tracking and modeling until convergence. We propose to perform joint face modeling and tracking with novel constraints to disambiguate the tracking parameters from the model.

3   Methodology

3.1   Overview

Our network architecture, as shown in Fig. 1, has two parts: (1) *ModelNet* which learns to capture the user-specific facial details and (2) *TrackNet* which learns to capture the user-independent facial motion. The networks are trained together in an end-to-end manner using multi-frame images of different identities, i.e., multiple images $\{I_1, \ldots, I_N\}$ of the same person sampled from a video in each mini-batch. We leverage the fact that the person's facial geometry and appearance remain unchanged across all the frames in a video, whereas the facial expression, head pose and scene illumination change on a per-frame basis. The *ModelNet* extracts a common feature from all the N images to learn a user-specific face shape, expression blendshapes and dynamic albedo maps (Sect. 3.2). The *TrackNet* processes each of the N images individually to learn the image-specific expression blendshape coefficients, pose and illumination parameters (Sect. 3.3). The predictions of *ModelNet* and *TrackNet* are combined to reconstruct the 3D faces and then projected to the 2D space using a differentiable renderer in order to train the network in a self-supervised manner using multi-image photometric consistency, landmark alignment and other constraints. During testing, the default settings can perform 3D face reconstruction. However, our network architecture and training strategy allow simultaneous tracking of one person's face using *TrackNet* and modeling another person's face using *ModelNet*, and then retarget the tracked person's facial motion to the modeled person or an external face model having similar topology as our face model.

3.2   Learning Personalized Face Model

Our template 3D face consists of a mean (neutral) face mesh S_0 having 12K vertices, per-vertex colors (converted to 2D mean albedo map R_0 using UV coordinates) and 56 expression blendshapes $\{S_1, \ldots, S_{56}\}$. Given a set of expression coefficients $\{w_1, \ldots, w_{56}\}$, the template face shape can be written as $\bar{S} = w_0 S_0 + \sum_{i=1}^{56} w_i S_i$ where $w_0 = (1 - \sum_{i=1}^{56} w_i)$. Firstly, we propose to learn an identity-specific corrective deformation Δ_0^S from the identity of the input images to convert $\bar{S}$ to identity-specific shape. Then, in order to better fit the facial expression of the input images, we learn corrective deformations Δ_i^S for each of the template blendshapes S_i to get identity-specific blendshapes. Similarly, we learn a corrective albedo map Δ_0^R to convert R_0 to identity-specific static albedo map. In addition, we also learn corrective albedo maps Δ_i^R corresponding to each S_i to get identity-specific dynamic (expression-specific) albedo maps.

In our *ModelNet*, we use a shared convolutional encoder E^{model} to extract features F_n^{model} from each image $I_n \in \{I_1, \ldots, I_N\}$ in a mini-batch. Since all the N images belong to the same person, we take an average over all the F_n^{model} features to get a common feature F^{model} for that person. Then, we pass F^{model} through two separate convolutional decoders, D_S^{model} to estimate the shape corrections Δ_0^S and Δ_i^S, and D_R^{model} to estimate the albedo corrections Δ_0^R and Δ_i^R.

We learn the corrections in the UV space instead of the vertex space to reduce the number of network parameters and preserve the contextual information.

User-Specific Expression Blendshapes. A naive approach to learn corrections on top of template blendshapes based on the user's identity would be to predict corrective values for all the vertices and add them to the template blendshapes. However, since blendshape deformation is local, we want to restrict the corrected deformation to a similar local region as the template deformation. To do this, we first apply an attention mask over the per-vertex corrections and then add it to the template blendshape. We compute the attention mask A_i corresponding to the blendshape S_i by calculating the per-vertex euclidean distances between S_i and S_0, thresholding them at 0.001, normalizing them by the maximum distance, and then converting them into the UV space. We also smooth the mask discontinuities using a small amount of Gaussian blur following [33]. Finally, we multiply A_i with Δ_i^S and add it to S_i to obtain a corrected S_i. Note that the masks are precomputed and then fixed during network operations. The final face shape is thus given by:

$$S = w_0 S_0 + \mathcal{F}(\Delta_0^S) + \sum_{i=1}^{56} w_i[S_i + \mathcal{F}(A_i \Delta_i^S)] \tag{1}$$

where $\mathcal{F}(\cdot)$ is a sampling function for UV space to vertex space conversion.

User-Specific Dynamic Albedo Maps. We use one static albedo map to represent the identity-specific neutral face appearance and 56 dynamic albedo maps, one for each expression blendshape, to represent the expression-specific face appearance. Similar to blendshape corrections, we predict 56 albedo correction maps in the UV space and add them to the static albedo map after multiplying the dynamic correction maps with the corresponding UV attention masks. Our final face albedo is thus given by:

$$R = R_0^t + \Delta_0^R + \sum_{i=1}^{56} w_i[A_i \Delta_i^R] \tag{2}$$

where R_0^t is the trainable mean albedo initialized with the mean albedo R_0 from our template face similar to [44].

3.3 Joint Modeling and Tracking

The *TrackNet* consists of a convolutional encoder E^{track} followed by multiple fully connected layers to regress the tracking parameters $\mathbf{p}_n = (\mathbf{w}_n, \mathbf{R}_n, \mathbf{t}_n, \gamma_n)$ for each image I_n. The encoder and fully connected layers are shared over all the N images in a mini-batch. Here $\mathbf{w}_n = (w_0^n, \ldots, w_{56}^n)$ is the expression coefficient vector and $\mathbf{R}_n \in SO(3)$ and $\mathbf{t}_n \in \mathbb{R}^3$ are the head rotation (in terms of Euler angles) and 3D translation respectively. $\gamma_n \in \mathbb{R}^{27}$ are the 27 Spherical Harmonics coefficients (9 per color channel) following the illumination model of [44].

Training Phase: We first obtain a face shape S_n and albedo R_n for each I_n by combining S (Eq. 1) and R (Eq. 2) from the *ModelNet* and the expression coefficient vector $\mathbf{w}_n$ from the *TrackNet*. Then, similar to [15,44], we transform the shape using head pose as $\tilde{S}_n = R_n S_n + \mathbf{t}_n$ and project it onto the 2D camera space using a perspective camera model $\Phi : \mathbb{R}^3 \rightarrow \mathbb{R}^2$. Finally, we use a differentiable renderer $\mathcal{R}$ to obtain the reconstructed 2D image as $\hat{I}_n = \mathcal{R}(\tilde{S}_n, \mathbf{n}_n, R_n, \gamma_n)$ where $\mathbf{n}_n$ are the per-vertex normals. We also mark 68 facial landmarks on our template mesh which we can project onto the 2D space using Φ to compare with the ground truth 2D landmarks.

Testing Phase: The *ModelNet* can take a variable number of input images of a person (due to our feature averaging technique) to predict a personalized face model. The *TrackNet* executes independently on one or more images of the same person given as input to *ModelNet* or a different person. For face reconstruction, we feed images of the same person to both the networks and combine their outputs as in the training phase to get the 3D faces. In order to perform facial motion retargeting, we first obtain the personalized face model of the target subject using *ModelNet*. We then predict the facial motion of the source subject on a per-frame basis using the *TrackNet* and combine it with the target face model. It is important to note that the target face model can be any external face model with semantically similar expression blendshapes.

3.4   Loss Functions

We train both the *TrackNet* and the *ModelNet* together in an end-to-end manner using the following loss function:

$$L = \lambda_{\mathrm{ph}} L_{\mathrm{ph}} + \lambda_{\mathrm{lm}} L_{\mathrm{lm}} + \lambda_{\mathrm{pa}} L_{\mathrm{pa}} + \lambda_{\mathrm{sd}} L_{\mathrm{sd}} + \lambda_{\mathrm{bg}} L_{\mathrm{bg}} + \lambda_{\mathrm{reg}} L_{\mathrm{reg}} \tag{3}$$

where the loss weights λ_* are chosen empirically and their values are given in the supplementary material[1].

Photometric and Landmark Losses: We use the $l_{2,1}$ [49] loss to compute the multi-image photometric consistency loss between the input images I_n and the reconstructed images $\hat{I}_n$. The loss is given by

$$L_{\mathrm{ph}} = \sum_{n=1}^{N} \frac{\sum_{q=1}^{Q} ||M_n(q) * [I_n(q) - \hat{I}_n(q)]||_2}{\sum_{q=1}^{Q} M_n(q)} \tag{4}$$

where M_n is the mask generated by the differentiable renderer (to exclude the background, eyeballs and mouth interior) and q ranges over all the pixels Q in the image. In order to further improve the quality of the predicted albedo by preserving high-frequency details, we add the image (spatial) gradient loss having

---

[1] https://homes.cs.washington.edu/~bindita/personalizedfacemodeling.html.

the same expression as the photometric loss with the images replaced by their gradients. Adding other losses as in [15] resulted in no significant improvement. The landmark alignment loss L_{lm} is computed as the l_2 loss between the ground truth and predicted 68 2D facial landmarks.

Face Parsing Loss: The photometric and landmark loss constraints are not strong enough to overcome the ambiguity between shape and albedo in the 2D projection of a 3D face. Besides, the landmarks are sparse and often unreliable especially for extreme poses and expressions which are difficult to model because of depth ambiguity. So, we introduce the face parsing loss given by:

$$L_{\text{pa}} = \sum_{n=1}^{N} ||I_n^{\text{pa}} - \hat{I}_n^{\text{pa}}||_2 \tag{5}$$

where I_n^{pa} is the ground truth parsing map generated using the method in [29] and $\hat{I}_n^{\text{pa}}$ is the predicted parsing map generated as $\mathcal{R}(\tilde{S}_n, \mathbf{n}_n, T)$ with a fixed precomputed UV parsing map T.

Shape Deformation Smoothness Loss: We employ Laplacian smoothness on the identity-specific corrective deformation to ensure that our predicted shape is locally smooth. The loss is given as:

$$L_{\text{sd}} = \sum_{v=1}^{V} \sum_{u \in \mathcal{N}_v} ||\Delta_0^S(v) - \Delta_0^S(u)||_2^2 \tag{6}$$

where V is the total number of vertices in our mesh and $\mathcal{N}_v$ is the set of neighboring vertices directly connected to vertex v.

Blendshape Gradient Loss: Adding free-form deformation to a blendshape, even after restricting it to a local region using attention masks, can change the semantic meaning of the blendshape. However, in order to retarget tracked facial motion of one person to the blendshapes of another person, the blendshapes of both the persons should have semantic correspondence. We introduce a novel blendshape gradient loss to ensure that the deformation gradients of the corrected blendshapes are similar to those of the template blendshapes. The loss is given by:

$$L_{\text{bg}} = \sum_{i=1}^{56} ||\mathbf{G}_{S_0 \to (S_i + \Delta_i^S)} - \mathbf{G}_{S_0 \to S_i}||_2^2 \tag{7}$$

where $\mathbf{G}_{a \to b}$ denotes the gradient of the deformed mesh b with respect to original mesh a. Details about how to compute $\mathbf{G}$ can be found in [26].

Tracking Parameter Regularization Loss: We use sigmoid activation at the output of the expression coefficients and regularize the coefficients using l_1 loss (L_{reg}^w) to ensure sparse coefficients in the range $[0, 1]$. In order to disentangle the albedo from the lighting, we use a lighting regularization loss given by:

$$L_{\text{reg}}^\gamma = ||\gamma - \gamma_{\text{mean}}||_2 + \lambda_\gamma ||\gamma||_2 \tag{8}$$

where the first term ensures that the predicted light is mostly monochromatic and the second term restricts the overall lighting. We found that regularizing the illumination automatically resulted in albedo consistency, so we don't use any additional albedo loss. Finally, $L_{\text{reg}} = L_{\text{reg}}^w + L_{\text{reg}}^\gamma$.

4   Experimental Setup

Datasets: We train our network using two datasets: (1) VoxCeleb2 [9] and (2) ExpressiveFaces. We set aside 10% of each dataset for testing. VoxCeleb2 has more than 140k videos of about 6000 identities collected from internet, but the videos are mostly similar. So, we choose a subset of 90k videos from about 4000 identities. The images in VoxCeleb2 vary widely in pose but lack variety in expressions and illumination, so we add a custom dataset (ExpressiveFaces) to our training, which contains 3600 videos of 3600 identities. The videos are captured by the users using a hand-held camera (typically the front camera of a mobile phone) as they perform a wide variety of expressions and poses in both indoor and outdoor environments. We sample the videos at 10fps to avoid multiple duplicate frames, randomly delete frames with neutral expression and

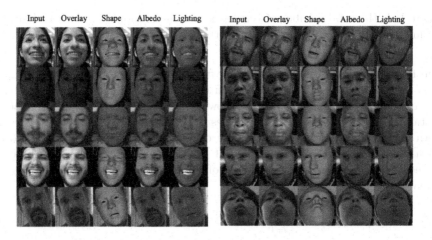

Fig. 2. Qualitative results of our method. Our modeling network accurately captures high-fidelity facial details specific to the user, thereby enabling the tracking network to learn user-independent facial motion. Our network can handle a wide variety of pose, expression, lighting conditions, facial hair and makeup etc. Refer to supplementary material for more results for images and videos.

pose based on a threshold on the expression and pose parameters predicted by [8], and then crop the face and extract ground truth 2D landmarks using [8]. The cropped faces are resized to 224×224 and grouped into mini-batches, each of N images chosen randomly from different parts of a video to ensure sufficient diversity in the training data. We set $N = 4$ during training and $N = 1$ during testing (unless otherwise mentioned) as evaluated to work best for real-time performance in [44].

Implementation Details: We implemented our networks in Tensorflow and used TF mesh renderer [16] for differentiable rendering. During the first stage of training, we train both *TrackNet* and *ModelNet* in an end-to-end manner using Eq. 3. During the second stage of training, we fix the weights of *TrackNet* and fine-tune the *ModelNet* to better learn the expression-specific corrections. The fine-tuning is done using the same loss function as before except the tracking parameter regularization loss since the tracking parameters are now fixed. This training strategy enables us to tackle the bilinear optimization problem of optimizing the blendshapes and the corresponding coefficients, which is generally solved through alternate minimization by existing optimization-based methods. For training, we use a batch size of 8, learning rates of 10^{-4} (10^{-5} during second stage) and Adam optimizer for loss minimization. Training takes $\sim$20 h on a single Nvidia Titan X for the first stage, and another $\sim$5 h for the second stage. The encoder and decoder together in *ModelNet* has an architecture similar to U-Net [39] and the encoder in *TrackNet* has the same architecture as ResNet-18 (details in the supplementary). Since our template mesh contains 12264 vertices, we use a corresponding UV map of dimensions 128×128.

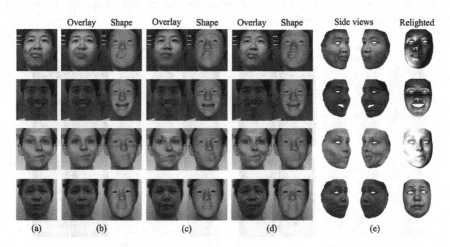

Fig. 3. Importance of personalization. (a) input image, (b) reconstruction using 3DMM prior only, (c) reconstruction after adding only identity-based corrections, i.e. Δ_0^S and Δ_0^R in Eqs. (2) and (3) respectively, (d) reconstruction after adding expression-specific corrections, (e) results of (d) with different viewpoints and illumination.

5   Results

We evaluate the effectiveness of our framework using both qualitative results and quantitative comparisons. Figure 2 shows the personalized face shape and albedo, scene illumination and the final reconstructed 3D face generated from monocular images by our method. Learning a common face shape and albedo from multiple images of a person separately from the image-specific facial motion helps in successfully decoupling the tracking parameters from the learned face model. As a result, our tracking network have the capacity to represent a wide range of expressions, head pose and lighting conditions. Moreover, learning a unified model from multiple images help to overcome issues like partial occlusion, self-occlusion, blur in one or more images. Figure 3 shows a gallery of examples that demonstrate the effectiveness of personalized face modeling for better reconstruction. Figure 7a shows that our network can be efficiently used to perform facial motion retargeting to another user or to an external 3D model of a stylized character in addition to face reconstruction.

5.1   Importance of Personalized Face Model

Importance of Personalized Blendshapes: Modeling the user-specific local geometry deformations while performing expressions enable our modeling network to accurately fit the facial shape of the input image. Figure 4a shows examples of how the same expression can look different on different identities and how the corrected blendshapes capture those differences for more accurate reconstruction than with the template blendshapes. In the first example, the extent of opening of the mouth in the *mouth open* blendshape is adjusted according to the user expression. In the second example, the mouth shape of the *mouth funnel* blendshape is corrected.

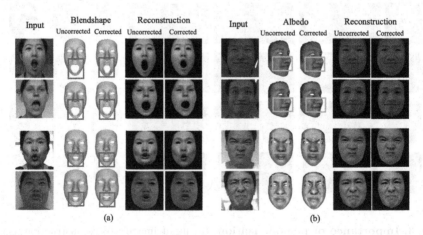

Fig. 4. Visualization of corrected blendshapes and albedo. The corrections are highlighted. (a) Learning user-specific blendshapes corrects the mouth shape of the blendshapes, (b) Learning user-specific dynamic albedo maps captures the high-frequency details like skin folds and wrinkles.

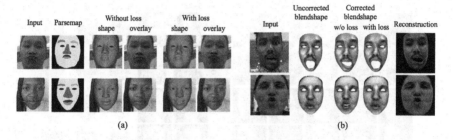

Fig. 5. Importance of novel training constraints. (a) importance of face parsing loss in obtaining accurate geometry decoupled from albedo, (b) importance of blendshape gradient loss in retaining the semantic meaning of *mouth open* (row 1) and *kiss* (row 2) blendshapes after correction.

Importance of Dynamic Textures: Modeling the user-specific local variations in skin reflectance while performing expressions enable our modeling network to generate a photorealistic texture for the input image. Figure 4b shows examples of how personalized dynamic albedo maps help in capturing the high-frequency expression-specific details compared to static albedo maps. In the first example, our method accurately captures the folds around the nose and mouth during smile expression. In the second example, the unique wrinkle patterns between the eyebrows of the two users are correctly modeled during a disgust expression.

5.2   Importance of Novel Training Constraints

Importance of Parsing Loss: The face parse map ensures that each face part of the reconstructed geometry is accurate as shown in [57]. This prevents the albedo to compensate for incorrect geometry, thereby disentangling the albedo from the geometry. However, the authors of [57] use parse map in a geometry fitting framework which, unlike our learning framework, does not generalize well to in-the-wild images. Besides, due to the dense correspondence of parse map compared to the sparse 2D landmarks, parsing loss (a) provides a stronger supervision on the geometry, and (b) is more robust to outliers. We demonstrate the effectiveness of face parsing loss in Fig. 5a. In the first example, the kiss expression is correctly reconstructed with the loss, since the 2D landmarks are not enough to overcome the depth ambiguity. In the second example, without parsing loss the albedo tries to compensate for the incorrect geometry by including the background in the texture. With loss, the nose shape, face contour and the lips are corrected in the geometry, resulting in better reconstruction.

Importance of Blendshape Gradient Loss: Even after applying attention masks to restrict the blendshape corrections to local regions, our method can distort a blendshape such that it loses its semantic meaning, which is undesirable for retargeting purposes. We prevent this by enforcing the blendshape gradient

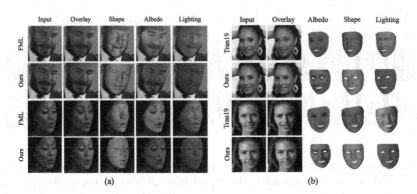

Fig. 6. Visual comparison with (a) FML [44], (b) Non-linear 3DMM [49].

loss, as shown in Fig. 5b. In the first example, without gradient loss, the *mouth open* blendshape gets combined with *jaw left* blendshape after correction in order to minimize the reconstruction loss. With gradient loss, the reconstruction is same but the *mouth open* blendshape retains its semantics after correction. Similarly in the second example, without gradient loss, the *kiss* blendshape gets combined with the *mouth funnel* blendshape, which is prevented by the loss.

5.3   Visual Comparison with State-of-the-art Methods

3D Face Reconstruction: We test the effectiveness of our method on Vox-Celeb2 test set to compare with FML results [44] as shown in Fig. 6a. In the first example, our method captures the mouth shape and face texture better. The second example shows that our personalized face modeling can efficiently model complex expressions like kissing and complex texture like eye shadow better than FML. We also show the visual comparisons between our method and Non-linear 3DMM [49] on the AFLW2000-3D dataset [58] in Fig. 6b. Similar to FML, Non-linear 3DMM fails to accurately capture the subtle facial details.

Face Tracking and Retargeting: By increasing the face modeling capacity and decoupling the model from the facial motion, our method performs superior face tracking and retargeting compared to a recent deep learning based retargeting approach [8]. Figure 7a shows some frames from a user video and how personalization helps in capturing the intensity of the expressions more accurately.

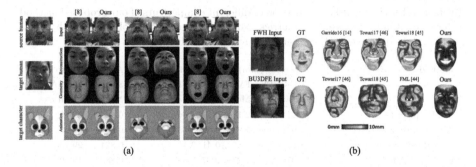

(a)   (b)

Fig. 7. (a) Tracking comparison with [8]. (b) 3D reconstruction error maps.

5.4   Quantitative Evaluation

3D Face Reconstruction: We compute 3D geometry reconstruction error (root mean squared error between a predicted vertex and ground truth correspondence point) to evaluate our predicted mesh on 3D scans from BU-3DFE [55] and Facewarehouse (FWH) [6]. For BU-3DFE we use both the views per scan as input, and for FWH we do not use any special template to start with, unlike Asian face template used by FML. Our personalized face modeling and novel constraints together result in lower reconstruction error compared to state-of-the-art methods (Table 2a and Fig. 7b). The optimization-based method [14] obtains 1.59 mm 3D error for FWH compared to our 1.68 mm, but is much slower (120 s/frames) compared to our method (15.4 ms/frame). We also show how each component of our method helps in improving the overall output in Table 1. For photometric error, we used 1000 images of CelebA [31] (referred as CelebA*) dataset to be consistent with [44].

Face Tracking: We evaluate the tracking performance of our method using two metrics: (1) Normalized Mean Error (NME), defined as an average Euclidean distance between the 68 predicted and ground truth 2D landmarks normalized by the bounding box dimension, on AFLW2000-3D dataset [58], and (2) Area under the Curve (AUC) of the cumulative error distribution curve for 2D landmark error [10] on 300VW video test set [42]. Table 2b shows that we achieve lower landmark error compared to state-of-the-art methods although our landmarks are generated by a third-party method. We also outperform existing methods on video data (Table 2c). For video tracking, we detect the face in the first frame and use the bounding box from previous frame for subsequent frames similar to [8]. However, the reconstruction is performed on a per-frame basis to avoid inconsistency due to the choice of random frames.

Facial Motion Retargeting: In order to evaluate whether our tracked facial expression gets correctly retargeted on the target model, we use the expression metric defined as the mean absolute error between the predicted and ground

Table 1. Ablation study. Evaluation of different components of our proposed method in terms of standard evaluation metrics. Note that B and C are obtained with all the loss functions other than the parsing loss and the gradient loss.

Method	3D error (mm)				NME	AUC	Photo error
	BU-3DFE		FWH		AFLW2000-3D	300 VW	CelebA*
	Mean	SD	Mean	SD	Mean	Mean	Mean
3DMM prior (A)	2.21	0.52	2.13	0.49	3.94	0.845	22.76
A + blendshape corrections (B)	2.04	0.41	1.98	0.44	3.73	0.863	22.25
B + albedo corrections (C)	1.88	0.39	1.85	0.41	3.68	0.871	20.13
C + parsing loss (D)	1.67	0.35	1.73	0.37	3.53	0.883	19.49
D + gradient loss (final)	1.61	0.32	1.68	0.35	3.49	0.890	18.91

Table 2. Quantitative evaluation with state-of-the-art methods. (a) 3D reconstruction error (mm) on BU-3DFE and FWH datasets, (b) NME (%) on AFLW2000-3D (divided into 3 groups based on yaw angles), (c) AUC for cumulative error distribution of the 2D landmark error for 300VW test set (divided into 3 scenarios by the authors). Note that higher AUC is better, whereas lower value is better for the other two metrics.

(a)	(b)	(c)

Method	BU-3DFE		FWH	
	Mean	SD	Mean	SD
[45]	1.83	0.39	1.84	0.38
[46]	3.22	0.77	2.19	0.54
[44]	1.74	0.43	1.90	0.40
Ours	**1.61**	**0.32**	**1.68**	**0.35**

Method	[0-30°]	[30-60°]	[60-90°]	Mean
[58]	3.43	4.24	7.17	4.94
[1]	3.15	4.33	5.98	4.49
[12]	2.75	3.51	4.61	3.62
[8]	2.91	3.83	4.94	3.89
Ours	**2.56**	**3.39**	**4.51**	**3.49**

Method	Sc. 1	Sc. 2	Sc. 3
[54]	0.791	0.788	0.710
[56]	0.748	0.760	0.726
[10]	0.847	0.838	0.769
[8]	0.901	0.884	0.842
Ours	**0.913**	**0.897**	**0.861**

Table 3. Quantitative evaluation of retargeting accuracy (measured by the expression metric) on [8] expression test set. Lower error means the model performs better for extreme expressions.

Model	Eye close	Eye wide	Brow raise	Brow anger	Mouth open	Jaw L/R	Lip Roll	Smile	Kiss	Avg
(1) Retargeting [8]	0.117	0.407	0.284	0.405	0.284	0.173	0.325	0.248	0.349	0.288
(2) Ours	0.140	0.389	0.259	0.284	0.208	0.394	0.318	0.121	0.303	**0.268**

truth blendshape coefficients as in [8]. Our evaluation results in Table 3 emphasize the importance of personalized face model in improved retargeting, since [8] uses a generic 3DMM as its face model.

6   Conclusion

We propose a novel deep learning based approach that learns a user-specific face model (expression blendshapes and dynamic albedo maps) and user-independent facial motion disentangled from each other by leveraging in-the-wild videos. Extensive evaluation have demonstrated that our personalized face modeling combined with our novel constraints effectively performs high-fidelity 3D face reconstruction, facial motion tracking, and retargeting of the tracked facial motion from one identity to another.

Acknowledgements:. We thank the anonymous reviewers for their constructive feedback, Muscle Wu, Wenbin Zhu and Zeyu Chen for helping, and Alex Colburn for valuable discussions.

References

1. Bhagavatula, C., Zhu, C., Luu, K., Savvides, M.: Faster than real-time facial alignment: a 3D spatial transformer network approach in unconstrained poses. In: IEEE International Conference on Computer Vision (ICCV) (2017)
2. Blanz, V., Vetter, T.: A morphable model for the synthesis of 3D faces. In: Proceedings SIGGRAPH, pp. 187–194 (1999)
3. Bouaziz, S., Wang, Y., Pauly, M.: Online modeling for realtime facial animation. ACM Trans. Graph. **32**(4), 1–10 (2013)
4. Cao, C., Chai, M., Woodford, O., Luo, L.: Stabilized real-time face tracking via a learned dynamic rigidity prior. ACM Trans. Graph. **37**(6), 1–11 (2018)
5. Cao, C., Hou, Q., Zhou, K.: Displaced dynamic expression regression for real-time facial tracking and animation. ACM Trans. Graph. **33**(4), 1–10 (2014)
6. Cao, C., Weng, Y., Lin, S., Zhou, K.: 3D shape regression for real-time facial animation. ACM Trans. Graph. **32**(4), 1–10 (2013)
7. Cao, C., Wu, H., Weng, Y., Shao, T., Zhou, K.: Real-time facial animation with image-based dynamic avatars. ACM Trans. Graph. **35**(4), 126:1–126:12 (2016)
8. Chaudhuri, B., Vesdapunt, N., Wang, B.: Joint face detection and facial motion retargeting for multiple faces. In: IEEE Conference on Computer Vision and Pattern Recognition (CVPR) (2019)
9. Chung, J.S., Nagrani, A., Zisserman, A.: Voxceleb2: deep speaker recognition. In: INTERSPEECH (2018)
10. Deng, J., Trigeorgis, G., Zhou, Y., Zafeiriou, S.: Joint multi-view face alignment in the wild. arXiv preprint arXiv:1708.06023 (2017)
11. Deng, Y., Yang, J., Xu, S., Chen, D., Jia, Y., Tong, X.: Accurate 3D face reconstruction with weakly-supervised learning: from single image to image set. In: IEEE Conference on Computer Vision and Pattern Recognition Workshop on Analysis and Modeling of Faces and Gestures (CVPRW) (2019)
12. Feng, Y., Wu, F., Shao, X., Wang, Y., Zhou, X.: Joint 3D face reconstruction and dense alignment with position map regression network. In: Ferrari, V., Hebert, M., Sminchisescu, C., Weiss, Y. (eds.) Computer Vision – ECCV 2018. LNCS, vol. 11218, pp. 557–574. Springer, Cham (2018). https://doi.org/10.1007/978-3-030-01264-9_33
13. Garrido, P., Valgaerts, L., Wu, C., Theobalt, C.: Reconstructing detailed dynamic face geometry from monocular video. ACM Trans. Graph. (Proc. SIGGRAPH Asia 2013) **32**(6), 1–10 (2013)
14. Garrido, P., et al.: Reconstruction of personalized 3D face rigs from monocular video. ACM Trans. Graph. **35**(3), 1–15 (2016)
15. Gecer, B., Ploumpis, S., Kotsia, I., Zafeiriou, S.: GANFIT: generative adversarial network fitting for high fidelity 3D face reconstruction. In: IEEE Conference on Computer Vision and Pattern Recognition (CVPR) (2019)

16. Genova, K., Cole, F., Maschinot, A., Sarna, A., Vlasic, D., Freeman, W.T.: Unsupervised training for 3D Morphable model regression. In: IEEE Conference on Computer Vision and Pattern Recognition (CVPR) (2018)
17. Gotardo, P., Riviere, J., Bradley, D., Ghosh, A., Beeler, T.: Practical dynamic facial appearance modeling and acquisition. ACM Trans. Graph. **37**(6), 1–13 (2018)
18. Guo, Y., Zhang, J., Cai, J., Jiang, B., Zheng, J.: CNN-based real-time dense face reconstruction with inverse-rendered photo-realistic face images. IEEE Trans. Pattern Anal. Mach. Intell. (TPAMI) **41**, 1294–1307 (2018)
19. Huynh, L., et al.: Mesoscopic facial geometry inference using deep neural networks. In: IEEE Conference on Computer Vision and Pattern Recognition (CVPR) (2018)
20. Ichim, A.E., Bouaziz, S., Pauly, M.: Dynamic 3D avatar creation from hand-held video input. ACM Trans. Graph. **34**(4), 1–14 (2015)
21. Jackson, A.S., Bulat, A., Argyriou, V., Tzimiropoulos, G.: Large pose 3D face reconstruction from a single image via direct volumetric CNN regression. In: IEEE International Conference on Computer Vision (ICCV) (2017)
22. Jiang, L., Zhang, J., Deng, B., Li, H., Liu, L.: 3D face reconstruction with geometry details from a single image. IEEE Trans. Image Process. **27**(10), 4756–4770 (2018)
23. Kim, H., et al.: Deep video portraits. ACM Trans. Graph. **37**(4), 163:1–163:14 (2018)
24. Kim, H., Zollöfer, M., Tewari, A., Thies, J., Richardt, C., Theobalt, C.: InverseFaceNet: deep single-shot inverse face rendering from a single image. In: IEEE Conference on Computer Vision and Pattern Recognition (CVPR) (2018)
25. Laine, S., et al.: Production-level facial performance capture using deep convolutional neural networks. In: Eurographics Symposium on Computer Animation (2017)
26. Li, H., Weise, T., Pauly, M.: Example-based facial rigging. ACM Trans. Graph. (Proc. SIGGRAPH) **29**(3), 1–6 (2010)
27. Li, H., Yu, J., Ye, Y., Bregler, C.: Realtime facial animation with on-the-fly correctives. ACM Trans. Graph. **32**(4), 1–10 (2013)
28. Li, T., Bolkart, T., Black, M.J., Li, H., Romero, J.: Learning a model of facial shape and expression from 4D scans. ACM Trans. Graph. **36**(6), 1–17 (2017)
29. Lin, J., Yang, H., Chen, D., Zeng, M., Wen, F., Yuan, L.: Face parsing with RoI Tanh-Warping. In: IEEE Conference on Computer Vision and Pattern Recognition (CVPR) (2019)
30. Liu, F., Zhu, R., Zeng, D., Zhao, Q., Liu, X.: Disentangling features in 3D face shapes for joint face reconstruction and recognition. In: IEEE Conference on Computer Vision and Pattern Recognition (CVPR) (2018)
31. Liu, Z., Luo, P., Wang, X., Tang, X.: Deep learning face attributes in the wild. In: International Conference on Computer Vision (ICCV) (2015)
32. Lombardi, S., Saragih, J., Simon, T., Sheikh, Y.: Deep appearance models for face rendering. ACM Trans. Graph. **37**(4), 1–13 (2018)
33. Nagano, K., et al.: paGAN: real-time avatars using dynamic textures. ACM Trans. Graph. **37**(6), 1–12 (2018)
34. Olszewski, K., et al.: Realistic dynamic facial textures from a single image using GANs. In: The IEEE International Conference on Computer Vision (ICCV) (2017)
35. Paysan, P., Knothe, R., Amberg, B., Romdhani, S., Vetter, T.: A 3D face model for pose and illumination invariant face recognition. In: IEEE International Conference on Advanced Video and Signal Based Surveillance (AVSS) (2009)
36. Ribera, R.B., Zell, E., Lewis, J.P., Noh, J., Botsch, M.: Facial retargeting with automatic range of motion alignment. ACM Trans. Graph. **36**(4), 1–12 (2017)

37. Richardson, E., Sela, M., Or-El, R., Kimmel, R.: Learning detailed face reconstruction from a single image. In: IEEE Conference on Computer Vision and Pattern Recognition (CVPR) (2017)
38. Romdhani, S., Vetter, T.: Estimating 3D shape and texture using pixel intensity, edges, specular highlights, texture constraints and a prior. In: IEEE Conference on Computer Vision and Pattern Recognition (CVPR) (2005)
39. Ronneberger, O., Fischer, P., Brox, T.: U-Net: convolutional networks for biomedical image segmentation. In: Navab, N., Hornegger, J., Wells, W.M., Frangi, A.F. (eds.) MICCAI 2015. LNCS, vol. 9351, pp. 234–241. Springer, Cham (2015). https://doi.org/10.1007/978-3-319-24574-4_28
40. Roth, J., Tong, Y., Liu, X.: Adaptive 3D face reconstruction from unconstrained photo collections. In: IEEE Conference on Computer Vision and Pattern Recognition (CVPR) (2016)
41. Sanyal, S., Bolkart, T., Feng, H., Black, M.: Learning to regress 3D face shape and expression from an image without 3D supervision. In: IEEE Conference on Computer Vision and Pattern Recognition (CVPR) (2019)
42. Shen, J., Zafeiriou, S., Chrysos, G.G., Kossaifi, J., Tzimiropoulos, G., Pantic, M.: The first facial landmark tracking in-the-wild challenge: benchmark and results. In: IEEE International Conference on Computer Vision Workshops (ICCVW) (2015)
43. Sumner, R.W., Popović, J.: Deformation transfer for triangle meshes. In: ACM SIGGRAPH (2004)
44. Tewari, A., et al.: FML: face model learning from videos. In: IEEE Conference on Computer Vision and Pattern Recognition (CVPR) (2019)
45. Tewari, A., et al.: Self-supervised multi-level face model learning for monocular reconstruction at over 250 Hz. In: IEEE Conference on Computer Vision and Pattern Recognition (CVPR) (2018)
46. Tewari, A., et al.: MoFA: model-based deep convolutional face autoencoder for unsupervised monocular reconstruction. In: IEEE International Conference on Computer Vision (ICCV) (2017)
47. Thies, J., Zollhöfer, M., Stamminger, M., Theobalt, C., Nießner, M.: Face2Face: real-time face capture and reenactment of RGB videos. In: IEEE Conference on Computer Vision and Pattern Recognition (CVPR) (2016)
48. Tran, A.T., Hassner, T., Masi, I., Medioni, G.: Regressing robust and discriminative 3D morphable models with a very deep neural network. In: IEEE Conference on Computer Vision and Pattern Recognition (CVPR) (2017)
49. Tran, L., Liu, F., Liu, X.: Towards high-fidelity nonlinear 3D face morphable model. In: IEEE Conference on Computer Vision and Pattern Recognition (CVPR) (2019)
50. Tran, L., Liu, X.: Nonlinear 3D face morphable model. In: IEEE Conference on Computer Vision and Pattern Recognition (CVPR) (2018)
51. Vlasic, D., Brand, M., Pfister, H., Popović, J.: Face transfer with multilinear models. In: ACM SIGGRAPH (2005)
52. Weise, T., Bouaziz, S., Li, H., Pauly, M.: Realtime performance-based facial animation. In: ACM SIGGRAPH (2011)
53. Wu, C., Shiratori, T., Sheikh, Y.: Deep incremental learning for efficient high-fidelity face tracking. ACM Trans. Graph. **37**(6), 1–12 (2018)
54. Yang, J., Deng, J., Zhang, K., Liu, Q.: Facial shape tracking via spatio-temporal cascade shape regression. In: IEEE International Conference on Computer Vision Workshops (ICCVW) (2015)
55. Yin, L., Wei, X., Sun, Y., Wang, J., Rosato, M.J.: A 3D facial expression database for facial behavior research. In: IEEE International Conference on Automatic Face and Gesture Recognition (FGR), pp. 211–216 (2006)

56. Zhang, K., Zhang, Z., Li, Z., Qiao, Y.: Joint face detection and alignment using multitask cascaded convolutional networks. IEEE Signal Process. Lett. **23**(10), 1499–1503 (2016)

57. Zhu, W., Wu, H., Chen, Z., Vesdapunt, N., Wang, B.: ReDA: reinforced differentiable attribute for 3D face reconstruction. In: IEEE Conference on Computer Vision and Pattern Recognition (CVPR) (2020)

58. Zhu, X., Lei, Z., Liu, X., Shi, H., Li, S.Z.: Face alignment across large poses: a 3D solution. In: IEEE Conference on Computer Vision and Pattern Recognition (CVPR) (2016)

59. Zollhöfer, M., et al.: State of the art on monocular 3D face reconstruction, tracking, and applications. Comput. Graph. Forum **37**, 523–550 (2018)

Entropy Minimisation Framework for Event-Based Vision Model Estimation

Urbano Miguel Nunes[(✉)] and Yiannis Demiris[(✉)]

Department of Electrical and Electronic Engineering,
Personal Robotics Laboratory, Imperial College London, London, UK
{um.nunes,y.demiris}@imperial.ac.uk

Abstract. We propose a novel Entropy Minimisation (EMin) framework for event-based vision model estimation. The framework extends previous event-based motion compensation algorithms to handle models whose outputs have arbitrary dimensions. The main motivation comes from estimating motion from events directly in 3D space (e.g. events augmented with depth), without projecting them onto an image plane. This is achieved by modelling the event alignment according to candidate parameters and minimising the resultant dispersion. We provide a family of suitable entropy loss functions and an efficient approximation whose complexity is only linear with the number of events (e.g. the complexity does not depend on the number of image pixels). The framework is evaluated on several motion estimation problems, including optical flow and rotational motion. As proof of concept, we also test our framework on 6-DOF estimation by performing the optimisation directly in 3D space.

Keywords: Event-based vision · Optimisation framework · Model estimation · Entropy Minimisation

1 Introduction

Event-based cameras asynchronously report pixel-wise brightness changes, denominated as *events*. This working principle allows them to have clear advantages over standard frame-based cameras, such as: very high dynamic ranges (>120 dB vs. ≈ 60 dB), high bandwidth in the order of millions of events per second, low latency in the order of microseconds. Thus, event-based cameras offer suitable and appealing traits to tackle a wide range of computer vision problems [3,6,7,14,18]. However, due to the different encoding of visual information, new algorithms need to be developed to properly process event-based data.

Significant research has been driven by the benefits of event-based cameras over frame-based ones, including high-speed motion estimation [3,6,7,28], depth estimation [2,10,18,24,26] and high dynamic range tracking [5,14,16,21], with applications in robotics [8,9] and SLAM [11,22], among others. In particular, event-based motion compensation approaches [6,7,16,26] have been successful

Electronic supplementary material The online version of this chapter (https://doi.org/10.1007/978-3-030-58558-7_10) contains supplementary material, which is available to authorized users.

© Springer Nature Switzerland AG 2020
A. Vedaldi et al. (Eds.): ECCV 2020, LNCS 12350, pp. 161–176, 2020.
https://doi.org/10.1007/978-3-030-58558-7_10

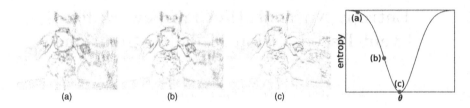

Fig. 1. Entropy measure examples of modelled events according to candidate parameters. (a)-(c) Projected events modelled according to candidate parameters (for visualisation purposes only) and respective entropy measures (right). As a particular case, our framework can also produce motion-corrected images, by minimising the events' entropy. Events generated from a moving DAVIS346B observing an iCub humanoid [15]

in tackling several estimation problems (*e.g.* optical flow, rotational motion and depth estimation). These methods seek to maximise the event alignment of point trajectories on the image plane, according to some loss function of the warped events. By solving this optimisation, the parameters that better compensate for the motion observed can be retrieved, whilst the resultant point trajectories produce sharp, motion-corrected and high contrast images (*e.g.* Fig. 1 (c)). All these event-based motion compensation methods require the warped events to be projected onto an image, where mature computer vision tools can then be applied. For instance, Gallego *et al.* [6] proposed to maximise the variance of the Image of Warped Events (IWE), which is a known measure of image contrast.

We propose a distinct design principle, where event alignment of point trajectories can still be achieved, but without projecting the events onto any particular sub-space. This can be useful in modelling point trajectories directly in 3D space, *e.g.* acquired by fusing event-based cameras with frame-based depth sensors [23]. Thus, instead of using a contrast measure of the IWE as the optimisation loss function, we propose to minimise a dispersion measure in the model's output space, where events are seen as nodes in a graph. Particularly, we will consider the entropy as a principled measure of dispersion. Figure 1 presents an example of events modelled according to candidate parameters. As shown, modelled events that exhibit a lower dispersion also produce motion-corrected images. Although the main motivation for developing the proposed method comes from being able to model events directly in 3D space, in principle, there is no constraint on the number of dimensions the framework can handle and we will use the term *features* throughout the paper to emphasise this. We note that this work is not focused on how to get such features, which is application/model dependent.

Main Contributions: We propose an EMin framework for arbitrary event-based model estimation that only requires event-based data, which can be augmented by other sensors (*e.g.* depth information). This is achieved by modelling the event alignment according to candidate parameters and minimising the resultant model outputs' dispersion according to an entropy measure, without explicitly establishing correspondences between events. In contrast to previous methods, our framework does not need to project the modelled events onto a particular subspace (*e.g.* image plane) and thus can handle model outputs of arbitrary dimensions. We present a family of entropy functions that are suitable for the optimisation framework, as well as efficient approximations whose complexity is only linear with the number of events, offering a valid trade-off between accuracy and computational complexity. We evaluate our framework on several estimation problems, including 6-DOF parameters estimation directly from the 3D spatial coordinates of augmented events.

2   Related Work

Event-based motion compensation algorithms have been proposed to tackle several motion estimation problems, by solving an optimisation procedure, whereby the event alignment of point trajectories is maximised. These estimation problems include rotation [7], depth [18,26], similarity transformations [16].

Gallego *et al.* [6] proposed a Contrast Maximisation (CMax) framework that unifies previous model estimation methods. The framework maximises a contrast measure, by warping a batch of events according to candidate model parameters and projecting them onto the IWE. It is unifying in the sense that the optimisation procedure is independent of the geometric model to be estimated (*e.g.* optical flow, rotation). Several new loss functions were recently added to this framework in [4]. However, the general working principle was not modified and, for brevity reasons, hereinafter, we will only refer to the CMax framework [6], which we review next.

Contrast Maximisation Framework: An event $e = (\mathbf{x}, t, p)$ is characterised by its coordinates $\mathbf{x} = (x, y)^{\mathsf{T}}$, the time-stamp it occurred t and its polarity $p \in \{-1, +1\}$, *i.e.* brightness change. Given a set of events $\mathcal{E} = \{e_k\}_{k=1}^{N_e}$, where N_e is the number of events considered, the CMax framework estimates the model parameters $\boldsymbol{\theta}^*$ that best fit the observed set of events along point trajectories. This is achieved by first warping all events to a reference time-stamp t_{ref} according to the candidate parameters $\boldsymbol{\theta}$

$$e_k \rightarrow e_k' : \quad \mathbf{x}_k' = \mathcal{W}(\mathbf{x}_k, t_k; \boldsymbol{\theta}), \tag{1}$$

where $\mathcal{W}$ is a function that warps the events according to the candidate parameters. Then, the warped events are projected onto the IWE and accumulated according to the Kronecker delta function $\delta_{\mathbf{x}_k}(\mathbf{x})$, where $\delta_{\mathbf{x}_k}(\mathbf{x}) = 1$ when $\mathbf{x} = \mathbf{x}_k$ and 0 otherwise:

$$\mathbf{I}'(\mathbf{x}; \boldsymbol{\theta}) = \sum_k^{N_e} b_k \delta_{\mathbf{x}_k'}(\mathbf{x}), \tag{2}$$

such that if $b_k = p_k$ the polarities are summed, whereas if $b_k = 1$ the number of events are summed instead. In [6,7], the variance of the IWE

$$f(\boldsymbol{\theta}) = \sigma^2 \left(\mathbf{I}'(\mathbf{x}; \boldsymbol{\theta}) \right) = \frac{1}{N_p} \sum_{i,j}^{N_p} \left(i_{ij}' - \mu_{\mathbf{I}'} \right)^2 \tag{3}$$

was proposed to be maximised, where N_p corresponds to the number of pixels of $\mathbf{I'} = (i'_{ij})$ and $\mu_{\mathbf{I'}} = \frac{1}{N_p} \sum_{i,j} i'_{ij}$ is the mean of the IWE. The variance of an image is a well known suitable contrast measure. Thus, the model parameters $\boldsymbol{\theta}^*$ that maximise the contrast of the IWE correspond to the parameters that best fit the event data $\mathcal{E}$, by compensating for the observed motion

$$\boldsymbol{\theta}^* = \arg\max_{\boldsymbol{\theta}} f(\boldsymbol{\theta}). \tag{4}$$

The proposed EMin framework can also be used to estimate arbitrary models, by minimising the entropy of the modelled events. As opposed to the CMax framework, whose optimisation is constrained to the image space, our framework can estimate models whose outputs have arbitrary dimensions, since the entropy measure can be computed on a space of arbitrary dimensions.

Related similarity measures were proposed by Zhu *et al.* [25] for feature tracking, where events were explicitly grouped into features based on the optical flow predicted as an expectation maximisation over all events' associations. Distinctively, we avoid the data association problem entirely, since the entropy already implicitly measures the similarity between event trajectories.

3   Entropy Minimisation Framework

The CMax framework [6] requires the warped events to be projected onto a subspace, *i.e.* the IWE. This means that the optimisation framework is constrained to geometric modelling and does not handle more than 2D features, *i.e.* IWE pixel coordinates. Thus, we propose an EMin framework for event-based model estimation that does not require projection onto any sub-space. The proposed framework relies on a family of entropy loss functions which measure the dispersion of features related to events in a feature space (Sect. 3.2). The EMin procedure complexity is quadratic with the number of events, which will motivate an efficient Approximate Entropy Minimisation (AEMin) solution, whose complexity is only linear with the number of events (Sect. 3.3).

3.1   Intuition from Pairwise Potentials

Events are optimally modelled if in the temporal-image space they are aligned along point trajectories that reveal strong edge structures [6,7]. This means that events become more concentrated (on the edges) and thus, their average pairwise distance is minimised (*e.g.* the average distance is zero if all events are projected onto the same point). This can be measured by using fully connected Conditional Random Field (CRF) models [13], which are a standard tool in computer vision, *e.g.* with applications to semantic image labelling [12]. An image is represented as a graph, each pixel is a node, and the aim is to obtain a labelling for each pixel that maximises the corresponding Gibbs energy

$$E_{\text{Gibbs}} = \sum_{i}^{N_p} \psi_u(l_i) + \sum_{i,j}^{N_p} \psi_p(l_i, l_j), \tag{5}$$

where ψ_u is the unary potential computed independently for each pixel and ψ_p is the pairwise potential. The unary potential incorporates descriptors of a pixel, *e.g.* texture and color, whereas the pairwise potential can be interpreted as a similarity measure between pixels, which can take the form

$$\psi_p(l_i, l_j) = \xi(l_i, l_j) \sum_{m}^{M} \omega_m \mathcal{K}_m(\mathbf{f}_i, \mathbf{f}_j), \tag{6}$$

where ξ is some label compatibility function, ω_m represents the weight of the contribution of the kernel $\mathcal{K}_m$ and $\mathbf{f}$ is a d-dimensional feature vector in an arbitrary feature space.

This model can be used to measure the similarity between events, by representing each event as a node of a graph, and then maximising the similarity measure. According to Eq. (6), this is equivalent to maximising the pairwise potentials between all events in the case where we consider one kernel ($M=1$) and the feature vector corresponds to the modelled event coordinates in the IWE ($\mathbf{f}_k = \mathbf{x}'_k$). This can be formalised by defining the *Potential* energy which we seek to minimise according to parameters $\boldsymbol{\theta}$ as

$$P(\mathbf{f}; \boldsymbol{\theta}) := -\frac{1}{N_e^2} \sum_{i,j}^{N_e} \mathcal{K}_{\boldsymbol{\Sigma}}(\mathbf{f}_i, \mathbf{f}_j; \boldsymbol{\theta}). \tag{7}$$

The *Potential* energy measures the average dispersion of events relative to each other in the feature space, *e.g.* if the events are more concentrated, then Eq. (7) is minimised. Based on this formulation, the modelled events are not required to be projected onto any particular space and we can handle feature vectors of arbitrary dimensions. For easy exposition, we will consider the d-dimensional multivariate Gaussian kernel parameterised by the covariance matrix $\boldsymbol{\Sigma}$

$$\mathcal{K}_{\boldsymbol{\Sigma}}(\mathbf{f}_i, \mathbf{f}_j; \boldsymbol{\theta}) = \frac{\exp\left(-\frac{1}{2}(\mathbf{f}_i - \mathbf{f}_j)^{\mathsf{T}} \boldsymbol{\Sigma}^{-1}(\mathbf{f}_i - \mathbf{f}_j)\right)}{(2\pi)^{\frac{d}{2}} |\boldsymbol{\Sigma}|^{\frac{1}{2}}}. \tag{8}$$

3.2 Description of Framework

Another well-known measure of dispersion is the entropy, which is at the core of the proposed optimisation framework. By interpreting the kernel $\mathcal{K}_{\boldsymbol{\Sigma}}(\mathbf{f}_i, \mathbf{f}_j; \boldsymbol{\theta})$ as a conditional probability distribution $p(\mathbf{f}_i | \mathbf{f}_j, \boldsymbol{\theta})$, we can consider the corresponding *Sharma-Mittal* entropies [20]

$$H_{\alpha,\beta}(\mathbf{f}; \boldsymbol{\theta}) := \frac{1}{1-\beta} \left[\left(\frac{1}{N_e^2} \sum_{i,j}^{N_e} \mathcal{K}_{\boldsymbol{\Sigma}}(\mathbf{f}_i, \mathbf{f}_j; \boldsymbol{\theta})^{\alpha} \right)^{\gamma} - 1 \right], \tag{9}$$

Algorithm 1. Entropy Minimisation Framework

Input: Set of events $\mathcal{E} = \{e_k\}_k^{N_e}$.
Output: Estimated model parameters $\boldsymbol{\theta}^*$.
Procedure:
1: Initialise the candidate model parameters $\boldsymbol{\theta}$.
2: Model events according to parameters $\boldsymbol{\theta}$, Eq. (13).
3: Compute the entropy $E(\mathbf{f}; \boldsymbol{\theta})$ (Sect. 3.2 and 3.3).
4: Find the best parameters $\boldsymbol{\theta}^*$ by minimising the entropy $f(\boldsymbol{\theta})$, Eq. (14).

---

where $\alpha > 0$, $\alpha, \beta \neq 1$ and $\gamma = \frac{1-\beta}{1-\alpha}$. This family of entropies tends in limit cases to *Rényi* R_α ($\beta \to 1$), *Tsallis* T_α ($\beta \to \alpha$) and *Shannon* S ($\alpha, \beta \to 1$) entropies:

$$R_\alpha(\mathbf{f}; \boldsymbol{\theta}) := \frac{1}{1-\alpha} \log \left(\frac{1}{N_e^2} \sum_{i,j}^{N_e} \mathcal{K}_{\boldsymbol{\Sigma}}(\mathbf{f}_i, \mathbf{f}_j; \boldsymbol{\theta})^\alpha \right), \tag{10}$$

$$T_\alpha(\mathbf{f}; \boldsymbol{\theta}) := \frac{1}{1-\alpha} \left(\frac{1}{N_e^2} \sum_{i,j}^{N_e} \mathcal{K}_{\boldsymbol{\Sigma}}(\mathbf{f}_i, \mathbf{f}_j; \boldsymbol{\theta})^\alpha - 1 \right), \tag{11}$$

$$S(\mathbf{f}; \boldsymbol{\theta}) := \frac{1}{N_e^2} \sum_{i,j}^{N_e} \mathcal{K}_{\boldsymbol{\Sigma}}(\mathbf{f}_i, \mathbf{f}_j; \boldsymbol{\theta}) \log \mathcal{K}_{\boldsymbol{\Sigma}}(\mathbf{f}_i, \mathbf{f}_j; \boldsymbol{\theta}). \tag{12}$$

Gallego *et al.* [4] proposed a loss function based on the image entropy, which measures the dispersion over the distribution of accumulated events in the IWE. Instead, we propose to measure the dispersion over the distribution of the modelled events in the feature space (by considering the distributions are given by each modelled event). The proposed measure is actually closer in spirit to the spatial autocorrelation loss functions proposed in [4].

The proposed framework then follows a similar flow to that of the CMax framework [6], which is summarised in Algorithm 1. A set of events $\mathcal{E}$ is modelled according to candidate parameters $\boldsymbol{\theta}$

$$\mathbf{f}_k = \mathcal{M}(e_k; \boldsymbol{\theta}), \tag{13}$$

where $\mathbf{f}_k$ is the resultant feature vector in a d-dimensional feature space associated with event e_k and $\mathcal{M}$ is a known model. Then, we find the best model parameters $\boldsymbol{\theta}^*$, by minimising the entropy in the feature space

$$\boldsymbol{\theta}^* = \arg\min_{\boldsymbol{\theta}} f(\boldsymbol{\theta}) = \arg\min_{\boldsymbol{\theta}} E(\mathbf{f}; \boldsymbol{\theta}), \tag{14}$$

where $E(\mathbf{f}; \boldsymbol{\theta})$ is one of the proposed loss functions.

3.3   Efficient Approximation

The complexity of the proposed framework is quadratic with the number of events and thus is more computationally demanding than the CMax framework [6], which is linear with the number of events and the number of pixels of the IWE. To overcome the increased complexity, we propose an efficient approximation to compute the entropy functions. This is achieved by approximating the kernel $\mathcal{K}_{\Sigma}$ (Eq. 8) with a truncated version $\mathbf{K}_{\Sigma}$, where values beyond certain standard deviations are set to zero, and then convolve each feature vector with $\mathbf{K}_{\Sigma}$. This idea is illustrated in Fig. 2. To achieve linear complexity with the number of events, we asynchronously convolve each feature vector using the event-based convolution method proposed by Scheerlinck et al. [19].

For simplicity, assuming the kernel $\mathbf{K}_{\Sigma}$ has size κ^d, evaluating the approximate functions has a complexity of $O(N_e \kappa^d)$, which is linear with the number of events. The complexity may still be exponential with the number of dimensions, if we do not consider efficient high-dimensional convolution operations that reduce the computational complexity to become linear with the number of dimensions, e.g. [1]. Although we need to discretise the feature space, the computational complexity of the AEMin approach is independent of the actual number of discretised bins. In contrast, the complexity of the CMax [6] framework is also linear with the number of discretised bins (e.g. number of image pixels N_p). This means that although directly extending the CMax framework to handle higher dimensions is possible, in practise it would be inefficient and not scalable, without considering sophisticated data structures.

The *Approximate Potential* energy can be expressed as

$$\tilde{P}(\mathbf{f}; \boldsymbol{\theta}) := -\frac{1}{N_e^2} \sum_{k,l}^{N_e} \mathbf{K}_{\Sigma} * \delta_{\mathbf{f}_k}(\mathbf{f}_l), \tag{15}$$

where each feature $\mathbf{f}_k$ is convolved with the truncated kernel $\mathbf{K}_{\Sigma}$ in the feature space. Similarly, the *Sharma-Mittal*, *Rényi* and *Tsallis* entropies can be approximately expressed based on the kernel $\mathbf{K}_{\Sigma}^{\alpha}$ and the *Shannon* entropy can be approximately expressed based on the kernel $\mathbf{K}_{\Sigma} \odot \log \mathbf{K}_{\Sigma}$:

$$\tilde{H}_{\alpha,\beta}(\mathbf{f}; \boldsymbol{\theta}) := \frac{1}{1-\beta} \left[\left(\frac{1}{N_e^2} \sum_{k,l}^{N_e} \mathbf{K}_{\Sigma}^{\alpha} * \delta_{\mathbf{f}_k}(\mathbf{f}_l) \right)^{\gamma} - 1 \right], \tag{16}$$

$$\tilde{R}_{\alpha}(\mathbf{f}; \boldsymbol{\theta}) := \frac{1}{1-\alpha} \log \left(\frac{1}{N_e^2} \sum_{k,l}^{N_e} \mathbf{K}_{\Sigma}^{\alpha} * \delta_{\mathbf{f}_k}(\mathbf{f}_l) \right), \tag{17}$$

$$\tilde{T}_{\alpha}(\mathbf{f}; \boldsymbol{\theta}) := \frac{1}{1-\alpha} \left(\frac{1}{N_e^2} \sum_{k,l}^{N_e} \mathbf{K}_{\Sigma}^{\alpha} * \delta_{\mathbf{f}_k}(\mathbf{f}_l) - 1 \right), \tag{18}$$

$$\tilde{S}(\mathbf{f}; \boldsymbol{\theta}) := \frac{1}{N_e^2} \sum_{k,l}^{N_e} (\mathbf{K}_\Sigma \odot \log \mathbf{K}_\Sigma) * \delta_{\mathbf{f}_k}(\mathbf{f}_l), \tag{19}$$

where $\odot$ represents the Hadamard product and log represents the natural logarithm applied element-wise.

4   Experiments and Results

In this section, we test our framework to estimate several models by providing qualitative and quantitative assessments. In every experiment, we assume that the camera is calibrated and lens distortion has been removed. We also assume that each batch of events spans a short time interval, such that the model parameters can be considered constants (which is reasonable, since events can be triggered with a microsecond temporal resolution). For brevity, we will only consider the *Tsallis* entropy (11) and respective approximation (18) where $\alpha = 2$. Note that α and β are not tunable parameters since each one specifies an entropy.

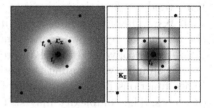

Fig. 2. In the EMin approach, all features pairwise distances are computed (left). In the AEMin approach, the feature space is discretised and each feature $\mathbf{f}_k$ is convolved with a truncated kernel $\mathbf{K}_\Sigma$ (right)

The proposed AEMin requires that we discretise the feature space and perform convolution operations. We use bilinear interpolation/voting for each feature vector $\mathbf{f}_k$ to update the nearest bins, as suggested in [7]. We asynchronously convolve each feature vector, using the event-based asynchronous convolution method proposed by Scheerlinck *et al.* [19]. This convolution method was also used in the custom implementation of the CMax framework [6] and the decay factor was set to 0, to emulate a synchronous convolution over the entire space. We use a 3×3 Gaussian truncated kernel with 1 bin as standard deviation.

Our optimisation framework was implemented in C++ and we used the CG-FR algorithm of the scientific library GNU-GSL for the optimisation and the Auto-Diff module of the Eigen library for the (automatic) derivatives computation.[1] Additional models are provided in the supplementary material, as well as additional results (for more entropies), and practical considerations.

4.1   Motion Estimation in the Image Plane

We tested our framework using sequences from the dataset provided by Mueggler *et al.* [17]. The dataset consists of real sequences acquired by a DAVIS240 camera, each with approximately one minute duration and increasing motion magnitude. A motion-capture system provides the camera's pose at 200 Hz and a built-in Inertial Measurement Unit (IMU) provides acceleration and angular velocity measurements at 1000 Hz.

---

[1] Code publicly available: www.imperial.ac.uk/personal-robotics.

Optical Flow: The optical flow model has 2-DOF and can be parameterised by the 2D linear velocity on the image plane $\boldsymbol{\theta} = (u_x, u_y)^\mathsf{T}$, being expressed as

$$\mathbf{f}_k = \mathcal{M}(e_k; \boldsymbol{\theta}) = \mathbf{x}_k - \Delta t_k \boldsymbol{\theta}, \tag{20}$$

where $\mathbf{f}_k = \mathbf{x}'_k$ represents the image coordinates of the warped events, according to the notation in [6], and $\Delta t_k = t_k - t_\text{ref}$. For our framework, this is just a special case, where the feature vector $\mathbf{f}_k$ is 2-dimensional.

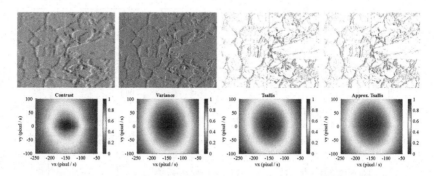

Fig. 3. Optical flow estimation between frames 17 and 18 of the `poster_translation` sequence [17]. Top row, from left to right: Original events projected onto the IWE, where the flow is dominated by the horizontal component $\left(\boldsymbol{\theta}^* \approx (-150, 0)^\mathsf{T}\right)$; Motion compensated events projected onto the IWE, according to the CMax framework [6] $\left(\boldsymbol{\theta}^* = (-150.3, 3.7)^\mathsf{T}\right)$, *Tsallis* entropy $\left(\boldsymbol{\theta}^* = (-150.8, 5.6)^\mathsf{T}\right)$ and corresponding approximate entropy $\left(\boldsymbol{\theta}^* = (-150.7, 3.8)^\mathsf{T}\right)$, respectively. Bottom row, from left to right: IWE contrast, Variance [6], *Tsallis* entropy and respective approximation profiles in function of the optical flow parameters, respectively

Figure 3 shows the results of the optical flow estimation between frames 17 and 18 of the `poster_translation` sequence [17]. Similar results are obtained if we use the CMax framework [6]. Distinctively, however, by using an entropy measure, our framework minimises the modelled events' dispersion in a 2D feature space. The profiles of the *Tsallis* entropy and corresponding approximation in function of the optical flow parameters are also presented (two most right plots in the bottom row). We can see that both profiles are similar and the optical flow parameters are correctly estimated.

Rotational Motion: The rotational model has 3-DOF and can be parameterised by the 3D angular velocity $\boldsymbol{\theta} = (w_x, w_y, w_z)^\mathsf{T}$, being expressed as

$$\mathbf{f}_k = \mathcal{M}(e_k; \boldsymbol{\theta}) \propto \mathcal{R}^{-1}(t_k; \boldsymbol{\theta}) \begin{pmatrix} \mathbf{x}_k \\ 1 \end{pmatrix}, \tag{21}$$

where the rotation matrix $\mathcal{R} \in SO(3)$ can be written as a matrix exponential map of the angular velocity parameters $\boldsymbol{\theta}$ as

$$\mathcal{R}(t_k; \boldsymbol{\theta}) = \exp\left(\Delta t_k \hat{\boldsymbol{\theta}}\right), \tag{22}$$

where $\hat{\boldsymbol{\theta}} \in \mathbb{R}^{3 \times 3}$ is the associated skew-symmetric matrix.

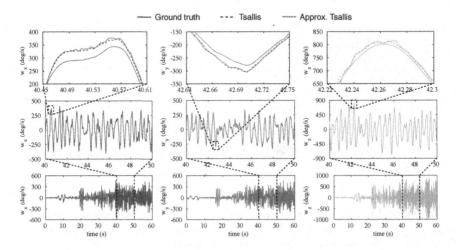

Fig. 4. Comparison of the angular velocity estimated against the ground truth from IMU measurements on the `poster_rotation` sequence [17], considering batches of 20000 events. Bottom row: Whole sequence. Middle and top rows: Zoomed-in plots of the corresponding bounded regions. Both angular velocity profiles are almost identical to the ground truth

Figure 4 presents a comparison of the angular velocity parameters estimated by the proposed framework using the *Tsallis* entropy and the corresponding approximation against the ground truth provided by IMU measurements on the `poster_rotation` sequence [17]. Qualitatively, both angular velocity profiles are identical to the ground truth, even during peak excursions of approximately ± 940 deg/s, which correspond to rotations of more than 2.5 turns per second.

A more detailed quantitative performance comparison is presented in Table 1. We compare the *Tsallis* and corresponding approximate entropy functions in terms of errors for each angular velocity component and the respective standard deviation and RMS, considering batches of 20000 events, on the `boxes_rotation` and `poster_rotation` sequences [17]. The exact entropy function achieves the best overall results, while the proposed efficient approximate entropy achieves competitive performance.

Table 1. Accuracy comparison on the `boxes_rotation` and `poster_rotation` sequences [17]. The angular velocity errors for each component $(e_{w_x}, e_{w_y}, e_{w_z})$, their standard deviation (σ_{e_w}) and RMS are presented in deg/s, w.r.t. IMU measurements, considering batches of 20000 events. The RMS error compared to the maximum excursions are also presented in percentage (RMS %), as well as the absolute and relative maximum errors. The best value per column is highlighted in bold

Sequence	Function	e_{w_x}	e_{w_y}	e_{w_z}	σ_{e_w}	RMS	RMS %	max	max %
boxes_rotation	Variance (3) [6]	7.49	6.87	**7.53**	10.73	10.80	1.15	65.05	6.92
	Tsallis (11)	**7.43**	**6.53**	8.21	**9.75**	**9.91**	**1.06**	**45.76**	**4.87**
	Approx. Tsallis (18)	7.93	7.57	7.87	10.25	10.33	1.10	47.40	5.04
poster_rotation	Variance (3) [6]	**13.26**	8.73	8.73	14.60	14.65	1.56	**62.41**	**6.64**
	Tsallis (11)	13.40	**8.66**	**7.88**	**13.39**	**13.65**	**1.45**	67.34	7.16
	Approx. Tsallis (18)	14.31	9.16	8.34	14.16	14.22	1.51	64.68	6.88

Motion in Planar Scenes: Motion estimation in planar scenes can be achieved by using the homography model. This model has 8-DOF and can be parameterised by the 3D angular velocity $\mathbf{w} = (w_x, w_y, w_z)^{\mathsf{T}}$, the up to scale 3D linear velocity $\bar{\mathbf{v}} = \mathbf{v}/s = (\bar{v}_x, \bar{v}_y, \bar{v}_z)^{\mathsf{T}}$ and the normalised 3D normal vector $\mathbf{n} = (n_x, n_y, n_z)^{\mathsf{T}}$ of the inducing plane $\boldsymbol{\pi} = (\mathbf{n}^{\mathsf{T}}, s)^{\mathsf{T}}$, being expressed as

Fig. 5. Motion estimation in planar scenes. (Left) Original events projected. (Right) Modelled events projected onto the image plane

$$\mathbf{f}_k = \mathcal{M}(e_k; \boldsymbol{\theta}) \propto \mathcal{H}^{-1}(t_k; \boldsymbol{\theta})\begin{pmatrix} \mathbf{x}_k \\ 1 \end{pmatrix}, \quad (23)$$

where $\boldsymbol{\theta} = (\mathbf{w}^{\mathsf{T}}, \bar{\mathbf{v}}^{\mathsf{T}}, \mathbf{n}^{\mathsf{T}})^{\mathsf{T}}$ are the model parameters and the homography matrix $\mathcal{H}$ can be written in function of the model parameters $\boldsymbol{\theta}$ as

$$\mathcal{H}(t_k; \boldsymbol{\theta}) = \mathcal{R}(t_k; \mathbf{w}) - \Delta t_k \bar{\mathbf{v}} \mathbf{n}^{\mathsf{T}}. \quad (24)$$

Figure 5 shows the results of motion estimation in a planar scene, according to the homography model, between frames 673 and 674 of the `poster_6dof` sequence [17]. The following parameters were obtained: $\mathbf{w} = (0.81, -0.11, -1.47)^{\mathsf{T}}$, $\bar{\mathbf{v}} = (-0.19, 0.12, 1.73)^{\mathsf{T}}$, $\mathbf{n} = (-0.10, -0.14, -0.99)^{\mathsf{T}}$.

4.2 Motion Estimation in 3D Space

As proof of concept, we synthetically generated events from the corners of a moving cube in 3D space and used our framework to estimate its 6-DOF motion parameters, without projecting the events onto the image plane. We also tested our framework on sequences from the dataset provided by Zhu et al. [27]. The dataset consists of real sequences acquired by a set of different sensors for event-based 3D perception. The set of sensors includes an event-based stereo system of two DAVIS346B cameras and a frame-based stereo system of two Aptina MT9V034 cameras at 20fps, which provides the depth of the events generated.

6-DOF: The 6-DOF can be parameterised by the 3D angular velocity $\mathbf{w} = (w_x, w_y, w_z)^\mathsf{T}$ and the 3D linear velocity $\mathbf{v} = (v_x, v_y, v_z)^\mathsf{T}$, being expressed as

$$\mathbf{f}_k = \mathcal{M}(e_k; \boldsymbol{\theta}) = \mathcal{S}^{-1}(t_k; \boldsymbol{\theta})\mathbf{z}_k, \tag{25}$$

where $\mathbf{z}_k$ and $\mathbf{f}_k$ represent the 3D coordinates of the original and modelled events, respectively. The matrix exponential map $\mathcal{S} \in SE(3)$ encodes the rigid body transformation and can be written in function of $\boldsymbol{\theta} = (\mathbf{w}^\mathsf{T}, \mathbf{v}^\mathsf{T})^\mathsf{T}$ as

$$\mathcal{S}(t_k; \boldsymbol{\theta}) = \exp\left[\Delta t_k \begin{pmatrix} \hat{\mathbf{w}} & \mathbf{v} \\ \mathbf{0}^\mathsf{T} & 0 \end{pmatrix}\right], \tag{26}$$

where $\hat{\mathbf{w}} \in \mathbb{R}^{3\times3}$ is the associated skew-symmetric matrix.

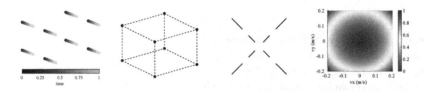

Fig. 6. 6-DOF estimation. From left to right: Events generated from the corners of a synthetic cube moving with $v_z = -1$ (m/s) in 3D space. Modelled events according to the optimal parameters $\mathbf{w} = (0,0,0)^\mathsf{T}$, $\mathbf{v} = (0,0,-1)^\mathsf{T}$ (dashed lines included for better visualisation only). Original events projected onto the image plane. Projected *Tsallis* entropy profile at $\mathbf{w} = (0,0,0)^\mathsf{T}$, $v_z = -1$, in function of v_x and v_y

Figure 6 illustrates a cube moving along the z-axis with $v_z = -1$ (m/s), whose corners generate events in 3D space. Our framework can accurately estimate the parameters of the moving cube, by modelling the events' trajectory according to the 6-DOF model. The CMax framework [6] can not estimate these parameters directly in 3D, because it requires the events to be projected onto the image plane. We also present the projected entropy profile in function of the velocities in the x and y direction, at $\mathbf{w} = (0,0,0)^\mathsf{T}$ and $v_z = -1$ (m/s). The profile exhibits a minimum at $\mathbf{v} = (0,0,-1)^\mathsf{T}$, corresponding to the motion parameters.

In Fig. 7, we compare the original and modelled events from a moving 3D scene of the `indoor_flying1` sequence [27] (for illustration purposes, we consider the first 75000 events at the 24 s timestamp). We can retrieve the 6-DOF parameters by minimising the modelled augmented events' dispersion directly in 3D space, while also aligning the events to a reference frame.

We also present a quantitative evaluation in Table 2 on three sequences from the dataset provided by Zhu *et al.* [27]. Both the exact and approximate entropies achieve similar errors. The 6-DOF parameters were estimated from highly corrupted data since the events' 3D coordinates were obtained from depth measurements at 20fps. Moreover, in the `outdoor_driving_night1` sequence, the events generated are significantly influenced by several other relative motions, *e.g.* due to other cars moving in the field-of-view. Our framework is capable of coping with noise, provided it is not the predominant factor.

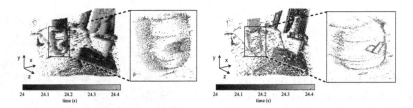

Fig. 7. 6-DOF estimation on the `indoor_flying1` sequence [27]. The trajectory of the original augmented events (left) can be modelled in 3D to retrieve the 6-DOF parameters of the moving camera, while also aligning the events (right)

Table 2. Accuracy comparison on the `indoor_flying1` and `indoor_flying4` and `outdoor_driving_night1` sequences [27]. The average angular and linear velocity errors (e_w, e_v), their standard deviation $(\sigma_{e_w}, \sigma_{e_v})$ and RMS $(\text{RMS}_{e_w}, \text{RMS}_{e_v})$ are presented in deg/s and m/s, respectively. The RMS errors compared to the maximum excursions are also presented in percentage $(\text{RMS}_{e_w} \%, \text{RMS}_{e_v} \%)$

Sequence	Function	e_w	e_v	σ_{e_w}	σ_{e_v}	RMS_{e_w}	RMS_{e_v}	RMS_{e_w} %	RMS_{e_v} %
indoor_flying1	*Tsallis* (11)	2.22	0.13	2.87	0.16	3.03	0.17	7.47	19.88
	Approx. Tsallis (18)	2.08	0.10	2.40	0.12	2.70	0.12	6.66	14.80
indoor_flying4	*Tsallis* (11)	4.08	0.25	5.16	0.30	5.12	0.30	20.53	16.38
	Approx. Tsallis (18)	4.43	0.23	5.30	0.29	5.56	0.30	22.28	16.35
outdoor_driving_night1	*Tsallis* (11)	4.18	1.73	14.73	1.71	14.85	2.22	3.98	21.85
	Approx. Tsallis (18)	7.27	1.82	17.97	1.94	17.97	2.41	4.81	23.71

In Table 3, we compare our framework to a deep learning method [28] that predicts the egomotion of a moving camera from events, in terms of relative pose error $(\text{RPE} = \arccos \frac{t_{\text{pred}} \cdot t_{\text{gt}}}{\|t_{\text{pred}}\| \cdot \|t_{\text{gt}}\|})$ and relative rotation error $(\text{RRE} = \left\| \text{logm} \left(\mathbf{R}^{\mathcal{T}}_{\text{pred}} \mathbf{R}_{\text{gt}} \right) \right\|$, where logm is the

Table 3. Quantitative evaluation of our framework compared to Zhu *et al.* [28] on the `outdoor_driving_day1` sequence [27]

	ARPE (deg)	ARRE (rad)
Approx. Tsallis (18)	**4.44**	**0.00768**
Zhu *et al.* [28]	7.74	0.00867

matrix log). Our framework compares favourably possibly because the deep learning method also estimates the depth.

4.3 Discussion and Limitations

We have demonstrated that the proposed EMin framework can be used to tackle several common computer vision estimation problems. The EMin approach can achieve better performance in rotational motion estimation. The proposed AEMin approach can still achieve competitive performance, whilst being computationally more efficient. Nevertheless, we consider that the framework's capability of handling models whose outputs have arbitrary dimensions is its most relevant property. As proof of concept, we showed that our framework can estimate the 6-DOF of a moving cube in 3D space. The quantitative tests on three sequences support the potential of the proposed framework to estimate the parameters of models with arbitrary output dimensions.

Our framework estimates the model parameters by minimising an entropy measure of the resultant modelled events. In the limit, the entropy is minimised if all events are mapped onto the same point, which in practise can occur by trying to estimate up-to scale 3D linear velocities (*e.g.* homography model). Figure 8 exemplifies this situation, where we show the original events and the modelled events according to the estimated parameters, in the first and second rows, respectively. The CMax framework [6] exhibits a similar limitation since the contrast of the IWE is also maximised if all events are warped onto a line.

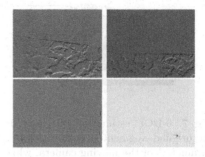

Fig. 8. Estimation failure cases. Original events (top row) and respective modelled events according to the best estimated parameters (bottom row), using (left) the variance [6] and (right) the *Approx. Tsallis* entropy, Eq. (18)

5 Conclusion

We have proposed a framework for event-based model estimation that can handle arbitrary dimensions. Our approach takes advantage of the benefits of the event-based cameras while allowing to incorporate additional sensory information, by augmenting the events, under the same framework. Additionally, since it can handle features of arbitrary dimensions, it can be readily applied to estimation problems that have output features of 4 or more dimensions, which we leave for future work. Thus, the proposed framework can be seen as an extension of previous motion compensation approaches. The exact EMin approach achieves the best performance, although its complexity is quadratic with the number of events. This motivated the proposed AEMin approach that achieves competitive performance while being computationally more efficient.

Acknowledgements. Urbano Miguel Nunes was supported by the Portuguese Foundation for Science and Technology under Doctoral Grant with reference SFRH/BD/130732/2017. Yiannis Demiris is supported by a Royal Academy of Engineering Chair in Emerging Technologies. This research was supported in part by EPRSC Grant EP/S032398/1. The authors thank the reviewers for their insightful feedback, and the members of the Personal Robotics Laboratory at Imperial College London for their support.

References

1. Adams, A., Baek, J., Davis, M.A.: Fast high-dimensional filtering using the permutohedral lattice. Comput. Graph. Forum **29**(2), 753–762 (2010)
2. Andreopoulos, A., Kashyap, H.J., Nayak, T.K., Amir, A., Flickner, M.D.: A low power, high throughput, fully event-based stereo system. In: IEEE Conference on Computer Vision and Pattern Recognition, pp. 7532–7542 (2018)

3. Bardow, P., Davison, A.J., Leutenegger, S.: Simultaneous optical flow and intensity estimation from an event camera. In: IEEE Conference on Computer Vision and Pattern Recognition, pp. 884–892 (2016)

4. Gallego, G., Gehrig, M., Scaramuzza, D.: Focus is all you need: loss functions for event-based vision. In: IEEE Conference on Computer Vision and Pattern Recognition, pp. 12280–12289 (2019)

5. Gallego, G., Lund, J.E., Mueggler, E., Rebecq, H., Delbruck, T., Scaramuzza, D.: Event-based, 6-DOF camera tracking from photometric depth maps. IEEE Trans. Pattern Anal. Mach. Intell. 40(10), 2402–2412 (2018)

6. Gallego, G., Rebecq, H., Scaramuzza, D.: A unifying contrast maximization framework for event cameras, with applications to motion, depth, and optical flow estimation. In: IEEE Conference on Computer Vision and Pattern Recognition, pp. 3867–3876 (2018)

7. Gallego, G., Scaramuzza, D.: Accurate angular velocity estimation with an event camera. IEEE Rob. Autom. Lett. 2(2), 632–639 (2017)

8. Glover, A., Bartolozzi, C.: Robust visual tracking with a freely-moving event camera. In: IEEE/RSJ International Conference on Intelligent Robots and Systems, pp. 3769–3776 (2017)

9. Glover, A., Vasco, V., Bartolozzi, C.: A controlled-delay event camera framework for on-line robotics. In: IEEE International Conference on Robotics and Automation, pp. 2178–2183 (2018)

10. Haessig, G., Berthelon, X., Ieng, S.H., Benosman, R.: A spiking neural network model of depth from defocus for event-based neuromorphic vision. Sci. Rep. 9(1), 3744 (2019)

11. Kim, H., Leutenegger, S., Davison, A.J.: Real-time 3D reconstruction and 6-DoF tracking with an event camera. In: Leibe, B., Matas, J., Sebe, N., Welling, M. (eds.) ECCV 2016. LNCS, vol. 9910, pp. 349–364. Springer, Cham (2016). https://doi.org/10.1007/978-3-319-46466-4_21

12. Krähenbühl, P., Koltun, V.: Efficient inference in fully connected CRFs with Gaussian edge potentials. In: Advances in Neural Information Processing Systems, pp. 109–117 (2011)

13. Lafferty, J., McCallum, A., Pereira, F.C.: Conditional random fields: probabilistic models for segmenting and labeling sequence data. In: International Conference on Machine Learning, pp. 282–289 (2001)

14. Manderscheid, J., Sironi, A., Bourdis, N., Migliore, D., Lepetit, V.: Speed invariant time surface for learning to detect corner points with event-based cameras. In: IEEE Conference on Computer Vision and Pattern Recognition, pp. 10245–10254 (2019)

15. Metta, G., et al.: The iCub humanoid robot: an open-systems platform for research in cognitive development. Neural Netw. 23(8–9), 1125–1134 (2010)

16. Mitrokhin, A., Fermüller, C., Parameshwara, C., Aloimonos, Y.: Event-based moving object detection and tracking. In: IEEE/RSJ International Conference on Intelligent Robots and Systems, pp. 1–9 (2018)

17. Mueggler, E., Rebecq, H., Gallego, G., Delbruck, T., Scaramuzza, D.: The event-camera dataset and simulator: Event-based data for pose estimation, visual odometry, and slam. Int. J. Rob. Res. 36(2), 142–149 (2017)

18. Rebecq, H., Gallego, G., Mueggler, E., Scaramuzza, D.: EMVS: event-based multi-view stereo-3D reconstruction with an event camera in real-time. Int. J. Comput. Vis. 126(12), 1394–1414 (2018)

19. Scheerlinck, C., Barnes, N., Mahony, R.: Asynchronous spatial image convolutions for event cameras. IEEE Rob. Autom. Lett. 4(2), 816–822 (2019)

20. Sharma, B.D., Mittal, D.P.: New non-additive measures of inaccuracy. J. Math. Sci. **10**, 120–133 (1975)
21. Valeiras, D.R., Lagorce, X., Clady, X., Bartolozzi, C., Ieng, S.H., Benosman, R.: An asynchronous neuromorphic event-driven visual part-based shape tracking. IEEE Trans. Neural Netw. Learn. Syst. **26**(12), 3045–3059 (2015)
22. Vidal, A.R., Rebecq, H., Horstschaefer, T., Scaramuzza, D.: Ultimate SLAM? Combining events, images, and IMU for robust visual SLAM in HDR and high-speed scenarios. IEEE Rob. Autom. Lett. **3**(2), 994–1001 (2018)
23. Weikersdorfer, D., Adrian, D.B., Cremers, D., Conradt, J.: Event-based 3D SLAM with a depth-augmented dynamic vision sensor. In: IEEE International Conference on Robotics and Automation, pp. 359–364 (2014)
24. Xie, Z., Chen, S., Orchard, G.: Event-based stereo depth estimation using belief propagation. Front. Neurosci. **11**, 535 (2017)
25. Zhu, A.Z., Atanasov, N., Daniilidis, K.: Event-based feature tracking with probabilistic data association. In: IEEE International Conference on Robotics and Automation, pp. 4465–4470 (2017)
26. Zhu, A.Z., Chen, Y., Daniilidis, K.: Realtime time synchronized event-based stereo. In: Ferrari, V., Hebert, M., Sminchisescu, C., Weiss, Y. (eds.) ECCV 2018. LNCS, vol. 11210, pp. 438–452. Springer, Cham (2018). https://doi.org/10.1007/978-3-030-01231-1_27
27. Zhu, A.Z., Thakur, D., Özaslan, T., Pfrommer, B., Kumar, V., Daniilidis, K.: The multivehicle stereo event camera dataset: an event camera dataset for 3d perception. IEEE Rob. Autom. Lett. **3**(3), 2032–2039 (2018)
28. Zhu, A.Z., Yuan, L., Chaney, K., Daniilidis, K.: Unsupervised event-based learning of optical flow, depth, and egomotion. In: IEEE Conference on Computer Vision and Pattern Recognition, pp. 989–997 (2019)

Reconstructing NBA Players

Luyang Zhu[(⊠)], Konstantinos Rematas, Brian Curless, Steven M. Seitz,
and Ira Kemelmacher-Shlizerman

University of Washington, Seattle, USA
`lyzhu@cs.washington.edu`

Abstract. Great progress has been made in 3D body pose and shape
estimation from a single photo. Yet, state-of-the-art results still suffer
from errors due to challenging body poses, modeling clothing, and self
occlusions. The domain of basketball games is particularly challenging, as
it exhibits all of these challenges. In this paper, we introduce a new approach for reconstruction of basketball players that outperforms the state-
of-the-art. Key to our approach is a new method for creating poseable,
skinned models of NBA players, and a large database of meshes (derived
from the NBA2K19 video game) that we are releasing to the research
community. Based on these models, we introduce a new method that
takes as input a single photo of a clothed player in any basketball pose
and outputs a high resolution mesh and 3D pose for that player. We
demonstrate substantial improvement over state-of-the-art, single-image
methods for body shape reconstruction. Code and dataset are available
at http://grail.cs.washington.edu/projects/nba_players/.

Keyword: 3D human reconstruction

1 Introduction

Given regular, broadcast video of an NBA basketball game, we seek a complete
3D reconstruction of the players, viewable from any camera viewpoint. This
reconstruction problem is challenging for many reasons, including the need to
infer hidden and back-facing surfaces, and the complexity of basketball poses,
e.g., reconstructing jumps, dunks, and dribbles.

Human body modeling from images has advanced dramatically in recent
years, due in large part to availability of 3D human scan datasets, e.g., CAESAR
[56]. Based on this data, researchers have developed powerful tools that enable
recreating realistic humans in a wide variety of poses and body shapes [41], and
estimating 3D body shape from single images [58,64]. These models, however,
are largely limited to the domains of the source data – people in underwear [56],
or clothed models of people in static, staged poses [4]. Adapting this data to a
domain such as basketball is extremely challenging, as we must not only match
the physique of an NBA player, but also their unique basketball poses.

Electronic supplementary material The online version of this chapter (https://
doi.org/10.1007/978-3-030-58558-7_11) contains supplementary material, which is
available to authorized users.

© Springer Nature Switzerland AG 2020
A. Vedaldi et al. (Eds.): ECCV 2020, LNCS 12350, pp. 177–194, 2020.
https://doi.org/10.1007/978-3-030-58558-7_11

Fig. 1. Single input photo (left), estimated 3D posed model that is viewed from **a new** camera position (middle), same model with video game texture for visualization purposes. The insets show the estimated shape from the input camera viewpoint. (Court and basketball meshes are extracted from the video game) *Photo Credit:* [5]

Sports video games, on the other hand, have become extremely realistic, with renderings that are increasingly difficult to distinguish from reality. The player models in games like NBA2K [6] are meticulously crafted to capture each player's physique and appearance (Fig. 3). Such models are ideally suited as a training set for 3D reconstruction and visualization of real basketball games.

In this paper, we present a novel dataset and neural networks that reconstruct high quality meshes of basketball players and retarget these meshes to fit frames of real NBA games. Given an image of a player, we are able to reconstruct the action in 3D, and apply new camera effects such as close-ups, replays, and bullet-time effects (Fig. 1).

Our new dataset is derived from the video game NBA2K (with approval from the creator, Visual Concepts), by playing the game for hours and intercepting rendering instructions to capture thousands of meshes in diverse poses. Each mesh provides detailed shape and texture, down to the level of wrinkles in clothing, and captures all sides of the player, not just those visible to the camera. Since the intercepted meshes are not rigged, we learn a mapping from pose parameters to mesh geometry with a novel *deep skinning* approach. The result of our skinning method is a detailed deep net basketball body model that can be retargeted to any desired player and basketball pose.

We also introduce a system to fit our retargetable player models to real NBA game footage by solving for 3D player pose and camera parameters for each frame. We demonstrate the effectiveness of this approach on synthetic and real NBA input images, and compare with the state of the art in 3D pose and human body model fitting. Our method outperforms the state-of-the-art methods when reconstructing basketball poses and players even when these methods, to the extent possible, are retrained on our new dataset. This paper focuses on basketball shape estimation, and leaves texture estimation as future work.

Our biggest contributions are, first, a deep skinning approach that produces high quality, pose-dependent models of NBA players. A key differentiator is that we leverage *thousands of poses* and capture detailed geometric variations as a function of pose (e.g., folds in clothing), rather than a small number of poses which is the norm for datasets like CAESAR (1–3 poses/person) and modeling methods like SMPL (trained on CAESAR and ∼45 poses/person). While our approach is applicable to any source of registered 3D scan data, we apply it to reconstruct models of NBA players from NBA2K19 game play screen captures.

As such, a second key contribution is pose-dependent models of different basketball players, and raw capture data for the research community. Finally, we present a system that fits these player models to images, enabling 3D reconstructions from photos of NBA players in real games. Both our skinning and pose networks are evaluated quantitatively and qualitatively, and outperform the current state of the art.

One might ask, why spend so much effort reconstructing mesh models that already exist (within the game)? NBA2K's rigged models and in-house animation tools are proprietary IP. By reconstructing a posable model from intercepted meshes (eliminating requirement of proprietary animation and simulation tools), we can provide these best-in-the-world models of basketball players to researchers for the first time (with the company's support). These models provide a number of advantages beyond existing body models such as SMPL. In particular, they capture not just static poses, but human body dynamics for running, walking, and many other challenging activities. Furthermore, the plentiful pose-dependent data enables robust reconstruction even in the presence of heavy occlusions. In addition to producing the first high quality reconstructions of basketball from regular photos, our models can facilitate synthetic data collection for ML algorithms. Just as simulation provides a critical source of data for many ML tasks in robotics, self-driving cars, depth estimation, etc., our derived models can generate much more simulated content under any desired conditions (we can render any pose, viewpoint, combination of players, against any background, etc.)

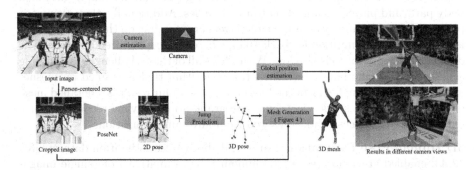

Fig. 2. Overview: Given a single basketball image (top left), we begin by detecting the target player using [15,59], and create a person-centered crop (bottom left). From this crop, our PoseNet predicts 2D pose, 3D pose, and jump information. The estimated 3D pose and the cropped image are then passed to mesh generation networks to predict the full, clothed 3D mesh of the target player. Finally, to globally position the player on the 3D court (right), we estimate camera parameters by solving the PnP problem on known court lines and predict global player position by combining camera, 2D pose, and jump information. Blue boxes represent novel components of our method. (Color figure online)

2   Related Work

Video Game Training Data. Recent works [38,53–55] have shown that, for some domains, data derived from video games can significantly reduce manual labor and labeling, since ground-truth labels can be extracted automatically while playing the game. E.g., [14,53] collected depth maps of soccer players by playing the FIFA soccer video game, showing generalization to images of real games. Those works, however, focused on low level vision data, e.g., optical flow and depth maps rather than full high quality meshes. In contrast, we collect data that includes 3D triangle meshes, texture maps, and detailed 3D body pose, which requires more sophisticated modeling of human body pose and shape.

Sports 3D Reconstruction. Reconstructing 3D models of athletes playing various sports from images has been explored in both academic research and industrial products. Most previous methods use multiple camera inputs rather than a single view. Grau *et al.* [20,21] and Guillemaut *et al.* [24,25] used multi-view stereo methods for free viewpoint navigation. Germann *et al.* [18] proposed an articulated billboard presentation for novel view interpolation. Intel demonstrated 360 degree viewing experiences[1], with their True View [2] technology by installing 38 synchronized 5k cameras around the venue and using this multi-view input to build a volumetric reconstruction of each player. This paper aims to achieve similar reconstruction quality but from a *single* image.

Rematas *et al.* [53] reconstructed soccer games from monocular YouTube videos. However, they predicted only depth maps, thus can not handle occluded body parts and player visualization from all angles. Additionally, they estimated players' global position by assuming all players are standing on the ground, which is not a suitable assumption for basketball, where players are often airborne. The detail of the depth maps is also low. We address all of these challenges by building a basketball specific player reconstruction algorithm that is trained on meshes and accounts for complex airborne basketball poses. Our result is a detailed mesh of the player from a single view, but comparable to multi-view reconstructions. Our reconstructed mesh can be viewed from any camera position.

3D Human Pose Estimation. Large scale body pose estimation datasets [31, 42,44] enabled great progress in 3D human pose estimation from single images [28,43,45,46,62]. We build on [45] but train on our new basketball pose data, use a more detailed skeleton (35 joints including fingers and face keypoints), and an explicit model of jumping and camera to predict global position. Accounting for jumping is an important step that allows our method outperform state of the art pose.

3D Human Body Shape Reconstruction. Parametric human body models [10,33,41,48,51,57] are commonly fit to images to derive a body skeleton, and provide a framework to optimize for shape parameters [12,30,33,39,48,65,69]. [64] further 2D warped the optimized parametric model to approximately account for clothing and create a rigged animated mesh from a single photo.

---

[1] https://www.intel.com/content/www/us/en/sports/technology/true-view.html.

[26,34–37,49,50,70] trained a neural network to directly regress body shape parameters from images. Most parametric model based methods reconstruct undressed humans, since clothing is not part of the parametric model.

Clothing can be modeled to some extent by warping SMPL [41] models, e.g., to silhouettes: Weng et al. [64] demonstrated 2D warping of depth and normal maps from a single photo silhouette, and Alldeick et al. [7–9] addressed multi-image fitting. Alternatively, given predefined garment models [11] estimated a clothing mesh layer on top of SMPL.

Non-parametric methods [47,52,58,63] proposed voxel [63] or implicit function [58] representations to model clothed humans by training on representative synthetic data. Xu et al. [67,68] and Habermann et al. [27] assumed a pre-captured multi-view model of the clothed human, retargeted based on new poses.

We focus on single-view reconstruction of players in NBA basketball games, producing a complete 3D model of the player pose and shape, viewable from any camera viewpoint. This reconstruction problem is challenging for many reasons, including the need to infer hidden and back-facing surfaces, and the complexity of basketball poses, e.g., reconstructing jumps, dunks, and dribbles. Unlike prior methods modeling undressed people in various poses or dressed people in a frontal pose, we focus on modeling clothed people under challenging basketball poses and provide a rigorous comparison with the state of the art.

3   The NBA2K Dataset

Fig. 3. Our novel NBA2K dataset examples, extracted from the NBA2K19 video game. Our NBA2K dataset captures 27,144 basketball poses spanning 27 subjects, extracted from the NBA2K19 video game.

Imagine having thousands of 3D body scans of NBA players, in every conceivable pose during a basketball game. Suppose that these models were extremely detailed and realistic, down to the level of wrinkles in clothing. Such a dataset would be instrumental for sports reconstruction, visualization, and analysis.

This section describes such a dataset, which we call *NBA2K*, after the video game from which these models derive. These models of course are not literally player scans, but are produced by professional modelers for use in the NBA2K19 video game, based on a variety of data including high resolution player photos, scanned models and mocap data of some players. While they do not exactly match each player, they are among the most accurate 3D renditions in existence (Fig. 3).

Our NBA2K dataset consists of body mesh and texture data for several NBA players, each in around 1000 widely varying poses. For each mesh (vertices, faces and texture) we also provide its 3D pose (35 keypoints including face and hand fingers points) and the corresponding RGB image with its camera parameters. While we used meshes of 27 real famous players to create many of figures in this paper, we do not have permission to release models of current NBA players. Instead, we additionally collected the same kind of data for 28 synthetic players and retrained our pipeline on this data. The synthetic player's have the same geometric and visual quality as the NBA models and their data along with trained models will be shared with the research community upon publication of this paper. Our released meshes, textures, and models will have the same quality as what's in the paper, and span a similar variety of player types, but not be named individuals. Visual Concepts [6] has approved our collection and sharing of the data.

The data was collected by playing the NBA2K19 game and intercepting calls between the game engine and the graphics card using RenderDoc [3]. The program captures all drawing events per frame, where we locate player rendering events by analyzing the hashing code of both vertex and pixel shaders. Next, triangle meshes and textures are extracted by reverse-engineering the compiled code of the vertex shader. The game engine renders players by body parts, so we perform a nearest neighbor clustering to decide which body part belongs to which player. Since the game engine optimizes the mesh for real-time rendering, the extracted meshes have different mesh topologies, making them harder to use in a learning framework. We register the meshes by resampling vertices in texture space based on a template mesh. After registration, the processed mesh has 6036 vertices and 11576 faces with fixed topology across poses and players (point-to-point correspondence), has multiple connected components (not a watertight manifold), and comes with no skinning information. We also extract the rest-pose skeleton and per-bone transformation matrix, from which we can compute forward kinematics to get full 3D pose.

4   From Single Images to Meshes

Figure 2 shows our full reconstruction system, starting from a single image of a basketball game, and ending with output of a complete, high quality mesh of the target player with pose and shape matching the image. Next, we describe the individual steps to achieve the final results.

4.1   3D Pose in World Coordinates

2D Pose, Jump, and 3D Pose Estimation. Since our input meshes are not rigged (no skeletal information or blending weights), we propose a neural network called *PoseNet* to estimate the 3D pose and other attributes of a player from a single image. This 3D pose information will be used later to facilitate shape reconstruction. PoseNet takes a single image as input and is trained to output 2D body pose, 3D body pose, a binary jump classification (is the person airborne or not), and the jump height (vertical height of the feet from ground). The two jump-related outputs are key for global position estimation and are our novel addition to existing generic body pose estimation.

From the input image, we first extract ResNet [66] features (from layer 4) and supply them to four separate network branches. The output of the 2D pose branch is a set of 2D heatmaps (one for each 2D keypoint) indicating where the particular keypoint is located. The output of the 3D pose branch is a set of XYZ location maps (one for each keypoint) [45]. The location map indicates the possible 3D location for every pixel. The 2D and 3D pose branches use the same architecture as [66]. The *jump branch* estimates a class label, and the *jump height branch* regresses the height of the jump. Both networks use a fully connected layer followed by two linear residual blocks [43] to get the final output.

The PoseNet model is trained using the following loss:

$$\mathcal{L}_{pose} = \omega_{2d}\mathcal{L}_{2d} + \omega_{3d}\mathcal{L}_{3d} + \omega_{bl}\mathcal{L}_{bl} + \omega_{jht}\mathcal{L}_{jht} + \omega_{jcls}\mathcal{L}_{jcls} \tag{1}$$

where $\mathcal{L}_{2d} = \|H - \hat{H}\|_1$ is the loss between predicted (H) and ground truth ($\hat{H}$) heatmaps, $\mathcal{L}_{3d} = \|L - \hat{L}\|_1$ is the loss between predicted (L) and ground truth ($\hat{L}$) 3D location maps, $\mathcal{L}_{bl} = \|B - \hat{B}\|_1$ is the loss between predicted (B) and ground truth ($\hat{B}$) bone lengths to penalize unnatural 3D poses (we pre-computed the ground truth bone length over the training data), $\mathcal{L}_{jht} = \|h - \hat{h}\|_1$ is the loss between predicted (h) and ground truth ($\hat{h}$) jump height, and $\mathcal{L}_{jcls}$ is the cross-entropy loss for the jump class. For all experiments, we set $\omega_{2d} = 10$, $\omega_{3d} = 10$, $\omega_{bl} = 0.5$, $\omega_{jht} = 0.4$, and $\omega_{jcls} = 0.2$.

Global Position. To estimate the global position of the player we need the camera parameters of the input image. Since NBA courts have known dimensions, we generate a synthetic 3D field and align it with the input frame. Similar to [16,53], we use a two-step approach. First, we provide four manual correspondences between the input image and the 3D basketball court to initialize the camera parameters by solving PnP [40]. Then, we perform a line-based camera optimization similar to [53], where the projected lines from the synthetic 3D court should match the lines on the image. Given the camera parameters, we can estimate a player's global position on (or above) the 3D court by the lowest keypoint and the jump height. We cast a ray from the camera center through the image keypoint; the 3D location of that keypoint is where the ray-ground height is equal to the estimated jump height.

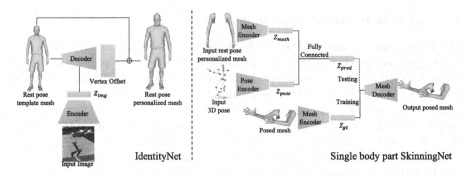

Fig. 4. Mesh generation contains two sub networks: IdentityNet and SkinningNet. IdentityNet deforms a rest pose template mesh (average rest pose over all players in the database), into a rest pose personalized mesh given the image. SkinningNet takes the rest pose personalized mesh and 3D pose as input and outputs the posed mesh. There is a separate SkinningNet per body part, here we illustrate the arms.

4.2   Mesh Generation

Reconstruction of a complete detailed 3D mesh (including deformation due to pose, cloth, fingers and face) from a single image is a key technical contribution of our method. To achieve this we introduce two sub-networks (Fig. 4): *IdentityNet* and *SkinningNet*. IdentityNet takes as input an image of a player whose rest mesh we wish to infer, and outputs the person's rest mesh by deforming a template mesh. The template mesh is the average of all training meshes and is the same starting point for any input. The main benefit of this network is that it allows us to estimate the body size and arm span of the player according to the input image. SkinningNet takes the rest pose personalized mesh and the 3D pose as input, and outputs the posed mesh. To reduce the learning complexity, we pre-segment the mesh into six parts: head, arms, shirt, pants, legs and shoes. We then train a SkinningNet on each part separately. Finally, we combine the six reconstructed parts into one, while removing interpenetration of garments with body parts. Details are described below.

IdentityNet. We propose a variant of 3D-CODED [22] to deform the template mesh. We first use ResNet [29] to extract features from input images. Then we concatenate template mesh vertices with image features and send them into an AtlasNet decoder [23] to predict per vertex offsets. Finally, we add this offset to the template mesh to get the predicted personalized mesh. We use the L1 loss between the prediction and ground truth to train IdentityNet.

SkinningNet. We propose a TL-embedding network [19] to learn an embedding space with generative capability. Specifically, the 3D keypoints $K_{pose} \in R^{35 \times 3}$ are processed by the pose encoder to produce a latent code $Z_{pose} \in R^{32}$. The rest pose personalized mesh vertices $V_{rest} \in R^{N \times 3}$ (where N is the number of vertices in a mesh part) are processed by the mesh encoder to produce a latent code $Z_{rest} \in R^{32}$. Then Z_{pose} and Z_{rest} are concatenated and fed into a fully connected layer to get $Z_{pred} \in R^{32}$. Similarly, the ground truth posed mesh vertices $V_{posed} \in R^{N \times 3}$ are processed by another mesh encoder to produce a latent code $Z_{gt} \in R^{32}$. Z_{gt} is sent into the mesh decoder during training while Z_{pred} is sent into the mesh decoder during testing.

The Pose encoder is comprised of two linear residual blocks [43] followed by a fully connected layer. The mesh encoders and shared decoder are built with spiral convolutions [13]. See supplementary material for detailed network architecture. SkinningNet is trained with the following loss:

$$\mathcal{L}_{skin} = \omega_Z \mathcal{L}_Z + \omega_{mesh} \mathcal{L}_{mesh} \qquad (2)$$

where $\mathcal{L}_Z = \|Z_{pred} - Z_{gt}\|_1$ forces the space of Z_{pred} and Z_{gt} to be similar, and $\mathcal{L}_{mesh} = \|V_{pred} - V_{posed}\|_1$ is the loss between decoded mesh vertices V_{pred} and ground truth vertices V_{posed}. The weights of different losses are set to $\omega_Z = 5$, $\omega_{mesh} = 50$. See supplementary for detailed training parameters.

Combining Body Part Meshes. Direct concatenation of body parts results in interpenetration between the garment and the body. Thus, we first detect all body part vertices in collision with clothing as in [48], and then follow [60,61] to deform the mesh by moving collision vertices inside the garment while preserving local rigidity of the mesh. This detection-deformation process is repeated until there is no collision or the number of iterations is above a threshold (10 in our experiments). See supplementary material for details of the optimization.

Table 1. Quantitative comparison of 3D pose estimation to state of the art. The metric is mean per joint position error with (MPJPE-PA) and without (MPJPE) Procrustes alignment. Baseline methods are fine-tuned on our NBA2K dataset.

	HMR [34]	CMR [37]	SPIN [36]	Ours (Reg+BL)	Ours (Loc)	Ours (Loc+BL)
MPJPE	115.77	82.28	88.72	81.66	66.12	**51.67**
MPJPE-PA	78.17	61.22	59.85	63.70	52.73	**40.91**

Table 2. Quantitative comparison of our mesh reconstruction to state of the art. We use Chamfer distance denoted by CD (scaled by 1000, lower is better), and Earth-mover distance denoted by EMD (lower is better) for comparison. Both distance metrics show that our method significantly outperforms state of the art for mesh estimation. All related works are retrained or fine-tuned on our data, see text.

	HMR [34]	SPIN [36]	SMPLify-X [48]	PIFu [58]	Ours
CD	22.411	14.793	47.720	23.136	**4.934**
EMD	0.137	0.125	0.187	0.207	**0.087**

5   Experiments

Dataset Preparation. We evaluate our method with respect to the state of the art on our NBA2K dataset. We collected 27,144 meshes spanning 27 subjects performing various basketball poses (about 1000 poses per player). PoseNet training requires generalization on real images. Thus, we augment the data to 265,765 training examples, 37,966 validation examples, and 66,442 testing examples. Augmentation is done by rendering and blending meshes into various random basketball courts. For IdentityNet and SkinningNet, we select 19,667 examples from 20 subjects as training data and test on 7,477 examples from 7 unseen players. To further evaluate generalization of our method, we also provide qualitative results on real images. Note that textures are extracted from the game and not estimated by our algorithm.

5.1   3D Pose, Jump, and Global Position Evaluation

We evaluate pose estimation by comparing to state of the art SMPL-based methods that released training code. Specifically we compare with HMR [34], CMR [37], and SPIN [36]. For fair comparison, we fine-tuned their models with 3D and 2D ground-truth NBA2K poses. Since NBA2K and SMPL meshes have different topology we do not use mesh vertices and SMPL parameters as part of the supervision. Table 1 shows comparison results for 3D pose. The metric is defined as mean per joint position error (MPJPE) with and without procrustes alignment. The error is computed on 14 joints as defined by the LSP dataset [32]. Our method outperforms all other methods even when they are fine-tuned on our NBA2K dataset (lower number is better).

To further evaluate our design choices, we compare the location-map-based representation (used in our network) with direct regression of 3D joints, and also evaluate the effect of bone length (BL) loss on pose prediction. A direct regression baseline is created by replacing our deconvolution network with fully connected layers [43]. The effectiveness of BL loss is evaluated by running the network with and without it. As shown in Table 1, both location maps and BL loss can boost the performance. In supplementary material, we show our results on global position estimation. We can see that our method can accurately place players (both airborne and on ground) on the court due to accurate jump estimation.

5.2   3D Mesh Evaluation

Quantitative Results. Table 2 shows results of comparing our mesh reconstruction method to the state of the art on NBA2K data. We compare to both undressed (HMR [34], SMPLify-X [48], SPIN [36]) and clothed (PIFu [58]) human reconstruction methods. For fair comparison, we retrain PIFU on our NBA2K meshes. SPIN and HMR are based on the SMPL model where we do not have groundtruth meshes, so we fine-tuned with NBA2K 2D and 3D pose. SMPLify-X is an optimization method, so we directly apply it to our testing examples. The meshes generated by baseline methods and the NBA2K meshes do not have

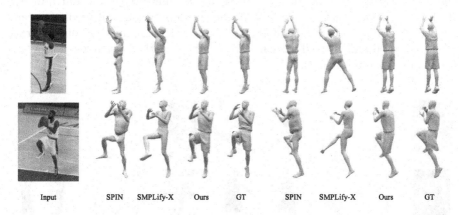

Input   SPIN   SMPLify-X   Ours   GT   SPIN   SMPLify-X   Ours   GT

Fig. 5. Comparison with SMPL-based methods. Column 1 is input, columns 2–5 are reconstructions in the image view, columns 6–9 are visualizations from a novel view. Note the significant difference in body pose between ours and SMPL-based methods.

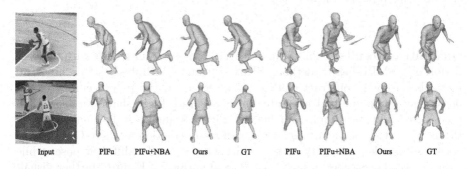

Input   PIFu   PIFu+NBA   Ours   GT   PIFu   PIFu+NBA   Ours   GT

Fig. 6. Comparison with PIFu [58]. Column 1 is input, columns 2–5 are reconstructions in the image viewpoint, columns 6–9 are visualizations from a novel view. PIFu significantly over-smooths shape details and produces lower quality reconstruction even when trained on our dataset (PIFu+NBA).

Input   Prediction   Garment details   Input   Prediction   Garment details

Fig. 7. Garment details at various poses. For each input image, we show the predicted shape, close-ups from two viewpoints.

one-to-one vertex correspondence, thus we use Chamfer (CD) and Earth-mover (EMD) as distance metrics. Prior to distance computations, all predictions are aligned to ground-truth using ICP. We can see that our method outperforms both undressed and clothed human reconstruction methods even when they are trained on our data.

Fig. 8. Results on real images. For each example, column 1 is the input image, 2–3 are reconstructions rendered in different views. 4–5 are corresponding renderings using texture from the video game, just for visualization. Our technical method is focused only on shape recovery. *Photo Credit:* [1]

Qualitative Results. Figure 5 qualitatively compares our results with the best performing SMPL-based methods SPIN [36] and SMPLify-X [48]. These two methods do not reconstruct clothes, so we focus on the pose accuracy of the body shape. Our method generates more accurate body shape for basketball poses, especially for hands and fingers. Figure 6 qualitatively compares with PIFu [58], a state-of-the-art *clothed* human reconstruction method. Our method generates detailed geometry such as shirt wrinkles under different poses while PIFu tends to over-smooth faces, hands, and garments. Figure 7 further visualizes garment details in our reconstructions. Figure 8 shows results of our method on real images, demonstrating robust generalization. Please also refer to the supplementary pdf and video for high quality reconstruction of real NBA players.

5.3   Ablative Study

Comparison with SMPL-NBA. We follow the idea of SMPL [41] to train a skinning model from NBA2K registered mesh sequences. The trained body model is called SMPL-NBA. Since we don't have rest pose meshes for thousands of different subjects, we cannot learn a meaningful PCA shape basis as SMPL did. Thus, we focus on the pose dependent part and fit the SMPL-NBA model to 2000 meshes of a single player. We use the same skeleton rig as SMPL to drive the mesh. Since our mesh is comprised of multiple connected parts, we initialize the skinning weights using a voxel-based heat diffusion method [17]. The whole training process of SMPL-NBA is the same as the pose parameter

Input   SMPL-NBA   Ours   GT   SMPL-NBA   Ours   GT

Fig. 9. Comparison with SMPL-NBA. Column 1 is input, columns 2–4 are reconstructions in the image view, columns 5–7 are visualizations from a novel viewpoint. SMPL-NBA fails to model clothing and the fitting process is unstable.

training of SMPL. We fit the learned model to predicted 2D keypoints and 3D keypoints from PoseNet following SMPLify [12]. Figure 9 compares SkinningNet with SMPL-NBA, showing that SMPL-NBA has severe artifacts for garment deformation – an inherent difficulty for traditional skinning methods. It also suffers from twisted joints which is a common problem when fitting per bone transformation to 3D and 2D keypoints.

Table 3. Quantitative comparison with 3D-CODED [22] and CMR [37]. The metric is mean per vertex position error in mm with (MPVPE-PA) and without (MPVPE) Procrustes alignment. All baseline methods are trained on the NBA2K data.

	CMR [37]	3D-CODED [22]	Ours
MPVPE	85.26	84.22	**76.41**
MPVPE-PA	64.32	63.13	**54.71**

Comparison with Other Geometry Learning Methods. Figure 10 compares SkinningNet with two state of the art mesh-based shape deformation networks: 3D-CODED [22] and CMR [37]. The baseline methods are retrained on the same data as SkinningNet for fair comparison. For 3D-CODED, we take 3D pose as input instead of a point cloud to deform the template mesh. For CMR, we only use their mesh regression network (no SMPL regression network) and replace images with 3D pose as input. Both methods use the same 3D pose encoder as SkinningNet. The input template mesh is set to the prediction of IdentityNet. Unlike baseline methods, SkinningNet does not suffer from substantial deformation errors when the target pose is far from the rest pose. Table 3 provides further quantitative results based on mean per vertex position error (MPVPE) with and without procrustes alignment.

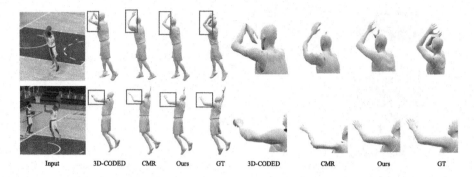

Fig. 10. Comparison with 3D-CODED [22] **and CMR** [37]. Column 1 is input, columns 2–5 are reconstructions in the image view, columns 6–9 are zoomed-in version of the red boxes. The baseline methods exhibit poor deformations for large deviations from the rest pose. (Color figure online)

6 Discussion

We have presented a novel system for state-of-the-art, detailed 3D reconstruction of complete basketball player models from single photos. Our method includes 3D pose estimation, jump estimation, an identity network to deform a template mesh to the person in the photo (to estimate rest pose shape), and finally a skinning network that retargets the shape from rest pose to the pose in the photo. We thoroughly evaluated our method compared to prior art; both quantitative and qualitative results demonstrate substantial improvements over the state-of-the-art in pose and shape reconstruction from single images. For fairness, we retrained competing methods to the extent possible on our new data. Our data, models, and code will be released to the research community.

Limitations and Future Work. This paper focuses solely on high quality shape estimation of basketball players, and does not estimate texture – a topic for future work. Additionally IdentityNet can not model hair and facial identity due to lack of details in low resolution input images. Finally, the current system operates on single image input only; a future direction is to generalize to video with temporal dynamics.

Acknowledgments. This work was supported by NSF/Intel Visual and Experimental Computing Award #1538618 and the UW Reality Lab funding from Facebook, Google and Futurewei. We thank Visual Concepts for allowing us to capture, process, and share NBA2K19 data for research.

References

1. Getty images. https://www.gettyimages.com
2. Intel true view. www.intel.com/content/www/us/en/sports/technology/true-view. html

3. RenderDoc. https://renderdoc.org
4. RenderPeople. https://renderpeople.com
5. USA TODAY network. https://www.commercialappeal.com
6. Visual Concepts. https://vcentertainment.com
7. Alldieck, T., Magnor, M., Bhatnagar, B.L., Theobalt, C., Pons-Moll, G.: Learning to reconstruct people in clothing from a single RGB camera. In: IEEE Conference on Computer Vision and Pattern Recognition (CVPR) (2019)
8. Alldieck, T., Magnor, M., Xu, W., Theobalt, C., Pons-Moll, G.: Video based reconstruction of 3D people models. In: IEEE Conference on Computer Vision and Pattern Recognition (CVPR) (2018)
9. Alldieck, T., Pons-Moll, G., Theobalt, C., Magnor, M.: Tex2Shape: detailed full human body geometry from a single image. In: IEEE International Conference on Computer Vision (ICCV). IEEE (2019)
10. Anguelov, D., Srinivasan, P., Koller, D., Thrun, S., Rodgers, J., Davis, J.: Scape: shape completion and animation of people. ACM Trans. Graph. (TOG) **24**, 408–416 (2005)
11. Bhatnagar, B.L., Tiwari, G., Theobalt, C., Pons-Moll, G.: Multi-garment net: learning to dress 3D people from images. In: IEEE International Conference on Computer Vision (ICCV). IEEE, October 2019
12. Bogo, F., Kanazawa, A., Lassner, C., Gehler, P., Romero, J., Black, M.J.: Keep it SMPL: automatic estimation of 3D human pose and shape from a single image. In: Leibe, B., Matas, J., Sebe, N., Welling, M. (eds.) ECCV 2016. LNCS, vol. 9909, pp. 561–578. Springer, Cham (2016). https://doi.org/10.1007/978-3-319-46454-1_34
13. Bouritsas, G., Bokhnyak, S., Ploumpis, S., Bronstein, M., Zafeiriou, S.: Neural 3D morphable models: spiral convolutional networks for 3D shape representation learning and generation. In: The IEEE International Conference on Computer Vision (ICCV) (2019)
14. Calagari, K., Elgharib, M., Didyk, P., Kaspar, A., Matuisk, W., Hefeeda, M.: Gradient-based 2-D to 3-D conversion for soccer videos. In: ACM Multimedia, pp. 605–619 (2015)
15. Cao, Z., Hidalgo, G., Simon, T., Wei, S.E., Sheikh, Y.: OpenPose: realtime multi-person 2D pose estimation using Part Affinity Fields. In: arXiv preprint arXiv:1812.08008 (2018)
16. Carr, P., Sheikh, Y., Matthews, I.: Pointless calibration: camera parameters from gradient-based alignment to edge images. In: WACV (2012)
17. Dionne, O., de Lasa, M.: Geodesic voxel binding for production character meshes. In: Proceedings of the 12th ACM SIGGRAPH/Eurographics Symposium on Computer Animation, pp. 173–180. ACM (2013)
18. Germann, M., Hornung, A., Keiser, R., Ziegler, R., Würmlin, S., Gross, M.: Articulated billboards for video-based rendering. In: Computer Graphics Forum, vol. 29, pp. 585–594. Wiley Online Library (2010)
19. Girdhar, R., Fouhey, D.F., Rodriguez, M., Gupta, A.: Learning a predictable and generative vector representation for objects. In: Leibe, B., Matas, J., Sebe, N., Welling, M. (eds.) ECCV 2016. LNCS, vol. 9910, pp. 484–499. Springer, Cham (2016). https://doi.org/10.1007/978-3-319-46466-4_29
20. Grau, O., Hilton, A., Kilner, J., Miller, G., Sargeant, T., Starck, J.: A free-viewpoint video system for visualization of sport scenes. SMPTE Motion Imaging J. **116**(5–6), 213–219 (2007)
21. Grau, O., Thomas, G.A., Hilton, A., Kilner, J., Starck, J.: A robust free-viewpoint video system for sport scenes. In: 2007 3DTV Conference, pp. 1–4. IEEE (2007)

22. Groueix, T., Fisher, M., Kim, V.G., Russell, B.C., Aubry, M.: 3D-CODED: 3D correspondences by deep deformation. In: Ferrari, V., Hebert, M., Sminchisescu, C., Weiss, Y. (eds.) ECCV 2018. LNCS, vol. 11206, pp. 235–251. Springer, Cham (2018). https://doi.org/10.1007/978-3-030-01216-8_15

23. Groueix, T., Fisher, M., Kim, V.G., Russell, B.C., Aubry, M.: A papier-mâché approach to learning 3D surface generation. In: Proceedings of the IEEE Conference on Computer Vision and Pattern Recognition, pp. 216–224 (2018)

24. Guillemaut, J.Y., Hilton, A.: Joint multi-layer segmentation and reconstruction for free-viewpoint video applications. IJCV **93**, 73–100 (2011)

25. Guillemaut, J.Y., Kilner, J., Hilton, A.: Robust graph-cut scene segmentation and reconstruction for free-viewpoint video of complex dynamic scenes. In: ICCV (2009)

26. Guler, R.A., Kokkinos, I.: Holopose: holistic 3D human reconstruction in-the-wild. In: The IEEE Conference on Computer Vision and Pattern Recognition (CVPR), June 2019

27. Habermann, M., Xu, W., Zollhoefer, M., Pons-Moll, G., Theobalt, C.: Livecap: real-time human performance capture from monocular video. ACM Trans. Graph. (Proc. SIGGRAPH) **38**(2), 1–17 (2019)

28. Habibie, I., Xu, W., Mehta, D., Pons-Moll, G., Theobalt, C.: In the wild human pose estimation using explicit 2D features and intermediate 3D representations. In: IEEE Conference on Computer Vision and Pattern Recognition (CVPR), June 2019

29. He, K., Zhang, X., Ren, S., Sun, J.: Deep residual learning for image recognition. In: Proceedings of the IEEE Conference on Computer Vision and Pattern Recognition, pp. 770–778 (2016)

30. Huang, Y., et al.: Towards accurate marker-less human shape and pose estimation over time. In: 2017 International Conference on 3D Vision (3DV), pp. 421–430. IEEE (2017)

31. Ionescu, C., Papava, D., Olaru, V., Sminchisescu, C.: Human3. 6m: large scale datasets and predictive methods for 3D human sensing in natural environments. IEEE Trans. Pattern Anal. Mach. Intell. **36**(7), 1325–1339 (2013)

32. Johnson, S., Everingham, M.: Clustered pose and nonlinear appearance models for human pose estimation. In: Proceedings of the British Machine Vision Conference (2010). https://doi.org/10.5244/C.24.12

33. Joo, H., Simon, T., Sheikh, Y.: Total capture: a 3D deformation model for tracking faces, hands, and bodies. In: Proceedings of the IEEE Conference on Computer Vision and Pattern Recognition, pp. 8320–8329 (2018)

34. Kanazawa, A., Black, M.J., Jacobs, D.W., Malik, J.: End-to-end recovery of human shape and pose. In: Computer Vision and Pattern Recognition (CVPR) (2018)

35. Kanazawa, A., Zhang, J.Y., Felsen, P., Malik, J.: Learning 3D human dynamics from video. In: Computer Vision and Pattern Recognition (CVPR) (2019)

36. Kolotouros, N., Pavlakos, G., Black, M.J., Daniilidis, K.: Learning to reconstruct 3D human pose and shape via model-fitting in the loop. In: Proceedings of the IEEE International Conference on Computer Vision (2019)

37. Kolotouros, N., Pavlakos, G., Daniilidis, K.: Convolutional mesh regression for single-image human shape reconstruction. In: CVPR (2019)

38. Krähenbühl, P.: Free supervision from video games. In: CVPR (2018)

39. Lassner, C., Romero, J., Kiefel, M., Bogo, F., Black, M.J., Gehler, P.V.: Unite the people: closing the loop between 3D and 2D human representations. In: Proceedings of the IEEE Conference on Computer Vision and Pattern Recognition, pp. 6050–6059 (2017)

40. Lepetit, V., Moreno-Noguer, F., Fua, P.: EPnP: an accurate o (n) solution to the PnP problem. Int. J. Comput. Vis. **81**(2), 155 (2009)
41. Loper, M., Mahmood, N., Romero, J., Pons-Moll, G., Black, M.J.: SMPL: a skinned multi-person linear model. ACM Trans. Graph. (TOG) **34**(6), 248 (2015)
42. von Marcard, T., Henschel, R., Black, M.J., Rosenhahn, B., Pons-Moll, G.: Recovering accurate 3D human pose in the wild using IMUs and a moving camera. In: Ferrari, V., Hebert, M., Sminchisescu, C., Weiss, Y. (eds.) ECCV 2018. LNCS, vol. 11214, pp. 614–631. Springer, Cham (2018). https://doi.org/10.1007/978-3-030-01249-6_37
43. Martinez, J., Hossain, R., Romero, J., Little, J.J.: A simple yet effective baseline for 3D human pose estimation. In: Proceedings of the IEEE International Conference on Computer Vision, pp. 2640–2649 (2017)
44. Mehta, D., et al.: Monocular 3D human pose estimation in the wild using improved CNN supervision. In: 2017 International Conference on 3D Vision (3DV), pp. 506–516. IEEE (2017)
45. Mehta, D., et al.: VNect: real-time 3d human pose estimation with a single RGB camera. ACM Trans. Graph. (TOG) **36**(4), 44 (2017)
46. Moon, G., Chang, J., Lee, K.M.: Camera distance-aware top-down approach for 3D multi-person pose estimation from a single RGB image. In: The IEEE Conference on International Conference on Computer Vision (ICCV) (2019)
47. Natsume, R., et al.: Siclope: silhouette-based clothed people. In: Proceedings of the IEEE Conference on Computer Vision and Pattern Recognition, pp. 4480–4490 (2019)
48. Pavlakos, G., et al.: Expressive body capture: 3D hands, face, and body from a single image. In: Proceedings of the IEEE Conference on Computer Vision and Pattern Recognition (2019)
49. Pavlakos, G., Kolotouros, N., Daniilidis, K.: Texturepose: supervising human mesh estimation with texture consistency. In: ICCV (2019)
50. Pavlakos, G., Zhu, L., Zhou, X., Daniilidis, K.: Learning to estimate 3D human pose and shape from a single color image. In: Proceedings of the IEEE Conference on Computer Vision and Pattern Recognition, pp. 459–468 (2018)
51. Pons-Moll, G., Romero, J., Mahmood, N., Black, M.J.: Dyna: a model of dynamic human shape in motion. ACM Trans. Actions Graph. (Proc. SIGGRAPH) **34**(4), 120:1–120:14 (2015)
52. Pumarola, A., Sanchez, J., Choi, G., Sanfeliu, A., Moreno-Noguer, F.: 3DPeople: modeling the geometry of dressed humans. In: ICCV (2019)
53. Rematas, K., Kemelmacher-Shlizerman, I., Curless, B., Seitz, S.: Soccer on your tabletop. In: Proceedings of the IEEE Conference on Computer Vision and Pattern Recognition, pp. 4738–4747 (2018)
54. Richter, S.R., Hayder, Z., Koltun, V.: Playing for benchmarks. In: ICCV (2017)
55. Richter, S.R., Vineet, V., Roth, S., Koltun, V.: Playing for data: ground truth from computer games. In: Leibe, B., Matas, J., Sebe, N., Welling, M. (eds.) ECCV 2016. LNCS, vol. 9906, pp. 102–118. Springer, Cham (2016). https://doi.org/10.1007/978-3-319-46475-6_7
56. Robinette, K.M., Blackwell, S., Daanen, H., Boehmer, M., Fleming, S.: Civilian American and European Surface Anthropometry Resource (CAESAR), final report. vol. 1. summary. Technical report, SYTRONICS INC DAYTON OH (2002)
57. Romero, J., Tzionas, D., Black, M.J.: Embodied hands: modeling and capturing hands and bodies together. ACM Trans. Graph. (Proc. SIGGRAPH Asia) **36**(6) (2017)

58. Saito, S., Huang, Z., Natsume, R., Morishima, S., Kanazawa, A., Li, H.: PIFu: pixel-aligned implicit function for high-resolution clothed human digitization. arXiv preprint arXiv:1905.05172 (2019)
59. Simon, T., Joo, H., Matthews, I., Sheikh, Y.: Hand keypoint detection in single images using multiview bootstrapping. In: CVPR (2017)
60. Sorkine, O., Alexa, M.: As-rigid-as-possible surface modeling. Symp. Geom. Process. **4**, 109–116 (2007)
61. Sorkine, O., Cohen-Or, D., Lipman, Y., Alexa, M., Rössl, C., Seidel, H.P.: Laplacian surface editing. In: Proceedings of the 2004 Eurographics/ACM SIGGRAPH Symposium on Geometry Processing, pp. 175–184 (2004)
62. Sun, X., Xiao, B., Wei, F., Liang, S., Wei, Y.: Integral human pose regression. In: Ferrari, V., Hebert, M., Sminchisescu, C., Weiss, Y. (eds.) ECCV 2018. LNCS, vol. 11210, pp. 536–553. Springer, Cham (2018). https://doi.org/10.1007/978-3-030-01231-1_33
63. Varol, G., et al.: BodyNet: volumetric inference of 3D human body shapes. In: Ferrari, V., Hebert, M., Sminchisescu, C., Weiss, Y. (eds.) ECCV 2018. LNCS, vol. 11211, pp. 20–38. Springer, Cham (2018). https://doi.org/10.1007/978-3-030-01234-2_2
64. Weng, C.Y., Curless, B., Kemelmacher-Shlizerman, I.: Photo wake-up: 3D character animation from a single photo. In: Proceedings of the IEEE Conference on Computer Vision and Pattern Recognition, pp. 5908–5917 (2019)
65. Xiang, D., Joo, H., Sheikh, Y.: Monocular total capture: posing face, body, and hands in the wild. In: Proceedings of the IEEE Conference on Computer Vision and Pattern Recognition (2019)
66. Xiao, B., Wu, H., Wei, Y.: Simple baselines for human pose estimation and tracking. In: Ferrari, V., Hebert, M., Sminchisescu, C., Weiss, Y. (eds.) ECCV 2018. LNCS, vol. 11210, pp. 472–487. Springer, Cham (2018). https://doi.org/10.1007/978-3-030-01231-1_29
67. Xu, F., et al.: Video-based characters: creating new human performances from a multi-view video database. ACM Trans. Graph. **30**(4), 32:1–32:10 (2011). https://doi.org/10.1145/2010324.1964927
68. Xu, W., et al.: Monoperfcap: human performance capture from monocular video. ACM Trans. Graph **37**(2), 1–15 (2018)
69. Zanfir, A., Marinoiu, E., Sminchisescu, C.: Monocular 3D pose and shape estimation of multiple people in natural scenes-the importance of multiple scene constraints. In: Proceedings of the IEEE Conference on Computer Vision and Pattern Recognition, pp. 2148–2157 (2018)
70. Zhu, H., Zuo, X., Wang, S., Cao, X., Yang, R.: Detailed human shape estimation from a single image by hierarchical mesh deformation. In: The IEEE Conference on Computer Vision and Pattern Recognition (CVPR), June 2019

PIoU Loss: Towards Accurate Oriented Object Detection in Complex Environments

Zhiming Chen[1,2], Kean Chen[2]($\boxtimes$), Weiyao Lin[2]($\boxtimes$), John See[3], Hui Yu[1], Yan Ke[1], and Cong Yang[1]($\boxtimes$)

[1] Clobotics, Shanghai, China
cong.yang@clobotics.com
[2] Department of Electronic Engineering,
Shanghai Jiao Tong University, Shanghai, China
wylin@sjtu.edu.cn
[3] Faculty of Computing and Informatics, Multimedia University, Cyberjaya, Malaysia

Abstract. Object detection using an oriented bounding box (OBB) can better target rotated objects by reducing the overlap with background areas. Existing OBB approaches are mostly built on horizontal bounding box detectors by introducing an additional angle dimension optimized by a distance loss. However, as the distance loss only minimizes the angle error of the OBB and that it loosely correlates to the IoU, it is insensitive to objects with high aspect ratios. Therefore, a novel loss, Pixels-IoU (PIoU) Loss, is formulated to exploit both the angle and IoU for accurate OBB regression. The PIoU loss is derived from IoU metric with a pixel-wise form, which is simple and suitable for both horizontal and oriented bounding box. To demonstrate its effectiveness, we evaluate the PIoU loss on both anchor-based and anchor-free frameworks. The experimental results show that PIoU loss can dramatically improve the performance of OBB detectors, particularly on objects with high aspect ratios and complex backgrounds. Besides, previous evaluation datasets did not include scenarios where the objects have high aspect ratios, hence a new dataset, Retail50K, is introduced to encourage the community to adapt OBB detectors for more complex environments.

Keywords: Orientated object detection · IoU loss

1 Introduction

Object detection is a fundamental task in computer vision and many detectors [17, 21, 25, 34] using convolutional neural networks have been proposed in recent years. In spite of their state-of-the-art performance, those detectors have inherent limitations on rotated and densely crowded objects. For example, bounding boxes (BB) of a rotated or perspective-transformed objects usually contain a significant amount of background that could mislead the classifiers. When bounding boxes have high overlapping areas, it is difficult to separate the densely crowded objects.

© Springer Nature Switzerland AG 2020
A. Vedaldi et al. (Eds.): ECCV 2020, LNCS 12350, pp. 195–211, 2020.
https://doi.org/10.1007/978-3-030-58558-7_12

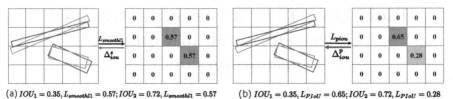

(a) $IOU_1 = 0.35, L_{smoothl1} = 0.57; IOU_2 = 0.72, L_{smoothl1} = 0.57$ (b) $IOU_1 = 0.35, L_{PIoU} = 0.65; IOU_2 = 0.72, L_{PIoU} = 0.28$

Fig. 1. Comparison between PIoU and SmoothL1 [34] losses. (a) Loss values between IoU and SmoothL1 are totally different while their SmoothL1 loss values are the same. (b) The proposed PIoU loss is consistent and correlated with IoU.

Because of these limitations, researchers have extended existing detectors with oriented bounding boxes (OBB). In particular, as opposed to the BB which is denoted by (c_x, c_y, w, h), an OBB is composed by (c_x, c_y, w, h, θ) where (c_x, c_y), (w, h) and θ are the center point, size and rotation of an OBB, respectively. As a result, OBBs can compactly enclose the target object so that rotated and densely crowded objects can be better detected and classified.

Existing OBB-based approaches are mostly built on anchor-based frameworks by introducing an additional angle dimension optimized by a distance loss [6,18, 19,24,41,43] on the parameter tuple (c_x, c_y, w, h, θ). While OBB has been primarily used for simple rotated target detection in aerial images [1,18,23,26,31,39,50], the detection performance in more complex and close-up environments is limited. One of the reasons is that the distance loss in those approaches, *e.g.* SmoothL1 Loss [34], mainly focus on minimizing the angle error rather than global IoU. As a result, it is insensitive to targets with high aspect ratios. An intuitive explanation is that object parts far from the center (c_x, c_y) are not properly enclosed even though the angle distance may be small. For example, [6,19] employ a regression branch to extract rotation-sensitive features and thereby the angle error of the OBB can be modelled in using a transformer. However, as shown in Fig. 1(a), the IoU of predicted boxes (green) and that of the ground truth (red) are very different while their losses are the same.

To solve the problem above, we introduce a novel loss function, named *Pixels-IoU (PIoU) Loss*, to increase both the angle and IoU accuracy for OBB regression. In particular, as shown in Fig. 1(b), the PIoU loss directly reflects the IoU and its local optimum compared to standard distance loss. The rationale behind this is that the IoU loss normally achieves better performance than the distance loss [35,45]. However, the IoU calculation between OBBs is more complex than BBs since the shape of intersecting OBBs could be any polygon of less than eight sides. For this reason, the PIoU, a continuous and derivable function, is proposed to jointly correlate the five parameters of OBB for checking the position (inside or outside IoU) and the contribution of each pixel. The PIoU loss can be easily calculated by accumulating the contribution of interior overlapping pixels. To demonstrate its effectiveness, the PIoU loss is evaluated on both anchor-based and anchor-free frameworks in the experiments.

To overcome the limitations of existing OBB-based approaches, we encourage the community to adopt more robust OBB detectors in a shift from conventional aerial imagery to more complex domains. We collected a new benchmark dataset,

Retail50K, to reflect the challenges of detecting oriented targets with high aspect ratios, heavy occlusions, and complex backgrounds. Experiments show that the proposed frameworks with PIoU loss not only have promising performances on aerial images, but they can also effectively handle new challenges in Retail50K.

The contributions of this work are summarized as follows: (1) We propose a novel loss function, PIoU loss, to improve the performance of oriented object detection in highly challenging conditions such as high aspect ratios and complex backgrounds. (2) We introduce a new dataset, Retail50K, to spur the computer vision community towards innovating and adapting existing OBB detectors to cope with more complex environments. (3) Our experiments demonstrate that the proposed PIoU loss can effectively improve the performances for both anchor-based and anchor-free OBB detectors in different datasets.

2 Related Work

2.1 Oriented Object Detectors

Existing oriented object detectors are mostly extended from generic horizontal bounding box detectors by introducing an additional angle dimension. For instance, [24] presented a rotation-invariant detector based on one-stage SSD [25]. [18] introduced a rotated detector based on two-stage Faster RCNN [34]. [6] designed an RoI transformer to learn the transformation from BB to OBB and thereafter, the rotation-invariant features are extracted. [12] formulated a generative probabilistic model to extract OBB proposals. For each proposal, the location, size and orientation are determined by searching the local maximum likelihood. Other possible ways of extracting OBB include, fitting detected masks [3,10] and regressing OBB with anchor-free models [49], two new concepts in literature. While these approaches have promising performance on aerial images, they are not well-suited for oriented objects with high aspect ratios and complex environments. For this reason, we hypothesize that a new kind of loss is necessary to obtain improvements under challenging conditions. For the purpose of comparative evaluation, we implement both anchor-based and anchor-free frameworks as baselines in our experiments. We later show how these models, when equipped with PIoU Loss, can yield better results in both retail and aerial data.

2.2 Regression Losses

For bounding box regression, actively used loss functions are Mean Square Error [29] (MSE, L2 loss, the sum of squared distances between target and predicted variables), Mean Absolute Error [38] (MAE, L1 loss, the sum of absolute differences between target and predicted variables), Quantile Loss [2] (an extension of MAE, predicting an interval instead of only point predictions), Huber Loss [13] (basically absolute error, which becomes quadratic when error is small) and Log-Cosh Loss (the logarithm of the hyperbolic cosine of the prediction error) [30]. In practise, losses in common used detectors [25,32,34] are extended from the base functions above. However, we can not directly use them since there is an additional angle dimension involved in the OBB descriptor.

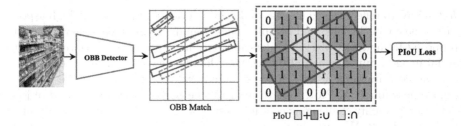

Fig. 2. Our proposed PIoU is a general concept that is applicable to most OBB-based frameworks. All possible predicted (green) and g/t (red) OBB pairs are matched to compute their PIoU. Building on that, the final PIoU loss is calculated using Eq. 14. (Color figure online)

Besides the base functions, there have been several works that introduce IoU losses for horizontal bounding box. For instance, [45] propose an IoU loss which regresses the four bounds of a predicted box as a whole unit. [35] extends the idea of [45] by introducing a Generalized Intersection over Union loss (GIoU loss) for bounding box regression. The main purpose of GIoU is to get rid of the case that two polygons do not have an intersection. [37] introduce a novel bounding box regression loss based on a set of IoU upper bounds. However, when using oriented bounding box, those approaches become much more complicated thus are hard to implement, while the proposed PIoU loss is much simpler and suitable for both horizontal and oriented box. It should be noted that the proposed PIoU loss is different from [48] in which the IoU is computed based on axis alignment and polygon intersection, our method is more straightforward, i.e. IoU is calculated directly by accumulating the contribution of interior overlapping pixels. Moreover, the proposed PIoU loss is also different from Mask Loss in Mask RCNN [10]. Mask loss is calculated by the average binary cross-entropy with per-pixel sigmoid (also called Sigmoid Cross-Entropy Loss). Different from it, our proposed loss is calculated based on positive IoU to preserve intersection and union areas between two boxes. In each area, the contribution of pixels are modeled and accumulated depending on their spatial information. Thus, PIoU loss is more general and sensitive to OBB overlaps.

3   Pixels-IoU (PIoU) Loss

In this section, we present in detail the PIoU Loss. For a given OBB b encoded by (c_x, c_y, w, h, θ), an ideal loss function should effectively guide the network to maximize the IoU and thereby the error of b can be minimized. Towards this goal, we first explain the IoU method. Generally speaking, an IoU function should accurately compute the area of an OBB as well as its intersection with another box. Since OBB and the intersection area are constructed by pixels in image space, their areas are approximated by the number of interior pixels. Specifically, as shown in Fig. 3(a), $t_{i,j}$ (the purple point) is the intersection point between the mid-vertical line and its perpendicular line to pixel $p_{i,j}$ (the green point). As a result, a triangle is constructed by OBB center c (the red point),

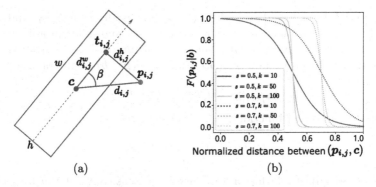

Fig. 3. General idea of the IoU function. (a) Components involved in determining the relative position (inside or outside) between a pixel p (green point) and an OBB b (red rectangle). Best viewed in color. (b) Distribution of the kernelized pixel contribution $F(p_{i,j}|b)$ with different distances between $p_{i,j}$ and box center c. We see that $F(p_{i,j}|b)$ is continuous and differentiable due to Eq. 9. Moreover, it approximately reflects the value distribution in Eq. 1 when the pixels $p_{i,j}$ are inside and outside b. (Color figure online)

$p_{i,j}$ and $t_{i,j}$. The length of each triangle side is denoted by $d_{i,j}^w$, $d_{i,j}^h$ and $d_{i,j}$. To judge the relative location (inside or outside) between $p_{i,j}$ and b, we define the binary constraints as follows:

$$\delta(p_{i,j}|b) = \begin{cases} 1, & d_{i,j}^w \le \dfrac{w}{2}, d_{i,j}^h \le \dfrac{h}{2} \\ 0, & otherwise \end{cases} \tag{1}$$

where d_{ij} denotes the L2-norm distance between pixel (i,j) and OBB center (c_x, c_y), d_w and d_h denotes the distance d along horizontal and vertical direction respectively:

$$d_{ij} = d(i,j) = \sqrt{(c_x - i)^2 + (c_y - j)^2} \tag{2}$$

$$d_{ij}^w = |d_{ij} \cos \beta| \tag{3}$$

$$d_{ij}^h = |d_{ij} \sin \beta| \tag{4}$$

$$\beta = \begin{cases} \theta + \arccos \dfrac{c_x - i}{d_{ij}}, & c_y - j \ge 0 \\ \theta - \arccos \dfrac{c_x - i}{d_{ij}}, & c_y - j < 0 \end{cases} \tag{5}$$

Let $B_{b,b'}$ denotes the smallest horizontal bounding box that covers both b and b'. We can then compute the intersection area $S_{b\cap b'}$ and union area $S_{b\cup b'}$ between two OBBs b and b' using the statistics of all pixels in $B_{b,b'}$:

$$S_{b\cap b'} = \sum_{p_{i,j} \in B_{b,b'}} \delta(p_{i,j}|b)\delta(p_{i,j}|b') \tag{6}$$

$$S_{b\cup b'} = \sum_{p_{i,j} \in B_{b,b'}} \delta(p_{i,j}|b) + \delta(p_{i,j}|b') - \delta(p_{i,j}|b)\delta(p_{i,j}|b') \tag{7}$$

The final IoU of b and b' can be calculated by dividing $S_{b \cap b'}$ and $S_{b \cup b'}$. However, we observe that Eq. 1 is not a continuous and differentiable function. As a result, back propagation (BP) cannot utilize an IoU-based loss for training. To solve this problem, we approximate Eq. 1 as $F(p_{i,j}|b)$ taking on the product of two kernels:

$$F(p_{i,j}|b) = K(d_{i,j}^w, w)K(d_{i,j}^h, h) \tag{8}$$

Particularly, the kernel function $K(d, s)$ is calculated by:

$$K(d, s) = 1 - \frac{1}{1 + e^{-k(d-s)}} \tag{9}$$

where k is an adjustable factor to control the sensitivity of the target pixel $p_{i,j}$. The key idea of Eq. 8 is to obtain the contribution of pixel $p_{i,j}$ using the kernel function in Eq. 9. Since the employed kernel is calculated by the relative position (distance and angle of the triangle in Fig. 3(a)) between $p_{i,j}$ and b, the intersection area $S_{b \cap b'}$ and union area $S_{b \cup b'}$ are inherently sensitive to both OBB rotation and size. In Fig. 3(b), we find that $F(p_{i,j}|b)$ is continuous and differentiable. More importantly, it functions similarly to the characteristics of Eq. 1 such that $F(p_{i,j}|b)$ is close to 1.0 when the pixel $p_{i,j}$ is inside and otherwise when $F(p_{i,j}|b) \sim 0$. Following Eq. 8, the intersection area $S_{b \cap b'}$ and union area $S_{b \cup b'}$ between b and b' are approximated by:

$$S_{b \cap b'} \approx \sum_{p_{i,j} \in B_{b,b'}} F(p_{i,j}|b)F(p_{i,j}|b') \tag{10}$$

$$S_{b \cup b'} \approx \sum_{p_{i,j} \in B_{b,b'}} F(p_{i,j}|b) + F(p_{i,j}|b') - F(p_{i,j}|b)F(p_{i,j}|b') \tag{11}$$

In practice, to reduce the computational complexity of Eq. 11, $S_{b \cup b'}$ can be approximated by a simpler form:

$$S_{b \cup b'} = w \times h + w' \times h' - S_{b \cap b'} \tag{12}$$

where (w, h) and (w', h') are the size of OBBs b and b', respectively. Our experiment in Sect. 5.2 shows that Eq. 12 can effectively reduce the complexity of Eq. 10 while preserving the overall detection performance. With these terms, our proposed Pixels-IoU ($PIoU$) is computed as:

$$PIoU(b, b') = \frac{S_{b \cap b'}}{S_{b \cup b'}} \tag{13}$$

Let b denotes the predicted box and b' denotes the ground-truth box. A pair (b, b') is regarded as positive if the predicted box b is based on a positive anchor and b' is the matched ground-truth box (an anchor is matched with a ground-truth box if the IoU between them is larger them 0.5). We use M to denote the set of all positive pairs. With the goal to maximize the PIoU between b and b', the proposed PIoU Loss is calculated by:

$$L_{piou} = \frac{-\sum_{(b,b') \in M} \ln PIoU(b, b')}{|M|} \tag{14}$$

Table 1. Comparison between different datasets with OBB annotations. $\approx$ indicate estimates based on selected annotated samples as full access was not possible.

Dataset	Scenario	Median ratio	Images	Instances
SZTAKI [1]	Aerial	$\approx$1:3	9	665
VEDAI [31]	Aerial	1:3	1268	2950
UCAS-AOD [50]	Aerial	1:1.3	1510	14596
HRSC2016 [26]	Aerial	1:5	1061	2976
Vehicle [23]	Aerial	1:2	20	14235
DOTA [39]	Aerial	1:2.5	2806	188282
SHIP [18]	Aerial	$\approx$1:5	640	-
OOP [12]	PASCAL	$\approx$1:1	4952	-
Proposed	**Retail**	**1:20**	**47000**	**48000**

Theoretically, Eq. 14 still works if there is no intersection between b and b'. This is because $PIoU(b, b') > 0$ based on Eq. 9 and the gradients still exist in this case. Moreover, the proposed PIoU also works for horizontal bounding box regression. Specifically, we can simply set $\theta = 0$ in Eq. 5 for this purpose. In Sect. 5, we experimentally validate the usability of PIoU for horizontal bounding box regression.

4   Retail50K Dataset

OBB detectors have been actively studied for many years and several datasets with such annotations have been proposed [1,12,18,23,26,31,39,50]. As shown in Table 1, most of them only focused on aerial images (Fig. 4(a) and (b)) while a few are annotated based on existing datasets such as MSCOCO [22], PASCAL VOC [7] and ImageNet [5]. These datasets are important to evaluate the detection performance with simple backgrounds and low aspect ratios. For example, aerial images are typically gray and texture-less. The statistics in [39] shows that most datasets of aerial images have a wide range of aspect ratios, but around 90% of these ratios are distributed between 1:1 and 1:4, and very few images contain OBBs with aspect ratios larger than 1:5. Moreover, aspect ratios of OBBs on PASCAL VOC are mostly close to square (1:1). As a result, it is hard to assess the capability of detectors on objects with high aspect ratios and complex backgrounds using existing datasets. Motivated by this, we introduce a new dataset, namely Retail50K, to advance the research of detection of rotated objects in complex environments. We intend to make this publicly available to the community (https://github.com/clobotics/piou).

Figure 4(c) illustrates a sample image from Retail50K dataset. Retail50K is a collection of 47,000 images from different supermarkets. Annotations on those images are the layer edges of shelves, fridges and displays. We focus on such retail environments for three reasons: (1) **Complex background.** Shelves and

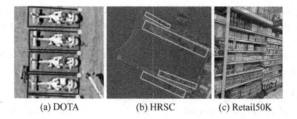

(a) DOTA   (b) HRSC   (c) Retail50K

Fig. 4. Sample images and their annotations of three datasets evaluated in our experiments: (a) DOTA [39] (b) HRSC2016 [26] (c) Retail50K. There are two unique characteristics of Retail50K: (1) Complex backgrounds such as occlusions (by price tags), varied colours and textures. (2) OBB with high aspect ratios.

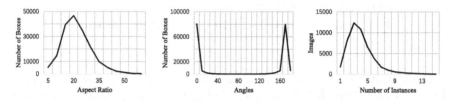

Fig. 5. Statistics of different properties of Retail50K dataset.

fridges are tightly filled with many different items with a wide variety of colours and textures. Moreover, layer edges are normally occluded by price tags and sale tags. Based on our statistics, the mean occlusion is around 37.5%. It is even more challenging that the appearance of price tags are different in different supermarkets. (2) **High aspect ratio.** Aspect ratio is one of the essential factors for anchor-based models [33]. Bounding boxes in Retail50K dataset not only have large variety in degrees of orientation, but also a wide range of aspect ratios. In particular, the majority of annotations in Retail50K are with high aspect ratios. Therefore, this dataset represents a good combination of challenges that is precisely the type we find in complex retail environments. (3) **Useful in practice.** The trained model based on Retail50K can be used for many applications in retail scenarios such as shelf retail tag detection, automatic shelf demarcation, shelf layer and image yaw angle estimation, etc. It is worth to note that although SKU-110K dataset [9] is also assembled from retail environment such as supermarket shelves, the annotations in this dataset are horizontal bounding boxes (HBB) of shelf products since it mainly focuses on object detection in densely packed scenes. The aspect ratios of its HBB are distributed between 1:1–1:3 and hence, it does not cater to the problem that we want to solve.

Images and Categories: Images in Retail50K were collected from 20 supermarket stores in China and USA. Dozens of volunteers acquired data using their personal cellphone cameras. To increase the diversity of data, images were collected in multiple cities from different volunteers. Image quality and view settings were unregulated and so the collected images represent different scales, viewing angles, lighting conditions, noise levels, and other sources of variability.

We also recorded the meta data of the original images such as capture time, volunteer name, shop name and MD5 [40] checksum to filter out duplicated images. Unlike existing datasets that contain multiple categories [5,7,22,39], there is only one category in Retail50K dataset. For better comparisons across datasets, we also employ DOTA [39] (15 categories) and HRSC2016 [26] (the aspect ratio of objects is between that of Retail50K and DOTA) in our experiments (Fig. 4).

Annotation and Properties: In Retail50K dataset, bounding box annotations were provided by 5 skilled annotators. To improve their efficiency, a handbook of labelling rules was provided during the training process. Candidate images were grouped into 165 labelling tasks based on their meta-data so that peer reviews can be applied. Finally, considering the complicated background and various orientations of layer edges, we perform the annotations using arbitrary quadrilateral bounding boxes (AQBB). Briefly, AQBB is denoted by the vertices of the bounding polygon in clockwise order. Due to high efficiency and empirical success, AQBB is widely used in many benchmarks such as text detection [15], object detection in aerial images [18], etc. Based on AQBB, we can easily compute the required OBB format which is denoted by (c_x, c_y, w, h, θ).

Since images were collected with personal cellphone cameras, the original images have different resolutions; hence they were uniformly resized into 600×800 before annotation took place. Figure 5 shows some statistics of Retail50K. We see that the dataset contains a wide range of aspect ratios and orientations (Fig. 5(a) and (b)). In particular, Retail50K is more challenging as compared to existing datasets [18,23,39] since it contains rich annotations with extremely high aspect ratios (higher than 1:10). Similar to natural-image datasets such as ImageNet (average 2) and MSCOCO (average 7.7), most images in our dataset contain around 2–6 instances with complex backgrounds (Fig. 5(c)). For experiments, we selected half of the original images as the training set, 1/6 as validation set, and 1/3 as the testing set.

5   Experiments

5.1   Experimental Settings

We evaluate the proposed PIoU loss with anchor-based and anchor-free OBB-detectors (RefineDet, CenterNet) under different parameters, backbones. We also compare the proposed method with other state-of-the-art OBB-detection methods in different benchmark datasets (*i.e.* DOTA [39], HRSC2016 [26], PASCAL VOC [7]) and the proposed Retail50K dataset. The training and testing tasks are accomplished on a desktop machine with Intel(R) Core(TM) i7-6850K CPU @ 3.60 GHzs, 64 GB installed memory, a GeForce GTX 1080TI GPU (11 GB global memory), and Ubuntu 16.04 LTS. With this machine, the batch size is set to 8 and 1 for training and testing, respectively.

Anchor-Based OBB Detector: For anchor-based object detection, we train RefineDet [46] by updating its loss using the proposed PIoU method. Since the detector is optimized by classification and regression losses, we can easily replace the regression one with PIoU loss L_{piou} while keeping the original Softmax Loss

L_{cls} for classification. We use ResNet [11] and VGG [36] as the backbone models. The oriented anchors are generated by rotating the horizontal anchors by $k\pi/6$ for $0 \leq k < 6$. We adopt the data augmentation strategies introduced in [25] except cropping, while including rotation (i.e. rotate the image by a random angle sampled in $[0, \pi/6]$). In training phase, the input image is resized to 512×512. We adopt the mini-batch training on 2 GPUs with 8 images per GPU. SGD is adopted to optimize the models with momentum set to 0.9 and weight decay set to 0.0005. All evaluated models are trained for 120 epochs with an initial learning rate of 0.001 which is then divided by 10 at 60 epochs and again at 90 epochs. Other experimental settings are the same as those in [46].

Anchor-Free OBB Detector: To extend anchor-free frameworks for detecting OBB, we modify CenterNet [49] by adding an angle dimension regressed by L1-Loss in its overall training objective as our baseline. To evaluate the proposed loss function, in similar fashion as anchor-based approach, we can replace the regression one with PIoU loss L_{piou} while keeping the other classification loss L_{cls} the same. Be noted that CenterNet uses a heatmap to locate the center of objects. Thus, we do not back-propagate the gradient of the object's center when computing the PIoU loss. We use DLA [44] and ResNet [11] as the backbone models. The data augmentation strategies is the same as those for RefineDet-OBB (shown before). In training phase, the input image is resized to 512×512. We adopt the mini-batch training on 2 GPUs with 16 images per GPU. ADAM is adopted to optimize the models. All evaluated models are trained for 120 epochs with an initial learning rate of 0.0005 which is then divided by 10 at 60 epochs and again at 90 epochs. Other settings are the same as those in [49].

5.2 Ablation Study

Comparison on Different Parameters: In Eq. 9, k is an adjustable factor in our kernel function to control the sensitivity of each pixel. In order to evaluate its influence as well as to find a proper value for the remaining experiments, we conduct a set of experiments by varying k values based on DOTA [39] dataset with the proposed anchor-based framework. To simplify discussions, results of $k = 5, 10, 15$ are detailed in Table 2 while their distributions can be visualized in Fig. 3(b). We finally select $k = 10$ for the rest of the experiments since it achieves the best accuracy.

Comparison for Oriented Bounding Box: Based on DOTA [39] dataset, we compare the proposed PIoU loss with the commonly used L1 loss, SmoothL1 loss as well as L2 loss. For fair comparisons, we fix the backbone to VGGNet [36] and build the network based on FPN [20]. Table 3 details the comparisons and we can clearly see that the proposed PIoU Loss improves the detection performance by around 3.5%. HPIoU (Hard PIoU) loss is the simplified PIoU loss using Eq. 12. Its performance is slightly reduced but still comparable to PIoU loss. Thus, HPIoU loss can be a viable option in practise as it has lower computational complexity. We also observe that the proposed PIoU costs 15–20% more time than other three loss functions, which shows that it is still acceptable in practice. We also observed

Table 2. Comparison between different sensitivity factor k in Eq. 9 for PIoU loss on DOTA dataset. RefineDet [46] is used as the detection model.

k	AP	AP_{50}	AP_{75}
5	46.88	59.03	34.73
10	54.24	67.89	40.59
15	53.41	65.97	40.84

Table 3. Comparison between different losses for oriented bounding box on DOTA dataset. RefineDet [46] is used as the detection model. HPIoU (Hard PIoU) loss refers to the PIoU loss simplified by Eq. 12. Training time is estimated in hours.

Loss	AP	AP_{50}	AP_{75}	Training time
L1 Loss	50.66	64.14	37.18	20
L2 Loss	49.70	62.74	36.65	20
SmoothL1 Loss	51.46	65.68	37.25	21.5
PIoU Loss	**54.24**	**67.89**	**40.59**	**25.7**
HPIoU Loss	**53.37**	**66.38**	**40.36**	**24.8**

Table 4. Comparison between different losses for horizontal bounding box on PASCAL VOC2007 dataset. SSD [25] is used as the detection model.

Loss	AP	AP_{50}	AP_{60}	AP_{70}	AP_{80}	AP_{90}
SmoothL1 Loss	48.8	79.8	72.9	60.6	40.3	10.2
GIoU Loss [35]	49.9	79.8	74.1	63.2	41.9	12.4
PIoU Loss	**50.3**	**80.1**	**74.9**	**63.0**	**42.5**	**12.2**

that HPIoU costs less training time than PIoU. Such observation verifies the theoretical analysis and usability of Eq. 12.

Comparison for Horizontal Bounding Box: Besides, we also compare the PIoU loss with SmoothL1 loss and GIoU loss [35] for horizontal bounding box on PASCAL VOC dataset [7]. In Table 4, we observe that the proposed PIoU loss is still better than SmoothL1 loss and GIoU loss for horizontal bounding box regression, particularly at those AP metrics with high IoU threshold. Note that the GIoU loss is designed only for horizontal bounding box while the proposed PIoU loss is more robust and well suited for both horizontal and oriented bounding box. Together with the results in Table 3, we observe the strong generalization ability and effectiveness of the proposed PIoU loss.

5.3   Benchmark Results

Retail50K: We evaluate our PIoU loss with two OBB-detectors (*i.e.* the OBB versions of RefineDet [46] and CenterNet [49]) on Retail50K dataset. The experimental results are shown in Table 5. We observe that, both detectors achieve

Table 5. Detection results on Retail50K dataset. The PIoU loss is evaluated on RefineDet [46] and CenterNet [49] with different backbone models.

Method	Backbone	AP	AP$_{50}$	AP$_{75}$	Time (ms)	FPS
RefineDet-OBB [46]	ResNet-50	53.96	74.15	33.77	142	7
RefineDet-OBB + PIoU	ResNet-50	**61.78**	**80.17**	**43.39**	**142**	**7**
RefineDet-OBB [46]	ResNet-101	55.46	77.05	33.87	167	6
RefineDet-OBB + PIoU	ResNet-101	**63.00**	**79.08**	**46.01**	**167**	**6**
CenterNet-OBB [49]	ResNet18	54.44	76.58	32.29	7	140
CenterNet-OBB + PIoU	ResNet18	**61.02**	**87.19**	**34.85**	**7**	**140**
CenterNet-OBB [49]	DLA-34	56.13	78.29	33.97	18.18	55
CenterNet-OBB + PIoU	DLA-34	**61.64**	**88.47**	**34.80**	**18.18**	**55**

Table 6. Detection results on HRSC2016 dataset. *Aug.* indicates data augmentation. *Size* means the image size that used for training and testing.

Method	Backbone	Size	Aug.	mAP	FPS
R^{2}CNN [14]	ResNet101	800 × 800	×	73.03	2
RC1 & RC2 [27]	VGG-16	-	-	75.7	<1fps
RRPN [28]	ResNet101	800 × 800	×	79.08	3.5
R^{2}PN [47]	VGG-16	-	√	79.6	<1fps
RetinaNet-H [41]	ResNet101	800 × 800	√	82.89	14
RetinaNet-R [41]	ResNet101	800 × 800	√	89.18	10
RoI-Transformer [6]	ResNet101	512 × 800	×	86.20	-
R^{3}Det [41]	ResNet101	300 × 300	√	87.14	18
	ResNet101	600 × 600	√	88.97	15
	ResNet101	800 × 800	√	89.26	12
CenterNet-OBB [49]	ResNet18	512 × 512	√	67.73	140
CenterNet-OBB + PIoU	**ResNet18**	**512 × 512**	√	**78.54**	**140**
CenterNet-OBB [49]	ResNet101	512 × 512	√	77.43	45
CenterNet-OBB + PIoU	**ResNet101**	**512 × 512**	√	**80.32**	**45**
CenterNet-OBB [49]	DLA-34	512 × 512	√	87.98	55
CenterNet-OBB + PIoU	**DLA-34**	**512 × 512**	√	**89.20**	**55**

significant improvements with the proposed PIoU loss (~7% improvement for RefineDet-OBB and ~6% improvement for CenterNet-OBB). One reason for obtaining such notable improvements is that the proposed PIoU loss is much better suited for oriented objects than the traditional regression loss. Moreover, the improvements from PIoU loss in Retail50K are more obvious than those in DOTA (*c.f.* Table 3), which could mean that the proposed PIoU loss is extremely

Table 7. Detection results on DOTA dataset. We report the detection results for each category to better demonstrate where the performance gains come from.

Method	Backbone	Size	PL	BD	BR	GTF	SV	LV	SH	TC	BC	ST	SBF	RA	HA	SP	HC	mAP
SSD [25]	VGG16	512	39.8	9.1	0.6	13.2	0.3	0.4	1.1	16.2	27.6	9.2	27.2	9.1	3.0	1.1	1.0	10.6
YOLOV2 [33]	DarkNet19	416	39.6	20.3	36.6	23.4	8.9	2.1	4.8	44.3	38.4	34.7	16.0	37.6	47.2	25.5	7.5	21.4
R-FCN [4]	ResNet101	800	37.8	38.2	3.6	37.3	6.7	2.6	5.6	22.9	46.9	66.0	33.4	47.2	10.6	25.2	18.0	26.8
FR-H [34]	ResNet101	800	47.2	61.0	9.8	51.7	14.9	12.8	6.9	56.3	60.0	57.3	47.8	48.7	8.2	37.3	23.1	32.3
FR-O [39]	ResNet101	800	79.1	69.1	17.2	63.5	34.2	37.2	36.2	89.2	69.6	59.0	49.	52.5	46.7	44.8	46.3	52.9
R-DFPN [42]	ResNet101	800	80.9	65.8	33.8	58.9	55.8	50.9	54.8	90.3	66.3	68.7	48.7	51.8	55.1	51.3	35.9	57.9
R²CNN [14]	ResNet101	800	80.9	65.7	35.3	67.4	59.9	50.9	55.8	90.7	66.9	72.4	55.1	52.2	55.1	53.4	48.2	60.7
RRPN [28]	ResNet101	800	88.5	71.2	31.7	59.3	51.9	56.2	57.3	90.8	72.8	67.4	56.7	52.8	53.1	51.9	53.6	61.0
RefineDet [46]	VGG16	512	80.5	26.3	33.2	28.5	63.5	75.1	78.8	90.8	61.1	65.9	12.1	23.0	50.9	50.9	22.6	50.9
RefineDet+PIoU	VGG16	512	**80.5**	**33.3**	**34.9**	**28.1**	**64.9**	**74.3**	**78.7**	**90.9**	**65.8**	**66.6**	**19.5**	**24.6**	**51.1**	**50.8**	**23.6**	**52.5**
RefineDet [46]	ResNet101	512	80.7	44.2	27.5	32.8	61.2	76.1	78.8	90.7	69.9	73.9	24.9	31.9	55.8	51.4	26.8	55.1
RefineDet + PIoU	ResNet101	512	**80.7**	**48.8**	**26.1**	**38.7**	**65.2**	**75.5**	**78.6**	**90.8**	**70.4**	**75.0**	**32.0**	**28.0**	**54.3**	**53.7**	**29.6**	**56.5**
CenterNet [49]	DLA-34	512	81.0	64.0	22.6	56.6	38.6	64.0	64.9	90.8	78.0	72.5	44.0	41.1	55.5	55.0	57.4	59.1
CenterNet + PIoU	DLA-34	512	**80.9**	**69.7**	**24.1**	**60.2**	**38.3**	**64.4**	**64.8**	**90.9**	**77.2**	**70.4**	**46.5**	**37.1**	**57.1**	**61.9**	**64.0**	**60.5**

useful for objects with high aspect ratios and complex environments. This verifies the effectiveness of the proposed method.

HRSC2016: The HRSC2016 dataset [26] contains 1070 images from two scenarios including ships on sea and ships close inshore. We evaluate the proposed PIoU with CenterNet [49] on different backbones, and compare them with several state-of-the-art detectors. The experimental results are shown in Table 6. It can be seen that the CenterNet-OBB+PIoU outperforms all other methods except R^3Det-800. This is because we use a smaller image size (512×512) than R^3Det-800 (800×800). Thus, our detector preserves a reasonably competitive detection performance, but with far better efficiency (55 fps $v.s$ 12 fps). This exemplifies the strength of the proposed PIoU loss on OBB detectors.

DOTA: The DOTA dataset [39] contains 2806 aerial images from different sensors and platforms with crowd-sourcing. Each image is of size about 4000×4000 pixels and contains objects of different scales, orientations and shapes. Note that image in DOTA is too large to be directly sent to CNN-based detectors. Thus, similar to the strategy in [39], we crop a series of 512×512 patches from the original image with the stride set to 256. For testing, the detection results are obtained from the DOTA evaluation server. The detailed performances for each category are reported so that deeper observations could be made. We use the same short names, benchmarks and forms as those existing methods in [41] to evaluate the effectiveness of PIoU loss on this dataset. The final results are shown in Table 7. We find that the performance improvements vary among different categories. However, it is interesting to find that the improvement is more plausible for some categories with high aspect ratios. For example, harbour (HA), ground track field (GTF), soccer-ball field (SBF) and basketball court (BC) all naturally have large aspect ratios, and they appear to benefit from the inclusion of PIoU. Such observations confirm that the PIoU can effectively improve the performance of OBB detectors, particularly on objects with high-aspect ratios. These verify again the effectiveness of the proposed PIoU loss on OBB detectors.

Fig. 6. Samples results using PIoU (red boxes) and SmoothL1 (yellow boxes) losses on Retail50K (first row), HRSC2016 (second row) and DOTA (last row) datasets. (Color figure online)

We also find that our baselines are relatively low than some state-of-the-art performances. We conjecture the main reason is that we use much smaller input size than other methods (512 vs 1024 on DOTA). However, note that the existing result (89.2 mAP) for HRSC2016 in Table 6 already achieves the state-of-the-art level performance with only 512 × 512 image size. Thus, the proposed loss function can bring gain in this strong baseline.

In order to visually verify these performance improvements, we employ the anchor-based model RefineDet [46] and conduct two independent experiments using PIoU and SmoothL1 losses. The experiments are applied on all three datasets (*i.e.* Retail50K, DOTA [39], HRSC2016 [26]) and selected visual results are presented in Fig. 6. We can observe that the OBB detector with PIoU loss (in red boxes) has more robust and accurate detection results than the one with SmoothL1 loss (in yellow boxes) on all three datasets, particularly on Retail50K, which demonstrates its strength in improving the performance for high aspect ratio oriented objects. Here, we also evaluate the proposed HPIoU loss with the same configuration of PIoU. In our experiments, the performances of HPIoU loss are slightly lower than those of PIoU loss (0.87, 1.41 and 0.18 mAP on DOTA, Retail50K and HRSC2016 respectively), but still better than smooth-L1 loss while having higher training speed than PIoU loss. Overall, the performances of HPIoU are consistent on all three datasets.

6   Conclusion

We introduce a simple but effective loss function, PIoU, to exploit both the angle and IoU for accurate OBB regression. The PIoU loss is derived from IoU metric with a pixel-wise form, which is simple and suitable for both horizontal and oriented bounding box. To demonstrate its effectiveness, we evaluate the PIoU loss on both anchor-based and anchor-free frameworks. The experimental results show that PIoU loss can significantly improve the accuracy of OBB detectors, particularly on objects with high-aspect ratios. We also introduce a new challenging dataset, Retail50K, to explore the limitations of existing OBB detectors as well as to validate their performance after using the PIoU loss. In the future, we will extend PIoU to 3D rotated object detection. Our preliminary results show that PIoU can improve PointPillars [16] on KITTI val dataset [8] by 0.65, 0.64 and 2.0 AP for car, pedestrian and cyclist in moderate level, respectively.

Acknowledgements. The paper is supported in part by the following grants: China Major Project for New Generation of AI Grant (No. 2018AAA0100400), National Natural Science Foundation of China (No. 61971277). The work is also supported by funding from Clobotics under the Joint Research Program of Smart Retail.

References

1. Benedek, C., Descombes, X., Zerubia, J.: Building development monitoring in multitemporal remotely sensed image pairs with stochastic birth-death dynamics. IEEE Trans. Pattern Anal. Mach. Intell. **34**(1), 33–50 (2012)
2. Cannon, A.J.: Quantile regression neural networks: implementation in R and application to precipitation downscaling. Comput. Geosci. **37**, 1277–1284 (2011)
3. Chen, B., Tsotsos, J.K.: Fast visual object tracking with rotated bounding boxes. In: IEEE International Conference on Computer Vision, pp. 1–9 (2019)
4. Dai, J., Li, Y., He, K., Sun, J.: R-FCN: Object detection via region-based fully convolutional networks. In: Advances in Neural Information Processing Systems, pp. 379–387 (2016)
5. Deng, J., Dong, W., Socher, R., Li, L.J., Li, K., Fei-Fei, L.: Imagenet: a large-scale hierarchical image database. In: IEEE Conference on Computer Vision and Pattern Recognition, pp. 248–255 (2009)
6. Ding, J., Xue, N., Long, Y., Xia, G.S., Lu, Q.: Learning ROI transformer for detecting oriented objects in aerial images. In: IEEE Conference on Computer Vision and Pattern Recognition, pp. 1–9 (2019)
7. Everingham, M., Eslami, S.A., Van Gool, L., Williams, C.K., Winn, J., Zisserman, A.: The pascal visual object classes challenge: a retrospective. Int. J. Comput. Vis. **111**(1), 98–136 (2015)
8. Geiger, A., Lenz, P., Stiller, C., Urtasun, R.: Vision meets robotics: the kitti dataset. Int. J. Robot. Res. **32**(11), 1231–1237 (2013)
9. Goldman, E., Herzig, R., Eisenschtat, A., Goldberger, J., Hassner, T.: Precise detection in densely packed scenes. In: IEEE Conference on Computer Vision and Pattern Recognition, pp. 1–9 (2019)
10. He, K., Gkioxari, G., Dollár, P., Girshick, R.: Mask R-CNN. In: IEEE International Conference on Computer Vision, pp. 2980–2988 (2017)

11. He, K., Zhang, X., Ren, S., Sun, J.: Deep residual learning for image recognition. In: IEEE Conference on Computer Vision and Pattern Recognition, pp. 770–778 (2016)
12. He, S., Lau, R.W.: Oriented object proposals. In: IEEE International Conference on Computer Vision, pp. 280–288 (2015)
13. Huber, P.J.: Robust estimation of a location parameter. Ann. Stat. **53**, 73–101 (1964)
14. Jiang, Y., et al.: R2CNN: rotational region CNN for orientation robust scene text detection. arXiv:1706.09579 (2017)
15. Karatzas, D., et al.: ICDAR competition on robust reading. In: International Conference on Document Analysis and Recognition, pp. 1156–1160 (2015)
16. Lang, A.H., Vora, S., Caesar, H., Zhou, L., Yang, J., Beijbom, O.: Pointpillars: fast encoders for object detection from point clouds. In: IEEE Conference on Computer Vision and Pattern Recognition, pp. 12697–12705 (2019)
17. Law, H., Deng, J.: CornerNet: detecting objects as paired keypoints. In: European Conference on Computer Vision, pp. 734–750 (2018)
18. Li, S., Zhang, Z., Li, B., Li, C.: Multiscale rotated bounding box-based deep learning method for detecting ship targets in remote sensing images. Sensors **18**(8), 1–14 (2018)
19. Liao, M., Zhu, Z., Shi, B., Xia, G.S., Bai, X.: Rotation-sensitive regression for oriented scene text detection. In: IEEE Conference on Computer Vision and Pattern Recognition, pp. 5909–5918 (2018)
20. Lin, T.Y., Dollár, P., Girshick, R., He, K., Hariharan, B., Belongie, S.: Feature pyramid networks for object detection. In: IEEE Conference on Computer Vision and Pattern Recognition, pp. 2117–2125 (2017)
21. Lin, T.Y., Goyal, P., Girshick, R., He, K., Dollár, P.: Focal loss for dense object detection. In: IEEE International Conference on Computer Vision, pp. 2980–2988 (2017)
22. Lin, T.Y., et al.: Microsoft COCO: common objects in context. In: European Conference on Computer Vision, pp. 740–755 (2014)
23. Liu, K., Mattyus, G.: Fast multiclass vehicle detection on aerial images. IEEE Geosci. Remote Sens. Lett. **12**(9), 1938–1942 (2015)
24. Liu, L., Pan, Z., Lei, B.: Learning a rotation invariant detector with rotatable bounding box. arXiv:1711.09405 (2017)
25. Liu, W., et al.: SSD: single shot multibox detector. In: European Conference on Computer Vision, pp. 21–37 (2016)
26. Liu, Z., Wang, H., Weng, L., Yang, Y.: Ship rotated bounding box space for ship extraction from high-resolution optical satellite images with complex backgrounds. IEEE Geosci. Remote Sens. Lett. **13**(8), 1074–1078 (2016)
27. Liu, Z., Yuan, L., Weng, L., Yang, Y.: A high resolution optical satellite image dataset for ship recognition and some new baselines. In: Proceedings of the 6th International Conference on Pattern Recognition Applications and Methods - Volume 2 (ICPRAM), pp. 324–331 (2017)
28. Ma, J., et al.: Arbitrary-oriented scene text detection via rotation proposals. IEEE Trans. Multimedia **20**, 3111–3122 (2018)
29. Introduction to the Theory of Statistics, p. 229. McGraw-Hill, New York (1974)
30. Muller, R.R., Gerstacker, W.H.: On the capacity loss due to separation of detection and decoding. IEEE Trans. Inf. Theor. **50**, 1769–1778 (2004)
31. Razakarivony, S., Jurie, F.: Vehicle detection in aerial imagery : a small target detection benchmark. J. Vis. Commun. Image Represent. **34**(1), 187–203 (2016)

32. Redmon, J., Divvala, S., Girshick, R., Farhadi, A.: You Only Look Once: Uified, Real-time Object Detection, pp. 779–788 (2016)
33. Redmon, J., Farhadi, A.: YOLO9000: better, faster, stronger. arXiv:1612.08242 (2016)
34. Ren, S., He, K., Girshick, R., Sun, J.: Faster R-CNN: Towards real-time object detection with region proposal networks. In: Advances in Neural Information Processing Systems, pp. 91–99 (2015)
35. Rezatofighi, H., et al.: Generalized intersection over union: a metric and a loss for bounding box regression. In: IEEE Conference on Computer Vision and Pattern Recognition, pp. 658–666 (2019)
36. Simonyan, K., Zisserman, A.: Very deep convolutional networks for large-scale image recognition. arXiv:1409.1556 (2014)
37. Tychsen-Smith, L., Petersson, L.: Lars petersson: Improving object localization with fitness NMS and bounded IoU loss. In: IEEE Conference on Computer Vision and Pattern Recognition, pp. 6877–6887 (2018)
38. Willmott, C.J., Matsuura, K.: Advantages of the mean absolute error (MAE) over the root mean square error (RMSE) in assessing average model performance. Clim. Res. **30**, 79–82 (2005)
39. Xia, G.S., et al.: DOTA: a large-scale dataset for object detection in aerial images. In: IEEE Conference on Computer Vision and Pattern Recognition, pp. 3974–3983 (2018)
40. Xie, T., Liu, F., Feng, D.: Fast collision attack on MD5. IACR Cryptol. ePrint Arch. **2013**, 170 (2013)
41. Yang, X., Liu, Q., Yan, J., Li, A.: R3DET: refined single-stage detector with feature refinement for rotating object. arXiv:1908.05612 (2019)
42. Yang, X., et al.: Automatic ship detection in remote sensing images from google earth of complex scenes based on multiscale rotation dense feature pyramid networks. Remote Sens. **10**(1), 132 (2018)
43. Yang, X., et al.: SCRDet: Towards more robust detection for small, cluttered and rotated objects. In: IEEE International Conference on Computer Vision, pp. 1–9 (2019)
44. Yu, F., Wang, D., Shelhamer, E., Darrell, T.: Deep layer aggregation. In: IEEE Conference on Computer Vision and Pattern Recognition, pp. 2403–2412 (2018)
45. Yu, J., Jiang, Y., Wang, Z., Cao, Z., Huang, T.: Unitbox: an advanced object detection network. ACM Int. Conf. Multimedia, 516–520 (2016)
46. Zhang, S., Wen, L., Bian, X., Lei, Z., Li, S.Z.: Single-shot refinement neural network for object detection. In: IEEE Conference on Computer Vision and Pattern Recognition, pp. 4203–4212 (2018)
47. Zhang, Z., Guo, W., Zhu, S., Yu, W.: Toward arbitrary-oriented ship detection with rotated region proposal and discrimination networks. IEEE Geosci. Remote Sens. Lett. **15**(11), 1745–1749 (2018)
48. Zhou, D., et al.: IoU loss for 2D/3D object detection. In: IEEE International Conference on 3D Vision, pp. 1–10 (2019)
49. Zhou, X., Wang, D., Krähenbühl, P.: Objects as points. arXiv:1904.07850 (2019)
50. Zhu, H., Chen, X., Dai, W., Fu, K., Ye, Q., Jiao, J.: Orientation robust object detection in aerial images using deep convolutional neural network. In: IEEE International Conference on Image Processing, pp. 3735–3739 (2015)

TENet: Triple Excitation Network for Video Salient Object Detection

Sucheng Ren[1], Chu Han[2], Xin Yang[3], Guoqiang Han[1], and Shengfeng He[1(✉)]

[1] School of Computer Science and Engineering, South China University of Technology, Guangzhou, China
hesfe@scut.edu.cn
[2] Guangdong Provincial People's Hospital, Guangdong Academy of Medical Sciences, Guangzhou, China
[3] Department of Computer Science and Technology, Dalian University of Technology, Dalian, China

Abstract. In this paper, we propose a simple yet effective approach, named Triple Excitation Network, to reinforce the training of video salient object detection (VSOD) from three aspects, spatial, temporal, and online excitations. These excitation mechanisms are designed following the spirit of curriculum learning and aim to reduce learning ambiguities at the beginning of training by selectively exciting feature activations using ground truth. Then we gradually reduce the weight of ground truth excitations by a curriculum rate and replace it by a curriculum complementary map for better and faster convergence. In particular, the spatial excitation strengthens feature activations for clear object boundaries, while the temporal excitation imposes motions to emphasize spatio-temporal salient regions. Spatial and temporal excitations can combat the saliency shifting problem and conflict between spatial and temporal features of VSOD. Furthermore, our semi-curriculum learning design enables the first online refinement strategy for VSOD, which allows exciting and boosting saliency responses during testing without re-training. The proposed triple excitations can easily plug in different VSOD methods. Extensive experiments show the effectiveness of all three excitation methods and the proposed method outperforms state-of-the-art image and video salient object detection methods.

1   Introduction

When humans look into an image or a video, our visual system will unconsciously focus on the most salient region. The importance of visual saliency has been demonstrated in a bunch of applications, *e.g.*, image manipulation [32,38], object tracking [22], person re-identification [54,61,62], and video understanding [44,45,47]. According to the slightly different goals, saliency detection can be further separated into two research interests, eye-fixation detection [20,48] which mimics the attention mechanism of the human visual system, and salient object detection (SOD) [27,36,58] which focuses on segmenting the salient objects. In this paper, we focus on the latter.

Image-based salient object detection [27,36] has been made great achievements recently. However, detecting salient objects in videos is a different story. This is because the human visual system is influenced not only by appearance but also by motion

© Springer Nature Switzerland AG 2020
A. Vedaldi et al. (Eds.): ECCV 2020, LNCS 12350, pp. 212–228, 2020.
https://doi.org/10.1007/978-3-030-58558-7_13

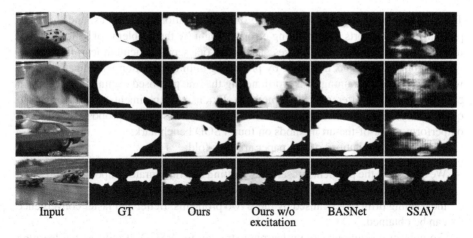

| Input | GT | Ours | Ours w/o excitation | BASNet | SSAV |

Fig. 1. We propose to manually excite feature activations from three aspects, spatial and temporal excitations during training, and online excitation during testing for video salient object detection. Our simple yet effective solution injects additional spatio-temporal guidance during training and even testing for better and faster convergence. In contrast, the image-based method BASNet [36] lacks temporal understanding (first row), while the video-based method SSAV [11] suffers from spatially coarse results.

stimulus. Therefore the severe saliency shifting problem [18,21,41] poses the challenge in video salient object detection (VSOD). Despite different cues, *e.g.*, optical flow [11,23,40] and eye-fixation [11], are used to deal with this problem, the sudden shift of ground truth label makes the training difficult to converge.

Another issue in VSOD training is the contradictory features in spatial and temporal domains. As motion stimulus is a key factor of the human visual system, humans may pay attention to a moving object that does not distinct in appearance. Although features fusion is typically applied, extracting temporal features is much more difficult than spatial ones, as motion blurring, object, and camera movements are involved. They cannot capture clear object boundaries as the spatial features do. As a consequence, VSOD methods (*e.g.*, the last column of Fig. 1) produce spatially coarse results in the scenarios with moving objects. We argue that a simple feature fusion strategy cannot solve this problem, and alternative guidance during training is desired.

To address the above two issues, we propose a Triple Excitation Network (TENet) for video salient object detection. Three types of excitations are tailored for VSOD, *i.e.*, spatial and temporal excitations during training, and online excitation during testing. We adopt a similar spirit with curriculum learning [2], that we aim to loosen the training difficulties at the beginning by exciting selective feature activations using ground truth, then gradually increase task difficulty by replacing such ground truth by our learnable complementary maps. This strategy simplifies the training process of VSOD and boosts the performance with faster convergence and we name it as semi-curriculum learning. In particular, spatial excitation learns spatial features to obtain a boundary-sharp segment. While the temporal excitation aims to leverage previous predictions and excites spatio-temporal salient regions from a spatial excitation map and an optical flow.

These excitations are directly performed on the activations of features, and therefore provide direct supports on mitigating errors brought by the problems of saliency shifting and inaccurate temporal features. Thanks to our semi-curriculum learning design, we can apply excitations in testing by proposing online excitation which can be done without any further training. It is worth noting that the proposed excitation mechanism can easily plug in different VSOD methods. Extensive experiments are performed to qualitatively and quantitatively evaluate the effectiveness of the proposed method, it outperforms state-of-the-art methods on four VSOD benchmarks.

The main contributions of this paper are four-fold:

- We delve into the problems of saliency shifting and inaccurate temporal features and tailor a triple excitation mechanism to excite spatio-temporal features for mitigating the training difficulties caused by these two problems. Better and faster convergence can be obtained.
- We present a semi-curriculum learning strategy for VSOD. It reduces the learning ambiguities by exciting certain activations during the beginning of training, then gradually reduces the curriculum rate to zero and transfers the weight of excitation from ground truth to our learnable complementary maps. This learning strategy progressively weans the network off dependence on ground truth, which is not only beneficial for training but also for testing.
- We propose an online excitation that allows the network to keep self-refining during the testing phase.
- We outperform state-of-the-art SOD and VSOD methods on four benchmarks.

2   Related Works

Image Salient Object Detection. Traditional image saliency detection methods [6,37, 43] usually rely on the hand-crafted features, *e.g.*, color contrast, brightness. It can be separated into two categories, bottom-up [3,13,19] and top-down [3,53] approaches. Driven by a large amount of labeled data, researchers attempt to consider saliency detection as a classification problem [24,60] by simply classifying the patches into non-salient or salient. However, the patches cropped from the original image are usually small and lack of global information. Recent approaches adopt FCN [30] as a basic architecture to detect saliency in an end-to-end manner. Based on that, edge information is incorporated to promote clear object boundaries, by a boundary-enhanced loss [12] or jointly trained with edge detection [27]. Attention mechanism [28,63] is also introduced to filter out a cluttered background. All these methods provide a useful guideline to handle spatial information.

Video Salient Object Detection. The involved temporal information makes video salient object detection much harder than image salient object detection. Some existing approaches try to fuse spatial and temporal information using graph cut [26,33], gradient flow [49], and low-rank coherence [6]. Researchers also try to associate spatial with temporal information using deep networks. Wang *et al.* [50] concatenate the current frame and saliency of the previous frame to process temporal information. Li *et al.* [23] propose to use optical flow to guide the recurrent neural encoder. They use ConvLSTM to process optical flow and warp latent features before feeding into another

ConvLSTM. To capture a wider range of temporal information, a deeper bi-directional ConvLSTM [40] has been proposed. Fan *et al.* [11] mitigate the saliency shifting problem by introducing the eye-fixation mechanism. Similar to VSOD, unsupervised video object segmentation [31,51,55] aims at segmenting primarily objects with temporal information. However, as discussed above, replying only to additional features cannot solve problems of saliency shifting and inaccurate temporal features well. We resolve them from the perspective of reducing training difficulties.

Extra Guidance for CNNs. Introducing extra guidance is a popular solution to aid the training of a deep network. For example, jointly training semantic segmentation and object detection [5,8,14] improves the performances for both tasks. One limitation of multi-task training is that it needs two types of annotations. Some other works introduce two different types of annotations from the same task, *e.g.*, box and segmentation labels for semantic segmentation [15,59], to boost the training performance. Different from these methods, we do not introduce extra task or annotation for training, but directly employ ground truth of the same task as well as pseudo-label for exciting features activations.

3   Triple Excitation Network

3.1   System Overview

Given a series of frames $\{T_n|n = 1, 2, ..., N\}$, we aim to predict the salient object in frame T_n. Figure 2 shows the pipeline of our proposed Triple Excitation Network. The basic network is an encoder-decoder architecture with skip connections (which are hidden in Fig. 2 for simplification). Our framework consists of three branches with respective purposes. The spatial excitation prediction branch is proposed to predict spatial complementary maps with rich spatial information for generating spatial excitation map. The temporal excitation prediction branch leverages the optical flow and spatial excitation maps to generate the temporal complementary maps. These two complementary maps are combined with ground truth to provide the additional guidance for the network training and testing. ConvLSTM [39] is introduced to inject temporal information on the feature maps extracted from the encoder in video saliency prediction branch. After the spatial and temporal excitations, the final saliency map of the current frame T_n is generated by the saliency decoder. During the testing phase, the final saliency map is further adopted for online excitation.

3.2   Excitation Module

Due to the difficulties of handling saliency shifting and the contradictory features in spatial and temporal domains, a simple feature fusion strategy is no longer sufficient for video salient object detection. Therefore, we propose an excitation mechanism shown in Fig. 3 as the additional guidance to reinforce certain feature activations during training.

It is worth noting that our proposed excitation mechanism is a super lightweight plug-in that does not waste computational power because it does not involve any convolution operation. Given an input tensor M, an excitation map E with the pixel values

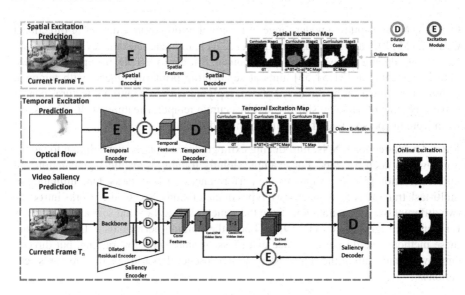

Fig. 2. Network architecture of TENet. The upper two branches provide spatial and temporal excitation with a curriculum learning strategy. In each curriculum stage, we balance the contribution of the ground truth and the complementary map to avoid the overdependency of ground truth during training phase. ConvLSTM is applied in the third branch to introduce the temporal features from the previous frames. During testing phase, an optional online excitation allows the network to keep refining the saliency prediction results by keeping updating the complementary maps with previous predictions recurrently. Note that, the structures of all the encoders (E) in the network are the exactly the same but with different parameters. We only show the saliency encoder for simplification.

under the range $[0, 1]$, we can obtain the output excited tensor M' by the following equation:

$$M' = \beta \odot E \odot M + (1 - \beta) \odot M, \tag{1}$$

where $\odot$ is element-wise multiplication. β is a learnable excitation rate which determines the intensity of excitation based on the feedback of the model itself. It learns an optimum balance between manual excitation and learned activations.

Semi-curriculum Learning. The excitation map can actually be any map that reflects the feature responses required excitation. It can be the ground truth saliency map in this application. However, directly utilizing the ground truth for excitation may let the network excessively rely on the ground truth itself. Therefore, we introduce a semi-curriculum learning strategy for our excitation mechanism. This strategy shares a similar spirit to the curriculum learning framework [2], in which they argue that training with an easy task at first then continues with more difficult tasks gradually may leads to better optimization. Therefore, as the training goes on, we update the excitation map by

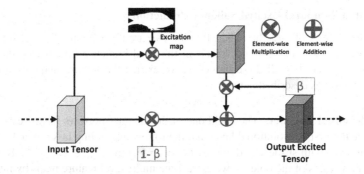

Fig. 3. The proposed excitation module. It strengthens saliency features responses by manually exciting certain feature activations, and the training is controlled by a learnable excitation rate β automatically adjusted according to the feedback of neural network.

trading off the intensity between the ground truth GT and a learnable complementary map S as follows:

$$E = \alpha \odot GT + (1 - \alpha) \odot S \tag{2}$$

where α is the curriculum rate which has been initially set as 1 and will automatically decay to 0 to transfer the contribution from the ground truth to the learnable complementary map. The gradually decreased curriculum rate α will increase the task difficulty and thus can effectively avoid the overdependence on the ground truth. In the meanwhile, this is also the key to enable online excitation.

In practice, we divide the training process into three curriculum stages along with the change of training epoch e. The excitation map E in Eq. 2 can be reformulated as follows:

$$E = \begin{cases} GT & \text{Stage1} : e \leq 2 \\ \alpha \odot GT + (1 - \alpha) \odot S & \text{Stage2} : 2 < e \leq 10 \\ S & \text{Stage3} : e > 10 \end{cases} \tag{3}$$

Stage 1: Due to the imbalance foreground and background pixels in VSOD, for example, the salient pixels take only 8.089% in the DAVIS Dataset [35], reinforcing the network to focus on salient region by using the ground truth as the excitation map provides a shortcut for a better optimization at the beginning of training.

Stage 2: However, models tend to rely on the perfect ground truth and it degrades the model performance once the ground truth is removed. As a result, we gradually replace the ground truth by our learned complementary map (controlled by the curriculum rate α). During this period, the predicted complementary map is to inject perturbation and prevent the model from too sensitive to the perfect ground truth.

Stage 3: When α decays to zero, our model is excited only by the complementary maps. This avoids our network overdependence on GT, more importantly, this is the key to enabling online excitation. We keep training the network in this stage to 15 epochs.

3.3   Spatial-Temporal Excited Saliency Prediction

Our model consists of three branches, the first two are for generating excitation maps, and we predict the video saliency result in the third branch by predicting the video frames one by one. For all these branches, we extract the feature maps by a dilated residual encoder as described below.

Dilated Residual Encoder. The backbone of the encoder borrows from ResNet [16]. We replace the first convolutional layer and the following pooling layer by a 3×3 convolutional layer with stride 1 and extract the deep features $\mathcal{X}_n \in \mathbb{R}^{w \times h \times c}$. To handle the uncertain scales of the objects, we extract the multi-level feature maps by introducing dilated convolution and keep increasing the dilation rates $\{2^k\}_{k=1}^{K}$. The k-th level features extracted from the dilated convolution with dilation rate 2^k is $\mathcal{D}^k \in \mathbb{R}^{w \times h \times c'}$. The output feature maps $\mathcal{F}$ is the concatenation of all the outputs from dilated residual encoder:

$$\mathcal{F}_n = \left[\mathcal{X}_n, \mathcal{D}_n^1, \mathcal{D}_n^2, ..., \mathcal{D}_n^K \right], \tag{4}$$

where $\mathcal{F} \in \mathbb{R}^{w \times h \times (c + K * c')}$. The feature maps $\mathcal{F}_n$ from dilated residual encoder not only keeps the original features $\mathcal{X}_n$ but also covers much larger receptive fields with local-global information. All the encoders in the three branches share the same structure but have different parameters.

Spatial Excitation Prediction Branch. We predict the spatial excitation map from a single frame in the spatial excitation branch which has an encoder-decoder structure. We use the dilated residual encoder as mentioned above, and the decoder has four convolutional stages. Each stage contains three convolutional blocks, each of which is a combination of a convolutional layer, a batch normalization layer, and a ReLU activation layer. With the spatial complementary map S_n^s generated by the spatial decoder, we can calculate the spatial excitation map E_n^s by Eq. 3.

Temporal Excitation Prediction Branch. The temporal branch is designed to tackle the human visual attention shifting problem. It takes an input optical flow, which is calculated by the state-of-the-art optical flow prediction method [29], and outputs a temporal excitation map. This branch has the same network structure with the spatial one. The difference is, we make a spatial excitation on the latent features in the temporal branch in order to associate the temporal excitation with the spatial excitation.

Given the input optical flow from the frame T_{n-1} to the frame T_n, we have the latent features $\mathcal{F}_n^t$ extracted from the temporal encoder. Then the spatial excitation map E_n^s is for spatial excitation. The excited temporal latent features $\mathcal{F}_n'^t$ is calculated as follows:

$$\mathcal{F}_n'^t = \beta^{s \rightarrow t} \odot E_n^s \odot \mathcal{F}_n^t + (1 - \beta^{s \rightarrow t}) \odot \mathcal{F}_n^t, \tag{5}$$

where $\beta^{s \rightarrow t}$ is a learnable temporal excitation rate. The optical flow reveals the moving objects explicitly and the predicted temporal complementary map covers temporally salient regions for governing training. With the temporal complementary map $\mathcal{S}_n^t$ generated by the temporal decoder, we can calculate the temporal excitation map E_n^t by Eq. 3.

Video Saliency Prediction Branch. In this branch, we aim to predict saliency maps with spatially sharpen the object boundaries by leveraging the spatio-temporal excitation mechanism. After the feature extraction with dilated residual encoder, We apply the bi-directional ConvLSTM on the extracted feature maps $\mathcal{F}_n^v$ to obtain long-short term spatial and temporal features:

$$\overrightarrow{\mathcal{H}}_n^v = ConvLSTM(\mathcal{F}_n^v, \overrightarrow{\mathcal{H}}_{n-1}^v), \tag{6}$$

$$\overleftarrow{\mathcal{H}}_n^v = ConvLSTM(\mathcal{F}_n^v, \overleftarrow{\mathcal{H}}_{n-1}^v). \tag{7}$$

We consolidate the features representations on two temporal directions by leveraging both spatial excitation and temporal excitation and obtain bi-directional feature maps $\overrightarrow{\mathcal{H}'}_n^v$ and $\overleftarrow{\mathcal{H}'}_n^v$ of frame T_n as follows:

$$
\begin{aligned}
\overrightarrow{\mathcal{H}'}_n^v = cat(&\overrightarrow{\beta}^{s\to v} \odot E_n^s \odot \overrightarrow{\mathcal{H}}_n^v + (1 - \overrightarrow{\beta}^{s\to v}) \odot \overrightarrow{\mathcal{H}}_n^v \\
&+ \overrightarrow{\beta}^{t\to v} \odot E_n^t \odot \overrightarrow{\mathcal{H}}_n^v + (1 - \overrightarrow{\beta}^{t\to v}) \odot \overrightarrow{\mathcal{H}}_n^v),
\end{aligned} \tag{8}
$$

$$
\begin{aligned}
\overleftarrow{\mathcal{H}'}_n^v = cat(&\overleftarrow{\beta}^{s\to v} \odot E_n^s \odot \overleftarrow{\mathcal{H}}_n^v + (1 - \overleftarrow{\beta}^{s\to v}) \odot \overleftarrow{\mathcal{H}}_n^v \\
&+ \overleftarrow{\beta}^{t\to v} \odot E_n^t \odot \overleftarrow{\mathcal{H}}_n^v + (1 - \overleftarrow{\beta}^{t\to v}) \odot \overleftarrow{\mathcal{H}}_n^v),
\end{aligned} \tag{9}
$$

where E_n^s and E_n^t are the spatial and temporal excitation maps respectively and $cat(\cdot, \cdot)$ is the concatenation operation. $\beta^\cdot$ are the learnable parameters. Since we perform the excitation strategy on the latent space, the excitation maps will be first downsampled to the same size with the feature maps.

The bi-directional excited hidden stage is then concatenated together to produce the final saliency result $\mathcal{S}_n^v$ of the frame T_n by the saliency decoder W^s:

$$\mathcal{S}_n^v = W^s \otimes cat(\overrightarrow{\mathcal{H}'}_n^v, \overleftarrow{\mathcal{H}'}_n^v). \tag{10}$$

Note that the saliency decoder W^s here has the same structure but different parameters with temporal and spatial decoders.

3.4 Loss Function

We borrow the loss function from BASNet [36]. It includes the cross entropy loss [9], SSIM loss [52], and IoU loss [56]. They measure the quality of saliency map in pixel-level, patch-level, and object-level respectively.

$$l = l_{bce} + l_{ssim} + l_{IoU}. \tag{11}$$

The cross entropy loss l_{bce} measures the distance between two probability distributions which is the most common loss function in binary classification and salient object detection.

$$l_{bce}(\mathcal{S}, GT) = -\sum_{i=1}^{w}\sum_{j=1}^{h}[GT(i,j)\log\mathcal{S}(i,j) + (1-GT(i,j))\log(1-\mathcal{S}(i,j))], \tag{12}$$

where S is the network predicted saliency map, S_{gt} is the ground truth saliency map.

The SSIM is originally designed for measuring the structural similarity of two images. When applying this loss into saliency detection, it helps the network pay more attention to the object boundary due to the higher SSIM activation around the boundary. Let $x = \{x_i : i = 1, \cdots, N^2\}, y = \{y_i : i = 1, \cdots, N^2\}$ be the corresponding N × N patches of predict saliency and ground truth label respectively, we have:

$$l_{ssim}(\mathcal{S}, GT) = 1 - \frac{(2\mu_x\mu_y + c_1)(2\sigma_{xy} + c_2)}{(\mu_x^2 + \mu_y^2 + c_1)(\sigma_x^2 + \sigma_y^2 + c_2)}, \qquad (13)$$

where μ_x, μ_y and σ_x, σ_y are the mean and variance, and σ_{xy} is the co-variance. $c_1 = 0.01^2$, $c_2 = 0.03^2$ are constants for maintaining stability.

Intersection over Union (IoU) is widely used in detection and segmentation for evaluation and also used as training loss. The IoU loss is defined as:

$$l_{iou}(\mathcal{S}, GT) = 1 - \frac{\sum_{i=1}^{w}\sum_{j=1}^{h} \mathcal{S}(i,j)GT(i,j)}{\sum_{i=1}^{w}\sum_{j=1}^{h} [\mathcal{S}(i,j) + GT(i,j) - \mathcal{S}(i,j)GT(i,j)]}. \qquad (14)$$

The above losses apply to three branches, and the total objective function of our network is the combination of the spatial excitation loss $l(\mathcal{S}^s, GT)$, temporal excitation loss $l(\mathcal{S}^t, GT)$, and the video saliency loss $l(\mathcal{S}^v, GT)$:

$$\mathcal{L} = l(\mathcal{S}^s, GT) + l(\mathcal{S}^t, GT) + l(\mathcal{S}^v, GT). \qquad (15)$$

3.5   Online Excitation

In our network design, the quality of the excitation map plays an important role in final saliency map prediction. During training, we use a predicted excitation map for highlighting salient activations in the features. Thanks to our semi-curriculum learning strategy, the excitation map does not rely on GT during the testing phase, and we can use a better excitation map to replace the initial guidance. In this way, we design an additional excitation strategy in the testing phase to refine the predicted saliency map without any further training, we call it online excitation. Users can refine the saliency result by recurrently replacing the excitation maps with previous video saliency prediction outputs for better guidance. It provides an additional option for the users to trade-off between the saliency prediction quality and the computational cost during testing. Theoretically, if the excitation map is the same as the ground truth saliency map, our network can give the optimal solution of saliency prediction. We have conducted an experiment in the ablation study to prove the effectiveness of our online excitation.

4   Experiments

4.1   Implementation Details

Our method is trained with three datasets DUTS [46], DAVIS [35] and DAVSOD [11]. Images are loaded into a batch according to their dataset, and we alternately train the Spatial Excitation branch with images from DUTS and DAVIS, Temporal Excitation branch with optical flow from DAVIS and DAVSOD, and the whole model with video from DAVIS and DAVSOD. The optimizer is SGD with momentum 0.9 and weight decay 0.0005. The learning rate starts from 5e−4 and decay to 1e−6. The curriculum rate initially set to 1, and decays following a cosine function.For data argumentation, all inputs are randomly flip horizontally and vertically. During testing every inputs will be resized to 256 × 256. It takes about 40 h to converge, which is one and a half times shorter (80 h) than training without excitation. It shows that our excitation not only boosts the performance as shown below but also accelerates the training process.

4.2 Datasets

We conduct the experiments on four most frequently used VSOD datasets, including Freiburg-Berkeley motion segmentation dataset (FBMS) [4], video salient object detection dataset (ViSal) [49], densely annotation video segmentation dataset (DAVIS) [35], and densely annotation video salient object detection dataset (DAVSOD) [11]. **FBMS** contains 59 videos with only 720 annotated frames. There are 29 videos for training and the rest of them are for testing. **DAVIS** is a high quality and high resolution densely annotated dataset under two resolutions, 480p and 1080p. There are 50 video sequences with 3455 densely annotated frames in pixel level. **ViSal** is the first dataset specially designed for video salient object detection which includes 17 videos and 193 manual annotated frames. **DAVSOD** is the latest and most challenging video salient detection dataset with pixel-wise annotations and eye-fixation labels. We follow the same setting of SSAV [11] and evaluate on 35 test videos.

4.3 Evaluation Metrics

We take three measurements to evaluate our methods: MAE [34], F-measure [1], Structural measurement (S-measure) [10]. MAE is the mean absolute value between predicted saliency map and the groundtruth.

F_β takes both precision and recall into consideration:

$$F_\beta = \frac{(1 + \beta^2) \times Precision \times Recall}{\beta^2 \times Precision + Recall}, \tag{16}$$

where β^2 is usually set to 0.3 and we use maximum F_β for evaluation.

S-measure takes both region and object structural similarity into consideration:

$$S = \mu * S_o + (1 - \mu) * S_r, \tag{17}$$

where S_0 and S_r denotes the region-aware structural similarity and object-aware structural similarity respectively. μ is set to 0.5.

4.4 Comparisons with State-of-the-Arts

We compare our method with 13 saliency methods, including image salient object detection methods (DSS [17], BMPM [57], BASNet [36]) and video salient object

Table 1. Quantitative comparison with image salient object detection methods (labeled as I) and the state-of-the-art VSOD methods (labelled as V) by three evaluation metrics. $Ours$ and $Ours^\star$ indicate the results without and with the online excitation. Top three performances are marked in Red, Green, and Blue respectively.

Method		FBMS			ViSal			DAVIS			DAVSOD		
		MAE↓	max F_β ↑	S ↑	MAE↓	max F_β ↑	S ↑	MAE↓	max F_β ↑	S ↑	MAE↓	max F_β ↑	S ↑
DSS [17]	I	0.080	0.760	0.831	0.024	0.917	0.925	0.059	0.720	0.791	0.112	0.545	0.630
BMPM [57]	I	0.056	0.791	0.844	0.022	0.925	0.930	0.046	0.769	0.834	0.089	0.599	0.704
BASNet [36]	I	0.051	0.817	0.861	0.011	0.949		0.029	0.818	0.862	0.110	0.597	0.670
SIVM [37]	V	0.233	0.416	0.551	0.199	0.521	0.611	0.211	0.461	0.551	0.291	0.299	0.491
MSTM [43]	V	0.177	0.501	0.617	0.091	0.681	0.744	0.166	0.437	0.588	0.210	0.341	0.529
SFLR [6]	V	0.119	0.665	0.690	0.059	0.782	0.815	0.055	0.726	0.781	0.132	0.477	0.627
SCOM [7]	V	0.078	0.796	0.789	0.110	0.829	0.761	0.048	0.789	0.836	0.217	0.461	0.603
SCNN [42]	V	0.091	0.766	0.799	0.072	0.833	0.850	0.066	0.711	0.785	0.129	0.533	0.677
FCNS [50]	V	0.095	0.745	0.788	0.045	0.851	0.879	0.055	0.711	0.781	0.121	0.545	0.664
FGRNE [23]	V	0.085	0.771	0.811	0.041	0.850	0.861	0.043	0.782	0.840	0.099	0.577	0.701
PDBM [40]	V	0.066	0.801	0.845	0.022	0.916	0.929	0.028	0.850	0.880	0.107	0.585	0.699
SSAV [11]	V	0.044	0.855	0.873	0.018	0.939	0.943	0.029	0.861	0.891	0.092	0.602	0.719
MGAN [25]	V	0.028		0.907	0.015	0.944	0.944	0.022			0.114	0.627	0.737
Ours	V	0.027	0.887	0.910	0.014	0.947	0.943	0.021	0.894	0.905	0.078	0.648	0.753
Ours*	V	0.026	0.897	0.915		0.949	0.946	0.019	0.904	0.916	0.074	0.664	0.780

detection methods (SIVM [37], MSTM [43], SFLR [6], SCOM [7], SCNN [42], FCNS [50], FGRNE [23], PDBM [40], SSAV [11], MGAN [25]).

Table 1 show the quantitative evaluation with existing methods. For image salient object detection methods, they perform well in some video salient datasets, because objects that distinct in appearance draws the most attention from the viewer if they are not moving dramatically in the video. On the other hand, although the most recent VSOD method SSAV [11] leverages eye-fixation information to guide the network, the imbalance of spatial and temporal domains harm the accuracy of their saliency results. Our method performs the best statistics results among all the methods in all datasets. Note that, '$Ours$' is the model with the excitation map generated by the excitation prediction branch, *i.e.*, without online excitation. '$Ours^\star$' indicates the results involve online excitation by recurrently applying the previous network outputs. We can find a significant improvement when we apply online excitation. It proves that a precise excitation map can give more accurate guidance for the network, even without further training.

Another interesting observation that can be found in Table 1 is that the datasets also affect the network performance a lot. Because when looking into a video, people tend to focus on moving objects. Some moving objects are not salient in the single frame will definitely distract the network. For some easy dataset whose salient objects are moving and occupy a large part of the image, like *ViSal*, the performance is good for both image-based methods and video-based methods. While in some complicated datasets like *DAVSOD* and *FBMS*, the salient objects in the temporal domain are unfortunately not salient in the spatial domain. The statistical results of them are much worse than those easier datasets. Since our proposed excitation mechanism governs both the

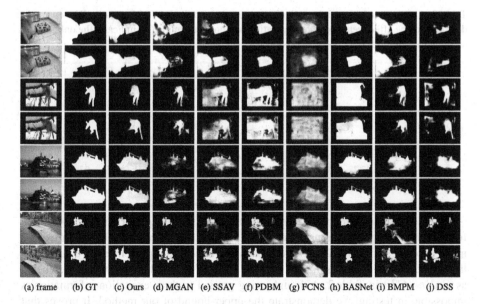

(a) frame   (b) GT   (c) Ours   (d) MGAN   (e) SSAV   (f) PDBM   (g) FCNS   (h) BASNet   (i) BMPM   (j) DSS

Fig. 4. Qualitative comparison with state-of-the-art methods. Our TENet produces clear object boundaries while capture temporally salient objects in the video.

spatial and temporal information, our methods gain a much higher performance than the existing methods.

Figure 4 shows the qualitative comparison. The results predicted by image-based methods shown in Fig. 4(h)–(j) fails to detect the object region accurately and sometimes cannot distinguish the foreground and background due to the lack of temporal information. VSOD methods shown in Fig. 4(d)–(g) provide visually more reasonable saliency maps. However, the boundary of the salient object is not clear and the inside region is blurry, due to the contradictory spatial and temporal features. In contrast, our results (without online excitation, see Fig. 4(c)) show clear boundaries as well as high-confidence interior salient regions.

4.5   Ablation Study

In this section, we explore the effectiveness of our proposed modules. We test the performance on DAVSOD which is the most challenging VSOD dataset.

Effectiveness of Triple Excitation. Table 2 shows an ablation study to evaluate the effectiveness of our triple excitation method. In this experiment, we choose 14 configurations of different excitation strategies. The checkmark in Table 2 indicates the activated excitation component. We can observe that both temporal and spatial excitations boost the detection performance by a large margin (comparing to the first column without checkmark).

Table 2. Ablation study for triple excitation mechanism on the DAVSOD dataset. We separately demonstrate the effectiveness of each excitation component. Online (N) indicates the online excitation with N iterations.

		1	2	3	4	5	6	7	8	9	10	11	12
	Temporal	✓	✓	✓	✓					✓	✓	✓	✓
	Spatial					✓	✓	✓	✓	✓	✓	✓	✓
	Online (1)		✓				✓				✓		
	Online (20)			✓				✓				✓	
	Online (GT)				✓				✓				✓
Results	MAE	0.092	0.090	0.091	0.069	0.084	0.081	0.080	0.062	0.078	0.075	0.074	0.053
	F_β	0.591	0.595	0.594	0.691	0.615	0.628	0.631	0.688	0.648	0.659	0.664	0.841
	S	0.693	0.702	0.708	0.738	0.715	0.733	0.741	0.764	0.753	0.772	0.780	0.862

We then demonstrate the proposed online excitation. We perform online excitation for one and multiple iterations, labeled as $Online(1)$ and $Online(20)$. In our experiment, applying more than 20 iterations cannot bring extra improvement. We also show the ideal case, in which uses the ground truth saliency map as the excitation map, labeled as $Online(GT)$ in Table 2. Although using ground truth as excitation information is impossible in testing, we demonstrate the upper-bound of our method. It proves that a more accuracy excitation map will bring more profits which also implicitly demonstrates the effectiveness of the excitation mechanism.

The statistical results reveal that our three excitations all together work well as we expected. Online excitation does not only introduces more precise excitation guidance but also provides additional optional to exchange the computational cost for the prediction accuracy, by iteratively running the online excitation. $Online(20)$ shows the results of 20 times online iterations. We can find an obvious improvement in three measurements. '$Ours^*$' in Table 1 is implemented with 20 iterations.

Effectiveness of Semi-curriculum Learning. Our semi-curriculum learning involves two main components, GT and learned complementary maps. We consider using the GT only as the traditional curriculum solution. As can be seen in Table 3, the traditional curriculum learning hugely reduces the convergence time. However, as the network relies too much on the perfect ground truth, once no guidance is provided during testing, the curriculum learning strategy does not bring too much improvement. On the other hand, using a learned complementary map for excitation can ease this problem. To provide initial supervision of the network, we pretrain the complementary map for static saliency detection. Using the learned complementary map can provide guidance during testing, which is the key to maintain consistent performances between training and testing phrases. The main limitation is that it requires a separated pretraining, which largely increases the convergence time. The proposed semi-curriculum learning strategy remedies the limitations of previous two methods, leading to faster and better convergence.

Table 3. Ablation study on the proposed semi-curriculum learning.

Method	DAVSOD			
	MAE↓	max F_β ↑	S ↑	Convergence Time
Baseline	0.112	0.579	0.694	80 h
Baseline + Curriculum	0.108	0.584	0.699	32 h
Baseline + Learned Excitation	0.080	0.641	0.743	46 (pre-training) + 38 h
Baseline + Semi-curriculum	0.078	0.648	0.753	40 h

Table 4. Running time comparison of existing methods.

Method	SIVM [37]	BMPM [57]	FCNS [50]	PMDB [40]	SSAV [11]	Ours
Time(s)	18.1	0.03	0.50	0.08	0.08	0.06

4.6 Timing Statistics

We also show the running time of different models in Table 4. All the methods are tested on the same platform: Intel(R) Xeon(R) CPU E5-2620v4 @2.10 GHz and GTX1080Ti. The timing statistics do not include the pre-/post-processing time. Due to our plug-and-play excitations, we can have a fast timing performance compared with most of the deep learning based VSOD methods.

5 Conclusion

This paper proposes a novel video salient object detection method equipped with a triple excitation mechanism. Spatial and temporal excitations are proposed during training phase to tackle the saliency shifting and contradictory spatio-temporal features problems. Besides, we introduce semi-curriculum learning during training to loosen the task difficulty at first and reach a better converage. Furthermore, we propose the first online excitation in testing phase to allow the network keep refining the saliency result by using the network output saliency map for excitation. Extensive experiments show that our results outperform all the competitors.

Acknowledgement. This project is supported by the National Natural Science Foundation of China (No. 61472145, No. 61972162, and No. 61702194), the Special Fund of Science and Technology Research and Development of Applications From Guangdong Province (SF-STRDA-GD) (No. 2016B010127003), the Guangzhou Key Industrial Technology Research fund (No. 201802010036), the Guangdong Natural Science Foundation (No. 2017A030312008), and the CCF-Tencent Open Research fund (CCF-Tencent RAGR20190112).

References

1. Achanta, R., Hemami, S., Estrada, F., Süsstrunk, S.: Frequency-tuned salient region detection. In: CVPR, pp. 1597–1604 (2009)
2. Bengio, Y., Louradour, J., Collobert, R., Weston, J.: Curriculum learning. In: ICML, pp. 41–48 (2009)
3. Borji, A.: Boosting bottom-up and top-down visual features for saliency estimation. In: CVPR, pp. 438–445. IEEE (2012)
4. Brox, T., Malik, J.: Object segmentation by long term analysis of point trajectories. In: Daniilidis, K., Maragos, P., Paragios, N. (eds.) Computer Vision – ECCV 2010. LNCS, vol. 6315, pp. 282–295. Springer, Berlin (2010). https://doi.org/10.1007/978-3-642-15555-0_21
5. Cao, J., Pang, Y., Li, X.: Triply supervised decoder networks for joint detection and segmentation. In: CVPR, pp. 7392–7401 (2019)
6. Chen, C., Li, S., Wang, Y., Qin, H., Hao, A.: Video saliency detection via spatial-temporal fusion and low-rank coherency diffusion. IEEE TIP 26(7), 3156–3170 (2017)
7. Chen, Y., et al.: Scom: spatiotemporal constrained optimization for salient object detection. IEEE TIP 27(7), 3345–3357 (2018)
8. Dai, J., He, K., Sun, J.: Instance-aware semantic segmentation via multi-task network cascades. In: CVPR, pp. 3150–3158 (2016)
9. De Boer, P.T., Kroese, D.P., Mannor, S., Rubinstein, R.Y.: A tutorial on the cross-entropy method. Ann. Oper. Res. 134(1), 19–67 (2005)
10. Fan, D.P., Cheng, M.M., Liu, Y., Li, T., Borji, A.: Structure-measure: a new way to evaluate foreground maps. In: ICCV, pp. 4548–4557 (2017)
11. Fan, D.P., Wang, W., Cheng, M.M., Shen, J.: Shifting more attention to video salient object detection. In: CVPR, pp. 8554–8564 (2019)
12. Feng, M., Lu, H., Ding, E.: Attentive feedback network for boundary-aware salient object detection. In: CVPR (2019)
13. Gao, D., Vasconcelos, N.: Bottom-up saliency is a discriminant process. In: ICCV, pp. 1–6 (2007)
14. Hariharan, B., Arbeláez, P., Girshick, R., Malik, J.: Simultaneous detection and segmentation. In: Fleet, D., Pajdla, T., Schiele, B., Tuytelaars, T. (eds.) Computer Vision – ECCV 2014. Lecture Notes in Computer Science, pp. 297–312. Springer, Cham (2014). https://doi.org/10.1007/978-3-319-10584-0_20
15. He, K., Gkioxari, G., Dollár, P., Girshick, R.: Mask r-cnn. In: ICCV, pp. 2961–2969 (2017)
16. He, K., Zhang, X., Ren, S., Sun, J.: Deep residual learning for image recognition. In: CVPR. pp. 770–778 (2016)
17. Hou, Q., et al.: Deeply supervised salient object detection with short connections. In: CVPR, pp. 3203–3212 (2017)
18. Itti, L., Koch, C., Niebur, E.: A model of saliency-based visual attention for rapid scene analysis. IEEE TPAMI 11, 1254–1259 (1998)
19. Jiang, H., et al.: Salient object detection: a discriminative regional feature integration approach. In: CVPR, pp. 2083–2090 (2013)
20. Jiang, M., Huang, S., Duan, J., Zhao, Q.: Salicon: saliency in context. In: CVPR (2015)
21. Koch, C.: Shifts in selective visual attention: towards the underlying neural circuitry. In: Vaina, L.M. (ed.) Matters of Intelligence. Synthese Library, vol. 188, pp. 115–141. Springer, Dordrecht (1987). https://doi.org/10.1007/978-94-009-3833-5_5
22. Lee, H., Kim, D.: Salient region-based online object tracking. In: WACV, pp. 1170–1177. IEEE (2018)
23. Li, G., Xie, Y., Wei, T., Wang, K., Lin, L.: Flow guided recurrent neural encoder for video salient object detection. In: ICCV, pp. 3243–3252 (2018)

24. Li, G., Yu, Y.: Visual saliency based on multiscale deep features. In: CVPR, pp. 5455–5463 (2015)
25. Li, H., Chen, G., Li, G., Yu, Y.: Motion guided attention for video salient object detection. In: ICCV (2019)
26. Li, S., Seybold, B., Vorobyov, A., Lei, X., Kuo, C.C.J.: Unsupervised video object segmentation with motion-based bilateral networks. In: Ferrari, V., Hebert, M., Sminchisescu, C., Weiss, Y. (eds.) Computer Vision – ECCV 2018. Lecture Notes in Computer Science, vol. 11207, pp. 215–231. Springer, Cham (2018). https://doi.org/10.1007/978-3-030-01219-9_13
27. Liu, J.J., Hou, Q., Cheng, M.M., Feng, J., Jiang, J.: A simple pooling-based design for real-time salient object detection. In: CVPR (2019)
28. Liu, N., Han, J., Yang, M.H.: PiCANet: learning pixel-wise contextual attention for saliency detection. In: CVPR, pp. 3089–3098 (2018)
29. Liu, P., Lyu, M., King, I., Xu, J.: Selflow: self-supervised learning of optical flow. In: CVPR, pp. 4571–4580 (2019)
30. Long, J., Shelhamer, E., Darrell, T.: Fully convolutional networks for semantic segmentation. In: CVPR, pp. 3431–3440 (2015)
31. Lu, X., et al.: See more, know more: Unsupervised video object segmentation with co-attention siamese networks. In: CVPR (2019)
32. Mechrez, R., Shechtman, E., Zelnik-Manor, L.: Saliency driven image manipulation. Mach. Vis. Appl. **30**(2), 189–202 (2019)
33. Papazoglou, A., Ferrari, V.: Fast object segmentation in unconstrained video. In: ICCV, pp. 1777–1784 (2013)
34. Perazzi, F., Krähenbühl, P., Pritch, Y., Hornung, A.: Saliency filters: Contrast based filtering for salient region detection. In: CVPR, pp. 733–740 (2012)
35. Perazzi, F., et al.: A benchmark dataset and evaluation methodology for video object segmentation. In: CVPR, pp. 724–732 (2016)
36. Qin, X., et al.: BASNet: boundary-aware salient object detection. In: CVPR. pp. 7479–7489 (2019)
37. Rahtu, E., Kannala, J., Salo, M., Heikkilä, J.: Segmenting salient objects from images and videos. In: Daniilidis, K., Maragos, P., Paragios, N. (eds.) Computer Vision – ECCV 2010. Lecture Notes in Computer Science, vol. 6315, pp. 366–379. Springer, Berlin, Heidelberg (2010). https://doi.org/10.1007/978-3-642-15555-0_27
38. Shafieyan, F., Karimi, N., Mirmahboub, B., Samavi, S., Shirani, S.: Image seam carving using depth assisted saliency map. In: ICIP, pp. 1155–1159. IEEE (2014)
39. Shi, X., et al.: Convolutional LSTM network: a machine learning approach for precipitation nowcasting. In: NeurIPS, pp. 802–810 (2015)
40. Song, H., Wang, W., Zhao, S., Shen, J., Lam, K.M.: Pyramid dilated deeper convLSTM for video salient object detection. In: Ferrari, V., Hebert, M., Sminchisescu, C., Weiss, Y. (eds.) Computer Vision – ECCV 2018. Lecture Notes in Computer Science, vol. 11215, pp. 744–760. Springer, Cham (2018). https://doi.org/10.1007/978-3-030-01252-6_44
41. Squire, L.R., Dronkers, N., Baldo, J.: Encyclopedia of Neuroscience. Elsevier, London (2009)
42. Tang, Y., et al.: Weakly supervised salient object detection with spatiotemporal cascade neural networks. IEEE Trans. Circuits Syst. Video Technol. (2018)
43. Tu, W.C., He, S., Yang, Q., Chien, S.Y.: Real-time salient object detection with a minimum spanning tree. In: CVPR, pp. 2334–2342 (2016)
44. Wang, H., Kläser, A., Schmid, C., Liu, C.L.: Dense trajectories and motion boundary descriptors for action recognition. IJCV **103**(1), 60–79 (2013)
45. Wang, H., Schmid, C.: Action recognition with improved trajectories. In: ICCV, pp. 3551–3558 (2013)

46. Wang, L., et al.: Learning to detect salient objects with image-level supervision. In: CVPR, pp. 136–145 (2017)
47. Wang, L., Qiao, Y., Tang, X.: Action recognition with trajectory-pooled deep-convolutional descriptors. In: CVPR, pp. 4305–4314 (2015)
48. Wang, W., Shen, J., Guo, F., Cheng, M.M., Borji, A.: Revisiting video saliency: a large-scale benchmark and a new model. In: CVPR (2018)
49. Wang, W., Shen, J., Shao, L.: Consistent video saliency using local gradient flow optimization and global refinement. IEEE TIP **24**(11), 4185–4196 (2015)
50. Wang, W., Shen, J., Shao, L.: Video salient object detection via fully convolutional networks. IEEE TIP **27**(1), 38–49 (2017)
51. Wang, W., et al.: Learning unsupervised video object segmentation through visual attention. In: CVPR (2019)
52. Wang, Z., et al.: Image quality assessment: from error visibility to structural similarity. IEEE TIP **13**(4), 600–612 (2004)
53. Yang, J., Yang, M.H.: Top-down visual saliency via joint CRF and dictionary learning. IEEE TPAMI **39**(3), 576–588 (2016)
54. Yang, Y., et al.: Salient color names for person re-identification. In: Fleet, D., Pajdla, T., Schiele, B., Tuytelaars, T. (eds.) Computer Vision – ECCV 2014. Lecture Notes in Computer Science, vol. 8689, pp. 536–551. Springer, Cham (2014). https://doi.org/10.1007/978-3-319-10590-1_35
55. Yang, Z., et al.: Anchor diffusion for unsupervised video object segmentation. In: ICCV (2019)
56. Yu, J., Jiang, Y., Wang, Z., Cao, Z., Huang, T.: UnitBox: an advanced object detection network. In: ACM MM, pp. 516–520 (2016)
57. Zhang, L., Dai, J., Lu, H., He, Y., Wang, G.: A bi-directional message passing model for salient object detection. In: CVPR, pp. 1741–1750 (2018)
58. Zhang, X., Wang, T., Qi, J., Lu, H., Wang, G.: Progressive attention guided recurrent network for salient object detection. In: CVPR (2018)
59. Zhang, Z., et al.: Single-shot object detection with enriched semantics. In: CVPR, pp. 5813–5821 (2018)
60. Zhao, R., Ouyang, W., Li, H., Wang, X.: Saliency detection by multi-context deep learning. In: CVPR, pp. 1265–1274 (2015)
61. Zhao, R., Ouyang, W., Wang, X.: Unsupervised salience learning for person re-identification. In: CVPR, pp. 3586–3593 (2013)
62. Zhao, R., Oyang, W., Wang, X.: Person re-identification by saliency learning. IEEE TPAMI **39**(2), 356–370 (2016)
63. Zhao, T., Wu, X.: Pyramid feature attention network for saliency detection. In: CVPR (2019)

Deep Feedback Inverse Problem Solver

Wei-Chiu Ma[1,2]([✉]), Shenlong Wang[1,3], Jiayuan Gu[1,4],
Sivabalan Manivasagam[1,3], Antonio Torralba[2], and Raquel Urtasun[1,3]

[1] Uber Advanced Technologies Group, Pittsburgh, USA
[2] Massachusetts Institute of Technology, Cambridge, USA
weichium@mit.edu
[3] University of Toronto, Toronto, Canada
[4] University of California San Diego, San Diego, USA

Abstract. We present an efficient, effective, and generic approach towards solving inverse problems. The key idea is to leverage the feedback signal provided by the forward process and learn an iterative update model. Specifically, at each iteration, the neural network takes the feedback as input and outputs an update on current estimation. Our approach does not have any restrictions on the forward process; it does not require any prior knowledge either. Through the feedback information, our model not only can produce accurate estimations that are coherent to the input observation but also is capable of recovering from early incorrect predictions. We verify the performance of our model over a wide range of inverse problems, including 6-DoF pose estimation, illumination estimation, as well as inverse kinematics. Comparing to traditional optimization-based methods, we can achieve comparable or better performance while being two to three orders of magnitude faster. Compared to deep learning-based approaches, our model consistently improves the performance on all metrics.

1 Introduction

Given a 3D model of an object, the light source(s), and their relevant pose to the camera, one can generate highly realistic images of the scene with one click. While such a *forward* rendering process is complicated and requires explicit modeling of interreflection, self-occlusion, as well as distortion, it is well-defined and can be computed effectively. However, if we were to recover the illumination parameters or predict the 6 DoF pose of the object from the image in an *inverse* fashion, the task becomes extremely challenging. This is because a lot of crucial information is lost during the forward (rendering) process. In fact, many complicated systems in natural science, signal processing, and robotics, all face similar challenges – the model parameters of interest cannot be measured directly and need to be estimated from limited observations. This family of problems are commonly referred to as **inverse problems**. Unfortunately, while there exists

---

Electronic supplementary material The online version of this chapter (https://doi.org/10.1007/978-3-030-58558-7_14) contains supplementary material, which is available to authorized users.

© Springer Nature Switzerland AG 2020
A. Vedaldi et al. (Eds.): ECCV 2020, LNCS 12350, pp. 229–246, 2020.
https://doi.org/10.1007/978-3-030-58558-7_14

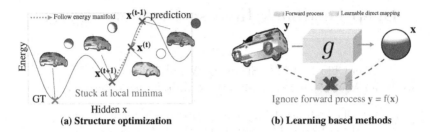

Fig. 1. Prior work on inverse problems: (a) Structured optimization approaches require hand-crafted energy/objective functions and are sensitive to initializations which makes them easy to get stuck in local optima. (b) Direct learning based methods do not utilize the available forward process as feedback to guarantee the quality of the solution. Without this feedback, the models cannot rectify the estimates effectively as shown above.

sophisticated theories on how to design the forward processes, how to address the inherent ambiguities of the inverse problem still remains an open question.

One popular strategy to disambiguate the problem is to model the inverse problem as a structured optimization task and incorporate human knowledge into the model [19,21,49,56]. For instance, the estimated solution should agree with the observation [48] and be smooth [3,47], or should follow a certain statistical distribution [2,33,64]. Through imposing carefully designed objectives, classic structure optimization methods are able to find a solution that not only agrees with the observation but also satisfies our prior knowledge about the solution. In practice, however, almost no hand-crafted priors can succeed in including all phenomena. To ensure that the optimization problem can be solved efficiently, there are multiple restrictions on the form of the priors as well as the forward process [4], both of which increases the difficulty of design. Furthermore, most optimization approaches require many iterations to converge and are sensitive to initialization (Fig. 1).

On the other hand, learning based methods propose to directly learn a mapping from observations to the model parameters [24,55,60,68,72]. They capitalize on powerful machine learning tools to extract task-specific priors in a data-driven fashion. With the help of large-scale datasets and the flourishing of deep learning, they are able to achieve state-of-the-art performance on a variety of inverse problems [13,23,57,61,69]. Unfortunately, these methods often ignore the fact that the forward model for inverse problems is available. Their systems remain open loop and do not have the capability to *update* their prediction based on the *feedback signal*. Consequently, the estimated parameters, while performing well in majority cases, may generate a result that is either incompatible with the real observation or not realistic.

With these challenges in mind, we develop a novel approach to solving inverse problems that takes the best of both worlds. The key idea is to *learn to iteratively update* the current estimation through the *feedback signal* from the forward process. Specifically, we design a neural network that takes the observation and the forward simulation result of the previous estimation as input, and outputs a steepest update towards the ideal model parameters. The advantages are

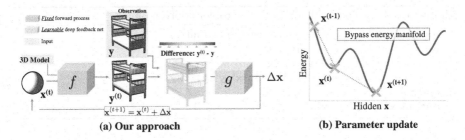

(a) Our approach   (b) Parameter update

Fig. 2. Overview: Our model iteratively updates the estimation based on the feedback signal from the forward process. We adopt a closed-loop scheme to ensure the consistency between the estimation and the observation. We neither require an objective at test time, nor have any restrictions on the forward process. Click here to watch an animated version of the update procedure.

four-fold: First, as each update is trained to aggressively move towards the ground truth, we can accelerate the update procedure and reach the target with much fewer iterations than classic optimization approaches. Second, our approach does not need to explicitly define the energy. Third, we do not have any restrictions on the forward process, such as differentiability, which greatly expands the applicable domains. Finally, in contrast to the conventional learning methods, our method incorporates feedback signals from the forward process so that the network is aware of how close the current estimation is to the ground truth and can react accordingly. The estimated parameters generally lead to results closer to the observation (Fig. 2).

We demonstrate the effectiveness of our approach on three different inverse problems in graphics and robotics: illumination estimation, 6 DoF pose estimation, and inverse kinematics. Compared to traditional optimization based methods, we are able to achieve comparable or better performance while being two to three orders of magnitude faster. Compared to deep learning based approaches, our model consistently improves the performance on all metrics.

2 Background

Let $\mathbf{x} \in \mathcal{X}$ be the hidden parameters of interest and let $\mathbf{y} \in \mathcal{Y}$ be the measurable observations. Denote $f : \mathbf{x} \to \mathbf{y}$ as the deterministic forward process. The aim of inverse problem is to recover $\mathbf{x}$ given the observation $\mathbf{y}$ and the forward mapping f. In the tasks that we consider, $\mathcal{X}$ is a group such as $\mathcal{X} = \mathrm{SE}(3)$ for 6 DoF pose estimation and $\mathcal{X} = \mathbb{R}^3$ when estimating the position of the light source.

2.1 Structured Optimization

Structured optimization methods generally formulate the inverse problem as an energy minimization task [8,11,12,21,29,48,52]:

$$\mathbf{x}^* = \arg\min_{\mathbf{x}} E(\mathbf{x}) = \arg\min_{\mathbf{x}} E_{\mathrm{data}}(f(\mathbf{x}), \mathbf{y}) + \lambda E_{\mathrm{prior}}(\mathbf{x}),$$

Algorithm 1. Deep Feedback Inverse Problem Solver

1: **input** observation $\mathbf{y}$, forward model $f(\cdot)$ and init $\mathbf{x}^0$
2: **for** $iter = 0, 1, \ldots, T - 1$ **do**
3:     Run forward model: $\mathbf{y}^t = f(\mathbf{x}^t)$
4:     Compute update: $\mathbf{x}^{t+1} = \mathbf{x}^t + g_{\mathbf{w}}(\mathbf{x}^t, \mathbf{y}^t, \mathbf{y})$
5: **end for**
6: **output** $\mathbf{x}^T$

---

where the data term E_{data} measures the similarity between the observation $\mathbf{y}$ and the forward simulated results $f(\mathbf{x})$ of the hidden parameters $\mathbf{x}$; and the prior term E_{prior} encodes humans' knowledge about the solution $\mathbf{x}$. As the energy function is often non-convex, iterative algorithms are used to refine the estimation. Without loss of generality, the update rule can be written as:

$$\mathbf{x}^{t+1} = \mathbf{x}^t + g_E(\mathbf{x}^t, \mathbf{y}^t, \mathbf{y}) \tag{1}$$

where $g_E(\mathbf{x}^t, \mathbf{y}^t, \mathbf{y})$ is an analytical update function derived from the energy function E, and $\mathbf{y}^t = f(\mathbf{x}^t)$. For instance, in continuous-valued inverse problems, $g_E = -A_E(\mathbf{x}^t)\nabla_E(\mathbf{x}^t)$, where $\nabla_E(\mathbf{x})$ is the first-order Jacobian and A_E is a warping matrix that depends on the optimization algorithm and the form of the energy. For instance, A_E is simply a (approximated) Hessian matrix in Newton method and is equivalent to the step size in first order gradient descent.

One major advantage of these approaches is that they explicitly take into account how close $f(\mathbf{x})$ and $\mathbf{y}$ are via the data term E_{data}, and exploit such *feedback* as a guidance for the update. This ensures that the result $f(\mathbf{x}^*)$ generated from the final estimation $\mathbf{x}^*$ is close to the observation $\mathbf{y}$. While impressive results have been achieved, there are several challenges remaining: first, they require both the forward process f as well as the prior E_{prior} to be optimization-friendly (*e.g.* differentiable) so that inference algorithms can be applied. Unfortunately this is not the case for many inverse problems and tailored approximations are required [26,39,42,54,67]. The performance may thus be affected. Second, they often require many updates to reach a decent solution (*e.g.* first-order methods). If higher order methods are exploited to speed up the process, the update may become expensive (*e.g.*, second-order methods). Third, carefully designed priors are necessary for identifying the true solution from multiple feasible answers. This is particularly true for ill-posed inverse problems, such as super-resolution and inverse kinematics, in which there exists infinite number of feasible solutions that could generate the observation. Additionally, the energy must be designed in a way that is easy to optimize, which is sometimes non-trivial. Finally, these optimization methods are typically sensitive to the initialization.

2.2   Learning Based Methods

Another line of work [10,30,32,37,71] has been devoted to directly learning a mapping from the observations $\mathbf{y}$ to the solution $\mathbf{x}$:

$$\mathbf{x}^* = g(\mathbf{y}; \mathbf{w}). \tag{2}$$

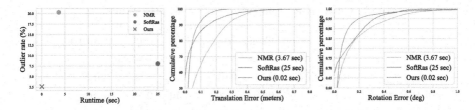

Fig. 3. Quantitative analysis on 6 DoF pose estimation. Our deep optimizer is robust, accurate, and significantly faster.

Here, $g(\cdot; \mathbf{w})$ is a learnable function parameterized by $\mathbf{w}$. These approaches try to capitalize on the feature learning capabilities of deep neural networks to extract statistical priors from data, and approximate the inverse process without the help of any hand-crafted energies. While these methods have achieved state-of-the-art performance in many challenging inverse tasks such as inverse kinematics [51, 74], super-resolution [32,63], compressive sensing [28], image inpainting [41,50], illumination estimation [35,43], reflection separation [73], and image deblurring [46], they ignore the fact that the forward process f is known (Fig. 3).

As a consequence, there is no *feedback* mechanism within the model that scores if $f(\mathbf{x}^*)$ is close to $\mathbf{y}$ after the inference, and the model cannot update the estimation accordingly. The whole system remains *open loop*.

3   Deep Feedback Inverse Problem Solver

In this paper we aim to develop an extremely efficient yet effective approach to solving structured inverse problems. We build our model based on the observation that traditional optimization approaches and current learning based methods are complementary – one is good at exploiting feedback signals as guidance and inducing human priors, while the other excels at learning data-driven inverse mapping. Towards this goal, we present a simple solution that takes the best of both worlds. We first describe our deep feedback network that iteratively updates the solution based on the feedback signal generated by the forward process. Then we demonstrate how to perform efficient inference as well as learning. Finally, we discuss our design choices and the relationships to related work.

3.1   Deep Feedback Network

As we have alluded to above, structured optimization and deep learning have very different yet complementary strengths. Our goal is to bring together the two paradigms, and develop a generic approach to inverse problems.

The key innovation of our approach is to replace the analytical function g_E defined in structured optimization approach at Eq. 1 with a neural network. Specifically, we design a neural network $g_{\mathbf{w}}$ that takes the same set of inputs as g_E and outputs the update. The hope is that the model can perceive the

Table 1. Quantitative comparison on 6 DoF pose estimation.

Methods	Optimization		Trans. error		Rot. error		Outlier (%)
	Step	Time	Mean	Median	Mean	Median	
NMR [26]	105	3.67 s	0.1	0.05	5.78	1.68	20.3
SoftRas [42]	157	25 s	0.05	0.003	4.14	0.5	8.03
Deep regression	1	0.004 s	0.07	0.06	10.07	7.68	5
Ours	5	0.02 s	0.02	0.009	2.64	1.02	2.6

difference between the observation $\mathbf{y}$ and the simulated forward results $\mathbf{y}^t$ and then predict a new solution based on the *feedback* signal. In practice, we employ a simple addition rule and fold the step size, parameter priors all into $g_{\mathbf{w}}$:

$$\mathbf{x}^{t+1} = \mathbf{x}^t + g_{\mathbf{w}}(\mathbf{x}^t, \mathbf{y}^t, \mathbf{y}), \quad \text{where } \mathbf{y}^t = f(\mathbf{x}^t). \tag{3}$$

The network architectures design depends on the form of observational data $\mathbf{y}$ and solution $\mathbf{x}$. For instance, for inverse graphics tasks, we utilize convolutional neural networks, since the observations are images. This not only allows us to sidestep all requirements imposed on f (*e.g.* differentiability), but also removes the need for explicitly defining energies. Unlike conventional learning based methods, we take both $\mathbf{y}^t$ and $\mathbf{y}$ as input to the update so that we incorporate the feedback signal through comparing the two.

We derive our final deep structured inverse problem solver by applying the aforementioned update functions in an iterative manner. The algorithm is summarized in Algorithm 1. At each step, the solver takes as input the current solution $\mathbf{x}^t$, the observation $\mathbf{y}$, and the forward simulated results $\mathbf{y}^t$, and predicts the next best solution as defined in Eq. 3. In practice, the stopping criteria could either be based on a predefined iteration number or on checking convergence by measuring the difference between solutions from two consecutive iterations.

3.2 Learning

The full deep structured inverse problem solver can be learned in an end-to-end fashion via back-propagation through time (BPTT). Yet in practice we find that applying loss function over each stage's intermediate solution $\mathbf{x}^t$ yields better results. Deep supervision greatly accelerates the speed of convergence.

However, it is non-trivial to design a learning procedure for each iterative update function $g_{\mathbf{w}}$, as there exist infinite paths towards the ideal solution. Ideally, we would like our solution to descend towards the ideal solution as quickly as possible. Thus, inspired by [70], at each iteration, we learn to aggressively predict the update required to reach the ideal solution. At each stage, the learning procedure finds the best $\mathbf{w}$ through minimizing the following loss function:

$$\arg\min_{\mathbf{w}} \sum_{(\mathbf{y}, \mathbf{x}_{\text{gt}})} \sum_t \ell(\mathbf{x}_{\text{gt}}, \mathbf{x}^t + g_{\mathbf{w}}(\mathbf{x}^t, \mathbf{y}^t, \mathbf{y})).$$

ℓ is a task-specific loss function; for instance, ℓ is l2-norm for inverse kinematics.

GT   NMR   SoftRas Regress.   Ours   GT   NMR   SoftRas Regress.   Ours

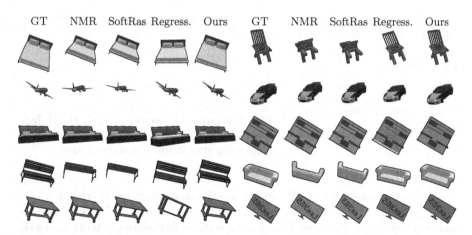

Fig. 4. Qualitative comparison on 6 DoF pose estimation: We infer the poses from only silhouette images. The rendered colored images in the figure are for visualization purpose. (Color figure online)

3.3   Discussions

Stage-Wise Network: In our standard approach described before, $g_\mathbf{w}$ is shared across all steps. However, the proximity to the ideal solution varies at different step. As a consequence, early iteration often takes inputs that are farther to the ideal solution than what a late iteration update step takes. This brings difficulties to the network as it needs to handle a variety of output scales across different iteration steps. This motivates us to train a separate update function per step $g_\mathbf{w}^t(\mathbf{x}^t, \mathbf{y}^t, \mathbf{y})$ that better captures the input data distribution at each iteration. To learn this non-shared weight network, we conduct a stage-wise training procedure. We start to train the $g_\mathbf{w}^0$ first. Then acquire $\mathbf{y}^0$ for all the training data, which allow us to train $g_\mathbf{w}^1$. We repeat this procedure until $g_\mathbf{w}^T$ is trained. In total T models $\{g_\mathbf{w}^t\}$ are trained. Please refer to the supp. material for the comparison between sharing weights and not sharing weights.

Adaptive Update: Our current update rule is simply an addition, yet it can be easily extended to more sophisticated settings to handle more complex scenarios. For instance, one can apply the classic momentum technique on top of the predicted gradient to stabilize the optimization trajectory. One can also learn another meta-network to dynamically adjust the output of our update network. While all of these options are feasible, we find that in practice a simple strategy suffices. Inspired by the Levenberg-Marquardt method [4], we exploit a damping factor λ to control the effectiveness of the update network, *i.e.*, $\mathbf{x}^{t+1} = \mathbf{x}^t + \lambda \cdot g_\mathbf{w}(\mathbf{x}^t, \mathbf{y}^t, \mathbf{y})$. Specifically, λ is initialized to 1 at the beginning of each update. If the new estimation results in a lower data energy than that of the original one, we update the estimation. Otherwise we reduce λ by half and re-compute. We only need to compute the update gradient once. The forward

Table 2. Runtime breakdown of a single optimization step for 6 DoF pose estimation.

Module	Forward rendering	Inverse update	Total
NMR [26]	28 ms	7 ms	35 ms
SoftRas [42]	76 ms	84 ms	160 ms
Ours	2.6 ms	0.9 ms	3.5 ms

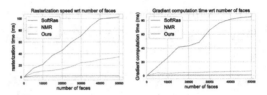

Fig. 5. Runtime vs number of faces. (Left) Forward rasterization time. (Right) Backward gradient computation (inverse update) time.

process is executed on the GPU and hence the computational overhead is negligible. Through this simple rule, we can guarantee that $E_{\text{data}}(\mathbf{x}^t, \mathbf{y})$ decreases after every iteration. Empirically $\mathbf{x}^t$ becomes closer to the ground truth $\mathbf{x}$ as well, since the ambiguity arising from the data term disappears when the estimation is already sufficiently close.

Relationship to Existing Work: Our model is closely related to the family of iterative networks [5,7,16,17,36,38,40,53,58,59,65,66], in particular the stacked inference machines [7,53,58,65,66]. Unlike previous methods that require the model to implicitly learn the relationship between the input and the preceding estimation, we leverage the forward process to explicitly establish the connection among them and close the loop. This is of crucial importance for inverse problems since the two spaces are very distinct (*e.g.* illumination parameters vs RGB image). The idea of learning to update is inspired by supervised descent methods [70]. However, unlike their approach we learn the mapping and the feature simultaneously. Furthermore, we focus on inverse problems and design a closed-loop scheme to incorporate feedback signals, while they simply perform iterative update in an open loop setting. Developed independently, Flynn *et al.* [14] propose a similar approach for view synthesis. Their model, however, relies on the analytical gradient components. They thus requires the system to be differentiable. In contrast, our approach directly predicts the update from the observation and the feedback signal. We do not require explicit gradient computation and thus do not have such a limitation. Please refer to the supp. material for more discussion on reinforcement learning and other prior art [7,34].

Applicability: Unlike previous work, our approach neither has restrictions on the forward process f, nor need to construct domain-specific objectives at test time. During inference, at each iteration, we simply adopt a feed-forward operation g on top of current estimate and predict the update. Our method is applicable to a wide range of tasks so long as the forward process function f is available. In the following sections, we showcase our approach on two different inverse graphics tasks (object pose estimation and illumination estimation from a single image) as well as one robotics task (inverse kinematics) (Table 3).

Table 3. Test on unseen rotations.

Training on 0°–40°	Trans. error		Rot. error (°)	
Evaluation Rot. Range	Mean	Median	Mean	Median
40°–45°	0.05	0.03	11.33	4.97
45°–50°	0.05	0.04	15.62	5.60
50°–55°	0.06	0.04	18.58	6.86
55°–60°	0.07	0.05	24.14	9.58

4 Application I: 6-DoF Object Pose Estimation

Problem Formulation: Assume that the 3D model of the object is given [6,20] and the camera intrinsic parameters are known. For a given object pose w.r.t. the camera, denoted as $x \in SE(3)$, we can generate the corresponding image observation y through a forward rendering function $f : x \rightarrow y$, powered by a graphics engine. The goal of 6 DoF pose estimation is to *invert* the process and recover the latent pose x from the observation image y. This problem is particularly important for problems such as robot grasping [34] and self-driving [44]. Unlike previous approaches that leverage RGB information or depth geometry to guide the pose estimation, we focus on a more challenging setting where the observation is *a single silhouette image* $y \in \{0,1\}^{H \times W}$. The object pose $x = (x_{quat}; x_{trans})$ is represented by a unit quaternion for rotation x_{quat} and a 3D translation vector x_{trans}.

Data: We use the 3D CAD models from ShapeNet [9] within 10 categories: cars, planes, chairs, bench, table, sofa, cabinet, bed, monitor, and couch. The dataset is split into training (70%), validation (10%) and testing (20%). For each object, we randomly sample an axis from the unit sphere and rotate the object around the axis by $\theta \sim [-40, 40]$ degrees. We further translate the object along each axis by a random offset within $[-0.2, 0.2]$ meters. Given the randomly generated ground truth object poses, we render 128×128 silhouette images with non-differentiable PyRender [1] as input observations. We refer the readers to the supp. material for the performance of our model on other image sizes.

Metrics: We measure the translation error with euclidean distance and the rotation error with angular difference. Inspired by [15], we also compute the *outlier ratio* as an indicator of the general quality of the output. Specifically, we define the prediction to be an outlier if the translation error is higher than 0.2 or the rotation error is larger than 30°.

Network Architecture: Our deep feedback network g_w is akin to the classic LeNet [31]. It takes as input the rendered image $y^t = f(x^t)$, the observed image y, as well as the difference image $\hat{y} - y^t$, and directly outputs the update Δx. We apply an additional normalization operator over the rotation component to correct it to a valid unit quaternion. We unroll our deep feedback network for

Table 4. Illumination estimation on ShapeNet.

Methods	Optimization		Directional light			Point light		
	Step	Time	Mean	Median	Outliers	Mean	Median	Outliers
NMR[a] [26]	166.7	58.3 s	0.099	0.037	19.2%	–	–	–
Deep regression [22]	1	0.043 s	0.067	0.022	24%	0.111	0.084	11%
Ours	7	0.183 s	0.052	0.008	8%	0.084	0.064	9%

[a]NMR does not support point light. Furthermore, its directional light is highly simplified and did not consider self-occlusion.

five steps. MSE is employed as the loss function for both rotation and translation since it produces the most stable results.

Baselines: For optimization methods, the energy function consists of a data term $E_{data}(f(\mathbf{x}), \mathbf{y})$ that favors agreement and a prior term $E_{prior}(\mathbf{x})$ that encourages the quaternion to remain on the manifold. To make the forward rendering procedure f differentiable, we utilize the state-of-the-art differentiable renderers for comparison, *i.e.* neural mesh renderer (NMR [26]) and soft rasterization (SoftRas [42]). We utilize the following stopping criteria for the optimizer: (i) 500 iterations, or (ii) the silhouette difference between the observation and the one generated by the renderer stops improving for 20 iterations. For the deep regression method, we use the same architecture as our deep feedback network except that no feedback is provided.

Results: As shown in Table 1, our method achieves a significantly lower outlier ratio compared to other approaches. This indicates that our model is more robust and less susceptible to becoming stuck in local optimum. It also has comparable performance to differentiable renderers in terms of mean translation and angular error, while being two to three orders of magnitude faster. On the other hand, our method has much better performance than the non-feedback deep regression method. For the category-wise performance, please refer to the supp. material. Figure 4 showcases some qualitative results. Our method is robust to extreme poses, whereas optimization based method is easy to get stuck in a local optimum.

Deep Feedback Network as Initialization: Due to the highly non-convex structure of the energy model, a good initialization is required for optimization methods to achieve good performance. One natural solution is to exploit our model as an initialization and employ classic solvers for the final optimization. By combining our approach with SoftRas, we can further reduce the error by more than 50%. We refer the readers to supp. material for detailed analysis.

Runtime Analysis: We show the runtime break down for a single update step in Table 2 and the runtime w.r.t. the number of faces in Fig. 5. As we neither need to construct the computation graph nor storing any activation value for gradient computation during the forward rasterization process, our rendering is significantly faster. For gradient computation, SoftRas is far slower as it needs to propagate the gradient to multiple faces. In contrast, our update model is

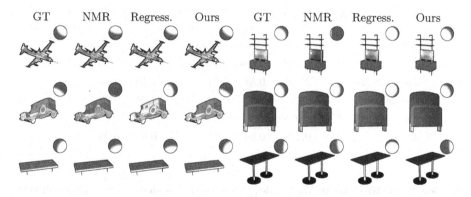

GT   NMR   Regress.   Ours   GT   NMR   Regress.   Ours

Fig. 6. Qualitative comparison on illumination estimation.

simply an efficient feed-forward neural net that takes as input the (difference) silhouette images. Its speed is invariant to the number of faces.

5   Application II: Illumination Estimation

Problem Formulation: We next evaluate our method on the task of illumination estimation. The goal is to recover the lighting parameter $\mathbf{x} \in \mathbb{R}^3$ from the observation RGB image $\mathbf{y} \in \mathbb{R}^{H \times W \times 3}$. It has critical applications in image relighting and photo-realistic rendering [25]. As in the 6-DoF pose estimation task, we assume the 3D model is given.

Data: We use the same dataset as the 6-DoF pose estimation experiment for the illumination estimation experiment. Specifically, we consider two types of light source: directional light and point light. The two light sources are complementary and can result in very different rendering effect. During training, we randomly sample the light position from the half unit sphere on the camera side [22,43]. If the light is directional, we point the light towards the origin. All the objects are set to have Lambertian surfaces. We ignore the scenario where the light source lies on the other side of the object, as it has no effect on the rendered image. For evaluation, we follow the same criteria. We perform rendering in pyrender and the image size is set to 256×256. Empirically we found this size provides the best balance between performance and the computational speed.

Metrics: Following [22], we use the standard mean-squared error (MSE) between the ground truth light and estimated light pose to measure the difference. We also compute the outlier rate as described in Sect. 4.

Table 5. Quantitative results on CMU MoCap.

Methods	Optimization		Position error (cm)		Rotation error (°)	
	Step	Time	Mean	Median	Mean	Median
L-BFGS [18]	73	27.9 s	0.38	0.01	7.19	4.68
Adam [27]	196	38.8 s	0.04	0.04	7.96	7.92
Deep6D [74]	1	0.012 s	1.9	1.6	–	–
Ours	4	0.12 s	0.64	0.36	0.88	0.03

Network Architecture: We employ an encoder-decoder architecture with skip connections as our deep feedback network. Since the 3D geometry of the object plays an important role during rendering, we adopt depth prediction as an auxiliary task. This allows the model to implicitly capture such notion and reason about its relationship with illumination. During training, our deep feedback network estimates both the depth of the object as well as the illumination parameters. We use MSE as the objective for both tasks. During inference, we simply discard the depth decoder and output only the illumination part. We unroll our network for 7 steps according to the validation performance.

Baselines: We exploit NMR [26] to minimize the energy $E_{\text{data}} + E_{\text{prior}}$. The data term is the ℓ_2 distance between the observation image and the rendered image, while the prior term constrains the light source to lie on the sphere. We adopt the same stopping criteria as in Sect. 4. The size of the rendered image is set to 256×256 based on the performance on the validation set. For deep regression method, we exploit the state-of-the-art model from Janner *et al.* [22].

Results: As shown in Table 4, our deep feedback network outperforms the baselines on both setup. The improvement is significant especially in the directional light case. We conjecture this is because the intensity of directional light does not decay w.r.t. the travel distance, and the signals from the image are weaker. Learning based approaches thus have to rely on feedback signals to refine the light direction. The performance of the optimization method is limited by the hand-crafted energy as well as the capability of renderer. NMR is sub-optimal as it approximates the gradient with a manually designed function and does not handle self-occlusion. In contrast, our method allows us to exploit complex rendering machines as the forward model as we do not require it to be differentiable. We note that we only report the optimization results on directional light since NMR does not support point light source. Figure 6 depicts the qualitative comparison against the baselines. It is clear that our deep feedback mechanism is able to recover accurate lighting information based on subtle difference between the forward results and the observations.

GT   Step 1   Step 3   GT   Step 1   Step 3

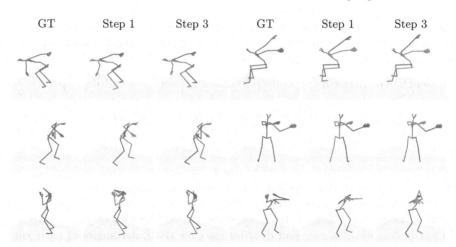

Fig. 7. Qualitative results on CMU MoCap: Our approach is able to accurately predict the joint rotations within a few steps. It can also correct wrong estimations through the feedback from the forward model (see the feet/toes in the right column). Bottom right shows an example where our model fails.

6 Application III: Inverse Kinematics

Problem Formulation. Finally we exploit how our proposed method to tackle the inverse kinematics problem. Given the 3D location of the joints of a reference pose $\mathbf{y}_{1:N}^{\text{ref}}$ and the desired joint rotations $\mathbf{x}_{1:N} \in SO(3)$, the forward kinematics function f rotates the joints and computes their 3D positions by recursively applying the follow update rule from parents to children: $\mathbf{y}_n = \mathbf{y}_{\text{parent}(n)} + \mathbf{x}_n(\mathbf{y}_n^{\text{ref}} - \mathbf{y}_{\text{parent}(n)}^{\text{ref}})$. The goal of inverse kinematics is to recover the $SO(3)$ rotations $\mathbf{x}_{1:N}$ that ensure the specific joints are placed at the desired 3D locations $\mathbf{y}_{1:N}$. Inverse kinematics has a wide range of applications, such as robot arm manipulation, legged robot motion planning and computer re-animation. The problem is inherently ill-posed as different rotations can result in the same observation through the forward kinematics function f, *i.e.*, $\mathbf{y} = f(\mathbf{x}_{1:N}) = f(\mathbf{x}'_{1:N})$. However, not all angles are feasible or natural due to the dynamic constraints. Therefore, in order to accurately recover the rotations, one has to either come up with a powerful prior or learn it from data. In this paper, we focus on inverse kinematics over human body skeletons.

Data: We validate our model on the CMU Motion Capture Dataset (CMU MoCap) as it contains complex human motions and a diverse range of joint rotations. Following Yi *et al.* [74], we select 865 motion clips from 37 motion categories and hold out 37 clips for testing. Each skeleton in the dataset has 57 joints. We fix the position of the hip to remove the effect of global motion.

Metrics: We evaluate the performance of our model with joint position error [74] and joint angular error [45,51]. The two metrics are complementary since a small rotation error may result in a large position error due to the recursive nature of the forward kinematics model, and small position error cannot guarantee correct joint rotation due to ambiguities.

Network Architecture: Our deep feedback network is a multilayer perception akin to [74]. Following [51,62], the network takes as input the estimated joint position, reference joint position, as well as the difference between the two, and outputs a rotation for each joint. We unroll our model three steps. We train the network with L2 loss on both position error and rotation error.

Baselines: We compare our model against two optimization-based approaches and one deep regression method. For optimization methods, we employ joint position error as our data term, *i.e.* $E_{\text{data}}(f(\mathbf{x}), \mathbf{y}) = \|f(\mathbf{x}) - \mathbf{y}\|_2^2$, and derive a prior energy term from data to alleviate the ambiguities of joint rotations. In particular, we fit a gaussian distribution over the Euler angles of each joint from training data and employ it as a regularization term during inference. We set the weight of the prior term to 0.001 and optimize both energies jointly. We exploit two different types of optimizers: a first-order method (*i.e.*, Adam [27]) and a quasi-Newton method (*i.e.*, L-BFGS [18]). For deep regression method, we compare with the current state of the art (Deep6D [74]).

Results: As shown in Table 5, our deep feedback network outperforms the baselines on the rotation metric and achieve comparable performance on the position error. By unrolling more steps and gathering feedback signals from the forward model, we are able to reduce incorrect estimation and improve the performance (see the Fig. 7). We refer the readers to the supp. material for detailed analysis. On average, a single step of L-BFGS, Adam, and our approach takes 383 ms, 198 ms, 30 ms respectively. L-BFGS takes longer to compute as it needs to conduct gradient evaluation multiple times to approximate the Hessian. Adam is faster in terms of computation, yet it takes far more steps to converge. Our approach, in comparison, is significantly faster and better.

7   Conclusions

In this paper, we propose a deep feedback inverse problem solver. Our method combines the strength of both learning-based approaches and optimization-based methods. Specifically, it learns to conduct an iterative update over the current solution based on the feedback signals provided from the forward process of the problem. Unlike prior work, it does not have any restrictions on the forward process. Further, it learns to conduct an update without explicitly define an objective function. Our results showcase that the proposed method is extremely effective, efficient, and widely applicable.

References

1. Pyrender (2020). https://github.com/mmatl/pyrender
2. Barron, J.T., Malik, J.: Shape, illumination, and reflectance from shading. TPAMI **37**, 1670–1687 (2014)
3. Bell, S., Bala, K., Snavely, N.: Intrinsic images in the wild. TOG **33**, 1–12 (2014)
4. Boyd, S., Vandenberghe, L.: Convex Optimization. Cambridge University Press, Cambridge (2004)
5. Byeon, W., Breuel, T.M., Raue, F., Liwicki, M.: Scene labeling with LSTM recurrent neural networks. In: CVPR (2015)
6. Cao, Z., Sheikh, Y., Banerjee, N.K.: Real-time scalable 6DOF pose estimation for textureless objects. In: ICRA (2016)
7. Carreira, J., Agrawal, P., Fragkiadaki, K., Malik, J.: Human pose estimation with iterative error feedback. In: CVPR (2016)
8. Chan, T.F., Shen, J., Zhou, H.M.: Total variation wavelet inpainting. J. Math. Imag. Vis. **25**, 107–125 (2006)
9. Chang, A.X., et al.: ShapeNet: an information-rich 3D model repository. arXiv (2015)
10. Dong, C., Loy, C.C., He, K., Tang, X.: Learning a deep convolutional network for image super-resolution. In: Fleet, D., Pajdla, T., Schiele, B., Tuytelaars, T. (eds.) ECCV 2014. LNCS, vol. 8692, pp. 184–199. Springer, Cham (2014). https://doi.org/10.1007/978-3-319-10593-2_13
11. Dong, W., Zhang, L., Shi, G., Wu, X.: Image deblurring and super-resolution by adaptive sparse domain selection and adaptive regularization. TIP **20**, 1838–1857 (2011)
12. Donoho, D.L.: De-noising by soft-thresholding. Trans. Inf. Theory **41**, 613–627 (1995)
13. Epstein, D., Chen, B., Vondrick, C.: Oops! Predicting unintentional action in video. arXiv (2019)
14. Flynn, J., et al.: DeepView: view synthesis with learned gradient descent. arXiv (2019)
15. Geiger, A., Lenz, P., Urtasun, R.: Are we ready for autonomous driving? The KITTI vision benchmark suite. In: Conference on Computer Vision and Pattern Recognition (CVPR) (2012)
16. Gkioxari, G., Toshev, A., Jaitly, N.: Chained predictions using convolutional neural networks. In: Leibe, B., Matas, J., Sebe, N., Welling, M. (eds.) ECCV 2016. LNCS, vol. 9908, pp. 728–743. Springer, Cham (2016). https://doi.org/10.1007/978-3-319-46493-0_44
17. Greff, K., Srivastava, R.K., Schmidhuber, J.: Highway and residual networks learn unrolled iterative estimation. arXiv (2016)
18. Grochow, K., Martin, S.L., Hertzmann, A., Popović, Z.: Style-based inverse kinematics. In: TOG (2004)
19. He, K., Sun, J., Tang, X.: Single image haze removal using dark channel prior. TPAMI **33**, 2341–2353 (2010)
20. Hinterstoisser, S., et al.: Model based training, detection and pose estimation of texture-less 3D objects in heavily cluttered scenes. In: Lee, K.M., Matsushita, Y., Rehg, J.M., Hu, Z. (eds.) ACCV 2012. LNCS, vol. 7724, pp. 548–562. Springer, Heidelberg (2013). https://doi.org/10.1007/978-3-642-37331-2_42
21. Huang, J.B., Singh, A., Ahuja, N.: Single image super-resolution from transformed self-exemplars. In: CVPR (2015)

22. Janner, M., Wu, J., Kulkarni, T.D., Yildirim, I., Tenenbaum, J.: Self-supervised intrinsic image decomposition. In: NeurIPS (2017)
23. Kanazawa, A., Black, M.J., Jacobs, D.W., Malik, J.: End-to-end recovery of human shape and pose. In: CVPR (2018)
24. Kanazawa, A., Tulsiani, S., Efros, A.A., Malik, J.: Learning category-specific mesh reconstruction from image collections. In: Ferrari, V., Hebert, M., Sminchisescu, C., Weiss, Y. (eds.) ECCV 2018. LNCS, vol. 11219, pp. 386–402. Springer, Cham (2018). https://doi.org/10.1007/978-3-030-01267-0_23
25. Karsch, K., Hedau, V., Forsyth, D., Hoiem, D.: Rendering synthetic objects into legacy photographs. TOG **30**, 1–12 (2011)
26. Kato, H., Ushiku, Y., Harada, T.: Neural 3D mesh renderer. In: CVPR (2018)
27. Kingma, D.P., Ba, J.: Adam: a method for stochastic optimization. arXiv (2014)
28. Kulkarni, K., Lohit, S., Turaga, P., Kerviche, R., Ashok, A.: ReconNet: non-iterative reconstruction of images from compressively sensed measurements. In: CVPR (2016)
29. Laffont, P.Y., Bazin, J.C.: Intrinsic decomposition of image sequences from local temporal variations. In: ICCV (2015)
30. Lai, W.S., Huang, J.B., Ahuja, N., Yang, M.H.: Deep Laplacian pyramid networks for fast and accurate super-resolution. In: CVPR (2017)
31. LeCun, Y., Bottou, L., Bengio, Y., Haffner, P., et al.: Gradient-based learning applied to document recognition. Proc. IEEE **86**, 2278–2324 (1998)
32. Ledig, C., et al.: Photo-realistic single image super-resolution using a generative adversarial network. In: CVPR (2017)
33. Levin, A.: Blind motion deblurring using image statistics. In: NeurIPS (2007)
34. Li, Y., Wang, G., Ji, X., Xiang, Yu., Fox, D.: DeepIM: deep iterative matching for 6D pose estimation. In: Ferrari, V., Hebert, M., Sminchisescu, C., Weiss, Y. (eds.) ECCV 2018. LNCS, vol. 11210, pp. 695–711. Springer, Cham (2018). https://doi.org/10.1007/978-3-030-01231-1_42
35. Li, Z., Snavely, N.: Learning intrinsic image decomposition from watching the world. In: CVPR (2018)
36. Liang, M., Hu, X.: Recurrent convolutional neural network for object recognition. In: CVPR (2015)
37. Lin, C.H., Kong, C., Lucey, S.: Learning efficient point cloud generation for dense 3D object reconstruction. In: AAAI (2018)
38. Lin, C.H., Lucey, S.: Inverse compositional spatial transformer networks. In: CVPR (2017)
39. Lin, C.H., et al.: Photometric mesh optimization for video-aligned 3D object reconstruction. In: CVPR (2019)
40. Lin, C.H., Yumer, E., Wang, O., Shechtman, E., Lucey, S.: ST-GAN: spatial transformer generative adversarial networks for image compositing. In: CVPR (2018)
41. Liu, G., Reda, F.A., Shih, K.J., Wang, T.-C., Tao, A., Catanzaro, B.: Image inpainting for irregular holes using partial convolutions. In: Ferrari, V., Hebert, M., Sminchisescu, C., Weiss, Y. (eds.) ECCV 2018. LNCS, vol. 11215, pp. 89–105. Springer, Cham (2018). https://doi.org/10.1007/978-3-030-01252-6_6
42. Liu, S., Chen, W., Li, T., Li, H.: Soft rasterizer: differentiable rendering for unsupervised single-view mesh reconstruction. arXiv (2019)
43. Ma, W.-C., Chu, H., Zhou, B., Urtasun, R., Torralba, A.: Single image intrinsic decomposition without a single intrinsic image. In: Ferrari, V., Hebert, M., Sminchisescu, C., Weiss, Y. (eds.) Computer Vision – ECCV 2018. LNCS, vol. 11218, pp. 211–229. Springer, Cham (2018). https://doi.org/10.1007/978-3-030-01264-9_13

44. Ma, W.C., Wang, S., Hu, R., Xiong, Y., Urtasun, R.: Deep rigid instance scene flow. In: CVPR (2019)
45. Martinez, J., Black, M.J., Romero, J.: On human motion prediction using recurrent neural networks. In: CVPR (2017)
46. Nah, S., Hyun Kim, T., Mu Lee, K.: Deep multi-scale convolutional neural network for dynamic scene deblurring. In: CVPR (2017)
47. Oh, B.M., Chen, M., Dorsey, J., Durand, F.: Image-based modeling and photo editing. In: SIGGRAPH (2001)
48. Pan, J., Hu, Z., Su, Z., Yang, M.H.: L_0-regularized intensity and gradient prior for deblurring text images and beyond. TPAMI **39**, 342–355 (2016)
49. Pan, J., Sun, D., Pfister, H., Yang, M.H.: Blind image deblurring using dark channel prior. In: CVPR (2016)
50. Pathak, D., Krahenbuhl, P., Donahue, J., Darrell, T., Efros, A.A.: Context encoders: feature learning by inpainting. In: CVPR (2016)
51. Pavllo, D., Grangier, D., Auli, M.: QuaterNet: a quaternion-based recurrent model for human motion. In: BMVS (2018)
52. Portilla, J., Strela, V., Wainwright, M.J., Simoncelli, E.P.: Image denoising using scale mixtures of Gaussians in the wavelet domain. TIP **12**, 1338–1351 (2003)
53. Ramakrishna, V., Munoz, D., Hebert, M., Andrew Bagnell, J., Sheikh, Y.: Pose machines: articulated pose estimation via inference machines. In: Fleet, D., Pajdla, T., Schiele, B., Tuytelaars, T. (eds.) ECCV 2014. LNCS, vol. 8690, pp. 33–47. Springer, Cham (2014). https://doi.org/10.1007/978-3-319-10605-2_3
54. Ravi, N., et al.: PyTorch3D (2020). https://github.com/facebookresearch/pytorch3d
55. Rick Chang, J., Li, C.L., Poczos, B., Vijaya Kumar, B., Sankaranarayanan, A.C.: One network to solve them all-solving linear inverse problems using deep projection models. In: ICCV (2017)
56. Rother, C., Kiefel, M., Zhang, L., Schölkopf, B., Gehler, P.V.: Recovering intrinsic images with a global sparsity prior on reflectance. In: NeurIPS (2011)
57. Shocher, A., Cohen, N., Irani, M.: "Zero-shot" super-resolution using deep internal learning. In: CVPR (2018)
58. Toshev, A., Szegedy, C.: DeepPose: human pose estimation via deep neural networks. In: CVPR (2014)
59. Tu, Z.: Auto-context and its application to high-level vision tasks. In: CVPR (2008)
60. Tung, H.Y., Tung, H.W., Yumer, E., Fragkiadaki, K.: Self-supervised learning of motion capture. In: NeurIPS (2017)
61. Tung, H.Y.F., Harley, A.W., Seto, W., Fragkiadaki, K.: Adversarial inverse graphics networks: learning 2D-to-3D lifting and image-to-image translation from unpaired supervision. In: ICCV (2017)
62. Villegas, R., Yang, J., Ceylan, D., Lee, H.: Neural kinematic networks for unsupervised motion retargetting. In: CVPR (2018)
63. Wang, X., et al.: ESRGAN: enhanced super-resolution generative adversarial networks. In: Leal-Taixé, L., Roth, S. (eds.) ECCV 2018. LNCS, vol. 11133, pp. 63–79. Springer, Cham (2019). https://doi.org/10.1007/978-3-030-11021-5_5
64. Wang, Z., Yang, Y., Wang, Z., Chang, S., Yang, J., Huang, T.S.: Learning super-resolution jointly from external and internal examples. TIP **24**, 4359–4371 (2015)
65. Wei, S.E., Ramakrishna, V., Kanade, T., Sheikh, Y.: Convolutional pose machines. In: CVPR (2016)
66. Weiss, D., Taskar, B.: Structured prediction cascades. In: AISTATS (2010)
67. Wiles, O., Gkioxari, G., Szeliski, R., Johnson, J.: SynSin: end-to-end view synthesis from a single image. arXiv (2019)

68. Wu, J., Lim, J.J., Zhang, H., Tenenbaum, J.B.: Physics 101: learning physical object properties from unlabeled videos
69. Wu, J., Yildirim, I., Lim, J.J., Freeman, B., Tenenbaum, J.: Galileo: perceiving physical object properties by integrating a physics engine with deep learning. In: NeurIPS (2015)
70. Xiong, X., De la Torre, F.: Supervised descent method and its applications to face alignment. In: CVPR (2013)
71. Xu, L., Ren, J.S., Liu, C., Jia, J.: Deep convolutional neural network for image deconvolution. In: NeurIPS (2014)
72. Yao, S., et al.: 3D-aware scene manipulation via inverse graphics. In: NeurIPS (2018)
73. Zhang, X., Ng, R., Chen, Q.: Single image reflection separation with perceptual losses. In: CVPR (2018)
74. Zhou, Y., Barnes, C., Lu, J., Yang, J., Li, H.: On the continuity of rotation representations in neural networks. In: CVPR (2019)

Learning From Multiple Experts: Self-paced Knowledge Distillation for Long-Tailed Classification

Liuyu Xiang[1], Guiguang Ding[1(✉)], and Jungong Han[2]

[1] Beijing National Research Center for Information Science and Technology (BNRist), School of Software, Tsinghua University, Beijing, China
xiangly17@mails.tsinghua.edu.cn, dinggg@tsinghua.edu.cn
[2] Computer Science Department, Aberystwyth University, Aberystwyth SY23 3FL, UK
jungonghan77@gmail.com

Abstract. In real-world scenarios, data tends to exhibit a long-tailed distribution, which increases the difficulty of training deep networks. In this paper, we propose a novel self-paced knowledge distillation framework, termed Learning From Multiple Experts (LFME). Our method is inspired by the observation that networks trained on less imbalanced subsets of the distribution often yield better performances than their jointly-trained counterparts. We refer to these models as 'Experts', and the proposed LFME framework aggregates the knowledge from multiple 'Experts' to learn a unified student model. Specifically, the proposed framework involves two levels of adaptive learning schedules: Self-paced Expert Selection and Curriculum Instance Selection, so that the knowledge is adaptively transferred to the 'Student'. We conduct extensive experiments and demonstrate that our method is able to achieve superior performances compared to state-of-the-art methods. We also show that our method can be easily plugged into state-of-the-art long-tailed classification algorithms for further improvements.

1 Introduction

Deep convolutional neural networks (CNNs) have achieved remarkable success in various computer vision applications such as image classification, object detection and face recognition. Training a CNN typically relies on carefully collected large-scale datasets, such as ImageNet [6] and MS COCO [30] with hundreds of examples for each class. However, collecting such a uniformly distributed dataset in real-world scenarios is usually difficult since the underlying natural data distribution tends to exhibit a long-tailed property with few majority classes (head) and large amount of minority classes (tail) [36,37,51]. When deep models are trained under such imbalanced distribution, they are unlikely to achieve the expected performances which necessitates developing relevant algorithms.

Recent approaches tackle this problem mainly from two aspects. First is via re-sampling schemes or cost-sensitive loss functions to alleviate the negative

© Springer Nature Switzerland AG 2020
A. Vedaldi et al. (Eds.): ECCV 2020, LNCS 12350, pp. 247–263, 2020.
https://doi.org/10.1007/978-3-030-58558-7_15

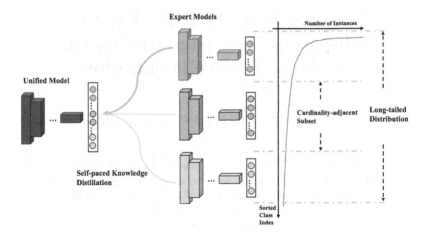

Fig. 1. Schematic illustration of our proposed method.

impact of data imbalance. Second is by head-to-tail knowledge transfer, where prior knowledge or induction bias is learned from the richly annotated classes and generalize to the minority ones.

Orthogonal to the above two perspectives, we propose a novel self-paced knowledge distillation method which can be easily plugged into previous methods. Our method is motivated by an interesting observation that learning a more uniform distribution with fewer samples is sometimes easier than learning a long-tailed distribution with more samples [37]. We first introduce four metrics to measure the *'longtailness'* of a long-tailed distribution. We then show that if we sort all the categories according to their cardinality, then splitting the entire long-tailed dataset into subsets will lead to a smaller *longtailness*, which indicates that they suffer a less severe data imbalance problem. Therefore training a CNN on these subsets is expected to perform better than their jointly-trained counterparts. For clarity, we refer to such a subset as **cardinality-adjacent subset**, and the CNN trained on these subsets as **Expert Models**.

Once we acquire the well-trained expert models, they can be utilized as guidance to train a unified student model. If we take a look at human learning process as students, we can conclude two characteristics: (1) the student often takes various courses from easy to hard, (2) as the learning proceeds, the student acquires more knowledge from self-learning than from teachers and he/she may even exceed his/her teachers. Inspired by these findings, we propose a *Learning From Multiple Experts (LFME)* framework with two levels of adaptive learning schemes, termed as self-paced expert selection and curriculum instance selection. Specifically, the self-paced expert selection automatically controls the impact of knowledge distillation from each expert, so that the learned student model will gradually acquire the knowledge from the experts, and finally exceed the expert. The curriculum instance selection, on the other hand, designs a curriculum for the unified model where the training samples are organized from easy to hard, so that the unified student model will receive a less challenging learning schedule, and gradually learns from easy to hard samples. A schematic illustration of our LFME framework is shown in Fig. 1.

To verify the effectiveness of our proposed framework, we conduct extensive experiments on three benchmark long-tailed classification datasets, and show that our method is able to yield superior performances compared to the state-of-the-art methods. It is worth noting that our method can be easily combined with other state-of-the-art methods and achieve further improvements. Moreover, we conduct extensive ablation studies to verify the contribution of each component.

Our contributions can be summarized as follows: (1) We introduce four metrics for evaluating the *'longtailness'* of a distribution and further propose a *Learning From Multiple Experts* knowledge distillation framework. (2) We propose two levels of adaptive learning schemes, i.e. model level and instance level, to learn a unified *Student* model. (3) Our proposed method achieves state-of-the-art performances on three benchmark long-tailed classification datasets, and can be easily combined with state-of-art methods for further improvements.

2   Related Work

Long-Tailed, Data-Imbalanced Learning. The long-tailed learning problem has been comprehensively studied due to the prevalence of data imbalance problem [17,37]. Most previous methods tackle this problem using either re-sampling, re-weighting or 'head-to-tail' knowledge transfer. Re-sampling methods either adopt over-sampling on tail classes [3,16] or use under-sampling [9,22,44] on head classes. On the other hand, various cost-sensitive loss functions have been proposed in the literature to re-weight majority and minority instances [2,5,8,20,26,29,51]. Among them, Range Loss [51] minimizes the range of each class to enhance the learning towards face recognition with long-tail while Focal Loss [29] down-weights the loss assigned to well-classified examples to deal with class imbalance in object detection. Label-Distribution-Aware Margin Loss (LDAM) [2], on the other hand, encourages minority classes to have larger margins.

Researchers also try to employ head-to-tail knowledge transfer for data imbalance. In [45,46] a transformation from minority classes to majority classes regressors/classifiers is learned progressively while in [31] a meta embedding equipped with a feature memory is proposed for such knowledge transfer.

Few-shot learning methods [10,15,47] also try to generalize knowledge from a richly annotated dataset to a low-shot dataset. This is often achieved by training a meta-learner [12,48] from the many-shot classes and then generalize to new few-shot classes. However, different from few-shot learning algorithms, we mainly focus on learning a continuous spectrum of data distribution jointly, rather than focus solely on the few-shot classes.

Knowledge Distillation. The idea of knowledge distillation was first introduced in [19] for the purpose of model compression where a student network is trained to mimic the behavior of a teacher network so that the knowledge is compressed to the compact student network. Then the distillation target is further extended to hidden layer features [41] and visual attention [50], where attention map from the teacher model is transferred to the student. Apart from distilling for model compression, knowledge distillation is also proved to be effective when the teacher and the student have identical architecture. i.e. self-distillation [11,49]. Knowledge distillation is also applied in other areas such as continual learning [28,39], semi-supervised learning [7,32] and neural style transfer [21].

Curriculum and Self-paced Learning. The basic idea of curriculum learning is to organize samples or tasks in ascending order of difficulty, and it has been widely adopted for weakly supervised learning [14,24,27] and reinforcement learning [33,34,43]. Apart from designing easy-to-hard curriculums based on prior knowledge, efforts have also been made to incorporate learning process to dynamically adjust the curriculum. In [23] a self-paced curriculum determined by both prior knowledge and learning dynamics is proposed. In [13] a multi-armed bandit algorithm is used to determine a syllabus, where the rate of increase in prediction accuracy and network complexity are utilized as reward signals. In [40], meta learning is employed to assign weights to training samples based on gradient directions.

Table 1. Comparison of entire distribution and subsets under four metrics. Larger values indicate more *longtailness*.

Metric	Accuracy			
	I_{Ratio}	I_{KL}	I_{Abs}	I_{Gini}
Entire	256.0	0.707	0.769	0.524
Many-shot	12.8	0.278	0.481	0.322
Medium-shot	4.7	0.122	0.356	0.235
Low-shot	4.0	0.099	0.320	0.209

3 Motivation and Metrics for Evaluating Data Imbalance

The problem we address in this work is to train a CNN on a long-tailed classification task. Our method is inspired by an interesting empirical finding that, training a CNN on a balanced dataset with fewer samples sometimes leads to superior performances than on a long-tailed dataset with more samples. As the experiment in [37] reveals that even when 40% of the positive samples are left out for the representation learning, the object detection performances can still be surprisingly improved a bit due to a more uniform distribution. This observation successfully emphasizes the importance of learning a balanced, uniform distribution. To learn a more balanced distribution, a natural question to ask would be, how to measure the data imbalance. Since almost every manually collected dataset more or less contains various number of samples per class. To this end, we first introduce four metrics for data imbalance measurement.

For a long-tailed dataset, if we denote N, N_i, C to be the total number of samples, number of samples in class i, and number of classes respectively, then the four metrics for measuring data imbalance is introduced as follows:

Imbalance Ratio [42] is defined as the ratio between the largest and the smallest number of samples:

$$I_{Ratio} = \frac{N_{i_{max}}}{N_{i_{min}}}.$$

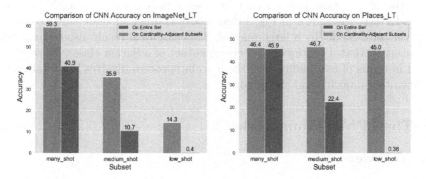

Fig. 2. Comparison of CNN performances trained on cardinality-adjacent subsets and the entire dataset.

Imbalance Divergence is defined as the KL-Divergence between the long-tailed distribution and the uniform distribution:

$$I_{KL} = D(P||Q) = \sum_i p_i log\frac{p_i}{q_i}$$

where $p_i = \frac{N_i}{N}$ is the class probability for class i, and $q_i = \frac{1}{C}$ denotes the uniform probability.

Imbalance Absolute Deviation [4] is defined as the sum of absolute distance between each long-tailed and uniform probability:

$$I_{Abs} = \sum_i |\frac{1}{C} - \frac{N_i}{N}|$$

Gini Coefficient is defined as

$$I_{Gini} = \frac{\sum_{i=1}^{C}(2i - C - 1)N_i}{C\sum_{i=1}^{C} N_i}$$

where i is the class index when all classes are sorted by cardinality in ascending order.

The last three metrics all measure the distance between the current long-tailed distribution and a uniform distribution. For all four metrics, smaller values indicate a more uniform distribution. Having these metrics at hand, we show that for a long-tailed dataset, if we sort the classes by their cardinality, i.e. number of samples, then a subset of the classes with adjacent cardinality will become less long-tailed under these imbalance measurements. As Table 1 shows that, if we split the long-tailed benchmark dataset ImageNet-LT into three splits (many-shot, medium-shot, few-shot) according to class cardinality (following [31]), then all four metrics become smaller, which indicates that these subsets become less imbalanced than the original. Then the CNNs trained on these subsets are expected to perform better than their jointly-trained counterparts. We verify this assumption on two long-tailed benchmark datasets ImageNet-LT and Places-LT [31]. As shown in Fig. 2 that CNNs trained on these subsets outperform the joint model by a large margin. This empirical result also accords with our intuition that a continuous spectrum subset of adjacent classes is more balanced in terms of cardinality, and the learning process involves less interference between the majority and the minority.

4   The LFME Framework

4.1   Overview

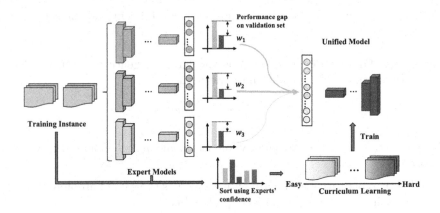

Fig. 3. Overview of the LFME framework.

In this section, we describe the proposed LFME framework in detail. Formally, given a long-tailed dataset D with C classes, we split the entire set of categories into L cardinality-adjacent subsets $S_1, S_2, \ldots, S_L$ using $L-1$ thresholds $T_1, T_2, \ldots, T_{L-1}$ such that $S_i = \{c | T_i \leq N_c \leq T_{i+1}\}$, where N_c is class c's cardinality. Then we train L expert models on each of the cardinality-adjacent subset and denote them as $\mathcal{M}_{E_1}, \mathcal{M}_{E_2}, \ldots, \mathcal{M}_{E_L}$. These expert models serve to (1) provide output logits' distribution for knowledge transfer (2) provide output confidence as instance difficulty cues. These information enables us to develop self-paced and curriculum learning schemes from both model level and instance level. From the perspective of **self-paced expert selection**, we adopt a weighted knowledge distillation between the output logits from the expert models and the student model. As the learning proceeds, the student will gradually approach the experts' performances. In such cases, we do not want the experts to limit the learning process of the student. To achieve this goal, we introduce a self-paced weighted scheme based on the performance gap on the validation set between the expert models and the student model. As the student model acquires knowledge from both data and the expert models, the importance weight of knowledge distillation will gradually decrease, and finally the unified student model is able to achieve comparable or even superior performance compared to the experts. From the perspective of **curriculum instance selection**, given the confidence scores from the Expert models, we re-organize the training data from easy to hard, i.e. from low-confidence to high-confidence. Then we exploit a soft weighted instance selection scheme to conduct such curriculum, so that easy samples are trained first, then harder samples are added to the training set gradually. This progressive learning curriculum has proved to be beneficial for training deep models [1]. Finally, with the two levels of self-paced and curriculum learning schemes, the knowledge from the expert models will be gradually transferred to the unified student model. An overview of the LFME framework is shown in Fig. 3.

4.2   Self-paced Expert Selection

Once we acquire the well-trained expert models, we feed the training data and obtain their output predictions. Then we employ knowledge distillation as an extra supervision signal to the student model. Specifically, for the expert $\mathcal{M}_{E_l}$ trained on l-th cardinality-adjacent subset S_l, if we denote $\mathbf{z}^{(l)}, \hat{\mathbf{z}}$ to be the output logits of the current expert model and the current student model respectively, then the knowledge distillation loss for expert $\mathcal{M}_{E_l}$ is given by:

$$L_{KD_l} = -H(\tau(\mathbf{z}^{(l)}), \tau(\hat{\mathbf{z}}^{(l)})) = -\sum_{i=1}^{|S_l|} \tau(z_i^{(l)}) \log(\tau(\hat{z}_i^{(l)}))$$

where $\hat{\mathbf{z}}^{(l)} = \hat{\mathbf{z}}_{c \in S_l}$ is the student logits for classes in S_l and

$$\tau(z_i^{(l)}) = \frac{\exp(z_i^{(l)}/T)}{\sum_j \exp(z_j^{(l)}/T)}, \quad \tau(\hat{z}_i^{(l)}) = \frac{\exp(\hat{z}_i^{(l)}/T)}{\sum_j \exp(\hat{z}_j^{(l)}/T)}$$

are the output probabilities using Softmax with temperature T. T is usually set to be greater than 1 to increase the weight for smaller probabilities. In this way, for each expert model $\mathcal{M}_{E_l}$ we have its knowledge distillation loss to transfer its knowledge to the student model, and there are L losses in total, corresponding to L experts trained on L cardinality-adjacent subsets. The most straightforward way to aggregate these losses would be simply summing them up. However, this could be problematic, since as the learning process goes on, the student model's performance will gradually approach or even exceed the expert's. In such cases, we do not want the expert models become performance ceilings for the student model, and we wish to gradually weaken the guidance from the experts as the data-driven learning proceeds.

To achieve this goal, we propose a Self-paced Expert Selection scheme based on the performance gap between the student and the experts. In the experiments, we use the Top-1 Accuracy on the validation set after each training epoch as the measurement for performance gap. If we denote the Top-1 Accuracy of the student model $\mathcal{M}$ and the expert model $\mathcal{M}_{E_l}$ at epoch e to be $Acc_{\mathcal{M}}$ and Acc_{E_l} respectively, then a weighting scheme is defined as follows:

$$w_l = \begin{cases} 1.0 & if \ Acc_{\mathcal{M}} \leq \alpha Acc_{E_l} \\ \dfrac{Acc_{E_l} - Acc_{\mathcal{M}}}{Acc_{E_l}(1 - \alpha)} & if \ Acc_{\mathcal{M}} > \alpha Acc_{E_l} \end{cases}$$

and w_l is updated at the end of each epoch. The weight scheduling threshold α controls the knowledge distillation to switch from ordinary to a self-paced decaying schedule. With the self-paced weight scheduling weight w_l, the knowledge transfer from the experts is automatically controlled by the student model's performance. The final knowledge distillation loss is the automatic weighted sum of knowledge distillation loss from all expert models:

$$L_{KD} = \sum_{l=1}^{L} w_l L_{KD_l}$$

4.3   Curriculum Instance Selection

Following the spirit of curriculum learning which mimics the human learning process, three questions need to be answered: (1) how to evaluate the difficulty of each instance, (2) how to *select* or *unselect* a sample, (3) how to design a curriculum so that samples are organized from easy to hard.

For the first question, we use the expert models' output confidence for each instance as an indication for instance difficulty. Given a training instance (x_i, y_i), suppose its ground-truth class y_i falls into the l-th subset S_l, i.e. $y_i \in S_l$, then we take the corresponding l-th expert model and use its output prediction for class y_i as confidence score, denoted as p_i. In this way, we can obtain the confidence score for all the instances in the training set.

For the second question, we adopt a soft selection method for instance selection. For instance (x_i, y_i), we replace the cross entropy loss with a weighted version:

$$L_{CE} = \sum_{i=1}^{N} v_i^{(k)} L_{CE}(x_i)$$

where $v_i^{(k)} \in [0, 1]$ is the selection weight at k-th epoch. A higher value of v_i (close to 1) indicates a soft selection of i-th instance, while a smaller value indicates a soft unselection of that instance.

Finally, to answer the third question, we design an automatic curriculum to determine the value of $v_i^{(k)}$, so that the instances are selected from easy to hard. The simplest approach is to sort the instances using their confidence score p_i

obtained by the expert models. However, different from traditional curriculum learning scenarios, our long-tailed classification problem involves both many-shot and low-shot categories, where low-shot instances tend to have lower confidence scores than many-shot instances. When sorted by the confidence score, the low-shot samples tend to be classified as hard examples and are not selected at first, which we do not wish to happen. To deal with such scenarios, instead of sorting across the whole training set, we sort instances according to their confidence scores **within each cardinality subset**. To be more specific, given the expert output confidence, v_i^k should be determined by three factors (1) the expert confidence p_i, (2) current epoch k, (3) the cardinality-adjacent subset $\mathcal{S}_l$ the i-th instance belongs to. Since the whole dataset is long-tailed, while we select samples from easy to hard, we also wish to select as uniform as possible across all subsets at the beginning of the training, and gradually add more hard samples as the epoch increases. In other words, at the first epoch we wish to select all the samples in the subset with lowest shots $\mathcal{S}_{min}$ (i.e. classes in $\mathcal{S}_{min}$ have the smallest number of samples) and same amount of samples in other subsets, and gradually add more samples until all the samples in all subsets are selected in the last epoch.

To achieve this goal, if we denote $N_{\mathcal{S}_l} = \frac{1}{|\mathcal{S}_l|} \sum_{i=1}^{|\mathcal{S}_l|} N_i$ as the average shot (average number of samples per class) in subset $\mathcal{S}_l$, then $v_i^{(k)}$ is determined by $p_i \frac{N_{\mathcal{S}_{min}}}{N_{\mathcal{S}_l}}$ at epoch 1, and grows gradually to 1 at the last epoch. Then at epoch 1, each subset softly selects its $N_{\mathcal{S}_{min}}$ easiest samples, and harder samples are gradually softly added to the training process. Formally, we use a monotonically increasing function f as scheduling function, so that $v_i^{(k)}$ will gradually grow from $p_i \frac{N_{\mathcal{S}_{min}}}{N_{\mathcal{S}_l}}$ to 1. For simplicity, we choose the linear function in the experiments and f is defined as

$$f(v_i^k) = (1 - v_i^{(1)})\frac{e}{E} + v_i^{(1)}$$

where $v_i^{(1)} = p_i \frac{N_{\mathcal{S}_{min}}}{N_{\mathcal{S}_l}}$ is the initial soft selection weight at epoch 1, and e, E are the current epoch and the total number of epochs respectively. It is worth noting that the scheduling function f can also be any convex or concave function as long as it is monotonically increasing within $[1, E]$. The impact of choosing different f is further analyzed in the experimental section. A schematic illustration of w_l and v_i can be found in Fig. 4.

4.4 Training

With the Self-paced Expert Selection and Curriculum Instance Selection, we obtain the final loss function:

$$L = \sum_{i=1}^{N} v_i L_{CE}(x_i, y_i) + \sum_{l=1}^{L} \sum_{i=1}^{N} w_l L_{KD_l}(\mathcal{M}, \mathcal{M}_{Exp}; x_i)$$

where N, L are the number of training instances and number of experts respectively, and v_i, w_i controls the two levels of adaptive learning schedules.

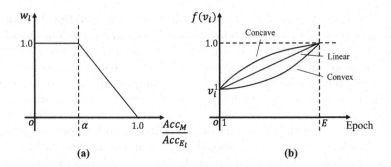

Fig. 4. Weight scheduling function for (a) model level selection w_l, and (b) instance level selection v_i.

In practice, we first train expert models using ordinary Instance-level Random Sampling, where each instance is sampled with equal probability. We then train the whole LFME using Class-level Random Sampling adopted in [25,31], where each class is sampled with equal number of samples and probability.

5   Experiments

5.1   Experimental Settings

Dataset. We evaluate our proposed method on three benchmark long-tailed classification datasets: ImageNet-LT, Places-LT proposed in [31] and CIFAR100-LT proposed in [2]. ImageNet-LT is created by sampling a subset of ImageNet [6] following the Pareto distribution with power value $\alpha = 6$. It contains 1000 categories with class cardinality ranging from 5 to 1280. Places-LT is created similarly from Places dataset [52] and contains 365 categories with class cardinality ranging from 5 to 4980. CIFAR100-LT is created with exponential decay imbalance and controllable imbalance ratio.

Baselines. For the first two datasets, similar to [31], our baseline methods include three re-weighting methods: Lifted Loss [35], Focal Loss [29] and Range Loss [51], and one SOTA few-shot learning method FSLwF [12], as well as the recent SOTA method OLTR [31]. For the CIFAR100-LT dataset, we mainly compare with the SOTA method LDAM proposed in [2].

Implementation Details. For the first two datasets, we choose the number of cardinality-adjacent subsets $L = 3$ with thresholds $\{20, 100\}$ following the splits in [31]. We refer to these subsets as many, medium and few-shot subsets. For CIFAR100-LT, We equally split the 100 classes into two subsets: many and few-shot. We use PyTorch [38] to implement all experiments. For the first two datasets, we first train the experts using SGD for 120 epochs. Then the LFME model is trained with SGD with momentum 0.9, weight decay 0.0005 for 90 epochs, batch size 256, learning rate 0.1 and divide by 0.1 every 40 epochs. We use ResNet-10 [18] training from scratch for ImageNet-LT and ImageNet pretrained Resnet-152 for Places-LT. During training, class-balanced sampling is adopted. For the

CIFAR100-LT experiments, we first train the experts for 200 epochs and then train the LFME model using SGD with momentum 0.9, weight decay 2×10^{-4}, batch size 128, epochs 200, initial learning rate 0.1 and decay by 0.01 at 160 and 180 epochs, as well as deferred class-balanced sampling, same as [2]. The backbone network is ResNet-32. The distillation temperature T is set to 2, and the expert weight scheduling threshold α is set to 0.6 during the experiments.

Table 2. Long-tailed classification results on ImageNet-LT and Place-LT. * denotes reproduced results, other results are from [31].

Method	Acc.							
	ImageNet-LT				Places-LT			
	Many	Med.	Few	All	Many	Med.	Few	All
Plain model	40.9	10.7	0.4	20.9	**45.9**	22.4	0.36	27.2
Lifted loss [35]	35.8	30.4	17.9	30.8	41.1	35.4	24	35.2
Focal loss [29]	36.4	29.9	16.0	30.5	41.1	34.8	22.4	34.6
Range loss [51]	35.8	30.3	17.6	30.7	41.1	35.4	23.2	35.1
FSLwF [12]	40.9	22.1	15.0	28.4	43.9	29.9	**29.5**	34.9
OLTR [31]	43.2	35.1	18.5	35.6	44.7	37.0	25.3	35.9
OLTR [31]*	40.7	33.2	17.4	33.8	42.2	38.1	17.8	35.3
Ours	**47.1**	35.0	17.5	37.2	38.4	39.1	21.7	35.2
Ours + Focal Loss	46.7	35.8	17.3	37.3	37.0	**39.6**	23.0	35.2
Ours + OLTR	47.0	**37.9**	**19.2**	**38.8**	39.3	**39.6**	24.2	**36.2**

5.2 Main Results on Long-Tailed Classification Benchmarks

Table 2 shows the long-tailed classification results on ImageNet-LT and Places-LT dataset. As can be found that our method is able to achieve superior or at least comparable results to the state-of-the-art methods such as OLTR. We found that many-shot categories benefit most from our LFME framework, while few-shot classes also demonstrate improvements and perform similarly with the re-weighting methods. Moreover, we also demonstrate that our method can be easily incorporated with other state-of-the-art methods, and we show the result of LFME + Focal Loss and LFME + OLTR (where LFME is added in the second stage of OLTR). We observe that both methods benefit from our expert model on all three subsets, and the combination of our method and OLTR outperforms previous methods by a large margin. It is also worth noting that our expert models are trained using vanilla CNNs, and utilizing other techniques will further lead to superior expert models, and assumably, superior student model.

To further demonstrate the statistical significance of the proposed method, we conduct experiments on CIFAR100-LT with imbalance ratio 100. The results in Table 3 show that LFME is able to achieve comparable performances with the SOTA method LDAM [2] and combining them will further improve LDAM on both many and few-shot subsets.

Table 3. Results on CIFAR100-LT.

Methods	Many	Few	All
Plain CNN	59.0	18.2	38.6
Ours	59.0	25.5	42.3
LDAM [2]	58.8	26.1	42.4
Ours + LDAM	**59.5**	**28.0**	**43.8**

Table 4. Effect of different scheduling functions.

Schedule	Many	Medium	Few	All
Linear	47.1	**35.0**	**17.5**	**37.2**
Convex	47.2	34.6	16.7	36.9
Concave	**47.5**	34.7	17.0	37.1

5.3 Ablation Study

Effectiveness of Each Component. We evaluate each part of our method and the result is shown in Table 5. We compare with the following variants: (1) Instance-level Random Sampling (Ins.Samp.), where each instance is sampled with equal probability. (2) Instance-level Random Sampling + Ordinary Knowledge distillation (Ins.Samp. + KD), where non-self-paced version knowledge distillation from the experts is added, i.e. $w_l = 1.0$. (3) Class-level Random Sampling (Cls.Samp.), where each class is sampled with equal number of samples and probability. (4) Class-level Random Sampling + Ordinary Knowledge Distillation (Cls.Samp. + KD). (5) Class-level Random Sampling + Knowledge Distillation + Self-paced Expert Selection (Cls.Samp. + KD + SpES). (6) Curriculum Instance Selection + Self-paced Expert Selection (CurIS + KD + SpES), which constitute our LFME framework.

From the results, we come up with the following observations: first, compared to the instance-random sampling, the adopted class-level random sampling is able to largely improve the few-shot performance while also decrease the many-shot performance slightly, since it samples more few-shot and less many-shot

Table 5. Effectiveness of each component.

Method	Accuracy			
	Many	Med.	Few	All
Ins.Samp	40.9	10.7	0.4	20.9
Ins.Samp. + KD	55.7	22.2	0.02	32.2
Cls.Samp	38.8	32.3	17.0	32.6
Cls.Samp. + KD	44.9	34.5	15.8	35.8
Cls.Samp. + KD + SpES	46.6	35.8	16.5	37.1
CurIS + KD + SpES	47.1	35.0	17.5	37.2

instances. Second, the introduction of knowledge distillation from experts can significantly improve the results, as it brings **11.3%** and **3.2%** to the Instance-level Sampling and Class-level Sampling baselines. However, while knowledge distillation brings improvements for many-shot classes, it will also decrease the few-shot accuracy slightly. Third, the Self-paced Expert Selection improves the knowledge distillation on all three subsets. It removes the performance ceiling brought from the experts and allows the student to exceed the experts. As the results show that SpES brings **0.4%** and **1.3%** overall performance gain respectively. Finally, the proposed Curriculum Instance Selection further improves on the few-shot categories with **1.0%** in accuracy, so that the decrease on few-shot subset caused by the knowledge distillation is compensated.

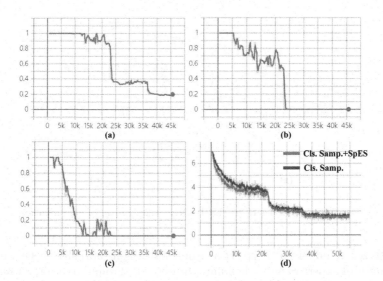

Fig. 5. (a)–(c): Visualization of self-paced expert selection scheduler w_l for many-shot, medium-shot, few-shot expert model. (d): Loss curves before and after adding Self-paced Expert Selection.

Visualization of Self-paced Expert Selection. Self-paced expert selection plays an important role in LFME for more efficient and effective knowledge transfer. Fig. 5(a)–(c) gives a visualization of the expert selection weights w_l for many-shot, medium-shot, few-shot model. From the visualization, we observe that w_l serves to automatically control the knowledge transfer, as for many-shot and medium-shot experts the knowledge is consistently distilled, while for few-shot experts, the student instantly exceeds the expert's performance, thus leading to a decay in $w_{fewshot}$. Moreover, we also visualize the impact of Self-paced Expert Selection in terms of cross-entropy loss curves, shown in Fig. 5(d), and we find that it leads to a lower cross-entropy loss, which also verifies the effectiveness of the proposed Self-paced Expert Selection.

Effect of Learning Scheduler. We also discuss the impact of different learning schedules for v_i as shown in Table 4. Given the initial instance confidence $v^{(1)}$, we test with the following scheduling functions:

- $f_{Linear} = (1 - v^{(1)})\frac{e}{E} + v^{(1)}$
- $f_{Convex} = 1 - (1 - v^{(1)})\cos(\frac{e}{E}\frac{\pi}{2})$
- $f_{Concave} = (1 - v^{(1)})\frac{\log(1+\frac{e}{E})}{\log 2}$

The result shows that the linear growing function yields the best result, while the concave the convex function f also produce similar performances. The convex function yields the worst performances as it selects fewest instances at the start of the training which may not be beneficial for the training dynamics.

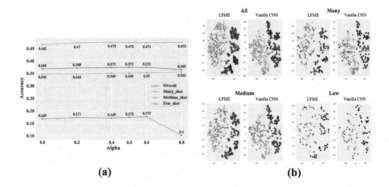

(a)   (b)

Fig. 6. (a) Sensitivity analysis of α. (b) T-SNE visualization of classification weights.

Sensitivity Analysis of Hyperparameter α. Figure 6(a) shows the sensitivity analysis of expert weight scheduling threshold α. From the result, we observe that our model is robust to most α values. When α grows to 1.0, the Self-paced Expert Selection becomes ordinary knowledge distillation, and result in a performance decline.

Visualization of Classification Weights. We visualize the classification weights of vanilla CNN and our LFME via T-SNE in Fig. 6(b). The results show that our method results in a more structured, compact feature manifold. It shows that without particular re-weighting, our method is also able to produce discriminative feature space and classifiers.

6   Conclusions

In this paper, we propose a Learning From Multiple Experts framework for long-tailed classification problem. By introducing the idea of cardinality-adjacent subset which is less long-tailed, we train several expert models and propose two levels of adaptive learning to distill the knowledge from the expert models to a unified student model. From the extensive experiments and visualizations, we verify the effectiveness of our proposed method as well as each of its component, and show that the LFME framework is able to achieve state-of-the-art performances on the long-tailed classification benchmarks.

Acknowledgement. This work is supported by the National Natural Science Foundation of China (No. U1936202, No. 61925107). We also thank anonymous reviewers for their constructive comments.

References

1. Bengio, Y., Louradour, J., Collobert, R., Weston, J.: Curriculum learning. In: Proceedings of the 26th Annual International Conference on Machine Learning, pp. 41–48. ACM (2009)

2. Cao, K., Wei, C., Gaidon, A., Arechiga, N., Ma, T.: Learning imbalanced datasets with label-distribution-aware margin loss. arXiv preprint arXiv:1906.07413 (2019)
3. Chawla, N.V., Bowyer, K.W., Hall, L.O., Kegelmeyer, W.P.: SMOTE: synthetic minority over-sampling technique. J. Artif. Intell. Res. **16**, 321–357 (2002)
4. Collins, E., Rozanov, N., Zhang, B.: Evolutionary data measures: understanding the difficulty of text classification tasks. arXiv preprint arXiv:1811.01910 (2018)
5. Cui, Y., Jia, M., Lin, T.Y., Song, Y., Belongie, S.: Class-balanced loss based on effective number of samples. In: Proceedings of the IEEE Conference on Computer Vision and Pattern Recognition, pp. 9268–9277 (2019)
6. Deng, J., Dong, W., Socher, R., Li, L.J., Li, K., Fei-Fei, L.: ImageNet: a large-scale hierarchical image database. In: 2009 IEEE Conference on Computer Vision and Pattern Recognition, pp. 248–255. IEEE (2009)
7. Ding, G., Guo, Y., Chen, K., Chu, C., Han, J., Dai, Q.: DECODE: deep confidence network for robust image classification. IEEE Trans. Image Process. **28**, 3752–3765 (2019)
8. Dong, Q., Gong, S., Zhu, X.: Class rectification hard mining for imbalanced deep learning. In: Proceedings of the IEEE International Conference on Computer Vision, pp. 1851–1860 (2017)
9. Drummond, C., Holte, R.C., et al.: C4.5, class imbalance, and cost sensitivity: why under-sampling beats over-sampling. In: Workshop on Learning from Imbalanced Datasets II, vol. 11, pp. 1–8. Citeseer (2003)
10. Finn, C., Abbeel, P., Levine, S.: Model-agnostic meta-learning for fast adaptation of deep networks. In: Proceedings of the 34th International Conference on Machine Learning, vol. 70, pp. 1126–1135. JMLR.org (2017)
11. Furlanello, T., Lipton, Z., Tschannen, M., Itti, L., Anandkumar, A.: Born-again neural networks. In: International Conference on Machine Learning, pp. 1602–1611 (2018)
12. Gidaris, S., Komodakis, N.: Dynamic few-shot visual learning without forgetting. In: Proceedings of the IEEE Conference on Computer Vision and Pattern Recognition, pp. 4367–4375 (2018)
13. Graves, A., Bellemare, M.G., Menick, J., Munos, R., Kavukcuoglu, K.: Automated curriculum learning for neural networks. In: Proceedings of the 34th International Conference on Machine Learning, vol. 70, pp. 1311–1320. JMLR.org (2017)
14. Guo, S., et al.: CurriculumNet: weakly supervised learning from large-scale web images. In: Ferrari, V., Hebert, M., Sminchisescu, C., Weiss, Y. (eds.) ECCV 2018. LNCS, vol. 11214, pp. 139–154. Springer, Cham (2018). https://doi.org/10.1007/978-3-030-01249-6_9
15. Guo, Y., Ding, G., Han, J., Gao, Y.: Zero-shot learning with transferred samples. IEEE Trans. Image Process. **26**, 3277–3290 (2017)
16. Han, H., Wang, W.-Y., Mao, B.-H.: Borderline-SMOTE: a new over-sampling method in imbalanced data sets learning. In: Huang, D.-S., Zhang, X.-P., Huang, G.-B. (eds.) ICIC 2005. LNCS, vol. 3644, pp. 878–887. Springer, Heidelberg (2005). https://doi.org/10.1007/11538059_91
17. He, H., Garcia, E.A.: Learning from imbalanced data. IEEE Trans. Knowl. Data Eng. **21**(9), 1263–1284 (2009)
18. He, K., Zhang, X., Ren, S., Sun, J.: Deep residual learning for image recognition. In: Proceedings of the IEEE Conference on Computer Vision and Pattern Recognition, pp. 770–778 (2016)
19. Hinton, G., Vinyals, O., Dean, J.: Distilling the knowledge in a neural network. arXiv preprint arXiv:1503.02531 (2015)

20. Huang, C., Li, Y., Chen, C.L., Tang, X.: Deep imbalanced learning for face recognition and attribute prediction. IEEE Trans. Pattern Anal. Mach. Intell. (2019)
21. Huang, Z., Wang, N.: Like what you like: knowledge distill via neuron selectivity transfer. arXiv preprint arXiv:1707.01219 (2017)
22. Jeatrakul, P., Wong, K.W., Fung, C.C.: Classification of imbalanced data by combining the complementary neural network and SMOTE algorithm. In: Wong, K.W., Mendis, B.S.U., Bouzerdoum, A. (eds.) ICONIP 2010. LNCS, vol. 6444, pp. 152–159. Springer, Heidelberg (2010). https://doi.org/10.1007/978-3-642-17534-3_19
23. Jiang, L., Meng, D., Zhao, Q., Shan, S., Hauptmann, A.G.: Self-paced curriculum learning. In: Twenty-Ninth AAAI Conference on Artificial Intelligence (2015)
24. Jiang, L., Zhou, Z., Leung, T., Li, L.J., Fei-Fei, L.: MentorNet: learning data-driven curriculum for very deep neural networks on corrupted labels. arXiv preprint arXiv:1712.05055 (2017)
25. Kang, B., et al.: Decoupling representation and classifier for long-tailed recognition. arXiv preprint arXiv:1910.09217 (2019)
26. Khan, S.H., Hayat, M., Bennamoun, M., Sohel, F.A., Togneri, R.: Cost-sensitive learning of deep feature representations from imbalanced data. IEEE Trans. Neural Netw. Learn. Syst. 29(8), 3573–3587 (2017)
27. Li, D., Huang, J.B., Li, Y., Wang, S., Yang, M.H.: Weakly supervised object localization with progressive domain adaptation. In: Proceedings of the IEEE Conference on Computer Vision and Pattern Recognition, pp. 3512–3520 (2016)
28. Li, Z., Hoiem, D.: Learning without forgetting. IEEE Trans. Pattern Anal. Mach. Intell. 40(12), 2935–2947 (2017)
29. Lin, T.Y., Goyal, P., Girshick, R., He, K., Dollár, P.: Focal loss for dense object detection. In: Proceedings of the IEEE International Conference on Computer Vision, pp. 2980–2988 (2017)
30. Lin, T.-Y., et al.: Microsoft COCO: common objects in context. In: Fleet, D., Pajdla, T., Schiele, B., Tuytelaars, T. (eds.) ECCV 2014. LNCS, vol. 8693, pp. 740–755. Springer, Cham (2014). https://doi.org/10.1007/978-3-319-10602-1_48
31. Liu, Z., Miao, Z., Zhan, X., Wang, J., Gong, B., Yu, S.X.: Large-scale long-tailed recognition in an open world. In: Proceedings of the IEEE Conference on Computer Vision and Pattern Recognition, pp. 2537–2546 (2019)
32. Lopez-Paz, D., Bottou, L., Schölkopf, B., Vapnik, V.: Unifying distillation and privileged information. arXiv preprint arXiv:1511.03643 (2015)
33. Narvekar, S.: Curriculum learning in reinforcement learning. In: IJCAI, pp. 5195–5196 (2017)
34. Narvekar, S., Sinapov, J., Stone, P.: Autonomous task sequencing for customized curriculum design in reinforcement learning. In: IJCAI, pp. 2536–2542 (2017)
35. Oh Song, H., Xiang, Y., Jegelka, S., Savarese, S.: Deep metric learning via lifted structured feature embedding. In: Proceedings of the IEEE Conference on Computer Vision and Pattern Recognition, pp. 4004–4012 (2016)
36. Oksuz, K., Cam, B.C., Kalkan, S., Akbas, E.: Imbalance problems in object detection: a review. arXiv preprint arXiv:1909.00169 (2019)
37. Ouyang, W., Wang, X., Zhang, C., Yang, X.: Factors in finetuning deep model for object detection with long-tail distribution. In: Proceedings of the IEEE Conference on Computer Vision and Pattern Recognition, pp. 864–873 (2016)
38. Paszke, A., et al.: Automatic differentiation in PyTorch (2017)
39. Rebuffi, S.A., Kolesnikov, A., Sperl, G., Lampert, C.H.: iCaRL: incremental classifier and representation learning. In: Proceedings of the IEEE Conference on Computer Vision and Pattern Recognition, pp. 2001–2010 (2017)

40. Ren, M., Zeng, W., Yang, B., Urtasun, R.: Learning to reweight examples for robust deep learning. arXiv preprint arXiv:1803.09050 (2018)
41. Romero, A., Ballas, N., Kahou, S.E., Chassang, A., Gatta, C., Bengio, Y.: FitNets: hints for thin deep nets. arXiv preprint arXiv:1412.6550 (2014)
42. Stamatatos, E.: Author identification: using text sampling to handle the class imbalance problem. Inf. Process. Manag. **44**(2), 790–799 (2008)
43. Svetlik, M., Leonetti, M., Sinapov, J., Shah, R., Walker, N., Stone, P.: Automatic curriculum graph generation for reinforcement learning agents. In: Thirty-First AAAI Conference on Artificial Intelligence (2017)
44. Tahir, M.A., Kittler, J., Yan, F.: Inverse random under sampling for class imbalance problem and its application to multi-label classification. Pattern Recogn. **45**(10), 3738–3750 (2012)
45. Wang, Y.-X., Hebert, M.: Learning to learn: model regression networks for easy small sample learning. In: Leibe, B., Matas, J., Sebe, N., Welling, M. (eds.) ECCV 2016. LNCS, vol. 9910, pp. 616–634. Springer, Cham (2016). https://doi.org/10.1007/978-3-319-46466-4_37
46. Wang, Y.X., Ramanan, D., Hebert, M.: Learning to model the tail. In: Advances in Neural Information Processing Systems, pp. 7029–7039 (2017)
47. Xiang, L., Jin, X., Ding, G., Han, J., Li, L.: Incremental few-shot learning for pedestrian attribute recognition. In: Proceedings of the 28th International Joint Conference on Artificial Intelligence, pp. 3912–3918. AAAI Press (2019)
48. Xiang, L., Jin, X., Yi, L., Ding, G.: Adaptive region embedding for text classification. In: Proceedings of the AAAI Conference on Artificial Intelligence, vol. 33, pp. 7314–7321 (2019)
49. Yim, J., Joo, D., Bae, J., Kim, J.: A gift from knowledge distillation: fast optimization, network minimization and transfer learning. In: Proceedings of the IEEE Conference on Computer Vision and Pattern Recognition, pp. 4133–4141 (2017)
50. Zagoruyko, S., Komodakis, N.: Paying more attention to attention: improving the performance of convolutional neural networks via attention transfer. arXiv preprint arXiv:1612.03928 (2016)
51. Zhang, X., Fang, Z., Wen, Y., Li, Z., Qiao, Y.: Range loss for deep face recognition with long-tailed training data. In: Proceedings of the IEEE International Conference on Computer Vision, pp. 5409–5418 (2017)
52. Zhou, B., Lapedriza, A., Khosla, A., Oliva, A., Torralba, A.: Places: a 10 million image database for scene recognition. IEEE Trans. Pattern Anal. Mach. Intell. **40**(6), 1452–1464 (2017)

Hallucinating Visual Instances in Total Absentia

Jiayan Qiu[1], Yiding Yang[2], Xinchao Wang[2], and Dacheng Tao[1(✉)]

[1] Faculty of Engineering, School of Computer Science, UBTECH Sydney AI Centre,
The University of Sydney, Darlington, NSW 2008, Australia
jqiu3225@uni.sydney.edu.au, dacheng.tao@sydney.edu.au
[2] Stevens Institute of Technology, Hoboken, NJ 07030, USA
{yyang99,xinchao.wang}@stevens.edu

Abstract. In this paper, we investigate a new visual restoration task, termed as hallucinating visual instances in total absentia (HVITA). Unlike conventional image inpainting task that works on images with only part of a visual instance missing, HVITA concerns scenarios where an object is completely absent from the scene. This seemingly minor difference in fact makes the HVITA a much challenging task, as the restoration algorithm would have to not only infer the category of the object in total absentia, but also hallucinate an object of which the appearance is consistent with the background. Towards solving HVITA, we propose an end-to-end deep approach that explicitly looks into the global semantics within the image. Specifically, we transform the input image to a semantic graph, wherein each node corresponds to a detected object in the scene. We then adopt a Graph Convolutional Network on top of the scene graph to estimate the category of the missing object in the masked region, and finally introduce a Generative Adversarial Module to carry out the hallucination. Experiments on COCO, Visual Genome and NYU Depth v2 datasets demonstrate that the proposed approach yields truly encouraging and visually plausible results.

1 Introduction

If we are given a masked image as the one shown in Fig. 1 (a) and are asked to hallucinate the object that is entirely absent, we can, without much effort, tell the type of the absent object and further imagine a fairly reasonable overall appearance of the image content in the masked region. In this regard, we identify this image to be an indoor scene featuring a desktop, on top of which we see a monitor, a keyboard, and a lamp, and then based on their relative spatial layouts, we may safely draw the conclusion that, the missing image content concerns about a wireless phone, which often appears in the vicinity of these visual instances.

We term this task, which has been barely explored in prior works, as *hallucinating visual instances in total absentia* (HVITA). Despite human brains are able to accomplish this task in a considerably effortless way, existing image

Electronic supplementary material The online version of this chapter (https://doi.org/10.1007/978-3-030-58558-7_16) contains supplementary material, which is available to authorized users.

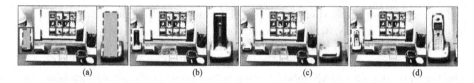

Fig. 1. Hallucinating instances in total absentia. Given a masked image with a scene object completely absent like (a), our approach hallucinates a contextually and visually plausible result, in this case the back cover of a cell phone shown in (b), by explicitly accounting for the global semantics. State-of-the-art inpainting approach [87], in this case where the target object is completely absent, fills the blank region with the background texture, as depicted in (c). The ground truth is shown in (d).

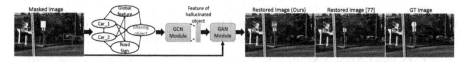

Fig. 2. Illustration of our framework. Given a masked image, we first detect the objects in the un-masked region and transform the image to a graph. We then use a GCN module to predict the semantic information of the masked part. Finally, conditioned on the masked image and the predicted semantic information of the missing region, we introduce a GAN module to produce our hallucination.

generation methods that omit the explicit reasoning on the inter-object semantics, produce visually inferior results. In the case of Fig. 1 (c), a state-of-the-art image inpainting approach [87], without analyzing the instance-level contexts, fills the masked region using the background texture. In fact, all existing image inpainting methods, to our best knowledge, rely on first observing a considerable fraction of the target instance before filling the absent part, and are thus incompetent of tackling HVITA.

Towards solving the proposed HVITA task, we introduce an interesting approach that, to some degree, imitates the human reasoning process for hallucinating the missing content. Unlike inpainting schemes that rely upon a partial instance being present, our approach works on scenarios where objects are in total absentia. This means our approach attempts to infer not only the category of the missing object but also a legitimate appearance consistent with the overall ambiance of the input image.

Specifically, our proposed approach follows a end-to-end trainable architecture that comprises three stages as illustrated in Fig. 2. In the first stage, we transform the input image to a semantic graph wherein each node corresponds to a detected object in the scene, obtained from an off-the-shelf detector. The masked region is also modeled as a node, but with an unknown label to be predicted. In the second stage, we introduce a Graph Convolutional Network (GCN) to regress the semantic information of the masked area, during which process the global-level inter-object semantics are explicitly accounted for. Finally, we stack a Generative Adversarial Network (GAN) to restore the masked region with the predicted semantic information and to generate visually realistic results.

Our contribution is therefore a novel framework designated to handle the proposed HVITA task, which is to our best knowledge the first attempt. This is accomplished by modeling each input image as a semantic graph, and then estimating the semantic information using a GCN module, followed by feeding the predictions into a GAN module so as to produce the final hallucination. The whole pipeline is end-to-end trainable. We evaluate our method on COCO [51], Visual Genome [41], and NYU Depth v2 [61], whose images comes from real world scenes with complex structural information, and obtain very encouraging qualitative and quantitative results.

2   Related Work

We briefly review here prior works related to ours, including graph convolutional network, image inpainting, image generation, and image insertion.

Graph Convolutional Network. Earlier works on graph-related tasks either assume the node features to be fixed [43,54,55,66,80,81], or apply iterative schemes to learn node representations, which are computationally expensive and at times unstable [17,21,71,75]. Inspired by the progress of deep learning [27,42,72], two categories graph convolutional neural networks are proposed: spectral-based approaches, which aims to develop graph convolution based on the spectral theory [9,12,28,39,45,47,79], and spatial-based ones, which investigates information mutual dependency [3,4,10,18,19,23,31,60,62,78,86]. More recently, the attention-based methods are introduced to GCN and have achieved very promising results [1,44,53,77,84,85,90].

Image Inpainting. Conventional image inpainting methods, both diffusion-based [5,7,46] and patch-based ones [6,8,11], use the intra-image information to fill the masked area. Thanks to the progress of deep learning especially generative adversarial networks [20], many deep approaches have been proposed [25,40,48,65,69,74,88]. These methods ensure the semantic continuity, yet at times yield artifacts around the border. The more recent methods make use of both intra-image information and learning from large datasets, and gain significant improvement in terms of semantic continuity and visual authenticity [34,73,82,83,87–89,92,93]. However, image inpainting methods handles only parts of a missing object, of which the semantic label is known, and therefore can not be exploited to hallucinate completely-absent objects.

Image Generation. Image generation has gained unprecedented improvement since the introduction of GAN [20]. Following works focus on improving the capacity of the generator and the learning capability of the discriminator [52, 58,68,70,89], designing more stable loss function [2,50,56,76], or introducing semantic information into the generation process [33,35,59,63,64,91]. In this work, we build our network based on the popular Conditional GAN [57] to produce the final hallucinated image.

Object Insertion. Object insertion aims to insert a human-chosen object into the target scene. Approaches of [14–16, 24, 29] focus on geometry for object modeling [67], and those of [22, 36] utilize user interactions for model generation. Methods like [37, 38, 49], on the other hand, model the generation process as a self-adaptive algorithm. Unlike our task, object insertion concerns only on the insertion process, rather than inferring and further hallucinating the missing object using scene semantics.

3   Method

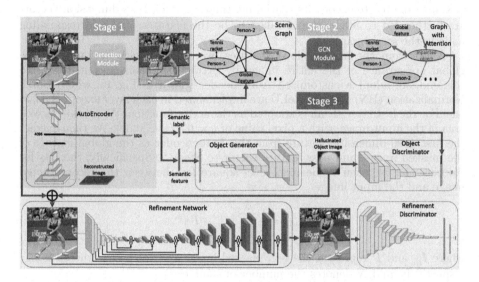

Fig. 3. Illustration of the proposed framework. The GCN module, object generator, and refinement network are trained end-to-end. $\oplus$ denotes the operation that fills the masked area with the restored object image, and $\otimes$ denotes concatenation. Note that, in the *Graph with Attention* of Stage 2, the thickness of an edge reflects its value of attention: a thicker edge indicates that the detected node is of higher importance to the hallucination inference.

In this section, we show the working scheme of the proposed framework in detail. As depicted in Fig. 3, our framework comprises three stages. In Stage 1, we utilize a detection module on the input image, where the object of interest is masked out, to detect the objects in the scene and extract their semantic and spatial information. Meanwhile, we adopt an Autoencoder to obtain the global feature of the masked image as a whole. In Stage 2, we model the masked image as a graph, in which each node corresponds to an object in the scene. We then insert two additional nodes, one for the absent object to be hallucinated and one for the global feature, into the graph. Afterwards, we apply a GCN module to estimate the features of the absent object. In Stage 3, we feed the hallucinated features of the absent object and the masked image into a GAN-based restoring module to achieve the restoration task.

3.1 Stage 1: Detection Module and Autoencoder

We adopt a detection module in Stage 1 to detect the objects in the masked image so as to derive *instance-level* features. Specifically, we train the Mask-RCNN model [26] on our training set, where masked images are used, and apply the trained detector on the test images. For each detected object, we construct a 1028-dimension feature vector that encodes both the semantic and spatial information. Features in first 1024 dimensions are taken directly from the last layer of Mask-RCNN to embrace the semantics, while features in the last four dimension, namely upper-left and lower-right coordinates of the detection bounding box, are adopted to encode its spatial information. Such 1028-dimension vectors are further fed to Stage 2, and taken to be the features of the corresponding node in the scene graph.

Apart from the instance-level features, we also explicitly account for the global appearance of the entire image by extracting *image-level* features. To this end, we adopt a customized Autoencoder shown in Fig. 4a, where Batch Normalization (BN) [32] and ReLU are implemented in each layer except the last one. For the last layer, Tanh is adopted as the active function. We fill the masked area in the input image with a pixel value of $v = 0.5$ ($v \in [0, 1]$). We then extract the first 4096-dimension feature from the Autoencoder as the global feature of the masked image, and pass feature vector to Stage 2. Specifically, the loss for Autoencoder is taken to be pixel-level square loss between the reconstructed image and the input image:

$$\mathcal{L}_{AE} = \frac{1}{N} \sum_{i=1}^{N} \|I_{in}^i - \hat{I}^i\|^2, \tag{1}$$

where I_{in}^i and $\hat{I}^i$ represent the i-th masked image and the reconstructed image respectively, and N denotes the number of samples.

3.2 Stage 2: GCN Module

The second stage of our framework takes as input both the instance- and image-level semantics obtained in Stage 1, and models the interplays between the scene objects using a scene graph. Let N denotes the number of detected objects in Stage 1. We construct a graph of $N+2$ nodes: N nodes that correspond to the N detected object instances, one node that encodes the image-level global features, and one node that denotes the missing object. We then link all the pairs of the $N + 2$ nodes to form a complete graph.

Each node in the graph holds a 1028-dimensions feature. For the N detected-object nodes, we directly take their *instance-level* features obtained in Stage 1 to be the node features. For the image node that accounts for the global semantics, we take the *image-level* feature of 4096 dimensions in Stage 1, and reduce it to 1024 dimensions using a learnable fully connected layer; we then pad 4 other dimensions, namely the origin coordinate $(0, 0)$ and the size of the image, to the 1024-dimension feature and form a 1028-dimension one. Finally for the absent-object node, we initialize its first 1024 dimensions using Gaussian noise and stack its 4-dimension location, i.e., the upper-left and lower-right coordinates of mask.

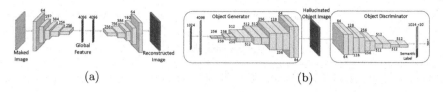

(a)   (b)

Fig. 4. (a) shows the architecture of our Autoenocoder for extracting image-level semantics, in which the first 4096-dimension feature is used as the global feature. (b) shows the network architecture of the object generative adversarial network (OGAN) for hallucinating the absent object in the mask.

Next, we adopt a graph convolutional network to reveal the intrinsic relationship between objects and derive the semantic features of the absent one to be hallucinated. Specifically, we adopt the popular graph attention network (GAT) model [77] to achieve this task. Given the input graph $\mathcal{G} = \{\mathcal{V}, \mathcal{E}\}$ and the corresponding features of nodes $X = \{x_1, x_2, ...x_{|\mathcal{V}|}\}, x_i \in \mathbb{R}^F$, one layer of the adopted GCN model can be formulated as

$$x_i' = \sum_{j:j \to i \in \mathcal{E}} \frac{e^{\mathcal{A}_\theta(h_\phi(x_i), h_\phi(x_j))}}{\sum_{j:j \to i \in \mathcal{E}} e^{\mathcal{A}_\theta(h_\phi(x_i), h_\phi(x_j))}} * h_\phi(x_j), \qquad (2)$$

where $\mathcal{A}$ is a function that takes a pair of nodes as input and outputs the attention score between them, and h is a non-linear mapping that takes the node features as input and maps them to a new space. The two functions $\mathcal{A}$ and h are controlled by parameters θ and ϕ, respectively. The new feature of the center node i is updated as the weighted sum of all the features from its neighbors. Thanks to the attention mechanism, the network learns to recover the feature of the absent object by interacting with other objects and global semantics of the entire image.

The loss of the GCN module comprises two parts, $\mathcal{L}_{CE}$ and $\mathcal{L}_{feat}$, defined as follows,

$$\mathcal{L}_{CE} = \frac{1}{N} \sum_{i=1}^{N} \mathcal{H}(f(x_i'), f(x_i)), \quad \mathcal{L}_{feat} = \frac{1}{N} \sum_{i=1}^{N} ||x_i' - x_i||_2, \qquad (3)$$

where f denotes the classifier of detector in Stage 1 and $f(\cdot)$ returns the classification probability, x_i denotes the ground truth feature of the hallucinated object in image i, and x_i' is the aggregated feature of the hallucinated object from GCN module. $\mathcal{H}$ denotes the cross entropy function. Specifically, $\mathcal{L}_{CE}$ accounts for the predicted label of the absent object, and $\mathcal{L}_{feat}$ enforces the similarity between the aggregated feature and the ground truth one. The loss for GCN module is thus taken as

$$\mathcal{L}_{GCN} = \mathcal{L}_{CE} + \mathcal{L}_{feat}. \qquad (4)$$

The semantic feature of the hallucinated object, aggregated from its neighbouring objects and the entire scene by the GCN module, is then passed to the next stage.

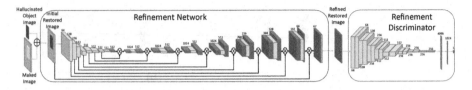

Fig. 5. Architecture of our global refinement adversarial network (GRAN). $\oplus$ denotes summation and the $\otimes$ denotes concatenation. GRAN tasks as input the initial restored image and outputs the refined one, ensuring the semantic continuity and visual authenticity.

3.3   Stage 3: GAN Module

Given the predicted class and features of the absent object, we conduct hallucination on the masked area to produce the final image output. This is achieved, in our implementation, via a GAN module. Specifically, our GAN module consists of two networks, an object generative adversarial network (OGAN) for generating the object and a global refinement adversarial network (GRAN) for refining restored image.

The structure of our OGAN is showed in Fig. 4b. It can be seen that the network is designed as a conditional GAN to generate the target object with the predicted class. In the Object Generator network, BN and Leaky ReLU are implemented in every layer except the last, where Tanh is used as the activation function. As for the Object Discriminator, BN and ReLU are applied to every layer but the last, where no activation function is applied and least squares loss [56] is implemented. To this end, we write,

$$\min_{OD} V_{OGAN}(OD) = \frac{1}{2}\mathbb{E}_{x \sim p_{data}(x)}[(OD(x,y) - b)^2] + \frac{1}{2}\mathbb{E}_{z \sim p_z(z)}[(OD(OG(z),y) - a)^2]$$

$$\min_{OG} V_{OGAN}(OG) = \frac{1}{2}\mathbb{E}_{z \sim p_z(z)}[(OD(OG(z),y) - c)^2],$$

$$(5)$$

where OD denotes the Object Discriminator network, and OG denotes the Object Generator network. a and b denote the ground truth fake and real labels, respectively. c denotes the value that OG wants OD to believe for the fake data, and y denotes the label of predicted class.

Once the image of the absent object is hallucinated, we resize and then insert it into the masked area of the input image, and obtain the initial restored image. Such restored images, despite semantically meaningful, turn out to be visually implausible, as the OGAN focuses on producing an object image of a specified class but overlooks the background content within the mask. This calls for another image generator that looks at the image as a whole and explicitly accounts for the visual continuity.

To this end, we introduce GRAN, a second GAN-based module that ensures the semantic continuity and visual authenticity of the restored image. The architecture of our GRAN is shown in Fig. 5. It comprises a Refinement Network and a Refinement Discriminator. For the Refinement Network, we implement it as an

U-net structure, which utilizes multi-scale features and avoids gradient vanishing. Specifically, in this network, BN and Leaky ReLU are implemented in every layer except the last, where again Tanh is adopted as active function. As for the Refinement Discriminator, we implement BN and ReLU in every layer except the last one. We also adopt the least square loss for the Refinement Disriminator:

$$\min_{RD} V_{GRAN}(RD) = \frac{1}{2}\mathbb{E}_{x \sim p_{data}(x)}[(RD(x) - b)^2] + \frac{1}{2}\mathbb{E}_{z \sim p_z(z)}[(RD(RN(z)) - a)^2]$$

$$\min_{RN} V_{GRAN}(RN) = \frac{1}{2}\mathbb{E}_{z \sim p_z(z)}[(RD(RN(z)) - c)^2],$$

$$(6)$$

where RD denotes the Refinement Discriminator network, and RN denotes the Refinement network. Moreover, we impose a l_1 loss on the Refinement network for the un-masked areas:

$$\mathcal{L}_{valid} = \frac{1}{N_{I_{valid}}}\|(1 - M) \odot (I_{out} - I_{is})\|_1, \qquad (7)$$

where $N_{I_{valid}}$ denotes the number of unmasked pixels, M denotes the binary mask (1 for holes), I_{out} denotes the output of the Refinement network, and I_{is} denotes the initial restored image.

In the first training step, our GAN module is pre-trained to ensure the ability of generating visual realistic images. The details will be explained in the experiment section.

3.4 End-to-End Training

The GCN and GAN module in our framework are end-to-end trainable. In our implementation, we adopt the end-to-end training scheme of the two modules, as the visual authenticity loss of the GAN in Stage 3 may facilitate the GCN in Stage 2 to aggregates and encode more discriminant information into its obtained semantic feature. We shown in Fig. 6 a comparison between the results obtained by end-to-end training and by separate training, where the former yields to the more semantically plausible result.

In sum, the losses for GCN module, OGAN, and GRAN are summarized as follows,

$$\mathcal{L}_{GCN} = \lambda_1\mathcal{L}_{RN_{GRAN}} + \lambda_2\mathcal{L}_{OG_{OGAN}} + \mathcal{L}_{CE} + \mathcal{L}_{feat},$$
$$\mathcal{L}_{OG} = \mathcal{L}_{RN_{GRAN}} + \lambda_3\mathcal{L}_{OG_{OGAN}}, \qquad (8)$$
$$\mathcal{L}_{RN} = \mathcal{L}_{RN_{GRAN}} + \lambda_4\mathcal{L}_{valid},$$

where λ_1, λ_2, λ_3 and λ_4 denotes the balancing weights.

4 Experiments

In this section, we provide our experimental setups and show the results. Since we are not aware of any existing work that performs exactly the same task as we do here, we mainly focus on showing the promise of the proposed framework. We also compare part of our framework with other popular models. Our goal

Fig. 6. Comparison between training Stage 2 and Stage 3 separately and end-to-end. (a), (b), (c) and (d) denote, respectively, the masked image, restored image with end-to-end training, restored image with separately training, and the ground truth.

Fig. 7. An example that our method produces visually and structurally reasonable, yet distorted image completion, in this case the distorted car. Image (a), (b) and (c) respectively denote the input, our restoration image, and the ground truth.

is, again, to show the possibility of hallucinating totally absent but reasonable objects in the masked image, rather than trying to beat the state-of-the-art GCN, GAN, and inpainting models. More complicated networks, as long as they are end-to-end trainable, can be adopted in our framework to achieve potentially better performances.

4.1 Datasets and Implementation Details

We adopt three datasets, COCO [51], Visual Genome [41], and NYU Depth v2 [61] to validate the proposed object hallucination approach. Image from these datasets feature complex scenes with rich contextual information, from which we can draw discriminant information to infer the type and appearance of the absent object. Other datasets, such as Pascal VOC [30] and ImageNet [13], do not fit our purpose well, since images from Pascal VOC typically contain only one object, while those from ImageNet often feature objects of the same class.

Microsoft COCO 2017 Dataset [51]. It is a detection dataset that comprises 123k complex scene images from 81 object classes for classification. We perform HVITA on 20 out of 81 classes: we use 40k images for training, 5k for validation, and 32k images for testing. We leave out the other classes, as the state-of-the-art object detector, GCN, and GAN are still short of the capability to detect and generate objects under in-the-wild higher-order physical constraints, such as the 3D orientation of a scene object. In the case of Fig. 7, for example, even though our approach successfully encodes the type of the missing object, in this case a car, and further generates a visually reasonable image, the 3D orientation of car, however, deviates from ground truth.

Table 1. Comparative results of our GCN module and GNN on COCO and NYU v2.

Term	Ours	GNN[39]	Ours-NYU	GNN-NYU	Ours-Genome	GNN-Genome
Accuracy (%)	87.93	79.23	80.41	75.54	85.01	80.47
Balanced Acc (%)	85.62	74.82	76.83	69.31	82.56	77.51

Visual Genome [41]. It comprises 110k images and 3.8 m object instances. We adopt this dataset because it features a large number of complex scenes. This dataset, however, includes a total number of 38k fine-grained object classes, making it infeasible to train a classifier and detector that can perform well on all classes with the limited samples. To this end, we group the finer-grained classes into coarser ones, such as clustering *blue suitcase* and *black suitcase* into a *suitcase* class. Then, we use the grouped classes as ground truth to train our detection and GCN module. In our experiment, we implement our framework on the same 20 classes as done for COCO. Thus, we use 30k images for training, 5k for validation, and 20k for testing. We use exactly same network structures and training parameters as for the COCO dataset.

NYU Depth v2 [61]. It comprises 1449 images from real-world indoor scenes. We adopt this dataset for testing the generalization capability of our method, as it provides complex scene with abundant structural information.

Implementation Details. Our networks are implemented using PyTorch and with 4 T V-100 SXM2 GPUs. In the pre-training stage, the batch size for Autoencoder, GCN module and GAN module are 192, 32 and 96, respectively. During the end-to-end training stage, the batch size is set to 16 for GCN and GAN module. The loss balancing weights λ_1, λ_2, λ_3 and λ_4 are set to be 0.01, 0.05, 5 and 0.1, respectively. We follow a two-step training strategy, which we find to be more efficient and effective than the single-step strategy that trans Stage 2 and Stage 3 all at once. In the first step, we supervisedly pre-training Stage 2 and Stage 3 independently. And in the second step, we end-to-end training these two stages with loss adjustment.

4.2 GCN Module

We compare our Graph Attention Network with GNN proposed by Kipf et al. [39] on the classification accuracy and balanced class accuracy, to demonstrate the its capability to infer the correct object class. It can be seen from Table 1 that our GCN model outperforms GNN on both metrics. This is because the GNN model aggregates the information of missing node from its neighbours averagely, thus unable to learn relationship importance embedded in the structural information.

For the GCN module, we use a three-layer neural network with residual connections. The number of attention heads for each layer are all set to be 6, and the feature dimension of the head for each layer are set to be 128, 256, 1024 respectively. For the last layer, we adopt a ReLU activation function to make the aggregated features comparable with the ground truth. The learning rate of GCN module is set to be 0.003 in the pre-training stage and 0.00001 in the end-to-end training stage.

Table 2. Results of the GAN networks trained separately (OG-only/RN-only) and with end-to-end training (Full-OG/Full-RN).

Term	OG-only	RN-only	Full-OG	Full-RN
PSNR	26.23	29.06	27.14	29.17
SSIM	0.9002	0.9144	0.9078	0.9165

4.3 GAN Module

We show here the pre-training process of our GAN module in detail. We feed the ground-truth object feature and class label to the GAN module, and use ground-truth object image and the scene image as the target output of OG and RN to compute a l_1 loss. An extra loss for OG is taken to be $\mathcal{L}_{OG}^{sup} = \frac{1}{N_{I_{Ogt}}}\|I_{OG}-I_{Ogt}\|_1$, where the $N_{I_{Ogt}}$ denotes the number of pixels in the ground truth, I_{OG} denotes the generated image from OG, and I_{Ogt} denotes the ground truth object image. We also define an extra loss for RN to be $\mathcal{L}_{RN}^{sup} = \frac{1}{N_{I_{hole}}}\|(M) \odot (I_{out} - I_{in})\|_1$. For the OG network, a squared RGB image with size 128 is generated. For the RN network, however, a squared RGB image with size 512 is required for input and output. To end this, for ground truth scene image, we first resize and then crop it centered at the masked area.

Moreover, in the pre-training process, we randomly add Gaussian noises on the background area in image generated by OG. This operation helps RN improve the performance on ensuring semantic continuity and visual authenticity. Here, the loss for OG and RN are computed as

$$\mathcal{L}_{OG}^{pre} = \mathcal{L}_{RN_{GRAN}} + \beta_1 \mathcal{L}_{OG_{OGAN}} + \beta_2 \mathcal{L}_{OG}^{sup},$$
$$\mathcal{L}_{RN}^{pre} = \mathcal{L}_{RN_{GRAN}} + \beta_3 \mathcal{L}_{valid} + \beta_4 \mathcal{L}_{RN}^{sup}, \tag{9}$$

where β_1, β_2, β_3 and β_4 denotes the loss balancing weights and are set to 5, 0.1, 0.1 and 0.5, respectively.

We compare the results of two training schemes for GAN module in pre-training step: training OGAN and GRAN separately, and training them jointly end-to-end. It can be seen from Table 2 that the end-to-end training improves the performance of OG network by a large margin. This can be in part explained by that, when training with GRAN, the supervision of background refinement are passed to OGAN, allowing for the OG network to ensure both the visual authenticity and the background continuity. The learning rate for GAN module is set to 10^{-5} during the pre-training stage. During the end-to-end training stage, the learning rate is reduced from 10^{-5} to 10^{-6}.

4.4 User Study

To validate the authenticity of the hallucinated image, we conduct two user-study experiments, where 166 users are involved to evaluate the visual quality of our produced images. In the first experiment, we send each user 30 randomly

Table 3. Statistical results of user study experiments. *std* here denotes the standard deviation of people.

Term	COCO-Uexp1	Visual Genome-Uexp1	NYU-Uexp1	COCO-Uexp2	Visual Genome-Uexp2	NYU-Uexp2
Score	0.406	0.399	0.373	3.38	3.31	3.17
Std	0.049	0.052	0.075	0.098	0.092	0.15

Table 4. Results of Ours and GNN under different setups. We compare the performances of the two networks trained with full settings in our paper (Ours-full/GNN-full), the performances of the two networks trained using Kullback–Leibler divergence loss L_{KL} to replace the cross-entropy loss L_{class} (Ours-L_{KL}/GNN-L_{KL}), the performance of the two networks trained without global feature from Autoencoder (Ours-without-G/GNN-without-G), and those of the two trainings without spatial information (Ours-without-S/GNN-without-S).

Term-full	Ours-full	GNN-full	Ours-L_{KL}	GNN-L_{KL}	Ours-without-G	GNN-without-G	Ours-without-S	GNN-without-S
Accuracy (%)	87.93	79.23	85.46	73.01	86.55	75.67	84.86	77.64
Balanced Acc (%)	85.62	74.82	80.12	70.56	84.19	71.78	81.21	72.58

selected image pairs, where one of them is the ground-truth image and the other is our hallucinated one, and ask the user to pick which image of the two is the real one. Finally, the proposed method achieves 40.6% real chosen on COCO, 39.9% on Visual Genome, and 37.3% on NYU v2.

For the second experiment, we send each user 30 randomly selected image triplets: the original image, the image with an instance masked out, and the hallucinated image. We then ask the user to give a grade for each hallucinated image: very visually real (4 points), fairly visually real (3 points), borderline (2 points), fairly visually fake (1 points), very visually fake (0 points). The obtained average score is 3.38 on COCO, 3.31 on Visual Genome and 3.17 on NYU V2. The results of these two experiments are summarized in Table 3, where our proposed approach indeed achieves promising and stable performances on the user study.

4.5 Ablation Studies

More Visual Samples. We show more visual examples in Fig. 8 for COCO and Visual Genome and NYU v2 datasets, where our methods generates visually pleasing results.

Results using Kullback–Leibler Divergence Loss in GCN Module. We replace the cross-entropy loss $\mathcal{L}_{CE}$ of GCN module by the Kullback–Leibler divergence loss $\mathcal{L}_{KL}$, which regresses the probabilities of the predicted classes. The supervised label is the ground-truth probability from the detection module. It can be seen from Table 4 that the $\mathcal{L}_{KL}$ decreases the prediction accuracies of both models. This shows when the hallucinated object's semantic feature

Fig. 8. Results on the COCO (Rows 1,2,4), NYU v2 (Row 3) and Visual Genome (Rows 5,6) dataset images. For each group of images, the first one is the masked image, the second one is our restored image and the third one is the original image.

contains multi-class information, its relationship with other objects is hard to determine, thus increasing the complexity of semantic understanding.

Results Without Using Global Feature in GCN Module. In this experiment, we train our GCN module and GNN without the global information from the Autoencoder. This means the GCN can only extract structural information from the remaining objects. As can be seen from Table 4, the performance decreased. This is because the missing objects are with strong relationship with the scene style. An visual example is shown in Fig. 9: when the semantic feature of the baseball is not trained with global feature, a tennis ball is hallucinated. One reason for this error is that, COCO contains both types of balls in the same class. After training with the global feature, however, a baseball is hallucinated, showing effect of the global feature for the image generating process.

Results Without Using Spatial Information in GCN Module. We train our GCN module and GNN without the spatial information, and thus the feature of an object is shortened to be 1024 dimensions. It can be seen from Table 4 that the performances of both methods decrease significantly, which demonstrates the important role played by the spatial coordinates of the defections.

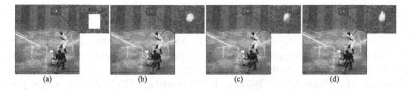

Fig. 9. Comparing training GAN module with and without global scene features from Autoencoder. The image (a), (b), (c) and (d) denotes the masked image, the restored image with global feature from Autoencoder, the restored image without global feature, and the ground truth image, respectively. Ours GCN module is used to hallucinate the semantic feature of the missing object.

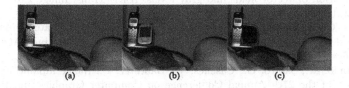

Fig. 10. An example that our framework produces visual realistic and structurally reasonable restoration but with a incorrectly predicted label.

An Interesting Case. We show a case in Fig. 10, where an incorrect object label is predicted, yet a visually plausible and contextually reasonable object is hallucinated. This highlights that, our method can utilize the structural information to produce realistic object hallucinations, even with a wrong label.

5 Conclusion

In this paper, we introduce a new image restoration task, named hallucinating visual instances in total absentia (HAVITA). Unlike image inpainting that aims to fill the missing part of a visual instance present in the scene, HAVITA concerns about hallucinating an completely-absent object. To this end, we propose an end-to-end network that models the input image as a semantic graph, and then predicting the semantic information using a GCN module, followed by feeding the semantic information into our GAN module to generate the final hallucination. Results on three datasets demonstrate the encouraging potential of our approach. In our future work, we will study hallucinating objects with more complex physical constraints, such as the 3D orientation of a running car following that of the road.

Acknowledgement. This research was supported by Australian Research Council Projects FL-170100117, DP-180103424, LE-200100049 and the startup funding of Stevens Institute of Technology.

References

1. Abu-El-Haija, S., Perozzi, B., Al-Rfou, R., Alemi, A.A.: Watch your step: Learning node embeddings via graph attention. In: Advances in Neural Information Processing Systems, pp. 9180–9190 (2018)
2. Arjovsky, M., Chintala, S., Bottou, L.: Wasserstein GAN (2017). arXiv preprint arXiv:1701.07875
3. Atwood, J., Towsley, D.: Diffusion-convolutional neural networks. In: Advances in Neural Information Processing Systems, pp. 1993–2001 (2016)
4. Bacciu, D., Errica, F., Micheli, A.: Contextual graph markov model: A deep and generative approach to graph processing. In: ICML (2018)
5. Ballester, C., Bertalmio, M., Caselles, V., Sapiro, G., Verdera, J.: Filling-in by joint interpolation of vector fields and gray levels. IEEE Trans. Image Process. **10**(8), 1200–1211 (2001)
6. Barnes, C., Shechtman, E., Finkelstein, A., Goldman, D.B.: Patchmatch: A randomized correspondence algorithm for structural image editing. ACM Trans. Graphics (ToG) **28**, 24 (2009). ACM
7. Bertalmio, M., Sapiro, G., Caselles, V., Ballester, C.: Image inpainting. In: Proceedings of the 27th Annual Conference on Computer Graphics and Interactive Techniques, pp. 417–424. ACM Press/Addison-Wesley Publishing Co. (2000)
8. Bertalmio, M., Vese, L., Sapiro, G., Osher, S.: Simultaneous structure and texture image inpainting. IEEE Trans. Image Process. **12**(8), 882–889 (2003)
9. Bruna, J., Zaremba, W., Szlam, A., LeCun, Y.: Spectral networks and locally connected networks on graphs (2013). arXiv preprint arXiv:1312.6203
10. Chen, J., Zhu, J., Song, L.: Stochastic training of graph convolutional networks with variance reduction (2017). arXiv preprint arXiv:1710.10568
11. Criminisi, A., Perez, P., Toyama, K.: Object removal by exemplar-based inpainting. In: 2003 IEEE Computer Society Conference on Computer Vision and Pattern Recognition, 2003, Proceedings, vol. 2, p. II. IEEE (2003)
12. Defferrard, M., Bresson, X., Vandergheynst, P.: Convolutional neural networks on graphs with fast localized spectral filtering. In: Advances in Neural Information Processing Systems, pp. 3844–3852 (2016)
13. Deng, J., Dong, W., Socher, R., Li, L.J., Li, K., Fei-Fei, L.: Imagenet: A large-scale hierarchical image database. In: 2009 IEEE Conference on Computer Vision and Pattern Recognition, pp. 248–255. IEEE (2009)
14. Furukawa, Y., Hernández, C., et al.: Multi-view stereo: A tutorial. Found. Trends® Comput. Graphics Vis. **9**(1–2), 1–148 (2015)
15. Fyffe, G., Jones, A., Alexander, O., Ichikari, R., Graham, P., Nagano, K., Busch, J., Debevec, P.: Driving high-resolution facial blendshapes with video performance capture. In: ACM SIGGRAPH 2013 Talks, p. 1 (2013)
16. Fyffe, G., Nagano, K., Huynh, L., Saito, S., Busch, J., Jones, A., Li, H., Debevec, P.: Multi-view stereo on consistent face topology. In: Computer Graphics Forum, vol. 36, pp. 295–309. Wiley Online Library (2017)
17. Gallicchio, C., Micheli, A.: Graph echo state networks. In: The 2010 International Joint Conference on Neural Networks (IJCNN), pp. 1–8. IEEE (2010)
18. Gao, H., Wang, Z., Ji, S.: Large-scale learnable graph convolutional networks. In: Proceedings of the 24th ACM SIGKDD International Conference on Knowledge Discovery & Data Mining, pp. 1416–1424. ACM (2018)
19. Gilmer, J., Schoenholz, S.S., Riley, P.F., Vinyals, O., Dahl, G.E.: Neural message passing for quantum chemistry. In: Proceedings of the 34th International Conference on Machine Learning, vol. 70, pp. 1263–1272. JMLR.org (2017)

20. Goodfellow, I., Pouget-Abadie, J., Mirza, M., Xu, B., Warde-Farley, D., Ozair, S., Courville, A., Bengio, Y.: Generative adversarial nets. In: Advances in Neural Information Processing Systems, pp. 2672–2680 (2014)
21. Gori, M., Monfardini, G., Scarselli, F.: A new model for learning in graph domains. In: Proceedings, 2005 IEEE International Joint Conference on Neural Networks, 2005, vol. 2, pp. 729–734. IEEE (2005)
22. Grosse, R., Johnson, M.K., Adelson, E.H., Freeman, W.T.: Ground truth dataset and baseline evaluations for intrinsic image algorithms. In: 2009 IEEE 12th International Conference on Computer Vision, pp. 2335–2342. IEEE (2009)
23. Hamilton, W., Ying, Z., Leskovec, J.: Inductive representation learning on large graphs. In: Advances in Neural Information Processing Systems, pp. 1024–1034 (2017)
24. Hartley, R., Zisserman, A.: Multiple View Geometry in Computer Vision. Cambridge University Press (2003)
25. Hays, J., Efros, A.A.: Scene completion using millions of photographs. Commun. ACM **51**(10), 87–94 (2008)
26. He, K., Gkioxari, G., Dollár, P., Girshick, R.: Mask R-CNN. In: Proceedings of the IEEE International Conference on Computer Vision, pp. 2961–2969 (2017)
27. He, K., Zhang, X., Ren, S., Sun, J.: Deep residual learning for image recognition. In: Proceedings of the IEEE Conference on Computer Vision and Pattern Recognition, pp. 770–778 (2016)
28. Henaff, M., Bruna, J., LeCun, Y.: Deep convolutional networks on graph-structured data (2015). arXiv preprint arXiv:1506.05163
29. Hernandez, C., Vogiatzis, G., Cipolla, R.: Multiview photometric stereo. IEEE Trans. Pattern Anal. Mach. Intell. **30**(3), 548–554 (2008)
30. Hoiem, D., Divvala, S.K., Hays, J.H.: Pascal VOC 2008 challenge. In: PASCAL Challenge Workshop in ECCV. Citeseer (2009)
31. Huang, W., Zhang, T., Rong, Y., Huang, J.: Adaptive sampling towards fast graph representation learning. In: Advances in Neural Information Processing Systems, pp. 4558–4567 (2018)
32. Ioffe, S., Szegedy, C.: Batch normalization: Accelerating deep network training by reducing internal covariate shift (2015). arXiv preprint arXiv:1502.03167
33. Isola, P., Zhu, J.Y., Zhou, T., Efros, A.A.: Image-to-image translation with conditional adversarial networks. In: Proceedings of the IEEE Conference on Computer Vision and Pattern Recognition, pp. 1125–1134 (2017)
34. Jaderberg, M., Simonyan, K., Zisserman, A., et al.: Spatial transformer networks. In: Advances in Neural Information Processing Systems, pp. 2017–2025 (2015)
35. Karras, T., Laine, S., Aila, T.: A style-based generator architecture for generative adversarial networks. In: Proceedings of the IEEE Conference on Computer Vision and Pattern Recognition, pp. 4401–4410 (2019)
36. Karsch, K., Hedau, V., Forsyth, D., Hoiem, D.: Rendering synthetic objects into legacy photographs. ACM Trans. Graph. (TOG) **30**(6), 1–12 (2011)
37. Karsch, K., Liu, C., Kang, S.B.: Depth transfer: Depth extraction from video using non-parametric sampling. IEEE Trans. Pattern Anal. Mach. Intell. **36**(11), 2144–2158 (2014)
38. Karsch, K., Sunkavalli, K., Hadap, S., Carr, N., Jin, H., Fonte, R., Sittig, M., Forsyth, D.: Automatic scene inference for 3d object compositing. ACM Trans. Graph. (TOG) **33**(3), 1–15 (2014)
39. Kipf, T.N., Welling, M.: Semi-supervised classification with graph convolutional networks (2016). arXiv preprint arXiv:1609.02907

40. Köhler, R., Schuler, C., Schölkopf, B., Harmeling, S.: Mask-specific inpainting with deep neural networks. In: Jiang, X., Hornegger, J., Koch, R. (eds.) GCPR 2014. LNCS, vol. 8753, pp. 523–534. Springer, Cham (2014). https://doi.org/10.1007/978-3-319-11752-2_43

41. Krishna, R., Zhu, Y., Groth, O., Johnson, J., Hata, K., Kravitz, J., Chen, S., Kalantidis, Y., Li, L.J., Shamma, D.A., et al.: Visual genome: Connecting language and vision using crowdsourced dense image annotations. Int. J. Comput. Vision **123**(1), 32–73 (2017)

42. Krizhevsky, A., Sutskever, I., Hinton, G.E.: Imagenet classification with deep convolutional neural networks. In: Advances in Neural Information Processing Systems, pp. 1097–1105 (2012)

43. Lan, L., Wang, X., Zhang, S., Tao, D., Gao, W., Huang, T.S.: Interacting tracklets for multi-object tracking. IEEE Trans. Image Process. **27**(9), 4585–4597 (2018)

44. Lee, J.B., Rossi, R., Kong, X.: Graph classification using structural attention. In: Proceedings of the 24th ACM SIGKDD International Conference on Knowledge Discovery & Data Mining, pp. 1666–1674. ACM (2018)

45. Levie, R., Monti, F., Bresson, X., Bronstein, M.M.: Cayleynets: Graph convolutional neural networks with complex rational spectral filters. IEEE Trans. Signal Process. **67**(1), 97–109 (2018)

46. Levin, A., Zomet, A., Weiss, Y.: Learning how to inpaint from global image statistics. In: Null, p. 305. IEEE (2003)

47. Li, R., Wang, S., Zhu, F., Huang, J.: Adaptive graph convolutional neural networks. In: Thirty-Second AAAI Conference on Artificial Intelligence (2018)

48. Li, Y., Liu, S., Yang, J., Yang, M.H.: Generative face completion. In: Proceedings of the IEEE Conference on Computer Vision and Pattern Recognition, pp. 3911–3919 (2017)

49. Liao, Z., Karsch, K., Zhang, H., Forsyth, D.: An approximate shading model with detail decomposition for object relighting. Int. J. Comput. Vision **127**(1), 22–37 (2019)

50. Lim, J.H., Ye, J.C.: Geometric GAN (2017). arXiv preprint arXiv:1705.02894

51. Lin, T.-Y., Maire, M., Belongie, S., Hays, J., Perona, P., Ramanan, D., Dollár, P., Zitnick, C.L.: Microsoft COCO: Common objects in context. In: Fleet, D., Pajdla, T., Schiele, B., Tuytelaars, T. (eds.) ECCV 2014. LNCS, vol. 8693, pp. 740–755. Springer, Cham (2014). https://doi.org/10.1007/978-3-319-10602-1_48

52. Liu, M.Y., Tuzel, O.: Coupled generative adversarial networks. In: Advances in Neural Information Processing Systems, pp. 469–477 (2016)

53. Liu, Z., Chen, C., Li, L., Zhou, J., Li, X., Song, L., Qi, Y.: Geniepath: Graph neural networks with adaptive receptive paths. Proc. AAAI Conf. Artif. Intell. **33**, 4424–4431 (2019)

54. Maksai, A., Wang, X., Fleuret, F., Fua, P.: Non-markovian globally consistent multi-object tracking. In: The IEEE International Conference on Computer Vision (ICCV) (2017)

55. Maksai, A., Wang, X., Fua, P.: What players do with the ball: A physically constrained interaction modeling. In: The IEEE Conference on Computer Vision and Pattern Recognition (CVPR) (2016)

56. Mao, X., Li, Q., Xie, H., Lau, R.Y., Wang, Z., Paul Smolley, S.: Least squares generative adversarial networks. In: Proceedings of the IEEE International Conference on Computer Vision, pp. 2794–2802 (2017)

57. Mirza, M., Osindero, S.: Conditional generative adversarial nets (2014). arXiv preprint arXiv:1411.1784

58. Miyato, T., Koyama, M.: Cgans with projection discriminator (2018). arXiv preprint arXiv:1802.05637
59. Mo, S., Cho, M., Shin, J.: Instagan: Instance-aware image-to-image translation (2018). arXiv preprint arXiv:1812.10889
60. Monti, F., Boscaini, D., Masci, J., Rodola, E., Svoboda, J., Bronstein, M.M.: Geometric deep learning on graphs and manifolds using mixture model CNNs. In: Proceedings of the IEEE Conference on Computer Vision and Pattern Recognition, pp. 5115–5124 (2017)
61. Silberman, N., Hoiem, D., Kohli, P., Fergus, R.: Indoor segmentation and support inference from RGBD images. In: ECCV (2012)
62. Niepert, M., Ahmed, M., Kutzkov, K.: Learning convolutional neural networks for graphs. In: International Conference on Machine Learning, pp. 2014–2023 (2016)
63. Park, E., Yang, J., Yumer, E., Ceylan, D., Berg, A.C.: Transformation-grounded image generation network for novel 3d view synthesis. In: Proceedings of the IEEE Conference on Computer Vision and Pattern Recognition, pp. 3500–3509 (2017)
64. Park, T., Liu, M.Y., Wang, T.C., Zhu, J.Y.: Gaugan: Semantic image synthesis with spatially adaptive normalization. In: ACM SIGGRAPH 2019 Real-Time Live! p. 2. ACM (2019)
65. Pathak, D., Krahenbuhl, P., Donahue, J., Darrell, T., Efros, A.A.: Context encoders: Feature learning by inpainting. In: Proceedings of the IEEE Conference on Computer Vision and Pattern Recognition, pp. 2536–2544 (2016)
66. Qiu, J., Wang, X., Fua, P., Tao, D.: Matching Seqlets: An unsupervised approach for locality preserving sequence matching. IEEE Trans. Pattern Anal. Mach. Intell. (2019)
67. Qiu, J., Wang, X., Maybank, S.J., Tao, D.: World from blur. In: The IEEE Conference on Computer Vision and Pattern Recognition (CVPR) (June 2019)
68. Radford, A., Metz, L., Chintala, S.: Unsupervised representation learning with deep convolutional generative adversarial networks (2015). arXiv preprint arXiv:1511.06434
69. Ren, J.S., Xu, L., Yan, Q., Sun, W.: Shepard convolutional neural networks. In: Advances in Neural Information Processing Systems, pp. 901–909 (2015)
70. Salimans, T., Goodfellow, I., Zaremba, W., Cheung, V., Radford, A., Chen, X.: Improved techniques for training GANs. In: Advances in Neural Information Processing Systems, pp. 2234–2242 (2016)
71. Scarselli, F., Gori, M., Tsoi, A.C., Hagenbuchner, M., Monfardini, G.: The graph neural network model. IEEE Trans. Neural Netw. **20**(1), 61–80 (2008)
72. Simonyan, K., Zisserman, A.: Very deep convolutional networks for large-scale image recognition (2014). arXiv preprint arXiv:1409.1556
73. Song, Y., Yang, C., Lin, Z., Liu, X., Huang, Q., Li, H., Jay Kuo, C.C.: Contextual-based image inpainting: Infer, match, and translate. In: Proceedings of the European Conference on Computer Vision (ECCV), pp. 3–19 (2018)
74. Song, Y., Yang, C., Shen, Y., Wang, P., Huang, Q., Kuo, C.C.J.: SPG-Net: Segmentation prediction and guidance network for image inpainting (2018). arXiv preprint arXiv:1805.03356
75. Sperduti, A., Starita, A.: Supervised neural networks for the classification of structures. IEEE Trans. Neural Netw. **8**(3), 714–735 (1997)
76. Tran, D., Ranganath, R., Blei, D.: Hierarchical implicit models and likelihood-free variational inference. In: Advances in Neural Information Processing Systems, pp. 5523–5533 (2017)
77. Veličković, P., Cucurull, G., Casanova, A., Romero, A., Lio, P., Bengio, Y.: Graph attention networks (2017). arXiv preprint arXiv:1710.10903

78. Veličković, P., Fedus, W., Hamilton, W.L., Liò, P., Bengio, Y., Hjelm, R.D.: Deep graph infomax (2018). arXiv preprint arXiv:1809.10341
79. Wang, X., Li, Z., Tao, D.: Subspaces indexing model on grassmann manifold for image search. IEEE Trans. Image Process. **20**(9), 2627–2635 (2011)
80. Wang, X., Türetken, E., Fleuret, F., Fua, P.: Tracking interacting objects using intertwined flows. IEEE Trans. Pattern Anal. Mach. Intell. **38**(11), 2312–2326 (2016)
81. Wang, X., Türetken, E., Fleuret, F., Fua, P.: Tracking interacting objects optimally using integer programming. In: European Conference on Computer Vision and Pattern Recognition (ECCV), pp. 17–32 (2014)
82. Yan, Z., Li, X., Li, M., Zuo, W., Shan, S.: Shift-net: Image inpainting via deep feature rearrangement. In: Proceedings of the European Conference on Computer Vision (ECCV), pp. 1–17 (2018)
83. Yang, C., Lu, X., Lin, Z., Shechtman, E., Wang, O., Li, H.: High-resolution image inpainting using multi-scale neural patch synthesis. In: Proceedings of the IEEE Conference on Computer Vision and Pattern Recognition, pp. 6721–6729 (2017)
84. Yang, Y., Qiu, J., Song, M., Tao, D., Wang, X.: Distilling knowledge from graph convolutional networks. In: IEEE Conference on Computer Vision and Pattern Recognition (CVPR) (2020)
85. Yang, Y., Wang, X., Song, M., Yuan, J., Tao, D.: SPAGAN: shortest path graph attention network. In: International Joint Conference on Artificial Intelligence (IJCAI) (2019)
86. Ying, Z., You, J., Morris, C., Ren, X., Hamilton, W., Leskovec, J.: Hierarchical graph representation learning with differentiable pooling. In: Advances in Neural Information Processing Systems, pp. 4800–4810 (2018)
87. Yu, J., Lin, Z., Yang, J., Shen, X., Lu, X., Huang, T.S.: Generative image inpainting with contextual attention. In: Proceedings of the IEEE Conference on Computer Vision and Pattern Recognition, pp. 5505–5514 (2018)
88. Yu, J., Lin, Z., Yang, J., Shen, X., Lu, X., Huang, T.S.: Free-form image inpainting with gated convolution. In: Proceedings of the IEEE International Conference on Computer Vision, pp. 4471–4480 (2019)
89. Zhang, H., Goodfellow, I., Metaxas, D., Odena, A.: Self-attention generative adversarial networks (2018). arXiv preprint arXiv:1805.08318
90. Zhang, J., Shi, X., Xie, J., Ma, H., King, I., Yeung, D.Y.: Gaan: Gated attention networks for learning on large and spatiotemporal graphs (2018). arXiv preprint arXiv:1803.07294
91. Zheng, C., Cham, T.J., Cai, J.: T2net: Synthetic-to-realistic translation for solving single-image depth estimation tasks. In: Proceedings of the European Conference on Computer Vision (ECCV), pp. 767–783 (2018)
92. Zheng, C., Cham, T.J., Cai, J.: Pluralistic image completion. In: Proceedings of the IEEE Conference on Computer Vision and Pattern Recognition, pp. 1438–1447 (2019)
93. Zhou, T., Tulsiani, S., Sun, W., Malik, J., Efros, A.A.: View synthesis by appearance flow. In: Leibe, B., Matas, J., Sebe, N., Welling, M. (eds.) ECCV 2016. LNCS, vol. 9908, pp. 286–301. Springer, Cham (2016). https://doi.org/10.1007/978-3-319-46493-0_18

Weakly-Supervised 3D Shape Completion in the Wild

Jiayuan Gu[1,2]([⊠]), Wei-Chiu Ma[1,3], Sivabalan Manivasagam[1,4],
Wenyuan Zeng[1,4], Zihao Wang[1], Yuwen Xiong[1,4], Hao Su[2],
and Raquel Urtasun[1,4]

[1] Uber Advanced Technologies Group, Pittsburgh, USA
{wichiu,manivasagam,wenyuan,yuwen,urtasun}@uber.com
[2] University of California, San Diego, USA
{jigu,haosu}@eng.ucsd.edu
[3] Massachusetts Institute of Technology, Cambridge, USA
[4] University of Toronto, Toronto, Canada

Abstract. 3D shape completion for real data is important but challenging, since partial point clouds acquired by real-world sensors are usually sparse, noisy and unaligned. Different from previous methods, we address the problem of learning 3D complete shape from unaligned and real-world partial point clouds. To this end, we propose a weakly-supervised method to estimate both 3D canonical shape and 6-DoF pose for alignment, given multiple partial observations associated with the same instance. The network jointly optimizes canonical shapes and poses with multi-view geometry constraints during training, and can infer the complete shape given a single partial point cloud. Moreover, learned pose estimation can facilitate partial point cloud registration. Experiments on both synthetic and real data show that it is feasible and promising to learn 3D shape completion through large-scale data without shape and pose supervision.

1 Introduction

We are interested in the problem of 3D shape completion, which estimates the complete geometry of objects from partial observations. This task is a prerequisite for many important real-world applications. For example, complete shapes can facilitate automated vehicles to track objects [12] and robots to figure out the best pose to grasp objects [29]. Previous works [8,14,33] have successfully applied deep learning methods to learn shape priors from large-scale synthetic data, which results in improvement of the 3D shape completion task. However, most these prior works have two major limitations: 1) they require the ground-truth shape for learning, and 2) they assume the input partial point clouds are aligned and normalized to the canonical frame, in which the object faces forward and are centered at the origin. In addition, models trained on synthetic data do not transfer well to the real world due to the domain gap.

Electronic supplementary material The online version of this chapter (https://doi.org/10.1007/978-3-030-58558-7_17) contains supplementary material, which is available to authorized users.

We aim to use real data for the 3D shape completion task. However, since there is a lack of real 3D data that comes with sufficient high-quality ground-truth 3D shapes, we cannot directly adopt these supervised learning methods developed in the synthetic domain. Although there are a few datasets containing real scans, such as KITTI [11] and ScanNet [7], no efforts are made to explore the possibility of learning 3D shape completion in a weakly-supervised fashion. There are three challenges to work on real 3D data, unique from the synthetic ones: 1) No or few ground-truth complete shapes are available for full supervision. Note that annotating 3D shapes are more difficult and expensive than annotating 2D images; 2) Partial point clouds acquired by real-world sensors like RGB-D cameras or LiDAR are sparse and noisy; 3) Poses and sizes of objects are more diverse, and partial observations may be occluded by other objects.

In this paper, we address the problem of learning 3D shape completion from real, unaligned partial point clouds without shape and pose supervision (Sect. 3). The proposed method is weakly-supervised by multi-view consistency of instances (Sect. 4). The key contributions of our work are as follows:

1. We propose a weakly-supervised[1] approach to learn 3D shape completion from unaligned point clouds. Our promising results show that it is feasible to learn 3D shape completion from large-scale 3D data without shape and pose supervision.
2. We showcase the extension of our method to tackle the challenging partial point cloud registration problem.

2   Related Work

3D Reconstruction from Single Images. 3D shape completion is highly related to 3D reconstruction from single images, since a partial point cloud can be obtained from a RGB-D image. Since the problem is ill-posed by nature, many learning-based approaches are developed to learn shape priors from large-scale data. [6] uses a recurrent 3D CNN to predict a 3D occupancy grid given one or more images of an object. [27] proposes a differentiable 'view consistency' loss and a probabilistic occupancy grid. [10] pioneers the representation of point clouds as output. However, they all require full supervision from synthetic images rendered from ShapeNet [4]. Performance on real datasets like Pascal 3D+ [31] suffer from unrealistic ground truth shapes from aligned CAD models.

Thus, [14,26,32,37] focus on reconstructing 3D shapes with weak supervision. Especially, [26] enforces geometric consistency between the independently predicted shape and pose from two views of the same instance. Differential point clouds (DPC) [14] uses a similar strategy to reconstruct point clouds and devises differentiable projection of point clouds. However, it is non-trivial to extend these methods to real-world data, which will be discussed in Sect. 5.3.

3D Reconstruction from Multiple Frames. By leveraging consecutive frames, 3D shapes can be reconstructed from RGB images [1,9,25] or depth

---

[1] We use the term "weakly-supervised" instead of "unsupervised learning of shape and pose" [14] to avoid confusion, which are in fact equivalent.

images [17]. The problem is also known as Structure-from-Motion (SfM). [23,28,35] are proposed to tackle it with deep learning. Although poses are estimated in both SfM and our 3D shape completion, the main difference is that unseen 3D points are hallucinated in our 3D shape completion while depths are estimated in the SfM.

KinectFusion [17] fuses all the depth data streamed from a Kinect sensor into a single global implicit surface model of the observed scene in real-time. It demonstrates the advantages of maintaining a full surface model compared to frame-to-frame tracking. Our method benefits from the similar idea, but differs from it in 2 aspects: 1) The proposed approach is a learning-based framework based on 3D point clouds only. 2) The trained model can predict the complete shape from a single point cloud and the relative pose between two distant views during inference, which is demonstrated in our experiments.

3D Shape Completion. 3D shape completion is usually performed on partial scans of individual objects. With the success of deep learning, learning-based approaches show more flexibility and better performance compared with geometry-based and alignment-based methods. [8] combines a data-driven shape predictor and analytic 3D shape synthesis. [33] proposes a variant of PointNet [19] to directly process point clouds and generate high-resolution outputs. [24] devises a tree-style neural network to generate structured point clouds.

3D shape completion without full supervision is of increasing interest to the community. [22] finetunes the encoder on the target dataset, like KITTI [11], with a fixed generator pretrained on the ground truth SDF representation of synthetic data, like ShapeNet [4]. [13] generates half-to-half sequence pairs from the ground truth complete point clouds of ShapeNet, and learns features by half-to-half prediction and self-reconstruction. [5] trains autoencoders to learn embedding features of shape on clean, complete synthetic data and noisy, partial target data. An adaption network is learned to transform the embedding space of noisy point clouds to that of clean point clouds with a GAN setup. However, none of those works deals with unaligned point clouds and relies on complete synthetic data to pretrain.

Deep Learning for Point Clouds. PointNet [19] is the pioneer to directly process point clouds with a deep neural network, followed by many variants [15,20,30]. It extracts features for each point with a shared multi-layer perceptron (MLP), and outputs with an aggregation function invariant to permutation. Any point-cloud-based neural network can work as the encoder of our method.

3   Problem

The goal of 3D shape completion is to predict a complete shape Y given a partial observation X. In this work, we represent the partial observation and the complete shape as point clouds: $X \in \mathcal{R}^{n \times 3}$ and $Y \in \mathcal{R}^{m \times 3}$, where n and m are the number of partial and complete points respectively.

Previous approaches [22,33] have assumed that partial observations are normalized according to ground-truth bounding boxes and transformed into a predefined canonical frame, *e.g.*, the forward-facing object centered at the origin.

Past works may also assume the ground-truth shape $Y^{gt} \in \mathcal{R}^{m_{gt} \times 3}$ is available and train a model in a supervised setting via a permutation-invariant loss function $L(Y, Y^{gt}; X)$, such as Chamfer Distance (CD) or Earth Mover Distance (EMD) [10] to evaluate reconstruction quality. While these ground truth information may be available on synthetic data, they may not be available in the real-world setting. Thus, we build on past works and propose a more general and challenging setting:

- We do not assume knowledge of the ground-truth canonical frame for normalizing and aligning the partial observations. Instead, we maintain the partial observations in the *sensor* coordinate system.
- We do not assume knowledge of the ground-truth shape.
- The point cloud observation is captured by a sensor (*e.g.*, a LiDAR) from the real world and can therefore be sparse and also noisy.
- Instead of ground truth, we have access to a set of unaligned partial observations of the instance, captured at different viewpoints by the sensor.

We call this more realistic setting "weakly-supervised shape completion in the wild". This setting is especially applicable to the real-world setting such as in autonomous driving or indoor scene navigation, where the robot may observe other moving agents from multiple viewpoints and needs to reason about the shape and pose to perform shape completion. In the next section, we propose our method for tackling weakly-supervised shape completion in the wild.

4 Method

4.1 Overview

We tackle weakly-supervised shape completion in the wild by jointly learning the canonical shape and pose of the object. The underlying idea is that predicting a complete shape Y_{sen} in the *sensor* coordinate system is equivalent to predicting a complete shape Y_{can} in the *canonical* coordinate system[2] and then transforming it according to a 6-DoF pose T_{can}^{sen}, where $Y_{sen} = T_{can}^{sen} Y_{can}$. But a key question remains: how do we learn the complete shape and pose when we do not have access to the ground-truth for either? We leverage the fact that, during training, we have access to multiple observations of the object from different viewpoints. We know that these observations, while noisy, accurately represent different views of the GT shape. By enforcing that predicted shapes and poses are consistent with recorded observations, we can train the network in a weakly-supervised fashion to estimate both the shape and pose from a single observation.

Our training approach is as follows: Given a set of sensor observations of the object of interest, $\{X_{sen}^1, X_{sen}^2, \cdots, X_{sen}^M\}$, we apply a deep autoencoder network

---

[2] The *canonical* frame in our method is not predefined, but emerges during training.

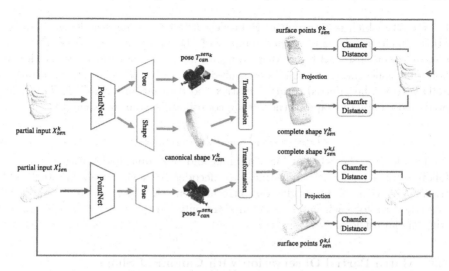

Fig. 1. Overview of our weakly-supervised shape completion pipeline at training time. This illustration is for a pair of partial inputs and can be extended to partial input of a shape from multiple views, by optimizing the averaged loss among selected pairs of inputs.

to each observation X^i_{sen} and predict a canonical shape Y_{can} and pose $T^{sen_i}_{can}$. We then apply two loss terms based on these outputs to guide the network to learn the correct shape and pose: (1) the partial observation points should match the completed shape transformed by the estimated pose (observation-matching-shape), and (2) the surface points of the completed shape as viewed by the estimated sensor pose should match the observation points with self-occlusion taken into consideration (shape-projection-matching-observation). Because we have access to multiple observations and multiple pose predictions, we can encourage the network to generate a completed shape estimate that minimizes these two loss terms with respect to all observations and poses. We call this *multi-view consistency*.

During inference, the trained network takes as input a single partial point cloud, and outputs the estimated pose and completed shape. Our pipeline is illustrated in Fig. 1. The following sections describe in detail our approach. Sect. 4.2 describes the network architecture to predict 3D canonical shapes and 6-DoF poses. Sect. 4.3 and Sect. 4.4 present the loss terms (observation-matching-shape and shape-silhouette-matching-observation). Sect. 4.5 describes how we extend the two loss terms to work on multiple observations.

4.2   Predict Canonical Shape and Pose

Given a partial observation X_{sen}, we employ a deep encoder-decoder network to predict both the 3D canonical shape and 6-DoF pose. The encoder encodes the input X_{sen} to a latent code z. We use the same encoder architecture as

PCN's [33], which is a variant of PointNet [19]. The shape decoder is a 3-layer MLP, which decodes z to a fixed number of 3D coordinates $Y_{can} \in R^{m \times 3}$. The pose decoder, also a 3-layer MLP, outputs a rotation $\hat{R}$ and a translation t. Following [36], the rotation is represented as a 6D vector, and the translation a 3D vector. The inferred rotation matrix and translation form T_{can}^{sen}. Thus, the predicted complete shape in the *sensor* coordinate system is calculated as:

$$Y_{sen} = T_{can}^{sen} Y_{can} = f_T(z) f_{shape}(z) \tag{1}$$

To alleviate the issue of local minima and overcome bad initialization, we follow prior art and have multiple pose decoder branches in our network and train them with the hindsight loss introduced in DPC [14]. In brief, hindsight loss is where, for each batch, gradients are only backpropagated to the branch with the lowest loss.

4.3 Match Partial Observation with Canonical Shape

We now describe the observation-matching-shape loss, which we implement as an asymmetric Chamfer-Distance (CD) between the observation point cloud and the completed shape point cloud in the sensor-coordinate space. The asymmetric CD (Eq. 2) between the input observation X_{sen} and the output shape Y_{sen} is

$$CD(X_{sen} \mapsto Y_{sen}) = \frac{1}{|X_{sen}|} \sum_{x \in X_{sen}} \min_{y \in Y_{sen}} ||x - y||_2 \tag{2}$$

This forces the output canonical shape to completely cover the input observation. However, it does not guarantee that Y_{sen} is close to X_{sen}—even a point cloud that fills the whole 3D space would minimize Eq. 2, which is not desired. Thus, we need to develop a more sophisticated loss term to enforce how the sensor acquires the point cloud and compute the distance between the input observation and the projection of the output, which is described next.

4.4 Project Canonical Shape to Partial Observation

We now describe the shape-projection-matching-observation term. Using our knowledge of how the sensor acquires observations, we can "simulate" which points on the surface are observed based on the estimated complete shape point cloud and the estimated pose. We can then force the "generated" point cloud to match the true observation. We tailor this loss term based on knowledge of how the LiDAR sensor works.

Given a subset point cloud $\hat{Y}_{sen}$ of the predicted point cloud Y_{sen}, which are on the surface as viewed from the sensor (the "simulated" observation), another asymmetric CD (Eq. 3) between the input X_{sen} and those *surface* points is optimized.

$$CD(\hat{Y}_{sen} \mapsto X_{sen}) = \frac{1}{|\hat{Y}_{sen}|} \sum_{\hat{y} \in \hat{Y}_{sen}} \min_{x \in X_{sen}} ||\hat{y} - x||_2 \tag{3}$$

We introduce a simple, flexible and efficient way to infer the *surface* points. The point cloud acquired by the LiDAR sensor can be projected to a range image, which is essentially the polar-coordinate system of the LiDAR sensor. The polar coordinate (ϕ, θ, r) of a cartesian sensor observation point (x, y, z) is calculated as Eq. 4:

$$r = \sqrt{x^2 + y^2 + z^2}, \phi = tan^{-1}\left(\frac{x}{y}\right), \theta = sin^{-1}\left(\frac{z}{r}\right) \tag{4}$$

where r is radial distance to the sensor and ϕ, θ are the azimuth and pitch angles, respectively, of the ray shot from the LiDAR sensor. According to the resolution of LiDAR (d_ϕ, d_θ), we can discretize (ϕ, θ) to $(\lfloor\frac{\phi}{d_\phi}\rfloor, \lfloor\frac{\theta}{d_\theta}\rfloor)$, which forms several bins. For each bin, the point with the smallest distance is considered to be on the *surface*. This is the depth buffer approach widely used for rasterization in the computer graphics literature.

This "projection" approach is differentiable and simple to implement, since we can just count the occupied bins and find the smallest distance in each. No voxelization or normalization of the points is needed like in DPC [14], which makes our approach more flexible, especially for real data without a normalized scale. Additionally, we can flexibly adjust the projection resolution to be coarser than the real resolution, which helps with noisy and occluded real-world data. Furthermore, this method is also efficient as the computation complexity is $O(m)$, where m is the number of predicted points.

4.5 Multi-view Consistency

Both loss terms, observation-matching-shape and shape-projection-matching-observation, work not only for the input observation X^i_{sen}, but also works for all other observations of the instance in the set. Inspired by [14,26], we leverage the consistency among multiple views associated with the same instance to supervise 3D shape prediction and 6-DoF pose estimation. During training, we sample M observations $\{X^1_{sen}, X^2_{sen}, \cdots, X^M_{sen}\}$ of one instance within a batch. One view is selected as the *reference*, denoted by index k. Intuitively, all observations share the same complete rigid shape in the *canonical* coordinate system. In other words, for any view i, Y^i_{can} should be close to Y^k_{can}. Therefore, we can replace Y^i_{sen} with $Y^{k,i}_{sen} = Y^k_{can}\hat{R}^T_i + t_i$, which forces the network to learn a complete canonical shape matching all the partial views.

The full loss for a given training example $\{X^i_{sen}, i \in 1...N\}$ with reference view k is calculated as Eq. 5:

$$L(\{X^i_{sen}\}) = \sum_{i=1}^{M} CD(X^i_{sen} \mapsto Y^{k,i}_{sen}) + \beta CD(\hat{Y}^{k,i}_{sen} \mapsto X^i_{sen}) \tag{5}$$

where β is a hyper-parameter, which can be adjusted according to the quality of data. While we could apply multi-view consistency between all possible pairs (*i.e.*, make each index in the observation set the reference index and sum all terms), we choose one randomly to reduce training complexity.

5   Experiments

In this section, we demonstrate our method and several baselines on this new setting of weakly-supervised shape completion in the wild. We first evaluate our method on the standard synthetic data benchmark, ShapeNet. We then showcase the performance of our method on two real-world self-driving datasets for which we construct ground-truth complete shapes. Furthermore, we demonstrate our method also works on the task of point cloud registration. Finally, we compare our approach against a fully-supervised oracle.

5.1   Datasets

ShapeNet [4]. ShapeNet is a richly-annotated, large-scale dataset of 3D synthetic shapes. We focus on 3 categories: chairs, cars, and airplanes. We use the same data and split provided by DPC [14], where the data available for each training example is 5 random RGB-D views of the model. We note that this data only has random viewpoint/orientation, and the translation component of the view is fixed. To acquire partial point clouds in the camera coordinate system, we backproject depth maps according to the intrinsic matrix. The average number of points of the partial point clouds for chairs, cars and airplanes is 3018, 2956, 756 respectively. For evaluation, 8192 ground-truth points are sampled from the surface of the CAD models.

3D Vehicle Dataset [16]. We build a collection of complete vehicle object point clouds from a large-scale LiDAR dataset for self-driving that contain bounding box instance annotations for over 1.2 million frames. We generate the ground-truth complete shape as follows: for each static object, we accumulate the LiDAR points inside the bounding box and determine the object relative coordinates for the LiDAR points based on the bounding box center. Since cars are usually symmetric, we postprocess data by mirroring the aggregate point cloud along the vehicle's heading axis, followed by Gaussian statistical outlier removal, to acquire complete shapes for annotated objects. Visualizations of the ground-truth shape can be seen in Fig. 3. There are 13700 annotated objects in total, splitted into 10000/700/3000 for train/val/test. On average, each object is associated with 80 scans, and each scan contains 1163 points. We filter observations to include at least 100 points to avoid overly sparse observations.

SemanticKITTI [2]. Instance and semantic annotations for the LiDAR point clouds are provided for all sequences of the Odometry Benchmark. We use SemanticKITTI's odometry localization to aggregate partial point clouds of the same parked vehicle instance (with at least 512 points on average) into a single vehicle frame and apply radius outlier removal. Following [2], we train our network on instances generated from sequences 00 to 10, except for sequence 08 instances which are used as test set. There are 659/229 instances and 51186/16299 observations for training/test. On average, each object is associated with 95 scans, and each scan contains 1377 points.

5.2   Tasks and Metrics

3D Shape Completion. For shape completion, the algorithm is required to predict shapes in the sensor coordinate system. Given the ground-truth canonical shape and pose, we can compute the ground-truth shape in the sensor coordinate system. Then, we adopt the standard metrics used in the literature [10,14,26,33]. The main metric to evaluate shape completion against ground truth point cloud Y_{sen}^{gt} is the Chamfer Distance (Eq. 6). The first term is called the *Precision* and the second term is called the *Coverage*.

$$
\begin{aligned}
CD(\hat{Y}_{sen} \leftrightarrow Y_{sen}^{gt}) = \frac{1}{|\hat{Y}_{sen}|} \sum_{\hat{y} \in \hat{Y}_{sen}} \min_{y \in Y_{sen}} ||\hat{y} - y^{gt}||_2 \\
+ \frac{1}{|Y_{sen}^{gt}|} \sum_{y^{gt} \in Y_{sen}^{gt}} \min_{\hat{y} \in \hat{Y}_{sen}} ||y^{gt} - \hat{y}||_2
\end{aligned}
\tag{6}
$$

Partial Point Cloud Registration. Given two partial observations, the algorithm is required to predict the relative pose from one to the other. This task is more challenging than common point cloud registration. The algorithms are evaluated by calculating the quaternion distance θ, or *angle difference*, between the estimated pose q_{pred} and the GT pose q_{gt} for all instances in the dataset: $\theta = 2 \arccos\langle q_{pred}, q_{gt}\rangle$. Following DPC [14], we report the median of angle difference and accuracy (the percentage of samples for which $\theta \leq 30°$). In addition, if the translation is predicted, we also report the median of mean-square-error between the prediction and the ground truth.

5.3   Baselines

To our knowledge, there are currently no weakly-supervised methods for shape completion that take as input a single unaligned partial point cloud. We instead compare our method against the state-of-the-art single-image 3D reconstruction method, and standard point cloud alignment methods.

DPC [14]. DPC is a weakly-supervised method, which is trained on image pairs to reconstruct 3D point clouds. We compare to their reported results of shape reconstruction on ShapeNet, and adapt their method to range images. We argue that it is non-trivial to adapt DPC to the "wild" setting, where both pose rotation and translation are unknown and the shape size is not bounded. We list their drawbacks as follows: 1) It is assumed that canonical shapes are normalized into a unit cube; 2) A fixed camera distance to the object is provided and only rotations are considered in the original paper; 3) The projection loss only is sensitive to density and occlusion. Despite these drawbacks, we modify DPC by replacing perspective transformation with polar transformation and scaling the normalized output by a fixed factor to match the real scale, denoted by DPC-LIDAR. We refer readers to the supplementary for more details.

ICP. Since our ground-truth complete shapes are acquired by accumulating partial point clouds given ground-truth transformations, we introduce a baseline based on Iterative Closest Point (ICP). For two consecutive frames, we calculate the rigid transformation between two partial point clouds by ICP. Thus, given a pair of partial point clouds, the transformation can be calculated by accumulating results from ICP. All the partial point clouds can be transformed into a certain frame and fused. To reduce the accumulated error, we choose the middle frame as the reference. For 3D shape completion, the fused point cloud in the reference frame is transformed according to the ground-truth pose to compare with the ground-truth complete shape. For point cloud registration, we compare the transformation from one frame to the reference one with the ground-truth transformation. Local ICP [3] and Global ICP [21] are the two ICP algorithms we compare against. We use the implementation of Open3D [34] and search the best hyper-parameters on the validation set (0.175 for the distance threshold in Local ICP and 0.125 for the voxel size in Global ICP).

5.4   Implementation Details

As our synthetic and real datasets are of different sizes and come very different distributions, we slightly modify our data input and our implementation of the model for each setting:

Input. To ease the learning requirements for our model, we preprocess input partial point clouds. Given a partial point cloud in the sensor coordinate system, we shift the point cloud to be centered at the origin without knowledge of the ground-truth shape or pose: For ShapeNet, an axis-aligned bounding box in the camera coordinate system is calculated, and its center is shifted to the origin; for real data, we first extract a bounding frustum of the input partial point cloud, and then centralize the frustum[3]. After converting to this origin-shifted coordinate system, the input point cloud is resampled with replacement to a fixed size, as done in the original PointNet [19]. For ShapeNet and for real LiDAR datasets, we uniformly resample 3096 and 1024 points, respectively, from the input partial point cloud.

Training. During training, 4 observations per instance are sampled in a batch. Adam is used as the optimizer. For synthetic data, models are trained with an initial learning rate of $1e^{-4}$ for 300 k iterations and a batch size of 32. The learning rate is decayed by 0.5 every 100 K iterations. For real data, models are trained with an initial learning rate of $1e^{-4}$ for 400 k iterations and a batch size of 32. The learning rate is decayed by 0.7 every 100K iterations. Especially, all the observations of one instance in a batch are within a window of 20 frames. It takes less than 16 h to train our model with a GTX 1080Ti. The loss weight β is set to 0.25 and 0.05 for synthetic and real data respectively.

5.5   Results of 3D Shape Completion

ShapeNet. We first demonstrate shape completion results on ShapeNet. The chamfer distance, precision and coverage are reported on the test set. We also

---

[3] The resulting coordinate system is similar to *3D mask coordinate* introduced in [18].

Table 1. Quantitative results of 3D shape completion on the test set of ShapeNet (airplane/car/chair). All the values are multiplied by 100. We also report the original numbers from [14]. Note that they align predicted shapes according to the validation set before evaluating on the test set.

	CD	Precision	Coverage
DPC	6.26/3.54/4.85	4.19/1.95/2.59	2.07/1.59/2.26
DPC†	13.17/4.73/7.32	8.17/2.52/3.81	5.00/2.20/3.52
DPC (pre-aligned [14])	3.91/3.47/4.30	-	-
DPC† (pre-aligned [14])	5.07/4.09/5.86	-	-
Ours	**1.95/2.68/3.33**	**0.91/1.27/1.69**	**1.05/1.41/1.64**

include the chamfer distance of DPC reported by [14]. Besides, we compare our approach with DPC†, which predicts both rotation and translation of the pose. Note that [14] evaluates *shape prediction*, rather than *shape completion*, by aligning the predicted canonical shape according to the ground truth of the validation set. We argue that this evaluation protocol assumes that all the objects share the same canonical space and it does not really disentangle shape and pose.

Table 1 shows the quantitative comparison between our method and the DPC variants. Despite not having access to the ground-truth translation or size of the object, our model is able to predict a more accurate complete shape. Fig. 2 shows the qualitative results. Since planes are usually flat and result in sparse observations, DPC fails to learn a clean shape while our approach is more robust. We refer readers to the supplementary for ablation studies and more details.

Real LiDAR Datasets. We now apply our method to real-world partial LiDAR scans of vehicles. Table 2a and 2b show the comparison between our method and ICP baselines on real LiDAR datasets. Figure 3 showcases the qualitative results. DPC-LIDAR does not converge and performs much worser than our approach, which implies it is better to process point clouds directly rather than project them into 2D planes and rely on existing 2D methods. Moreover, compared to strong ICP baselines, our method shows higher precision and comparable coverage. More results and analysis are provided in the supplementary.

Table 2. 3D shape completion results on the test sets of real LiDAR datasets.

	CD	Precision	Coverage
DPC-LIDAR	0.928	0.489	0.439
Local-ICP	0.315	0.170	0.145
Global-ICP	0.309	0.174	**0.135**
Ours	**0.255**	**0.083**	0.172

(a) 3D vehicle dataset

	CD	Precision	Coverage
Local-ICP	0.246	0.152	0.094
Global-ICP	0.213	0.138	**0.075**
Ours	**0.194**	**0.087**	0.107

(b) SemanticKITTI

Input(Ours)   Input(DPC)   GT   DPC   DPC†   Ours

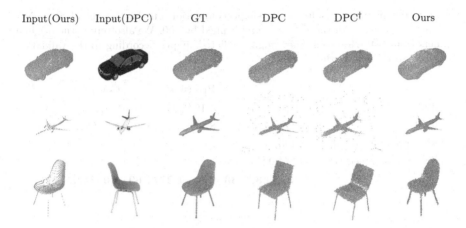

Fig. 2. Qualitative results of 3D shape completion on the test set of ShapeNet. All the (input and predicted) point clouds are transformed to the ground-truth canonical frame and visualized at a fixed viewpoint. *Note that the input of our method is the point cloud lifted from the depth map input of DPC.*

5.6   Partial Point Cloud Registration

In this section, we showcase that our method can be applied to point cloud registration. It is challenging to align real-world partial point clouds for traditional methods like ICP, due to incompleteness, noise, and sparsity. Furthermore, even if the transformation between two consecutive frames is accurate, the error may accumulate across frames. However, our approach alleviates these issues since it implicitly encodes a complete canonical shape. To evaluate the performance of point cloud registration, we select the middle frame in a sequence as the target, and align other frames to the reference according to estimated poses. Given a source and a target point cloud, our method predicts T_{can}^{src} and T_{can}^{tgt}. Thus, the transformation from the source to the target is calculated as $T_{src}^{tgt} = T_{can}^{tgt}(T_{can}^{src})^{-1}$. We report the accuracy, median angle difference, and median MSE between the groundtruth vs. our method, Local-ICP, Global-ICP in Table 3a and 3b. Figure 3 demonstrates fused point clouds according to estimated poses. Our method outperforms standard alignment methods.

5.7   Comparison with Fully-Supervised Counterparts

Our method does not rely on any ground-truth shape and pose, or any prior knowledge of where the object is located, *e.g.*, the bounding box. Yet, it can still reconstruct the 3D shape reliably and accurately. In order to understand the upper bound of our method for shape completion, we include an *oracle* baseline, where our model is trained with ground-truth complete shapes. Concretely, during training, given a partial point cloud in the *sensor* coordinate system as input, we employ the same network to encode the input and decode the canonical

Input   GT   Local-ICP   Global-ICP   Ours(shape)   Ours(fusion)

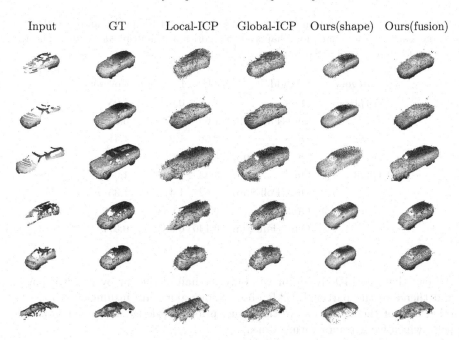

Fig. 3. Qualitative results of our method compared against ground-truth and ICP on the real datasets (row 1–3: 3D vehicle dataset; row 4–6: SemanticKITTI). All the point clouds are transformed to the ground-truth canonical frame and visualized at a fixed viewpoint. We denote our approach for 3D shape completion and point cloud registration by *Ours(shape)* and *Ours(fusion)*.

Table 3. Point cloud registration results on the test sets of real LiDAR datasets. We report the median of angle difference and accuracy (the percentage of samples for which $\Delta\theta \leq 30°$, as well as the median of translation error Δt.)

	Acc	Rot $\Delta\theta$	Trans Δt		Acc	Rot $\Delta\theta$	Trans Δt
Local-ICP	84.09	11.33	0.30	Local-ICP	85.29	13.04	0.31
Global-ICP	83.83	10.69	0.26	Global-ICP	85.28	10.59	0.23
Ours	**97.68**	**2.37**	**0.13**	Ours	**89.37**	**2.86**	**0.17**
(a) 3D vehicle dataset				(b) SemanticKITTI			

complete shape without estimating the pose. The Chamfer Distance is calculated between the output canonical shape and the ground-truth canonical complete shape. Note that our method with full supervision is identical to *PCN-FC* [33], except for unaligned point clouds as input.

Table 4 shows that there exists a gap between our weakly-supervised approach and its fully-supervised counterpart. However, the gap is even smaller than that between ours and DPC [14]. The Chamfer Distance of the fully-supervised oracle is almost half of that of our weakly-supervised approach. This may be due to

Table 4. Shape completion results on the test sets of ShapeNet and our 3D vehicle dataset. All the values for ShapeNet categories are multiplied by 100. The full-supervision oracle is denoted by *Ours(Full-Sup)*.

Category	Method	CD	Precision	Coverage
Airplane	Ours	1.95	0.91	1.05
	Ours(Full-Sup)	1.65	0.77	0.88
Car	Ours	2.68	1.27	1.41
	Ours(Full-Sup)	2.04	1.01	1.03
Chair	Ours	3.33	1.69	1.64
	Ours(Full-Sup)	2.82	1.47	1.35
Real vehicle	Ours	0.255	0.083	0.172
	Ours (Full-Sup)	0.140	0.064	0.077

the fact that the LiDAR sensor will only see half of the car by which it passes, and therefore the partial LiDAR observations alone are insufficient to see the other side of the car. To solve this issue, prior knowledge of the category may help, which we leave for future work.

6   Discussion and Future Work

We have proposed a new setting, weakly-supervised 3D shape completion in the wild, which better captures the realistic scenario of being able to infer unknown shape from real world scans of objects. We demonstrate that this challenging problem can be tackled by jointly learning both shape and pose with multi-view consistency. However, there remains much space to improve and explore. From visualization, we observe that the model tends to generate coarse shapes and miss details, due to the noise of pose estimation. It is also observed in training that the loss calculated on point clouds is more sensitive to the density compared to that using 2D projection. More efforts can be made to improve visual quality and narrow the gap with fully-supervised methods. Besides, we use PointNet as the backbone for simplicity and efficiency in this work. Differently designed networks can be applied to predict shapes and poses separately. Furthermore, our approach currently requires knowing multiple views of a single rigid object. Such "annotations" can be acquired by a 3D detector and tracker. Thus, one future direction is to study self-supervised or weakly-supervised 3D detection and tracking.

References

1. Agarwal, S., Snavely, N., Seitz, S.M., Szeliski, R.: Bundle adjustment in the large. In: Daniilidis, K., Maragos, P., Paragios, N. (eds.) ECCV 2010. LNCS, vol. 6312, pp. 29–42. Springer, Heidelberg (2010). https://doi.org/10.1007/978-3-642-15552-9_3

2. Behley, J., Garbade, M., Milioto, A., Quenzel, J., Behnke, S., Stachniss, C., Gall, J.: SemanticKITTI: A dataset for semantic scene understanding of LiDAR sequences. In: Proceedings of the IEEE/CVF International Conference on Computer Vision (ICCV) (2019)
3. Besl, P.J., McKay, N.D.: Method for registration of 3-d shapes. In: Sensor fusion IV: control paradigms and data structures, vol. 1611, pp. 586–606. International Society for Optics and Photonics (1992)
4. Chang, A.X., Funkhouser, T., Guibas, L., Hanrahan, P., Huang, Q., Li, Z., Savarese, S., Savva, M., Song, S., Su, H., et al.: Shapenet: An information-rich 3d model repository (2015). arXiv preprint arXiv:1512.03012
5. Chen, X., Chen, B., Mitra, N.J.: Unpaired point cloud completion on real scans using adversarial training (2019). arXiv preprint arXiv:1904.00069
6. Choy, C.B., Xu, D., Gwak, J.Y., Chen, K., Savarese, S.: 3D-R2N2: A unified approach for single and multi-view 3D object reconstruction. In: Leibe, B., Matas, J., Sebe, N., Welling, M. (eds.) ECCV 2016. LNCS, vol. 9912, pp. 628–644. Springer, Cham (2016). https://doi.org/10.1007/978-3-319-46484-8_38
7. Dai, A., Chang, A.X., Savva, M., Halber, M., Funkhouser, T., Nießner, M.: ScanNet: Richly-annotated 3D reconstructions of indoor scenes. In: Proceedings of the IEEE Conference on Computer Vision and Pattern Recognition, pp. 5828–5839 (2017)
8. Dai, A., Ruizhongtai Qi, C., Nießner, M.: Shape completion using 3d-encoder-predictor CNNs and shape synthesis. In: Proceedings of the IEEE Conference on Computer Vision and Pattern Recognition, pp. 5868–5877 (2017)
9. Eigen, D., Puhrsch, C., Fergus, R.: Depth map prediction from a single image using a multi-scale deep network. In: Advances in Neural Information Processing Systems, pp. 2366–2374 (2014)
10. Fan, H., Su, H., Guibas, L.J.: A point set generation network for 3d object reconstruction from a single image. In: Proceedings of the IEEE Conference on Computer Vision and Pattern Recognition, pp. 605–613 (2017)
11. Geiger, A., Lenz, P., Stiller, C., Urtasun, R.: Vision meets robotics: The KITTI dataset. Int. J. Robot. Res. (IJRR) (2013)
12. Giancola, S., Zarzar, J., Ghanem, B.: Leveraging shape completion for 3d siamese tracking. In: Proceedings of the IEEE Conference on Computer Vision and Pattern Recognition, pp. 1359–1368 (2019)
13. Han, Z., Wang, X., Liu, Y.S., Zwicker, M.: Multi-angle point cloud-VAE: Unsupervised feature learning for 3d point clouds from multiple angles by joint self-reconstruction and half-to-half prediction. arXiv (2019)
14. Insafutdinov, E., Dosovitskiy, A.: Unsupervised learning of shape and pose with differentiable point clouds. In: Advances in Neural Information Processing Systems, pp. 2802–2812 (2018)
15. Li, Y., Bu, R., Sun, M., Wu, W., Di, X., Chen, B.: PointCNN: Convolution on x-transformed points. In: Advances in Neural Information Processing Systems, pp. 820–830 (2018)
16. Manivasagam, S., Wang, S., Wong, K., Zeng, W., Sazanovich, M., Tan, S., Yang, B., Ma, W.C., Urtasun, R.: Lidarsim: Realistic LiDAR simulation by leveraging the real world. In: Proceedings of the IEEE/CVF Conference on Computer Vision and Pattern Recognition, pp. 11167–11176 (2020)
17. Newcombe, R.A., Izadi, S., Hilliges, O., Molyneaux, D., Kim, D., Davison, A.J., Kohi, P., Shotton, J., Hodges, S., Fitzgibbon, A.: Kinectfusion: Real-time dense surface mapping and tracking. In: 2011 10th IEEE International Symposium on Mixed and Augmented Reality, pp. 127–136. IEEE (2011)

18. Qi, C.R., Liu, W., Wu, C., Su, H., Guibas, L.J.: Frustum pointnets for 3d object detection from RGB-D data. In: Proceedings of the IEEE Conference on Computer Vision and Pattern Recognition, pp. 918–927 (2018)
19. Qi, C.R., Su, H., Mo, K., Guibas, L.J.: Pointnet: Deep learning on point sets for 3d classification and segmentation. In: Proceedings of the IEEE Conference on Computer Vision and Pattern Recognition, pp. 652–660 (2017)
20. Qi, C.R., Yi, L., Su, H., Guibas, L.J.: Pointnet++: Deep hierarchical feature learning on point sets in a metric space. In: Advances in Neural Information Processing Systems, pp. 5099–5108 (2017)
21. Rusu, R.B., Blodow, N., Beetz, M.: Fast point feature histograms (FPFH) for 3d registration. In: 2009 IEEE International Conference on Robotics and Automation, pp. 3212–3217. IEEE (2009)
22. Stutz, D., Geiger, A.: Learning 3d shape completion from laser scan data with weak supervision. In: Proceedings of the IEEE Conference on Computer Vision and Pattern Recognition, pp. 1955–1964 (2018)
23. Tang, C., Tan, P.: BA-Net: Dense bundle adjustment network (2018). arXiv preprint arXiv:1806.04807
24. Tchapmi, L.P., Kosaraju, V., Rezatofighi, H., Reid, I., Savarese, S.: TopNet: Structural point cloud decoder. In: Proceedings of the IEEE Conference on Computer Vision and Pattern Recognition, pp. 383–392 (2019)
25. Triggs, B., McLauchlan, P.F., Hartley, R.I., Fitzgibbon, A.W.: Bundle adjustment — A modern synthesis. In: Triggs, B., Zisserman, A., Szeliski, R. (eds.) IWVA 1999. LNCS, vol. 1883, pp. 298–372. Springer, Heidelberg (2000). https://doi.org/10.1007/3-540-44480-7_21
26. Tulsiani, S., Efros, A.A., Malik, J.: Multi-view consistency as supervisory signal for learning shape and pose prediction. In: Proceedings of the IEEE Conference on Computer Vision and Pattern Recognition, pp. 2897–2905 (2018)
27. Tulsiani, S., Zhou, T., Efros, A.A., Malik, J.: Multi-view supervision for single-view reconstruction via differentiable ray consistency. In: Proceedings of the IEEE Conference on Computer Vision and Pattern Recognition, pp. 2626–2634 (2017)
28. Ummenhofer, B., Zhou, H., Uhrig, J., Mayer, N., Ilg, E., Dosovitskiy, A., Brox, T.: Demon: Depth and motion network for learning monocular stereo. In: Proceedings of the IEEE Conference on Computer Vision and Pattern Recognition, pp. 5038–5047 (2017)
29. Varley, J., DeChant, C., Richardson, A., Ruales, J., Allen, P.: Shape completion enabled robotic grasping. In: 2017 IEEE/RSJ International Conference on Intelligent Robots and Systems (IROS), pp. 2442–2447. IEEE (2017)
30. Wang, Y., Sun, Y., Liu, Z., Sarma, S.E., Bronstein, M.M., Solomon, J.M.: Dynamic graph CNN for learning on point clouds. ACM Trans. Graphics (TOG), **1**, 1–13 (2019)
31. Xiang, Y., Mottaghi, R., Savarese, S.: Beyond PASCAL: A benchmark for 3d object detection in the wild. In: IEEE Winter Conference on Applications of Computer Vision, pp. 75–82. IEEE (2014)
32. Yan, X., Yang, J., Yumer, E., Guo, Y., Lee, H.: Perspective transformer nets: Learning single-view 3d object reconstruction without 3d supervision. In: Advances in Neural Information Processing Systems, pp. 1696–1704 (2016)
33. Yuan, W., Khot, T., Held, D., Mertz, C., Hebert, M.: PCN: Point completion network. In: 2018 International Conference on 3D Vision (3DV), pp. 728–737. IEEE (2018)
34. Zhou, Q.Y., Park, J., Koltun, V.: Open3D: A modern library for 3D data processing (2018). arXiv:1801.09847

35. Zhou, T., Brown, M., Snavely, N., Lowe, D.G.: Unsupervised learning of depth and ego-motion from video. In: CVPR (2017)
36. Zhou, Y., Barnes, C., Lu, J., Yang, J., Li, H.: On the continuity of rotation representations in neural networks. In: Proceedings of the IEEE Conference on Computer Vision and Pattern Recognition, pp. 5745–5753 (2019)
37. Zhu, R., Kiani Galoogahi, H., Wang, C., Lucey, S.: Rethinking reprojection: Closing the loop for pose-aware shape reconstruction from a single image. In: Proceedings of the IEEE International Conference on Computer Vision, pp. 57–65 (2017)

DTVNet: Dynamic Time-Lapse Video Generation via Single Still Image

Jiangning Zhang[1], Chao Xu[1], Liang Liu[1], Mengmeng Wang[1], Xia Wu[1],
Yong Liu[1(✉)], and Yunliang Jiang[2]

[1] APRIL Lab, College of Control Science and Engineering, Zhejiang University,
Hangzhou, Zhejiang, China
{186368,21832066,leonliuz,mengmengwang,xiawu}@zju.edu.cn
[2] Huzhou University, Huzhou, Zhejiang, China
jyl@zjhu.edu.cn

Abstract. This paper presents a novel end-to-end dynamic time-lapse video generation framework, named DTVNet, to generate diversified time-lapse videos from a single landscape image, which are conditioned on normalized motion vectors. The proposed DTVNet consists of two submodules: *Optical Flow Encoder* (OFE) and *Dynamic Video Generator* (DVG). The OFE maps a sequence of optical flow maps to a *normalized motion vector* that encodes the motion information inside the generated video. The DVG contains motion and content streams that learn from the motion vector and the single image respectively, as well as an encoder and a decoder to learn shared content features and construct video frames with corresponding motion respectively. Specifically, the *motion stream* introduces multiple *adaptive instance normalization* (AdaIN) layers to integrate multi-level motion information that are processed by linear layers. In the testing stage, videos with the same content but various motion information can be generated by different *normalized motion vectors* based on only one input image. We further conduct experiments on Sky Time-lapse dataset, and the results demonstrate the superiority of our approach over the state-of-the-art methods for generating high-quality and dynamic videos, as well as the variety for generating videos with various motion information (https://github.com/zhangzjn/DTVNet).

Keywords: Generative adversarial network · Optical flow encoding · Time-Lapse video generation

---

J. Zhang and C. Xu—Indicates equal contributions.

Electronic supplementary material The online version of this chapter (https://doi.org/10.1007/978-3-030-58558-7_18) contains supplementary material, which is available to authorized users.

A. Vedaldi et al. (Eds.): ECCV 2020, LNCS 12350, pp. 300–315, 2020.
https://doi.org/10.1007/978-3-030-58558-7_18

1   Introduction

Video generation is a task to generate video sequences from the noise or special conditions such as images and masks, which has promising application capabilities, *e.g.* video dataset expansion, texture material generation, and film production. However, this is a very challenging task where the model has to learn the content, motion, and relationship between objects simultaneously, especially when objects have no special shapes such as cloud and fog. In this work, we focus on a more challenging task that generates realistic and dynamic time-lapse videos based on a single natural landscape image.

Thanks to the development of *generative adversarial network* [4,10,35], many excellent models have been proposed that achieve generating realistic and dynamic videos. Vondrick *et al.* [28] explore how to leverage a random noise to learn the dynamic video in line with a moving foreground and a static background pathways from large amounts of unlabeled video by capitalizing on adversarial learning methods. However, this kind of noise-based method usually suffers from a low-quality video generation and a hard training, because of its mapping from a vector to a high-resolution feature map. Villegas *et al.* [27] first adopt the LSTM [6] to predict futureexploits two different high-level poses in a sequence to sequence manner, and then use an analogy-based encoder-decoder model to generate future images. The method can well generate high-quality and reasonable images, but it acquires extra label information such as human pose and several past reference images that is unpractical. Subsequent MoCoGAN [26] apply LSTM model to generate a sequence of random vectors that consist of content and motion parts, and then map them to a sequence of video frames. This method can generate diversified dynamic videos that contain the same content in theory, because we can fix the content part while change the motion part of the random vectors. Nevertheless, the resolution of the generated video is still low.

To obtain high-resolution and high-quality time-lapse videos, Aigner *et al.* [1] propose the FutureGAN that uses spatio-temporal 3d convolutions in all encoder and decoder modules, aiming at capturing both the spatial and temporal components of a video sequence. But FutureGAN is time-consuming during the training stage for utilizing concepts of the progressively growing GAN [11]. Xiong *et al.* [32] recently propose a two-stage 3DGAN-based model named MDGAN, which learns to generate long-term realistic time-lapse videos of high resolution given the first frame. Though MDGAN is capable of generating vivid motion and realistic video, it is hard to train such a two stage model and lacks of diverse video generation capabilities, which limits the practicality of the method. We follow the same time-lapse video generation task with MDGAN and design a practical one stage model.

Recently, some researches [13,14] introduce optical flow maps that contain the motion information during the training and testing stages, aiming at explicitly supplying motion signals to the network. Liang *et al.* [14] design a dual motion GAN that simultaneously solves the primal future-frame prediction and future-flow prediction tasks. Though it can generate high-quality video frames,

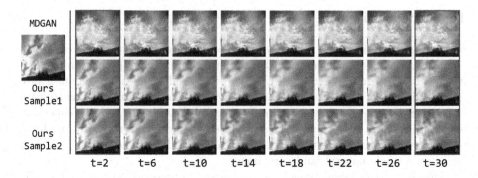

Fig. 1. The first row shows generated frames by MDGAN [32], and other two rows are from our method with random motion vectors. The left image is in-the-wild start landscape frame, and other frames are generation results at different times. Our method could generate more photorealistic and diversified video frames. Please zoom in the red rectangles for a more clear comparison. (Color figure online)

but it is time-consuming for taking more computational effort on future-flow predictions. Besides the quality and the resolution, the diversification of the generated time-lapse videos is likewise important for practical application. Li *et al.* [13] first map a sampled noise to consecutive flows, and then use proposed video prediction algorithm to synthesize a set of likely future frames in multiple time steps from one single still image. During the testing, the method can directly use sample points from the distribution for predictions. Different from aforementioned methods, our method is designed in an end-to-end manner and we introduce a normalized motion vector to control the generation process, which can generate high-quality and diversified time-lapse videos.

In this work, we propose an end-to-end dynamic time-lapse video generation framework, *i.e.* DTVNet, to generate diversified time-lapse videos from a single landscape image. The time-lapse landscape videos generally contain still objects, *e.g.* house, earth, and tree, as well as unspecific objects, *e.g.* cloud and fog, which is challenging to understand the motion relationship between objects. Specifically, the proposed DTVNet consists of two submodules: *Optical Flow Encoder* (OFE) and *Dynamic Video Generator* (DVG). OFE introduces unsupervised optical flow estimation method to get the motion maps among consecutive images and encoders them to a normalized motion vector. During the testing stage, we exclude the process of flow estimation as well as flow encoding, and directly sample from a normalization distribution as the motion vector, which reduces the network computing overhead and supply diversified motion information simultaneously. DVG contains motion and content streams that learn from the motion vector and the single image respectively, as well as an encoder and a decoder to learn shared content features and construct video frames respectively. In detail, the normalized motion vector is integrated into the motion stream by multiple *adaptive instance normalization* (AdaIN) layers [7]. During the training stage, we apply content loss, motion loss, and adversarial loss that ensures high-quality, dynamic, and diversified video generation, as shown in Fig. 1.

Specifically, we make the following four contributions:

- An *optical flow encoder* (OFE) is designed to supply normalized motion information in the training stage that are used to guide diversified video frames generation.
- A new *dynamic video generator* (DVG) is proposed to first learn disentangling content and motion features separately, and then use integrated features to generate target video.
- We apply content loss, motion loss, and adversarial loss during the training stage, which ensures high-quality, dynamic, and diversified video generation.
- Experimental results on Sky Time-lapse dataset indicate that the proposed DTVNet can generate high-quality and dynamic video frames in an end-to-end one stage network.

2 Related Work

Optical Flow Estimation. Optical flow is a reliable representation to characterize motion between frames. Starting from Flownet [3], many supervised methods for optical flow estimation are proposed, *e.g.* FlowNet2 [9], PWC-Net [24], IRR-PWC [8], etc. Though these methods are in high accuracy and efficiency, they heavily depend on the labeled dataset, while it it hard to acquire the ground truth in reality, which reduces the practicality of these methods.

As an alternative, some researchers focus on studying unsupervised methods [22,34] and have achieved great success. Liu *et al.* [17] propose the SelFlow that distills reliable flow estimations from non-occluded pixels, and uses these predictions as ground truth to learn optical flow for hallucinated occlusions. DDFlow [16] further improve the model performance by distilling unlabeled data. In this paper, we first apply unsupervised method [15] to estimate the optical flow map, and then encoder the flow information to a motion vector that is used as a condition when generating the video.

Generative Adversarial Networks. Since Goodfellow *et al.* [4] first introduces the generative adversarial network (GAN) that contains a generator and a discriminator, many GAN-based approaches are proposed and have achieved impressive results in various aspects, *e.g.* image inpainting, style translation, super-resolution, etc. Mehdi *et al.* [19] propose the cGAN that controls the mode of generated samples by adding extra conditional variable to the network. Pix2Pix [10] uses ℓ_1 and adversarial loss for paired image translation tasks, and Zhu *et al.* [35] further introduces a cycle consistency loss to deal with unpaired image-to-image translation tasks. ProGAN [11] describes a new training methodology that grows both the generator and discriminator progressively, which is capable of generating up to 1024 resolution images. Besides 2D-based GAN methods, Wu *et al.* [31] apply 3D convolution to generate 3D objects from a probabilistic space. MDGAN [32] presents a 3D convolutional based two-stage approach to generate realistic time-lapse videos of high resolution. Our model follows the GAN idea and takes extra motion information into consideration when generating videos.

Video Generation. Video generation aims at generating image sequences from a noise, image(s), or with extra condition such as human pose, semantic label map, and optical flow. Mathieu et al. [18] first adopt the GAN idea to mitigate the inherently blurry predictions obtained from the standard mean squared error loss function. Subsequently, VGAN [28] is proposed to untangle the foreground from the background of the scene with a spatio-temporal convolutional architecture, and many follow-up works borrowed the idea of disentangling. Saito et al. [23] exploits two different types of generators, i.e. a temporal generator and an image generator, to generate videos and achieve good performance, while MoCoGAN [26] maps a sequence of random vectors that consists of content and motion parts to a sequence of video frames.

However, these noise-input and progressively growing methods generally suffer from a low-quality generation or a difficult training process, so following methods that use a single image as input are proposed. MDGAN [32] adopts a two-stage network to generate long-term future frames. It generates videos of realistic content for each frame in the first stage and then refines the generated video from the first stage. Nam et al. [20] learns the correlation between the illumination change of an outdoor scene and the time of the day by a multi-frame joint conditional generation framework. Yang et al. [33] propose a pose guided method to synthesize human videos in a disentangled way: plausible motion prediction and coherent appearance generation. Similarly, Cai et al. [2] design a skeleton-to-image network to generate human action videos. Researches [21,29] take one semantic label map as input to synthesize a sequence of photo-realistic video frames. Recently, some flow based methods have made great success. Liang et al. [14] design a dual motion GAN that simultaneously solves the primal future-frame prediction and future-flow prediction tasks. Li et al. [13] propose a video prediction algorithm that synthesizes a set of likely future frames in multiple time steps from one single still image. Considering the resolution and the motion of generated frames, we design our model as two streams: motion and content stream for solving the motion vector and the image information, respectively, and then fuse features from both streams to generate time-lapse video frames. Our model neither has to map random vectors to a sequence of video frames from scratch, nor generates video sequences frame by frame.

3   Method

In this paper, a novel end-to-end dynamic time-lapse video generation framework named DTVNet is proposed to generate diversified time-lapse videos from a single landscape image. Formally, given a single landscape image I_0, the model learns to generate a time-lapse video sequences $\hat{V} = \{\hat{I}_1, \hat{I}_2, \ldots, \hat{I}_T\}$. As depicted in Fig. 2, DTVNet consists of two submodules: *Optical Flow Encoder* ψ and *Dynamic Video Generator* ϕ, and we will explain our approach as follows.

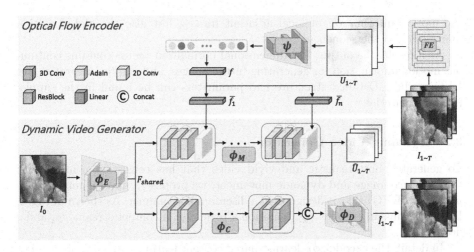

Fig. 2. Overview of the proposed DTVNet that consists of a *Optical Flow Encoder* ψ and a *Dynamic Video Generator* ϕ. Given the first landscape image I_0 and subsequent landscape images $I_{1\sim32}$, the *flow estimator* (FE) first estimates the consecutive flows $U_{1\sim T}$ and then ψ encodes the flows to the *normalized motion vector* $\bar{f}$. Dynamic video predictor ϕ successively apply encoder ϕ_E to learn shared content feature, motion stream ϕ_M and content stream ϕ_C to learn motion and content information from $\bar{f}$ and I_0 respectively, and a decoder ϕ_D to construct consecutive video frames $\hat{I}_{1\sim32}$.

3.1 Optical Flow Encoder

Considering the motion and diversity of the generated video, we design an *optical flow encoder* (OFE) module to encode the motion information to a *normalized motion vector*, thus not only can it supply motion information but also we could sample various motion vectors from a normalization distribution to generate diversified videos in the testing stage.

As shown in Fig. 2, we first apply unsupervised optical flow estimator which is more practical for not requiring label information (we use ARFlow [15] in this paper) to estimate consecutive flows $U_{1\sim T} = \{U_{0\rightarrow1}, U_{1\rightarrow2}, \ldots, U_{(T-1)\rightarrow T}\}$ from real video frames $V_{0\sim T} = \{I_0, I_1, \ldots, I_T\}$, where T indicates maximum frame number. We formulate this process as:

$$U_{1\sim T} = FE(V_{0\sim T}). \tag{1}$$

Then OFE module ψ encodes consecutive flows $U_{1\sim T}$ to a normalized motion vector f that contains the motion information, denoted as:

$$f = \psi(U_{1\sim T}). \tag{2}$$

Specifically, OFE employs a 3D encoder architecture, which is proven to be more suitable for learning spatial-temporal features than 2D convolution [25]. With the reduction of the time dimension, the 3D encoder can not only model the motion information of local adjacent frames, but also the global motion information of the sequence.

As a result, the output 512-dimensional normalized vector contains continuous motion information for generating future images, and will be integrated into DVG module. Detailed structure and parameters can be found in the supplementary material.

3.2 Dynamic Video Generator

To generate photo-realistic and vivid video that has consistent content with the reference image and dynamic movement, we propose a novel *dynamic video generator* (DVG) that is designed in a disentangling manner. As shown in Fig. 2, DVG contains an encoder ϕ_E, a motion stream ϕ_M, a content stream ϕ_C, and a decoder ϕ_D.

In detail, the encoder ϕ_E learns shared content feature F_{shared} from the first landscape image I_0, denoted as:

$$F_{shared} = \phi_E(I_0). \tag{3}$$

Considering the mismatch between the normalized motion vector f and motion stream features in semantic level that could inevitably increase the training difficulty, we introduce multiple linear layers to learn the adaptive motion vectors $\{\bar{f}_1, \ldots, \bar{f}_n\}$. Then adaptive motion vectors are integrated into the motion stream ϕ_M by multiple adaptive instance normalization (AdaIN) layers. Specifically, adaptive motion vector $\bar{f}_i$ is first specialized to motion styles $z_i = (z_i^{scale}, z_i^{shift})$ for i_{th} AdaIn layer with the input feature map F_i^{in}. Then we can calculate the output feature map F_i^{out} in the following formula:

$$F_i^{out} = z_i^{scale} \frac{F_i^{in} - \mu(F_i^{in})}{\sigma(F_i^{in})} + z_i^{shift}, \tag{4}$$

where $\mu(\cdot)$ and $\sigma(\cdot)$ calculate mean and variance respectively. Complete formula for motion stream is as follows:

$$\hat{U}_{1\sim T} = \phi_M(F_{shared}, \bar{f}_1, \ldots, \bar{f}_n), \tag{5}$$

where $\hat{U}_{1\sim T}$ indicates adapted low-resolution flows. During the training stage, $\hat{U}_{1\sim T}$ is supervised by real optical flows, aiming at adapting the motion to the input landscape image.

Analogously, the content stream ϕ_C also use F_{shared} as input to further learn deeper features. Subsequent decoder ϕ_D synthesize target video by combined motion and content information:

$$\hat{I}_{1\sim T} = \phi_D(\hat{U}_{1\sim T}, \phi_C(F_{shared})), \tag{6}$$

3.3 Objective Function

During the training stage of the DTVNet, we adopt content loss to monitor image quality at the pixel level, motion loss to ensure reasonable movements of the generated video, and adversarial loss to further boost video quality and authenticity. The full loss function $\mathcal{L}_{all}$ is defined as follow:

$$\mathcal{L}_{all} = \lambda_C \mathcal{L}_C + \lambda_M \mathcal{L}_M + \lambda_{adv} \mathcal{L}_{adv}, \tag{7}$$

where λ_C, λ_M, and λ_{adv} represent weight parameters to balance different terms.

Content Loss. The first term $\mathcal{L}_C$ calculates ℓ_1 errors between generated images $\hat{I}_{1\sim T}$ and real images $I_{1\sim T}$.

$$\mathcal{L}_C = \sum_{i=1}^{T} ||\hat{I}_i - I_i||_1. \tag{8}$$

Motion Loss. The second term $\mathcal{L}_M$ calculates ℓ_1 errors between adapted low-resolution flows $\hat{U}_{1\sim T}$ and real optical flows $U_{1\sim T}^{LR}$. Note that $U_{1\sim T}^{LR}$ are reconstructed low-resolution optical flow maps from $U_{1\sim T}$.

$$\mathcal{L}_M = \sum_{i=1}^{T} ||\hat{U}_i - U_i^{LR}||_1. \tag{9}$$

Adversarial Loss. The third term $\mathcal{L}_{adv}$ employ the improved WGAN with a gradient penalty for adversarial training [5]. Specifically, the discriminator D consists of six (3D-Conv)-(3D-InNorm)-(LeakyReLU) blocks that can capture discriminative spacial and temporal features.

$$\mathcal{L}_{GAN} = \mathbb{E}_{\tilde{V} \sim p_g}[D(\tilde{V})] - \mathbb{E}_{V \sim p_r}[D(V)] \\ + \lambda \mathbb{E}_{\hat{V} \sim p_{\hat{V}}}[(||\nabla_{\hat{V}} D(\hat{V})||_2 - 1)^2], \tag{10}$$

where p_r and p_g are real and generated video distribution respectively, and $p_{\hat{x}}$ is implicitly defined by sampling uniformly along straight lines between pairs of points sampled from p_r and p_g.

3.4 Training Scheme

We first train the unsupervised optical flow estimator (FE) under the instruction of ARFlow [15], and fix its parameters once the training is complete in all experiments. When training the DTVNet, we set loss weights λ_C, λ_M, and λ_{adv} to 100, 1, and 1 respectively. The layer number n of AdaIN in the DVG module is set to 6 in the paper.

4   Experiments

In this section, many experiments are conducted to evaluate the effectiveness of the method in various aspects on the Sky Time-lapse dataset. We first qualitatively and quantitatively compare our approach with two state-of-the-art methods, *i.e.* MoCoGAN [26] and MDGAN [32], and then conduct ablation studies to illustrate the effects of the structure and loss functions of our approach. Furthermore, we make a human study to demonstrate that our method can generate high-quality and dynamic video frames. Finally, we analyze the diversified generation ability of the proposed DTVNet.

4.1   Datasets and Implementations Details

Sky Time-Lapse. The Sky Time-lapse [32] dataset includes over 5000 time-lapse videos that are cut into short clips from Youtube, which contain dynamic sky scenes such as the cloudy sky with moving clouds and the starry sky with moving stars. Across the entire dataset, there are 35,392 training video clips and 2,815 testing video clips, each containing 32 frames. The original size of each frame is $3 \times 640 \times 360$, and we resize it into a square image of size $3 \times 128 \times 128$ as well as normalize the color values to $[-1, 1]$.

Evaluation Metric. We use *Peak Signal-to-Noise Ratio* (PSNR) to evaluate the frame quality in pixel level, and *Structural Similarity* (SSIM) [30] to measure the structural similarity between synthesized and real video frames. However, these two metrics could not well evaluate the motion information of video sequences, so we introduce another metric, named Flow-MSE [13], to calculate difference of the optical flow between generated video sequences and ground truth sequences. Furthermore, we conduct a *Human Study* (HS) to evaluate the visual quality of generated video frames by real persons.

Implementation Details. We follow the training scheme described in Sect. 3.4. The OFE module inputs the flow sequence $U_{1 \sim T}$ ($\in \mathbb{R}^{2 \times 32 \times 128 \times 128}$) and produces the normalized motion vector $\bar{f}$ ($\in \mathbb{R}^{512}$). The DVG module inputs the start frame I_0 ($\in \mathbb{R}^{3 \times 128 \times 128}$) and $\bar{f}$ to generate the flow sequence $\hat{U}_{1 \sim T}$ ($\in \mathbb{R}^{2 \times 32 \times 64 \times 64}$) and synthesized video frames $\hat{I}_{1 \sim T}$ ($\in \mathbb{R}^{3 \times 32 \times 128 \times 128}$). During the training stage, we use Adam [12] optimizer for all modules and set $\beta_1 = 0.99$, $\beta_2 = 0.999$. The initial learning rate is set to $3e^{-4}$, and it decays by ten every 150 epochs. We train the DTVNet for 200 epochs and the batch size is 12. We further test the inference speed of the DTVNet that can run 31 FPS with a single 1080 Ti GPU. Our model volume is 8.42M that is nearly a tenth of MDGAN [32] (78.14M). The details of the network architectures are given in the supplementary material.

4.2   Comparison with State-of-the-Arts

Qualitative Results. We conduct and discuss a series of qualitative results, compared with MoCoGAN and MDGAN, on Sky Time-Lapse dataset. As shown

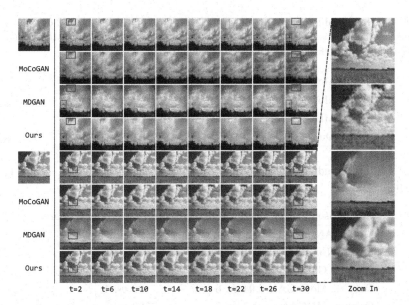

Fig. 3. Generative results compared with MoCoGAN [26] and MDGAN [32] on the Sky Time-lapse dataset. The first column lists two different landscape images as the start frames, the middle eight columns are generated video frames by different methods at different times. Right four enlarged images are long-term results at t = 30 for a better visual comparison. Please zoom in red and blue rectangles for a more clear comparison. (Color figure online)

in Fig. 3, we randomly sample two videos from the test dataset with different dynamic speed (The cloud in the top half moves much faster than the bottom), and the first column shows the start frames of two videos while the second to ninth columns are generated video frames by different methods at different times. Note that the first and fifth rows are ground truth frames, and the test models of other methods are supplied by official codes.

Results show that our method can not only keep better content information than other SOTA methods (Comparing the generated results in column at a special time), but also capture the dynamic motion (Viewing the generated results in row). In detail, the generated sequences produced by MoCoGAN (second and sixth rows) become more and more distorted over time, thus the quality and motion cannot be well identified. The results produced by MDGAN suffer from the distortion in color and cannot well keep the content. Specifically, we mark some dynamic and still details in red and blue rectangles respectively, and results show that our method can well keep the content of the still objects while predict reasonable dynamic details, which obviously outperforms all other state-of-the-art methods.

Quantitative Results. We choose PSNR, SSIM, and Flow-MSE metrics to quantitatively evaluate the effectiveness of our proposed method on Sky

Table 1. Metric evaluation results of MoCoGAN [26], MDGAN [32], and out approach on the Sky Time-lapse dataset. The up arrow indicates that the larger the value, the better the model performance, and vice versa.

Method	PSNR ↑	SSIM ↑	Flow-MSE ↓
MoCoGAN [26]	23.867	0.849	1.365
MDGAN [32]	23.042	0.822	1.406
Ours	**29.917**	**0.916**	**1.275**

Table 2. Human study about video frames and video quality evaluation on Sky Time-lapse dataset.

Comparison methods	Frame quality score	Video quality score
MoCoGAN *vs.* GT	3 *vs.* 97	2 *vs.* 98
MDGAN *vs.* GT	2 *vs.* 98	1 *vs.* 99
Ours *vs.* GT	**12 *vs.* 88**	**7 *vs.* 93**
Ours *vs.* MoCoGAN	94 *vs.* 6	96 *vs.* 4
Ours *vs.* MDGAN	97 *vs.* 3	99 *vs.* 1

Time-lapse dataset. In detail, we use all start frames in the test dataset to generate corresponding videos by different methods, and then calculate metric scores with ground truth videos.

As shown in Table 1, our approach gains +6.05 and +6.875 improvements for PSNR as well as 0.067 and 0.094 for SSIM compared to MoCoGAN and MDGAN, respectively. For the Flow-MSE metric, our method achieves the lowest value, which means that our generated video sequences are the closest to the ground truth videos in terms of the motion. On the whole, evaluation results indicate that our approach outperforms other two baselines in all three metrics, which illustrates that our model can generate more high-quality and dynamic videos than other methods.

Human Study. We further perform a human study for artificially evaluating the video frames quality. In detail, we first random choose 100 start images from the test dataset and generate corresponding videos by different methods. Then the generated videos by two different methods are simultaneously shown to a real person and we can get a result indicating "which one is better". Totally, the aforementioned experiment is conduct 30 times by 30 real workers, and we take the average of all results as the final result. Specifically, we split video quality evaluation to two aspects: the single frame quality evaluation (selecting frame 16 for each video sequence) and the dynamic video evaluation, which evaluates content and motion information separately. Note that the image is shown for 2 s and the video is played only once for a worker.

From the comparison results shown in Table 2, our method and other two baselines have a lower quality score than ground truth images, which means

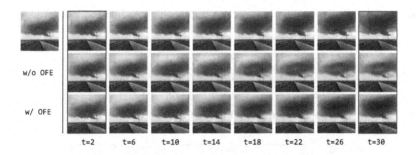

Fig. 4. A toy experiment for testing the effect of the *flow stream* of our approach. Images in the first row are ground truth frames at different times. Second and third rows are generated video frames without and with flow stream of our approach from the same start frame (the image in the first column). Please zoom in the red rectangles for a more clear comparison. (Color figure online)

that there are still many challenges for the video generation task. Nevertheless, our method has a better quality score than others: 12 *vs.* 3/2 in frame and 7 *vs.* 2/1 in video. To make a more intuitive comparison between our method and the others, we conduct another experiments in the bottom two rows. Results demonstrate that our approach outperforms other two SOTA methods.

4.3 Ablation Study

In this section, we conduct several ablation studies on the Sky Time-Lapse dataset to analyze the contribution of the *optical flow encoder* submodule and the effectiveness of different loss terms.

Influence of OFE. To evaluate the effectiveness of the optical flow encoder module, we conduct an ablation experiment that with or without OFE module of our proposed DTVNet. As show in Fig. 4, images in the first row are ground truth frames, and the second and third rows illustrate generation results without and with OFE module respectively.

By analyzing the results, we find that the generated frames loss the motion information and could not capture the clear movement and texture of the cloud without OFE module (as shown in the second row of Fig. 4). In detail, the generated frames remains almost stationary and becomes more obscure as time goes on. When adding the OFE module, the model can generate high-quality and long-term dynamic video (as shown in the third row of Fig. 4), which demonstrates that the OFE is critical for synthesizing photorealistic and dynamic videos. Please zoom in the red rectangles to compare long-term generation results of different structures.

Table 3. Quantitative comparisons of our approach with different loss terms on Sky Time-lapse dataset.

Method	PSNR ↑	SSIM ↑	Flow-MSE ↓
$\mathcal{L}_C$	28.768	0.897	1.624
$\mathcal{L}_C + \mathcal{L}_M$	29.364	0.906	1.509
$\mathcal{L}_C + \mathcal{L}_M + \mathcal{L}_{adv}$	**29.917**	**0.916**	**1.275**

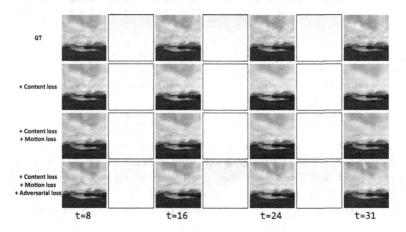

Fig. 5. Qualitative comparisons of our approach with different loss terms on Sky Time-lapse dataset. GT indicates the ground truth of the video frames. The images in even columns are the visualization results of the optical flow between two adjacent images.

Influence of Loss Functions. To further illustrate the effectiveness of different loss functions, *i.e.* content loss, motion loss, and adversarial loss, we conduct qualitative and quantitative experiments with different loss functions on Sky Time-lapse dataset. As shown in Fig. 5, the first row indicates the ground truth of the video frames, and the other three rows are generation results under the supervision of different loss functions. Generated video frames in odd columns indicate that *motion loss* and *adversarial loss* can greatly improve the model performance, where the generated video frames have more clear details and a higher quality when gradually adding loss terms. *i.e.* the results in the third row are better than the second row and the results in the fourth row are better than the third row.

We further visualize the optical flow between two adjacent images in even columns, and the result shows that our approach can well learn the motion information when generating video frames. Also, we quantitatively evaluate the effectiveness of different loss functions, and obtain a similar result. As shown in Table 3, the model obtains the best metric results, *i.e.* PSNR = 29.917, SSIM = 0.916, and Flow-MSE = 1.275, when all the three loss functions are used.

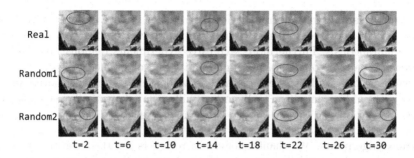

Fig. 6. Diversified video generation experiment. Given the same start frame, our approach can generate different videos under the different motion vectors, *e.g.* encoded from the real video (denoted as **Real**) and sampled from the normalization distribution (denoted as **Random1** and **Random2**). Please zoom in the red rectangles to compare each generated video in temporal and blue ellipses to compare different video frames generated by different motion vectors. (Color figure online)

4.4 Diversified Video Generation

Diversity is a key factor in content creation. Besides generating a fixed target video from a single start frame, diversified video generation from one still landscape image is critical for practical application. In this section, we conduct an additional diversified video generation experiment to illustrate the advantage of our approach for using *normalized motion vector* to provide motion information.

In detail, different video frames are generated by DTVNet under the different motion vectors, *e.g.* encoded from the real video (denoted as **Real**) and sampled from the normalization distribution (denoted as **Random1** and **Random2**), as shown in Fig. 6. We can obviously observe that our approach can generated diversified video frames by different motion vectors, which have the same content information but different motion information.

5   Conclusions

In this paper, we propose a novel end-to-end one-stage dynamic time-lapse video generation framework, *i.e.* DTVNet, to generate diversified time-lapse videos from a single landscape image. The *Optical Flow Encoder* submodule maps a sequence of optical flow maps to a *normalized motion vector* that encodes the motion information inside the generated video. The *Dynamic Video Generator* submodule contains motion and content streams that learn the movement and the texture of the generated video separately, as well as an encoder and a decoder to learn shared content features and construct target video frames respectively. During the training stage, we design three loss functions, *i.e.* content loss, motion loss, and adversarial loss, to the network, in order to generate high-quality and diversified dynamic videos. During the testing stage, we exclude the OFE module and directly sample from normalization distribution as the motion vector, which reduces the network computing overhead and supply diversified motion information simultaneously. Furthermore, extensive experiments demonstrate that our approach is capable of generating high-quality and diversified videos.

We hope our study to help researchers and users to achieve more effective works in the video generation task, and we will explore how to efficiently produce higher resolution and higher quality videos in the future.

Acknowledgements. We thank anonymous reviewers for their constructive comments. This work is partially supported by the National Natural Science Foundation of China (NSFC) under Grant No. 61836015 and the Fundamental Research Funds for the Central Universities (2020XZA205).

References

1. Aigner, S., Körner, M.: FutureGAN: anticipating the future frames of video sequences using spatio-temporal 3D convolutions in progressively growing GANs. arXiv preprint arXiv:1810.01325 (2018)
2. Cai, H., Bai, C., Tai, Y.W., Tang, C.K.: Deep video generation, prediction and completion of human action sequences. In: ECCV, pp. 366–382 (2018)
3. Dosovitskiy, A., et al.: FlowNet: learning optical flow with convolutional networks. In: ICCV, pp. 2758–2766 (2015)
4. Goodfellow, I., et al.: Generative adversarial nets. In: NeurIPS, pp. 2672–2680 (2014)
5. Gulrajani, I., Ahmed, F., Arjovsky, M., Dumoulin, V., Courville, A.C.: Improved training of Wasserstein GANs. In: NeurIPS, pp. 5767–5777 (2017)
6. Hochreiter, S., Schmidhuber, J.: Long short-term memory. Neural Comput. **9**(8), 1735–1780 (1997)
7. Huang, X., Belongie, S.: Arbitrary style transfer in real-time with adaptive instance normalization. In: ICCV, pp. 1501–1510 (2017)
8. Hur, J., Roth, S.: Iterative residual refinement for joint optical flow and occlusion estimation. In: CVPR, pp. 5754–5763 (2019)
9. Ilg, E., Mayer, N., Saikia, T., Keuper, M., Dosovitskiy, A., Brox, T.: FlowNet 2.0: evolution of optical flow estimation with deep networks. In: CVPR, pp. 2462–2470 (2017)
10. Isola, P., Zhu, J.Y., Zhou, T., Efros, A.A.: Image-to-image translation with conditional adversarial networks. In: CVPR, pp. 1125–1134 (2017)
11. Karras, T., Aila, T., Laine, S., Lehtinen, J.: Progressive growing of GANs for improved quality, stability, and variation. arXiv preprint arXiv:1710.10196 (2017)
12. Kingma, D.P., Ba, J.: Adam: a method for stochastic optimization. arXiv preprint arXiv:1412.6980 (2014)
13. Li, Y., Fang, C., Yang, J., Wang, Z., Lu, X., Yang, M.H.: Flow-grounded spatial-temporal video prediction from still images. In: ECCV, pp. 600–615 (2018)
14. Liang, X., Lee, L., Dai, W., Xing, E.P.: Dual motion GAN for future-flow embedded video prediction. In: ICCV, pp. 1744–1752 (2017)
15. Liu, L., et al.: Learning by analogy: reliable supervision from transformations for unsupervised optical flow estimation. In: CVPR, pp. 6489–6498 (2020)
16. Liu, P., King, I., Lyu, M.R., Xu, J.: DDFlow: learning optical flow with unlabeled data distillation. In: AAAI, vol. 33, pp. 8770–8777 (2019)

17. Liu, P., Lyu, M., King, I., Xu, J.: SelFlow: self-supervised learning of optical flow. In: CVPR, pp. 4571–4580 (2019)
18. Mathieu, M., Couprie, C., LeCun, Y.: Deep multi-scale video prediction beyond mean square error. arXiv preprint arXiv:1511.05440 (2015)
19. Mirza, M., Osindero, S.: Conditional generative adversarial nets. arXiv preprint arXiv:1411.1784 (2014)
20. Nam, S., Ma, C., Chai, M., Brendel, W., Xu, N., Kim, S.J.: End-to-end time-lapse video synthesis from a single outdoor image. In: CVPR, pp. 1409–1418 (2019)
21. Pan, J., et al.: Video generation from single semantic label map. In: CVPR, pp. 3733–3742 (2019)
22. Ranjan, A., et al.: Competitive collaboration: joint unsupervised learning of depth, camera motion, optical flow and motion segmentation. In: CVPR, pp. 12240–12249 (2019)
23. Saito, M., Matsumoto, E., Saito, S.: Temporal generative adversarial nets with singular value clipping. In: ICCV, pp. 2830–2839 (2017)
24. Sun, D., Yang, X., Liu, M.Y., Kautz, J.: PWC-Net: CNNs for optical flow using pyramid, warping, and cost volume. In: CVPR, pp. 8934–8943 (2018)
25. Tran, D., Bourdev, L., Fergus, R., Torresani, L., Paluri, M.: Learning spatiotemporal features with 3D convolutional networks. In: ICCV, pp. 4489–4497 (2015)
26. Tulyakov, S., Liu, M.Y., Yang, X., Kautz, J.: MoCoGAN: decomposing motion and content for video generation. In: CVPR, pp. 1526–1535 (2018)
27. Villegas, R., Yang, J., Zou, Y., Sohn, S., Lin, X., Lee, H.: Learning to generate long-term future via hierarchical prediction. In: ICML, pp. 3560–3569 (2017)
28. Vondrick, C., Pirsiavash, H., Torralba, A.: Generating videos with scene dynamics. In: NeurIPS, pp. 613–621 (2016)
29. Wang, T.C., et al.: Video-to-video synthesis. arXiv preprint arXiv:1808.06601 (2018)
30. Wang, Z., Bovik, A.C., Sheikh, H.R., Simoncelli, E.P.: Image quality assessment: from error visibility to structural similarity. IEEE Trans. Image Process. 13(4), 600–612 (2004)
31. Wu, J., Zhang, C., Xue, T., Freeman, B., Tenenbaum, J.: Learning a probabilistic latent space of object shapes via 3D generative-adversarial modeling. In: NeurIPS, pp. 82–90 (2016)
32. Xiong, W., Luo, W., Ma, L., Liu, W., Luo, J.: Learning to generate time-lapse videos using multi-stage dynamic generative adversarial networks. In: CVPR, pp. 2364–2373 (2018)
33. Yang, C., Wang, Z., Zhu, X., Huang, C., Shi, J., Lin, D.: Pose guided human video generation. In: ECCV, pp. 201–216 (2018)
34. Zhong, Y., Ji, P., Wang, J., Dai, Y., Li, H.: Unsupervised deep epipolar flow for stationary or dynamic scenes. In: CVPR, pp. 12095–12104 (2019)
35. Zhu, J.Y., Park, T., Isola, P., Efros, A.A.: Unpaired image-to-image translation using cycle-consistent adversarial networks. In: ICCV, pp. 2223–2232 (2017)

CLIFFNet for Monocular Depth Estimation with Hierarchical Embedding Loss

Lijun Wang[1] (ID), Jianming Zhang[2] (ID), Yifan Wang[1(✉)] (ID), Huchuan Lu[1,3] (ID), and Xiang Ruan[4] (ID)

[1] Dalian University of Technology, Dalian, China
{ljwang,wyfan,lhchuan}@dlut.edu.cn
[2] Adobe Research, San Jose, USA
jianmzha@adobe.com
[3] Peng Cheng Lab, Shenzhen , China
[4] tiwaki Co., Ltd., Kusatsu, Japan
ruanxiang@tiwaki.com

Abstract. This paper proposes a hierarchical loss for monocular depth estimation, which measures the differences between the prediction and ground truth in hierarchical embedding spaces of depth maps. In order to find an appropriate embedding space, we design different architectures for hierarchical embedding generators (HEGs) and explore relevant tasks to train their parameters. Compared to conventional depth losses manually defined on a per-pixel basis, the proposed hierarchical loss can be learned in a data-driven manner. As verified by our experiments, the hierarchical loss even learned without additional labels can capture multi-scale context information, is more robust to local outliers, and thus delivers superior performance. To further improve depth accuracy, a cross level identity feature fusion network (CLIFFNet) is proposed, where low-level features with finer details are refined using more reliable high-level cues. Through end-to-end training, CLIFFNet can learn to select the optimal combinations of low-level and high-level features, leading to more effective cross level feature fusion. When trained using the proposed hierarchical loss, CLIFFNet sets a new state of the art on popular depth estimation benchmarks.

Keywords: Monocular depth estimation · Hierarchical loss · Hierarchical embedding space · Feature fusion

1 Introduction

Depth estimation is traditionally tackled by shallow models [20, 22] with hand-crafted features. More recent works [6, 14] have shown that the success of deep convolutional neural networks (CNNs) in other computer vision areas can also be

Electronic supplementary material The online version of this chapter (https://doi.org/10.1007/978-3-030-58558-7_19) contains supplementary material, which is available to authorized users.

© Springer Nature Switzerland AG 2020
A. Vedaldi et al. (Eds.): ECCV 2020, LNCS 12350, pp. 316–331, 2020.
https://doi.org/10.1007/978-3-030-58558-7_19

transferred to monocular depth estimation. The hierarchical structure of trainable CNN features provides stronger representation capabilities, yielding more accurate monocular depth estimation.

To train CNNs, a well defined loss function is required in the first place to provide supervision by measuring the differences between predictions and targets. A wide range of loss functions have been explored in the literature of depth estimation. For instance, the reverse Huber loss [14] and depth aware loss [11] are used to address the heavy-tailed distribution of depth values in some existing datasets, while the scale invariant loss [6] and depth gradient loss [18] are designed to balance depth relations and scales. Although good performance has been achieved, these losses are manually designed, which require rich domain knowledge. As such, their generalization ability across different datasets cannot be guaranteed. Besides, the existing depth losses are mostly defined on a per pixel basis, which fail to capture context information. Therefore, they may be over sensitive to label noise and outliers, leading to unstable network training. In order to address the above issues, it is interesting to investigate alternative representation spaces where training losses can be more effective and robust for depth supervision.

In light of the above observations, we propose to leverage a loss function computed in a hierarchical embedding space for training depth estimation models. To this purpose, we devise different hierarchical embedding generators (HEGs) which take depth maps as input and generate their corresponding hierarchical embedding spaces, which in our cases are hierarchical convolutional feature maps extracted from the input depth maps. The loss function for training depth estimation networks is then computed on both the original depth space and the generated embedding space, giving rise to a *hierarchical embedding loss*. In order to seek desired hierarchical embeddings, we design multiple tasks to train HEGs. It is found that training on relevant tasks even without additional annotations can effectively improve depth estimation performance. It can also be shown that the widely adopted gradient loss is a special form of our hierarchical loss computed by a HEG with hand-designed network parameters. However, our experiments confirm that properly trained HEGs can significantly outperform either hand-designed ones or those trained on irrelevant tasks.

Another contribution of this paper is a cross level identity feature fusion (CLIFF) module acting as a basic building block of our depth estimation network. Fully convolutional networks with multi-level feature pyramids have become the *de facto* technique for solving pixel-level prediction tasks [23,32] A number of evidences [19,23,28] suggest that high-level features with more semantic and global context information is able to facilitate more reliable and accurate predictions. In comparison, low-level features with higher resolutions contain more detailed local information, which may benefit high-resolution predictions. Nonetheless, the low-level features also carry more noise which may reduce the reliability of the predictions. In light of the above observations, given features of two different levels the proposed CLIFF module first enhances low-level features using high-level ones through an attention scheme. In addition, the proposed architecture allows our CLIFF module to learn to select optimal features from the combination of high-level, original and enhanced low-level features. Finally, an identity mapping path connecting the high-level input feature and output is built to preserve the reliable semantic information. By applying CLIFF modules recursively, we obtain a new depth estimation network termed as CLIFFNet.

Our main contribution can be summarized into three folds.

- A new form of hierarchical loss computed in depth embedding spaces is proposed for depth estimation.
- Different architectures and training schemes of hierarchical embedding generators are investigated to find desirable hierarchical losses.
- A new CLIFFNet architecture is designed with more effective cross level feature fusion mechanism.

When trained using the proposed hierarchical losses, our CLIFFNet sets new state-of-the-art performance on popular depth estimation benchmarks.

2   Related Work

Monocular depth estimation is a long standing problem in computer vision [10, 20, 22, 26]. Recent years have witnessed tremendous progress achieved by deep learning based depth estimation methods. In the seminal work by Eigen *et al.* [6], a multi-scale deep network based method is proposed, where a global network is used to predict coarse-scale depth and a local network further refines the prediction with finer details. This network is extended by [5] into three levels, and is successfully applied to depth prediction, normal estimation and segmentation. Later on, Laina *et al.* [14] propose one of the earliest fully convolutional network architectures for monocular depth estimation, which significantly boosts the estimation accuracy. Motivated by [14], convolutional architectures have been intensively studied for depth estimation. For instance, a two-stream convolutional network is proposed in [17], which simultaneously predicts depth and depth gradients to restore fine depth details. Fu *et al.* [7] discretize depth values and propose a deep ordinal regression network. In contrast, [16] decomposes metric depth prediction into relative depth prediction and recombination, where a new convolutional network is proposed for relative depth estimation. Recently, Zhi *et al.* [35] proposes a new type of convolution which considers the camera parameters to learn calibration-aware patterns for monocular depth estimation. In addition, different training strategies have been explored to benefit monocular depth estimation, including multi-task training [30, 33, 35], self-supervised learning with photometric losses [8, 31], and those with sparse ordinal [2] or relative depth [29, 32] supervisions.

Although, the above deep learning based methods have significantly improved depth prediction accuracy, the scheme of deep feature fusion across levels is not thoroughly studied for depth estimation. Nonetheless, our experiments show that effective multi-level feature fusion can yield considerable performance boost.

Another line of work which correlates to ours is the design of loss functions for training depth estimation networks. Among others, [6] proposes a scale invariant loss, which enforces the network to learn depth relations rather than scales. In a similar spirit, Li *et al.* [18] propose depth gradient loss, which computes the L1 losses in the gradient space of the predicted and ground truth depth. Meanwhile, the heavy-tailed distribution of depth values have been observed in both [14] and [11]. They propose to address this issue using the reverse Huber loss and depth-aware loss, respectively, both of which attach higher weight towards samples with large residuals. Our hierarchical losses differ from the above works mainly in two aspects. First, most of the above losses are manually designed, whereas ours can be learned in a data-driven manner on relevant tasks. Second, the above losses are mostly defined in the original depth space on a per pixel basis. In comparison, ours are defined in hierarchial embedding spaces of the depth. Our experiments show that the hierarchical loss can capture contextual information and is more robust to local noises, leading to significant performance gain. Our hierarchical loss is also related to perceptual losses [12,25]. However, the methods in [12,25] aim to improve visual quality of image generation/reconstruction by directly applying perceptual losses, while our focus is on architecture design and relevant task exploration to achieve more superior hierarchical embeddings to compute hierarchical losses.

3   Method

3.1   Hierarchical Embedding Loss for Depth Estimation

For monocular depth estimation, a deep network takes a single image as input and estimates its depth map $\hat{d}$. Given the corresponding ground truth depth d and a loss function $L(d, \hat{d})$ measuring the differences between the prediction and ground truth, the parameters of the network can then be learned by minimizing the loss function. Instead of directly comparing the differences in the original depth space, some existing works demonstrate that loss functions defined on some manually designed embeddings (*eg.*, vertical and horizontal gradients) of the original depth may embody more appealing properties, leading to considerable accuracy gains. Motivated by this fact, we aim to design an embedding generator $G(d, \theta)$ parameterized by θ to map the input depth into an embedding space. As such, the parameter θ can be learned in a data-driven manner rather than through hand-engineering.

Inspired by the impressive performance of hierarchical structures in deep networks, we propose to transfer their success to the supervision domain by defining loss functions on hierarchical embedding spaces. To this end, we implement the hierarchical embedding generator (HEG) G using multi-layer CNNs[1]. By feeding a depth map d into G, we obtain a set of K hierarchical convolutional feature maps $\{G_1(d), G_2(d), \ldots, G_K(d)\}$, which are treated as an embedding hierarchy of the input depth. The final loss function can then be computed as:

$$L_D(d, \hat{d}) = \sum_{k=0}^{K} w_k L(G_k(d), G_k(\hat{d})), \tag{1}$$

---

[1] We drop the parameter θ for notational simplicity.

Table 1. Architecture details of HEG-S. Conv, Max, FC, BN, NS, SM, and α denote convolutional layers with kernel size 3×3, adaptive max pooling with output size 2×2, fully connected layer, batch normalization, number of scenes, softmax layer and negative slop of leaky ReLUs, respectively.

#Layer	1	2	3	4	5	6	7	8	9	10
Type	Conv	Conv	Conv	Conv	Conv	Conv	Max+ Flatten	FC	FC	FC
Output Channel	16	16	32	32	64	64	256	256	256	NS
Stride	1×1	1×1	2×2	1×1	2×2	1×1	–	–	–	–
α	0.01	0.01	0.01	0.01	0.01	0.01	–	0.0	0.0	–
Normal.	BN	BN	BN	BN	BN	BN	–	BN	BN	SM

where G_0 denotes the identify mapping, *ie.*, $G_0(\boldsymbol{d}) = \boldsymbol{d}$; w_k indicates the loss weight. As a result, the above loss function combines the supervision from both the original depth space and its embedding spaces of different levels.

A essential problem remaining is how to learn the parameters of HEGs. In this paper, we identify appropriate tasks for training HEGs according to the following two standards.

- The task, including both the input and output target, should be relevant to depth estimation. Otherwise, the learned HEGs can hardly benefit depth estimation. Consider a HEG pre-trained on image classification, which can also be adopted for training depth estimation. However, our experiments show that its performance in terms of depth accuracy gain is similar to a randomly initialized HEG.
- Although additional annotations maybe beneficial, we focus on tasks that require limited additional manual annotations. As a result, the idea of learning hierarchical embedding losses can be more easily applied across different datasets, and the comparison against baseline approaches trained without a hierarchical loss is more fair.

According to the above standards, we mainly select depth-based scene classification and depth reconstruction as two tasks for training HEGs. We further design appropriate HEG network architectures for the two tasks and study their impact on depth estimation.

HEG-S Trained on Depth-Based Scene Classification. The image and depth sample pairs in existing datasets are collected in various locations and scenes. For instance, the NYU-Depth V2 dataset [27] contains 464 scenes, while the data of Cityscape [3] belong to 50 scenes. The scene name of each sample can be easily recorded as meta data when collecting the depth data (*eg.*, in many datasets the depth samples recorded under the same scene are stored in one folder), and therefore does not require heavy manual labour for additional annotations. Motivated by this observation, we propose a depth-based scene

Table 2. Architecture details of HEG-R encoder. Conv, Max, FC, and α denote convolutional layers with kernel size 3×3, adaptive max pooling with output size 2×2, fully connected layer, and negative slop of leaky ReLUs, respectively.

#Layer	1	2	3	4	5	6	7	8
Type	Conv	Conv	Conv	Conv	Conv	Conv	Max+ Flatten	FC
Output Channel	16	16	32	32	64	64	256	256
Stride	1×1	1×1	2×2	1×1	2×2	1×1	–	–
α	0.01	0.01	0.01	0.01	0.01	0.01	–	0.0

classification task to train HEG. Technically, the HEG takes as input a depth map, rather than an RGB image, and is trained to infer its corresponding scene label from a pre-defined label set. It is very likely that depth maps taken from the same scene share similar properties, *eg.*, depth scales and structures. By learning to identify the correlation between depth and scenes, we hope the embeddings generated by the trained HEG are able to capture the key properties of the input depth map, and further benefit depth estimation training in the subsequent stage.

We design a CNN termed as HEG-S for depth-based scene classification. Table 1 illustrates the detailed network architecture. The first 6 trainable layers are 3×3 convolutional layers. The output feature maps are spatially downsampled to 2×2 using an adaptive max pooling layer and reshaped into a feature vector, which is then consumed by 3 additional fully connected layers. A batch normalization and leaky ReLU layer are appended to each intermediate trainable layer. The final fully connected layer generates a score for each scene class, which is further normalized into a probability via a softmax layer. Given the inferred scene class probabilities and the ground truth labels, HEG-S is trained by optimizing a cross-entropy loss. After training, the output feature maps of the intermediate convolutional layers can be adopted as hierarchical embeddings to compute supervisions for training the depth estimation network.

HEG-R Trained on Depth Reconstruction. Depth reconstruction aims to extract representative features from the input depth and restore the depth information from the extracted features. For one thing, it can be learned without additional labels. For another, it is highly relevant to depth estimation since both the input and the target output are depth maps. As a result, we propose to explore depth reconstruction as the second task for training HEGs.

We design a new HEG network with an encoder-decoder architecture for depth reconstruction. The encoder network consists of 6 convolutional layers and 1 fully connected layer. For a fair comparison, the detailed architecture of the encoder (as shown in Table 2) mostly follows that of HEG-S, except that batch normalization after each convolutional layer is discarded due to the reconstruction purpose. The decoder architecture is symmetric to that of the encoder,

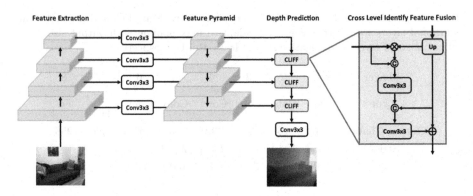

Fig. 1. Overview of the proposed CLIFFNet.

where only 2×2 strided convolutional layers are replaced by transpose convolutional layers with a $\times 2$ upsampling factor. One of the key ingredients in the proposed network is the 256 dimensional feature vector generated by the encoder, which serves as a bottleneck connecting the encoder and decoder. As the bottleneck structure significantly squeezes the feature dimension, it forces the convolutional layers of the encoder to capture the most representative features from the input depth map, preventing the reconstruction network degenerating into a trivial identity mapping.

We name the above HEG trained on depth reconstruction as HEG-R. The multi-level convolutional feature maps generated by the encoder of HEG-R are investigated as an embedding hierarchy for training depth estimation.

Discussion. The proposed hierarchical loss is reminiscent of the perceptual losses which are mainly adopted by generative models to produce photo-realistic results. It has been shown that the perceptual losses can effectively improve the visual quality but may hinder the quantitative performance [12]. In comparison, we focus on analyzing different training tasks and HEG architectures to compute hierarchical depth losses. Our experiment shows that the proposed hierarchical loss can not only benefit the perceptual quality but also significantly improves the quantitative performance in terms of depth metrics. It should also be noted that although our current training strategies are selected according to the proposed two standards, they are not directly coupled with our ultimate goal of finding an optimal embedding space for a hierarchical loss. In our future work, we will explore meta-learning techniques to learn optimal hierarchical embedding spaces for depth supervision.

3.2 CLIFFNet for Depth Estimation

Following most existing works [14,33], the proposed CLIFFNet performs depth estimation with a fully convolutional architecture, which consists of three components: a feature extraction sub-network, a feature pyramid sub-network, and a depth prediction sub-network. The feature extraction sub-network takes a single RGB image as input and extracts a collection of multi-level convolutional feature maps of various resolutions. The generated feature maps are then fed into the feature pyramid sub-network through lateral connections, which propagates the semantic information from high-level to low-level feature maps, producing a feature pyramid. The depth prediction sub-network make the final prediction based on the feature pyramid. Figure 1 provides an overview of the architecture.

In order to take full advantage of the feature pyramid, some prior methods adopt a direct fusion strategy. They first upsample all feature maps in the pyramid into the same resolution, which are then combined through concatenation and used to estimate the depth map. Although high-level features with rich semantic information are used to benefit robust predictions, they are directly upsampled from very low-resolutions, leading to blurry depth prediction. An alternative idea is the progressive fusion strategy, where high-level features are gradually upsampled (*e.g.*, by ×2 each time) and combined with lower level features of the same resolution. Though the blurry prediction issue can be alleviated, the output features are dominated by low-level cues which are not robust to challenging scenarios. To address this issue, we propose the cross level identity feature fusion (CLIFF) module, which not only enhances the visual quality but also preserves high-level features to facilitate more robust depth estimation.

CLIFF Module. The CLIFF module takes a high-level and low-level feature map as input. We first upsample the high-level feature map using bilinear interpolation to ensure that two input feature maps have the same spatial resolution. Since high-level feature is more reliable with less noise, we refine the low-level feature through an attention mechanism by multiplying it with the high-level feature. As such, accurate responses in the low-level feature are further strengthened, while noisy responses are weakened. In order to achieve the optimal combination of the high-level feature, original and refined low-level feature, these features are further selected through two convolutional layers. Specifically, the first convolutional layer learns to select and aggregate low-level features by taking the concatenation of the original and refined low-level feature as input. Its output is then concatenated with the high-level feature and serves as the input to the second convolutional layer, further allowing feature selection between low-level and high-level feature maps. Finally, to facilitate gradients back-propagation and to preserve high-level semantic cues, an identity mapping from the high-level feature to the output feature is added. Denoting the low-level feature as F^l, the upsampled high-level feature as F^h, and the output as F^o, the above operations can be formally described as:

$$
\begin{cases}
F^o = F_2^c + F^h, \\
F_2^c = W_2 * [F_1^c, F^h] + b_2, \\
F_1^c = W_1 * [F^l, F^a] + b_1, \\
F^a = F^l \odot F^h,
\end{cases}
\tag{2}
$$

where F_i^c denotes the selected feature using convolutional layers parameterized by weight W_i and bias b_i. $[\cdot, \cdot]$ indicates the concatenation of two feature maps along the channel dimension. The operators $*$ and $\odot$ indicate convolution and element-wise multiplication, respectively.

In the proposed depth prediction sub-network, the CLIFF modules are repeatedly applied to gradually perform feature fusion from high-level to low-level features. The fused feature generated by the last CLIFF module is then fed into a convolutional layer to produce the final depth prediction.

4   Experiments

4.1   Implementation Details

We compute L1 losses on embedding spaces for training depth estimation and exhaustively search the optimal combination of embedding spaces generated by the proposed HEGs. Our empirical results (See supplementary materials for details) show that the combination of the original depth space and the embedding spaces generated by the 2nd and 4th layer of HEGs delivers the best performance when used for hierarchical loss computation. The result is consistent to both HEG-S and HEG-R, giving rise to our final loss function as below:

$$L_D(d, \hat{d}) = \sum_{k=\in\{0,2,4\}} w_k \|G_k(d) - G_k(\hat{d})\|_1, \tag{3}$$

where the loss weights are determined through grid-search and fixed as $w_0 = 1.0, w_2 = 10.0, w_4 = 15.0$.

For the proposed CLIFFNet, we adopt the first 5 residual block of a pre-trained ResNet-50 network [9] as the feature extraction sub-network. The feature pyramid sub-network are designed closely following [19] (See supplementary materials for architecture details). We resize each input image to have a minimum side of 228 pixels by maintaining its aspect ratio. All the networks are trained using Adam optimizer [13] with a batch size of 8 images and initial learning rates $1e-3$, $1e-4$, and $1e-4$ for HEG-S, HEG-R, and CLIFFNet, respectively. Source code will be made publicly available[2].

Our experiments are conducted on NYU-Depth V2 [27] and Cityscapes [3] dataset. The NYU-Depth V2 dataset contains 464 indoor scenes, where 249 of them are for training and the rest for testing. 40K image-depth pairs are sampled from all the 120 K training samples. We first use the sampled depth to train HEG-S for 249 scene classification and HEG-R for depth reconstruction, then use the trained HEGs to compute losses to learn CLIFFNet for depth estimation. On one NVIDIA 1080Ti GPU, the training processes of HEG-S and HEG-R take around 4 h, respectively, while the depth network is trained for around 22 h. The depth network with 36.89M parameters runs at 37.86 FPS during inference. As in [6,7], we evaluate the proposed method using 7 widely adopted metrics defined

---

[2] https://github.com/scott89/CLIFFNet.

Table 3. Adopted evaluation metrics for depth estimation. d_i and $\hat{d}_i$ denote the ground truth and estimated depth value of pixel i. N denotes the total number of pixels.

Metric	Definition	Metric	Definition
RMSE	$(\frac{1}{N}\sum_i(\hat{d}_i - d_i)^2)^{(\frac{1}{2})}$	RMSE (log)	$(\frac{1}{N}\sum_i(\log\hat{d}_i - \log d_i)^2)^{(\frac{1}{2})}$
Abs Rel	$\frac{1}{N}\sum_i |\hat{d}_i - d_i|/d_i$	Sq Rel	$\frac{1}{N}\sum_i(\hat{d}_i - d_i)^2/d_i$
Pn	Percentage of d_i such that $\max\{\frac{d_i}{\hat{d}_i}, \frac{\hat{d}_i}{d_i}\} < 1.25^n$		

Table 4. Comparison with state-of-the-art methods on NYU-Depth V2 dataset [27]. The best and second best results are in **bold** font and underlined, respectively.

Method	Error				Accuracy		
	RMSE	RMSE (log)	Abs Rel	Sq Rel	P1	P2	P3
Eigen *et al.* [6]	0.874	0.284	0.218	0.207	0.616	0.889	0.971
Liu *et al.* [21]	0.756	0.261	0.209	0.180	0.662	0.913	0.979
Eigen and Fergus [5]	0.874	0.284	0.218	0.207	0.616	0.889	0.971
Laina *et al.* [14]	0.584	0.198	0.136	0.101	0.822	0.956	0.989
Chakrabarti *et al.* [1]	0.620	0.205	0.149	0.118	0.806	0.958	0.987
Xu *et al.* [34]	0.593	–	0.125	–	0.806	0.952	0.986
Qi *et al.* [24]	0.569	–	0.128	–	0.834	0.960	0.990
Lee *et al.* [15]	0.572	0.193	0.139	0.096	0.815	0.963	**0.991**
Fu *et al.* [7]	0.509	0.188	**0.116**	**0.089**	0.828	**0.965**	0.986
Xu *et al.* [33]	0.582	–	0.120	–	0.817	0.954	0.987
CLIFFNet-R	0.497	0.180	0.129	**0.089**	0.841	0.963	**0.991**
CLIFFNet-S	**0.493**	**0.171**	0.128	**0.089**	**0.844**	0.964	**0.991**

in Table 3. We compute these metrics using the implementation provided by [7]. Due to page limits, we present the results on Cityscapes in the supplementary materials.

4.2   Comparison to State of the Arts

On NYU-Depth V2, we compare with 10 state-of-the-art methods. The quantitative results are reported in Table 4, where CLIFFNet-R and CLIFFNet-S denote the proposed CLIFFNet trained with hierarchical losses computed using HEG-R and HEG-S, respectively. Both CLIFFNet-R and CLIFFNet-S compare favorably against state-of-the-art methods. Among others, CLIFFNet-S consistently outperforms the other methods and achieves the top performance in terms of 5 metrics. Though the performance of CLIFFNet-R is slightly worse than CLIFFNet-S, it also delivers state-of-the-art performance in terms of all the 7 metrics. In particular, its performance in terms of RMSE, RMSE (log) and P2 is also comparable to CLIFFNet-R. It should be noted that some of the compared

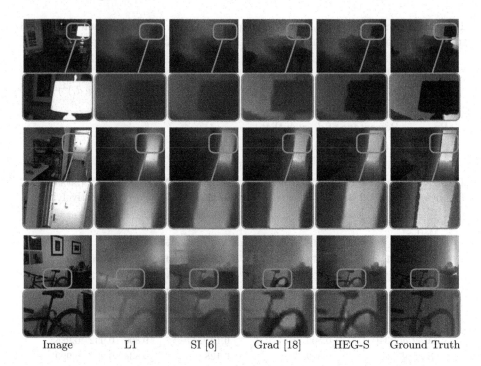

Image   L1   SI [6]   Grad [18]   HEG-S   Ground Truth

Fig. 2. Depth maps predicted using different loss functions.

Table 5. Comparison of different losses on NYU-Depth V2 dataset [27]. The best results are in **bold** font.

Methods	L1	Grad	SI	DA	MT	HEG-Rn	HEG-Im	HEG-R	HEG-S
RMS	0.529	0.513	0.520	0.511	0.530	0.517	0.523	0.497	**0.493**
Abs Rel	0.135	0.132	0.134	0.130	0.134	0.134	0.132	0.129	**0.128**
P1	0.817	0.830	0.820	0.835	0.815	0.815	0.829	0.841	**0.844**
P2	0.961	0.964	0.963	0.964	0.960	0.961	0.963	**0.963**	**0.964**

approaches [6,7] use all the 120 K training images, while our proposed methods use only a subset of the training data.

4.3 Ablation Study

Effectiveness of Hierarchical Loss. To further verify the effectiveness of the proposed hierarchical losses, we evaluate the performance of our CLIFFNet variants trained with different losses. Since the network architectures are the same, we refer to different variants using the name of the adopted loss function. Among them, L1 represents only using the L1 loss computed on the original depth space, while all the other variants combine the depth space L1 loss with other form of loss functions. Specifically, Grad indicates the combination of depth space L1 loss and depth gradient loss [18]. SI denotes the scale invariant loss [6]. DA indicates the depth aware loss. HEG-S and HEG-R represent the hierarchical losses proposed in Sect. 3.1. HEG-Rn denotes the proposed hierarchical loss computed using a randomly initialized HEG. HEG-Im indicates the hierarchical loss computed by a HEG with the same architecture as HEG-S trained on ImageNet classification task [4]. The input channels of the kernels on the first convolutional layer are averaged in order to take depth map as input.

Table 5 shows the comparison results of different variants on NYU-Depth V2 dataset. The comparison of L1 against SI and DA confirms the advantage of loss functions defined on additional embeddings over those computed only on the original depth space. However, compared with the hand-designed losses SI and DA, the proposed hierarchical embeddings generated by HEG-S and HEG-R are learned in a data-driven manner, leading to more superior performance. The proposed HEG-S and HEG-R trained on carefully designed tasks significantly outperform the randomly initialized HEG-Rn and HEG-Im trained on irrelevant image classification. Figure 2 shows the predicted depth maps of our CLIFFNet trained with different losses. It can be observed that the predictions using HEG-S are perceptually more realistic than other losses. The performance of HEG-Rn and HEG-Im further justifies the importance of seeking relevant tasks for learning loss embeddings.

One may wonder that the advantages of HEG-S may be caused by using additional scene labels. To verify this, we design another variant model named MT, which adds an additional scene classification module on top of the Res-5 feature map generated by the feature extraction sub-network. It consists of a global average pooling followed by two fully connected layers. We then train MT on both depth estimation (using depth space L1 loss) and scene classification (using cross-entropy loss) in a multi-task training manner. As illustrated in Table 5, the depth estimation performance of MT trained on additional scene classification task is similar to that of the baseline L1, and is worse to our proposed HEG-R and HEG-S by a considerable margin. The results suggest that compared to muti-task learning the proposed HEG-S can serve as a more superior strategy to benefit depth estimation with additional scene classification annotations.

Visualization of Loss Gradients. To gain more intuitive understanding of the hierarchical losses, we perform additional visualization experiments to analyze the impact of different loss functions on network training. During one intermediate training iteration, we forward-propagate an input image through CLIFFNet, compute different losses using the predicted and ground truth depth maps, and then back-propagate the gradients of the loss functions to the predicted depth space. Figure 3 provides the visualization of depth space gradients back-propagated from different loss functions. It can be observed that the gradient magnitude of L1 is almost uniform in each pixels, while DA [11] attaches a higher weight on distant regions with larger depth values. The Grad loss focuses more on the low-level boundary regions. In comparison, the gradients of HEG-S demonstrates more clear hierarchical patterns. The behavior of HEG-S2 (with

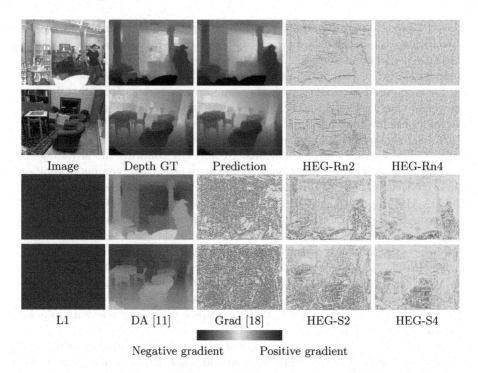

Fig. 3. Gradients backpropagated to the predicted depth space from different losses. L1: L1 loss computed on the original depth space. DA: depth aware loss [11]. Grad: depth gradient loss [18]. HEG-Rn: L1 loss on the embedding produced by the i-th layer of a randomly initialized HEG. HEG-Si: L1 loss on the embedding produced by the i-th layer of a HEG pre-trained on scene classification.

loss computed in the 2nd layer of HEG-S) is similar to Grad, but seems to be more robust to noisy depth edges. Meanwhile, HEG-S4 (with loss computed in the 4th layer of HEG-S) focuses on more on interior of object regions with semantic meaning. Compared with HEG-S, the gradients of randomly initialized HEG-Rn fail to exhibit such hierarchical patterns. The above observations on HEG-S also hold for HEG-R. According to their behaviors, we conjecture that the proposed hierarchical losses is able to capture multi-scale contexts and therefore more robust to local noise labels and outliers.

Impact of CLIFF Module. The core architecture designs of the proposed CLIFF module include a) attention based low-level feature refinement, b) multi-level feature selection, and c) identity mapping of high-level features. We ablate these core designs by comparing 4 variants of CLIFF module for cross level feature fusion. Among them, CLIFF-w/o-att discards attention based feature refinement. It select the input features by applying two convolutional layers on their concatenation. An identity mapping of high-level features is then added

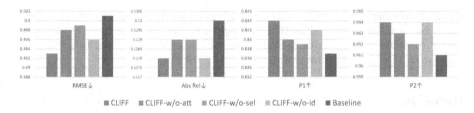

Fig. 4. Comparison results of different CLIFF variants on NYU-Depth V2 in terms of errors (the first row) and accuracy (the second row).

to the selected feature to produce the output. CLIFF-w/o-sel discards feature selection, where two convolutional features are directly applied to the sum of the original and refined low-level features and high-level features. CLIFF-w/o-id removes the identity mapping. Finally, a baseline module that does not contain any of the above 3 architecture design is developed. It combines two input feature maps through addition and then produces the output features with two convolutional layers.

We apply the above 4 variants to the depth prediction module in the same way as the proposed CLIFF module, leading to 4 variants of the proposed method. We then train the 4 variants as well as the proposed CLIFFNet using hierarchical embedding losses computed by HEG-S. Figure 4 demonstrates the comparison results on NYU-Depth V2 dataset. It can be observed that each of the three core architecture designs can effectively improve depth estimation performance. By combining all the architecture designs, CLIFF outperforms the baseline for a large margin, suggesting the contribution of each design is relative orthogonal to the others. We also performs additional ablation studies to investigate the performance of intermediate output from CLIFF modules. We leave the detailed results in the supplementary materials due to page limits.

5   Conclusion

We propose hierarchical losses for monocular depth estimation. Rather than defined on a per pixel basis, they are computed in hierarchical embedding spaces and can be automatically learned from training data. To obtain superior hierarchical embeddings, we design two embedding generators, named as HEG-S and HEG-R, which are trained on scene classification and depth reconstruction, respectively. Experiments show that learned hierarchical losses can capture multi-scale contexts and are more robust to outliers, leading to significant performance gain. In addition, we further propose CLIFFNet for depth estimation, which provides a more effective manner for cross level feature fusion. CLIFFNet trained with hierarchical losses sets new record on popular benchmarks.

Acknowledgements. This work is supported by National Key R&D Program of China (2018AAA0102001), National Natural Science Foundation of China (61725202,

U1903215, 61829102, 91538201, 61771088, 61751212, 61906031), Fundamental Research Funds for the Central Universities (DUT19GJ201), Dalian Innovation Leader's Support Plan (2018RD07), China Postdoctoral Science Foundation (2019M661095), National Postdoctoral Program for Innovative Talent (BX20190055).

References

1. Chakrabarti, A., Shao, J., Shakhnarovich, G.: Depth from a single image by harmonizing over complete local network predictions. In: Lee, D.D., Sugiyama, M., von Luxburg, U., Guyon, I., Garnett, R. (eds.) NIPS, pp. 2658–2666 (2016)
2. Chen, W., Fu, Z., Yang, D., Deng, J.: Single-image depth perception in the wild. In: NIPS, pp. 730–738 (2016)
3. Cordts, M., et al.: The cityscapes dataset for semantic urban scene understanding. In: CVPR, pp. 3213–3223 (2016)
4. Deng, J., Dong, W., Socher, R., Li, L., Li, K., Li, F.: ImageNet: a large-scale hierarchical image database. In: CVPR, pp. 248–255 (2009)
5. Eigen, D., Fergus, R.: Predicting depth, surface normals and semantic labels with a common multi-scale convolutional architecture. In: ICCV, pp. 2650–2658 (2015)
6. Eigen, D., Puhrsch, C., Fergus, R.: Depth map prediction from a single image using a multi-scale deep network. In: NIPS, pp. 2366–2374 (2014)
7. Fu, H., Gong, M., Wang, C., Batmanghelich, K., Tao, D.: Deep ordinal regression network for monocular depth estimation. In: CVPR, pp. 2002–2011 (2018)
8. Godard, C., Aodha, O.M., Firman, M., Brostow, G.J.: Digging into self-supervised monocular depth estimation. In: ICCV, pp. 3827–3837 (2019)
9. He, K., Zhang, X., Ren, S., Sun, J.: Deep residual learning for image recognition. In: CVPR, pp. 770–778 (2016)
10. Hoiem, D., Efros, A.A., Hebert, M.: Geometric context from a single image. In: ICCV, pp. 654–661 (2005)
11. Jiao, J., Cao, Y., Song, Y., Lau, R.: Look deeper into depth: monocular depth estimation with semantic booster and attention-driven loss. In: ECCV, pp. 53–69 (2018)
12. Johnson, J., Alahi, A., Fei-Fei, L.: Perceptual losses for real-time style transfer and super-resolution. In: Leibe, B., Matas, J., Sebe, N., Welling, M. (eds.) ECCV 2016. LNCS, vol. 9906, pp. 694–711. Springer, Cham (2016). https://doi.org/10.1007/978-3-319-46475-6_43
13. Kingma, D.P., Ba, J.: Adam: a method for stochastic optimization. arXiv preprint arXiv:1412.6980 (2014)
14. Laina, I., Rupprecht, C., Belagiannis, V., Tombari, F., Navab, N.: Deeper depth prediction with fully convolutional residual networks. In: 3DV, pp. 239–248 (2016)
15. Lee, J.H., Heo, M., Kim, K.R., Kim, C.S.: Single-image depth estimation based on Fourier domain analysis. In: CVPR, pp. 330–339 (2018)
16. Lee, J.H., Kim, C.S.: Monocular depth estimation using relative depth maps. In: CVPR, pp. 9729–9738 (2019)
17. Li, J., Klein, R., Yao, A.: A two-streamed network for estimating fine-scaled depth maps from single RGB images. In: ICCV, pp. 3372–3380 (2017)
18. Li, Z., Snavely, N.: MegaDepth: learning single-view depth prediction from internet photos. In: CVPR, pp. 2041–2050 (2018)
19. Lin, T.Y., Dollár, P., Girshick, R., He, K., Hariharan, B., Belongie, S.: Feature pyramid networks for object detection. In: CVPR, pp. 2117–2125 (2017)

20. Liu, B., Gould, S., Koller, D.: Single image depth estimation from predicted semantic labels. In: CVPR, pp. 1253–1260 (2010)
21. Liu, F., Shen, C., Lin, G.: Deep convolutional neural fields for depth estimation from a single image. In: CVPR, pp. 5162–5170 (2015)
22. Liu, M., Salzmann, M., He, X.: Discrete-continuous depth estimation from a single image. In: CVPR, pp. 716–723 (2014)
23. Long, J., Shelhamer, E., Darrell, T.: Fully convolutional networks for semantic segmentation. In: CVPR, pp. 3431–3440 (2015)
24. Qi, X., Liao, R., Liu, Z., Urtasun, R., Jia, J.: GeoNet: geometric neural network for joint depth and surface normal estimation. In: CVPR, pp. 283–291 (2018)
25. Rad, M.S., Bozorgtabar, B., Marti, U.V., Basler, M., Ekenel, H.K., Thiran, J.P.: SROBB: targeted perceptual loss for single image super-resolution. In: ICCV, pp. 2710–2719 (2019)
26. Saxena, A., Sun, M., Ng, A.Y.: Make3D: learning 3D scene structure from a single still image. TPAMI **31**(5), 824–840 (2008)
27. Silberman, N., Hoiem, D., Kohli, P., Fergus, R.: Indoor segmentation and support inference from RGBD images. In: Fitzgibbon, A., Lazebnik, S., Perona, P., Sato, Y., Schmid, C. (eds.) ECCV 2012. LNCS, vol. 7576, pp. 746–760. Springer, Heidelberg (2012). https://doi.org/10.1007/978-3-642-33715-4_54
28. Wang, L., Ouyang, W., Wang, X., Lu, H.: Visual tracking with fully convolutional networks. In: ICCV, pp. 3119–3127 (2015)
29. Wang, L., et al.: DeepLens: shallow depth of field from a single image. ACM Trans. Graph. **37**(6), 245:1–245:11 (2018)
30. Wang, L., Zhang, J., Wang, O., Lin, Z., Lu, H.: SDC-Depth: semantic divide-and-conquer network for monocular depth estimation. In: CVPR, June 2020
31. Watson, J., Firman, M., Brostow, G.J., Turmukhambetov, D.: Self-supervised monocular depth hints. In: ICCV, pp. 2162–2171 (2019)
32. Xian, K., et al.: Monocular relative depth perception with web stereo data supervision. In: CVPR, pp. 311–320 (2018)
33. Xu, D., Ouyang, W., Wang, X., Sebe, N.: PAD-Net: multi-tasks guided prediction-and-distillation network for simultaneous depth estimation and scene parsing. In: CVPR, pp. 675–684 (2018)
34. Xu, D., Wang, W., Tang, H., Liu, H., Sebe, N., Ricci, E.: Structured attention guided convolutional neural fields for monocular depth estimation. In: CVPR, pp. 3917–3925 (2018)
35. Zhi, S., Bloesch, M., Leutenegger, S., Davison, A.J.: SceneCode: monocular dense semantic reconstruction using learned encoded scene representations. In: CVPR, pp. 11776–11785 (2019)

Collaborative Video Object Segmentation by Foreground-Background Integration

Zongxin Yang[1,2], Yunchao Wei[2], and Yi Yang[2(✉)]

[1] Baidu Research, Beijing, China
zongxin.yang@student.uts.edu.au
[2] ReLER, Centre for Artificial Intelligence, University of Technology Sydney,
Ultimo, SYD, Australia
{yunchao.wei,yi.yang}@uts.edu.au

Abstract. This paper investigates the principles of embedding learning to tackle the challenging semi-supervised video object segmentation. Different from previous practices that only explore the embedding learning using pixels from foreground object (s), we consider background should be equally treated and thus propose Collaborative video object segmentation by Foreground-Background Integration (CFBI) approach. Our CFBI implicitly imposes the feature embedding from the target foreground object and its corresponding background to be contrastive, promoting the segmentation results accordingly. With the feature embedding from both foreground and background, our CFBI performs the matching process between the reference and the predicted sequence from both pixel and instance levels, making the CFBI be robust to various object scales. We conduct extensive experiments on three popular benchmarks, *i.e.*, DAVIS 2016, DAVIS 2017, and YouTube-VOS. Our CFBI achieves the performance ($\mathcal{J}\&\mathcal{F}$) of 89.4%, 81.9%, and 81.4%, respectively, outperforming all the other state-of-the-art methods. Code: https://github.com/z-x-yang/CFBI.

Keywords: Video Object Segmentation · Metric learning

1 Introduction

Video Object Segmentation (VOS) is a fundamental task in computer vision with many potential applications, including augmented reality [25] and self-driving cars [44]. In this paper, we focus on semi-supervised VOS, which targets on segmenting a particular object across the entire video sequence based on the object mask given at the first frame. The development of semi-supervised VOS can benefit many related tasks, such as video instance segmentation [13,41] and interactive video object segmentation [21,24,26].

This work was done when Zongxin Yang interned at Baidu Research.

Electronic supplementary material The online version of this chapter (https://doi.org/10.1007/978-3-030-58558-7_20) contains supplementary material, which is available to authorized users.

© Springer Nature Switzerland AG 2020
A. Vedaldi et al. (Eds.): ECCV 2020, LNCS 12350, pp. 332–348, 2020.
https://doi.org/10.1007/978-3-030-58558-7_20

Early VOS works [2,23,35] rely on fine-tuning with the first frame in evaluation, which heavily slows down the inference speed. Recent works (*e.g.*, [27,34,42]) aim to avoid fine-tuning and achieve better run-time. In these works, STMVOS [27] introduces memory networks to learn to read sequence information and outperforms all the fine-tuning based methods. However, STMVOS relies on simulating extensive frame sequences using large image datasets [7,12,15,22,32] for training. The simulated data significantly boosts the performance of STMVOS but makes the training procedure elaborate. Without simulated data, FEELVOS [34] adopts a semantic pixel-wise embedding together with a global (between the first and current frames) and a local (between the previous and current frames) matching mechanism to guide the prediction. The matching mechanism is simple and fast, but the performance is not comparable with STMVOS.

Even though the efforts mentioned above have made significant progress, current state-of-the-art works pay little attention to the feature embedding of background region in videos and only focus on exploring robust matching strategies for the foreground object (s). Intuitively, it is easy to extract the foreground region from a video when precisely removing all the background. Moreover, modern video scenes commonly focus on many similar objects, such as the cars in car racing, the people in a conference, and the animals on a farm. For these cases,

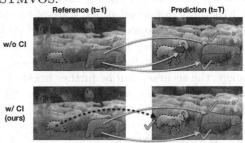

Fig. 1. CI means collaborative integration. There are two foreground sheep (pink and blue). In the top line, the contempt of background matching leads to a confusion of sheep's prediction. In the bottom line, we relieve the confusion problem by introducing background matching (dot-line arrow). (Color figure online)

the contempt of integrating foreground and background embeddings traps VOS in an unexpected background confusion problem. As shown in Fig. 1, if we focus on only the foreground matching like FEELVOS, a similar and same kind of object (sheep here) in the background is easy to confuse the prediction of the foreground object. Such an observation motivates us that the background should be equally treated compared with the foreground so that better feature embedding can be learned to relieve the background confusion and promote the accuracy of VOS.

We propose a novel framework for Collaborative video object segmentation by Foreground-Background Integration (CFBI) based on the above motivation. Different from the above methods, we not only extract the embedding and do match for the foreground target in the reference frame, but also for the background region to relieve the background confusion. Besides, our framework extracts two types of embedding (*i.e.*, pixel-level, and instance-level embedding) for each video frame to cover different scales of features. Like FEELVOS, we employ pixel-level embedding to match all the objects' details with the same global & local mechanism. However, the pixel-level matching is not sufficient and robust to match those objects with larger scales and may bring unexpected noises due to the pixel-wise diversity. Thus we introduce instance-level embedding to help the segmentation of large-scale objects by using attention mechanisms. Moreover, we propose a collaborative ensembler to aggregate the foreground & background and pixel-level & instance-level information and learn the collaborative relationship among them implicitly. For better convergence, we take a balanced random-crop scheme in training to avoid learned attributes being biased to the background attributes. All these proposed strategies can significantly improve the quality of the learned collaborative embeddings for conducting VOS while keeping the network simple yet effective simultaneously.

We perform extensive experiments on DAVIS [30,31], and YouTube-VOS [40] to validate the effectiveness of the proposed CFBI approach. Without any bells and whistles (such as the use of simulated data, fine-tuning or post-processing), CFBI outperforms all other state-of-the-art methods on the validation splits of DAVIS 2016 (ours, $\mathcal{J}\&\mathcal{F}$ **89.4%**), DAVIS 2017 (**81.9%**) and YouTube-VOS (**81.4%**) while keeping a competitive single-object inference speed of about 5 FPS. By additionally applying multi-scale & flip augmentation at the testing stage, the accuracy can be further boosted to **90.1%**, **83.3%** and **82.7%**, respectively. We hope our simple yet effective CFBI will serve as a solid baseline and help ease VOS's future research.

2   Related Work

Semi-supervised Video Object Segmentation. Many previous methods for semi-supervised VOS rely on fine-tuning at test time. Among them, OSVOS [2] and MoNet [39] fine-tune the network on the first-frame ground-truth at test time. OnAVOS [35] extends the first-frame fine-tuning by an online adaptation mechanism, *i.e.*, online fine-tuning. MaskTrack [29] uses optical flow to propagate the segmentation mask from one frame to the next. PReMVOS [23] combines four different neural networks (including an optical flow network [11]) using extensive fine-tuning and a merging algorithm. Despite achieving promising results, all these methods are seriously slowed down by fine-tuning during inference.

Some other recent works (*e.g.*, [6,42]) aim to avoid fine-tuning and achieve a better run-time. OSMN [42] employs two networks to extract the instance-level information and make segmentation predictions, respectively. PML [5] learns a pixel-wise embedding with the nearest neighbor classifier. Similar to PML, VideoMatch [18] uses a soft matching layer that maps the pixels of the current frame to the first frame in a learned embedding space. Following PML and Video-Match, FEELVOS [34] extends the pixel-level matching mechanism by additionally matching between the current frame and the previous frame. Compared to the methods with fine-tuning, FEELVOS achieves a much higher speed, but there is still a gap inaccuracy. Like FEELVOS, RGMP [38] and STMVOS [27] does not require any fine-tuning. STMVOS, which leverages a memory network to store and read the information from past frames, outperforms all the previous methods. However, STMVOS relies on an elaborate training procedure using extensive simulated data generated from multiple datasets. Moreover, the above methods do not focus on background matching.

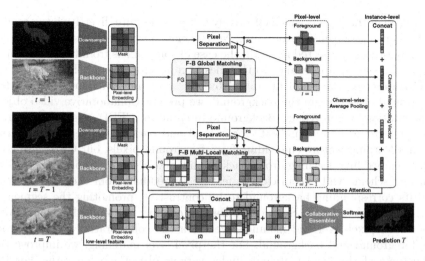

Fig. 2. An **overview** of CFBI. F-G denotes Foreground-Background. We use red and blue to indicate foreground and background separately. The deeper the red or blue color, the higher the confidence. Given the first frame ($t = 1$), previous frame ($t = T - 1$), and current frame ($t = T$), we firstly extract their pixel-wise embedding by using a backbone network. Second, we separate the first and previous frame embeddings into the foreground and background pixels based on their masks. After that, we use F-G pixel-level matching and instance-level attention to guide our collaborative ensembler network to generate a prediction. (Color figure online)

Our CFBI utilizes both the pixel-level and instance-level embeddings to guide prediction. Furthermore, we propose a collaborative integration method by additionally learning background embedding.

Attention Mechanisms. Recent works introduce the attention mechanism into convolutional networks (*e.g.*, [9,14]). Following them, SE-Nets [17] introduced a lightweight gating mechanism that focuses on enhancing the representational power of the convolutional network by modeling channel attention. Inspired by SE-Nets, CFBI uses an instance-level average pooling method to embed collaborative instance information from pixel-level embeddings. After that, we conduct a channel-wise attention mechanism to help guide prediction. Compared to OSMN, which employs an additional convolutional network to extract instance-level embedding, our instance-level attention method is more efficient and lightweight.

3   Method

Overview. Learning foreground feature embedding has been well explored by previous practices (*e.g.*, [34,42]). OSMN proposed to conduct an instance-level matching, but such a matching scheme fails to consider the feature diversity among the details of the target's appearance and results in coarse predictions. PML and FEELVOS alternatively adopt the pixel-level matching by matching each pixel of the target, which effectively takes the feature diversity into account and achieves promising performance. Nevertheless, performing pixel-level matching may bring unexpected noises in the case of some pixels from the background are with a similar appearance to the ones from the foreground (Fig. 1).

To overcome the problems raised by the above methods and promote the foreground objects from the background, we present Collaborative video object segmentation by Foreground-Background Integration (CFBI), as shown in Fig. 2. We use red and blue to indicate foreground and background separately. First, beyond learning feature embedding from foreground pixels, our CFBI also considers embedding learning from background pixels for collaboration. Such a learning scheme will encourage the feature embedding from the target object and its corresponding background to be contrastive, promoting the segmentation results accordingly. Second, we further conduct the embedding matching from both pixel-level and instance-level with the collaboration of pixels from the foreground and background. For the pixel-level matching, we improve the robustness of the local matching under various object moving rates. For the instance-level matching, we design an instance-level attention mechanism to augment the pixel-level matching efficiently. Moreover, to implicitly aggregate the learned foreground & background and pixel-level & instance-level information, we employ a collaborative ensembler to construct large receptive fields and make precise predictions.

3.1 Collaborative Pixel-Level Matching

For the pixel-level matching, we adopt a global and local matching mechanism similar to FEELVOS for introducing the guided information from the first and previous frames, respectively. Unlike previous methods [5,34], we additionally incorporate background information and apply multiple windows in the local matching, which is shown in the middle of Fig. 2.

For incorporating background information, we firstly redesign the pixel distance of [34] to further distinguish the foreground and background. Let B_t and F_t denote the pixel sets of background and all the foreground objects of frame t, respectively. We define a new distance between pixel p of the current frame T and pixel q of frame t in terms of their corresponding embedding, e_p and e_q, by

$$D_t(p,q) = \begin{cases} 1 - \frac{2}{1+exp(||e_p-e_q||^2+b_B)} & \text{if } q \in B_t \\ 1 - \frac{2}{1+exp(||e_p-e_q||^2+b_F)} & \text{if } q \in F_t \end{cases}, \tag{1}$$

where b_B and b_F are trainable background bias and foreground bias. We introduce these two biases to make our model be able further to learn the difference between foreground distance and background distance.

Foreground-Background Global Matching. Let $\mathcal{P}_t$ denote the set of all pixels (with a stride of 4) at time t and $\mathcal{P}_{t,o} \subseteq \mathcal{P}_t$ is the set of pixels at time t which belongs to the foreground object o. The global foreground matching between one pixel p of the current frame T and the pixels of the first reference frame (*i.e.*, $t = 1$) is,

$$G_{T,o}(p) = \min_{q \in \mathcal{P}_{1,o}} D_1(p,q). \tag{2}$$

Similarly, let $\overline{\mathcal{P}}_{t,o} = \mathcal{P}_t \backslash \mathcal{P}_{t,o}$ denote the set of relative background pixels of object o at time t, and the global background matching is,

$$\overline{G}_{T,o}(p) = \min_{q \in \overline{\mathcal{P}}_{1,o}} D_1(p,q). \tag{3}$$

Foreground-Background Multi-Local Matching.
In FEELVOS, the local matching is limited in only one fixed extent of neighboring pixels, but the offset of objects across two adjacent frames in VOS is variable, as shown in Fig. 3. Thus, we propose to apply the local matching mechanism on different scales and let the network learn how to select an appropriate local scale, which makes our framework more robust to various moving rates of objects. Notably, we use the intermediate results of the local matching with the largest window to calculate on other windows. Thus, the increase of computational resources of our multi-local matching is negligible.

Fig. 3. The moving rate of objects across two adjacent frames is largely variable for different sequences. Examples are from YouTube-VOS [40].

Formally, let $K = \{k_1, k_2, ..., k_n\}$ denote all the neighborhood sizes and $H(p,k)$ denote the neighborhood set of pixels that are at most k pixels away from p in both x and y directions, our foreground multi-local matching between the current frame T and its previous frame $T - 1$ is

$$ML_{T,o}(p,K) = \{L_{T,o}(p,k_1), L_{T,o}(p,k_2), ..., L_{T,o}(p,k_n)\}, \tag{4}$$

where

$$L_{T,o}(p,k) = \begin{cases} \min_{q \in \mathcal{P}^{p,k}_{T-1,o}} D_{T-1}(p,q) & \text{if } \mathcal{P}^{p,k}_{T-1,o} \neq \emptyset \\ 1 & \text{otherwise} \end{cases}. \tag{5}$$

Here, $\mathcal{P}^{p,k}_{T-1,o} := \mathcal{P}_{T-1,o} \cap H(p,k)$ denotes the pixels in the local window (or neighborhood). And our background multi-local matching is

$$\overline{ML}_{T,o}(p,K) = \{\overline{L}_{T,o}(p,k_1), \overline{L}_{T,o}(p,k_2), ..., \overline{L}_{T,o}(p,k_n)\}, \tag{6}$$

where

$$\overline{L}_{T,o}(p,k) = \begin{cases} \min_{q \in \overline{\mathcal{P}}^{p,k}_{T-1,o}} D_{T-1}(p,q) & \text{if } \overline{\mathcal{P}}^{p,k}_{T-1,o} \neq \emptyset \\ 1 & \text{otherwise} \end{cases}. \tag{7}$$

Here similarly, $\overline{\mathcal{P}}^{p,k}_{T-1,o} := \overline{\mathcal{P}}_{T-1,o} \cap H(p,k)$.

In addition to the global and multi-local matching maps, we concatenate the pixel-level embedding feature and mask of the previous frame with the current frame feature. FEELVOS demonstrates the effectiveness of concatenating the previous mask. Following this, we empirically find that introducing the previous embedding can further improve the performance ($\mathcal{J}\&\mathcal{F}$) by about 0.5%.

In summary, the output of our collaborative pixel-level matching is a concatenation of (1) the pixel-level embedding of the current frame, (2) the pixel-level embedding and mask of the previous frame, (3) the multi-local matching map and (4) the global matching map, as shown in the bottom box of Fig. 2.

3.2 Collaborative Instance-Level Attention

As shown in the right of Fig. 2, we further design a Collaborative instance-level attention mechanism to guide the segmentation for large-scale objects.

After getting the pixel-level embeddings of the first and previous frames, we separate them into foreground and background pixels (i.e., $\mathcal{P}_{1,o}$, $\overline{\mathcal{P}}_{1,o}$, $\mathcal{P}_{T-1,o}$, and $\overline{\mathcal{P}}_{T-1,o}$) according to their masks. Then, we apply channel-wise average pooling on each group of pixels to generate a total of four instance-level embedding vectors and concatenate these vectors into one collaborative instance-level guidance vector. Thus, the guidance vector contains the information from both the first and previous frames, and both the foreground and background regions.

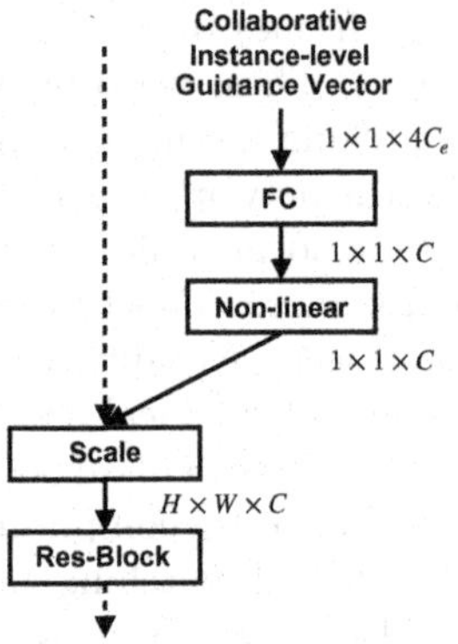

In order to efficiently utilize the instance-level information, we employ an attention mechanism to adjust our Collaborative Ensembler (CE). We show a detailed illustration in Fig. 4. Inspired by SE-Nets [17], we leverage a fully-connected (FC) layer (we found this setting is better than using two FC layers as adopted by SE-Net) and a non-linear activation function to construct a gate for the input of each Res-Block in the CE. The gate will adjust the scale of the input feature channel-wisely.

Fig. 4. The trainable part of the instance-level attention. C_e denotes the channel dimension of pixel-wise embedding. H, W, C denote the height, width, channel dimension of CE features.

By introducing collaborative instance-level attention, we can leverage a full scale of foreground-background information to guide the prediction further. The information with a large (instance-level) receptive field is useful to relieve local ambiguities [33], which is inevitable with a small (pixel-wise) receptive field.

3.3 Collaborative Ensembler (CE)

In the lower right of Fig. 2, we design a collaborative ensembler for making large receptive fields to aggregate pixel-level and instance-level information and implicitly learn the collaborative relationship between foreground and background.

Inspired by ResNets [16] and Deeplabs [3,4], which both have shown significant representational power in image segmentation tasks, our CE uses a downsample-upsample structure, which contains three stages of Res-Blocks [16] and an Atrous Spatial Pyramid Pooling (ASPP) [4] module. The number of Res-Blocks in Stage 1, 2, and 3 are 2, 3, 3 in order. Besides, we employ dilated convolutional layers to improve the receptive fields efficiently. The dilated rates of the 3×3 convolutional layer of Res-Blocks in one stage are separately 1, 2, 4 (or 1, 2 for Stage 1). At the beginning of Stage 2 and Stage 3, the feature maps will be downsampled by the first Res-Block with a stride of 2. After these three stages, we employ an ASPP and a Decoder [4] module to increase the receptive fields further, upsample the scale of feature and fine-tune the prediction collaborated with the low-level backbone features.

4   Implementation Details

For better convergence, we modify the random-crop augmentation and the training method in previous methods [27,34].

Balanced Random-Crop. As shown in Fig. 5, there is an apparent imbalance between the foreground and the background pixel number on VOS datasets. Such an issue usually makes the models easier to be biased to background attributes.

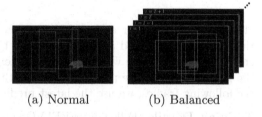

(a) Normal   (b) Balanced

Fig. 5. When using normal random-crop, some red windows contain few or no foreground pixels. For relieving this problem, we propose balanced random-crop.

In order to relieve this problem, we take a balanced random-crop scheme, which crops a sequence of frames (*i.e.*, the first frame, the previous frame, and the current frame) by using a same cropped window and restricts the cropped region of the first frame to contain enough foreground information. The restriction method is simple yet effective. To be specific, the balanced random-crop will decide on whether the randomly cropped frame contains enough pixels from foreground objects or not. If not, the method will continually take the cropping operation until we obtain an expected one.

Sequential Training. In the training stage, FEELVOS predicts only one step in one iteration, and the guidance masks come from the ground-truth data. RGMP and STMVOS uses previous guidance information (mask or feature memory) in training, which is more consistent with the inference stage and performs better. In the evaluation stage, the previous guidance masks are always generated by the network in the previous inference steps.

Following RGMP, we train the network using a sequence of consecutive frames in each SGD iteration. In each iteration, we randomly sample a batch

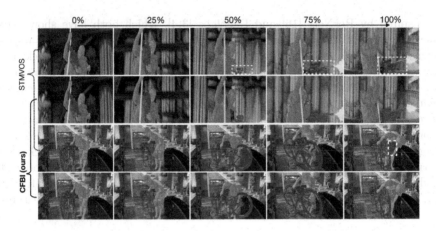

Fig. 6. Qualitative comparison with STMVOS on DAVIS 2017. In the first video, STMVOS fails in tracking the gun after occlusion and blur. In the second video, STMVOS is easier to partly confuse with bicycle and person.

of video sequences. For each video sequence, we randomly sample a frame as the reference frame and a continuous $N + 1$ frames as the previous frame and current frame sequence with N frames. When predicting the first frame, we use the ground-truth of the previous frame as the previous mask. When predicting the following frames, we use the latest prediction as the previous mask.

Training Details. Following FEELVOS, we use the DeepLabv3+ [4] architecture as the backbone for our network. However, our backbone is based on the dilated Resnet-101 [4] instead of Xception-65 [8] for saving computational resources. We apply batch normalization (BN) [19] in our backbone and pre-train it on ImageNet [10] and COCO [22]. The backbone is followed by one depth-wise separable convolution for extracting pixel-wise embedding with a stride of 4.

We initialize b_B and b_F to 0. For the multi-local matching, we further downsample the embedding feature to a half size using bi-linear interpolation for saving GPU memory. Besides, the window sizes in our setting are $K = \{2, 4, 6, 8, 10, 12\}$. For the collaborative ensembler, we apply group normalization (GN) [37] and gated channel transformation [43] to improving training stability and performance when using a small batch size. For sequential training, the current sequence's length is $N = 3$, which makes a better balance between computational resources and network performance.

We use the DAVIS 2017 [31] training set (60 videos) and the YouTube-VOS [40] training set (3471 videos) as the training data. We downsample all the videos to 480P resolution, which is same as the default setting in DAVIS. We adopt SGD with a momentum of 0.9 and apply a bootstrapped cross-entropy loss, which only considers the 15% hardest pixels. During the training stage, we freeze the parameters of BN in the backbone. For the experiments on YouTube-VOS, we use a learning rate of 0.01 for $100,000$ steps with a batch size of 4 videos (*i.e.*, 20 frames in total) per GPU using 2 Tesla V100 GPUs. The training time on YouTube-VOS is about 5 d. For DAVIS, we use a learning rate of 0.006 for $50,000$ steps with a batch size of 3 videos (*i.e.*, 15 frames in total) per GPU using 2 GPUs. We apply flipping, scaling, and balanced random-crop as data augmentations. The cropped window size is 465 × 465. For the multi-scale testing, we apply the scales of {1.0, 1.15, 1.3, 1.5} and {2.0, 2.15, 2.3} for YouTube-VOS and DAVIS, respectively. CFBI achieves similar results in PyTorch [28] and PaddlePaddle [1].

5   Experiments

Following the previous state-of-the-art method [27], we evaluate our method on YouTube-VOS [40], DAVIS 2016 [30] and DAVIS 2017 [31]. For the evaluation on YouTube-VOS, we train our model on the YouTube-VOS training set [40] (3471 videos). For DAVIS, we train our model on the DAVIS-2017 training set [31] (60 videos). Both DAVIS 2016 and 2017 are evaluated using an identical model trained on DAVIS 2017 for a fair comparison with the previous works [27,34]. Furthermore, we provide DAVIS results using both DAVIS 2017 and YouTube-VOS for training following some latest works [27,34].

The evaluation metric is the $\mathcal{J}$ score, calculated as the average IoU between the prediction and the ground truth mask, and the $\mathcal{F}$ score, calculated as an average

Table 1. The quantitative evaluation on YouTube-VOS [40]. F, S, and * separately denote fine-tuning at test time, using simulated data in the training process and performing model ensemble in evaluation. CFBI^{MS} denotes using a multi-scale and flip strategy in evaluation.

Methods	F	S	Avg	Seen		Unseen	
				$\mathcal{J}$	$\mathcal{F}$	$\mathcal{J}$	$\mathcal{F}$
Validation 2018 split							
AG [20]			66.1	67.8	-	60.8	-
PReM [23]	✓		66.9	71.4	75.9	56.5	63.7
BoLT [36]	✓		71.1	71.6	-	64.3	-
STM$^-$ [27]			68.2	-	-	-	-
STM [27]		✓	79.4	79.7	84.2	72.8	80.9
CFBI			**81.4**	**81.1**	**85.8**	**75.3**	**83.4**
CFBI^{MS}			**82.7**	**82.2**	**86.8**	**76.9**	**85.0**
Testing 2019 split							
MST* [45]		✓	81.7	80.0	83.3	**77.9**	85.5
EMN* [46]		✓	81.8	**80.7**	**84.7**	77.3	84.7
CFBI			81.5	79.6	84.0	77.3	85.3
CFBI^{MS}			**82.2**	80.4	**84.7**	**77.9**	**85.7**

boundary similarity measure between the boundary of the prediction and the ground truth, and their average value ($\mathcal{J}$&$\mathcal{F}$). We evaluate our results on the official evaluation server or use the official tools.

5.1   Compare with the State-of-the-art Methods

YouTube-VOS. [40] is the latest large-scale dataset for multi-object video segmentation. Compared to the popular DAVIS benchmark that consists of 120 videos, YouTube-VOS is about 37 times larger. In detail, the dataset contains 3471 videos in the training set (65 categories), 507 videos in the validation set (additional 26 unseen categories), and 541 videos in the test set (additional 29 unseen categories). Due to the existence of unseen object categories, the

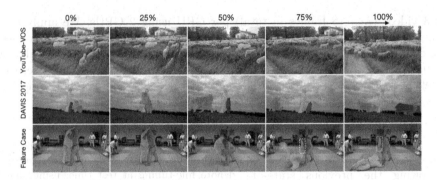

Fig. 7. Qualitative results on DAVIS 2017 and YouTube-VOS. In the first video, we succeed in tracking many similar-looking sheep. In the second video, our CFBI tracks the person and the dog with a red mask after occlusion well. In the last video, CFBI fails to segment one hand of the right person (the white box). A possible reason is that the two persons are too similar and close. (Color figure online)

YouTube-VOS validation set is much suitable for measuring the generalization ability of different methods.

As shown in Table 1, we compare our method to existing methods on both Validation 2018 and Testing 2019 splits. Without using any bells and whistles, like fine-tuning at test time [2,35] or pre-training on larger augmented simulated data [27,38], our method achieves an average score of **81.4%**, which significantly outperforms all other methods in every evaluation metric. Particularly, the 81.4% result is 2.0% higher than the previous state-of-the-art method, STMVOS, which uses extensive simulated data from [7,12, 15,22,32] for training. Without simulated data, the performance of STMVOS will drop from 79.4% to 68.2%. More-

Table 2. The quantitative evaluation on DAVIS 2016 [30] validation set. (**Y**) denotes using YouTube-VOS for training.

Methods	F	S	Avg	$\mathcal{J}$	$\mathcal{F}$	t/s
OSMN [42]	-		-		74.0	0.14
PML [5]			77.4	75.5	79.3	0.28
VideoMatch [18]			80.9	81.0	80.8	0.32
RGMP⁻ [38]			68.8	68.6	68.9	0.14
RGMP [38]		✓	81.8	81.5	82.0	0.14
A-GAME [20] (**Y**)			82.1	82.2	82.0	**0.07**
FEELVOS [34] (**Y**)			81.7	81.1	82.2	0.45
OnAVOS [35]	✓		85.0	85.7	84.2	13
PReMVOS [23]	✓		86.8	84.9	88.6	32.8
STMVOS [27]		✓	86.5	84.8	88.1	0.16
STMVOS [27] (**Y**)		✓	**89.3**	**88.7**	89.9	0.16
CFBI			86.1	85.3	86.9	0.18
CFBI (**Y**)			**89.4**	88.3	**90.5**	0.18
CFBI^{MS} (**Y**)			**90.7**	**89.6**	**91.7**	9

over, we further boost our performance to **82.7%** by applying a multi-scale and flip strategy during the evaluation.

We also compare our method with two of the best results on the Testing 2019 split, i.e., *Rank 1* (EMN [46]) and *Rank 2* (MST [45]) results in the 2nd Large-scale Video Object Segmentation Challenge. Without applying model ensemble, our single-model result (**82.2%**) outperforms the *Rank 1* result (81.8%) in the unseen and average metrics, which further demonstrates our generalization ability and effectiveness.

DAVIS 2016. [30] contains 20 videos annotated with high-quality masks each for a single target object. We compare our CFBI method with state-of-the-art methods in Table 2. On the DAVIS-2016 validation set, our method trained with an additional YouTube-VOS training set achieves an average score of **89.4%**, which is slightly better than STMVOS (89.3%), a method using simulated data as mentioned before. The accuracy gap between CFBI and STMVOS on DAVIS is smaller than the gap on YouTube-VOS. A possible reason is that DAVIS is too small and easy to over-fit. Compare to a much fair baseline (*i.e.*, FEELVOS) whose setting is same to ours, the proposed CFBI not only achieves much better accuracy (**89.4%** *vs.* 81.7%) but also maintains a comparable fast inference speed (0.18*s* *vs.* 0.45*s*). After applying multi-scale and flip for evaluation, we can improve the performance from **89.4%** to **90.1%**. However, this strategy will cost much more inference time (9*s*).

DAVIS 2017. [31] is a multi-object extension of DAVIS 2016. The validation set of DAVIS 2017 consists of 59 objects in 30 videos. Next, we evaluate the generalization ability of our model on the popular DAVIS-2017 benchmark.

As shown in Table 3, our CFBI makes significantly improvement over FEELVOS (**81.9%** *vs.* 71.5%). Besides, our CFBI without using simulated data is slightly better than the previous state-of-the-art method, STMVOS (**81.9%** *vs.* 81.8%). We show some examples compared with STMVOS in Fig. 6. Same as previous experiments, the augmentation in evaluation can further boost the results to a higher score of **83.3%**. We also evaluate our method on the testing split of DAVIS 2017, which is much more challenging than the validation split. As shown in Table 3, we significantly outperforms STMVOS (72.2%) by **2.6%**. By applying augmentation, we can further boost the result to **77.5%**. The strong results prove that our method has the best generalization ability among the latest methods.

Table 3. The quantitative evaluation on DAVIS-2017 [31].

Methods	F	S	Avg	$\mathcal{J}$	$\mathcal{F}$
Validation split					
OSMN [42]			54.8	52.5	57.1
VideoMatch [18]			62.4	56.5	68.2
OnAVOS [35]	✓		63.6	61.0	66.1
RGMP [38]		✓	66.7	64.8	68.6
A-GAME [20] (**Y**)			70.0	67.2	72.7
FEELVOS [34] (**Y**)			71.5	69.1	74.0
PReMVOS [23]	✓		77.8	73.9	81.7
STMVOS [27]		✓	71.6	69.2	74.0
STMVOS [27] (**Y**)		✓	**81.8**	**79.2**	**84.3**
CFBI			74.9	72.1	77.7
CFBI (**Y**)			**81.9**	**79.1**	**84.6**
CFBI^{MS} (**Y**)			**83.3**	**80.5**	**86.0**
Testing split					
OSMN [42]			41.3	37.7	44.9
OnAVOS [35]	✓		56.5	53.4	59.6
RGMP [38]		✓	52.9	51.3	54.4
FEELVOS [34] (**Y**)			57.8	55.2	60.5
PReMVOS [23]	✓		71.6	67.5	75.7
STMVOS [27] (**Y**)		✓	72.2	69.3	75.2
CFBI (**Y**)			**74.8**	**71.1**	**78.5**
CFBI^{MS} (**Y**)			**77.5**	**73.8**	**81.1**

Qualitative Results. We show more results of CFBI on the validation set of DAVIS 2017 (**81.9%**) and YouTube-VOS (**81.4%**) in Fig. 7. It can be seen that CFBI is capable of producing accurate segmentation under challenging situations, such as large motion, occlusion, blur, and similar objects. In the *sheep* video, CFBI succeeds in tracking five selected sheep inside a crowded flock. In the *judo* video, CFBI fails to segment one hand of the right person. A possible reason is that the two persons are too similar in appearance and too close in position. Besides, their hands are with blur appearance due to the fast motion.

5.2 Ablation Study

We analyze the ablation effect of each component proposed in CFBI on the DAVIS-2017 validation set. Following FEELVOS, we only use the DAVIS-2017 training set as training data for these experiments.

Table 4. Ablation of background embedding. P and I separately denote the pixel-level matching and instance-level attention. * denotes removing the foreground and background bias.

P	I	Avg	$\mathcal{J}$	$\mathcal{F}$
✓	✓	74.9	72.1	77.7
✓*	✓	72.8	69.5	76.1
✓		73.0	69.9	76.0
	✓	72.3	69.1	75.4
		70.9	68.2	73.6

Background Embedding. As shown in Table 4, we first analyze the influence of removing the background embedding while keeping the foreground only as [34,42]. Without any background mechanisms, the result of our method heavily drops from 74.9% to 70.9%. This result shows that it is significant to embed both foreground and background features collaboratively. Besides, the missing of background information in the pixel-level matching or the instance-level attention will decrease the result to 73.0% or 72.3% separately. Thus, compared to instance-level attention, the pixel-level matching performance is more sensitive to the effect of background embedding. A possible reason for this phenomenon is that the possibility of existing some background pixels similar to the foreground is higher than some background instances. Finally, we remove the foreground and background bias, b_F and b_B, from the distance metric and the result drops to 72.8%, which further shows that the distance between foreground pixels and the distance between background pixels should be separately considered.

Other Components. The ablation study of other proposed components is shown in Table 5. Line 0 (74.9%) is the result of proposed CFBI, and Line 6 (68.3%) is our baseline method reproduced by us. Under the same setting, our CFBI significantly outperforms the baseline.

In line 1, we use only one local neighborhood window to conduct

Table 5. Ablation of other components.

	Ablation	Avg	$\mathcal{J}$	$\mathcal{F}$
0	Ours (CFBI)	74.9	72.1	77.7
1	w/o multi-local windows	73.8	70.8	76.8
2	w/o sequential training	73.3	70.8	75.7
3	w/o collaborative ensembler	73.3	70.5	76.1
4	w/o balanced random-crop	72.8	69.8	75.8
5	w/o instance-level attention	72.7	69.8	75.5
6	Baseline (FEELVOS)	68.3	65.6	70.9

the local matching following the setting of FEELVOS, which degrades the result from 74.9% to 73.8%. It demonstrates that our multi-local matching module is more robust and effective than the single-local matching module of FEELVOS. Notably, the computational complexity of multi-local matching dominantly depends on the biggest local window size because we use the intermediate results of the local matching of the biggest window to calculate on smaller windows.

In line 2, we replace our sequential training by using ground-truth masks instead of network predictions as the previous mask. By doing this, the performance of CFBI drops from 74.9% to 73.3%, which shows the effectiveness of our sequential training under the same setting.

In line 3, we replace our collaborative ensembler with 4 depth-wise separable convolutional layers. This architecture is the same as the dynamic segmentation head of [34]. Compared to our collaborative ensembler, the dynamic segmentation head has much smaller receptive fields and performs 1.6% worse.

In line 4, we use normal random-crop instead of our balanced random-crop during the training process. In this situation, the performance drops by 2.1% to 72.8% as well. As expected, our balanced random-crop is successful in relieving the model form biasing to background attributes.

In line 5, we disable the use of instance-level attention as guidance information to the collaborative ensembler, which means we only use pixel-level information to guide the prediction. In this case, the result deteriorates even further to 72.7, which proves that instance-level information can further help the segmentation with pixel-level information.

In summary, we explain the effectiveness of each proposed component of CFBI. For VOS, it is necessary to embed both foreground and background features. Besides, the model will be more robust by combining pixel-level information and instance-level information, and by using more local windows in the matching between two continuous frames. Apart from this, the proposed balanced random-crop and sequential training are useful but straightforward in improving training performance.

6   Conclusion

This paper proposes a novel framework for video object segmentation by introducing collaborative foreground-background integration and achieves new state-of-the-art results on three popular benchmarks. Specifically, we impose the feature embedding from the foreground target and its corresponding background to be contrastive. Moreover, we integrate both pixel-level and instance-level embeddings to make our framework robust to various object scales while keeping the network simple and fast. We hope CFBI will serve as a solid baseline and help ease the future research of VOS and related areas, such as video object tracking and interactive video editing.

Acknowledgements. This work is partly supported by ARC DP200100938 and ARC DECRA DE190101315.

References

1. Parallel distributed deep learning: Machine learning framework from industrial practice. https://www.paddlepaddle.org.cn/
2. Caelles, S., Maninis, K.K., Pont-Tuset, J., Leal-Taixé, L., Cremers, D., Van Gool, L.: One-shot video object segmentation. In: CVPR, pp. 221–230 (2017)
3. Chen, L.C., Papandreou, G., Kokkinos, I., Murphy, K., Yuille, A.L.: DeepLab: semantic image segmentation with deep convolutional nets, atrous convolution, and fully connected CRFs. TPAMI **40**(4), 834–848 (2017)
4. Chen, L.-C., Zhu, Y., Papandreou, G., Schroff, F., Adam, H.: Encoder-decoder with atrous separable convolution for semantic image segmentation. In: Ferrari, V., Hebert, M., Sminchisescu, C., Weiss, Y. (eds.) ECCV 2018, Part VII. LNCS, vol. 11211, pp. 833–851. Springer, Cham (2018). https://doi.org/10.1007/978-3-030-01234-2_49
5. Chen, Y., Pont-Tuset, J., Montes, A., Van Gool, L.: Blazingly fast video object segmentation with pixel-wise metric learning. In: CVPR, pp. 1189–1198 (2018)
6. Cheng, J., Tsai, Y.H., Hung, W.C., Wang, S., Yang, M.H.: Fast and accurate online video object segmentation via tracking parts. In: CVPR, pp. 7415–7424 (2018)
7. Cheng, M.M., Mitra, N.J., Huang, X., Torr, P.H., Hu, S.M.: Global contrast based salient region detection. TPAMI **37**(3), 569–582 (2014)
8. Chollet, F.: Xception: deep learning with depthwise separable convolutions. In: CVPR, pp. 1251–1258 (2017)
9. Dauphin, Y.N., Fan, A., Auli, M., Grangier, D.: Language modeling with gated convolutional networks. In: ICML (2017)
10. Deng, J., Dong, W., Socher, R., Li, L.J., Li, K., Fei-Fei, L.: ImageNet: a large-scale hierarchical image database. In: CVPR, pp. 248–255. IEEE (2009)
11. Dosovitskiy, A., et al.: FlowNet: learning optical flow with convolutional networks. In: ICCV, pp. 2758–2766 (2015)
12. Everingham, M., Van Gool, L., Williams, C.K., Winn, J., Zisserman, A.: The pascal visual object classes (VOC) challenge. IJCV **88**(2), 303–338 (2010)
13. Feng, Q., Yang, Z., Li, P., Wei, Y., Yang, Y.: Dual embedding learning for video instance segmentation. In: ICCV Workshops (2019)
14. Gehring, J., Auli, M., Grangier, D., Yarats, D., Dauphin, Y.N.: Convolutional sequence to sequence learning. In: ICML (2017)
15. Hariharan, B., Arbeláez, P., Bourdev, L., Maji, S., Malik, J.: Semantic contours from inverse detectors. In: ICCV, pp. 991–998. IEEE (2011)
16. He, K., Zhang, X., Ren, S., Sun, J.: Deep residual learning for image recognition. In: CVPR (2016)
17. Hu, J., Shen, L., Sun, G.: Squeeze-and-excitation networks. In: CVPR (2018)
18. Hu, Y.-T., Huang, J.-B., Schwing, A.G.: VideoMatch: matching based video object segmentation. In: Ferrari, V., Hebert, M., Sminchisescu, C., Weiss, Y. (eds.) ECCV 2018, Part VIII. LNCS, vol. 11212, pp. 56–73. Springer, Cham (2018). https://doi.org/10.1007/978-3-030-01237-3_4
19. Ioffe, S., Szegedy, C.: Batch normalization: accelerating deep network training by reducing internal covariate shift. In: ICML (2015)
20. Johnander, J., Danelljan, M., Brissman, E., Khan, F.S., Felsberg, M.: A generative appearance model for end-to-end video object segmentation. In: CVPR, pp. 8953–8962 (2019)
21. Liang, C., Yang, Z., Miao, J., Wei, Y., Yang, Y.: Memory aggregated CFBI+ for interactive video object segmentation. In: CVPR Workshops (2020)

22. Lin, T.-Y., et al.: Microsoft COCO: common objects in context. In: Fleet, D., Pajdla, T., Schiele, B., Tuytelaars, T. (eds.) ECCV 2014, Part V. LNCS, vol. 8693, pp. 740–755. Springer, Cham (2014). https://doi.org/10.1007/978-3-319-10602-1_48

23. Luiten, J., Voigtlaender, P., Leibe, B.: PReMVOS: proposal-generation, refinement and merging for video object segmentation. In: Jawahar, C.V., Li, H., Mori, G., Schindler, K. (eds.) ACCV 2018, Part IV. LNCS, vol. 11364, pp. 565–580. Springer, Cham (2019). https://doi.org/10.1007/978-3-030-20870-7_35

24. Miao, J., Wei, Y., Yang, Y.: Memory aggregation networks for efficient interactive video object segmentation. In: CVPR (2020)

25. Ngan, K.N., Li, H.: Video Segmentation and Its Applications. Springer Science & Business Media, New York (2011). https://doi.org/10.1007/978-1-4419-9482-0

26. Oh, S.W., Lee, J.Y., Xu, N., Kim, S.J.: Fast user-guided video object segmentation by interaction-and-propagation networks. In: CVPR, pp. 5247–5256 (2019)

27. Oh, S.W., Lee, J.Y., Xu, N., Kim, S.J.: Video object segmentation using space-time memory networks. In: ICCV (2019)

28. Paszke, A., et al.: Automatic differentiation in PyTorch (2017)

29. Perazzi, F., Khoreva, A., Benenson, R., Schiele, B., Sorkine-Hornung, A.: Learning video object segmentation from static images. In: CVPR, pp. 2663–2672 (2017)

30. Perazzi, F., Pont-Tuset, J., McWilliams, B., Van Gool, L., Gross, M., Sorkine-Hornung, A.: A benchmark dataset and evaluation methodology for video object segmentation. In: CVPR, pp. 724–732 (2016)

31. Pont-Tuset, J., Perazzi, F., Caelles, S., Arbeláez, P., Sorkine-Hornung, A., Van Gool, L.: The 2017 davis challenge on video object segmentation. arXiv preprint arXiv:1704.00675 (2017)

32. Shi, J., Yan, Q., Xu, L., Jia, J.: Hierarchical image saliency detection on extended CSSD. TPAMI **38**(4), 717–729 (2015)

33. Torralba, A.: Contextual priming for object detection. IJCV **53**(2), 169–191 (2003)

34. Voigtlaender, P., Chai, Y., Schroff, F., Adam, H., Leibe, B., Chen, L.C.: FEELVOS: fast end-to-end embedding learning for video object segmentation. In: CVPR, pp. 9481–9490 (2019)

35. Voigtlaender, P., Leibe, B.: Online adaptation of convolutional neural networks for video object segmentation. In: BMVC (2017)

36. Voigtlaender, P., Luiten, J., Leibe, B.: BoLTVOS: Box-level tracking for video object segmentation. arXiv preprint arXiv:1904.04552 (2019)

37. Wu, Y., He, K.: Group normalization. In: Ferrari, V., Hebert, M., Sminchisescu, C., Weiss, Y. (eds.) ECCV 2018, Part XIII. LNCS, vol. 11217, pp. 3–19. Springer, Cham (2018). https://doi.org/10.1007/978-3-030-01261-8_1

38. Wug Oh, S., Lee, J.Y., Sunkavalli, K., Joo Kim, S.: Fast video object segmentation by reference-guided mask propagation. In: CVPR, pp. 7376–7385 (2018)

39. Xiao, H., Feng, J., Lin, G., Liu, Y., Zhang, M.: MoNet: deep motion exploitation for video object segmentation. In: CVPR, pp. 1140–1148 (2018)

40. Xu, N., et al.: YouTube-VOS: A large-scale video object segmentation benchmark. arXiv preprint arXiv:1809.03327 (2018)

41. Yang, L., Fan, Y., Xu, N.: Video instance segmentation. In: ICCV, pp. 5188–5197 (2019)

42. Yang, L., Wang, Y., Xiong, X., Yang, J., Katsaggelos, A.K.: Efficient video object segmentation via network modulation. In: CVPR, pp. 6499–6507 (2018)

43. Yang, Z., Zhu, L., Wu, Y., Yang, Y.: Gated channel transformation for visual recognition. In: CVPR (2020)

44. Zhang, Z., Fidler, S., Urtasun, R.: Instance-level segmentation for autonomous driving with deep densely connected MRFs. In: CVPR, pp. 669–677 (2016)
45. Zhou, Q., et al.: Motion-guided spatial time attention for video object segmentation. In: ICCV Workshops (2019)
46. Zhou, Z., et al.: Enhanced memory network for video segmentation. In: ICCV Workshops (2019)

Adaptive Margin Diversity Regularizer for Handling Data Imbalance in Zero-Shot SBIR

Titir Dutta, Anurag Singh, and Soma Biswas[✉]

Indian Institute of Science, Bangalore, India
{titird,anuragsingh2,somabiswas}@iisc.ac.in

Abstract. Data from new categories are continuously being discovered, which has sparked significant amount of research in developing approaches which generalize to previously unseen categories, i.e. zero-shot setting. Zero-shot sketch-based image retrieval (ZS-SBIR) is one such problem in the context of cross-domain retrieval, which has received lot of attention due to its various real-life applications. Since most real-world training data have a fair amount of imbalance; in this work, for the first time in literature, we extensively study the effect of training data imbalance on the generalization to unseen categories, with ZS-SBIR as the application area. We evaluate several state-of-the-art data imbalance mitigating techniques and analyze their results. Furthermore, we propose a novel framework AMDReg (Adaptive Margin Diversity Regularizer), which ensures that the embeddings of the sketches and images in the latent space are not only semantically meaningful, but they are also separated according to their class-representations in the training set. The proposed approach is model-independent, and it can be incorporated seamlessly with several state-of-the-art ZS-SBIR methods to improve their performance under imbalanced condition. Extensive experiments and analysis justify the effectiveness of the proposed AMDReg for mitigating the effect of data imbalance for generalization to unseen classes in ZS-SBIR.

1 Introduction

Sketches-based image retrieval (SBIR) [15,35], which deals with retrieving natural images, given a hand-drawn sketches query, has gained significant traction because of its potential applications in e-commerce, forensics, etc. Since new categories of data are continuously being added to the system, it is important for algorithms to generalize well to unseen classes, which is termed as Zero-Shot Sketches-Based Image Retrieval (ZS-SBIR) [5–7,16]. Majority of ZS-SBIR approaches learn a shared latent-space representation for both sketch and image, where sketches and images from same category come closer to each other and also incorporate additional techniques to facilitate generalization to unseen classes.

One important factor that has been largely overlooked in this task of generalization to unseen classes is the distribution of the training data. Real-world data, used to train the model, is not always class-wise or domain-wise well-balanced.

© Springer Nature Switzerland AG 2020
A. Vedaldi et al. (Eds.): ECCV 2020, LNCS 12350, pp. 349–364, 2020.
https://doi.org/10.1007/978-3-030-58558-7_21

When training and test categories are same, as expected, class imbalance in the training data results in severe degradation in testing performance, specially for the minority classes. Many seminal approaches have been proposed to mitigate this effect for the task of image classification [2,4,11,14], but the effect of data imbalance on generalization to unseen classes is relatively unexplored, both for single and cross-domain applications. In fact, both of the two large-scale datasets, widely used for SBIR/ZS-SBIR, namely Sketchy Extended [25] and TU-Berlin Extended [8] have data imbalance. In cross-domain data, there can be two types of imbalance: 1) domain imbalance - where the number of data samples in one domain is significantly different compared to the other domain; 2) class imbalance - where there is a significant difference in the number of data samples per class. TU-Berlin Ext. exhibits imbalance of both types. Although a recent paper [5] has attributed poor retrieval performance for TU-Berlin Ext. to data imbalance, no measures have been proposed to handle this.

Here, we aim to study the effect of class imbalance in the training data on the retrieval performance of unseen classes in the context of ZS-SBIR, but interestingly we observe that the proposed framework works well even when both types of imbalances are present. We analyze several state-of-the-art approaches for mitigating the effect of training data imbalance on the final retrieval performance. To this end, we propose a novel regularizer termed **AMDReg - Adaptive Margin Diversity Regularizer**, which ensures that the embeddings of the data samples in the latent space account for the distribution of classes in the training set. To facilitate generalization to unseen classes for ZS-SBIR, majority of the ZS-SBIR approaches impose a direct or indirect semantic constraint on the latent-space which ensures that the sketch and image samples from *unseen* classes during testing are embedded in the neighborhood of its related *seen* classes. But merely imposing a semantic constraint does not account for the training class imbalance. The proposed AMDReg, which is computed from the class-wise training data distribution present in sketch and image domains, helps to appropriately position the semantic embeddings. It tries to enforce a broader margin/spread for the classes for which less number of training samples are available as compared to the classes which have larger number of samples. Extensive analysis and evaluation on two benchmark datasets validate the effectiveness of the proposed approach. The contributions of this paper have been summarized below.

1. We analyze the effect of class-imbalance on generalization to unseen classes for the ZS-SBIR task. To the best of our knowledge, this is the first work in literature which addresses the data-imbalance problem in the context of cross-domain retrieval.
2. We analyze the performance of several state-of-the-art techniques for handling data imbalance problem for this task.
3. We propose a novel regularizer termed **AMDReg**, which can seamlessly be used with several ZS-SBIR methods to improve their performance. We have observed significant improvement in the performance of three state-of-the-art ZS-SBIR methods.
4. We obtain state-of-the-art performance for ZS-SBIR and generalized ZS-SBIR for two large-scale benchmark datasets.

2   Related Work

Here, we discuss relevant work in the literature for this study. We include recent papers for sketch-based image retrieval (SBIR), zero-shot sketch-based image retrieval (ZS-SBIR), as well as the class-imbalance problems in classification.

Sketch-Based Image Retrieval (SBIR): The primary goal of these approaches is to bridge the domain-gap between natural images and hand-drawn sketches. Early methods for SBIR, such as HOG [12], LKS [24] aim to extract hand-crafted features from the sketches as well as from the edge-maps obtained from natural images, which are then directly used for retrieval. The advent of deep networks have advanced the state-of-the-art significantly. Siamese network [22] with triplet-loss or contrastive-loss, GoogleNet [25] with triplet loss, etc. are some of the initial architectures. Recently a number of hashing-based methods, such as [15,35] have achieved significant success. [15] uses a heterogeneous network, which employs the edge maps from images, along with the sketch-image training data to learn a shared representation space. In contrast, GDH [35] exploits a generative model to learn the equivalent image representation from a given sketch and performs the final retrieval in the image space.

Zero-Shot Sketch-Based Image Retrieval (ZS-SBIR): The knowledge-gap encountered by the retrieval model, when a sketch query or database image is from a previously unseen class makes ZS-SBIR extremely challenging. ZSIH [26], generative-model based ZS-SBIR [32] are some of the pioneering works in this direction. However, as identified by [6], ZSIH [26] requires a fusion-layer for learning the model, which shoots up the learning cost and [32] requires strictly paired sketch-image data for training. Some of the recent works, [5–7,16] have reported improved performance for ZS-SBIR over the early techniques. [6] introduces a further generalization in the evaluation protocol for ZS-SBIR, termed as generalized ZS-SBIR; where the search set contains images from both the sets of seen and unseen classes. This poses even greater challenge to the algorithm, and the performances degrade significantly for this evaluation protocol [6,7]. Few of the ZS-SBIR approaches are discussed in more details later.

Handling Data Imbalance for Classification: Since real-world training data are often imbalanced, recently, a number of works [2,4,11,14] have been proposed to address this problem. [14] mitigates the problem of foreground background class imbalance problem in the context of object recognition and proposes a modification to the traditional cross-entropy based classification loss. [4] introduces an additional cost-sensitive term to be included with any classification loss, designed on the basis of effective number of samples in a particular class. [2] and [11] both propose a modification in the margin of the class-boundary learned via minimizing intra-class variations and maximizing inter-class margin. [17] discusses a dynamic meta-embedding technique to address classification problem under long-tailed training data scenario.

Equipped with the knowledge of recent algorithms for both ZS-SBIR and single domain class imbalance mitigating techniques, we now move forward to discuss the problem of imbalanced training data for cross-domain retrieval.

3   Does Imbalanced Training Data Effect ZS-SBIR?

First, we analyze what is the effect of training data imbalance on generalization to unseen classes in the context of ZS-SBIR. Here, for ease of analysis, we consider only class imbalance, but our approach is effective for the mixed imbalance too, as justified by the experimental results later. Since both the standard datasets for this task, namely Sketchy Ext. [25] and TU-Berlin Ext. [8] are already imbalanced, to systematically study the effect of imbalance, we create a smaller balanced dataset, which is a subset of Sketchy Ext. dataset. This is termed as *mini*-**Sketchy Dataset** and contains sketches and images from 60 classes, with 500 images and sketches per class. Among them, randomly selected 10 classes are used as *unseen* classes and the rest 50 classes are used for training.

To study the effect of imbalance, motivated by the class-imbalance literature in image classification [11,14], we introduce two different types of class imbalance: 1) Step imbalance - where few of the classes in the training set contains less amount of samples compared to other classes; 2) Long-tailed imbalance - where the number of samples across the classes decrease gradually following the rule, $n_k^{lt} = n_k \mu^{\frac{k}{C_{seen}-1}}$; where n_k^{lt} is the available samples for k^{th} class under long-tailed distribution and n_k is the number of original samples of that class ($=500$ here). Here, $k \in \{1, 2, ..., C_{seen}\}$, i.e. C_{seen} is the number of training classes and $\mu = \frac{1}{p}$. We define imbalance factor p for a particular data-distribution to be the ratio of the highest number of samples in any class to the lowest number of samples in any class in that data and higher value of p implies more severe training class imbalance. Since the analysis is with class-imbalance, we assume that the data samples in image and sketch domain are the same.

As mentioned earlier, the proposed regularizer is generic and can be used with several baseline approaches to improve their performance in presence of data imbalance. For this analysis, we choose one recent auto-encoder based approach [7]. We term this as *Baseline Model* for this discussion, since the analysis is equally applicable for other approaches as well. We systematically introduce both the step and long-tailed imbalances for two different values of p and observe the performance for each of them. The results are reported in Table 1.

As compared to the balanced setting, we observe significant degradation in performance of the baseline whenever any kind of imbalance is present in the training data. This implies that training data imbalance not only effects the test performance when the classes remain the same, it also adversely effects the generalization performance significantly. This is due to the fact that unseen classes are recognized by embedding them close to their semantically relevant seen classes. Data imbalance results in (1) latent embedding space which is not sufficiently discriminative and (2) improperly learnt embedding functions, both of which negatively affects the embeddings of the unseen classes. The goal of the proposed AMDReg is to mitigate these limitations, which in turn will help in better generalization to unseen classes (Table 1 bottom row). Thus we see, that if the imbalance is handled properly, it may reduce the need for collecting large-scale balanced training samples.

Table 1. Evaluation (MAP@200) of Baseline Model [7] for ZS SBIR on *mini-Sketchy* dataset. Results for long-tailed and step imbalance with different imbalance factors are reported. The final performance using the proposed AMDReg is also compared.

Experimental protocol	Balanced data	Imbalanced data			
		Long-tailed		Step	
		$p = 10$	$p = 100$	$p = 10$	$p = 100$
Baseline [7]	0.395	0.234	0.185	0.241	0.156
Baseline [7] + AMDReg		**0.332**	**0.240**	**0.315**	**0.218**

4 Proposed Approach

Here, we describe the proposed Adaptive Margin Diversity Regularizer (AMDReg), which when used with existing ZS-SBIR approaches can help to mitigate the adverse effect of training data imbalance. We observe that majority of the state-of-the-art ZS-SBIR [6,7,16] approaches have two objectives: (1) projecting the sketches and images to a common discriminative latent space, where retrieval can be performed; (2) to ensure that the latent space is semantically meaningful so that the approach generalizes to unseen classes. For the first objective, a classification loss is used while learning the shared latent-space, which constraints the latent-space embeddings of both sketches and images from same classes to be clustered together, and samples from different classes to be well-separated. For the second objective, different direct or indirect techniques are utilized to make the embeddings semantically meaningful to ensure better generalization.

Semantically Meaningful Class Prototypes: Without loss of generality, we again chose the same baseline [7] to explain how to incorporate the proposed AMDReg into an existing ZS-SBIR approach. Let us consider that there are C_{seen} number of classes present in the dataset, and d is the latent space dimension. The baseline model has two parallel branches $F_{im}(\theta_{im})$ and $F_{sk}(\theta_{sk})$ for extracting features from images and sketches, $\{f^{(m)}\}$, where $m \in \{im, sk\}$, respectively. These features are then passed through corresponding content encoder networks to learn the shared latent-space embeddings for the same, i.e. $z^{(m)} = E_m(f^{(m)})$. In [7], a distance-based cross-entropy loss is used to learn these latent embeddings such that the embeddings is close to the semantic information. As is widely used, the class-name embeddings $h(y)$ of the *seen*-class labels $y \in \{1, 2, ..., C_{seen}\}$ are used as the semantic information. These embeddings are extracted from a pre-trained language model, such as, word2vec [18] or GloVe [20]. Please refer to Fig. 1 for illustration of the proposed AMDReg with respect to this baseline model.

The last fully connected (fc) layer of the encoders is essentially the classification layer and the weights of this layer, $\mathbf{P} = [\mathbf{p}_1, \mathbf{p}_2, ..., \mathbf{p}_{C_{seen}}], \mathbf{p}_i \in \mathbb{R}^d$ can be considered as the shared class-prototypes or the representatives of the corresponding class [21]. To ensure a semantically meaningful latent representation, one can learn the prototypes ($\mathbf{p}_i$'s) such that they are close to the class-name embeddings, or the prototypes can themselves be set equal to the semantic embeddings, i.e. $\mathbf{p}_i = h(y)$ and kept fixed. If the training data is imbalanced, just ensuring that the prototypes are semantically meaningful is not sufficient, we should also ensure that they take into account the label distribution of the training data. In our modification, to be able to adjust the prototypes properly, instead of fixing them as the class-embeddings, we initialize them using these attributes. Since the output of this fc layer is given by $\mathbf{z}^{(m)} = [z_1^{(m)}, z_2^{(m)}, ..., z_{C_{seen}}^{(m)}]$; the encoder with the prototypes is learnt using standard cross-entropy loss as,

$$\mathcal{L}_{CE}(\mathbf{z}^{(m)}, y) = -\log\frac{exp(z_y^{(m)})}{\sum\limits_{j=1}^{C_{seen}} exp(z_j^{(m)})} \tag{1}$$

Now, with this as the background, we will describe the proposed regularizer, AMDReg, which ensures that the prototypes are modified in such a way that they are spread out according to their class representation in the training set.

Adaptive Margin Diversity Regularizer: Our proposed AMDReg is inspired from the recently proposed Diversity Regularizer [11], which addresses data imbalance in image classification by adjusting the classifier weights (here prototypes) so that they are uniformly spread out in the feature space. In our context, it can be enforced by the following regularizer

$$\mathcal{R}(\mathbf{P}) = \frac{1}{C_{seen}}\sum_{i<j}[\|\mathbf{p}_i - \mathbf{p}_j\|_2^2 - d_{mean}]^2, \ \forall j \in \{1, 2, ..., C_{seen}\} \tag{2}$$

Here d_{mean} is the mean distance between all the class prototypes and is computed as

$$d_{mean} = \frac{2}{C_{seen}^2 - C_{seen}}\sum_{i<j}\|\mathbf{p}_i - \mathbf{p}_j\|_2^2, \ \forall j \in \{1, 2, ..., C_{seen}\} \tag{3}$$

The above regularizer tries to spread out all the class prototypes, without considering the amount of imbalance present in the training data. As has been observed in many recent works [2], due to insufficient number of samples of the minority classes, it is more likely that their test samples will have a wider spread instead of being clustered around the class prototype during testing. For our problem, this implies greater uncertainty for samples of unseen classes, which are semantically similar to the minority classes in the training set.

Towards this end, we propose to adjust the class prototypes adaptively, which takes into account the data imbalance. Since there can be both class and domain imbalance in the cross-domain retrieval problem, we propose to use the total number of sketches and image samples per class in the training set, and we refer to this combined number for k^{th}-class as the *effective* number of samples, n_k^{eff}, in this work. We then define the imbalance-based margin for the kth class as,

$$\Delta_k = \frac{K}{n_k^{eff}} \tag{4}$$

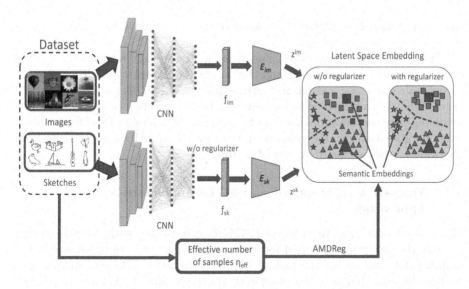

Fig. 1. Illustration of the proposed Adaptive Margin Diversity Regularizer (AMDReg). The AMDReg ensures that the embeddings of the shared prototypes of the images and sketcheses are not only placed away from each other, but also accounts for the increased uncertainty when the training class distribution is imbalanced. This results in better generalization to unseen classes.

This is similar to the inverse frequency of occurrence, except for the experimental hyper-parameter K. Thus the final AMDReg is given by

$$\mathcal{R}_\Delta(\mathbf{P}) = \frac{1}{C_{seen}} \sum_{i<j} [\|\mathbf{p}_i - \mathbf{p}_j\|_2^2 - (d_{mean} + \Delta_j)]^2, \ \forall j \in \{1, 2, ..., C_{seen}\} \quad (5)$$

Thus, we adjust the relative distance between $\mathbf{p}_i$'s such that they are atleast separated by a distance which is more than the mean-distance by the class imbalance margin. This ensures that the prototypes for the minority classes have more margin around them, which will reduce the chances of confusion for the semantically similar unseen classes during testing. Finally, the encoder with the prototypes are learnt using the CE loss along with the AMDReg as

$$\mathcal{L}_{CE}^{AMDReg} = \mathcal{L}_{CE} + \lambda \mathcal{R}_\Delta \quad (6)$$

where λ is an experimental hyper-parameter, which controls the contribution of the regularizer towards the learning.

Difference with Related Work: Even though the proposed AMDReg is inspired from [11], there are significant differences, namely (1) [11] addresses the imbalanced classification task for a single domain, while our work addresses generalization to unseen classes in the context of cross-domain retrieval (ZS-SBIR); (2) While [11] ensures that the weight vectors are equally spread out, AMDReg accounts for the training data distribution while designing the relative distances between the semantic embeddings; (3) Finally, [11] works with the max-margin loss, but AMDReg is used with the standard CE loss while learning the semantic class prototypes.

The proposed approach also differs from another closely related work LDAM [2]. LDAM loss is a modification on the standard cross-entropy or Hinge-loss to incorporate class-wise margin. In contrast, proposed AMDReg is a margin-based regularizer with adaptive margins between class-prototypes, based on the corresponding representation of classes in the training set. Thus, while [2] is inspired from margin-based generalization bound, the proposed AMDReg is inspired from the widely used inverse frequency of occurance.

4.1 Analysis with Standard and SOTA Imbalance-Aware Approaches

Here, we analyze how the proposed AMDReg compares with several existing state-of-the-art techniques used for addressing the problem of imbalance in the training data mainly for the task of image classification. These techniques can be broadly classified into two categories, (1) re-sampling techniques to balance the existing imbalanced dataset and (2) cost-sensitive learning or modification of the classifier. For this analysis also, we use the same retrieval backbone [7]. In this context, we first compute the average number of samples in the dataset. Any class which has lesser number of samples than the average are considered *minority* classes, and the remaining are considered *majority* classes.

1) **Re-balancing the Dataset:** Re-sampling is a standard and effective technique used to balance out the dataset distribution bias. The most common methods are under-sampling of the majority classes [1] or over-sampling of minority classes [3]. We systematically use such imbalance data-sampling techniques on the training data to address the class imbalance for ZS-SBIR as discussed below. Here, the re-sampled/balanced dataset created by individual re-sampling operations described below is used for training the baseline network and reporting the retrieval performance.

1. *Naive Under-Sampling:* Here, we randomly select $1/p$-th of total samples per class for the majority classes and discard their remaining samples. Naturally, we loose a significant amount of important samples with such random sampling technique.

2. *Selective Decontamination* [1]: This technique is used to intelligently under-sample the majority classes instead of randomly throwing away excess samples. As per [1], we also modify the Euclidean distance function $d_E(\mathbf{x}_i, \mathbf{x}_j)$ between two samples of cth class, $\mathbf{x}_i$ and $\mathbf{x}_j$ as,

$$d_{modified}(\mathbf{x}_i, \mathbf{x}_j) = \left(\frac{n_c}{N}\right)^{(1/m)} d_E(\mathbf{x}_i, \mathbf{x}_j) \tag{7}$$

where n_c and N are the number of samples in c^{th} class and in all classes, respectively. m represents the dimension of the feature space. We retain only those samples in the majority classes for which the classes of majority of samples in top-K nearest neighbors agree completely.

3. **Naive Over-Sampling:** Here, the minority classes are augmented by repeating the instances (as in [35]) and using the standard image augmentation techniques (such as, rotation, translation, flipping etc.).

4. **SMOTE** [3]: In this intelligent over-sampling technique, instead of replacing the samples, the minority classes are augmented by generating *synthetic* features along the line-segment joining each minority class sample with its K-nearest neighbors.

5. **GAN Based Augmentation** [29]: Finally, we propose to augment the minority classes by generating features with the help of generative models, which have been very successful for zero-shot [29]/few-shot [19]/any-shot [30] image classification. Towards that goal, we use f-GAN [29] model to generate synthetic features for the minority classes using their attributes and augment those features with the available training dataset to reduce the imbalance.

2) **Cost-Sensitive Learning of Classifier:** The goal of cost-sensitive learning based methods is to learn a better classifier using the original imbalanced training data, but with a more suitable loss function which can account for the data imbalance. To observe the effect of the different kinds of losses, we modify the distance-based CE-loss in the baseline model to the following ones, keeping the rest of the network fixed.

1. **Focal Loss:** This loss [14] was proposed to address foreground-background class imbalance issue in the context of object detection. It is based on a simple yet effective modification of standard cross-entropy loss, such that while computation, the *easy* or well-classified samples are given less weights compared to the difficult samples.

2. **Class-Balanced Focal Loss:** It is a variant of focal loss, recently proposed in [4], which incorporates the effective number of samples for a class in the imbalanced dataset.

3. **Diversity Regularizer:** This recently proposed regularizer [11] ensures that both the majority and minority classes are at equal distance from each other in the latent-space and reported significant performance improvement for imbalanced training data for image classification.

4. **LDAM:** [2] proposes a margin-based modification of standard cross-entropy loss or hinge loss, to ensure that the classes are well-separated from each other.

The retrieval performance obtained with these imbalance-handling methods are reported in Table 2. We observe that all the techniques result in varying degree of improvement over the base model. Among the data augmentation techniques, GAN-based augmentation outperforms the other approaches. In general, all the cost-sensitive learning techniques perform quite well, specially the recently proposed diversity regularizer and the LDAM cross-entropy loss. However, the proposed AMDReg outperforms both the data balancing and cost-sensitive learning approaches, giving the best performance across all types and degrees of imbalance.

Table 2. ZS-SBIR performance (MAP@200) of different kinds of imbalance handling techniques applied on the Baseline Model [7] for the *mini-Sketchy* dataset. Results of the original Baseline Model is also reported for reference.

Imbalance handler	Methods	Long-tailed		Step	
		$p = 10$	$p = 100$	$p = 10$	$p = 100$
	Baseline model [7]	0.234	0.185	0.241	0.156
Data balancing methods	Naive under-sampling	0.235	0.191	0.256	0.159
	Naive over-sampling	0.269	0.219	0.258	0.155
	Selective decontamination [1]	0.268	0.221	0.251	0.164
	SMOTE [3]	0.269	0.217	0.269	0.183
	GAN-based augmentation [29]	0.305	0.229	0.274	0.188
Loss-modification techniques	Focal loss [14]	0.273	0.228	0.289	0.195
	Class-balanced focal loss [4]	0.299	0.236	0.296	0.210
	Diversity-regularizer [11]	0.296	0.222	0.285	0.207
	LDAM-CE loss [2]	0.329	0.234	0.310	0.213
	Proposed AMDReg	**0.332**	**0.240**	**0.315**	**0.218**

5 Experimental Evaluation on ZS-SBIR

Here, we provide details of the extensive experiments performed to evaluate the effectiveness of the proposed AMDReg for handing data imbalance in ZS-SBIR.

Datasets Used and Experimental Protocol: We have used two large-scale standard benchmarks for evaluating ZS-SBIR approaches, namely, Sketchy Ext. [25] and TU-Berlin Ext. [8].

Sketchy Ext. [25] dataset originally contained approximately 75,000 sketches and 12,500 images from 125 object categories. Later, [15] collected and added additional 60,502 images to this dataset. Following the standard protocol [6,16], we randomly choose 25 classes as *unseen*-classes (sketches as query and images in the search set) and the rest 100 classes for training.

TU-Berlin Ext. [8] originally contained 80 hand-drawn sketches per class from total 250 classes. To make it a better fit for large-scale experiments, [34] included additional 2,04,489 images. As followed in literature [6,7], we randomly select 30-classes as *unseen*, while the rest 220-classes are used for training.

The dataset statistics are shown in Fig. 2, which depicts data imbalance in both the datasets. This is specially evident in TU-Berlin Ext., which has huge domain-wise imbalance as well as class-wise imbalance. These real-world datasets reinforce the importance of handling data imbalance for the ZS-SBIR task.

5.1 State-of-the-Art ZS-SBIR Approaches Integrated with AMDReg

As already mentioned, the proposed AMDReg is generic and can be seamlessly integrated with most state-of-the-art ZS-SBIR approaches for handling the

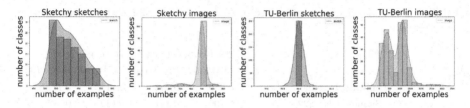

Fig. 2. Dataset statistics of sketches and images of Sketchy-extended and TU-Berlin-extended are shown in the first two and last two plots respectively in that order.

training data imbalance. Here, we have integrated AMDReg with three state-of-the-art approaches, namely (1) Semantically-tied paired cycle-consistency based network (SEM-PCYC) [6]; (2) Semantic-aware knowledge preservation for ZS-SBIR (SAKE) [16]. (3) Style-guided network for ZS-SBIR [7]. Now, we briefly describe the three approaches along with the integration of AMDReg.

SEM-PCYC [6] with AMDReg: SEM-PCYC is a generative model with two separate branches for image and sketch; for visual-to-semantic mapping along with cyclic consistency loss. Further, to ensure that the semantic output of the generators is also class-discriminative, a classification loss is used. This classifier is pre-trained on *seen*-class training data and kept frozen while the whole retrieval model is trained. We modify the training methodology by enabling the classifier to train along with the rest of the model, by including the AMDReg with the CE-loss. Here, the semantic information is enforced through an auto-encoder, which uses a hierarchical and a text-based model as input, and thus the weights are randomly initialized. Please refer to [6] for more details.

SAKE [16] with AMDReg: This ZS-SBIR method extends the concept of domain-adaptation for fine-tuning a pre-trained model on ImageNet [23] for the specific ZS-SBIR datasets. The network contains a shared branch to extract features from both sketches and images, which are later used for the categorical classification task using the soft-max CE-loss. Simultaneously, the semantic structure with respect to the ImageNet [23] classes are maintained. Here also, we modify the CE-loss using the proposed AMDReg to mitigate the adverse effect of training data imbalance. The rest of the branches and the proposed SAKE-loss remain unchanged. Please refer to [16] for more details of the base algorithm.

Style-Guide [7] with AMDReg: This is a two-step process, where the shared latent-space is learnt first. Then, the latent-space content extracted from the sketch query is combined with the styles of the relevant images to obtain the final retrieval in the image-space. While learning the latent-space, a distance-based cross-entropy loss is used, which is modified as explained in details earlier. Please refer to [7] for more details of the base algorithm.

Implementation Details: The proposed regularizer is implemented using Pytorch. We use a single Nvidia GeForce GTX TITAN X for all our experiments. For all the experiments, we set $\lambda = 10^3$ and $K = 1$. Adam optimizer has been used with $\beta_1 = 0.5$, $\beta_2 = 0.999$ and a learning rate of $lr = 10^{-3}$. The

Table 3. Performance of several state-of-the-art approaches for ZS-SBIR and generalized ZS-SBIR.

	Algorithms	TU-Berlin extended		Sketchy-extended	
		MAP@all	Prec@100	MAP@all	Prec@100
SBIR	Softmax baseline	0.089	0.143	0.114	0.172
	Siamese CNN [22]	0.109	0.141	0.132	0.175
	SaN [33]	0.089	0.108	0.115	0.125
	GN triplet [25]	0.175	0.253	0.204	0.296
	3D shape [28]	0.054	0.067	0.067	0.078
	DSH (binary) [15]	0.129	0.189	0.171	0.231
	GDH (binary) [35]	0.135	0.212	0.187	0.259
ZSL	CMT [27]	0.062	0.078	0.087	0.102
	DeViSE [10]	0.059	0.071	0.067	0.077
	SSE [36]	0.089	0.121	0.116	0.161
	JLSE [37]	0.109	0.155	0.131	0.185
	SAE [13]	0.167	0.221	0.216	0.293
	FRWGAN [9]	0.110	0.157	0.127	0.169
	ZSH [31]	0.141	0.177	0.159	0.214
Zero-Shot	ZSIH (binary) [26]	0.223	0.294	0.258	0.342
SBIR	ZS-SBIR [32]	0.005	0.001	0.196	0.284
	SEM-PCYC [6]	0.297	0.426	0.349	0.463
	SEM-PCYC + AMDReg	**0.330**	**0.473**	**0.397**	**0.494**
	Style-guide [7]	0.254	0.355	0.375	0.484
	Style-guide + AMDReg	**0.291**	**0.376**	**0.410**	**0.512**
	SAKE [16]	0.428*	0.534*	0.547	0.692
	SAKE + AMDReg	**0.447**	**0.574**	**0.551**	**0.715**
Generalized	Style-guide [7]	0.149	0.226	**0.330**	0.381
Zero-shot	SEM-PCYC [6]	0.192	0.298	0.307	0.364
SBIR	**SEM-PCYC + AMDReg**	**0.245**	**0.303**	0.320	**0.398**

implementation of different baselines and the choice of hyper-parameters for their implementation has been done as described in the corresponding papers.

5.2   Evaluation for ZS-SBIR

Here, we report the results of the modifications to the state-of-the-art approaches for ZS-SBIR. We first train all the three original models (as described before) to replicate the results reported in the respective papers. We use the codes given by the authors and are able to replicate all the results for SEM-PCYC and Style-guide as reported. However, for SAKE, in two cases, the results we obtained are slightly different from that reported in the paper. So we report the results as we obtained, for fair evaluation of proposed improvement (marked with a star to indicate that they are different from the reported numbers in the paper).

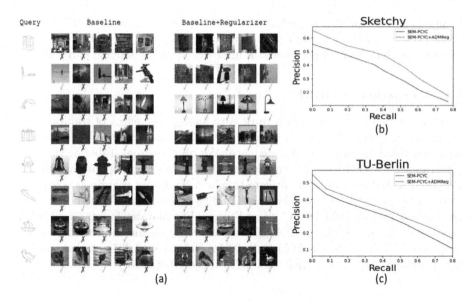

Fig. 3. Performance comparison of the base model (SEM-PCYC) and the modified base-model using proposed AMDReg: (a) Few examples of top-5 retrieved images against the given unseen sketch query from TU-Berlin dataset; (b) P-R curve on Sketchy dataset; (c) P-R curve on TU-Berlin dataset.

We incorporate the proposed modifications for AMDReg in all three approaches and retrained the models. The results are reported in Table 3. All the results of the other approaches are taken directly from [6]. We observe significant improvement in the performance of all the state-of-the-art approaches, when trained using the proposed regularizer. This experiment throws insight that by handling the data-imabalance, which is inherently present in the collected data, it is possible to gain siginificant improvement in the final performance. Since AMDReg is generic, it can potentially be incorporated with other approaches, developed for the ZS-SBIR task, to handle the training data imbalance problem.

Figure 3 shows top-5 retrieved results for a few unseen queries (first column), using SEM-PCYC as the baseline model, without and with AMDReg, respectively. We observe significant improvement when AMDReg is used, justifying its effectiveness. We make similar observations from the P-R curves in Fig. 3.

5.3 Evaluation for Generalized ZS-SBIR

In real scenarios, the search set may consist of both the seen and unseen image samples, which makes the problem much more challenging. This is termed as the generalized ZS-SBIR. To evaluate the effectiveness of proposed AMDReg for this scenario, we follow the experimental protocol in [6] and SEM-PCYC [6] as the base model. From the results in Table 3, we observe that AMDReg is able to significantly improve the performance of the base model and yields state-of-the-art results for three out of the four cases. Only for Sketchy Ext., it performs slightly less than Style-Guide, but still improves upon its baseline performance.

5.4   Evaluation for SBIR

Though the main purpose of this work is to analyze the effect of training data imbalance on generalization to unseen classes, this approach should also benefit standard SBIR in presence of imbalance. We observe from Table 4, that the performance of SBIR indeed decreases drastically with training data imbalance. Proposed AMDReg is able to mitigate this by a significant margin as compared to the other state-of-the-art imbalance handing techniques. We further analyze the performance of SEM-PCYC [6] on Sketchy Ext. dataset for standard SBIR protocol with and without AMDReg. We observe significant improvement when proposed AMDReg is used (MAP@all: 0.811; Prec@100: 0.897) as compared to the baseline SEM-PCYC (MAP@all: 0.771; Prec@100: 0.871).

Table 4. SBIR evaluation (MAP@200) of Baseline Model [7] on *mini-Sketchy*.

Balanced data	Step imb $(p = 100)$	GAN-based Aug. [29]	CB focal loss [4]	Diversity regularizer [11]	Proposed AMDReg
0.839	0.571	0.580	0.613	0.636	**0.647**

6   Conclusion

In this work, for the first time in literature, we analyzed the effect of training data imbalance for the task of generalization to unseen classes in context of ZS-SBIR. We observe that most real-world SBIR datasets are in-fact imbalanced, and that this imbalance does effect the generalization adversely. We systematically evaluate several state-of-the-art imbalanced mitigating approaches (for classification) for this problem. Additionally, we propose a novel adaptive margin diversity regularizer (AMDReg), which ensures that the shared latent space embeddings of the images and sketches account for the data imbalance in the training set. The proposed regularizer is generic, and we show how it can be seamlessly incorporated in three existing state-of-the-art ZS-SBIR approaches with slight modifications. Finally, we show that the proposed AMDReg results in significant improvement in both ZS-SBIR and generalized ZS-SBIR protocols, setting the new state-of-the-art result.

Acknowledgement. This work is partly supported through a research grant from SERB, Department of Science and Technology, Government of India.

References

1. Barandela, R., Rangel, E., Sanchez, J.S., Ferri, F.J.: Restricted decontamination for the imbalanced training sample problem. Springer, Iberoamerican Congress on Pattern Recognition (2003)
2. Cao, K., Wei, C., Gaidon, A., Arechiga, N., Ma, T.: Learning imbalanced datasets with label-distribution-aware margin loss. In: NeurIPS (2019)
3. Chawla, N.V., Bowyer, K.W., Hall, L.O., Kegelmeyer, W.P.: Smote: synthetic minority over-sampling technique. J. Artifi. Intell. Res. **16**, 321–357 (2002)
4. Cui, Y., Jia, M., Lin, T.Y., Song, Y.: Class-balanced loss based on effective number of samples. In: CVPR (2019)
5. Dey, S., Riba, P., Dutta, A., Llados, J., Song, Y.Z.: Doodle to search: practical zero-shot sketch-based image retrieval. In: CVPR (2019)
6. Dutta, A., Akata, Z.: Sematically tied paired cycle consistency for zero-shot sketch-based image retrieval. In: CVPR (2019)
7. Dutta, T., Biswas, S.: Style-guided zero-shot sketch-based image retrieval. In: BMVC (2019)
8. Eitz, M., Hays, J., Alexa, M.: How do humans sketch objects? ACM TOG **31**(4), 1–10 (2012)
9. Felix, R., Kumar, V.B., Reid, I., Carneiro, G.: Multi-modal cycle-consistent generalized zero-shot learning. In: ECCV (2018)
10. Frome, A., et al.: Devise: A deep visual-semantic embedding model. In: NeurIPS (2013)
11. Hayat, M., Khan, S., Zamir, S.W., Shen, J., Shao, L.: Gaussian affinity for max-margin class imbalanced learning. In: ICCV (2019)
12. Hu, R., Collomosse, J.: A performance evaluation of gradient field hog descriptor for sketch based image retrieval. CVIU **117**(7), 790–806 (2013)
13. Kodirov, E., Xiang, T., Gong, S.: Semantic autoencoder for zero-shot learning. In: CVPR (2017)
14. Lin, T.Y., Goyal, P., Girshiick, R., He, K., Dollar, P.: Focal loss for dense object detection. arXiv:1708.02002 [cs.CV] (2018)
15. Liu, L., Shen, F., Shen, Y., Liu, X., Shao, L.: Deep sketch hashing: fast free-hand sketch-based image retrieval. In: CVPR (2017)
16. Liu, Q., Xie, L., Wang, H., Yuille, A.: Semantic-aware knowledge preservation for zero-shot sketch-based image retrieval. In: ICCV (2019)
17. Liu, Z., Miao, Z., Zhan, X., Wang, J., Gong, B., Yu, S.X.: Large-scale long-tailed recognition in an open world. In: CVPR (2019)
18. Mikolov, T., Sutskever, I., Chen, K., Corrado, G.S., Dean, J.: Distributed representations of words and phrases and their compositionality. NeurIPS (2013)
19. Mishra, A., Reddy, S.K., Mittal, A., Murthy, H.A.: A generative model for zero-shot learning using conditional variational auto-encoders. CVPR-W (2018)
20. Pennington, J., Socher, R., Manning, C.D.: Glove: global vectors for word representation. In: EMNLP (2014)
21. Qi, H., Brown, M., Lowe, D.G.: Low-shot learning with imprinted weights. In: CVPR (2018)
22. Qi, Y., Song, Y.Z., Zhang, H., Liu, J.: Sketch-based image retrieval via siamese convolutional neural network. In: ICIP (2016)
23. Russakovsky, O., Deng, J., Su, H., Krause, J., Satheesh, S., Ma, S., Huang, Z., Karpathy, A., Khosla, A., Bernstein, M., Berg, A.C., Li, F.F.: Imagenet: large-scale visual recognition challenge. IJCV **115**(3), 211–252 (2015)

24. Saavedra, J.M., Barrios, J.M.: Sketch-based image retrieval using learned keyshapes (lks). In: BMVC (2015)
25. Sangkloy, P., Burnell, N., Ham, C., Hays, J.: The sketchy database: learning to retrieve badly drawn bunnies. ACM TOG **35**(4), 1–12 (2016)
26. Shen, Y., Liu, L., Shen, F., Shao, L.: Zero-shot sketch-image hashing. In: CVPR (2018)
27. Socher, R., Ganjoo, M., Manning, C.D., Ng, A.: Zero-shot learning through cross-modal transfer. In: NeurIPS (2013)
28. Wang, M., Wang, C., Wu, J.X., Zhang, J.: Community detection in social networks: an in-depth benchmarking study with a procedure-oriented framework. In: VLDB (2015)
29. Xian, Y., Lorenz, T., Schiele, B., Akata, Z.: Feature generating networks for zero-shot learning. In: CVPR (2018)
30. Xian, Y., Sharma, S., Schiele, B., Akata, Z.: f-vaegan-d2: A feature generating framework for any-shot learning. In: CVPR (2019)
31. Yang, Z., Cohen, W.W., Salakhutdinov, R.: Revisiting semi-supervised learning with graph embeddings. arXiv preprint arXiv:1603.08861 (2016)
32. Yelamarthi, S.K., Reddy, S.K., Mishra, A., Mittal, A.: A zero-shot framework for sketch-based image retrieval. In: ECCV (2018)
33. Yu, Q., Yang, Y., Liu, F., Song, Y.Z., Xiang, T., Hospedales, T.M.: Sketch-a-net that beats humans. In: BMVC (2015)
34. Zhang, J., Liu, S., Zhang, C., Ren, W., Wang, R., Cao, X.: Sketchnet: sketch classification with web images. In: CVPR (2016)
35. Zhang, J., et al.: Generative domain-migration hashing for sketch-to-image retrieval. In: ECCV (2018)
36. Zhang, R., Lin, L., Zhang, R., Zuo, W., Zhang, L.: Bit-scalable deep hashing with regularized similarity learning for image retrieval and person re-identification. IEEE Trans. Image Process. **24**(12), 4766–4779 (2015)
37. Zhang, Z., Saligrama, V.: Zero-shot learning via joint latent similarity embedding. In: CVPR (2016)

ETH-XGaze: A Large Scale Dataset for Gaze Estimation Under Extreme Head Pose and Gaze Variation

Xucong Zhang[1($\boxtimes$)], Seonwook Park[1], Thabo Beeler[2], Derek Bradley[3], Siyu Tang[1], and Otmar Hilliges[1]

[1] Department of Computer Science, ETH Zurich, Zürich, Switzerland
{xucong.zhang,spark,siyu.tang,otmar.hilliges}@inf.ethz.ch
[2] Google Inc., Zürich, Switzerland
tbeeler@google.com
[3] Zürich, Switzerland

Abstract. Gaze estimation is a fundamental task in many applications of computer vision, human computer interaction and robotics. Many state-of-the-art methods are trained and tested on custom datasets, making comparison across methods challenging. Furthermore, existing gaze estimation datasets have limited head pose and gaze variations, and the evaluations are conducted using different protocols and metrics. In this paper, we propose a new gaze estimation dataset called ETH-XGaze, consisting of over one million high-resolution images of varying gaze under extreme head poses. We collect this dataset from 110 participants with a custom hardware setup including 18 digital SLR cameras and adjustable illumination conditions, and a calibrated system to record ground truth gaze targets. We show that our dataset can significantly improve the robustness of gaze estimation methods across different head poses and gaze angles. Additionally, we define a standardized experimental protocol and evaluation metric on ETH-XGaze, to better unify gaze estimation research going forward. The dataset and benchmark website are available at https://ait.ethz.ch/projects/2020/ETH-XGaze.

1 Introduction

Estimating eye-gaze from monocular images alone has recently received significant interest in computer vision [9,35,38] due to its significance in many application domains ranging from the cognitive sciences and HCI to robotics and semi-autonomous driving [7,23,32]. Many arising computing paradigms such as smart-home appliances, autonomous cars and robots, as well as body-worn cameras will rely on understanding the attention and intent of humans without directly interacting with the observed person. We argue that in order to be more robust to a larger variety of environmental conditions, future methods should be

Electronic supplementary material The online version of this chapter (https://doi.org/10.1007/978-3-030-58558-7_22) contains supplementary material, which is available to authorized users.

© Springer Nature Switzerland AG 2020
A. Vedaldi et al. (Eds.): ECCV 2020, LNCS 12350, pp. 365–381, 2020.
https://doi.org/10.1007/978-3-030-58558-7_22

able to accurately estimate the gaze of humans in a broader range of settings, including variation of viewpoint, extreme gaze angles, lighting variation, input image resolutions, and in the presence of occluders such as glasses.

Unfortunately, existing gaze datasets do not cater to such use-cases and are mostly limited to the frontal setting, covering a relatively narrow range of head poses and gaze directions. These are typically collected via laptops [42], mobile devices [13,19] or in stationary settings [10]. Recent work has moved towards more unconstrained environmental conditions in particular with respect to lighting but the coverage of head pose and gaze direction ranges remains limited [9,16,40].

In this paper we detail a new dataset, dubbed ETH-XGaze, to facilitate research into robust gaze estimation methods. The dataset exhaustively samples large variations in head poses, up to the limit of where both eyes are still visible (maximum $\pm70°$ from directly facing the camera) as well as comprehensive gaze directions (maximum $\pm50°$ in the head coordinate system) [27]. The dataset will allow for the development of new methods that can robustly estimate gaze direction without requiring a quasi-frontal camera placement. We show experimentally that i) the data distribution of ETH-XGaze is more comprehensive than other datasets (e.g., our dataset broadens the scope for eye-gaze research), and ii) that training on our dataset significantly improves robustness towards head pose and gaze direction variations. Beyond extending the gaze and head-pose ranges, the proposed dataset allocates considerably more pixels to the periocular region compared to existing datasets (e.g. refer to Fig. 3). This allows to train gaze estimators that can take advantage of the high-resolution imagery of modern camera hardware to improve gaze prediction. We collect data from 110 participants with different ethnicity, age, and gender – some with glasses and some without – in order to provide a rich and diverse dataset. For each of the participants we capture over 500 gaze directions with full-on illumination, plus an additional 90 samples under 15 different illumination conditions. This results in a total of over 1 million labeled samples. For all samples, the ground-truth gaze direction is known since the gaze is guided by stimuli displayed on a large screen in front of the participant, ensuring good label quality even under extreme view angles. The capture setup is depicted in Fig. 1 (left).

To ensure fair and systematic comparisons between future methods that leverage this new large-scale dataset, we also propose a standardized evaluation protocol. Unlike other fields in computer vision that have benefited from such benchmark frameworks (i.e. image classification [28], face recognition [24], full-body [15], hand pose estimation [43] and multiview stereo reconstruction [29]), the gaze estimation community has so far relied on a heterogeneous environment where many papers employ custom data pre-processing and evaluation protocols, rendering direct comparisons challenging. Motivated by the benchmarking approaches in adjacent areas we create a website open to the public to submit, evaluate and compare gaze estimation methods based on ETH-XGaze.

Finally, in order to provide initial insights into the value of our dataset, we provide results from a simple gaze estimation method that can serve as a baseline. Our estimation approach leverages a standard CNN architecture (i.e., ResNet-50 [11]), trained with the task of estimating gaze from a monocular face patch. We present

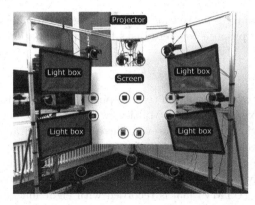 

Fig. 1. Our data collection device includes 18 high-resolution Canon 250D digital SLR cameras (marked with red circles), a projector to project the stimuli on the screen, and four Walimex Daylight 250 light boxes. A chair with a head rest is positioned approximately one meter away from the screen. Captured samples under different head poses and lighting conditions are shown on the right. (Color figure online)

the estimation results as well as an ablation study of training on different subsets of our dataset, indicating the importance of all sampled dimensions (e.g. head pose and gaze angles, number of subjects, lighting conditions and input image resolution). We hope this baseline method and evaluations will inspire future research in gaze estimation using our ETH-XGaze dataset.

In summary, our contribution is three-fold:

- A large scale dataset (over 1 Mio samples) for gaze estimation covering a large head pose and gaze range from 110 participants of different age, gender and ethnicity with consistent label quality and high-resolution images.
- Standardized experimental protocol and evaluation metrics including a new robustness evaluation.
- Detailed analysis on different factors for gaze estimation training.

2   Related Work

2.1   Gaze Estimation Algorithms

Initial learning-based gaze estimation methods often assume a static head pose [1,21], with later works allowing for gradually more head pose freedom [22,33]. In parallel, gaze estimation errors on public datasets have improved rapidly in recent years, through the use of domain adaptation [30], Bayesian networks [34], adversarial approaches [35], coarse-to-fine [6], and multi-region CNNs [9, 19]. Recent development in the person-specific adaptation of gaze estimators [20,25,38] are quickly reducing error metrics on public datasets even further. However, gaze-estimation is studied mostly in the frontal setting which does not apply to many emerging application domains. There is hence a need for a systematic method to understanding the robustness of a model with regards to gaze direction and head orientation ranges. We thus propose our gaze estimation dataset to cover these factors and propose concrete tasks for their evaluation.

2.2   Gaze Datasets

Newly introduced datasets in any area of research tend to push the limits of the data distribution represented in existing datasets. Multi-view cameras have been used to cover head poses in previous works. However, there are limited range of head poses [31], or limited effective resolution on face region using machine vision cameras [33] or wide-angle cameras [40]. The Columbia dataset uses five high-resolution camera while only 5,880 samples with discrete gaze directions are recorded [31]. UT Multi-view (UTMV) [33] is recorded with eight machine version cameras, and HUMBI is recorded with multiple wide-angle cameras, however, their resolution of eye region is small in the captured image. Capturing different head poses with a single camera can be achieved by asking participants to explicitly move their head during recording as in EYEDIAP [10], moving the camera and gaze target around the participant as in Gaze360 [16], or both as in RT-GENE [9]. Some of these approaches do result in lower resolution images, and as such are not informative in the development of generative models [30] or gaze redirection methods [12,38]. Therefore, these methods had to revert to the synthetic data from UnityEye [37] or the relatively small Columbia datasets [31]. In addition, it is more challenging to aim for the acquisition of a balanced dataset in terms of head pose and gaze estimation ranges when capturing in the wild (cf. [16,19,42]), as is later shown in this paper in parameter range comparisons between our proposed dataset and existing ones. Our high resolution dataset tackles the mentioned challenge of limited head pose and gaze direction ranges in existing datasets, taking meaningful steps towards constructing a balanced set of training data for learning high performance and robust gaze estimation models. Furthermore, we see potential in leveraging the high quality imagery to enable future work in areas adjacent to gaze-estimation such as generative modeling of the eye-region, Computer Graphics and facial reconstruction.

A comprehensive summary of current gaze estimation datasets in relationship to ours is shown in Table 1.

2.3   Evaluation Protocols

Having public benchmark frameworks for evaluation of popular algorithms is common for many computer vision tasks such as image classification [28], face recognition [17], pedestrian detection [8] and hand pose estimation [43]. Unfortunately, there is neither a unified evaluation protocol for gaze estimation nor an existing dataset that can serve as a general evaluation platform. Despite existing best practices, most previous work relies on their own data pre-processing and sometimes uses different training-test splits for evaluation. To provide a platform for gaze estimation evaluation, we share our dataset ETH-XGaze and define a set of clearly defined evaluation procedures. Furthermore, an online evaluation system and public leader-board are released along with the dataset).

Table 1. Overview of popular gaze estimation datasets showing the number of participants, the maximum head poses and gaze in horizontal (around yaw axis) and vertical (around pitch axis) directions in the camera coordinate system, amount of data (number of images or duration of video), and image resolution.

	# Peo.	Maximum head pose	Maximum gaze	# Data	Resolution
Columbia [31]	56	0°, ±30°	±15°, ±10°	5,880	5184 × 3456
UTMV [33]	50	±36°, ±36°	±50°, ±36°	64,000	1280 × 1024
EYEDIAP [10]	16	±15°, 30°	±25°, 20°	237 min	HD & VGA
MPIIGaze [42]	15	±15°, 30°	±20°, ±20°	213,659	1280 × 720
GazeCapture [19]	1,474	±30°, 40°	±20°, ±20°	2,445,504	640 × 480
RT-GENE [9]	15	±40°, ±40°	±40°, −40°	122,531	1920 × 1080
Gaze360 [16]	238	±90°, unknown	±140°, −50°	172,000	4096 × 3382
ETH-XGaze	110	**±80°, ±80°**	**±120°, ±70°**	1,083,492	**6000 × 4000**

3   ETH-XGaze Dataset

There are several parameters that define a comprehensive gaze estimation dataset, including: head pose, gaze direction, subject appearance, illumination condition, and image resolution. We design the ETH-XGaze data collection procedure with the main objective to maximize the parameter range along each of those dimensions as much as possible.

3.1   Acquisition Setup

The setup used for data collection is shown in the left of Fig. 1. We capture the subject with 18 Canon 250D digital SLR cameras from different viewpoints to cover a large range of head poses. There are five paired cameras for geometry capture and eight cameras for texture acquisition, such as to enable 3D face reconstruction in the future. The resolution of the captured images is very high (6000 × 4000 pixels). All cameras are connected via ESPER trigger boxes[1] to a Raspberry Pi, and a wireless mouse is used to send the triggering signal to the Raspberry Pi. The delay between mouse click and triggering the camera is below 0.05 s. A large screen (120 × 100 cm) is placed in the center of the cameras to show the stimuli controlled by the Raspberry Pi and projected by a projector. Since some cameras are placed behind the screen, we create cutout holes for their lenses. There are four light boxes (Walimex Daylight 250) surrounding the screen and each of them is equipped with a light bulb that emits ∼4500lm. The Raspberry Pi can turn each of the light boxes on or off to simulate different illumination conditions. We mount polarization filters in front of both the light box and camera and carefully adjust the filter angle to attenuate specular reflection off the face of the participants. During recording, the participants are sitting at approximately one meter distance in front of the screen, with the head placed in a head rest to reduce unintentional head motion.

---

[1] https://www.esperhq.com.

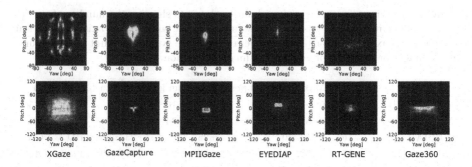

Fig. 2. Head pose (top row) and gaze direction (bottom row) distributions of different datasets. The head pose of Gaze360 is not shown here since it is not provided by the dataset.

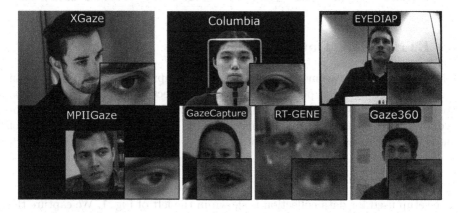

Fig. 3. Data examples and corresponding cropped eye images from different gaze estimation datasets. ETH-XGaze has the highest image resolution and quality.

3.2 Collection Procedure

During data collection, the participant focuses on a shrinking circle and clicks the mouse when the circle becomes a dot, providing the gaze point. The position of gaze points are randomly distributed on the screen. We have three methods to ensure the participant is looking at the dot when clicking the mouse. First, the participant has a short time window of 0.5 s to click the mouse to successfully collect one sample. Second, the shrinking time of the circle is random such that the participant has to focus on the shrinking circle to avoid missing the triggering time window. Third, the participant is told to collect a fixed amount of samples and any missing mouse click will increase the collection time. For most of the data collection, all four light boxes are fully on, in order to provide the maximum brightness, but we additionally simulate 15 illumination conditions by switching on and off the four light boxes.

3.3   Data Characteristics

In total, we collect data from 110 participants (47 female and 63 male), aged between 19 and 41 years. 17 of them wore contact lenses and 17 of them wore eye glasses during recording. The ethnicities of the participants includes Caucasian, Middle Eastern, East Asian, South Asian and African. Each participant collected 525 gaze points under the full-lighting condition, and 90 gaze points under the varying lighting conditions - six gaze points for each of the 15 lighting conditions. For each gaze point, a total of 18 images was collected by the 18 different cameras. We manually removed samples for which the participant was not looking at the ground-truth point-of-regard due to blinking, motion blur etc. This results in total 1,083,492 images samples for whole ETH-XGaze dataset.

A comparison between the proposed and existing datasets can be found in Table 1. Our dataset surpasses existing datasets regarding the following aspects.

Head Pose. ETH-XGaze has the largest range of head poses compared to existing datasets, as shown in the first row of Fig. 2. Examples from ETH-XGaze with different head poses are shown in Fig. 5. In [16], it is stated that the effective head pose range of Gaze360 is ±90° in horizontal direction and limited head poses in vertical direction. However, head pose annotations are not provided in their dataset and hence we cannot visualize it here.

Gaze Direction. ETH-XGaze has the largest range of gaze directions compared to existing datasets. The second row of Fig. 2 compares the gaze direction distributions. Although Gaze360 reports ±140° coverage on the horizontal gaze direction, the dataset contains only very few samples beyond ±70°. ETH-XGaze is evenly sampled across a large range of horizontal and vertical gaze directions.

Image Resolution. ETH-XGaze has the highest image resolution compared to existing datasets, especially the effective resolution on the face region. We show some examples and corresponding cropped eye images from different datasets in Fig 3. The Columbia dataset al.so has high image resolution, however, the dataset is comprised of only 5,880 samples. While EYEDIAP, MPIIGaze, RT-GENE and Gaze360 have fairly high resolution imagery as well, the participant is far away from the camera which results in low effective eye region resolution.

Controlled Illumination Conditions. ETH-XGaze provides a set of controlled illumination conditions. Although uncontrolled in-the-wild illumination conditions are important for gaze estimation [19,42], controlled illumination conditions provide complementary information to better understand illumination impact and enable light synthesis. As shown in Fig. 4, we record 16 different illumination conditions.

3.4   ETH-XGaze Utility

ETH-XGaze makes it possible to *train* gaze estimators that cover large ranges of head poses and gaze directions. This allows to better estimate gaze from oblique viewpoints, such as overhead cameras. ETH-XGaze can also be used to *evaluate*

Fig. 4. Samples of the 16 illumination conditions created by switching on and off the four light boxes. The first row are the original samples, and the second row employs histogram equalization. The first column is the full-lighting setting.

the robustness of a gaze estimation method with respect to these factors. In our dataset the head pose remains fixed and thus does not follow the traditional *head-pose-following-gaze* pattern. However, by imaging from 18 viewpoints we densely sample all natural pose-gaze combinations with respect to the camera, suitable for varied applications like gaze estimation from a personal laptop or attention measurement inside a smart home.

Our dataset allows future gaze prediction methods to train on high-resolution imagery, which is critical for generative methods [12,25,34,38]. Since the generated image quality highly depends on the training image quality, Columbia and the synthetic UnityEYE dataset have been used during training in the past. Our ETH-XGaze provides high-resolution images (6000 × 4000 pixels), and more importantly face region occupies a large portion of the image.

Since the data in ETH-XGaze has been captured to allow for 3D geometry reconstruction using multi-view photogrammetry methods (i.e. [2]), it provides the potential of synthesizing high-quality gaze estimation data in the future. Parametric eye models [3,36] can be fit to the data to build a controllable rig of the eye [4]. Such a rig can then be used to re-render novel images of different lighting conditions, gaze directions, and head poses with state-of-the-art rendering techniques, providing additional training data for gaze estimation task.

3.5 Data Pre-processing

We crop the face patch out of the original image as input for gaze estimation model training. For each input image sample, we first perform face and facial landmark detection using a state-of-the-art method [5]. We then fit a 3D morphable model of the face to the detected landmarks to estimate the 3D head pose [14]. The 3D head pose along with camera calibration information is used to perform data normalization [41]. In a nutshell, the data normalization method maps the input image to a normalized space where a virtual camera is used to warp the face patch out of the original input image according to 3D head pose. It rotates a virtual camera to cancel the head rotation around the row axis, and moves the virtual camera to a fixed distance from the face center to warp the face patch of fixed size. More details can be found in the original paper [41]. During data normalization, we define the face center as the center of the four eye corners and two nose corners, we set the focal length of the virtual camera to be 960 mm, the normalized distance to be 300 mm, and the cropped face image is 448 × 448 pixels. Examples of face patches after data normalization are shown in Fig. 5. The processed data along with original imagery are released to public.

Fig. 5. Data examples captured by 18 different camera views. The red arrow is the gaze direction. The face patch images shown are after data normalization. (Color figure online)

4   Evaluation Protocol

One goal of this paper is to establish a benchmark to evaluate gaze estimation algorithms. For this purpose, we define four evaluations on ETH-XGaze. The first three evaluations - cross-dataset, within-dataset, and person-specific evaluations - are popular evaluations found in the current gaze estimation literature. In addition, we propose to also assess robustness over head poses and gaze directions as a fourth evaluation criteria, which is made possible by ETH-XGaze.

4.1   Baseline Method

We provide a baseline gaze estimation method using an off-the-shelf ResNet-50 network [11]. This baseline takes the full-face patch covering 224×224 pixels as input and outputs the horizontal and vertical gaze angles. We used the ADAM [18] optimizer with an initial learning rate of 0.0001, and the batch size is set to be 50. We trained the baseline model for 25 epochs and decay the learning rate by a factor of 0.1 every 10 epochs.

4.2   Dataset Preparation

We split ETH-XGaze into three parts: a training set $\mathbb{TR}$ comprised of 80 participants, a test set for within-dataset evaluation $\mathbb{TE}$ containing 15 participants, and a test set for person-specific evaluation $\mathbb{TES}$ consisting of another 15 participants. Splitting the test data into two disjoint sets allows us to release ground truth gaze required for the person-specific evaluation (Sect. 4.5). We ensured that the subjects in both training and test sets exhibit diverse gender, age, and ethnicity, some with and some without glasses. While we release both ground-truth gaze and imagery for the training set, we withhold the ground-truth gaze for the test sets. Authors are encouraged to submit gaze predictions on test samples to the benchmark website, and the performance will be evaluated and reported. This enables future research to compare to existing methods on neutral grounds.

Aside from the proposed ETH-XGaze dataset, we also evaluated other existing datasets with our baseline method. These datasets were pre-processed as we described in Sect. 3.5. For the *EYEDIAP* dataset, we used both screen sequence

Table 2. Gaze estimation errors in degrees on cross-dataset evaluations. The last column shows the average ranking on each test sets, and all other numbers are gaze estimation error in degrees.

Train	Test						
	MPIIGaze	EYEDIAP	Gaze capture	RT-GENE	Gaze360	ETH-XGaze	Ave. rank
MPIIGaze	–	17.9	**6.3**	14.9	31.7	34.9	2.6
EYEDIAP	16.9	–	14.2	15.6	33.7	41.7	4.2
GazeCapture	**4.5**	13.7	–	**14.7**	30.2	29.4	**1.8**
RT-GENE	12.0	21.2	13.2	–	34.7	42.6	4.6
Gaze360	10.3	11.3	12.9	26.6	–	**17.0**	2.8
ETH-XGaze	7.5	**11.0**	10.5	31.2	**27.3**	–	2.0

and floating target sequences and sampled the video sequences every 15 frames. For the *GazeCapture* dataset, we used the pre-processing pipeline from [25] to obtain 3D head poses since the dataset does not provide camera parameters. For the *Gaze360* dataset, we used the face bounding box provided by the dataset to crop the face patch, alongside the 3D gaze ground-truth. We will ask authors of these datasets for permission to release the processed data such that the community can use it for evaluations on ETH-XGaze.

4.3   Cross-Dataset Evaluation

Cross-dataset evaluation has gained popularity since it indicates the generalization capabilities of a gaze estimation method. We define the cross-dataset evaluation as training the model on ETH-XGaze and testing on other datasets, as well as training on other datasets and testing on ETH-XGaze.

We conducted the pair-wise cross-dataset evaluations on different datasets and show results achieved by the baseline in Table 2. The results exhibit rather large gaze estimation errors when testing on our ETH-XGaze, indicating that there is a big domain gap between ETH-XGaze and previous datasets. This stems from the fact that ETH-XGaze exhibits much larger variation in head pose and gaze direction compared to other datasets. Therefore, the gaze estimator has to extrapolate to those unseen head poses and gaze directions which is known to be a difficult machine learning task.

Training on GazeCapture achieves the best overall ranking since it contains similar head pose and gaze ranges compared to MPIIGaze, RT-GENE and EYE-DIAP. However, it performs poorly on test datasets that exhibit large variation in head pose and gaze direction such as Gaze360 and our ETH-XGaze. In contrast, ETH-XGaze enables thorough benchmarking of generalization capabilities of future gaze estimation approaches.

The model trained on Gaze360 achieves the best cross-dataset performance on ETH-XGaze since they contain similar head pose and gaze direction ranges. However, Gaze360 has been collected "in the wild" setting and can suffer from low-quality images and gaze labels (see Fig. 6). Our dataset, despite the lab setting, still allows for good performance (the best on EYEDIAP and Gaze360) without any data augmentation.

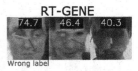

Fig. 6. Test samples from different datasets. We show results from training on Gaze360 and testing on ETH-XGaze (left), and training on ETH-XGaze and testing on Gaze360 (middle) and RT-GENE (right). The green arrow denotes ground truth and the red arrow is the prediction. The numbers give the respective gaze estimation errors in degrees. (Color figure online)

Table 3. Comparison of the baseline with current state-of-the-art on within dataset evaluations. Numbers are gaze estimation errors in degrees.

	ETH-XGaze	MPIIGaze	EYEDIAP	GazeCapture	RT-GENE
[25]	–	5.2	–	3.5	–
[9]	–	4.8	—	–	**8.7**
[26]	–	**4.5**	10.3	–	–
[39]	–	–	6.8	–	–
Baseline	4.5	4.8	**6.5**	**3.3**	12.0

4.4 Within-Dataset Evaluation

Within-dataset evaluation is another popular means of evaluating gaze estimation methods. Here the method is trained on $\mathbb{TR}$ and evaluated on $\mathbb{TE}$. Table 3 shows performances of the baseline alongside comparisons to recent state-of-the-art methods. The baseline achieves an error of 4.7° on average on ETH-XGaze, which is reasonably low given the large ranges of head poses and gaze directions. On other datasets, the baseline exhibits an accuracy comparable to current state-of-the-art methods, indicating that it is a strong baseline. The results of the other methods are taken from the respective publications.

4.5 Person-Specific Evaluation

Person-specific gaze estimation has gained a lot of attention in recent years [20, 25,38] due to the huge improvements that can be achieved from even just a few personal calibration samples. We randomly selected 200 samples from each participant in $\mathbb{TES}$ as the personal calibration samples. The protocol is to train the model with $\mathbb{TR}$ and up to 200 personal calibration samples, and to test on the remaining samples of $\mathbb{TES}$ – for each of the 15 test subjects. We pre-trained the model on $\mathbb{TR}$ and then fine-tune it using the 200 samples with 25 epochs.

Results from the baseline in Fig. 7 show that personal calibration improves the gaze estimates by a large margin. The goal of this evaluation is not only to achieve good results but also to rely on as few calibration samples as possible.

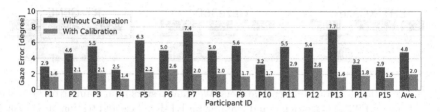

Fig. 7. Gaze estimation errors for person-specific evaluation of our baseline. We show the gaze estimation errors with and without training with 200 calibration samples. The number above each bar is the gaze estimation error in degrees.

4.6 Robustness Evaluation

Previous gaze estimation works usually only report the mean gaze estimation errors without detailed analysis across head poses and gaze directions. This is partly due to the lack of sufficient data samples to cover a wide range. Knowing the performance of an algorithm with respect to these factors is important, since a method with a higher overall error might have lower error within a specific range of interest. Hence we introduce a detailed evaluation to show the robustness across head poses and gaze directions. Figure 8 shows the performance of the baseline on TE over horizontal and vertical axes of the head pose and gaze direction. The different colors represent the different training sets. While these plots evaluate the performance of the different training sets, the benchmark will compare different algorithms instead. A flat curve across the entire graph, as in the case of training on ETH-XGaze, indicates robustness to head pose and gaze direction variation.

5   Demonstration Of ETH-XGaze

In this section, we evaluate the importance of different factors during training. Previous gaze estimation datasets cannot serve as the evaluation set for an ablation study of different factors such as head poses, gaze directions and illumination conditions due to the limited coverage. In contrast, the proposed ETH-XGaze is an ideal dataset for these evaluations.

Head Pose and Gaze Direction. We created several training subsets from TR by constraining the head poses and/or gaze directions angle ranges to be ± 80, ± 60, ± 40, and ± 20 in both horizontal and vertical directions. To keep the same amount of training samples for each subsets, we randomly re-sampled each training subset to have the same amount of samples as the minimal training set, i.e. the training set of ± 20 in both head poses and gaze directions. The results of testing on TE are shown in the left of Fig. 9. As we can see from the figure, constraining the head pose and gaze direction results in worse performance in general, especially when we constrain both head pose and gaze direction ranges. Constraining the gaze directions achieves worse results than constraining head

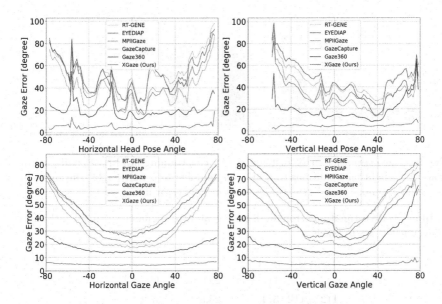

Fig. 8. Gaze estimation error distribution across head poses (first row) and gaze directions (second row) in horizontal and vertical directions respectively. The colored curves represent results with different training sets tested on TE.

poses, which indicates gaze directions have more impact than the head poses. Specifically, when we constrained the angle range to be ±40∘, the performance decrease caused by constraining head poses is 34.6%, constraining gaze directions is 82.1%, and constraining both head poses and gaze directions is 206.4%.

Illumination Condition. In the center of Fig. 9, we show results by training the baseline with all lighting conditions or only with the full-lighting condition. The performance drop (9% from 7.8° to 8.5°) indicates the impact of lighting conditions on gaze estimation performance.

Personal Appearance. In [19], the authors show gaze estimation performance with different numbers of participants. Our repeated experiment with our baseline on ETH-XGaze shows the same trend as increasing number of participants improves the performance (see Fig. 9, right).

Input Resolution. The image resolution analysis in [42] was only for eye images and the highest resolution was 60×36. The default input face patch image size to ResNet is 224×224 which we used in our baseline. We resized the input image to be 112×112 and 448×448 and then fed them into the baseline. Since there is an average pooling layer at the end of the ResNet convolutional layers, we do not need to modify the architecture with respect to different resolutions.

The results of resolution variation are shown in Table 4. The performance is improved when training and testing on higher resolutions, which indicates the potential of high-resolution gaze estimation. However, different with results

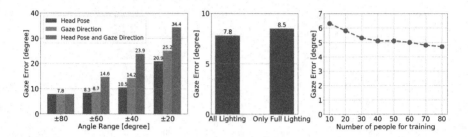

Fig. 9. Gaze estimation error distribution by constraining head poses and gaze directions (left), lighting conditions (middle), and number of people (right) during training. The number above each bar is the gaze estimation error in degrees.

Table 4. Gaze estimation errors in degrees generated by models trained with different input image sizes in pixels.

Train	Test		
	112×112	224×224	448×448
112×112	**5.4**	25.3	37.2
224×224	20.2	**4.5**	42.1
448×448	65.1	54.7	**4.2**

in [42], the model trained on one size achieves much worse results on other sizes. This can be caused by the much higher image resolution in ETH-XGaze with large appearance differences compared to the MPIIGaze in [42]. We did not specifically develop the method to handle cross-resolution input images and expect future works can properly deal with cross-resolution training.

6 Conclusion

We present a new large-scale gaze estimation dataset ETH-XGaze, featuring large variation in head pose and gaze direction, high-resolution imagery, varied subject appearance, systematic illumination conditions, as well as accurate ground-truth gaze vectors. Evaluation using a baseline method shows that training on ETH-XGaze significantly improves robustness towards variation in head pose and gaze direction compared to existing datasets, adding a very valuable resource for future work on gaze estimation. In addition, we propose a standardized experimental protocol and evaluation framework that will be made available via the benchmark website alongside the dataset, allowing for fair comparison of gaze estimation algorithms on neutral ground.

Acknowledgements. We thank the participants of our dataset for their contributions, our reviewers for helping us improve the paper, and Jan Wezel for helping with the hardware setup. This project has received funding from the European Research Council (ERC) under the European Union's Horizon 2020 research and innovation programme grant agreement No. StG-2016–717054.

Fig. 10. ERC logo.

References

1. Baluja, S., Pomerleau, D.: Non-intrusive gaze tracking using artificial neural networks. In: Advances in Neural Information Processing Systems, pp. 753–760 (1994)
2. Beeler, T., Bickel, B., Beardsley, P., Sumner, B., Gross, M.: High-quality single-shot capture of facial geometry. In: ACM Transactions on Graphics (TOG), pp. 1–9 (2010)
3. Bérard, P., Bradley, D., Gross, M., Beeler, T.: Lightweight eye capture using a parametric model. ACM Trans. Graph. (TOG) **35**(4), 1–12 (2016)
4. Bérard, P., Bradley, D., Gross, M., Beeler, T.: Practical person-specific eye rigging. In: Computer Graphics Forum, vol. 38, pp. 441–454. Wiley Online Library (2019)
5. Bulat, A., Tzimiropoulos, G.: How far are we from solving the 2D & 3D face alignment problem?(and a dataset of 230,000 3D facial landmarks. In: Proceedings of the IEEE International Conference on Computer Vision, pp. 1021–1030 (2017)
6. Cheng, Y., Huang, S., Wang, F., Qian, C., Lu, F.: A coarse-to-fine adaptive network for appearance-based gaze estimation. In: Proceedings of the AAAI Conference on Artificial Intelligence, pp. 10623–10630 (2020)
7. Demiris, Y.: Prediction of intent in robotics and multi-agent systems. Cogn. Process. **8**(3), 151–158 (2007)
8. Dollar, P., Wojek, C., Schiele, B., Perona, P.: Pedestrian detection: an evaluation of the state of the art. IEEE Trans. Pattern Anal. Mach. Intell. **34**(4), 743–761 (2011)
9. Fischer, T., Chang, H.J., Demiris, Y.: RT-GENE: real-time eye gaze estimation in natural environments. In: Ferrari, V., Hebert, M., Sminchisescu, C., Weiss, Y. (eds.) ECCV 2018. LNCS, vol. 11214, pp. 339–357. Springer, Cham (2018). https://doi.org/10.1007/978-3-030-01249-6_21
10. Funes Mora, K.A., Monay, F., Odobez, J.M.: Eyediap: a database for the development and evaluation of gaze estimation algorithms from RGB and RGB-D cameras. In: Proceedings of the ACM Symposium on Eye Tracking Research & Applications, pp. 255–258. ACM (2014)
11. He, K., Zhang, X., Ren, S., Sun, J.: Deep residual learning for image recognition. In: Proceedings of the IEEE Conference on Computer Vision and Pattern Recognition, pp. 770–778 (2016)
12. He, Z., Spurr, A., Zhang, X., Hilliges, O.: Photo-realistic monocular gaze redirection using generative adversarial networks. In: Proceedings of the IEEE International Conference on Computer Vision, pp. 6932–6941 (2019)
13. Huang, Q., Veeraraghavan, A., Sabharwal, A.: Tabletgaze: dataset and analysis for unconstrained appearance-based gaze estimation in mobile tablets. Mach. Vis. Appl. **28**(5–6), 445–461 (2017)

14. Huber, P., et al.: A multiresolution 3D morphable face model and fitting framework. In: Proceedings of the 11th International Joint Conference on Computer Vision, Imaging and Computer Graphics Theory and Applications (2016)
15. Ionescu, C., Papava, D., Olaru, V., Sminchisescu, C.: Human3.6m: large scale datasets and predictive methods for 3D human sensing in natural environments. IEEE Trans. Pattern Anal. Mach. Intell. **36**(7), 1325–1339 (2014)
16. Kellnhofer, P., Recasens, A., Stent, S., Matusik, W., Torralba, A.: Gaze360: physically unconstrained gaze estimation in the wild. In: Proceedings of the IEEE International Conference on Computer Vision, pp. 6912–6921 (2019)
17. Kemelmacher-Shlizerman, I., Seitz, S.M., Miller, D., Brossard, E.: The megaface benchmark: 1 million faces for recognition at scale. In: Proceedings of the IEEE Conference on Computer Vision and Pattern Recognition, pp. 4873–4882 (2016)
18. Kingma, D.P., Ba, J.: Adam: a method for stochastic optimization. arXiv preprint arXiv:1412.6980 (2014)
19. Krafka, K., Khosla, A., Kellnhofer, P., Kannan, H., Bhandarkar, S., Matusik, W., Torralba, A.: Eye tracking for everyone. In: Proceedings of the IEEE Conference on Computer Vision and Pattern Recognition, pp. 2176–2184 (2016)
20. Liu, G., Yu, Y., Mora, K.A.F., Odobez, J.M.: A differential approach for gaze estimation with calibration. In: British Machine Vision Conference, vol. 2, p. 6 (2018)
21. Lu, F., Sugano, Y., Okabe, T., Sato, Y.: Inferring human gaze from appearance via adaptive linear regression. In: Proceedings of the IEEE International Conference on Computer Vision, pp. 153–160. IEEE (2011)
22. Lu, F., Sugano, Y., Okabe, T., Sato, Y.: Adaptive linear regression for appearance-based gaze estimation. IEEE Trans. Pattern Anal. Mach. Intell. **36**(10), 2033–2046 (2014)
23. Majaranta, P., Bulling, A.: Eye tracking and eye-based human–computer interaction. In: Fairclough, S.H., Gilleade, K. (eds.) Advances in Physiological Computing. HIS, pp. 39–65. Springer, London (2014). https://doi.org/10.1007/978-1-4471-6392-3_3
24. Nech, A., Kemelmacher-Shlizerman, I.: Level playing field for million scale face recognition. In: Proceedings of the IEEE Conference on Computer Vision and Pattern Recognition, pp. 7044–7053 (2017)
25. Park, S., Mello, S.D., Molchanov, P., Iqbal, U., Hilliges, O., Kautz, J.: Few-shot adaptive gaze estimation. In: Proceedings of the IEEE International Conference on Computer Vision, pp. 9368–9377 (2019)
26. Park, S., Spurr, A., Hilliges, O.: Deep pictorial gaze estimation. In: Ferrari, V., Hebert, M., Sminchisescu, C., Weiss, Y. (eds.) ECCV 2018. LNCS, vol. 11217, pp. 741–757. Springer, Cham (2018). https://doi.org/10.1007/978-3-030-01261-8_44
27. Ruch, T.C., Fulton, J.F.: Medical physiology and biophysics. Acad. Med. **35**(11), 1067 (1960)
28. Russakovsky, O., et al.: Imagenet large scale visual recognition challenge. Int. J. Comput. Vis. **115**(3), 211–252 (2015)
29. Seitz, S.M., Curless, B., Diebel, J., Scharstein, D., Szeliski, R.: A comparison and evaluation of multi-view stereo reconstruction algorithms. In: Proceedings of the IEEE Conference on Computer Vision and Pattern Recognition, vol. 1, pp. 519–528. IEEE (2006)
30. Shrivastava, A., Pfister, T., Tuzel, O., Susskind, J., Wang, W., Webb, R.: Learning from simulated and unsupervised images through adversarial training. In: Proceedings of the IEEE Conference on Computer Vision and Pattern Recognition, pp. 2107–2116 (2017)

31. Smith, B.A., Yin, Q., Feiner, S.K., Nayar, S.K.: Gaze locking: passive eye contact detection for human-object interaction. In: Proceedings of the 26th Annual ACM Symposium on User Interface Software and Technology, pp. 271–280 (2013)
32. Soo Park, H., Jain, E., Sheikh, Y.: Predicting primary gaze behavior using social saliency fields. In: Proceedings of the IEEE International Conference on Computer Vision, pp. 3503–3510 (2013)
33. Sugano, Y., Matsushita, Y., Sato, Y.: Learning-by-synthesis for appearance-based 3D gaze estimation. In: Proceedings of the IEEE Conference on Computer Vision and Pattern Recognition, pp. 1821–1828 (2014)
34. Wang, K., Zhao, R., Ji, Q.: A hierarchical generative model for eye image synthesis and eye gaze estimation. In: Proceedings of the IEEE Conference on Computer Vision and Pattern Recognition, pp. 440–448 (2018)
35. Wang, K., Zhao, R., Su, H., Ji, Q.: Generalizing eye tracking with Bayesian adversarial learning. In: Proceedings of the IEEE Conference on Computer Vision and Pattern Recognition, pp. 11907–11916 (2019)
36. Wood, E., Baltrušaitis, T., Morency, L.-P., Robinson, P., Bulling, A.: A 3D morphable eye region model for gaze estimation. In: Leibe, B., Matas, J., Sebe, N., Welling, M. (eds.) ECCV 2016. LNCS, vol. 9905, pp. 297–313. Springer, Cham (2016). https://doi.org/10.1007/978-3-319-46448-0_18
37. Wood, E., Baltrušaitis, T., Morency, L.P., Robinson, P., Bulling, A.: Learning an appearance-based gaze estimator from one million synthesised images. In: Proceedings of the ACM Symposium on Eye Tracking Research & Applications, pp. 131–138 (2016)
38. Yu, Y., Liu, G., Odobez, J.M.: Improving few-shot user-specific gaze adaptation via gaze redirection synthesis. In: Proceedings of the IEEE Conference on Computer Vision and Pattern Recognition, pp. 11937–11946 (2019)
39. Yu, Y., Odobez, J.M.: Unsupervised representation learning for gaze estimation. In: Proceedings of the IEEE Conference on Computer Vision and Pattern Recognition, pp. 7314–7324 (2020)
40. Yu, Z., Yoon, J.S., Venkatesh, P., Park, J., Yu, J., Park, H.S.: Humbi 1.0: Human Multiview Behavioral Imaging Dataset, June 2020
41. Zhang, X., Sugano, Y., Bulling, A.: Revisiting data normalization for appearance-based gaze estimation. In: Proceedings of the ACM Symposium on Eye Tracking Research & Applications, p. 12. ACM (2018)
42. Zhang, X., Sugano, Y., Fritz, M., Bulling, A.: Mpiigaze: real-world dataset and deep appearance-based gaze estimation. IEEE Trans. Pattern Anal. Mach. Intell. **41**(1), 162–175 (2019)
43. Zimmermann, C., Ceylan, D., Yang, J., Russell, B., Argus, M., Brox, T.: Freihand: a dataset for markerless capture of hand pose and shape from single RGB images. In: Proceedings of the IEEE International Conference on Computer Vision, pp. 813–822 (2019)

Calibration-Free Structure-from-Motion with Calibrated Radial Trifocal Tensors

Viktor Larsson[1]([✉]), Nicolas Zobernig[2], Kasim Taskin[3], and Marc Pollefeys[1,4]

[1] Department of Computer Science, ETH Zürich, Zürich, Switzerland
vlarsson@inf.ethz.ch
[2] Department of Information Technology and Electrical Engineering, ETH Zürich, Zürich, Switzerland
[3] KTH Royal Institute of Technology, Stockholm, Sweden
[4] Microsoft Mixed Reality & AI Zurich Lab, Zürich, Switzerland

Abstract. In this paper we consider the problem of Structure-from-Motion from images with unknown intrinsic calibration. Instead of estimating the internal camera parameters through some self-calibration procedure, we propose to use a subset of the reprojection constraints that is invariant to radial displacement. This allows us to recover metric 3D reconstructions without explicitly estimating the cameras' focal length or radial distortion parameters. The weaker projection model makes initializing the reconstruction especially difficult. To handle this additional challenge we propose two novel minimal solvers for radial trifocal tensor estimation. We evaluate our approach on real images and show that even for extreme optical systems, such as fisheye or catadioptric, we are able to get accurate reconstructions without performing any calibration.

1 Introduction

In this paper we revisit the classical Structure-from-Motion problem [19], which is to recover the camera poses (the motion) and the 3D scene geometry (the structure) from a set of images. Structure-from-Motion pipelines generally fall into one of two categories; incremental or global. Incremental SfM methods (see e.g. [42,45,53]) work by incrementally growing an initial reconstruction by alternating posing in new images and triangulating new 3D points. Global SfM methods (see e.g. [9,37,38,56]) instead first estimate pairwise epipolar geometries. In a second step the relative poses are then fused into a single reconstruction, typically using some form of rotation averaging [7,14,18]. There are also SfM methods which combine the two approaches (e.g. [8,33]).

In all of the above methods it is necessary to know the cameras' internal parameters (camera intrinsics and lens-distortion) to achieve accurate reconstruction results. These parameters can either be found during an offline calibration procedure (e.g. using some calibration object with known structure such as checkerboards, see [44,55]), or they are estimated during the reconstruction.

Electronic supplementary material The online version of this chapter (https://doi.org/10.1007/978-3-030-58558-7_23) contains supplementary material, which is available to authorized users.

© Springer Nature Switzerland AG 2020
A. Vedaldi et al. (Eds.): ECCV 2020, LNCS 12350, pp. 382–399, 2020.
https://doi.org/10.1007/978-3-030-58558-7_23

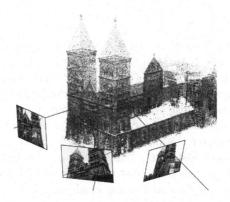

Fig. 1. Structure-from-Motion using radial alignment constraints. Instead of requiring the 3D point to project onto the 2D-point, we only require the projection to lie on the radial lines going through the image point. This makes projection equations invariant to focal length and radial distortion.

The second set of methods can be further divided into two approaches. Methods which first perform a projective reconstruction followed by a self-calibration step (see e.g. [20,21,34]). The self-calibration step entails estimating the Dual Absolute Quadric (see [6,19]) by adding assumptions on the camera intrinsics, such as unit aspect ratio and zero skew. The other approach is to estimate the camera intrinsics during the initial pose estimation process. This can be done either by using solvers which also estimate the internal camera parameters (e.g. [28–30,54]) or the camera parameters are initialized with some heuristic guess (e.g. using EXIF tags) followed by bundle adjustment. This approach is used in the open-source framework COLMAP [42,43] that uses focal length sampling [41] and zero-initialized distortion parameters, which are then refined in the bundle-adjustment step. For global SfM with unknown calibration, Sweeney et al. [47] proposed a method which optimizes the consistency of fundamental matrices in order to estimate a consistent focal length for each camera.

While the above approaches can work well in practice, they typically only work reliably for cameras with no or negligible radial distortion. Methods based on finding the calibration during reconstruction are prone to failure for cameras with severe distortion, especially for images where most point correspondences are in regions with high distortion (e.g. close to the image borders).

In this paper we propose a Structure-from-Motion pipeline that does not require knowing or estimating the camera calibration. We only make the assumptions that the camera has square pixels and approximately centered principal point (which is satisfied for essentially all consumer cameras today). The main idea is to use a subset of the geometric constraints which are invariant to any radial change in the projection (such as focal length and most lens-distortion). We show that it is possible to recover high quality reconstructions from this weaker set of constraints even for images from very extreme distortions (e.g. fisheye and catadioptric cameras). In contrast to previous work we do not estimate any distortion model or perform self-calibration.

1.1 Background

The *Radial Alignment Constraint* (RAC) was first introduced by Tsai [51] for camera calibration. The RAC simply states that the projection of a 3D point should lie on the radial line[1] passing through the image point (see Fig. 1). This constraint has the nice property that it does not depend on the camera's focal length or any purely radial distortion, since these only move the projections along the lines. However, since forward motion also moves the projections radially it is only possible to recover the pose of the camera up to an unknown forward translation using these constraints. This constraint has been used for absolute camera pose estimation with radial distortion, see Kukelova et al. [25] and more recently Camposeco et al. [5] and Larsson et al. [30].

1.2 1D Radial Camera Model

The idea in the RAC later gave rise to the *1D-Radial* camera model which considers the mapping from 3D points to radial lines in the image. Formally, this can be modelled as a projective mapping from $\mathbb{P}^3$ to $\mathbb{P}^1$. Similarly to pinhole cameras, we can describe this mapping with a matrix acting on homogeneous coordinates, i.e. $\mathbf{x} \sim P\mathbf{X}$, where $\mathbf{x} \in \mathbb{P}^1$, $\mathbf{X} \in \mathbb{P}^3$, $P \in \mathbb{R}^{2 \times 4}$. Note that in this case the camera matrix P is a 2×4 matrix instead of 3×4. The camera matrix can be thought of as the first two rows of the pinhole camera; giving only the direction of the pinhole projection and not the radial scaling.

As for pinhole cameras we can consider *calibrated* cameras. In the pinhole camera setting we require the first 3×3 block to be a rotation matrix; for radial cameras we require the first 2×3 block to consist of two orthonormal vectors,

$$P = \begin{bmatrix} \mathbf{r}_1^T & t_1 \\ \mathbf{r}_2^T & t_2 \end{bmatrix}, \quad \mathbf{r}_1^T \mathbf{r}_2 = 0, \quad \|\mathbf{r}_1\| = \|\mathbf{r}_2\| = 1. \tag{1}$$

It is important to note that this is not an approximation (like e.g. weak or para-perspective), but instead we essentially consider a subset of the geometric constraints which are independent. This means that for any perspective reconstruction (possibly with non-linear radial distortion), there exists a corresponding 1D radial reconstruction found by just taking the first two rows from each camera. In this paper we show that we can recover this reconstruction without ever estimating the focal length or radial distortion. Note also that the 1D radial camera model is not only valid for central cameras, but any optical system satisfying the RAC, e.g. spherical mirrors (chromeball images) or in general any radially-symmetric mirror, axial cameras, etc. For more details about the 1D-radial camera model see the supplementary material.

1.3 Multiple View Geometry of 1D Radial Cameras

The multiple view geometry of 1D radial cameras was studied by Thirthala and Pollefeys [48] in the framework of multi-focal tensors [49]. Since the radial model

---

[1] Radial lines are lines passing through the image center.

only provides a single constraint from each projection, it was shown in [48] that there does not exist any bi- or trifocal tensors for radial cameras in general position, and that it is first in four views that constraints appear. Furthermore, [48] showed that the quadrifocal tensor itself has two internal constraints. Ignoring these constraints the quadrifocal tensor can be linearly estimated from 15 quadruplet point matches. However, as mentioned in [48], this solver is mostly of theoretical interest and not usable for practical purposes due to the high-number of points required. There is currently no known minimal solver for the radial quadrifocal tensor which enforces the internal constraints. In [48] they also consider three special camera configurations where trifocal tensors exist: 1) three principal axes intersect, 2) the scene points are planar and 3) one pinhole camera and two radial cameras. For the tensors they only consider the projective setting, i.e. there are no constraints enforcing that the tensors they estimate can be factorized into calibrated cameras, as in (1).

The tensors above describe projective mappings from $\mathbb{P}^3$ to $\mathbb{P}^1$. There has also been a series of works which consider the multiple view geometry of cameras in lower dimensional spaces, i.e. $\mathbb{P}^2$ to $\mathbb{P}^1$. The trifocal tensor in this setting was first investigated by Quan and Kanade [39]. Faugeras et al. [16] showed that cameras undergoing planar motion can be modeled with 1D cameras by projecting the image measurements onto the ground plane, allowing for estimation with the radial trifocal tensor from [39]. Later, Åström and Oskarsson [4] derived the internal constraint for calibration for this tensor. In this simpler setting the calibration constraint turns out to be linear. These lower dimensional radial trifocal tensors were then used in [3, 11, 40] for localization of robotic platforms.

Calibrated Multiple View Geometry. In general, enforcing constraints for calibration on multi-focal tensors is very difficult for higher order tensors. For the two-view case (i.e. fundamental vs. essential matrix), these constraints are the well-known trace-constraints[2], $2EE^T E - \mathrm{tr}(EE^T)E = 0$. The constraints for the perspective trifocal tensor have received much attention recently ([13,15,22,32,35]), though currently the minimal solvers are based on homotopy continuation or other iterative methods and have far from practical runtimes, especially compared to their two-view counterparts. In this paper we will show that there exist analogous calibration constraints for the radial trifocal tensor as well as the mixed trifocal tensor. We also show that these constraints can be used to develop fast minimal solvers for calibrated radial reconstruction.

Related work by Kim et al. [23]. Structure-from-Motion with the 1D radial camera model was previously considered by Kim et al. [23]. In [23] the authors presented a method for performing projective reconstruction with 1D radial cameras based on matrix factorization techniques, similar to previous work on projective-factorization for perspective cameras [2,10,50]. In a post-processing step, the method attempts to upgrade the reconstruction to metric by estimating the dual absolute quadric. However, their approach does not handle outlier measurements which limits the applicability on real image sequences. Additionally,

---

[2] Also known as the Demazure constraints [12].

due to the matrix factorization based approach, the method does not scale to larger image collections, e.g. the largest reconstruction presented in [23] has 189 3D points from 79 images. For comparison, in Sect. 4.4 we present 3D reconstructions from over a thousand images and more than 400k 3D points.

2   Calibrated Radial Trifocal Tensors

In this section we will present two new minimal solvers for calibrated radial trifocal tensors. These will be used for initializing our incremental Structure-from-Motion pipeline in Sect. 3. Next we show that there exists one additional internal constraint for each of the two tensors we consider; the purely radial trifocal tensor with intersecting principal axes, and the mixed trifocal tensor with one central camera and two radial cameras in general position. In the supplementary material we also discuss the third case considered by Thirthala and Pollefeys [48] where the scene is planar.

2.1   Intersecting Principal Axes

First we consider the case of intersecting principal axes. Choosing the world-coordinate frame such that the point of intersection is the origin, then the 2×4 camera matrices will be of the form, $P_k = [A_k \ \mathbf{0}]$, $k = 1, 2, 3$, $A_k \in \mathbb{R}^{2 \times 3}$. The projection equations $\lambda \mathbf{x} = A_1 \mathbf{X}$, $\lambda' \mathbf{x}' = A_2 \mathbf{X}$ and $\lambda'' \mathbf{x}'' = A_3 \mathbf{X}$ can be rewritten

$$\begin{bmatrix} A_1 & \mathbf{x} & 0 & 0 \\ A_2 & 0 & \mathbf{x}' & 0 \\ A_3 & 0 & 0 & \mathbf{x}'' \end{bmatrix} \begin{pmatrix} \mathbf{X} \\ -\lambda \\ -\lambda' \\ -\lambda'' \end{pmatrix} = 0. \tag{2}$$

This 6×6 matrix must thus be rank deficient and its determinant yields an equation which depend on the image points, $\sum_{i,j,k} T_{ijk} \mathbf{x}_i \mathbf{x}'_j \mathbf{x}''_k = 0$ where $\mathbf{x}_i$ denotes the ith image coordinate. The coefficients T_{ijk} can be interpreted as the $2 \times 2 \times 2$ multi-focal tensor [49] corresponding to this camera configuration. This is the *radial trifocal tensor* from [48]. In the uncalibrated setting this camera configuration has $3 \cdot (2 \cdot 3 - 1) - (3 \cdot 3 - 1) = 7$ degrees of freedom. Since the corresponding multi-focal tensor is a homogeneous $2 \times 2 \times 2$ tensor (which also has 7 degrees of freedom), the radial trifocal tensor does not have any internal constraint, as was also stated in [48].

Now if we consider the calibrated setting we require each matrix A_k to have orthonormal rows, i.e. $P_k = [R_k \ \mathbf{0}]$ where $R_k R_k^T = I_2$, $R_k \in \mathbb{R}^{2 \times 3}$. In this case each camera only has 3 degrees of freedom and similarly the gauge freedom in the coordinate system is also reduced to 3 (since the projections are scale invariant) resulting in $3 \cdot 3 - 3 = 6$ degrees of freedom. This means that there must exist $7 - 6 = 1$ internal constraint on the corresponding trifocal tensor.

Internal Constraint for Calibration. Using techniques from numerical linear algebra we found that the internal constraint is a homogeneous quartic polynomial in the tensor elements. See the supplementary material for details on the

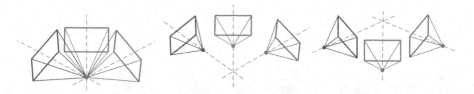

Fig. 2. The radial trifocal tensor describes three views with intersecting principal axes, e.g. from pure rotation (*Left*), panoramic motion (*Middle*) and orbital motion (*Right*).

constraint and how we found it. We have verified the validity of the constraint both empirically and symbolically using computer algebra software.

Estimation from Minimal Point Sets. As shown above, each triplet correspondence $(\mathbf{x}, \mathbf{x}', \mathbf{x}'')$ in the images yields one linear constraint on the elements of the radial trifocal tensor (see also [48]). To get a minimal problem we therefore need we need six triplet correspondences in total. From the six correspondences we can then find the two-dimensional linear subspace of possible $2 \times 2 \times 2$ tensors that satisfy the trifocal constraints, i.e.

$$T = \alpha_1 N_1 + \alpha_2 N_2, \tag{3}$$

where N_1 and N_2 are basis vectors to the nullspace. Since the tensor is homogeneous we can fix the scale by setting $\alpha_2 = 1$. To solve for the remaining unknown we insert (3) into the internal constraint from the previous section, yielding a single univariate quartic polynomial in α_1 that can be solved in closed form. In Sect. 4.1 we evaluate the proposed minimal solver.

2.2 Mixed Trifocal Tensor

Now we consider heterogeneous camera setups with both radial and pinhole cameras. The different minimal problems for heterogeneous camera setups were listed in Kozuka and Sato [24], though only in the projective setting. For the trifocal case there are two possibilities: 1) one pinhole and two radial cameras, 2) two pinhole and one radial camera. The second case becomes trivial since the minimal problem decouples into independent relative pose estimation between the pinhole cameras followed by pose estimation of the radial camera.

One Pinhole and Two Radial. The minimal solution for this camera case was first presented in [48] in the uncalibrated setting. In this case there are $11 + 7 + 7 - 15 = 10$ degrees of freedom[3]. Since the corresponding tensor is a homogeneous $3 \times 2 \times 2$ tensor with 11 degrees of freedom, there exist a single internal constraint. This constraint was derived in [48] and is a degree 6 polynomial in the tensor.

In the calibrated setting we have 6 d.o.f. in the calibrated pinhole camera, 5 in each of the calibrated 1D radial cameras and the coordinate system has

---

[3] The projective coordinate system has 15 d.o.f.

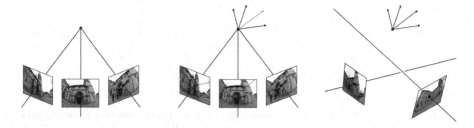

Fig. 3. Initialization for Structure-from-Motion with 1D radial cameras. *Left:* First we estimate a calibrated radial tensor describing the relative motion of three cameras with intersecting principal axes. *Middle:* Intersecting the backprojected feature correspondences of the three cameras we synthesize the image of a central camera with the intersection point as the projection center. *Right:* Finally we estimate a mixed trifocal tensor describing the relative motion of the synthesized central camera and two additional radial cameras.

7 d.o.f., yielding $6 + 5 + 5 - 7 = 9$ degrees of freedom. Thus there must exist one additional internal constraint in the case of calibrated cameras. Similarly to Sect. 2.1 we used numerical techniques to recover the internal constraint. For this case it was more difficult to recover the constraint, both due to its higher degree, and having to consider the multiples of the original internal constraint from [48]. The internal constraint is a homogeneous degree 8 polynomial in the elements of the tensor. For space reasons we do not print the full polynomial here (it has 3357 monomials). See the supplementary material for more details.

Estimation from Minimal Point Sets. Each point correspondence yields a single linear constraint on the 12 elements of the mixed trifocal tensor. From the minimal sample of nine point correspondences we get a three dimensional nullspace where the tensor must lie, $T = \alpha_1 N_1 + \alpha_2 N_2 + \alpha_3 N_3$. Fixing $\alpha_3 = 1$ and inserting into the two internal constraints we get two polynomials in two unknowns of degree 6 and 8. Empirically we found that the coefficients of the two polynomials are completely generic which means that we have 48 solutions in general. Note that in practice many of these solutions end up being complex and only a small subset needs to be verified in the end. Using the generator from Larsson et al. [27] we created a Groebner basis solver for this polynomial system, but other techniques such as resultants could have been used as well.

3   Calibration-Free Structure-from-Motion

In this section we present our incremental pipeline for Structure-from-Motion based on the 1D radial camera model (see Sect. 1.2). We base our method on the incremental SfM pipeline COLMAP [42]. The main steps in our pipeline are: Initialization (Sect. 3.1), Triangulation (Sect. 3.2), Camera Resectioning (Sect. 3.3) and Bundle Adjustment (Sect. 3.4). The main difference to traditional SfM frameworks is that each point-observation now only gives a single

constraint on the reconstruction instead of two. This makes the geometric estimation problems significantly harder, e.g. 3D points require at least three views
to triangulate. The benefit of this camera model is that we can perform reconstructions which are invariant to focal length or radial distortion. Note that at
no point in our reconstruction pipeline do we estimate these parameters. We
only make the assumption of square pixels and centered principal point. The
next sections detail the different parts of our framework.

3.1 Initialization

Initializing Structure-from-Motion is significantly harder for the 1D radial camera model compared to normal pinhole-like camera models. Without additional
assumptions on the camera motion, the first constraints on the reconstruction
appear in four views, i.e. it is (in general) impossible to estimate the structure
and motion from only two or three views. The projective four-view case was
investigated by Thirthala and Pollefeys in [48], but due to the high number of
points required (15 for the linear solver presented in [48]), it is not useful in
practice where we need to deal with outlier-contaminated data.

Now we present our approach for finding the initial reconstruction for the
incremental Structure-from-Motion pipeline. It is based on the assumption that
we can find three images where the principal axes are (approximately) intersecting (see Fig. 2). Note that while a purely rotating camera satisfies this assumption, intersecting principal axes is a weaker constraint since the camera centers
are not required to coincide. This also covers the spherical type of motion common in handheld panoramic image capture (see e.g. [46,52]) as well as orbital
motion. This camera configuration is also common in photo collections where
the cameras are often pointed towards some object of interest. The initialization
consists of three stages and is performed using a combination of the minimal
solvers described in Sect. 2.1 and 2.2. See Fig. 3 for an overview.

a) Estimate Calibrated Radial Trifocal Tensor. Using the 6 point minimal
solver from Sect. 2.1 in a RANSAC framework [31] we estimate a calibrated radial
trifocal tensor for the first three images (which we assume have approximately
intersecting principal axes). In Sect. 3.5 we present a simple heuristic we use for
finding such image triplets in an image collection and in Sect. 4.3 evaluate the
quality of the estimated camera poses on real images.

b) Create Synthetic Central Camera. From the three images with intersecting principal axes it is not possible to triangulate any 3D points. Each 2D
observation backprojects to a 3D plane which contain the 3D point as well as
the principal axis of the camera. If we intersect all three backprojected planes,
the intersection will contain both the true 3D point and the intersection point
of the three principal axes, and thus also the entire line between them. Thus
we can only determine the direction towards the 3D point from the principal
axes' intersection point. The idea is now that we can interpret these directions
as the viewing rays from a central camera with projection center at the intersection point. Note that this automatically becomes a *calibrated* central image,

since the directions were triangulated in the coordinate system defined by the calibrated radial trifocal tensor from the previous step. Note that we only triangulate the directions of the sparse set of correspondences we have and not generate a full synthetic image. The idea of generating synthetic pinhole images from the radial trifocal tensor was also used in [36] to create undistorted images from three views.

c) **Estimate Calibrated Mixed Trifocal Tensor.** Finally we use the 9 point solver from Sect. 2.2 in RANSAC [31] to estimate the calibrated mixed trifocal tensor between the synthetic central camera and two additional views (which are modeled as radial cameras and can be in general position). Once we have a reconstruction with the five radial cameras in the same coordinate system we perform bundle adjustment (Sect. 3.4). For the refinement we remove the constraint that the first three views have intersecting principal axes.

3.2   Triangulation

Each 2D-3D correspondence yields a single constraint,

$$(-y, x) \begin{bmatrix} \mathbf{r}_1^T & t_1 \\ \mathbf{r}_2^T & t_2 \end{bmatrix} \begin{pmatrix} \mathbf{X} \\ 1 \end{pmatrix} = 0 \tag{4}$$

Geometrically this can be interpreted as restricting the 3D point $\mathbf{X}$ to lie on the plane $\mathbf{n}^T\mathbf{X} + d = 0$, where $\mathbf{n} = x\mathbf{r}_2 - y\mathbf{r}_1$ and $d = xt_2 - yt_1$. Given at least three correspondences (for cameras in general position) we can find the intersection point of the planes by solving the corresponding linear system of equations (possibly in a least squares sense for overconstrained problems). Note that the triangulation problem is minimal with three views which means that the triangulated point will always have zero reprojection error. Therefore it is not possible to determine if the matches used were correct or not. To avoid this problem we only triangulate points seen in at least four views. For pinhole cameras the same number of constraints is achieved from two views.

3.3   Camera Resectioning

Resectioning is the problem of estimating the camera pose given 2D-3D correspondences. For calibrated radial cameras each camera has five degrees of freedom and thus we require at least five correspondences for estimation. The minimal solver for this problem was proposed by Kukelova et al. [25], where it was used in a two-step approach for radial distortion estimation. Note that for the case where the principal point is not known, the 1D radial solver from [29] which also estimates principal point, could in principle be used as a drop-in replacement. However, this solver has significantly larger runtime and requires two additional correspondences. We did not use this solver and found that the method is stable for the principal point offsets observed in practice.

3.4 Bundle Adjustment

We measure the reprojection error as the orthogonal distance from the projected radial line to the 2D point correspondence, i.e. for a camera $[R \ t] \in \mathbb{R}^{2 \times 4}$, 3D point $\mathbf{X} \in \mathbb{R}^3$ and 2D-observation $\mathbf{x} \in \mathbb{R}^2$, we measure

$$\varepsilon = \left\| \left(\frac{\mathbf{nn}^T}{\mathbf{n}^T \mathbf{n}} - I \right) \mathbf{x} \right\|, \quad \text{where} \quad \mathbf{n} = R\mathbf{X} + \mathbf{t} \tag{5}$$

For the Bundle-Adjustment step in our pipeline we minimize the squared reprojection errors using the Ceres Solver [1]. If we have multiple images from the same camera we also refine the principal point.

3.5 Implementation Details

We have implemented our Structure-from-Motion pipeline by extending the open-source framework COLMAP [42]. The trifocal tensors estimated by the solvers in Sect. 2.1 and 2.2 can be factorized into the respective camera matrices. To perform this factorization we use the methods from [17,39], see the supplementary material for more detail. The runtimes of the solvers are $3.6 \, \mu s$ (radial trifocal) and 0.8 ms (mixed trifocal). In the synthetic experiments the solvers returned 3.09/4 and 8.76/48 real solutions in average.

Initialization Image Selection. The proposed initialization method (Sect. 3.1) requires three images with intersecting principal axes. These images can either be manually selected by the user, or we use a simple heuristic for finding suitable image triplets to initialize from. We restrict ourselves to the case where the camera is undergoing purely rotational motion. For normal Structure-from-Motion this is a degenerate case for initialization which is avoided. In [42] this is detected by checking if a homography fits the image pair. We use this to identify potential image triplets for initialization. For a triplet we can then geometrically verify if the image triplet has intersecting principal axes by fitting a radial trifocal tensor. With this simple heuristic we could find good initialization images for all datasets used in the evaluation in Sect. 4.4.

4 Experimental Evaluation

4.1 Solver Stability, Robustness and Runtime

In this section we evaluate the numerical stability of the two proposed minimal solvers. Figure 4 (Left) shows the $\log_{10}$-residuals for 10,000 synthetically generated instances. For the residuals we compute the ℓ_2-distance to the ground truth tensor after normalizing each tensor to unit length (i.e. $\|\text{vec}(T)\|_2 = 1$). In the experiment 0.03% (radial trifocal) and 4.25% (mixed trifocal) instances had errors larger than 10^{-8}. The mixed trifocal tensor is slightly less numerically stable and had a few failures as can be seen in the figure.

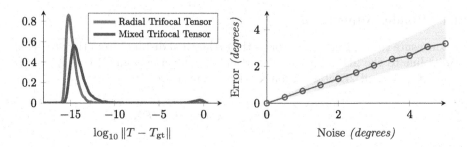

Fig. 4. *Left:* **Numerical stability.** The figure shows the distribution of the errors for 10,000 synthetically generated scenes. *Right:* **Stability to non-intersecting principal axes.** The graph shows the median errors in the relative rotations (in degrees) for the estimated calibrated radial trifocal tensor as the intersection constraint is violated. The shaded regions show the quartiles.

We also performed an experiment where we evaluate how the solutions for the radial trifocal tensor degrade as the assumption of intersecting principal axes is violated. We generate randomized synthetic scenes with three pinhole cameras looking at the origin from unit-distance. We then perturb the cameras by rotating each camera around a random axis with the camera center being fixed. Figure 4 (Right) shows how the rotation estimates from the radial trifocal tensor degrades as the rotation angle increases.

4.2   Comparison with Thirthala and Pollefeys [48]

In [48] the authors propose minimal solvers for estimating the radial trifocal tensor (intersecting principal axes) and the mixed radial trifocal tensor (perspective + two radial cameras) in the projective setting. These solvers do not enforce the additional constraint that ensures the tensors can be factorized into calibrated cameras (see Sect. 2.1 and Sect. 2.2). Since they use less constraints on the tensor itself, they also require one additional point correspondence. We generated synthetic scenes with varying levels of noise and compared how close the resulting cameras were to calibrated after factorizing the tensor and attempting metric upgrade (see supplementary material). Figure 5 shows the error in the rotation matrix constraint, $\|R_i R_i^T - I_2\|$, for varying levels of noise added to the image coordinates. Even for low noise levels the solvers from [48] yields solutions which are quite far from calibrated.

4.3   Evaluation of the Initialization on Real Data

The initialization pipeline we propose requires five images where three of them have close to intersecting principal axes. Intersecting principal axes can e.g. come from a purely rotating camera. In Sect. 3.5 we proposed a simple heuristic for finding this type of motion. To evaluate the initialization method we use the aforementioned method to find potential image triplets to initialize from.

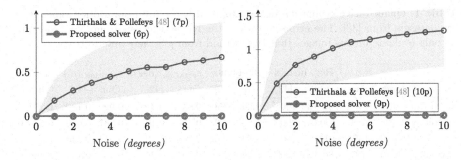

Fig. 5. Comparison with projective solvers from [48]. The graphs show the median error of the rotation matrix constraint (shadowed region shows quartiles) for 10,000 random instances. *Left:* Radial trifocal tensor. (Sect. 2.1) *Right:* Mixed trifocal tensor. (Sect. 2.2)

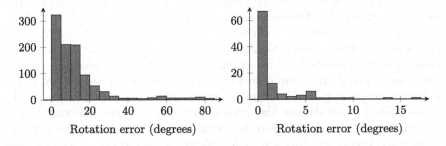

Fig. 6. Rotation errors (in degrees) for the initialization. *Left:* Three-view radial trifocal tensor estimation. *Right:* Full initialization pipeline (five views).

We select 1000 triplets from the *Lund Cathedral* dataset from [38]. For each triplet we estimate the trifocal tensor and compute the errors in the relative rotations w.r.t. the reconstruction provided in [38]. This is shown in Fig. 6 (Left). Note that some of the selected triplets do not satisfy the assumption of intersecting principal axes, leading to large errors. For the 100 best triplets (highest inlier ratio) we try to further initialize by selecting the two additional images with the most matches. Figure 6 (Right) shows the distribution of the rotation errors for all five cameras used to initialize.

4.4   Structure-from-Motion Evaluation

For the quantitative evaluation of our SfM pipeline we consider five datasets from Olsson et al. [38]. We compare with vanilla COLMAP [42] using the ground truth camera intrinsics. Since we use a subset of the geometric constraints used in COLMAP, this provides an upper bound on the reconstruction quality we can achieve. The reconstructions from [38] are used as a pseudo-ground truth. Table 1 shows the camera pose errors and statistics after robustly aligning the coordinate systems to the ground truth. Since we only recover the camera position up to an

Table 1. Quantitative evaluation of the proposed Structure-from-Motion pipeline on the datasets from [38]. The errors are w.r.t. the reconstructions from [38]. Note that we only evaluate on the images that the method from [38] were able to register.

		Reg. Images		3D Points		$\varepsilon_{\text{rotation}}$ (deg)		$\varepsilon_{\text{position}}$ (m)		$\varepsilon_{\text{reproj}}$ (px)	
Dataset	Images	Our	[42]	Our	[42]	Our	[42]	Our	[42]	Our	[42]
Lund Cathedral	1208	99.6%	100%	422k	535k	0.93	0.39	0.929	0.180	0.287	0.578
Orebro Castle	761	100.0%	100%	197k	246k	0.15	0.10	0.387	0.089	0.276	0.532
San Marco	1498	100.0%	100%	293k	325k	0.50	0.29	0.614	0.140	0.443	0.751
Spilled Blood	781	80.3%	100%	285k	328k	0.72	0.26	0.231	0.134	0.409	0.571
Doge Palace	241	100.0%	100%	74k	93k	0.20	0.20	0.154	0.110	0.293	0.605

unknown forward translation, the position error measures the distance from the ground truth camera center to the principal axis for both our and the baseline method [42]. The scales of the reconstructions from [38] were manually corrected. Image are considered correctly registered if the rotation error is below 5° and it has at least 100 inliers. The table shows that we are able to achieve comparable reconstructions to the state-of-the-art pipeline [42] without knowing the intrinsic calibration. As expected, our reprojection errors are lower since they ignore the radial component of the errors. Figure 9 shows some qualitative results. For the *Spilled Blood* dataset the scene is highly symmetric and some images are being incorrectly registered to the wrong side of the building. Since we use weaker projection constraints it is more difficult to disambiguate these incorrect matches.

Reconstruction with Severe Radial Distortion. Next we present qualitative results of our method applied to highly distorted images and show that we achieve accurate reconstruction without directly modeling the non-linear distortion. Figure 7 shows the reconstruction results from images taken with fisheye camera (from Camposeco et al. [5]) and Fig. 8 from a GoPro camera (from Kukelova et al. [26]). In Fig. 10 we show a reconstruction from 148 fisheye images. For comparison we also show the result of running COLMAP [42] without providing it intrinsic/distortion parameters, which fails to reconstruct the scene. This experiment shows that COLMAP [42] is not always able to converge to the correct intrinsic/distortion parameters during the bundle adjustment which motivates our method. More results can be found in the supplementary material.

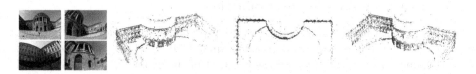

Fig. 7. *Building dataset* from [5]. 60 images, 8984 3D-points, 0.38 px average reprojection error.

Fig. 8. *Rotunda dataset* from [26]. 62 images, 16292 3D-points, 0.41 px average reprojection error.

Fig. 9. Qualitative results for *Lund Cathedral* from Sect. 4.4. 1226 images, 422939 3D-points, 0.29 px average reprojection error.

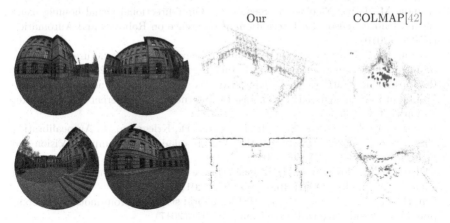

Fig. 10. *Fisheye Dataset*, 148 images, 14893 points, 0.51 px average reprojection error. Without known intrinsic/distortion parameters COLMAP fails to reconstruct the scene, while the proposed method successfully reconstructs it.

5   Conclusions

We have presented an incremental Structure-from-Motion pipeline using the 1D radial camera model. Since the model is invariant to radial displacements in the image, we can directly perform reconstruction from heavily distorted images without any offline calibration or even explicitly modelling the type of distortion.

In this paper we deliberately focused on the most difficult setup where every camera is modeled as a radial camera, making the initialization more complex. In practice, for heterogeneous image collections it is possible to only model the cameras with high distortion effects as radial cameras and use a pinhole-like model for the others. This would allow for an easier and more general initialization procedure; either from two pinhole cameras, or from one pinhole camera together with two radial cameras (Sect. 2.2). In principle it is possible to use the reconstructions we recover to calibrate the cameras, e.g. using [25, 30] for parametric distortion models or [5] for non-parametric. Even without this post-calibration step we have shown that we can achieve accurate 3D reconstruction.

Acknowledgements. Viktor Larsson was supported by an ETH Zurich Postdoctoral Fellowship.

References

1. Agarwal, S., Mierle, K., Others: Ceres solver. http://ceres-solver.org
2. Angst, R., Zach, C., Pollefeys, M.: The generalized trace-norm and its application to structure-from-motion problems. In: International Conference on Computer Vision (ICCV) (2011)
3. Aranda, M., López-Nicolás, G., Sagüés, C.: Omnidirectional visual homing using the 1D trifocal tensor. In: International Conference on Robotics and Automation (ICRA) (2010)
4. Åström, K., Oskarsson, M.: Solutions and ambiguities of the structure and motion problem for 1D retinal vision. J. Math. Imaging Vis. (JMIV) **12**(2), 121–135 (2000)
5. Camposeco, F., Sattler, T., Pollefeys, M.: Non-parametric structure-based calibration of radially symmetric cameras. In: International Conference on Computer Vision (ICCV) (2015)
6. Chandraker, M., Agarwal, S., Kahl, F., Nistér, D., Kriegman, D.: Autocalibration via rank-constrained estimation of the absolute quadric. In: Computer Vision and Pattern Recognition (CVPR) (2007)
7. Chatterjee, A., Govindu, V.M.: Robust relative rotation averaging. Trans. Pattern Anal. Mach. Intell. (PAMI) **40**(4), 958–972 (2017)
8. Cui, H., Gao, X., Shen, S., Hu, Z.: HSFM: hybrid structure-from-motion. In: Computer Vision and Pattern Recognition (CVPR) (2017)
9. Cui, Z., Tan, P.: Global structure-from-motion by similarity averaging. In: International Conference on Computer Vision (ICCV) (2015)
10. Dai, Y., Li, H., He, M.: Projective multiview structure and motion from element-wise factorization. Trans. Pattern Anal. Mach. Intell. (PAMI) **35**(9), 2238–2251 (2013)
11. Dellaert, F., Stroupe, A.W.: Linear 2D localization and mapping for single and multiple robot scenarios. In: International Conference on Robotics and Automation (ICRA) (2002)
12. Demazure, M.: Sur deux problemes de reconstruction. Technical report, RR-0882, INRIA, July 1988. https://hal.inria.fr/inria-00075672
13. Duff, T., Kohn, K., Leykin, A., Pajdla, T.: PLMP-point-line minimal problems in complete multi-view visibility. In: International Conference on Computer Vision (ICCV) (2019)

14. Eriksson, A., Olsson, C., Kahl, F., Chin, T.J.: Rotation averaging and strong duality. In: Computer Vision and Pattern Recognition (CVPR) (2018)
15. Fabbri, R., et al.: Trifocal relative pose from lines at points and its efficient solution. arXiv preprint arXiv:1903.09755 (2019)
16. Faugeras, O., Quan, L., Strum, P.: Self-calibration of a 1D projective camera and its application to the self-calibration of a 2D projective camera. Trans. Pattern Anal. Mach. Intell. (PAMI) **22**(10), 1179–1185 (2000)
17. Hartley, R., Schaffalitzky, F.: Reconstruction from projections using grassmann tensors. Int. J. Comput. Vis. (IJCV) **83**(3), 274–293 (2009)
18. Hartley, R., Trumpf, J., Dai, Y., Li, H.: Rotation averaging. Int. J. Comput. Vis. (IJCV) **103**(3), 267–305 (2013)
19. Hartley, R., Zisserman, A.: Multiple View Geometry in Computer Vision. Cambridge University Press, Cambridge (2003)
20. Hong, J.H., Zach, C., Fitzgibbon, A., Cipolla, R.: Projective bundle adjustment from arbitrary initialization using the variable projection method. In: Leibe, B., Matas, J., Sebe, N., Welling, M. (eds.) ECCV 2016. LNCS, vol. 9905, pp. 477–493. Springer, Cham (2016). https://doi.org/10.1007/978-3-319-46448-0_29
21. Hyeong Hong, J., Zach, C.: pose: Pseudo object space error for initialization-free bundle adjustment. In: Computer Vision and Pattern Recognition (CVPR) (2018)
22. Kileel, J.: Minimal problems for the calibrated trifocal variety. SIAM J. Appl. Algebra Geom. **1**(1), 575–598 (2017)
23. Kim, J.H., Dai, Y., Li, H., Du, X., Kim, J.: Multi-view 3D reconstruction from uncalibrated radially-symmetric cameras. In: International Conference on Computer Vision (ICCV) (2013)
24. Kozuka, K., Sato, J.: Multiple view geometry for mixed dimensional cameras. In: International Conference on Computer Vision Theory and Applications (VISAPP) (2008)
25. Kukelova, Z., Bujnak, M., Pajdla, T.: Real-time solution to the absolute pose problem with unknown radial distortion and focal length. In: International Conference on Computer Vision (ICCV) (2013)
26. Kukelova, Z., Heller, J., Bujnak, M., Fitzgibbon, A., Pajdla, T.: Efficient solution to the epipolar geometry for radially distorted cameras. In: International Conference on Computer Vision (ICCV) (2015)
27. Larsson, V., Astrom, K., Oskarsson, M.: Efficient solvers for minimal problems by syzygy-based reduction. In: Computer Vision and Pattern Recognition (CVPR) (2017)
28. Larsson, V., Kukelova, Z., Zheng, Y.: Making minimal solvers for absolute pose estimation compact and robust. In: International Conference on Computer Vision (ICCV) (2017)
29. Larsson, V., Kukelova, Z., Zheng, Y.: Camera pose estimation with unknown principal point. In: Computer Vision and Pattern Recognition (CVPR) (2018)
30. Larsson, V., Sattler, T., Kukelova, Z., Pollefeys, M.: Revisiting radial distortion absolute pose. In: International Conference on Computer Vision (ICCV) (2019)
31. Lebeda, K., Matas, J., Chum, O.: Fixing the locally optimized RANSAC. In: British Machine Vision Conference (BMVC) (2012)
32. Leonardos, S., Tron, R., Daniilidis, K.: A metric parametrization for trifocal tensors with non-colinear pinholes. In: Computer Vision and Pattern Recognition (CVPR) (2015)
33. Locher, A., Havlena, M., Van Gool, L.: Progressive structure from motion. In: European Conference on Computer Vision (ECCV) (2018)

34. Magerand, L., Del Bue, A.: Revisiting projective structure for motion: a robust and efficient incremental solution. Trans. Pattern Anal. Mach. Intell. (PAMI) **42**(2), 430–443 (2018)
35. Martyushev, E.: On some properties of calibrated trifocal tensors. J. Math. Imaging Vis. (JMIV) **58**(2), 321–332 (2017)
36. Molana, R., Daniilidis, K.: A single-perspective novel panoramic view from radially distorted non-central images. In: British Machine Vision Conference (BMVC) (2007)
37. Moulon, P., Monasse, P., Marlet, R.: Global fusion of relative motions for robust, accurate and scalable structure from motion. In: International Conference on Computer Vision (ICCV) (2013)
38. Olsson, C., Enqvist, O.: Stable structure from motion for unordered image collections. In: Scandinavian Conference on Image Analysis (SCIA) (2011)
39. Quan, L., Kanade, T.: Affine structure from line correspondences with uncalibrated affine cameras. Trans. Pattern Anal. Mach. Intell. (PAMI) **19**(8), 834–845 (1997)
40. Sagues, C., Murillo, A., Guerrero, J.J., Goedemé, T., Tuytelaars, T., Van Gool, L.: Localization with omnidirectional images using the radial trifocal tensor. In: International Conference on Robotics and Automation (ICRA) (2006)
41. Sattler, T., Sweeney, C., Pollefeys, M.: On sampling focal length values to solve the absolute pose problem. In: Fleet, D., Pajdla, T., Schiele, B., Tuytelaars, T. (eds.) ECCV 2014. LNCS, vol. 8692, pp. 828–843. Springer, Cham (2014). https://doi.org/10.1007/978-3-319-10593-2_54
42. Schonberger, J.L., Frahm, J.M.: Structure-from-motion revisited. In: Computer Vision and Pattern Recognition (CVPR) (2016)
43. Schönberger, J.L., Zheng, E., Frahm, J.-M., Pollefeys, M.: Pixelwise view selection for unstructured multi-view stereo. In: Leibe, B., Matas, J., Sebe, N., Welling, M. (eds.) ECCV 2016. LNCS, vol. 9907, pp. 501–518. Springer, Cham (2016). https://doi.org/10.1007/978-3-319-46487-9_31
44. Schops, T., Larsson, V., Pollefeys, M., Sattler, T.: Why having 10,000 parameters in your camera model is better than twelve. In: Computer Vision and Pattern Recognition (CVPR) (2020)
45. Snavely, N., Seitz, S.M., Szeliski, R.: Modeling the world from internet photo collections. Int. J. Comput. Vis. (IJCV) **80**(2), 189–210 (2008)
46. Sweeney, C., Holynski, A., Curless, B., Seitz, S.M.: Structure from motion for panorama-style videos. arXiv preprint arXiv:1906.03539 (2019)
47. Sweeney, C., Sattler, T., Hollerer, T., Turk, M., Pollefeys, M.: Optimizing the viewing graph for structure-from-motion. In: International Conference on Computer Vision (ICCV) (2015)
48. Thirthala, S., Pollefeys, M.: Radial multi-focal tensors. Int. J. Comput. Vis. (IJCV) **96**(2), 195–211 (2012)
49. Triggs, B.: Matching constraints and the joint image. In: International Conference on Computer Vision (ICCV) (1995)
50. Triggs, B.: Factorization methods for projective structure and motion. In: Computer Vision and Pattern Recognition (CVPR) (1996)
51. Tsai, R.: A versatile camera calibration technique for high-accuracy 3D machine vision metrology using off-the-shelf TV cameras and lenses. J. Robot. Autom. **3**(4), 323–344 (1987)
52. Ventura, J.: Structure from motion on a sphere. In: Leibe, B., Matas, J., Sebe, N., Welling, M. (eds.) ECCV 2016. LNCS, vol. 9907, pp. 53–68. Springer, Cham (2016). https://doi.org/10.1007/978-3-319-46487-9_4

53. Wu, C.: Towards linear-time incremental structure from motion. In: International Conference on 3D Vision (3DV) (2013)
54. Wu, C.: P3.5P: pose estimation with unknown focal length. In: Computer Vision and Pattern Recognition (CVPR) (2015)
55. Zhang, Z., et al.: Flexible camera calibration by viewing a plane from unknown orientations. In: International Conference on Computer Vision (ICCV) (1999)
56. Zhu, S., et al.: Very large-scale global SFM by distributed motion averaging. In: Computer Vision and Pattern Recognition (CVPR) (2018)

Occupancy Anticipation for Efficient Exploration and Navigation

Santhosh K. Ramakrishnan[1,2(✉)], Ziad Al-Halah[1], and Kristen Grauman[1,2]

[1] The University of Texas at Austin, Austin, TX 78712, USA
{srama,grauman}@cs.utexas.edu,
ziadlhlh@gmail.com
[2] Facebook AI Research, Austin, TX 78701, USA

Abstract. State-of-the-art navigation methods leverage a spatial memory to generalize to new environments, but their occupancy maps are limited to capturing the geometric structures directly observed by the agent. We propose *occupancy anticipation*, where the agent uses its egocentric RGB-D observations to infer the occupancy state beyond the visible regions. In doing so, the agent builds its spatial awareness more rapidly, which facilitates efficient exploration and navigation in 3D environments. By exploiting context in both the egocentric views and top-down maps our model successfully anticipates a broader map of the environment, with performance significantly better than strong baselines. Furthermore, when deployed for the sequential decision-making tasks of exploration and navigation, our model outperforms state-of-the-art methods on the Gibson and Matterport3D datasets. Our approach is the winning entry in the 2020 Habitat PointNav Challenge. Project page: http://vision.cs. utexas.edu/projects/occupancy_anticipation/.

1 Introduction

In visual navigation, an agent must move intelligently through a 3D environment in order to reach a goal. Visual navigation has seen substantial progress in the past few years, fueled by large-scale datasets and photo-realistic 3D environments [4,9,62,67], simulators [3,31,35,67], and public benchmarks [3,12,35]. Whereas traditionally navigation was attempted using purely geometric representations (i.e., SLAM), recent work shows the power of *learned* approaches to navigation that integrate both geometry and semantics [11,19,38,50,70,72]. Learned approaches operating directly on pixels and/or depth as input can be robust to noise [10,11] and can generalize well on unseen environments [10,19,35,70]—even outperforming pure SLAM given sufficient experience [35].

One of the key factors for success in navigation has been the movement towards complex map-based architectures [10,11,19,41] that capture both geometry [10,11,19] and semantics [18,19,22,41], thereby facilitating efficient policy

Electronic supplementary material The online version of this chapter (https:// doi.org/10.1007/978-3-030-58558-7_24) contains supplementary material, which is available to authorized users.

A. Vedaldi et al. (Eds.): ECCV 2020, LNCS 12350, pp. 400–418, 2020.
https://doi.org/10.1007/978-3-030-58558-7_24

learning and planning. These learned maps allow an agent to exploit prior knowledge from training scenes when navigating in novel test environments.

Despite such progress, state-of-the-art approaches to navigation are limited to encoding *what the agent actually sees in front of it*. In particular, they build maps of the environment using only the *observed* regions, whether via geometry [11,22] or learning [10,18,19,41]. Thus, while promising, today's models suffer from an important inefficiency: to map a space in the 3D environment as free or occupied, the agent must directly see evidence thereof in its egocentric camera.

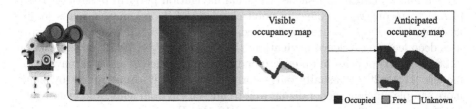

Fig. 1. Occupancy anticipation: A robot's perception of the 3D world is limited by its field-of-view and obstacles (the visible map). We propose to anticipate occupancy for unseen regions (anticipated map) by exploiting the context from egocentric views. We then train a deep reinforcement learning agent to move intelligently in a 3D environment, rewarding movements that improve the anticipated map. (Color figure online)

Our key idea is to *anticipate occupancy*. Rather than wait to directly observe a more distant or occluded region of the 3D environment to declare its occupancy status, the proposed agent infers occupancy for unseen regions based on the visual context in its egocentric views. For example, in Fig. 1, with only the partial observation of the scene, the agent could infer that it is quite likely that the wall extends to its right, a corridor is present on its left, and the region immediately in front of it is free space. Such intelligent extrapolation beyond the observed space would lead to more efficient exploration and navigation. To achieve this advantage, we introduce a model that anticipates occupancy maps from normal field-of-view RGB(D) observations, while aggregating its predictions over time in tight connection with learning a navigation policy. Furthermore, we incorporate the anticipation objective directly into the agent's exploration policy, encouraging movements in the 3D space that will efficiently yield broader and more accurate inferred occupancy maps.

We validate our approach on Gibson [67] and Matterport3D [9], two 3D environment datasets spanning over 170 real-world spaces with a variety of obstacles and floor plans. Using only RGB(D) inputs to anticipate occupancy, the proposed agent learns to explore intelligently, achieving faster and more accurate maps compared to a state-of-the-art approach for neural SLAM [10], and navigating more efficiently than strong baselines. Furthermore, for navigation under noisy actuation and sensing, our agent improves the state of the art, winning the 2020 Habitat PointNav Challenge [1] by a margin of 6.3 SPL points.

Our main contributions are: (1) a novel occupancy anticipation framework that leverages visual context from egocentric RGB(D) views; (2) a novel exploration approach that incorporates intelligent anticipation for efficient environment mapping, providing better maps in less time; and (3) successful navigation results that improve the state of the art.

2   Related Work

Navigation. Classical approaches to visual navigation perform passive or active SLAM to reconstruct geometric point-clouds [21,64] or semantic maps [5,49], facilitated by loop closures or learned odometry [7,8,36]. More recent work uses deep learning to learn navigation [19,38,50,53,57,68,70,72] or exploration [6,26,42,45,51] policies in an end-to-end fashion. Explicit *map-based* navigation models [11,18,20,41] usually outperform their implicit counterparts by being more sample-efficient, generalizing well to unseen environments, and even transferring from simulation to real robots [10,19]. However, existing approaches only encode *visible* regions for mapping (i.e., the ground plane projection of the observed or inferred depth). In contrast, our model goes beyond the visible cues and anticipates maps for unseen regions to accelerate navigation.

Layout Estimation. Recent work predicts 3D Manhattan layouts of indoor scenes given 360 panoramas [14,63,66,69,73]. These methods predict structured outputs such as layout boundaries [63,73], corners [73], and floor/ceiling probability maps [69]. However, they do not extrapolate to unseen regions. FloorNet [33] and Floor-SP [27] use walkthroughs of previously scanned buildings to reconstruct detailed floorplans that may include predictions for the room type, doors, objects, etc. However, they assume that the layouts are polygonal, the scene is fully explored, and that detailed human annotations are available. Our occupancy map representation can be seen as a new way for the agent to infer the layout of its surroundings. Unlike any of the above approaches, our model does not make strict assumptions on the scene structure, nor does it require detailed semantic annotations. Furthermore, the proposed anticipation model is learned jointly with the exploration policy and without human guidance. Finally, unlike prior work, our goal is to accelerate navigation and map creation.

Scene Completion. Past work in scene completion focuses on pixelwise reconstruction of 360 panoramas with limited glimpses [26,44,45,55], inpainting [24,32,43], and inferring unseen 3D structure and semantics [61,71]. While some methods allow pixelwise extrapolation outside the current field of view (FoV) [25,45,61,71], they do not permit inferences about occluded regions in the scene. Our results show that this limitation is detrimental to successful occupancy estimation (cf. our view extrapolation baseline). SSCNet [60] performs voxelwise geometric and semantic predictions for unseen 3D structures; however, it is computationally expensive, requires voxelwise semantic labels, limits predictions to the agent's FoV, and needs carefully curated viewpoints for training. In contrast, our approach predicts 2D occupancy from egocentric RGB(D) views, and it learns to do so in an active perception setting. Since the agent controls its own camera, the viewpoints tend to be more challenging than those in curated datasets of human-taken photos used in the scene completion literature [26,44,60,61,71].

Occupancy Maps. In robotics, methods for occupancy focus on building continuous representations of the world [40,47,56], mapping for autonomous driving [23,34,37,39,59], and indoor robot navigation [15,29,58]. Prior extrapolation methods assume wide FoV LIDAR inputs, only exploit geometric cues from top-down views, and demonstrate results in relatively simple 2D floorplans devoid of non-wall obstacles [15,29,30,58]. In contrast, our approach does not require expensive LIDAR sensors. It operates with standard RGB(D) camera inputs, and it exploits both semantic and geometric context from those egocentric views to perform accurate occupancy anticipation. Furthermore, we demonstrate efficient navigation in visually rich 3D environments with challenging obstacles other than walls. Finally, unlike prior work, our anticipation models are learned jointly with a navigation policy that rewards accurate anticipatory mapping.

3 Approach

We propose an occupancy anticipation approach for efficient exploration and navigation. Our model anticipates areas not directly visible to the agent because of occlusion (e.g., behind a table, around a corner) or due to being outside its FoV. The agent's first-person view is provided in the form of RGB-D images (see Fig. 2 left). The goal is to anticipate the occupancy for a fixed region in front of the agent, and integrate those predictions over time as the agent moves about.

Next, we define the task setup and notation, followed by our approach for occupancy anticipation (Sect. 3.1) and a new formulation for exploration that rewards correctly anticipated regions (Sect. 3.2). Then, we explain how our occupancy anticipation model can be integrated into a state-of-the-art approach [10] for autonomous exploration and navigation in 3D environments (Sect. 3.3).

3.1 Occupancy Anticipation Model

We formulate occupancy anticipation as a pixelwise classification task. The egocentric occupancy is represented as a two-channel top-down map $p \in [0,1]^{2 \times V \times V}$ which comprises a local area of $V \times V$ cells in front of the camera. Each cell in the map represents a $25\,\mathrm{mm} \times 25\,\mathrm{mm}$ region. The two channels contain the probabilities (confidence values) of the cell being occupied and explored, respectively. A cell is considered to be occupied if there is an obstacle, and it is explored if we know whether it is occupied or free. For training, we use the 3D meshes of indoor environments (Sect. 4.1) to obtain the ground-truth local occupancy of a $V \times V$ region in front of the camera, which includes parts that may be occluded or outside the field of view (Fig. 2, bottom right).

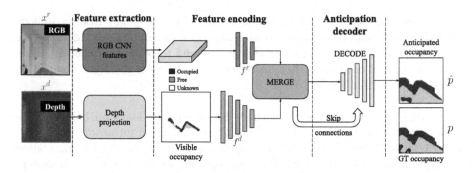

Fig. 2. Our occupancy anticipation model uses RGB(D) inputs to extract features, and processes them using a UNet to anticipate the occupancy. The depth map is projected to the ground plane to obtain the preliminary visible occupancy map. See text. (Color figure online)

Our occupancy anticipation model consists of three main components (Fig. 2):

(1) Feature Extraction: Given egocentric RGB-D inputs, we compute:

RGB CNN features: We encode the RGB images using blocks 1 and 2 of a ResNet-18 that is pre-trained on ImageNet, followed by three additional convolution layers that prepare these features to be passed forward with the visible occupancy map. This step extracts a mixture of textural and semantic features.

Depth Projection: We estimate a map of occupied, free, and unknown space by setting height thresholds on the point cloud obtained from depth and camera intrinsics [11]. Consistent with past work [10,11], we restrict the projection-based estimates to points within $\sim$ 3m, the range at which modern depth sensors would provide reliable results. This yields the initial visible occupancy map.

(2) Feature Encoding: Given the RGB-D features, we independently encode them using UNet [48] encoders and project them to a common feature space. We encode the depth projection features using a stack of five convolutional blocks which results in features $\boldsymbol{f}^d = f_{1:5}^d$. Since the RGB features are already at a lower resolution, we use only three convolutional blocks to encode them, which results in features $\boldsymbol{f}^r = f_{3:5}^r$. We then combine these features using the MERGE module which contains layer-specific convolution blocks to merge each $[\boldsymbol{f}_i^r, \boldsymbol{f}_i^d]$:

$$\boldsymbol{f} = \text{MERGE}(\boldsymbol{f}^d, \boldsymbol{f}^r). \tag{1}$$

For experiments with only the depth modality, we skip the RGB feature extractor and MERGE layer and directly use the occupancy features obtained from the depth image. For experiments with only the RGB modality, we learn a model to infer the visible occupancy features from RGB (to be defined at the end of Sect. 4.1) and use that instead of the features computed from the depth image.

(3) Anticipation Decoding: Given the encoded features f, we use a UNet decoder that outputs a $2 \times V \times V$ tensor of probabilities:

$$\hat{p} = \sigma(\text{DECODE}(f)), \tag{2}$$

where $\hat{p} \in [0,1]^{2 \times V \times V}$ is the estimated egocentric occupancy and σ is the sigmoid activation function. For training the occupancy anticipation model, we use binary cross entropy loss per pixel and per channel:

$$L = \sum_{i=1}^{V^2} \sum_{j=1}^{2} -\left[p_{ij} \log \hat{p}_{ij} + (1 - p_{ij}) \log(1 - \hat{p}_{ij}) \right], \tag{3}$$

where p is the ground-truth (GT) occupancy map that is derived from the 3D mesh of training environments (see Sect. S5 in Supp. for details).

So far, we have presented our occupancy anticipation approach supposing a single RGB-D observation as input. However, our model is ultimately used in the context of an embodied agent that moves in the environment and actively collects a sequence of RGB-D views to build a complete map of the environment. Next, we introduce a new reward function that utilizes the agent's anticipation performance to guide its exploration during training.

3.2   Anticipation Reward for Exploration Policy Learning

In *visual exploration*, an agent must quickly map a new environment without having a specified target. Prior work on exploration [10,11,16,46] often uses area-coverage—the area seen in the environment during navigation—as a reward function to guide exploration. However, the traditional area-coverage approach is limited to rewarding the agent only for *directly seeing* areas. Arguably, an ideal exploration agent would obtain an accurate and complete map of the environment *without* necessarily directly observing all areas.

Thus, we propose to encourage exploratory behaviors that yield a correctly *anticipated* map. In this case, the occupancy entries in the map need not be obtained via direct agent observations to register a reward; it is sufficient to correctly infer them. In particular, we reward agent actions that yield accurate occupancy predictions for the global environment map, i.e., the number of grid cells where the predicted occupancy matches the layout of the environment.

More concretely, let $\hat{m}_t \in [0,1]^{2 \times G \times G}$ be the global environment map obtained by anticipating occupancy for the RGB-D observations $\{x_{1:t}^r, x_{1:t}^d\}$ from time 1 to t, and then geometrically registering the predictions to a single global map based on the agent's pose estimates at each time step (see Fig. 3). Note $G > V$. Let m be the ground-truth layout of the environment. Then, the unnormalized accuracy of a map prediction $\hat{m}$ is measured as follows:

$$\text{Accuracy}(\hat{m}, m) = \sum_{i=1}^{G^2} \sum_{j=1}^{2} \mathbb{1}[\hat{m}_{ij} = m_{ij}], \tag{4}$$

where $\mathbb{1}[\hat{m}_{ij} = m_{ij}]$ is an indicator function that returns one if $\hat{m}_{ij} = m_{ij}$ and zero otherwise. We reward the increase in map accuracy from time $t - 1$ to t:

$$R_t^{anticp} = \text{Accuracy}(\hat{m}_t, m) - \text{Accuracy}(\hat{m}_{t-1}, m). \tag{5}$$

This function rewards actions leading to correct global map predictions, irrespective of whether the agent actually *observed* those locations. For example, if the agent correctly anticipates free space behind a table and is rewarded for that, it then learns to avoid spending additional time around tables in the future to observe that space directly. Resources can be instead allocated to visiting more interesting regions that are harder to anticipate. Additionally, this reward provides a better learning signal while training under noisy conditions by accounting for mapping errors arising from noisy pose and map predictions. Thus, our approach encourages more intelligent exploration behavior by injecting our anticipated occupancy idea directly into the agent's sequential decision-making.

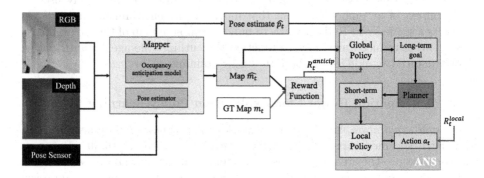

Fig. 3. Exploration with occupancy anticipation: We introduce two key upgrades to the original Active Neural SLAM (ANS) model [10] (see text): (1) We replace the projection unit in the mapper with our occupancy anticipation model (see Fig. 2). (2) We replace the area-coverage reward function with the proposed reward (Eq. 5), which encourages the agent to efficiently explore and build accurate maps through occupancy anticipation. Note that the reward signals (in red) are provided only during training. (Color figure online)

3.3   Exploration and Navigation with Occupancy Anticipation

Having defined the core occupancy anticipation components, we now demonstrate how our model can be used to benefit embodied navigation in 3D environments. We consider both exploration (discussed above) and *PointGoal navigation* [2,52], a.k.a PointNav, where the agent must navigate efficiently to a target specified by a displacement vector from the agent's starting position.

For both tasks, we adapt the state-of-the-art Active Neural SLAM (ANS) architecture [10] that previously achieved the best exploration results in the literature and was the winner of the 2019 Habitat PointNav challenge. However, our anticipation model is generic and can be easily integrated with most map-based embodied navigation models [11,17,19].

The ANS model is a hierarchical, modular policy for exploration that consists of a mapper, a planner, a local policy, and a global policy (shown in Fig. 3). Given RGB images, the mapper estimates the egocentric occupancy and agent pose, and then temporally aggregates the maps into a global top-down map using the pose estimates. At regular time intervals Δ, the global policy picks a location on the global map to explore. A shortest-path planner decides what trajectory to take from the current position to the target and picks an intermediate goal (within 1.25 m) to navigate to. The local policy then selects actions that lead to the intermediate goal; it gets another intermediate goal upon reaching the current goal. See [10] for details. Critically, and like other prior work, the model of [10] is supervised to generate occupancy estimates based solely on the *visible* occupancy obtained from the egocentric views.

We adapt ANS by modifying the mapper and the reward function. For the mapper, we replace the projection unit from ANS with our anticipation model (see Fig. 3). Additionally, we account for incorrect occupancy estimates in two ways: (1) we filter out high entropy predictions and (2) we maintain a moving average estimate of occupancy at each location in the global map (see Sect. S7 in Supp.). For the reward function, we use the anticipation-based reward presented in Sect. 3.2.

We train the exploration policy with our anticipation model end-to-end, as this allows adapting to the changing distribution of the agent's inputs. Both the local and the global reinforcement learning policies are trained with Proximal Policy Optimization (PPO) [54]. In our model, the reward of the global policy is our anticipation-based reward defined in Eq. 5. This replaces the traditional area-coverage reward used in ANS and other current models [10,11,46], which rewards the increment in the actual area seen, not the correctly registered area in the map. The reward for the local policy is simply based on the reduction in the distance to the local goal: $R_t^{local} = d_{t-1} - d_t$, where d is the Euclidean distance between the current position and the local goal.

4   Experiments

In the following experiments we demonstrate that 1) our occupancy anticipation module can successfully infer unseen parts of the map (Sect. 4.2) and 2) trained together with an exploration and navigation policy, it accelerates active mapping and navigation in new environments (Sect. 4.3 and Sect. 4.4).

4.1   Experimental Setup

We use the Habitat [35] simulator along with Gibson [67] and Matterport3D [9] environments. Each dataset contains around 90 challenging large-scale photo-realistic 3D indoor environments such as houses and office buildings. On average, the Matterport3D environments are larger. Our observation space consists of 128×128 RGB-D observations and odometry sensor readings that denote the change in the agent's pose x, y, θ. Our action space consists of three actions: MOVE-FORWARD by 25 cm, TURN-LEFT by $10°$, TURN-RIGHT by $10°$. For navigation, we add a STOP action, which the agent emits when it believes it has reached the goal. We simulate noisy actuation and odometer readings for realistic evaluation (see Sect. S6 in Supp.).

We train our exploration models on Gibson, and then transfer them to Point-Goal navigation on Gibson and exploration on Matterport3D. We use the default train/val/test splits provided for both datasets [35] with disjoint environments across the splits. For evaluation on Gibson, we divide the validation environments into small (area less than $36 \, \mathrm{m}^2$) and large (area greater than $36 \, \mathrm{m}^2$) to observe the influence of environment size on results. For policy learning, we use the Adam optimizer and train on episodes of length 1000 for $1.5 - 2$ million frames of experience. Please see Sect. S8 in Supp. for more details.

Baselines: We define baselines based on prior work:

- **ANS(rgb)** [10]: This is the state-of-the-art Active Neural SLAM approach for exploration and navigation. We use the original mapper architecture [10], which infers the visible occupancy from RGB.[1]
- **ANS(depth)**: We use depth projection to infer the visible occupancy (similar to [11]) instead of predicting it from RGB.
- **View-extrap.**: We extrapolate an $180°$ FoV depth map from $90°$ FoV RGB-D and project it to the top-down view. This is representative of scene completion approaches [61,71]. See Sect. S11 in Supp. for network details.
- **OccAnt(GT)**: This is an upper bound that cheats by using the ground-truth anticipation maps for exploration and navigation.

We implement all baselines on top of the ANS framework. Our goal is to show the impact of our occupancy model, while fixing the backbone navigation architecture and policy learning approach across methods for a fair comparison. We consider three versions of our models based on the input modality:

- **OccAnt(depth)**: anticipate occupancy given the visible occupancy map.
- **OccAnt(rgb)**: anticipate occupancy given only the RGB image. We replace the depth projections in Fig. 2 with the pre-trained ANS(rgb) estimates (kept frozen throughout training).
- **OccAnt(rgbd)**: anticipate occupancy given the full RGB-D inputs.

By default, our methods use the proposed anticipation reward from Sect. 3.2. We denote ablations without this reward as "w/o AR".

4.2 Occupancy Anticipation Results

First we evaluate the per-frame prediction accuracy of the mapping models trained during exploration. We evaluate on a separate dataset of images sampled from validation environments in Gibson at uniform viewpoints from discrete

---

[1] We use our own implementation of ANS since authors' code was unavailable at the time of our experiments. See Sect. S7 in Supp. for details.

Table 1. Occupancy anticipation results on the Gibson validation set. Our models, OccAnt($\cdot$), substantially improve the map quality and extent, showing the advantage of learning to anticipate 3D structures beyond those directly observed.

Method	IoU %			F1 score %		
	Free	Occ	Mean	Free	Occ	Mean
All-free	30.1	0	15.1	43.6	0	21.8
All-occupied	0	25.1	12.6	0	39.2	19.6
ANS(rgb)	12.1	14.9	13.5	19.6	24.9	22.5
ANS(depth)	14.5	24.1	19.3	23.1	37.6	30.4
View-extrap	15.5	26.4	21.0	25.0	40.4	32.7
OccAnt(rgb)	44.4	47.9	46.1	58.2	62.9	60.6
OccAnt(depth)	50.4	**61.9**	56.1	63.8	**75.0**	69.4
OccAnt(rgbd)	**51.5**	61.5	**56.5**	**64.9**	74.8	**69.8**

locations on a 1m grid, a total of 1,034 (input, output) samples. This allows standardized evaluation of the mapper, independent of the exploration policy.

To quantify the local occupancy maps' accuracy, we compare the predicted maps to the ground truth. We report the Intersection over Union (IoU) and F1 scores for the "free" and "occupied" classes independently. In addition to the baselines from Sect. 4.1, we add two naive baselines that classify all locations as free (all-free), or occupied (all-occupied).

Table 1 shows the results. Our anticipation models OccAnt are substantially better than all the baselines. Comparing different modalities, OccAnt(depth) is much better than OccAnt(rgb) under all the metrics. This makes sense, as visible occupancy is directly computable from the depth input, but must be inferred for RGB (see Fig. 4). Interestingly, the rgbd models are not better than the depth-only models, likely because (1) geometric cues are more easily learned from depth than RGB, and (2) the RGB encoder contains significantly more parameters and could lead to overfitting. See Table S5 in Supp. for network sizes. Overall, Table 1 demonstrates our occupancy anticipation models successfully broaden the coverage of the map beyond the visible regions.

4.3   Exploration Results

Next we deploy our models for visual exploration. The agent is given a limited time budget (T=1000) to intelligently explore and build a 2D top-down occupancy map of a previously unseen environment.

To quantify exploration, we measure both map quality and speed (number of agent actions): (1) **Map accuracy (m^2):** the area in the global map built during exploration (both free and occupied) that matches with the ground-truth layout of the environment. The map is built using predicted occupancy maps which are registered using estimated pose (may be noisy). Note that this is an unnormalized accuracy measure (see Eq. 4). (2) **IoU:** the intersection over union between that same global map and the ground-truth layout of the environment. (3) **Area seen** (m^2): the amount of free and occupied regions *directly seen* during exploration. The map for this metric is built using ground-truth pose and depth-projections (similar to [10,11]). (4) **Episode steps:** the number of actions taken by the agent. While the first two metrics measure the quality of the created map, the latter two are a function of how (and how long) the agent moved to get that map. Higher accuracy in fewer steps or lower area-seen is better.

All agents are trained on 72 scenes from Gibson under noisy odometry and actuation (see Sect. 4.1), and evaluated on Gibson and Matterport3D under both noisy and noise-free conditions.

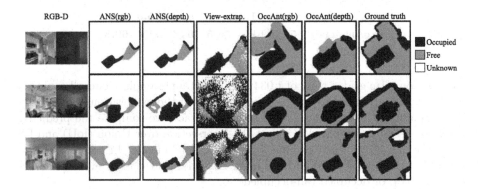

Fig. 4. Per-frame local occupancy predictions: First and last columns show the RGB-D input and anticipation ground-truth, respectively. ANS(*) are restricted to only predicting occupancy for visible regions. View-extrap. extrapolates, but is unable to predict occupancy for occluded regions (first row) and struggles to make correct predictions in cluttered scenes (second row). Our model successfully anticipates with either RGB or depth. For example, in the first row, we successfully predict the presence of a corridor and another room on the left. In the second row, we successfully predict the presence of navigable space behind the table. In the third row, we are able to correctly anticipate the free space behind the chair and the corridor to the right. (Color figure online)

Figure 6 shows the exploration results. Our approach generally outperforms the baselines, improving the map quality more rapidly, whether in terms of time (top row) or area seen (bottom row). When compared on a same-modality basis, we see that OccAnt(rgb) converges much faster than ANS(rgb). Similarly, OccAnt(depth) is able to rapidly improve the map quality and outperforms ANS(depth) on all cases. This apples-to-apples comparison shows that anticipating occupancy leads to much more efficient mapping in unseen environments. Again, using depth generally provides more reliable mapping than pure RGB.

Furthermore, the proposed anticipation reward generally provides significant benefits to map accuracy in the noisy setting (compare our full model to the

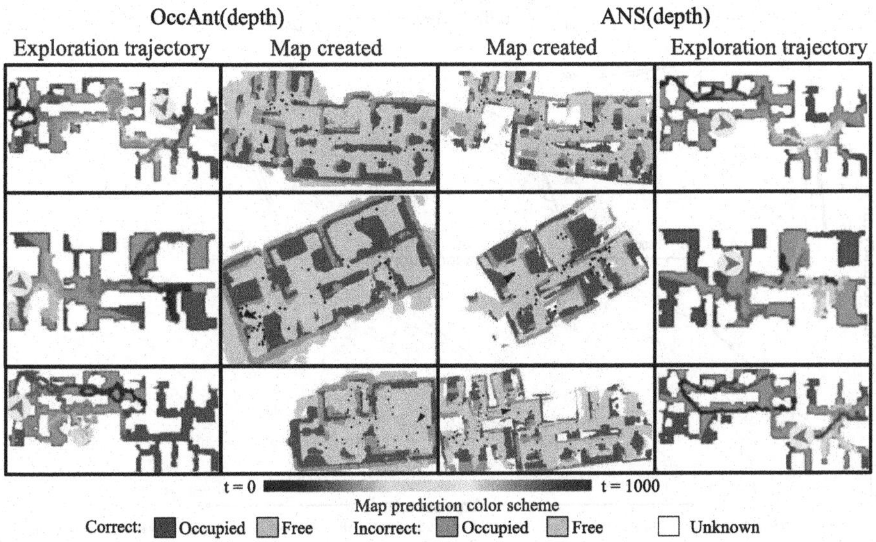
OccAnt(depth)
ANS(depth)
Exploration trajectory
Map created
Map created
Exploration trajectory
t = 0
t = 1000
Map prediction color scheme
Correct: Occupied Free
Incorrect: Occupied Free
Unknown

Fig. 5. Exploration examples: We compare OccAnt with ANS [10] in Gibson under noisy actuation and odometry. The exploration trajectories and the corresponding maps are shown at the extremes and center, respectively. **Row 1:** Both methods cover similar area, but our method better anticipates the unseen parts with fewer registration errors. **Row 2:** Our method achieves better area coverage and mapping quality whereas the baseline gets stuck in a small room for extended periods of time. **Row 3:** A failure case for our method, where it gets stuck in one part of the house after anticipating that a narrow corridor leading to a different room was occupied. (Color figure online)

Table 2. Timed exploration results: Map quality at $T = 500$ for all models and datasets. See text for details.

	Noisy test conditions						Noise-free test conditions					
	Gibson small		Gibson large		Matterport3D		Gibson small		Gibson large		Matterport3D	
Method	Map acc.	IoU	Map acc.	IoU	Map acc.	IoU	Map acc.	IoU	Map acc.	IoU	Map acc.	IoU
ANS(rgb) [10]	18.5	55	35.0	47	44.7	18	22.4	76	43.4	64	53.4	23
ANS(depth)	18.5	56	39.4	53	72.5	26	21.4	74	48.0	72	85.9	34
View-extrap.	12.0	26	28.1	27	39.4	14	12.1	27	26.5	27	33.9	13
OccAnt(rgb) w/o AR	21.8	66	44.2	57	65.8	23	22.6	71	45.2	60	64.4	24
OccAnt(depth) w/o AR	20.2	58	44.2	54	92.7	29	**24.9**	**84**	**54.1**	**75**	**104.7**	**38**
OccAnt(rgbd) w/o AR	16.9	45	35.6	40	76.3	23	24.8	**84**	52.0	71	98.7	34
OccAnt(rgb)	20.9	62	42.1	54	66.2	22	22.3	70	43.5	58	64.4	22
OccAnt(depth)	**22.7**	**71**	**50.3**	**67**	94.1	**33**	24.8	83	53.1	74	96.5	35
OccAnt(rgbd)	**22.7**	**71**	48.4	62	**99.9**	32	24.5	82	51.0	69	100.3	34
OccAnt(GT)	21.7	67	51.9	63	-	-	26.1	93	65.4	91	-	-

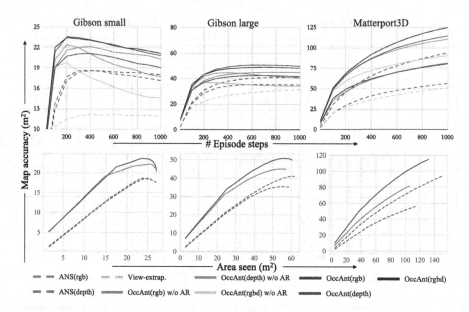

Fig. 6. Exploration results: Map accuracy (m²) as a function of episode duration (top row) and area seen (bottom row) for Gibson (small and large splits) and Matterport3D under noisy conditions (see Sect. S1 in Supp. for noise-free). Higher and steeper curves are better. **Top:** Our OccAnt approach rapidly attains higher map accuracy than the baselines (dotted lines). **Bottom:** OccAnt achieves higher map accuracy for the same area seen (we show the best variants here to avoid clutter). These results show the agent *actively moves better* to explore the environment with occupancy anticipation. (Color figure online)

Table 3. PointNav results: Our approach provides more efficient navigation.

Method	SPL %	Success %	Time taken
ANS(rgb) [10]	66.8	87.9	254.109
ANS(depth)	76.8	86.6	226.161
View-extrap.	10.4	33.3	835.556
OccAnt(rgb)	71.2	88.2	223.411
OccAnt(depth)	77.8	91.3	194.751
OccAnt(rgbd)	**80.0**	**93.0**	**171.874**
OccAnt(GT)	89.5	96.0	125.018

"w/o AR" models in Fig. 6). While map accuracy generally increases over time for noise-free conditions (see Sect. S1 in Supp.), it sometimes saturates early or even declines slightly over time in the noisy setting as noisy pose estimates accumulate and hurt map registration accuracy. This is most visible in Gibson small (top left plot). However, our anticipatory reward alleviates this decline.

Table 2 summarizes the map accuracy and IoU for all methods at $T=500$. Our method obtains significant improvements, supporting our claim that occupancy anticipation accelerates exploration and mapping. Additionally, perfect anticipation with the OccAnt(GT) model gives comparably good noisy exploration, and good gains in noise-free exploration (+10–20% IoU). This shows that there is indeed a lot of mileage in anticipating occupancy; our model moves the state-of-the-art towards this ceiling. Figure 5 shows example exploration trajectories and the final global map predictions on Gibson.

4.4   Navigation Results

Next we evaluate the utility of occupancy anticipation for quickly reaching a target. In PointNav [2,52], the agent is given a 2D coordinate (relative to its position) and needs to reach that target as quickly as possible. Following [10], we use noise-free evaluation and directly transfer the mapper, planner, and local policy learned during exploration to this task. In this way, instead of navigating to a point specified by the global policy, the agent has to navigate to a fixed goal location. To evaluate navigation, we use the standard metrics—success rate, success rate normalized by inverse path length (SPL) [2], and time taken. The agent succeeds if it stops within 0.2m of the target under a time budget of $T = 1000$.

Table 3 shows the navigation results on the Gibson validation set. Our approach outperforms the baselines. Thus, not only does occupancy anticipation successfully map the environment, but it also allows the agent to move to a specified goal more quickly by modeling the navigable spaces. This apples-to-apples comparison shows that our idea improves the state of the art for PointNav. As with exploration, using ground truth (GT) anticipation leads to good gains in the navigation performance, and our methods bridge the gap between the prior state of the art and perfect anticipation.

Table 4. Habitat Challenge 2020 results: Our approach is the winning entry.

		Test standard			Test challenge		
Rank	Team	SPL %	Success %	Team	SPL %	Success %
1	**Occupancy Anticipation**	19.2	24.8	**Occupancy Anticipation**	20.9	27.5
2	ego-localization [13]	10.4	13.6	ego-localization [13]	14.6	19.2
3	Information Bottleneck	5.0	7.5	DAN [28]	13.2	25.3
4	cogmodel_team	0.8	1.3	Information Bottleneck	6.0	8.8
5	UCULab	0.5	0.8	cogmodel_team	0.7	1.2
6	Habitat Team (DD-PPO) [65]	0.0	0.2	UCULab	0.1	0.2

In concurrent work, the DD-PPO approach [65] obtains 0.96 SPL for Point-Nav, but it requires 2.5 billion frames of experience to do so (and it fails for noisy conditions; see below). To achieve the performance of our method (0.8 SPL in 2M frames), DD-PPO requires more than 50× the experience. Our sample efficiency can be attributed to explicit mapping along with occupancy anticipation.

Finally, we validate our approach on the 2020 Habitat PointNav Challenge [1], which requires the agent to adapt to noisy RGB-D sensors and noisy actuators, and to operate without an odometer. This presents a much more difficult evaluation setup than past work which assumes perfect odometry as well as noise-free sensing and actuation [10,35,65]. See Sect. S13 in Supp. for more details. Table 4 shows the results. Our method won the challenge, outperforming the competing approaches by large margins. While our approach generalizes well to this setting, DD-PPO [65] fails (0 SPL) due to its reliance on perfect odometry.

5   Conclusion

We introduced the idea of occupancy anticipation from egocentric views in 3D environments. By learning to anticipate the navigable areas beyond the agent's actual field of view, we obtain more accurate maps more efficiently in novel environments. We demonstrate our idea both for individual local maps, as well as integrated within sequential models for exploration and navigation, where the agent continually refines its (anticipated) map of the world. Our results clearly demonstrate the advantages on multiple datasets, including improvements to the state-of-the-art embodied AI model for exploration and navigation.

Acknowledgements. UT Austin is supported in part by DARPA Lifelong Learning Machines and the GCP Research Credits Program. We thank Devendra Singh Chaplot for clarifying the implementation details for ANS.

References

1. The Habitat Challenge 2020. https://aihabitat.org/challenge/2020/
2. Anderson, P., et al.: On evaluation of embodied navigation agents. arXiv preprint arXiv:1807.06757 (2018)
3. Anderson, P., et al.: Vision-and-language navigation: interpreting visually-grounded navigation instructions in real environments. In: Proceedings of the IEEE Conference on Computer Vision and Pattern Recognition (CVPR) (2018)
4. Armeni, I., Sax, A., Zamir, A.R., Savarese, S.: Joint 2D–3D-semantic data for indoor scene understanding. ArXiv e-prints, February 2017
5. Bao, S.Y., Bagra, M., Chao, Y.W., Savarese, S.: Semantic structure from motion with points, regions, and objects. In: 2012 IEEE Conference on Computer Vision and Pattern Recognition, pp. 2703–2710. IEEE (2012)
6. Burda, Y., Edwards, H., Pathak, D., Storkey, A., Darrell, T., Efros, A.A.: Large-scale study of curiosity-driven learning. arXiv:1808.04355 (2018)
7. Cadena, C., et al.: Past, present, and future of simultaneous localization and mapping: toward the robust-perception age. IEEE Trans. Rob. **32**(6), 1309–1332 (2016)
8. Carrillo, H., Reid, I., Castellanos, J.A.: On the comparison of uncertainty criteria for active slam. In: 2012 IEEE International Conference on Robotics and Automation, pp. 2080–2087. IEEE (2012)
9. Chang, A., et al.: Matterport3D: learning from RGB-D data in indoor environments. In: Proceedings of the International Conference on 3D Vision (3DV), MatterPort3D dataset license (2017). http://kaldir.vc.in.tum.de/matterport/MP_TOS.pdf

10. Chaplot, D.S., Gupta, S., Gandhi, D., Gupta, A., Salakhutdinov, R.: Learning to explore using active neural mapping. In: 8th International Conference on Learning Representations, ICLR 2020 (2020)
11. Chen, T., Gupta, S., Gupta, A.: Learning exploration policies for navigation. In: 7th International Conference on Learning Representations, ICLR 2019 (2019)
12. Das, A., Datta, S., Gkioxari, G., Lee, S., Parikh, D., Batra, D.: Embodied question answering. In: Proceedings of the IEEE Conference on Computer Vision and Pattern Recognition Workshops, pp. 2054–2063 (2018)
13. Datta, S., Maksymets, O., Hoffman, J., Lee, S., Batra, D., Parikh, D.: Integrating egocentric localization for more realistic pointgoal navigation agents. In: CVPR 2020 Embodied AI Workshop (2020)
14. Dhamo, H., Navab, N., Tombari, F.: Object-driven multi-layer scene decomposition from a single image. In: The IEEE International Conference on Computer Vision (ICCV), October 2019
15. Elhafsi, A., Ivanovic, B., Janson, L., Pavone, M.: Map-predictive motion planning in unknown environments. arXiv preprint arXiv:1910.08184 (2019)
16. Fang, K., Toshev, A., Fei-Fei, L., Savarese, S.: Scene memory transformer for embodied agents in long-horizon tasks. In: Proceedings of the IEEE Conference on Computer Vision and Pattern Recognition, pp. 538–547 (2019)
17. Gan, C., Zhang, Y., Wu, J., Gong, B., Tenenbaum, J.B.: Look, listen, and act: towards audio-visual embodied navigation. arXiv preprint arXiv:1912.11684 (2019)
18. Gordon, D., Kembhavi, A., Rastegari, M., Redmon, J., Fox, D., Farhadi, A.: IQA: visual question answering in interactive environments. In: Proceedings of the IEEE Conference on Computer Vision and Pattern Recognition, pp. 4089–4098 (2018)
19. Gupta, S., Davidson, J., Levine, S., Sukthankar, R., Malik, J.: Cognitive mapping and planning for visual navigation. In: Proceedings of the IEEE Conference on Computer Vision and Pattern Recognition, pp. 2616–2625 (2017)
20. Gupta, S., Fouhey, D., Levine, S., Malik, J.: Unifying map and landmark based representations for visual navigation. arXiv preprint arXiv:1712.08125 (2017)
21. Hartley, R., Zisserman, A.: Multiple View Geometry in Computer Vision. Cambridge University Press, Cambridge (2003)
22. Henriques, J.F., Vedaldi, A.: MapNet: an allocentric spatial memory for mapping environments. In: proceedings of the IEEE Conference on Computer Vision and Pattern Recognition, pp. 8476–8484 (2018)
23. Hoermann, S., Bach, M., Dietmayer, K.: Dynamic occupancy grid prediction for urban autonomous driving: a deep learning approach with fully automatic labeling. In: 2018 IEEE International Conference on Robotics and Automation (ICRA), pp. 2056–2063. IEEE (2018)
24. Iizuka, S., Simo-Serra, E., Ishikawa, H.: Globally and locally consistent image completion. ACM Trans. Graph. (ToG) **36**(4), 1–14 (2017)
25. Jayaraman, D., Gao, R., Grauman, K.: ShapeCodes: self-supervised feature learning by lifting views to viewgrids. In: Proceedings of the European Conference on Computer Vision (ECCV), pp. 120–136 (2018)
26. Jayaraman, D., Grauman, K.: Learning to look around: intelligently exploring unseen environments for unknown tasks. In: 2018 IEEE Conference on Computer Vision and Pattern Recognition (2018)
27. Chen, J., Liu, C., Wu, J., Furukawa, Y.: Floor-SP: inverse CAD for floorplans by sequential room-wise shortest path. In: The IEEE International Conference on Computer Vision (ICCV) (2019)

28. Karkus, P., Ma, X., Hsu, D., Kaelbling, L.P., Lee, W.S., Lozano-Pérez, T.: Differentiable algorithm networks for composable robot learning. arXiv preprint arXiv:1905.11602 (2019)
29. Katyal, K., Popek, K., Paxton, C., Burlina, P., Hager, G.D.: Uncertainty-aware occupancy map prediction using generative networks for robot navigation. In: 2019 International Conference on Robotics and Automation (ICRA), pp. 5453–5459. IEEE (2019)
30. Katyal, K., et al.: Occupancy map prediction using generative and fully convolutional networks for vehicle navigation. arXiv preprint arXiv:1803.02007 (2018)
31. Kolve, E., et al.: AI2-THOR: An Interactive 3D Environment for Visual AI. arXiv (2017)
32. Li, Y., Liu, S., Yang, J., Yang, M.H.: Generative face completion. In: Proceedings of the IEEE Conference on Computer Vision and Pattern Recognition, pp. 3911–3919 (2017)
33. Liu, C., Wu, J., Furukawa, Y.: FloorNet: a unified framework for floorplan reconstruction from 3D scans. In: Proceedings of the European Conference on Computer Vision (ECCV), pp. 201–217 (2018)
34. Lu, C., Dubbelman, G.: Hallucinating beyond observation: learning to complete with partial observation and unpaired prior knowledge (2019)
35. Savva, M., et al.: Habitat: a platform for embodied AI research. In: Proceedings of the IEEE/CVF International Conference on Computer Vision (ICCV) (2019)
36. Martinez-Cantin, R., De Freitas, N., Brochu, E., Castellanos, J., Doucet, A.: A Bayesian exploration-exploitation approach for optimal online sensing and planning with a visually guided mobile robot. Auton. Rob. **27**(2), 93–103 (2009)
37. Mohajerin, N., Rohani, M.: Multi-step prediction of occupancy grid maps with recurrent neural networks. In: Proceedings of the IEEE Conference on Computer Vision and Pattern Recognition, pp. 10600–10608 (2019)
38. Mousavian, A., Toshev, A., Fišer, M., Košecká, J., Wahid, A., Davidson, J.: Visual representations for semantic target driven navigation. In: 2019 International Conference on Robotics and Automation (ICRA), pp. 8846–8852. IEEE (2019)
39. Müller, M., Dosovitskiy, A., Ghanem, B., Koltun, V.: Driving policy transfer via modularity and abstraction. arXiv preprint arXiv:1804.09364 (2018)
40. O'Callaghan, S.T., Ramos, F.T.: Gaussian process occupancy maps. Int. J. Robot. Res. **31**(1), 42–62 (2012)
41. Parisotto, E., Salakhutdinov, R.: Neural map: structured memory for deep reinforcement learning. arXiv preprint arXiv:1702.08360 (2017)
42. Pathak, D., Agrawal, P., Efros, A.A., Darrell, T.: Curiosity-driven exploration by self-supervised prediction. In: International Conference on Machine Learning (2017)
43. Pathak, D., Krahenbuhl, P., Donahue, J., Darrell, T., Efros, A.A.: Context encoders: feature learning by inpainting. In: The IEEE Conference on Computer Vision and Pattern Recognition (CVPR), June 2016
44. Ramakrishnan, S.K., Grauman, K.: Sidekick policy learning for active visual exploration. In: Proceedings of the European Conference on Computer Vision (ECCV), pp. 413–430 (2018)
45. Ramakrishnan, S.K., Jayaraman, D., Grauman, K.: Emergence of exploratory look-around behaviors through active observation completion. Sci. Robot. **4**(30) (2019). https://doi.org/10.1126/scirobotics.aaw6326, https://robotics.sciencemag.org/content/4/30/eaaw6326
46. Ramakrishnan, S.K., Jayaraman, D., Grauman, K.: An exploration of embodied visual exploration. arXiv preprint arXiv:2001.02192 (2020)

47. Ramos, F., Ott, L.: Hilbert maps: scalable continuous occupancy mapping with stochastic gradient descent. Int. J. Robot. Res. **35**(14), 1717–1730 (2016)
48. Ronneberger, O., Fischer, P., Brox, T.: U-net: convolutional networks for biomedical image segmentation. In: Navab, N., Hornegger, J., Wells, W.M., Frangi, A.F. (eds.) MICCAI 2015. LNCS, vol. 9351, pp. 234–241. Springer, Cham (2015). https://doi.org/10.1007/978-3-319-24574-4_28
49. Salas-Moreno, R.F., Newcombe, R.A., Strasdat, H., Kelly, P.H., Davison, A.J.: Slam++: simultaneous localisation and mapping at the level of objects. In: Proceedings of the IEEE Conference on Computer Vision and Pattern Recognition, pp. 1352–1359 (2013)
50. Savinov, N., Dosovitskiy, A., Koltun, V.: Semi-parametric topological memory for navigation. arXiv preprint arXiv:1803.00653 (2018)
51. Savinov, N., et al.: Episodic curiosity through reachability. arXiv preprint arXiv:1810.02274 (2018)
52. Savva, M., Chang, A.X., Dosovitskiy, A., Funkhouser, T., Koltun, V.: MINOS: multimodal indoor simulator for navigation in complex environments. arXiv preprint arXiv:1712.03931 (2017)
53. Sax, A., Emi, B., Zamir, A.R., Guibas, L., Savarese, S., Malik, J.: Mid-level visual representations improve generalization and sample efficiency for learning visuomotor policies. arXiv preprint arXiv:1812.11971 (2018)
54. Schulman, J., Wolski, F., Dhariwal, P., Radford, A., Klimov, O.: Proximal policy optimization algorithms. arXiv preprint arXiv:1707.06347 (2017)
55. Seifi, S., Tuytelaars, T.: Where to look next: unsupervised active visual exploration on 360 {\deg} input. arXiv preprint arXiv:1909.10304 (2019)
56. Senanayake, R., Ganegedara, T., Ramos, F.: Deep occupancy maps: a continuous mapping technique for dynamic environments (2017)
57. Shen, W.B., Xu, D., Zhu, Y., Guibas, L.J., Fei-Fei, L., Savarese, S.: Situational fusion of visual representation for visual navigation. In: Proceedings of the IEEE International Conference on Computer Vision, pp. 2881–2890 (2019)
58. Shrestha, R., Tian, F.P., Feng, W., Tan, P., Vaughan, R.: Learned map prediction for enhanced mobile robot exploration. In: 2019 International Conference on Robotics and Automation (ICRA), pp. 1197–1204. IEEE (2019)
59. Sless, L., Cohen, G., Shlomo, B.E., Oron, S.: Self supervised occupancy grid learning from sparse radar for autonomous driving. arXiv preprint arXiv:1904.00415 (2019)
60. Song, S., Yu, F., Zeng, A., Chang, A.X., Savva, M., Funkhouser, T.: Semantic scene completion from a single depth image. In: Proceedings of 30th IEEE Conference on Computer Vision and Pattern Recognition (2017)
61. Song, S., Zeng, A., Chang, A.X., Savva, M., Savarese, S., Funkhouser, T.: Im2pano3D: extrapolating 360 structure and semantics beyond the field of view. In: Proceedings of the IEEE Conference on Computer Vision and Pattern Recognition, pp. 3847–3856 (2018)
62. Straub, J., et al.: The replica dataset: a digital replica of indoor spaces. arXiv preprint arXiv:1906.05797 (2019)
63. Sun, C., Hsiao, C.W., Sun, M., Chen, H.T.: HorizonNet: learning room layout with 1D representation and pano stretch data augmentation. In: The IEEE Conference on Computer Vision and Pattern Recognition (CVPR), June 2019
64. Thrun, S.: Probabilistic robotics. Commun. ACM **45**(3), 52–57 (2002)
65. Wijmans, E., Kadian, A., Morcos, A., Lee, S., Essa, I., Parikh, D., Savva, M., Batra, D.: DD-PPO: learning near-perfect pointgoal navigators from 2.5 billion frames (2020)

66. Wu, W., Fu, X.M., Tang, R., Wang, Y., Qi, Y.H., Liu, L.: Data-driven interior plan generation for residential buildings. ACM Trans. Graph. **38**(6), 1–2 (2019). https://doi.org/10.1145/3355089.3356556
67. Xia, F., et al.: Gibson Env: real-world perception for embodied agents. In: Proceedings of the IEEE Conference on Computer Vision and Pattern Recognition. pp. 9068–9079, Gibson dataset license agreement (2018). https://storage.googleapis.com/gibson_material/Agreement%20GDS%2006-04-18.pdf
68. Yang, J., et al.: Embodied amodal recognition: learning to move to perceive objects. In: ICCV (2019)
69. Yang, S.T., Wang, F.E., Peng, C.H., Wonka, P., Sun, M., Chu, H.K.: DuLa-Net: a dual-projection network for estimating room layouts from a single RGB panorama. In: Proceedings of the IEEE Conference on Computer Vision and Pattern Recognition, pp. 3363–3372 (2019)
70. Yang, W., Wang, X., Farhadi, A., Gupta, A., Mottaghi, R.: Visual semantic navigation using scene priors. arXiv preprint arXiv:1810.06543 (2018)
71. Yang, Z., Pan, J.Z., Luo, L., Zhou, X., Grauman, K., Huang, Q.: Extreme relative pose estimation for RGB-D scans via scene completion. In: The IEEE Conference on Computer Vision and Pattern Recognition (CVPR), June 2019
72. Zhu, Y., et al.: Visual semantic planning using deep successor representations. In: 2017 IEEE International Conference on Computer Vision (2017)
73. Zou, C., Colburn, A., Shan, Q., Hoiem, D.: LayoutNet: reconstructing the 3D room layout from a single RGB image. In: Proceedings of the IEEE Conference on Computer Vision and Pattern Recognition, pp. 2051–2059 (2018)

Unified Image and Video Saliency Modeling

Richard Droste[✉], Jianbo Jiao, and J. Alison Noble

University of Oxford, Oxford, UK
{richard.droste,jianbo.jiao,alison.noble}@eng.ox.ac.uk

Abstract. Visual saliency modeling for images and videos is treated as two independent tasks in recent computer vision literature. While image saliency modeling is a well-studied problem and progress on benchmarks like SALICON and MIT300 is slowing, video saliency models have shown rapid gains on the recent DHF1K benchmark. Here, we take a step back and ask: Can image and video saliency modeling be approached via a unified model, with mutual benefit? We identify different sources of domain shift between image and video saliency data and between different video saliency datasets as a key challenge for effective joint modelling. To address this we propose four novel domain adaptation techniques—Domain-Adaptive Priors, Domain-Adaptive Fusion, Domain-Adaptive Smoothing and Bypass-RNN—in addition to an improved formulation of learned Gaussian priors. We integrate these techniques into a simple and lightweight encoder-RNN-decoder-style network, UNISAL, and train it jointly with image and video saliency data. We evaluate our method on the video saliency datasets DHF1K, Hollywood-2 and UCF-Sports, and the image saliency datasets SALICON and MIT300. With one set of parameters, UNISAL achieves state-of-the-art performance on all video saliency datasets and is on par with the state-of-the-art for image saliency datasets, despite faster runtime and a 5 to 20-fold smaller model size compared to all competing deep methods. We provide retrospective analyses and ablation studies which confirm the importance of the domain shift modeling. The code is available at https://github.com/rdroste/unisal.

Keywords: Visual saliency · Video saliency · Domain adaptation

1 Introduction

When processing static scenes (images) and dynamic scenes (videos), humans direct their visual attention towards important information, which can be measured by recording eye fixations. The task of predicting the fixation distribution

---

R. Droste and J. Jiao—Contributed equally to this work.

Electronic supplementary material The online version of this chapter (https://doi.org/10.1007/978-3-030-58558-7_25) contains supplementary material, which is available to authorized users.

© Springer Nature Switzerland AG 2020
A. Vedaldi et al. (Eds.): ECCV 2020, LNCS 12350, pp. 419–435, 2020.
https://doi.org/10.1007/978-3-030-58558-7_25

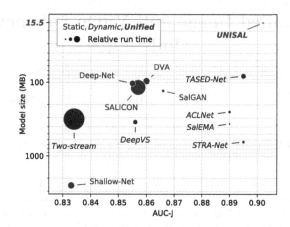

Fig. 1. Comparison of the proposed model with current state-of-the-art methods on the DHF1K benchmark [47]. The proposed model is more accurate (as measured by the official ranking metric AUC-J [5]) despite a model size reduction of 81% or more.

is referred to as *(visual) saliency prediction/modeling*, and the predicted distributions as *saliency maps.* Convolutional neural networks (CNNs) have emerged as the most performant technique for saliency modeling due to their capacity to learn complex feature hierarchies from large-scale datasets [2,20].

While most prior work focuses on image data, interest in video saliency modeling was recently accelerated through ACLNet, a dynamic saliency model that outperforms static models on the large-scale, diverse DHF1K benchmark [47]. However, as methods for video saliency modeling progress, it is usually considered a separate task to image saliency prediction [1,19,25,29,35,48] although both strive to model human visual attention. Current dynamic models use image data only for pre-training [1,19,25,29,35] or auxilliary loss functions [47]. In addition, many dynamic models are incompatible with image inputs since they require optical flow [1,25] or fixed-length video clips for spatio-temporal convolutions [19,35]. In this paper, we ask the question: *Is it possible to model static and dynamic saliency via one unified framework, with mutual benefit?*

First, we present experiments that identify the domain shift between image and video saliency data and between different video saliency datasets as a crucial hurdle for joint modelling. Consequently, we propose suitable domain adaptation techniques for the identified sources of domain shift. To study the benefit of the proposed techniques, we introduce the UNISAL neural network, which is designed to model visual saliency on image and video data coequally while aiming for simplicity and low computational complexity. The network is simultaneously trained on three video datasets—DHF1K [47], Hollywood-2 and UCF-Sports [34]—and one image saliency dataset, SALICON [20].

We evaluate our method on the four training datasets, among which DHF1K and SALICON have held-out test sets. In addition, we evaluate on the established MIT300 image saliency benchmark [21]. We find that our model significantly outperforms current state-of-the-art methods on all video saliency datasets and achieves competitive performance for the image saliency datasets, with a fraction of the model size and faster runtime than competing models. The performance of UNISAL on the challenging DHF1K benchmark is shown in Fig 1. In summary, our contributions are as follows:

- To the best of our knowledge, we make the first attempt to model image and video visual saliency with one unified framework.
- We identify different sources of domain shift as the main challenge for joint image and video saliency modeling and propose four novel domain adaptation techniques to enable strong shared features: Domain-Adaptive Priors, Domain-Adaptive Fusion, Domain-Adaptive Smoothing, and Bypass-RNN.
- Our method achieves state-of-the-art performance on all video saliency datasets and is on par with the state-of-the-art for all image saliency datasets. At the same time, the model achieves a 5 to 20-fold reduction in model size and faster runtime compared to all existing deep saliency models.

2   Related Work

Image Saliency Modeling. Most visual saliency modeling literature aims to predict human visual attention mechanisms on static scenes. Early saliency models [3,13,17,22,26,42] focus on low-level image features such as intensity/contrast, color, edges, *etc.*, and are are therefore referred to as *bottom-up* methods. Recently, the field has achieved significant performance gains through deep neural networks and their capacity to learn high-level, *top-down* features, starting with Vig *et al.* [45] who propose the first neural network-based approach. Jiang *et al.* [20] collect a large-scale saliency dataset, SALICON, to facilitate the exploration of deep learning-based saliency modeling. Zheng *et al.* [51] investigate the impact of high-level observer tasks on saliency modeling. Other papers mainly focus on network architecture design with increasing model sizes. For instance, Pan *et al.* [37] evaluate shallow and deep CNNs for saliency prediction, and Kruthiventi *et al.* [23] introduce dilated convolutions and Gaussian priors into the VGG network architecture. Kuemmerer *et al.* [24] propose a simplified VGG-based network while Wang *et al.* [46] add skip connections to fuse multiple scales and Cornia *et al.* [7] add an attentive convolutional LSTM and learned Gaussian priors. Yang *et al.* [50] expand on the idea of dilated convolutions based on the inception network architecture. While exploration is still ongoing for image saliency modeling, dynamic scenes are arguably at least as relevant to human visual experience, but have received less attention in the literature to date.

Video Saliency Modeling. Similar to image saliency models, early dynamic models [15,32,33,39] predict video saliency based on low-level visual statistics, with additional temporal features (*e.g.*, optical flow). Marat *et al.* [33] use video frame pairs to compute a static and a dynamic saliency map, which are fused for the final prediction. Marat *et al.* [33] and Zhong *et al.* [52] combine spatial and temporal saliency features and fuse the predictions. By extending the center-surround saliency in static scenes, Mahadevan *et al.* [32] use dynamic textures to model video saliency. The performance of these early models is limited by the ability of the low-level features to represent temporal information. Consequently, deep learning based methods have been introduced for dynamic saliency modeling in recent years. Gorji et al. [10] propose to incorporate attentional push for video saliency prediction, via a multi-stream convolutional long short-term memory network (ConvLSTM). Jiang et al. [19] show that human attention is attracted to moving objects and propose a saliency-structured ConvLSTM to generate video saliency. A recent work [48] presents a new large-scale video saliency dataset, DHF1K, and propose an attention mechanism with ConvLSTM to achieve better performance than static deep models. The DHF1K dataset, sparked advances [25,29,35] in video saliency prediction, exploring different strategies to extract temporal features (optical flow, 3D convolutions, different recurrences). However, the above methods either extend prior image saliency models or focus on video data alone with limited applicability to static scenes. Guo et al. [11] present a spatio-temporal model that predicts image and video saliency through the phase spectrum of the Quaternion Fourier Transform but the model lacks the necessary high-level information for accurate saliency prediction. While a recent learning-based approach [30] extends the image domain to the spatio-temporal domain by using LSTMs, such models are specialized for video data, rendering them unable to simultaneously model image saliency.

Domain Adaptation. We focus on domain specific learning, a form of domain adaptation which enables a learning system to process data from different domains by separating domain-invariant (shared) and domain-specific (private) parameters [6]. Domain Separation Networks (DSN) [4], for instance, are autoencoders with additional private encoders. Instead of an autoencoder, Tsai et al. [43] introduce an adversarial loss that enforces shared and private encoders networks. Xiao et al. [49] propose Domain Guided Dropout that results in different sub-networks for each domain, and Rozantev et al. [38] train entirely separate networks for each domain, coupled through a similarity loss. In contrast to using separate networks, the AdaBN method [28] adjusts the batch-normalization (BN) parameters of a shared network based on samples from a given target domain. The DSBN method [6] generalizes this idea by training a separate set of BN parameters for each domain. In general, these existing methods result in a large proportion of domain-specific parameters. In contrast, we propose domain-adaptation techniques that are aimed to bridge the domain gap of saliency datasets with a maximum proportion of shared parameters.

3   Unified Image and Video Saliency Modeling

3.1   Domain-Shift Modeling

In this section we present analyses to examine the domain shift between image and video data and between different video saliency datasets. We use the insights to design corresponding domain adaptation methods. Following

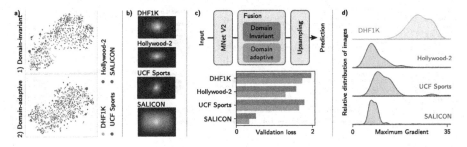

Fig. 2. Experiments to examine the domain shift between the saliency datasets. **a)** t-SNE visualization of MNet V2 features after domain-invariant and domain-adaptive normalization. **b)** Average ground truth saliency maps. **c)** Comparison of validation losses when training a simple saliency model with domain-invariant and domain-adaptive fusion. **d)** Distributions of ground truth saliency map sharpness.

Wang *et al.* [48], we select the video saliency datasets DHF1K [48], Hollywood-2 and UCF Sports [34], and the image saliency dataset SALICON [20].

Domain-Adaptive Batch Normalization. Batch normalization (BN) aims to reduce the internal covariate shift of neural network activations by transforming their distribution to zero mean and unit variance for each training batch. Simultaneously, it computes running estimates of the distribution mean and variance for inference. However, estimating these statistics across different domains results in inaccurate intra-domain statistics, and therefore a performance trade-off. In order to examine the domain shift between the datasets, we conduct a simple experiment: We randomly sample 256 images/frames from each dataset and compute their average pooled MobileNet V2 (MNet V2) features. We then visualize the distribution of the feature vectors via t-SNE [31] after normalizing them with the mean and variance of 1) all samples (domain-invariant) or 2) the samples from the respective dataset (domain-adaptive). The results, shown in Fig 2 a), reveal a significant domain shift among the different datasets, which is mitigated by the domain-adaptive normalization. Consequently, we employ *Domain-Adaptive Batch Normalization* (DABN), *i.e.*, a different set of BN modules for each dataset. During training and inference, each batch is constructed with data from one dataset and passed through the corresponding BN modules.

Domain-Adaptive Priors. Figure 2b) shows the average ground truth saliency map for each training dataset. Among the video datasets, Hollywood-2 and UCF Sports exhibit the strongest center bias, which is plausible since they are biased towards certain content (movies and sports) while DHF1K is more diverse. SALICON has a much weaker center bias than the video saliency datasets, which can potentially be explained by the longer viewing time of each image/frame (5 s *vs.* 30 ms to 42 ms) that allows secondary stimuli to be fixated. Accordingly, we propose to learn a separate set of Gaussian prior maps for each dataset.

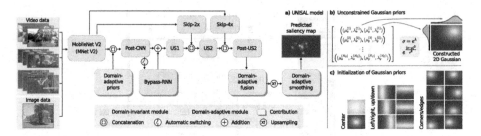

Fig. 3. a) Overview of the proposed framework. The model consists of a MobileNet V2 (MNet V2) encoder, followed by concatenation with learned Gaussian prior maps, a *Bypass-RNN*, a decoder network with skip connections, and *Fusion* and *Smoothing* layers. The prior maps, fusion, smoothing and batch-normalization modules are domain-adaptive in order to account for domain-shift between the image and video saliency datasets and enable high-quality shared features. **b)** Construction of the prior maps from learned Gaussian parameters. **c)** Prior maps initialization.

Domain-Adaptive Fusion. We hypothesize that similar image features can have varying visual saliency for images/frames from different training datasets. For example, the Hollywood-2 and UCF Sports datasets are *task-driven, i.e.*, the viewer is instructed to identify the main action shown. On the other hand, the DHF1K and SALICON datasets contains *free-viewing* fixations. To test the hypothesis, we design a simple saliency predictor (see Fig. 2 c): The outputs of the MNet V2 model are fused to a single map by a *Fusion* layer (1×1 convolution) and upsampled through bilinear interpolation. We train the *Fusion* layer until convergence with 1) one set of weights (domain-invariant) or 2) different weights for each dataset (domain-adaptive). We find that the validation loss is lower for all datasets for setting 2), where the network can weigh the importance of the feature maps differently for each dataset. Consequently, we propose to learn a different set of *Fusion* layer weights for each dataset.

Domain-Adaptive Smoothing. The size of the blurring filter which is used to generate the ground truth saliency maps from fixation maps can vary between datasets, especially since the images/frames are resized by different amounts. To examine this effect, we compute the distribution of the ground truth saliency map sharpness for each dataset. Sharpness is computed as the maximum image gradient magnitude after resizing to the model input resolution. The results in Fig. 2 d) confirm the heterogeneous distributions across datasets, revealing the highest sharpness for DHF1K. Therefore, we propose to blur the network output with a different learned *Smoothing* kernel for each dataset.

3.2 UNISAL Network Architecture

We introduce a simple and lightweight neural network architecture termed *UNISAL* that is designed to model image and video saliency coequally and implements the proposed domain-adaptation techniques. The architecture, illustrated in Fig. 3, follows an encoder-RNN-decoder design tailored for saliency modeling.

Encoder Network. We use MobileNet-V2 (MNet V2) [40] as our backbone encoder for three reasons: First, its small memory footprint enables training with sufficiently large sequence length and batch size; second, its small number of floating point operations allows for real time inference; and third, we expect the relatively small number of parameters to mitigate overfitting on smaller datasets like UCF Sports. The main building blocks of MNet V2 are *inverted residuals*, *i.e.*, sequences of pointwise convolutions that decompress and compress the feature space, interleaved with depthwise separable 3×3 convolutions. Overall, for an input resolution of $[r_x, r_y]$, MNet V2 computes feature maps at resolutions of $\frac{1}{2^\alpha}[r_x, r_y]$ with $\alpha \in \{1, 2, 3, 4, 5\}$. The output has 1280 channels and scale $\alpha = 5$. Domain-Adaptive Batch Normalization is not used in MNet V2 since we initialize it with ImageNet-pretrained parameters.

Gaussian Prior Maps. The domain-adaptive Gaussian prior maps are constructed at runtime from learned means and standard deviations. The map with index $i = 1, \ldots, N_G$ is computed as

$$g^{(i)}(x, y) = \gamma \exp\left(-\frac{(x - \mu_x^{(i)})^2}{(\sigma_x^{(i)})^2} - \frac{(y - \mu_y^{(i)})^2}{(\sigma_y^{(i)})^2}\right), \tag{1}$$

where $\gamma = 6$ is a scaling factor since the maps are concatenated with the ReLU6 activations of MNet V2. In this formulation, if the standard deviation $\sigma_{xy}^{(i)}$ is optimized over $\mathbb{R}$, then the resulting variance $(\sigma_{xy}^{(i)})^2$ has the domain $\mathbb{R}_{\geq 0}$, which can lead to division by zero. Prior work which uses non-adaptive prior maps [7] addresses this by clipping $\sigma_{xy}^{(i)}$ to a predefined interval $[a, b]$ with $a > 0$ and clipping $\mu_{xy}^{(i)}$ to an interval around the center of the map. However, these constraints potentially limit the ability to learn the optimal parameters. Here, we propose *unconstrained Gaussian prior maps* by substituting $\sigma_{xy}^{(i)} = e^{\lambda_{xy}^{(i)}}$ and optimizing $\lambda_{xy}^{(i)}$ and $\mu_{xy}^{(i)}$ over $\mathbb{R}$. Moreover, instead of drawing the initial Gaussian parameters from a normal distribution, which results in highly correlated maps, we initialize $N_G = 16$ maps as shown in Fig. 3c), covering a broad range of priors. Finally, previous work usually introduces prior maps at the second to last layer in order to model the static center bias. Here, we concatenate the prior maps with the encoder output before the RNN and decoder, in order to leverage the prior maps in higher-level features.

Bypass-RNN. Modeling video saliency data requires a strategy to extract temporal features, such as an RNN, optical flow or 3D convolutions. However, none of these techniques are generally suitable to process static inputs, whereas our goal is to process images and videos with one model. Therefore, we introduce a *Bypass-RNN*, *i.e.*, a RNN whose output is added to its input features via a residual connection that is automatically omitted (bypassed) for static batches. during training and inference. Thus, the RNN only models the residual variations in visual saliency that are caused by temporal features.

Table 1. Network modules and corresponding operations. $ConvDW(c)$ denotes a depthwise separable convolution with c channels and kernel size 3×3, followed by batch normalization and ReLU6 activation. $ConvPW(c_{in}, c_{out})$ is a pointwise 1×1 convolution with c_{in} input and c_{out} output channels, followed by batch normalization and, if $c_{in} \leq c_{out}$, by ReLU6 activation. $DO(p)$ denotes 2D dropout with probability p. $Up(c, n)$ denotes n-fold upsampling with bilinear interpolation of feature maps with c channels.

Module	Operations
Post-CNN	$ConvDW(1280)$, $ConvPW(1280, 256)$
Skip-4x	$ConvPW(64, 128)$, $DO(0.6)$, $ConvPW(128, 64)$
Skip-2x	$ConvPW(160, 256)$, $DO(0.6)$, $ConvPW(256, 128)$
US1	$Bilinear(256, 2)$
US2	$ConvPW(384, 768)$, $ConvDW(768)$, $ConvPW(768, 128)$, $Up(128, 2)$
Post-US2	$ConvPW(200, 400)$, $ConvDW(400)$, $ConvPW(400, 64)$
Fusion	$ConvPW(64, 1)$

In the UNISAL model, the *Bypass-RNN* is preceded by a *post-CNN* module, which compresses the concatenated MNet V2 outputs and Gaussian prior maps to 256 channels. For the Bypass-RNN, we use a convolutional GRU (*cGRU*) RNN [44] due to its relative simplicity, followed by a pointwise convolution. The cGRU has 256 hidden channels, 3×3 kernel size, recurrent dropout [9] with probability $p = 0.2$, and MobileNet-style convolutions, *i.e.*, depthwise separable convolutions followed by pointwise convolutions.

Decoder Network and Smoothing. The details of the decoder modules are listed in Table 1. First, the Bypass-RNN features are upsampled to scale $\alpha = 4$ by *US1* and concatenated with the output of *Skip-2x*. Next, the concatenated feature maps are upsampled to scale $\alpha = 3$ by *US2* and concatenated with the output of *Skip-4x*. The *Post-US2* features are reduced to a single channel by an *Domain-Adaptive Fusion* layer (1×1 convolution) and upsampled to the input resolution via nearest-neighbor interpolation. The upsampling is followed by a *Domain-Adaptive Smoothing* layer with 41×41 convolutional kernels that explicitly models the dataset-dependent blurring of the ground-truth saliency maps. Finally, following Jetley *et al.* [18], we transform the output into a generalized Bernoulli distribution by applying a softmax operation across all output values.

3.3   Domain-Aware Optimization

Domain-Adaptive Input Resolution. The images/frames have different aspect ratios for each dataset, specifically 4:3 for SALICON, 16:9 for DHF1K, 1.85:1 (median) for Hollywood-2, and 3:2 (median) for UCF Sports. Our network architecture is fully-convolutional, and therefore agnostic to exact the input resolution. Moreover, each mini-batch is constructed from one dataset due to

DABN. Therefore, we use input resolutions of 288×384, 224×384, 224×416 and 256×384 for SALICON, DHF1K, Hollywood-2 and UCF Sports, respectively.

Assimilated Frame Rate. The frame rate of the DHF1K videos is 30 fps compared to 24 fps for Hollywood-2 and UCF Sports. In order to assimilate the frame rates during training, and to train on longer time intervals, we construct clips using every 5th frame for DHF1K and every 4th frame for all others, yielding 6 fps overall. During inference, the predictions are interleaved.

4 Experiments

In this section, we compare the proposed method with current state-of-the-art image and video saliency models and provide detailed analyses are presented to gain an understanding of the proposed approach.

4.1 Experimental Setup

Datasets and Evaluation Metrics. To evaluate our proposed unified image and video saliency modeling framework, we jointly train UNISAL on datasets from both modalities. For fair comparison, we use the same training data as [47], i.e., the SALICON [20] image saliency dataset and the Hollywood-2 [34], UCF Sports [34], and DHF1K [47] video saliency datasets. For SALICON, we use the official training/validation/testing split of 10,000/5,000/5,000. For Hollywood-2 and UCF Sports, we use the training and testing splits of 823/884 and 103/47 videos, and the corresponding validation sets are randomly sampled 10% from the training sets, following [47]. Hollywood-2 videos are divided into individual shots. For DHF1K, we use the official training/validation/testing splits of 600/100/300 videos. We compare against the state-of-the-art methods listed in [47] and add newer models with available implementations [7,25,29,35,50]. Moreover, test on the MIT300 benchmark [21], after fine-tuning with the MIT1003 dataset as suggested by the benchmark authors. As in prior work [3,47], we use the evaluation metrics AUC-Judd (AUC-J), Similarity Metric (SIM), shuffled AUC (s-AUC), Linear Correlation Coefficient (CC), and Normalized Scanpath Saliency (NSS) [5].

Implementation Details. We optimize the network via Stochastic Gradient Descent with momentum of 0.9 and weight decay of 10^{-4}. Gradients are clipped to ± 2. The learning rate is set to 0.04 and exponentially decayed by a factor of 0.8 after each epoch. The batch size is set to 4 for video data and 32 for SALICON. The video clip length is set to 12 frames that are sampled as described in Sect. 3.3. Videos that are too short are discarded for training, which applies to Hollywood-2. For comparability, we use the same loss formulation as Wang et al. [48]. The model is trained for 16 epochs and with early stopping on the DHF1K validation set. To prevent overfitting, the weights of MNet V2 are frozen for the first two epochs

Table 2. Quantitative performance on the video saliency datasets. The training settings (i) to (vi) denote training with: (i) DHF1K, (ii) Hollywood-2, (iii) UCF Sports, (iv) SALICON, (v) DHF1K+Hollywood-2+UCF Sports, and (vi) DHF1K+Hollywood-2+UCF Sports+SALICON. Best performance is shown in **bold** while the second best is underlined. The * symbol denotes training under setting (vi), while † indicates that the method is fine-tuned for each dataset.

Dataset / Method	DHF1K					Hollywood-2					UCF Sports				
	AUC-J	SIM	s-AUC	CC	NSS	AUC-J	SIM	s-AUC	CC	NSS	AUC-J	SIM	s-AUC	CC	NSS
Dynamic models															
PQFT [12]	0.699	0.139	0.562	0.137	0.749	0.723	0.201	0.621	0.153	0.755	0.825	0.250	0.722	0.338	1.780
Seo et al. [41]	0.635	0.142	0.499	0.070	0.334	0.652	0.155	0.530	0.076	0.346	0.831	0.308	0.666	0.336	1.690
Rudoy et al. [39]	0.769	0.214	0.501	0.285	1.498	0.783	0.315	0.536	0.302	1.570	0.763	0.271	0.637	0.344	1.619
Hou et al. [15]	0.726	0.167	0.545	0.150	0.847	0.731	0.202	0.580	0.146	0.684	0.819	0.276	0.674	0.292	1.399
Fang et al. [8]	0.819	0.198	0.537	0.273	1.539	0.859	0.272	0.659	0.358	1.667	0.845	0.307	0.674	0.395	1.787
OBDL [14]	0.638	0.171	0.500	0.117	0.495	0.640	0.170	0.541	0.106	0.462	0.759	0.193	0.634	0.234	1.382
AWS-D [27]	0.703	0.157	0.513	0.174	0.940	0.694	0.175	0.637	0.146	0.742	0.823	0.228	0.750	0.306	1.631
OM-CNN [19]	0.856	0.256	0.583	0.344	1.911	0.887	0.356	0.693	0.446	2.313	0.870	0.321	0.691	0.405	2.089
Two-stream [1]	0.834	0.197	0.581	0.325	1.632	0.863	0.276	0.710	0.382	1.748	0.832	0.264	0.685	0.343	1.753
*ACLNet [48]	0.890	0.315	0.601	0.434	2.354	0.913	<u>0.542</u>	0.757	0.623	3.086	0.897	0.406	0.744	0.510	2.567
TASED-Net [35]	0.895	0.361	0.712	0.470	2.667	0.918	0.507	0.768	0.646	3.302	0.899	0.469	0.752	0.582	2.920
STRA-Net [25]	0.895	0.355	0.663	0.458	2.558	0.923	0.536	<u>0.774</u>	0.662	3.478	0.910	0.479	0.751	0.593	3.018
†SalEMA [29]	0.890	**0.465**	0.667	0.449	2.573	0.919	0.487	0.708	0.613	3.186	0.906	0.431	0.740	0.544	2.638
*SalEMA [29]	0.895	0.283	**0.739**	0.414	2.285	0.875	0.371	0.663	0.456	2.214	0.899	0.381	0.769	0.521	2.503
Static models															
ITTI [17]	0.774	0.162	0.553	0.233	1.207	0.788	0.221	0.607	0.257	1.076	0.847	0.251	0.725	0.356	1.640
GBVS [13]	0.828	0.186	0.554	0.283	1.474	0.837	0.257	0.633	0.308	1.336	0.859	0.274	0.697	0.396	1.818
SALICON [16]	0.857	0.232	0.590	0.327	1.901	0.586	0.321	0.711	0.425	2.013	0.848	0.304	0.738	0.375	1.838
Shallow-Net [37]	0.833	0.182	0.529	0.295	1.509	0.851	0.276	0.694	0.423	1.680	0.846	0.276	0.691	0.382	1.789
Deep-Net [37]	0.855	0.201	0.592	0.331	1.775	0.884	0.300	0.736	0.451	2.066	0.861	0.282	0.719	0.414	1.903
*Deep-Net [37]	0.874	0.288	0.610	0.374	1.983	0.901	0.482	0.740	0.597	2.834	0.880	0.365	0.729	0.475	2.448
DVA [46]	0.860	0.262	0.595	0.358	2.013	0.886	0.372	0.727	0.482	2.459	0.872	0.339	0.725	0.439	2.311
*DVA [46]	0.883	0.297	0.623	0.397	2.237	0.907	0.497	0.753	0.607	2.942	0.892	0.387	0.740	0.492	2.503
SalGAN [36]	0.866	0.262	0.709	0.370	2.043	0.901	0.393	**0.789**	0.535	2.542	0.876	0.332	0.762	0.470	2.238
UNISAL (ours)															
Training setting (i)	<u>0.899</u>	0.378	0.686	0.481	2.707	0.920	0.496	0.710	0.612	3.279	0.896	0.443	0.717	0.553	2.689
Training setting (ii)	0.881	0.313	0.690	0.422	2.352	<u>0.932</u>	0.534	0.762	0.672	3.803	0.892	0.440	0.735	0.566	2.768
Training setting (iii)	0.869	0.286	0.664	0.375	2.056	0.890	0.392	0.683	0.475	2.350	0.908	0.502	0.764	0.614	3.076
Training setting (iv)	0.883	0.288	<u>0.715</u>	0.410	2.259	0.912	0.432	0.750	0.565	2.897	0.892	0.428	<u>0.776</u>	0.561	2.740
Training setting (v)	**0.901**	0.384	0.692	<u>0.488</u>	<u>2.739</u>	**0.934**	**0.544**	0.758	**0.675**	3.909	<u>0.917</u>	<u>0.514</u>	**0.786**	0.642	<u>3.260</u>
Training setting (vi)	**0.901**	0.390	0.691	**0.490**	**2.776**	**0.934**	<u>0.542</u>	0.759	<u>0.673</u>	3.901	**0.918**	**0.523**	0.775	**0.644**	**3.381**

and afterwards trained with a learning rate that is reduced by a factor of 10. The pretrained BN statistics of MNet V2 are frozen throughout training. To account for dataset imbalance, the learning rate for SALICON batches is reduced by a factor of 2. Our model is implemented using the PyTorch framework and trained on a NVIDIA GTX 1080 Ti GPU.

4.2 Quantitative Evaluation

The results of the quantitative evaluation are shown in Table 2 for the video saliency datasets and in Tables 3 and 4 for the image datasets. For video saliency prediction, in order to analyze the impact of—and generalization across—different datasets, we evaluate six training settings: i) DHF1K, ii) Hollywood-2, iii) UCF Sports, iv) SALICON, v) DHF1K, Hollywood-2, and UCF Sports, vi) DHF1K, Hollywood-2, UCF Sports and SALICON. For fair comparison, we include state-of-the-art methods that are trained on our best-performing training setting (iv): The ACLNet [48] video saliency model and the Deep-Net [37] and DVA [46] image saliency models. In addition, we provide the performance of

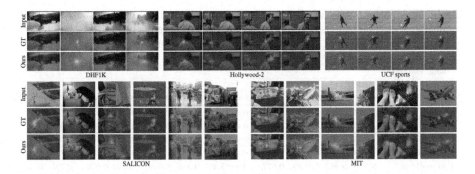

Fig. 4. Qualitative performance of the proposed approach on video (top part) and image (bottom part) saliency prediction.

Table 3. performance on the SALICON and MIT300 benchmarks. Best performance is shown in **bold** while the second best is <u>underlined</u>. Training setting (vi) is used for UNISAL (see supplementary material for other settings).

Method	Dataset									
	SALICON					MIT300				
	AUC-J	SIM	s-AUC	CC	NSS	AUC-J	SIM	s-AUC	CC	NSS
ITTI [17]	0.667	0.378	0.610	0.205	-	0.75	0.44	0.63	0.37	0.97
GBVS [13]	0.790	0.446	0.630	0.421	-	0.81	0.48	0.63	0.48	1.24
SALICON [16]	-	-	-	-	-	**0.87**	0.60	**0.74**	<u>0.74</u>	<u>2.12</u>
Shallow-Net [37]	0.836	<u>0.520</u>	0.670	0.596	-	0.80	0.46	0.64	0.53	-
Deep-Net [37]	-	-	0.724	0.609	1.859	0.83	0.52	0.69	0.58	1.51
SAM-ResNet [7]	**0.886**	-	**0.787**	0.844	**3.260**	**0.87**	**0.68**	0.70	**0.78**	**2.34**
DVA [46]	-	-	-	-	-	0.85	0.58	0.71	0.68	1.98
DINet [50]	**0.884**	-	<u>0.782</u>	0.860	<u>3.249</u>	0.86	-	0.71	**0.79**	2.33
SalGAN [36]	-	-	0.772	0.781	2.459	<u>0.86</u>	0.63	<u>0.72</u>	0.73	2.04
UNISAL (ours)	<u>0.864</u>	**0.775**	0.739	**0.879**	1.952	<u>0.872</u>	<u>0.674</u>	0.743	0.784	2.322

Table 4. Comparison for dynamic models on the static SALICON benchmark. Best performance is shown in **bold** while the second best is <u>underlined</u>. Training setting (vi) is used for all methods.

Method	AUC-J	SIM	s-AUC	CC	NSS
SalEMA [29]	0.732	0.470	0.519	0.411	0.760
ACLNet [48]	0.843	0.688	<u>0.698</u>	0.771	1.618
UNISAL (w/o DA)	<u>0.848</u>	<u>0.690</u>	0.676	<u>0.799</u>	<u>1.654</u>
UNISAL (final)	**0.864**	**0.775**	**0.739**	**0.879**	**1.952**

SalEMA [29], which is based on SalGAN [36], after fine-tuning the model with training setting (vi). Other state-of-the-art video saliency models [19,25,35] are not suitable for training with image data as discussed in Sect. 1. We observe that the proposed UNISAL model significantly outperforms previous static and dynamic methods, across almost all metrics. We obtain the following additional findings: 1) Training with all video saliency datasets (setting (v)) *always* improves performance compared to individual video saliency datasets (settings (i) to (iii)). This has not been the case for UCF Sports in a previous cross-dataset evaluation study [48]. 2) Additionally including image saliency data (setting (vi)) further improves performance for most metrics for DHF1K and UCF Sports. The exception is Hollywood-2, but the performance decrease is less than 1%.

For image saliency prediction, UNISAL performs on par with state-of-the-art image saliency models both on the SALICON and MIT300 benchmark as shown in Table 3. In addition, we evaluate state-of-the-art video saliency models on SALICON dataset as shown in Table 4. For ACLNet [48] we use the auxiliary output which is trained on SALICON (using the LSTM output yielded worse performance). For SalEMA [29], we fine-tuned their best performing model with training setting (vi). A large performance jump can be observed for the domain-adaptive UNISAL model.

Table 5. Ablation study of the proposed approach on the DHF1K and SALICON validation sets. The proposed components are added incrementally to the baseline to quantify their contribution. Training setting (vi) is used for this study.

Config. Dataset	DHF1K						SALICON					
	KLD ↓	AUC-J ↑	SIM ↑	s-AUC ↑	CC ↑	NSS ↑	KLD ↓	AUC-J ↑	SIM ↑	s-AUC ↑	CC ↑	NSS ↑
Baseline	1.877	0.863	0.282	0.659	0.372	2.057	0.551	0.824	0.607	0.633	0.711	1.415
+ Gaussian	1.776	0.879	0.300	0.668	0.411	2.273	0.394	0.848	0.675	0.685	0.801	1.634
+ RNNRes	1.754	0.881	0.302	0.666	0.411	2.274	0.450	0.843	0.648	0.665	0.770	1.531
+ SkipConnect	1.749	0.884	0.308	0.658	0.412	2.301	0.404	0.841	0.673	0.664	0.777	1.600
+ Smoothing	1.770	0.882	0.295	0.677	0.416	2.305	0.369	0.848	0.690	0.676	0.799	1.654
+ DomainAdaptive	1.526	0.907	0.373	0.685	0.482	2.731	0.231	0.867	0.768	0.712	0.877	1.925
Final	1.531	0.907	0.381	0.691	0.487	2.755	0.226	0.867	0.771	0.725	0.880	1.923

4.3 Qualitative Evaluation

In Fig. 4, we show randomly selected saliency predictions for both images and videos. It is visible that the proposed unified model performs well on both modalities. For challenging dynamic scenes with complete occlusion (DHF1K, left), the model correctly memorizes the salient object location, indicating that long-term temporal dependencies are effectively modeled. Moreover, the model correctly predicts shifting observer focus in the presence of multiple salient objects, as evident from the Hollywood-2 and UCF Sports samples. The results on static scenes (bottom part of Fig. 4) confirm that the proposed unified model indeed generalizes to static scenes.

4.4 Ablation Study

We analyze the contribution of each proposed component: 1) Gaussian prior maps; 2) RNN residual connection; 3) skip connections; 4) *Smoothing* layer; 5) domain-adaptive operations (incl. Bypass-RNN); and 6) domain-aware optimization. We perform the ablation on the representative DHF1K and SALICON validation sets. The results in Table 5 show that each of the proposed components contributes a considerable performance increase. Overall, the domain-adaptive operations contribute the most, both for DHF1K and SALICON. This indicates that mitigating the domain shift between datasets is a crucial component of UNISAL, confirming our initial studies in Sect. 3.1. The Gaussian prior maps yield the second largest gain, indicating the effectiveness of their proposed unconstrained optimization and early position in the model.

4.5 Inter-Dataset Domain Shift

Figure 5 shows the retrospective analysis of the four domain-adaptive modules. The DABN estimated means in Fig. 5 a) are correlated among video datasets with Pearson correlation coefficients r between 82% to 83%, but not correlated

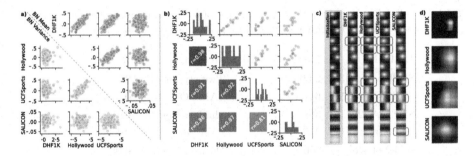

Fig. 5. Retrospective analysis of the domain-adaptive modules. a) Correlation of the batch normalization statistics between datasets (*US2* module, representative). The upper-right plots correlate the estimated means and the lower-left plots the estimated variances. b) Correlation of the *Fusion* layer weights between datasets. The plots on the diagonal show the distribution of weights of the respective dataset. The lower-left part shows Pearson's correlation coefficients. c) Gaussian prior maps. Significant deviations from the initialization are highlighted. d) *Smoothing* kernel of each dataset.

between SALICON and the video datasets ($r < 3\%$). For the estimated variances, only Hollywood-2 and UCF Sports are correlated ($r = 82\%$). This confirms the shift of the feature distributions between datasets, especially between SALICON and the video data. The domain-adaptive *Fusion* layer weights shown in Fig. 5 b) are generally correlated across datasets, with $r > 81\%$. However, as for the DABN, SALICON is the least correlated with the other datasets. Moreover, many of the SALICON *Fusion* weights lie near zero compared to the video datasets, which indicates that only a subset of the video saliency features is relevant for image saliency. The *Domain-Adaptive Fusion* layer models these differences while the remaining network weights are shared. The domain-adaptive Gaussian prior maps shown in Fig. 5 c) are successfully learned with our proposed unconstrained parametrization, as observed by the deviations from the initialization. Some prior maps are similar across datasets while others vary visibly, indicating that the different domains have different optimal priors. Finally, the learned *Smoothing* kernels shown in Fig. 5 d) vary significantly across datasets. As expected, the DHF1K dataset, which has the least blurry training targets, results in the most narrow *Smoothing* filter.

4.6 Computational Load

With the design of ever more complex network architectures, few studies evaluate the model size, although performance gains can often be traced back to more parameters. We compare the size of UNISAL to the state-of-the-art video saliency predictors in the left column of Table 6. Our model is the most lightweight by a significant margin, with over 5× smaller size than TASED-Net, which is the current state-of-the-art on the DHF1K benchmark (see also Fig. 1). The same result applies when comparing to the deep image saliency methods from Table 3, whose sizes range from 92 MB for DVA to 2.5 GB for Shallow-Net.

Table 6. Model size and runtime comparison of video saliency prediction methods (based on the DHF1K benchmark [48]). Best performance is shown in **bold**.

Method	Model size (MB)	Method	Runtime (s)
Shallow-Net [37]	2,500	Two-stream [1]	20
STRA-Net [25]	641	SALICON [16]	0.5
SalEMA [29]	364	Shallow-Net [37]	0.1
Two-stream [1]	315	DVA [46]	0.1
ACLNet [48]	250	Deep-Net [37]	0.08
SalGAN [36]	130	TASED-Net [35]	0.06
SALICON [16]	117	ACLNet [48]	0.02
Deep-Net [37]	103	SalGAN [36]	0.02
DVA [46]	96	STRA-Net [25]	0.02
TASED-Net [35]	82	SalEMA [29]	0.01
UNISAL (ours)	**15.5**	UNISAL (ours)	**0.009**

Another key issue for real-world applications is the model efficiency. Consequently, we present a GPU runtime comparison (processing time per frame) of video saliency models in the right column of Table 6. Our model is the most efficient compared to previous state-of-the-art methods. In addition, we observe a CPU (Intel Xeon W-2123 at 3.60 GHz) runtime of 0.43 s (2.3 fps), which is faster than some models' GPU runtime. Considering both the model size and the runtime, the proposed saliency modeling approach achieves state-of-the-art performance in terms of real-world applicability. While the MNet V2 encoder makes a large contribution to low model size and runtime, other contributing factors are: Separable convolutions throughout the cGRU and decoder; cGRU at the low-resolution bottleneck; bilinear upsampling. Without these measures the model size and runtime increase to 59.4 MB and 0.017 s, respectively.

5   Discussion and Conclusion

In this paper, we have presented a simple yet effective approach to unify static and dynamic saliency modeling. To bridge the domain gap, we found it crucial to account for different sources of inter-dataset domain shift through corresponding novel domain-adaptive modules. We integrated the domain-adaptive modules into the new, lightweight and simple UNISAL architecture which is designed to model both data modalities coequally. We observed state-of-the-art performance on video saliency datasets, and competitive performance on image saliency datasets, with a 5 to 20-fold reduction in model size compared to the *smallest* previous deep model, and faster runtime. We found that the domain-adaptive modules capture the differences between image and video saliency data, resulting in improved performance on each individual dataset through joint training. We presented preliminary and retrospective experiments which explain the merit of the domain-adaptive modules. To our knowledge, this is the first attempt towards unifying image and video saliency modeling in a single framework. We believe that our work can serve as a basis for further research into joint modeling of these modalities.

Acknowledgements. We acknowledge the EPSRC (Project Seebibyte, reference EP/M013774/1) and the NVIDIA Corporation for the donation of GPU.

References

1. Bak, C., Kocak, A., Erdem, E., Erdem, A.: Spatio-temporal saliency networks for dynamic saliency prediction. IEEE TMM **20**(7), 1688–1698 (2017)
2. Borji, A.: Saliency prediction in the deep learning era: an empirical investigation. arXiv:1810.03716 (2018)
3. Borji, A., Itti, L.: State-of-the-art in visual attention modeling. IEEE TPAMI **35**(1), 185–207 (2012)
4. Bousmalis, K., Trigeorgis, G., Silberman, N., Krishnan, D., Erhan, D.: Domain separation networks. In: NeurIPS (2016)
5. Bylinskii, Z., Judd, T., Oliva, A., Torralba, A., Durand, F.: What do different evaluation metrics tell us about saliency models? IEEE TPAMI **41**(3), 740–757 (2019)
6. Chang, W.G., You, T., Seo, S., Kwak, S., Han, B.: Domain-specific batch normalization for unsupervised domain adaptation. In: CVPR (2019)
7. Cornia, M., Baraldi, L., Serra, G., Cucchiara, R.: Predicting human eye fixations via an LSTM-based saliency attentive model. IEEE TIP **27**(10), 5142–5154 (2016)
8. Fang, Y., Wang, Z., Lin, W., Fang, Z.: Video saliency incorporating spatiotemporal cues and uncertainty weighting. IEEE TIP **23**(9), 3910–3921 (2014)
9. Gal, Y., Ghahramani, Z.: A Theoretically grounded application of dropout in recurrent neural networks. In: NeurIPS (2016)
10. Gorji, S., Clark, J.J.: Going from image to video saliency: augmenting image salience with dynamic attentional push. In: CVPR (2018)
11. Guo, C., Ma, Q., Zhang, L.: Spatio-temporal saliency detection using phase spectrum of quaternion fourier transform. In: CVPR (2008)
12. Guo, C., Zhang, L.: A novel multiresolution spatiotemporal saliency detection model and its applications in image and video compression. IEEE TIP **19**(1), 185–198 (2009)
13. Harel, J., Koch, C., Perona, P.: Graph-based visual saliency. In: NeurIPS (2007)
14. Hossein Khatoonabadi, S., Vasconcelos, N., Bajic, I.V., Shan, Y.: How many bits does it take for a stimulus to be salient? In: CVPR (2015)
15. Hou, X., Zhang, L.: Dynamic visual attention: searching for coding length increments. In: NeurIPS (2009)
16. Huang, X., Shen, C., Boix, X., Zhao, Q.: SALICON: reducing the semantic gap in saliency prediction by adapting deep neural networks. In: ICCV (2015)
17. Itti, L., Koch, C., Niebur, E.: A model of saliency-based visual attention for rapid scene analysis. IEEE TPAMI **20**(11), 1254–1259 (1998)
18. Jetley, S., Murray, N., Vig, E.: End-to-end saliency mapping via probability distribution prediction. In: CVPR (2016)
19. Jiang, L., Xu, M., Liu, T., Qiao, M., Wang, Z.: DeepVS: a deep learning based video saliency prediction approach. In: ECCV (2018)

20. Jiang, M., Huang, S., Duan, J., Zhao, Q.: SALICON: saliency in context. In: CVPR (2015)
21. Judd, T., Durand, F., Torralba, A.: A Benchmark of computational models of saliency to predict human fixations. In: MIT-CSAIL-TR-2012, vol. 1, pp. 1–7 (2012)
22. Judd, T., Ehinger, K., Durand, F., Torralba, A.: Learning to predict where humans look. In: ICCV (2009)
23. Kruthiventi, S.S.S., Ayush, K., Babu, R.V.: DeepFix: a fully convolutional neural network for predicting human eye fixations. IEEE TIP **26**(9), 4446–4456 (2015)
24. Kümmerer, M., Wallis, T.S.A., Bethge, M.: DeepGaze II: reading fixations from deep features trained on object recognition. arXiv:1610.01563 (2016)
25. Lai, Q., Wang, W., Sun, H., Shen, J.: Video saliency prediction using spatiotemporal residual attentive networks. IEEE TIP **26**, 1113–1126 (2019)
26. Le Meur, O., Le Callet, P., Barba, D., Thoreau, D.: A coherent computational approach to model bottom-up visual attention. IEEE TPAMI **28**(5), 802–817 (2006)
27. Leboran, V., Garcia-Diaz, A., Fdez-Vidal, X.R., Pardo, X.M.: Dynamic whitening saliency. IEEE TPAMI **39**(5), 893–907 (2016)
28. Li, Y., Wang, N., Shi, J., Liu, J., Hou, X.: Revisiting batch normalization for practical domain adaptation. In: ICLR (2016)
29. Linardos, P., Mohedano, E., Nieto, J.J., McGuinness, K., Giro-i Nieto, X., O'Connor, N.E.: Simple vs complex temporal recurrences for video saliency prediction. In: BMVC (2019)
30. Liu, J., Shahroudy, A., Xu, D., Wang, G.: Spatio-temporal LSTM with trust gates for 3D human action recognition. In: Leibe, B., Matas, J., Sebe, N., Welling, M. (eds.) ECCV 2016. LNCS, vol. 9907, pp. 816–833. Springer, Cham (2016). https://doi.org/10.1007/978-3-319-46487-9_50
31. Maaten, L.V.D., Hinton, G.: Visualizing data using t-SNE. J. Mach. Learn. Res. **9**(Nov), 2579–2605 (2008)
32. Mahadevan, V., Vasconcelos, N.: Spatiotemporal saliency in dynamic scenes. IEEE TPAMI **32**(1), 171–177 (2009)
33. Marat, S., Phuoc, T.H., Granjon, L., Guyader, N., Pellerin, D., Guérin-Dugué, A.: Modelling spatio-temporal saliency to predict gaze direction for short videos. Int. J. Comput. Vis. **82**(3), 231 (2009)
34. Mathe, S., Sminchisescu, C.: Actions in the eye: dynamic gaze datasets and learnt saliency models for visual recognition. IEEE TPAMI **37**(7), 1408–1424 (2015)
35. Min, K., Corso, J.J.: TASED-net: temporally-aggregating spatial encoder-decoder network for video saliency detection. In: ICCV (2019)
36. Pan, J., et al.: SalGAN: visual saliency prediction with generative adversarial networks. arXiv:1701.01081 (2017)
37. Pan, J., Sayrol, E., Giro-i Nieto, X., McGuinness, K., O'Connor, N.E.: Shallow and deep convolutional networks for saliency prediction. In: CVPR (2016)
38. Rozantsev, A., Salzmann, M., Fua, P.: Beyond sharing weights for deep domain adaptation. IEEE TPAMI **41**(4), 801–814 (2019)
39. Rudoy, D., Goldman, D.B., Shechtman, E., Zelnik-Manor, L.: Learning video saliency from human gaze using candidate selection. In: CVPR (2013)
40. Sandler, M., Howard, A., Zhu, M., Zhmoginov, A., Chen, L.C.: MobileNetV2: inverted residuals and linear bottlenecks. In: CVPR (2018)
41. Seo, H.J., Milanfar, P.: Static and space-time visual saliency detection by self-resemblance. J. Vis. **9**(12), 15–15 (2009)
42. Sun, Y., Fisher, R.: Object-based visual attention for computer vision. Artif. Intell. **146**(1), 77–123 (2003)

43. Tsai, J.C., Chien, J.T.: Adversarial domain separation and adaptation. In: 2017 IEEE 27th International Workshop on Machine Learning for Signal Processing (MLSP), pp. 1–6 (2017)

44. Valipour, S., Siam, M., Jagersand, M., Ray, N.: Recurrent fully convolutional networks for video segmentation. In: IEEE WACV, pp. 29–36 (2017)

45. Vig, E., Dorr, M., Cox, D.: Large-scale optimization of hierarchical features for saliency prediction in natural images. In: CVPR (2014)

46. Wang, W., Shen, J.: Deep visual attention prediction. IEEE TIP **27**(5), 2368–2378 (2017)

47. Wang, W., Shen, J., Guo, F., Cheng, M.M., Borji, A.: Revisiting video saliency: a large-scale benchmark and a new model. In: CVPR (2018)

48. Wang, W., Shen, J., Xie, J., Cheng, M.M., Ling, H., Borji, A.: Revisiting video saliency prediction in the deep learning era. IEEE TPAMI (2019, early access)

49. Xiao, T., Li, H., Ouyang, W., Wang, X.: Learning deep feature representations with domain guided dropout for person re-identification. arXiv:1604.07528 (2016)

50. Yang, S., Lin, G., Jiang, Q., Lin, W.: A dilated inception network for visual saliency prediction. IEEE TMM **22**(8), 2163–2176 (2020)

51. Zheng, Q., Jiao, J., Cao, Y., Lau, R.W.: Task-driven webpage saliency. In: ECCV (2018)

52. Zhong, S.H., Liu, Y., Ren, F., Zhang, J., Ren, T.: Video saliency detection via dynamic consistent spatio-temporal attention modelling. In: AAAI (2013)

TAO: A Large-Scale Benchmark
for Tracking Any Object

Achal Dave[1]([⊠]), Tarasha Khurana[1], Pavel Tokmakov[1], Cordelia Schmid[2],
and Deva Ramanan[1,3]

[1] Carnegie Mellon University, Pittsburgh, USA
achald@cs.cmu.edu
[2] Inria, Montbonnot, France
[3] Argo AI, Pittsburgh, USA

Abstract. For many years, multi-object tracking benchmarks have
focused on a handful of categories. Motivated primarily by surveillance
and self-driving applications, these datasets provide tracks for people,
vehicles, and animals, ignoring the vast majority of objects in the world.
By contrast, in the related field of object detection, the introduction
of large-scale, diverse datasets (e.g., COCO) have fostered significant
progress in developing highly robust solutions. To bridge this gap, we
introduce a similarly diverse dataset for Tracking Any Object (TAO)
(http://taodataset.org/). It consists of 2,907 high resolution videos, captured in diverse environments, which are half a minute long on average. Importantly, we adopt a bottom-up approach for discovering a large
vocabulary of 833 categories, an order of magnitude more than prior
tracking benchmarks. To this end, we ask annotators to label objects
that move at any point in the video, and give names to them post factum. Our vocabulary is both significantly larger and qualitatively different from existing tracking datasets. To ensure scalability of annotation,
we employ a federated approach that focuses manual effort on labeling
tracks for those relevant objects in a video (e.g., those that move). We
perform an extensive evaluation of state-of-the-art trackers and make a
number of important discoveries regarding large-vocabulary tracking in
an open-world. In particular, we show that existing single- and multi-
object trackers struggle when applied to this scenario in the wild, and
that detection-based, multi-object trackers are in fact competitive with
user-initialized ones. We hope that our dataset and analysis will boost
further progress in the tracking community.

Keywords: Datasets · Video object detection · Tracking

Electronic supplementary material The online version of this chapter (https://
doi.org/10.1007/978-3-030-58558-7_26) contains supplementary material, which is
available to authorized users.

A. Vedaldi et al. (Eds.): ECCV 2020, LNCS 12350, pp. 436–454, 2020.
https://doi.org/10.1007/978-3-030-58558-7_26

1   Introduction

A key component in the success of modern object detection methods was the introduction of large-scale, diverse benchmarks, such as MS COCO [38] and LVIS [27]. By contrast, multi-object tracking datasets tend to be small [40,56], biased towards short videos [65], and, most importantly, focused on a very small vocabulary of categories [40,56,60] (see Table 1). As can be seen from Fig. 1, they predominantly target people and vehicles. Due to the lack of proper benchmarks, the community has shifted towards solutions tailored to the few videos used for evaluation. Indeed, Bergmann et al. [5] have recently and convincingly demonstrated that simple baselines perform on par with state-of-the-art (SOTA) multi-object trackers.

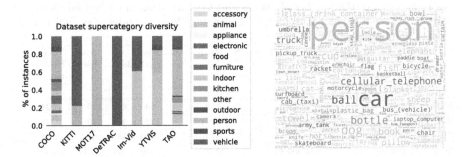

Fig. 1. (left) Super-category distribution in existing multi-object tracking datasets compared to TAO and COCO [38]. Previous work focused on people, vehicles and animals. By contrast, our bottom-up category discovery results in a more diverse distribution, covering many small, hand-held objects that are especially challenging from the tracking perspective. (right) Wordcloud of TAO categories, weighted by number of instances, and colored according to their supercategory.

In this work we introduce a large-scale benchmark for Tracking Any Object (TAO). Our dataset features 2,907 high resolution videos captured in diverse environments, which are 30 s long on average, and has tracks labeled for 833 object categories. We compare the statistics of TAO to existing multi-object tracking benchmarks in Table 1 and Fig. 1, and demonstrate that it improves upon them both in terms of complexity and in terms of diversity (see Fig. 2 for representative frames from TAO). Collecting such a dataset presents three main challenges: (1) how to select a large number of diverse, long, high-quality videos; (2) how to define a set of categories covering all the objects that might be of interest for tracking; and (3) how to label tracks for these categories at a realistic cost. Below we summarize our approach for addressing these challenges. A detailed description of dataset collection is provided in Sect. 4.

Existing datasets tend to focus on one or just a few domains when selecting the videos, such as outdoor scenes in MOT [40], or road scenes in KITTI [24]. This results in methods that fail when applied in the wild. To avoid this bias, we construct TAO with videos from as many environments as possible. We include indoor videos from Charades [52], movie scenes from AVA [26], outdoor videos from LaSOT [21], road-scenes from ArgoVerse [14], and a diverse sample of videos from HACS [68] and YFCC100M [54]. We ensure all videos are of high quality, with the smallest dimension larger or equal to 480px, and contain at least 2 moving objects. Table 1 reports the full statistics of the collected videos, showing that TAO provides an evaluation suite that is significantly larger, longer, and more diverse than prior work. Note that TAO contains fewer training videos than recent tracking datasets, as we intentionally dedicate the majority of videos for in-the-wild *benchmark* evaluation, the focus of our effort.

Table 1. Statistics of major multi-object tracking datasets. TAO is by far the largest dataset in terms of the number of classes and total duration of evaluation videos. In addition, we ensure that each video is challenging (long, containing several moving objects) and high quality.

Dataset	Classes	Videos		Avg length (s)	Tracks/video	Min resolution	Ann. fps	Total Eval length (s)
		Eval	Train					
MOT17 [40]	1	7	7	35.4	112	640 × 480	30	248
KITTI [24]	2	29	21	12.6	52	1242 × 375	10	365
UA-DETRAC [60]	4	40	60	56	57.6	960 × 540	5	2,240
ImageNet-Vid [48]	30	1,314	4,000	10.6	2.4	480 × 270	~25	13,928
YTVIS [65]	40	645	2,238	4.6	1.7	320 × 240	5	2,967
TAO (ours)	833	2,407	500	36.8	5.9	640 × 480	1	88,605

Given the selected videos, we must choose *what* to annotate. Most datasets are constructed with a top-down approach, where categories of interest are predefined by benchmark curators. That is, curators first select the subset of categories deemed relevant for the task, and then collect images or videos expressly for these categories [19,38,55]. This approach naturally introduces curator bias. An alternative strategy is bottom-up, open-world *discovery* of what objects are present in the data. Here, the vocabulary emerges post factum [26,27,69], an approach that dates back to LabelMe [49]. Inspired by this line of work, we devise the following strategy to discover an ontology of objects relevant for tracking: first annotators are asked to label *all* objects that either move by themselves or are moved by people. They then give names to the labeled objects, resulting in a vocabulary that is not only significantly larger, but is also qualitatively different from that of any existing tracking dataset (see Fig. 1). To facilitate training of object detectors, that can be later used by multi-object trackers on our dataset, we encourage annotators to choose categories that exists in the LVIS dataset [27]. If no appropriate category can be found in the LVIS vocabulary, annotators can provide free-form names (see Sect. 4.2 for details).

Exhaustively labeling tracks for such a large collection of objects in 2,907 long videos is prohibitively expensive. Instead, we extend the federated annotation approach proposed in [27] to the tracking domain. In particular, we ask the annotators to label tracks for up to 10 objects in every video. We then separately collect exhaustive labels for every category for a subset of videos, indicating whether all the instances of the category have been labeled in the video. During evaluation of a particular category, we use only videos with exhaustive labels for computing precision and all videos for computing recall. This allows us to reliably measure methods' performance at a fraction of the cost of exhaustively annotating the videos. We use the LVIS federated mAP metric [27] for evaluation, replacing 2D IoU with 3D IoU [65]. For detailed comparisons, we further report the standard MOT challenge [40] metrics in supplementary.

Fig. 2. Representative frames from TAO, showing videos sourced from multiple domains with annotations at two different timesteps.

Equipped with TAO, we set out to answer several questions about the state of the tracking community. In particular, in Sect. 5 we report the following discoveries: (1) SOTA trackers struggle to generalize to a large vocabulary of objects, particularly for infrequent object categories in the tail; (2) while trackers work significantly better for the most-explored category of people, tracking people in diverse scenarios (e.g., frequent occlusions or camera motion) remains challenging; (3) when scaled to a large object vocabulary, multi-object trackers become competitive with user-initialized trackers, despite the latter being provided with a ground truth initializations. We hope that these insights will help to define the most promising directions for future research.

2 Related Work

The domain of object tracking is subdivided based on how tracks are initialized. Our work falls into the multi-object tracking category, where all objects out of a fixed vocabulary of classes must be detected and tracked. Other formulations include user-initialized and saliency-based tracking. In this section, we first review the most relevant benchmarks datasets in each of these areas, and then discuss SOTA methods for multi-object and user-initialized tracking.

2.1   Benchmarks

Multi-object tracking (MOT) is the task of tracking an unknown number of objects from a known set of categories. Most MOT benchmarks [23, 24, 40, 60] focus on either people or vehicles (see Fig. 1), motivated by surveillance and self-driving applications. Moreover, they tend to include only a few dozen videos, captured in outdoor or road environments, encouraging methods that are overly adapted to the benchmark and do not generalize to different scenarios (see Table 1). In contrast, TAO focuses on diversity both in the category and visual domain distribution, resulting in a realistic benchmark for tracking *any* object.

Several works have attempted to extend the MOT task to a wider vocabulary of categories. In particular, the ImageNet-Vid [48] benchmark provides exhaustive trajectories annotations for objects of 30 categories in 1314 videos. While this dataset is both larger and more diverse than standard MOT benchmarks, videos tend to be relatively short and the categories cover only animals and vehicles. The recent YTVIS dataset [65] has the most broad vocabulary to date, covering 40 classes, but the majority of the categories still correspond to people, vehicles and animals. Moreover, the videos are 5 s long on average, making the tracking problem considerably easier in many cases. Unlike previous work, we take a bottom-up approach for defining the vocabulary. This results in not only the largest set of categories among MOT datasets to date, but also in a qualitatively different category distribution. In addition, our dataset is over 7 times larger than YTVIS in the number of frames. The recent VidOR dataset [51] explores Video Object Relations, including tracks for a large vocabulary of objects. But, since ViDOR focuses on relations rather than tracks, object trajectories tend to be missing or incomplete, making it hard to repurpose for tracker benchmarking. In contrast, we ensure TAO maintains high quality for both accuracy and completeness of labels (see supplementary for a quantitative analysis).

Finally, several recent works have proposed to label masks instead of bounding boxes for benchmarking multi-object tracking [56, 65]. In collecting TAO we made a conscious choice to prioritize scale and diversity of the benchmark over pixel-accurate labeling. Instance mask annotations are significantly more expensive to collect than bounding boxes, and we show empirically that tracking at the box level is already a challenging task that current methods fail to solve.

User-initialized tracking forgoes a fixed vocabulary of categories and instead relies on the user to provide bounding box annotations for objects at need to be tracked at test time [21, 30, 34, 55, 61] (in particular, the VOT challenge [34] has driven the progress in this field for many years). The benchmarks in this category tend to be larger and more diverse than their MOT counterparts, but most still offer a tradeoff between the number of videos and the average length of the videos (see supplementary). Moreover, even if the task itself is category-agnostic, empirical distribution of categories in the benchmarks tends to be heavily skewed towards a few common objects. We study whether this bias in category selection results in methods failing to generalize to more challenging objects by evaluating state-of-the-art user-initialized trackers on TAO in Sect. 5.2.

Semi-supervised video object segmentation differs from user-initialized tracking in that both the input to the tracker and the output are object masks, not boxes [43,64]. As a result, such datasets are a lot more expensive to collect, and videos tend to be extremely short. The main focus of the works in this domain [12,33,57] is on accurate mask propagation, not solving challenging identity association problems, thus their effort is complementary to ours.

Saliency-based tracking is an intriguing direction towards open-world tracking, where the objects of interest are defined not with a fixed vocabulary of categories, or manual annotations, but with bottom-up, motion- [42,43] or appearance-based [13,59] saliency cues. Our work similarly uses motion-based saliency to define a comprehensive vocabulary of categories, but presents a significantly larger benchmark with class labels for each object, enabling the use and evaluation of large-vocabulary object recognition approaches.

2.2 Algorithms

Multi-object trackers for people and other categories have historically been studied by separate communities. The former have been mainly developed on the MOT benchmark [40] and follow the tracking-by-detection paradigm, linking outputs of person detectors in an offline, graph-based framework [3,4,10,20]. These methods mainly differ in the way they define the edge cost in the graph. Classical approaches use overlap between detections in consecutive frames [31,44,67]. More recent methods define edge costs based on appearance similarity [41,47], or motion-based models [1,15,16,35,45,50]. Very recently, Bergmann et al. [5] proposed a simple baseline approach for tracking people that performs on par with SOTA by repurposing an object detector's bounding box regression capability to predict the position of an object in the next frame. All these methods have been developed and evaluated on the relatively small MOT dataset, containing 14 videos captured in very similar environments. By contrast, TAO provides a much richer, more diverse set of videos, encouraging trackers more robust to tracking challenges such as occlusion and camera motion.

The more general multi-object tracking scenario is usually studied using ImageNet-Vid [48]. Methods in this group also use offline, graph-based optimization to link frame-level detections into tracks. To define the edge potentials, in addition to box overlap, Feichtenhofer et al. [22] propose a similarity embedding, which is learned jointly with the detector. Kang et al. [32] directly predict short tubelets, and Xiao et al. [63] incorporate a spatio-temporal memory module inside a detector. Inspired by [5], we show that a simple baseline relying on the Viterbi algorithm for linking detections [22,25] performs on par with the aforementioned methods on ImageNet-Vid. We then use this baseline for evaluating generic multi-object tracking on TAO in Sect. 5.2, and demonstrate that it struggles when faced with a large vocabulary and a diverse data distribution.

User-initialized trackers tend to rely on a Siamese network architecture that was first introduced for signature verification [11], and later adapted for tracking [7,18,29,53]. They learn a patch-level distance embedding and find the closest patch to the one annotated in the first frame in the following frames. To simplify the matching problem, state-of-the-art approaches limit the search space to the region in which the object was localized in the previous frame. Recently there have been several attempts to introduce some ideas from CNN architectures for object detection into Siamese trackers. In particular, Li et al. [37] use the similarity map obtained by matching the object template to the test frame as input to an RPN-like module adapted from Faster-RCNN [46]. Later this architecture was extended by introducing hard negative mining and template updating [71], as well as mask prediction [58]. In another line of work, Siamese-based trackers have been augmented with a target discrimination module to improve their robustness to distractors [9,17]. We evaluate several state-of-the-art methods in this paradigm for which public implementation is available [9,17,18,36,58] on TAO, and demonstrate that they achieve only a moderate improvement over our multi-object tracking baseline, despite being provided with a ground truth initialization for each track (see Sect. 5.2 for details).

3   Dataset Design

Our primary goal is a large-scale video dataset with a diverse vocabulary of labeled objects to evaluate trackers in the wild. This requires designing a strategy for (1) video collection, (2) vocabulary discovery, (3) scalable annotation, and (4) evaluation. We detail our strategies for (2–4) below, and defer (1) to Sect. 4.1.

Category Discovery. Rather than manually defining a set of categories, we discover an object vocabulary from unlabeled videos which span diverse operating domains. Our focus is on *dynamic* objects in the world. Towards this end, we ask annotators to mark all objects that *move* in our collection of videos, without any object vocabulary in mind. We then construct a vocabulary by giving names for all the discovered objects, following the recent trend for open-world dataset collection [27,69]. In particular, annotators are asked to provide a free-form name for every object, but are encouraged to select a category from the LVIS [27] vocabulary whenever possible. We detail this process further in Sect. 4.2.

Federation. Given this vocabulary, one option might be to exhaustively label all instances of each category in all videos. Unfortunately, exhaustive annotation of a large vocabulary is expensive, even for images [27]. We choose to use our labeling budget instead on collecting a large-scale, diverse dataset, by extending the federated annotation protocol [27] from image datasets to videos. Rather than labeling every video v with every category c, we define three subsets of our dataset for each category: P_c, containing videos where all instances of c are labeled, N_c, videos with no instance of c present in the video, and U_c, videos where *some* instances of c are annotated. Videos not belonging to any of these subsets are ignored when evaluating category c. For each category c, we only use videos in P_c and N_c to measure the *precision* of trackers, and videos in P_c and U_c to measure recall. We describe how to define P_c, N_c, and U_c in Sect. 4.2.

Granularity of Annotations. To collect TAO, we choose to prioritize scale and diversity of the data at the cost of annotation granularity. In particular, we label tracks at 1 frame per second with bounding box labels but don't annotate segmentation masks. This allows us to label 833 categories in 2,907 videos at a relatively modest cost. Our decision is motivated by the observation of [55] that dense frame labeling does not change the relative performance of the methods.

Evaluation and Metric. Traditionally, multi-object tracking datasets use either the CLEAR MOT metrics [6,24,40] or a 3D intersection-over-union (IoU) based metric [48,65]. We report the former in supplementary (with modifications for large-vocabularies of classes, including multi-class aggregation and federation), but focus our experiments on the latter. To formally define 3D IoU, let $G = \{g_1, \ldots, g_T\}$ and $D = \{d_1, \ldots, d_T\}$ be a groundtruth and predicted track for a video with T frames. 3D IoU is defined as: $\text{IoU}_{3d}(D, G) = \frac{\sum_{t=1}^{T} g_t \cap d_t}{\sum_{t=1}^{T} g_t \cup d_t}$. If an object is not present at time t, we assign g_t to an empty bounding box, and similarly for a missing detection. We choose 3D IoU (with a threshold of 0.5) as the default metric for TAO, and provide further analysis in supplementary.

Similar to standard object detection metrics, (3D) IoU together with (track) confidence can be used to compute mean average precision across categories. For methods that provide a score for each frame in a track, we use the average frame score as the track score. Following [27], we measure precision for a category c in video v only if all instances of the category are verified to be labeled in it.

4   Dataset Collection

4.1   Video Selection

Most video datasets focus on one or a few domains. For instance, MOT benchmarks [40] correspond to urban, outdoor scenes featuring crowds, while AVA [26] contains produced films, typically capturing actors with close shots in carefully staged scenes. As a result, methods developed on any single dataset (and hence domain) fail to generalize in the wild. To avoid this bias, we constructed TAO by selecting videos from a variety of sources to ensure scene and object diversity.

Diversity. In particular, we used datasets for action recognition, self-driving cars, user-initialized tracking, and in-the-wild Flickr videos. In the action recognition domain we selected 3 datasets: Charades [52], AVA [26], and HACS [68]. Charades features complex human-human and human-object interactions, but all videos are indoor with limited camera motion. By contrast, AVA has a much wider variety of scenes and cinematographic styles but is scripted. HACS provides unscripted, in-the-wild videos. These action datasets are naturally focused on people and objects with which people interact. To include other animals and

vehicles, we source clips from LaSOT [21] (a benchmark for user-initialized tracking), BDD [66] and ArgoVerse [14] (benchmarks for self-driving cars). LaSOT
is a diverse collection whereas BDD and ArgoVerse consist entirely of outdoor,
urban scenes. Finally we sample in-the-wild videos from the YFCC100M [54]
Flickr collection.

Quality. The videos are automatically filtered to remove short videos and videos
with a resolution below 480p. For longer videos, as in AVA, we use [39] to extract
scenes without shot changes. In addition, we manually reviewed each sampled
video to ensure it is high quality: i.e., we removed grainy videos as well as
videos with excessive camera motion or shot changes. Finally, to focus on the
most challenging tracking scenarios, we only kept videos that contain at least 2
moving objects. The full statistics of the collected videos are provided in Table 1.
We point out that many prior video datasets tend to limit one or more quality
dimensions (in terms of resolution, length, or number of videos) in order to
keep evaluation and processing times manageable. In contrast, we believe that
in order to truly enable tracking in the open-world, we need to appropriately
scale benchmarks.

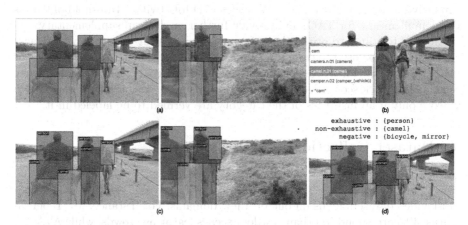

Fig. 3. Our federated video annotation pipeline. First (a), annotators mine and track
moving objects. Second (b), annotators categorize tracks using the LVIS vocabulary
or free-form text, producing the labeled tracks (c). Finally, annotators identify categories that are exhaustively annotated or verified to be absent. In (d), 'person's are
identified as being exhaustively annotated, 'camel's are present but not exhaustively
annotated and 'bicycle's and 'mirror's are verified as absent. These labels allow accurately penalizing false-positives and missed detections for exhaustively annotated and
verified categories.

4.2   Annotation Pipeline

Our annotation pipeline is illustrated in Fig. 3. We designed it to separate low-
level tracking from high-level semantic labeling. As pointed out by others [2],

semantic labeling can be subtle and error-prone because of ambiguities and corner-cases that arise in category boundaries. By separating tasks into low vs high-level, we are able to take advantage of unskilled annotators for the former and highly-vetted workers for the latter.

Object Mining and Tracking. We combine object mining and track labeling into a single stage. Given the videos described above, we ask annotators to mark *objects that move at any point in the video*. To avoid overspending our annotation budget on a few crowded videos, we limited the number of labeled objects per video to 10. Note that this stage is *category-agnostic*: annotators are not instructed to look for objects from any specific vocabulary, but instead to use motion as a *saliency* cue for mining relevant objects. They are then asked to label these objects throughout the video with bounding boxes at 1 frame-per-second. Finally, the tracks are verified by one independent annotator. This process is illustrated in Fig. 3, where we can see that 6 objects are discovered and tracked.

Object Categorization. Next, we collected category labels for objects discovered in the previous stage and simultaneously constructed the dataset vocabulary. We focus on the large vocabulary from the LVIS [27] object detection dataset, which contains 1,230 synsets discovered in a bottom-up manner similar to ours. Doing so also allows us to make use of LVIS as a training set of relevant object detectors (which we later use within a tracking pipeline to produce strong baselines - Sect. 5.1). Because maintaining a mental list of 1,230 categories is challenging even for expert annotators, we use an auto-complete annotation interface to suggest categories from the LVIS vocabulary (Fig. 3(b)). The auto-complete interface displays classes with a matching synset (e.g., "person.n.01"), name, synonym, and finally those with a matching definition. Interestingly, we find that some objects discovered in TAO, such as "door" or "marker cap", do not exist in LVIS. To accommodate such important exceptions, we allow annotators to label objects with free-form text if they do not fit in the LVIS vocabulary. Overall, annotators labeled 16,144 objects (95%) with 488 LVIS categories, and 894 objects (5%) with 345 free-form categories. We use the 488 LVIS categories for MOT experiments (because detectors can be trained on LVIS), but use all categories for user-initialized tracking experiments in supplementary.

Federated "Exhaustive" Labeling. Finally, we ask annotators to verify which categories are exhaustively labeled for each video. For each category c labeled in video v, we ask annotators whether all instances of c are labeled. In Fig. 3, after this stage, annotators marked that 'person' is exhaustively labeled, while 'camel' is not. Next, we show annotators a sampled subset of categories that are not labeled in the video, and ask them to indicate which categories are absent in the video. In Fig. 3, annotators indicated that 'bicycle' and 'mirror' are absent.

4.3   Dataset Splits

We split TAO into three subsets: train, validation and test, containing 500, 988 and 1,419 videos respectively. Typically, train splits tend to be larger than val and test. We choose to make TAO train small for several reasons. First, our primary goal is to reliably benchmark trackers in-the-wild. Second, most MOT systems are modularly trained using image-based detectors with hyper-parameter tuning of the overall tracking system. We ensure TAO train is sufficiently large for tuning, and that our large-vocabulary is aligned with the LVIS image dataset. This allows devoting most of our annotation budget for large-scale val and held-out test sets. We ensure that the videos in train, val and test are well-separated (e.g., each Charades subject appears in only one split); see supp. for details.

5   Analysis of State-of-the-Art Trackers

We now use TAO to analyze how well existing multi- and single-object trackers perform in the wild and when they fail. We tune the hyperparameters of each tracking approach on the 'train' set, and report results on the 'val' set. To capitalize on existing object detectors, we evaluate using the 488 LVIS categories in TAO. We begin by shortly describing the methods used in our analysis.

5.1   Methods

Detection. We analyze how well state-of-the-art object detectors perform on our dataset. To this end, we present results using a standard Mask R-CNN [46] detector trained using [62] in Sect. 5.2.

Multi-object Tracking. We analyze SOTA multi-object tracking methods on ImageNet-Vid. We first clarify whether such approaches improve detection or tracking. Table 2 reports the standard ImageNet-Vid Detection and Track mAP. The 'Detection' row corresponds to a detection-only baseline widely reported by prior work [22,63,70]. D&T [22] and

Table 2. ImageNet-Vid detection and track mAP; see text (left) for details.

	Viterbi	Det mAP	Track mAP
Detection		73.4 [63]	–
D&T [22]	✓	79.8	–
STMN [63]	✓	79.0	60.4
Detection	✓	79.2	60.3

STMN [63] are spatiotemporal architectures that produce 6–7% detection mAP improvements over a per-frame detector. However, both D&T and STMN post-process their per-frame outputs using the Viterbi algorithm, which iteratively links and re-weights the confidences of per-frame detections (see [25]). *When the same post-processing is applied to a single-frame detector, one achieves nearly the same performance gain (Table 2, last row).*

Our analysis reinforces the bleak view of multi-object tracking progress suggested by [5]: while ever-more complex approaches have been proposed for the task, their improvements are often attributable to simple, baseline strategies. To foster meaningful progress on TAO, we evaluate a number of strong baselines. We evaluate a per-frame detector trained on LVIS [27] and COCO [38], followed by two linking methods: SORT [8], a simple, online linker initially proposed for tracking people, and the Viterbi post-processing step used by [22,63], in Sect. 5.2.

Person Detection and Tracking. Detecting and tracking people has been a distinct focus in the multi-object tracking community. Sect. 5.2 compares the above baselines to a recent SOTA people-tracker [5].

User-Initialized Tracking. We evaluate several recent user-initialized trackers for which public implementation is available [9,17,18,36,58]. Unfortunately, these trackers do not classify tracked objects, and cannot directly be compared to multi-object trackers which simultaneously detect and track objects. However, these trackers *can* be evaluated with an oracle classifier, enabling direct comparisons.

Oracles. Finally, to disentangle the complexity of classification and tracking, we use two oracles. The first, a class oracle, computes the best matching between predicted and groundtruth tracks. Predicted tracks that match to a groundtruth track with 3D IoU > 0.5 are assigned the corresponding groundtruth category. Tracks that do not match to a groundtruth track are not modified, and count as false positives. This allows us to evaluate the performance of trackers assuming the semantic *classification* task is solved. The second oracle computes the best possible assignment of per-frame detections to tracks, by comparing them with groundtruth. When doing so, class predictions for each detection are held constant. Any detections that are not matched are discarded. This oracle allows us to analyze the best performance we could expect given a fixed set of detections.

5.2   Results

How Hard is Object Detection on TAO? We start by assessing the difficulty of detection on TAO by evaluating the SOTA object detector [28] using detection mAP. We train this model on LVIS and COCO, as training on LVIS alone led to low accuracy in detecting people. The final model achieves 27.1 mAP on TAO val at IoU 0.5, suggesting that single-frame detection is challenging on TAO.

Do Multi-object Trackers Generalize to TAO? Table 3 reports results using tracking mAP on TAO. As a sanity check, we first evaluate a per-frame detector by assigning each detection to its own track. As expected, this achieves an mAP of nearly 0 (which isn't quite 0 due to the presence of short tracks).

Next, we evaluate two multi-object tracking approaches. We compare the Viterbi linking method to an online SORT tracker [8]. We tune SORT on our diverse 'train' set, which is key for good accuracy. As the offline Viterbi algorithm takes over a month to run on TAO train, we only tune the post-processing score threshold for reporting a detection at each frame. See supplementary for tuning details. Surprisingly, we find that the simpler, online SORT approach outperforms Viterbi, perhaps because the latter has been heavily tuned for ImageNet-Vid. Because of its scalablity (to many categories and long videos) and relatively better performance, we focus on SORT for the majority of our

Table 3. SORT and Viterbi linking provide strong baselines on TAO, but detection and tracking remain challenging. Relabeling and linking detections from current detectors using the class and track oracles leads to high performance, suggesting a pathway for progress on TAO.

Method	Oracle		Track mAP
	Class	Track	
Detection			0.6
Viterbi [22,25]			6.3
SORT [8]			13.2
Detection		✓	31.5
Viterbi [22,25]	✓		15.7
SORT [8]	✓		30.2
Detection	✓	✓	83.6

Fig. 4. SORT qualitative results, showing (left) a successful tracking result, and (right) a failure case due to semantic flicker between similar classes, suggesting that large-vocabulary tracking on TAO requires additional machinery.

experiments. However, the performance of both methods remains low, suggesting TAO presents a major challenge for the tracking community, requiring principled novel approaches.

To better understand the nature of the complexity of TAO, we separately measure the challenges of tracking and classification. To this end, we first evaluate the "track" oracle that perfectly links per-frame detections. It achieves a stronger mAP of 31.5, compared to 13.2 for SORT. Interestingly, providing SORT tracks with an oracle class label provides a similar improvement, boosting mAP to 30.2. We posit that these improvements are orthogonal, and verify this by combining them; we link detections with oracle tracks and assign these tracks oracle class labels. This provides the largest delta, dramatically improving mAP to 83.6%. This suggests that *large-vocabulary tracking requires jointly improving tracking and classification accuracy (e.g., reducing semantic flicker as shown in Fig. 4).*

How Well Can We Track People? We now evaluate tracking on one particularly important category: people. Measuring AP for individual categories in a federated dataset can be noisy [27], so we emphasize *relative* performance of trackers rather than their absolute AP. We evaluate Tracktor++ [5], the state-of-the-art method designed specifically for people tracking, and compare it to the SORT and Viterbi baselines in Table 4. We update Tracktor++ to use the same detector used

Table 4. Person-tracking results on TAO. See text (left).

Method	Person AP
Viterbi [22,25]	16.5
SORT [8]	18.5
Tracktor++ [5]	36.7

by SORT and Viterbi, using only the 'person' predictions. We tune the score threshold on TAO 'train', but find the method is largely robust to this parameter (see supp.). Tracktor++ strongly outperforms other approaches (36.7 AP), while SORT modestly outperforms Viterbi (18.6 vs 16.5 AP). It is interesting to note that SORT, which can scale to all object categories, performs noticeably worse on all categories on average (13.2 mAP). This delta between 'person' and overall is even more dramatic using the MOTA metric (6.7 overall vs 54.8 for 'person'; see supp.). We attribute the higher accuracy for 'person' to two factors: (1) a rich history of focused research on this one category, which has led to more accurate detectors and trackers, and (2) more complex categories present significant challenges, such as hand-held objects which exhibit frequent occlusions during interactions.

To further investigate Tracktor++'s performance, we evaluate a simpler variant of the method from [5], which does not use appearance-based re-identification nor pixel-level frame alignment. We find that removing these components reduces AP from 36.7 to 25.9, suggesting these components contribute to a majority of improvements over the baselines. Our results contrast [5], which suggests that re-id and frame alignment are not particularly helpful. *Compared to prior benchmarks, the diversity of TAO results in a challenging testbed for person tracking which encourages trackers robust to occlusion and camera jitter.*

Do User-Initialized Trackers Generalize Better? Next, we evaluate recent user-initialized trackers in Table 5. We provide the tracker with the groundtruth box for each object from its first visible frame. As these trackers do not report when an object is *absent* [55], we modify them to report an object as absent when the confidence drops below a threshold tuned on TAO 'train' (see supp).

We compare these trackers to SORT, supplying both with a class oracle. As expected, the use of a ground-truth initialization allows the best user-initialized methods to outperform the multi-object tracker. However, even with this oracle box initialization and an oracle classifier, tracking remains challenging on TAO. Indeed, most user-initialized trackers provide at most modest improvements over SORT. We provide further analysis in supplementary, showing that (1) while a more informative initialization

Table 5. User-initialized trackers on 'val'. We re-train some trackers on their train set with TAO videos removed, denoted *.

Method	Oracle		Track mAP
	Init	Class	
SORT		✓	30.2
ECO [18]	✓	✓	23.7
SiamMask [58]	✓	✓	30.8
SiamRPN++ LT [36]	✓	✓	27.2
SiamRPN++ [36]	✓	✓	29.7
ATOM* [17]	✓	✓	30.9
DIMP* [9]	✓	✓	**33.2**

frame improves user-initialized tracker accuracy, SORT remains competitive, and (2) user-initialized trackers accurately track for a few frames after init, leading to improvements in MOTA, but provide little benefits in long-term tracking. We hypothesize that the small improvement of user-initialized trackers over SORT is due to the fact that the former are trained on a small vocabulary of objects with limited occlusions, leading to methods that do not generalize to the most challenging cases in TAO. One goal of user-initialized trackers is open-world tracking of objects without good detectors. TAO's large vocabulary allows us to analyze progress towards this goal, indicating that *large-vocabulary multi-object trackers may now address the open-world of objects as well as category-agnostic, user-initialized trackers.*

6   Discussion

Developing tracking approaches that can be deployed in-the-wild requires being able to reliably measure their performance. With nearly 3,000 videos, TAO provides a robust evaluation benchmark. Our analysis provides new conclusions about the state of tracking, while raising important questions for future work.

The Role of User-Initialized Tracking. User-initialized trackers aim to track *any* object, without requiring category-specific detectors. In this work, we raise a provocative question: with the advent of large vocabulary object detectors [27], to what extent can (detection-based) multi-object trackers perform generic tracking *without* user initialization? Table 5, for example, shows that large-vocabulary datasets (such as TAO and LVIS) now allow multi-object trackers to match or outperform user-initialization for a number of categories.

Specialized Tracking Approaches. TAO aims to measure progress in tracking in-the-wild. A valid question is whether progress may be better achieved by building trackers for *application-specific* scenarios. An indoor robot, for example, has little need for tracking elephants. However, success in many computer vision fields has been driven by the pursuit of *generic* approaches, that can then be tailored for specific applications. We do not build one class of object detectors for indoor scenes, and another for outdoor scenes, and yet another for surveillance videos. We believe that tracking will similarly benefit from targeting diverse scenarios. Of course, due to its size, TAO also lends itself to use for evaluating trackers for specific scenarios or categories, as in Sect. 5.2 for 'person.'

Video Object Detection. Although image-based object detectors have significantly improved in recent years, our analysis in Sect. 5.1 suggests that simple post-processing of detection outputs remains a strong baseline for detection in videos. While we do not emphasize it in this work, TAO can also be used to measure progress in video object *detection*, where the goal is not to maintain the identity of objects, but only to reliably detect them in every video frame. TAO's large vocabulary particularly provides avenues for incorporating temporal information to resolve classification errors, which remain challenging (see Fig. 4).

Acknowledgements. We thank Jonathon Luiten and Ross Girshick for detailed feedback, and Nadine Chang and Kenneth Marino for reviewing early drafts. Annotations for this dataset were provided by Scale.ai. This work was supported in part by the CMU Argo AI Center for Autonomous Vehicle Research, the Inria associate team GAYA, and by the Intelligence Advanced Research Projects Activity (IARPA) via Department of Interior/Interior Business Center (DOI/IBC) contract number D17PC00345. The U.S. Government is authorized to reproduce and distribute reprints for Governmental purposes not withstanding any copyright annotation theron. Disclaimer: The views and conclusions contained herein are those of the authors and should not be interpreted as necessarily representing the official policies or endorsements, either expressed or implied of IARPA, DOI/IBC or the U.S. Government.

References

1. Alahi, A., Goel, K., Ramanathan, V., Robicquet, A., Fei-Fei, L., Savarese, S.: Social LSTM: human trajectory prediction in crowded spaces. In: CVPR (2016)
2. Barriuso, A., Torralba, A.: Notes on image annotation. arXiv preprint arXiv:1210.3448 (2012)
3. Berclaz, J., Fleuret, F., Fua, P.: Robust people tracking with global trajectory optimization. In: CVPR (2006)
4. Berclaz, J., Fleuret, F., Turetken, E., Fua, P.: Multiple object tracking using k-shortest paths optimization. IEEE Trans. Pattern Anal. Mach. Intell. 33(9), 1806–1819 (2011)
5. Bergmann, P., Meinhardt, T., Leal-Taixe, L.: Tracking without bells and whistles. In: ICCV (2019)
6. Bernardin, K., Stiefelhagen, R.: Evaluating multiple object tracking performance: the CLEAR MOT metrics. J. Image Video Process. **2008**, 1 (2008). https://doi.org/10.1155/2008/246309
7. Bertinetto, L., Valmadre, J., Henriques, J.F., Vedaldi, A., Torr, P.H.S.: Fully-convolutional siamese networks for object tracking. In: Hua, G., Jégou, H. (eds.) ECCV 2016. LNCS, vol. 9914, pp. 850–865. Springer, Cham (2016). https://doi.org/10.1007/978-3-319-48881-3_56
8. Bewley, A., Ge, Z., Ott, L., Ramos, F., Upcroft, B.: Simple online and realtime tracking. In: ICIP (2016)
9. Bhat, G., Danelljan, M., Gool, L.V., Timofte, R.: Learning discriminative model prediction for tracking. In: CVPR (2019)
10. Breitenstein, M.D., Reichlin, F., Leibe, B., Koller-Meier, E., Van Gool, L.: Robust tracking-by-detection using a detector confidence particle filter. In: ICCV (2009)
11. Bromley, J., Guyon, I., LeCun, Y., Säckinger, E., Shah, R.: Signature verification using a "siamese" time delay neural network. In: NIPS (1994)
12. Caelles, S., Maninis, K.K., Pont-Tuset, J., Leal-Taixé, L., Cremers, D., Van Gool, L.: One-shot video object segmentation. In: CVPR (2017)
13. Caelles, S., Pont-Tuset, J., Perazzi, F., Montes, A., Maninis, K.K., Van Gool, L.: The 2019 DAVIS challenge on VOS: unsupervised multi-object segmentation. arXiv preprint arXiv:1905.00737 (2019)
14. Chang, M.F., et al.: Argoverse: 3D tracking and forecasting with rich maps. In: CVPR (2019)
15. Chen, B., Wang, D., Li, P., Wang, S., Lu, H.: Real-time 'actor-critic' tracking. In: Ferrari, V., Hebert, M., Sminchisescu, C., Weiss, Y. (eds.) ECCV 2018. LNCS, vol. 11211, pp. 328–345. Springer, Cham (2018). https://doi.org/10.1007/978-3-030-01234-2_20
16. Choi, W., Savarese, S.: Multiple target tracking in world coordinate with single, minimally calibrated camera. In: Daniilidis, K., Maragos, P., Paragios, N. (eds.) ECCV 2010. LNCS, vol. 6314, pp. 553–567. Springer, Heidelberg (2010). https://doi.org/10.1007/978-3-642-15561-1_40
17. Danelljan, M., Bhat, G., Khan, F.S., Felsberg, M.: ATOM: accurate tracking by overlap maximization. In: CVPR (2019)

18. Danelljan, M., Bhat, G., Shahbaz Khan, F., Felsberg, M.: ECO: efficient convolution operators for tracking. In: CVPR (2017)
19. Deng, J., Dong, W., Socher, R., Li, L.J., Li, K., Fei-Fei, L.: ImageNet: a large-scale hierarchical image database. In: CVPR (2009)
20. Ess, A., Leibe, B., Schindler, K., Van Gool, L.: A mobile vision system for robust multi-person tracking. In: CVPR (2008)
21. Fan, H., et al.: LaSOT: a high-quality benchmark for large-scale single object tracking. In: CVPR (2019)
22. Feichtenhofer, C., Pinz, A., Zisserman, A.: Detect to track and track to detect. In: ICCV (2017)
23. Fisher, R., Santos-Victor, J., Crowley, J.: Context aware vision using image-based active recognition. EC's Information Society Technology's Programme Project IST2001-3754 (2001)
24. Geiger, A., Lenz, P., Urtasun, R.: Are we ready for autonomous driving? The KITTI vision benchmark suite. In: CVPR (2012)
25. Gkioxari, G., Malik, J.: Finding action tubes. In: CVPR (2015)
26. Gu, C., et al.: AVA: a video dataset of spatio-temporally localized atomic visual actions. In: CVPR (2018)
27. Gupta, A., Dollar, P., Girshick, R.: LVIS: a dataset for large vocabulary instance segmentation. In: CVPR (2019)
28. He, K., Gkioxari, G., Dollár, P., Girshick, R.: Mask R-CNN. In: ICCV (2017)
29. Held, D., Thrun, S., Savarese, S.: Learning to track at 100 FPS with deep regression networks. In: Leibe, B., Matas, J., Sebe, N., Welling, M. (eds.) ECCV 2016. LNCS, vol. 9905, pp. 749–765. Springer, Cham (2016). https://doi.org/10.1007/978-3-319-46448-0_45
30. Huang, L., Zhao, X., Huang, K.: GOT-10k: a large high-diversity benchmark for generic object tracking in the wild. arXiv preprint arXiv:1810.11981 (2018)
31. Jiang, H., Fels, S., Little, J.J.: A linear programming approach for multiple object tracking. In: CVPR (2007)
32. Kang, K., et al.: Object detection in videos with tubelet proposal networks. In: CVPR (2017)
33. Khoreva, A., Benenson, R., Ilg, E., Brox, T., Schiele, B.: Lucid data dreaming for video object segmentation. Int. J. Comput. Vis. 127(9), 1175–1197 (2019). https://doi.org/10.1007/s11263-019-01164-6
34. Kristan, M., et al.: A novel performance evaluation methodology for single-target trackers. TPAMI 38(11), 2137–2155 (2016)
35. Leal-Taixé, L., Fenzi, M., Kuznetsova, A., Rosenhahn, B., Savarese, S.: Learning an image-based motion context for multiple people tracking. In: CVPR (2014)
36. Li, B., Wu, W., Wang, Q., Zhang, F., Xing, J., Yan, J.: SiamRPN++: evolution of siamese visual tracking with very deep networks. In: CVPR (2019)
37. Li, B., Yan, J., Wu, W., Zhu, Z., Hu, X.: High performance visual tracking with siamese region proposal network. In: CVPR (2018)
38. Lin, T.-Y., et al.: Microsoft COCO: common objects in context. In: Fleet, D., Pajdla, T., Schiele, B., Tuytelaars, T. (eds.) ECCV 2014. LNCS, vol. 8693, pp. 740–755. Springer, Cham (2014). https://doi.org/10.1007/978-3-319-10602-1_48
39. Lokoč, J., Kovalčík, G., Souček, T., Moravec, J., Čech, P.: A framework for effective known-item search in video. In: ACMM (2019). https://doi.org/10.1145/3343031.3351046
40. Milan, A., Leal-Taixé, L., Reid, I., Roth, S., Schindler, K.: MOT16: a benchmark for multi-object tracking. arXiv preprint arXiv:1603.00831 (2016)

41. Milan, A., Rezatofighi, S.H., Dick, A., Reid, I., Schindler, K.: Online multi-target tracking using recurrent neural networks. In: Thirty-First AAAI Conference on Artificial Intelligence (2017)
42. Ochs, P., Malik, J., Brox, T.: Segmentation of moving objects by long term video analysis. IEEE Trans. Pattern Anal. Mach. Intell. **36**(6), 1187–1200 (2013)
43. Perazzi, F., Pont-Tuset, J., McWilliams, B., Van Gool, L., Gross, M., Sorkine-Hornung, A.: A benchmark dataset and evaluation methodology for video object segmentation. In: CVPR (2016)
44. Pirsiavash, H., Ramanan, D., Fowlkes, C.C.: Globally-optimal greedy algorithms for tracking a variable number of objects. In: CVPR (2011)
45. Ren, L., Lu, J., Wang, Z., Tian, Q., Zhou, J.: Collaborative deep reinforcement learning for multi-object tracking. In: Ferrari, V., Hebert, M., Sminchisescu, C., Weiss, Y. (eds.) ECCV 2018. LNCS, vol. 11207, pp. 605–621. Springer, Cham (2018). https://doi.org/10.1007/978-3-030-01219-9_36
46. Ren, S., He, K., Girshick, R., Sun, J.: Faster R-CNN: towards real-time object detection with region proposal networks. In: NIPS (2015)
47. Ristani, E., Tomasi, C.: Features for multi-target multi-camera tracking and re-identification. In: CVPR (2018)
48. Russakovsky, O., et al.: ImageNet large scale visual recognition challenge. Int. J. Comput. Vis. **115**(3), 211–252 (2015). https://doi.org/10.1007/s11263-015-0816-y
49. Russell, B.C., Torralba, A., Murphy, K.P., Freeman, W.T.: LabelMe: a database and web-based tool for image annotation. Int. J. Comput. Vis. **77**(1–3), 157–173 (2008). https://doi.org/10.1007/s11263-007-0090-8
50. Scovanner, P., Tappen, M.F.: Learning pedestrian dynamics from the real world. In: ICCV (2009)
51. Shang, X., Di, D., Xiao, J., Cao, Y., Yang, X., Chua, T.S.: Annotating objects and relations in user-generated videos. In: ICMR (2019)
52. Sigurdsson, G.A., Varol, G., Wang, X., Farhadi, A., Laptev, I., Gupta, A.: Holly-wood in homes: crowdsourcing data collection for activity understanding. In: Leibe, B., Matas, J., Sebe, N., Welling, M. (eds.) ECCV 2016. LNCS, vol. 9905, pp. 510–526. Springer, Cham (2016). https://doi.org/10.1007/978-3-319-46448-0_31
53. Tao, R., Gavves, E., Smeulders, A.W.: Siamese instance search for tracking. In: CVPR (2016)
54. Thomee, B., et al.: YFCC100M: the new data in multimedia research. arXiv preprint arXiv:1503.01817 (2015)
55. Valmadre, J., et al.: Long-term tracking in the wild: a benchmark. In: Ferrari, V., Hebert, M., Sminchisescu, C., Weiss, Y. (eds.) ECCV 2018. LNCS, vol. 11207, pp. 692–707. Springer, Cham (2018). https://doi.org/10.1007/978-3-030-01219-9_41
56. Voigtlaender, P., et al.: MOTS: multi-object tracking and segmentation. In: CPVR (2019)
57. Voigtlaender, P., Leibe, B.: Online adaptation of convolutional neural networks for video object segmentation. In: BMVC (2017)
58. Wang, Q., Zhang, L., Bertinetto, L., Hu, W., Torr, P.H.: Fast online object tracking and segmentation: a unifying approach. In: CVPR (2019)
59. Wang, W., et al.: Learning unsupervised video object segmentation through visual attention. In: CVPR (2019)
60. Wen, L., et al.: UA-DETRAC: a new benchmark and protocol for multi-object detection and tracking. arXiv preprint arXiv:1511.04136 (2015)
61. Wu, Y., Lim, J., Yang, M.H.: Online object tracking: a benchmark. In: CVPR (2013)

62. Wu, Y., Kirillov, A., Massa, F., Lo, W.Y., Girshick, R.: Detectron2 (2019). https:// github.com/facebookresearch/detectron2

63. Xiao, F., Lee, Y.J.: Video object detection with an aligned spatial-temporal memory. In: Ferrari, V., Hebert, M., Sminchisescu, C., Weiss, Y. (eds.) ECCV 2018. LNCS, vol. 11212, pp. 494–510. Springer, Cham (2018). https://doi.org/10.1007/978-3-030-01237-3_30

64. Xu, N., et al.: YouTube-VOS: a large-scale video object segmentation benchmark. arXiv preprint arXiv:1809.03327 (2018)

65. Yang, L., Fan, Y., Xu, N.: Video instance segmentation. In: ICCV (2019)

66. Yu, F., et al.: BDD100K: a diverse driving dataset for heterogeneous multitask learning. In: CVPR, June 2020

67. Zhang, L., Li, Y., Nevatia, R.: Global data association for multi-object tracking using network flows. In: CVPR (2008)

68. Zhao, H., Torralba, A., Torresani, L., Yan, Z.: HACS: human action clips and segments dataset for recognition and temporal localization. In: ICCV (2019)

69. Zhou, B., Zhao, H., Puig, X., Fidler, S., Barriuso, A., Torralba, A.: Scene parsing through ADE20K dataset. In: CVPR (2017)

70. Zhu, X., Wang, Y., Dai, J., Yuan, L., Wei, Y.: Flow-guided feature aggregation for video object detection. In: 2017 IEEE International Conference on Computer Vision (ICCV), pp. 408–417 (2017)

71. Zhu, Z., Wang, Q., Li, B., Wu, W., Yan, J., Hu, W.: Distractor-aware siamese networks for visual object tracking. In: Ferrari, V., Hebert, M., Sminchisescu, C., Weiss, Y. (eds.) ECCV 2018. LNCS, vol. 11213, pp. 103–119. Springer, Cham (2018). https://doi.org/10.1007/978-3-030-01240-3_7

A Generalization of Otsu's Method and Minimum Error Thresholding

Jonathan T. Barron[✉]

Google Research, San Francisco, USA
barron@google.com

Abstract. We present Generalized Histogram Thresholding (GHT), a simple, fast, and effective technique for histogram-based image thresholding. GHT works by performing approximate maximum a posteriori estimation of a mixture of Gaussians with appropriate priors. We demonstrate that GHT subsumes three classic thresholding techniques as special cases: Otsu's method, Minimum Error Thresholding (MET), and weighted percentile thresholding. GHT thereby enables the continuous interpolation between those three algorithms, which allows thresholding accuracy to be improved significantly. GHT also provides a clarifying interpretation of the common practice of coarsening a histogram's bin width during thresholding. We show that GHT outperforms or matches the performance of all algorithms on a recent challenge for handwritten document image binarization (including deep neural networks trained to produce per-pixel binarizations), and can be implemented in a dozen lines of code or as a trivial modification to Otsu's method or MET.

1 Introduction

Histogram-based thresholding is a ubiquitous tool in image processing, medical imaging, and document analysis: The grayscale intensities of an input image are used to compute a histogram, and some algorithm is then applied to that histogram to identify an optimal threshold (corresponding to a bin location along the histogram's x-axis) with which the histogram is to be "split" into two parts. That threshold is then used to binarize the input image, under the assumption that this binarized image will then be used for some downstream task such as classification or segmentation. Thresholding has been the focus of a half-century of research, and is well documented in several survey papers [20,27]. One might assume that a global thresholding approach has little value in a modern machine learning context: preprocessing an image via binarization discards potentially-valuable information in the input image, and training a neural network to perform binarization is a straightforward alternative. However, training requires significant amounts of training data, and in many contexts (particularly medical imaging) such data may be prohibitively expensive or difficult to obtain. As

Electronic supplementary material The online version of this chapter (https://doi.org/10.1007/978-3-030-58558-7_27) contains supplementary material, which is available to authorized users.

such, there continues to be value in "automatic" thresholding algorithms that do not require training data, as is evidenced by the active use of these algorithms in the medical imaging literature.

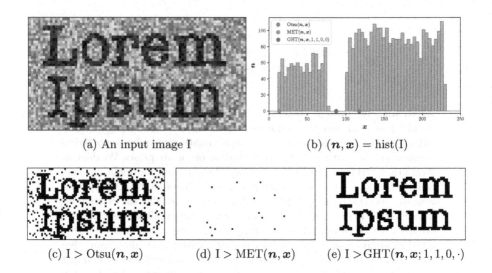

(a) An input image I   (b) $(\boldsymbol{n}, \boldsymbol{x}) = \text{hist}(\text{I})$

(c) I > Otsu$(\boldsymbol{n}, \boldsymbol{x})$ (d) I > MET$(\boldsymbol{n}, \boldsymbol{x})$ (e) I > GHT$(\boldsymbol{n}, \boldsymbol{x}; 1, 1, 0, \cdot)$

Fig. 1. Despite the visually apparent difference between foreground and background pixels in the image (a) and its histogram (b), both Otsu's method (c) and MET (d) fail to correctly binarize this image. GHT (e), which includes both Otsu's method and MET as special cases, produces the expected binarization.

2   Preliminaries

Given an input image I (Fig. 1a) that is usually assumed to contain 8-bit or 16-bit quantized grayscale intensity values, a histogram-based automatic thresholding technique first constructs a histogram of that image consisting of a vector of histogram bin locations $\boldsymbol{x}$ and a corresponding vector of histogram counts or weights $\boldsymbol{n}$ (Fig. 1b). With this histogram, the algorithm attempts to find some threshold t that separates the two halves of the histogram according to some criteria (*e.g.* maximum likelihood, minimal distortion, *etc.*), and this threshold is used to produce a binarized image I > t (Figs. 1c–1e). Many prior works assume that $x_i = i$, which addresses the common case in which histogram bins exactly correspond to quantized pixel intensities, but we will make no such assumption. This allows our algorithm (and our descriptions and implementations of baseline algorithms) to be applied to arbitrary sets of sorted points, and to histograms whose bins have arbitrary locations. For example, it is equivalent to binarize some quantized image I using a histogram:

$$(n, x) = \text{hist}(I), \tag{1}$$

or using a vector of sorted values each with a "count" of 1:

$$n = \vec{1}, \qquad x = \text{sort}(I). \tag{2}$$

This equivalence is not used in any result presented here, but is useful when binarizing a set of continuous values.

Most histogram-based thresholding techniques work by considering each possible "split" of the histogram: each value in x is considered as a candidate for t, and two quantities are produced that reflect the surface statistics of $n_{x \leq t}$ and $n_{x > t}$. A critical insight of many classic histogram-based thresholding techniques is that some of these quantities can be computed recursively. For example, the sum of all histogram counts up through some index i ($w_i^{(0)} = \sum_{j=0}^{i} n_j$) need not be recomputed from scratch if that sum has already been previously computed for all the previous histogram element, and can instead just be updated with a single addition ($w_i^{(0)} = w_{i-1}^{(0)} + n_i$). Here we construct pairs of vectors of intermediate values that measure surface statistics of the histogram above and below each "split", to be used by GHT and our baselines:

$$
\begin{aligned}
w^{(0)} &= \text{csum}(n) & w^{(1)} &= \text{dsum}(n) \\
\pi^{(0)} &= w^{(0)} / \|n\|_1 & \pi^{(1)} &= w^{(1)} / \|n\|_1 = 1 - \pi^{(0)} \\
\mu^{(0)} &= \text{csum}(nx) / w^{(0)} & \mu^{(1)} &= \text{dsum}(nx) / w^{(1)} \\
d^{(0)} &= \text{csum}(nx^2) - w^{(0)} \left(\mu^{(0)}\right)^2 & d^{(1)} &= \text{dsum}(nx^2) - w^{(1)} \left(\mu^{(1)}\right)^2
\end{aligned}
\tag{3}
$$

All multiplications, divisions, and exponentiations are element-wise, and csum() and dsum() are cumulative and "reverse cumulative" sums respectively. $\|n\|_1$ is the sum of all histogram counts in n, while $w^{(0)}$ and $w^{(1)}$ are the sums of histogram counts in n below and above each split of the histogram, respectively. Similarly, each $\pi^{(k)}$ respectively represent normalized histogram counts (or mixture weights) above and below each split. Each $\mu^{(k)}$ and $d^{(k)}$ represent the mean and distortion of all elements of x below and above each split, weighted by their histogram counts n. Here "distortion" is used in the context of k-means, where it refers to the sum of squared distances of a set of points to their mean. The computation of $d^{(k)}$ follows from three observations: First, csum($\cdot$) or dsum($\cdot$) can be used (by weighting by n and normalizing by w) to compute a vector of expected values. Second, the sample variance of a quantity is a function of the expectation of its values and its values squared ($\text{Var}(X) = \text{E}[X^2] - \text{E}[X]^2$). Third, the sample variance of a set of points is proportional to the total distortion of those points.

Formally, $\text{csum}(\cdot)_i$ is the inclusive sum of every vector element at or before index i, and $\text{dsum}(\cdot)_i$ is the exclusive sum of everything after index i:

$$\text{csum}(\boldsymbol{n})_i = \sum_{j=0}^{i} n_j = \text{csum}(\boldsymbol{n})_{i-1} + n_i, \tag{4}$$

$$\text{dsum}(\boldsymbol{n})_i = \sum_{j=i+1}^{\text{len}(\boldsymbol{n})-1} n_j = \text{dsum}(\boldsymbol{n})_{i+1} + n_{i+1}.$$

Note that the outputs of these cumulative and decremental sums have a length that is one less than that of $\boldsymbol{x}$ and $\boldsymbol{n}$, as these values measure "splits" of the data, and we only consider splits that contain at least one histogram element. The subscripts in our vectors of statistics correspond to a threshold *after* each index, which means that binarization should be performed by a greater-than comparison with the returned threshold ($\boldsymbol{x} > t$). In practice, $\text{dsum}(\boldsymbol{n})$ can be computed using just the cumulative and total sums of $\boldsymbol{n}$:

$$\text{dsum}(\boldsymbol{n}) = \|\boldsymbol{n}\|_1 - \text{csum}(\boldsymbol{n}). \tag{5}$$

As such, these quantities (and all other vector quantities that will be described later) can be computed efficiently with just one or two passes over the histogram.

3 Algorithm

Our Generalized Histogram Thresholding (GHT) algorithm is motivated by a straightforward Bayesian treatment of histogram thresholding. We assume that each pixel in the input image is generated from a mixture of two probability distributions corresponding to intensities below and above some threshold, and we maximize the joint probability of all pixels (which are assumed to be IID):

$$\prod_{x,y} \left(p^{(0)}(\mathrm{I}_{x,y}) + p^{(1)}(\mathrm{I}_{x,y}) \right). \tag{6}$$

Our probability distributions are parameterized as:

$$p^{(k)}(x) = \pi^{(k)} \mathcal{N}\left(\mathrm{I}_{x,y} \mid \mu^{(k)}, \sigma^{(k)}\right) \chi^2_{\text{SI}}\left(\sigma^{(k)} \mid \pi^{(k)}\nu, \tau^2\right) \text{Beta}\left(\pi^{(k)} \mid \kappa, \omega\right). \tag{7}$$

This is similar to a straightforward mixture of Gaussians: we have a mixture probability ($\pi^{(0)}$, $\pi^{(1)} = 1 - \pi^{(0)}$), two means ($\mu^{(0)}$, $\mu^{(1)}$), and two standard deviations ($\sigma^{(0)}$, $\sigma^{(1)}$). But unlike a standard mixture of Gaussians, we place conjugate priors on the model parameters. We assume each standard deviation is the mode (*i.e.*, the "updated" τ value) of a scaled inverse chi-squared distribution, which is the conjugate prior of a Gaussian with a known mean.

$$\chi^2_{\text{SI}}(x \mid \nu, \tau^2) = \frac{(\tau^2\nu/2)^{\nu/2}}{\Gamma(\nu/2)} \frac{\exp\left(\frac{-\nu\tau^2}{2x}\right)}{x^{1+\nu/2}}. \tag{8}$$

And we assume the mixture probability is drawn from a beta distribution (the conjugate prior of a Bernoulli distribution).

$$\text{Beta}(x \mid \kappa, \omega) = \frac{x^{\alpha-1}(1-x)^{\beta-1}}{\text{B}(\alpha, \beta)}, \quad \alpha = \kappa\omega + 1, \quad \beta = \kappa(1-\omega) + 1. \quad (9)$$

where B is the beta function. Note that our beta distribution uses a slightly unconventional parameterization in terms of its concentration κ and mode ω, which will be explained later. Also note that in Eq. 7 the degrees-of-freedom parameter ν of the scaled inverse chi-squared distribution is rescaled according to the mixture probability, as will also be discussed later. This Bayesian mixture of Gaussians has four hyperparameters that will determine the behavior of GHT: $\nu \geq 0, \tau \geq 0, \kappa \geq 0, \omega \in [0, 1]$.

From this probabilistic framework we can derive our histogram thresholding algorithm. Let us rewrite Eq. 6 to be in terms of bin locations and counts:

$$\prod_i \left(p^{(0)}(x_i) + p^{(1)}(x_i)\right)^{n_i}. \quad (10)$$

Taking its logarithm gives us this log-likelihood:

$$\sum_i n_i \log\left(p^{(0)}(x_i) + p^{(1)}(x_i)\right). \quad (11)$$

Now we make a simplifying model assumption that enables a single-pass histogram thresholding algorithm. We will assume that, at each potential "split" of the data, each of the two splits of the histogram is generated entirely by one of the two Gaussians in our mixture. More formally, we consider all possible sorted assignments of each histogram bin to either of the two Gaussians and maximize the expected complete log-likelihood (ECLL) of that assignment. Jensen's inequality ensures that this ECLL is a lower bound on the marginal likelihood of the data that we would actually like to maximize. This is the same basic approach (though not the same justification) as in "Minimum Error Thresholding" (MET) [10] and other similar approaches, but where our technique differs is in how the likelihood and the parameters of these Gaussians are estimated. For each split of the data, we assume the posterior distribution of the latent variables determining the ownership of each histogram bin by each Gaussian (the missing "z" values) are wholly assigned to one of the two Gaussians according to the split. This gives us the following ECLL as a function of the assumed split location's array index i:

$$\mathcal{L}_i = \sum_{j=0}^{i} n_j \log\left(p^{(0)}(x_j)\right) + \sum_{j=i+1}^{\text{len}(n)-1} n_j \log\left(p^{(1)}(x_j)\right). \quad (12)$$

Our proposed algorithm is to simply iterate over all possible values of i (of which there are 255 in our experiments) and return the value that maximizes $\mathcal{L}_i$. As such, our algorithm can be viewed as a kind of inverted expectation-maximization in which the sweep over threshold resembles an M-step and the assignment of latent variables according to that threshold resembles an E-step [21]. Our GHT algorithm will be defined as:

$$\text{GHT}(\boldsymbol{x}, \boldsymbol{n}; \nu, \tau, \kappa, \omega) = x_{\text{argmax}_i(\mathcal{L}_i)}. \tag{13}$$

This definition of GHT is a bit unwieldy, but using the preliminary math of Eq. 3 and ignoring global shifts and scales of $\mathcal{L}_i$ that do not affect the argmax it can be simplified substantially:

$$\boldsymbol{v}^{(0)} = \frac{\boldsymbol{\pi}^{(0)}\nu\tau^2 + \boldsymbol{d}^{(0)}}{\boldsymbol{\pi}^{(0)}\nu + \boldsymbol{w}^{(0)}} \qquad \boldsymbol{v}^{(1)} = \frac{\boldsymbol{\pi}^{(1)}\nu\tau^2 + \boldsymbol{d}^{(1)}}{\boldsymbol{\pi}^{(1)}\nu + \boldsymbol{w}^{(1)}}$$

$$\boldsymbol{f}^{(0)} = -\frac{\boldsymbol{d}^{(0)}}{\boldsymbol{v}^{(0)}} - \boldsymbol{w}^{(0)} \log\left(\boldsymbol{v}^{(0)}\right) + 2\left(\boldsymbol{w}^{(0)} + \kappa\omega\right) \log\left(\boldsymbol{w}^{(0)}\right)$$

$$\boldsymbol{f}^{(1)} = -\frac{\boldsymbol{d}^{(1)}}{\boldsymbol{v}^{(1)}} - \boldsymbol{w}^{(1)} \log\left(\boldsymbol{v}^{(1)}\right) + 2\left(\boldsymbol{w}^{(1)} + \kappa(1-\omega)\right) \log\left(\boldsymbol{w}^{(1)}\right)$$

$$\text{GHT}(\boldsymbol{x}, \boldsymbol{n}; \nu, \tau, \kappa, \omega) = x_{\text{argmax}_i\left(\boldsymbol{f}^{(0)} + \boldsymbol{f}^{(1)}\right)} \tag{14}$$

Each $\boldsymbol{f}^{(k)}$ can be thought of as a log-likelihood of the data for each Gaussian at each split, and each $\boldsymbol{v}^{(k)}$ can be thought of as an estimate of the variance of each Gaussian at each split. GHT simply scores each split of the histogram and returns the value of $\boldsymbol{x}$ that maximizes the $\boldsymbol{f}^{(0)} + \boldsymbol{f}^{(1)}$ score (in the case of ties we return the mean of all $\boldsymbol{x}$ values that maximize that score). Because this only requires element-wise computation using the previously-described quantities in Eq. 3, GHT requires just a single linear scan over the histogram.

The definition of each $\boldsymbol{v}^{(k)}$ follows from our choice to model each Gaussian's variance with a scaled inverse chi-squared distribution: the ν and τ^2 hyperparameters of the scaled inverse chi-squared distribution are updated according to the scaled sample variance (shown here as distortion $\boldsymbol{d}^{(k)}$) to produce an updated posterior hyperparameter that we use as the Gaussian's estimated variance $\boldsymbol{v}^{(k)}$. Using a conjugate prior update instead of the sample variance has little effect when the subset of $\boldsymbol{n}$ being processed has significant mass (*i.e.*, when $\boldsymbol{w}^{(k)}$ is large) but has a notable effect when a Gaussian is being fit to a sparsely populated subset of the histogram. Note that $\boldsymbol{v}^{(k)}$ slightly deviates from the traditional conjugate prior update, which would omit the $\boldsymbol{\pi}^{(k)}$ scaling on ν in the numerator and denominator (according to our decision to rescale each degrees-of-freedom parameter by each mixture probability). This additional scaling counteracts the fact that, at different splits of our histogram, the total histogram count on either side of the split will vary. A "correct" update would assume a constant number of degrees-of-freedom regardless of where the split is being performed, which would result in the conjugate prior here having a substantially different effect at each split, thereby making the resulting $\boldsymbol{f}^{(0)} + \boldsymbol{f}^{(1)}$ scores not comparable to each other and making the argmax over these scores not meaningful.

The beta distribution used in our probabilistic formulation (parameterized by a concentration κ and a mode ω) has a similar effect as the "anisotropy coefficients" used by other models [8,19]: setting ω near 0 biases the algorithm towards thresholding near the start of the histogram, and setting it near 1 biases the algorithm towards thresholding near the end of the histogram. The concentration κ parameter determines the strength of this effect, and setting $\kappa = 0$ disables this regularizer entirely. Note that our parameterization of concentration κ differs slightly from the traditional formulation, where setting $\kappa = 1$ has no effect and setting $\kappa < 1$ moves the predicted threshold *away* from the mode. We instead assume the practitioner will only want to consider biasing the algorithm's threshold *towards* the mode, and parameterize the distribution accordingly. These hyperparameters allow for GHT to be biased towards outputs where a particular fraction of the image lies above or below the output threshold. For example, in an OCR/digit recognition context, it may be useful to inform the algorithm that the majority of the image is expected to not be ink, so as to prevent output thresholds that erroneously separate the non-ink background of the page into two halves.

Note that, unlike many histogram based thresholding algorithms, we do not require or assume that our histogram counts n are normalized. In all experiments, we use raw counts as our histogram values. This is important to our approach, as it means that the behavior of the magnitude of n and the behavior induced by varying our ν and κ hyperparameters is consistent:

$$\forall a \in \mathbb{R}_{>0} \ \mathrm{GHT}(x, an; a\nu, \tau, a\kappa, \omega) = \mathrm{GHT}(n, x; \nu, \tau, \kappa, \omega). \tag{15}$$

This means that, for example, doubling the number of pixels in an image is equivalent to halving the values of the two hyperparameters that control the "strength" of our two Bayesian model components.

Additionally, GHT's output and hyperparameters (excluding τ, which exists solely to encode absolute scale) are invariant to positive affine transformations of the histogram bin centers:

$$\forall a \in \mathbb{R}_{>0} \ \forall b \in \mathbb{R} \ \mathrm{GHT}(ax + b, n; \nu, a\tau, \kappa, \omega) = \mathrm{GHT}(n, x; \nu, \tau, \kappa, \omega). \tag{16}$$

One could extend GHT to be sensitive to absolute intensity values by including conjugate priors over the means of the Gaussians, but we found little experimental value in that level of control. This is likely because the dataset we evaluate on does not have a standardized notion of brightness or exposure (as is often the case for binarization tasks).

In the following subsections we demonstrate how GHT generalizes three standard approaches to histogram thresholding by including them as special cases of our hyperparameter settings: Minimum Error Thresholding [10], Otsu's method [17], and weighted percentile thresholding.

3.1 Special Case: Minimum Error Thresholding

Because Minimum Error Thresholding (MET) [10], like our approach, works by maximizing the expected complete log-likelihood of the input histogram under a mixture of two Gaussians, it is straightforward to express it using the already-defined quantities in Eq. 3:

$$\ell = 1 + w^{(0)} \log\left(\frac{d^{(0)}}{w^{(0)}}\right) + w^{(1)} \log\left(\frac{d^{(1)}}{w^{(1)}}\right) - 2\left(w^{(0)} \log\left(w^{(0)}\right) + w^{(1)} \log\left(w^{(1)}\right)\right)$$

$$\mathrm{MET}(x, n) = x_{\arg\min_i(\ell)} \tag{17}$$

Because GHT is simply a Bayesian treatment of this same process, it includes MET as a special case. If we set $\nu = 0$ and $\kappa = 0$ (under these conditions the values of τ and ω are irrelevant) in Eq. 14, we see that $v^{(k)}$ reduces to $d^{(k)}/w^{(k)}$, which causes each score to simplify dramatically:

$$\nu = \kappa = 0 \implies f^{(k)} = -w^{(k)} - w^{(k)} \log\left(\frac{d^{(k)}}{w^{(k)}}\right) + 2w^{(k)} \log\left(w^{(k)}\right). \tag{18}$$

Because $w^{(0)} + w = \|n\|_1$, the score maximized by GHT can be simplified:

$$\nu = \kappa = 0 \implies f^{(0)} + f^{(1)} = 1 - \|n\|_1 - \ell. \tag{19}$$

We see that, for these particular hyperparameter settings, the total score $f^{(0)} + f^{(1)}$ that GHT maximizes is simply an affine transformation (with a negative scale) of the score ℓ that MET maximizes. This means that the index that optimizes either quantity is identical, and so the algorithms (under these hyperparameter settings) are equivalent:

$$\mathrm{MET}(n, x) = \mathrm{GHT}(n, x; 0, \cdot, 0, \cdot). \tag{20}$$

3.2   Special Case: Otsu's Method

Otsu's method for histogram thresholding [17] works by directly maximizing inter-class variance for the two sides of the split histogram, which is equivalent to indirectly minimizing the total intra-class variance of those two sides [17]. Otsu's method can also be expressed using the quantities in Eq. 3:

$$o = w^{(0)} w^{(1)} \left(\mu^{(0)} - \mu^{(1)}\right)^2, \quad \mathrm{Otsu}(n, x) = x_{\mathrm{argmax}_i(o)}. \tag{21}$$

To clarify the connection between Otsu's method and GHT, we rewrite the score o of Otsu's method as an explicit sum of intra-class variances:

$$o = \|n\|_1 \langle n, x^2 \rangle - \langle n, x \rangle^2 - \|n\|_1 \left(d^{(0)} + d^{(1)}\right). \tag{22}$$

Now let us take the limit of the $f^{(0)} + f^{(1)}$ score that is maximized by GHT as ν approaches infinity:

$$\lim_{\nu \to \infty} f^{(0)} + f^{(1)} = -\frac{d^{(0)} + d^{(1)}}{\tau^2} + 2w^{(0)} \log\left(\frac{w^{(0)}}{\tau}\right) + 2w^{(1)} \log\left(\frac{w^{(1)}}{\tau}\right). \tag{23}$$

With this we can set the τ hyperparameter to be infinitesimally close to zero, in which case the score in Eq. 23 becomes dominated by its first term. Therefore, as ν approaches infinity and τ approaches zero, the score maximized by GHT is proportional to the (negative) score that is indirectly minimized by Otsu's method:

$$\nu \gg 0, \tau = \epsilon \implies f^{(0)} + f^{(1)} \approx -\frac{d^{(0)} + d^{(1)}}{\tau^2}, \tag{24}$$

where ϵ is a small positive number. This observation fits naturally with the well-understood equivalence of k-means (which works by minimizing distortion) and the asymptotic behavior of maximum likelihood under a mixture of Gaussians model as the variance of each Gaussian approaches zero [12]. Given this equivalence between the scores that are optimized by MET and GHT (subject to these specific hyperparameters) we can conclude that Otsu's method is a special case of GHT:

$$\text{Otsu}(\boldsymbol{n}, \boldsymbol{x}) = \lim_{(\nu, \tau) \to (\infty, 0)} \text{GHT}(\boldsymbol{n}, \boldsymbol{x}; \nu, \tau, 0, \cdot). \tag{25}$$

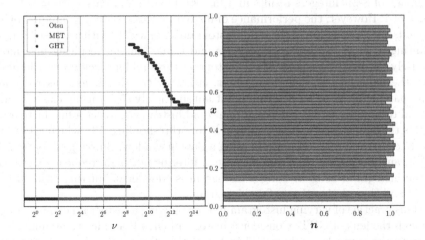

Fig. 2. On the right we have a toy input histogram (shown rotated), and on the left we have a plot showing the predicted threshold (y-axis) of GHT ($\tau = 0.01$, $\kappa = 0$, shown in black) relative to the hyperparameter value ν (x-axis). Note that the y-axis is shared across the two plots. Both MET and Otsu's method predict incorrect thresholds (shown in red and blue, respectively) despite the evident gap in the input histogram. GHT, which includes MET and Otsu's method as special cases ($\nu = 0$ and $\nu = \infty$, respectively) allows us to interpolate between these two extremes by varying ν, and produces the correct threshold for a wide range of values $\nu \in [\sim 2^2, \sim 2^8]$. (Color figure online)

We have demonstrated that GHT is a generalization of Otsu's method (when $\nu = \infty$) and MET (when $\nu = 0$), but for this observation to be of any practical importance, there must be value in GHT when ν is set to some value in between those two extremes. To demonstrate this, in Fig. 2 we visualize the effect of ν on GHT's behavior. Similar to the example shown in Fig. 1, we see that both Otsu's method and MET both perform poorly when faced with a simple histogram that contains two separate and uneven uniform distributions: MET prefers to express all histogram elements using a single Gaussian by selecting a threshold at one end of the histogram, and Otsu's method selects a threshold that splits the larger of the two modes in half instead of splitting the two modes apart. But by varying ν from 0 to ∞ (while also setting τ to a small value) we see that a wide range of values of ν results in GHT correctly separating the two modes of this histogram. Additionally, we see that the range of hyperparameters that reproduces this behavior is wide, demonstrating the robustness of performance with respect to the specific value of this parameter.

3.3   Relationship to Histogram Bin Width

Most histogram thresholding techniques (including GHT as it is used in this paper) construct a histogram with as many bins as there are pixel intensities—an image of 8-bit integers results in a histogram with 256 bins, each with a bin width of 1. However, the performance of histogram thresholding techniques often depends on the selection of the input histogram's bin width, with a coarse binning resulting in more "stable" performance and a fine binning resulting in a more precisely localized threshold. Many histogram thresholding techniques therefore vary the bin width of their histograms, either by treating it as a user-defined hyperparameter [2] or by determining it automatically [7,16] using classical statistical techniques [3,24]. Similarly, one could instead construct a fine histogram that is then convolved with a Gaussian before thresholding [13], which is equivalent to constructing the histogram using a Parzen window. Blurring or coarsening histograms both serve the same purpose of filtering out high frequency variation in the histogram, as coarsening a histogram is equivalent blurring (with a box filter) and then decimating a fine histogram.

This practice of varying histogram bins or blurring a histogram can be viewed through the lens of GHT. Consider a histogram (n, x), and let us assume (contrary to the equivalence described in Eq. 2) that the spacing between the bins centers in x is constant. Consider a discrete Gaussian blur filter $f(\sigma)$ with a standard deviation of σ, where that filter is normalized ($\|f(\sigma)\|_1 = 1$). Let us consider $n * f(\sigma)$, which is the convolution of histogram counts with our Gaussian blur (assuming a "full" convolution of n and x with zero boundary conditions). This significantly affects the sample variance of the histogram v, which (because convolution is linear) is:

$$v = \frac{\sum_i (n * f(\sigma))_i (x_i - \mu)^2}{\|n\|_1} = \frac{\|n\|_1 \sigma^2 + d}{\|n\|_1}, \qquad d = \sum_i n_i (x_i - \mu)^2. \quad (26)$$

We use v and d to describe sample variance and distortion as before, though here these values are scalars as we compute a single estimate of both for the entire histogram. This sample variance v resembles the definition of $v^{(k)}$ in Eq. 14, and we can make the two identical in the limit by setting GHT's hyperparameters to $\nu = \epsilon \|n\|_1$ and $\tau = \sigma/\sqrt{\epsilon}$, where ϵ is a small positive number. With this equivalence we see that GHT's τ hyperparameter serves a similar purpose as coarsening/blurring the input histogram—blurring the input histogram with a Gaussian filter with standard deviation of σ is roughly equivalent to setting GHT's τ parameter proportionally to σ. Or more formally:

$$\text{MET}(n, x * f(\sigma)) \approx \text{GHT}\left(n, x; \|n\|_1 \epsilon, \cdot, \sigma/\sqrt{\epsilon}, \cdot\right). \tag{27}$$

Unlike the other algorithmic equivalences stated in this work, this equivalence is approximate: changing σ is not exactly equivalent to blurring the input histogram. Instead, it is equivalent to, at each split of n, blurring the left and right halves of n independently—the histogram cannot be blurred "across" the split. Still, we observed that these two approaches produce identical threshold results in the vast majority of instances.

This relationship between GHT's τ hyperparameter and the common practice of varying a histogram's bin width provides a theoretical grounding of both GHT and other work: GHT may perform well because varying τ implicitly blurs the input histogram, or prior techniques based on variable bin widths may work well because they are implicitly imposing a conjugate prior on sample variance. This relationship may also shed some light on other prior work exploring the limitations of MET thresholding due to sample variance estimates not being properly "blurred" across the splitting threshold [1].

3.4 Special Case: Weighted Percentile

A simple approach for binarizing some continuous signal is to use the nth percentile of the input data as the threshold, such as its median. This simple baseline is also expressible as a special case of GHT. If we set κ to a large value we see that the score $f^{(0)} + f^{(1)}$ being maximized by GHT becomes dominated by the terms of the score related to $\log(\pi^{(k)})$:

$$\kappa \gg 0 \implies \frac{f^{(0)} + f^{(1)}}{2} \approx (\|n\|_1 + \kappa) \log(\|n\|_1) + \kappa \left(\omega \log\left(\pi^{(0)}\right) + (1-\omega) \log\left(1 - \pi^{(0)}\right)\right). \tag{28}$$

By setting its derivative to zero we see that this score is maximized at the split location i where $\pi_i^{(0)}$ is as close as possible to ω. This condition is also satisfied by the (100ω)th percentile of the histogrammed data in I, or equivalently, by taking the (100ω)th weighted percentile of x (where each bin is weighted by n). From this we can conclude that the weighted percentile of the histogram (or equivalently, the percentile of the image) is a special case of GHT:

$$\text{wprctile}(x, n, 100\omega) = \lim_{\kappa \to \infty} \text{GHT}(n, x, 0, \cdot, \kappa, \omega). \tag{29}$$

This also follows straightforwardly from how GHT uses a beta distribution: if this beta distribution's concentration κ is set to a large value, the mode of the posterior distribution will approach the mode of the prior ω.

In Fig. 3 we demonstrate the value of this beta distribution regularization. We see that Otsu's method and MET behave unpredictably when given multimodal inputs, as the metrics they optimize have no reason to prefer a split that groups more of the histogram modes above the threshold versus one that groups more of the modes below the threshold. This can be circumvented by using the weighted percentile of the histogram as a threshold, but this requires that the target percentile ω be precisely specified beforehand: For any particular input histogram, it is possible to identify a value for ω that produces the desired threshold, but this percentile target will not generalize to another histogram with the same overall appearance but with a slightly different distribution of mass across that desired threshold level. GHT, in contrast, is able to produce sensible thresholds for all valid values of ω—when the value is less than $1/2$ the recovered threshold precisely separates the first mode from the latter two, and when the value is greater than $1/2$ it precisely separates the first two modes from the latter one. As such, we see that this model component provides an effective means to bias GHT towards specific kinds of thresholds, while still causing it to respect the relative spread of histogram bin counts.

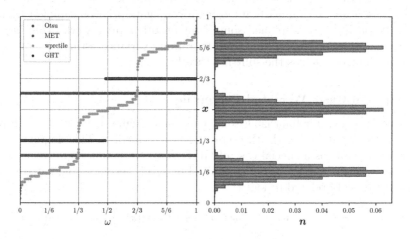

Fig. 3. On the right we have a toy input histogram (shown rotated), and on the left we have a plot showing the predicted threshold (y-axis) of GHT ($\nu = 200$, $\tau = 0.01$, $\kappa = 0.1$, shown in black) relative to the hyperparameter value ω (x-axis). Note that the y-axis is shared across the two plots. Both MET and Otsu's method (shown in red and blue, respectively) predict thresholds that separate two of the three modes from each other arbitrarily. Computing the weighted percentile of the histogram (shown in yellow) can produce *any* threshold, but reproducing the desired split requires exactly specifying the correct value of ω, which likely differs across inputs. GHT (which includes these three other algorithms as special cases) produces thresholds that accurately separate the three modes, and the location of that threshold is robust to the precise value of ω and depends only on it being below or above $1/2$. (Color figure online)

Table 1. Results on the 2016 Handwritten Document Image Binarization Contest (H-DIBCO) challenge [18], in terms of the arithmetic mean and standard deviation of F1 score (multiplied by 100 to match the conventions used by [18]), PSNR, and Distance-Reciprocal Distortion (DRD) [15]. The scores for all baseline algorithms are taken from [18]. Our GHT algorithm and it's ablations and special cases (some of which correspond to other algorithms) are indicated in bold. The settings of the four hyperparameters governing GHT's behavior are also provided.

Algorithm	ν	τ	κ	ω	$F_1 \times 100$ ↑	PSNR ↑	DRD [15] ↓
GHT (MET Case)	-	-	-	-	60.40 ± 20.65	11.21 ± 3.50	45.32 ± 41.35
Kefali *et al.* [18,22]					76.10 ± 13.81	15.35 ± 3.19	9.16 ± 4.87
Raza [18]					76.28 ± 9.71	14.21 ± 2.21	15.14 ± 9.42
GHT (wprctile Case)	-	-	10^{60}	$2^{-3.75}$	76.77 ± 14.50	15.44 ± 3.40	12.91 ± 17.19
Sauvola [18,23]					82.52 ± 9.65	16.42 ± 2.87	7.49 ± 3.97
Khan and Mollah [18]					84.32 ± 6.81	16.59 ± 2.99	6.94 ± 3.33
Tensmeyer and Martinez [14,25,26]					85.57 ± 6.75	17.50 ± 3.43	5.00 ± 2.60
de Almeida and de Mello [18]					86.24 ± 5.79	17.52 ± 3.42	5.25 ± 2.88
Otsu's Method [17,18]					86.61 ± 7.26	17.80 ± 4.51	5.56 ± 4.44
GHT (No wprctile)	$2^{50.5}$	$2^{0.125}$	-	-	87.16 ± 6.32	17.97 ± 4.00	5.04 ± 3.17
GHT (Otsu Case)	10^{60}	10^{-15}	-	-	87.19 ± 6.28	17.97 ± 4.01	5.04 ± 3.16
Otsu's Method (Our Impl.) [17]					87.19 ± 6.28	17.97 ± 4.01	5.04 ± 3.16
Nafchi *et al.* - 1 [18,28]					87.60 ± 4.85	17.86 ± 3.51	4.51 ± 1.62
Kligler [6,9,11]					87.61 ± 6.99	18.11 ± 4.27	5.21 ± 5.28
Roe and de Mello [18]					87.97 ± 5.17	18.00 ± 3.68	4.49 ± 2.65
Nafchi *et al.* - 2 [18,28]					88.11 ± 4.63	18.00 ± 3.41	4.38 ± 1.65
Hassaïne *et al.* - 1 [5,18]					88.22 ± 4.80	18.22 ± 3.41	4.01 ± 1.49
Hassaïne *et al.* - 2 [4,18]					88.47 ± 4.45	18.29 ± 3.35	3.93 ± 1.37
Hassaïne *et al.* - 3 [4,5,18]					88.72 ± 4.68	18.45 ± 3.41	3.86 ± 1.57
GHT	$2^{29.5}$	$2^{3.125}$	$2^{22.25}$	$2^{-3.25}$	88.77 ± 4.99	18.55 ± 3.46	3.99 ± 1.77
Oracle Global Threshold					90.69 ± 3.92	19.17 ± 3.29	3.57 ± 1.84

4   Experiments

To evaluate GHT, we use the 2016 Handwritten Document Image Binarization Contest (H-DIBCO) challenge [18]. This challenge consists of images of handwritten documents alongside ground-truth segmentations of those documents, and algorithms are evaluated by how well their binarizations match the ground truth. We use the 2016 challenge because this is the most recent instantiation of this challenge where a global thresholding algorithm (one that selects a single threshold for use on all pixels) can be competitive: Later versions of this challenge use input images that contain content outside of the page being binarized which, due to the background of the images often being black, renders any algorithm that produces a single global threshold for binarization ineffective. GHT's four hyperparameters were tuned using coordinate descent to maximize the arithmetic mean of F_1 scores (a metric used by the challenge) over a small training dataset (or perhaps more accurately, a "tuning" dataset). For training data we use the 8 hand-written images from the 2013 version of the same challenge, which is consistent with the tuning procedure advocated by the challenge (any training data from past challenges may be used, though we found it sufficient to use only relevant data from the most recent challenge). Input images are processed by taking the per-pixel max() across color channels to produce a grayscale image, computing a single 256-bin histogram of the resulting values, applying GHT to that histogram to produce a single global threshold, and binarizing the grayscale input image according to that threshold.

In Table 1 we report GHT's performance against all algorithms evaluated in the challenge [18], using the F_1, PSNR, and Distance-Reciprocal Distortion [15] metrics used by the challenge [18]. We see that GHT produces the lowest-error of all entrants to the H-DIBCO 2016 challenge for two of the three metrics used. GHT outperforms Otsu's method and MET by a significant margin, and also outperforms or matches the performance of significantly more complicated techniques that rely on large neural networks with thousands or millions of learned parameters, and that produce an arbitrary per-pixel binary mask instead of GHT's single global threshold. This is despite GHT's simplicity: it requires roughly a dozen lines of code to implement (far fewer if an implementation of MET or Otsu's method is available), requires no training, and has only four parameters that were tuned on a small dataset of only 8 training images.

We augment Table 1 with some additional results not present in [18]. We present an "Oracle Global Threshold" algorithm, which shows the performance of an oracle that selects the best-performing (according to F_1) global threshold individually for each test image. We present the special cases of GHT that correspond to MET and Otsu's method, to verify that the special case corresponding to Otsu's method performs identically to our own implementation of Otsu's method and nearly identically to the implementation presented in [18]. We present additional ablations of GHT to demonstrate the contribution of each algorithm component: The "wprctile Case" model sets κ to an extremely large value and tunes ω on our training set, and performs poorly. The "No wprctile" model sets $\kappa = 0$ and exhibits worse performance than complete GHT.

See the supplement for reference implementations of GHT, as well as reference implementations of the other algorithms that were demonstrated to be special cases of GHT in Sects. 3.1–3.4.

5   Conclusion

We have presented Generalized Histogram Thresholding, a simple, fast, and effective technique for histogram-based image thresholding. GHT includes several classic techniques as special cases (Otsu's method, Minimum Error Thresholding, and weighted percentile thresholding) and thereby serves as a unifying framework for those discrete algorithms, in addition to providing a theoretical grounding for the common practice of varying a histogram's bin width when thresholding. GHT is exceedingly simple: it can be implemented in just a dozen lines of python (far fewer if an implementation of MET or Otsu's method is available) and has just four tunable hyperparameters. Because it requires just a single sweep over a histogram of image intensities, GHT is fast to evaluate: its computational complexity is comparable to Otsu's method. Despite its simplicity and speed (and its inherent limitations as a *global* thresholding algorithm) GHT outperforms or matches the performance of all submitted techniques on the 2016H-DIBCO image binarization challenge—including deep neural networks that have been trained to produce arbitrary per-pixel binarizations.

References

1. Cho, S., Haralick, R., Yi, S.: Improvement of Kittler and Illingworth's minimum error thresholding. Pattern Recogn. **22**(5), 609–617 (1989)
2. Coudray, N., Buessler, J.L., Urban, J.P.: A robust thresholding algorithm for unimodal image histograms (2010)
3. Doane, D.P.: Aesthetic frequency classifications. Am. Stat. **30**(4), 181–183 (1976)
4. Hassaïne, A., Al-Maadeed, S., Bouridane, A.: A set of geometrical features for writer identification. In: Huang, T., Zeng, Z., Li, C., Leung, C.S. (eds.) ICONIP 2012. LNCS, vol. 7667, pp. 584–591. Springer, Heidelberg (2012). https://doi.org/10.1007/978-3-642-34500-5_69
5. Hassaïne, A., Decencière, E., Besserer, B.: Efficient restoration of variable area soundtracks. Image Anal. Stereol. **28**(2), 113–119 (2009)
6. Howe, N.R.: Document binarization with automatic parameter tuning. Int. J. Doc. Anal. Recogn. **16**, 247–258 (2013). https://doi.org/10.1007/s10032-012-0192-x
7. Kadhim, N., Mourshed, M.: A shadow-overlapping algorithm for estimating building heights from VHR satellite images. IEEE Geosci. Remote Sens. Lett. **15**(1), 8–12 (2018)
8. Kapur, J.N., Sahoo, P.K., Wong, A.K.: A new method for gray-level picture thresholding using the entropy of the histogram. Comput. Vis. Graph. Image Process. **29**(3), 273–285 (1985)
9. Katz, S., Tal, A., Basri, R.: Direct visibility of point sets. In: SIGGRAPH (2007)
10. Kittler, J., Illingworth, J.: Minimum error thresholding. Pattern Recogn. **19**(1), 41–47 (1986)
11. Kligler, N., Tal, A.: On visibility and image processing. Ph.D. thesis, Computer Science Department, Technion (2017)
12. Kulis, B., Jordan, M.I.: Revisiting K-means: new algorithms via Bayesian nonparametrics. In: ICML (2012)
13. Lindblad, J.: Histogram thresholding using kernel density estimates. In: SSAB Symposium on Image Analysis (2000)
14. Long, J., Shelhamer, E., Darrell, T.: Fully convolutional networks for semantic segmentation. In: CVPR (2015)
15. Lu, H., Kot, A.C., Shi, Y.Q.: Distance-reciprocal distortion measure for binary document images. IEEE Sig. Process. Lett. **11**(2), 228–231 (2004)
16. Onumanyi, A., Onwuka, E., Aibinu, A., Ugweje, O., Salami, M.J.E.: A modified Otsu's algorithm for improving the performance of the energy detector in cognitive radio. AEU-Int. J. Electron. Commun. **79**, 53–63 (2017)
17. Otsu, N.: A threshold selection method from gray-level histograms. IEEE Trans. Syst. Man Cybern. **9**(1), 62–66 (1979)
18. Pratikakis, I., Zagoris, K., Barlas, G., Gatos, B.: ICFHR 2016 handwritten document image binarization contest (H-DIBCO 2016). ICFHR (2016)

19. Pun, T.: Entropic thresholding, a new approach. Comput. Graph. Image Process. **16**(3), 210–239 (1981)
20. Sahoo, P.K., Soltani, S., Wong, A.K.: A survey of thresholding techniques. Comput. Vis. Graph. Image Process. **41**(2), 233–260 (1988)
21. Salakhutdinov, R., Roweis, S.T., Ghahramani, Z.: Optimization with EM and expectation-conjugate-gradient. In: ICML (2003)
22. Sari, T., Kefali, A., Bahi, H.: Text extraction from historical document images by the combination of several thresholding techniques. Adv. Multimed. (2014). https://doi.org/10.1155/2014/934656
23. Sauvola, J.J., Pietikäinen, M.: Adaptive document image binarization. Pattern Recogn. **33**(2), 225–236 (2000)
24. Sturges, H.A.: The choice of a class interval. J. Am. Stat. Assoc. **21**(153), 65–66 (1926)
25. Tensmeyer, C., Martinez, T.: Document image binarization with fully convolutional neural networks. In: ICDAR (2017)
26. Wolf, C., Jolion, J.M., Chassaing, F.: Text localization, enhancement and binarization in multimedia documents. In: Object Recognition Supported by User Interaction for Service Robots (2002)
27. Zhang, H., Fritts, J.E., Goldman, S.A.: Image segmentation evaluation: a survey of unsupervised methods. Comput. Vis. Image Underst. **110**(2), 260–280 (2008)
28. Ziaei Nafchi, H., Farrahi Moghaddam, R., Cheriet, M.: Historical document binarization based on phase information of images. In: Park, J.-I., Kim, J. (eds.) ACCV 2012. LNCS, vol. 7729, pp. 1–12. Springer, Heidelberg (2013). https://doi.org/10.1007/978-3-642-37484-5_1

A Cordial Sync: Going Beyond Marginal Policies for Multi-agent Embodied Tasks

Unnat Jain[1]([✉]), Luca Weihs[2], Eric Kolve[2], Ali Farhadi[3], Svetlana Lazebnik[1], Aniruddha Kembhavi[2,3], and Alexander Schwing[1]

[1] University of Illinois at Urbana-Champaign, Champaign, USA
unnatjain@gmail.com
[2] Allen Institute for AI, Seattle, USA
[3] University of Washington, Seattle, USA

Abstract. Autonomous agents must learn to collaborate. It is not scalable to develop a new centralized agent every time a task's difficulty outpaces a single agent's abilities. While multi-agent collaboration research has flourished in gridworld-like environments, relatively little work has considered visually rich domains. Addressing this, we introduce the novel task FurnMove in which agents work together to move a piece of furniture through a living room to a goal. Unlike existing tasks, FurnMove requires agents to coordinate at every timestep. We identify two challenges when training agents to complete FurnMove: existing decentralized action sampling procedures do not permit expressive joint action policies and, in tasks requiring close coordination, the number of failed actions dominates successful actions. To confront these challenges we introduce SYNC-policies (synchronize your actions coherently) and CORDIAL (coordination loss). Using SYNC-policies and CORDIAL, our agents achieve a 58% completion rate on FurnMove, an impressive absolute gain of 25% points over competitive decentralized baselines. Our dataset, code, and pretrained models are available at https://unnat.github.io/cordial-sync.

Keywords: Embodied agents · Multi-agent reinforcement learning · Collaboration · Emergent communication · AI2-THOR

1 Introduction

Progress towards enabling artificial embodied agents to learn collaborative strategies is still in its infancy. Prior work mostly studies collaborative agents in gridworld like environments. Visual, multi-agent, collaborative tasks have not been studied until very recently [24,41]. While existing tasks are well designed to study some aspects of collaboration, they often don't require agents to closely collaborate *throughout* the task. Instead such tasks either require initial coordination (distributing tasks) followed by almost independent execution, or collaboration at

---

U. Jain and L. Weihs—Equal contribution.

Electronic supplementary material The online version of this chapter (https://doi.org/10.1007/978-3-030-58558-7_28) contains supplementary material, which is available to authorized users.

© Springer Nature Switzerland AG 2020
A. Vedaldi et al. (Eds.): ECCV 2020, LNCS 12350, pp. 471–490, 2020.
https://doi.org/10.1007/978-3-030-58558-7_28

Fig. 1. Two agents communicate and synchronize their actions to move a heavy object through an indoor environment towards a goal. (a) Agents begin holding the object in a randomly chosen location. (b) Given only egocentric views, successful navigation requires agents to communicate their intent to reposition themselves, and the object, while contending with collisions, mutual occlusion, and partial information. (c) Agents successfully moved the object above the goal

a task's end (*e.g.*, verifying completion). Few tasks require frequent coordination, and we are aware of none within a visual setting.

To study our algorithmic ability to address tasks which require close and frequent collaboration, we introduce the furniture moving (FurnMove) task (see Fig. 1), set in the AI2-THOR environment. Agents hold a lifted piece of furniture in a living room scene and, given only egocentric visual observations, must collaborate to move it to a visually distinct goal location. As a piece of furniture cannot be moved without both agents agreeing on the direction, agents must explicitly *coordinate at every timestep*. Beyond coordinating actions, FurnMove requires agents to visually anticipate possible collisions, handle occlusion due to obstacles and other agents, and estimate free space. Akin to the challenges faced by a group of roommates relocating a widescreen television, this task necessitates extensive and ongoing coordination amongst all agents at every time step.

In prior work, collaboration between multiple agents has been enabled primarily by (i) sharing observations or (ii) learning low-bandwidth communication. (i) is often implemented using a *centralized* agent, *i.e.*, a single agent with access to all observations from all agents [9,70,89]. While effective it is also unrealistic: the real world poses restrictions on communication bandwidth, latency, and modality. We are interested in the more realistic *decentralized* setting enabled via option (ii). This is often implemented by one or more rounds of message passing between agents before they choose their actions [27,41,57]. Training decentralized agents when faced with FurnMove's requirement of coordination at each timestep leads to two technical challenges. Challenge 1: as each agent independently samples an action from its policy at every timestep, the joint probability tensor of all agents' actions at any given time is rank-one. This severely limits which multi-agent policies are representable. Challenge 2: the number of possible mis-steps or failed actions increases dramatically when requiring that agents closely coordinate with each other, complicating training.

Addressing challenge 1, we introduce SYNC (**S**ynchronize **Y**our actio**N**s **C**oherently) policies which permit expressive (*i.e.*, beyond rank-one) joint policies for decentralized agents while using interpretable communication. To ameliorate challenge 2 we introduce the **Coordination Loss** (**CORDIAL**) that replaces the standard entropy loss in actor-critic algorithms and guides agents away from actions that are mutually incompatible. A 2-agent system using SYNC and CORDIAL obtains a 58% success rate on test scenes in FURNMOVE, an impressive absolute gain of 25% points over the baseline from [41] (76% relative gain). In a 3-agent setting, this difference is even more extreme.

In summary, our contributions are: (i) FURNMOVE, a new multi-agent embodied task that demands ongoing coordination, (ii) SYNC, a collaborative mechanism that permits expressive joint action policies for decentralized agents, (iii) CORDIAL, a training loss for multi-agent setups which, when combined with SYNC, leads to large gains, and (iv) open-source improvements to the AI2-THOR environment including a 16× faster gridworld equivalent for prototyping.

2 Related Work

Single-Agent Embodied Systems: Single-agent embodied systems have been considered extensively in the literature. For instance, literature on visual navigation, *i.e.*, locating an object of interest given only visual input, spans geometric and learning based methods. Geometric approaches have been proposed separately for mapping and planning phases of navigation. Methods entailing structure-from-motion and SLAM [13,25,71,76,77,87] were used to build maps. Planning algorithms on existing maps [14,45,51] and combined mapping & planning [6,26,30,48,49] are other related research directions.

While these works propose geometric approaches, the task of navigation can be cast as a reinforcement learning (RL) problem, mapping pixels to policies in an end-to-end manner. RL approaches [1,20,33,43,61,67,82,88,92] have been proposed to address navigation in synthetic layouts like mazes, arcade games, and other visual environments [8,42,46,53,80,98]. Navigation within photo-realistic environments [5,11,15,35,47,58,75,94,99,100] led to the development of *embodied* AI agents. The early work [105] addressed object navigation (find an object given an image) in AI2-THOR. Soon after, [35] showed how imitation learning permits agents to learn to build a map from which they navigate. Methods also investigate the utility of topological and latent memory maps [35,37,74,97], graph-based learning [97,101], meta-learning [96], unimodal baselines [86], 3D point clouds [95], and effective exploration [16,72,74,91] to improve embodied navigational agents. Extensions of embodied navigation include instruction following [3,4,38,78,91], city navigation [19,62,63,90], question answering [21–23,34,95], and active visual recognition [102,103]. Recently, with visual and acoustic rendering, agents have been trained for audio-visual embodied navigation [18,31].

In contrast to the above single-agent embodied tasks and approaches, we focus on collaboration between multiple embodied agents. Extending the above single-agent architectural novelties (or a combination of them) to multi-agent systems such as ours is an interesting direction for future work.

Non-visual MARL: Multi-agent reinforcement learning (MARL) is challenging due to non-stationarity when learning. Several methods have been proposed to address such issues [29,83–85]. For instance, permutation invariant critics have been developed recently [54]. In addition, for MARL, cooperation and competition between agents has been studied [12,28,36,50,54,57,59,68,69]. Similarly, communication and language in the multi-agent setting has been investigated [7,10,27,32,44,52,60,66,79] in maze-based setups, tabular tasks, and Markov games. These algorithms mostly operate on low-dimensional observations (*e.g.*, position, velocity, *etc.*) and top-down occupancy grids. For a survey of centralized and decentralized MARL methods, kindly refer to [104]. Our work differs from the aforementioned MARL works in that we consider complex visual environments. Our contribution of SYNC-Policies is largely orthogonal to RL loss function or method. For a fair comparison to [41], we used the same RL algorithm (A3C) but it is straightforward to integrate SYNC into other MARL methods [28,57,73] (for details, see the supplement).

Visual MARL: Jain *et al.* [41] introduced a collaborative task for two embodied visual agents, which we refer to as FURNLIFT. In this task, two agents are randomly initialized in an AI2-THOR living room scene, must visually navigate to a TV, and, in a singe coordinated PICKUP action, work to lift that TV up. FURNLIFT doesn't demand that agents coordinate their actions at each timestep. Instead, such coordination only occurs at the last timestep of an episode. Moreover, as success of an action executed by an agent is independent (with the exception of the PICKUP action), a high performance joint policy need not be complex, *i.e.*, it may be near low-rank. More details on this analysis and the complexity of our proposed FURNMOVE task are provided in Sect. 3. Similarly, Chen *et al.* [17] proposes a visual hide-and-seek task, where agents can move independently. Das *et al.* [24] enable agents to learn who to communicate with, on predominantly 2D tasks. In visual environments they study the task where multiple agents jointly navigate to the same object. Jaderberg *et al.* [40] recently studied the game of Quake III and Weihs *et al.* [93] develop agents to play an adversarial hiding game in AI2-THOR. Collaborative perception for semantic segmentation and recognition classification have also been investigated [55,56].

To the best of our knowledge, all prior work in decentralized MARL uses a single marginal probability distribution per agent, *i.e.*, a rank-1 joint distribution. Moreover, FURNMOVE is the first multi-agent collaborative task in a visually rich domain requiring close coordination between agents at every timestep.

3 The Furniture Moving Task (FURNMOVE)

We describe our new multi-agent task FURNMOVE, grounded in the real-world experience of moving furniture. We begin by introducing notation.

RL Background and Notation. Consider $N \geq 1$ collaborative agents $A^1, \ldots, A^N$. At every timestep $t \in \mathbb{N} = \{0, 1, \ldots\}$ the agents, and environment, are in some state $s_t \in \mathcal{S}$ and each agent A^i obtains an observation o_t^i recording some partial information about s_t. For instance, o_t^i might be the egocentric visual view of an agent A^i embedded in some simulated environment. From observation o_t^i and history h_{t-1}^i, which records prior observations and decisions made by the agent, each agent A^i forms a policy $\pi_t^i : \mathcal{A} \to [0, 1]$ where $\pi_t^i(a)$ is the probability that agent A^i chooses to take action $a \in \mathcal{A}$ from a finite set of options $\mathcal{A}$ at time t. After the agents execute their respective actions $(a_t^1, \dots, a_t^N)$, which we call a *multi-action*, they enter a new state s_{t+1} and receive individual rewards $r_t^1, \dots, r_t^N \in \mathbb{R}$. For more on RL see [64,65,81].

Task Definition. FURNMOVE is set in the near-photorealistic and physics-enabled simulated environment AI2-THOR [47]. In FURNMOVE, N agents collaborate to move a lifted object through an indoor environment with the goal of placing this object above a visually distinct target as illustrated in Fig. 1. Akin to humans moving large items, agents must navigate around other furniture and frequently walk in-between obstacles on the floor.

In FURNMOVE, each agent at every timestep receives an egocentric observation (a $3 \times 84 \times 84$ RGB image) from AI2-THOR. In addition, agents are allowed to communicate with other agents at each timestep via a low-bandwidth communication channel. Based on their local observation and communication, each agent executes an action from the set $\mathcal{A}$. The space of actions $\mathcal{A} = \mathcal{A}^{\text{NAV}} \cup \mathcal{A}^{\text{MWO}} \cup \mathcal{A}^{\text{MO}} \cup \mathcal{A}^{\text{RO}}$ available to an agent is comprised of the four single-agent navigational actions $\mathcal{A}^{\text{NAV}} = \{$MOVEAHEAD, ROTATELEFT, ROTATERIGHT, PASS$\}$ used to move the agent independently, four actions $\mathcal{A}^{\text{MWO}} = \{$MOVEWITHOBJECTX $\mid X \in \{$AHEAD, RIGHT, LEFT, BACK$\}\}$ used to move the lifted object and the agents simultaneously in the same direction, four actions $\mathcal{A}^{\text{MO}} = \{$MOVEOBJECTX$\mid X \in \{$AHEAD, RIGHT, LEFT, BACK$\}\}$ used to move the lifted object while the agents stay in place, and a single action used to rotate the lifted object clockwise $\mathcal{A}^{\text{RO}} = \{$ROTATEOBJECTRIGHT$\}$. We assume that all movement actions for agents and the lifted object result in a displacement of 0.25 m (similar to [41,58]) and all rotation actions result in a rotation of $90°$ (counter-)clockwise when viewing the agents from above.

Close and on-going collaboration is required in FURNMOVE due to restrictions on the set of actions which can be successfully completed jointly by all the agents. These restrictions reflect physical constraints: for instance, if two people attempt to move in opposite directions while carrying a heavy object they will either fail to move or drop the object. For two agents, we summarize these restrictions using the *coordination matrix* shown in Fig. 2a. For comparison, we include a similar matrix in Fig. 2b corresponding to the FURNLIFT task from [41]. We defer a more detailed discussion of these restrictions to the supplement. Generalizing the coordination matrix shown in Fig. 2a, at every timestep t we let S_t be the $\{0, 1\}$-valued $|\mathcal{A}|^N$-dimensional tensor where $(S_t)_{i_1, \dots, i_N} = 1$ if and only if the agents are configured such that multi-action $(a^{i_1}, \dots, a^{i_N})$ satisfies the restrictions detailed in the supplement. If $(S_t)_{i_1, \dots, i_N} = 1$ we say the actions $(a^{i_1}, \dots, a^{i_N})$ are *coordinated*.

3.1 Technical Challenges

As we show in our experiments in Sect. 6, standard communication-based models similar to the ones proposed in [41] perform rather poorly when trained to

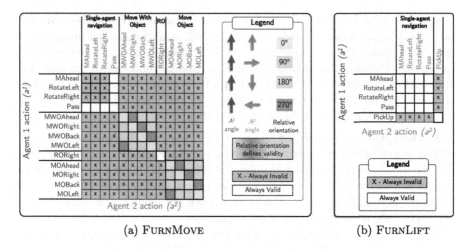

(a) FurnMove   (b) FurnLift

Fig. 2. Coordination matrix for tasks. The matrix S_t records the validity of multi-action (a^1, a^2) for different relative orientations of agents A^1 & A^2. (a) S_t for all 4 relative orientations of two agents, for FurnMove. Only $16/169 = 9.5\%$ of multi-actions are coordinated at any given relative orientation, (b) FurnLift where single agent actions are always valid and coordination is needed only for PickUp action, i.e. at least $16/25 = 64\%$ actions are always valid.

complete the FurnMove task. In the following we identify two key challenges that contribute to this poor performance.

Challenge 1: Rank-One Joint Policies. In classical multi-agent work [12,57,69], each agent A^i samples its action $a_t^i \sim \pi_t^i$ independently of all other agents. Due to this independent sampling, at time t, the probability of the agents taking multi-action $(a^1, \ldots, a^N)$ equals $\prod_{i=1}^N \pi_t^i(a^i)$. This means that the joint probability tensor of all actions at time t can be written as the rank-one tensor $\Pi_t = \pi_t^1 \otimes \cdots \otimes \pi_t^N$. This rank-one constraint limits the joint policy that can be executed by the agents, which has real impact. In the supplement, we consider two agents playing rock-paper-scissors with an adversary: the rank-one constraint reduces the expected reward achieved by an optimal policy from 0 to -0.657 (minimal reward being -1). Intuitively, a high-rank joint policy is not well approximated by a rank-one probability tensor obtained via independent sampling.

Challenge 2: Exponential Failed Actions. The number of possible multi-actions $|\mathcal{A}|^N$ increases exponentially as the number of agents N grows. While this is not problematic if agents act relatively independently, it's a significant obstacle when the agents are *tightly coupled*, i.e., when the success of agent A^i's action a^i is highly dependent on the actions of the other agents. Just consider a randomly initialized policy (the starting point of almost all RL problems): agents stumble upon positive rewards with an extremely low probability which leads to slow learning. We focus on small N, nonetheless, the proportion of coordinated action tuples is small (9.5% when $N = 2$ and 2.1% when $N = 3$).

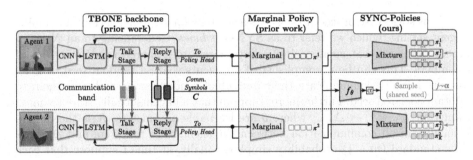

Fig. 3. Model overview for 2 agents in the decentralized setting. *Left*: all decentralized methods in this paper have the same TBONE [41] backbone. *Right*: marginal vs SYNC-policies. With marginal policies, the standard in prior work, each agent constructs its own policy and independently samples from it. With SYNC-policies, agents communicate to construct a distribution α over multiple "strategies" which they then sample from using a shared random seed

4   A Cordial Sync

To address the above challenges we develop: (a) a novel action sampling procedure named **S**ynchronize **Y**our actio**N**s **C**oherently (SYNC) and (b) an intuitive & effective multi-agent training loss named the **Cord**ination L**o**ss (CORDIAL).

Addressing Challenge 1: SYNC. For concreteness we let $N = 2$, so the joint probability tensor $\varPi_t$ is matrix of size $|\mathcal{A}| \times |\mathcal{A}|$, and provide an overview in Fig. 3. Recall our goal: using little communication, multiple agents should sample their actions from a high-rank joint policy. This is difficult as (i) little communication means that, except in degenerate cases, no agent can form the full joint policy and (ii) even if all agents had access to the joint policy it is not obvious how to ensure that the decentralized agents will sample a valid coordinated action.

To achieve our goal recall that, for any rank $m \leq |\mathcal{A}|$ matrix $L \in \mathbb{R}^{|\mathcal{A}| \times |\mathcal{A}|}$, there are vectors $v_1, w_1, \ldots, v_m, w_m \in \mathbb{R}^{|\mathcal{A}|}$ such that $L = \sum_{j=1}^{m} v_j \otimes w_j$. Here, $\otimes$ denotes the outer product. Also, the *non-negative rank* of a matrix $L \in \mathbb{R}_{\geq 0}^{|\mathcal{A}| \times |\mathcal{A}|}$ equals the smallest integer s such that L can be written as the sum of s non-negative rank-one matrices. A non-negative matrix $L \in \mathbb{R}_{\geq 0}^{|\mathcal{A}| \times |\mathcal{A}|}$ has non-negative rank bounded above by $|\mathcal{A}|$. Since $\varPi_t$ is a $|\mathcal{A}| \times |\mathcal{A}|$ joint probability matrix, *i.e.*, $\varPi_t$ is non-negative and its entries sum to one, it has non-negative rank $m \leq |\mathcal{A}|$, *i.e.*, there exist non-negative vectors $\alpha \in \mathbb{R}_{\geq 0}^{m}$ and $p_1, q_1, \ldots, p_m, q_m \in \mathbb{R}_{\geq 0}^{|\mathcal{A}|}$ whose entries sum to one such that $\varPi_t = \sum_{j=1}^{m} \alpha_j \cdot p_j \otimes q_j$. We call a sum of the form $\sum_{j=1}^{m} \alpha_j \cdot p_j \otimes q_j$ a *mixture-of-marginals*. With this decomposition at hand, randomly sampling action pairs (a^1, a^2) from $\sum_{j=1}^{m} \alpha_j \cdot p_j \otimes q_j$ can be interpreted as a two-step process: first sample an index $j \sim \text{Multinomial}(\alpha)$ and then sample $a^1 \sim \text{Multinomial}(p_j)$ and $a^2 \sim \text{Multinomial}(q_j)$.

This stage-wise procedure suggests a strategy for sampling actions in a multi-agent setting, which we refer to as SYNC-*policies*. Generalizing to an N agent setup, suppose that agents $(A^i)_{i=1}^N$ have access to a shared random stream of numbers. This can be accomplished if all agents share a random seed or if all agents initially communicate their individual random seeds and sum them to obtain a shared seed. Furthermore, suppose that all agents locally store a shared function $f_\theta : \mathbb{R}^K \to \Delta_{m-1}$ where θ are learnable parameters, K is the dimensionality of all communication between the agents in a timestep, and Δ_{m-1} is the standard $(m-1)$-probability simplex. Finally, at time t suppose that each agent A^i produces not a single policy π_t^i but instead a collection of policies $\pi_{t,1}^i, \ldots, \pi_{t,m}^i$. Let $C_t \in \mathbb{R}^K$ be all communication sent between agents at time t. Each agent A^i then samples its action as follows: (i) compute the shared probabilities $\alpha_t = f_\theta(C_t)$, (ii) sample an index $j \sim \text{Multinomial}(\alpha_t)$ using the shared random number stream, (iii) sample, independently, an action a^i from the policy $\pi_{t,j}^i$. Since both f_θ and the random number stream are shared, the quantities in (i) and (ii) are equal across all agents despite being computed individually. This sampling procedure is equivalent to sampling from the tensor $\sum_{j=1}^m \alpha_j \cdot \pi_{t,j}^1 \otimes \ldots \otimes \pi_{t,j}^N$ which, as discussed above, may have rank up to m. Intuitively, SYNC enables decentralized agents to have a more expressive joint policy by allowing them to agree upon a strategy by sampling from α_t.

Addressing Challenge 2: CORDIAL. We encourage agents to rapidly learn to choose coordinated actions via a new loss. In particular, letting Π_t be the joint policy of our agents, we propose the *coordination loss* (CORDIAL)

$$\text{CL}_\beta(S_t, \Pi_t) = -\beta \cdot \langle S_t, \log(\Pi_t) \rangle \, / \, \langle S_t, S_t \rangle, \tag{1}$$

where log is applied element-wise, $\langle *, * \rangle$ is the usual Frobenius inner product, and S_t is defined in Sect. 3. CORDIAL encourages agents to have a near uniform policy over the actions which are coordinated. We use this loss to replace the standard entropy encouraging loss in policy gradient algorithms (*e.g.*, the A3C algorithm [65]). Similarly to the parameter for the entropy loss in A3C, β is chosen to be a small positive constant so as to not overly discourage learning.

The coordination loss is less meaningful when $\Pi_t = \pi^1 \otimes \cdots \otimes \pi^N$, *i.e.*, when Π_t is rank-one. For instance, suppose that S_t has ones along the diagonal, and zeros elsewhere, so that we wish to encourage the agents to all take the same action. In this case it is straightforward to show that $\text{CL}_\beta(S_t, \Pi_t) = -\beta \sum_{i=1}^N \sum_{j=1}^M (1/M) \log \pi_t^i(a^j)$ so that $\text{CL}_\beta(S_t, \Pi_t)$ simply encourages each agent to have a uniform distribution over its actions and thus actually encourages the agents to place a large amount of probability mass on uncoordinated actions. Indeed, Table 4 shows that using CORDIAL without SYNC leads to poor results.

5   Models

We study four distinct policy types: *central, marginal, marginal w/o comm,* and *SYNC. Central* samples actions from a joint policy generated by a central agent with access to observations from all agents. While often unrealistic in practice due to communication bottlenecks, *central* serves as an informative baseline. *Marginal* follows prior work, *e.g.*, [41]: each agent independently samples its actions from its individual policy after communication. *Marginal w/o comm* is identical to *marginal* but does not permit agents to communicate explicitly (agents may still see each other). Finally, *SYNC* is our newly proposed policy described in Sect. 4. For a fair comparison, all decentralized agents (*i.e.*, *SYNC, marginal,* and *marginal w/o comm*), use the same TBONE backbone architecture from [41], see Fig. 3. We have ensured that parameters are fairly balanced so that our proposed *SYNC* has close to (and never more) parameters than the *marginal* and *marginal w/o comm* nets. We train *central* and *SYNC* with CORDIAL, and the *marginal* and *marginal w/o comm* without it. This choice is mathematically explained in Sect. 4 and empirically validated in Sect. 6.3.

Architecture: For clarity, we describe the policy and value net for the 2 agent setup, extending to any number of agents is straightforward. Decentralized agents use the TBONE backbone from [41]. Our primary architectural novelty extends TBONE to SYNC-policies. An overview of the TBONE backbone and differences between sampling with *marginal* and *SYNC* policies is shown in Fig. 3.

As a brief summary of TBONE, agent i observes at time t inputs o_t^i, *i.e.*, a $3 \times 84 \times 84$ RGB image returned from AI2-THOR which represents the i-th agent's egocentric view. Each o_t^i is encoded by a 4-layer CNN and combined with an agent-specific learned embedding (encoding the agent's ID) along with the history embedding h_{t-1}^i. The resulting vector is fed into an LSTM [39] unit to produce a 512-dimensional embedding $\tilde{h}_t^i$ corresponding to the ith agent. The agents then undergo two rounds of communication resulting in two final hidden states h_t^1, h_t^2 and messages $c_{t,j}^i \in \mathbb{R}^{16}$, $1 \leq i,j \leq 2$ with message $c_{t,j}^i$ being produced by agent i in round j and then sent to the other agent in that round. In [41], the value of the agents' state as well as logits corresponding to the policy of the agents are formed by applying linear functions to h_t^1, h_t^2.

We now show how SYNC can be integrated into TBONE to allow our agents to represent high-rank joint distributions over multi-actions (see Fig. 3). First each agent computes the logits corresponding to α_t. This is done using a 2-layer MLP applied to the messages sent between the agents, at the second stage. In particular, $\alpha_t = \mathbf{W}_3 \text{ReLU}(\mathbf{W}_2 \text{ReLU}(\mathbf{W}_1 [c_{t,2}^1; c_{t,2}^2] + \mathbf{b}_1) + \mathbf{b}_2) + \mathbf{b}_3$ where $\mathbf{W}_1 \in \mathbb{R}^{64 \times 32}, \mathbf{W}_2 \in \mathbb{R}^{64 \times 64}, \mathbf{W}_3 \in \mathbb{R}^{m \times 64}, \mathbf{b}_1 \in \mathbb{R}^{32}, \mathbf{b}_2 \in \mathbb{R}^{64}$, and $\mathbf{b}_3 \in \mathbb{R}^m$ are a learnable collection of weight matrices and biases. After computing α_t we compute a collection of policies $\pi_{t,1}^i, \ldots, \pi_{t,m}^i$ for $i \in \{1,2\}$. Each of these policies is computed following the TBONE architecture but using $m-1$ additional, and learnable, linear layers per agent.

6   Experiments

6.1   Experimental Setup

Simulator. We evaluate our models in the AI2-THOR environment [47] with several novel upgrades including support for initializing lifted furniture and a top-down gridworld version of AI2-THOR for faster prototyping (16× faster than [41]). For details about framework upgrades, see the supplement.

Tasks. We compare against baselines on FurnMove, Gridworld-FurnMove, and FurnLift [41]. FurnMove is the novel task introduced in this work (Sect. 3): agents observe egocentric visual views (90° field-of-view). In Gridworld-FurnMove the agents are provided a top-down egocentric 3D tensor as observations. The third dimension of the tensor contains semantic information such as, if the location is navigable by an agent or navigable by the lifted object, or whether the location is occupied by another agent, the lifted object, or the goal object. Hence, Gridworld-FurnMove agents do not need visual understanding, but face other challenges of the FurnMove task – coordinating actions and planning trajectories. We consider only the harder variant of FurnLift, where communication was shown to be most important ('constrained' with no implicit communication in [41]). In FurnLift, agents observe egocentric visual views.

Data. As in [41], we train and evaluate on a split of the 30 living room scenes. As FurnMove is already quite challenging, we only consider a single piece of lifted furniture (a television) and a single goal object (a TV-stand). Twenty rooms are used for training, 5 for validation, and 5 for testing. The test scenes have very different lighting conditions, furniture, and layouts. For evaluation our test set includes 1000 episodes equally distributed over the five scenes.

Training. For training we augment the A3C algorithm [65] with CORDIAL. For our studies in the visual domain, we use 45 workers and 8 GPUs. Models take around two days to train. For more details, see the supplement.

6.2   Metrics

For completeness, we consider a variety of metrics. We adapt SPL, *i.e.*, Success weighted by (normalized inverse) Path Length [2], so that it doesn't require shortest paths but still provides similar semantic information[1]: We define a Manhattan Distance based SPL as $MD\text{-}SPL = N_{\mathrm{ep}}^{-1} \sum_{i=1}^{N_{\mathrm{ep}}} S_i \frac{m_i/d_{\mathrm{grid}}}{\max(p_i, m_i/d_{\mathrm{grid}})}$, where i denotes an index over episodes, N_{ep} equals the number of test episodes, and S_i is a binary indicator for success of episode i. Also p_i is the number of actions taken per agent, m_i is the Manhattan distance from the lifted object's start location to the goal, and d_{grid} is the distance between adjacent grid points, for us 0.25 m. We also report other metrics capturing complementary information. These include mean number of actions in an episode per agent (*Ep len*), success rate (*Success*), and distance to target at the end of the episode (*Final dist*).

---

[1] For FurnMove, each location of the lifted furniture corresponds to 404,480 states, making shortest path computation intractable (more details in the supplement).

Table 1. Quantitative results on three tasks. $\uparrow$ (or $\downarrow$) indicates that higher (or lower) value of the metric is desirable while $\updownarrow$ denotes that no value is, a priori, better than another. †denotes that a centralized agent serves only as an upper bound to decentralized methods and cannot be fairly compared with. Among decentralized agents, our SYNC model has the best metric values across all reported metrics (**bolded** values). Values are highlighted in green if their 95% confidence interval has no overlap with the confidence intervals of other values

Methods	MD-SPL $\uparrow$	Success $\uparrow$	Ep len $\downarrow$	Final dist $\downarrow$	Invalid prob. $\downarrow$	TVD $\updownarrow$
FURNMOVE (ours)						
Marginal w/o comm [41]	0.032	0.164	224.1	2.143	0.815	0
Marginal [41]	0.064	0.328	194.6	1.828	0.647	0
SYNC	**0.114**	**0.587**	**153.5**	**1.153**	**0.31**	0.474
Central†	0.161	0.648	139.8	0.903	0.075	0.543
Gridworld-FURNMOVE (ours)						
Marginal w/o comm [41]	0.111	0.484	172.6	1.525	0.73	0
Marginal [41]	0.218	0.694	120.1	0.960	0.399	0
SYNC	**0.228**	**0.762**	**110.4**	**0.711**	**0.275**	0.429
Central†	0.323	0.818	87.7	0.611	0.039	0.347
Gridworld-FURNMOVE-3Agents (ours)						
Marginal [41]	0	0	250.0	3.564	0.823	0
SYNC	**0.152**	**0.578**	**149.1**	**1.05**	**0.181**	0.514
Central†	0.066	0.352	195.4	1.522	0.138	0.521

Table 2. Quantitative results on the FURNLIFT task. For legend, see Table 1

Methods	MD-SPL $\uparrow$	Success $\uparrow$	Ep len $\downarrow$	Final dist $\downarrow$	Invalid prob. $\downarrow$	TVD $\updownarrow$	Failed pickups $\downarrow$	Missed pickups $\downarrow$
FURNLIFT [41] ('constrained' setting with no implicit communication)								
Marginal w/o comm [41]	0.029	0.15	229.5	2.455	0.11	0	25.219	6.501
Marginal [41]	**0.145**	**0.449**	**174.1**	2.259	0.042	0	8.933	1.426
SYNC	0.139	0.423	176.9	**2.228**	0	0.027	**4.873**	1.048
Central†	0.145	0.453	172.3	2.331	0	0.059	5.145	0.639

We also introduce two metrics unique to coordinating actions: TVD, the mean total variation distance between Π_t and its best rank-one approximation, and $Invalid\ prob$, the average probability mass allotted to uncoordinated actions, $i.e.$, the dot product between $1 - S_t$ and Π_t. By definition, TVD is zero for the $marginal$ model, and higher values indicate divergence from independent marginal sampling. Without measuring TVD we would have no way of knowing if our SYNC model was actually using the extra expressivity we've afforded it. Lower $Invalid\ prob$ values imply an improved ability to avoid uncoordination actions as detailed in Sect. 3 and Fig. 2.

6.3 Quantitative Evaluation

We conduct four studies: (a) performance of different methods and relative difficulty of the three tasks, (b) effect of number of components on SYNC performance, (c) effect of CORDIAL (ablation), and (d) effect of number of agents.

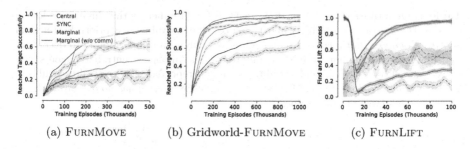

(a) FURNMOVE   (b) Gridworld-FURNMOVE   (c) FURNLIFT

Fig. 4. Success rate during training. Train (solid lines) and validation (dashed lines) performance of our agents for FURNMOVE, Gridworld-FURNMOVE, and FURN-LIFT (with 95% confidence intervals). For additional plots, see supplement

Table 3. Effect of number of mixture components m on *SYNC*'s performance (in FURNMOVE). Generally, larger m means better performance and larger *TVD*.

K in SYNC	MD-SPL ↑	Success ↑	Ep len ↓	Final dist ↓	Invalid prob. ↓	TVD ↕
			FURNMOVE			
1 component	0.064	0.328	194.6	1.828	0.647	0
2 components	0.084	0.502	175.5	1.227	**0.308**	0.206
4 components	0.114	0.569	154.1	**1.078**	0.339	0.421
13 components	**0.114**	**0.587**	**153.5**	1.153	0.31	0.474

Comparing Methods and Tasks. We compare models detailed in Sect. 5 on tasks of varying difficulty, report metrics in Table 1, and show the progress of metrics over training episodes in Fig. 4. In our FURNMOVE experiments, *SYNC* performs better than the best performing method of [41] (*i.e.*, *marginal*) on all metrics. Success rate increases by 25.9% and 6.8% absolute percentage points on FURNMOVE and Gridworld-FURNMOVE respectively. Importantly, SYNC is significantly better at allowing agents to coordinate their actions: for FURNMOVE, the joint policy of SYNC assigns, on average, 0.31 probability mass to invalid actions pairs while the *marginal* and *marginal w/o comm* models assign 0.647 and 0.815 probability mass to invalid action pairs. Additionally, SYNC goes beyond rank-one *marginal* methods by capturing a more expressive joint policy using the mixture of marginals. This is evidenced by the high TVD of 0.474 *vs.* 0 for *marginal*. In Gridworld-FURNMOVE, oracle-perception of a 2D gridworld helps raise performance of all methods, though the trends are similar. Table 2 shows similar trends for FURNLIFT but, perhaps surprisingly, the *Success* of *SYNC* is somewhat lower than the *marginal* model (2.6% lower, within statistical error). As is emphasized in [41] however, *Success* alone is a poor measure of model performance: equally important are the *failed pickups* and *missed pickups* metrics (for details, see the supplement). For these metrics, *SYNC* outperforms the *marginal* model. That *SYNC* does not completely outperform *marginal* in FURNLIFT is intuitive, as FURNLIFT does not require continuous close coordination the benefits of *SYNC* are less pronounced.

Table 4. Ablation study of CORDIAL on *marginal* [41], *SYNC*, and *central* methods. *Marginal* performs better without CORDIAL whereas *SYNC* and *central* show improvement with CORDIAL. For legend, see Table 1

Method	CORDIAL	MD-SPL ↑	Success ↑	Ep len ↓	Final dist ↓	Invalid prob. ↓	TVD ↕
			FurnMove				
Marginal	✗	0.064	0.328	194.6	1.828	0.647	0
Marginal	✓	0.015	0.099	236.9	2.134	0.492	0
SYNC	✗	0.091	0.488	170.3	1.458	0.47	0.36
SYNC	✓	**0.114**	**0.587**	**153.5**	**1.153**	**0.31**	0.474
Central[†]	✗	0.14	0.609	146.9	1.018	0.155	0.6245
Central[†]	✓	0.161	0.648	139.8	0.903	0.075	0.543

While the difficulty of a task is hard to quantify, we will consider the relative test-set metrics of agents on various tasks as an informative proxy. Replacing the complex egocentric vision in FurnMove with the semantic 2D gridworld in Gridworld-FurnMove, we see that all agents show large gains in *Success* and *MD-SPL*, suggesting that Gridworld-FurnMove is a dramatic simplification of FurnMove. Comparing FurnMove to FurnLift is particularly interesting: the *MD-SPL* and *Success* metrics for the *central* agent do not provide a clear picture of relative task difficulty. However, the much higher *TVD* for the *central* agent for FurnMove and the superior *MD-SPL* and *Success* of the *Marginal* agents for FurnLift suggest that FurnMove requires more coordination and more expressive joint policies than FurnLift.

Effect of Number of Mixture Components in SYNC. Recall (Sect. 4) that the number of mixture components m in SYNC is a hyperparameter controlling the maximal rank of the joint policy. *SYNC* with $m = 1$ is equivalent to *marginal*. In Table 3 we see TVD increase from 0.206 to 0.474 when increasing m from 2 to 13. This suggests that SYNC learns to use the additional expressivity. Moreover, we see that this increased expressivity results in better performance. A success rate jump of 17.4% from $m = 1$ to $m = 2$ demonstrates that substantial benefits are obtained by even small increases in expressitivity. Moreover with more components, *i.e.*, $m = 4$ & $m = 13$ we obtain more improvements. There are, however, diminishing returns with the $m = 4$ model performing nearly as well as the $m = 13$ model. This suggests a trade-off between the benefits of expressivity and the increasing complexities in optimization.

Effect of CORDIAL. In Table 4 we quantify the effect of CORDIAL. When adding CORDIAL to *SYNC* we obtain a 9.9% improvement in success rate. This is accompanied by a drop in *Invalid prob.* from 0.47 to 0.31, which signifies better coordination of actions. Similar improvements are seen for the *central* model. In 'Challenge 2' (Sect. 4) we mathematically laid out why *marginal* models gain little from CORDIAL. We substantiate this empirically with a 22.9% drop in success rate when training the *marginal* model with CORDIAL.

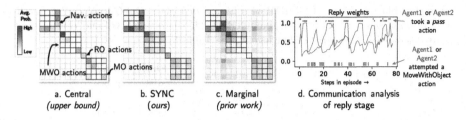

Fig. 5. Qualitative results. (a,b,c) joint policy summary (Π_t averaged over steps in test episodes in FURNMOVE) and (d) communication analysis.

Effect of More Agents. The final three rows of Table 1 show the test-set performance of *SYNC*, *marginal*, and *central* models trained to accomplish a 3-agent variant of our Gridworld-FURNMOVE task. In this task the *marginal* model fails to train at all, achieving a 0% success rate. *SYNC*, on the other hand, successfully completes the task 57.8% of the time. *SYNC*'s success rate drops by nearly 20% points when moving from the 2- to the 3-agent variant of the task: clearly increasing the number of agents substantially increases the task's difficult. Surprisingly, the *central* model performs worse than *SYNC* in this setting. A discussion of this phenomenon is deferred to supplement.

6.4 Qualitative Evaluation

We present three qualitative results on FURNMOVE: joint policy summaries, analysis of learnt communication, and visualizations of agent trajectories.

Joint Policy Summaries. In Fig. 5 we show summaries of the joint policy captured by the *central*, *SYNC*, and *marginal* models. These matrices average over action steps in the test-set episodes for FURNMOVE. Other tasks show similar trends, see the supplement. In Fig. 5a, the sub-matrices corresponding to $\mathcal{A}^{\mathrm{MWO}}$ and $\mathcal{A}^{\mathrm{MO}}$ are diagonal-dominant, indicating that agents are looking in the same direction ($0_{\circ}$ relative orientation in Fig. 2). Also note the high probability associated to (PASS, ROTATEX) and (ROTATEX, PASS), within the $\mathcal{A}^{\mathrm{NAV}}$ block. Together, this means that the *central* method learns to coordinate single-agent navigational actions to rotate one of the agents (while the other holds the TV by executing PASS) until both face the same direction. They then execute the same action from $\mathcal{A}^{\mathrm{MO}}$ ($\mathcal{A}^{\mathrm{MWO}}$) to move the lifted object. Comparing Fig. 5b *vs.* Fig. 5c, shows the effect of CORDIAL. Recall that the *marginal* model doesn't support CORDIAL and thus suffers by assigning probability to invalid action pairs (color outside the block-diagonal submatrices). The banded nature of Fig. 5c suggesting agents frequently fail to coordinate.

Communication Analysis. A qualitative discussion of communication follows. Agent are colored red and green. We defer a quantitative treatment to the supplement. As we apply SYNC on the TBONE backbone introduced by Jain *et al.* [41], we use similar tools to understand the communication emerging with

SYNC policy heads. In line with [41], we plot the weight assigned by each agent to the first communication symbol in the reply stage. Fig. 5d strongly suggests that the reply stage is directly used by the agents to coordinate the modality of actions they intend to take. In particular, the large weight being assigned to the first reply symbol is consistently associated with the other agent taking a PASS action. Similarly, we see that small reply weights coincide with agents taking a MOVEWITHOBJECT action. The talk weights' interpretation is intertwined with the reply weights, and is deferred to supplement.

Agent Trajectories. Our supplementary video includes examples of policy roll-outs. These clips include both agents' egocentric views and a top-down trajectory visualization. This enables direct comparisons of *marginal* and SYNC on the same test episode. We also allow for hearing patterns in agents' communication: we convert scalar weights (associated with reply symbols) to audio.

7   Conclusion

We introduce FURNMOVE, a collaborative, visual, multi-agent task requiring close coordination between agents and develop novel methods that allow for moving beyond existing marginal action sampling procedures, these methods lead to large gains across a diverse suite of metrics.

Acknowledgements. This material is based upon work supported in part by the National Science Foundation under Grants No. 1563727, 1718221, 1637479, 165205, 1703166, Samsung, 3M, Sloan Fellowship, NVIDIA Artificial Intelligence Lab, Allen Institute for AI, Amazon, AWS Research Awards, and Siebel Scholars Award. We thank M. Wortsman and K.-H. Zeng for their insightful comments.

References

1. Abel, D., Agarwal, A., Diaz, F., Krishnamurthy, A., Schapire, R.E.: Exploratory gradient boosting for reinforcement learning in complex domains. arXiv preprint arXiv:1603.04119 (2016)
2. Anderson, P., et al.: On evaluation of embodied navigation agents. arXiv preprint arXiv:1807.06757 (2018)
3. Anderson, P., Shrivastava, A., Parikh, D., Batra, D., Lee, S.: Chasing ghosts: instruction following as bayesian state tracking. In: NeurIPS (2019)
4. Anderson, P., et al.: Vision-and-language navigation: interpreting visually-grounded navigation instructions in real environments. In: CVPR (2018)
5. Armeni, I., Sax, S., Zamir, A.R., Savarese, S.: Joint 2D–3D-semantic data for indoor scene understanding. arXiv preprint arXiv:1702.01105 (2017)
6. Aydemir, A., Pronobis, A., Göbelbecker, M., Jensfelt, P.: Active visual object search in unknown environments using uncertain semantics. IEEE Trans. Robot. **29**, 986–1002 (2013)
7. Baker, B., et al.: Emergent tool use from multi-agent autocurricula. arXiv preprint arXiv:1909.07528 (2019)

8. Bellemare, M.G., Naddaf, Y., Veness, J., Bowling, M.: The arcade learning environment: an evaluation platform for general agents. J. Artif. Intell. Res. **47**, 253–279 (2013)

9. Boutilier, C.: Sequential optimality and coordination in multiagent systems. In: IJCAI (1999)

10. Bratman, J., Shvartsman, M., Lewis, R.L., Singh, S.: A new approach to exploring language emergence as boundedly optimal control in the face of environmental and cognitive constraints. In: Proceedings of International Conference on Cognitive Modeling (2010)

11. Brodeur, S., et al.: HoME: a household multimodal environment. arXiv preprint arXiv:1711.11017 (2017)

12. Busoniu, L., Babuska, R., Schutter, B.D.: A comprehensive survey of multiagent reinforcement learning. IEEE Trans. Syst. Man Cybern. **38**, 156–172 (2008)

13. Cadena, C., et al.: Past, present, and future of simultaneous localization and mapping: toward the robust-perception age. IEEE Trans. Robot. **32**, 1309–1332 (2016)

14. Canny, J.: The Complexity of Robot Motion Planning. MIT Press, Cambridge (1988)

15. Chang, A., et al.: Matterport3D: learning from RGB-D data in indoor environments. In: 3DV (2017)

16. Chaplot, D.S., Gupta, S., Gupta, A., Salakhutdinov, R.: Learning to explore using active neural mapping. In: ICLR (2020)

17. Chen, B., Song, S., Lipson, H., Vondrick, C.: Visual hide and seek. arXiv preprint arXiv:1910.07882 (2019)

18. Chen, C., et al.: Audio-visual embodied navigation. arXiv preprint arXiv:1912.11474 (2019). First two authors contributed equally

19. Chen, H., Suhr, A., Misra, D., Snavely, N., Artzi, Y.: Touchdown: natural language navigation and spatial reasoning in visual street environments. In: CVPR (2019)

20. Daftry, S., Bagnell, J.A., Hebert, M.: Learning transferable policies for monocular reactive MAV control. In: Kulić, D., Nakamura, Y., Khatib, O., Venture, G. (eds.) ISER 2016. SPAR, vol. 1, pp. 3–11. Springer, Cham (2017). https://doi.org/10.1007/978-3-319-50115-4_1

21. Das, A., Datta, S., Gkioxari, G., Lee, S., Parikh, D., Batra, D.: Embodied question answering. In: CVPR (2018)

22. Das, A., Gkioxari, G., Lee, S., Parikh, D., Batra, D.: Neural modular control for embodied question answering. In: ECCV (2018)

23. Das, A., et al.: Probing emergent semantics in predictive agents via question answering. In: ICML (2020). First two authors contributed equally

24. Das, A., et al.: TarMAC: targeted multi-agent communication. In: ICML (2019)

25. Dellaert, F., Seitz, S., Thorpe, C., Thrun, S.: Structure from motion without correspondence. In: CVPR (2000)

26. Elfes, A.: Using occupancy grids for mobile robot perception and navigation. Computer **22**, 46–57 (1989)

27. Foerster, J.N., Assael, Y.M., de Freitas, N., Whiteson, S.: Learning to communicate with deep multi-agent reinforcement learning. In: NeurIPS (2016)

28. Foerster, J.N., Farquhar, G., Afouras, T., NArdelli, N., Whiteson, S.: Counterfactual multi-agent policy gradients. In: AAAI (2018)

29. Foerster, J.N., Nardelli, N., Farquhar, G., Torr, P.H.S., Kohli, P., Whiteson, S.: Stabilising experience replay for deep multi-agent reinforcement learning. In: ICML (2017)

30. Fraundorfer, F., et al.: Vision-based autonomous mapping and exploration using a quadrotor MAV. In: IROS (2012)

31. Gao, R., Chen, C., Al-Halah, Z., Schissler, C., Grauman, K.: VisualEchoes: spatial image representation learning through echolocation. In: ECCV (2020)

32. Giles, C.L., Jim, K.-C.: Learning communication for multi-agent systems. In: Truszkowski, W., Hinchey, M., Rouff, C. (eds.) WRAC 2002. LNCS (LNAI), vol. 2564, pp. 377–390. Springer, Heidelberg (2003). https://doi.org/10.1007/978-3-540-45173-0_29

33. Giusti, A., et al.: A machine learning approach to visual perception of forest trails for mobile robots. IEEE Robot. Autom. Lett. 1, 661–667 (2015)

34. Gordon, D., Kembhavi, A., Rastegari, M., Redmon, J., Fox, D., Farhadi, A.: IQA: visual Question Answering in Interactive Environments. In: CVPR (2018)

35. Gupta, A., Johnson, J., Fei-Fei, L., Savarese, S., Alahi, A.: Social GAN: socially acceptable trajectories with generative adversarial networks. In: CVPR (2018)

36. Gupta, J.K., Egorov, M., Kochenderfer, M.: Cooperative multi-agent control using deep reinforcement learning. In: Sukthankar, G., Rodriguez-Aguilar, J.A. (eds.) AAMAS 2017. LNCS (LNAI), vol. 10642, pp. 66–83. Springer, Cham (2017). https://doi.org/10.1007/978-3-319-71682-4_5

37. Henriques, J.F., Vedaldi, A.: MapNet: an allocentric spatial memory for mapping environments. In: CVPR (2018)

38. Hill, F., Hermann, K.M., Blunsom, P., Clark, S.: Understanding grounded language learning agents. arXiv preprint arXiv:1710.09867 (2017)

39. Hochreiter, S., Schmidhuber, J.: Long short-term memory. Neural Comput. 9, 1735–1780 (1997)

40. Jaderberg, M., et al.: Human-level performance in 3D multiplayer games with population-based reinforcement learning. Science 364, 859–865 (2019)

41. Jain, U., et al.: Two body problem: collaborative visual task completion. In: CVPR (2019), first two authors contributed equally

42. Johnson, M., Hofmann, K., Hutton, T., Bignell, D.: The malmo platform for artificial intelligence experimentation. In: IJCAI (2016)

43. Kahn, G., Zhang, T., Levine, S., Abbeel, P.: Plato: policy learning using adaptive trajectory optimization. In: ICRA (2017)

44. Kasai, T., Tenmoto, H., Kamiya, A.: Learning of communication codes in multi-agent reinforcement learning problem. In: Proceedings of IEEE Soft Computing in Industrial Applications (2008)

45. Kavraki, L.E., Svestka, P., Latombe, J.C., Overmars, M.H.: Probabilistic roadmaps for path planning in high-dimensional configuration spaces. IEEE Trans. Robot. Autom. 12, 566–580 (1996)

46. Kempka, M., Wydmuch, M., Runc, G., Toczek, J., Jakowski, W.: ViZDoom: a doom-based AI research platform for visual reinforcement learning. In: Proceedings of IEEE Conference on Computational Intelligence and Games (2016)

47. Kolve, E., et al.: AI2-THOR: an interactive 3D environment for visual AI. arXiv preprint arXiv:1712.05474 (2019)

48. Konolige, K., et al.: View-based maps. Int. J. Robot. Res. 29, 941–957 (2010)

49. Kuipers, B., Byun, Y.T.: A robot exploration and mapping strategy based on a semantic hierarchy of spatial representations. Robot. Auton. Syst. 8, 47–63 (1991)

50. Lauer, M., Riedmiller, M.: An algorithm for distributed reinforcement learning in cooperative multi-agent systems. In: ICML (2000)

51. Lavalle, S.M., Kuffner, J.J.: Rapidly-exploring random trees: progress and prospects. Algorithmic Comput. Robot.: New Direct (2000)

52. Lazaridou, A., Peysakhovich, A., Baroni, M.: Multi-agent cooperation and the emergence of (natural) language. In: arXiv preprint arXiv:1612.07182 (2016)
53. Lerer, A., Gross, S., Fergus, R.: Learning physical intuition of block towers by example. In: ICML (2016)
54. Liu, I.J., Yeh, R., Schwing, A.G.: PIC: permutation invariant critic for multi-agent deep reinforcement learning. In: CoRL (2019). First two authors contributed equally
55. Liu, Y.C., Tian, J., Glaser, N., Kira, Z.: When2com: multi-agent perception via communication graph grouping. In: CVPR (2020)
56. Liu, Y.C., Tian, J., Ma, C.Y., Glaser, N., Kuo, C.W., Kira, Z.: Who2com: collaborative perception via learnable handshake communication. In: ICRA (2020)
57. Lowe, R., Wu, Y., Tamar, A., Harb, J., Abbeel, P., Mordatch, I.: Multi-agent actor-critic for mixed cooperative-competitive environments. In: NeurIPS (2017)
58. Savva, M., et al.: Habitat: a platform for embodied AI research. In: ICCV (2019)
59. Matignon, L., Laurent, G.J., Fort-Piat, N.L.: Hysteretic Q-learning: an algorithm for decentralized reinforcement learning in cooperative multi-agent teams. In: IROS (2007)
60. Melo, F.S., Spaan, M.T.J., Witwicki, S.J.: QueryPOMDP: POMDP-based communication in multiagent systems. In: Cossentino, M., Kaisers, M., Tuyls, K., Weiss, G. (eds.) EUMAS 2011. LNCS (LNAI), vol. 7541, pp. 189–204. Springer, Heidelberg (2012). https://doi.org/10.1007/978-3-642-34799-3_13
61. Mirowski, P., et al.: Learning to navigate in complex environments. In: ICLR (2017)
62. Mirowski, P., et al.: The streetlearn environment and dataset. arXiv preprint arXiv:1903.01292 (2019)
63. Mirowski, P., et al.: Learning to navigate in cities without a map. In: NeurIPS (2018)
64. Mnih, V., et al.: Human-level control through deep reinforcement learning. Nature **518**, 529–533 (2015)
65. Mnih, V., et al.: Asynchronous methods for deep reinforcement learning. In: ICML (2016)
66. Mordatch, I., Abbeel, P.: Emergence of grounded compositional language in multi-agent populations. In: AAAI (2018)
67. Oh, J., Chockalingam, V., Singh, S., Lee, H.: Control of memory, active perception, and action in minecraft. In: ICML (2016)
68. Omidshafiei, S., Pazis, J., Amato, C., How, J.P., Vian, J.: Deep decentralized multi-task multi-agent reinforcement learning under partial observability. In: ICML (2017)
69. Panait, L., Luke, S.: Cooperative multi-agent learning: the state of the art. Autonom. Agents Multi-Agent Syst. AAMAS **11**, 387–434 (2005)
70. Peng, P., et al.: Multiagent bidirectionally-coordinated nets: emergence of human-level coordination in learning to play starcraft combat games. arXiv preprint arXiv:1703.10069 (2017)
71. Smith, R.C., Cheeseman, P.: On the representation and estimation of spatial uncertainty. Int. J. Robot. Res. **5**, 56–68 (1986)
72. Ramakrishnan, S.K., Jayaraman, D., Grauman, K.: An exploration of embodied visual exploration. arXiv preprint arXiv:2001.02192 (2020)
73. Rashid, T., Samvelyan, M., Schroeder, C., Farquhar, G., Foerster, J., Whiteson, S.: QMIX: monotonic value function factorisation for deep multi-agent reinforcement learning. In: ICML (2018)

74. Savinov, N., Dosovitskiy, A., Koltun, V.: Semi-parametric topological memory for navigation. In: ICLR (2018)
75. Savva, M., Chang, A.X., Dosovitskiy, A., Funkhouser, T., Koltun, V.: MINOS: multimodal indoor simulator for navigation in complex environments. arXiv preprint arXiv:1712.03931 (2017)
76. Schönberger, J.L., Frahm, J.M.: Structure-from-motion revisited. In: CVPR (2016)
77. Smith, R.C., Self, M., Cheeseman, P.: Estimating uncertain spatial relationships in robotics. In: UAI (1986)
78. Suhr, A., et al.: Executing instructions in situated collaborative interactions. In: EMNLP (2019)
79. Sukhbaatar, S., Szlam, A., Fergus, R.: Learning multiagent communication with backpropagation. In: NeurIPS (2016)
80. Sukhbaatar, S., Szlam, A., Synnaeve, G., Chintala, S., Fergus, R.: MazeBase: a sandbox for learning from games. arXiv preprint arXiv:1511.07401 (2015)
81. Sutton, R.S., Barto, A.G.: Reinforcement Learning: An Introduction. MIT Press, Cambridge (1998)
82. Tamar, A., Wu, Y., Thomas, G., Levine, S., Abbeel, P.: Value iteration networks. In: NeurIPS (2016)
83. Tampuu, A., et al.: Multiagent cooperation and competition with deep reinforcement learning. PloS 12, e0172395 (2017)
84. Tan, M.: Multi-agent reinforcement learning: independent vs. cooperative agents. In: ICML (1993)
85. Tesauro, G.: Extending Q-learning to general adaptive multi-agent systems. In: NeurIPS (2004)
86. Thomason, J., Gordon, D., Bisk, Y.: Shifting the baseline: Single modality performance on visual navigation & QA. In: NAACL (2019)
87. Tomasi, C., Kanade, T.: Shape and motion from image streams under orthography: a factorization method. IJCV 9, 137–154 (1992)
88. Toussaint, M.: Learning a world model and planning with a self-organizing, dynamic neural system. In: NeurIPS (2003)
89. Usunier, N., Synnaeve, G., Lin, Z., Chintala, S.: Episodic exploration for deep deterministic policies: an application to starcraft micromanagement tasks. In: ICLR (2016)
90. de Vries, H., Shuster, K., Batra, D., Parikh, D., Weston, J., Kiela, D.: Talk the walk: navigating new York city through grounded dialogue. arXiv preprint arXiv:1807.03367 (2018)
91. Wang, X., et al.: Reinforced cross-modal matching and self-supervised imitation learning for vision-language navigation. In: CVPR (2019)
92. Weihs, L., Jain, U., Salvador, J., Lazebnik, S., Kembhavi, A., Schwing, A.: Bridging the imitation gap by adaptive insubordination. arXiv preprint arXiv:2007.12173 (2020). The first two authors contributed equally
93. Weihs, L., et al.: Artificial agents learn flexible visual representations by playing a hiding game. arXiv preprint arXiv:1912.08195 (2019)
94. Weihs, L., et al.: AllenAct: a framework for embodied AI research. arXiv (2020)
95. Wijmans, E., et al.: Embodied question answering in photorealistic environments with point cloud perception. In: CVPR (2019)
96. Wortsman, M., Ehsani, K., Rastegari, M., Farhadi, A., Mottaghi, R.: Learning to learn how to learn: self-adaptive visual navigation using meta-learning. In: CVPR (2019)

97. Wu, Y., Wu, Y., Tamar, A., Russell, S., Gkioxari, G., Tian, Y.: Bayesian relational memory for semantic visual navigation. In: ICCV (2019)
98. Wymann, B., Espié, E., Guionneau, C., Dimitrakakis, C., Coulom, R., Sumner, A.: TORCS, the open racing car simulator (2013). http://www.torcs.org
99. Xia, F., et al.: Interactive Gibson: a benchmark for interactive navigation in cluttered environments. arXiv preprint arXiv:1910.14442 (2019)
100. Xia, F., Zamir, A.R., He, Z., Sax, A., Malik, J., Savarese, S.: Gibson ENv: real-world perception for embodied agents. In: CVPR (2018)
101. Yang, J., Lu, J., Lee, S., Batra, D., Parikh, D.: Visual curiosity: learning to ask questions to learn visual recognition. In: CoRL (2018)
102. Yang, J., et al.: Embodied amodal recognition: learning to move to perceive objects. In: ICCV (2019)
103. Yang, W., Wang, X., Farhadi, A., Gupta, A., Mottaghi, R.: Visual semantic navigation using scene priors. In: ICLR (2018)
104. Zhang, K., Yang, Z., Başar, T.: Multi-agent reinforcement learning: a selective overview of theories and algorithms. arXiv preprint arXiv:1911.10635 (2019)
105. Zhu, Y., et al.: Target-driven visual navigation in indoor scenes using deep reinforcement learning. In: ICRA (2017)

Big Transfer (BiT): General Visual Representation Learning

Alexander Kolesnikov, Lucas Beyer, Xiaohua Zhai$^{(\boxtimes)}$, Joan Puigcerver,
Jessica Yung, Sylvain Gelly, and Neil Houlsby

Google Research, Brain Team, Zürich, Switzerland
akolesnikov@google.com, lbeyer@google.com, xzhai@google.com,
jpuigcerver@google.com, jessicayung@google.com, sylvaingelly@google.com,
neilhoulsby@google.com

Abstract. Transfer of pre-trained representations improves sample efficiency and simplifies hyperparameter tuning when training deep neural
networks for vision. We revisit the paradigm of pre-training on large
supervised datasets and fine-tuning the model on a target task. We scale
up pre-training, and propose a simple recipe that we call Big Transfer
(BiT). By combining a few carefully selected components, and transferring using a simple heuristic, we achieve strong performance on over
20 datasets. BiT performs well across a surprisingly wide range of data
regimes—from 1 example per class to 1M total examples. BiT achieves
87.5% top-1 accuracy on ILSVRC-2012, 99.4% on CIFAR-10, and 76.3%
on the 19 task Visual Task Adaptation Benchmark (VTAB). On small
datasets, BiT attains 76.8% on ILSVRC-2012 with 10 examples per class,
and 97.0% on CIFAR-10 with 10 examples per class. We conduct detailed
analysis of the main components that lead to high transfer performance.

1 Introduction

Strong performance using deep learning usually requires a large amount of task-
specific data and compute. These per-task requirements can make new tasks
prohibitively expensive. Transfer learning offers a solution: task-specific data
and compute are replaced with a pre-training phase. A network is trained once
on a large, generic dataset, and its weights are then used to initialize subsequent
tasks which can be solved with fewer data points, and less compute [9, 34, 37].

We revisit a simple paradigm: pre-train on a large supervised source dataset,
and fine-tune the weights on the target task. Numerous improvements to deep network training have recently been introduced, e.g. [1, 17, 21, 29, 46, 47, 52, 54, 56, 59].
We aim not to introduce a new component or complexity, but to provide a recipe
that uses the minimal number of tricks yet attains excellent performance on many
tasks. We call this recipe "Big Transfer" (BiT).

A. Kolesnikov, L. Beyer and X. Zhai—Equal contribution.

Electronic supplementary material The online version of this chapter (https://
doi.org/10.1007/978-3-030-58558-7_29) contains supplementary material, which is
available to authorized users.

A. Vedaldi et al. (Eds.): ECCV 2020, LNCS 12350, pp. 491–507, 2020.
https://doi.org/10.1007/978-3-030-58558-7_29

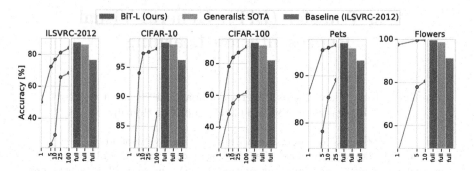

Fig. 1. Transfer performance of our pre-trained model, BiT-L, the previous state-of-the-art (SOTA), and a ResNet-50 baseline pre-trained on ILSVRC-2012 to downstream tasks. Here we consider only methods that are pre-trained independently of the final task (generalist representations), like BiT. The bars show the accuracy when fine-tuning on the full downstream dataset. The curve on the left-hand side of each plot shows that BiT-L performs well even when transferred using only few images (1 to 100) per class.

We train networks on three different scales of datasets. The largest, BiT-L is trained on the JFT-300M dataset [43], which contains 300M noisily labelled images. We transfer BiT to many diverse tasks; with training set sizes ranging from 1 example per class to 1M total examples. These tasks include ImageNet's ILSVRC-2012 [6], CIFAR-10/100 [22], Oxford-IIIT Pet [35], Oxford Flowers-102 [33] (including few-shot variants), and the 1000-sample VTAB-1k benchmark [58], which consists of 19 diverse datasets. BiT-L attains state-of-the-art performance on many of these tasks, and is surprisingly effective when very little downstream data is available (Fig. 1). We also train BiT-M on the public ImageNet-21k dataset, and attain marked improvements over the popular ILSVRC-2012 pre-training.

Importantly, BiT only needs to be pre-trained once and subsequent fine-tuning to downstream tasks is cheap. By contrast, other state-of-the-art methods require extensive training on support data conditioned on the task at hand [32,53,55]. Not only does BiT require a short fine-tuning protocol for each new task, but BiT also does not require extensive hyperparameter tuning on new tasks. Instead, we present a heuristic for setting the hyperparameters for transfer, which works well on our diverse evaluation suite.

We highlight the most important components that make Big Transfer effective, and provide insight into the interplay between scale, architecture, and training hyperparameters. For practitioners, we will release the performant BiT-M model trained on ImageNet-21k.

2   Big Transfer

We review the components that we found necessary to build an effective network for transfer. *Upstream* components are those used during pre-training, and *downstream* are those used during fine-tuning to a new task.

2.1 Upstream Pre-training

The first component is scale. It is well-known in deep learning that larger networks perform better on their respective tasks [10,40]. Further, it is recognized that larger datasets require larger architectures to realize benefits, and vice versa [20,38]. We study the effectiveness of scale (during pre-training) in the context of transfer learning, including transfer to tasks with very few datapoints. We investigate the interplay between computational budget (training time), architecture size, and dataset size. For this, we train three BiT models on three large datasets: ILSVRC-2012 [39] which contains 1.3M images (BiT-S), ImageNet-21k [6] which contains 14M images (BiT-M), and JFT [43] which contains 300M images (BiT-L).

The second component is Group Normalization (GN) [52] and Weight Standardization (WS) [28]. Batch Normalization (BN) [16] is used in most state-of-the-art vision models to stabilize training. However, we find that BN is detrimental to Big Transfer for two reasons. First, when training large models with small per-device batches, BN performs poorly or incurs inter-device synchronization cost. Second, due to the requirement to update running statistics, BN is detrimental for transfer. GN, when combined with WS, has been shown to improve performance on small-batch training for ImageNet and COCO [28]. Here, we show that the combination of GN and WS is useful for training with large batch sizes, and has a significant impact on transfer learning.

2.2 Transfer to Downstream Tasks

We propose a cheap fine-tuning protocol that applies to many diverse downstream tasks. Importantly, we avoid expensive hyperparameter search for every new task and dataset size; we try only one hyperparameter per task. We use a heuristic rule—which we call BiT-HyperRule—to select the most important hyperparameters for tuning as a simple function of the task's intrinsic image resolution and number of datapoints. We found it important to set the following hyperparameters per-task: training schedule length, resolution, and whether to use MixUp regularization [59]. We use BiT-HyperRule for over 20 tasks in this paper, with training sets ranging from 1 example per class to over 1M total examples. The exact settings for BiT-HyperRule are presented in Sect. 3.3.

During fine-tuning, we use the following standard data pre-processing: we resize the image to a square, crop out a smaller random square, and randomly horizontally flip the image at training time. At test time, we only resize the image to a fixed size. In some tasks horizontal flipping or cropping destroys the label semantics, making the task impossible. An example is if the label requires predicting object orientation or coordinates in pixel space. In these cases we omit flipping or cropping when appropriate.

Recent work has shown that existing augmentation methods introduce inconsistency between training and test resolutions for CNNs [49]. Therefore, it is common to scale up the resolution by a small factor at test time. As an alternative, one can add a step at which the trained model is fine-tuned to the test resolution [49]. The latter is well-suited for transfer learning; we include the resolution change during our fine-tuning step.

We found that MixUp [59] is not useful for pre-training BiT, likely due to the abundance of data. However, it is sometimes useful for transfer. Interestingly, it is most useful for mid-sized datasets, and not for few-shot transfer, see Sect. 3.3 for where we apply MixUp.

Surprisingly, we do not use any of the following forms of regularization during downstream tuning: weight decay to zero, weight decay to initial parameters [25], or dropout. Despite the fact that the network is very large—BiT has 928 million parameters—the performance is surprisingly good without these techniques and their respective hyperparameters, even when transferring to very small datasets. We find that setting an appropriate schedule length, i.e. training longer for larger datasets, provides sufficient regularization.

3   Experiments

We train three upstream models using three datasets at different scales: BiT-S, BiT-M, and BiT-L. We evaluate these models on many downstream tasks and attain very strong performance on high and low data regimes.

3.1   Data for Upstream Training

BiT-S is trained on the ILSVRC-2012 variant of ImageNet, which contains 1.28 million images and 1000 classes. Each image has a single label. BiT-M is trained on the full ImageNet-21k dataset [6], a public dataset containing 14.2 million images and 21k classes organized by the WordNet hierarchy. Images may contain multiple labels. BiT-L is trained on the JFT-300M dataset [32,43,53]. This dataset is a newer version of that used in [4,13]. JFT-300M consists of around 300 million images with 1.26 labels per image on average. The labels are organized into a hierarchy of 18 291 classes. Annotation is performed using an automatic pipeline, and are therefore imperfect; approximately 20% of the labels are noisy. We remove all images present in downstream test sets from JFT-300M. We provide details in supplementary material. Note: the "-S/M/L" suffix refers to the pre-training datasets size and schedule, not architecture. We train BiT with several architecture sizes, the default (largest) being ResNet152x4.

3.2   Downstream Tasks

We evaluate BiT on long-standing benchmarks: ILSVRC-2012 [6], CIFAR-10/100 [22], Oxford-IIIT Pet [35] and Oxford Flowers-102 [33]. These datasets differ in the total number of images, input resolution and nature of their categories, from general object categories in ImageNet and CIFAR to fine-grained ones in Pets and Flowers. We fine-tune BiT on the official training split and report results on the official *test* split if publicly available. Otherwise, we use the *val* split.

Table 1. Top-1 accuracy for BiT-L on many datasets using a single model and single hyperparameter setting per task (BiT-HyperRule). The entries show median ± standard deviation across 3 fine-tuning runs. Specialist models are those that condition pre-training on each task, while generalist models, including BiT, perform task-independent pre-training. (*Concurrent work.)

	BiT-L	Generalist SOTA	Specialist SOTA
ILSVRC-2012	**87.54 ± 0.02**	86.4 [49]	88.4 [53]*
CIFAR-10	**99.37 ± 0.06**	99.0 [14]	-
CIFAR-100	**93.51 ± 0.08**	91.7 [47]	-
Pets	**96.62 ± 0.23**	95.9 [14]	97.1 [32]
Flowers	**99.63 ± 0.03**	98.8 [47]	97.7 [32]
VTAB (19 tasks)	**76.29 ± 1.70**	70.5 [50]	-

To further assess the generality of representations learned by BiT, we evaluate on the Visual Task Adaptation Benchmark (VTAB) [58]. VTAB consists of 19 diverse visual tasks, each of which has 1000 training samples (VTAB-1k variant). The tasks are organized into three groups: *natural, specialized* and *structured*. The VTAB-1k score is top-1 recognition performance averaged over these 19 tasks. The *natural* group of tasks contains classical datasets of natural images captured using standard cameras. The *specialized* group also contains images captured in the real world, but through specialist equipment, such as satellite or medical images. Finally, the *structured* tasks assess understanding of the structure of a scene, and are mostly generated from synthetic environments. Example structured tasks include object counting and 3D depth estimation.

3.3   Hyperparameter Details

Upstream Pre-training. All of our BiT models use a vanilla ResNet-v2 architecture [11], except that we replace all Batch Normalization [16] layers with Group Normalization [52] and use Weight Standardization [36] in all convolutional layers. See Sect. 4.3 for analysis. We train ResNet-152 architectures in all datasets, with every hidden layer widened by a factor of four (ResNet152x4). We study different model sizes and the coupling with dataset size in Sect. 4.1.

We train all of our models upstream using SGD with momentum. We use an initial learning rate of 0.03, and momentum 0.9. During image preprocessing stage we use image cropping technique from [45] and random horizontal mirroring followed by 224×224 image resize. We train both BiT-S and BiT-M for 90 epochs and decay the learning rate by a factor of 10 at 30, 60 and 80 epochs. For BiT-L, we train for 40 epochs and decay the learning rate after 10, 23, 30 and 37 epochs. We use a global batch size of 4096 and train on a Cloud TPUv3-512 [19], resulting in 8 images per chip. We use linear learning rate warm-up for 5000 optimization steps and multiply the learning rate by $\frac{\text{batch size}}{256}$

Table 2. Improvement in accuracy when pre-training on the public ImageNet-21k dataset over the "standard" ILSVRC-2012. Both models are ResNet152x4.

	ILSVRC-2012	CIFAR-10	CIFAR-100	Pets	Flowers	VTAB-1k
BiT-S (ILSVRC-2012)	81.30	97.51	86.21	93.97	89.89	66.87
BiT-M (ImageNet-21k)	85.39	98.91	92.17	94.46	99.30	70.64
Improvement	+4.09	+1.40	+5.96	+0.49	+9.41	+3.77

following [7]. During pre-training we use a weight decay of 0.0001, but as discussed in Sect. 2, we do not use any weight decay during transfer.

Downstream Fine-Tuning. To attain a low per-task adaptation cost, we do not perform any hyperparameter sweeps downstream. Instead, we present BiT-HyperRule, a heuristic to determine all hyperparameters for fine-tuning. Most hyperparameters are fixed across all datasets, but schedule, resolution, and usage of MixUp depend on the task's image resolution and training set size.

For all tasks, we use SGD with an initial learning rate of 0.003, momentum 0.9, and batch size 512. We resize input images with area smaller than 96×96 pixels to 160×160 pixels, and then take a random crop of 128×128 pixels. We resize larger images to 448×448 and take a 384×384-sized crop.[1] We apply random crops and horizontal flips for all tasks, except those for which cropping or flipping destroys the label semantics, we provide details in supplementary material.

For schedule length, we define three scale regimes based on the number of examples: we call *small* tasks those with fewer than 20k labeled examples, *medium* those with fewer than 500k, and any larger dataset is a *large* task. We fine-tune BiT for 500 steps on small tasks, for 10k steps on medium tasks, and for 20k steps on large tasks. During fine-tuning, we decay the learning rate by a factor of 10 at 30%, 60% and 90% of the training steps. Finally, we use MixUp [59], with $\alpha = 0.1$, for medium and large tasks.

3.4 Standard Computer Vision Benchmarks

We evaluate BiT-L on standard benchmarks and compare its performance to the current state-of-the-art results (Table 1). We separate models that perform task-independent pre-training ("general" representations), from those that perform task-dependent auxiliary training ("specialist" representations). The specialist methods condition on a particular task, for example ILSVRC-2012, then train using a large support dataset, such as JFT-300M [32] or Instagram-1B [55]. See discussion in Sect. 5. Specialist representations are highly effective, but require a large training cost *per task*. By contrast, generalized representations require large-scale training *only once*, followed by a cheap adaptation phase.

---

[1] For our largest R152x4, we increase resolution to 512×512 and crop to 480×480.

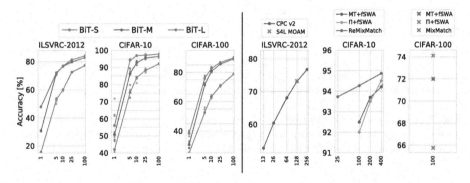

Fig. 2. Experiments in the low data regime. **Left:** Transfer performance of BiT-L. Each point represents the result after training on a balanced random subsample of the dataset (5 subsamples per dataset). The median across runs is highlighted by the curves. The variance across data samples is usually low, with the exception of 1-shot CIFAR-10, which contains only 10 images. **Right:** We summarize the state-of-the-art in semi-supervised learning as reference points. Note that a direct comparison is not meaningful; unlike BiT, semi-supervised methods have access to extra unlabelled data from the training distribution, but they do not make use of out-of-distribution labeled data.

BiT-L outperforms previously reported generalist SOTA models as well as, in many cases, the SOTA specialist models. Inspired by strong results of BiT-L trained on JFT-300M, we also train models on the public ImageNet-21k dataset. This dataset is more than 10 times bigger than ILSVRC-2012, but it is mostly overlooked by the research community. In Table 2 we demonstrate that BiT-M trained on ImageNet-21k leads to substantially improved visual representations compared to the same model trained on ILSVRC-2012 (BiT-S), as measured by all our benchmarks. In Sect. 4.2, we discuss pitfalls that may have hindered wide adoption of ImageNet-21k as a dataset model for pre-training and highlight crucial components of BiT that enabled success on this large dataset.

For completeness, we also report top-5 accuracy on ILSVRC-2012 with median ± standard deviation format across 3 runs: 98.46% ± 0.02% for BiT-L, 97.69% ± 0.02% for BiT-M and 95.65% ± 0.03% for BiT-S.

3.5 Tasks with Few Datapoints

We study the number of *downstream* labeled samples required to transfer BiT-L successfully. We transfer BiT-L using subsets of ILSVRC-2012, CIFAR-10, and CIFAR-100, down to 1 example per class. We also evaluate on a broader suite of 19 VTAB-1k tasks, each of which has 1000 training examples.

Figure 2 (left half) shows BiT-L using few-shots on ILSVRC-2012, CIFAR-10, and CIFAR-100. We run multiple random subsamples, and plot every trial. Surprisingly, even with very few samples per class, BiT-L demonstrates strong performance and quickly approaches performance of the full-data regime.

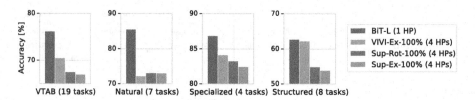

Fig. 3. Results on VTAB (19 tasks) with 1000 examples/task, and the current SOTA. It compares methods that sweep few hyperparameters per task: either four hyperparameters in previous work ("4 HPs") or the single BiT-HyperRule.

In particular, with just 5 labeled samples per class it achieves top-1 accuracy of 72.0% on ILSVRC-2012 and with 100 samples the top-1 accuracy goes to 84.1%. On CIFAR-100, we achieve 82.6% with just 10 samples per class.

Semi-supervised learning also tackles learning with few labels. However, such approaches are not directly comparable to BiT. BiT uses extra labelled out-of-domain data, whereas semi-supervised learning uses extra unlabelled in-domain data. Nevertheless, it is interesting to observe the relative benefits of transfer from out-of-domain labelled data versus in-domain semi-supervised data. In Fig. 2 we show state-of-the-art results from the semi-supervised learning.

Figure 3 shows the performance of BiT-L on the 19 VTAB-1k tasks. BiT-L with BiT-HyperRule substantially outperforms the previously reported state-of-the-art. When looking into performance of VTAB-1k task subsets, BiT is the best on *natural*, *specialized* and *structured* tasks. The recently-proposed VIVI-Ex-100% [50] model that employs video data during upstream pre-training shows very similar performance on the *structured* tasks.

We investigate heavy per-task hyperparameter tuning in supplementary material and conclude that this further improves performance.

3.6   Object Detection

Finally, we evaluate BiT on object detection. We use the COCO-2017 dataset [28] and train a top-performing object detector, RetinaNet [27], using pre-trained BiT models as backbones. Due to memory constraints, we use the ResNet-101x3 architecture for all of our BiT models. We fine-tune the detection models on the COCO-2017 train split and report results on the

Table 3. Object detection performance on COCO-2017 [28] validation data of RetinaNet models with pre-trained BiT backbones and the literature baseline.

Model	Upstream data	AP
RetinaNet [27]	ILSVRC-2012	40.8
RetinaNet (BiT-S)	ILSVRC-2012	41.7
RetinaNet (BiT-M)	ImageNet-21k	43.2
RetinaNet (BiT-L)	JFT-300M	**43.8**

validation split using the standard metric [28] in Table 3. Here, we do not use BiT-HyperRule, but stick to the standard RetinaNet training protocol, we provide details in supplementary material. Table 3 demonstrates that BiT models outperform standard ImageNet pre-trained models. We can see clear benefits of pre-training on large data beyond ILSVRC-2012: pre-training on ImageNet-21k results in a 1.5 point improvement in Average Precision (AP), while pre-training on JFT-300M further improves AP by 0.6 points.

4 Analysis

We analyse various components of BiT: we demonstrate the importance of model capacity, discuss practical optimization caveats and choice of normalization layer.

4.1 Scaling Models and Datasets

The general consensus is that larger neural networks result in better performance. We investigate the interplay between model capacity and upstream dataset size on downstream performance. We evaluate the BiT models of different sizes (ResNet-50x1, ResNet-50x3, ResNet-101x1, ResNet-101x3, and ResNet-152x4) trained on ILSVRC-2012, ImageNet-21k, and JFT-300M on various downstream benchmarks. These results are summarized in Fig. 4.

When pre-training on ILSVRC-2012, the benefit from larger models diminishes. However, the benefits of larger models are more pronounced on the larger two datasets. A similar effect is observed when training on Instagram hashtags [30] and in language modelling [20].

Not only is there limited benefit of training a large model size on a small dataset, but there is also limited (or even negative) benefit from training a small model on a larger dataset. Perhaps surprisingly, the ResNet-50x1 model trained on the JFT-300M dataset can even performs worse than the same architecture trained on the smaller ImageNet-21k. Thus, if one uses only a ResNet50x1, one may conclude that scaling up the dataset does not bring any additional benefits. However, with larger architectures, models pre-trained on JFT-300M significantly outperform those pre-trained on ILSVRC-2012 or ImageNet-21k.

Figure 2 shows that BiT-L attains strong results even on tiny downstream datasets. Figure 5 ablates few-shot performance across different pre-training datasets and architectures. In the extreme case of one example per class, larger architectures outperform smaller ones when pre-trained on large upstream data.

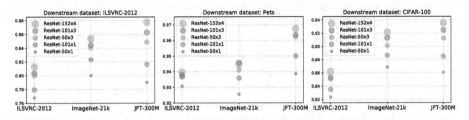

Fig. 4. Effect of upstream data (shown on the x-axis) and model size on downstream performance. Note that exclusively using more data or larger models may hurt performance; instead, both need to be increased in tandem.

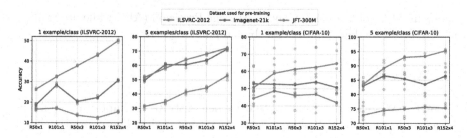

Fig. 5. Performance of BiT models in the low-data regime. The x-axis corresponds to the architecture, where R is short for ResNet. We pre-train on the three upstream datasets and evaluate on two downstream datasets: ILSVRC-2012 (left) and CIFAR-10 (right) with 1 or 5 examples per class. For each scenario, we train 5 models on random data subsets, represented by the lighter dots. The line connects the medians of these five runs.

Interestingly, on ILSVRC-2012 with few shots, BiT-L trained on JFT-300M outperforms the models trained on the entire ILSVRC-2012 dataset itself. Note that for comparability, the classifier head is re-trained from scratch during fine-tuning, even when transferring ILSVRC-2012 full to ILSVRC-2012 few shot.

4.2 Optimization on Large Datasets

For standard computer vision datasets such as ILSVRC-2012, there are well-known training procedures that are robust and lead to good performance. Progress in high-performance computing has made it feasible to learn from much larger datasets, such as ImageNet-21k, which has 14.2M images compared to ILSVRC-2012's 1.28M. However, there are no established procedures for training from such large datasets. In this section we provide some guidelines.

Sufficient computational budget is crucial for training performant models on large datasets. The standard ILSVRC-2012 training schedule processes roughly 100 million images (1.28M images × 90 epochs). However, if the same computational budget is applied to ImageNet-21k, the resulting model performs worse on ILSVRC-2012, see Fig. 6, left. Nevertheless, as shown in the same figure, by increasing the computational budget, we not only recover ILSVRC-2012 performance, but significantly outperforms it. On JFT-300M the validation error may not improve over a long time—Fig. 6 middle plot, "8 GPU weeks" zoom-in—although the model is still improving as evidenced by the longer time window.

Another important aspect of pre-training with large datasets is the weight decay. Lower weight decay can result in an apparent acceleration of convergence, Fig. 6 rightmost plot. However, this setting eventually results in an underperforming final model. This counter-intuitive behavior stems from the interaction of weight decay and normalization layers [23, 26]. Low weight decay results in growing weight norms, which in turn results in a diminishing effective learning rate. Initially this effect creates an impression of faster convergence, but it

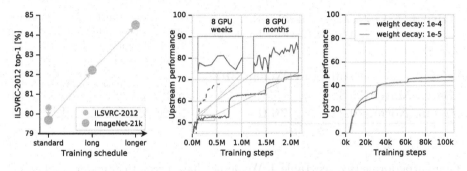

Fig. 6. Left: Applying the "standard" computational budget of ILSVRC-2012 to the larger ImageNet-21k seems detrimental. Only when we train longer (3x and 10x) do we see the benefits of training on the larger dataset. **Middle:** The learning progress of a ResNet-101x3 on JFT-300M seems to be flat even after 8 GPU-weeks, but after 8 GPU-months progress is clear. If one decays the learning rate too early (dashed curve), final performance is significantly worse. **Right:** Faster initial convergence with lower weight decay may trick the practitioner into selecting a sub-optimal value. Higher weight decay converges more slowly, but results in a better final model.

eventually prevents further progress. A sufficiently large weight decay is required to avoid this effect, and throughout we use 10^{-4}.

Finally, we note that in all of our experiments we use stochastic gradient descent with momentum without any modifications. In our preliminary experiments we did not observe benefits from more involved adaptive gradient methods.

4.3 Large Batches, Group Normalization, Weight Standardization

Currently, training on large datasets is only feasible using many hardware accelerators. Data parallelism is the most popular distribution strategy, and this naturally entails large batch sizes. Many known algorithms for training with large batch sizes use Batch Normalization (BN) [16] as a component [7] or even highlight it as the key instrument required for large batch training [5].

Our larger models have a high memory requirement for any single accelerator chip, which necessitates small per-device batch sizes. However, BN performs worse when the number of images on each accelerator is too low [15]. An alternative strategy is to accumulate BN statistics across all of the accelerators. However, this has two major drawbacks. First, computing BN statistics across large batches has been shown to harm generalization [5]. Second, using global BN requires many aggregations across accelerators which incurs significant latency.

We investigated Group Normalization (GN) [52] and Weight Standardization (WS) [36] as alternatives to BN. We tested large batch training using 128 accelerator chips and a batch size of 4096. We found that GN alone does not scale to large batches; we observe a performance drop of 5.4% on ILSVRC-2012 top-1 accuracy compared to using BN in a ResNet-50x1, and less stable training. The addition of WS enables GN to scale to such large batches, stabilizes training, and

Table 4. Top-1 accuracy of ResNet-50 trained from scratch on ILSVRC-2012 with a batch-size of 4096.

	Plain Conv	Weight Std.
Batch Norm.	75.6	75.8
Group Norm.	70.2	**76.0**

Table 5. Transfer performance of the corresponding models from Table 4 fine-tuned to the 19 VTAB-1k tasks.

	Plain Conv	Weight Std.
Batch Norm.	67.72	66.78
Group Norm.	68.77	**70.39**

even outperforms BN, see Table 4. We do not have theoretical understanding of this empirical finding.

We are not only interested in upstream performance, but also how models trained with GN and WS transfer. We thus transferred models with different combinations of BN, GN, and WS pre-trained on ILSVRC-2012 to the 19 tasks defined by VTAB. The results in Table 5 indicate that the GN/WS combination transfers better than BN, so we use GN/WS in all BiT models.

5   Related Work

Large-Scale Weakly Supervised Learning of Representations. A number of prior works use large supervised datasets for pre-training visual representations [18,24,30,43]. In [18,24] the authors use a dataset containing 100M Flickr images [48]. This dataset appears to transfer less well than JFT-300M. While studying the effect of dataset size, [43] show good transfer performance when training on JFT-300M, despite reporting a large degree of noise (20% precision errors) in the labels. An even larger, noisily labelled dataset of 3.5B Instagram images is used in [30]. This increase in dataset size and an improved model architecture [54] lead to better results when transferring to ILSVRC-2012. We show that we can attain even better performance with ResNet using JFT-300M with appropriate adjustments presented in Sect. 2. The aforementioned papers focus on transfer to ImageNet classification, and COCO or VOC detection and segmentation. We show that transfer is also highly effective in the low data regime, and works well on the broader set of 19 tasks in VTAB [58].

Specialized Representations. Rather than pre-train generic representations, recent works have shown strong performance by training task-specific representations [32,53,55]. These papers condition on a particular task when training on a large support dataset. [53,55] train student networks on a large unlabelled support dataset using the predictions of a teacher network trained on the target task. [32] compute importance weights on the a labelled support dataset by conditioning on the target dataset. They then train the representations on the re-weighted source data. Even though these approaches may lead to superior results, they require knowing the downstream dataset in advance and substantial computational resources for each downstream dataset.

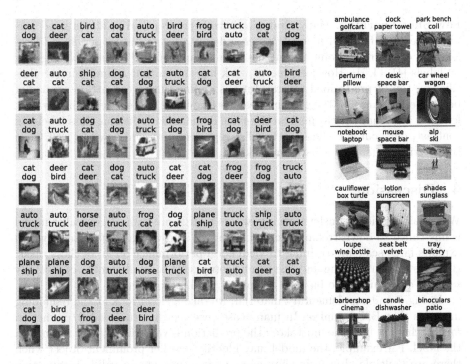

Fig. 7. Cases where BiT-L's predictions (top word) do not match the ground-truth labels (bottom word), and hence are counted as top-1 errors. **Left:** *All* mistakes on CIFAR-10, colored by whether five human raters agreed with BiT-L's prediction (green), with the ground-truth label (red) or were unsure or disagreed with both (yellow). **Right:** Selected representative mistakes of BiT-L on ILSVRC-2012. Top group: The model's prediction is more representative of the primary object than the label. Middle group: According to top-1 accuracy the model is incorrect, but according to top-5 it is correct. Bottom group: The model's top-10 predictions are incorrect. (Color figure online)

Unsupervised and Semi-Supervised Representation Learning. Self-supervised methods have shown the ability to leverage unsupervised datasets for downstream tasks. For example, [8] show that unsupervised representations trained on 1B unlabelled Instagram images transfer comparably or better than supervised ILSVRC-2012 features. Semi-supervised learning exploits unlabelled data drawn from the same domain as the labelled data. [2,42] used semi-supervised learning to attain strong performance on CIFAR-10 and SVHN using only 40 or 250 labels. Recent works combine self-supervised and semi-supervised learning to attain good performance with fewer labels on ImageNet [12,57]. [58] study many representation learning algorithms (unsupervised, semi-supervised, and supervised) and evaluate their representation's ability to generalize to novel tasks, concluding that a combination of supervised and self-supervised signals works best. However, all models were trained on ILSVRC-2012. We show that supervised pre-training on larger datasets continues to be an effective strategy.

Few-Shot Learning. Many strategies have been proposed to attain good performance when faced with novel classes and only a few examples per class. Meta-learning or metric-learning techniques have been proposed to learn with few or no labels [41, 44, 51]. However, recent work has shown that a simple linear classifier on top of pre-trained representations or fine-tuning can attain similar or better performance [3, 31]. The upstream pre-training and downstream few-shot learning are usually performed on the same domain, with disjoint class labels. In contrast, our goal is to find a generalist representation which works well when transferring to many downstream tasks.

6 Discussion

We revisit classical transfer learning, where a large pre-trained generalist model is fine-tuned to downstream tasks of interest. We provide a simple recipe which exploits large scale pre-training to yield good performance on all of these tasks. BiT uses a clean training and fine-tuning setup, with a small number of carefully selected components, to balance complexity and performance.

In Fig. 7 and supplementary material, we take a closer look at the remaining mistakes that BiT-L makes. In many cases, we see that these label/prediction mismatches are not true 'mistakes': the prediction is valid, but it does not match the label. For example, the model may identify another prominent object when there are multiple objects in the image, or may provide an valid classification when the main entity has multiple attributes. There are also cases of label noise, where the model's prediction is a better fit than the ground-truth label. In a quantitative study, we found that around half of the model's mistakes on CIFAR-10 are due to ambiguity or label noise (see Fig. 7, left), and in only 19.21% of the ILSVRC-2012 mistakes do human raters clearly agree with the label over the prediction. Overall, by inspecting these mistakes, we observe that performance on the standard vision benchmarks seems to approach a saturation point.

We therefore explore the effectiveness of transfer to two classes of more challenging tasks: classical image recognition tasks, but with very few labelled examples to adapt to the new domain, and VTAB, which contains more diverse tasks, such as spatial localization in simulated environments, and medical and satellite imaging tasks. These benchmarks are much further from saturation; while BiT-L performs well on them, there is still substantial room for further progress.

References

1. Athiwaratkun, B., Finzi, M., Izmailov, P., Wilson, A.G.: There are many consistent explanations of unlabeled data: why you should average. In: ICLR (2019)
2. Berthelot, D., et al.: ReMixMatch: Semi-supervised learning with distribution alignment and augmentation anchoring. arXiv preprint arXiv:1911.09785 (2019)
3. Chen, W., Liu, Y., Kira, Z., Wang, Y.F., Huang, J.: A closer look at few-shot classification. In: ICLR (2019)
4. Chollet, F.: Xception: deep learning with depthwise separable convolutions. In: CVPR (2017)

5. De, S., Smith, S.L.: Batch normalization has multiple benefits: an empirical study on residual networks (2020). https://openreview.net/forum?id=BJeVklHtPr
6. Deng, J., Dong, W., Socher, R., Li, L.J., Li, K., Fei-Fei, L.: Imagenet: a large-scale hierarchical image database. In: CVPR (2009)
7. Goyal, P., et al.: Accurate, large minibatch sgd: training imagenet in 1 h. arXiv preprint arXiv:1706.02677 (2017)
8. He, K., Fan, H., Wu, Y., Xie, S., Girshick, R.: Momentum contrast for unsupervised visual representation learning. arXiv preprint arXiv:1911.05722 (2019)
9. He, K., Girshick, R., Dollár, P.: Rethinking imagenet pre-training. In: ICCV (2019)
10. He, K., Zhang, X., Ren, S., Sun, J.: Deep residual learning for image recognition. In: CVPR (2016)
11. He, K., Zhang, X., Ren, S., Sun, J.: Identity mappings in deep residual networks. In: Leibe, B., Matas, J., Sebe, N., Welling, M. (eds.) ECCV 2016. LNCS, vol. 9908, pp. 630–645. Springer, Cham (2016). https://doi.org/10.1007/978-3-319-46493-0_38
12. Hénaff, O.J., Razavi, A., Doersch, C., Eslami, S., van den Oord, A.: Data-efficient image recognition with contrastive predictive coding. arXiv preprint arXiv:1905.09272 (2019)
13. Hinton, G., Vinyals, O., Dean, J.: Distilling the knowledge in a neural network. arXiv preprint arXiv:1503.02531 (2015)
14. Huang, Y., et al.: GPipe: Efficient training of giant neural networks using pipeline parallelism. arXiv preprint arXiv:1811.06965 (2018)
15. Ioffe, S.: Batch renormalization: towards reducing minibatch dependence in batch-normalized models. In: Guyon, I., et al. (eds.) Advances in Neural Information Processing Systems 30, pp. 1945–1953. Curran Associates, Inc. (2017)
16. Ioffe, S., Szegedy, C.: Batch normalization: accelerating deep network training by reducing internal covariate shift. In: ICML (2015)
17. Izmailov, P., Podoprikhin, D., Garipov, T., Vetrov, D., Wilson, A.G.: Averaging weights leads to wider optima and better generalization. arXiv preprint arXiv:1803.05407 (2018)
18. Joulin, A., van der Maaten, L., Jabri, A., Vasilache, N.: Learning visual features from large weakly supervised data. In: Leibe, B., Matas, J., Sebe, N., Welling, M. (eds.) ECCV 2016. LNCS, vol. 9911, pp. 67–84. Springer, Cham (2016). https://doi.org/10.1007/978-3-319-46478-7_5
19. Jouppi, N.P., et al.: In-datacenter performance analysis of a tensor processing unit. In: International Symposium on Computer Architecture (ISCA) (2017)
20. Kaplan, J., et al.: Scaling laws for neural language models. arXiv preprint arXiv:2001.08361 (2020)
21. Kingma, D.P., Ba, J.: Adam: A method for stochastic optimization. arXiv preprint arXiv:1412.6980 (2014)
22. Krizhevsky, A.: Learning multiple layers of features from tiny images. University of Toronto, Technical report (2009)
23. van Laarhoven, T.: L2 regularization versus batch and weight normalization. CoRR (2017)
24. Li, A., Jabri, A., Joulin, A., van der Maaten, L.: Learning visual n-grams from web data. In: ICCV (2017)
25. Li, X., Grandvalet, Y., Davoine, F.: Explicit inductive bias for transfer learning with convolutional networks. In: ICML (2018)
26. Li, Z., Arora, S.: An exponential learning rate schedule for deep learning. arXiv preprint arXiv:1910.07454 (2019)
27. Lin, T.Y., Goyal, P., Girshick, R., He, K., Dollár, P.: Focal loss for dense object detection. In: ICCV (2017)

28. Lin, T.-Y., et al.: Microsoft COCO: common objects in context. In: Fleet, D., Pajdla, T., Schiele, B., Tuytelaars, T. (eds.) ECCV 2014. LNCS, vol. 8693, pp. 740–755. Springer, Cham (2014). https://doi.org/10.1007/978-3-319-10602-1_48

29. Loshchilov, I., Hutter, F.: Sgdr: Stochastic gradient descent with warm restarts. arXiv preprint arXiv:1608.03983 (2016)

30. Mahajan, D., et al.: Exploring the limits of weakly supervised pretraining. In: Ferrari, V., Hebert, M., Sminchisescu, C., Weiss, Y. (eds.) ECCV 2018. LNCS, vol. 11206, pp. 185–201. Springer, Cham (2018). https://doi.org/10.1007/978-3-030-01216-8_12

31. Nakamura, A., Harada, T.: Revisiting fine-tuning for few-shot learning. arXiv preprint arXiv:1910.00216 (2019)

32. Ngiam, J., Peng, D., Vasudevan, V., Kornblith, S., Le, Q.V., Pang, R.: Domain adaptive transfer learning with specialist models. arXiv:1811.07056 (2018)

33. Nilsback, M.E., Zisserman, A.: Automated flower classification over a large number of classes. In: Indian Conference on Computer Vision, Graphics and Image Processing (2008)

34. Pan, S.J., Yang, Q.: A survey on transfer learning. IEEE Trans. Knowl. Data Eng (2009)

35. Parkhi, O.M., Vedaldi, A., Zisserman, A., Jawahar, C.V.: Cats and dogs. In: CVPR (2012)

36. Qiao, S., Wang, H., Liu, C., Shen, W., Yuille, A.: Weight standardization. arXiv preprint arXiv:1903.10520 (2019)

37. Raghu, M., Zhang, C., Kleinberg, J., Bengio, S.: Transfusion: Understanding transfer learning with applications to medical imaging. arXiv:1902.07208 (2019)

38. Rosenfeld, J.S., Rosenfeld, A., Belinkov, Y., Shavit, N.: A constructive prediction of the generalization error across scales. In: ICLR (2020)

39. Russakovsky, O., et al.: ImageNet large scale visual recognition challenge. Int. J. Comput. Vis. **115**(3), 211–252 (2015). https://doi.org/10.1007/s11263-015-0816-y

40. Simonyan, K., Zisserman, A.: Very deep convolutional networks for large-scale image recognition. arXiv preprint arXiv:1409.1556 (2014)

41. Snell, J., Swersky, K., Zemel, R.: Prototypical networks for few-shot learning. In: NIPS (2017)

42. Sohn, K., et al.: Fixmatch: Simplifying semi-supervised learning with consistency and confidence. arXiv preprint arXiv:2001.07685 (2020)

43. Sun, C., Shrivastava, A., Singh, S., Gupta, A.: Revisiting unreasonable effectiveness of data in deep learning era. In: ICCV (2017)

44. Sung, F., Yang, Y., Zhang, L., Xiang, T., Torr, P.H., Hospedales, T.M.: Learning to compare: Relation network for few-shot learning. In: CVPR (2018)

45. Szegedy, C., et al.: Going deeper with convolutions. In: CVPR (2015)

46. Szegedy, C., Vanhoucke, V., Ioffe, S., Shlens, J., Wojna, Z.: Rethinking the inception architecture for computer vision. In: CVPR (2016)

47. Tan, M., Le, Q.: Efficientnet: rethinking model scaling for convolutional neural networks. In: ICML (2019)

48. Thomee, B., et al.: Yfcc100m: The new data in multimedia research. arXiv preprint arXiv:1503.01817 (2015)

49. Touvron, H., Vedaldi, A., Douze, M., Jégou, H.: Fixing the train-test resolution discrepancy. In: NeurIPS (2019)

50. Tschannen, M., et al.: Self-supervised learning of video-induced visual invariances. In: Proceedings of the IEEE/CVF Conference on Computer Vision and Pattern Recognition (CVPR), June 2020

51. Vinyals, O., Blundell, C., Lillicrap, T., Wierstra, D., et al.: Matching networks for one shot learning. In: NIPS (2016)
52. Wu, Y., He, K.: Group normalization. In: Ferrari, V., Hebert, M., Sminchisescu, C., Weiss, Y. (eds.) ECCV 2018. LNCS, vol. 11217, pp. 3–19. Springer, Cham (2018). https://doi.org/10.1007/978-3-030-01261-8_1
53. Xie, Q., Hovy, E., Luong, M.T., Le, Q.V.: Self-training with noisy student improves imagenet classification. arXiv preprint arXiv:1911.04252 (2019)
54. Xie, S., Girshick, R., Dollár, P., Tu, Z., He, K.: Aggregated residual transformations for deep neural networks. In: CVPR (2017)
55. Yalniz, I.Z., Jégou, H., Chen, K., Paluri, M., Mahajan, D.: Billion-scale semi-supervised learning for image classification. arXiv preprint arXiv:1905.00546 (2019)
56. Yun, S., Han, D., Oh, S.J., Chun, S., Choe, J., Yoo, Y.: Cutmix: Regularization strategy to train strong classifiers with localizable features. arXiv preprint arXiv:1905.04899 (2019)
57. Zhai, X., Oliver, A., Kolesnikov, A., Beyer, L.: S^4L: self-supervised semi-supervised learning. In: ICCV (2019)
58. Zhai, X., et al.: A large-scale study of representation learning with the visual task adaptation benchmark. arXiv preprint arXiv:1910.04867 (2019)
59. Zhang, H., Cisse, M., Dauphin, Y.N., Lopez-Paz, D.: mixup: Beyond empirical risk minimization. In: ICLR (2017)

VisualCOMET: Reasoning About the Dynamic Context of a Still Image

Jae Sung Park[1,2]([⊠]), Chandra Bhagavatula[2], Roozbeh Mottaghi[1,2],
Ali Farhadi[1], and Yejin Choi[1,2]

[1] Paul G. Allen School of Computer Science and Engineering, Seattle, WA, USA
jspark96@cs.washington.edu
[2] Allen Institute for Artificial Intelligence, Seattle, WA, USA

Abstract. Even from a single frame of a still image, people can reason about the dynamic story of the image *before*, *after*, and *beyond* the frame. For example, given an image of a man struggling to stay afloat in water, we can reason that the man fell into the water sometime in the past, the intent of that man at the moment is to stay alive, and he will need help in the near future or else he will get washed away. We propose **VisualCOMET**, (**Visual Com**monsense Reasoning in **T**ime.) the novel framework of visual commonsense reasoning tasks to predict events that might have happened before, events that might happen next, and the intents of the people at present. To support research toward visual commonsense reasoning, we introduce the first large-scale repository of **Visual Commonsense Graphs** that consists of over **1.4 million** textual descriptions of visual commonsense inferences carefully annotated over a diverse set of 59,000 images, each paired with short video summaries of before and after. In addition, we provide person-grounding (i.e., co-reference links) between people appearing in the image and people mentioned in the textual commonsense descriptions, allowing for tighter integration between images and text. We establish strong baseline performances on this task and demonstrate that integration between visual and textual commonsense reasoning is the key and wins over non-integrative alternatives.

1 Introduction

Given a still image, people can reason about the rich dynamic story underlying the visual scene that goes far beyond the frame of the image. For example, in Fig. 1, given the image of a desperate man holding onto a statue in water, we

visualcomet.xyz.

Electronic supplementary material The online version of this chapter (https://doi.org/10.1007/978-3-030-58558-7_30) contains supplementary material, which is available to authorized users.

A. Vedaldi et al. (Eds.): ECCV 2020, LNCS 12350, pp. 508–524, 2020.
https://doi.org/10.1007/978-3-030-58558-7_30

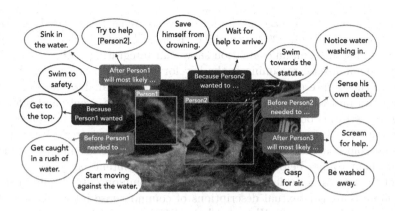

Fig. 1. Given a person in the image, **VisualCOMET** provides a *graph* of common sense inferences about 1) what needed to happen before, 2) intents of the people at present, and 3) what will happen next.

can reason far beyond what are immediately visible in that still frame; sometime in the past, an accident might have happened and a ship he was on might have started sinking. Sometime in the future, he might continue struggling and eventually be washed away. In the current moment, his intent and motivation must be that he wants to save himself from drowning. This type of visual understanding requires a major leap from recognition-level understanding to cognitive-level understanding, going far beyond the scope of image classification, object detection, activity recognition, or image captioning. An image caption such as "a man in a black shirt swimming in water", for example, while technically correct, falls far short of understanding the dynamic situation captured in the image that requires reasoning about the context that spans before, after, and beyond the frame of this image. Key to this rich cognitive understanding of visual scenes is visual commonsense reasoning, which in turn, requires rich background knowledge about how the visual world works, and how the social world works.

In this paper, we propose **VisualCOMET**, a new framework of task formulations to reason about the rich visual context that goes beyond the immediately visible content of the image, ranging from events that might have happened before, to events that might happen next, and to the intents of the people at present. To support research toward visual commonsense reasoning, we introduce the first large-scale repository of **Visual Commonsense Graphs** that consists of **1.4 million** textual descriptions of visual commonsense inferences that are carefully annotated over a diverse set of about 59,000 people-centric images from VCR [47]. In addition, we provide person-grounding (i.e., co-reference links) between people appearing in the image and people mentioned in the textual commonsense descriptions, allowing for tighter integration between images and text. The resulting Visual Commonsense Graphs are rich, enabling a number of task formulations with varying levels of difficulties for future research.

We establish strong baseline performances on such tasks based on GPT-2 transformer architecture [29] to combine visual and textual information. Quantitative results and human evaluation show that integrating both the visual and textual commonsense reasoning is the key for enhanced performance. Furthermore, when the present eventual description is not available and only image is given, we find that the model trained to predict both events and inferential sentences performs better than the one trained to predict only inferences.

In summary, our contributions are as follows. (1) We introduce a new task of visual commonsense reasoning for cognitive visual scene understanding, to reason about events before and after and people's intents at present. (2) We present the first large-scale repository of Visual Commonsense Graphs that contains more than 1M textual descriptions of commonsense inferences over 60 K complex visual scenes. (3) We extend the GPT-2 model to incorporate visual information and allow direct supervision for grounding people in images. (4) Empirical results and human evaluations show that model trained jointly with visual and textual cues outperform models with single modality, and can generate meaningful inferences from still images.

2   Related Work

Visual Understanding with Language: Various tasks have been introduced for joint understanding of visual information and language, such as image captioning [7,33,42], visual question answering [1,16,23] and referring expressions [17,22,28]. These works, however, perform inference about only the current content of images and fall short of understanding the dynamic situation captured in the image, which is the main motivation of our work. There is also a recent body of work addressing representation learning using vision and language cues [21,35,37]. We propose a baseline for our task, which is inspired by these techniques.

Visual Commonsense Inference: Prior works have tried to incorporate commonsense knowledge in the context of visual understanding. [39] use human-generated abstract scenes made from clipart to learn common sense, but not on real images. [27] try to infer the motivation behind the actions of people from images. Visual Commonsense Reasoning (VCR) [47] tests if the model can answer questions with rationale using commonsense knowledge. While [47] includes rich visual common sense information, their question answering setup makes it difficult to have models to generate commonsense inferences. ATOMIC [32] provides a commonsense knowledge graph containing if-then inferential textual descriptions in generative setting; however, it relies on generic, textual events and does not consider visually contextualized information. In this work, we are interested in extending [47] and [32] for general visual commonsense by building a large-scale repository of visual commonsense graphs and models that can explicitly generate commonsense inferences for given images.

Visual Future Prediction: There is a large body of work on future prediction in different contexts such as future frame generation [5, 24, 30, 34, 41, 43, 46], prediction of the trajectories of people and objects [2, 25, 44], predicting human pose in future frames [6, 12, 45] and semantic future action recognition [19, 36, 50]. In contrast to all these approaches, we provide a compact description for the future events using language.

3   Task: Cognitive Image Understanding via Visual Commonsense Graphs

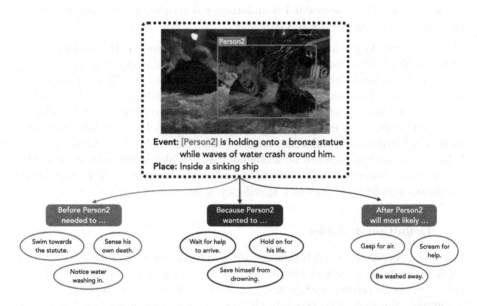

Fig. 2. Task Overview: Our proposed task is to generate commonsense inferences of **events before**, **events after** and **intents at present**, given an image, a description of an **event at present** in the image and a plausible scene/location of the image.

3.1   Definition of Visual Commonsense Graphs

The ultimate goal is to generate the entire visual commonsense graph illustrated in Fig. 1 that requires reasoning about the dynamic story underlying the input image. This graph consists of four major components (Fig. 2):

- (1) a set of textual descriptions of **events at present**,
- (2) a set of commonsense inferences on **events before**,
- (3) a set of commonsense inferences on **events after**, and
- (4) a set of commonsense inferences on people's **intents at present**.

The events before and after can broadly include any of the following: (a) *actions* people might take before and after (e.g., people jumping to the water), (b) *events* that might happen before and after (e.g., a ship sinking), and (c) *mental states* of people before and after (e.g., people scared and tired). Our design of the commonsense graph representation is inspired by ATOMIC [32], a text-only atlas of machine commonsense knowledge for *if-then* reasoning, but tailored specifically for cognitive understanding of visual scenes in images.

Location and Person Grounding: In addition, the current event descriptions are accompanied by additional textual descriptions of the **place** or the overall scene of the image, e.g., "at a bar" or "at a party". We also provide person-grounding (i.e., co-reference links) between people appearing in the image and people mentioned in the textual commonsense descriptions, allowing for tighter integration between images and text.

Dense Event Annotations with Visual Commonsense Reasoning: Generally speaking, the first component of the visual commonsense graph, *"events at present"*, is analogous to dense image captioning in that it focuses on the immediate visual content of the image, while components (2)–(4), *events before and after* and *intents at present*, correspond to visual commonsense reasoning.

Importantly, in an image that depicts a complex social scene involving multiple people engaged in different activities simultaneously, the inferences about before, after, and intents can be ambiguous as to which exact current event the inferences are based upon. Therefore, in our graph representation, we link up all the commonsense inferences to a specific event at present.

3.2 Definition of Tasks

Given the complete visual commonsense graph representing an image, we can consider multiple task formulations of varying degrees of difficulties. In this paper, we focus on two such tasks: (1) Given an image and one of the events at present, the task is to generate the rest of visual commonsense graph that is connected to the specific current event. (2) Given an image, the task is to generate the complete set of commonsense inferences from scratch.

4 Dataset Overview

We present the first large-scale dataset of Visual Commonsense Graphs for images with person grounding (i.e., multimodal co-reference chains). We collect a dataset of 1.4 million commonsense inferences over 59,356 images and 139,377 distinct events at present (Table 1). Figure 3 gives an overview of our Visual Commonsense Graphs including a diverse set of images, connected with the inference sentences[1].

---

[1] Larger figure available in the Appendix.

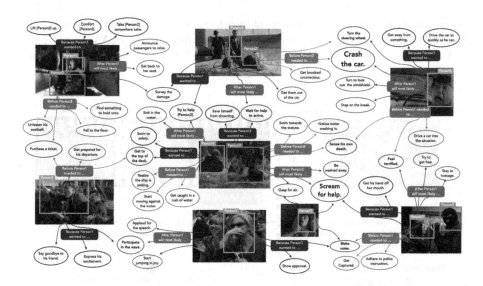

Fig. 3. Overview of our **Visual Commonsense Graphs**. We see that a diverse set of images are covered and connected with inference sentences. Red bubbles indicate if inference sentences shared by two or more images. (Color figure online)

Table 1. Statistics of our Visual Commonsense Graph repository: there are in total 139,377 distinct Visual Commonsense Graphs over 59,356 images involving 1,465,704 commonsense inferences.

	Train	Dev	Test	Total
# Images/Places	47,595	5,973	5,968	**59,356**
# Events at Present	111,796	13,768	13,813	**139,377**
# Inferences on Events Before	467,025	58,773	58,413	584,211
# Inferences on Events After	469,430	58,665	58,323	586,418
# Inferences on Intents at Present	237,608	28,904	28,568	295,080
# Total inferences	1,174,063	146,332	145,309	**1,465,704**

4.1   Source of Images

As the source of the images, we use the VCR [47] dataset that consists of images corresponding to complex visual scenes with multiple people and activities. The dataset also includes automatically detected object bounding boxes, and each person in the image uniquely identified with a referent tag (e.g. Person1 and Person2 in Fig. 1).

4.2   Crowdsourcing Visual Commonsense Graphs

Annotating the entire commonsense graph solely from an image is a daunting task even for humans. We design a two-stage crowdsourcing pipeline to make the annotation task feasible and to obtain focused and consistent annotations. We run our annotation pipeline on Amazon Mechanical Turk (AMT) platform and maintain the ethical pay rate of at least \$15/hr. This amounts to \$4 per image on average. Figure 4 shows an overview of our annotation pipeline.

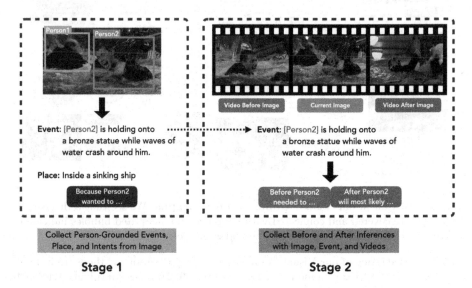

Fig. 4. Annotation Pipeline: Our two-stage crowdsourcing annotation pipeline used for collecting our high-quality Visual Commonsense Graphs.

Stage 1: Grounded Event Descriptions with Locations and Intents

In the first stage, we show crowdworkers an image along with tags identifying each person in the image. Crowdworkers select a person and author a description for the event involving that person. One key concern during event annotation is to encourage crowdworkers to annotate informative, interesting events as opposed to low-level events like standing, sitting, looking, etc. While technically correct, such descriptions do not contribute to higher-level understanding of the image. To obtain more meaningful events, we ask crowdworkers to write an event description and intent inferences at the same time. Finally, we ask crowdworkers to annotate the plausible location of the scene depicted in the image. In addition to priming workers, such location information provides more contextualized information for the task. The location information is not just a physical place, but can also include occasions, e.g., in a business meeting. At the end of stage 1, we collect (i) the location of an image, (ii) two to three events for each image, and (iii) two to four intents for each event of each image.

Stage 2: Collecting Before and After Inferences

In stage 2, we collect visual commonsense inferences of what might have happened *before* and what might happen *after* for each event description for each image annotated in stage 1 above. Images in our dataset were originally part of movie scenes. Based on the timestamp of the image being annotated, we show crowdworkers two short, fast-forwarded clips of events that happen before and after the image. This allows crowdworkers to author inferences that are more meaningful, rather than authoring correct but trivial inferences – e.g. "before, Person1 needed to be born", "after, Person1 will be dead", etc.

We assign two workers for each event and ask each to annotate between two and four *before* and *after* inferences. At the end of the two stages in our annotation pipeline, we have up to ten (2 intent, 4 before, 4 after) inferences for each pair of image and a textual description of event at present.

5   Our Approach

Our task assumes the following inputs for each image: a sequence of visual embeddings $\mathcal{V}$ representing the image and people detected in the image, grounded event description e, scene's location information p, and inference type r. Then, we wish to generate a set of possible inferences $H = \{s_1^r, s_2^r, ...s_{|H|}^r\}$.

5.1   Visual Features

The sequence of visual representations $\mathcal{V}$ consists of a representation of the whole image and an additional representations for each person detected in the image. We use Region of Interest (RoI) Align features [13] from Faster RCNN [31] as our visual embedding and pass it through a non-linear layer to obtain the final representation for an image or each detected person. The final sequence of representations $\mathcal{V} = \{v_0, v_1, ..v_k\}$ where k is the number of people detected.

As described in Sect. 4.2, we provide special tags identifying each person in the image (e.g. Person1 in Fig. 4) in our dataset. To use these tags, we introduce new person tokens, e.g. [Person1], in the vocabulary and create additional word embedding for these tokens. Then, we sum the visual representation for a person with the word embedding of the token referencing the person in text. This way, our model has visually grounded information about the image. We refer to this approach as "Person Grounding" (PG) input.

5.2   Text Representation

Transformer models used for language tasks [11,29] use special separator tokens to enable better understanding of the input structure. Since our task involves textual information of different kinds (event, place, and relation), we follow [3,4,48] to include special tokens for our language representation as well. Specifically, we append special token indicating the start and end of image (e.g. s_img, e_img), event, place, and inference fields. To generate inference statements, we use one of the three inference types (*before, intent, after*) as the start token, depending on the desired dimension.

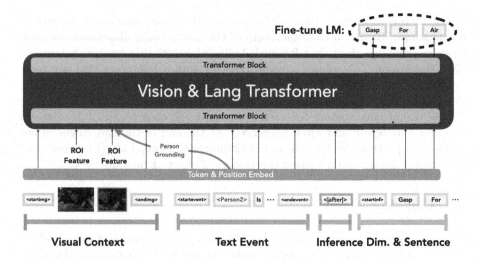

Fig. 5. Model Overview. Vision-Language Transformer for our approach. Our sequence of inputs uses special tokens indicating the start and end of image, event, place, and inference. We only show the start token in the figure for simplicity.

5.3   Single Stream Vision-Language Transformer

We fix the model architecture as GPT-2 [29], a strong Transformer model [38] for natural language generation, conditioned on $\mathcal{V}, e, p$. Our model is a single stream transformer that encodes visual and language representations with a single transformer model, which has been shown to be more effective in vision and language tasks [8,49] compared to designing separate transformer models for each modality [21].

For each inference $s_h^r \in H$, our objective is to maximize $P(s_h^r|v, e, p, r)$. Suppose $s_h^r = \{w_{h1}^r, w_{h2}^r, ...w_{hl}^r\}$ is a sequence of l tokens. Then, we minimize the negative log-likelihood loss over inference instances in dataset (Fig. 5):

$$\mathcal{L} = -\sum_{i=1}^{l} \log P(w_{hi}^r|w_{h<i}^r, r, e, p, v) \tag{1}$$

While our dataset provides events associated with each image, it is impractical to assume the availability of this information on new images. We experiment with a more general version of our model which does not take e and p as input. Nonetheless, we can supervise such models to generate e and p in the training phase. If we denote the *event at present* $\{e\} = \{w_1^e, w_2^e, ...w_n^e\}$ and place $\{p\} = \{w_1^p, w_2^p, ...w_m^p\}$ as a sequence of tokens, we apply the seq2seq loss on e, p

(EP Loss in Sect. 6) as follows:

$$\mathcal{L} = -\sum_{i=1}^{n} \log P(w_i^e | w_{<i}^e, v)) - \sum_{i=1}^{m} \log P(w_i^p | w_{<i}^p, e, v))$$
$$-\sum_{i=1}^{l} \log P(w_{hi}^r | w_{h<i}^r, r, e, p, v) \tag{2}$$

6   Experiments and Results

6.1   Implementation Details

We use Adam optimizer [18] with a learning rate of 5e-5 and batch size of 64. Visual features for image and person embeddings use ResNet101 [14] backbone pretrained on ImageNet [10]. We set the maximum number of visual features to 15. We use pre-trained GPT2-base model [29] as our model architecture with maximum total sequence length as 256. For decoding, we use nucleus sampling [15] with $p = 0.9$, which has shown to be effective generating text that is diverse and coherent. We have found beam search, which is a popular decoding scheme for generating multiple candidates, to be repetitive and produce uninteresting inferences. We report the effect of different decoding schemes in the supplementary material.

6.2   Experimental Setup

Baselines Based on Different Inputs
In our experiments, we fix the same model architecture but ablate on the inputs available, e.g. place, event, and image. We also measure the effect of Person Grounding (PG) trick stated in Sect. 5.1. The models are trained with the same seq2seq objective in Eq. 1, and we mask out the visual and/or textual input based on the ablation of interest. We additionally experiment if learning to generate the event at present and place can improve the performance of generating the inferences using the objective in Eq. 2. For simplicity, we denote the loss on the two textual input as [+ EP. Loss]. Thus, we test two settings when generating the inferences: 1) one that uses event, place, and image, and 2) one that uses only image. We mark the two options in the Text Given column.[2]

Automatic Evaluation
Here, we describe the automatic evaluation measuring the quality of inference sentences. We first report the automatic metrics used in image captioning [7], such as BLEU-2 [26], METEOR [20], and CIDER [40] across the 5 inferences. Inspired by the metric in visual dialog [9], we also use perplexity score to rank

---

[2] We have tried running inferences on predicted events at present, but have gotten worse results than using no events. We report the results on predicted events in the supplemental.

Table 2. Ablation Results. Ablations of our baseline model on the Validation set. We use nucleus sampling with $p = 0.9$ to generate 5 sentences for all models. Automatic metrics used are BLEU-2 (B-2) [26], METEOR (M) [20], and CIDER (C) [40]. Acc@50 is the accuracy of correctly retrieved inference sentences with 50 candidates to choose from. Unique is the number of inference sentences that are unique within the generated sentences, divided by the total number of sentences. Novel refers to the number of generated sentences that are not in the training data, divided by the total number of sentences. Text Given is when model is given any textual input during test time to generate the inferences. We bold the models based on the following order: 1) Best Text only model, 2) Best Image + Text model given visual and text input, 3) Best Image only model, and 4) Best Image + Text model given just visual input.

Modalities	Text given	B-2	M	C	Acc@50	Unique	Novel
Place	Yes	6.26	6.25	4.59	14.55	8.08	47.88
Event	Yes	10.37	9.58	14.48	31.95	44.12	47.37
Event + Place	Yes	10.75	9.82	15.42	33.06	46.22	47.90
Image + Place	Yes	7.40	7.33	6.55	20.39	31.36	46.90
Image + Event	Yes	11.47	10.73	16.14	37.00	49.31	47.39
Image + Event + Place	Yes	11.74	10.87	16.79	38.25	50.39	48.37
Image + Event + Place + EP Loss	Yes	10.4	10.02	13.32	33.23	49.00	51.55
Image + Event + Place + PG	Yes	**12.33**	**11.55**	**17.94**	**38.72**	**50.57**	49.24
Image + Event + Place + PG + EP Loss	Yes	11.12	10.74	14.70	34.07	47.9	**52.02**
No Input	No	4.96	5.23	0.02	6.87	0.00	33.33
Image	No	6.97	7.13	5.50	18.22	29.88	47.16
Image + PG	No	8.09	8.44	7.43	21.5	34.25	45.83
Image + Event + Place	No	7.03	7.55	5.85	16.81	31.09	45.49
Image + Event + Place + EP Loss	No	7.12	7.77	6.22	20.02	39.36	50.67
Image + Event + Place + PG	No	8.58	9.19	8.57	17.35	33.56	47.75
Image + Event + Place + PG + EP Loss	No	**9.71**	**10.66**	**11.60**	**22.7**	**44.20**	**50.02**
GT	–	–	–	–	–	81.67	56.05

the ground truth inferences and inferences from the different image. We append negatives such that there are 50 candidates to choose from, rank each candidate using perplexity score, and get the average accuracy of retrieved ground truth inferences (Acc@50 in Table 2). Note that perplexity is not necessarily the perfect measure to rank the sentences, but good language models should still be able to filter out inferences that do not match the content in image and event at present. Lastly, we measure the diversity of sentences, so that we do not reward the model for being conservative and making the same predictions. We report the number of inference sentences that are unique within the generated sentences divided by the total number of sentences (Unique in Table 2), and the number of generated sentences that are not in the training data divided by the total number of sentences (Novel in Table 2). To capture the semantic diversity, we replace the predicted person tags with the same tag when calculating the above diversity scores.

Table 3. Generated Inference Results. BLEU-2 (B-2) [26], METEOR (M) [20], CIDER (C), [40] and Human scores for the generated inferences on the Test split. We select 200 random images and generate 5 sentences for each of the three inference type (3000 sentences total). Then, we assign three annotators to determine if each inference sentence is correct, and take the majority vote. The models are chosen based on their best performance on the validation set when visual and/or textual modalities are available (bolded models in Table 2).

Modalities	B-2	M	C	Human before	Human intent	Human after	Human avg
With text input							
Event + Place	10.49	9.65	14.83	54.9	52.6	42.9	50.1
Image + Event + Place + PG	**11.76**	**11.13**	**17.05**	**63.36**	**63.5**	**56.0**	**61.0**
Without text input							
No Input	4.88	5.20	1.77	5.3	4.9	3.5	4.6
Image + PG	7.84	8.17	7.14	38.2	34.8	30.3	34.4
Image + Event + Place + PG + EP Loss	**9.07**	**10.12**	**10.59**	**42.9**	**36.8**	**34.8**	**38.2**
GT	–	–	–	83.8	84.5	76	81.4

6.3 Results

Table 2 shows our experimental results testing multiple training schemes and input modalities. We make the following observations: 1) Adding PG trick gives a boost for model over all metrics. 2) Model trained with both visual and textual (Image + Event + Place + PG) modalities outperform models trained with only one of modality (Event + Place; Image + PG) in every metric, including retrieval accuracy and diversity scores. This indicates that the task needs visual information to get higher quality inferences. 3) Adding place information helps in general. 4) Models with access to textual event and place information during test time, generate higher quality sentences than the same models without them (Text Given Yes vs No). This is not surprising as our dataset was collected with workers looking at the event, and the event already gives a strong signal understanding the content in the image. 5) Lastly, adding the EP Loss boosts the performance if only the image content is available in the test time. This indicates that training the model to recognize events at present helps the performance, when the model has to generate inferences directly from image.

Human Evaluation

While the numbers in automatic evaluation give favorable results to our Image + Text model, they are not sufficient enough to evaluate the quality of generated inferences. We choose the best performing model when only image, text, or both inputs are available (model trained with no input and bolded models in Table 2). We take 200 random images and the generated inferences, and ask the humans to evaluate their quality based on just the image content. Even for models that use ground truth inferences, we do not show the events to the workers and make them rely on image to make the decision. Specifically, we ask three different workers to evaluate if each inference is likely (1) or unlikely (0) to happen based on the image. We then take the majority out of three and calculate the average across all the inferences.

Table 3 shows automatic metrics and human evaluation scores on the test split. We notice a similar pattern based on our automatic metric results: Image + Text model outperforms the Text only model (61.0 vs 50.1 on average) when text input is given in test time, and Image + Text model outperforms Image only model when text input is not given (38.2 vs 34.4 on average). We see that Text only model performs better than the Image + Text model without text input in test time, as the event sentence already describes the relevant details in the image and is a strong signal itself. Note that there is still a 20 point gap between our best model and ground truth inferences, meaning there is more room to improve our best model.

6.4   Qualitative Examples

Figure 6 presents some qualitative examples comparing the outputs of the various systems with the human annotated ground truth inferences. Overall, models that integrate information from both the visual and textual modalities generate more consistent and better contextualized predictions than models that only use either visual or textual information.

Specifically, the first example (on the top) illustrates that in the absence of the event description, a model that solely relies on visual information generates incorrect predictions like "order a drink at a bar", "dance and have fun" etc. – none of which are reasonable in the context of the event description. Similarly, a model that solely relies on the textual description, but not the visual information, generates "get off of the stage" and even predicts "her job as a scientist". This inference could be true in the absence of the visual features, but the image clearly shows that the person is in the audience, and not the one giving a presentation, nor she is portrayed as a scientist.

This pattern continues in the bottom example. [Person2] clearly looks worried but the Text only model predicts that he wants to "alleviate his boredom", and does not incorporate this visual detail. Image only model again hallucinates wrong objects like "have grabbed the wire". On the other hand, Image + Text model has the appropriate balance between the two models by stating there is possibly a criminal nearby as Person 2 is making an urgent call, and still predicts relevant visual details in the image. Thus, we see that both visual and textual features contribute to generating coherent inferences.

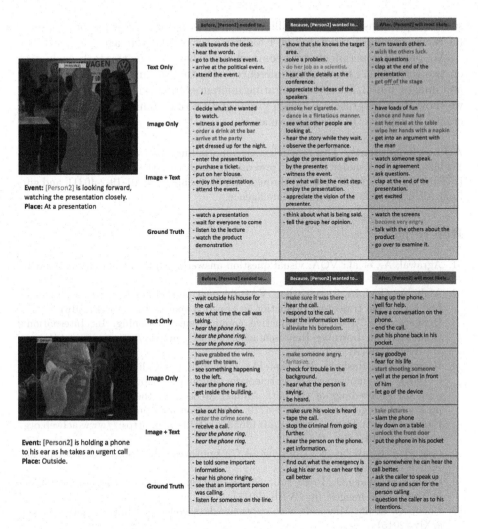

Event: [Person2] is looking forward, watching the presentation closely.
Place: At a presentation

Event: [Person2] is holding a phone to his ear as he takes an urgent call
Place: Outside.

	Before, [Person2] needed to...	Because, [Person2] wanted to...	After, [Person2] will most likely...
Text Only	- walk towards the desk. - hear the words. - go to the business event. - arrive at the political event. - attend the event.	- show that she knows the target area. - solve a problem. - do her job as a scientist. - hear all the details at the conference. - appreciate the ideas of the speakers	- turn towards others. - wish the others luck. - ask questions - clap at the end of the presentation - get off of the stage
Image Only	- decide what she wanted to watch. - witness a good performer - order a drink at the bar - arrive at the party - get dressed up for the night.	- smoke her cigarette. - dance in a flirtatious manner. - see what other people are looking at. - hear the story while they wait. - observe the performance.	- have loads of fun - dance and have fun - eat her meal at the table - wipe her hands with a napkin - get into an argument with the man
Image + Text	- enter the presentation. - purchase a ticket. - put on her blouse. - enjoy the presentation. - attend the event.	- judge the presentation given by the presenter. - witness the event. - see what will be the next step. - enjoy the presentation. - appreciate the vision of the presenter.	- watch someone speak. - nod in agreement - ask questions. - clap at the end of the presentation. - get excited
Ground Truth	- watch a presentation - wait for everyone to come - listen to the lecture - watch the product demonstration	- think about what is being said. - tell the group her opinion.	- watch the screens - become very angry - talk with the others about the product - go over to examine it.

	Before, [Person2] needed to...	Because, [Person2] wanted to...	After, [Person2] will most likely...
Text Only	- wait outside his house for the call. - see what time the call was taking. - hear the phone ring. - hear the phone ring. - hear the phone ring.	- make sure it was there - hear the call. - respond to the call. - hear the information better. - alleviate his boredom.	- hang up the phone. - yell for help. - have a conversation on the phone. - end the call. - put his phone back in his pocket
Image Only	- have grabbed the wire. - gather the team. - see something happening to the left. - hear the phone ring. - get inside the building.	- make someone angry. - fantasize. - check for trouble in the background. - hear what the person is saying. - be heard.	- say goodbye - fear for his life - start shooting someone - yell at the person in front of him - let go of the device
Image + Text	- take out his phone. - enter the crime scene. - receive a call. - hear the phone ring. - hear the phone ring.	- make sure his voice is heard - tape the call. - stop the criminal from going further. - hear the person on the phone. - get information.	- take pictures - slam the phone - lay down on a table - unlock the front door - put the phone in his pocket
Ground Truth	- be told some important information. - hear his phone ringing. - see that an important person was calling. - listen for someone on the line.	- find out what the emergency is - plug his ear so he can hear the call better	- go somewhere he can hear the call better. - ask the caller to speak up - stand up and scan for the person calling - question the caller as to his intentions.

Fig. 6. Qualitative Results. Qualitative Examples comparing our best Text only, Image only, and Image + Text only model. Red highlights inference statements that are incorrect. Orange highlights if the sentences are plausible, but not expected. We see that our Image + Text model gives more consistent and contextualized predictions than the baseline models. (Color figure online)

7   Conclusion

We present **VisualCOMET**, a novel framework of visual commonsense reasoning tasks to predict events that might have happened *before*, events that that might happen *after*, and the intents of people at *present*. To support research in this direction, we introduce the first large-scale dataset of Visual Commonsense Graphs consisting of 1.4 million textual descriptions of visual commonsense inferences carefully annotated over a diverse set of 59,000 images.

We present experiments with comprehensive baselines on this task, evaluating on two settings: 1) Generating inferences with textual input (event and place) and images, and 2) Directly generating inferences from images. For both setups, we show that integration between visual and textual commonsense reasoning is crucial to achieve the best performance.

Acknowledgements. This research was supported in part by NSF (IIS1524371, IIS-1714566), DARPA under the CwC program through the ARO (W911NF-15-1-0543), DARPA under the MCS program through NIWC Pacific (N66001-19-2-4031), and gifts from Allen Institute for Artificial Intelligence.

References

1. Agrawal, A., et al.: VQA: visual question answering. Int. J. Comput. Vision **123**, 4–31 (2015)
2. Alahi, A., Goel, K., Ramanathan, V., Robicquet, A., Fei-Fei, L., Savarese, S.: Social LSTM: human trajectory prediction in crowded spaces. In: CVPR (2016)
3. Bhagavatula, C., et al.: Abductive commonsense reasoning. In: International Conference on Learning Representations (2020). https://openreview.net/forum?id=Byg1v1HKDB
4. Bosselut, A., Rashkin, H., Sap, M., Malaviya, C., Celikyilmaz, A., Choi, Y.: COMET: commonsense transformers for automatic knowledge graph construction. In: Proceedings of the 57th Annual Meeting of the Association for Computational Linguistics, pp. 4762–4779. Association for Computational Linguistics, Florence (2019). https://doi.org/10.18653/v1/P19-1470. https://www.aclweb.org/anthology/P19-1470
5. Castrejón, L., Ballas, N., Courville, A.C.: Improved VRNNs for video prediction. In: ICCV (2019)
6. Chao, Y.W., Yang, J., Price, B.L., Cohen, S., Deng, J.: Forecasting human dynamics from static images. In: CVPR (2017)
7. Chen, X., et al.: Microsoft coco captions: data collection and evaluation server. arXiv (2015)
8. Chen, Y.C., et al.: Uniter: learning universal image-text representations. arXiv (2019)
9. Das, A., et al.: Visual dialog. In: CVPR (2017)
10. Deng, J., Dong, W., Socher, R., Li, L.J., Li, K., Fei-Fei, L.: Imagenet: a large-scale hierarchical image database. In: Proceedings of the IEEE Conference on Computer Vision and Pattern Recognition (CVPR) (2009)
11. Devlin, J., Chang, M.W., Lee, K., Toutanova, K.: Bert: pre-training of deep bidirectional transformers for language understanding. arXiv (2018)
12. Fragkiadaki, K., Levine, S., Felsen, P., Malik, J.: Recurrent network models for human dynamics. In: ICCV (2015)
13. He, K., Gkioxari, G., Dollar, P., Girshick, R.: Mask R-CNN. In: Proceedings of the IEEE International Conference on Computer Vision (ICCV) (2017)
14. He, K., Zhang, X., Ren, S., Sun, J.: Deep residual learning for image recognition. In: Proceedings of the IEEE Conference on Computer Vision and Pattern Recognition (CVPR) (2016)

15. Holtzman, A., Buys, J., Forbes, M., Choi, Y.: The curious case of neural text degeneration. arXiv (2019)
16. Johnson, J.E., Hariharan, B., van der Maaten, L., Fei-Fei, L., Zitnick, C.L., Girshick, R.B.: Clevr: a diagnostic dataset for compositional language and elementary visual reasoning. In: CVPR (2017)
17. Kazemzadeh, S., Ordonez, V., Matten, M., Berg, T.: ReferItGame: referring to objects in photographs of natural scenes. In: EMNLP (2014)
18. Kingma, D.P., Ba, J.: Adam: a method for stochastic optimization. arXiv (2014)
19. Lan, T., Chen, T.-C., Savarese, S.: A hierarchical representation for future action prediction. In: Fleet, D., Pajdla, T., Schiele, B., Tuytelaars, T. (eds.) ECCV 2014. LNCS, vol. 8691, pp. 689–704. Springer, Cham (2014). https://doi.org/10.1007/978-3-319-10578-9_45
20. Lavie, M.D.A.: Meteor universal: language specific translation evaluation for any target language. In: Proceedings of the Annual Meeting of the Association for Computational Linguistics (ACL) (2014)
21. Lu, J., Batra, D., Parikh, D., Lee, S.: Vilbert: pretraining task-agnostic visiolinguistic representations for vision-and-language tasks. In: NeurIPS (2019)
22. Mao, J., Huang, J., Toshev, A., Camburu, O., Murphy, K.: Generation and comprehension of unambiguous object descriptions. In: CVPR (2016)
23. Marino, K., Rastegari, M., Farhadi, A., Mottaghi, R.: OK-VQA: a visual question answering benchmark requiring external knowledge. In: CVPR (2019)
24. Mathieu, M., Couprie, C., LeCun, Y.: Deep multi-scale video prediction beyond mean square error. In: Bengio, Y., LeCun, Y. (eds.) ICLR (2016)
25. Mottaghi, R., Rastegari, M., Gupta, A., Farhadi, A.: "What happens if..." learning to predict the effect of forces in images. In: Leibe, B., Matas, J., Sebe, N., Welling, M. (eds.) ECCV 2016. LNCS, vol. 9908, pp. 269–285. Springer, Cham (2016). https://doi.org/10.1007/978-3-319-46493-0_17
26. Papineni, K., Roukos, S., Ward, T., Zhu, W.J: BLEU: a method for automatic evaluation of machine translation. In: Proceedings of the Annual Meeting of the Association for Computational Linguistics (ACL) (2002)
27. Pirsiavash, H., Vondrick, C., Torralba, A.: Inferring the why in images. arXiv (2014)
28. Plummer, B.A., Wang, L., Cervantes, C.M., Caicedo, J.C., Hockenmaier, J., Lazebnik, S.: Flickr30k entities: collecting region-to-phrase correspondences for richer image-to-sentence models. In: IJCV (2015)
29. Radford, A., Wu, J., Child, R., Luan, D., Amodei, D., Sutskever, I.: Language models are unsupervised multitask learners. OpenAI Blog **1**(8) (2019)
30. Ranzato, M., Szlam, A., Bruna, J., Mathieu, M., Collobert, R., Chopra, S.: Video (language) modeling: a baseline for generative models of natural videos. arXiv (2014)
31. Ren, S., He, K., Girshick, R., Sun, J.: Faster R-CNN: towards real-time object detection with region proposal networks. In: Advances in Neural Information Processing Systems (NIPS) (2015)
32. Sap, M., et al.: Atomic: an atlas of machine commonsense for if-then reasoning. In: Proceedings of the Conference on Artificial Intelligence (AAAI) (2019)
33. Sharma, P., Ding, N., Goodman, S., Soricut, R.: Conceptual captions: a cleaned, hypernymed, image alt-text dataset for automatic image captioning. In: Proceedings of the Annual Meeting of the Association for Computational Linguistics (ACL) (2018)
34. Srivastava, N., Mansimov, E., Salakhudinov, R.: Unsupervised learning of video representations using LSTMS. In: ICML (2015)

35. Su, W., et al.: Vl-bert: pre-training of generic visual-linguistic representations. In: ICLR (2020)
36. Sun, C., Shrivastava, A., Vondrick, C., Sukthankar, R., Murphy, K., Schmid, C.: Relational action forecasting. In: CVPR (2019)
37. Tan, H., Bansal, M.: Lxmert: learning cross-modality encoder representations from transformers. In: EMNLP (2019)
38. Vaswani, A., et al.: Attention is all you need. In: Advances in Neural Information Processing Systems (NIPS) (2017)
39. Vedantam, R., Lin, X., Batra, T., Zitnick, C.L., Parikh, D.: Learning common sense through visual abstraction. In: ICCV (2015)
40. Vedantam, R., Zitnick, C.L., Parikh, D.: Cider: consensus-based image description evaluation. In: Proceedings of the IEEE Conference on Computer Vision and Pattern Recognition (CVPR) (2015)
41. Villegas, R., Pathak, A., Kannan, H., Erhan, D., Le, Q.V., Lee, H.: High fidelity video prediction with large stochastic recurrent neural networks. In: NeurIPS (2019)
42. Vinyals, O., Toshev, A., Bengio, S., Erhan, D.: Show and tell: a neural image caption generator. In: CVPR (2015)
43. Vondrick, C., Pirsiavash, H., Torralba, A.: Generating videos with scene dynamics. In: NeurIPS (2016)
44. Walker, J., Gupta, A., Hebert, M.: Patch to the future: unsupervised visual prediction. In: CVPR (2014)
45. Walker, J., Marino, K., Gupta, A., Hebert, M.: The pose knows: video forecasting by generating pose futures. In: ICCV (2017)
46. Xue, T., Wu, J., Bouman, K., Freeman, B.: Visual dynamics: probabilistic future frame synthesis via cross convolutional networks. In: NeurIPS (2016)
47. Zellers, R., Bisk, Y., Farhadi, A., Choi, Y.: From recognition to cognition: visual commonsense reasoning. In: CVPR (2019)
48. Zellers, R., et al.: Defending against neural fake news. In: Advances in Neural Information Processing Systems (NIPS) (2019)
49. Zhou, L., Hamid, P., Zhang, L., Hu, H., Corso, J., Gao, J.: Unified vision-language pre-training for image captioning and question answering. In: Proceedings of the Conference on Artificial Intelligence (AAAI) (2020)
50. Zhou, Y., Berg, T.L.: Temporal perception and prediction in ego-centric video. In: ICCV (2015)

Few-Shot Action Recognition
with Permutation-Invariant Attention

Hongguang Zhang[1,2,3,5(✉)], Li Zhang[2], Xiaojuan Qi[2,4], Hongdong Li[1,5],
Philip H. S. Torr[2], and Piotr Koniusz[1,3]

[1] Australian National University, Canberra, Australia
hongguang.zhang@anu.edu.au
[2] University of Oxford, Oxford, UK
[3] Data61/CSIRO, Canberra, Australia
[4] The University of Hong Kong, Pokfulam, Hong Kong, China
[5] Australian Centre for Robotic Vision, Brisbane City, Australia

Abstract. Many few-shot learning models focus on recognising images. In contrast, we tackle a challenging task of few-shot action recognition from videos. We build on a C3D encoder for spatio-temporal video blocks to capture short-range action patterns. Such encoded blocks are aggregated by permutation-invariant pooling to make our approach robust to varying action lengths and long-range temporal dependencies whose patterns are unlikely to repeat even in clips of the same class. Subsequently, the pooled representations are combined into simple relation descriptors which encode so-called query and support clips. Finally, relation descriptors are fed to the comparator with the goal of similarity learning between query and support clips. Importantly, to re-weight block contributions during pooling, we exploit spatial and temporal attention modules and self-supervision. In naturalistic clips (of the same class) there exists a temporal distribution shift–the locations of discriminative temporal action hotspots vary. Thus, we permute blocks of a clip and align the resulting attention regions with similarly permuted attention regions of non-permuted clip to train the attention mechanism invariant to block (and thus long-term hotspot) permutations. Our method outperforms the state of the art on the HMDB51, UCF101, *mini*MIT datasets.

1 Introduction

Few-shot learning is an open problem with the goal to design algorithms that learn in the low-sample regime. Examples include meta-learning [1,11,32,41,42], robust feature representations by relation learning [44,47,51,54], gradient-based [39,43,59] and hallucination strategies for insufficient data [18,55].

Electronic supplementary material The online version of this chapter (https://doi.org/10.1007/978-3-030-58558-7_31) contains supplementary material, which is available to authorized users.

A. Vedaldi et al. (Eds.): ECCV 2020, LNCS 12350, pp. 525–542, 2020.
https://doi.org/10.1007/978-3-030-58558-7_31

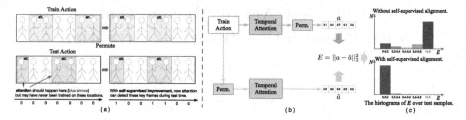

Fig. 1. Augmentation-guided attention by alignment. Figure 1a shows that discriminative action blocks (in red, top left) may be misaligned with discriminative action blocks of test clip (in blue, bottom left). If the attention unit observes different distributions of locations of discriminative blocks at training and testing time, it fails. With the right approach, one may overcome the distribution shift (top and bottom right panels). Figure 1b shows how the augmentation-guided attention by alignment works for permutation-based augmentations: (i) we shuffle training blocks of a clip to train an attention on permuted blocks, (ii) we shuffle in the same way coefficients of the attention vector from the non-permuted blocks. Both attention vectors are then encouraged to align by a dedicated loss term during training. Figure 1c shows histograms of alignment-errors on test data. The top histogram shows larger errors (no alignment loss used in training) while the bottom histogram shows small errors (alignment during testing improves). (Color figure online)

However, very few papers address video-based few-shot learning. As annotating large video datasets is prohibitive, this makes the problem we study particularly valuable. While results are far from satisfactory on *Kinetics* [3], the largest action recognition dataset, its size of 300,000 video clips with hundreds of frames each exceeds the size of large-scale image ImageNet [38] and Places205 [56] datasets.

There exist few limited works on few-shot learning for action recognition [17,35,52,58]. However, they focus on modeling 3D body joints with graphs [17], attribute-based learning in generative models [35], network design for low-sample classification [52] and salient memory approach [58]. In contrast, we focus on robust relation/similarity, spatial and temporal modeling of short- and long-term range action patterns via permutation-invariant pooling and attention.

To obtain a robust few-shot action recognition approach, we investigate how to: (i) represent discriminative short- and non-repetitive long-term action patterns for relation/similarity learning, (ii) localize temporally discriminative action blocks with limited number of training samples, and (iii) deal with long-term temporal distribution shift of such discriminative patterns (these patterns never re-appear at the same temporal locations even for clips of the same class).

To address the first point, our early experiments indicated that short-term discriminative action patterns can be captured by an encoder with C3D convolutional blocks. Thus, resulting features from a clip undergo permutation-invariant pooling which discards long-term non-repetitive dependencies. Finally, pooled query/support representations form relation descriptors are fed into a comparator.

Regarding the second point, aggregating spatio-temporal blocks with equal weights is suboptimal. Thus, we devise spatial/temporal attention units to emphasize discriminative blocks. Under the low-sample regime, self-supervision by *jigsaw* and *rotation*[1] helps train a more robust encoder, comparator and attention. However, vanilla attention (and/or self-supervision) cannot fully promote the invariance to temporal (or spatial) permutations as described next.

To address the third point, we note that long-term dependencies in clips are non-repetitive *e.g.*, videos of the same class often contain relevant action blocks at different temporal locations. Figure 1a shows that discriminative blocks of training and testing clips of *dance* class do not align (top left *vs.* bottom left corner). By permuting the blocks of training (top right), one can make them align with the most discriminative test samples (bottom right). Figure 1b shows that for a given clip, we (i) shuffle its blocks and feed them to the attention mechanism (shuffling pass), (ii) we shuffle accordingly the attention coefficients from a non-shuffled pass through attention, and (iii) we force attention coefficients from both passes to align. Such an attention by alignment deals with the distribution shift of discriminative temporal (and spatial) patterns via *jigsaw* augmentation (but applies also to *rotation, zoom, etc.*) To summarize, our contributions include:

i. A robust pipeline with a C3D-based encoder capturing short-term dependencies which yields block representations subsequently aggregated by permutation-invariant pooling into fixed-length representations which form relation descriptors for relational/similarity learning in an episodic setting [47].

ii. Spatial and temporal attention units which re-weight block contributions during the aggregation step. To improve training of the encoder, comparator and the attention unit under the low-sample regime, we introduce spatial and temporal self-supervision by *rotations*, and *spatial* and *temporal jigsaws*.

iii. An improved self-supervised attention unit by applying augmentation patterns such as *jigsaws* and/or *rotations* on the input of the attention unit and aligning the output with augmented the same way attention vector coefficients from non-augmented data passed by the attention unit (See footnote 1). Thus, the attention unit becomes invariant to a given augmentation action by design.

iv. We propose new data splits for a systematic comparison of few-shot action recognition algorithms and we make them available as existing approaches use each different pipeline concepts, data modality, data splits and protocols[2].

---

[1] Self-supervision assumes generating cheap-to-obtain data (*e.g.*, augmentation by rotations) from the original data and imposing an auxiliary task whose goal is to predict the label of an augmentation with the goal of robust representation learning [4,5,16,19,20]. We are first to apply the self-supervision by alignment paradigm to the problem of robust attention training (we devise an augmentation-guided attention).

[2] Section 2.1 explains that existing works do not specify class/validation splits which yields ∼6% variations in accuracy rendering their protocols highly inaccurate. Section 2.2 explains this issue and how we compare our method to existing works.

2 Related Work

Below, we discuss zero-, one- and few-shot learning models followed by a discussion on self-supervised learning and second-order pooling.

One- and Few-shot learning models have been widely studied in both the shallow [2,8,10,33,34] and deep learning pipelines [11,22,44,44,47,49]. Motivated by the human ability to learn new concepts from few samples, early works [8,31] employ generative models with an iterative inference. Siamese Network [22] is a two-stream convolutional neural network which generates image descriptors and learns the similarity between them. Matching Network [49] proposes query-support episodic learning and L-way Z-shot learning protocols[3]. The similarity between a query and support images is learnt for each episode. At the testing time, each test query (of novel class) is compared against a handful of annotated test support images for rapid recognition. Prototypical Networks [44] compute distances between a datapoint and class-wise prototypes. Model-Agnostic Meta-Learning (MAML) [11] is trained on multiple learning tasks. Relation Net [47] learns relations between query and support images, and it leverages a similarity learning neural network to compare query-support pairs. SalNet [55] uses saliency-guided end-to-end sample hallucination to grow the training set. Graph Neural Networks (GNN) have also been used in few-shot learning [13,15,21].

Self-supervised learning leverages free supervision signals residing in images and videos to promote robust representation learning in image recognition [4,5,16], video recognition [9,12,40], video object segmentation [30,57] and few-shot image classification [14,46]. Approaches [4,5,16] learn to predict random image rotations, relative pixel positions, and surrogate classes, respectively. Finally, [14,46] improve few-shot results by predicting image rotations/jigsaw patterns.

Second-order statistics are used by us for permutation-invariant pooling. They are also used for texture recognition [37,48] by so-called Region Covariance Descriptors (RCD), object and action recognition [25,26,50]. Second-order pooling has also been used in fine-grained image classification [7,28,50], domain adaptation [24] and the fine-grained few-shot learning [27,51,54].

Few-shot action recognition approaches [17,35,52] use a generative model, graph matching on 3D coordinates and a dilated networks with class-wise classifiers, respectively. Approach [58] proposes a so-called compound memory network using key-value memory associations. ProtoGAN [6] proposes a GAN model to generate action prototypes to address few-shot action recognition.

2.1 Contrast with Existing Works

Unlike [17], we use video clips rather than 3D skeletal coordinates. In contrast to [52], we use relation/similarity learning and our training/testing class concepts

---

[3] Kindly see [44,47,49] for the concept of query, support and episodic learning, and the evaluation protocols which differ from traditional recognition and low-shot learning.

are disjoint. While [58] memorizes key values/frames, we model short- and long-term dependencies. While [6] forms action prototypes by GAN, we focus on self-supervised attention learning and permutation-invariant aggregation.

In contrast to self-supervision by rotations and jigsaw [14,46], we use a sophisticated self-supervision on the attention unit for which a dedicated loss performs the alignment between the attention vector of augmented attention unit and the augmented in the same way attention vector from the non-augmented attention unit. Thus, we train a permutation-invariant attention to deal with the distribution shift of discriminative action locations.

Finally, we use second-order pooling [28,54] for a permutation-invariant aggregation of temporal blocks while [28,54] work with images. We develop a theory explaining why Power Normalization helps episodic learning.

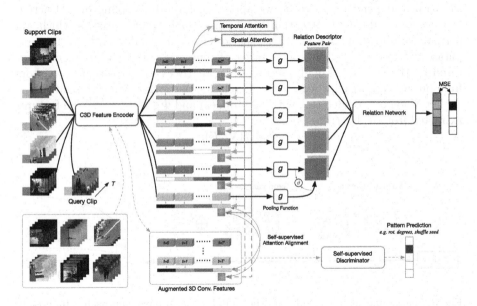

Fig. 2. Our few-shot Action Relation Network (ARN) contains: feature encoder with 4-layer 3D conv. blocks, relation network with 2D conv. blocks, and spatial and temporal attention units which refine the aggregation step. Specifically, we apply second-order pooling (operator g) over encoder outputs (re-weighted by attention vectors) per clip to obtain a Power Normalized Autocorrelation Matrix (AM). Query and support AMs per episode form relation descriptors (by operator ϑ) from which the relation network learns to capture relations. The block (blue dashed line) is the self-supervised learning module which encourages our pipeline to learn auxiliary tasks *e.g.*, *jigsaws*, *rotations*. (Color figure online)

2.2 Issues with Fair Comparisons

Each few-shot action recognition method from Sect. 2 uses different datasets and evaluation protocols making fair comparisons impossible. Class-wise splits/

validation sets are unavailable *ie.*, model [35] uses a random split. Figures 5c and 5d of Sect. 4 show that the random choice of the split set yields up to ~6% deviation in accuracy rendering such a protocol problematic. Thus, we propose a new protocol with class splits and validation sets made publicly available.

3   Approach

3.1   Pipeline

Figure 2 shows our Action Relation Network (ARN). In contrast to the *Conv-4-64* backbone in few-shot image classification [47,54,55], we adopt a C3D-based *Conv-4-64* backbone to extract spatio-temporal features which capture short-range dependencies. Next, we apply second-order pooling on 3D action features re-weighted by attention to obtain second-order statistics which are permutation-invariant [28] w.r.t. the spatio-temporal order of features. To paraphrase, we discard the long-range order of temporal (and spatial) blocks captured by the encoder. Finally, second-order matrices form relation descriptors from query/support clips fed into a 2D relation network to capture relations.

Let $\mathbf{V}$ denote a video (*ie.*, with ~20 frames) and $\boldsymbol{\Phi} \in \mathbb{R}^{C \times T \times H \times W}$ be features extracted from $\mathbf{V}$ by f:

$$\boldsymbol{\Phi} = f(\mathbf{V}; \mathcal{F}). \tag{1}$$

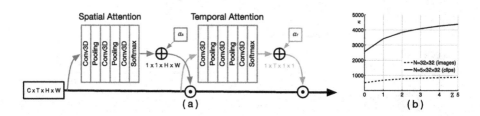

Fig. 3. Spatial and temporal units are shown in Fig. 3a. A naive approach is to directly extract the temporal and spatial attention, whose size is $1 \times T \times H \times W$. However, this is computationally expensive and results in overparametrization. Thus, we split the attention block into separate spatial and temporal branches whose impact is adjusted by α_s and α_t. Figure 3b is the κ ratio w.r.t. the Z-shot value (see Eq. (6)). The dashed curve shows that as Z grows (0 denotes the regular classification), the memorization burden of co-occurrence (i, j) on the comparator grows κ times for second-order pooling without Power Normalization (as opposed to Power Normalization). The solid line shows that as we use larger N (video clips *vs.* images), not using PN is even more detrimental.

To aggregate $\boldsymbol{\Phi}$ per clip into $\boldsymbol{\Psi}$, we apply a pooling operator g over the support and query features, resp. For g, we use pooling operators from Sect. 3.2:

$$\boldsymbol{\Psi} = g(\boldsymbol{\Phi}). \tag{2}$$

Once $\boldsymbol{\Psi}$ are computed for query/support clips, they form relation descriptors (via operator ϑ) passed to the relation network r to obtain the relation score ζ_{sq}:

$$\zeta_{sq} = r(\vartheta(\boldsymbol{\Psi}_s, \boldsymbol{\Psi}_q); \mathcal{R}), \tag{3}$$

where $\mathcal{R}$ are parameters of network r, and ϑ forms relation descriptors *e.g.*, we use the concatenation along the channel mode.

We use the Mean Square Error (MSE) loss over support and query pairs:

$$L = \sum_{s \in S} \sum_{q \in Q} (\zeta_{sq} - \delta(l_s - l_q))^2, \tag{4}$$

where $\delta(l_s - l_q) = 1$ if $l_s = l_q$, $\delta(l_s - l_q) = 0$ otherwise. Class labels of support and query action clips are denoted as l_s and l_q.

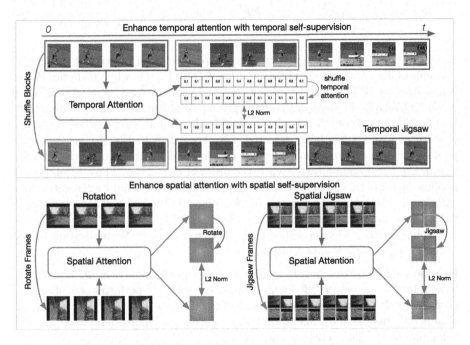

Fig. 4. Augmentation-guided attention by alignment. We firstly collect the encoded representations of original and augmented data, then we extract the temporal or spatial attention vectors from them. We apply the same augmentation(s) on the temporal or spatial attention vectors of the original data resulting in the augmented attention vectors which we align with attention vectors of the augmented data.

3.2 Pooling of Encoded Representations

For permutation-invariant pooling of temporal (and spatial) blocks, we investigate three pooling mechanisms discussed below.

Average and max pooling are two widely-used pooling functions which can be used for aggregation of $N = T \times W \times H$ fibers (channel-wise vectors) of feature map $\boldsymbol{\Phi}$ defined in Eq. (1). The average pooling is given as $\boldsymbol{\psi} = \frac{1}{N} \sum_{n=1}^{N} \boldsymbol{\phi}_n$ where $\boldsymbol{\psi} \in \mathbb{R}^C$, and $\boldsymbol{\phi}_n \in \mathbb{R}^C$ are N fibers. Similarly, max pooling is given by $\psi_c = \max_{n=1,...,N} \phi_{cn}$, $c = 1, ..., C$, and $\boldsymbol{\psi} = [\psi_1, ..., \psi_C]^T$. Average and max pooling are commutative w.r.t. the input fibers, thus being permutation-invariant. However, first-order pooling is less informative than second-order [23] discussed next.

Second-order pooling captures correlations (or co-occurrences) between pairs of features in N fibres of feature map $\boldsymbol{\Phi}$, which is reshaped such that $\boldsymbol{\Phi} \in \mathbb{R}^{C \times N}$, $N = T \times H \times W$. Such an operator proved robust in classification [23] and few-shot learning [27,51,54]. Specifically, we define:

$$\boldsymbol{\Psi} = \eta\Big(\frac{1}{N}\sum_{n=1}^{N}\boldsymbol{\phi}_n\boldsymbol{\phi}_n^T\Big) = \eta\Big(\frac{1}{N}\boldsymbol{\Phi}\boldsymbol{\Phi}^T\Big) \quad \text{where} \quad \eta(\mathbf{X}) = \frac{1 - \exp(\sigma\mathbf{X})}{1 + \exp(\sigma\mathbf{X})}. \quad (5)$$

Matrix $\boldsymbol{\Psi} \in \mathbb{R}^{C \times C}$ is a Power Normalized autocorrelation matrix capturing correlations of fiber features $\boldsymbol{\phi}_n$ of feature map $\boldsymbol{\Phi}$ from Eq. (1) while η applies Power Normalization (PN): we use the zero-centered element-wise Sigmoid [23,54] on $\mathbf{X}$, and σ controls the slope of PN. For a given pair of features i and j in matrix $\boldsymbol{\Psi}$, that is Ψ_{ij}, the role of PN is to detect the likelihood if at least one co-occurrence of features i and j has been detected [23]. According to Eq. (5), second-order pooling is permutation-invariant w.r.t. the order of input fibers as the summation in Eq. (5) is commutative w.r.t. the order of $\boldsymbol{\phi}_1, ..., \boldsymbol{\phi}_N$. Thus, second-order pooling factors out the spatial and temporal modes of $\boldsymbol{\Phi}$ and aggregates clips with varying numbers of temporal blocks (discards the order of long-range spatial/temporal dependencies) into a fixed length representation $\boldsymbol{\Psi} \in \mathbb{R}^{C \times C}$. Below we explain further why second-order pooling with Power Normalization is well suited for episodic few-shot learning.

Relation descriptors between query/support pooled matrices $\boldsymbol{\Psi}_q$ and $\boldsymbol{\Psi}_s$ are formed by operation $\vartheta(\boldsymbol{\Psi}_q, \boldsymbol{\Psi}_s)$ which, in our case, simply performs concatenation of $\boldsymbol{\Psi}_q$ with $\boldsymbol{\Psi}_s$ along the channel mode by $\text{cat}(\boldsymbol{\Psi}_q, \boldsymbol{\Psi}_s) \in \mathbb{R}^{2 \times C \times C}$, and $\boldsymbol{\Psi}_s$ is obtained by the mean (or maximum) along the channel mode between $\boldsymbol{\Psi}_s^1, ..., \boldsymbol{\Psi}_s^Z$ belonging to the same episode and class ($Z > 1$ for the few-shot case).

It is known from [28] that the Power Normalization in Eq. (5) performs a co-occurrence detection rather than counting (correlation). For classification problems, assume a probability mass function $p_{X_{ij}}(x) = 1/(N+1)$ if $x = 0, ..., N$, $p_{X_{ij}}(x) = 0$ otherwise, that tells the probability that co-occurrence between Φ_{in} and Φ_{jn} happened $x = 0, ..., N$ times (given some clip). Note that classification often depends on detecting a co-occurrence (e.g., is there a flower co-occurring with a pot?) rather than counts (e.g., how many flowers and pots co-occur?). Using second-order pooling without PN requires a classifier to observe $N+1$ training samples of *flower and pot* co-occurring in quantities $0, ..., N$ to memorise all possible co-occurrence count configurations. For relation learning, our ϑ stacks pairs of samples to compare, thus a comparator now has to deal with a probability mass function of $R_{ij} = X_{ij} + Y_{ij}$ depicting *flowers and pots* whose support$(p_{R_{ij}}) = 2N+1 > $ support$(p_{X_{ij}}) = N+1$ if random variable $X = Y$ (same class). The same is shown by variances var$(p_{R_{ij}}) > $ var$(p_{X_{ij}})$. For Z-shot learning, the growth of variance and support equal $(Z+1)N+1$ indicates that the comparator has to memorize more configurations of co-occurrence (i, j) as Z grows.

However, this situation is alleviated by Power Normalization (operator η) whose probability mass function can be modeled as $p_{X_{ij}^\eta}(x) = 1/2$ if $x = \{0, 1\}$, $p_{X_{ij}^\eta}(x) = 0$ otherwise, as PN detects a co-occurrence (or its lack). For Z-shot learning, support$(p_{R_{ij}^\eta}) = Z+2 \ll $ support$(p_{R_{ij}}) = (Z+1)N+1$. The ratio given as

$$\kappa = \frac{\text{support}(p_{R_{ij}})}{\text{support}(p_{R_{ij}^\eta})} = \frac{(Z+1)N+1}{Z+2} \tag{6}$$

shows that the comparator has to memorize many more count configurations of co-occurrence (i, j) for naive pooling compared to PN as Z and/or N grow (N depends on the number of temporal and spatial blocks T, H and W). Figure 3 shows how κ varies w.r.t. Z and N. Our modeling assumptions are simple *e.g.*, the assumption on mass functions with uniform probabilities, the use of the support of mass functions rather than variances to describe variability of co-occurrence (i, j). Yet, substituting these modeling choices with more sophisticated ones does not affect theoretical conclusions that: (i) PN benefits few-shot learning ($Z \geq 1$) more than the regular classification ($Z = 0$) in terms of reducing possible count configurations of (i, j), and (ii) for videos (large N) PN reduces the number of count configurations of (i, j) more rapidly than for images (smaller N). While classifiers and comparators do not learn exhaustively all count configurations of co-occurrence (i, j) as they have some generalization ability, they learn quicker and better if the number of count configurations of (i, j) is limited.

3.3 Temporal and Spatial Attention

Figure 3a introduces decoupled spatial/temporal attention units consisting of three 3D Convolutional blocks and a Sigmoid output layer. Let t and s denote the temporal and spatial attention modules, and the attention be applied ahead of second-order pooling. We obtain temporal and spatial attention maps $\mathbf{T} \in \mathbb{R}^{1 \times T \times 1 \times 1}$ and $\mathbf{S} \in \mathbb{R}^{1 \times 1 \times H \times W}$, and attentive action features $\boldsymbol{\Phi}^*$ by:

$$\mathbf{T} = t(\boldsymbol{\Phi}; \boldsymbol{\mathcal{T}}), \quad \mathbf{S} = s(\boldsymbol{\Phi}; \boldsymbol{\mathcal{S}}), \tag{7}$$

$$\boldsymbol{\Phi}^* = (\alpha_t + \mathbf{T}) \cdot (\alpha_s + \mathbf{S}) \cdot \boldsymbol{\Phi}, \tag{8}$$

where $\boldsymbol{\mathcal{T}}$ and $\boldsymbol{\mathcal{S}}$ are network parameters of temporal/spatial attention units while α_t and α_s control the impact of attention vectors.

Using attention helps spot discriminative temporal/spatial blocks, and suppress uninformative regions. However, the attention should be robust to varying distributions of locations of discriminative blocks in clips as proposed below.

3.4 Temporal and Spatial Self-supervision

Self-supervised Learning (SsL) helps learn representations without using manually-labeled annotations. We impose self-supervision both on encoders and attention units. For temporal self-supervision, we augment clips by shuffling the order of temporal blocks, which primes our network to become robust to long-term non-repetitive temporal dependencies in clips. Self-supervision also helps overcome the low-sample by encouraging network to learn auxiliary tasks. In contrast, previous works shuffled frames which breaks the highly discriminative short-term temporal dependencies. We use the following self-supervision strategies:

i. **Temporal jigsaw.** Jigsaw, a popular self-supervisory task breaks the object location bias and teaches the network to recognize shuffling. As in [53], we split clips into non-overlapping fixed-length temporal blocks and shuffle them.
ii. **Spatial jigsaw.** We split frames into four non-overlapping regions, then randomly permute them.
iii. **Rotation.** As the most popular self-supervisory task are rotations, we uniformly rotate all frames per clip by a random angle ($0°$, $90°$, $180°$, $270°$).

Figure 2 (blue frame) shows how we apply and recognize the self-supervision patterns *e.g.*, shuffling and rotation angles. Below, we illustrate self-supervision via rotations. Consider the objective function L_{rot} for self-supervised learning with a self-supervision discriminator d, where $\mathcal{D}$ are parameters of d. Thus:

$$\hat{\boldsymbol{\Phi}}_i = f(\ rot(\mathbf{V}_i, \theta)\ ; \mathcal{F}), \tag{9}$$

$$\mathbf{p}_{rot_i} = d(\hat{\boldsymbol{\Phi}}_i; \mathcal{D}), \tag{10}$$

$$L_{rot} = -\sum_i log\left(\frac{\exp(\mathbf{p}_{rot_i}[l_{\theta i}])}{\sum_s \exp(\mathbf{p}_{rot_i}[l_s])} \right), \tag{11}$$

where $\mathbf{V}_i$ is a randomly sampled clip, $\theta \in \{0°, 90°, 180°, 270°\}$ is a randomly selected rotation angle of a frame, $l_{\theta_i} \in \{0, 1, 2, 3\}$ is the rot. label for sample i.

Combining the original loss function L with such a self-supervision term L_{rot} results in a self-supervised few-shot action recognition pipeline. However, this objective does not make the attention to be invariant to augmentations per se.

3.5 Augmentation-Guided Attention by Alignment

Figure 4 presents a strategy in which we extract the attention vector for an augmented clip, then we apply the same augmentation to the attention vector obtained from the original non-augmented clip, and we encourage such a pair of augmentation vectors to align by a dedicated MSE loss. This encourages the attention unit to be invariant w.r.t. a given augmentation type. Figure 1a explains why the temporal permutation strategy benefits few-shot learning while

Figure 4 shows how to apply permutations and rotations. As an example, for a rotation-guided spatial-attention we define the alignment loss L_{att}:

$$\hat{\boldsymbol{\Phi}}_i = f(\ rot(\mathbf{V}_i, \theta) \ ; \mathcal{F}), \tag{12}$$

$$S_i = s(\boldsymbol{\Phi}_i; \boldsymbol{\mathcal{S}}), \ \hat{S}_i = s(\hat{\boldsymbol{\Phi}}_i; \boldsymbol{\mathcal{S}}), \tag{13}$$

$$L_{att} = \sum_i || \ | \ rot(S_i, \theta) - \hat{S}_i \ | - \lambda \ ||_F^2. \tag{14}$$

where λ controls the strictness of alignment. The final objective then becomes:

$$\underset{\mathcal{F}, \mathcal{D}, \mathcal{T}, \mathcal{S}}{\arg\min} \quad L + \beta L_{ss} + \gamma L_{att} \tag{15}$$

where β and γ are the hyper-parameters adjusted by cross-validation, L_{ss} is a chosen type of self-supervision $e.g.$, via rotations as introduced in Eq. (11).

4 Experiments

4.1 Experimental Setup

Below, we describe our setup and evaluations in detail. To exclude complicated data pre-processing and frame sampling steps typically used in action recognition, we sample uniformly 20 frames along the temporal mode for each dataset.

Table 1. Comparisons between our ARN model and existing works on HMDA51 and UCF101 splits proposed in [35] and a Kinetics split from [58] (given 5-way acc.)

Model	HMDB51 [35]		UCF101 [35]		Kinetics [6]	
	1-shot	5-shot	1-shot	5-shot	1-shot	5-shot
$GenApp$ [35]	–	52.5 ± 3.10	–	78.6 ± 2.1	-	-
$ProtoGAN$ [6]	34.7 ± 9.20	54.0 ± 3.90	57.8 ± 3.0	80.2 ± 1.3	-	-
CMN [58]	-	-	-	-	60.5	78.9
$Ours$	$\mathbf{44.6 \pm 0.9}$	$\mathbf{59.1 \pm 0.8}$	$\mathbf{62.1 \pm 1.0}$	$\mathbf{84.8 \pm 0.8}$	$\mathbf{63.7}$	$\mathbf{82.4}$

HMDB51 [29] contains 6849 clips divided into 51 action categories, each with at least 101 clips, 31, 10 and 10 classes selected for training, validation and testing.

Mini Moments in Time (miniMIT) [36] contains 200 classes and 550 videos per class. We select 120, 40 and 40 classes for training, validation and testing.

UCF101 [45], action videos from Youtube, has 13320 video clips and 101 action classes. We randomly select 70 training, 10 validation and 21 testing classes.

Table 2. Ablations of different modules of ARN given our proposed HMDB51 protocol (given 5-way acc.) We used *spatial-jigsaw* for self-supervision.

Baseline	Spatial attention	Self-supervision	Alignment	1-shot	5-shot
✓				40.83	55.18
✓	✓			41.27	56.12
✓		✓		44.19	58.50
✓	✓	✓		44.61	59.71
✓	✓		✓	43.11	57.35
✓	✓	✓	✓	**45.17**	**60.56**

Kinetics, used by [58] to select a subset for few-shot learning, consists of 64, 12 and 24 training, validation and testing classes. We use it for comparisons.

Training, validation and testing splits on the first three datasets are detailed in our supplementary material while authors of [58] provide the split on Kinetics. The frames of action clips from all datasets are resized to 128 × 128. All models are trained on training splits. Validation set is only used for cross-validation.

4.2   Comparison with Previous Works

Section 2.2 explains the issues with existing methods, protocols, and the lack of publicly available codes. For a fair comparison, we use firstly the protocol of [35] (HMDB51 and UCF101 datasets) but we chose 5 splits at random according to their protocol to average results over multiple runs: we report an average-case result not the best case or a single run result (in contrast to [35]). We also use the Kinetics split of [58], and compare our approach with [6,35,58].

Table 1 shows that our ARN (best variant) outperforms GenApp [35], ProtoGAN [6], CMN [58] by a large margin of 3% to 10% on the three protocols. Our standard errors are low as they result from 5 runs on 5 splits (average case) while a large deviation of ProtoGAN [6] was obtained w.r.t. episodes on a single split.

The Weakness of Protocol [35]. Evaluation protocols in [35] rely on randomly selecting training/testing classes with 50–50 ratio from all classes to form training/testing splits on HMDB51 and UCF101. The performance of that protocol varies heavily due to the randomness. Moreover, results of [35] are reported on a single run. Figures 5c and 5d show up to 6% variations due to the randomness, making a fair comparison between models difficult on such a protocol. The lack of validation set makes cross-validation also a random process affecting results.

We rectify all this by providing standardized training, validation, and testing splits on HMDB51, *mini*MIT and UCF101. In what follows, we use our new splits with our few-shot ARN. We equip the Prototypical Net [44], Relation Net [47] and SoSN [54] with a 3D conv. feature encoder (*C3D*) for baselines used below.

ARN Modules (Ablations). We start by studying ARN modules on HMDB51. Table 2 shows that combining attention with the baseline C3D SoSN

Table 3. Evaluations on HMDB51 (5-way acc.) Attention: Temporal (TA), Spatial (SA). Self-super.: Temp. (TS), Spat. (SS), Self-Super. & Alignment: Temp. (TSA), Spat. (SSA).

Model	1-shot	5-shot
Baseline		
C3D Prototypical Net [44]	38.05 ± 0.89	53.15 ± 0.90
C3D RelationNet [47]	38.23 ± 0.97	53.17 ± 0.86
C3D SoSN [54]	40.83 ± 0.96	55.18 ± 0.86
Temporal/Spatial Attention only (TA *vs.* SA)		
ARN+TA	41.97 ± 0.97	57.67 ± 0.88
ARN+SA	41.27 ± 0.98	56.12 ± 0.89
ARN+SA+TA	42.41 ± 0.99	56.81 ± 0.87
Temporal/Spatial Self-supervision only (TS *vs.* SS)		
ARN+TS (temp. jigsaw)	43.79 ± 0.96	58.13 ± 0.88
ARN+SS (spat. jigsaw)	44.19 ± 0.96	58.50 ± 0.86
ARN+SS (rotation)	43.90 ± 0.92	57.20 ± 0.90
Temp./Spat. Self-super. & Att. by alignment (TSA *vs.* SSA)		
ARN+TSA (temp. jigsaw)	45.20 ± 0.98	59.11 ± 0.86
ARN+SSA (spat. jigsaw)	45.15 ± 0.96	60.56 ± 0.86
ARN+SSA (rotation)	45.52 ± 0.96	58.96 ± 0.87

pipeline brings ∼1% gain. Switching to the attention by alignment brings 1.2–1.9% gain over the naive attention unit. Combining self-supervision with (i) the baseline and (ii) the baseline with attention brings ∼3% and ∼3.5% gain, resp. Combining all units together (attention, self-supervision and alignment) yields **∼5%** gain. The computational cost is similar to running either self-supervision or alignment.

Thus, in what follows we will report results for the most distinct four variants: (i) baseline (*C3D SoSN*), (ii) Temporal/Spatial Attention only (*TA & SA*), Temporal/Spatial Self-supervision w/o attention (*TS & SS*), and Temporal/Spatial Self-supervision with attention by alignment (*TSA & SSA*).

Pooling (Ablations). Section 3.2 discusses pooling variants from Table 5. Second-order pooling (with PN) outperforms second-order pooling (w/o PN) followed by average and max pooling. Combining average pooling with PN boosts its results which is consistent with the theoretical analysis in Figure 3. In what follows, we use the best pooling variant only, that is second-order pooling with PN.

Main Evaluations. Tables 3 and 4 present main evaluations on the proposed by us protocols. Notably, our approaches outperform all baselines (known approaches enhanced by us with the C3D-based encoder). Below, we detail the results.

Table 4. Evaluations on *mini*MIT and UCF101 datasets (given 5-way acc.) See the legend at the top of Table 3 for the description of abbreviations.

Model	miniMIT		UCF101	
	1-shot	5-shot	1-shot	5-shot
C3D Prototypical net [44]	33.65 ± 1.01	45.11 ± 0.90	57.05 ± 1.02	78.25 ± 0.73
C3D RelationNet [47]	35.71 ± 1.02	47.32 ± 0.91	58.21 ± 1.02	78.35 ± 0.72
C3D SoSN [54]	40.83 ± 0.99	52.16 ± 0.95	62.57 ± 1.03	81.51 ± 0.74
Temporal/Spatial Attention only (TA *vs.* SA)				
ARN+TA	41.65 ± 0.97	56.75 ± 0.93	63.35 ± 1.03	80.59 ± 0.77
ARN+SA	41.27 ± 0.98	55.69 ± 0.92	63.73 ± 1.08	82.19 ± 0.70
ARN+TA+SA	41.85 ± 0.99	56.43 ± 0.87	64.48 ± 1.06	82.37 ± 0.72
Temporal/Spatial Self-supervision only (TS *vs.* SS)				
ARN+TS (temp. jigsaw)	42.45 ± 0.96	54.67 ± 0.87	63.79 ± 1.02	82.14 ± 0.77
ARN+SS (spat. jigsaw)	42.68 ± 0.95	54.46 ± 0.88	63.75 ± 0.98	80.92 ± 0.72
ARN+SS (rotation)	42.01 ± 0.94	56.83 ± 0.86	63.95 ± 1.03	81.09 ± 0.76
Temp./Spat. Self-super. & Att. by alignment (TSA *vs.* SSA)				
ARN+TSA (temp. jigsaw)	42.65 ± 0.94	57.35 ± 0.85	65.46 ± 1.05	82.97 ± 0.71
ARN+SSA (spat. jigsaw)	42.92 ± 0.95	56.21 ± 0.85	66.04 ± 1.01	82.68 ± 0.72
ARN+SSA (rotation)	43.05 ± 0.97	56.71 ± 0.87	66.32 ± 0.99	83.12 ± 0.70

Attention. Tables 3 and 4 investigate the Temporal and Spatial Attention denoted as (TA) and (SA) on our few-shot ARN. TA on the 1-shot and 5-shot protocols improves the accuracy by 1.0% and 2.5% while SA boosts the 1- and 5-shot accuracy by 0.5% and 1.0%, respectively. For the combined Temporal and Spatial Attention ($TA+SA$), the Eq. (8) is used with $\alpha_s = 1.0$ and $\alpha_t = 0.5$ (HMDB51) and $\alpha_s = 1.5$ and $\alpha_t = 1.0$ (UCF101) chosen on the validation split. Tables 3 and 4 show that SA+TA achieves a further improvement of up to 1.1% for 1-shot learning but for 5-shot learning it may suffer an 0.8% drop in accuracy compared to the best score of TA and SA while still achieving between an 0.8 and 4.2% gain over the baseline C3D SoSN. This is consistent with our argument that vanilla attention units can be further improved for a better performance.

Table 5. Comparison of pooling functions on our HMDB51 split (5-way 1-shot).

No Pooling	Average	Average+PN	Max	Second-order (w/o PN)	Second-order (with PN)
35.71	39.51	40.02	38.95	39.97	**40.83**

Temporal/Spatial Self-supervision. In this experiment, we disable attention units. Tables 3 and 4 show that self-supervision w.r.t. either temporal or spatial mode boosts performance of 1-shot and 5-shot learning over the C3D SoSN baseline on HMDB51 up to 3.2%. On miniMIT, we observe gains between 1.2 and 4.6%. On UCF101, we see gains between 0.6 and 1.4%. However, for UCF101 dataset, self-supervision by the spatial jigsaw and rotation lead to a marginal performance drop on 5-shot learning compared to C3D SoSN.

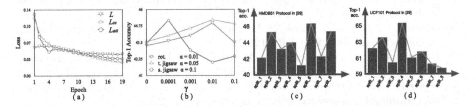

(a)   (b)   (c)   (d)

Fig. 5. In Fig. 5a are the loss curves for Eq. (15). Figure 5b shows the validation score w.r.t. γ (1-shot prot.) Applying Spatial Self-super. & Attention by alignment (SSA) $\gamma > 0$ outperforms the Spatial Self-super. & Attention only ($\gamma = 0$). Figure 5c and 5d show the performance variation on random splits of HMDB51 and UCF101 proposed by [35].

Temporal/Spatial Self-supervision and Attention by Alignment. According to Tables 3 and 4, the gains are in 2–5% range compared to the baseline C3D SoSN. Figure 5a shows the training loss w.r.t. epoch (HMDB51) (*temp. jigsaw*). Figure 5b shows the validation accuracy (HMDB51) w.r.t. γ for SS (*rot.*) As can be seen, Self-supervision combined with Attention by alignment (any curve for $\gamma > 0$) scores higher than Self-supervision with Attention only ($\gamma = 0$).

5   Conclusions

We have proposed a new few-shot Action Recognition Network (ARN) which comprises an encoder, comparator and an attention mechanism to model short- and long-range temporal patterns. We have investigated the role of self-supervision via spatial and temporal augmentations/auxiliary tasks. Moreover, we have proposed a novel mechanism dubbed *attention by alignment* which tackles the so-called distribution shift of temporal positions of discriminative long-range blocks. By combining losses of self-supervision and attention by alignment, we see gains of up to 6% accuracy. We make our dataset splits publicly available to facilitate fair comparisons of few-shot action recognition pipelines.

Acknowledgements. This research is supported in part by the Australian Research Council through Australian Centre for Robotic Vision (CE140100016), Australian Research Council grants (DE140100180), the China Scholarship Council (CSC Student ID 201603170283). Hongdong Li is funded in part by ARC-DP (190102261) and ARC-LE (190100080). We thank CSIRO Scientific Computing, NVIDIA (GPU grant) and the National University of Defense Technology.

References

1. Antoniou, A., Edwards, H., Storkey, A.: How to train your MAML. arXiv preprint (2018)
2. Bart, E., Ullman, S.: Cross-generalization: Learning novel classes from a single example by feature replacement. In: CVPR (2005)
3. Carreira, J., Zisserman, A.: Quo Vadis, action recognition? a new model and the kinetics dataset. In: CVPR (2018)
4. Doersch, C., Gupta, A., Efros, A.A.: Unsupervised visual representation learning by context prediction. In: CVPR (2015)
5. Dosovitskiy, A., Springenberg, J.T., Riedmiller, M., Brox, T.: Discriminative unsupervised feature learning with convolutional neural networks. In: NeurIPS (2014)
6. Dwivedi, S.K., Gupta, V., Mitra, R., Ahmed, S., Jain, A.: Protogan: towards few shot learning for action recognition. arXiv preprint (2019)
7. Engin, M., Wang, L., Zhou, L., Liu, X.: DeepKSPD: learning kernel-matrix-based SPD representation for fine-grained image recognition. In: Ferrari, V., Hebert, M., Sminchisescu, C., Weiss, Y. (eds.) ECCV 2018. LNCS, vol. 11206, pp. 629–645. Springer, Cham (2018). https://doi.org/10.1007/978-3-030-01216-8_38
8. Fei-Fei, L., Fergus, R., Perona, P.: One-shot learning of object categories. In: TPAMI (2006)
9. Fernando, B., Bilen, H., Gavves, E., Gould, S.: Self-supervised video representation learning with odd-one-out networks. In: CVPR (2017)
10. Fink, M.: Object classification from a single example utilizing class relevance metrics. In: NeurIPS (2005)
11. Finn, C., Abbeel, P., Levine, S.: Model-agnostic meta-learning for fast adaptation of deep networks. In: ICML (2017)
12. Gan, C., Gong, B., Liu, K., Su, H., Guibas, L.J.: Geometry guided convolutional neural networks for self-supervised video representation learning. In: CVPR (2018)
13. Garcia, V., Bruna, J.: Few-shot learning with graph neural networks. In: ICLR (2018)
14. Gidaris, S., Bursuc, A., Komodakis, N., Pérez, P., Cord, M.: Boosting few-shot visual learning with self-supervision. In: ICCV (2019)
15. Gidaris, S., Komodakis, N.: Generating classification weights with GNN denoising autoencoders for few-shot learning. In: CVPR (2019)
16. Gidaris, S., Singh, P., Komodakis, N.: Unsupervised representation learning by predicting image rotations. arXiv preprint (2018)
17. Guo, M., Chou, E., Huang, D.-A., Song, S., Yeung, S., Fei-Fei, L.: Neural graph matching networks for fewshot 3D action recognition. In: Ferrari, V., Hebert, M., Sminchisescu, C., Weiss, Y. (eds.) ECCV 2018. LNCS, vol. 11205, pp. 673–689. Springer, Cham (2018). https://doi.org/10.1007/978-3-030-01246-5_40
18. Hariharan, B., Girshick, R.: Low-shot visual recognition by shrinking and hallucinating features. In: ICCV (2017)
19. Jian, S., Hu, L., Cao, L., Lu, K.: Representation learning with multiple Lipschitz-constrained alignments on partially-labeled cross-domain data. In: AAAI, pp. 4320–4327 (2020)
20. Jian, S., Hu, L., Cao, L., Lu, K., Gao, H.: Evolutionarily learning multi-aspect interactions and influences from network structure and node content. In: Proceedings of the AAAI Conference on Artificial Intelligence, vol. 33, pp. 598–605 (2019)
21. Kim, J., Kim, T., Kim, S., Yoo, C.D.: Edge-labeling graph neural network for few-shot learning. In: CVPR (2019)

22. Koch, G., Zemel, R., Salakhutdinov, R.: Siamese neural networks for one-shot image recognition. In: ICML Deep Learning Workshop (2015)
23. Koniusz, P., Cherian, A., Porikli, F.: Tensor representations via kernel linearization for action recognition from 3D skeletons. In: Leibe, B., Matas, J., Sebe, N., Welling, M. (eds.) ECCV 2016. LNCS, vol. 9908, pp. 37–53. Springer, Cham (2016). https://doi.org/10.1007/978-3-319-46493-0_3
24. Koniusz, P., Tas, Y., Porikli, F.: Domain adaptation by mixture of alignments of second-or higher-order scatter tensors. In: CVPR (2017)
25. Koniusz, P., Wang, L., Cherian, A.: Tensor representations for action recognition. TPAMI (2020)
26. Koniusz, P., Yan, F., Gosselin, P.H., Mikolajczyk, K.: Higher-order occurrence pooling for bags-of-words: Visual concept detection. TPAMI (2017)
27. Koniusz, P., Zhang, H.: Power normalizations in fine-grained image, few-shot image and graph classification. TPAMI (2020)
28. Koniusz, P., Zhang, H., Porikli, F.: A deeper look at power normalizations. In: CVPR (2018)
29. Kuehne, H., Jhuang, H., Garrote, E., Poggio, T., Serre, T.: HMDB: a large video database for human motion recognition. In: ICCV (2011)
30. Lai, Z., Lu, E., Xie, W.: Mast: a memory-augmented self-supervised tracker. In: CVPR (2020)
31. Lake, B.M., Salakhutdinov, R., Gross, J., Tenenbaum, J.B.: One shot learning of simple visual concepts. CogSci (2011)
32. Lee, K., Maji, S., Ravichandran, A., Soatto, S.: Meta-learning with differentiable convex optimization. In: CVPR (2019)
33. Li, F.F., VanRullen, R., Koch, C., Perona, P.: Rapid natural scene categorization in the near absence of attention. Proc. Natl. Acad. Sci. 99, 9596–9601 (2002)
34. Miller, E.G., Matsakis, N.E., Viola, P.A.: Learning from one example through shared densities on transforms. In: CVPR (2000)
35. Mishra, A., Verma, V.K., Reddy, M.S.K., Arulkumar, S., Rai, P., Mittal, A.: A generative approach to zero-shot and few-shot action recognition. In: WACV (2018)
36. Monfort, M., et al.: Moments in time dataset: one million videos for event understanding. TPAMI (2019)
37. Romero, A., Terán, M.Y., Gouiffès, M., Lacassagne, L.: Enhanced local binary covariance matrices (ELBCM) for texture analysis and object tracking. MIRAGE (2013)
38. Russakovsky, O.: ImageNet largescale visual recognition challenge. IJCV 115, 211–252 (2015). https://doi.org/10.1007/s11263-015-0816-y
39. Rusu, A.A., Rao, D., Sygnowski, J., Vinyals, O., Pascanu, R., Osindero, S., Hadsell, R.: Meta-learning with latent embedding optimization. In: ICLR (2019)
40. Sermanet, P., Lynch, C., Chebotar, Y., Hsu, J., Jang, E., Schaal, S., Levine, S.: Time-contrastive networks: self-supervised learning from pixels. In: ICRA (2017)
41. Simon, C., Koniusz, P., Nock, R., Harandi, M.: Deep subspace networks for few-shot learning. In: NeurIPS Workshops (2019)
42. Simon, C., Koniusz, P., Nock, R., Harandi, M.: Adaptive subspaces for few-shot learning. In: CVPR (2020)
43. Simon, C., Koniusz, P., Nock, R., Harandi, M.: On modulating the gradient formeta-learning. In: ECCV (2020)
44. Snell, J., Swersky, K., Zemel, R.: Prototypical networks for few-shot learning. In: NeurIPS (2017)
45. Soomro, K., Zamir, A.R., Shah, M.: Ucf101: a dataset of 101 human actions classes from videos in the wild. arXiv preprint (2012)

46. Su, J.C., Maji, S., Hariharan, B.: Boosting supervision with self-supervision for few-shot learning. arXiv preprint (2019)
47. Sung, F., Yang, Y., Zhang, L., Xiang, T., Torr, P.H., Hospedales, T.M.: Learning to compare: Relation network for few-shot learning. In: CVPR (2018)
48. Tuzel, O., Porikli, F., Meer, P.: Region covariance: a fast descriptor for detection and classification. In: Leonardis, A., Bischof, H., Pinz, A. (eds.) ECCV 2006. LNCS, vol. 3952, pp. 589–600. Springer, Heidelberg (2006). https://doi.org/10.1007/11744047_45
49. Vinyals, O., Blundell, C., Lillicrap, T., Wierstra, D., et al.: Matching networks for one shot learning. In: NeurIPS (2016)
50. Wang, L., Zhang, J., Zhou, L., Tang, C., Li, W.: Beyond covariance: Feature representation with nonlinear kernel matrices. In: ICCV, pp. 4570–4578 (2015). https://doi.org/10.1109/ICCV.2015.519
51. Wertheimer, D., Hariharan, B.: Few-shot learning with localization in realistic settings. In: CVPR (2019)
52. Xu, B., Ye, H., Zheng, Y., Wang, H., Luwang, T., Jiang, Y.G.: Dense dilated network for few shot action recognition. In: ICMR (2018)
53. Xu, D., Xiao, J., Zhao, Z., Shao, J., Xie, D., Zhuang, Y.: Self-supervised spatiotemporal learning via video clip order prediction. In: CVPR (2019)
54. Zhang, H., Koniusz, P.: Power normalizing second-order similarity network for few-shot learning. In: WACV (2019)
55. Zhang, H., Zhang, J., Koniusz, P.: Few-shot learning via saliency-guided hallucination of samples. In: CVPR (2019)
56. Zhou, B., Lapedriza, A., Xiao, J., Torralba, A., Oliva, A.: Learning deep features for scene recognition using places database. In: NeurIPS (2014)
57. Zhu, F., Zhang, L., Fu, Y., Guo, G., Xie, W.: Self-supervised video object segmentation. arXiv preprint (2020)
58. Zhu, L., Yang, Y.: Compound memory networks for few-shot video classification. In: Ferrari, V., Hebert, M., Sminchisescu, C., Weiss, Y. (eds.) ECCV 2018. LNCS, vol. 11211, pp. 782–797. Springer, Cham (2018). https://doi.org/10.1007/978-3-030-01234-2_46
59. Zintgraf, L., Shiarli, K., Kurin, V., Hofmann, K., Whiteson, S.: Fast context adaptation via meta-learning. In: ICML (2019)

Character Grounding
and Re-identification in Story
of Videos and Text Descriptions

Youngjae Yu[1,2], Jongseok Kim[1], Heeseung Yun[1], Jiwan Chung[1],
and Gunhee Kim[1,2(✉)]

[1] Seoul National University, Seoul, Korea
{yj.yu,js.kim,heeseung.yung,jiwanchung}@vision.snu.ac.kr,
gunhee@snu.ac.kr
[2] Ripple AI, Seoul, Korea
http://vision.snu.ac.kr/projects/CharacterReid/

Abstract. We address character grounding and re-identification in multiple story-based videos like movies and associated text descriptions. In order to solve these related tasks in a mutually rewarding way, we propose a model named *Character in Story Identification Network* (CiSIN). Our method builds two semantically informative representations via joint training of multiple objectives for character grounding, video/text re-identification and gender prediction: Visual Track Embedding from videos and Textual Character Embedding from text context. These two representations are learned to retain rich semantic multimodal information that enables even simple MLPs to achieve the state-of-the-art performance on the target tasks. More specifically, our CiSIN model achieves the best performance in the Fill-in the Characters task of LSMDC 2019 challenges [35]. Moreover, it outperforms previous state-of-the-art models in M-VAD Names dataset [30] as a benchmark of multimodal character grounding and re-identification.

1 Introduction

Searching persons in videos accompanying with free-form natural language descriptions is a challenging problem in computer vision and natural language research [30,32,34,35]. For example, in the story-driven videos such as movies and TV series, distinguishing *who is who* is a prerequisite to understanding the relationships between characters in the storyline. Thanks to the recent rapid progress of deep neural network models for joint visual-language representation [19,28,42,46], it has begun to be an achievable goal to understand interactions of characters that reside in the complicated storyline of videos and associate text.

In this work, we tackle the problem of character grounding and re-identification in consecutive pairs of movie video clips and corresponding language descriptions. The character grounding indicates the task of locating the character mentioned in the text within videos. The re-identification can be done

A. Vedaldi et al. (Eds.): ECCV 2020, LNCS 12350, pp. 543–559, 2020.
https://doi.org/10.1007/978-3-030-58558-7_32

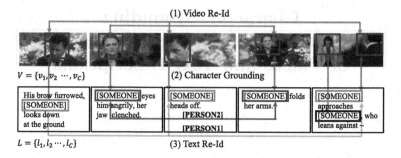

Fig. 1. The problem statement. Given consecutive C pairs of video clips and corresponding language descriptions, our CiSIN model aims at solving three multimodal tasks in a mutually rewarding way. Character grounding is the identity matching between person tracks and [SOMEONE] tokens. Video/text re-identification is the identity matching between person tracks in videos and [SOMEONE] tokens in text, respectively.

in both text and image domain; it groups the tracks of the same person across video clips or identifies tokens of the identical person across story sentences. As the main testbed of our research, we choose the recently proposed Fill-in the Characters task of the Large Scale Movie Description Challenge (LSMDC) 2019 [35], since it is one of the most large-scale and challenging datasets for character matching in videos and associated text. Its problem statement is as follows. Given five pairs of video clips and text descriptions that include the [SOMEONE] tokens for characters, the goal is to identify which [SOMEONE] tokens are identical to one another. This task can be tackled minimally with text re-identification but can be synergic to jointly solve with video re-identification and character grounding.

We propose a new model named *Character-in-Story Identification Network* (CiSIN) to jointly solve the character grounding and video/text re-identification in a mutually rewarding way, as shown in Fig. 1. The character grounding, which connects the characters between different modalities, complements both visual and linguistic domains to improve re-identification performance. In addition, each character's grounding can be better solved by closing the loop between both video/text re-identification and neighboring character groundings.

Our method proposes two semantically informative representations. First, Visual Track Embedding (VTE) involves motion, face and body-part information of the tracks from videos. Second, Textual Character Embedding (TCE) learns rich information of characters and their actions from text using BERT [6]. They are trained together to share various multimodal information via multiple objectives, including character grounding, video/text re-identification and attribute prediction. The two representations are powerful enough for simple MLPs to achieve state-of-the-art performance on the target tasks.

We summarize the contributions of this work as follows.

1. We propose the CiSIN model that can jointly tackle character grounding and re-identification in both video and text narratives. To the best of our knowledge, our work is the first to jointly solve these three tasks, each of which has been addressed separately in previous research. Our model is jointly trained via multi-task objectives of these complementary tasks in order to create synergetic effects from one domain to another and vice versa.
2. Our CiSIN model achieves the best performance so far in two benchmarks datasets: LSMDC 2019 challenge [35] and M-VAD Names [30] dataset. The CiSIN model attains the best accuracy in the Fill-in the Characters task of LSMDC 2019 challenges. Moreover, it achieves the new state-of-the-art results on grounding and re-identification tasks in M-VAD Names.

2   Related Work

Linking Characters with Visual Tracks. There has been a long line of work that aims to link character names in movie or TV scripts with their corresponding visual tracks [2,7,14,27,29,32,38,40]. However, this line of research deals with more constrained problems than ours; for example, having the templates of main characters or knowing which characters appear in a clip. On the other hand, our task requires person grounding and re-identification in more free-formed multiple sentences, without ever seeing characters before.

Human Retrieval with Natural Language. Visual content retrieval with natural language queries has been mainly addressed by joint visual-language embedding models [12,13,18,21,22,24,26,42,43], and extended to the video domain [28,42,46]. Such methodologies have also been applied to movie description datasets such as MPII-MD [34] and M-VAD Names [30]. Pini *et al.* [30] propose a neural network that learns a joint multimodal embedding for human tracking in videos and verb embedding in text. Rohrbach *et al.* [34] develop an attention model that aligns human face regions and mentioned characters in the description. However, most previous approaches tend to focus on the retrieval of human bounding boxes with sentence queries in a single clip while ignoring story coherence. On the other hand, our model understands the context within consecutive videos and sentences; thereby can achieve the best result reported so far in the human retrieval in the video story.

Person Re-identification in Multiple Videos. The goal of person re-identification (re-id) is to associate with identical individuals across multiple camera views [1,4,8–10,15,20,23,47,51,52]. Some methods exploit part information for better re-identification against occlusion and pose variation [37,39,44]. However, this line of work often has dealt with visual information only and exploited the videos with less diverse human appearances (*e.g.* standing persons taken from CCTV) than movies taken by various camerawork (*e.g.* head shots, half body shots and two shots). Moreover, our work takes a step further by considering the video description context of story-coherent multiple clips for character search and matching.

Visual Co-reference Resolution. Visual co-reference resolves pronouns and linked character mentions in text with the character appearances in video clips. Ramanathan *et al.* [32] address the co-reference resolution in TV show description with the aid of character visual models and linguistic co-reference resolution features. Rohrbach *et al.* [34] aim generate video descriptions with grounded and co-referenced characters. In the visual dialogue domain [19,36], the visual co-reference problem is addressed to identify the same entity/object instances in an image for answering questions with pronouns. On the other hand, we focus on identifying all character mentions blanked as [SOMEONE] tokens in multiple descriptions. This is inherently different from the co-reference resolution that can use pronouns or explicit gender information (*e.g.* she, him) as clues.

3   Approach

Problem Statement. We address the problem of character identity matching in the movie clips and associated text descriptions. To be specific, we follow the Fill-in the Characters task of the Large Scale Movie Description Challenge (LSMDC) 2019[1]. As shown in Fig. 1, the input is a sequence of C pairs of video clips and associated descriptions that may include the [SOMEONE] tokens for characters (*i.e.* $C = 5$ in LSMDC 2019). The [SOMEONE] tokens correspond to proper nouns or pronouns of characters in the original description. The goal of the task is to replace the [SOMEONE] tokens with *local* character IDs that are consistent within the input set of clips, not globally in the whole movie.

We decompose the Fill-in the Characters task into three subtasks: (*i*) *character grounding* finds the character tracks of each [SOMEONE] token in the video, (*ii*) *video re-identification* that groups visual tracks of the identical character in videos and (*iii*) *text re-identification* that connects between the [SOMEONE] tokens of the same person. We define these three subtasks because each of them is an important research problem with its own applications, and jointly solving the problems is mutually rewarding. Thereby we can achieve the best result reported so far on LSMDC 2019 and M-VAD Names [30] (See the details in Sect. 4).

In the following, we first present how to define person tracks from videos (Sect. 3.1). We then discuss two key representations of our model: Visual Track Embedding (VTE) for person tracks (Sect. 3.2) and Textual Character Embedding (TCE) for [SOMEONE] tokens (Sect. 3.3). We then present the details of our approach to character grounding (Sect. 3.4) and video/text re-identification (Sect. 3.5). Finally, we discuss how to train all the components of our model via multiple objectives so that the solution to each subtask helps one another (Sect. 3.6). Figure 2 illustrates the overall architecture of our model.

3.1   Video Preprocessing

For each of C video clips, we first resize it to 224×224 and uniformly sample 24 frames per second. We then detect multiple person tracks as basic character

---

[1] https://sites.google.com/site/describingmovies/lsmdc-2019.

instances in videos for grounding and re-identification tasks. That is, the tasks reduce to the matching problems between person tracks and [SOMEONE] tokens.

Person Tracks. We obtain person tracks $\{\mathbf{t}_m\}_{m=1}^M$ as follows. M denotes the number of person tracks throughout C video clips. First, we detect bounding boxes (bbox) of human bodies using the rotation robust CenterNet [53] and group them across consecutive frames using the DeepSORT tracker [45]. The largest bbox in each track is regarded as its representative image $\{\mathbf{h}_m\}_{m=1}^M$, which can be used as a simple representation of the person track.

Face Regions. Since person tracks are not enough to distinguish "who is who", we detect the faces for better identification. In every frame, we obtain face regions using MTCNN [49] and resize them to 112×112. We then associate each face detection with the person track that has the highest IoU score.

Track Metadata. We extract track metadata $\{\mathbf{m}_m\}_{m=1}^M$, which include the (x, y) coordinates, the track length and the size of the representative image $\mathbf{h}_m$. All of them are normalized with respect to the original clip size and duration.

3.2 Visual Track Embedding

For better video re-identification, it is critical to correctly measure the similarity between track i and track j. As rich semantic representation of tracks, we build Visual Track Embedding (VTE) as a set of motion, face and body-part features.

Motion Embedding. We apply the I3D network [3] to each video clip and obtain the last CONV feature of the final Inception module, whose dimension is (time, width, height, feature) $= (\lfloor \frac{t}{8} \rfloor, 7, 7, 1024)$. That is, each set of 8 frames is represented by a 7×7 feature map whose dimension is 1024. Then, we extract motion embedding $\{\mathbf{i}_m\}_{m=1}^M$ of each track by mean-pooling over the cropped spatio-temporal tensor of this I3D feature.

Face Embedding. We obtain the face embedding $\{\mathbf{f}_m\}_{m=1}^M$ of a track, by applying ArcFace [5] to all face regions of the track and mean-pooling over them.

Body-Part Embedding. To more robustly represent visual tracks against the ill-posed views or cameraworks (*e.g.* no face is shown), we also utilize the body parts of a character in a track (*e.g.* limb and torso). We first extract the pose keypoints using the pose detector of [53], and then obtain the body-part embedding $\{\{\mathbf{p}_{m,k}\}_{k=1}^K\}_{m=1}^M$ by selecting keypoint-corresponding coordinates from the last CONV layer ($7 \times 7 \times 2048$) of ImageNet pretrained ResNet-50 [11]:

$$\mathbf{p}_{m,k} = \text{ResNet}(\mathbf{h}_m)[x_k, y_k], \tag{1}$$

where $\mathbf{p}_{m,k}$ is the representation of the k-th keypoint in a pose, $\mathbf{h}_m$ is the representative image of the track, and (x_i, y_i) is a relative position of the keypoint. To ensure the quality of pose estimation, we only consider keypoints whose confidences are above a certain threshold ($\tau = 0.3$).

In summary, VTE of each track m includes three sets of embeddings for motion, face and body parts: $\text{VTE} = \{\mathbf{i}_m, \mathbf{f}_m, \mathbf{p}_m\}$.

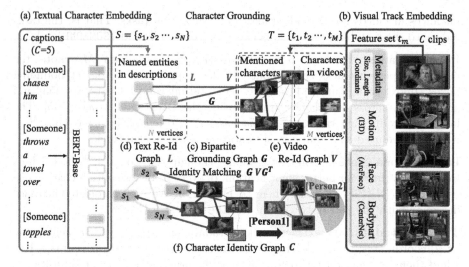

Fig. 2. Overview of the proposed *Character-in-Story Identification Network* (CiSIN) model. Using (a) Textual Character Embedding (TCE) (Sect. 3.3) and (b) Visual Track Embedding (VTE) (Sect. 3.2), we obtain the (c) bipartite Character Grounding Graph **G** (Sect. 3.4). We build (d)–(e) Video/Text Re-Id Graph **V** and **L**, from which (f) Character Identity Graph **C** is created. Based on the graphs, we can perform the three subtasks jointly (Sect. 3.5).

3.3   Textual Character Embedding

Given C sentences, we make a unified language representation of the characters (*i.e.* [SOMEONE] tokens), named Textual Character Embedding (TCE) as follows.

Someone Embedding. We use the BERT model [6] to embed the [SOMEONE] tokens in the context of C consecutive sentences. We load pretrained BERT-Base-Uncased with initial weights. We denote each sentence $\{\mathbf{w}_l^c\}_{l=1}^{W_c}$ where W_c is the number of words in the c-th sentence. As an input to the BERT model, we concatenate the C sentences $\{\{\mathbf{w}_l^c\}_{l=1}^{W_c}\}_{c=1}^C$ by placing the [SEP] token in each sentence boundary. We obtain the embedding of each [SOMEONE] token as

$$\mathbf{b}_n = \text{BERT}(\{\{\mathbf{w}_l^c\}_{l=1}^{W_c}\}_{c=1}^C, k), \tag{2}$$

where k is the position of the [SOMEONE] token. We let this word representation $\{\mathbf{b}_n\}_{n=1}^N$ (*i.e.* the BERT output at position k), where N is the number of [SOMEONE] tokens in the C sentences.

Action Embedding. For a better representation of characters, we also consider the actions of [SOMEONE] described in the sentences. We build a dependency tree for each sentence with the Stanford Parser [31], and find the ROOT word associated with [SOMEONE], which is generally the verb of the sentence. We then obtain the ROOT word representation $\{\mathbf{a}_n\}_{n=1}^N$ as done in Eq. (2).

In summary, TCE of each character n includes two sets of embeddings for someone and action. Since they have the same dimension as the output of BERT, we simply concatenate them as a single embedding: $\text{TCE} = \mathbf{s}_n = [\mathbf{b}_n; \mathbf{a}_n]$.

3.4   Character Grounding

For each clip c, we perform the character grounding using $\text{VTE} = \{\mathbf{i}_m, \mathbf{f}_m, \mathbf{p}_m\}$ of person tracks $\mathbf{t}_m$ (Sect. 3.2) and $\text{TCE} = \mathbf{s}_n = [\mathbf{b}_n; \mathbf{a}_n]$ of characters (Sect. 3.3). Due to the heterogeneous nature of our VTE (*i.e.* $\mathbf{f}_m$ contains appearance information such as facial expression while $\mathbf{i}_m$ does motion information), we separately fuse VTE with TCE and then concatenate to form fused ground representation $\mathbf{g}_{n,m}$:

$$\mathbf{g}_{n,m}^{\text{face}} = \text{MLP}(\mathbf{s}_n) \odot \text{MLP}(\mathbf{f}_m), \quad \mathbf{g}_{n,m}^{\text{motion}} = \text{MLP}(\mathbf{s}_n) \odot \text{MLP}(\mathbf{i}_m), \qquad (3)$$

$$\mathbf{g}_{n,m} = [\text{MLP}(\mathbf{m}_m); \mathbf{g}_{n,m}^{\text{face}}; \mathbf{g}_{n,m}^{\text{motion}}], \qquad (4)$$

where $\odot$ is a Hadamard product, $[;]$ is concatenation, $\mathbf{m}_m$ is the track metadata, and the MLP is made up of two FC layers with the identical output dimension. Finally, we calculate the bipartite Grounding Graph $\mathbf{G} \in \mathbb{R}^{N \times M}$ as

$$\mathbf{G}_{nm} = g(\mathbf{s}_n, \mathbf{t}_m) = \text{MLP}(\mathbf{g}_{n,m}), \qquad (5)$$

where the MLP consists of two FC layers with the scalar output.

Attributes. In addition to visual and textual embeddings, there are some attributes of characters that may be helpful for grounding. For example, as in MPII-MD Co-ref [34], gender information can be an essential factor for matching characters. However, most [SOMEONE] tokens have vague context to hardly infer gender information, although some of them may be clear like gendered pronouns (*e.g.* she, he) or nouns (*e.g.* son, girl). Thus, we add an auxiliary attribute module that predicts the gender of [SOMEONE] using $\mathbf{b}_n$ as input:

$$attr_{logit} = \sigma(\text{MLP}(\mathbf{b}_n)), \qquad (6)$$

where the MLP has two FC layers with scalar output, and σ is a sigmoid function.

The predicted attribute is neither explicitly used for any grounding or re-identification tasks. Instead, the module is jointly trained with other components to encourage the shared character representation to learn the gender context implicitly. Moreover, it is straightforward to extend this attribute module to other information beyond gender like the characters' roles.

3.5   Re-identification

We present our method to solve video/text re-identification tasks, and explain how to aggregate all subtasks to achieve the Fill-in the Characters task.

Video Re-id Graph. We calculate the Video Re-Id Graph $\mathbf{V} \in \mathbb{R}^{M \times M}$ that indicates the pairwise similarity scores between person tracks. We use the face and body-part embeddings of VTE. We first calculate the face matching score of track i and j, namely $f_{i,j}$, by applying a 2-layer MLP to face embedding $\mathbf{f}_m$ followed by a dot product and a scaling function:

$$f_{i,j} = w_f^{scale} \mathbf{f}_i' \cdot \mathbf{f}_j' + b_f^{scale}, \quad \mathbf{f}_m' = \text{MLP}(\mathbf{f}_m), \tag{7}$$

where $w_f^{scale}, b_f^{scale} \in \mathbb{R}$ are learnable scalar weights for scaling.

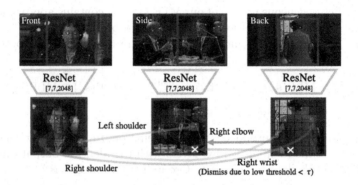

Fig. 3. Pose-based body-part matching. We only consider keypoints that appear across scenes with confidences scores above a certain threshold $\tau(=0.3)$ such as right/left shoulder and right elbow, while ignoring right wrist as it falls short of the threshold.

We next calculate the pose-guided body-part matching score $q_{i,j}$. Unlike conventional re-identification tasks, movies may not contain full keypoints of a character due to camerawork like upper body close up shots. Therefore, we only consider the pairs of keypoints that are visible in both tracks (Fig. 3):

$$q_{i,j} = \frac{w_p^{scale}}{Z_{i,j}} \sum_k \delta_{i,k} \delta_{j,k} \mathbf{p}_{i,k} \cdot \mathbf{p}_{j,k} + b_p^{scale}, \tag{8}$$

where $\mathbf{p}_{i,k}$ is the body-part embedding of VTE for keypoint k of track i, $\delta_{i,k}$ is its binary visibility value (1 if visible otherwise 0), $Z_{i,j} = \sum_k \delta_{i,k} \delta_{j,k}$ is a normalizing constant and w_p^{scale}, b_p^{scale} are scalar weights.

Finally, we obtain Video Re-Id Graph $\mathbf{V}$ by summing both face matching score and body-part matching score:

$$\mathbf{V}_{i,j} = f_{i,j} + q_{i,j}. \tag{9}$$

Text Re-id Graph. The Text Re-Id Graph $\mathbf{L} \in \mathbb{R}^{N \times N}$ measure the similarity between every pair of [SOMEONE] token as

$$\mathbf{L}_{i,j} = \sigma(\text{MLP}(\mathbf{b}_i \odot \mathbf{b}_j)), \tag{10}$$

where $\odot$ is a Hadamard product, σ is a sigmoid and the MLP has two FC layers.

Solutions to Three Subtasks. Since we have computed all pairwise similarity between and within person tracks and [SOMEONE] tokens in Eq. (5 and Eq. (9)–(10), we can achieve the three tasks by thresholding. We perform the character grounding by finding the column with the maximum value for each row in the bipartite Grounding Graph $\mathbf{G} \in \mathbb{R}^{N \times M}$. The video re-identification is carried out by finding the pair whose score in Video Re-Id Graph $\mathbf{V}$ is positive. Finally, the text re-identification can be done as will be explained below for the Fill-in the Characters task since they share the same problem setting.

Fill-in the Characters Task. We acquire the Character Identity Graph $\mathbf{C}$:

$$\mathbf{C} = \mathrm{avg}(\mathbf{L}, \mathbf{R}) \quad \text{where } \mathbf{R} = \sigma(\mathbf{GVG}^T), \ \mathbf{G}_n = \mathrm{argmax}_{m \in M_c} \, g(\mathbf{s}_n, \mathbf{t}_m), \quad (11)$$

where σ is a sigmoid, $\mathbf{G} \in \mathbb{R}^{N \times M}$ is the bipartite Grounding Graph in Eq. (5), $\mathbf{V} \in \mathbb{R}^{M \times M}$ is the Video Re-Id Graph in Eq. (9), and $\mathbf{L} \in \mathbb{R}^{N \times N}$ is the Text Re-Id Graph in Eq. (10). We perform the Fill-in the Characters task by finding $\mathbf{C}_{ij} \geq 0.5$, for which we decide [SOMEONE] token i and j as the same character.

Although we can solve the task using only the Text Re-Id graph $\mathbf{L}$, the key idea of Eq. (11) is that we also consider the other loop leveraging character grounding $\mathbf{G}$ and the Video Re-Id graph $\mathbf{V}$. That is, we find the best matching track for each token in the same clip c, where M_c is the candidate tracks in the clip c. We then use the Video Re-Id graph to match tracks and apply $\mathbf{G}$ again to find their character tokens. Finally, we average the scores from these two loops of similarity between [SOMEONE] tokens.

3.6 Joint Training

We perform joint training of all components in the CiSIN model so that both VTE and TCE representation share rich multimodal semantic information, and subsequently help solve character grounding and video/text re-identification in a mutually rewarding way. We first introduce the loss functions.

Losses. For character grounding, we use a triplet loss where a positive pair maximizes the ground matching score g in Eq. (5) while a negative one minimizes

$$\mathcal{L}(\mathbf{s}, \mathbf{t}, \mathbf{t}^-) = \max(0, \ \alpha - g(\mathbf{s}, \mathbf{t}) + g(\mathbf{s}, \mathbf{t}^-)), \quad (12)$$

$$\mathcal{L}(\mathbf{s}, \mathbf{t}, \mathbf{s}^-) = \max(0, \ \alpha - g(\mathbf{s}, \mathbf{t}) + g(\mathbf{s}^-, \mathbf{t})), \quad (13)$$

where α is a margin, $(\mathbf{s}, \mathbf{t})$ is a positive pair, $\mathbf{t}^-$ is a negative track, and $\mathbf{s}^-$ is a negative token. For video re-identification, we also use a triplet loss:

$$\mathcal{L}(\mathbf{t}_0, \mathbf{t}_+, \mathbf{t}_-) = \max(0, \ \beta - \mathbf{V}_{0,+} + \mathbf{V}_{0,-}) \quad (14)$$

where β is a margin and $\mathbf{V}$ is the score in Video Re-Id Graph. $(\mathbf{t}_0, \mathbf{t}_+)$ and $(\mathbf{t}_0, \mathbf{t}_-)$ are positive and negative track pair, respectively. For text re-identification, we train parameters to make Character Identity Graph $\mathbf{C}$ in Eq. (11) closer to the ground truth $\mathbf{C}^{gt}$ with a binary cross-entropy (BCE) loss. When computing $\mathbf{C}$

in Eq. (11), we replace the argmax of $\mathbf{G}_n$ with the softmax for differentiability: $\mathbf{G}_n = \text{softmax}_{m \in M} g(\mathbf{s}_n, \mathbf{t}_m)$. Additionally, the attribute module is trained with the binary cross-entropy loss for gender class (*i.e.* female, male).

Training. We use all losses to train the model jointly. While fixing the parameters in the motion embedding (I3D) and the face embedding (ArcFace), we update all the other parameters in all MLPs, ResNets and BERTs during training. Notably, BERT models in TCE are trained by multiple losses (*i.e.* character grounding, text re-identification and the attribute loss) to learn better multimodal representation.

In Eq. (11), the Grounding Graph $\mathbf{G}$ is a bipartite graph for cross-domain retrieval between the Text Re-Id Graph $\mathbf{L}$ and the Video Re-Id Graph $\mathbf{V}$. The bipartite graph $\mathbf{G}$ identifies a subgraph of $\mathbf{V}$ for the characters mentioned in the text such that the subgraph is topologically similar to $\mathbf{L}$. By joint training of multiple losses, the similarity metric between the visual and textual representation of the same character increases, consequently improving both character grounding and re-identification performance.

Details. We unify the hidden dimension size of all MLPs as 1024 and use the leaky ReLU activation for every FC layer in our model. The whole model is trained with the Adam optimizer [17] with a learning rate of 10^{-5} and a weight decay of 10^{-8}. We train for 86 K iterations for 15 epochs. We set the margin $\alpha = 3.0$ for the triplet loss in Eq. (12)–(13) and $\beta = 2.0$ for Eq. (14).

4   Experiments

We evaluate the proposed CiSIN model in two benchmarks of character grounding and re-identification: M-VAD Names [30] dataset and LSMDC 2019 challenge [35], on which we achieve new state-of-the-art performance.

4.1   Experimental Setup

M-VAD Names. We experiment three tasks with M-VAD Names [30] dataset: character grounding and video/text re-identification. We group five successive clips into a single set, which is the same setting with LSMDC 2019, to evaluate the model's capability to understand the story in multiple videos and text. M-VAD Names dataset is annotated with persons' name and their face tracks, which we use as ground truth for training.

For evaluation of character grounding, we measure the accuracy by checking whether the positive track is correctly identified. For evaluation of video re-identification, we predict the Video Re-Id Graph and then calculate the accuracy by comparing it with the ground truth. For evaluation of text re-identification, we calculate the accuracy with the ground truth character matching graph.

LSMDC 2019. We report experimental results for the Fill-in the Characters task of LSMDC 2019 [35], which is the superset of M-VAD dataset [41]. Contrary to M-VAD Names, LSMDC 2019 only provides local ID ground truths of [SOMEONE] tokens with no other annotation (*e.g.* face bbox annotation and characters' names). We exactly follow the evaluation protocol of the challenge.

Table 1. Results of video re-identification on the validation set of M-VAD names [30].

Video Re-id	Accuracy
ResNet-50 (ImageNet pretrained)	0.607
Strong-ReID-baseline [25]	0.617
Face+fullbody+poselets [50]	0.776
Face+head+body+upperbody [16]	0.779
RANet visual context [15]	0.787
CiSIN Face only	0.783
+ Human bbox	0.781
+ Body parts	0.799
+ Body parts + Text	**0.806**

Table 2. Results of character grounding (**left**) and text re-identification (**right**) on the validation set of M-VAD Names [30]. Attr, Text, Visual and Grounding means joint training with the attribute module, Text Re-Id Graph, Video Re-Id Graph and Character Grounding Graph, respectively.

Character Grounding	Accuracy	Text Re-Id	Accuracy
M-VAD Names [30]	0.621	BERT baseline [6]	0.734
MPII-MD Co-ref [33]	0.622	JSFusion [48] with BERT	0.744
CiSIN w/o motion	0.606	CiSIN w/o Grounding & Visual	0.743
CiSIN w/o Attr & Text	0.651	CiSIN w/o Visual	0.754
CiSIN w/o Text	0.673	CiSIN w/o Text	0.737
CiSIN	**0.684**	CiSIN	**0.767**

4.2 Quantitative Results

M-VAD Names. Table 1 summarizes the quantitative results of video re-identification tasks. We compare with state-of-the-art strong-ReID-baseline [25] using the official implementation[2] which is pretrained on market-1501 [51]. For fair comparison, we choose the same ResNet-50 as the CNN backbone. We also test other visual cue-based baselines [15,16,50] using the same face and body features with CiSIN. We then fine-tune the model with M-VAD Names dataset.

---

[2] https://github.com/michuanhaohao/reid-strong-baseline.

Despite its competence in person re-identification research, the strong-ReID-baseline significantly underperforms our model by 18.9% of accuracy drop. M-VAD Names dataset contains much more diverse person appearances in terms of postures, sizes and ill-posedness (*e.g.* extremely wide shots, partial and back views), which makes the existing re-identification model hard to attain competitive scores. In ablation studies, our model with only face embedding achieves a considerable gain of 16.6%, which implies that utilizing facial information is crucial for re-identification in movies. However, adding simple average-pooled human bbox embedding (+ Human bbox in Table 1) does not improve the performance. Instead, combining with our body-part embedding ($\mathbf{p}_m$) yields better accuracy, meaning that it can convey full-body information better than simple bbox embedding especially when human appearances are highly diverse. The variant (+ Text) in Table 1 uses the grounding and text matching to further improve video re-identification accuracy, as done in Eq. (11). We update the Video Re-Id Graph as $\mathbf{V} = \mathbf{V} + \lambda \mathbf{G}^T \mathbf{L} \mathbf{G}$, where λ is a trainable parameter, $\mathbf{G}$ is the bipartite Grounding Graph, and L is the Text Re-Id Graph. The result hints that contextual information inside the text helps improve re-identification in videos.

Table 2 presents the results of character grounding and text re-identification in M-VAD Names dataset. For character grounding, we report the results of two state-of-the-art models and different variants of our CiSIN model. Originally, the MPII-MD Co-ref [34] is designed to link the given character's name and gender to corresponding visual tracks. For fair comparison, we use the supervised version in [33] that utilizes ground truth track information, but do not provide exact character name nor gender information at test time. Our model outperforms existing models with large margins. Moreover, joint training enhances our grounding performance by 3.3%p than naïvely trained CiSIN w/o Attr & Text.

For the text-re identification task, we use JSFusion [48] and the BERT of our model with no other component as the baselines. JSFusion is the model that won the LSMDC 2017 and we modify its Fill-in-the blank model to be applicable to the new task of Fill-in the Characters. The CiSIN w/o Grounding & Visual indicates this BERT variant with additional joint training with the attribute loss. It enhances accuracy by 0.9%p and additional training with the Character Grounding graph further improves 1.1%p. Also, we report the score of using only Video Re-Id Graph and bipartite Grounding Graph, which is 73.7%. As expected, jointly training of the whole model increases the accuracy to 76.7%.

LSMDC 2019. Table 3 shows the results of the Fill-in the Characters task in LSMDC 2019 challenge. The baseline and human performance are referred to LSMDC 2019 official website[3]. Our model trained with both textual and visual context achieves 0.673, which is the best score for the benchmark. For the variant of CiSIN (Separate training), each component is separately trained with its own loss function. This model without joint training shows the performance drop by 2%p. Obviously, our model that lacks the text Re-Id graph significantly underperforms, meaning the association with text is critical for the task.

---

[3] https://sites.google.com/site/describingmovies/lsmdc-2019.

Table 3. Quantitative results of the "Fill-in the Characters in Description" task for blinded test dataset in LSMDC 2019 challenge. * denotes the scores from the final official scores reported in LSMDC 2019 challenge slides.

Fill-in the characters	Accuracy
Human median*	0.870
Human (w/o video) median*	0.700
Official baseline*	0.639
YASA*	0.648
CiSIN (Visual-only)	0.620
CiSIN (Text-only)	0.639
CiSIN (Separate training)	0.653
CiSIN	**0.673**

4.3 Qualitative Results

Fill-in the Characters. Figure 4 illustrates the results of CiSIN model for the Fill-in the Characters task with correct (left) and near-miss (right) examples. Figure 4(a) is a correct example where our model identifies characters in conversation. Most frames expose only the upper body of characters, where clothing and facial features become decisive clues. Figure 4(b) shows a failure case to distinguish two main characters [PERSON1,2] in black coats since the characters are seen in a distant side view. Figure 4(c) is another successful case; in the first and fourth video clips, the Video Re-Id Graph alone hardly identifies the character because he is too small (in the 1st clip) or only small parts (hands) appear (in the 4th clip). Nonetheless, our model can distinguish character identity from the story descriptions in text. In Fig. 4(d), although the model can visually identify the same characters in neighboring video clips, the character grounding module repeatedly pays false attention to the other character. In such cases, our model often selects the candidate that stands out the most.

Character Grounding. Figure 5 shows character grounding results with top six visual tracks. Figure 5(a) depicts three character grounding examples where our model uses action (*e.g.* shaving, showing a photo, climbing) and facial (*e.g.* shaving cream, cringe) information to pick the right character. Figure 5(b) shows the effect of the Text Re-Id Graph. In the first clip, two people are standing still in front of the lobby, and our grounding module would select [PERSON2] as an incorrect prediction without the Text Re-Id Graph. However, it can later capture the coherence of the text that the mentioned person is the same.

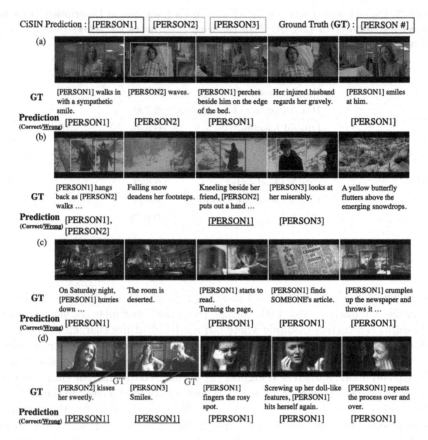

Fig. 4. Qualitative results of our CiSIN model. (a)–(d) are the Fill-in the Characters examples with median frames of the character grounding trajectories. Green examples indicate correct inferences by our model, while red ones with underlines in (b) and (d) are incorrect. Note that the bounding boxes are generated by our model.

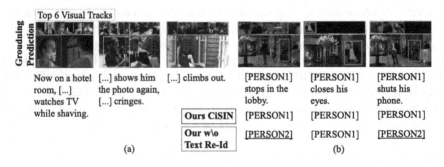

Fig. 5. Qualitative results of character grounding by the CiSIN model. (a) Examples of character grounding where [...] denotes a [SOMEONE] token. (b) Examples of Fill-in the Characters with and without the Text Re-Id Graph.

5 Conclusion

We proposed the Character-in-Story Identification Network (CiSIN) model for character grounding and re-identification in a sequence of videos and descriptive sentences. The two key representations of the model, Visual Track Embedding and Textualized Character Embedding, are easily adaptable in many video retrieval tasks, including person retrieval with free-formed language queries. We demonstrated that our method significantly improved the performance of video character grounding and re-identification in multiple clips; our method achieved the best performance in a challenge track of LSMDC 2019 and outperformed existing baselines for both tasks on M-VAD Names dataset.

Moving forward, there may be some interesting future works to expand the applicability of the CiSIN model. First, we can explore human retrieval and re-identification tasks in other domains of videos. Second, as this work only improved the re-identification of those mentioned in descriptions, we can integrate with a language generation or summarization module to better understand the details of a specific target person in the storyline.

Acknowledgement. We thank SNUVL lab members for helpful comments. This research was supported by Seoul National University, Brain Research Program by National Research Foundation of Korea (NRF) (2017M3C7A1047860), and AIR Lab (AI Research Lab) in Hyundai Motor Company through HMC-SNU AI Consortium Fund.

References

1. Ahmed, E., Jones, M., Marks, T.: An improved deep learning architecture for person re-identification. In: CVPR (2015)
2. Bojanowski, P., Bach, F., Laptev, I., Ponce, J., Schmid, C., Sivic, J.: Finding actors and actions in movies. In: ICCV (2013)
3. Carreira, J., Zisserman, A.: Quo vadis, action recognition? A new model and the kinetics dataset. In: CVPR (2017)
4. Cheng, D., Gong, Y., Zhou, S., Wang, J., Zheng, N.: Person re-identification by multi-channel parts-based CNN with improved triplet loss function. In: CVPR (2016)
5. Deng, J., Guo, J., Niannan, X., Zafeiriou, S.: ArcFace: additive angular margin loss for deep face recognition. In: CVPR (2019)
6. Devlin, J., Chang, M.W., Lee, K., Toutanova, K.: BERT: pre-training of deep bidirectional transformers for language understanding. In: NAACL-HLT (2019)
7. Everingham, M., Sivic, J., Zisserman, A.: "Hello! my name is... buffy"-automatic naming of characters in TV video. In: BMVC (2006)
8. Farenzena, M., Bazzani, L., Perina, A., Murino, V., Cristani, M.: Person re-identification by symmetry-driven accumulation of local features. In: CVPR (2010)
9. Gheissari, N., Sebastian, T.B., Hartley, R.: Person reidentification using spatiotemporal appearance. In: CVPR (2006)
10. Gray, D., Tao, H.: Viewpoint invariant pedestrian recognition with an ensemble of localized features. In: Forsyth, D., Torr, P., Zisserman, A. (eds.) ECCV 2008. LNCS, vol. 5302, pp. 262–275. Springer, Heidelberg (2008). https://doi.org/10.1007/978-3-540-88682-2_21

11. He, K., Zhang, X., Ren, S., Sun, J.: Deep residual learning for image recognition. In: CVPR (2016)
12. Hodosh, M., Young, P., Hockenmaier, J.: Framing image description as a ranking task: data, models and evaluation metrics. JAIR **47**, 853–899 (2013)
13. Hu, R., Xu, H., Rohrbach, M., Feng, J., Saenko, K., Darrell, T.: Natural language object retrieval. In: CVPR (2016)
14. Huang, Q., Liu, W., Lin, D.: Person search in videos with one portrait through visual and temporal links. In: Ferrari, V., Hebert, M., Sminchisescu, C., Weiss, Y. (eds.) ECCV 2018. LNCS, vol. 11217, pp. 437–454. Springer, Cham (2018). https://doi.org/10.1007/978-3-030-01261-8_26
15. Huang, Q., Xiong, Y., Lin, D.: Unifying identification and context learning for person recognition. In: CVPR (2018)
16. Joon Oh, S., Benenson, R., Fritz, M., Schiele, B.: Person recognition in personal photo collections. In: ICCV (2015)
17. Kingma, D., Ba, J.: Adam: a method for stochastic optimization. In: ICLR (2015)
18. Kiros, R., Salakhutdinov, R., Zemel, R.S.: Unifying visual-semantic embeddings with multimodal neural language models. TACL (2014)
19. Kottur, S., Moura, J.M.F., Parikh, D., Batra, D., Rohrbach, M.: Visual coreference resolution in visual dialog using neural module networks. In: Ferrari, V., Hebert, M., Sminchisescu, C., Weiss, Y. (eds.) ECCV 2018. LNCS, vol. 11219, pp. 160–178. Springer, Cham (2018). https://doi.org/10.1007/978-3-030-01267-0_10
20. Li, S., Bak, S., Carr, P., Wang, X.: Diversity regularized spatiotemporal attention for video-based person re-identification. In: CVPR (2018)
21. Li, S., Xiao, T., Li, H., Yang, W., Wang, X.: Identity-aware textual-visual matching with latent co-attention. In: ICCV (2017)
22. Li, S., Xiao, T., Li, H., Zhou, B., Yue, D., Wang, X.: Person search with natural language description. In: CVPR (2017)
23. Li, W., Zhao, R.R., Xiao, T., Wang, X.: DeepReID: deep filter pairing neural network for person re-identification. In: CVPR (2014)
24. Lin, D., Fidler, S., Kong, C., Urtasun, R.: Visual semantic search: retrieving videos via complex textual queries. In: CVPR (2014)
25. Luo, H., Gu, Y., Liao, X., Lai, S., Jiang, W.: Bag of tricks and a strong baseline for deep person re-identification. In: CVPR Workshop (2019)
26. Mao, J., Huang, J., Toshev, A., Camburu, O., Yuille, A.L., Murphy, K.: Generation and comprehension of unambiguous object descriptions. In: CVPR (2016)
27. Nagrani, A., Zisserman, A.: From benedict cumberbatch to sherlock holmes: character identification in TV series without a script. In: BMVC (2017)
28. Otani, M., Nakashima, Y., Rahtu, E., Heikkilä, J., Yokoya, N.: Learning joint representations of videos and sentences with web image search. In: Hua, G., Jégou, H. (eds.) ECCV 2016. LNCS, vol. 9913, pp. 651–667. Springer, Cham (2016). https://doi.org/10.1007/978-3-319-46604-0_46
29. Parkhi, O.M., Rahtu, E., Zisserman, A.: It's in the bag: stronger supervision for automated face labelling. In: ICCV Workshop (2015)
30. Pini, S., Cornia, M., Bolelli, F., Baraldi, L., Cucchiara, R.: M-VAD names: a dataset for video captioning with naming. MTA **78**, 14007–14027 (2019)
31. Qi, P., Dozat, T., Zhang, Y., Manning, C.D.: Universal dependency parsing from scratch. In: CoNLL 2018 UD Shared Task (2018)
32. Ramanathan, V., Joulin, A., Liang, P., Fei-Fei, L.: Linking people in videos with "their" names using coreference resolution. In: Fleet, D., Pajdla, T., Schiele, B., Tuytelaars, T. (eds.) ECCV 2014. LNCS, vol. 8689, pp. 95–110. Springer, Cham (2014). https://doi.org/10.1007/978-3-319-10590-1_7

33. Rohrbach, A., Rohrbach, M., Hu, R., Darrell, T., Schiele, B.: Grounding of textual phrases in images by reconstruction. In: Leibe, B., Matas, J., Sebe, N., Welling, M. (eds.) ECCV 2016. LNCS, vol. 9905, pp. 817–834. Springer, Cham (2016). https://doi.org/10.1007/978-3-319-46448-0_49
34. Rohrbach, A., Rohrbach, M., Tang, S., Oh, S.J., Schiele, B.: Generating descriptions with grounded and co-referenced people. In: CVPR (2017)
35. Rohrbach, A., et al.: Movie description. IJCV **123**, 94–120 (2017)
36. Seo, P.H., Lehrmann, A., Han, B., Sigal, L.: Visual reference resolution using attention memory for visual dialog. In: NIPS (2017)
37. Shen, Y., Lin, W., Yan, J., Xu, M., Wu, J., Wang, J.: Person re-identification with correspondence structure learning. In: ICCV (2015)
38. Sivic, J., Everingham, M., Zisserman, A.: "Who are you?"-learning person specific classifiers from video. In: CVPR (2009)
39. Su, C., Li, J., Zhang, S., Xing, J., Gao, W., Tian, Q.: Pose-driven deep convolutional model for person re-identification. In: ICCV (2017)
40. Tapaswi, M., Bäuml, M., Stiefelhagen, R.: "Knock! Knock! Who is it?" Probabilistic person identification in TV-series. In: CVPR (2012)
41. Torabi, A., Pal, C., Larochelle, H., Courville, A.: Using descriptive video services to create a large data source for video annotation research. arXiv:1503.01070 (2015)
42. Torabi, A., Tandon, N., Sigal, L.: Learning language-visual embedding for movie understanding with natural-language. arXiv:1609.08124 (2016)
43. Vendrov, I., Kiros, R., Fidler, S., Urtasun, R.: Order-embeddings of images and language. In: ICLR (2016)
44. Wei, L., Zhang, S., Yao, H., Gao, W., Tian, Q.: GLAD: global-local-alignment descriptor for pedestrian retrieval. In: ACM MM (2017)
45. Wojke, N., Bewley, A., Paulus, D.: Simple online and realtime tracking with a deep association metric. In: ICIP (2017)
46. Xu, R., Xiong, C., Chen, W., Corso, J.J.: Jointly modeling deep video and compositional text to bridge vision and language in a unified framework. In: AAAI (2015)
47. Yan, Y., Zhang, Q., Ni, B., Zhang, W., Xu, M., Yang, X.: Learning context graph for person search. In: CVPR (2019)
48. Yu, Y., Kim, J., Kim, G.: A joint sequence fusion model for video question answering and retrieval. In: Ferrari, V., Hebert, M., Sminchisescu, C., Weiss, Y. (eds.) ECCV 2018. LNCS, vol. 11211, pp. 487–503. Springer, Cham (2018). https://doi.org/10.1007/978-3-030-01234-2_29
49. Zhang, K., Zhang, Z., Li, Z., Qiao, Y.: Joint face detection and alignment using multitask cascaded convolutional networks. In: IEEE Signal Proceedings (2016)
50. Zhang, N., Paluri, M., Taigman, Y., Fergus, R., Bourdev, L.: Beyond frontal faces: improving person recognition using multiple cues. In: CVPR (2015)
51. Zheng, L., Shen, L., Tian, L., Wang, S., Wang, J., Tian, Q.: Scalable person re-identification: a benchmark. In: ICCV (2015)
52. Zheng, L.: MARS: a video benchmark for large-scale person re-identification. In: Leibe, B., Matas, J., Sebe, N., Welling, M. (eds.) ECCV 2016. LNCS, vol. 9910, pp. 868–884. Springer, Cham (2016). https://doi.org/10.1007/978-3-319-46466-4_52
53. Zhou, X., Wang, D., Krähenbühl, P.: Objects as points. arXiv:1904.07850 (2019)

AABO: Adaptive Anchor Box Optimization for Object Detection via Bayesian Sub-sampling

Wenshuo Ma[1], Tingzhong Tian[1], Hang Xu[2(✉)], Yimin Huang[2],
and Zhenguo Li[2]

[1] Tsinghua University, Beijing, China
[2] Huawei Noah's Ark Lab, Hong Kong, China
xbjxh@live.com

Abstract. Most state-of-the-art object detection systems follow an anchor-based diagram. Anchor boxes are densely proposed over the images and the network is trained to predict the boxes position offset as well as the classification confidence. Existing systems pre-define anchor box shapes and sizes and ad-hoc heuristic adjustments are used to define the anchor configurations. However, this might be sub-optimal or even wrong when a new dataset or a new model is adopted. In this paper, we study the problem of automatically optimizing anchor boxes for object detection. We first demonstrate that the number of anchors, anchor scales and ratios are crucial factors for a reliable object detection system. By carefully analyzing the existing bounding box patterns on the feature hierarchy, we design a flexible and tight hyper-parameter space for anchor configurations. Then we propose a novel hyper-parameter optimization method named AABO to determine more appropriate anchor boxes for a certain dataset, in which Bayesian Optimization and sub-sampling method are combined to achieve precise and efficient anchor configuration optimization. Experiments demonstrate the effectiveness of our proposed method on different detectors and datasets, e.g. achieving around 2.4% mAP improvement on COCO, 1.6% on ADE and 1.5% on VG, and the optimal anchors can bring 1.4%–2.4% mAP improvement on SOTA detectors by only optimizing anchor configurations, e.g. boosting Mask RCNN from 40.3% to 42.3%, and HTC detector from 46.8% to 48.2%.

Keywords: Object detection · Hyper-parameter optimization · Bayesian optimization · Sub-sampling

1 Introduction

Object detection is a fundamental and core problem in many computer vision tasks and is widely applied on autonomous vehicles [3], surveillance camera [21],

Electronic supplementary material The online version of this chapter (https://doi.org/10.1007/978-3-030-58558-7_33) contains supplementary material, which is available to authorized users.

© Springer Nature Switzerland AG 2020
A. Vedaldi et al. (Eds.): ECCV 2020, LNCS 12350, pp. 560–575, 2020.
https://doi.org/10.1007/978-3-030-58558-7_33

facial recognition [2], to name a few. Object detection aims to recognize the location of objects and predict the associated class labels in an image. Recently, significant progress has been made on object detection tasks using deep convolution neural network [17,20,25,27]. In many of those deep learning based detection techniques, anchor boxes (or default boxes) are the fundamental components, serving as initial suggestions of object's bounding boxes. Specifically, a large set of densely distributed anchors with pre-defined scales and aspect ratios are sampled uniformly over the feature maps, then both shape offsets and position offsets relative to the anchors, as well as classification confidence, are predicted using a neural network.

While anchor configurations are rather critical hyper-parameters of the neural network, the design of anchors always follows straight-forward strategies like handcrafting or using statistical methods such as clustering. Taking some widely used detection frameworks for instance, Faster R-CNN [27] uses pre-defined anchor shapes with 3 scales ($128^2, 256^2, 512^2$) and 3 aspect ratios ($1 : 1$, $1 : 2$, $2 : 1$), and YOLOv2 [26] models anchor shapes by performing k-means clustering on the ground-truth of bounding boxes. And when the detectors are extended to a new certain problem, anchor configurations must be manually modified to adapt the property and distribution of this new domain, which is difficult and inefficient, and could be sub-optimal for the detectors.

While it is irrational to determine hyper-parameters manually, recent years have seen great development in hyper-parameter optimization (HPO) problems and a great quantity of HPO methods are proposed. The most efficient methods include Bayesian Optimization (BO) and bandit-based policies. BO iterates over the following three steps: a) Select the point that maximizes the acquisition function. b) Evaluate the objective function. c) Add the new observation to the data and refit the model, which provides an efficacious method to select promising hyper-parameters with sufficient resources. Different from BO, Bandit-based policies are proposed to efficiently measure the performance of hyper-parameters. Among them, Hyperband [16] (HB) makes use of cheap-to-evaluate approximations of the acquisition function on smaller budgets, which calls SuccessiveHalving [13] as inner loop to identify the best out of n randomly-sampled configurations. Bayesian Optimization and Hyperband (BOHB) introduced in [9] combined these two methods to deal with HPO problems in a huge search space, and it is regarded as a very advanced HPO method. However, BOHB is less applicable to our anchor optimization problems, because the appropriate anchors for small objects are always hard to converge, then the optimal anchor configurations could be early-stopped and discarded by SuccesiveHalving.

In this paper, we propose an adaptive anchor box optimization method named AABO to automatically discover optimal anchor configurations, which can fully exploit the potential of the modern object detectors. Specifically, we illustrate that anchor configurations such as the number of anchors, anchor scales and aspect ratios are crucial factors for a reliable object detector, and demonstrate that appropriate anchor boxes can improve the performance of object detection systems. Then we prove that anchor shapes and distributions vary distinctly across different feature maps, so it is irrational to share identical anchor settings through all those feature maps. So we design a tight and adaptive search space for feature map pyramids after meticulous analysis of the distribution and pattern of the bounding boxes in existing datasets, to make full use of the search resources. After optimizing the anchor search space, we propose a novel hyper-parameter optimization method combining the benefits of both Bayesian Optimization and sub-sampling method. Compared with existing HPO methods, our proposed approach uses sub-sampling method to estimate acquisition function as accurately as possible, and gives opportunity to the configuration to be assigned with more budgets if it has chance to be the best configuration, which can ensure that the promising configurations will not be discarded too early. So our method can efficiently determine more appropriate anchor boxes for a certain dataset using limited computation resources, and achieves better performance than previous HPO methods such as random search and BOHB.

We conduct extensive experiments to demonstrate the effectiveness of our proposed approach. Significant improvements over the default anchor configurations are observed on multiple benchmarks. In particular, AABO achieves 2.4% mAP improvement on COCO, 1.6% on ADE and 1.5% on VG by only changing the anchor configurations, and consistently improves the performance of SOTA detectors by 1.4%–2.4%, e.g. boosts Mask RCNN [10] from 40.3% to 42.3% and HTC [6] from 46.8% to 48.2% in terms of mAP.

2   Related Work

Anchor-Based Object Detection. Modern object detection pipelines based on CNN can be categorized into two groups: One-stage methods such as SSD [20] and YOLOv2 [26], and two-stage methods such as Faster R-CNN [27] and R-FCN [8]. Most of those methods make use of a great deal of densely distributed anchor boxes. In brief, those modern detectors regard anchor boxes as initial references to the bounding boxes of objects in an image. The anchor shapes in those methods are typically determined by manual selection [8,20,27] or naive clustering methods [26]. Different from the traditional methods, there are several works focusing on utilizing anchors more effectively and efficiently [31,33]. MetaAnchor [31] introduces meta-learning to anchor generation, which models anchors using an extra neural network and computes anchors from customized priors. However, the network becomes more complicated. Zhong et al. [33] tries to learn the anchor shapes during training via a gradient-based method while the continuous relaxation may be not appropriate.

Hyper-paramter Optimization. Although deep learning has achieved great successes in a wide range, the performance of deep learning models depends strongly on the correct setting of many internal hyper-parameters, which calls for an effective and practical solution to the hyper-parameter optimization (HPO) problems. Bayesian Optimization (BO) has been successfully applied to many HPO works. For example, [29] obtained state-of-the-art performance on CIFAR-10 using BO to search out the optimal hyper-parameters for convolution

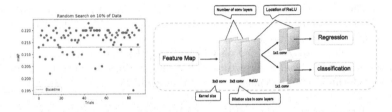

Fig. 1. (Left) The performance of Faster-R-CNN [27] under different anchor configurations on 10% data of COCO [19]. Randomly-sampled anchors significantly influence the performance of the detector. (Right) The search space for RPN Head in FPN [17] consists of the number of convolution layers, kernel size, dilation size, and the location of nonlinear activation functions ReLU, which are illustrated in the dialogue boxes.

neural networks. And [22] won 3 datasets in the 2016 AutoML challenge by automatically finding the proper architecture and hyper-parameters via BO methods. While BO approach can converge to the best configurations theoretically, it requires an awful lot of resources and is typically computational expensive. Compared to Bayesian method, there exist bandit-based configuration evaluation approaches based on random search such as Hyperband [16], which could dynamically allocate resources and use SuccessiveHalving [13] to stop poorly performing configurations. Recently, some works combining Bayesian Optimization with Hyperband are proposed like BOHB [9], which can obtain strong performance as well as fast convergence to optimal configurations. Other non-parametric methods that have been proposed include ε-greedy and Boltzmann exploration [30]. [4] proposed an efficient non-parametric solution and proved optimal efficiency of the policy which would be extended in our work. However, there exist some problems in those advanced HPO methods such as expensive computation in BO and early-stop in BOHB.

3   The Proposed Approach

3.1   Preliminary Analysis

As mentioned before, mainstream detectors, including one-stage and two-stage detectors, rely on anchor boxes to provide initial guess of the object's bounding box. And most detectors pre-define anchors and manually modify them when applied on new datasets. We believe that those manual methods can hardly find optimal anchor configurations and sub-optimal anchors will prevent the detectors from obtaining the optimal performance. To confirm this assumption, we construct two preliminary experiments.

Default Anchors Are Not Optimal. We randomly sample 100 sets of different anchor settings, each with 3 scales and 3 ratios. Then we examine the performance of Faster-RCNN [27] under those anchor configurations. The results are shown in Fig. 1.

Table 1. The respective contributions of RPN head architecture and anchor configurations. Anchor optimization could obviously improve the mAP of the detectors while architecture optimization produces very little positive effect, or even negative effect. All the experiments are conducted on COCO, using FPN as detector. Note that the search space for RPN head architecture and anchor settings are both relatively small, then the performance improvements are not that significant.

Performance	mAP	AP_{50}	AP_{75}	AP_S	AP_M	AP_L
Baseline	36.4	58.2	39.1	21.3	40.1	46.5
Only Anchor	$37.1^{+0.7}$	58.4	40.0	20.6	40.8	49.5
Only Architecture	$36.3^{-0.1}$	58.3	38.9	21.6	40.2	46.1
Anchor+Architecture	$\mathbf{37.2^{+0.8}}$	**58.7**	**40.2**	**20.9**	**41.3**	**49.1**

It's obvious that compared with default anchor setting (3 anchor scales: $128^2, 256^2, 512^2$ and 3 aspect ratios: $1:1$, $1:2$, and $2:1$), randomly-sampled anchor settings could significantly influence the performance of the detector, which clearly demonstrates that the default anchor settings may be less appropriate and the optimization of anchor boxes is necessary.

Anchors Influence More than RPN Structure. Feature Pyramid Networks [17] (FPN) introduces a top-down pathway and lateral connections to enhance the semantic representation of low-level features and is a widely used feature fusion scheme in modern detectors. In this section, we use BOHB [9] to search RPN head architecture in FPN as well as anchor settings simultaneously. The search space of RPN Head is illustrated in Fig. 1. The searched configurations, including RPN head architecture and anchor settings, are reported in the appendix. Then we analyze the respective contributions of RPN head architecture and anchor settings, and the results are shown in Table 1. Here, mean Average Precision is used to measure the performance and is denoted by mAP.

The results in Table 1 illustrate that searching for proper anchor configurations could bring more performance improvement than searching for RPN head architecture to a certain extent.

Thus, the conclusion comes clearly that anchor settings affect detectors substantially, and proper anchor settings could bring considerable improvement than doing neural architecture search (NAS) on the architecture of RPN head. Those conclusions indicate that the optimization of anchor configurations is essential and rewarding, which motivates us to view anchor configurations as hyperparameters and propose a better HPO method for our anchor optimization case.

3.2   Search Space Optimization for Anchors

Since we have decided to search appropriate anchor configurations to increase the performance of detectors over a certain dataset, one critical problem is how to design the search space. In the preliminary analysis, we construct two experiments whose search space is roughly determined regardless of the distribution

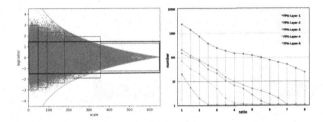

Fig. 2. (Left) Bounding boxes in COCO [19] dataset apparently only distribute in a certain area determined by the yellow curves. The region inside the black rectangle represents the previous search space which is coarse and unreasonable as analyzed. The intersecting regions between the 5 red rectangles and the anchor distribution bounds are our designed feature-map-wise search space, which is more accurate and adaptive. (Right) Numbers and shapes of bounding boxes vary a lot across different feature maps. In this figure, X-axis is the anchor ratio while Y-axis is the number of bounding boxes. It's obvious that the number of bounding boxes decreases rapidly, and the range of anchor ratios also becomes rather smaller in higher feature map. (Color figure online)

of bounding boxes. In this section, we will design a much tighter search space by analyzing the distribution characteristics of object's bounding boxes.

For a certain detection task, we find that anchor distribution satisfies some properties and patterns as follows.

Upper and Lower Limits of the Anchors. Note that the anchor scale and anchor ratio are calculated from the *width* and *height* of the anchor, which are not independent. Besides, we discover that both anchor *width* and *height* are limited within fixed values, denoted by W and H. Then the anchor ratio and scale must satisfy constraints as follows:

$$\begin{cases} scale = \sqrt{width * height} \\ ratio = height/width \\ width \leq W, \ height \leq H. \end{cases} \quad (1)$$

From the formulas above, we calculate the upper bound and lower bound of the ratio for the anchor boxes:

$$\frac{scale^2}{W^2} \leq ratio \leq \frac{H^2}{scale^2}. \quad (2)$$

Figure 2 shows an instance of the distribution of bounding boxes (the blue points) as well as the upper and lower bounds of anchors (the yellow curves) in COCO [19] dataset. And the region inside the black rectangle is the previous search space using in preliminary experiments. We can observe that there exists an area which is beyond the upper and lower bounds so that bounding boxes won't appear, while the search algorithm will still sample anchors here. So it's necessary to limit the search space within the upper and lower bounds.

Adaptive Feature-Map-Wised Search Space. We then study the distribution of anchor boxes in different feature maps of Feature Pyramid Networks

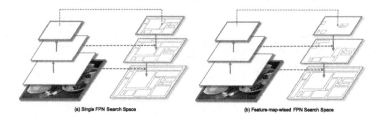

(a) Single FPN Search Space   (b) Feature-map-wised FPN Search Space

Fig. 3. We design an adaptive feature-map-wised search space for anchor configuration optimization. Compared to the former single search space illustrated in (a), where there exist exactly identical anchors among different feature maps, feature-map-wised search space illustrated in (b) takes bounding-box distribution into consideration, so extreme anchors are fewer in higher feature maps. That is, in lower feature layers, there are more diverse, larger and more anchors, while in higher feature layers, there are less diverse, smaller and fewer anchors.

(FPN) [17] and discover that the numbers, scales and ratios of anchor boxes vary a lot across different feature maps, shown in the right subgraph of Fig. 2. There are more and bigger bounding boxes in lower feature maps whose receptive fields are smaller, less and tinier bounding boxes in higher feature maps whose respective fields are wider.

As a result, we design an adaptive search space for FPN [17] as shown in the left subgraph of Fig. 2. The regions within 5 red rectangles and the anchor distribution bounds represent the search space for each feature map in FPN. As feature maps become higher and smaller, the numbers of anchor boxes are less, as well as the anchor scales and ratios are limited to a narrower range.

Compared with the initial naive search space, we define a tighter and more adaptive search space for FPN [17], as shown in Fig. 3. Actually, the new feature-map-wised search space is much bigger than the previous one, which makes it possible to select more flexible anchors and cover more diverse objects with different sizes and shapes. Besides, the tightness of the search space can help HPO algorithms concentrate limited resources on more meaningful areas and avoid wasting resources in sparsely distributed regions of anchors.

3.3   Bayesian Anchor Optimization via Sub-sampling

As described before, we regard anchor configurations as hyper-parameters and try to use HPO method to choose optimal anchor settings automatically. However, existing HPO methods are not suitable for solving our problems. For random search or grid search, it's unlikely to find a good solution because the search space is too big for those methods. For Hyperband or BOHB, the proper configurations for the small objects could be discarded very early since the anchors of small objects are always slowly-converged. So we propose a novel method which combines Bayesian Optimization and sub-sampling method, to search out the optimal configurations as quickly as possible.

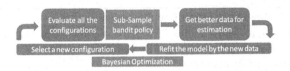

Fig. 4. Our proposed method iterates over the following four steps: (a) Select the point that maximizes the acquisition function. (b) Evaluate the objective function on the whole configurations with Sub-Sampling policy. (c) Get more appropriate data for estimating the densities in the model. (d) Add the new observation to the data and refit the model.

Specifically, our proposed approach makes use of BO to select potential configurations, which estimates the acquisition function based on the configurations already evaluated, and then maximizes the acquisition function to identify promising new configurations. Meanwhile, sub-sampling method is employed to determine which configurations should be allocated more budgets, and explore more configurations in the search space. Figure 4 illustrates the process of our proposed method. In conclusion, our approach can achieve good performance as well as better speed, and take full advantage of models built on previous budgets.

Bayesian Optimization. Bayesian Optimization (BO) is a sequential design strategy for parameter optimization of black-box functions. In hyper-parameter optimization problems, the validation performance of machine learning algorithms is regarded as a function $f : \chi \to \mathbb{R}$ of hyper-parameters $x \in \chi$, and hyper-parameter optimization problem aims to determine the optimal $x_* \in argmin_\chi f(x)$. In most machine learning problems, $f(x)$ is unobservable, so Bayesian Optimization approach treats it as a random function with a prior over it. Then some data points are sampled and BO updates the prior and models the function based on those gathering data points and evaluations. Then new data points are selected and observed to refit the model function.

In our approach, we use Tree Parzen Estimator (TPE) [1] which uses a kernel density estimator to model the probability density functions $l(x) = p(y < \alpha|x, D)$ and $g(x) = p(y > \alpha|x, D)$ instead of modeling function $p(f|D)$ directly, where $D = \{(x_0, y_0), \ldots, (x_n, y_n)\}$ and $\alpha = \min\{y_0, \ldots, y_n\}$, as seen in BOHB [9]. Note that there exists a serious problem that the theory of Hyperband only guarantees to return the best x_i which has the smallest y_i among all these configurations, while the quality of other configurations may be very poor. Consequently, larger responses lead to an inaccurate estimation of $l(x)$, which plays an important role in TPE. Thus, we need to propose a policy that has a better sequence of responses y_i to solve this problem.

Sub-sample Method. To better explain the sub-sampling method used in our proposed approach, we first introduce the standard multi-armed bandit problem with K arms in this section. Recall the traditional setup for the classic multi-armed bandit problem. Let $\mathcal{I} = \{1, 2, \ldots, K\}$ be a given set of $K \geq 2$ arms. Consider a sequential procedure based on past observations for selecting an arm to pull. Let N_k be the number of observations from the arm k, and $N = \sum_{k=1}^{K} N_k$

Algorithm 1. Sub-sample Mean Comparisons.

Input: The set of configurations $\mathcal{I} = \{1, \ldots, K\}$, parameter c_n, minimum budget b.
Output: $\hat{\pi}_1, \ldots, \hat{\pi}_N \in \mathcal{I}$.
  1: $r = 1$, evaluate each configuration with budget b.
  2: **for** $r = 2, 3, \ldots$ **do**
  3:      The configuration with the most budgets is denoted by ζ^r and called the leader;
  4:      **for** $k \neq \zeta^r$ **do**
  5:         Evaluate the k-th configuration with one more budget b if it is "better" than
      the ζ^r-th configuration.
  6:      **end for**
  7:      If there is no configuration "better" than the leader, evaluate the leader with
      one more budget b.
  8: **end for**

---

is the number of total observations. Observations $Y_1^{(k)}, Y_2^{(k)}, \ldots$, $1 \leq k \leq K$ are also called rewards from the arm k. In each arm, rewards $\{Y_t^{(k)}\}_{t \geq 1}$ are assumed to be independent and identically distributed with expectation given by $\mathbb{E}(Y_t^{(k)}) = \mu_k$ and $\mu_* = \max_{1 \leq k \leq K} \mu_k$. For simplicity, assume without loss of generality that the best arm is *unique* which is also assumed in [24] and [4].

A *policy* $\pi = \{\pi_t\}$ is a sequence of random variables $\pi_t \in \{1, 2, \ldots, K\}$ denoting that at each time $t = 1, 2, \ldots, N$, the arm π_t is selected to pull. Note that π_t depends only on previous $t - 1$ observations. The objective of a good policy π is to minimize the *regret*

$$R_N(\pi) = \sum_{k=1}^{K} (\mu_* - \mu_k) \mathbb{E} N_k = \sum_{t=1}^{N} (\mu_* - \mu_{\pi_t}). \tag{3}$$

Note that for a data-driven policy $\hat{\pi}$, the regret monotonically increases with respect to N. Hence, minimizing the growth rate of R_N becomes an important criterion which will be considered later.

Then we introduce an efficient nonparametric solution to the multi-armed bandit problem. First, we revisit the Sub-sample Mean Comparisons (SMC) introduced in [4] for the HPO case. The set of configurations $\mathcal{I} = \{1, \ldots, K\}$, minimum budget b and parameter c_n are inputs. Output is the sequence of the configurations $\hat{\pi}_1, \ldots, \hat{\pi}_N \in \mathcal{I}$ to be evaluated in order.

First, it is defined that the k-th configuration is "better" than the k'-th configuration, if one of the following conditions holds:

1. $n_k < n_{k'}$ and $n_k < c_n$.
2. $c_n \leq n_k < n_{k'}$ and $\bar{Y}_{1:n_k}^{(k)} \geq \bar{Y}_{j:(j+n_k-1)}^{(k')}$, for some $1 \leq j \leq n_{k'} - n_k + 1$, where $\bar{Y}_{l:u}^{(k)} = \sum_{v=l}^{u} Y_v^{(k)} / (u - l + 1)$.

In SMC, let r denotes the round number. In the first round, all configurations are evaluated since there is no information about them. In round $r \geq 2$, we define the leader of configurations which has been evaluated with the most budgets.

And, the k-th configuration will be evaluated with one more budget b, if it is "better" than the leader. Otherwise, if there is no configuration "better" than the leader, the leader will be evaluated again. Hence, in each round, there are at most $K - 1$ configurations and at least one configuration to be evaluated. Let n^r be the total number of evaluations at the beginning of round r, and n_k^r be the corresponding number from the k-th configuration. Then, we have $K + r - 2 \leq n \leq K + (K - 1)(r - 2)$.

The Sub-sample Mean Comparisons(SMC) is shown in Algorithm 1. In SMC, the parameter c_n is a non-negative monotone increasing threshold for SMC satisfied that $c_n = o(\log n)$ and $c_n / \log \log n \to \infty$ as $n \to \infty$. In [4], they set $c_n = \sqrt{\log n}$ for efficiency of SMC.

Note that when the round r ends, the number of evaluations n^r usually doesn't equal to N exactly, i.e., $n^r < N < n^{r+1}$. For this case, $N - n^r$ configurations are randomly chosen from the $n^{r+1} - n^r$ configurations selected by SMC in the r-th round. A main advantage of SMC is that unlike the UCB-based procedures, underlying probability distributions need not be specified. Still, it remains asymptotic optimal efficiency. The detailed discussion about theoretical results refers to [12].

4 Experiments

4.1 Datasets, Metrics and Implementation Details

We conduct experiments to evaluate the performance of our proposed method AABO on three object detection datasets: COCO 2017 [19], Visual Genome (VG) [15], and ADE [34]. COCO is a common object detection dataset with 80 object classes, containing 118K training images (*train*), 5K validation images (*val*) and 20K unannotated testing images (*test-dev*). VG and ADE are two large-scale object detection benchmarks with thousands of object classes. For COCO, we use the *train* split for training and *val* split for testing. For VG, we use release v1.4 and synsets [28] instead of raw names of the categories due to inconsistent annotations. Specifically, we consider two sets containing different target classes: VG_{1000} and VG_{3000}, with 1000 most frequent classes and 3000 most frequent classes respectively. In both VG_{1000} and VG_{3000}, we use 88K images for training, 5K images for testing, following [7,14]. For ADE, we consider 445 classes, use 20K images for training and 1K images for testing, following [7,14]. Besides, the ground-truths of ADE are given as segmentation masks, so we first convert them to bounding boxes for all instances before training.

As for evaluation, the results of the detection tasks are estimated with standard COCO metrics, including mean Average Precision (mAP) across IoU thresholds from 0.5 to 0.95 with an interval of 0.05 and AP_{50}, AP_{75}, as well as AP_S, AP_M and AP_L, which respectively concentrate on objects of small size $(32 \times 32-)$, medium size $(32 \times 32-96 \times 96)$ and large size $(96 \times 96+)$.

During anchor configurations searching, we sample anchor configurations and train Faster-RCNN [27] (combined with FPN [17]) using the sampled anchor configurations, then compare the performance of these models (regrading mAP

Table 2. The results of our proposed method on some large-scale methods. We use Faster-RCNN [27] combined with FPN [17] as detectors and ResNet-50 as backbones.

Dataset	Method	mAP	AP_{50}	AP_{75}	AP_S	AP_M	AP_L
COCO	Faster-RCNN w FPN	36.4	58.2	39.1	21.3	40.1	46.5
	Search via AABO	$\mathbf{38.8^{+2.4}}$	$\mathbf{60.7^{+2.5}}$	$\mathbf{41.6^{+2.5}}$	$\mathbf{23.7^{+2.4}}$	$\mathbf{42.5^{+2.4}}$	$\mathbf{51.5^{+5.0}}$
VG_{1000}	Faster-RCNN w FPN	6.5	12.0	6.4	3.7	7.2	9.5
	Search via AABO	$\mathbf{8.0^{+1.5}}$	$\mathbf{13.2^{+1.2}}$	$\mathbf{8.2^{+1.8}}$	$\mathbf{4.2^{+0.5}}$	$\mathbf{8.3^{+1.1}}$	$\mathbf{12.0^{+2.5}}$
VG_{3000}	Faster-RCNN w FPN	3.7	6.5	3.6	2.3	4.9	6.8
	Search via AABO	$\mathbf{4.2^{+0.5}}$	$\mathbf{6.9^{+0.4}}$	$\mathbf{4.6^{+1.0}}$	$\mathbf{3.0^{+0.7}}$	$\mathbf{5.8^{+0.9}}$	$\mathbf{7.9^{+1.1}}$
ADE	Faster-RCNN w FPN	10.3	19.1	10.0	6.1	11.2	16.0
	Search via AABO	$\mathbf{11.9^{+1.6}}$	$\mathbf{20.7^{+1.6}}$	$\mathbf{11.9^{+1.9}}$	$\mathbf{7.4^{+1.3}}$	$\mathbf{12.2^{+1.0}}$	$\mathbf{17.5^{+1.5}}$

as evaluation metrics) to reserve better anchor configurations and stop the poorer ones, following the method we proposed before. The search space is feature-map-wised as introduced. All the experiments are conducted on 4 servers with 8 Tesla V100 GPUs, using the Pytorch framework [5,23]. ResNet-50 [11] pre-trained on ImageNet [28] is used as the shared backbone networks. We use SGD (momentum = 0.9) with batch size of 64 and train 12 epochs in total with an initial learning rate of 0.02, and decay the learning rate by 0.1 twice during training.

4.2 Anchor Optimization Results

We first evaluate the effectiveness of AABO over 3 large-scale detection datasets: COCO [19], VG [15] (including VG_{1000} and VG_{3000}) and ADE [34]. We use Faster-RCNN [27] combined with FPN [17] as our detector, and the baseline model is FPN with default anchor configurations. The results are shown in Table 2 and the optimal anchors searched out by AABO are reported in the appendix.

It's obvious that AABO outperforms Faster-RCNN with default anchor settings among all the 3 datasets. Specifically, AABO improves mAP by 2.4% on COCO, 1.5% on VG_{1000}, 0.5% on VG_{3000}, and 1.6% on ADE. The results illustrate that the pre-defined anchor used in common-used detectors are not optimal. Treat anchor configurations as hyper-parameters and optimize them using AABO can assist to determine better anchor settings and improve the performance of the detectors without increasing the complexity of the network.

Note that the searched anchors increase all the AP metrics, and the improvements on AP_L are always more significant than AP_S and AP_M: AABO boosts AP_L by 5% on COCO, 2.5% on VG_{1000}, 1.1% on VG_{3000}, and 1.5% on ADE. These results indicate that anchor configurations determined by AABO concentrate better on all objects, especially on the larger ones. It can also be found that AABO is especially useful for the large-scale object detection dataset such as VG_{3000}. We conjecture that this is because the searched anchors can better capture the various sizes and shapes of objects in a large number of categories.

Table 3. Improvements on SOTA detectors over COCO *val.* The optimal anchors are applied on several SOTA detectors with different backbones.

Model		mAP	AP_{50}	AP_{75}	AP_S	AP_M	AP_L
FPN [17] w r101	Default	38.4	60.1	41.7	21.6	42.7	50.1
	Searched	$40.5^{+2.1}$	**61.8**	**43.3**	**23.4**	**43.6**	**51.3**
FPN [17] w x101	Default	40.1	62.0	43.8	24.0	44.8	51.7
	Searched	$42.0^{+1.9}$	**63.9**	**65.1**	**25.2**	**46.3**	**54.4**
Mask RCNN [10] w r101	Default	40.3	61.5	44.1	22.2	44.8	52.9
	Searched	$42.3^{+2.0}$	**63.6**	**46.3**	**26.1**	**46.3**	**55.3**
RetinaNet [18] w r101	Default	38.1	58.1	40.6	20.2	41.8	50.8
	Searched	$39.5^{+1.4}$	**60.2**	**41.9**	**21.7**	**42.7**	**53.7**
DCNv2 [35] w x101	Default	43.4	61.3	47.0	24.3	46.7	58.0
	Searched	$45.8^{+2.4}$	**67.5**	**49.7**	**28.9**	**49.4**	**60.9**
HTC [6] w x101	Default	46.8	66.2	51.2	28.0	50.6	62.0
	Searched	$48.2^{+1.4}$	**67.3**	**52.2**	**28.6**	**51.9**	**62.7**

4.3 Benefit of the Optimal Anchor Settings on SOTA Methods

After searching out optimal anchor configurations via AABO, we apply them on several other backbones and detectors to study the generalization property of the anchor settings. For backbone, we change ResNet-50 [11] to ResNet-101 [11] and ResNeXt-101 [32], with detector (FPN) and other conditions constant. For detectors, we apply our searched anchor settings on several state-of-the-art detectors: a) Mask RCNN [10], b) RetinaNet [18], which is a one-stage detector, c) DCNv2 [35], and d) Hybrid Task Cascade (HTC) [6], with different backbones: ResNet-101 and ResNeXt-101. All the experiments are conducted on COCO.

The results are reported in Table 3. We can observe that the optimal anchors can consistently boost the performance of SOTA detectors, whether one-stage methods or two-stage methods. Concretely, the optimal anchors bring 2.1% mAP improvement on FPN with ResNet-101, 1.9% on FPN with ResNeXt-101, 2.0% on Mask RCNN, 1.4% on RetinaNet, 2.4% on DCNv2, and 1.4% improvement on HTC. The results demonstrate that our optimal anchors can be widely applicable across different network backbones and SOTA detection algorithms, including both one-stage and two-stage detectors. We also evaluate these optimized SOTA detectors on COCO *test-dev*, and the results are reported in the appendix. The performance improvements on *val* split and *test-dev* are consistent.

4.4 Comparison with Other Optimization Methods

Comparison with Other Anchor Initialization Methods. In this section, we compare AABO with several existing anchor initialization methods: a) Pre-define anchor settings, which is used in most modern detectors. b) Use k-means to obtain clusters and treat them as default anchors, which is used in YOLOv2 [26].

Table 4. Comparison with other anchor initialization methods. Here HB denotes Hyperband while SS denotes sub-sampling. The experiments are conducted on COCO, using FPN (with ResNet-50) as detector. Note that feature-map-wised search space is much huger than the single one, so random search fails to converge.

Methods to determine anchor		mAP	AP_{50}	AP_{75}	AP_S	AP_M	AP_L
Manual methods	Pre-defined	36.4	58.2	39.1	21.3	40.1	46.5
Statistical methods	K-Means	$37.0^{+0.6}$	58.9	39.8	21.9	40.5	48.5
HPO	Random search	$11.5^{-24.9}$	19.1	8.2	4.5	10.1	13.6
	AABO w HB	$38.2^{+1.8}$	59.3	40.7	22.6	42.1	50.1
	AABO w SS	$\mathbf{38.8^{+2.4}}$	**60.7**	**41.6**	**23.7**	**42.5**	**51.5**

Table 5. The search efficiency of some HPO methods on COCO with FPN (with ResNet-50). HB denotes Hyperband while SS denotes sub-sampling.

Search space	Search method	mAP of optimal anchor	Number of searched parameters
Single	Random	37.2	100
	AABO w HB	$37.8^{+0.6}$	64
	AABO w SS	$38.3^{+1.1}$	64
Feature-map-wised	Random	11.5	100
	AABO w HB	$38.2^{+26.7}$	64
	AABO w SS	$\mathbf{38.8^{+27.3}}$	64

c) Use random search to determine anchors. d) Use AABO combined with Hyperband [16] to determine anchors. e) Use AABO (combined with sub-sampling) to determine anchors. Among all these methods, the latter three use HPO methods to select anchor boxes automatically, while a) and b) use naive methods like handcrafting and k-means. The results are recorded in Table 4.

Among all these anchor initialization methods, our proposed approach can boost the performance most significantly, bring 2.4% improvement than using default anchors, while the improvements of other methods including statistical methods and previous HPO methods are less remarkable. The results illustrate that the widely used anchor initialization approaches might be sub-optimal, while AABO can fully utilize the ability of advanced detection systems.

Comparison with Other HPO Methods. As Table 5 shows, our proposed method could find better anchor configurations in fewer trials and can improve the performance of the detector significantly: With single search space, AABO combined with HB and SS boosts the mAP of FPN [17] from 36.4% to 37.8% and 38.3% respectively, while random search only boosts 36.4% to 37.2%. With feature-map-wised search space, AABO combined with HB and SS can obtain 38.2% and 38.8% mAP respectively, while random search fails to converge due to the huge and flexible search space. The results illustrate the effectiveness and the high efficiency of our proposed approach.

Table 6. Comparison with previous anchor optimization methods: Zhong's method [33] and MetaAnchor [31]. The results are extracted from their papers respectively. Note that we search optimal anchors for Faster-RCNN originally, then directly apply them to RetinaNet. Therefore, the performance improvement on RetinaNet is not as significant as Faster-RCNN, but still better than the other methods.

| Methods | YOLOv2 | RetinaNet | Faster-RCNN | RetinaNet |
	w Zhong's Method [33]	w MetaAnchor [31]	w AABO	w AABO
mAP of Baseline	23.5	36.9	36.4	38.1
mAP after Optimization	$24.5^{+1.0}$	$37.9^{+1.0}$	$38.8^{+2.4}$	$\mathbf{39.5^{+1.4}}$

Table 7. Regard Bayesian Optimization (BO), Sub-sampling (SS), Feature-map-wised search space as key components of AABO, we study the effectiveness of all these components. HB denotes Hyperband and SS denotes sub-sampling. The experiments are conducted on COCO, using FPN (with ResNet-50) as detector. Note that random search fails to converge due to the huge feature-map-wised search space.

Model	Search	BO	SS	Feature-map-wised	mAP	AP_{50}	AP_{75}	AP_S	AP_M	AP_L
Default					36.4	58.2	39.1	21.3	40.1	46.5
Random search	✓				$37.2^{+0.8}$	58.8	39.9	21.7	40.6	48.1
	✓			✓	$11.5^{-24.9}$	19.1	8.2	4.5	10.1	13.6
AABO w HB	✓	✓			$37.8^{+1.4}$	58.9	40.4	22.7	41.3	49.9
	✓	✓		✓	$38.2^{+1.8}$	59.3	40.7	22.6	42.1	50.1
AABO w SS	✓	✓	✓		$38.3^{+1.9}$	59.6	40.9	22.9	42.2	50.8
	✓	✓	✓	✓	$\mathbf{38.8^{+2.4}}$	**60.7**	**41.6**	**23.7**	**42.5**	**51.5**

Comparison with Other Anchor Optimization Methods.

We also compare AABO with some previous anchor optimization methods like [33] and MetaAnchor [31]. As shown in Table 6, all these methods can boost the performance of detectors, while our method brings 2.4% mAP improvement on Faster-RCNN and 1.4% on RetinaNet, and the other two methods only bring 1.0% improvement, which demonstrates the superiority of our proposed AABO.

4.5 Ablation Study

In this section, we study the effects of all the components used in AABO: a) Treat anchor settings as hyper-parameters, then use HPO methods to search them automatically. b) Use Bayesian Optimization method. c) Use sub-sampling method to determine the reserved anchors. d) Feature-map-wised search space.

As shown in Table 7, using HPO methods such as random search to optimize anchors can bring 0.8% performance improvement, which indicates the default anchors are sub-optimal. Using single search space (not feature-map-wised), AABO combined with HB brings 1.4% mAP improvement, and AABO combined with SS brings 1.9% mAP improvement, which demonstrates the advantage of BO and accurate estimation of acquisition function. Besides, our tight and adaptive feature-map-wised search space can give a guarantee to search out better anchors with limited computation resources, and brings about 0.5% mAP improvement as well. Our method AABO can increase mAP by 2.4% overall.

5 Conclusion

In this work, we propose AABO, an adaptive anchor box optimization method for object detection via Bayesian sub-sampling, where optimal anchor configurations for a certain dataset and detector are determined automatically without manually adjustment. We demonstrate that AABO outperforms both hand-adjusted methods and HPO methods on popular SOTA detectors over multiple datasets, which indicates that anchor configurations play an important role in object detection frameworks and our proposed method could help exploit the potential of detectors in a more effective way.

References

1. Bergstra, J.S., Bardenet, R., Bengio, Y., Kégl, B.: Algorithms for hyper-parameter optimization. In: NIPS (2011)
2. Bhagavatula, C., Zhu, C., Luu, K., Savvides, M.: Faster than real-time facial alignment: a 3D spatial transformer network approach in unconstrained poses. In: ICCV (2017)
3. Chabot, F., Chaouch, M., Rabarisoa, J., Teuliere, C., Chateau, T.: Deep manta: a coarse-to-fine many-task network for joint 2D and 3D vehicle analysis from monocular image. In: CVPR (2017)
4. Chan, H.P.: The multi-armed bandit problem: an efficient non-parametric solution. Ann. Stat. **48**, 346–373 (2019)
5. Chen, K., et al.: mmdetection (2018). https://github.com/open-mmlab/mmdetection
6. Chen, K., et al.: Hybrid task cascade for instance segmentation. In: IEEE Conference on Computer Vision and Pattern Recognition (2019)
7. Chen, X., Li, L.J., Fei-Fei, L., Gupta, A.: Iterative visual reasoning beyond convolutions. In: CVPR (2018)
8. Dai, J., Li, Y., He, K., Sun, J.: R-FCN: object detection via region-based fully convolutional networks. In: NIPS (2016)
9. Falkner, S., Klein, A., Hutter, F.: BOHB: robust and efficient hyperparameter optimization at scale. arXiv preprint arXiv:1807.01774 (2018)
10. He, K., Gkioxari, G., Dollar, P., Girshick, R.: Mask R-CNN. In: 2017 IEEE International Conference on Computer Vision (ICCV) (2017)
11. He, K., Zhang, X., Ren, S., Sun, J.: Deep residual learning for image recognition. In: CVPR (2016)
12. Huang, Y., Li, Y., Li, Z., Zhang, Z.: An asymptotically optimal multi-armed bandit algorithm and hyperparameter optimization. arXiv e-prints arXiv:2007.05670 (2020)
13. Jamieson, K., Talwalkar, A.: Non-stochastic best arm identification and hyperparameter optimization. In: Artificial Intelligence and Statistics, pp. 240–248 (2016)
14. Jiang, C., Xu, H., Liang, X., Lin, L.: Hybrid knowledge routed modules for large-scale object detection. In: NIPS (2018)
15. Krishna, R., et al.: Visual genome: connecting language and vision using crowd-sourced dense image annotations. Int. J. Comput. Vis. **123**, 32–73 (2016)
16. Li, L., Jamieson, K., Desalvo, G., Rostamizadeh, A., Talwalkar, A.: Hyperband: a novel bandit-based approach to hyperparameter optimization. J. Mach. Learn. Res. **18**, 1–52 (2016)

17. Lin, T.Y., Dollár, P., Girshick, R., He, K., Hariharan, B., Belongie, S.: Feature pyramid networks for object detection. In: CVPR (2017)
18. Lin, T.Y., Goyal, P., Girshick, R., He, K., Dollár, P.: Focal loss for dense object detection. In: ICCV, pp. 2980–2988 (2017)
19. Lin, T.Y., et al.: Microsoft COCO: common objects in context. In: Fleet, D., Pajdla, T., Schiele, B., Tuytelaars, T. (eds.) ECCV 2014. LNCS, vol. 8693, pp. 740–755. Springer, Cham (2014). https://doi.org/10.1007/978-3-319-10602-1_48
20. Liu, W., et al.: SSD: single shot multibox detector. In: Leibe, B., Matas, J., Sebe, N., Welling, M. (eds.) ECCV 2016. LNCS, vol. 9905, pp. 21–37. Springer, Cham (2016). https://doi.org/10.1007/978-3-319-46448-0_2
21. Luo, P., Tian, Y., Wang, X., Tang, X.: Switchable deep network for pedestrian detection. In: CVPR (2014)
22. Mendoza, H., Klein, A., Feurer, M., Springenberg, J.T., Hutter, F.: Towards automatically-tuned neural networks. In: Workshop on Automatic Machine Learning, pp. 58–65 (2016)
23. Paszke, A., et al.: Automatic differentiation in pytorch. In: NIPS Workshop (2017)
24. Perchet, V., Rigollet, P.: The multi-armed bandit problem with covariates. Ann. Stat. 41(2), 693–721 (2013)
25. Redmon, J., Divvala, S., Girshick, R., Farhadi, A.: You only look once: unified, real-time object detection. In: CVPR (2016)
26. Redmon, J., Farhadi, A.: Yolo9000: better, faster, stronger. In: CVPR (2017)
27. Ren, S., He, K., Girshick, R., Sun, J.: Faster R-CNN: towards real-time object detection with region proposal networks. In: NIPS (2015)
28. Russakovsky, O., et al.: ImageNet large scale visual recognition challenge. IJCV 115(3), 211–252 (2015)
29. Snoek, J., Larochelle, H., Adams, R.P.: Practical Bayesian optimization of machine learning algorithms. In: NIPS (2012)
30. Sutton, R.S., Barto, A.G.: Reinforcement Learning: An Introduction. MIT Press, Cambridge (2018)
31. Tong, Y., Zhang, X., Zhang, W., Jian, S.: Metaanchor: learning to detect objects with customized anchors (2018)
32. Xie, S., Girshick, R., Dollár, P., Tu, Z., He, K.: Aggregated residual transformations for deep neural networks. In: CVPR, pp. 1492–1500 (2017)
33. Zhong, Y., Wang, J., Peng, J., Zhang, L.: Anchor box optimization for object detection. arXiv preprint arXiv:1812.00469 (2018)
34. Zhou, B., Zhao, H., Puig, X., Fidler, S., Barriuso, A., Torralba, A.: Scene parsing through ade20k dataset. In: CVPR (2017)
35. Zhu, X., Hu, H., Lin, S., Dai, J.: Deformable convnets v2: more deformable, better results. arXiv preprint arXiv:1811.11168 (2018)

Learning Visual Context by Comparison

Minchul Kim[1], Jongchan Park[1], Seil Na[1], Chang Min Park[2],
and Donggeun Yoo[1(✉)]

[1] Lunit Inc., Seoul, Republic of Korea
{minchul.kim,jcpark,seil.na,dgyoo}@lunit.io
[2] Seoul National University Hospital, Seoul, Republic of Korea
cmpark.morphius@gmail.com

Abstract. Finding diseases from an X-ray image is an important yet highly challenging task. Current methods for solving this task exploit various characteristics of the chest X-ray image, but one of the most important characteristics is still missing: the necessity of comparison between related regions in an image. In this paper, we present Attend-and-Compare Module (ACM) for capturing the difference between an object of interest and its corresponding context. We show that explicit difference modeling can be very helpful in tasks that require direct comparison between locations from afar. This module can be plugged into existing deep learning models. For evaluation, we apply our module to three chest X-ray recognition tasks and COCO object detection & segmentation tasks and observe consistent improvements across tasks. The code is available at https://github.com/mk-minchul/attend-and-compare.

Keywords: Context modeling · Attention mechanism · Chest X-ray

1 Introduction

Among a variety of medical imaging modalities, chest X-ray is one of the most common and readily available examinations for diagnosing chest diseases. In the US, more than 35 million chest X-rays are taken every year [20]. It is primarily used to screen diseases such as lung cancer, pneumonia, tuberculosis and pneumothorax to detect them at their earliest and most treatable stage. However, the problem lies in the heavy workload of reading chest X-rays. Radiologists usually read tens or hundreds of X-rays every day. Several studies regarding radiologic errors [9,28] have reported that 20–30% of exams are misdiagnosed. To compensate for this shortcoming, many hospitals equip radiologists with computer-aided

---

M. Kim and J. Park—The authors have equally contributed.

Electronic supplementary material The online version of this chapter (https:// doi.org/10.1007/978-3-030-58558-7_34) contains supplementary material, which is available to authorized users.

© Springer Nature Switzerland AG 2020
A. Vedaldi et al. (Eds.): ECCV 2020, LNCS 12350, pp. 576–592, 2020.
https://doi.org/10.1007/978-3-030-58558-7_34

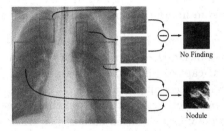

Fig. 1. An example of a comparison procedure for radiologists. Little differences indicate no disease (blue), the significant difference is likely to be a lesion (red). (Color figure online)

diagnosis systems. The recent developments of medical image recognition models have shown potentials for growth in diagnostic accuracy [26].

With the recent presence of large-scale chest X-ray datasets [3,18,19,37], there has been a long line of works that find thoracic diseases from chest X-rays using deep learning [12,23,29,41]. Most of the works attempt to classify thoracic diseases, and some of the works further localize the lesions. To improve recognition performance, Yao et al. [41] handles varying lesion sizes and Mao et al. [25] takes the relation between X-rays of the same patient into consideration. Wang et al. [35] introduces an attention mechanism to focus on regions of diseases.

While these approaches were motivated by the characteristics of chest X-rays, we paid attention to how radiology residents are trained, which led to the following question: why don't we model the way radiologists read X-rays? When radiologists read chest X-rays, they compare zones [1], paying close attention to any asymmetry between left and right lungs, or any changes between semantically related regions, that are likely to be due to diseases. This comparison process provides contextual clues for the presence of a disease that local texture information may fail to highlight. Figure 1 illustrates an example of the process. Previous studies [4,15,36,38] proposed various context models, but none addressed the need for the explicit procedure to *compare* regions in an image.

In this paper, we present a novel module, called *Attend-and-Compare Module* (ACM), that extracts features of an object of interest and a corresponding context to explicitly compare them by subtraction, mimicking the way radiologists read X-rays. Although motivated by radiologists' practices, we pose no explicit constraints for symmetry, and ACM learns to compare regions in a data-driven way. ACM is validated over three chest X-ray datasets [37] and object detection & segmentation in COCO dataset [24] with various backbones such as ResNet [14], ResNeXt [40] or DenseNet [16]. Experimental results on chest X-ray datasets and natural image datasets demonstrate that the explicit comparison process by ACM indeed improves the recognition performance.

Contributions. To sum up, our major contributions are as follows:

1. We propose a novel context module called ACM that explicitly compares different regions, following the way radiologists read chest X-rays.
2. The proposed ACM captures multiple comparative self-attentions whose difference is beneficial to recognition tasks.
3. We demonstrate the effectiveness of ACM on three chest X-ray datasets [37] and COCO detection & segmentation dataset [24] with various architectures.

2 Related Work

2.1 Context Modeling

Context modeling in deep learning is primarily conducted with the self-attention mechanism [15, 22, 30, 33]. Attention related works are broad and some works do not explicitly pose themselves in the frame of context modeling. However, we include them to highlight different methods that make use of global information, which can be viewed as context.

In the visual recognition domain, recent self-attention mechanisms [7, 15, 22, 34, 38] generate dynamic attention maps for recalibration (e.g., emphasize salient regions or channels). Squeeze-and-Excitation network (SE) [15] learns to model channel-wise attention using the spatially averaged feature. A Style-based Recalibration Module (SRM) [22] further explores the global feature modeling in terms of style recalibration. Convolutional block attention module (CBAM) [38] extends SE module to the spatial dimension by sequentially attending the important location and channel given the feature. The attention values are computed with global or larger receptive fields, and thus, more contextual information can be embedded in the features. However, as the information is aggregated into a single feature by average or similar operations, spatial information from the relationship among multiple locations may be lost.

Works that explicitly tackle the problem of using context stem from using pixel-level pairwise relationships [4, 17, 36]. Such works focus on long-range dependencies and explicitly model the context aggregation from dynamically chosen locations. Non-local neural networks (NL) [36] calculate pixel-level pairwise relationship weights and aggregate (weighted average) the features from all locations according to the weights. The pairwise modeling may represent a more diverse relationship, but it is computationally more expensive. As a result, Global-Context network (GC) [4] challenges the necessity of using all pairwise relationships in NL and suggests to softly aggregate a single distinctive context feature for all locations. Criss-cross attention (CC) [17] for semantic segmentation reduces the computation cost of NL by replacing the pairwise relationship attention maps with criss-cross attention block which considers only horizontal and vertical directions separately. NL and CC explicitly model the pairwise relationship between regions with affinity metrics, but the qualitative results in [17, 36] demonstrate a tendency to aggregate features only among foreground objects or among pixels with similar semantics.

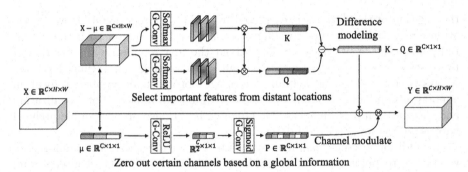

Fig. 2. Illustration of the ACM module. It takes in an input feature and uses the mean-subtracted feature to calculate two feature vectors (K, Q). Each feature vector $(K$ or $Q)$ contains multiple attention vectors from multiple locations calculated using grouped convolutions and normalizations. The difference of the vectors is added to the main feature to make the information more distinguishable. The resulting feature is modulated channel-wise, by the global information feature.

Sharing a similar philosophy, there have been works on contrastive attention [31,42]. MGCAM [31] uses the contrastive feature between persons and backgrounds, but it requires extra mask supervision for persons. C-MWP [42] is a technique for generating more accurate localization maps in a contrastive manner, but it is not a learning-based method and uses pretrained models.

Inspired by how radiologists diagnose, our proposed module, namely, ACM explicitly models a comparing mechanism. The overview of the module can be found in Fig. 2. Unlike the previous works proposed in the natural image domain, our work stems from the precise need for incorporating difference operation in reading chest radiographs. Instead of finding an affinity map based attention as in NL [36], ACM explicitly uses direct comparison procedure for context modeling; instead of using extra supervision to localize regions to compare as in MGCAM [31], ACM automatically learns to focus on meaningful regions to compare. Importantly, the efficacy of our explicit and data-driven contrastive modeling is shown by the superior performance over other context modeling works.

2.2 Chest X-ray as a Context-Dependent Task

Recent releases of the large-scale chest X-ray datasets [3,18,19,37] showed that commonly occurring diseases can be classified and located in a weakly-supervised multi-label classification framework. ResNet [14] and DenseNet [16,29] pre-trained on ImageNet [8] have set a strong baseline for these tasks, and other studies have been conducted on top of them to cover various issues of recognition task in the chest X-ray modality.

To address the issue of localizing diseases using only class-level labels, Guendel *et al.* [12] propose an auxiliary localization task where the ground truth of the location of the diseases is extracted from the text report. Other works use attention module to indirectly align the class-level prediction with the potentially abnormal location [10,32,35] without the text reports on the location of the disease. Some works observe that although getting annotations for chest X-rays is costly, it is still helpful to leverage both a small number of location annotations and a large number of class-level labels to improve both localization and classification performances [23]. Guan *et al.* [11] also proposes a hierarchical hard-attention for cascaded inference.

In addition to such characteristics inherent in the chest X-ray image, we would like to point out that the difference between an object of interest and a corresponding context could be the crucial key for classifying or localizing several diseases as it is important to compare semantically meaningful locations. However, despite the importance of capturing the semantic difference between regions in chest X-ray recognition tasks, no work has dealt with it yet. Our work is, to the best of our knowledge, the first to utilize this characteristic in the Chest X-ray image recognition setting.

3   Attend-and-Compare Module

3.1   Overview

Attend-and-Compare Module (ACM) extracts an object of interest and the corresponding context to compare, and enhances the original image feature with the comparison result. Also, ACM is designed to be light-weight, self-contained, and compatible with popular backbone architectures [14,16,40]. We formulate ACM comprising three procedures as

$$Y = f_{\mathrm{ACM}}(X) = P(X + (K - Q)), \tag{1}$$

where f_{ACM} is a transformation mapping an input feature $X \in \mathbb{R}^{C \times H \times W}$ to an output feature $Y \in \mathbb{R}^{C \times H \times W}$ in the same dimension. Between $K \in \mathbb{R}^{C \times 1 \times 1}$ and $Q \in \mathbb{R}^{C \times 1 \times 1}$, one is intended to be the object of interest and the other is the corresponding context. ACM compares the two by subtracting one from the other, and add the comparison result to the original feature X, followed by an additional channel re-calibration operation with $P \in \mathbb{R}^{C \times 1 \times 1}$. Figure 2 illustrates Eq. (1). These three features K, Q and P are conditioned on the input feature X and will be explained in details below.

3.2   Components of ACM

Object of Interest and Corresponding Context. To fully express the relationship between different spatial regions of an image, ACM generates two features (K, Q) that focus on two spatial regions of the input feature map X. At first, ACM normalizes the input feature map as $X := X - \mu$ where μ is a C-dimensional mean vector of X. We include this procedure to make training more stable as K and Q will be generated by learnable parameters (W_K, W_Q) that are shared by all input features. Once X is normalized, ACM then calculates K with W_K as

$$K = \sum_{i,j \in H,W} \frac{\exp(W_K X_{i,j})}{\sum_{H,W} \exp(W_K X_{h,w})} X_{i,j}, \tag{2}$$

where $X_{i,j} \in \mathbb{R}^{C \times 1 \times 1}$ is a vector at a spatial location (i,j) and $W_K \in \mathbb{R}^{C \times 1 \times 1}$ is a weight of a 1×1 convolution. The above operation could be viewed as applying 1×1 convolution on the feature map X to obtain a single-channel attention map in $\mathbb{R}^{1 \times H \times W}$, applying softmax to normalize the attention map, and finally weighted averaging the feature map X using the normalized map. Q is also modeled likewise, but with W_Q. K and Q serve as features representing important regions in X. We add $K - Q$ to the original feature so that the comparative information is more distinguishable in the feature.

Channel Re-calibration. In light of the recent success in self-attention modules which use a globally pooled feature to re-calibrate channels [15,22,38], we calculate the channel re-calibrating feature P as

$$P = \sigma \circ \text{conv}_2^{1 \times 1} \circ \text{ReLU} \circ \text{conv}_1^{1 \times 1}(\mu), \tag{3}$$

where σ and $\text{conv}^{1 \times 1}$ denote a sigmoid function and a learnable 1×1 convolution function, respectively. The resulting feature vector P will be multiplied to $X + (K - Q)$ to scale down certain channels. P can be viewed as marking which channels to attend with respect to the task at hand.

Group Operation. To model a relation of multiple regions from a single module, we choose to incorporate group-wise operation. We replace all convolution operations with grouped convolutions [21,40], where the input and the output are divided into G number of groups channel-wise, and convolution operations are performed for each group separately. In our work, we use the grouped convolution to deliberately represent multiple important locations from the input. Here, we compute G different attention maps by applying grouped convolution to X, and then obtain the representation $K = [K^1, \cdots, K^G]$ by aggregating each group in X with each attention as follows:

$$K^g = \sum_{i,j \in H,W} \frac{\exp(W_K^g X_{i,j}^g)}{\sum_{H,W} \exp(W_K^g X_{h,w}^g)} X_{i,j}^g, \tag{4}$$

where g refers to g-th group.

Loss Function. ACM learns to utilize comparing information within an image by modeling $\{K, Q\}$ whose difference can be important for the given task. To further ensure diversity between them, we introduce an orthogonal loss. Based on a dot product. It is defined as

$$\ell_{\text{orth}}(K, Q) = \frac{K \cdot Q}{C}, \tag{5}$$

where C refers to the number of channels. Minimizing this loss can be viewed as decreasing the similarity between K and Q. One trivial solution to minimizing the term would be making K or Q zeros, but they cannot be zeros as they come from the weighted averages of X. The final loss function for a target task can be written as

$$\ell_{task} + \lambda \sum_{m}^{M} \ell_{orth}(K_m, Q_m), \tag{6}$$

where ℓ_{task} refers to a loss for the target task, and M refers to the number of ACMs inserted into the network. λ is a constant for controlling the effect of the orthogonal constraint.

Placement of ACMs. In order to model contextual information in various levels of feature representation, we insert multiple ACMs into the backbone network. In ResNet, following the placement rule of SE module [15], we insert the module at the end of every Bottleneck block. For example, a total of 16 ACMs are inserted in ResNet-50. Since DenseNet contains more number of DenseBlocks than ResNet's Bottleneck block, we inserted ACM in DenseNet every other three DenseBlocks. Note that we did not optimize the placement location or the number of placement for each task. While we use multiple ACMs, the use of grouped convolution significantly reduces the computation cost in each module.

4   Experiments

We evaluate ACM in several datasets: internally-sourced Emergency-Pneumothorax (Em-Ptx) and Nodule (Ndl) datasets for lesion localization in chest X-rays, Chest X-ray14 [37] dataset for multi-label classification, and COCO 2017 [24] dataset for object detection and instance segmentation. The experimental results show that ACM outperforms other context-related modules, in both chest X-ray tasks and natural image tasks.

Experimental Setting. Following the previous study [2] on multi-label classification with chest X-Rays, we mainly adopt ResNet-50 as our backbone network. To show generality, we sometimes adopt DenseNet [16] and ResNeXt [40] as backbone networks. In classification tasks, we report class-wise Area Under the Receiver Operating Characteristics (AUC-ROC) for classification performances. For localization tasks, we report the jackknife free-response receiver operating characteristic (JAFROC) [5] for localization performances. JAFROC is a metric widely used for tracking localization performance in radiology. All chest X-ray tasks are a weakly-supervised setting [27] in which the model outputs a probability map for each disease, and final classification scores are computed by global maximum or average pooling. If any segmentation annotation is available, extra map losses are given on the class-wise confidence maps. For all experiments, we initialize the backbone weights with ImageNet-pretrained weights and randomly initialize context-related modules. Experiment details on each dataset are elaborated in the following section.

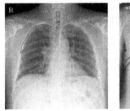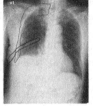

(a) Emergency   (b) Non-emergency

Fig. 3. Examples of pneumothorax cases and annotation maps in Em-Ptx dataset. Lesions are drawn in red. (a) shows a case with pneumothorax, and (b) shows a case which is already treated with a medical tube marked as blue. (Color figure online)

4.1 Localization on Em-Ptx Dataset

Task Overview. The goal of this task is to localize emergency-pneumothorax (Em-Ptx) regions. Pneumothorax is a fatal thoracic disease that needs to be treated immediately. As a treatment, a medical tube is inserted into the pneumothorax affected lung. It is often the case that a treated patient repeatedly takes chest X-rays over a short period to see the progress of the treatment. Therefore, a chest X-ray with pneumothorax is categorized as an emergency, but a chest X-ray with both pneumothorax and a tube is not an emergency. The goal of the task is to correctly classify and localize emergency-pneumothorax. To accurately classify the emergency-pneumothorax, the model should exploit the relationship between pneumothorax and tube within an image, even when they are far apart. In this task, utilizing the context as the presence/absence of a tube is the key to accurate classification.

We internally collected 8,223 chest X-rays, including 5,606 pneumothorax cases, of which 3,084 cases are emergency. The dataset is from a real-world cohort and contains cases with other abnormalities even if they do not have pneumothorax. We received annotations for 10 major x-ray findings (Nodule, Consolidation, etc) and the presence of medical devices (EKG, Endotracheal tube, Chemoport, etc). Their labels were not used for training. The task is a binary classification and localization of emergency-pneumothorax. All cases with pneumothorax and tube together and all cases without pneumothorax are considered a non-emergency. We separated 3,223 cases as test data, of which 1,574 cases are emergency-pneumothorax, 1,007 cases are non-emergency-pneumothorax, and 642 cases are pure normal. All of the 1,574 emergency cases in the test data were annotated with coarse segmentation maps by board-certified radiologists. Of the 1,510 emergency-pneumothorax cases in the training data, only 930 cases were annotated, and the rest were used with only the class label. An example of a pneumothorax case and an annotation map created by board-certified radiologists is provided in Fig. 3.

Table 1. Results on Em-Ptx dataset. Average of 5 random runs are reported for each setting with standard deviation. RN stands for ResNet [14].

Method	AUC-ROC	JAFROC	Method	AUC-ROC	JAFROC
RN-50	86.78 ± 0.58	81.84 ± 0.64	RN-101	89.75 ± 0.49	85.36 ± 0.44
RN-50 + SE [15]	93.05 ± 3.63	89.19 ± 4.38	RN-101 + SE [15]	90.36 ± 0.83	85.54 ± 0.85
RN-50 + NL [36]	94.63 ± 0.39	91.93 ± 0.68	RN-101 + NL [36]	94.24 ± 0.34	91.70 ± 0.83
RN-50 + CC [17]	87.73 ± 8.66	83.32 ± 10.36	RN-101 + CC [17]	92.57 ± 0.89	89.75 ± 0.89
RN-50 + ACM	**95.35 ± 0.12**	**94.16 ± 0.21**	RN-101 + ACM	**95.43 ± 0.14**	**94.47 ± 0.10**

Table 2. Performance with respect to varying module architectures and hyperparameters on Em-Ptx dataset. All the experiments are based on ResNet-50.

Module	AUC-ROC	JAFROC
None	86.78±0.58	81.84±0.64
$X + (K - Q)$	94.25±0.31	92.94±0.36
PX	87.16±0.42	82.05±0.30
$P(X + K)$	94.96±0.15	93.59±0.24
$P(X + (K - Q))$	**95.35±0.12**	**94.16±0.21**

(a) Ablations on K, Q and P.

#groups	AUC-ROC	JAFROC
8	90.96±1.88	88.79±2.23
32	**95.35±0.12**	**94.16±0.21**
64	95.08±0.25	93.73±0.31
128	94.89±0.53	92.88±0.53

(b) Ablations on number of groups.

λ	AUC-ROC	JAFROC
0.00	95.11±0.20	93.87±0.20
0.01	95.29±0.34	94.09±0.41
0.10	**95.35±0.12**	**94.16±0.21**
1.00	95.30±0.17	94.04±0.11

(c) Ablations on orthogonal loss weight λ.

Training Details. We use Binary Cross-Entropy loss for both classification and localization, and SGD optimizer with momentum 0.9. The initial learning rate is set to 0.01. The model is trained for 35 epochs in total and the learning rate is dropped by the factor of 10 at epoch 30. For each experiment setting, we report the average AUC-ROC and JAFROC of 5 runs with different initialization.

Result. The experimental result is summarized in Table 1. Compared to the baseline ResNet-50 and ResNet-101, all the context modules have shown performance improvements. ACM outperforms all other modules in terms of both AUC-ROC and JAFROC. The result supports our claim that the contextual information is critical to Em-Ptx task, and our module, with its explicit feature-comparing design, shows the biggest improvement in terms of classification and localization.

4.2 Analysis on ACM with Em-Ptx Dataset

In this section, we empirically validate the efficacy of each component in ACM and search for the optimal hyperparameters. For the analysis, we use the

Em-Ptx dataset. The training setting is identical to the one used in Sect. 4.1. The average of 5-runs is reported.

Effect of Sub-modules. As described in Sect. 3, our module consists of 2 sub-modules: difference modeling and channel modulation. We experiment with the two sub-modules both separately and together. The results are shown in Table 2. Each sub-module brings improvements over the baseline, indicating the context modeling in each sub-module is beneficial to the given task. Combining the sub-modules brings extra improvement, showing the complementary effect of the two sub-modules. The best performance is achieved when all sub-modules are used.

Number of Groups. By dividing the features into groups, the module can learn to focus on multiple regions, with only negligible computational or parametric costs. On the other hand, too many groups can result in too few channels per group, which prevents the module from exploiting correlation among many channels. We empirically find the best setting for the number of groups. The results are summarized in Table 2. Note that, training diverges when the number of groups is 1 or 4. The performance improves with the increasing number of groups and saturates after 32. We set the number of groups as 32 across all other experiments, except in DenseNet121 (Sect. 4.4) where we set the number of channel per group to 16 due to channel divisibility.

Orthogonal Loss Weight. We introduce a new hyperparameter λ to balance between the target task loss and the orthogonal loss. Although the purpose of the orthogonal loss is to diversify the compared features, an excessive amount of λ can disrupt the trained representations. We empirically determine the magnitude of λ. Results are summarized in Table 2. From the results, we can empirically verify the advantageous effect of the orthogonal loss on the performance, and the optimal value of λ is 0.1. In the validation set, the average absolute similarities between K and Q with $\lambda = 0, 0.1$ are 0.1113 and 0.0394, respectively. It implies that K and Q are dissimilar to some extent, but the orthogonal loss further encourages it. We set λ as 0.1 across all other experiments.

4.3   Localization on Ndl Dataset

Task Overview. The goal of this task is to localize lung nodules (Ndl) in chest X-ray images. Lung nodules are composed of fast-growing dense tissues and thus are displayed as tiny opaque regions. Due to inter-patient variability, view-point changes and differences in imaging devices, the model that learns to find nodular patterns with respect to the normal side of the lung from the same image (context) may generalize better. We collected 23,869 X-ray images, of which 3,052 cases are separated for testing purposes. Of the 20,817 training cases, 5,817 cases have nodule(s). Of the 3,052 test cases, 694 cases are with nodules. Images without nodules may or may not contain other lung diseases. All cases

Table 3. Results on Ndl dataset. Average of 5 random runs are reported for each setting with standard deviation.

Method	AUC-ROC	JAFROC
ResNet-50	87.34 ± 0.34	77.35 ± 0.50
ResNet-50 + SE [15]	87.66 ± 0.40	77.57 ± 0.44
ResNet-50 + NL [36]	88.35 ± 0.35	80.51 ± 0.56
ResNet-50 + CC [17]	87.72 ± 0.18	78.63 ± 0.40
ResNet-50 + ACM	**88.60** ± 0.23	**83.03** ± 0.24

with nodule(s) are annotated with coarse segmentation maps by board-certified radiologists. We use the same training procedure for the nodule localization task as for the Em-Ptx localization. We train for 25 epochs with the learning rate dropping once at epoch 20 by the factor of 10.

Results. The experimental result is summarized in Table 3. ACM outperforms all other context modeling methods in terms of both AUC-ROC and JAFROC. The results support our claim that the comparing operation, motivated by how radiologists read X-rays, provides a good contextual representation that helps with classifying and localizing lesions in X-rays. Note that the improvements in this dataset may seem smaller than in the Em-Ptx dataset. In the usage of context modules in general, the bigger increase in performance in Em-Ptx dataset is because emergency classification requires knowing both the presence of the tube and the presence of Ptx even if they are far apart. So the benefit of the contextual information is directly related to the performance. However, nodule classification can be done to a certain degree without contextual information. Since not all cases need contextual information, the performance gain may be smaller.

4.4   Multi-label Classification on Chest X-ray14

Task Overview. In this task, the objective is to identify the presence of 14 diseases in a given chest X-ray image. Chest X-ray14 [37] dataset is the first large-scale dataset on 14 common diseases in chest X-rays. It is used as a benchmark dataset in previous studies [6,10,12,23,25,29,32,41]. The dataset contains 112,120 images from 30,805 unique patients. Image-level labels are mined from image-attached reports using natural language processing techniques (each image can have multi-labels). We split the dataset into training (70%), validation (10%), and test (20%) sets, following previous works [37,41].

Training Details. As shown in Table 4, previous works on CXR14 dataset vary in loss, input size, etc. We use CheXNet [29] implementation[1] to conduct

---

[1] https://github.com/jrzech/reproduce-chexnet.

Table 4. Reported performance of previous works on CXR14 dataset. Each work differs in augmentation schemes and some even in the usage of the dataset. We choose CheXNet [29] as the baseline model for adding context modules.

Method	Backbone Arch	Loss Family	Input size	Reported AUC (%)
Wang et al. [37]	ResNet-50	CE	1,024	74.5
Yao et al. [41]	ResNet+DenseNet	CE	512	76.1
Wang and Xia [35]	ResNet-151	CE	224	78.1
Li et al. [23]	ResNet-v2-50	BCE	299	80.6
Guendel et al. [12]	DenseNet121	BCE	1,024	80.7
Guan et al. [10]	DenseNet121	BCE	224	81.6
ImageGCN [25]	Graph Convnet	CE	224	82.7
CheXNet [29]	DenseNet121	BCE	224	84.1

Table 5. Performance in average AUC of various methods on CXR14 dataset. The numbers in the bracket after model names are the input sizes.

Modules	DenseNet121(448)	ResNet-50(448)
None	(CheXNet [29]) 84.54	84.19
SE [15]	84.95	84.53
NL [36]	84.49	85.08
CC [17]	84.43	85.11
ACM	**85.03**	**85.39**

context module comparisons, and varied with the backbone architecture and the input size to find if context modules work in various settings. We use BCE loss with the SGD optimizer with momentum 0.9 and weight decay 0.0001. Although ChexNet uses the input size of 224, we use the input size of 448 as it shows a better result than 224 with DenseNet121. More training details can be found in the supplementary material.

Results. Table 5 shows test set performance of ACM compared with other context modules in multiple backbone architectures. ACM achieves the best performance of 85.39 with ResNet-50 and 85.03 with DenseNet121. We also observe that Non-local (NL) and Cross-Criss Attention (CC) does not perform well in DenseNet architecture, but attains a relatively good performance of 85.08 and 85.11 in ResNet-50. On the other hand, a simpler SE module performs well in DenseNet but does poorly in ResNet50. However, ACM shows consistency across all architectures. One of the possible reasons is that it provides a context based on a contrasting operation, thus unique and helpful across different architectures.

Table 6. Results on COCO dataset. All experiments are based on Mask-RCNN [13].

Method	AP^{bbox}	AP$^{bbox}_{50}$	AP$^{bbox}_{75}$	AP^{mask}	AP$^{mask}_{50}$	AP$^{mask}_{75}$
ResNet-50	38.59	59.36	42.23	35.24	56.24	37.66
ResNet-50+SE [15]	39.10	60.32	42.59	35.72	57.16	38.20
ResNet-50+NL [36]	39.40	60.60	43.02	35.85	57.63	38.15
ResNet-50+CC [17]	39.82	60.97	42.88	36.05	57.82	38.37
ResNet-50+ACM	**39.94**	**61.58**	**43.30**	**36.40**	**58.40**	**38.63**
ResNet-101	40.77	61.67	44.53	36.86	58.35	39.59
ResNet-101+SE [15]	41.30	62.36	45.26	37.38	59.34	40.00
ResNet-101+NL [36]	41.57	62.75	45.39	37.39	59.50	40.01
ResNet-101+CC [17]	**42.09**	63.21	**45.79**	**37.77**	59.98	**40.29**
ResNet-101+ACM	41.76	**63.38**	45.16	37.68	**60.16**	40.19
ResNeXt-101	43.23	64.42	47.47	39.02	61.10	42.11
ResNeXt-101+SE [15]	43.44	64.91	47.66	39.20	61.92	42.17
ResNeXt-101+NL [36]	43.93	65.44	48.20	39.45	61.99	42.33
ResNeXt-101+CC [17]	43.86	65.28	47.74	39.26	62.06	41.97
ResNeXt-101+ACM	**44.07**	**65.92**	**48.33**	**39.54**	**62.53**	**42.44**

4.5 Detection and Segmentation on COCO

Task Overview. In this experiment, we validate the efficacy of ACM in the natural image domain. Following the previous studies [4,17,36,38], we use COCO dataset [24] for detection and segmentation tasks. Specifically, we use COCO Detection 2017 dataset, which contains more than 200,000 images and 80 object categories with instance-level segmentation masks. We train both tasks simultaneously using Mask-RCNN architecture [13] in Detectron2 [39].

Training Details. Basic training details are identical to the default settings in Detectron2 [39]: learning rate of 0.02 with the batch size of 16. We train for 90,000 iterations, drop the learning rate by 0.1 at iterations 60,000 and 80,000. We use COCO2017-train for training and use COCO2017-val for testing.

Results. The experimental results are summarized in Table 6. Although originally developed for chest X-ray tasks, ACM significantly improves the detection and segmentation performance in the natural image domain as well. In ResNet-50 and ResNeXt-101, ACM outperforms all other modules [4,15,36]. The result implies that the comparing operation is not only crucial for X-ray images but is also generally helpful for parsing scene information in the natural image domain.

4.6 Qualitative Results

To analyze ACM, we visualize the attention maps for objects of interest and the corresponding context. The network learns to attend different regions, in such a

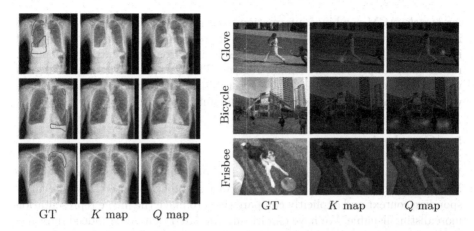

Fig. 4. Left: The visualized attention maps for the localization task on Em-Ptx dataset. The 11th group in the 16th module is chosen. Em-Ptx annotations are shown as red contours on the chest X-ray image. Right: The visualization on COCO dataset. Ground-truth segmentation annotations for each category are shown as red contours. (Color figure online)

way to maximize the performance of the given task. We visualize the attention maps to see if maximizing performance is aligned with producing attention maps that highlight interpretable locations. Ground-truth annotation contours are also visualized.

We use the Em-Ptx dataset and COCO dataset for the analysis. Since there are many attention maps to check, we sort the maps by the amount of overlap between each attention map and the ground truth location of lesions. We visualize the attention map with the most overlap. Qualitative results of other tasks are included in the supplementary material due to a space limit.

Pneumothorax is a collapsed lung, and on the X-ray image, it is often portrayed as a slightly darker region than the normal side of the lung. A simple way to detect pneumothorax is to find a region slightly darker than the normal side. ACM learns to utilize pneumothorax regions as objects of interest and normal lung regions as the corresponding context. The attention maps are visualized in Fig. 4. It clearly demonstrates that the module attends to both pneumothorax regions and normal lung regions and compare the two sets of regions. The observation is coherent with our intuition that comparing can help recognize, and indicates that ACM automatically learns to do so.

We also visualize the attended regions in the COCO dataset. Examples in Fig. 4 shows that ACM also learns to utilize the object of interest and the corresponding context in the natural image domain; for *baseball glove*, ACM combines the corresponding context information from the players' heads and feet;

for *bicycle*, ACM combines information from roads; for *frisbee*, ACM combines information from dogs. We observe that the relationship between the object of interest and the corresponding context is mainly from co-occurring semantics, rather than simply visually similar regions. The learned relationship is aligned well with the design principle of ACM; selecting features whose semantics differ, yet whose relationship can serve as meaningful information.

5   Conclusion

We have proposed a novel self-contained module, named Attend-and-Compare Module (ACM), whose key idea is to extract an object of interest and a corresponding context and explicitly compare them to make the image representation more distinguishable. We have empirically validated that ACM indeed improves the performance of visual recognition tasks in chest X-ray and natural image domains. Specifically, a simple addition of ACM provides consistent improvements over baselines in COCO as well as Chest X-ray14 public dataset and internally collected Em-Ptx and Ndl dataset. The qualitative analysis shows that ACM automatically learns dynamic relationships. The objects of interest and corresponding contexts are different yet contain useful information for the given task.The qualitative analysis shows that ACM automatically learns dynamic relationships. The objects of interest and corresponding contexts are different yet contain useful information for the given task.

References

1. Armato III, S.G., Giger, M.L., MacMahon, H.: Computerized detection of abnormal asymmetry in digital chest radiographs. Med. Phys. **21**(11), 1761–1768 (1994)
2. Baltruschat, I.M., Nickisch, H., Grass, M., Knopp, T., Saalbach, A.: Comparison of deep learning approaches for multi-label chest X-ray classification. Sci. Rep. **9**(1), 6381 (2019)
3. Bustos, A., Pertusa, A., Salinas, J.M., de la Iglesia-Vayá, M.: PadChest: a large chest X-ray image dataset with multi-label annotated reports. arXiv (2019)
4. Cao, Y., Xu, J., Lin, S., Wei, F., Hu, H.: GCNet: non-local networks meet squeeze-excitation networks and beyond. In: ICCV (2019)
5. Chakraborty, D.P.: Recent advances in observer performance methodology: jackknife free-response ROC (JAFROC). Radiat. Prot. Dosim. **114**(1–3), 26–31 (2005)
6. Chen, B., Li, J., Guo, X., Lu, G.: DualCheXNet: dual asymmetric feature learning for thoracic disease classification in chest X-rays. Biomed. Signal Process. Control **53**, 101554 (2019)
7. Chen, L., et al.: SCA-CNN: spatial and channel-wise attention in convolutional networks for image captioning. In: CVPR (2017)
8. Deng, J., Dong, W., Socher, R., Li, L.J., Li, K., Fei-Fei, L.: ImageNet: a large-scale hierarchical image database. In: CVPR (2009)
9. Forrest, J.V., Friedman, P.J.: Radiologic errors in patients with lung cancer. West. J. Med. **134**(6), 485 (1981)
10. Guan, Q., Huang, Y.: Multi-label chest X-ray image classification via category-wise residual attention learning. Pattern Recogn. Lett. **130**, 259–266 (2018)

11. Guan, Q., Huang, Y., Zhong, Z., Zheng, Z., Zheng, L., Yang, Y.: Thorax disease classification with attention guided convolutional neural network. Pattern Recogn. Lett. **131**, 38–45 (2020). https://doi.org/10.1016/j.patrec.2019.11.040. http://www.sciencedirect.com/science/article/pii/S0167865519303617

12. Gündel, S., Grbic, S., Georgescu, B., Liu, S., Maier, A., Comaniciu, D.: Learning to recognize abnormalities in chest X-rays with location-aware dense networks. In: Vera-Rodriguez, R., Fierrez, J., Morales, A. (eds.) CIARP 2018. LNCS, vol. 11401, pp. 757–765. Springer, Cham (2019). https://doi.org/10.1007/978-3-030-13469-3_88

13. He, K., Gkioxari, G., Dollár, P., Girshick, R.: Mask R-CNN. In: ICCV (2017)

14. He, K., Zhang, X., Ren, S., Sun, J.: Deep residual learning for image recognition. In: CVPR (2016)

15. Hu, J., Shen, L., Sun, G.: Squeeze-and-excitation networks. In: CVPR (2018)

16. Huang, G., Liu, Z., Van Der Maaten, L., Weinberger, K.Q.: Densely connected convolutional networks. In: CVPR (2017)

17. Huang, Z., Wang, X., Huang, L., Huang, C., Wei, Y., Liu, W.: CCNet: criss-cross attention for semantic segmentation. In: ICCV (2019)

18. Irvin, J., et al.: CheXpert: a large chest radiograph dataset with uncertainty labels and expert comparison. In: AAAI (2019)

19. Johnson, A.E., et al.: MIMIC-CXR, a de-identified publicly available database of chest radiographs with free-text reports. Scientific Data (2019)

20. Kamel, S.I., Levin, D.C., Parker, L., Rao, V.M.: Utilization trends in noncardiac thoracic imaging, 2002–2014. J. Am. Coll. Radiol. **14**(3), 337–342 (2017)

21. Krizhevsky, A., Sutskever, I., Hinton, G.E.: ImageNet classification with deep convolutional neural networks. In: NIPS (2012)

22. Lee, H., Kim, H.E., Nam, H.: SRM: a style-based recalibration module for convolutional neural networks. In: ICCV (2019)

23. Li, Z., et al.: Thoracic disease identification and localization with limited supervision. In: CVPR (2018)

24. Lin, T.Y., et al.: Microsoft COCO: common objects in context. In: Fleet, D., Pajdla, T., Schiele, B., Tuytelaars, T. (eds.) ECCV 2014. LNCS, vol. 8693, pp. 740–755. Springer, Cham (2014). https://doi.org/10.1007/978-3-319-10602-1_48

25. Mao, C., Yao, L., Luo, Y.: ImageGCN: multi-relational image graph convolutional networks for disease identification with chest X-rays. arXiv (2019)

26. Nam, J.G., et al.: Development and validation of deep learning-based automatic detection algorithm for malignant pulmonary nodules on chest radiographs. Radiology **290**(1), 218–228 (2019). https://doi.org/10.1148/radiol.2018180237. pMID: 30251934

27. Oquab, M., Bottou, L., Laptev, I., Sivic, J.: Is object localization for free?-weakly-supervised learning with convolutional neural networks. In: CVPR (2015)

28. Quekel, L.G., Kessels, A.G., Goei, R., van Engelshoven, J.M.: Miss rate of lung cancer on the chest radiograph in clinical practice. Chest **115**(3), 720–724 (1999)

29. Rajpurkar, P., et al.: CheXNet: radiologist-level pneumonia detection on chest X-rays with deep learning. arXiv (2017)

30. Roy, A.G., Navab, N., Wachinger, C.: Concurrent spatial and channel 'squeeze & excitation' in fully convolutional networks. In: Frangi, A.F., Schnabel, J.A., Davatzikos, C., Alberola-López, C., Fichtinger, G. (eds.) MICCAI 2018. LNCS, vol. 11070, pp. 421–429. Springer, Cham (2018). https://doi.org/10.1007/978-3-030-00928-1_48

31. Song, C., Huang, Y., Ouyang, W., Wang, L.: Mask-guided contrastive attention model for person re-identification. In: Proceedings of the IEEE Conference on Computer Vision and Pattern Recognition, pp. 1179–1188 (2018)
32. Tang, Y., Wang, X., Harrison, A.P., Lu, L., Xiao, J., Summers, R.M.: Attention-guided curriculum learning for weakly supervised classification and localization of thoracic diseases on chest radiographs. In: Shi, Y., Suk, H.-I., Liu, M. (eds.) MLMI 2018. LNCS, vol. 11046, pp. 249–258. Springer, Cham (2018). https://doi.org/10.1007/978-3-030-00919-9_29
33. Vaswani, A., et al.: Attention is all you need. In: NIPS (2017)
34. Wang, F., et al.: Residual attention network for image classification. In: CVPR (2017)
35. Wang, H., Xia, Y.: ChestNet: a deep neural network for classification of thoracic diseases on chest radiography. arXiv (2018)
36. Wang, X., Girshick, R., Gupta, A., He, K.: Non-local neural networks. In: CVPR (2018)
37. Wang, X., Peng, Y., Lu, L., Lu, Z., Bagheri, M., Summers, R.M.: ChestX-ray8: hospital-scale chest X-ray database and benchmarks on weakly-supervised classification and localization of common thorax diseases. In: CVPR (2017)
38. Woo, S., Park, J., Lee, J.-Y., Kweon, I.S.: CBAM: convolutional block attention module. In: Ferrari, V., Hebert, M., Sminchisescu, C., Weiss, Y. (eds.) ECCV 2018. LNCS, vol. 11211, pp. 3–19. Springer, Cham (2018). https://doi.org/10.1007/978-3-030-01234-2_1
39. Wu, Y., Kirillov, A., Massa, F., Lo, W.Y., Girshick, R.: Detectron2 (2019). https://github.com/facebookresearch/detectron2
40. Xie, S., Girshick, R., Dollár, P., Tu, Z., He, K.: Aggregated residual transformations for deep neural networks. In: CVPR (2017)
41. Yao, L., Prosky, J., Poblenz, E., Covington, B., Lyman, K.: Weakly supervised medical diagnosis and localization from multiple resolutions. arXiv (2018)
42. Zhang, J., Bargal, S.A., Lin, Z., Brandt, J., Shen, X., Sclaroff, S.: Top-down neural attention by excitation backprop. Int. J. Comput. Vis. **126**(10), 1084–1102 (2018)

Large Scale Holistic Video Understanding

Ali Diba[1,5(✉)], Mohsen Fayyaz[2], Vivek Sharma[3], Manohar Paluri[1,2,3,4,5],
Jürgen Gall[2], Rainer Stiefelhagen[3], and Luc Van Gool[1,4,5]

[1] KU Leuven, Leuven, Belgium
{ali.diba,luc.gool}@kuleuven.be, Balamanohar@gmail.com
[2] University of Bonn, Bonn, Germany
{fayyaz,gall}@iai.uni-bonn.de
[3] KIT, Karlsruhe, Karlsruhe, Germany
{vivek.sharma,rainer.stiefelhagen}@kit.edu
[4] ETH Zürich, Zürich, Switzerland
[5] Sensifai, Brussels, Belgium

Abstract. Video recognition has been advanced in recent years by benchmarks with rich annotations. However, research is still mainly limited to human action or sports recognition - focusing on a highly specific video understanding task and thus leaving a significant gap towards describing the overall content of a video. We fill this gap by presenting a large-scale "Holistic Video Understanding Dataset" (HVU). HVU is organized hierarchically in a semantic taxonomy that focuses on multi-label and multi-task video understanding as a comprehensive problem that encompasses the recognition of multiple semantic aspects in the dynamic scene. HVU contains approx. 572k videos in total with 9 million annotations for training, validation and test set spanning over 3142 labels. HVU encompasses semantic aspects defined on categories of scenes, objects, actions, events, attributes and concepts which naturally captures the real-world scenarios.

We demonstrate the generalisation capability of HVU on three challenging tasks: 1) Video classification, 2) Video captioning and 3) Video clustering tasks. In particular for video classification, we introduce a new spatio-temporal deep neural network architecture called "Holistic Appearance and Temporal Network" (HATNet) that builds on fusing 2D and 3D architectures into one by combining intermediate representations of appearance and temporal cues. HATNet focuses on the multi-label and multi-task learning problem and is trained in an end-to-end manner. Via our experiments, we validate the idea that holistic representation learning is complementary, and can play a key role in enabling many real-world applications. https://holistic-video-understanding.github.io/.

A. Diba, M. Fayyaz and V. Sharma—Contributed equally to this work and listed in alphabetical order.

Electronic supplementary material The online version of this chapter (https://doi.org/10.1007/978-3-030-58558-7_35) contains supplementary material, which is available to authorized users.

A. Vedaldi et al. (Eds.): ECCV 2020, LNCS 12350, pp. 593–610, 2020.
https://doi.org/10.1007/978-3-030-58558-7_35

1   Introduction

Video understanding is a comprehensive problem that encompasses the recognition of multiple semantic aspects that include: a scene or an environment, objects, actions, events, attributes, and concepts. Even if considerable progress is made in video recognition, it is still rather limited to action recognition - this is due to the fact that there is no established video benchmark that integrates joint recognition of multiple semantic aspects in the dynamic scene. While Convolutional Networks (ConvNets) have caused several sub-fields of computer vision to leap forward, one of the expected drawbacks of training the ConvNets for video understanding with a single label per task is insufficiency to describe the content of a video. This issue primarily impedes the ConvNets to learn a generic feature representation towards challenging holistic video analysis. To this end, one can easily overcome this issue by recasting the video understanding problem as multi-task classification, where multiple labels are assigned to a video from multiple semantic aspects. Furthermore, it is possible to learn a generic feature representation for video analysis and understanding. This is in line with image classification ConvNets trained on ImageNet that facilitated the learning of generic feature representation for several vision tasks. Thus, training ConvNets on a multiple semantic aspects dataset can be directly applied for holistic recognition and understanding of concepts in video data, which makes it very useful to describe the content of a video (Fig. 1).

To address the above drawbacks, this work presents the "Holistic Video Understanding Dataset" (**HVU**). HVU is organized hierarchically in a semantic taxonomy that aims at providing a multi-label and multi-task large-scale video benchmark with a comprehensive list of tasks and annotations for video analysis and understanding. HVU dataset consists of 476k, 31k and 65k samples in train, validation and test set, and is a sufficiently large dataset, which means that the scale of dataset approaches that of image datasets. HVU contains approx. 572k videos in total, with ~7.5M annotations for training set, ~600K for validation set, and ~1.3M for test set spanning over 3142 labels. A full spectrum encompasses the recognition of multiple semantic aspects defined on them including 248 categories for scenes, 1678 for objects, 739 for actions, 69 for events, 117 for attributes and 291 for concepts, which naturally captures the long tail

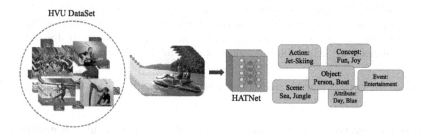

Fig. 1. Holistic Video Understanding Dataset: A multi-label and multi-task fully annotated dataset and HATNet as a new deep ConvNet for video classification.

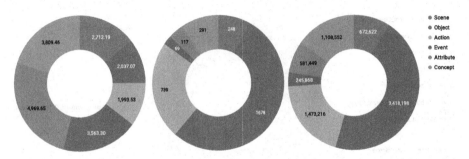

Fig. 2. Left: Average number of samples per label in each of main categories. Middle: Number of labels for each main category. Right: Number of samples per main category.

distribution of visual concepts in the real world problems. All these tasks are supported by rich annotations with an average of 2112 annotations per label. The HVU action categories builds on action recognition datasets [23,27,29,47,64] and further extend them by incorporating labels of scene, objects, events, attributes, and concepts in a video. The above thorough annotations enable developments of strong algorithms for a holistic video understanding to describe the content of a video. Table 1 shows the dataset statistics.

In order to show the importance of holistic representation learning, we demonstrate the influence of HVU on three challenging tasks: video classification, video captioning and video clustering. Motivated by holistic representation learning, for the task of video classification, we introduce a new spatio-temporal architecture called "Holistic Appearance and Temporal Network" (HATNet) that focuses on the multi-label and multi-task learning for jointly solving multiple spatio-temporal problems simultaneously. HATNet fuses 2D and 3D architectures into one by combining intermediate representations of appearance and temporal cues, leading to a robust spatio-temporal representation. Our HATNet is evaluated on challenging video classification datasets, namely HMDB51, UCF101 and Kinetics. We experimentally show that our HATNet achieves outstanding results. Furthermore, we show the positive effect of training models using more semantic concepts on transfer learning. In particular, we show that pre-training the model on HVU with more semantic concepts improves the fine-tuning results on other datasets and tasks compared to pre-training on single semantic category datasets such as, Kinetics. This shows the richness of our dataset as well as the importance of multi-task learning. Furthermore, our experiments on video captioning and video clustering demonstrates the generalisation capability of HVU on other tasks by showing promising results in comparison to the state-of-the-art.

2   Related Work

Video Recognition with ConvNets: As to prior hand-engineered [8,28, 30,39,55,61] and low-level temporal structure [18,19,35,58] descriptor learning there is a vast literature and is beyond the scope of this paper.

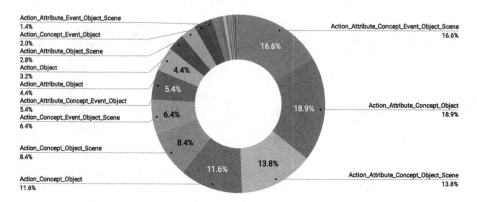

Fig. 3. Coverage of different subsets of the 6 main semantic categories in videos. 16.6% of the videos have annotations of all categories.

Recently ConvNets-based action recognition [16,26,46,50,59] has taken a leap to exploit the appearance and the temporal information. These methods operate on 2D (individual image-level) [12,14,20,48,49,59,63] or 3D (video-clips or snippets of K frames) [16,50,51,53]. The filters and pooling kernels for these architectures are 3D (x, y, time) i.e. 3D convolutions $(s \times s \times d)$ [63] where d is the kernel's temporal depth and s is the kernel's spatial size. These 3D ConvNets are intuitively effective because such 3D convolution can be used to directly extract spatio-temporal features from raw videos. Carreira et al.proposed inception [25] based 3D CNNs, which they referred to as I3D [6]. More recently, some works introduced temporal transition layer that models variable temporal convolution kernel depths over shorter and longer temporal ranges, namely T3D [11]. Further Diba et al. [10] propose spatio-temporal channel correlation that models correlations between channels of a 3D ConvNets wrt. both spatial and temporal dimensions. In contrast to these prior works, our work differs substantially in scope and technical approach. We propose an architecture, HATNet, that exploits both 2D ConvNets and 3D ConvNets to learn an effective spatio-temporal feature representation. Finally, it is worth noting the self-supervised ConvNet training works from unlabeled sources [21,42,44], such as Fernando et al. [17] and Mishra et al. [33] generate training data by shuffling the video frames; Sharma et al. [37,40,41,43] mines labels using a distance matrix or clustering based on similarity although for video face clustering; Wei et al. [60] predict the ordering task; Ng et al. [34] estimates optical flow while recognizing actions; Diba et al. [13] predicts short term future frames while recognizing actions. Self-supervised and unsupervised representation learning is beyond the scope of this paper.

The closest work to ours is by Ray et al. [36]. Ray et al. concatenate pre-trained deep features, learned independently for the different tasks, scenes, object and actions aiming to the recognition, in contrast our HATNet is trained end-to-end for multi-task and multi-label recognition in videos.

Table 1. Statistics of the HVU training set for different categories. The category with the highest number of labels and annotations is the object category.

Task category	Scene	Object	Action	Event	Attribute	Concept	Total
#Labels	248	1678	739	69	117	291	3142
#Annotations	672,622	3,418,198	1,473,216	245,868	581,449	1,108,552	7,499,905
#Videos	251,794	471,068	479,568	164,924	316,040	410,711	481,417

Table 2. Comparison of the HVU dataset with other publicly available video recognition datasets in terms of #labels per category. Note that SOA is not publicly available.

Dataset	Scene	Object	Action	Event	Attribute	Concept	#Videos	Year
HMDB51 [29]	–	–	51	–	–	–	7K	'11
UCF101 [47]	–	–	101	–	–	–	13K	'12
ActivityNet [5]	–	–	200	–	–	–	20K	'15
AVA [23]	–	–	80	–	–	–	57.6K	'18
Something-Something [22]	–	–	174	–	–	–	108K	'17
HACS [64]	–	–	200	–	–	–	140K	'19
Kinetics [27]	–	–	600	–	–	–	500K	'17
EPIC-KITCHEN [9]	–	323	149	–	–	–	39.6K	'18
SOA [36]	49	356	148	–	–	–	562K	'18
HVU (**Ours**)	248	1678	739	69	117	291	572K	'20

Video Classification Datasets: Over the last decade, several video classification datasets [4,5,29,38,47] have been made publicly available with a focus on action recognition, as summarized in Table 2. We briefly review some of the most influential action datasets available. The HMDB51 [29] and UCF101 [47] has been very important in the field of action recognition. However, they are simply not large enough for training deep ConvNets from scratch. Recently, some large action recognition datasets were introduced, such as ActivityNet [5] and Kinetics [27]. ActivityNet contains 849 h of videos, including 28,000 action instances. Kinetics-600 contains 500k videos spanning 600 human action classes with more than 400 examples for each class. The current experimental strategy is to first pre-train models on these large-scale video datasets [5,26,27] from scratch and then fine-tune them on small-scale datasets [29,47] to analyze their transfer behavior. Recently, a few other action datasets have been introduced with more samples, temporal duration and the diversity of category taxonomy, they are HACS [64], AVA [23], Charades [45] and Something-Something [22]. Sports-1M [26] and YouTube-8M [3] are the video datasets with million-scale samples. They consist quite longer videos rather than the other datasets and their annotations are provided in video-level and not temporally stamped. YouTube-8M labels are machine-generated without any human verification in the loop and Sports-1M is just focused on sport activities.

A similar spirit of HVU is observed in SOA dataset [36]. SOA aims to recognize visual concepts, such as scenes, objects and actions. In contrast, HVU has several orders of magnitude more semantic labels (6 times larger than SOA) and not just limited to scenes, objects, actions only, but also including events, attributes, and concepts. Our HVU dataset can help the computer vision community and bring more attention to holistic video understanding as a comprehensive, multi-faceted problem. Noticeably, the SOA paper was published in 2018, however the dataset is not released while our dataset is ready to become publicly available.

Motivated by efforts in large-scale benchmarks for object recognition in static images, i.e. the Large Scale Visual Recognition Challenge (ILSVRC) to learn a generic feature representation is now a back-bone to support several related vision tasks. We are driven by the same spirit towards learning a generic feature representation at the video level for holistic video understanding.

3   HVU Dataset

The HVU dataset is organized hierarchically in a semantic taxonomy of holistic video understanding. Almost all real-wold conditioned video datasets are targeting human action recognition. However, a video is not only about an action which provides a human-centric description of the video. By focusing on human-centric descriptions, we ignore the information about scene, objects, events and also attributes of the scenes or objects available in the video. While SOA [36] has categories of scenes, objects, and actions, to our knowledge it is not publicly available. Furthermore, HVU has more categories as it is shown in Table 2. One of the important research questions which is not addressed well in recent works on action recognition, is leveraging the other contextual information in a video. The HVU dataset makes it possible to assess the effect of learning and knowledge transfer among different tasks, such as enabling transfer learning of object recognition in videos to action recognition and vice-versa. In summary, HVU can help the vision community and bring more interesting solutions to holistic video understanding. Our dataset focuses on the recognition of scenes, objects, actions, attributes, events, and concepts in user generated videos. Scene, object, action and event categories definition is the same and standard as in other image and datasets. For attribute labels, we target attributes describing scenes, actions, objects or events. The concept category refers to any noun and label which present a grouping definition or related higher level in the taxonomy tree for labels of other categories.

3.1   HVU Statistics

HVU consists of **572k** videos. The number of video-clips for train, validation, and test set are **481k**, **31k** and **65k** respectively. The dataset consists of trimmed video clips. In practice, the duration of the videos are different with a maximum of 10 s length. HVU has 6 main categories: scene, object, action, event, attribute, and concept. In total, there are 3142 labels with approx. 7.5M annotations for the training, validation and test set. On average, there are ~2112 annotations per label. We depict the distribution of categories with respect to the number of annotations, labels, and annotations per label in Fig. 2. We can observe that the object category has the highest quota of labels and annotations, which is due to the abundance of objects in video. Despite having the highest quota of the labels and annotations, the object category does not have the highest annotations per label ratio. However, the average number of ~2112 annotations per label is a reasonable amount of training data for each label. The scene category does not have a large amount of labels and annotations which is due to two reasons: the trimmed videos of the dataset and the short duration of the videos. This distribution is somewhat the same for the action category. The dataset statistics for each category are shown in Table 1 for the training set.

3.2   Collection and Annotation

Building a large-scale video understanding dataset is a time-consuming task. In practice, there are two main tasks which are usually most time consuming for creating a large-scale video dataset: (a) data collection and (b) data annotation. Recent popular datasets, such as ActivityNet, Kinetics, and YouTube-8M are collected from Internet sources like YouTube. For the annotation of these datasets, usually a semi-automatic crowdsourcing strategy is used, in which a human manually verifies the crawled videos from the web. We adopt a similar strategy with difference in the technical approach to reduce the cost of data collection and annotation. Since, we are interested in the user generated videos, thanks to the taxonomy diversity of YouTube-8M [3], Kinetics-600 [27] and HACS [64], we use these datasets as main source of the HVU. By using these datasets as the source, we also do not have to deal with copyright or privacy issues so we can publicly release the dataset. Moreover, this ensures that none of the test videos of existing datasets is part of the training set of HVU. Note that, all of the aforementioned datasets are action recognition datasets.

Manually annotating a large number of videos with multiple semantic categories (i.e thousands of concepts and tags) has two major shortcomings, (a) manual annotations are error-prone because a human cannot be attentive to every detail occurring in the video that leads to mislabeling and are difficult to eradicate; (b) large scale video annotation in specific is a very time consuming task due to the amount and temporal duration of the videos. To overcome these issues, we employ a two-stage framework for the HVU annotation. In the first stage, we utilize the Google Vision API [1] and Sensifai Video Tagging API [2] to get rough annotations of the videos. The APIs predict 30 tags per video. We keep the probability threshold of the APIs relative low (~30%) as a guarantee to avoid false rejects of tags in the video. The tags were chosen from a dictionary with almost 8K words. This process resulted in almost 18 million tags for the whole dataset. In the second stage, we apply human verification to remove any possible mislabeled noisy tags and also add possible missing tags missed by the APIs from some recommended tags of similar videos. The human annotation step resulted in 9 million tags for the whole dataset with ~3500 different tags.

We provide more detailed statistics and discussion regarding the annotation process in the supplementary materials.

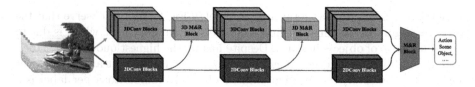

Fig. 4. HATNet: A new 2D/3D deep neural network with 2DConv, 3DConv blocks and merge and reduction (M&R) block to fuse 2D and 3D feature maps in intermediate stages of the network. HATNet combines the appearance and temporal cues with the overall goal to compress them into a more compact representation.

3.3 Taxonomy

Based on the predicted tags from the Google and the Sensifai APIs, we found that the number of obtained tags is approximately ∼8K before cleaning. The services can recognize videos with tags spanning over categories of scenes, objects, events, attributes, concepts, logos, emotions, and actions. As mentioned earlier, we remove tags with imbalanced distribution and finally, refine the tags to get the final taxonomy by using the WordNet [32] ontology. The refinement and pruning process aims to preserve the true distribution of labels. Finally, we ask the human annotators to classify the tags into 6 main semantic categories, which are scenes, objects, actions, events, attributes and concepts.

In fact, each video can be assigned to multiple semantic categories. Almost 100K of the videos have all of the semantic categories. In comparison to SOA, almost half of HVU videos have labels for scene, object and action together. Figure 3 shows the percentage of the different subsets of the main categories.

4 Holistic Appearance and Temporal Network

We first briefly discuss state-of-the-art 3D ConvNets for video classification and then propose our new proposed "Holistic Appearance and Temporal Network" (HATNet) for multi-task and multi-label video classification.

4.1 3D-ConvNets Baselines

3D ConvNets are designed to handle temporal cues available in video clips and are shown to be efficient performance-wise for video classification. 3D ConvNets exploit both spatial and temporal information in one pipeline. In this work, we chose 3D-ResNet [51] and STCnet [10] as our 3D CNNs baseline which have competitive results on Kinetics and UCF101. To measure the performance on the multi-label HVU dataset, we use mean average precision (mAP) over all labels. We also report the individual performance on each category separately. The comparison between all of the methods can be found in Table 3. These networks are trained with binary cross entropy loss.

4.2 Multi-task Learning 3D-ConvNets

Another approach which is studied in this work to tackle the HVU dataset is to have the problem solved with multi-task learning or a joint training method. As we know the HVU dataset consists of high-level categories like objects, scenes, events, attributes, and concepts, so each of these categories can be dealt like separate tasks. In our experiments, we have defined six tasks, scene, object, action, event, attribute, and concept classification. So our multi-task learning network is trained with six objective functions, that is with multi-label classification for each task. The trained network is a 3D-ConvNet which has separate Conv layers as separate heads for each of the tasks at the end of the network.

For each head we use the binary cross entropy loss since it is a multi-label classification for each of the categories.

4.3 2D/3D HATNet

Our "Holistic Appearance and Temporal Network" (HATNet) is a spatio-temporal neural network, which extracts temporal and appearance information in a novel way to maximize engagement of the two sources of information and also the efficiency of video recognition. The motivation of proposing this method is deeply rooted in a need of handling different levels of concepts in holistic video recognition. Since we are dealing with still objects, dynamic scenes, different attributes and also different human activities, we need a deep neural network that is able to focus on different levels of semantic information. We propose a flexible method to use a 2D pre-trained model on a large image dataset like ImageNet and a 3D pre-trained model on video datasets like Kinetics to fasten the process of training but the model can be trained from scratch as it is shown in our experiments as well. The proposed HATNet is capable of learning a hierarchy of spatio-temporal feature representation using appearance and temporal neural modules.

Appearance Neural Module. In HATNet design, we use 2D ConvNets with 2D Convolutional (2DConv) blocks to extract static cues of individual frames in a video-clip. Since we aim to recognize objects, scenes and attributes alongside of actions, it is necessary to have this module in the network which can handle these concepts better. Specifically, we use 2DConv to capture the spatial structure in the frame.

Temporal Neural Module. In HATNet architecture, the 3D Convolutions (3DConv) module handles temporal cues dealing with interaction in a batch of frames. 3DConv aims to capture the relative temporal information between frames. It is crucial to have 3D convolutions in the network to learn relational motion cues for efficiently understanding dynamic scenes and human activities. We use ResNet18/50 for both the 3D and 2D modules, so that they have the same spatial kernel sizes, and thus we can combine the output of the appearance and temporal branches at any intermediate stage of the network.

Table 3. MAP (%) performance of different architecture on the HVU dataset. The backbone ConvNet for all models is ResNet18.

Model	Scene	Object	Action	Event	Attribute	Concept	HVU overall %
3D-ResNet	50.6	28.6	48.2	35.9	29	22.5	35.8
3D-STCNet	51.9	30.1	50.3	35.8	29.9	22.7	36.7
HATNet	**55.8**	**34.2**	**51.8**	**38.5**	**33.6**	**26.1**	**40**

Table 4. Multi-task learning performance (mAP (%) comparison of 3D-ResNet18 and HATNet, when trained on HVU with all categories in the multi-task pipeline. The backbone ConvNet for all models is ResNet18.

Model	Scene	Object	Action	Event	Attribute	Concept	Overall
3D-ResNet (Standard)	50.6	28.6	48.2	35.9	29	22.5	35.8
HATNet (Standard)	55.8	34.2	51.8	38.5	33.6	26.1	40
3D-ResNet (Multi-Task)	51.7	29.6	48.9	36.6	31.1	24.1	37
HATNet (Multi-Task)	**57.2**	**35.1**	**53.5**	**39.8**	**34.9**	**27.3**	**41.3**

Figure 4 shows how we combine the 2DConv and 3DConv branches and use merge and reduction blocks to fuse feature maps at the intermediate stages of HATNet. Intuitively, combining the appearance and temporal features are complementary for video understanding and this fusion step aims to compress them into a more compact and robust representation. In the experiment section, we discuss in more detail about the HATNet design and how we apply merge and reduction modules between 2D and 3D neural modules. Supported by our extensive experiments, we show that HATNet complements the holistic video recognition, including understanding the dynamic and static aspects of a scene and also human action recognition. In our experiments, we have also performed tests on HATNet based multi-task learning similar to 3D-ConvNets based multi-task learning discussed in Sect. 4.2. HATNet has some similarity to the SlowFast [15] network but there are major differences. SlowFast uses two 3D-CNN networks for a slow and a fast branch. HATNet has one 3D-CNN branch to handle motion and dynamic information and one 2D-CNN to handle static information and appearance. HATNet also has skip connections with M&R blocks between 3D and 2D convolutional blocks to exploit more information.

2D/3D HATNet Design. The HATNet includes two branches: first is the 3D-Conv blocks with merging and reduction block and second branch is 2D-Conv blocks. After each 2D/3D blocks, we merge the feature maps from each block and perform a channel reduction by applying a $1 \times 1 \times 1$ convolution. Given the feature maps of the first block of both 2DConv and 3DConv, that have 64 channels each. We first concatenate these maps, resulting in 128 channels, and then apply $1 \times 1 \times 1$ convolution with 64 kernels for channel reduction, resulting in an output with 64 channels. The merging and reduction is done in the 3D

Table 5. Performance (mAP (%)) comparison of HVU and Kinetics datasets for transfer learning generalization ability when evaluated on different action recognition dataset. The trained model for all of the datasets is 3D-ResNet18.

Pre-training dataset	UCF101	HMDB51	Kinetics
From scratch	65.2	33.4	65.6
Kinetics	89.8	62.1	–
HVU	**90.5**	**65.1**	**67.8**

Table 6. State-of-the-art performance comparison on UCF101, HMDB51 test sets and Kinetics validation set. The results on UCF101 and HMDB51 are average mAP over three splits, and for Kinetics(400,600) is Top-1 mAP on validation set. For a fair comparison, here we report the performance of methods which utilize only RGB frames as input. *SlowFast uses multiple branches of 3D-ResNet with bigger backbones.

Method	Pre-trained dataset	CNN Backbone	UCF101	HMDB51	Kinetics-400	Kinetics-600
Two Stream (spatial stream) [46]	Imagenet	VGG-M	73	40.5	–	
RGB-I3D [6]	Imagenet	Inception v1	84.5	49.8	–	
C3D [50]	Sport1M	VGG11	82.3	51.6	–	
TSN [59]	Imagenet, Kinetics	Inception v3	93.2	–	72.5	
RGB-I3D [6]	Imagenet, Kinetics	Inception v1	95.6	74.8	72.1	
3D ResNext 101 (16 frames) [24]	Kinetics	ResNext101	90.7	63.8	65.1	
STC-ResNext 101 (64 frames) [10]	Kinetics	ResNext101	96.5	74.9	68.7	
ARTNet [57]	Kinetics	ResNet18	93.5	67.6	69.2	
R(2+1)D [53]	Kinetics	ResNet50	96.8	74.5	72	
ir-CSN-101 [52]	Kinetics	ResNet101	–	–	76.7	
DynamoNet [13]	Kinetics	ResNet101	–	–	76.8	
SlowFast 4×16 [15]	Kinetics	ResNet50	–	–	75.6	78.8
SlowFast 16×8* [15]	Kinetics	ResNet101	–	–	78.9*	81.1
HATNet (32 frames)	Kinetics	ResNet50	96.8	74.8	77.2	80.2
HATNet (32 frames)	HVU	ResNet18	96.9	74.5	74.2	77.4
HATNet (16 frames)	HVU	ResNet50	96.5	73.4	76.3	79.4
HATNet (32 frames)	HVU	ResNet50	**97.8**	**76.5**	**79.3**	**81.6**

and 2D branches, and continues independently until the last merging with two branches.

We employ 3D-ResNet and STCnet [10] with ResNet18/50 as the HATNet backbone in our experiments. The STCnet is a model of 3D networks with spatio-temporal channel correlation modules which improves 3D networks performance significantly. We also had to make a small change to the 2D branch and remove pooling layers right after the first 2D Conv to maintain a similar feature map size between the 2D and 3D branches since we use 112×112 as input-size.

5   Experiments

In this section, we demonstrate the importance of HVU on three different tasks: video classification, video captioning and video clustering. First, we introduce the implementation details and then show the results of each mentioned method on multi-label video recognition. Following, we compare the transfer learning ability of HVU against Kinetics. Next, as an additional experiment, we show the importance of having more categories of tags such as scenes and objects for video classification. Finally, we show the generalisation capability of HVU for video captioning and clustering tasks. For each task, we test and compare our method with the state-of-the-art on benchmark datasets. For all experiments, we use RGB frames as input to the ConvNet. For training, we use 16 or 32 frames long video clips as single input. We use PyTorch framework for implementation and all the networks are trained on a machine with 8 V100 NVIDIA GPUs.

5.1   HVU Results

In Table 3, we report the overall performance of different simpler or multi-task learning baselines and HATNet on the HVU validation set. The reported performance is mean average precision on all of the labels/tags. HATNet that exploits both appearance and temporal information in the same pipeline achieves the best performance, since recognizing objects, scenes and attributes need an appearance module which other baselines do not have. With HATNet, we show that combining the 3D (temporal) and 2D (appearance) convolutional blocks one can learn a more robust reasoning ability.

5.2   Multi-task Learning on HVU

Since the HVU is a multi-task classification dataset, it is interesting to compare the performance of different deep neural networks in the multi-task learning paradigm as well. For this, we have used the same architecture as in the previous experiment, but with different last layer of convolutions to observe multi-task learning performance. We have targeted six tasks: scene, object, action, event, attribute, and concept classification. In Table 4, we have compared standard training without multi-task learning heads versus multi-task learning networks.

The simple baseline multi-task learning methods achieve higher performance on individual tasks as expected, in comparison to standard networks learning for all categories as a single task. Therefore this initial result on a real-world multi-task video dataset motivates the investigation of more efficient multi-task learning methods for video classification.

Table 7. Captioning performance comparisons of [54] with different models and pre-training datasets. M denotes the motion features from optical flow extracted as in the original paper.

Model	Pre-training dataset	BLEU@4
SA (VGG+C3D) [62]	ImageNet+Sports1M	36.6
M3 (VGG+C3D) [56]	ImageNet+Sports1M	38.1
SibNet (GoogleNet) [31]	ImageNet	40.9
MGSA (Inception+C3D) [7]	ImageNet+Sports1M	42.4
I3D+M [54]	Kinetics	41.7
3D-ResNet50+M	Kinetics	41.8
3D-ResNet50+M	HVU	**42.7**

5.3 Transfer Learning: HVU vs Kinetics

Here, we study the ability of transfer learning with the HVU dataset. We compare the results of pre-training 3D-ResNet18 using Kinetics versus using HVU and then fine-tuning on UCF101, HMDB51 and Kinetics. Obviously, there is a large benefit from pre-training of deep 3D-ConvNets and then fine-tune them on smaller datasets (i.e. HVU, Kinetics ⇒ UCF101 and HMDB51). As it can be observed in Table 5, models pre-trained on our HVU dataset performed notably better than models pre-trained on the Kinetics dataset. Moreover, pre-training on HVU can improve the results on Kinetics also.

5.4 Benefit of Multiple Semantic Categories

Here, we study the effect of training models with multiple semantic categories, in comparison to using only a single semantic category, such as Kinetics which covers only action category. In particular, we designed an experiment by having the model trained in multiple steps by adding different categories of tags one by one. Specifically, we first train 3D-ResNet18 with action tags of HVU, following in second step we add tags from object category and in the last step we add tags from the scene category. For performance evaluation, we consider action category of HVU. In the first step the gained performance was 43.6% accuracy and after second step it was improved to 44.5% and finally in the last step it raised to 45.6%. The results show that adding high-level categories to the training, boosts the performance for action recognition in each step. As it was also shown in Table 4, training all the categories together yields 47.5% for the action category which is ~4% gain over action as single category for training. Thus we can infer from this that an effective feature representation can be learned by adding additional categories, and also acquire knowledge for an in-depth understanding of the video in holistic sense.

Table 8. Video clustering performance: evaluation based on extracted features from networks pre-trained on Kinetics and HVU datasets.

Model	Pre-training dataset	Clustering accuracy (%)
3D-ResNet50	Kinetics	50.3
3D-ResNet50	HVU	53.5
HATNet	HVU	**54.8**

5.5   Comparison on UCF, HMDB, Kinetics

In Table 6, we compare the HATNet performance with the state-of-the-art on UCF101, HMDB51 and Kinetics. For our baselines and HATNet, we employ pre-training in two separate setups: one with HVU and another with Kinetics, and then fine-tune on the target datasets. For UCF101 and HMDB51, we report the average accuracy over all three splits. We have used ResNet18/50 as backbone model for all of our networks with 16 and 32 input-frames. HATNet pre-trained on HVU with 32 frames input achieved superior performance on all three datasets with standard network backbones. Note that on Kinetics, HATNet even with ResNet18 as a backbone ConvNet performs almost comparable to SlowFast which is trained by dual 3D-ResNet50. In Table 6, however while SlowFast has better performance using dual 3D-ResNet101 architecture, HATNet obtains comparable results with much smaller backbone.

5.6   Video Captioning

We present a second task that showcases the effectiveness of our HVU dataset, we evaluate the effectiveness of HVU for video captioning task. We conduct experiments on a large-scale video captioning dataset, namely MSR-VTT [62]. We follow the standard training/testing splits and protocols provided originally in [62]. For video captioning, the performance is measured using the BLEU metric.

Method and Results: Most of the state-of-the-art video captioning methods use models pre-trained on Kinetics or other video recognition datasets. With this experiment, we intend to show another generalisation capability of HVU dataset where we evaluate the performance of pre-trained models trained on HVU against Kinetics. For our experiment, we use the Controllable Gated Network [54] method, which is to the best of our knowledge the state-of-the-art for captioning task.

For comparison, we considered two models of 3D-ResNet50, pre-trained on (i) Kinetics and (ii) HVU. Table 7 shows that the model trained on HVU obtained better gains in comparison to Kinetics. This shows HVU helps to learn more generic video representation to achieve better performance in other tasks.

5.7   Video Clustering

With this experiment, we evaluate the effectiveness of generic features learned using HVU when compared to Kinetics.

Dataset: We conduct experiments on ActivityNet-100 [5] dataset. For this experiment we provide results when considering 20 action categories with 1500 test videos. We have selected ActivityNet dataset to make sure there are no same videos in HVU and Kinetics training set. For clustering, the performance is measured using clustering accuracy [41].

Method and Results: We extract features using 3D-ResNet50 and HATNet pre-trained on Kinetics-600 and HVU for the test videos and then cluster them with KMeans clustering algorithm with the given number of action categories. Table 8 clearly shows that the features learned using HVU is far more effective compared to features learned using Kinetics.

6   Conclusion

This work presents the "Holistic Video Understanding Dataset" (HVU), a large-scale multi-task, multi-label video benchmark dataset with comprehensive tasks and annotations. It contains 572k videos in total with 9M annotations, which is richly labeled over 3142 labels encompassing scenes, objects, actions, events, attributes and concepts categories. Through our experiments, we show that the HVU can play a key role in learning a generic video representation via demonstration on three real-world tasks: video classification, video captioning and video clustering. Furthermore, we present a novel network architecture, HATNet, that combines 2D and 3D ConvNets in order to learn a robust spatio-temporal feature representation via multi-task and multi-label learning in an end-to-end manner. We believe that our work will inspire new research ideas for holistic video understanding. For the future plan, we are going to expand the dataset to 1 million videos with similar rich semantic labels and also provide annotations for other important tasks like activity and object detection and video captioning.

Acknowledgements. This work was supported by DBOF PhD scholarship & GC4 Flemish AI project, and the ERC Starting Grant ARCA (677650). We also would like to thank Sensifai for giving us access to the Video Tagging API for dataset preparation.

References

1. Google Vision AI API. cloud.google.com/vision
2. Sensifai Video Tagging API. www.sensifai.com
3. Abu-El-Haija, S., et al.: Youtube-8m: a large-scale video classification benchmark. arXiv:1609.08675 (2016)
4. Andriluka, M., Pishchulin, L., Gehler, P., Schiele, B.: 2D human pose estimation: new benchmark and state of the art analysis. In: CVPR (2014)

5. Caba Heilbron, F., Escorcia, V., Ghanem, B., Carlos Niebles, J.: ActivityNet: a large-scale video benchmark for human activity understanding. In: CVPR (2015)
6. Carreira, J., Zisserman, A.: Quo Vadis, action recognition? A new model and the kinetics dataset. In: CVPR (2017)
7. Chen, S., Jiang, Y.G.: Motion guided spatial attention for video captioning. In: AAAI (2019)
8. Dalal, N., Triggs, B., Schmid, C.: Human detection using oriented histograms of flow and appearance. In: Leonardis, A., Bischof, H., Pinz, A. (eds.) ECCV 2006. LNCS, vol. 3952, pp. 428–441. Springer, Heidelberg (2006). https://doi.org/10.1007/11744047_33
9. Damen, D., et al.: Scaling egocentric vision: the Epic-Kitchens dataset. In: Ferrari, V., Hebert, M., Sminchisescu, C., Weiss, Y. (eds.) ECCV 2018. LNCS, vol. 11208, pp. 753–771. Springer, Cham (2018). https://doi.org/10.1007/978-3-030-01225-0_44
10. Diba, A., et al.: Spatio-temporal channel correlation networks for action classification. In: Ferrari, V., Hebert, M., Sminchisescu, C., Weiss, Y. (eds.) ECCV 2018. LNCS, vol. 11208, pp. 299–315. Springer, Cham (2018). https://doi.org/10.1007/978-3-030-01225-0_18
11. Diba, A., et al.: Temporal 3D convnets using temporal transition layer. In: CVPR Workshops (2018)
12. Diba, A., Sharma, V., Van Gool, L.: Deep temporal linear encoding networks. In: CVPR (2017)
13. Diba, A., Sharma, V., Van Gool, L., Stiefelhagen, R.: DynamoNet: dynamic action and motion network. In: ICCV (2019)
14. Donahue, J., et al.: Long-term recurrent convolutional networks for visual recognition and description. In: CVPR (2015)
15. Feichtenhofer, C., Fan, H., Malik, J., He, K.: SlowFast networks for video recognition. In: ICCV (2019)
16. Feichtenhofer, C., Pinz, A., Zisserman, A.: Convolutional two-stream network fusion for video action recognition. In: CVPR (2016)
17. Fernando, B., Bilen, H., Gavves, E., Gould, S.: Self-supervised video representation learning with odd-one-out networks. In: CVPR (2017)
18. Fernando, B., Gavves, E., Oramas, J.M., Ghodrati, A., Tuytelaars, T.: Modeling video evolution for action recognition. In: CVPR (2015)
19. Gaidon, A., Harchaoui, Z., Schmid, C.: Temporal localization of actions with actoms. PAMI **35**, 2782–2795 (2013)
20. Girdhar, R., Ramanan, D., Gupta, A., Sivic, J., Russell, B.: ActionVLAD: learning spatio-temporal aggregation for action classification. In: CVPR (2017)
21. Girdhar, R., Tran, D., Torresani, L., Ramanan, D.: Distinit: learning video representations without a single labeled video. In: ICCV (2019)
22. Goyal, R., et al.: The "something something" video database for learning and evaluating visual common sense. In: ICCV (2017)
23. Gu, C., et al.: AVA: a video dataset of spatio-temporally localized atomic visual actions. In: CVPR (2018)
24. Hara, K., Kataoka, H., Satoh, Y.: Learning spatio-temporal features with 3D residual networks for action recognition. In: ICCV (2017)
25. Ioffe, S., Szegedy, C.: Batch normalization: accelerating deep network training by reducing internal covariate shift. In: ICML (2015)
26. Karpathy, A., Toderici, G., Shetty, S., Leung, T., Sukthankar, R., Fei-Fei, L.: Large-scale video classification with convolutional neural networks. In: CVPR (2014)

27. Kay, W., et al.: The kinetics human action video dataset. arXiv:1705.06950 (2017)
28. Klaser, A., Marszałek, M., Schmid, C.: A spatio-temporal descriptor based on 3d-gradients. In: BMVC (2008)
29. Kuehne, H., Jhuang, H., Stiefelhagen, R., Serre, T.: HMDB51: a large video database for human motion recognition. In: High Performance Computing in Science and Engineering (2013)
30. Laptev, I., Marszalek, M., Schmid, C., Rozenfeld, B.: Learning realistic human actions from movies. In: CVPR (2008)
31. Liu, S., Ren, Z., Yuan, J.: SibNet: sibling convolutional encoder for video captioning. In: ACMM (2018)
32. Miller, G.A.: Wordnet: a lexical database for English. Commun. ACM **38**, 39–41 (1995)
33. Misra, I., Zitnick, C.L., Hebert, M.: Shuffle and learn: unsupervised learning using temporal order verification. In: Leibe, B., Matas, J., Sebe, N., Welling, M. (eds.) ECCV 2016. LNCS, vol. 9905, pp. 527–544. Springer, Cham (2016). https://doi.org/10.1007/978-3-319-46448-0_32
34. Ng, J.Y.H., Choi, J., Neumann, J., Davis, L.S.: ActionFlowNet: learning motion representation for action recognition. In: WACV (2018)
35. Niebles, J.C., Chen, C.-W., Fei-Fei, L.: Modeling temporal structure of decomposable motion segments for activity classification. In: Daniilidis, K., Maragos, P., Paragios, N. (eds.) ECCV 2010. LNCS, vol. 6312, pp. 392–405. Springer, Heidelberg (2010). https://doi.org/10.1007/978-3-642-15552-9_29
36. Ray, J., et al.: Scenes-objects-actions: a multi-task, multi-label video dataset. In: Ferrari, V., Hebert, M., Sminchisescu, C., Weiss, Y. (eds.) Computer Vision – ECCV 2018. LNCS, vol. 11218, pp. 660–676. Springer, Cham (2018). https://doi.org/10.1007/978-3-030-01264-9_39
37. Roethlingshoefer, V., Sharma, V., Stiefelhagen, R.: Self-supervised face-grouping on graph. In: ACM MM (2019)
38. Schuldt, C., Laptev, I., Caputo, B.: Recognizing human actions: a local SVM approach. In: ICPR (2004)
39. Scovanner, P., Ali, S., Shah, M.: A 3-dimensional sift descriptor and its application to action recognition. In: ACM MM (2007)
40. Sharma, V., Sarfraz, S., Stiefelhagen, R.: A simple and effective technique for face clustering in TV series. In: CVPR workshop on Brave New Motion Representations (2017)
41. Sharma, V., Tapaswi, M., Sarfraz, M.S., Stiefelhagen, R.: Self-supervised learning of face representations for video face clustering. In: International Conference on Automatic Face and Gesture Recognition (2019)
42. Sharma, V., Tapaswi, M., Sarfraz, M.S., Stiefelhagen, R.: Video face clustering with self-supervised representation learning. IEEE Trans. Biometrics Behav. Identity Sci. **2**, 145–157 (2019)
43. Sharma, V., Tapaswi, M., Sarfraz, M.S., Stiefelhagen, R.: Clustering based contrastive learning for improving face representations. In: International Conference on Automatic Face and Gesture Recognition (2020)
44. Sharma, V., Tapaswi, M., Stiefelhagen, R.: Deep multimodal feature encoding for video ordering. In: ICCV Workshop on Holistic Video Understanding (2019)
45. Sigurdsson, G.A., Varol, G., Wang, X., Farhadi, A., Laptev, I., Gupta, A.: Hollywood in homes: crowdsourcing data collection for activity understanding. In: Leibe, B., Matas, J., Sebe, N., Welling, M. (eds.) ECCV 2016. LNCS, vol. 9905, pp. 510–526. Springer, Cham (2016). https://doi.org/10.1007/978-3-319-46448-0_31

46. Simonyan, K., Zisserman, A.: Two-stream convolutional networks for action recognition in videos. In: NIPS (2014)
47. Soomro, K., Zamir, A.R., Shah, M.: Ucf101: a dataset of 101 human actions classes from videos in the wild. arXiv:1212.0402 (2012)
48. Sun, L., Jia, K., Yeung, D.Y., Shi, B.E.: Human action recognition using factorized spatio-temporal convolutional networks. In: ICCV (2015)
49. Tang, P., Wang, X., Shi, B., Bai, X., Liu, W., Tu, Z.: Deep fishernet for object classification. arXiv:1608.00182 (2016)
50. Tran, D., Bourdev, L., Fergus, R., Torresani, L., Paluri, M.: Learning spatiotemporal features with 3D convolutional networks. In: ICCV (2015)
51. Tran, D., Ray, J., Shou, Z., Chang, S.F., Paluri, M.: Convnet architecture search for spatiotemporal feature learning. arXiv:1708.05038 (2017)
52. Tran, D., Wang, H., Torresani, L., Feiszli, M.: Video classification with channel-separated convolutional networks. In: ICCV (2019)
53. Tran, D., Wang, H., Torresani, L., Ray, J., LeCun, Y., Paluri, M.: A closer look at spatiotemporal convolutions for action recognition. In: CVPR (2018)
54. Wang, B., Ma, L., Zhang, W., Jiang, W., Wang, J., Liu, W.: Controllable video captioning with POS sequence guidance based on gated fusion network. In: ICCV (2019)
55. Wang, H., Schmid, C.: Action recognition with improved trajectories. In: ICCV (2013)
56. Wang, J., Wang, W., Huang, Y., Wang, L., Tan, T.: M3: multimodal memory modelling for video captioning. In: CVPR (2018)
57. Wang, L., Li, W., Li, W., Van Gool, L.: Appearance-and-relation networks for video classification. In: CVPR (2018)
58. Wang, L., Qiao, Yu., Tang, X.: Video action detection with relational dynamic-poselets. In: Fleet, D., Pajdla, T., Schiele, B., Tuytelaars, T. (eds.) ECCV 2014. LNCS, vol. 8693, pp. 565–580. Springer, Cham (2014). https://doi.org/10.1007/978-3-319-10602-1_37
59. Wang, L., et al.: Temporal segment networks: towards good practices for deep action recognition. In: Leibe, B., Matas, J., Sebe, N., Welling, M. (eds.) ECCV 2016. LNCS, vol. 9912, pp. 20–36. Springer, Cham (2016). https://doi.org/10.1007/978-3-319-46484-8_2
60. Wei, D., Lim, J., Zisserman, A., Freeman, W.T.: Learning and using the arrow of time. In: CVPR (2018)
61. Willems, G., Tuytelaars, T., Van Gool, L.: An efficient dense and scale-invariant spatio-temporal interest point detector. In: Forsyth, D., Torr, P., Zisserman, A. (eds.) ECCV 2008. LNCS, vol. 5303, pp. 650–663. Springer, Heidelberg (2008). https://doi.org/10.1007/978-3-540-88688-4_48
62. Xu, J., Mei, T., Yao, T., Rui, Y.: MSR-VTT: a large video description dataset for bridging video and language. In: CVPR (2016)
63. Yue-Hei Ng, J., Hausknecht, M., Vijayanarasimhan, S., Vinyals, O., Monga, R., Toderici, G.: Beyond short snippets: deep networks for video classification. In: CVPR (2015)
64. Zhao, H., Yan, Z., Torresani, L., Torralba, A.: HACS: human action clips and segments dataset for recognition and temporal localization. arXiv:1712.09374 (2019)

Indirect Local Attacks for Context-Aware Semantic Segmentation Networks

Krishna Kanth Nakka[1]([✉]) and Mathieu Salzmann[1,2]

[1] CVLab, EPFL, Lausanne, Switzerland
[2] ClearSpace, Ecublens, Switzerland
{krishna.nakka,mathieu.salzmann}@epfl.ch

Abstract. Recently, deep networks have achieved impressive semantic segmentation performance, in particular thanks to their use of larger contextual information. In this paper, we show that the resulting networks are sensitive not only to global adversarial attacks, where perturbations affect the entire input image, but also to indirect local attacks, where the perturbations are confined to a small image region that does not overlap with the area that the attacker aims to fool. To this end, we introduce an indirect attack strategy, namely adaptive local attacks, aiming to find the best image location to perturb, while preserving the labels at this location and producing a realistic-looking segmentation map. Furthermore, we propose attack detection techniques both at the global image level and to obtain a pixel-wise localization of the fooled regions. Our results are unsettling: Because they exploit a larger context, more accurate semantic segmentation networks are more sensitive to indirect local attacks. We believe that our comprehensive analysis will motivate the community to design architectures with contextual dependencies that do not trade off robustness for accuracy.

Keywords: Adversarial attacks · Semantic segmentation

1 Introduction

Deep Neural Networks (DNNs) are highly expressive models and achieve state-of-the-art performance in many computer vision tasks. In particular, the powerful backbones originally developed for image recognition have now be recycled for semantic segmentation, via the development of fully convolutional networks (FCNs) [29]. The success of these initial FCNs, however, was impeded by their limited understanding of surrounding context. As such, recent techniques have focused on exploiting contextual information via dilated convolutions [50], pooling operations [26,53], or attention mechanisms [12,54].

Despite this success, recent studies have shown that DNNs are vulnerable to adversarial attacks. That is, small, dedicated perturbations to the input images can make a network produce virtually arbitrarily incorrect predictions. While

Electronic supplementary material The online version of this chapter (https://doi.org/10.1007/978-3-030-58558-7_36) contains supplementary material, which is available to authorized users.

© Springer Nature Switzerland AG 2020
A. Vedaldi et al. (Eds.): ECCV 2020, LNCS 12350, pp. 611–628, 2020.
https://doi.org/10.1007/978-3-030-58558-7_36

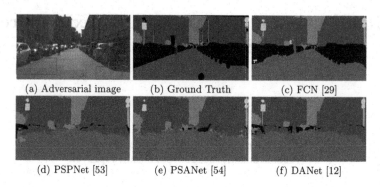

(a) Adversarial image   (b) Ground Truth   (c) FCN [29]

(d) PSPNet [53]   (e) PSANet [54]   (f) DANet [12]

Fig. 1. Indirect Local Attacks. An adversarial input image **(a)** is attacked with an imperceptible noise in local regions, shown as red boxes, to fool the dynamic objects. Such *indirect* local attacks barely affect an FCN [29] **(c)**. By contrast, modern networks that leverage context to achieve higher accuracy on clean (unattacked) images, such as PSPNet [53] **(d)**, PSANet [54] **(e)** and DANet [12] **(f)** are more strongly affected, even in regions far away from the perturbed area.

this has been mostly studied in the context of image recognition [9,23,34,35,39], a few recent works have nonetheless discussed such adversarial attacks for semantic segmentation [2,18,49]. These methods, however, remain limited to global perturbations to the entire image. Here, we argue that local attacks are more realistic, in that, in practice, they would allow an attacker to modify the physical environment to fool a network. This, in some sense, was the task addressed in [11], where stickers were placed on traffic poles so that an image recognition network would misclassify the corresponding traffic signs. In this scenario, however, the attack was directly performed on the targeted object.

In this paper, by contrast, we study the impact of *indirect local* attacks, where the perturbations are performed on regions outside the targeted objects. This, for instance, would allow one to place a sticker on a building to fool a self-driving system such that all nearby dynamic objects, such as cars and pedestrians, become mislabeled as the nearest background class. We choose this setting not only because it allows the attacker to perturb only a small region in the scene, but also because it will result in realistic-looking segmentation maps. By contrast, untargeted attacks would yield unnatural outputs, which can much more easily be detected by a defense mechanism. However, designing such targeted attacks that are effective is much more challenging than untargeted ones.

To achieve this, we first investigate the general idea of *indirect* attacks, where the perturbations can occur anywhere in the image except on the targeted objects. We then switch to the more realistic case of *localized* indirect attacks, and design a group sparsity-based strategy to confine the perturbed region to a small area outside of the targeted objects. For our attacks to remain realistic and imperceptible, we perform them without ground-truth information about the dynamic objects and in a norm-bounded setting. In addition to these indirect attacks, we evaluate the robustness of state-of-the-art networks to a single universal fixed-size local perturbation that can be learned from all training images to attack an unseen image in an untargeted manner.

The conclusions of our experiments are disturbing: In short, more accurate semantic segmentation networks are more sensitive to indirect local attacks. This is illustrated by Fig. 1, where perturbing a few patches in static regions has much larger impact on the dynamic objects for the context-aware PSPNet [53], PSANet [54] and DANet [12] than for a simple FCN [29]. This, however, has to be expected, because the use of context, which improves segmentation accuracy, also increases the network's receptive field, thus allowing the perturbation to be propagated to more distant image regions.

Motivated by this unsettling sensitivity of segmentation networks to indirect local attacks, we then turn our focus to adversarial attack detection. In contrast to the only two existing works that have tackled attack detection for semantic segmentation [25,48], we perform detection not only at the global image level, but locally at the pixel level. Specifically, we introduce an approach to localizing the regions whose predictions were affected by the attack, i.e., not the image regions that were perturbed. In an autonomous driving scenario, this would allow one to focus more directly on the potential dangers themselves, rather than on the image regions that caused them.

To summarize, our contributions are as follows. We introduce the idea of indirect local adversarial attacks for semantic segmentation networks, which better reflects potential real-world dangers. We design an adaptive, image-dependent local attack strategy to find the minimal location to perturb in the static image region. We show the vulnerability of modern networks to a universal, image-independent adversarial patch. We study the impact of context on a network's sensitivity to our indirect local attacks. We introduce a method to detect indirect local attacks at both image level and pixel level. Our code is available at https://github.com/krishnakanthnakka/Indirectlocalattacks/.

2   Related Work

Context in Semantic Segmentation Networks. While context has been shown to improve the results of traditional semantic segmentation methods [13,17,21,22], the early deep fully-convolutional semantic segmentation networks [15,29] only gave each pixel a limited receptive field, thus encoding relatively local relationships. Since then, several solutions have been proposed to account for wider context. For example, UNet [42] uses contracting path to capture larger context followed by a expanding path to upsample the intermediate low-resolution representation back to the input resolution. ParseNet [26] relies on global pooling of the final convolutional features to aggregate context information. This idea was extended to using different pooling strides in PSPNet [53], so as to encode different levels of context. In [50], dilated convolutions were introduced to increase the size of the receptive field. PSANet [54] is designed so that each local feature vector is connected to all the other ones in the feature map, thus learning contextual information adaptively. EncNet [52] captures context via a separate network branch that predicts the presence of the object categories in the scene without localizing them. DANet [12] uses a dual attention mechanism to attend to the most important spatial and channel locations in the final feature map. In particular, the DANet position attention module selectively aggregates the features at all positions using a weighted sum. In practice, all of these strategies to use larger contextual information have been shown to outperform simple FCNs on clean samples. Here, however, we show that this makes the resulting networks more vulnerable to indirect local adversarial attacks, even when the perturbed region covers less than 1% of the input image.

Adversarial Attacks on Semantic Segmentation: Adversarial attacks aim to perturb an input image with an imperceptible noise so as to make a DNN produce erroneous predictions. So far, the main focus of the adversarial attack literature has been image classification, for which diverse attack and defense strategies have been proposed [6,9,14,23,34,35,39]. In this context, it was shown that deep networks can be attacked even when one does not have access to the model weights [28,37], that attacks can be transferred across different networks [45], and that universal perturbations that can be applied to any input image exist [32,33,40].

Motivated by the observations made in the context of image classification, adversarial attacks were extended to semantic segmentation. In [2], the effectiveness of attack strategies designed for classification was studied for different segmentation networks. In [49], a dense adversary generation attack was proposed, consisting of projecting the gradient in each iteration with minimal distortion. In [18], a universal perturbation was learnt using the whole image dataset. Furthermore, [4] demonstrated the existence of perturbations that are robust over chosen distributions of transformations.

None of these works, however, impose any constraints on the location of the attack in the input image. As such, the entire image is perturbed, which, while effective when the attacker has access to the image itself, would not allow one to physically modify the scene so as to fool, e.g., autonomous vehicles. This, in essence, was the task first addressed in [5], where a universal targeted patch was shown to fool a recognition system to a specific target class. Later, patch attacks were studied in a more realistic setting in [11], where it was shown that placing a small, well-engineered patch on a traffic sign was able to fool a classification network into making wrong decisions. While these works focused on classification, patch attacks have been extended to object detection [27,30,43,44] and optical flow [41]. Our work differs fundamentally from these methods in the following ways. First, none of these approaches optimize the location of the patch perturbation. Second, [5,27,43] learn a separate perturbation for every target class, which, at test time, lets the attacker change the predictions to one class only. While this is suitable for recognition, it does not apply to our segmentation setup, where we seek to misclassify the dynamic objects as different background classes so as to produce a realistic segmentation map. Third, unlike [5,11,41], our perturbations are imperceptible. Finally, while the perturbations in [11,41,44] cover the regions that should be misclassified, and in [27,43] affect the predictions in the perturbed region, we aim to design an attack that affects only targeted locations outside the perturbed region.

In other words,we study the impact of *indirect* local attacks, where the perturbation is outside the targeted area. This would allow one to modify static portions of the scene so as to, e.g., make dynamic objects disappear to fool the self-driving system. Furthermore, we differ from these other patch-based attacks in that we study local attacks for semantic segmentation to understand the impact of the contextual information exploited by different networks, and introduce detection strategies at both image- and pixel-level.Similarly to most of the existing literature [5,11,18,25,43,44], we focus on the white-box setting for three reasons: (1) Developing effective defense mechanisms for semantic segmentation, which are currently lacking, requires assessing the sensitivity of semantic segmentation networks to the strongest attacks, i.e., white-box ones; (2) Recent model extraction methods [7,38,46] allow an attacker to obtain a close approximation of the deployed model. (3) While effective in classification [37], black-box attacks were observed to transfer poorly across semantic segmentation architectures [48], particularly in the targeted scenario. We nonetheless evaluate black-box attacks in the supplementary material.

When it comes to detecting attacks to semantic segmentation networks, only two techniques have been proposed [25,48]. In [48], detection is achieved by checking the consistency of predictions obtained from overlapping image patches. In [25], the attacked label map is passed through a pix2pix generator [19] to resynthesize an image, which is then compared with the input image to detect the attack. In contrast to these works that need either multiple passes through the network or an auxiliary detector, we detect the attack by analyzing the internal subspaces of the segmentation network. To this end, inspired by the algorithm of [24] designed for image classification, we compute the Mahalanobis distance of the features to pre-trained class conditional distributions. In contrast to [25,48], which study only global image-level detection, we show that our approach is applicable at both the image and the pixel level, yielding the first study on localizing the regions fooled by the attack.

3   Indirect Local Segmentation Attacks

Let us now introduce our diverse strategies to attack a semantic segmentation network. In semantic segmentation, given a clean image $\mathbf{X} \in \mathbb{R}^{W \times H \times C}$, where W, H and C are the width, height, and number of channels, respectively, a network is trained to minimize a loss function of the form $L(\mathbf{X}) = \sum_{j=1}^{W \times H} J(y_j^{true}, f(\mathbf{X})_j)$, where J is typically taken as the cross-entropy between the true label y_j^{true} and the predicted label $f(\mathbf{X})_j$ at spatial location j. In this context, an adversarial attack is carried out by optimizing for a perturbation that forces the network to output wrong labels for some (or all) of the pixels. Below, we denote by $\mathbf{F} \in \{0,1\}^{W \times H}$ the fooling mask such that $\mathbf{F}_j = 1$ if the j-th pixel location is targeted by the attacker to be misclassified and $\mathbf{F}_j = 0$ is the predicted label should be preserved. We first present our different local attack strategies, and finally introduce our attack detection technique.

3.1 Indirect Local Attacks

To study the sensitivity of segmentation networks, we propose to perform local perturbations, confined to predefined regions such as class-specific regions or patches, and to fool other regions than those perturbed. For example, in the context of automated driving, we may aim to perturb only the regions belonging to the road to fool the car regions in the output label map. This would allow one to modify the physical, static scene while targeting dynamic objects.

Formally, given a clean image $\mathbf{X} \in \mathbb{R}^{W \times H \times C}$, we aim to find an additive perturbation $\delta \in \mathbb{R}^{W \times H \times C}$ within a perturbation mask $\mathbf{M}$ that yields erroneous labels within the fooling mask $\mathbf{F}$. To achieve this, we define the perturbation mask $\mathbf{M} \in \{0, 1\}^{W \times H}$ such that $\mathbf{M}_j = 1$ if the j-th pixel location can be perturbed and $\mathbf{M}_j = 0$ otherwise. Let $\mathbf{y}_j^{pred}$ be the label obtained from the clean image at pixel j. An untargeted attack can then be expressed as the solution to the optimization problem

$$\delta^* = \arg\min_{\delta} \sum_{j | \mathbf{F}_j = 1} -J(\mathbf{y}_j^{pred}, f(\mathbf{X} + \mathbf{M} \odot \delta)_j) + \sum_{j | \mathbf{F}_j = 0} J(\mathbf{y}_j^{pred}, f(\mathbf{X} + \mathbf{M} \odot \delta)_j) \quad (1)$$

which aims to minimize the probability of $\mathbf{y}_j^{pred}$ in the targeted regions while maximizing it in the rest of the image. By contrast, for a targeted attack whose goal is to misclassify any pixel j in the fooling region to a pre-defined label $\mathbf{y}_j^t$, we write the optimization problem

$$\delta^* = \arg\min_{\delta} \sum_{j | \mathbf{F}_j = 1} J(\mathbf{y}_j^t, f(\mathbf{X} + \mathbf{M} \odot \delta)_j) + \sum_{i | \mathbf{F}_j = 0} J(\mathbf{y}_j^{pred}, f(\mathbf{X} + \mathbf{M} \odot \delta)_j) . \quad (2)$$

We solve (1) and (2) via the iterative projected gradient descent algorithm [3] with an ℓ_p-norm perturbation budget $\|\mathbf{M} \odot \delta\|_p < \epsilon$, where $p \in \{2, \infty\}$.

Note that the formulations above allow one to achieve any local attack. To perform *indirect* local attacks, we simply define the masks $\mathbf{M}$ and $\mathbf{F}$ so that they do not intersect, i.e., $\mathbf{M} \odot \mathbf{F} = \mathbf{0}$, where $\odot$ is the element-wise product.

3.2 Adaptive Indirect Local Attacks

The attacks described in Sect. 3.1 assume the availability of a fixed, predefined perturbation mask $\mathbf{M}$. In practice, however, one might want to find the best location for an attack, as well as make the attack as local as possible. In this section, we introduce an approach to achieving this by enforcing structured sparsity on the perturbation mask.

To this end, we first re-write the previous attack scheme under an ℓ_2 budget as an optimization problem. Let $J_t(\mathbf{X}, \mathbf{M}, \mathbf{F}, \delta, f, \mathbf{y}^{pred}, \mathbf{y}^t)$ denote the objective function of either (1) or (2), where $\mathbf{y}^t$ can be ignored in the untargeted case. Following [6], we write an adversarial attack under an ℓ_2 budget as the solution to the optimization problem

$$\delta^* = \arg\min_{\delta} \lambda_1 \|\delta\|_2^2 + J_t(\mathbf{X}, \mathbf{M}, \mathbf{F}, \delta, f, \mathbf{y}^{pred}, \mathbf{y}^t) , \quad (3)$$

where λ_1 balances the influence of the term aiming to minimize the magnitude of the perturbation. While solving this problem, we further constrain the resulting adversarial image $\mathbf{X} + \mathbf{M} \odot \delta$ to lie in the valid pixel range $[0, 1]$.

To identify the best location for an attack together with confining the perturbations to as small an area as possible, we divide the initial perturbation mask $\mathbf{M}$ into T non-overlapping patches. This can be achieved by defining T masks $\{\mathbf{M}_t \in \mathbb{R}^{W \times H}\}$ such that, for any s, t, with $s \neq t$, $\mathbf{M}_s \odot \mathbf{M}_t = \mathbf{0}$, and $\sum_{t=1}^{T} \mathbf{M}_t = \mathbf{M}$. Our goal then becomes that of finding a perturbation that is non-zero in the smallest number of such masks. This can be achieved by modifying (3) as

$$\delta^* = \arg\min_\delta \; \lambda_2 \sum_{t=1}^{T} \|\mathbf{M}_t \odot \delta\|_2 + \lambda_1 \|\delta\|_2^2 + J_t(\mathbf{X}, \mathbf{M}, \mathbf{F}, \delta, f, \mathbf{y}^{pred}, \mathbf{y}^t) \,, \quad (4)$$

whose first term encodes an $\ell_{2,1}$ group sparsity regularizer encouraging complete groups to go to zero. Such a regularizer has been commonly used in the sparse coding literature [36,51], and more recently in the context of deep networks for compression purposes [1,47]. In our context, this regularizer encourages as many as possible of the $\{\mathbf{M}_t \odot \delta\}$ to go to zero, and thus confines the perturbation to a small number of regions that most effectively fool the targeted area $\mathbf{F}$. λ_2 balances the influence of this term with the other ones.

3.3 Universal Local Attacks

The strategies discussed in Sects. 3.1 and 3.2 are image-specific. However, [5] showed the existence of a universal perturbation patch that can fool an image classification system to output any target class. In this section, instead of finding optimal locations for a specific image, we aim to learn a single fixed-size local perturbation that can fool any unseen image in an untargeted manner. This will allow us to understand the contextual dependencies of a fixed size universal patch on the output of modern networks. Unlike in the above-mentioned adaptive local attacks, but as in [5,28,41,43], such a universal patch attack will require a larger perturbation norm. Note also that, because it is image-independent, the resulting attack will typically not be indirect. While [5] uses a different patch for different target classes, we aim to learn a single universal local perturbation that can fool all classes in the image. This will help to understand the propagation of the attack in modern networks using different long-range contextual connections. As will be shown in our experiments in Sect. 4.3, such modern networks are the most vulnerable to universal attacks, while their effect on FCNs is limited to the perturbed region. To find a universal perturbation effective across all images, we write the optimization problem

$$\delta^* = \arg\min_\delta \frac{1}{N} \sum_{i=1}^{N} J_u(\mathbf{X}^i, \mathbf{M}, \mathbf{F}^i, \delta, f, \mathbf{y}_i^{pred}) \,, \quad (5)$$

where $J_u(\cdot)$ is the objective function for a single image, as in the optimization problem (1), N is the number of training images, $\mathbf{X}^i$ is the i-th image with fooling mask $\mathbf{F}^i$, and the mask $\mathbf{M}$ is the global perturbation mask used for all images. In principle, $\mathbf{M}$ can be obtained by sampling patches over all possible image locations. However, we observed such a strategy to be unstable during learning. Hence, in our experiments, we confine ourselves to one or a few fixed patch positions. Note that, to give the attacker more flexibility, we take the universal attack defined in (5) to be an untargeted attack.

3.4   Adversarial Attack Detection

To understand the strength of the attacks discussed above, we introduce a detection method that can act either at the global image level or at the pixel level. The latter is particularly interesting in the case of indirect attacks, where the perturbation regions and the fooled regions are different. In this case, our goal is to localize the pixels that were fooled, which is more challenging than finding those that were perturbed, since their intensity values were not altered. To this end, we use a score based on the Mahalanobis distance defined on the intermediate feature representations. This is because, as discussed in [24,31] in the context of image classification, the attacked samples can be better characterized in the representation space than in the output label space. Specifically, we use a set of training images to compute class-conditional Gaussian distributions, with class-specific means μ_c^ℓ and covariance $\mathbf{\Sigma}^\ell$ shared across all C classes, from the features extracted at every intermediate layer ℓ of the network within locations corresponding to class label c. We then define a confidence score for each spatial location j in layer ℓ as $C(\mathbf{X}_j^\ell) = \max_{c \in [1,C]} - \left(\mathbf{X}_j^\ell - \mu_c^\ell\right)^\top \mathbf{\Sigma}_\ell^{-1} \left(\mathbf{X}_j^\ell - \mu_c^\ell\right)$, where $\mathbf{X}_j^\ell$ denotes the feature vector at location j in layer ℓ. We handle the different spatial feature map sizes in different layers by resizing all of them to a fixed spatial resolution. We then concatenate the confidence scores in all layers at every spatial location and use the resulting L-dimensional vectors, with L being the number of layers, as input to a logistic regression classifier with weights $\{\alpha_\ell\}$. We then train this classifier to predict whether a pixel was fooled or not. At test time, we compute the prediction for an image location j as $\sum_\ell \alpha_\ell C(\mathbf{X}_j^\ell)$. To perform detection at the global image level, we sum over the confidence scores of all spatial positions. That is, for layer ℓ, we compute an image-level score as $C(\mathbf{X}^\ell) = \sum_j C(\mathbf{X}_j^\ell)$. We then train another logistic regression classifier using these global confidence scores as input.

4   Experiments

Datasets. In our experiments, we use the Cityscapes [8] and Pascal VOC [10] datasets, the two most popular semantic segmentation benchmarks. Specifically, for Cityscapes, we use the complete validation set, consisting of 500 images, for untargeted attacks, but use a subset of 150 images containing dynamic object instances of vehicle classes whose combined area covers at least 8% of the image for targeted attacks. This lets us focus on fooling sufficiently large regions, because reporting results on too small one may not be representative of the true behavior of our algorithms. For Pascal VOC, we use 250 randomly selected images from the validation set because of the limited resources we have access to relative to the number of experiments we performed.

Models. We use publicly-available state-of-the-art models, namely FCN [29], DRNet [50], PSPNet [53], PSANet [54], DANet [12] on Cityscapes, and FCN [29] and PSANet [54] on PASCAL VOC. FCN, PSANet, PSPNet and DANet share the same ResNet [16] backbone network. We perform all experiments at the image resolution of 512×1024 for Cityscapes and 512×512 for PASCAL VOC.

Adversarial Attacks. We use the iterative projected gradient descent (PGD) method with ℓ_∞ and ℓ_2 norm budgets, as described in Sect. 3. Following [2], we set the number of iterations for PGD to a maximum of 100, with an early termination criterion of 90% of attack success rate on the targeted objects. We evaluate ℓ_∞ attacks with a step size $\alpha \in \{$1e-5, 1e-4, 1e-3, 5e-3$\}$. For ℓ_2 attacks, we set $\alpha \in \{$8e-3, 4e-2, 8e-2$\}$. We set the maximum ℓ_p-norm of the perturbation ϵ to $100 \cdot \alpha$ for ℓ_∞ attacks, and to 100 for ℓ_2 attacks. For universal attacks, we use a higher ℓ_∞ ϵ bound of 0.3, with a step size $\alpha = 0.001$. We perform two types of attacks; targeted and untargeted. The untargeted attacks focus on fooling the network to move away from the predicted label. For the targeted attacks, we chose a safety-sensitive goal, and thus aim to fool the dynamic object regions to be misclassified as their (spatially) nearest background label. We do not use ground-truth information in any of the experiments but perform attacks based on the predicted labels only.

Evaluation Metric. Following [2,18,49], we report the mean Intersection over Union (mIoU) and Attack Success Rate (ASR) computed over the entire dataset. The mIoU of FCN [29], DRNet [50], PSPNet [53], PSANet [54], and DANet [12] on clean samples at full resolution are $0.66, 0.64, 0.73, 0.72$, and 0.67, respectively. For targeted attacks, we report the average ASR_t, computed as the percentage of pixels that were predicted as the target label. We additionally report the $mIoU_u$, which is computed between the adversarial and normal sample predictions. For untargeted attacks, we report the ASR_u, computed as the percentage of pixels that were assigned to a different class than their normal label prediction. Since, in most of our experiments, the fooling region is confined to local objects, we compute the metrics only within the fooling mask. We observed that the non-targeted regions retain their prediction label more than 98% of the time, and hence we report the metrics at non-targeted regions in the supplementary material. To evaluate detection, we report the Area under the Receiver Operating Characteristics (AUROC), both at image level, as in [25,48], and at pixel level.

4.1 Indirect Local Attacks

Let us study the sensitivity of the networks to indirect local attacks. In this setting, we first perform a targeted attack, formalized in (2), to fool the dynamic object areas by allowing the attacker to perturb any region belonging to the static object classes. This is achieved by setting the perturbation mask $\mathbf{M}$ to 1 at all

Table 1. Indirect attacks on Cityscapes to fool dynamic classes while perturbing static ones. The numbers indicate $mIoU_u/ASR_t$, obtained using different step sizes α for ℓ_∞ and ℓ_2 attacks. The most robust network in each case is underlined and the most vulnerable models are highlighted in **bold**.

Network	$\alpha=0.00001$	$\alpha=0.0001$	$\alpha=0.001$	$\alpha=0.005$
FCN [29]	0.64 / 5.0%	0.28 / 29%	0.13 / 55%	0.11 / 61%
PSPNet [53]	0.70 / 12%	0.05 / 85%	0.00 / 89%	0.00 / **90%**
PSANet [54]	0.59 / 14%	0.03 / 85%	0.01 / **90%**	0.00 / **90%**
DANet [12]	0.80 / 5.0%	0.11 / 79%	0.01 / **90%**	0.00 / **90%**
DRN [50]	0.64 / 6.0%	0.15 / 56%	0.03 / 84%	0.02 / 86%

(a) ℓ_∞ attack

Network	$\alpha=0.008$	$\alpha=0.04$	$\alpha=0.08$
FCN [29]	0.60 / 10%	0.56 / 26%	0.27 / 36%
PSPNet [53]	0.67 / **19%**	0.23 / 67%	0.06 / **84%**
PSANet [54]	0.59 / 14%	0.21 / 63%	0.06 / 82%
DANet [12]	0.79 / 11%	0.43 / 49%	0.13 / 79%
DRN [50]	0.63 / 10%	0.24 / 47%	0.13 / 64%

(b) ℓ_2 attack

Table 2. Impact of local attacks by perturbing pixels that are at least d pixels away from any dynamic class. We report $mIoU_u/ASR_t$ for different values of d.

Network	$d=0$	$d=50$	$d=100$	$d=150$
FCN [29]	0.11 / 64%	0.77 / 2.0%	0.98 / 0%	1.00 / 0.0%
PSPNet [53]	0.00 / **90%**	0.14 / 73%	0.24 / 60%	0.55 / 23%
PSANet [54]	0.00 / **90%**	0.11 / 71%	0.13 / **65%**	0.29 / 47%
DANet [12]	0.00 / **90%**	0.13 / **81%**	0.48 / 43%	0.80 / 10%
DRN [50]	0.02 / 86%	0.38 / 22%	0.73 / 3%	0.94 / 1.0%

(a) ℓ_∞ attack

Network	$d=0$	$d=50$	$d=100$	$d=150$
FCN [29]	0.27 / **36%**	0.79 / 2.0%	0.98 / 2.0%	0.99 / 1.0%
PSPNet [53]	0.06 / **84%**	0.18 / 73%	0.55 / 23%	0.99 / 0.0%
PSANet [54]	0.06 / 82%	0.10 / **75%**	0.14 / **66 %**	0.31 / 44%
DANet [12]	0.13 / 79%	0.27 / 71%	0.67 / 26%	0.85 / 7.0%
DRN [50]	0.13 / 64%	0.44 / 17%	0.76 / 3.0%	0.95 / 0.0%

(b) ℓ_2 attack

the static class pixels and the fooling mask $\mathbf{F}$ to 1 at all the dynamic class pixels. We report the $mIoU_u$ and ASR_t metrics in Tables 1a and 1b on Cityscapes for ℓ_∞ and ℓ_2 attacks, respectively. As evidenced by the tables, FCN is more robust to such indirect attacks than the networks that leverage contextual information. In particular, PSANet, which uses long range contextual dependencies, and PSPNet are highly sensitive to these attacks.

To further understand the impact of indirect *local* attacks, we constrain the perturbation region to a subset of the static class regions. To do this in a systematic manner, we perturb the static class regions that are at least d pixels away from any dynamic object, and vary the value d. The results of this experiment using ℓ_2 and ℓ_∞ attacks are provided in Table 2. Here, we chose a step size $\alpha = 0.005$ for ℓ_∞ and $\alpha = 0.08$ for ℓ_2. Similar conclusions to those in the previous non-local scenario can be drawn: Modern networks that use larger receptive fields are extremely vulnerable to such perturbations, even when they are far away from the targeted regions. By contrast, FCN is again more robust. For example, as shown in Fig. 2, while an adversarial attack occurring 100 pixels away from the nearest dynamic object has a high success rate on the context-aware networks, the FCN predictions remain accurate.

4.2   Adaptive Indirect Local Attacks

We now study the impact of our approach to adaptively finding the most sensitive context region to fool the dynamic objects. To this end, we use the group sparsity

(a) Adversarial (b) Perturbation (c) FCN [29] (d) PSPNet [53] (e) PSANet [54] (f) DANet [12]

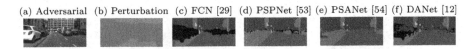

Fig. 2. Indirect Local attack on different networks with perturbations at least $d =$ 100 pixels away from any dynamic class.

Table 3. Adaptive indirect local attacks on Cityscapes and PASCAL VOC. We report mIoU$_u$/ASR$_t$ for different sparsity levels S.

Network	$S = 75\%$	$S = 85\%$	$S = 90\%$	$S = 95\%$
FCN [29]	0.52 / 12%	0.66 / 6%	0.73 / 4%	0.84 / 1.0%
PSPNet [53]	0.19 / 70%	0.31 / 54%	0.41 / 42%	0.53 / 21%
PSANet [54]	0.10 / **78%**	0.16 / **71%**	0.20 / **64%**	0.35 / **44%**
DANet [12]	0.30 / 64%	0.52 / 43%	0.64 / 30%	0.71 / 21%
DRN [50]	0.42 / 23%	0.55 / 13%	0.63 / 9%	0.77 / 4.5%

(a) Cityscapes

Network	$S = 75\%$	$S = 85\%$	$S = 90\%$	$S = 95\%$
FCN [29]	0.50 / 32%	0.59 / 27%	0.66 / 22%	0.80 / 12%
PSANet [54]	0.28 / **68%**	0.21 / **77%**	0.20 / **80%**	0.30 / **69%**

(b) PASCAL VOC

based optimization given in (4) and find the minimal perturbation region to fool all dynamic objects to their nearest static label. Specifically, we achieve this in two steps. First, we divide the perturbation mask $\mathbf{M}$ corresponding to all static class pixels into uniform patches of size $h \times w$, and find the most sensitive ones by solving (4) with a relatively large group sparsity weight $\lambda_2 = 100$ for Cityscapes and $\lambda_2 = 10$ for PASCAL VOC. Second, we limit the perturbation region by selecting the n patches that have the largest values $\|\mathbf{M}_t \odot \delta\|_2$, choosing n so as to achieve a given sparsity level $S \in \{75\%, 85\%, 90\%, 95\%\}$. Specifically, S is computed as the percentage of pixels that are not perturbed relative to the initial perturbation mask. We then re-optimize (4) with $\lambda_2 = 0$. In both steps, we set $\lambda_1 = 0.01$ and use the Adam optimizer [20] with a learning rate of 0.01. For Cityscapes, we use patch dimensions $h = 60$, $w = 120$, and, for PASCAL VOC, $h = 60$, $w = 60$. We clip the perturbation values below 0.005 to 0 at each iteration. This results in very local perturbation regions, active only in the most sensitive areas, as shown for PSANet in Fig. 3 on Cityscapes and in Fig. 4 for PASCAL VOC. As shown in Table 3, all context-aware networks are significantly affected by such perturbations, even when they are confined to small background regions. For instance, on Cityscapes, at a high sparsity level of 95%, PSANet yields an ASR$_t$ of 44% compared to 1% for FCN. This means that, in the physical world, an attacker could add a small sticker at a static position to essentially make dynamic objects disappear from the network's view.

4.3 Universal Local Attacks

In this section, instead of considering image-dependent perturbations, we study the existence of universal local perturbations and their impact on semantic segmentation networks. In this setting, we perform untargeted *local* attacks by placing a fixed-size patch at a predetermined position. While the patch location can in principle be sampled at any location, we found learning its position to

(a) Adversarial image (b) Perturbation (c) Normal Seg. (d) Adversarial Seg.

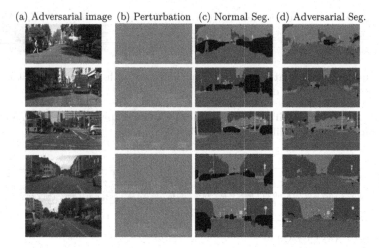

Fig. 3. Adaptive indirect local attacks on Cityscapes with PSANet [54]. An adversarial input image (**a**) when attacked at positions shown as red boxes with a perturbation (**b**) is misclassified within the dynamic object areas of the normal segmentation map (**c**) to result in (**d**).

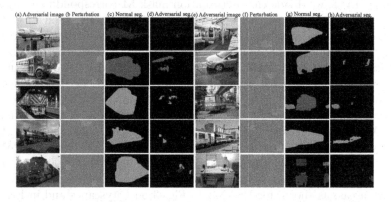

Fig. 4. Adaptive indirect local attacks on PASCAL VOC with PSANet [54]. An adversarial input image (**a**),(**e**) when attacked at positions shown as red boxes with a perturbation (**b**),(**f**) is misclassified within the foreground object areas of the normal segmentation map (**c**), (**g**) to result in (**d**), (**h**), respectively.

be unstable to due to the large number of possible patch locations in the entire dataset. Hence, here, we consider the scenario where the patch is located at the center of the image. We then learn a local perturbation that can fool the entire dataset of images for a given network by optimizing the objective given in (5). Specifically, the perturbation mask **M** is active only at the patch location and the fooling mask **F** at all image positions, i.e., at both static and dynamic classes. For Cityscapes, we learn the universal local perturbation using 100 images and use the remaining 400 images for evaluation purposes. For PASCAL VOC, we

(a) Adversarial image (b) Ground truth (c) FCN [29] (d) PSANet [54] (e) PSPNet [53]

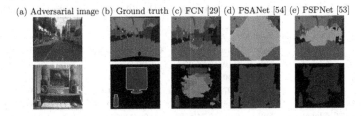

Fig. 5. Universal local attacks on Cityscapes and PASCAL VOC. In both datasets, the degradation in FCN [29] is limited to the attacked area, whereas for context-aware networks, such as PSPNet [53], PSANet [54], DANet [12], it extends to far away regions.

Table 4. Universal local attacks. We show the impact of the patch size $h \times w$ (area%) on different networks and report $\text{mIoU}_u / \text{ASR}_u$.

Network	51 × 102(1.0%)	76 × 157(2.3%)	102 × 204(4.0%)	153 × 306(9.0%)
FCN [29]	0.85 / 2.0%	0.78 / 4.0%	0.73 / 9.0%	0.58 / 18%
PSPNet [53]	0.79 / 3.0%	0.63 / 11%	0.44 / 27%	0.08 / 83%
PSANet [54]	0.41 / 37%	0.22 / 60%	0.14 / 70%	0.10 / 90%
DANet [12]	0.79 / 4.0%	0.71 / 10%	0.65 / 15%	0.40 / 42%
DRN [50]	0.82 / 3.0%	0.78 / 8.0%	0.71 / 14%	0.55 / 28%

(a) Cityscapes

Network	51 × 51(1.0%)	76 × 76(2.3%)	102 × 102(4.0%)	153 × 153(9.0%)
FCN [29]	0.70 / 6%	0.70 / 7%	0.63 / 10%	0.52 / 20%
PSANet [54]	0.83 / 4%	0.76 / 8 %	0.56 / 28%	0.35/ 56%

(b) PASCAL VOC

perform training on 100 images and evaluate on the remaining 150 images. We use ℓ_∞ optimization with $\alpha = 0.001$ for 200 epochs on the training set. We report the results of such universal patch attacks in Tables 4a and 4b on Cityscapes and PASCAL VOC for different patch sizes. As shown in the table, PSANet and PSPNet are vulnerable to such universal attacks, even when only 2.3% of the image area is perturbed. From Fig. 5, we can see that the fooling region propagates to a large area far away from the perturbed one. While these experiments study untargeted universal local attacks, we report additional results on single-class targeted universal local attacks in the supplementary material.

4.4 Attack Detection

We now turn to studying the effectiveness of the attack detection strategies described in Sect. 3.4. We also compare our approach to the only two detection techniques that have been proposed for semantic segmentation [25, 48]. The method in [48] uses the spatial consistency of the predictions obtained from $K = 50$ random overlapping patches of size 256×256. The one in [25] compares an image re-synthesized from the predicted labels with the input image. Both methods were designed to handle attacks that fool the entire label map, unlike our work where we aim to fool local regions. Furthermore, both methods perform detection at the image level, and thus, in contrast to ours, do not localize the fooled regions at the pixel level.

We study detection in four perturbation settings: Global image perturbations (Global) to fool the entire image; Universal patch perturbations (UP) at a fixed

Table 5. Attack detection on Cityscapes with different perturbation settings.

Networks	Perturbation region	Fooling region	ℓ_∞/ℓ_2 norm	Mis. pixels %	Global AUROC SC [48]/Re-Syn [25]/**Ours**	Local AUROC **Ours**
FCN [29]	Global	Full	0.10/17.60	90%	**1.00/1.00**/0.94	0.90
	UP	Full	0.30/37.60	4%	0.71/0.63/**1.00**	0.94
	FS	Dyn	0.07/2.58	13%	0.57/0.71/**1.00**	0.87
	AP	Dyn	0.14/3.11	1.7%	0.51/0.65/**0.87**	0.89
PSPNet [53]	Global	Full	0.06/10.74	83%	0.90/**1.00**/0.99	0.85
	UP	Full	0.30/38.43	11%	0.66/0.70/ **1.00**	0.96
	FS	Dyn	0.03/1.78	14%	0.57/0.75/**0.90**	0.87
	AP	Dyn	0.11/5.25	11%	0.57/0.75/**0.90**	0.82
PSANet [54]	Global	Full	0.05/8.26	92%	0.90/1.00/1.00	0.67
	UP	Full	0.30/38.6	60%	0.65/1.00/1.00	0.98
	FS	Dyn	0.02/1.14	12%	0.61/0.76/**1.00**	0.92
	AP	Dyn	0.10/5.10	10%	0.50/0.82/**1.00**	0.94
DANet [12]	Global	Full	0.06/12.55	82%	0.89/1.00/1.00	0.68
	UP	Full	0.30/37.20	10%	0.67/0.63/**0.92**	0.89
	FS	Dyn	0.05/1.94	13%	0.57/0.69/**0.94**	0.88
	AP	Dyn	0.14/6.12	43%	0.59/0.68/**0.98**	0.82

location to fool the entire image; Full static (FS) class perturbations to fool the dynamic classes; Adaptive patch (AP) perturbations in the static class regions to fool the dynamic objects. As shown in Table 5, while the state-of-the-art methods [25,48] have high Global AUROC in the first setting where the entire image is targeted, our detection strategy outperforms them by a large margin in the other scenarios. We believe this to be due to the fact that, with local attacks, the statistics obtained by studying the consistency across local patches, as in [48], are much closer to the clean image statistics. Similarly, the image re-synthesized by a pix2pix generator, as used in [25], will look much more similar to the input one in the presence of local attacks instead of global ones. For all the perturbation settings, we also report the mean percentage of pixels misclassified relative to the number of pixels in the image.

5   Conclusion

In this paper, we have studied the impact of indirect local image perturbations on the performance of modern semantic segmentation networks. We have observed that the state-of-the-art segmentation networks, such as PSANet and PSPNet, are more vulnerable to local perturbations because their use of context, which improves their accuracy on clean images, enables the perturbations to be propagated to distant image regions. As such, they can be attacked by perturbations that cover as little as 2.3% of the image area. We have then proposed a Mahalanobis distance-based detection strategy, which has proven effective for both image-level and pixel-level attack detection. Nevertheless, the performance at localizing the fooled regions in a pixel-wise manner can still be improved, which will be our goal in the future.

Acknowledgments. This work was funded in part by the Swiss National Science Foundation.

References

1. Alvarez, J.M., Salzmann, M.: Learning the number of neurons in deep networks. In: Advances in Neural Information Processing Systems, pp. 2270–2278 (2016)
2. Arnab, A., Miksik, O., Torr, P.H.: On the robustness of semantic segmentation models to adversarial attacks. In: Proceedings of the IEEE Conference on Computer Vision and Pattern Recognition, pp. 888–897 (2018)
3. Athalye, A., Carlini, N., Wagner, D.: Obfuscated gradients give a false sense of security: circumventing defenses to adversarial examples. arXiv preprint arXiv:1802.00420 (2018)
4. Athalye, A., Engstrom, L., Ilyas, A., Kwok, K.: Synthesizing robust adversarial examples. In: International Conference on Machine Learning, pp. 284–293 (2018)
5. Brown, T.B., Mané, D.: Aurko roy, martín abadi, and justin gilmer. Adversarial patch. CoRR, abs/1712.09665 (2017)
6. Carlini, N., Wagner, D.: Towards evaluating the robustness of neural networks. In: 2017 IEEE Symposium on Security and Privacy (SP), pp. 39–57. IEEE (2017)
7. Chen, P.Y., Zhang, H., Sharma, Y., Yi, J., Hsieh, C.J.: Zoo: zeroth order optimization based black-box attacks to deep neural networks without training substitute models. In: Proceedings of the 10th ACM Workshop on Artificial Intelligence and Security, pp. 15–26 (2017)
8. Cordts, M., et al.: The cityscapes dataset for semantic urban scene understanding. In: Proceedings of the IEEE Conference on Computer Vision and Pattern Recognition, pp. 3213–3223 (2016)
9. Dong, Y., et al.: Boosting adversarial attacks with momentum. In: Proceedings of the IEEE Conference on Computer Vision and Pattern Recognition, pp. 9185–9193 (2018)
10. Everingham, M., Van Gool, L., Williams, C.K., Winn, J., Zisserman, A.: The pascal visual object classes challenge 2007 (voc2007) results (2007)
11. Eykholt, K., et al.: Robust physical-world attacks on deep learning visual classification. In: Proceedings of the IEEE Conference on Computer Vision and Pattern Recognition, pp. 1625–1634 (2018)
12. Fu, J., et al.: Dual attention network for scene segmentation. In: Proceedings of the IEEE Conference on Computer Vision and Pattern Recognition, pp. 3146–3154 (2019)
13. Gonfaus, J.M., Boix, X., Van de Weijer, J., Bagdanov, A.D., Serrat, J., Gonzalez, J.: Harmony potentials for joint classification and segmentation. In: 2010 IEEE Computer Society Conference on Computer Vision and Pattern Recognition, pp. 3280–3287. IEEE (2010)
14. Goodfellow, I.J., Shlens, J., Szegedy, C.: Explaining and harnessing adversarial examples. arXiv preprint arXiv:1412.6572 (2014)
15. Hariharan, B., Arbeláez, P., Girshick, R., Malik, J.: Hypercolumns for object segmentation and fine-grained localization. In: Proceedings of the IEEE Conference on Computer Vision and Pattern Recognition, pp. 447–456 (2015)
16. He, K., Zhang, X., Ren, S., Sun, J.: Deep residual learning for image recognition. In: Proceedings of the IEEE Conference on Computer Vision and Pattern Recognition, pp. 770–778 (2016)

17. He, X., Zemel, R.S., Carreira-Perpiñán, M.Á.: Multiscale conditional random fields for image labeling. In: Proceedings of the 2004 IEEE Computer Society Conference on Computer Vision and Pattern Recognition, CVPR 2004, vol. 2, p. II. IEEE (2004)

18. Hendrik Metzen, J., Chaithanya Kumar, M., Brox, T., Fischer, V.: Universal adversarial perturbations against semantic image segmentation. In: Proceedings of the IEEE International Conference on Computer Vision, pp. 2755–2764 (2017)

19. Isola, P., Zhu, J.Y., Zhou, T., Efros, A.A.: Image-to-image translation with conditional adversarial networks. In: Proceedings of the IEEE Conference on Computer Vision and Pattern Recognition, pp. 1125–1134 (2017)

20. Kingma, D.P., Ba, J.: Adam: a method for stochastic optimization. arXiv preprint arXiv:1412.6980 (2014)

21. Kohli, P., Torr, P.H., et al.: Robust higher order potentials for enforcing label consistency. Int. J. Comput. Vision $82(3)$, 302–324 (2009)

22. Krähenbühl, P., Koltun, V.: Efficient inference in fully connected CRFs with gaussian edge potentials. In: Advances in Neural Information Processing Systems, pp. 109–117 (2011)

23. Kurakin, A., Goodfellow, I., Bengio, S.: Adversarial machine learning at scale. arXiv preprint arXiv:1611.01236 (2016)

24. Lee, K., Lee, K., Lee, H., Shin, J.: A simple unified framework for detecting out-of-distribution samples and adversarial attacks. In: Advances in Neural Information Processing Systems, pp. 7167–7177 (2018)

25. Lis, K., Nakka, K., Salzmann, M., Fua, P.: Detecting the unexpected via image resynthesis. arXiv preprint arXiv:1904.07595 (2019)

26. Liu, W., Rabinovich, A., Berg, A.C.: Parsenet: looking wider to see better. arXiv preprint arXiv:1506.04579 (2015)

27. Liu, X., Yang, H., Liu, Z., Song, L., Li, H., Chen, Y.: Dpatch: an adversarial patch attack on object detectors. arXiv preprint arXiv:1806.02299 (2018)

28. Liu, Y., Chen, X., Liu, C., Song, D.: Delving into transferable adversarial examples and black-box attacks. arXiv preprint arXiv:1611.02770 (2016)

29. Long, J., Shelhamer, E., Darrell, T.: Fully convolutional networks for semantic segmentation. In: Proceedings of the IEEE Conference on Computer Vision and Pattern Recognition, pp. 3431–3440 (2015)

30. Lu, J., Sibai, H., Fabry, E., Forsyth, D.: No need to worry about adversarial examples in object detection in autonomous vehicles. arXiv preprint arXiv:1707.03501 (2017)

31. Ma, X., et al.: Characterizing adversarial subspaces using local intrinsic dimensionality. arXiv preprint arXiv:1801.02613 (2018)

32. Moosavi-Dezfooli, S.M., Fawzi, A., Fawzi, O., Frossard, P.: Universal adversarial perturbations. In: Proceedings of the IEEE Conference on Computer Vision and Pattern Recognition, pp. 1765–1773 (2017)

33. Moosavi-Dezfooli, S.M., Fawzi, A., Fawzi, O., Frossard, P., Soatto, S.: Analysis of universal adversarial perturbations. arXiv preprint arXiv:1705.09554 (2017)

34. Moosavi-Dezfooli, S.M., Fawzi, A., Frossard, P.: Deepfool: a simple and accurate method to fool deep neural networks. In: Proceedings of the IEEE Conference on Computer Vision and Pattern Recognition, pp. 2574–2582 (2016)

35. Nguyen, A., Yosinski, J., Clune, J.: Deep neural networks are easily fooled: high confidence predictions for unrecognizable images. In: Proceedings of the IEEE Conference on Computer Vision and Pattern Recognition, pp. 427–436 (2015)

36. Nie, F., Huang, H., Cai, X., Ding, C.H.: Efficient and robust feature selection via joint l2, 1-norms minimization. In: Advances in Neural Information Processing Systems, pp. 1813–1821 (2010)
37. Papernot, N., McDaniel, P., Goodfellow, I.: Transferability in machine learning: from phenomena to black-box attacks using adversarial samples. arXiv preprint arXiv:1605.07277 (2016)
38. Papernot, N., McDaniel, P., Goodfellow, I., Jha, S., Celik, Z.B., Swami, A.: Practical black-box attacks against machine learning. In: Proceedings of the 2017 ACM on Asia Conference on Computer and Communications Security, pp. 506–519 (2017)
39. Papernot, N., McDaniel, P., Jha, S., Fredrikson, M., Celik, Z.B., Swami, A.: The limitations of deep learning in adversarial settings. In: 2016 IEEE European Symposium on Security and Privacy (EuroS&P), pp. 372–387. IEEE (2016)
40. Poursaeed, O., Katsman, I., Gao, B., Belongie, S.: Generative adversarial perturbations. In: Proceedings of the IEEE Conference on Computer Vision and Pattern Recognition, pp. 4422–4431 (2018)
41. Ranjan, A., Janai, J., Geiger, A., Black, M.J.: Attacking optical flow. In: Proceedings of the IEEE International Conference on Computer Vision, pp. 2404–2413 (2019)
42. Ronneberger, O., Fischer, P., Brox, T.: U-Net: convolutional networks for biomedical image segmentation. In: Navab, N., Hornegger, J., Wells, W.M., Frangi, A.F. (eds.) MICCAI 2015. LNCS, vol. 9351, pp. 234–241. Springer, Cham (2015). https://doi.org/10.1007/978-3-319-24574-4_28
43. Saha, A., Subramanya, A., Patil, K., Pirsiavash, H.: Adversarial patches exploiting contextual reasoning in object detection. arXiv preprint arXiv:1910.00068 (2019)
44. Thys, S., Van Ranst, W., Goedemé, T.: Fooling automated surveillance cameras: adversarial patches to attack person detection. In: Proceedings of the IEEE Conference on Computer Vision and Pattern Recognition Workshops (2019)
45. Tramèr, F., Kurakin, A., Papernot, N., Goodfellow, I., Boneh, D., McDaniel, P.: Ensemble adversarial training: Attacks and defenses. arXiv preprint arXiv:1705.07204 (2017)
46. Tramèr, F., Zhang, F., Juels, A., Reiter, M.K., Ristenpart, T.: Stealing machine learning models via prediction APIs. In: 25th {USENIX} Security Symposium ({USENIX} Security 2016), pp. 601–618 (2016)
47. Wen, W., Wu, C., Wang, Y., Chen, Y., Li, H.: Learning structured sparsity in deep neural networks. In: Advances in Neural Information Processing Systems, pp. 2074–2082 (2016)
48. Xiao, C., Deng, R., Li, B., Yu, F., Liu, M., Song, D.: Characterizing adversarial examples based on spatial consistency information for semantic segmentation. In: Proceedings of the European Conference on Computer Vision (ECCV), pp. 217–234 (2018)
49. Xie, C., Wang, J., Zhang, Z., Zhou, Y., Xie, L., Yuille, A.: Adversarial examples for semantic segmentation and object detection. In: Proceedings of the IEEE International Conference on Computer Vision, pp. 1369–1378 (2017)
50. Yu, F., Koltun, V., Funkhouser, T.: Dilated residual networks. In: Proceedings of the IEEE Conference on Computer Vision and Pattern Recognition, pp. 472–480 (2017)
51. Yuan, M., Lin, Y.: Model selection and estimation in regression with grouped variables. J. Roy. Stat. Soc.: Ser. B (Stat. Methodol.) 68(1), 49–67 (2006)
52. Zhang, H., et al.: Context encoding for semantic segmentation. In: Proceedings of the IEEE Conference on Computer Vision and Pattern Recognition, pp. 7151–7160 (2018)

53. Zhao, H., Shi, J., Qi, X., Wang, X., Jia, J.: Pyramid scene parsing network. In: Proceedings of the IEEE Conference on Computer Vision and Pattern Recognition, pp. 2881–2890 (2017)
54. Zhao, H., et al.: Psanet: point-wise spatial attention network for scene parsing. In: Proceedings of the European Conference on Computer Vision (ECCV), pp. 267–283 (2018)

Predicting Visual Overlap of Images Through Interpretable Non-metric Box Embeddings

Anita Rau[1]([✉]), Guillermo Garcia-Hernando[2], Danail Stoyanov[1],
Gabriel J. Brostow[1,2], and Daniyar Turmukhambetov[2]

[1] University College London, London, UK
`a.rau.16@ucl.ac.uk`
[2] Niantic, San Francisco, USA
`http://www.github.com/nianticlabs/image-box-overlap`

Abstract. To what extent are two images picturing the same 3D surfaces? Even when this is a known scene, the answer typically requires an expensive search across scale space, with matching and geometric verification of large sets of local features. This expense is further multiplied when a query image is evaluated against a gallery, *e.g.* in visual relocalization. While we don't obviate the need for geometric verification, we propose an interpretable image-embedding that cuts the search in scale space to essentially a lookup.

Our approach measures the **asymmetric** relation between two images. The model then learns a scene-specific measure of similarity, from training examples with known 3D visible-surface overlaps. The result is that we can quickly identify, for example, which test image is a close-up version of another, and by what scale factor. Subsequently, local features need only be detected at that scale. We validate our scene-specific model by showing how this embedding yields competitive image-matching results, while being simpler, faster, and also interpretable by humans.

Keywords: Image embedding · Representation learning · Image localization · Interpretable representation

1 Introduction

Given two images of the same scene, which one is the close-up? This question is relevant for many tasks, such as image-based rendering and navigation [52], because the close-up has higher resolution details [30] for texturing a generated

---

A. Rau—Work done during an internship at Niantic.

Electronic supplementary material The online version of this chapter (https://doi.org/10.1007/978-3-030-58558-7_37) contains supplementary material, which is available to authorized users.

A. Vedaldi et al. (Eds.): ECCV 2020, LNCS 12350, pp. 629–646, 2020.
https://doi.org/10.1007/978-3-030-58558-7_37

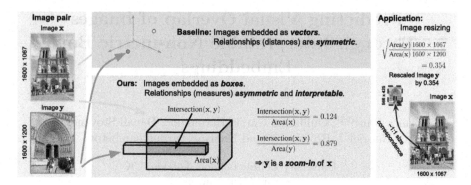

Fig. 1. Given two images from the test set, can we reason about their relationship? How much of the scene is visible in both images? Is one image a zoomed-in version of another? We propose a CNN model to predict box embeddings that approximate the visible surface overlap measures between two images. The surface overlap measures give an interpretable relationship between two images and their poses. Here, we extract the relative scale difference between two test images, without expensive geometric analysis.

3D model, while the other image has the context view of the scene. Related tasks include 3D scene reconstruction [48], robot navigation [5] and relocalization [3,40,42,44], which all need to reason in 3D about visually overlapping images.

For these applications, exhaustive searches are extremely expensive, so efficiency is needed in two related sub-tasks. First, it is attractive to cheaply identify which image(s) from a corpus have substantial overlap with a query image. Most images are irrelevant [53], so progress on whole-image descriptors like [50] narrows that search. Second, for every two images that are expected to have matchable content, the geometric verification of relative pose involves two-view feature matching and pose estimation [17], which can range from being moderately to very expensive [12], depending on design choices. RANSAC-type routines are more accurate and faster if the two images have matching scales [10,64,66], and if the detector and descriptor work well with that scene to yield low outlier-ratios.

For scenes where training data is available, we efficiently address both sub-tasks, relevant image-finding and scale estimation, with our new embedding. Our proposed model projects whole images to our custom asymmetric feature space. There, we can compare the non-Euclidean measure between image encoding $\mathbf{x}$ to image encoding $\mathbf{y}$, which is different from comparing $\mathbf{y}$ to $\mathbf{x}$ – see Fig. 1. Ours is distinct from previous methods which proposed a) learning image similarity with metric learning, *e.g.* [1,3,50], and b) estimation of relative scale or image overlap using geometric validation of matched image features, *e.g.* [29,30]. Overall, we

- advocate that normalized surface overlap (NSO) serves as an interpretable real-world measure of how the same geometry is pictured in two images,
- propose a new box embedding for images, that approximates the surface overlap measure while preserving interpretability, and
- show that the predicted surface overlap allows us to pre-scale images for same-or-better feature matching and pose estimation in a localization task.

The new representation borrows from box embeddings [26,56,60], that were designed to represent hierarchical relations in words. We are the first to adapt them for computer vision tasks, and hereby propose *image* box embeddings. We qualitatively validate their ability to yield *interpretable* image relationships that don't impede localization, and quantitatively demonstrate that they help with scale estimation.

2   Related Work

Ability to quickly determine if two images observe the same scene is useful in multiple computer vision pipelines, especially for SLAM, Structure-from-Motion (SfM), image-based localization, landmark recognition, *etc.*

For example, Bag-of-Words (BoW) models [13] are used in many SLAM systems, *e.g.* ORB-SLAM2 [33], to detect when the camera has revisited a mapped place, *i.e.* a loop closure detection, to re-optimize the map of the scene. The BoW model allows to quickly search images of a mapped area and find a match to the current frame for geometric verification. A similar use case for BoW model can be found in SfM pipelines [12,20,48], where thousands or even millions of images are downloaded from the Internet and two-view matching is impractical to do between all pairs of images. Once a 3D model of a scene is built, a user might want to browse the collection of images [53]. Finding images that are zoom-ins to parts of the query and provide detailed, high-resolution images of interesting parts of the scene was addressed in [29,30], where they modified a BoW model to also store local feature geometry information [36] that can be exploited for Document at a Time (DAAT) scoring [55] and further geometric verification. However, that approach relies on iterative query expansion to find zoomed-in matches, and re-ranking of retrieved images still requires geometric verification, even when choosing images for query expansion. Similarly the loss of detail in SfM reconstructions was identified to be a side-effect of retrieval approaches in [47] and solution based on modification of query expansion and DAAT scoring was proposed, that favors retrieval of images with scale and viewing directions changes. Again, each image query retrieves a subset of images from the image collection in a certain ranking order and geometric verification is performed to establish the pairwise relation of the views. Weyand and Leibe [62] build a hierarchy of iconic views of a scene by iteratively propagating and verifying homography between pairs of images and generating clusters of views. Later, they extended [63] it to find details and scene structure with hierarchical clustering over scale space. Again, the relation between views is estimated with geometric verification.

Schönberger *et al.* [46] propose to learn to identify image pairs with scene overlap without geometric verification for large-scale SfM reconstruction applications to decrease the total computation time. However, their random forest classifier is trained with binary labels, a pair of images either overlap or not, and no additional output about the relative pose relation is available without further geometric verification. Shen *et al.* [50] proposed to learn an embedding which can be used for quick assessment of image overlap for SfM reconstructions. They trained a neural network with triplet loss with ground-truth overlap triplets, however their embedding models minimal overlap between two images, which can be used to rank images according to predicted mutual overlap, but does not provide additional information about relative scale and pose of the image pairs.

Neural networks can be used to directly regress a homography between a pair of images [9,11,25,34], however a homography only explains a single planar surface co-visible between views.

Finally, image retrieval is a common technique in localization [3,43,44]: first, a database of geo-tagged images is searched to find images that are similar to the query, followed by image matching and pose estimation to estimate the pose of the query. Again, the image retrieval reduces the search space for local feature matching, and it is done using a compact representation such as BoW [38], VLAD [2] or Fisher vectors [37]. Recently proposed methods use neural networks to compute the image representation and learn the image similarity using geometric information available for the training images, e.g. GPS coordinates for panoramic images [1], camera poses in world coordinate system [3], camera pose and 3D model of the environment [50]. Networks trained to directly regress the camera pose [22,24] were also shown to be similar to image retrieval methods [45]. The image embeddings are learned to implicitly encapsulate the geometry of a scene so that similar views of a scene are close in the embedding space. However, these representations are typically learned with metric learning, and so they only allow to rank the database images with respect to the query, and no additional camera pose information is encoded in the representations. We should also mention that localization can also be tackled by directly finding correspondences from 2D features in the query to 3D points in the 3D model without explicit image retrieval, e.g. [6,40–42].

Image Matching. If a pair of images have a large viewing angle and scale difference, then estimating relative pose using standard correspondence matching becomes challenging. MODS [32] addresses the problem by iteratively generating and matching warped views of the input image pair, until a sufficient confidence in the estimated relative pose is reached. Zhou *et al.* [66] tackles the problem of very large scale difference between the two views by exploiting the consistent scale ratio between matching feature points. Thus, their two-stage algorithm first does exhaustive scale level matching to estimate the relative scale difference between the two views and then does feature matching only in corresponding scale levels. The first stage is done exhaustively, as there is no prior information about the scale difference, which our embeddings provide an estimate for.

Metric Learning. Many computer vision applications require learning an embedding, such that the relative distances between inputs can be predicted. For example, face recognition [49], descriptor learning [4,19,31,39,57], image retrieval [14], person re-identification [21], *etc.* The common approach is metric learning: models are learned to encode the input data (*e.g.* image, patch, *etc.*) as a vector, such that a desirable distance between input data points is approximated by distances in corresponding vector representations. Typically, the setup

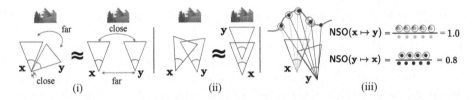

Fig. 2. Illustrations of world-space measures and their properties. (i) Left and right camera configurations result in the same value for weighted sum of rotation translation errors. (ii) Left and right camera configurations result in the same frustum overlap value. (iii) Illustration of our surface overlap. Images $\mathbf{x}$ and $\mathbf{y}$ have resolution of 5 pixels. All 5 pixels of $\mathbf{x}$ are visible in $\mathbf{y}$, meaning the corresponding 3D points of $\mathbf{x}$ are sufficiently close to 3D points backprojected from $\mathbf{y}$. Hence $\mathrm{NSO}(\mathbf{x} \mapsto \mathbf{y}) = 1.0$. However, only 80% of pixels of $\mathbf{y}$ are visible by $\mathbf{x}$, thus, $\mathrm{NSO}(\mathbf{y} \mapsto \mathbf{x}) = 0.8$.

uses siamese neural networks, and contrastive loss [16], triplet loss [49], average concentration loss [19] and ranking loss [8] is used to train the network. The distances in the embedding space can be computed as Euclidean distance between two vectors, an inner product of two vectors, a cosine of the angle between two vectors, *etc.* However, learned embeddings can be used to estimate the order of data points, but not other types relations, due to symmetric distance function used for similarity measure. Other relations, for example, hierarchies require asymmetric similarity measures, which is often encountered in word embeddings [26,35,56,59–61].

Some methods learn representations that are disentangled, so that they correspond to meaningful variations of the data. For example [23,54,65] model geometric transformations of the data in the representation space. However these representations are typically learned for a single class of objects, and it is not straight-forward how to use these embeddings for retrieval.

3   Method

Our aim is to i) interpret images according to a geometric world-space measure, and ii) to devise an embedding that, once trained for a specific reconstructed 3D scene, lends itself to easily compute those interpretations for new images. We now explain each, in turn.

3.1   World-Space Measures

We want to have interpretable world-space relationship between a pair of images $\mathbf{x}$ and $\mathbf{y}$, and their corresponding camera poses (orientations and positions).

A straightforward world-space measure is the distance between camera centers [1]. This world-space measure is useful for localizing omni-directional or panoramic cameras that observe the scene in 360°. However, most cameras are not omni-directional, so one needs to incorporate the orientation of the cameras.

But how can we combine the relative rotation of the cameras and their relative translation into a world-space measure? One could use a weighted sum of the rotation and translation differences [7,22,52,53], but there are many camera configurations where this measure is not satisfactory, *e.g.* Fig. 2(i).

Another example of a world-space measure is frustum overlap [3]. Indeed, if we extend the viewing frustum of each camera up to a cutoff distance D, we can assume that the amount of frustum overlap correlates with the positions and orientation of the cameras. However, the two cameras can be placed multiple ways and have the same frustum overlap – see Fig. 2(ii). So, the normalized frustum overlap value does not provide interpretable information about the two camera poses.

We propose to use normalized surface overlap for the world-space measure. See Fig. 2(iii) for an illustration. Formally, normalized surface overlap is defined as

$$\mathsf{NSO}(\mathbf{x} \mapsto \mathbf{y}) = \mathsf{overlap}(\mathbf{x} \mapsto \mathbf{y})/N_{\mathbf{x}}, \text{ and} \tag{1}$$

$$\mathsf{NSO}(\mathbf{y} \mapsto \mathbf{x}) = \mathsf{overlap}(\mathbf{y} \mapsto \mathbf{x})/N_{\mathbf{y}}, \tag{2}$$

where $\mathsf{overlap}(\mathbf{x} \mapsto \mathbf{y})$ is the number of pixels in image $\mathbf{x}$ that are visible in image $\mathbf{y}$, and $\mathsf{NSO}(\mathbf{x} \mapsto \mathbf{y})$ is $\mathsf{overlap}(\mathbf{x} \mapsto \mathbf{y})$ normalized by the number of pixels in $\mathbf{x}$ (denoted by $N_{\mathbf{x}}$), hence it is a number between 0 and 1 and it can be represented as a percentage. To compute it, we need to know camera poses and the depths of pixels for both image $\mathbf{x}$ and image $\mathbf{y}$. The pixels in both images are backprojected into 3D. The $\mathsf{overlap}(\mathbf{x} \mapsto \mathbf{y})$ are those 3D points in $\mathbf{x}$ that have a neighbor in the point cloud of image $\mathbf{y}$ within a certain radius.

The normalized surface overlap is not symmetric, $\mathsf{NSO}(\mathbf{x} \mapsto \mathbf{y}) \neq \mathsf{NSO}(\mathbf{y} \mapsto \mathbf{x})$, because only a few pixels in image $\mathbf{x}$ could be viewed in image $\mathbf{y}$, but all the pixels in $\mathbf{y}$ could be viewed in image $\mathbf{x}$. This asymmetric measure can have an interpretation that image $\mathbf{y}$ is a *close-up view of a part the scene* observed in image $\mathbf{x}$. So, the normalized surface overlap provides an interpretable relation between cameras; please see Fig. 4 for different cases.

In addition to the visibility of pixels in $\mathbf{x}$ by image $\mathbf{y}$, one could also consider the angle at which the overlapping surfaces are observed. Thus, we weight each point in $\mathsf{overlap}(\mathbf{x} \mapsto \mathbf{y})$ with $\cos(\mathbf{n_i}, \mathbf{n_j})$, $\mathbf{n_i}$ denotes the normal of a pixel $\mathbf{i}$ and $\mathbf{n_j}$ is the normal of the nearest 3D point of $\mathbf{i}$ in image $\mathbf{y}$. This will reduce the surface overlap between images observing the same scene from very different angles. Two images are difficult to match if there is substantial perspective distortion due to a difference in viewing angle. Incorporating the angle difference into the world-space measure captures this information.

3.2   Embeddings

We aim to learn a representation and embedding of RGB images. The values computed on those representations should approximate the normalized surface overlap measures, but without access to pose and depth data. Such a representation and embedding should provide estimates of the surface overlap measure cheaply and generalize to test images for downstream tasks such as localization.

Vector Embeddings. A common approach in computer vision to learn embeddings is to use metric learning. Images are encoded as vectors using a CNN, and the network is trained to approximate the relation in world-space measure by distances in vector space.

However, there is a fundamental limitation of vector representations and metric learning: the distance between vectors corresponding to images $\mathbf{x}$ and $\mathbf{y}$ can only be symmetric. In our case, the normalized surface overlaps are not symmetric. Hence they cannot be represented with vector embeddings. We could compromise and compute a symmetric value, for example the average of normalized surface overlaps computed in both directions,

$$\text{NSO}^{sym}(\mathbf{x},\mathbf{y}) = \text{NSO}^{sym}(\mathbf{y},\mathbf{x}) = \frac{1}{2}(\text{NSO}(\mathbf{x} \mapsto \mathbf{y}) + \text{NSO}(\mathbf{y} \mapsto \mathbf{x})). \quad (3)$$

Or, only consider the least amount of overlap in one of the directions [50], so $\text{NSO}^{min}(\mathbf{x},\mathbf{y}) = \text{NSO}^{min}(\mathbf{y},\mathbf{x}) = \min(\text{NSO}(\mathbf{x} \mapsto \mathbf{y}), \text{NSO}(\mathbf{y} \mapsto \mathbf{x}))$. As shown in Fig. 3 any such compromise would result in the loss of interpretability.

Different strategies to train vector embeddings are valid. One could consider training a symmetric version using NSO^{sym} or an asymmetric one using NSO; we hypothesize both are equivalent for large numbers of pairs. We empirically found the best performance with the loss function $\mathcal{L}_{\text{vector}} = ||\text{NSO}^{sym}(\mathbf{x},\mathbf{y}) - 1 - ||f(\mathbf{x}) - f(\mathbf{y})||_2||_2$, where f is a network that predicts a vector embedding.

Box Embeddings. We propose to use box representations to embed images with non-metric world-space measures. Our method is adapted from geometrically-inspired word embeddings for natural language processing [26,56,60]. The box representation of image $\mathbf{x}$ is a D-dimensional orthotope (hyperrectangle) parameterized as a $2D$-dimensional array $\mathbf{b_x}$. The values of this array are split into $m_{\mathbf{b_x}}$ and $M_{\mathbf{b_x}}$, which are the lower and upper bounds of the box in D-dimensional space. Crucially, the box representation allows us to compute the *intersection* of two boxes, $\mathbf{b_x} \wedge \mathbf{b_y}$, as

$$\mathbf{b_x} \wedge \mathbf{b_y} = \prod_d^D \sigma\left(\min(M_{\mathbf{b_x}}^d, M_{\mathbf{b_y}}^d) - \max(m_{\mathbf{b_x}}^d, m_{\mathbf{b_y}}^d)\right), \quad (4)$$

and the *volume* of a box,

$$A(\mathbf{b_x}) = \prod_d^D \sigma(M_{\mathbf{b_x}}^d - m_{\mathbf{b_x}}^d), \quad (5)$$

where $\sigma(v) = \max(0, v)$. This definition of $\sigma()$ has a problem of zero gradient to v for non-overlapping boxes that should be overlapping. As suggested by Li *et al.* [26], we train with a smoothed box intersection: $\sigma_{\text{smooth}}(v) = \rho \ln(1 + \exp(v/\rho))$, which is equivalent to $\sigma()$ as ρ approaches 0. Hence, we can approximate world-space surface overlap values with normalized box overlaps as

$$NBO(b_x \mapsto b_y) = \frac{b_x \wedge b_y}{A(b_x)} \approx NSO(x \mapsto y) \text{ and} \qquad (6)$$

$$NBO(b_y \mapsto b_x) = \frac{b_y \wedge b_x}{A(b_y)} \approx NSO(y \mapsto x). \qquad (7)$$

For training our embeddings, we can minimize the squared error between the ground truth surface overlap and the predicted box overlap of a random image pair (x, y) from the training set, so the loss is

$$\mathcal{L}_{box} = ||NSO(x \mapsto y) - NBO(b_x \mapsto b_y)||_2^2. \qquad (8)$$

It is important to note that computing the volume of a box is computationally very efficient. Given two boxes, intersection can be computed with just min, max and multiplication operations. Once images are embedded, one could build an efficient search system using R-trees [15] or similar data structures.

3.3   Implementation Details

We use a pretrained ResNet50 [18] as a backbone. The output of the 5-th layer is average pooled across spatial dimensions to produce a 2048-dimensional vector, followed by two densely connected layers with feature dimensions 512 and $2D$ respectively, where D denotes the dimensionality of the boxes. The first D values do not have a non-linearity and represent the position of the center of the box, and the other D values have softplus activation, and represent the size of the box in each dimension. These values are trivially manipulated to represent m_{b_x} and M_{b_x}. We found $D = 32$ to be optimal (ablation in supp. material). Input images are anisotropically rescaled to 256×456 resolution. We fix $\rho = 5$ when computing smoothed box intersection and volume. We train with batch size $b = 32$ for $20 - 60k$ steps, depending on the scene. The ground-truth overlap both for training and evaluation is computed over the original resolution of images, however the 3D point cloud is randomly subsampled to 5000 points for efficiency. Surface normals are estimated by fitting a plane to a 3×3 neighborhood of depths. We used 0.1 as the distance threshold for two 3D points to overlap.

4   Experiments

We conduct four main experiments. First, we validate how our box embeddings of images capture the surface overlap measure. Do these embeddings learn the asymmetric surface overlap and preserve interpretability? Do predicted relations of images match camera pose relations?

Second, we demonstrate that the proposed world measure is superior to alternatives like frustum overlap for the task of image localization. We evaluate localization quality on a small-scale indoor dataset and a medium-scale outdoor dataset. We also show that images retrieved with our embeddings can be ranked with metric distances, and are "backwards-compatible" with existing localization pipelines, despite also being interpretable.

Third, we demonstrate how the interpretability of our embeddings can be exploited for the task of localization, specifically for images that have large scale differences to the retrieved gallery images. Finally, we show how our embeddings give scale estimates that help with feature extraction.

Datasets. Since our proposed world-space measure requires depth information during training, we conduct our experiments on MegaDepth [27] and 7Scenes [51] datasets.

MegaDepth is a large scale dataset including 196 scenes, each of which consists of images with camera poses and camera intrinsics. However, only a subset of images have a corresponding semi-dense depth map. The dataset was originally proposed for the task of SfM reconstruction and to learn single-image depth estimation methods. The scenes have varying numbers of images, and large scenes have camera pose variations with significant zoom-in and zoom-out relationships. Most images do not have corresponding semi-dense depth maps, and may have only two-pixel ordinal labels or no labels at all. The low number of images with suitable depth also necessitates the generation of our own train and test splits, replacing the provided sets that are sampled across images with and without depths. As depth is needed for the computation of the ground-truth surface overlap, we only consider four scenes that provide enough data for training, validation and testing. So, we leave 100 images for validation and 100 images for testing, and the remaining images with suitable depth maps are used for training: 1336, 2130, 2155 and 1471 images for Big Ben, Notre-Dame, Venice, and Florence scenes, respectively.

7Scenes is an established small-scale indoor localization benchmark consisting of 3 to 10 sequences of a thousand images (per scene) with associated depth maps Kinect-acquired ground-truth poses. To evaluate our localization pipeline we follow the training and test splits from [51] and compare to methods such as RelocNet [3], which uses frustum overlap as a world-space measure.

4.1 Learning Surface Overlap

Figure 5 shows qualitative results of ground-truth surface overlaps and the predicted box overlaps between a random test image as query and training images as gallery. See supplementary materials for more examples.

In Table 1 we compare vector embeddings against box embeddings to experimentally validate that asymmetric surface overlap cannot be learned with a symmetric vector embedding. We evaluate the predictions against ground truth surface overlap on $1,000$ random pairs of test images. For each pair of images, we can compute ground truth $\mathsf{NSO}(\mathbf{x} \mapsto \mathbf{y})$ and $\mathsf{NSO}(\mathbf{y} \mapsto \mathbf{x})$. We report results on three metrics: L_1-Norm[1], the root mean square error (RMSE)[2] and the prediction accuracy. The prediction accuracy is defined as the percentage of individual overlaps that is predicted with an absolute error of less than 10%. Note, that

---

[1] L_1-Norm: $\frac{1}{N} \sum |\mathsf{NSO}(\mathbf{x} \mapsto \mathbf{y}) - \mathsf{NBO}(\mathbf{b_x} \mapsto \mathbf{b_y})| + |\mathsf{NSO}(\mathbf{y} \mapsto \mathbf{x}) - \mathsf{NBO}(\mathbf{b_y} \mapsto \mathbf{b_x})|$.

[2] RMSE: $\sqrt{\frac{1}{N} \sum [(\mathsf{NSO}(\mathbf{x} \mapsto \mathbf{y}) - \mathsf{NBO}(\mathbf{b_x} \mapsto \mathbf{b_y}))^2 + (\mathsf{NSO}(\mathbf{y} \mapsto \mathbf{x}) - \mathsf{NBO}(\mathbf{b_y} \mapsto \mathbf{b_x}))^2]}$.

the ground-truth depth information for the images may be incomplete, so there is inherent noise in the training signal and ground-truth measurements which makes smaller percentage thresholds not meaningful.

These results confirm that box embeddings are better than vectors at capturing the asymmetric world-space overlap relationships between images.

4.2 Interpreting Predicted Relations

For these experiments, we learn box embeddings for images in different scenes of the Megadepth dataset [27]. A trained network can be used to predict box representations for any image. We re-use our training data (which incidentally also have ground-truth depths) to form the gallery. Each query image q comes from the test-split. We compute its box representation b_q and compute surface overlap approximations using box representations as specified in Eq. (1) and (2).

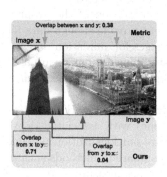

x	y	NSO $(x \mapsto y)$	NSO $(y \mapsto x)$	Interpretation
	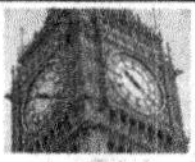	15.2% (15.3%)	83.1% (83.5%)	y is zoom-in on x
		80.8% (87.7%)	88.7% (89.2%)	x and y are clone-like
		85.3% (86.9%)	5.3% (4.8%)	y is zoom-out of x
		23.7% (20.1%)	22.3% (28.5%)	y is oblique-out or crop-out of x

Fig. 3. Top: Metric learning can only represent a single distance between two images of Big Ben. Bottom: While 4% of image y is visible in image x, 71% of image x is visible vice versa. The average of overlaps is 38%.

Fig. 4. Interpretation of image relationship based on the normalized surface overlap (NSO) between two images (predicted from just RGB with our approach and ground-truth using privileged 3D data). Four different relationships between image pairs can be observed. When NSO($x \mapsto y$) is low but NSO($y \mapsto$ x) is high, this indicates that most pixels of y are visible in x. Therefore y must be a close-up of x.

First, we analyze predicted relations qualitatively. We introduce two terms to discuss the relations between query image q and retrieved image r: *enclosure* and *concentration*. Enclosure refers to the predicted surface overlap from query to retrieved image, *i.e.* NBO($b_q \mapsto b_r$). Concentration refers to the predicted surface overlap from retrieved image to query, *i.e.* NBO($b_r \mapsto b_q$). Thus, if a retrieved image has a large value for enclosure, then it observes a large number of the query's pixels. See Fig. 4 for other interpretations.

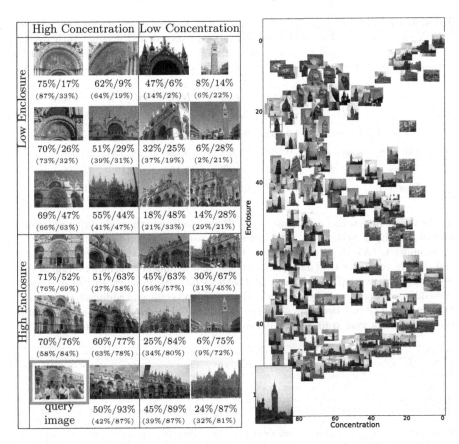

Fig. 5. Left: Results of predicted and ground-truth enclosure and concentration (defined in Sect. 4.2) relative to the query image indicated by a green frame. The numbers below each image indicate the predicted and ground-truth concentration/enclosure. Right: Results of predicting enclosure and concentration between a query image from the test set and test images (including test images without depth maps). Images are plotted at the coordinates of predicted (enclosure, concentration). Note in both plots how results in the upper right quadrant are oblique views, and bottom-right show zoomed-out views. The upper left quadrants show zoomed-in views, depending on the range of (low) enclosure selected.

To demonstrate the interpretability of the predicted relations between the images, we retrieve and show gallery images in different ranges of enclosure and concentration. Figure 5 shows results on two different scenes. On the left we see qualitative and quantitative results for the query image from the test data, and the images retrieved from the training set. As can be seen, the images in different quadrants of enclosure and concentration ranges are interpretable with different amounts of zoom-in, or zoom-out, or looking clone-like, or exhibiting crop-out/oblique-out. On the right, we retrieve images from the larger test set

Table 1. Evaluation of box *vs.* vector embeddings, trained on normalized surface overlap. We measure the discrepancy between the predicted and ground-truth overlaps on the test set. Across all measures and scenes, the non-metric embeddings with box representations outperform metric learning with vector representations

	NSO($\mathbf{x} \mapsto \mathbf{y}$) with boxes			NSO^{sym}($\mathbf{x}, \mathbf{y}$) with vectors		
	L_1-Norm	RMSE	Acc.< 0.1	L1-Norm	RMSE	Acc.< 0.1
Notre-Dame	0.070	0.092	93.3%	0.244	0.249	60.9%
Big Ben	0.126	0.138	82.6%	0.429	0.350	24.8%
Venice	0.066	0.112	89.9%	0.164	0.193	75.1%
Florence	0.063	0.094	89.7%	0.145	0.162	76.5%

Table 2. Results on 7Scenes dataset [51]. "Repr." indicates the embedding representation as box (B) or vector (V). Q specifies if the world space measure is symmetric (S) or asymmetric (A). Reported numbers show translation and rotation errors in meters and degrees respectively. The results indicate that symmetric surface overlap is superior to frustum overlap when represented with vectors. Asymmetric surface overlap box embeddings are similar to symmetric surface overlap vector embeddings, except for the Stairs scene. The last two rows show the generalization ability of our embeddings: the two embeddings were trained on Kitchen and used for retrieval on other scenes

Method	Repr.	Q	Chess	Fire	Heads	Office	Pumpkin	Kitchen	Stairs
DenseVLAD [58]	V	S	0.03, 1.40°	0.04, 1.62°	0.03, 1.21°	0.05, 1.37°	0.07, 2.02°	0.05, 1.63°	0.16, 3.85°
NetVLAD [1]	V	S	0.04, 1.29°	0.04, 1.85°	0.03, 1.14°	0.05, 1.45°	0.08, 2.16°	0.05, 1.77°	0.16, 4.00°
PoseNet [22]			0.32, 8.12°	0.47, 14.4°	0.29, 12.0°	0.48, 7.68°	0.47, 8.42°	0.59, 8.64°	0.47, 13.6°
RelocNet [3]			0.12, 4.14°	0.26, 10.4°	0.14, 10.5°	0.18, 5.32°	0.26, 4.17°	0.23, 5.08°	0.28, 7.53°
Active Search [42]			0.04, 1.96°	0.03, 1.53°	0.02, 1.45°	0.09, 3.61°	0.08, 3.10°	0.07, 3.37°	0.03, 2.22°
DSAC++ [6]			0.02, 0.50°	0.02, 0.90°	0.01, 0.80°	0.03, 0.70°	0.04, 1.10°	0.04, 1.10°	0.09, 2.60°
Ours NSO()	B	A	0.05, 1.47°	0.05, 1.91°	0.05, 2.54°	0.06, 1.60°	0.10, 2.46°	0.07, 1.73°	0.50, 9.18°
Ours NSO^{sym}()	V	S	0.04, 1.19°	0.05, 2.05°	0.05, 2.84°	0.06, 1.46°	0.10, 2.28°	0.06, 1.61°	0.22, 5.28°
Ours Frustum	V	S	0.05, 1.25°	0.05, 2.02°	0.04, 1.86°	0.07, 1.73°	0.10, 2.40°	0.07, 1.71°	1.82, 12.0°
Box (Kitchen)	B	A	0.06, 2.19°	0.08, 2.94°	0.08, 5.42°	0.17, 4.87°	0.13, 3.21°	*	1.83, 50.1°
Vector (Kitchen)	V	S	0.04, 1.54°	0.06, 2.26°	0.04, 1.90°	0.06, 1.88°	0.10, 2.37°	*	0.55, 9.22°

(without depth maps) plotted according to the estimated enclosure and concentration. This qualitative result demonstrates how box embeddings generalize to the test set. Please see supplementary materials for more examples.

4.3   Querying Box Embeddings for Localization

We now compare the differences between surface overlap and frustum overlap measures for retrieving images in a localization task. The task is to find the camera pose of a query image $\mathbf{q}$ with respect to the images in the training set, where the latter have known camera poses and semi-dense depths. The image embeddings are used to retrieve the top k-th image ($k = 1$) from the training data that are closest according to each embedding measure. After retrieval, 2D−3D correspondences are found between the query image and k-th retrieved image's 3D point cloud. We use SIFT [28] features, and correspondences are filtered with

Lowe's ratio test and matched using OpenCV's FLANN-based matcher. The pose is solved with RANSAC and PnP, and measured against the ground truth pose. We report median translation and rotation errors, with all test images as queries.

We compare three embeddings for this task, each trained with a different world-space measure: i) vector embeddings trained with frustum overlap, ii) vector embeddings trained with $\mathsf{NSO}^{sym}()$, and box embeddings trained with $\mathsf{NSO}()$. Ranking for (i) and (ii) is easy, and for (iii) we rank retrieved images according to $0.5(\mathsf{NBO}(\mathbf{b_x} \mapsto \mathbf{b_y}) + \mathsf{NBO}(\mathbf{b_y} \mapsto \mathbf{b_x}))$. This query function is used to show backwards compatibility with traditional metric embeddings.

7Scenes. Table 2 shows the results of using (i) frustum overlap and both (ii) symmetric and (iii) asymmetric surface overlap. We also report the results of state-of-the-art work [3,6,22,42] and two SOTA baselines, DenseVLAD and NetVLAD, that we generated by swapping our retrieval system with the ones of [58] and [1] respectively and leaving the rest of the pose estimation pipeline intact. Both surface overlap-based results are generally better than frustum overlap. Note, absolute differences between most recent methods on this dataset are relatively minor. For example, due to the use of PnP, we get better results using our frustum overlap than RelocNet. Ours-box is similar in performance to Ours-vector, except for the Stairs dataset, where we perform poorly because stairs have many repeated structures. This is a positive result, showing on-par performance while giving the benefit of an interpretable embedding. Although the localization task includes having a 3D reconstruction of the gallery/training images by its nature, we also compare two embeddings that were trained on Kitchen and tested generalization for retrieval on other scenes. Please see the supplementary material for further experimental results.

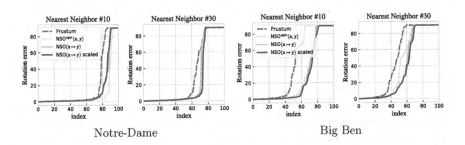

Notre-Dame   Big Ben

Fig. 6. Each plot shows (sorted) rotation error (capped at 90°) when each test image is matched against 10-th and 30-th closest retrieved image for pose estimation. As we can see, box embeddings with surface overlap measure tend to outperform alternatives, especially when rescaling images according to estimated relative scale.

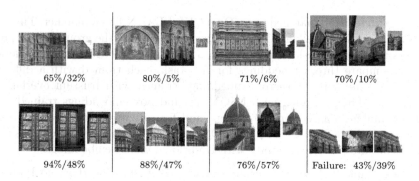

65%/32%   80%/5%   71%/6%   70%/10%

94%/48%   88%/47%   76%/57%   Failure: 43%/39%

Fig. 7. Qualitative results of relative scale estimation on Florence test set. For each pair, the enclosure and concentration are calculated from which the relative estimated scaled can be derived. Based on that scale, the first image is resized and shown in the third position. If the relative scale is accurate, the content in the second and third images should match in size (number of pixels). The resized images are sometimes small, so the reader is encouraged to zoom into the images. The two numbers below each image pair show the estimated enclosure and concentration values.

4.4 Predicting Relative Scale Difference for Image Matching

Given a pair of images $\mathbf{x}$ and $\mathbf{y}$, we can compare the predicted $\mathrm{NBO}(\mathbf{b_x} \mapsto \mathbf{b_y})$ and $\mathrm{NBO}(\mathbf{b_y} \mapsto \mathbf{b_x})$ to estimate the relative scale of the images, so

$$\frac{\mathrm{NBO}(\mathbf{b_x} \mapsto \mathbf{b_y})}{\mathrm{NBO}(\mathbf{b_y} \mapsto \mathbf{b_x})} \approx \frac{\mathrm{NSO}(\mathbf{x} \mapsto \mathbf{y})}{\mathrm{NSO}(\mathbf{y} \mapsto \mathbf{x})} = \frac{\mathrm{overlap}(\mathbf{x} \mapsto \mathbf{y})/N_\mathbf{x}}{\mathrm{overlap}(\mathbf{y} \mapsto \mathbf{x})/N_\mathbf{y}} \approx \frac{1}{s^2} \frac{N_\mathbf{y}}{N_\mathbf{x}}. \quad (9)$$

This estimates the scale factor s to be applied to image $\mathbf{x}$, so that overlaps $\mathrm{overlap}(\mathbf{x} \mapsto \mathbf{y})$ and $\mathrm{overlap}(\mathbf{y} \mapsto \mathbf{x})$ occupy approximately the same number of pixels in each of the two images.

Figure 7 shows qualitative examples of pairs of images from the test set, with predicted normalized box overlaps and estimated scale factor applied to one of the images. We observed that this relative scale estimate is in general accurate if the images have zoom-in/zoom-out relationships. However, if the images are in crop-out/oblique-out relation to one-another, then the rescaling may not necessarily make matching easier, as the overlap is already quite small. However, we can detect if images are in such a relationship by looking at predicted box overlaps (see Sect. 4.2). The estimated scale factor can be applied to the images before local feature detection and description extraction, to improve relative pose accuracy for pairs of images with large scale differences.

We simulate this scenario by retrieving $k = 10$ and $k = 30$ closest matches for each embedding, and solving for the pose. Additionally, for box embeddings, we do feature matching with and without pre-scaling of images according to the predicted overlaps. Figure 6 shows results on Notre-Dame and Big Ben scenes.

Please see supplementary materials for further details.

5   Conclusions

We found surface overlap to be a superior measure compared to frustum overlap. We have shown that normalized surface overlap can be embedded in our new box embedding. The benefit is that we can now *easily* compute interpretable relationships between pairs of images, without hurting localization. Further, this can help with pre-scaling of images for feature-extraction, and hierarchical organization of images.

Limitations and Future Work. An obvious limitation is that we rely on expensive depth information for training. This could be addressed in two ways: either approximate the visual overlap with sparse 3D points and their covisibility in two views, or train box embeddings with homography overlap of single images. Both of these approaches could also help with learning box embeddings that generalize across scenes, for image and object retrieval applications, as larger datasets could potentially be used for training.

Acknowledgements. Thanks to Carl Toft for help with normal estimation, to Michael Firman for comments on paper drafts and to the anonymous reviewers for helpful feedback.

References

1. Arandjelovic, R., Gronat, P., Torii, A., Pajdla, T., Sivic, J.: NetVLAD: CNN architecture for weakly supervised place recognition. In: CVPR (2016)
2. Arandjelovic, R., Zisserman, A.: All about VLAD. In: CVPR (2013)
3. Balntas, V., Li, S., Prisacariu, V.: RelocNet: continuous metric learning relocalisation using neural nets. In: ECCV (2018)
4. Balntas, V., Riba, E., Ponsa, D., Mikolajczyk, K.: Learning local feature descriptors with triplets and shallow convolutional neural networks. In: BMVC (2016)
5. Bonin-Font, F., Ortiz, A., Oliver, G.: Visual navigation for mobile robots: a survey. J. Intell. Robot. Syst. **53**(3), 263 (2008)
6. Brachmann, E., Rother, C.: Learning less is more-6D camera localization via 3D surface regression. In: CVPR (2018)
7. Buehler, C., Bosse, M., McMillan, L., Gortler, S., Cohen, M.: Unstructured lumigraph rendering. In: Computer Graphics and Interactive Techniques (2001)
8. Cakir, F., He, K., Xia, X., Kulis, B., Sclaroff, S.: Deep metric learning to rank. In: CVPR (2019)
9. DeTone, D., Malisiewicz, T., Rabinovich, A.: Deep image homography estimation. arXiv preprint arXiv:1606.03798 (2016)
10. Dufournaud, Y., Schmid, C., Horaud, R.: Image matching with scale adjustment. Comput. Vis. Image Underst. **93**(2), 175–194 (2004)
11. Erlik Nowruzi, F., Laganiere, R., Japkowicz, N.: Homography estimation from image pairs with hierarchical convolutional networks. In: ICCVW (2017)
12. Frahm, J.M., et al.: Building rome on a cloudless day. In: ECCV (2010)
13. Gálvez-López, D., Tardós, J.D.: Bags of binary words for fast place recognition in image sequences. IEEE Trans. Robot. **28**(5), 1188–1197 (2012)
14. Gordo, A., Almazán, J., Revaud, J., Larlus, D.: Deep image retrieval: learning global representations for image search. In: ECCV (2016)

15. Guttman, A.: R-trees: a dynamic index structure for spatial searching. In: ACM SIGMOD International Conference on Management of Data (1984)
16. Hadsell, R., Chopra, S., LeCun, Y.: Dimensionality reduction by learning an invariant mapping. In: CVPR (2006)
17. Hartley, R.I.: In defense of the eight-point algorithm. TPAMI **19**(6), 580–593 (1997)
18. He, K., Zhang, X., Ren, S., Sun, J.: Deep residual learning for image recognition. In: CVPR (2016)
19. He, K., Lu, Y., Sclaroff, S.: Local descriptors optimized for average precision. In: CVPR (2018)
20. Heinly, J., Schönberger, J.L., Dunn, E., Frahm, J.M.: Reconstructing the world* in six days. In: CVPR (2015)
21. Hermans, A., Beyer, L., Leibe, B.: In defense of the triplet loss for person re-identification. arXiv:1703.07737 (2017)
22. Kendall, A., Grimes, M., Cipolla, R.: PoseNet: a convolutional network for real-time 6-DOF camera relocalization. In: ICCV (2015)
23. Kulkarni, T.D., Whitney, W.F., Kohli, P., Tenenbaum, J.: Deep convolutional inverse graphics network. In: NeurIPS (2015)
24. Laskar, Z., Melekhov, I., Kalia, S., Kannala, J.: Camera relocalization by computing pairwise relative poses using convolutional neural network. In: ICCV (2017)
25. Le, H., Liu, F., Zhang, S., Agarwala, A.: Deep homography estimation for dynamic scenes. In: CVPR (2020)
26. Li, X., Vilnis, L., Zhang, D., Boratko, M., McCallum, A.: Smoothing the geometry of probabilistic box embeddings. In: ICLR (2019)
27. Li, Z., Snavely, N.: Megadepth: learning single-view depth prediction from internet photos. In: CVPR (2018)
28. Lowe, D.G.: Distinctive image features from scale-invariant keypoints. IJCV **60**(2), 91–110 (2004)
29. Mikulik, A., Chum, O., Matas, J.: Image retrieval for online browsing in large image collections. In: International Conference on Similarity Search and Applications (2013)
30. Mikulík, A., Radenović, F., Chum, O., Matas, J.: Efficient image detail mining. In: ACCV (2014)
31. Mishchuk, A., Mishkin, D., Radenovic, F., Matas, J.: Working hard to know your neighbor's margins: Local descriptor learning loss. In: NeurIPS (2017)
32. Mishkin, D., Matas, J., Perdoch, M.: MODS: fast and robust method for two-view matching. Comput. Vis. Image Underst. **141**, 81–93 (2015)
33. Mur-Artal, R., Tardós, J.D.: ORB-SLAM2: an open-source SLAM system for monocular, stereo and RGB-D cameras. IEEE Trans. Robot. **33**(5), 1255–1262 (2017)
34. Nguyen, T., Chen, S.W., Shivakumar, S.S., Taylor, C.J., Kumar, V.: Unsupervised deep homography: a fast and robust homography estimation model. IEEE Robot. Autom. Lett. **3**(3), 2346–2353 (2018)
35. Nickel, M., Kiela, D.: Poincaré embeddings for learning hierarchical representations. In: NeurIPS (2017)
36. Perd'och, M., Chum, O., Matas, J.: Efficient representation of local geometry for large scale object retrieval. In: CVPR (2009)
37. Perronnin, F., Liu, Y., Sánchez, J., Poirier, H.: Large-scale image retrieval with compressed Fisher vectors. In: CVPR (2010)
38. Philbin, J., Chum, O., Isard, M., Sivic, J., Zisserman, A.: Object retrieval with large vocabularies and fast spatial matching. In: CVPR (2007)

39. Revaud, J., et al.: R2D2: repeatable and reliable detector and descriptor. In: NeurIPS (2019)

40. Sattler, T., Leibe, B., Kobbelt, L.: Fast image-based localization using direct 2D-to-3D matching. In: ICCV (2011)

41. Sattler, T., Leibe, B., Kobbelt, L.: Improving image-based localization by active correspondence search. In: ECCV (2012)

42. Sattler, T., Leibe, B., Kobbelt, L.: Efficient & effective prioritized matching for large-scale image-based localization. TPAMI **39**(9), 1744–1756 (2016)

43. Sattler, T., et al.: Are large-scale 3D models really necessary for accurate visual localization? In: CVPR (2017)

44. Sattler, T., Weyand, T., Leibe, B., Kobbelt, L.: Image retrieval for image-based localization revisited. In: BMVC (2012)

45. Sattler, T., Zhou, Q., Pollefeys, M., Leal-Taixe, L.: Understanding the limitations of CNN-based absolute camera pose regression. In: CVPR (2019)

46. Schönberger, J.L., Berg, A.C., Frahm, J.M.: Paige: pairwise image geometry encoding for improved efficiency in structure-from-motion. In: CVPR (2015)

47. Schönberger, J.L., Radenovic, F., Chum, O., Frahm, J.M.: From single image query to detailed 3D reconstruction. In: CVPR (2015)

48. Schönberger, J.L., Frahm, J.M.: Structure-from-motion revisited. In: CVPR (2016)

49. Schroff, F., Kalenichenko, D., Philbin, J.: Facenet: a unified embedding for face recognition and clustering. In: CVPR (2015)

50. Shen, T., et al.: Matchable image retrieval by learning from surface reconstruction. In: ACCV (2018)

51. Shotton, J., Glocker, B., Zach, C., Izadi, S., Criminisi, A., Fitzgibbon, A.: Scene coordinate regression forests for camera relocalization in RGB-D images. In: CVPR (2013)

52. Snavely, N., Garg, R., Seitz, S.M., Szeliski, R.: Finding paths through the world's photos. ACM Trans. Graph. **27**(3), 1–11 (2008)

53. Snavely, N., Seitz, S.M., Szeliski, R.: Photo tourism: exploring photo collections in 3D. In: SIGGRAPH (2006)

54. Sohn, K., Lee, H.: Learning invariant representations with local transformations. In: ICML (2012)

55. Stewénius, H., Gunderson, S.H., Pilet, J.: Size matters: exhaustive geometric verification for image retrieval. In: ECCV (2012)

56. Subramanian, S., Chakrabarti, S.: New embedded representations and evaluation protocols for inferring transitive relations. In: SIGIR Conference on Research & Development in Information Retrieval (2018)

57. Tian, Y., Yu, X., Fan, B., Wu, F., Heijnen, H., Balntas, V.: Sosnet: second order similarity regularization for local descriptor learning. In: CVPR (2019)

58. Torii, A., Arandjelovic, R., Sivic, J., Okutomi, M., Pajdla, T.: 24/7 place recognition by view synthesis. In: CVPR (2015)

59. Vendrov, I., Kiros, R., Fidler, S., Urtasun, R.: Order-embeddings of images and language. In: ICLR (2016)

60. Vilnis, L., Li, X., Murty, S., McCallum, A.: Probabilistic embedding of knowledge graphs with box lattice measures. In: ACL (2018)

61. Vilnis, L., McCallum, A.: Word representations via gaussian embedding. In: ICLR (2015)

62. Weyand, T., Leibe, B.: Discovering favorite views of popular places with iconoid shift. In: ICCV (2011)

63. Weyand, T., Leibe, B.: Discovering details and scene structure with hierarchical iconoid shift. In: ICCV (2013)

64. Witkin, A.P.: Scale-space filtering. In: IJCAI (1983)
65. Worrall, D.E., Garbin, S.J., Turmukhambetov, D., Brostow, G.J.: Interpretable transformations with encoder-decoder networks. In: ICCV (2017)
66. Zhou, L., Zhu, S., Shen, T., Wang, J., Fang, T., Quan, L.: Progressive large scale-invariant image matching in scale space. In: ICCV (2017)

Connecting Vision and Language with Localized Narratives

Jordi Pont-Tuset[1(✉)], Jasper Uijlings[1], Soravit Changpinyo[2], Radu Soricut[2], and Vittorio Ferrari[1]

[1] Google Research, Zurich, USA
jponttuset@google.com
[2] Google Research, Venice, CA, USA

Abstract. We propose Localized Narratives, a new form of multimodal image annotations connecting vision and language. We ask annotators to describe an image with their voice while simultaneously hovering their mouse over the region they are describing. Since the voice and the mouse pointer are synchronized, we can localize every single word in the description. This dense visual grounding takes the form of a mouse trace segment per word and is unique to our data. We annotated 849k images with Localized Narratives: the whole COCO, Flickr30k, and ADE20K datasets, and 671k images of Open Images, all of which we make publicly available. We provide an extensive analysis of these annotations showing they are diverse, accurate, and efficient to produce. We also demonstrate their utility on the application of controlled image captioning.

1 Introduction

Much of our language is rooted in the visual world around us. A popular way to study this connection is through Image Captioning, which uses datasets where images are paired with human-authored textual captions [8,46,59]. Yet, many researchers want deeper visual grounding which links specific words in the caption to specific regions in the image [32,33,44,45]. Hence Flickr30k Entities [40] enhanced Flickr30k [59] by connecting the nouns from the captions to bounding boxes in the images. But these connections are still sparse and important aspects remain ungrounded, such as words capturing relations between nouns (as "holding" in "a woman holding a balloon"). Visual Genome [27] provides short descriptions of regions, thus words are not individually grounded either.

In this paper we propose Localized Narratives, a new form of multimodal image annotations in which we ask annotators to describe an image with their voice while simultaneously hovering their mouse over the region they are describing. Figure 1 illustrates the process: the annotator says "woman" while using their mouse to indicate her spatial extent, thus providing visual grounding for

Electronic supplementary material The online version of this chapter (https://doi.org/10.1007/978-3-030-58558-7_38) contains supplementary material, which is available to authorized users.

A. Vedaldi et al. (Eds.): ECCV 2020, LNCS 12350, pp. 647–664, 2020.
https://doi.org/10.1007/978-3-030-58558-7_38

Image and Trace: Caption: Voice:

In the front portion of
the picture we can see
a dried grass area with
dried twigs. There is a
woman standing wearing
light blue jeans and
ash colour long sleeve
length shirt. This
woman is holding a
black jacket in her
hand. On the other hand
she is holding a balloon
which is peach in
colour. On the top of
the picture we see a
clear blue sky with
clouds. The hair colour
of the woman is
brownish.

Fig. 1. Localized Narrative example: Caption, voice, and mouse trace synchronization represented by a color gradient ▬▬▬ . The project website [53] contains a visualizer with many live examples.

this noun. Later they move the mouse from the woman to the balloon following its string, saying "holding". This provides direct visual grounding of this relation. They also describe attributes like "clear blue sky" and "light blue jeans". Since voice is *synchronized* to the mouse pointer, we can determine the image location of every single word in the description. This provides dense visual grounding in the form of a mouse trace segment for each word, which is unique to our data.

In order to obtain written-word grounding, we additionally need to transcribe the voice stream. We observe that automatic speech recognition [1,15,41] typically results in imperfect transcriptions. To get data of the highest quality, we ask annotators to transcribe their own speech, immediately after describing the image. This delivers an accurate transcription, but without temporal synchronization between the mouse trace and the written words. To address this issue, we perform a sequence-to-sequence alignment between automatic and manual transcriptions, which leads to accurate *and* temporally synchronized captions. Overall, our annotation process tightly connects four modalities: the image, its spoken description, its textual description, and the mouse trace. Together, they provide dense grounding between language and vision.

Localized Narratives is an efficient annotation protocol. Speaking and pointing to describe things comes naturally to humans [22,38]. Hence this step takes little time (40.4 s on average). The manual transcription step takes 104.3 s, for a total of 144.7 s. This is lower than the cost of related grounded captioning datasets Flickr30k Entities and Visual Genome [27,40], which were made by more complicated annotation processes and involved manually drawing bounding boxes (Sect. 4.1 – Annotation Cost). Moreover, if automatic speech recognition improves in the future it might be possible to skip the manual transcription step, making our approach even more efficient.

Table 1. Tasks enabled by Localized Narratives. Each row represents different uses of the four elements in a Localized Narrative: image, textual caption, speech, and grounding (mouse trace); labeled as being input (In) or output (Out) for each task.

Image	Text	Speech	Grounding	Task
In	Out	–	–	Image captioning [51,52,55], Paragraph generation [26,56,64]
Out	In	–	–	Text-to-image Generation [42,47,57]
In	Out	–	Out	Dense image captioning [21,25,58], Dense relational captioning [25]
In	Out	–	In	Controllable and Grounded Captioning [10]
In	In	–	Out	Phrase grounding [13]
In	In + Out	–	–	Visual Question Answering [4,20,34]
In	In + Out	–	Out	Referring Expression Recognition [9,24,35]
In	–	In	Out	Discover visual objects and spoken words from raw sensory input [18]
–	In	Out	–	Speech recognition [1,15,41]
–	Out	In	–	Speech synthesis [23,36,37]
Out	In	–	In	Image generation/retrieval from traces
In	Out	In	In	Grounded speech recognition
In	–	In	Out	Voice-driven environment navigation

We collected Localized Narratives at scale: we annotated the whole COCO [31] (123k images), ADE20K [62] (20k) and Flickr30k [59] (32k) datasets, as well as 671k images of Open Images [29]. We make the Localized Narratives for these 848,749 images publicly available [53]. We provide an extensive analysis (Sect. 4) and show that: (i) Our data is rich: we ground all types of words (nouns, verbs, prepositions, etc.), and our captions are substantially longer than in most previous datasets [8,27,46,59]. (ii) Our annotations are diverse both in the language modality (e.g. caption length varies widely with the content of the image) and in the visual domain (different pointing styles and ways of grounding relationships). (iii) Our data is of high quality: the mouse traces match well the location of the objects, the words in the captions are semantically correct, and the manual transcription is accurate. (iv) Our annotation protocol is more efficient than for related grounded captioning datasets [27,40].

Since Localized Narratives provides four synchronized modalities, it enables many applications (Table 1). We envision that having each word in the captions grounded, beyond the sparse set of boxes of previous datasets [24,27,35,40], will enable richer results in many of these tasks and open new doors for tasks and research directions that would not be possible with previously existing annotations. As a first example, we show how to use the mouse trace as a fine-grained control signal for a user to request a caption on a particular image (Sect. 5). Mouse traces are a more natural way for humans to provide a sequence of grounding locations, compared to drawing a list of bounding boxes [10]. We therefore envision its use as assistive technology for people with imperfect vision.

Table 2. Datasets connecting vision and language via image captioning, compared with respect to their type of grounding, scale, and caption length. Num. captions is typically higher than num. images because of replication (i.e. several annotators writing a caption for the same image).

Dataset	Grounding	Num. captions	Num. images	Words/capt
COCO Captions [8]	Whole capt. → Whole im	616,767	123,287	10.5
Conceptual Capt. [46]	Whole capt. → Whole im	3,334,173	3,334,173	10.3
Stanford Vis. Par. [26]	Whole capt. → Whole im.	19,561	19,561	67.5
ReferIt [24]	Short phrase ⟶ Region	130,525	19,894	3.6
Google Refexp [35]	Short phrase ⟶ Region	104,560	26,711	8.4
Visual Genome [27]	Short phrase ⟶ Region	5,408,689	108,077	5.1
Flickr30k Ent. [40]	Nouns ⟶ Region	158,915	31,783	12.4
Loc. Narr. (Ours)	Each word ⟶ Region	873,107	848,749	36.5

In future work, the mouse trace in our Localized Narratives could be used as additional attention supervision at training time, replacing or complementing the self-supervised attention mechanisms typical of modern systems [3,7,46,55,60]. This might train better systems and improve captioning performance at test time, when only the image is given as input. Alternatively, our mouse traces could be used at test time only, to inspect whether current spatial attention models activate on the same image regions that humans associate with each word.

Besides image captioning, Localized Narratives are a natural fit for: (i) image generation: the user can describe which image they want to generate by talking and moving their mouse to indicate the position of objects (demonstration in supp. material, Sect. 1); (ii) image retrieval: the user naturally describes the content of an image they are looking for, in terms of both what and where; (iii) grounded speech recognition: considering the content of an image would allow better speech transcription, e.g. 'plant' and 'planet' are easier to distinguish in the visual than in the voice domain; (iv) voice-driven environment navigation: the user describes where they want to navigate to, using relative spatial language.

To summarize, our paper makes the following contributions: (i) We introduce Localized Narratives, a new form of multimodal image annotations where every word is localized in the image with a mouse trace segment; (ii) We use Localized Narratives to annotate 848,749 images and provide a thorough analysis of the data. (iii) We demonstrate the utility of our data for controlled image captioning.

2   Related Work

Captioning Datasets. Various annotation efforts connect vision and language via captioning (Table 2). We focus on how their captions are grounded, as this is the key differentiating factor of Localized Narratives from these works. As a

The man at bat readies to swing at the pitch while the umpire looks on

A man **with** pierced ears **is wearing** glasses **and** an orange hat.

Man jumping for a picture with a skateboard
Light brown shoe with red strip
Green shirt with logo across front
The Eiffel Tower in the background

There is a kid standing on the bed, holding one of the railing, with a hand under his chin. He is wearing a blue jacket. Behind him there is a pillow and bed sheets.

Fig. 2. Sample annotations from (a) COCO Captions [8], (b) Flickr30k Entities [40], (c) Visual Genome [27], and (d) Localized Narratives (Ours). For clarity, (b) shows a subset of region descriptions and (d) shows a shorter-than-average Localized Narrative.

starting point, classical image captioning [8,46,59] and visual paragraph generation [26] simply provide a whole caption for the whole image (Fig. 2(a)). This lack of proper grounding was shown to be problematic [32,33,44,45].

Flickr30k Entities [40] annotated the nouns mentioned in the captions of Flickr30k [59] and drew their bounding box in the image (Fig. 2(b)): the grounding is therefore from nouns to regions (including their attached adjectives, Table 2). Visual Genome [27] and related previous efforts [24,35] provide short phrases describing regions in the images (Fig. 2(c)): grounding is therefore at the phrase level (Table 2). While Visual Genome uses these regions as a seed to generate a scene graph, where each node is grounded in the image, the connection between the region descriptions and the scene graph is not explicit.

In Localized Narratives, in contrast, *every word* is grounded to a specific region in the image represented by its trace segment (Fig. 2(d)). This includes all types of words (nouns, verbs, adjectives, prepositions, etc.), in particular valuable spatial-relation markers ("above", "behind", etc.) and relationship indicators ("riding", "holding", etc.). Another disadvantage of Flickr30k Entities and Visual Genome is that their annotation processes require manually drawing many bounding boxes a posteriori, which is unnatural and time-consuming compared to our simpler and more natural protocol (Sect. 4.1 – Annotation Cost).

SNAG [48] is a proof of concept where annotators describe images using their voice while their gaze is tracked using specialized hardware. This enables inferring the image location they are looking at. As a consequence of the expensive and complicated setup, only 100 images were annotated. In our proposed Localized Narratives, instead, we collect the data using just a mouse, a keyboard, and a microphone as input devices, which are commonly available. This allows us to annotate a much larger set of images (848,749 to date).

In the video domain, ActivityNet-Entities [63] adds visual grounding to the ActivityNet Captions, also in two stages where boxes were drawn a posteriori.

Annotation Using Voice. A few recent papers use voice as an input modality for computer vision tasks [11,16–18,48,49]. The closest work to ours [16] uses voice to simultaneously annotate the class name and the bounding box of an object instance in an image. With Localized Narratives we bring it to the next level by producing richer annotations both in the language and vision domains with long free-form captions associated to synchronized mouse traces.

In the video domain, EPIC-KITCHENS [12] contains videos of daily kitchen activities collected with a head-mounted camera. The actions were annotated with voice, manually transcribed, and time-aligned using YouTube's automatic closed caption alignment tool.

3   Annotation Process

The core idea behind the Localized Narratives annotation protocol is to ask the annotators to describe the contents of the image using their voice while hovering their mouse over the region being described. Both voice and mouse location signals are timestamped, so we know where the annotators are pointing while they are speaking every word.

Figure 3 shows voice (a) and mouse trace data (b), where the color gradient represents temporal synchronization. We summarize how to process this data to produce a Localized Narrative example. First, we apply an Automatic Speech Recognition (ASR) algorithm and get a synchronized, but typically imperfect, transcription (c). After finishing a narration, the annotators transcribe their own recording, which gives us an accurate caption, but without synchronization with the mouse trace (d). Finally, we obtain a correct transcription with timestamps by performing sequence-to-sequence alignment between the manual and automatic transcriptions (e). This time-stamped transcription directly reveals which trace segment corresponds to each word in the caption (f), and completes the creation of a Localized Narrative instance. Below we describe each step in detail.

Annotation Instructions. One of the advantages of Localized Narratives is that it is a natural task for humans to do: speaking and pointing at what we are describing is a common daily-life experience [22,38]. This makes it easy for annotators to understand the task and perform as expected, while increasing the pool of qualified annotators for the task. The instructions we provide are concise:

Use the mouse to point at the objects in the scene.
Simultaneously, use your voice to describe what you are pointing at.

– Focus on concrete objects (e.g. cow, grass, person, kite, road, sky).
– Do not comment on things you cannot directly see in the image (e.g. feelings that the image evokes, or what might happen in the future).
– Indicate an object by moving your mouse over the whole object, roughly specifying its location and size.
– Say the relationship between two objects while you move the mouse between them, e.g. "a man *is flying* a kite", "a bottle *is on* the table".
– If relevant, also mention attributes of the objects (e.g. *old* car).

Automatic and Manual Transcriptions. We apply an ASR algorithm [14] to obtain an automatic transcription of the spoken caption, which is times-tamped but typically contains transcription errors. To fix these errors, we ask

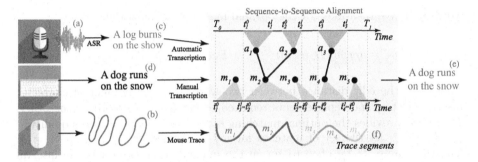

Fig. 3. Localized Narratives annotation: We align the automatic transcription (c) to the manual one (d) to transfer the timestamps from the former to the latter, resulting in a transcription that is both accurate and timestamped (e). To do so, we perform a sequence-to-sequence alignment (gray box) between a_i and m_j (black thick lines). The timestamps of matched words m_j are defined as the segment (green) containing the original timestamps (red) of the matched words a_i. Unmatched words m_j get assigned the time segments in between matched neighboring words (blue). These timestamps are transferred to the mouse trace and define the trace segment for each word m_j. (Color figure online)

the annotators to manually transcribe their own recorded narration. Right after they described an image, the annotation tool plays their own voice recording accompanied by the following instructions:

Type literally what you just said.

– Include filler words if you said them (e.g. "I think", "alright") but not filler sounds (e.g. "um", "uh", "er").
– Feel free to separate the text in multiple sentences and add punctuation.

The manual transcription is accurate but not timestamped, so we cannot associate it with the mouse trace to recover the grounding of each word.

Transcription Alignment. We obtain a correct transcription with timestamps by performing a sequence-to-sequence alignment between the manual and automatic transcriptions (Fig. 3).

Let $\mathbf{a} = \{(a_1, \ldots, a_{|\mathbf{a}|}\}$ and $\mathbf{m} = \{m_1, \ldots, m_{|\mathbf{m}|}\}$ be the automatic and manual transcriptions of the spoken caption, where a_i and m_j are individual words. a_i is timestamped: let $[t_i^0, t_i^1]$ be the time segment during which a_i was spoken. Our goal is to align $\mathbf{a}$ and $\mathbf{m}$ to transfer the timestamps from the automatically transcribed words a_i to the manually provided m_j.

To do so, we apply Dynamic Time Warping [28] between $\mathbf{a}$ and $\mathbf{m}$. Intuitively, we look for a matching function μ that assigns each word a_i to a word $m_{\mu(i)}$, such that if $i_2 > i_1$ then $\mu(i_2) \geq \mu(i_1)$ (it preserves the order of the words). Note that μ assigns each a_i to exactly one m_j, but m_j can match to zero or multiple words in **a**. We then look for the optimal matching μ^* such that:

$$\mu^* = \arg\min_\mu D_\mu(\mathbf{a}, \mathbf{m}) \qquad D_\mu(\mathbf{a}, \mathbf{m}) = \sum_{i=1}^{|\mathbf{a}|} d(a_i, m_{\mu(i)}) \qquad (1)$$

where d is the edit distance between two words, i.e. the number of character inserts, deletes, and replacements required to get from one word to the other. $D_{\mu^*}(\mathbf{a}, \mathbf{m})$ provides the optimal matching score (used below to assess quality).

Given μ^*, let the set of matches for m_j be defined as $A_j = \{i \mid \mu^*(i) = j\}$. The timestamp $[\bar{t}_j^0, \bar{t}_j^1]$ of word m_j in the manual transcription is the interval spanned by its matching words (if any) or to the time between neighboring matching words (if none). Formally:

$$\bar{t}_j^0 = \begin{cases} \min\{t_i^0 \mid i \in A_j\} & \text{if } A_j \neq \emptyset, \\ \max\{t_i^1 \mid i \in A_k \mid k < j\} & \text{if } \exists k < j \text{ s.t. } A_k \neq \emptyset \\ T^0 & \text{otherwise,} \end{cases} \qquad (2)$$

$$\bar{t}_j^1 = \begin{cases} \max\{t_i^1 \mid i \in A_j\} & \text{if } A_j \neq \emptyset, \\ \min\{t_i^0 \mid i \in A_k \mid k > j\} & \text{if } \exists k > j \text{ s.t. } A_k \neq \emptyset \\ T^1 & \text{otherwise,} \end{cases}$$

where T^0 is the first time the mouse pointer *enters* the image and T^1 is the last time it *leaves* it. Finally, we define the *trace segment* associated with a word m_j as the segment of the mouse trace spanned by the time interval $[\bar{t}_j^0, \bar{t}_j^1]$ (Fig. 3).

Automatic Quality Control. To ensure high-quality annotations, we devise an automatic quality control mechanism by leveraging the fact that we have a double source of voice transcriptions: the manual one given by the annotators (**m**) and the automatic one given by the ASR system (**a**, Fig. 3). We take their distance $D_{\mu^*}(\mathbf{a}, \mathbf{m})$ in the optimal alignment μ^* as a quality control metric (Eq. (1)). A high value of D_{μ^*} indicates large discrepancy between the two transcriptions, which could be caused by the annotator having wrongly transcribed the text, or due to the ASR failing to recognize the annotators' spoken words. In contrast, a low value of D_{μ^*} indicates that the transcription is corroborated by two sources. In practice, we manually analyzed a large number of annotations at different values of D_{μ^*} and choose a specific threshold below which essentially all transcriptions were correct. We discarded all annotations above this threshold.

In addition to this automatic quality control, we also evaluate the quality of the annotations in terms of semantic accuracy, visual grounding accuracy, and quality of manual voice transcription (Sect. 4.2).

4   Dataset Collection, Quality, and Statistics

4.1   Dataset Collection

Image Sources and Scale. We annotated a total of 848,749 images with Localized Narratives over 4 datasets: (i) COCO [8,31] (train and validation, 123k images); (ii) Flickr30k [59] (train, validation, and test, 32k); (iii) ADE20K [62] (train and validation, 20k); (iv) Open Images (full validation and test, 167k, and part of train, 504k). For Open Images, to enable cross-modal applications, we selected images for which object segmentations [5], bounding boxes or visual relationships [29] are already available. We annotated 5,000 randomly selected COCO images with replication 5 (i.e. 5 different annotators annotated each image). Beyond this, we prioritized having a larger set covered, so the rest of images were annotated with replication 1. All analyses in the remainder of this section are done on the full set of 849k images, unless otherwise specified.

Annotation Cost. Annotating one image with Localized Narratives takes 144.7 s on average. We consider this a relatively low cost given the amount of information harvested, and it allows data collection at scale. Manual transcription takes up the majority of the time (104.3 s, 72%), while the narration step only takes 40.4 s (28%). In the future, when ASR systems improve further, manual transcription could be skipped and Localized Narratives could become even faster thanks to our core idea of using speech.

To put our timings into perspective, we can roughly compare to Flickr30k Entities [40], which is the only work we are aware of that reports annotation times. They first manually identified which words constitute entities, which took 235 seconds per image. In a second stage, annotators drew bounding boxes for these selected entities, taking 408 s (8.7 entities per image on average). This yields a total of 643 seconds per image, without counting the time to write the actual captions (not reported). This is 4.4× slower than the total annotation cost of our method, which includes the grounding of 10.8 nouns per image and the writing of the caption. The Visual Genome [27] dataset was also annotated by a complex multi-stage pipeline, also involving drawing a bounding box for each phrase describing a region in the image.

4.2   Dataset Quality

To ensure high quality, Localized Narratives was made by 156 professional annotators working full time on this project. Annotator managers did frequent manual inspections to keep quality consistently high. In addition, we used an automatic quality control mechanism to ensure that the spoken and written transcriptions match (Sect. 3 – Automatic quality control). In practice, we placed a high quality bar, which resulted in discarding 23.5% of all annotations. Below we analyze the quality of the annotations that remained after this automatic discarding step (all dataset statistics reported in this paper are after this step too).

Semantic and Transcription Accuracy. In this section we quantify (i) how well the noun phrases and verbs in the caption correctly represent the objects in the image (Semantic accuracy) and (ii) how well the manually transcribed caption matches the voice recording (Transcription accuracy). We manually check every word in 100 randomly selected Localized Narratives on COCO and log each of these two types of errors. This was done carefully by experts (authors of this paper), not by the annotators themselves (hence an independent source).

In terms of semantic accuracy, we check every noun and verb in the 100 captions and assess whether that object or action is indeed present in the image. We allow generality up to a base class name (e.g. we count either "dog" or "Chihuahua" as correct for a Chihuahua in the image) and we strictly enforce correctness (e.g. we count "skating" as incorrect when the correct term is "snowboarding" or "bottle" in the case of a "jar"). Under these criteria, semantic accuracy is very high: 98.0% of the 1,582 nouns and verbs are accurate.

In terms of transcription accuracy, we listen to the voice recordings and compare them to the manual transcriptions. We count every instance of (i) a missing word in the transcription, (ii) an extra word in the transcription, and (iii) a word with typographical errors. We normalize these by the total number of words in the 100 captions (4,059). This results in 3.3% for type (i), 2.2% for (ii), and 1.1% for (iii), showing transcription accuracy is high.

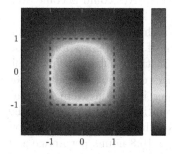

Fig. 4. Mouse trace segment locations on COCO with respect to the closest box of the relevant class (‑ ‑ ‑).

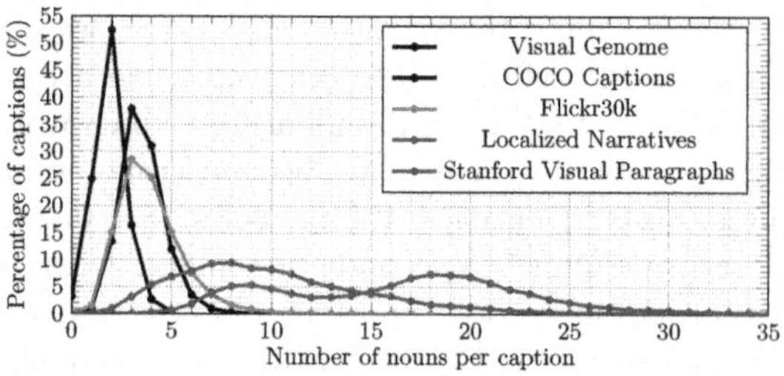

Fig. 5. Distribution of number of nouns per caption. As in Table 3, these counts are per individual caption.

Localization Accuracy. To analyze how well the mouse traces match the location of actual objects in the image, we extract all instances of any of the 80 COCO object classes in our captions (exact string matching, 600 classes in the case of Open Images). We recover 146,723 instances on COCO and 374,357 on Open Images train. We then associate each mouse trace segment to the closest ground-truth box of its corresponding class. Figure 4 displays the 2D histogram of the positions of all trace segment points with respect to the closest box (‑ ‑ ‑), normalized by box size for COCO. We observe that most of the trace points are within the correct bounding box (the figure for Open Images is near-identical, see supp. material Sect. 3).

We attribute the trace points that fall outside the box to two different effects. First, circling around the objects is commonly used by annotators (Fig. 1 and Fig. 6). This causes the mouse traces to be close to the box, but not inside it. Second, some annotators sometimes start moving the mouse before they describe the object, or vice versa. We see both cases as a research opportunity to better understand the connection between vision and language.

4.3   Dataset Statistics

Richness. The mean length of the captions we produced is 36.5 words (Table 2), substantially longer than all previous datasets, except Stanford Visual Paragraphs [26] (e.g. 3.5× longer than the individual COCO Captions [8]). Both Localized Narratives and Stanford Visual Paragraphs describe an image with a whole paragraph, as opposed to one sentence [8,24,27,35,46,59]. However, Localized Narratives additionally provide dense visual grounding via a mouse trace segment for each word, and has annotations for 40× more images than Stanford Visual Paragraphs (Table 2).

We also compare in terms of the average number of nouns, pronouns, adjectives, verbs, and adpositions (prepositions and postpositions, Table 3). We determined this using the spaCy [19] part-of-speech tagger. Localized Narratives has a higher occurrence per caption for each of these categories compared to most previous datasets, which indicates that our annotations provide richer use of natural language in connection to the images they describe.

Table 3. Richness of individual captions of Localized Narratives versus previous works. Please note that since COCO Captions and Flickr30K have replication 5 (and Visual Genome also has a high replication), counts *per image* would be higher in these datasets. However, many of them would be duplicates. We want to highlight the richness of captions as units and thus we show word counts averaged over *individual captions*.

Dataset	Words	Nouns	Pronouns	Adjectives	Adpositions	Verbs
Visual Genome [27]	5.1	1.9	0.0	0.6	0.7	0.3
COCO Captions [8]	10.5	3.6	0.2	0.8	1.7	0.9
Flickr30k [59]	12.4	3.9	0.3	1.1	1.8	1.4
Localized Narratives	36.5	10.8	3.6	1.6	4.7	4.2
Stanford Visual Paragraphs [26]	61.9	17.0	2.7	6.6	8.0	4.1

Diversity. To illustrate the diversity of our captions, we plot the distribution of the number of nouns per caption, and compare it to the distributions obtained over previous datasets (Fig. 5). We observe that the range of number of nouns is significantly higher in Localized Narratives than in COCO Captions, Flickr30k, Visual Genome, and comparable to Stanford Visual Paragraphs. This poses an additional challenge for captioning methods: automatically adapting the length of the descriptions to each image, as a function of the richness of its content. Beyond nouns, Localized Narratives provide visual grounding for every word (verbs, prepositions, etc.). This is especially interesting for relationship words, e.g. "woman holding ballon" (Fig. 1) or "with a hand under his chin" (Fig. 2(d)). This opens the door to a new venue of research: understanding how humans naturally ground visual relationships.

Ship   Open land with some grass on it   Main stairs

Fig. 6. Examples of mouse trace segments and their corresponding word(s) in the caption with different pointing styles: circling, scribbling, and underlining.

Diversity in Localized Narratives is present not only in the language modality, but also in the visual modality, such as the different ways to indicate the spatial location of objects in an image. In contrast to previous works, where the grounding is in the form of a bounding box, our instructions lets the annotator hover the mouse over the object in any way they feel natural. This leads to diverse styles of creating trace segments (Fig. 6): circling around an object (sometimes without even intersecting it), scribbling over it, underlining in case of text, etc. This diversity also presents another challenge: detect and adapt to different trace styles in order to make full use of them.

5   Controlled Image Captioning

We now showcase how localized narratives can be used for controlled image captioning. Controlled captioning was first proposed in [10] and enables a user to specify which parts of the image they want to be described, and in which order. In [10] the user input was in the form of user-provided bounding boxes. In this paper we enable controllability through a mouse trace, which provides a more intuitive and efficient user interface. One especially useful application for controlled captioning is assistive technology for people with imperfect vision [6,54,61], who could utilize the mouse to express their preferences in terms of how the image description should be presented.

Task Definition. Given both an image and a mouse trace, the goal is to produce an image caption which matches the mouse trace, i.e. it describes the image regions covered by the trace, and in the order of the trace. This task is illustrated by several qualitative examples of our controlled captioning system in Fig. 7. In both the image of the skiers and the factory, the caption correctly matches the given mouse trace: it describes the objects which were indicated by the user, in the order which the user wanted.

Method. We start from a state-of-the-art, transformer-based encoder-decoder image captioning model [3,7]. This captioning model consumes Faster-RCNN features [43] of the top 16 highest scored object proposals in the image. The Faster-RCNN module is pre-trained on Visual Genome [27] (excluding its intersection with COCO). The model uses these features to predict an image caption

 In this image there are **doughnuts** kept on the **grill**. In the front there is a **white color paper attached to the machine**. On the right side there is a **machine** which is kept on the floor. In the background there are group of people standing near the **table**. On the left side there is a person standing on the **floor**. In the background there is a wall on which there are different types of doughnuts. At the top there are **lights**.

 As we can see in the image there is a white color wall, few people here and there and there are food items.

 In this picture we can see a **person skiing on ski boards**, in the bottom there is snow, we can see some people standing and **sitting here**, at the bottom there is snow, we can see a flag here.

 In this image I can see ground full of snow and on it I can see few people are standing. Here I can see a flag and on it I can see something is written. I can also see something is written over here.

Fig. 7. Qualitative results for controlled image captioning. Gradient ▬▬■ indicates time. Captioning controlled by mouse traces (left) and without traces (right). The latter misses important objects: e.g. skiers in the sky, doughnuts – all in bold.

based on an attention model, inspired by the Bottom-Up Top-Down approach of [3]. This model is state of the art for standard image captioning, i.e. it produces captions given images alone as input (Fig. 8(a)).

We modify this model to also input the mouse trace, resulting in a model that consumes four types of features both at training time and test time: (i) Faster R-CNN features of the automatically-detected top object proposals, representing their semantic information; (ii) the coordinate and size features of these proposals, representing the location of the detected objects. (iii) the total time duration of the mouse trace, capturing information about the expected length of the full image description. (iv) the position of the mouse trace as it moves over the image, representing the visual grounding. To create this representation, we first divide the mouse trace evenly into pseudo-segments based on the prior median word duration (0.4 s over the whole training set). We then represent each pseudo-segment by its encapsulating bounding box, resulting in a set of features which take the same form as (ii). This new model takes an image *plus* a mouse trace as input and produces the caption that the user is interested in. More technical details in the supp. material, Sect. 4.

Evaluation. Our first metric is the standard ROUGE-L [30]. This metric determines the longest common subsequence (LCS) of words between the predicted caption and the reference caption, and calculates the F1-score (harmonic mean over precision and recall of words in the LCS). This means ROUGE-L explicitly measures word order. We also measure the F1 score of ROUGE-1, which we term ROUGE-1-F1. This measures the F1-score w.r.t. co-occurring words. Hence ROUGE-1-F1 is the orderless counterpart of ROUGE-L and enables us to separate the effects of caption completeness (the image parts which the user wanted to be described) and word order (the order in which the user wanted the image to be described). For completeness we also report other standard captioning metrics: BLEU-1, BLEU-4 [39], CIDEr-D [50], and SPICE [2].

Output 1: Caption
There is a fire hydrant and this is road.

Output 2: Controlled caption
In this image, we can see a platform, there is a yellow color pot on it. Left side, there is a road, car, few trees we can see on the right side.

Output 3: Controlled caption
This image consists of a car parked on the road. To the top left, there is a car. To the right, there is a footpath on which a fire hydrant is kept.

Fig. 8. Qualitative results for controlled image captioning. Standard (a) versus controlled captioning (b) and (c). (a) misses important objects such as the car or the footpath. In (b) and (c) the controlled output captions adapt to the order of the objects defined by the trace. Gradient ■■■■ indicates time. More in the supp. mat. Sect. 2.

For all measures, a higher number reflects a better agreement between the caption produced by the model and the ground-truth caption written by the annotator.

We observe that in standard image captioning tasks there typically are multiple reference captions to compare to [8,50,59], since that task is ambiguous: it is unclear what image parts should be described and in which order. In contrast, our controlled image captioning task takes away both types of ambiguity, resulting in a much better defined task. As such, in this evaluation we compare to a single reference only: given an image plus a human-provided mouse trace, we use its corresponding human-provided caption as reference.

Results. We perform experiments on the Localized Narratives collected on COCO images, using the standard 2017 training and validation splits. To get a feeling of what a trace can add to image captioning, we first discuss the qualitative examples in Fig. 7 and 8. First of all, the trace focuses the model attention on specific parts of the image, leading it to mention objects which would otherwise be missed: In the top-left image of Fig. 7, the trace focuses attention on the skiers, which are identified as such (in contrast to the top-right). Similarly, the top-left and right of Fig. 7, using the trace results in focusing on specific details which leads to more complete and more fine-grained descriptions (e.g. doughnuts, grill, machine, lights). Finally in Fig. 8a, the standard captioning model misses the car since it is not prominent in the image. In Fig. 8b and c instead, the augmented model sees both traces going over the car and produces a caption including it. In this same figure, we can also see that different traces

Table 4. Controlled image captioning results on the COCO validation set, versus standard (non-controlled) captioning, and two ablations.

Method	features	ROUGE-L	ROUGE-1-F1	BLEU-1	BLEU-4	CIDEr-D	SPICE
Standard captioning [3,7]	i	0.317	0.479	0.322	0.081	0.293	0.257
+ proposal locations	i+ii	0.318	0.482	0.323	0.082	0.295	0.257
+ mouse trace duration	i+ii+iii	0.334	0.493	0.372	0.097	0.373	0.265
Controlled captioning	i+ii+iii+iv	0.483	0.607	0.522	0.246	1.065	0.365

lead to different captions. These results suggests that conditioning on the trace helps with covering the image more completely and highlighting specific objects within it. At the same time, we can see in all examples that the trace order maps nicely to the word order in the caption, which is the order the user wanted.

Table 4 shows quantitative results. Compared to standard captioning [3,7], all metrics improve significantly when doing controlled captioning using the mouse trace. BLEU-4 and CIDEr-D are particularly affected and improve by more than 3×. ROUGE-1-F1 increased from 0.479 for standard captioning to 0.607 for controlled captioning using the full mouse trace. Since ROUGE-1-F1 ignores word order, this increase is due to the *completeness* of the caption only: it indicates that using the mouse trace enables the system to better describe those parts of the image which were indicated by the user.

Switching from ROUGE-1-F1 to ROUGE-L imposes a word order. The standard captioning model yields a ROUGE-L of 0.317, a drop of 34% compared to ROUGE-1-F1. Since standard captioning does not input any particular order within the image (but does use a linguistically plausible ordering), this drop can be seen as a baseline for not having information on the order in which the image should be described. When using the mouse trace, the controlled captioning model yields a ROUGE-L of 0.483, which is a much smaller drop of 20%. This demonstrates quantitatively that our controlled captioning model successfully exploits the input trace to determine the order in which the user wanted the image to be described. Overall, the controlled captioning model outperforms the standard captioning model by 0.166 ROUGE-L on this task (0.483 vs 0.317).

Ablations. We perform two ablations to verify whether most improvements indeed come from the mouse trace itself, as opposed to the other features we added. Starting from standard captioning, we add the locations of the object proposals from which the model extracts visual features (Table 4, "+ proposal locations", feature (ii)). This has negligible effects on performance, suggesting that this model does not benefit from knowing where in the image its appearance features (i) came from. Next, we add the trace time duration (Table 4, "+ mouse trace duration"). This gives an indication of how long the caption the user wants should be. This brings minor improvements only. Hence, most improvements come when using the full mouse trace, demonstrating that most information comes from the location and order of the mouse trace (Table 4, controlled captioning).

Summary. To summarize, we demonstrated that using the mouse trace leads to large improvements when compared to a standard captioning model, for the task of controlled captioning. Importantly, we do not claim the resulting captions are better in absolute terms. Instead, they are better fitting what the user wanted, in terms of which parts of the image are described and in which order.

6   Conclusions

This paper introduces Localized Narratives, an efficient way to collect image captions in which every single word is visually grounded by a mouse trace. We annotated 849k images with Localized Narratives. Our analysis shows that our data is rich and provides accurate grounding. We demonstrate the utility of our data through controlled image captioning using the mouse trace.

References

1. Amodei, D., et al.: Deep speech 2: end-to-end speech recognition in English and Mandarin. In: ICML (2016)
2. Anderson, P., Fernando, B., Johnson, M., Gould, S.: SPICE: semantic propositional image caption evaluation. In: ECCV (2016)
3. Anderson, P., et al.: Bottom-up and top-down attention for image captioning and visual question answering. In: CVPR (2018)
4. Antol, S., et al.: VQA: visual question answering. In: ICCV (2015)
5. Benenson, R., Popov, S., Ferrari, V.: Large-scale interactive object segmentation with human annotators. In: CVPR (2019)
6. Bigham, J.P., et al.: VizWiz: nearly real-time answers to visual questions. In: Proceedings of the 23nd Annual ACM Symposium on User Interface Software and Technology (2010)
7. Changpinyo, S., Pang, B., Sharma, P., Soricut, R.: Decoupled box proposal and featurization with ultrafine-grained semantic labels improve image captioning and visual question answering. In: EMNLP-IJCNLP (2019)
8. Chen, X., et al.: Microsoft COCO captions: data collection and evaluation server. arXiv (2015)
9. Cirik, V., Morency, L.P., Berg-Kirkpatrick, T.: Visual referring expression recognition: what do systems actually learn? In: NAACL (2018)
10. Cornia, M., Baraldi, L., Cucchiara, R.: Show, control and tell: a framework for generating controllable and grounded captions. In: CVPR (2019)
11. Dai, D.: Towards cost-effective and performance-aware vision algorithms. Ph.D. thesis, ETH Zurich (2016)
12. Damen, D., et al.: The EPIC-KITCHENS dataset: collection, challenges and baselines. IEEE Trans. PAMI (2020)
13. Dogan, P., Sigal, L., Gross, M.: Neural sequential phrase grounding (seqground). In: CVPR (2019)
14. Google cloud speech-to-text API. https://cloud.google.com/speech-to-text/
15. Graves, A., Mohamed, A.R., Hinton, G.: Speech recognition with deep recurrent neural networks. In: ICASSP (2013)
16. Gygli, M., Ferrari, V.: Efficient object annotation via speaking and pointing. In: IJCV (2019)

17. Gygli, M., Ferrari, V.: Fast object class labelling via speech. In: CVPR (2019)
18. Harwath, D., Recasens, A., Surís, D., Chuang, G., Torralba, A., Glass, J.: Jointly discovering visual objects and spoken words from raw sensory input. In: ECCV (2018)
19. Honnibal, M., Montani, I.: spaCy 2: natural language understanding with Bloom embeddings, convolutional neural networks and incremental parsing (2017). spacy.io
20. Hudson, D.A., Manning, C.D.: GQA: a new dataset for real-world visual reasoning and compositional question answering. In: CVPR (2019)
21. Johnson, J., Karpathy, A., Fei-Fei, L.: Densecap: fully convolutional localization networks for dense captioning. In: CVPR (2016)
22. Kahneman, D.: Attention and effort. Citeseer (1973)
23. Kalchbrenner, N., et al.: Efficient neural audio synthesis. In: ICML (2018)
24. Kazemzadeh, S., Ordonez, V., Matten, M., Berg, T.: Referitgame: referring to objects in photographs of natural scenes. In: EMNLP (2014)
25. Kim, D.J., Choi, J., Oh, T.H., Kweon, I.S.: Dense relational captioning: triple-stream networks for relationship-based captioning. In: CVPR (2019)
26. Krause, J., Johnson, J., Krishna, R., Fei-Fei, L.: A hierarchical approach for generating descriptive image paragraphs. In: CVPR (2017)
27. Krishna, R., et al.: Visual genome: connecting language and vision using crowd-sourced dense image annotations. IJCV **123**(1), 32–73 (2017)
28. Kruskal, J.B., Liberman, M.: The symmetric time-warping problem: from continuous to discrete. In: Time Warps, String Edits, and Macromolecules - The Theory and Practice of Sequence Comparison, chap. 4. CSLI Publications (1999)
29. Kuznetsova, A., et al.: The Open Images Dataset V4: Unified image classification, object detection, and visual relationship detection at scale. arXiv preprint arXiv:1811.00982 (2018)
30. Lin, C.Y.: ROUGE: a package for automatic evaluation of summaries. In: Text Summarization Branches Out (2004)
31. Lin, T.Y., et al.: Microsoft COCO: common objects in context. In: ECCV (2014)
32. Liu, C., Mao, J., Sha, F., Yuille, A.: Attention correctness in neural image captioning. In: AAAI (2017)
33. Lu, J., Yang, J., Batra, D., Parikh, D.: Neural baby talk. In: CVPR (2018)
34. Malinowski, M., Rohrbach, M., Fritz, M.: Ask your neurons: a neural-based approach to answering questions about images. In: ICCV (2015)
35. Mao, J., Huang, J., Toshev, A., Camburu, O., Yuille, A.L., Murphy, K.: Generation and comprehension of unambiguous object descriptions. In: CVPR (2016)
36. Mehri, S., et al.: Samplernn: an unconditional end-to-end neural audio generation model. In: ICLR (2017)
37. Oord, A.V.D., et al.: Wavenet: a generative model for raw audio. arXiv 1609.03499 (2016)
38. Oviatt, S.: Multimodal interfaces. In: The Human-Computer Interaction Handbook: Fundamentals, Evolving Technologies and Emerging Applications (2003)
39. Papineni, K., Roukos, S., Ward, T., Zhu, W.J.: Bleu: a method for automatic evaluation of machine translation. In: ACL (2002)
40. Plummer, B.A., Wang, L., Cervantes, C.M., Caicedo, J.C., Hockenmaier, J., Lazebnik, S.: Flickr30k entities: collecting region-to-phrase correspondences for richer image-to-sentence models. IJCV **123**(1), 74–93 (2017)
41. Ravanelli, M., Parcollet, T., Bengio, Y.: The Pytorch-Kaldi speech recognition toolkit. In: ICASSP (2019)

42. Reed, S.E., Akata, Z., Mohan, S., Tenka, S., Schiele, B., Lee, H.: Learning what and where to draw. In: NeurIPS, pp. 217–225 (2016)
43. Ren, S., He, K., Girshick, R., Sun, J.: Faster R-CNN: towards real-time object detection with region proposal networks. In: NeurIPS (2015)
44. Rohrbach, A., Hendricks, L.A., Burns, K., Darrell, T., Saenko, K.: Object hallucination in image captioning. In: EMNLP (2018)
45. Selvaraju, R.R., et al.: Taking a HINT: leveraging explanations to make vision and language models more grounded. In: ICCV (2019)
46. Sharma, P., Ding, N., Goodman, S., Soricut, R.: Conceptual captions: a cleaned, hypernymed, image alt-text dataset for automatic image captioning. In: ACL (2018)
47. Tan, F., Feng, S., Ordonez, V.: Text2scene: generating compositional scenes from textual descriptions. In: CVPR (2019)
48. Vaidyanathan, P., Prud, E., Pelz, J.B., Alm, C.O.: SNAG : spoken narratives and gaze dataset. In: ACL (2018)
49. Vasudevan, A.B., Dai, D., Van Gool, L.: Object referring in visual scene with spoken language. In: CVPR (2017)
50. Vedantam, R., Lawrence Zitnick, C., Parikh, D.: CIDEr: consensus-based image description evaluation. In: CVPR (2015)
51. Vinyals, O., Toshev, A., Bengio, S., Erhan, D.: Show and tell: a neural image caption generator. In: CVPR (2015)
52. Vinyals, O., Toshev, A., Bengio, S., Erhan, D.: Show and tell: lessons learned from the 2015 MSCOCO image captioning challenge. IEEE Trans. PAMI **39**(4), 652–663 (2016)
53. Website: Localized Narratives Data and Visualization (2020). https://google.github.io/localized-narratives
54. Wu, S., Wieland, J., Farivar, O., Schiller, J.: Automatic alt-text: computer-generated image descriptions for blind users on a social network service. In: Conference on Computer Supported Cooperative Work and Social Computing (2017)
55. Xu, K., et al.: Show, attend and tell: neural image caption generation with visual attention. In: ICML (2015)
56. Yan, S., Yang, H., Robertson, N.: ParaCNN: visual paragraph generation via adversarial twin contextual CNNs. arXiv (2020)
57. Yin, G., Liu, B., Sheng, L., Yu, N., Wang, X., Shao, J.: Semantics disentangling for text-to-image generation. In: CVPR (2019)
58. Yin, G., Sheng, L., Liu, B., Yu, N., Wang, X., Shao, J.: Context and attribute grounded dense captioning. In: CVPR (2019)
59. Young, P., Lai, A., Hodosh, M., Hockenmaier, J.: From image descriptions to visual denotations: new similarity metrics for semantic inference over event descriptions. TACL **2**, 67–78 (2014)
60. Yu, J., Li, J., Yu, Z., Huang, Q.: Multimodal transformer with multi-view visual representation for image captioning. arXiv 1905.07841 (2019)
61. Zhao, Y., Wu, S., Reynolds, L., Azenkot, S.: The effect of computer-generated descriptions on photo-sharing experiences of people with visual impairments. ACM Hum.-Comput. Interact. **1** (2017)
62. Zhou, B., Zhao, H., Puig, X., Fidler, S., Barriuso, A., Torralba, A.: Semantic understanding of scenes through the ADE20K dataset. IJCV **127**(3), 302–321 (2019)
63. Zhou, L., Kalantidis, Y., Chen, X., Corso, J.J., Rohrbach, M.: Grounded video description. In: CVPR (2019)
64. Ziegler, Z.M., Melas-Kyriazi, L., Gehrmann, S., Rush, A.M.: Encoder-agnostic adaptation for conditional language generation. arXiv (2019)

Adversarial T-Shirt!
Evading Person Detectors in a Physical World

Kaidi Xu[1], Gaoyuan Zhang[2], Sijia Liu[2], Quanfu Fan[2], Mengshu Sun[1], Hongge Chen[3], Pin-Yu Chen[2], Yanzhi Wang[1], and Xue Lin[1(✉)]

[1] Northeastern University, Boston, USA
xue.lin@northeastern.edu
[2] MIT-IBM Watson AI Lab, IBM Research, Cambridge, USA
[3] Massachusetts Institute of Technology, Cambridge, USA

Abstract. It is known that deep neural networks (DNNs) are vulnerable to adversarial attacks. The so-called *physical adversarial examples* deceive DNN-based decision makers by attaching adversarial patches to real objects. However, most of the existing works on physical adversarial attacks focus on static objects such as glass frames, stop signs and images attached to cardboard. In this work, we propose *Adversarial T-shirts*, a robust physical adversarial example for evading person detectors even if it could undergo non-rigid deformation due to a moving person's pose changes. To the best of our knowledge, this is the first work that models the effect of deformation for designing physical adversarial examples with respect to non-rigid objects such as T-shirts. We show that the proposed method achieves 74% and 57% attack success rates in the digital and physical worlds respectively against YOLOv2. In contrast, the state-of-the-art physical attack method to fool a person detector only achieves 18% attack success rate. Furthermore, by leveraging min-max optimization, we extend our method to the ensemble attack setting against two object detectors YOLO-v2 and Faster R-CNN simultaneously.

Keywords: Physical adversarial attack · Object detection · Deep learning

1 Introduction

The vulnerability of deep neural networks (DNNs) against adversarial attacks (namely, perturbed inputs deceiving DNNs) has been found in applications spanning from image classification to speech recognition [2,6,21,32–34,37]. Early works studied adversarial examples only in the digital space. Recently, some

Electronic supplementary material The online version of this chapter (https://doi.org/10.1007/978-3-030-58558-7_39) contains supplementary material, which is available to authorized users.

works showed that it is possible to create adversarial perturbations on physical objects and fool DNN-based decision makers under a variety of real-world conditions [1,5,7,14,15,20,25,28,30]. The design of *physical adversarial attacks* helps to evaluate the robustness of DNNs deployed in real-life systems, e.g., autonomous vehicles and surveillance systems. However, most of the studied physical adversarial attacks encounter two limitations: a) the physical objects are usually considered being *static*, and b) the possible *deformation* of adversarial pattern attached to a moving object (e.g., due to pose change of a moving person) is commonly neglected. In this paper, we propose a new type of physical adversarial attack, *adversarial T-shirt*, to evade DNN-based person detectors when a person wears the adversarial T-shirt; see the second row of Fig. 1 for illustrative examples.

Related Work. Most of the existing physical adversarial attacks are generated against image classifiers and object detectors. In [28], a face recognition system is fooled by a real eyeglass frame designed under a crafted adversarial pattern. In [14], a stop sign is misclassified by adding black or white stickers on it against the image classification system. In [20], an image classifier is fooled by placing a crafted sticker at the lens of a camera. In [1], a so-called Expectation over Transformation (EoT) framework was proposed to synthesize adversarial examples robust to a set of physical transformations such as rotation, translation, contrast, brightness, and random noise. Moreover, the crafted adversarial examples on the rigid objects can be designed in camouflage style [35] or natural style [11] that appear legitimate to human observers in the real world. Compared to attacking image classifiers, generating physical adversarial attacks against object detectors is more involved. For example, the adversary is required to mislead the bounding box detector of an object when attacking YOLOv2 [26] and SSD [24]. A well-known success of such attacks in the physical world is the generation of adversarial stop sign [15], which deceives state-of-the-art object detectors such as YOLOv2 and Faster R-CNN [27].

The most relevant approach to ours is the work of [30], which demonstrates that a person can evade a detector by holding a cardboard with an adversarial patch. However, such a physical attack restricts the adversarial patch to be attached to a *rigid* carrier (namely, cardboard), and is different from our setting here where the generated adversarial pattern is directly printed on a T-shirt. We show that the attack proposed by [30] becomes ineffective when the adversarial patch is attached to a T-shirt (rather than a cardboard) and worn by a moving person (see the fourth row of Fig. 1). At the technical side, different from [30] we propose a thin plate spline (TPS) based transformer to model deformation of non-rigid objects, and develop an ensemble physical attack that fools object detectors YOLOv2 and Faster R-CNN simultaneously. We highlight that our proposed adversarial T-shirt is not just a T-shirt with printed adversarial patch for clothing fashion, it is a physical adversarial wearable designed for evading person detectors in the real world.

Our work is also motivated by the importance of person detection on intelligent surveillance. DNN-based surveillance systems have significantly advanced the

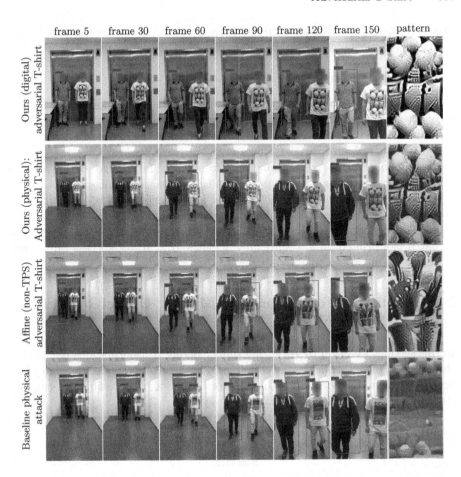

Fig. 1. Evaluation of the effectiveness of adversarial T-shirts to evade person detection by YOLOv2. Each row corresponds to a specific attack method while each column except the last one shows an individual frame in a video. The last column shows the adversarial patterns applied to the T-shirts. At each frame, there are two persons, one of whom wears the adversarial T-shirt. First row: digital adversarial T-shirt generated using TPS. Second row: physical adversarial T-shirt generated using TPS. Third row: physical adversarial T-shirt generated using affine transformation (namely, in the absence of TPS). Fourth row: T-shirt with physical adversarial patch considered in [30] to evade person detectors.

field of object detection [17,18]. Efficient object detectors such as Faster R-CNN [27], SSD [24], and YOLOv2 [26] have been deployed for human detection. Thus, one may wonder whether or not there exists a security risk for intelligent surveillance systems caused by adversarial human wearables, e.g., adversarial T-shirts. However, paralyzing a person detector in the physical world requires substantially more challenges such as low resolution, pose changes and occlusions. The success of our adversarial T-shirt against real-time person detectors offers new insights for designing practical physical-world adversarial human wearables.

Contributions. We summarize our contributions as follows:

- We develop a TPS-based transformer to model the temporal deformation of an adversarial T-shirt caused by pose changes of a moving person. We also show the importance of such non-rigid transformation to ensuring the effectiveness of adversarial T-shirts in the physical world.
- We propose a general optimization framework for design of adversarial T-shirts in both single-detector and multiple-detector settings.
- We conduct experiments in both digital and physical worlds and show that the proposed adversarial T-shirt achieves 74% and 57% attack success rates respectively when attacking YOLOv2. By contrast, the physical adversarial patch [30] printed on a T-shirt only achieves 18% attack success rate. Some of our results are highlighted in Fig. 1.

2   Modeling Deformation of a Moving Object by Thin Plate Spline Mapping

In this section, we begin by reviewing some existing transformations required in the design of physical adversarial examples. We then elaborate on the Thin Plate Spline (TPS) mapping we adopt in this work to model the possible deformation encountered by a moving and non-rigid object.

Let $\mathbf{x}$ be an original image (or a video frame), and $t(\cdot)$ be the physical transformer. The transformed image $\mathbf{z}$ under t is given by

$$\mathbf{z} = t(\mathbf{x}). \tag{1}$$

Existing Transformations. In [1], the parametric transformers include scaling, translation, rotation, brightness and additive Gaussian noise; see details in [1, Appendix D]. In [23], the geometry and lighting transformations are studied via parametric models. Other transformations including perspective transformation, brightness adjustment, resampling (or image resizing), smoothing and saturation are considered in [9,29]. All the existing transformations are included in our library of physical transformations. However, they are not sufficient to model the cloth deformation caused by pose change of a moving person. For example, the second and third rows of Fig. 1 show that adversarial T-shirts designed against only existing physical transformations yield low attack success rates.

TPS Transformation for Cloth Deformation. A person's movement can result in significantly and constantly changing wrinkles (aka deformations) in her clothes. This makes it challenging to develop an adversarial T-shirt effectively in the real world. To circumvent this challenge, we employ TPS mapping [4] to model the cloth deformation caused by human body movement. TPS has been widely used

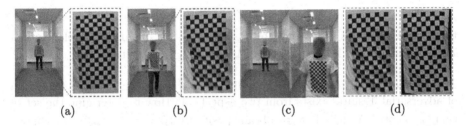

$$\text{(a)} \qquad\qquad \text{(b)} \qquad\qquad \text{(c)} \qquad\qquad \text{(d)}$$

Fig. 2. Generation of TPS. (a) and (b): Two frames with checkerboard detection results. (c): Anchor point matching process between two frames (d): Real-world close deformation in (b) versus the synthesized TPS transformation (right plot).

as the non-rigid transformation model in image alignment and shape matching [19]. It consists of an affine component and a non-affine warping component. We will show that the non-linear warping part in TPS can provide an effective means of modeling cloth deformation for learning adversarial patterns of non-rigid objects.

TPS learns a parametric deformation mapping from an original image $\mathbf{x}$ to a target image $\mathbf{z}$ through a set of control points with given positions. Let $\mathbf{p} := (\phi, \psi)$ denote the 2D location of an image pixel. The deformation from $\mathbf{x}$ to $\mathbf{z}$ is then characterized by the *displacement* of every pixel, namely, how a pixel at $\mathbf{p}^{(x)}$ on image $\mathbf{x}$ changes to the pixel on image $\mathbf{z}$ at $\mathbf{p}^{(z)}$, where $\phi^{(z)} = \phi^{(x)} + \Delta_\phi$ and $\psi^{(z)} = \psi^{(x)} + \Delta_\psi$, and Δ_ϕ and Δ_ψ denote the pixel displacement on image $\mathbf{x}$ along ϕ direction and ψ direction, respectively.

Given a set of n control points with locations $\{\hat{\mathbf{p}}_i^{(x)} := (\hat{\phi}_i^{(x)}, \hat{\psi}_i^{(x)})\}_{i=1}^n$ on image $\mathbf{x}$, TPS provides a parametric model of pixel displacement when mapping $\mathbf{p}^{(x)}$ to $\mathbf{p}^{(z)}$ [8]

$$\Delta(\mathbf{p}^{(x)}; \boldsymbol{\theta}) = a_0 + a_1 \phi^{(x)} + a_2 \psi^{(x)} + \sum_{i=1}^{n} c_i U(\|\hat{\mathbf{p}}_i^{(x)} - \mathbf{p}^{(x)}\|_2), \qquad (2)$$

where $U(r) = r^2 \log(r)$ and $\boldsymbol{\theta} = [\mathbf{c}; \mathbf{a}]$ are the TPS parameters, and $\Delta(\mathbf{p}^{(x)}; \boldsymbol{\theta})$ represents the displacement along either ϕ or ψ direction.

Moreover, given the locations of control points on the transformed image $\mathbf{z}$ (namely, $\{\hat{\mathbf{p}}_i^{(z)}\}_{i=1}^n$), TPS resorts to a regression problem to determine the parameters $\boldsymbol{\theta}$ in (2). The regression objective is to minimize the distance between $\{\Delta_\phi(\mathbf{p}_i^{(x)}; \boldsymbol{\theta}_\phi)\}_{i=1}^n$ and $\{\hat{\Delta}_{\phi,i} := \hat{\phi}_i^{(z)} - \hat{\phi}_i^{(x)}\}_{i=1}^n$ along the ϕ direction, and the distance between $\{\Delta_\psi(\mathbf{p}_i^{(x)}; \boldsymbol{\theta}_\psi)\}_{i=1}^n$ and $\{\hat{\Delta}_{\psi,i} := \hat{\psi}_i^{(z)} - \hat{\psi}_i^{(x)}\}_{i=1}^n$ along the ψ direction, respectively. Thus, TPS (2) is applied to coordinate ϕ and ψ separately (corresponding to parameters $\boldsymbol{\theta}_\phi$ and $\boldsymbol{\theta}_\psi$). The regression problem can be solved by the following linear system of equations [10]

$$\begin{bmatrix} \mathbf{K} & \mathbf{P} \\ \mathbf{P}^T & \mathbf{0}_{3\times3} \end{bmatrix} \boldsymbol{\theta}_\phi = \begin{bmatrix} \hat{\Delta}_\phi \\ \mathbf{0}_{3\times1} \end{bmatrix}, \quad \begin{bmatrix} \mathbf{K} & \mathbf{P} \\ \mathbf{P}^T & \mathbf{0}_{3\times3} \end{bmatrix} \boldsymbol{\theta}_\psi = \begin{bmatrix} \hat{\Delta}_\psi \\ \mathbf{0}_{3\times1} \end{bmatrix}, \qquad (3)$$

where the (i, j)th element of $\mathbf{K} \in \mathbb{R}^{n \times n}$ is given by $K_{ij} = U(\|\hat{\mathbf{p}}_i^{(x)} - \hat{\mathbf{p}}_j^{(x)}\|_2)$, the ith row of $\mathbf{P} \in \mathbb{R}^{n \times 3}$ is given by $P_i = [1, \hat{\phi}_i^{(x)}, \hat{\psi}_i^{(x)}]$, and the ith elements of $\hat{\boldsymbol{\Delta}}_\phi \in \mathbb{R}^n$ and $\hat{\boldsymbol{\Delta}}_\psi \in \mathbb{R}^n$ are given by $\hat{\Delta}_{\phi,i}$ and $\hat{\Delta}_{\psi,i}$, respectively.

Non-trivial Application of TPS. The difficulty of implementing TPS for design of adversarial T-shirts exists from two aspects: 1) How to determine the set of control points? And 2) how to obtain positions $\{\hat{\mathbf{p}}_i^{(x)}\}$ and $\{\hat{\mathbf{p}}_i^{(z)}\}$ of control points aligned between a pair of video frames $\mathbf{x}$ and $\mathbf{z}$?

To address the first question, we print a *checkerboard* on a T-shirt and use the camera calibration algorithm [16,36] to detect points at the intersection between every two checkerboard grid regions. These successfully detected points are considered as the control points of one frame. Figure 2-(a) shows the checkerboard-printed T-shirt, together with the detected intersection points. Since TPS requires a set of control points *aligned* between two frames, the second question on point matching arises. The challenge lies in the fact that the control points detected at one video frame are different from those at another video frame (e.g., due to missing detection). To address this issue, we adopt a 2-stage procedure, *coordinate system alignment* followed by *point aliment*, where the former refers to conducting a perspective transformation from one frame to the other, and the latter finds the matched points at two frames through the nearest-neighbor method. We provide an illustrative example in Fig. 2-(c). We refer readers to Appendix A for more details about our method.

3   Generation of Adversarial T-Shirt: An Optimization Perspective

In this section, we begin by formalizing the problem of adversarial T-shirt and introducing notations used in our setup. We then propose to design a *universal* perturbation used in our adversarial T-shirt to deceive a *single* object detector. We lastly propose a min-max (robust) optimization framework to design the universal adversarial patch against *multiple* object detectors.

Let $\mathcal{D} := \{\mathbf{x}_i\}_{i=1}^M$ denote M video frames extracted from one or multiple given videos, where $\mathbf{x}_i \in \mathbb{R}^d$ denotes the ith frame. Let $\boldsymbol{\delta} \in \mathbb{R}^d$ denote the universal adversarial perturbation applied to $\mathcal{D}$. The adversarial T-shirt is then characterized by $M_{c,i} \circ \boldsymbol{\delta}$, where $M_{c,i} \in \{0,1\}^d$ is a bounding box encoding the position of the cloth region to be perturbed at the ith frame, and $\circ$ denotes element-wise product. *The goal of adversarial T-shirt is to design $\boldsymbol{\delta}$ such that the perturbed frames of $\mathcal{D}$ are mis-detected by object detectors.*

Fooling a Single Object Detector. We generalize the Expectation over Transformation (EoT) method in [3] for design of adversarial T-shirts. Note that different from the conventional EoT, a transformers' composition is required for generating an adversarial T-shirt. For example, a perspective transformation on the bounding box of the T-shirt is composited with an TPS transformation

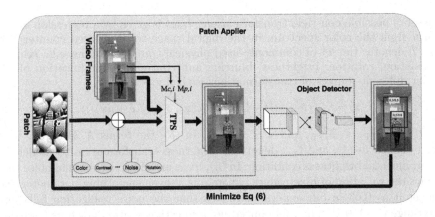

Fig. 3. Overview of the pipeline to generate adversarial T-shirts. First, the video frames containing a person whom wears the T-shirt with printed checkerboard pattern are used as training data. Second, the universal adversarial perturbation (to be designed) applies to the cloth region by taking into account different kinds of transformations. Third, the adversarial perturbation is optimized through problem (6) by minimizing the largest bounding-box probability belonging to the 'person' class. The optimization procedure is performed as a closed loop through back-propagation.

applied to the cloth region. Let us begin by considering two video frames, an anchor image $\mathbf{x}_0$ (e.g., the first frame in the video) and a target image $\mathbf{x}_i$ for $i \in [M]^1$. Given the bounding boxes of the person ($M_{p,0} \in \{0,1\}^d$) and the T-shirt ($M_{c,0} \in \{0,1\}^d$) at $\mathbf{x}_0$, we apply the perspective transformation from $\mathbf{x}_0$ to $\mathbf{x}_i$ to obtain the bounding boxes $M_{p,i}$ and $M_{c,i}$ at image $\mathbf{x}_i$. In the *absence* of physical transformations, the perturbed image $\mathbf{x}'_i$ with respect to (w.r.t.) $\mathbf{x}_i$ is given by

$$\mathbf{x}'_i = \underbrace{(1 - M_{p,i}) \circ \mathbf{x}_i}_{A} + \underbrace{M_{p,i} \circ \mathbf{x}_i}_{B} - \underbrace{M_{c,i} \circ \mathbf{x}_i}_{C} + \underbrace{M_{c,i} \circ \boldsymbol{\delta}}_{D}, \tag{4}$$

where the term A denotes the background region outside the bouding box of the person, the term B is the person-bounded region, the term C erases the pixel values within the bounding box of the T-shirt, and the term D is the newly introduced additive perturbation. In (4), the prior knowledge on $M_{p,i}$ and $M_{c,i}$ is acquired by person detector and manual annotation, respectively. Without taking into account physical transformations, Eq. (4) simply reduces to the conventional formulation of adversarial example $(1 - M_{c,i}) \circ \mathbf{x}_i + M_{c,i} \circ \boldsymbol{\delta}$.

Next, we consider *three main types* of physical transformations: a) TPS transformation $t_{\mathrm{TPS}} \in \mathcal{T}_{\mathrm{TPS}}$ applying to the adversarial perturbation $\boldsymbol{\delta}$ for modeling the effect of cloth deformation, b) physical color transformation t_{color} which converts digital colors to those printed and visualized in the physical world, and c) conventional physical transformation $t \in \mathcal{T}$ applying to the region within the person's bounding box, namely, $(M_{p,i} \circ \mathbf{x}_i - M_{c,i} \circ \mathbf{x}_i + M_{c,i} \circ \boldsymbol{\delta})$. Here $\mathcal{T}_{\mathrm{TPS}}$ denotes

---

[1] $[M]$ denotes the integer set $\{1, 2, \ldots, M\}$.

the set of possible non-rigid transformations, t_{color} is given by a regression model learnt from the color spectrum in the digital space to its printed counterpart, and $\mathcal{T}$ denotes the set of commonly-used physical transformations, e.g., scaling, translation, rotation, brightness, blurring and contrast. A modification of (4) under different sources of transformations is then given by

$$\mathbf{x}_i' = t_{\text{env}} \left(\text{A} + t \left(\text{B} - \text{C} + t_{\text{color}}(M_{c,i} \circ t_{\text{TPS}}(\boldsymbol{\delta} + \mu\mathbf{v}))\right)\right) \tag{5}$$

for $t \in \mathcal{T}$, $t_{\text{TPS}} \in \mathcal{T}_{\text{TPS}}$, and $\mathbf{v} \sim \mathcal{N}(0,1)$. In (5), the terms A, B and C have been defined in (4), and t_{env} denotes a brightness transformation to model the environmental brightness condition. In (5), $\mu\mathbf{v}$ is an additive Gaussian noise that allows the variation of pixel values, where μ is a given smoothing parameter and we set it as 0.03 in our experiments such that the noise realization falls into the range $[-0.1, 0.1]$. The randomized noise injection is also known as Gaussian smoothing [12], which makes the final objective function smoother and benefits the gradient computation during optimization.

Different with the prior works, e.g. [13,28], established a non-printability score (NPS) to measure the distance between the designed perturbation vector and a library of printable colors, we propose to model the color transformer t_{color} using a quadratic polynomial regression. The detailed color mapping is showed in Appendix B.

With the aid of (5), the EoT formulation to fool a single object detector is cast as

$$\underset{\boldsymbol{\delta}}{\text{minimize}} \ \frac{1}{M} \sum_{i=1}^{M} \mathbb{E}_{t,t_{\text{TPS}},\mathbf{v}} \left[f(\mathbf{x}_i') \right] + \lambda g(\boldsymbol{\delta}) \tag{6}$$

where f denotes an attack loss for misdetection, g is the total-variation norm that enhances perturbations' smoothness [15], and $\lambda > 0$ is a regularization parameter. We further elaborate on our attack loss f in problem (6). In YOLOv2, a probability score associated with a bounding box indicates whether or not an object is present within this box. Thus, we specify the attack loss as the largest bounding-box probability over all bounding boxes belonging to the 'person' class. For Faster R-CNN, we attack all bounding boxes towards the class 'background'. The more detailed derivation on the attack loss is provided in Appendix C. Figure 3 presents an overview of our approach to generate adversarial T-shirts.

Min-Max Optimization for Fooling Multiple Object Detectors. Unlike digital space, the transferability of adversarial attacks largely drops in the physical environment, thus we consider a *physical ensemble attack* against multiple object detectors. It was recently shown in [31] that the ensemble attack can be designed from the perspective of min-max optimization, and yields much higher worst-case attack success rate than the averaging strategy over multiple models. Given N object detectors associated with attack loss functions $\{f_i\}_{i=1}^{N}$, the physical ensemble attack is cast as

$$\underset{\boldsymbol{\delta} \in \mathcal{C}}{\text{minimize}} \ \underset{\mathbf{w} \in \mathcal{P}}{\text{maximize}} \ \sum_{i=1}^{N} w_i \phi_i(\boldsymbol{\delta}) - \frac{\gamma}{2} \|\mathbf{w} - 1/N\|_2^2 + \lambda g(\boldsymbol{\delta}), \tag{7}$$

where $\mathbf{w}$ are known as domain weights that adjust the importance of each object detector during the attack generation, $\mathcal{P}$ is a probabilistic simplex given by $\mathcal{P} = \{\mathbf{w} | \mathbf{1}^T\mathbf{w} = 1, \mathbf{w} \geq \mathbf{0}\}$, $\gamma > 0$ is a regularization parameter, and $\phi_i(\boldsymbol{\delta}) := \frac{1}{M}\sum_{i=1}^{M}\mathbb{E}_{t\in\mathcal{T}, t_{\mathrm{TPS}}\in\mathcal{T}_{\mathrm{TPS}}}[f(\mathbf{x}'_i)]$ following (6). In (7), if $\gamma = 0$, then the adversarial perturbation $\boldsymbol{\delta}$ is designed over the *maximum* attack loss (worst-case attack scenario) since $\mathrm{maximize}_{\mathbf{w}\in\mathcal{P}}\sum_{i=1}^{N} w_i\phi_i(\boldsymbol{\delta}) = \phi_{i^*}(\boldsymbol{\delta})$, where $i^* = \arg\max_i \phi_i(\boldsymbol{\delta})$ at a fixed $\boldsymbol{\delta}$. Moreover, if $\gamma \to \infty$, then the inner maximization of problem (7) implies $\mathbf{w} \to 1/N$, namely, an averaging scheme over M attack losses. Thus, the regularization parameter γ in (7) strikes a balance between the max-strategy and the average-strategy.

4   Experiments

In this section, we demonstrate the effectiveness of our approach (we call *advT-TPS*) for design of the adversarial T-shirt by comparing it with 2 attack baseline methods, a) adversarial patch to fool YOLOv2 proposed in [30] and its printed version on a T-shirt (we call *advPatch*[2]), and b) the variant of our approach in the absence of TPS transformation, namely, $\mathcal{T}_{\mathrm{TPS}} = \emptyset$ in (5) (we call *advT-Affine*). We examine the convergence behavior of proposed algorithms as well as its Attack Success Rate[3] (ASR) in both digital and physical worlds. We clarify our algorithmic parameter setting in Appendix D.

Prior to detailed illustration, we briefly summarize the attack performance of our proposed adversarial T-shirt. When attacking YOLOv2, our method achieves 74% ASR in the digital world and 57% ASR in the physical world, where the latter is computed by averaging successfully attacked video frames over all different scenarios (i.e., indoor, outdoor and unforeseen scenarios) listed in Table 2. When attacking Faster R-CNN, our method achieves 61% and 47% ASR in the digital and the physical world, respectively. By contrast, the baseline advPatch only achieves around 25% ASR in the best case among all digital and physical scenarios against either YOLOv2 or Faster R-CNN (e.g., 18% against YOLOv2 in the physical case).

4.1   Experimental Setup

Data Collection. We collect two datasets for learning and testing our proposed attack algorithm in digital and physical worlds. The training dataset contains 40 videos (2003 video frames) from 4 different scenes: one outdoor and three indoor scenes. each video takes 5–10 s and was captured by a moving person wearing a T-shirt with printed checkerboard. The desired adversarial pattern is then learnt from the training dataset. The test dataset in the digital space contains 10 videos

---

[2] For fair comparison, we modify the perturbation size same as ours and execute the code provided in [30] under our training dataset.

[3] ASR is given by the ratio of successfully attacked testing frames over the total number of testing frames.

captured under the same scenes as the training dataset. This dataset is used to evaluate the attack performance of the learnt adversarial pattern in the digital world. In the physical world, we customize a T-shirt with the printed adversarial pattern learnt from our algorithm. Another 24 test videos (Sect. 4.3) are then collected at a different time capturing two or three persons (one of them wearing the adversarial T-shirt) walking a) side by side or b) at different distances. An additional control experiment in which actors wearing adversarial T-shirts walk in an exaggerated way is conducted to introduce large pose changes in the test data. In addition, we also test our adversarial T-shirt by unforeseen scenarios, where the test videos involve different locations and different persons which are never covered in the training dataset. All videos were taken using an iPhone X and resized to 416×416. In Table A2 of the Appendix F, we summarize the collected dataset under all circumstances.

Object Detectors. We use two state-of-the-art object detectors: Faster R-CNN [27] and YOLOv2 [26] to evaluate our method. These two object detectors are both pre-trained on COCO dataset [22] which contains 80 classes including 'person'. The minimum detection threshold are set as 0.7 for both Faster R-CNN and YOLOv2 by default. The sensitivity analysis of this threshold is performed in Fig. A4 Appendix D.

4.2   Adversarial T-Shirt in the Digital World

Convergence Performance of Our Proposed Attack Algorithm. In Fig. 4, we show ASR against the epoch number used by our proposed algorithm to solve problem (6). Here the success of our attack at one testing frame is required to meet two conditions, a) misdetection of the person who wears the adversarial T-shirt, and b) successful detection of the person whom dresses a normal cloth. As we can see, the proposed attack method covnerges well for attacking both YOLOv2 and Faster R-CNN. We also note that attacking Faster R-CNN is more difficult than attacking YOLOv2. Furthermore, if TPS is not applied during training, then ASR drops around 30% compared to our approach by leveraging TPS.

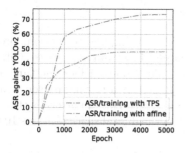 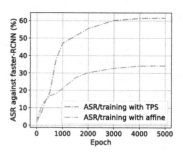

Fig. 4. ASR v.s. epoch numbers against YOLOv2 (left) and Faster R-CNN (right).

ASR of Adversarial T-shirts in Various Attack Settings. We perform a more comprehensive evaluation on our methods by digital simulation. Table 1 compares the ASR of adversarial T-shirts generated w/ or w/o TPS transformation in 4 attack settings: a) *single-detector attack* referring to adversarial T-shirts designed and evaluated using the same object detector, b) *transfer single-detector attack* referring to adversarial T-shirts designed and evaluated using different object detectors, c) *ensemble attack (average)* given by (7) but using the average of attack losses of individual models, and d) *ensemble attack (min-max)* given by (7). As we can see, it is crucial to incorporate TPS transformation in the design of adversarial T-shirts: without TPS, the ASR drops from 61% to 34% when attacking Faster R-CNN and drops from 74% to 48% when attacking YOLOv2 in the single-detector attack setting. We also note that the transferability of single-detector attack is weak in all settings. And Faster R-CNN is consistently more robust than YOLOv2, similar to the results in Fig. 4. Compared to our approach and *advT-Affine*, the baseline method *advPatch* yields the worst ASR when attacking a single detector. Furthermore, we evaluate the effectiveness of the proposed min-max ensemble attack (7). As we can see, when attacking Faster R-CNN, the min-max ensemble attack significantly outperforms its counterpart using the averaging strategy, leading to 15% improvement in ASR. This improvement is at the cost of 7% degradation when attacking YOLOv2.

Table 1. The ASR (%) of adversarial T-shirts generated from our approach, *advT-Affine* and the baseline *advPatch* in digital-world against Faster R-CNN and YOLOv2.

Method	Model	Target	Transfer	Ensemble (average)	Ensemble (min-max)
advPatch[30]	Faster R-CNN	22%	10%	N/A	N/A
advT-Affine		34%	11%	16%	32%
advT-TPS(ours)		**61%**	10%	**32%**	**47%**
advPatch[30]	YOLOv2	24%	10%	N/A	N/A
advT-Affine		48%	13%	31%	27%
advT-TPS(ours)		**74%**	13%	**60%**	**53%**

4.3 Adversarial T-Shirt in the Physical World

We next evaluate our method in the physical world. First, we generate an adversarial pattern by solving problem (6) against YOLOv2 and Faster R-CNN, following Sect. 4.2. We then print the adversarial pattern on a white T-shirt, leading to the adversarial T-shirt. For fair comparison, we also print adversarial patterns generated by the *advPatch* [30] and *advT-Affine* in Sect. 4.2 on white T-shirts of the same style. It is worth noting that different from evaluation by taking static photos of physical adversarial examples, our evaluation is conducted at a more practical and challenging setting. That is because we record videos to track a moving person wearing adversarial T-shirts, which could encounter multiple environment effects such as distance, deformation of the T-shirt, poses and angles of the moving person.

In Table 2, we compare our method with *advPatch* and *advT-Affine* under 3 specified scenarios, including the indoor, outdoor, and unforeseen scenarios[4], together with the overall case of all scenarios. We observe that our method achieves 64% ASR (against YOLOv2), which is much higher than *advT-Affine* (39%) and *advPatch* (19%) in the indoor scenario. Compared to the indoor scenario, evading person detectors in the outdoor scenario becomes more challenging. The ASR of our approach reduces to 47% but outperforms *advT-Affine* (36%) and *advPatch* (17%). This is not surprising since the outdoor scenario suffers more environmental variations such as lighting change. Even considering the unforeseen scenario, we find that our adversarial T-shirt is robust to the change of person and location, leading to 48% ASR against Faster R-CNN and 59% ASR against YOLOv2. Compared to the digital results, the ASR of our adversarial T-shirt drops around 10% in all tested physical-world scenarios; see specific video frames in Fig. A5 in Appendix.

Table 2. The ASR (%) of adversarial T-shirts generated from our approach, *advT-Affine* and *advPatch* under different physical-world scenes.

Method	Model	Indoor	Outdoor	New scenes	Average ASR
advPatch [30]	Faster R-CNN	15%	16%	12%	14%
advT-Affine		27%	25%	25%	26%
advT-TPS (ours)		**50%**	**42%**	**48%**	**47%**
advPatch [30]	YOLOv2	19%	17%	17%	18%
advT-Affine		39%	36%	34%	37%
advT-TPS (ours)		**64%**	**47%**	**59%**	**57%**

4.4   Ablation Study

In this section, we conduct more experiments for better understanding the robustness of our adversarial T-shirt against various conditions including angles and distances to camera, camera view, person's pose, and complex scenes that include crowd and occlusion. Since the baseline method (*advPatch*) performs poorly in most of these scenarios, we focus on evaluating our method (*advT-TPS*) against *advT-Affine* using YOLOv2. We refer readers to Appendix E for details on the setup of our ablation study.

---

[4] Unforeseen scenarios refer to test videos that involve different locations and actors that never seen in the training dataset.

Angles and Distances to Camera. In Fig. 5, we present ASRs of *advT-TPS* and *advT-Affine* when the actor whom wears the adversarial T-shit at different angles and distances to the camera. As we can see, *advT-TPS* works well within the angle 20° and the distance 4 m. And *advT-TPS* consistently outperforms *advT-Affine*. We also note that ASR drops significantly at the angle 30° since it induces occlusion of the adversarial pattern. Further, if the distance is greater than 7 m, the pattern cannot clearly be seen from the camera.

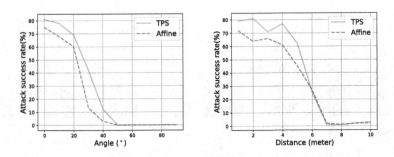

Fig. 5. Average ASR v.s. different angles (left) and distance (right).

Human Pose. In Table 3 (left), we evaluate the effect of pose change on *advT-TPS*, where videos are taken for an actor with some distinct postures including crouching, siting and running in place; see Fig. 6 for specific examples. To alleviate other latent effects, the camera was made to look straight at the person at a fixed distance of about 1–2 m away from the person. As we can see, *advT-TPS* consistently outperforms *advT-Affine*. In additional, we study the effect of occlusion on *advT-Affine* and *advT-TPS* in Appendix F.

Complex Scenes. In Table 3 (right), we test our adversarial T-shirt in several complex scenes with cluttered backgrounds, including a) an office with multiple objects and people moving around; b) a parking lot with vehicles and pedestrians; and c) a crossroad with busy traffic and crowd. We observe that compared to *advT-Affine*, *advT-TPS* is reasonably effective in complex scenes without suffering a significant loss of ASR. Compared to the other factors such as camera angle and occlusion, cluttered background and even crowd are probably the least of a concern for our approach. This is explainable, as our approach works on object proposals directly to suppress the classifier (Fig. 7).

Table 3. The ASR (%) of adversarial T-shirts generated from our approach, *advT-Affine* and *advPatch* under different physical-world scenarios.

Method \ Pose	crouching	siting	running	Method \ Scenario	office	parking lot	crossroad
advT-Affine	27%	26%	52%	*advT-Affine*	69%	53%	51%
advT-TPS	**53%**	**32%**	**63%**	*advT-TPS*	**73%**	**65%**	**54%**

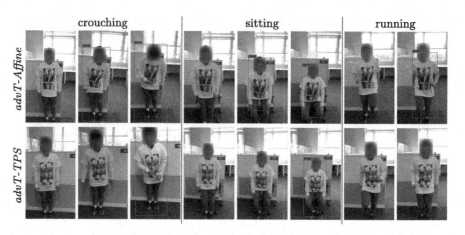

Fig. 6. Some video frames of person who wears adversarial T-shirt generated by *advT-Affine* (first row) and *advT-TPS* (second row) with different poses.

Fig. 7. The person who wear our adversarial T-shirt generate by TPS in three complex scenes: office, parking lot and crossroad.

5   Conclusion

In this paper, we propose *Adversarial T-shirt*, the first successful adversarial wearable to evade detection of moving persons. Since T-shirt is a non-rigid object, its deformation induced by a person's pose change is taken into account when generating adversarial perturbations. We also propose a min-max ensemble attack algorithm to fool multiple object detectors simultaneously. We show that our attack against YOLOv2 can achieve 74% and 57% attack success rate in the digital and physical world, respectively. By contrast, the *advPatch* method can only achieve 24% and 18% ASR. Based on our studies, we hope to provide some implications on how the adversarial perturbations can be implemented in physical worlds.

Acknowledgement. This work is partly supported by the National Science Foundation CNS-1932351. We would also like to extend our gratitude to MIT-IBM Watson AI Lab.

References

1. Athalye, A., Engstrom, L., Ilyas, A., Kwok, K.: Synthesizing robust adversarial examples. In: Dy, J., Krause, A. (eds.) Proceedings of the 35th International Conference on Machine Learning, vol. 80, pp. 284–293 (2018)
2. Athalye, A., Carlini, N., Wagner, D.: Obfuscated gradients give a false sense of security: circumventing defenses to adversarial examples. arXiv preprint arXiv:1802.00420 (2018)
3. Athalye, A., Engstrom, L., Ilyas, A., Kwok, K.: Synthesizing robust adversarial examples. In: International Conference on Machine Learning, pp. 284–293 (2018)
4. Bookstein, F.L.: Principal warps: thin-plate splines and the decomposition of deformations. IEEE Trans. Pattern Anal. Mach. Intell. **11**(6), 567–585 (1989)
5. Cao, Y., et al.: Adversarial objects against lidar-based autonomous driving systems. arXiv preprint arXiv:1907.05418 (2019)
6. Carlini, N., Wagner, D.: Audio adversarial examples: targeted attacks on speech-to-text. In: 2018 IEEE Security and Privacy Workshops (SPW), pp. 1–7. IEEE (2018)
7. Chen, S.-T., Cornelius, C., Martin, J., Chau, D.H.P.: ShapeShifter: robust physical adversarial attack on faster R-CNN object detector. In: Berlingerio, M., Bonchi, F., Gärtner, T., Hurley, N., Ifrim, G. (eds.) ECML PKDD 2018. LNCS (LNAI), vol. 11051, pp. 52–68. Springer, Cham (2019). https://doi.org/10.1007/978-3-030-10925-7_4
8. Chui, H.: Non-rigid point matching: algorithms, extensions and applications. Citeseer (2001)
9. Ding, G.W., Lui, K.Y.C., Jin, X., Wang, L., Huang, R.: On the sensitivity of adversarial robustness to input data distributions. In: International Conference on Learning Representations (2019)
10. Donato, G., Belongie, S.: Approximate thin plate spline mappings. In: Heyden, A., Sparr, G., Nielsen, M., Johansen, P. (eds.) ECCV 2002. LNCS, vol. 2352, pp. 21–31. Springer, Heidelberg (2002). https://doi.org/10.1007/3-540-47977-5_2
11. Duan, R., Ma, X., Wang, Y., Bailey, J., Qin, A.K., Yang, Y.: Adversarial camouflage: hiding physical-world attacks with natural styles. In: Proceedings of the IEEE/CVF Conference on Computer Vision and Pattern Recognition, pp. 1000–1008 (2020)
12. Duchi, J.C., Bartlett, P.L., Wainwright, M.J.: Randomized smoothing for stochastic optimization. SIAM J. Optim. **22**(2), 674–701 (2012)

13. Evtimov, I., et al.: Robust physical-world attacks on machine learning models. arXiv preprint arXiv:1707.08945 (2017)
14. Eykholt, K., et al.: Robust physical-world attacks on deep learning visual classification. In: Proceedings of the IEEE Conference on Computer Vision and Pattern Recognition, pp. 1625–1634 (2018)
15. Eykholt, K., et al.: Physical adversarial examples for object detectors. In: 12th USENIX Workshop on Offensive Technologies (WOOT 2018) (2018)
16. Geiger, A., Moosmann, F., Car, Ö., Schuster, B.: Automatic camera and range sensor calibration using a single shot. In: 2012 IEEE International Conference on Robotics and Automation, pp. 3936–3943. IEEE (2012)
17. Girshick, R.: Fast R-CNN. In: Proceedings of the IEEE International Conference on Computer Vision, pp. 1440–1448 (2015)
18. Girshick, R., Donahue, J., Darrell, T., Malik, J.: Rich feature hierarchies for accurate object detection and semantic segmentation. In: Proceedings of the IEEE Conference on Computer Vision and Pattern Recognition, pp. 580–587 (2014)
19. Jaderberg, M., Simonyan, K., Zisserman, A., et al.: Spatial transformer networks. In: Advances in Neural Information Processing Systems, pp. 2017–2025 (2015)
20. Li, J., Schmidt, F., Kolter, Z.: Adversarial camera stickers: a physical camera-based attack on deep learning systems. In: International Conference on Machine Learning, pp. 3896–3904 (2019)
21. Lin, J., Gan, C., Han, S.: Defensive quantization: when efficiency meets robustness. In: International Conference on Learning Representations (2019). https://openreview.net/forum?id=ryetZ20ctX
22. Lin, T.-Y., et al.: Microsoft COCO: common objects in context. In: Fleet, D., Pajdla, T., Schiele, B., Tuytelaars, T. (eds.) ECCV 2014. LNCS, vol. 8693, pp. 740–755. Springer, Cham (2014). https://doi.org/10.1007/978-3-319-10602-1_48
23. Liu, H.T.D., Tao, M., Li, C.L., Nowrouzezahrai, D., Jacobson, A.: Beyond pixel norm-balls: parametric adversaries using an analytically differentiable renderer. In: International Conference on Learning Representations (2019). https://openreview.net/forum?id=SJl2niR9KQ
24. Liu, W., et al.: SSD: single shot MultiBox detector. In: Leibe, B., Matas, J., Sebe, N., Welling, M. (eds.) ECCV 2016. LNCS, vol. 9905, pp. 21–37. Springer, Cham (2016). https://doi.org/10.1007/978-3-319-46448-0_2
25. Lu, J., Sibai, H., Fabry, E.: Adversarial examples that fool detectors. arXiv preprint arXiv:1712.02494 (2017)
26. Redmon, J., Farhadi, A.: Yolo9000: better, faster, stronger. In: Proceedings of the IEEE Conference on Computer Vision and Pattern Recognition, pp. 7263–7271 (2017)
27. Ren, S., He, K., Girshick, R., Sun, J.: Faster R-CNN: towards real-time object detection with region proposal networks. In: Advances in Neural Information Processing Systems, pp. 91–99 (2015)
28. Sharif, M., Bhagavatula, S., Bauer, L., Reiter, M.K.: Accessorize to a crime: real and stealthy attacks on state-of-the-art face recognition. In: Proceedings of the 2016 ACM SIGSAC Conference on Computer and Communications Security, pp. 1528–1540. ACM (2016)
29. Sitawarin, C., Bhagoji, A.N., Mosenia, A., Mittal, P., Chiang, M.: Rogue signs: deceiving traffic sign recognition with malicious ads and logos. arXiv preprint arXiv:1801.02780 (2018)
30. Thys, S., Van Ranst, W., Goedemé, T.: Fooling automated surveillance cameras: adversarial patches to attack person detection. In: Proceedings of the IEEE Conference on Computer Vision and Pattern Recognition Workshops (2019)

31. Wang, J., et al.: Beyond adversarial training: min-max optimization in adversarial attack and defense. arXiv preprint arXiv:1906.03563 (2019)
32. Xu, K., et al.: Topology attack and defense for graph neural networks: an optimization perspective. In: International Joint Conference on Artificial Intelligence (IJCAI) (2019)
33. Xu, K., et al.: Interpreting adversarial examples by activation promotion and suppression. arXiv preprint arXiv:1904.02057 (2019)
34. Xu, K., et al.: Structured adversarial attack: towards general implementation and better interpretability. In: International Conference on Learning Representations (2019)
35. Zhang, Y., Foroosh, H., David, P., Gong, B.: CAMOU: learning physical vehicle camouflages to adversarially attack detectors in the wild. In: International Conference on Learning Representations (2019). https://openreview.net/forum?id=SJgEl3A5tm
36. Zhang, Z.: A flexible new technique for camera calibration. IEEE Trans. Pattern Anal. Mach. Intell. **22**, 1330–1334 (2000)
37. Zhao, P., Xu, K., Liu, S., Wang, Y., Lin, X.: Admm attack: an enhanced adversarial attack for deep neural networks with undetectable distortions. In: Proceedings of the 24th Asia and South Pacific Design Automation Conference, pp. 499–505. ACM (2019)

Bounding-Box Channels for Visual Relationship Detection

Sho Inayoshi[1(✉)], Keita Otani[1], Antonio Tejero-de-Pablos[1],
and Tatsuya Harada[1,2]

[1] The University of Tokyo, Tokyo, Japan
{inayoshi,otani,antonio-t,harada}@mi.t.u-tokyo.ac.jp
[2] RIKEN, Wako, Japan

Abstract. Recognizing the relationship between multiple objects in an image is essential for a deeper understanding of the meaning of the image. However, current visual recognition methods are still far from reaching human-level accuracy. Recent approaches have tackled this task by combining image features with semantic and spatial features, but the way they relate them to each other is weak, mostly because the spatial context in the image feature is lost. In this paper, we propose the bounding-box channels, a novel architecture capable of relating the semantic, spatial, and image features strongly. Our network learns bounding-box channels, which are initialized according to the position and the label of objects, and concatenated to the image features extracted from such objects. Then, they are input together to the relationship estimator. This allows retaining the spatial information in the image features, and strongly associate them with the semantic and spatial features. This way, our method is capable of effectively emphasizing the features in the object area for a better modeling of the relationships within objects. Our evaluation results show the efficacy of our architecture outperforming previous works in visual relationship detection. In addition, we experimentally show that our bounding-box channels have a high generalization ability.

Keywords: Bounding-box channels · Visual relationship detection · Scene graph generation

1 Introduction

Although research on image understanding has been actively conducted, its focus has been on single object recognition and object detection. With the success of deep learning, the recognition and detection accuracy for a single object has improved significantly, becoming comparable to human recognition accuracy [17]. Previous works extract features from an input image using Convolutional Neural Networks (CNNs), and then the extracted features are used for recognition. However, single-object recognition and detection tasks for a single object cannot estimate the relationship between objects, which is essential for understanding the scene.

© Springer Nature Switzerland AG 2020
A. Vedaldi et al. (Eds.): ECCV 2020, LNCS 12350, pp. 682–697, 2020.
https://doi.org/10.1007/978-3-030-58558-7_40

person	on	skis	laptop	on	table
person	wear	helmet	laptop	below	sky
person	above	ramp	table	in front of	sky
person	stand on	ramp	sky	above	laptop

Fig. 1. Visual relationship detection (VRD) is the task of detecting relationships between objects in an image. Visual-relationship instances, or triplets, follow the subject-predicate-object pattern.

Visual relationship detection (VRD) is a very recent task that tackles this problem. The purpose of VRD is to recognize predicates that represent the relationship between two objects (e.g., *person wears helmet*), in addition to recognizing and detecting single objects, as shown in Fig. 1. In other words, the purpose of VRD is detecting *subject-predicate-object* triplets in images. When thinking about solving the VRD task, humans leverage the following three features: the image features, which represent the visual attributes of objects, the semantic features, which represent the combination of *subject-object* class labels, and the spatial features, which represent object positions in the image. Therefore, previous research in VRD [1,6,9,15,24,26,27] employs these three types of features. Previous works have attempted to extract these features in a variety of ways. In [6,9,13,15,25,26], semantic features are extracted from the label of the detected objects. For spatial features, in [6,15,22,24,26,27] the coordinate values of the object candidate area have been used. Alternatively, previous works [1,9,23] proposed extracting the spatial features from a binary mask filled with zeros except for the area of the object pairs. In spite to the efforts of the aforementioned approaches, the task of recognizing the relationship between objects it is still far from achieving human-level recognition accuracy. One of the main reasons is that the spatial information contained in the image features has not been successfully leveraged yet. For example, several image recognition approaches flatten image features, but this eliminates the spatial information contained in the image features.

In this paper, in order to improve the accuracy of VRD, we propose a novel feature fusion method, the bounding-box channels, which are capable of modeling together image features with semantic and spatial features without discarding spatial information (i.e., feature flattening). In our bounding-box channels, spatial information such as the location and overlap of the objects, is represented by adding channels to the image features. This allows for a strong association between the image features of each subject and object and their respective spatial information. Semantic features are also employed in the construction of the bounding-box channels to achieve a strong binding of all three features. Consequently, the relationship estimation network can learn a better model that leads to a better accuracy.

The contributions of this research are as follows:

- We propose the bounding-box channels, a new feature fusion method for visual relationship detection. It allows strongly combining semantic and spatial features with the image features for a boost in performance, without discarding spatial information.
- Our bounding-box channels follow a clear and straightforward implementation, and can be potentially used as replacements of the semantic and spatial feature extractors of other VRD methods.
- We provide extensive experimentation to show the generalization ability of our bounding-box channels, outperforming previous methods.

2   Related Work

Before the actual visual relationship detection, it is necessary to detect the image areas that contain objects and classify them. This is performed by object detection methods.

2.1   Object Detection

In recent years, the performance of object detectors has improved significantly through the use of deep learning. R-CNN [3] was a pioneer method in using convolutional neural networks (CNNs) for object detection. R-CNN uses a sliding window approach to input image patches to a CNN and performs object classification and regression of the bounding box (rectangular area containing the object). This method allowed for a significant improvement in object detection accuracy, but it has some limitations. First, the computational cost is huge because all windows are processed by the CNN. Second, since the size and position of the sliding window are static, some objects are misdetected. To solve this, Fast R-CNN [2] was proposed. In Fast R-CNN, image segmentation is applied to the whole image, and windows are generated dinamically by selecting regions of interest (RoI). Then, the CNN processes the whole image and classification and regression are performed only in the generated windows. This way, the computation cost is reduced and the accuracy is improved. Later, Faster R-CNN [17] was proposed, further improving computation time and accuracy by using a CNN to generate RoI.

Apart from the aforementioned methods, SSD [12] and YOLO [16], which are faster than the above methods, and FPN [10], which can detect objects of various scales, were also proposed. However, for the sake of comparison with previous VRD methods, directly using the detection results of R-CNN is a common practice [13,27].

2.2   Visual Relationship Detection

VRD aims to a deeper understanding of the meaning of an image by estimating the relationship between pairs of detected objects. A large number of VRD methods [1,6,9,13,15,19,22–27] have been recently proposed, which share the same three type of features: image, semantic and spatial.

Needless to say, image features play an important role when considering the relationship between objects. For example, in the case of the *subject-object* pair *man-chair*, several possible relationships can be considered (e.g., *"sitting"*, *"next to"*), but if an image is given, the possibilities are reduced to one. Previous works in image feature extraction proposed using multiple feature maps of a CNN, such as VGG16 [18]. The image feature map can be also obtained by cropping the feature map of the whole image by the smallest rectangular region containing subject and object using RoI Pooling [2] or RoI Alignment [4]. These feature extraction methods are widely used in various image recognition tasks including object detection, and their effectiveness has been thoroughly validated.

Semantic features are also an important element in VRD. For example, when considering the relationship between a man and a horse, our prior knowledge tells us that the relation *"man rides on a horse"* is more probable than *"man wears a horse"*. In order to provide such information, previous works extracted semantic features from the predicted class label of detected objects [9,13,15,22,25,27]. Previous works proposed two semantic feature extraction approaches. The first approach uses the word2vec embedding [14]. The class labels of the object pair were embedded in word2vec and processed by a fully-connected layer network, whose output is the semantic features. The second approach used the posterior probability distribution expressed by Eq. 1 [1,20,23,26].

$$P(p|s,o) = N_{spo}/N_{so} \qquad (1)$$

In Eq. 1, p, s, and o represents the labels of predicate, subject, and object respectively. N_{spo} is the number of *subject-predicate-object* triplets, and N_{so} is the number of *subject-object* pairs, both are emerged in the training set. This posterior probability distribution was used as the semantic features.

Last but not least, spatial features are crucial to detect spatial relationships like *"on"* and *"under"*. For example, when estimating the relationship between a bottle and a table, if the image area of the bottle is located above the area of the table, the relationship is probably *"a bottle on a table"*. In order to model these spatial relationships, two spatial feature extraction approaches were mainly proposed in previous work. The first approach used spatial scalar values (e.g., the distance between the centers of the pair of bounding boxes, the relative size of objects), concatenated them into a vector and passed them through fully-connected layers, to output the spatial features [6,15,22,24,26,27]. The second approach created binary images whose pixels are nonzero in the bounding box and zero otherwise, and feed it to a CNN to extract the spatial features [1,9,23].

As described above, a lot of approaches for extracting each feature were proposed. However, there was no deep consideration on how to effectively associate them together to learn a better model. Most of previous approaches either modeled each feature separately [13,15,25,26] or simply flattened and concatenated each feature [1,9,19,22,24]. However, the former approach could not effectively combine the spatial information in the image features and the spatial features, because they process the image features and the spatial features separately. Similarly, the latter approach loses the spatial information contained in the image features due to flattening.

In this work, we propose a novel VRD method capable of effectively modeling image features together with spatial and semantic features. Modeling together instead of separately and concatenating them before flattening the features increases the accuracy and adequacy of the estimated relationship.

3   Bounding-Box Channels

This section describes our novel proposed bounding-box channels, which effectively combines image, semantic, and spatial features. Due to the structure of CNNs, the image features retain the spatial information to some extent unless they are flattened. For example, the upper right area in an image is mapped to the upper right area in the corresponding image features of the CNN. Contrary to previous methods, we avoid flattening the image features in order to preserve spatial information. In our bounding-box channels, the spatial and semantic information of the objects is modeled in the form of channels that are directly added to the image features. This allows for a strong association between the image features of each subject and object and their respective spatial information, hence the relationship estimation network can learn a better model that leads to a better accuracy.

Image Feature: In our method, the image features $F_i \in \mathbb{R}^{H \times W \times n_i}$ are obtained from the feature map extracted from the backbone CNN such as VGG16 [18] aligned with the smallest area containing the bounding boxes of the subject-object pair by RoIAlign [4]. H and W are the height and width of the image features, and n_i is the number of channels of the image features.

Spatial Feature: As shown in Fig. 2, we encode the positions of the objects by leveraging the bounding box of the *subject-object* pair in the image used to extract the image features. We build two tensors $C_s, C_o \in \mathbb{R}^{H \times W \times n_c}$ with the same height H and width W as the image features F_i, and a number of channels n_c.

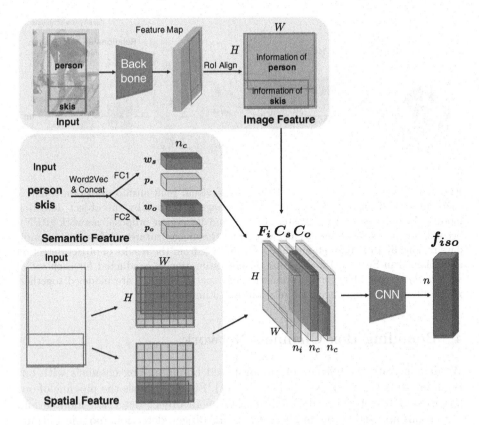

Fig. 2. Overview of how to construct the bounding-box channels. F_i is the aligned image feature. The words are transformed into the semantic features w_s and w_o via word2vec and fully-connected layers. In C_s, the inner region of the bounding box is filled with w_s, and the outer region is filled with a learnable parameter p_s. C_o is filled with w_o and p_o in the same way. Finally, we concatenate F_i, C_s and C_o in the channel direction, and fed into CNN to create our bounding-box channels f_{iso}.

Semantic Feature: As our semantic features, the words of the detected *subject-object* pair classes are embedded in word2vec [14]. In our implementation, we use the pretrained word2vec model and fix its weights. The obtained pair of word vectors are concatenated and mapped to n_c dimensions by fully-connected layers; we denote them w_s and w_o. Also, learn two n_c dimensional vectors, which are p_s and p_o, respectively.

Aggregation: For C_s, we fill the inner and outer regions representing the bounding box of the subject with w_s and p_s respectively. Similarly, for C_o, we fill the inner and outer regions representing the bounding box of the object with w_o and p_o respectively. Finally, we concatenate F_i, C_s, and C_o in the channel direction, and fed into CNN to create our bounding-box channels $f_{iso} \in \mathbb{R}^n$.

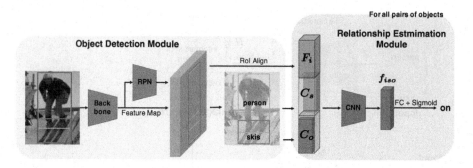

Fig. 3. Overview of the proposed BBCNet for visual relationship detection. First, a candidate set of *subject-object* pairs in the image is output by the object detection module. In our case, we use Faster R-CNN and its region proposal network (RPN). Second, for each *subject-object* pair, we extract the image features F_i from these candidate regions by RoI Align [4] and make C_s and C_o from the results of object detection as explained in Sect. 3. Finally, relationship estimation is conducted for each set of *subject-object* pair. The image, semantic, and spatial features are modeled together, which allows for a more accurate relationship estimation.

4   Bounding-Box Channels Network

We demonstrate the efficacy of our proposed bounding-box channels with our Bounding-Box Channels Network (BBCNet). Figure 3 shows the pipeline of our proposed BBCNet. The BBCNet consists of an object detection module and a relationship estimation module. First, an object detection module outputs the candidate set of *subject-object* pairs in the image. Second, we extract the image features F_i from the smallest area containing the bounding boxes of these candidate regions by RoI Align [4] and make C_s and C_o from the results of object detection as explained in Sect. 3. Finally, relationship estimation is conducted for each set of *subject-object* pair.

In previous VRD works, three types of features are leveraged for relationship estimation: image, semantic and spatial features. Our bounding-box channel module computes these three types of features for each candidate set of *subject-object* pair. The bounding-box channel module concatenates the image features extracted from the smallest rectangular region containing subject and object, the semantic features, and the spatial features (see Sect. 3). This way, the bounding-box channels are built. Our bounding-box channels are fed to a single layer CNN and two fully-connected layers to attain logit scores for each predicate class. As illustrated in Fig. 3, we obtain the probability distribution for each predicate class via sigmoid normalization. In multi-class classification tasks, classification is generally performed using softmax normalization. However, in VRD task, multiple predicate may be correct for one *subject-object* pair. For example, in Fig. 3, not only *person on bike* but also *person rides bike* can be correct. For such problem settings, not softmax normalization but sigmoid normalization is appropriate.

Fig. 4. We evaluate the performance of our method for visual relationship detection in the three tasks proposed in [13]: predicate detection, phrase detection and relationship detection. For predicate detection, classes and bounding boxes of objects are given in addition to an image, and the output is the predicate. Phrase detection and relationship detection take a single image, and output a set containing a pair of related objects or the individual related objects, respectively. In predicate detection, the given pair of objects is always related.

5 Experiments

5.1 Dataset

For our experiments, we used the Visual Relationship Detection (VRD) dataset [13] and the Visual Genome dataset [8], which are widely used to evaluate the performance of VRD methods. The VRD dataset contains 5000 images, with 100 object categories, and 70 predicate (relationship) categories among pairs of objects. Besides the categories, images are annotated with bounding boxes surrounding the objects. In total, VRD dataset contains 37993 subject-predicate-object triplets, of which 6672 are unique. We evaluate our method using the default splits, which contain 4000 training images and 1000 test images. The Visual Genome dataset contains 108073 images, with 150 object categories, and 50 predicate categories among pairs of objects. It is labeled in the same way as the VRD dataset. Our experiments follow the same train/test splits as [19].

5.2 Experimental Settings

VRD: We evaluate the proposed method in three relevant VRD tasks: predicate detection, phrase detection and relationship detection. The outline of each task is shown in Fig. 4.

The predicate detection task (left) aims to estimate the relationship between object pairs in an input image, given the object class labels and the surrounding bounding boxes. In other words, this task assumes an object detector with perfect accuracy, and thus, the predicate detection accuracy is not influenced by the object detection accuracy. Next, the phrase detection task (middle) aims to localize set boxes that include a pair of related objects in an input image, and then predict the corresponding predicate for each box. Lastly, the relationship detection task (right) aims to localize individual objects boxes in the input image, and then predict the relevant predicates between each pair of them. In phrase detection and relationship detection, not all object pairs have a relationship and, in contrast to the predicate detection task, the performance of the phrase detection and the relationship detection is largely influenced by the performance of the object detection.

Following the original paper that proposed the VRD dataset [13], we use Recall@50 (R@50) and Recall@100 (R@100) as our evaluation metrics. R@K computes the fraction of true positive relationships over the total relevant relationships among the top K predictions with the highest confidence (probability). Another reason for using recall is that, since the annotations do not contain all possible objects and relationships, the mean average precision (mAP) metric is usually low and not representative of the actual performance, as some predicted relationships, even if correct, they may not be included in the ground truth. Evaluating only the top prediction per object pair may mistakenly penalize correct predictions since annotators have bias over several plausible predicates. So we treat the number of chosen predictions per object pair (k) as a hyper-parameter, and report R@n for different k's for comparison with other methods. Since the number of predicates is 70, $k = 70$ is equivalent to evaluating all predictions w.r.t. the two detected objects.

Visual Genome: We evaluate the proposed method in two relevant VRD tasks: predicate classification (PRDCLS) and scene graph detection (SGDET) [19]. PRDCLS is equivalent to predicate detection, and SGDET is equivalent to relationship detection.

5.3   Implementation Details

In our method, the image, semantic, and spatial features are extracted from an input image (Fig. 3).

For object detection, we used the Faster R-CNN [17] structure[1] with ResNet50-FPN backbone [5,10] pretrained with MSCOCO [11]. First, we input the image into the backbone CNN, and get the feature map of the whole image and the object detection results. Next, as described in Sect. 3, we create the bounding-box channels. When extracting the semantic features, we leverage the word2vec model [14] pretrained with the Google News corpus and fixed weights. We embed the object class names of subject and object separately using the

---

[1] For the sake of comparison, some experiments replace Faster R-CNN with R-CNN [3].

word2vec model, which generates a 300-dimensional vector per word, and concatenate them into a 600-dimensional vector. Then, we feed this vector to two separate fully-connected layers, and denominate the output w_s and w_o as our semantic features. In this paper, we set $H = 7, W = 7, n_i = 512, n_c = 256$, and $n = 256$. Our implementation is partially based on the architecture proposed in the work of [27], but our BBCNet results outperform theirs.

We apply binary cross entropy loss to each predicted predicate class. We train our BBCNet using the Adam optimizer [7]. We set the initial learning rate to 0.0002 for backbone, and 0.002 for the rest of the network. We train the proposed model for 10 epochs and divide the learning rate by a factor of 5 after the 6th and 8th epochs. We set the weight decay to 0.0005. During training, from the training set, we sample all the positive triplets and the same number of negative triplets. This is due to the highly imbalance nature of the problem (only a few objects in the images are actually related).

During testing, we rank relationship proposals by multiplying the predicted subject, object, and predicate probabilities as follows:

$$p^{total} = p^{det}(s) \cdot p^{det}(o) \cdot p^{pred}(pred) \tag{2}$$

where p^{total} is the probability of subject-predicate-object triplets, $p^{det}(s), p^{det}(o)$ are the probability of subject and object classes respectively, and $p^{pred}(pred)$ is the probability of the predicted predicate class (i.e., the output of BBCNet). This reranking allows a fairer evaluation of the relationship detector, by giving preference to objects that are more likely to exist in the image.

Table 1. Performance comparison of the phrase detection and relationship detection tasks on the entire VRD dataset. "-" indicates performances that have not been reported in the original paper. The best performances are marked **in bold**. In the upper half, we compare our performance with four state-of-the-art methods that use the same object detection proposals. In the lower half, we compare with three state-of-the-art methods that use more sophisticated detectors. Our method achieves the state-of-the-art performance in almost all evaluation metrics.

Recall at	Phrase detection						Relationship detection					
	k = 1		k = 10		k = 70		k = 1		k = 10		k = 70	
	100	50	100	50	100	50	100	50	100	50	100	50
w/proposals from [13]												
CAI [27]	–	–	–	–	19.24	17.60	–	–	–	–	17.39	15.63
Language cues [15]	–	–	20.70	16.89	–	–	–	–	18.37	15.08	–	–
VRD-Full [13]	17.03	16.17	25.52	20.42	24.90	20.04	14.70	13.86	22.03	17.43	21.51	17.35
LSVR [25]	19.78	18.32	25.92	21.69	25.65	21.39	**17.07**	**16.08**	22.64	19.18	22.35	18.89
Ours	**20.95**	**19.72**	**28.33**	**24.46**	**28.38**	**24.47**	16.63	15.87	**22.79**	**19.90**	**22.86**	**19.91**
w/better proposals												
LK distillation [22]	24.03	23.14	29.76	26.47	29.43	26.32	21.34	19.17	29.89	22.56	31.89	22.68
LSVR [25]	32.85	28.93	39.66	32.90	39.64	32.90	26.67	23.68	32.63	26.98	32.59	26.98
GCL [26]	36.42	31.34	42.12	34.45	42.12	34.45	28.62	25.29	33.91	28.15	33.91	28.15
Ours	**40.72**	**34.25**	**46.18**	**36.71**	**46.18**	**36.71**	**33.36**	**28.21**	**38.50**	**30.61**	**38.50**	**30.61**

Table 2. Performance comparison of the phrase detection and relationship detection tasks on the zero-shot data (i.e., subject-predicate-object combinations not present in the training split) in the VRD dataset. We compare our performance with four methods which use the object detection results reported in [13] using R-CNN [3]. Our method outperforms the other related works in all evaluation metrics, and demonstrates high generalization ability.

Recall at	Phrase detection						Relationship detection					
	$k=1$		$k=10$		$k=70$		$k=1$		$k=10$		$k=70$	
	100	50	100	50	100	50	100	50	100	50	100	50
w/proposals from [13]												
CAI [27]	–	–	–	–	6.59	5.99	–	–	–	–	5.99	5.47
Language cues [15]	–	–	15.23	10.86	–	–	–	–	13.43	9.67	–	–
VRD-Full [13]	3.75	3.36	12.57	7.56	12.92	7.96	3.52	3.13	11.46	7.01	11.70	7.13
Ours	**8.81**	**8.13**	**16.51**	**12.57**	**16.60**	**12.66**	**6.42**	**5.99**	**13.77**	**10.09**	**13.94**	**10.27**

5.4   Quantitative Evaluation

VRD: As explained in Sect. 5.2, we compare the proposed method and related works via three evaluation metrics. For phrase detection and relationship detection, we compare our performance with four state-of-the-art methods [13,15,25,27], that use the same object detection proposals reported in [13] using R-CNN [3]. Also, we compare our performance with three state-of-the-art methods [21,25,26] using the object detection proposals of a more complex object detector (Faster R-CNN in our case). The phrase detection and the relationship detection performances are reported in Table 1 and Table 2. Also, the predicate detection performance is reported in Table 3. These results show that our BBCNet achieves state-of-the-art performance, outperforming previous works in almost all evaluation metrics on entire VRD dataset.

In particular, Table 2 and the zero-shot part in Table 3 show the results of the performance comparison when using combinations of triplets that exist in the test split but not in the training split (zero-shot data). A high generalization ability for zero-shot data has an important meaning in model evaluation. A poor generalization ability requires including all combinations of subject-predicate-object in the training data, which is unrealistic in terms of computational complexity and dataset creation. Our BBCNet achieves the highest performance in the zero-shot VRD dataset for all evaluation metrics, which shows its high generalization ability.

Visual Genome: As explained in Sect. 5.2, we compare the proposed method with four state-of-the-art methods [19,23,26] via two evaluation metrics. The scene graph detection and the predicate classification performances are reported in Table 4, in which graph constraint means that there is only one relationship between each object pair (that is, $k=1$ in VRD dataset). These results show that our BBCNet achieves the state-of-the-art performance in almost all evaluation metrics for the Visual Genome dataset as well.

Table 3. Performance comparison of the predicate detection task on the entire VRD and zero-shot VRD data sets. Our method outperforms the other state-of-the-art related works in all evaluation metrics on both entire set and zero-shot set. This result shows that when the object detection is perfectly conducted, our method is the most accurate for estimating the relationships between *subject-object* pairs.

Recall at	Predicate detection					
	Entire set			Zero-shot		
	$k = 1$	$k = 70$		$k = 1$	$k = 70$	
	100/50	100	50	100/50	100	50
VRD-Full [13]	47.87	–	–	8.45	–	–
VTransE [24]	44.76	–	–	–	–	–
LK distillation [22]	54.82	90.63	83.97	19.17	76.42	56.81
DSR [9]	–	93.18	86.01	–	79.81	60.90
Zoom-Net [21]	50.69	90.59	84.25	–	–	–
CAI + SCA-M [21]	55.98	94.56	89.03	–	–	–
Ours	**57.87**	**95.98**	**89.43**	**27.54**	**86.06**	**68.78**

Table 4. Comparison with the state-of-the-art methods on Visual Genome dataset. Graph constraint means that only one relationship is considered between each object pair. Our method achieves the state-of-the-art performance in most evaluation metrics.

Recall at	Graph constraint						No graph constraint			
	SGDET			PRDCLS			SGDET		PRDCLS	
	100	50	20	100	50	20	100	50	100	50
w/better proposals										
Message Passing [19]	4.2	3.4	–	53.0	44.8	–	–	–	–	–
Message Passing+ [23]	24.5	20.7	14.6	61.3	59.3	52.7	27.4	22.0	83.6	75.2
MotifNet-LeftRight [23]	30.3	27.2	**21.4**	67.1	65.2	58.5	35.8	**30.5**	88.3	81.1
RelDN [26]	32.7	28.3	21.1	68.4	68.4	66.9	36.7	30.4	97.8	93.8
Ours	**34.3**	**28.5**	20.4	**69.9**	**69.9**	**68.5**	**37.2**	29.9	**98.2**	**94.7**

Section 6 offers a more detailed discussion on the cause of the obtained results.

5.5   Qualitative Evaluation

In order to understand the contribution of the bounding box channels (BBC), we performed a comparison with the baseline in [27], whose architecture resembles ours, but without the BBC. Figure 5 shows a the VRD results of both our method and the baseline. Whereas the baseline outputs the relationship between a person and different person's belongings, our method outputs the relationship between a person and their own belongings. Similarly, in the other example, the relationship estimated by our method is more adequate than that of the baseline.

jacket on person : 0.88   person wear pants : 0.74

Baseline Model   Ours

Fig. 5. Qualitative comparison of our method with the baseline (no bounding box channels, as in [27]). The number over the image represents confidence of the output triplet. Thanks to our better modeling of the image-semantic-spatial features, the relationships detected by our method (right column) are more adequate than those of the baseline (left column).

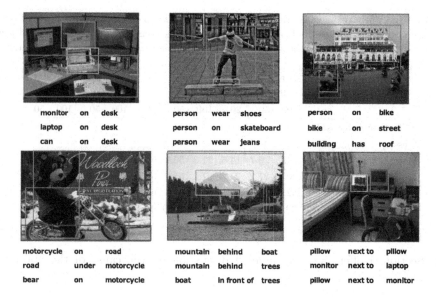

monitor	on	desk		person	wear	shoes		person	on	bike
laptop	on	desk		person	on	skateboard		bike	on	street
can	on	desk		person	wear	jeans		building	has	roof

motorcycle	on	road		mountain	behind	boat		pillow	next to	pillow
road	under	motorcycle		mountain	behind	trees		monitor	next to	laptop
bear	on	motorcycle		boat	in front of	trees		pillow	next to	monitor

Fig. 6. Qualitative examples of our proposed VRD method. The color of the text corresponds to the color of the bounding box of the object (same color means same object). "*bear on motorcycle*" in the lower-left image is an example of zero-shot data (i.e., *subject-predicate-object* combinations not present in the training split).

The reason is that, although the baseline is able to combine the semantic features with the image features, the spatial features do not work for removing inadequate relationships with respect to the objects location. On the other hand, the proposed

BBC allows considering the objects position properly when estimating their relationship. Figure 6 shows supplementary results of our method. Our method is able to estimate relationships not present in the training split (zero-shot data), as in the case of *"bear on motorcycle"* in the lower-left image.

6 Discussion

6.1 Quantitative Evaluation

Our method outperforms previous works in the vast majority of the conducted experiments. We can draw some conclusions from these results. First, our bounding-box channels can model more discriminative features for VRD than previous methods. The reason is that our BBCNet does not lose the spatial information in the image features, and effectively combines image, semantic and spatial features. Second, the word2vec based semantic feature extraction method has high generalization ability, because similar object names are projected on neighbor areas of the feature space. Therefore, the relationships not present in the training split but similar with the relationships present in the training split can be detected. For example, If the *dog under chair* triplet present in the training split, the *cat under chair* triplet is likely to be detected even it is not present in the training split. In contrast, the generalization ability is lower in methods whose semantic feature extraction uses the posterior probability distribution in Eq. 1. This occurs because restricting the semantic features to the posterior probability of the triplets included in the training set, worsens robustness against unseen samples (i.e., zero-shot data).

6.2 Qualitative Evaluation

As explained in Sect. 5.5, our BBCNet without the bounding-box channels (BBC) resembles the architecture of CAI [27]. Thus, these results can also be interpreted as an ablation study that shows the improvement in performance of a previous method by applying our BBC. But our bounding-box channels are potentially applicable not only to the architecture of CAI [27] adopted in this paper but also to other architectures. First, as far as the task of VRD is concerned, the image, semantic, and spatial features can be effectively combined by simply replacing the feature fusion modules with our bounding-box channels. In addition, not limited to VRD, if the task uses an object candidate area and an image, our bounding-box channels can be used to effectively combine both, expecting an improvement in performance.

7 Conclusion

In this paper, we proposed the bounding-box channels, a feature fusion method capable of successfully modeling together spatial and semantic features along with image features without discarding spatial information. Our experiments show that our architecture is beneficial for VRD, and outperforms the previous state-of-the-art works. As our future work, we plan to apply our bounding-box channels to a variety of network architectures, not limited to the VRD task, to further explore the combination of the image, semantic and spatial features.

Acknowledgements. This work was partially supported by JST AIP Acceleration Research Grant Number JPMJCR20U3, and partially supported by JSPS KAKENHI Grant Number JP19H01115. We would like to thank Akihiro Nakamura and Yusuke Mukuta for helpful discussions.

References

1. Dai, B., Zhang, Y., Lin, D.: Detecting visual relationships with deep relational networks. In: IEEE Conference on Computer Vision and Pattern Recognition (CVPR) (2017)
2. Girshick, R.: Fast R-CNN. In: IEEE International Conference on Computer Vision (ICCV) (2015)
3. Girshick, R.B., Donahue, J., Darrell, T., Malik, J.: Rich feature hierarchies for accurate object detection and semantic segmentation. In: IEEE Conference on Computer Vision and Pattern Recognition (CVPR) (2014)
4. He, K., Gkioxari, G., Dollár, P., Girshick, R.B.: Mask R-CNN. In: IEEE International Conference on Computer Vision (ICCV), pp. 2980–2988 (2017)
5. He, K., Zhang, X., Ren, S., Sun, J.: Deep residual learning for image recognition. In: IEEE Conference on Computer Vision and Pattern Recognition (CVPR), pp. 770–778 (2015)
6. Hu, R., Rohrbach, M., Andreas, J., Darrell, T., Saenko, K.: Modeling relationships in referential expressions with compositional modular networks. In: IEEE Conference on Computer Vision and Pattern Recognition (CVPR) (2017)
7. Kingma, D., Ba, J.: Adam: a method for stochastic optimization. In: International Conference on Learning Representations (ICLR) (2014)
8. Krishna, R., et al.: Visual genome: connecting language and vision using crowdsourced dense image annotations. Int. J. Comput. Vis. (IJCV) **123**(1), 32–73 (2017)
9. Liang, K., Guo, Y., Chang, H., Chen, X.: Visual relationship detection with deep structural ranking. In: Association for the Advancement of Artificial Intelligence (AAAI) (2018)
10. Lin, T.Y., Dollár, P., Girshick, R.B., He, K., Hariharan, B., Belongie, S.J.: Feature pyramid networks for object detection. In: IEEE Conference on Computer Vision and Pattern Recognition (CVPR) (2017)
11. Lin, T.Y., et al.: Microsoft coco: common objects in context. In: European Conference on Computer Vision (ECCV), pp. 740–755 (2014)
12. Liu, W., et al.: SSD: single shot multibox detector. In: European Conference on Computer Vision (ECCV) (2016)
13. Lu, C., Krishna, R., Bernstein, M., Fei-Fei, L.: Visual relationship detection with language priors. In: European Conference on Computer Vision (ECCV) (2016)
14. Mikolov, T., Corrado, G., Chen, K., Dean, J.: Efficient estimation of word representations in vector space. In: International Conference on Learning Representations (ICLR) (2013)
15. Plummer, B.A., Mallya, A., Cervantes, C.M., Hockenmaier, J., Lazebnik, S.: Phrase localization and visual relationship detection with comprehensive image-language cues. In: IEEE International Conference on Computer Vision (ICCV) (2017)

16. Redmon, J., Divvala, S.K., Girshick, R.B., Farhadi, A.: You only look once: unified, real-time object detection. In: IEEE Conference on Computer Vision and Pattern Recognition (CVPR) (2016)
17. Ren, S., He, K., Girshick, R., Sun, J.: Faster R-CNN: towards real-time object detection with region proposal networks. In: Neural Information Processing Systems (NIPS) (2015)
18. Simonyan, K., Zisserman, A.: Very deep convolutional networks for large-scale image recognition. In: International Conference on Learning Representations (ICLR) (2015)
19. Xu, D., Zhu, Y., Choy, C., Fei-Fei, L.: Scene graph generation by iterative message passing. In: IEEE Conference on Computer Vision and Pattern Recognition (CVPR) (2017)
20. Yang, J., Lu, J., Lee, S., Batra, D., Parikh, D.: Graph R-CNN for scene graph generation. In: European Conference on Computer Vision (ECCV), pp. 670–685 (2018)
21. Yin, G., et al.: Zoom-net: mining deep feature interactions for visual relationship recognition. In: European Conference on Computer Vision (ECCV), September 2018
22. Yu, R., Li, A., Morariu, V.I., Davis, L.S.: Visual relationship detection with internal and external linguistic knowledge distillation. In: IEEE International Conference on Computer Vision (ICCV), pp. 1068–1076 (2017)
23. Zellers, R., Yatskar, M., Thomson, S., Choi, Y.: Neural motifs: scene graph parsing with global context. In: IEEE Conference on Computer Vision and Pattern Recognition (CVPR) (2018)
24. Zhang, H., Kyaw, Z., Chang, S.F., Chua, T.S.: Visual translation embedding network for visual relation detection. In: IEEE Conference on Computer Vision and Pattern Recognition (CVPR) (2017)
25. Zhang, J., Kalantidis, Y., Rohrbach, M., Paluri, M., Elgammal, A., Elhoseiny, M.: Large-scale visual relationship understanding. In: Association for the Advancement of Artificial Intelligence (AAAI) (2019)
26. Zhang, J., Shih, K.J., Elgammal, A., Tao, A., Catanzaro, B.: Graphical contrastive losses for scene graph parsing. In: IEEE Conference on Computer Vision and Pattern Recognition (CVPR) (2019)
27. Zhuang, B., Liu, L., Shen, C., Reid, I.: Towards context-aware interaction recognition for visual relationship detection. In: IEEE International Conference on Computer Vision (ICCV) (2017)

Minimal Rolling Shutter Absolute Pose with Unknown Focal Length and Radial Distortion

Zuzana Kukelova[1]($\boxtimes$), Cenek Albl[2], Akihiro Sugimoto[3], Konrad Schindler[2], and Tomas Pajdla[4]

[1] Visual Recognition Group (VRG), Faculty of Electrical Engineering, Czech Technical University in Prague, Prague, Czechia
`kukelova@fel.cvut.cz`
[2] Photogrammetry and Remote Sensing, ETH Zurich, Zurich, Switzerland
[3] National Institute of Informatics, Tokyo, Japan
[4] Czech Institute of Informatics, Robotics and Cybernetics, Czech Technical University in Prague, Prague, Czechia

Abstract. The internal geometry of most modern consumer cameras is not adequately described by the perspective projection. Almost all cameras exhibit some radial lens distortion and are equipped with electronic rolling shutter that induces distortions when the camera moves during the image capture. When focal length has not been calibrated offline, the parameters that describe the radial and rolling shutter distortions are usually unknown. While, for global shutter cameras, minimal solvers for the absolute camera pose and unknown focal length and radial distortion are available, solvers for the rolling shutter were missing. We present the first minimal solutions for the absolute pose of a rolling shutter camera with unknown rolling shutter parameters, focal length, and radial distortion. Our new minimal solvers combine iterative schemes designed for calibrated rolling shutter cameras with fast generalized eigenvalue and Gröbner basis solvers. In a series of experiments, with both synthetic and real data, we show that our new solvers provide accurate estimates of the camera pose, rolling shutter parameters, focal length, and radial distortion parameters. The implementation of our solvers is available at *github.com/CenekAlbl/RnP*.

1 Introduction

Estimating the six degree-of-freedom (6DOF) pose of a camera is a fundamental problem in computer vision with applications in camera calibration [7], Structure-from-Motion (SfM) [36,37], augmented reality (AR) [32], and visual localization [33]. The task is to compute the camera pose in the world coordinate system from 3D points in the world and their 2D projections in an image.

Electronic supplementary material The online version of this chapter (https:// doi.org/10.1007/978-3-030-58558-7_41) contains supplementary material, which is available to authorized users.

A. Vedaldi et al. (Eds.): ECCV 2020, LNCS 12350, pp. 698–714, 2020.
https://doi.org/10.1007/978-3-030-58558-7_41

Fig. 1. Removing RS and radial distortion simultaneously using our minimal absolute pose solver. The original image (left) with tentative correspondences (black) the inliers captured by P4Pfr+R6P [4] (blue) and subsequent local optimization (red) compared to inliers captured by the proposed R7Pfr (cyan) and subsequent local optimization (green). The correction using R7Pfr without local optimization (right) is better than by P4Pfr+R6P with local optimization (middle). (Color figure online)

Solvers for the camera pose are usually used inside RANSAC-style hypothesis-and-test frameworks [12]. For efficiency it is therefore important to employ *minimal* solvers that generate the solution with a minimal number of point correspondences. The minimal number of 2D-to-3D correspondences necessary to solve the absolute pose problem is three for a calibrated perspective camera. The earliest solver dates back to 1841 [14]. Since then, the problem has been revisited several times [6,12,15,16,21]. In many situations, however, the internal camera calibration is unavailable, e.g.. when working with crowd-sourced images. Consequently, methods have been proposed to jointly estimate the camera pose together with focal length [8,27,38,39]. These methods have been extended to include also estimation of an unknown principal point [28], and unknown radial distortion [20,27]. The latter is particularly important for the wide-angle lenses commonly used in mobile phones and GoPro-style action cameras. The absolute pose of fully uncalibrated perspective camera without radial distortion can be estimated from six correspondence using the well-known DLT solver [1]. All these solutions assume a perspective camera model and are not suitable for cameras with rolling shutter (RS), unless the camera and the scene can be kept static.

Rolling shutter is omnipresent from consumer phones to professional SLR cameras. Besides technical advantages, like higher frame-rate and longer exposure time per pixel, it is also cheaper to produce. The price to pay is that the rows of an "image" are no longer captured synchronously, leading to motion-induced distortions and in general to a more complicated imaging geometry.

Motivation: While several minimal solutions have been proposed for the absolute pose of an RS camera with calibrated intrinsics [3–5,22], minimal solutions for uncalibrated RS cameras are missing. One obvious way to circumvent that problem is to first estimate the intrinsic and radial distortion parameters while ignoring the rolling shutter effect, then recover the 3D pose and rolling shutter parameters with an absolute pose solver for calibrated RS cameras [3,4,22]. Ignoring the deviation from the perspective projection in the first step can, however, lead to wrong estimates. For example, if the image point distribution is unfavourable, it may happen that RS distortion is compensated by an (incorrect) change of radial distortion, see Fig. 1.

Contribution: We present the first minimal solutions for two rolling shutter absolute pose problems:

1. absolute pose estimation of an RS camera with unknown focal length from 7 point correspondences; and
2. absolute pose estimation of an RS camera with unknown focal length *and* unknown radial distortion, also from 7 point correspondences.

The new minimal solvers combine two ingredients: a recent, iterative approach introduced for pose estimation of calibrated RS cameras [22]; and fast polynomial eigenvalue [11] and Gröbner basis solvers for comparatively simple, tractable systems of polynomial equations [10,26]. In experiments with synthetic and real data, we show that for uncalibrated RS cameras our new solvers find good estimates of camera pose, RS parameters, focal length, and radial distortion. We demonstrate that the new all-in-one solvers outperform alternatives that sequentially estimate first perspective intrinsics, then RS correction and extrinsics.

2   Related Work

The problem of estimating the absolute pose of a camera from a minimal number of 2D-to-3D point correspondences is important in geometric computer vision. Minimal solvers are often the main building blocks for SfM [36,37] and localization pipelines [33]. Therefore, during the last two decades a large variety of minimal absolute pose solvers for perspective cameras with or without radial distortion have been proposed.

For estimating the absolute pose of a calibrated camera, three points are necessary and the resulting system of polynomial equations can be solved in a closed form [21]. If the camera intrinsics and radial distortion are unknown, more point correspondences are required and the resulting systems of polynomial equations become more complex. The most common approach to solve such systems of polynomial equations is to use the Gröbner basis method [10] and automatic generators of efficient polynomial solvers [23,26].

Most of the minimal absolute pose solvers have been developed using the Gröbner basis method. These include four or 3.5 point minimal solvers (P4Pf or P3.5Pf solvers) for the perspective camera with unknown focal length, and known or zero radial distortion [8,27,38,39], four point (P4Pfr) solvers for perspective cameras with unknown focal length and unknown radial distortion [9,20,27], and P4.5Pfuv solver for unknown focal length and unknown principal point [28].

Recently, as RS cameras have become omnipresent, the focus has turned to estimating the camera absolute pose from images affected by RS distortion. RS cameras motion models [31] result in more complex systems of polynomial equations than perspective cameras models. Therefore, most of the existing RS absolute pose solvers use some model relaxations [3,4,22,35], scene assumptions such as planarity [2], additional information e.g. from IMU [5] or a video sequence [18], and a non-minimal number of point correspondences [2,25,30]. Moreover, all the existing solutions assume calibrated RS cameras, i.e., they assume that the camera intrinsic as well as radial distortion are known.

The first minimal solution to the absolute pose problem for a calibrated RS camera was presented in [3]. The proposed solver uses the minimal number of six 2D-to-3D point correspondences and the Gröbner basis method to generate an efficient solver. The proposed R6P is based on the constant linear and angular velocity model as in [2,19,30], but it uses the first order approximation to both the camera orientation and angular velocity, and, therefore, it requires an initialization of the camera orientation, e.g., from P3P [12]. It is shown in [3] that the proposed R6P solver significantly outperforms the P3P solver in terms of camera pose precision and the number of inliers captured in the RANSAC loop. The R6P solver was extended in [4] by linearizing only the angular velocity and also by proposing a faster solution to the "double-linearized" model. The model that linearizes only the angular velocity does not require any initialization of the camera orientation, however it results in a slower and more complicated solver. Moreover, it has been shown in [4] that such solver usually produces similar results as the "double-linearized" solver initialized with P3P [12,21].

The double-linearized model [3,4] results in a quite complex system of six quadratic equations in six unknowns with 20 solutions. The fastest solver to this problem, presented in [4], runs 0.3 ms and is not suitable for real-time applications. Therefore, a further simplification of the double-linearized model was proposed [22]. The model in [22] is based on the assumption that after the initialization with the P3P solver, the camera rotation is already close to the identity, and that in real applications, the rolling shutter rotation during the capture is usually small. Therefore, some nonlinear terms (monomials) in the double-linearized model are usually small, sometimes even negligible. Based on this assumption, a linear iterative algorithm was proposed in [22]. In the first iteration, the algorithm substitutes negligible monomials with zeros. In each subsequent iteration, it substitutes these monomials with the estimates from the previous iteration. In this way, the original, complicated system of polynomial equations is approximated with a system of linear equations. This new linear iterative algorithm usually converges to the solutions of the original system in no more than five iterations and is an order of magnitude faster than [4].

Different from the above mentioned methods for calibrated RS cameras, we combine the iterative scheme designed for calibrated RS cameras [22] with fast generalized eigenvalue and Gröbner basis solvers [26] for specific polynomial equation systems to solve the previously unsolved problem of estimating the absolute pose of an *uncalibrated* RS camera (i.e., unknown RS parameters, focal length, and radial distortion) from a minimal number of point correspondences.

3   Problem Formulation

For perspective cameras with radial distortion, the projection equation can be written as

$$\alpha_i u(\mathbf{x}_i, \boldsymbol{\lambda}) = \mathbf{K}[\,\mathbf{R} \mid \mathbf{C}\,]\mathbf{X}_i, \tag{1}$$

where $\mathbf{R} \in SO(3)$ and $\mathbf{C} \in \mathbb{R}^3$ are the rotation and translation bringing a 3D point $\mathbf{X}_i = [x_i, y_i, z_i, 1]^\top$ from the world coordinate system to the camera coordinate system, $\mathbf{x}_i = [r_i, c_i, 1]^\top$ are the homogeneous coordinates of a measured distorted image point, $u(\cdot, \boldsymbol{\lambda})$ is an image undistortion function with parameters $\boldsymbol{\lambda}$, and $\alpha_i \in \mathbb{R}$ is a scalar.

Matrix K is a 3×3 matrix known as the calibration matrix containing the intrinsic parameters of a camera. Natural constraints satisfied by most consumer cameras with modern CCD or CMOS sensor are zero skew and the unit aspect ratio [17]. The principal point [17] is usually also close to the image center ($[p_x, p_y]^\top = [0, 0]^\top$). Thus the majority of existing absolute pose solvers adhere to those assumptions, and we do so, too. Hence, we adopt calibration matrix

$$K = \mathrm{diag}(f, f, 1). \tag{2}$$

For cameras with lens distortion, measured image coordinates $\mathbf{x}_i$ have to be transformed into "pinhole points" with an undistortion function $u(\cdot, \boldsymbol{\lambda})$. For standard cameras, the radial component of the lens distortion is dominant, whereas the tangential component is negligible at this stage. Therefore, most camera models designed for minimal solvers consider only radial distortion[1]. A widely used model represents radial lens distortion with a one-parameter division [13]. This model is especially popular with absolute pose solvers thanks to its compactness and expressive power: it can capture even large distortions of wide-angle lenses (e.g., GoPro-type action cams) with a single parameter. Assuming that the distortion center is in the image center, the division model is

$$u(\mathbf{x}_i, \lambda) = u\left(\begin{bmatrix} r_i \\ c_i \\ 1 \end{bmatrix}, \lambda \right) = \begin{bmatrix} r_i \\ c_i \\ 1 + \lambda(r_i^2 + c_i^2) \end{bmatrix}. \tag{3}$$

Unlike perspective cameras, RS cameras capture every image row (or column) at a different time, and consequently, reveal the presence of relative motion between the camera and the scene at a different position. Camera rotation R and translation C are, therefore, functions of the image row r_i (or column). Together with the calibration matrix K of (2) and the distortion model (3), the projection equation of RS cameras is

$$\alpha_i u(\mathbf{x}_i, \lambda) = \alpha_i \begin{bmatrix} r_i \\ c_i \\ 1 + \lambda(r_i^2 + c_i^2) \end{bmatrix} = \begin{bmatrix} f & 0 & 0 \\ 0 & f & 0 \\ 0 & 0 & 1 \end{bmatrix} [R(r_i) \mid C(r_i)]\mathbf{X}_i. \tag{4}$$

Let R_0 and C_0 be the unknown rotation and translation of the camera at time $\tau = 0$, which denotes the acquisition time of the middle row $r_0 \in \mathbb{R}$. Then, for the short time-span required to record all rows of a frame (typically < 50 ms), the translation $C(r_i)$ can be approximated by a constant velocity model [2,3,19, 30,31,34]:

$$C(r_i) = C_0 + (r_i - r_0)\mathbf{t}, \tag{5}$$

---

[1] For maximum accuracy, tangential distortion can be estimated in a subsequent non-linear refinement.

with the translational velocity $\mathbf{t}$.

The rotation $\mathsf{R}(r_i)$, on the other hand, can be decomposed into two parts: the initial orientation R_0 of r_0, and the change of the orientation relative to it: $\mathsf{R}_\mathbf{w}(r_i - r_0)$. In [3,30], it was established that for realistic motions it is usually sufficient to linearize $\mathsf{R}_\mathbf{w}(r_i - r_0)$ around the initial rotation R_0 via the first-order Taylor expansion. Thereby the RS projection (4) becomes

$$\alpha_i \begin{bmatrix} r_i \\ c_i \\ 1 + \lambda(r_i^2 + c_i^2) \end{bmatrix} = \mathsf{K}\left[(\mathsf{I} + (r_i - r_0)[\mathbf{w}]_\times)\,\mathsf{R}_0 \mid \mathsf{C}_0 + (r_i - r_0)\mathbf{t}\right]\mathsf{X}_i, \qquad (6)$$

where $[\mathbf{w}]_\times$ is the skew-symmetric matrix for the vector $\mathbf{w} = [w_1, w_2, w_3]^\top$.

The linearized model (6) is sufficient for all scenarios except for the most extreme motions (which anyway present a problem due to motion blur that compromises keypoint extraction).

Unfortunately, the system of polynomial Eq. (6) is rather complex even with the linearized rolling shutter rotation. Already for calibrated RS camera and assuming Cayley parametrization of R_0, this model results in six equations of degree three in six unknowns and 64 solutions [4].

Therefore, following [3,4], we employ another linear approximation to the camera orientation R_0 to have the double-linearized model:

$$\alpha_i \begin{bmatrix} r_i \\ c_i \\ 1 + \lambda(r_i^2 + c_i^2) \end{bmatrix} = \mathsf{K}\left[(\mathsf{I} + (r_i - r_0)[\mathbf{w}]_\times)\,(\mathsf{I} + [\mathbf{v}]_\times) \mid \mathsf{C}_0 + (r_i - r_0)\mathbf{t}\right]\mathsf{X}_i. \qquad (7)$$

This model leads to a simpler way of solving the calibrated RS absolute pose from $\geq$ six 2D-3D point correspondences than the model in [4]. However, the drawback of this further simplification is the fact that, other than the relative intra-frame rotation, the absolute rotation R_0 can be of arbitrary magnitude, and therefore far from the linear approximation. A practical solution for calibrated cameras is to compute a rough approximate pose with a standard P3P solver [12,21], align the object coordinate system to it so that the remaining rotation is close enough to identity, and then run the RS solver [3,4,22].

The double-linearized model (7) is simpler than the original one (6), but still leads to a complex polynomial system (for calibrated RS cameras a system of six quadratic equations in six unknowns with up to 20 real solutions), and is rather slow for practical use. Therefore, further simplification of the double-linearized model was proposed in [22]. That model uses the fact that both the absolute rotation (after P3P initialisation) *and* the rolling shutter rotation $\mathbf{w}$ are small. As a consequence [22] assumes that the nonlinear term $[\mathbf{w}]_\times[\mathbf{v}]_\times$ in (7) is sufficiently small (sometimes even negligible). With this assumption, one can further linearize the nonlinear term $[\mathbf{w}]_\times[\mathbf{v}]_\times$ in (7) by approximating $[\mathbf{v}]_\times$ with $[\hat{\mathbf{v}}]_\times$, while keeping the remaining linear terms as they are; which leads to an efficient iterative solution of the original system: solve a resulting linearized system to estimate all unknowns including $[\mathbf{v}]_\times$, and iterate with updated $[\hat{\mathbf{v}}]_\times \leftarrow [\mathbf{v}]_\times$ until convergence. As initial approximation one can set $[\hat{\mathbf{v}}]_\times = 0$.

Algorithm 1. Iterative absolute pose solver for uncalibrated RS camera [with unknown radial distortion]

Input: $x_i, X_i, \{i = 1, \ldots, 7\}, k_{max}, \epsilon_{err}$
Output: $v, C_0, w, t, f, [\lambda]$

$v^0 \leftarrow 0, k \leftarrow 1$
$R_{GS}, C_{GS}, f_{GS} \leftarrow P4Pf(x_i, X_j)$ [27]
$[R_{GS}, C_{GS}, f_{GS}, \lambda_{GS} \leftarrow P4Pfr(x_i, X_j)$ [27]]
$X_i \leftarrow R_{GS}X_i$
while $k < k_{max}$ **do**
$\quad \hat{v} \leftarrow v^{k-1}$
$\quad err^k_{max} \leftarrow \infty$
$\quad [\lambda_{RS}], v_{RS}, C_{ORS}, w_{RS}, t_{RS}, f_{RS}, \leftarrow$ solve Eq. (8)
$\quad$**for** $j = 1$ **to** #solutions of Eq. (8) **do**
$\quad\quad err_j \leftarrow$ Residual error of Eq. (7) evaluated on $\{v_{RSj}, C_{ORSj}, w_{RSj}, t_{RSj}, f_{RSj}, [\lambda_{RSj}]\}$

$\quad\quad$**if** $err_j < err_{max}$ **then**
$\quad\quad\quad \{v^k, C^k, w^k, t^k, f^k, [\lambda^k]\} \leftarrow \{v_{RSj}, C_{ORSj}, w_{RSj}, t_{RSj}, f_{RSj}, [\lambda_{RSj}]\}$
$\quad\quad\quad err^k_{max} \leftarrow err_j$
$\quad\quad$**end if**
$\quad$**end for**
$\quad$**if** $err^k_{max} < \epsilon_{err}$ **or** $\left(|err^k_{max} - err^{k-1}_{max}| < \epsilon_{err} \ \& \ k > 1\right)$ **then**
$\quad\quad$**return** $\{v^k, C^k, w^k, t^k, f^k, [\lambda^k]\}$
$\quad$**end if**
$\quad k \leftarrow k + 1$
end while
return $\{v^{k-1}, C^{k-1}, w^{k-1}, t^{k-1}, f^{k-1}, [\lambda^{k-1}]\}$

Here we are interested in RS cameras with *unknown* focal length and radial distortion. In that setting (7) leads to a much more complicated system of polynomial equations that exceeds the capabilities of existing algebraic methods such as Gröbner bases [10,23,26]. We adopt a similar relaxation as in [22] and linearize $[w]_\times [v]_\times$ by substituting with the preliminary value $[v]_\times \leftarrow [\hat{v}]_\times$. Without loss of generality, let us assume that $r_0 = 0$. Then, the projection equation for this relaxed model is

$$\alpha_i \begin{bmatrix} r_i \\ c_i \\ 1 + \lambda(r_i^2 + c_i^2) \end{bmatrix} = K\left[I + r_i[w]_\times + [v]_\times + r_i[w]_\times[\hat{v}]_\times \mid C_0 + r_i t\right] X_i. \quad (8)$$

4   Minimal Solvers

To develop efficient minimal solvers for uncalibrated RS cameras, we advance the idea of the calibrated iterative RS solver of [22] by combining it with a generalized eigenvalue and efficient Gröbner basis solvers for specific polynomial equation systems.

We develop two new solvers. They both first pre-rotate the scene with a rotation estimated using efficient perspective absolute pose solvers for uncalibrated cameras, i.e. the P3.5Pf [27] and P4Pfr [27]/P5Pfr [24]. Then, they iterate two steps: (i) solve the system of polynomial equations derived from (8), with fixed preliminary $\hat{\mathbf{v}}$. (ii) update $\hat{\mathbf{v}}$ with the current estimates of the unknown parameters. The iteration is initialised with $\hat{\mathbf{v}} = 0$. A compact summary in the form of pseudo-code is given in Algorithm 1. Note that after solving the polynomial system (8), we obtain, in general, more than one feasible solution (where "feasible" means real and geometrically meaningful values, e.g., $f > 0$). To identify the correct one among them, we evaluate the (normalized) residual error of the original equations (7), and choose the one with the smallest error.

The described computational scheme of Algorithm 1 covers both the case where the radial distortion is known and only the pose, focal length and RS parameters must be found, and the case where also radial distortion is unknown. In the following, we separately work out the R7Pf solver for known radial distortion and the R7Pfr solver for unknown radial distortion. Both cases require seven point correspondences.

4.1 R7Pf - RS Absolute Pose with Unknown Focal Length

In the first solver, we assume that the camera has a negligible radial distortion (since known, non-zero distortion can be removed by warping the image point coordinates). This R7Pf solver follows the iterative procedure of Algorithm 1. What remains to be specified is how to efficiently solve the polynomial system (8) with $\lambda = 0$.

The R7Pf solver first eliminates the scalar values α_i by left-multiplying equation (8) with the skew-symmetric matrix $[\mathbf{x}_i]_\times$ for the vector $\mathbf{x}_i = \begin{bmatrix} r_i & c_i & 1 \end{bmatrix}^\top$. Since the projection equation (8) is defined only up to scale, we multiply the whole equation with $q = \frac{1}{f}$ ($f \neq 0$), resulting in

$$\begin{bmatrix} 0 & -1 & c_i \\ 1 & 0 & -r_i \\ -c_i & r_i & 0 \end{bmatrix} \begin{bmatrix} 1 & 0 & 0 \\ 0 & 1 & 0 \\ 0 & 0 & q \end{bmatrix} [\mathbf{I} + r_i[\mathbf{w}]_\times + [\mathbf{v}]_\times + r_i[\mathbf{w}]_\times[\hat{\mathbf{v}}]_\times \mid \mathbf{C}_0 + r_i\mathbf{t}] \mathbf{X}_i = 0. \quad (9)$$

(9) has 13 degrees of freedom (corresponding to 13 unknowns): $\mathbf{v}, \mathbf{w}, \mathbf{C}_0, \mathbf{t}$, and $q = \frac{1}{f}$. Since each 2D–3D point correspondence gives two linearly independent equations (only two equations in (9) are linearly independent due to the singularity of the skew-symmetric matrix), we need $6\frac{1}{2}$ point correspondences for a minimal solution.

Since half-points for which only one coordinate is known normally do not occur, we present a 7-point solver and just drop out one of the constraints in computing the camera pose, RS parameters, and focal length. The dropped constraint can be further used to filter out geometrically incorrect solutions.

After eliminating the scalar values α_i, the R7Pf solver starts with equations corresponding to the 3rd row of (9) for $i = 1, \ldots, 7$. These equations are linear in ten unknowns and do not contain the unknown $q = \frac{1}{f}$, indicating that they are independent of focal length. Let us denote the elements of unknown vectors by $\mathbf{v} = [v_1, v_2, v_3]^\top$, $\mathbf{w} = [w_1, w_2, w_3]^\top$, $\mathbf{C}_0 = [C_{0x}, C_{0y}, C_{0z}]^\top$, and $\mathbf{t} = [t_x, t_y, t_z]^\top$.

Then, the equations corresponding to the 3rd row of (9) for $i = 1, \ldots, 7$ can be written as

$$My = 0 \qquad (10)$$

where M is a 7×11 coefficient matrix and y is a 11×1 vector of monomials: $y = [v_1, v_2, v_3, w_1, w_2, w_3, C_{0x}, C_{0y}, t_x, t_y, 1]^{\top}$. For points in the general configuration, the matrix M in (10) has a 4-dimensional null-space, so we can write the unknown vector y as a linear combination of four 11×1 basis vectors y_1, y_2, y_3, and y_4 of that null-space:

$$y = \beta_1 y_1 + \beta_2 y_2 + \beta_3 y_3 + \beta_4 y_4, \qquad (11)$$

where β_j $(j = 1, \ldots, 4)$ are new unknowns. One of these unknowns, e.g.. β_4, can be eliminated (expressed as a linear combination of the remaining three unknowns $\beta_1, \beta_2, \beta_3$), using the constraint on the last element of y, which by construction is 1.

In the next step, the parameterization (11) is substituted into the equations corresponding to the 1st (or 2nd) row of (9) for $i = 1, \ldots, 6$. Note that here we use only six of seven available equations. The substitution results in six polynomial equations in six unknowns $\beta_1, \beta_2, \beta_3, C_{0z}, t_z, q$, and 10 monomials $m = [\beta_1 q, \beta_1, \beta_2 q, \beta_2, \beta_3 q, \beta_3, C_{0z} q, t_z q, q, 1]$. This is a system of six quadratic equations in six unknowns, which could be solved using standard algebraic methods based on Gröbner bases [10] and automatic Gröbner basis solver generators [23,26]. However, in this specific case, it is more efficient to transform it to a generalized eigenvalue problem (GEP) of size 6×6, by rewriting it as

$$q A_1 [\beta_1, \beta_2, \beta_3, C_{0z}, t_z, 1]^{\top} = A_0 [\beta_1, \beta_2, \beta_3, C_{0z}, t_z, 1]^{\top}, \qquad (12)$$

where A_0 and A_1 are 6×6 coefficient matrices. Equation (12) can be solved using standard efficient eigenvalue methods [11]. Alternatively, one can simplify even further by eliminating monomials $C_{0z} q$ and $t_z q$, thereby also eliminating two unknowns C_{0z} and t_z, and then solving a GEP of size 4×4. The remaining unknowns are obtained by back-substitution into (11).

4.2   R7Pfr - RS Absolute Pose with Unknown Focal Length and Unknown Radial Distortion

The R7Pfr solver finds the solution of the minimal problem with unknown absolute pose, RS parameters, focal length, and radial distortion. Compared to the first solver, there is one additional degree of freedom (14 unknowns in total); hence, we need seven 2D-to-3D point correspondences. R7Pfr follows the same iterative approach.

After eliminating the scalar values α_i by left-multiplying Eq. (8) with the skew-symmetric matrix $[u(x_i, \lambda)]_{\times}$ for $u(x_i, \lambda) = [r_i, c_i, 1 + \lambda(r_i^2 + c_i^2)]^{\top}$, and multiplying the complete system with $q = \frac{1}{f}(\neq 0)$, we obtain

$$\begin{bmatrix} 0 & d_i & c_i \\ d_i & 0 & -r_i \\ -c_i & r_i & 0 \end{bmatrix} \begin{bmatrix} 1 & 0 & 0 \\ 0 & 1 & 0 \\ 0 & 0 & q \end{bmatrix} [I + r_i[w]_{\times} + [v]_{\times} + r_i[w]_{\times}[\hat{v}]_{\times} \mid C_0 + r_i t] X_i = 0, \qquad (13)$$

with $d_i = 1 + \lambda(r_i^2 + c_i^2)$ for $i = 1, \ldots, 7$. The polynomial system (13) is more complicated than that without radial distortion, but the third row remains unchanged, as it is independent not only of focal length, but also of radial distortion. We can therefore proceed in the same way: find the 4-dimensional null space of matrix M, eliminate β_4, and substitute the parametrization (11) back into the 1st (or 2nd) row of (13), to obtain a system of seven quadratic polynomial equations in seven unknowns $\beta_1, \beta_2, \beta_3, C_{0z}, t_z, q, \lambda$, and 14 monomials $m = [\beta_1 q, \beta_1 \lambda, \beta_1, \beta_2 q, \beta_2 \lambda, \beta_2, \beta_3 q, \beta_3 \lambda, \beta_3, C_{0z} q, t_z q, q, \lambda, 1]$.

In the next step, we eliminate the monomials $C_{0z} q$ and $t_z q$ (and, consequently, also two unknowns C_{0z} and t_z) by simple Gauss-Jordan elimination. The resulting system of five quadratic equations in five unknowns $\beta_1, \beta_2, \beta_3, q, \lambda$, and 12 monomials has ten solutions. Different from the R7Pf case, this system does not allow a straight-forward transformation to a GEP. Instead, we solve it with the Gröbner basis method using the automatic solver generator [26].

To find a solver that is as efficient as possible, we follow the recent heuristic [29]. We generate solvers for 1000 different candidate bases and select the most efficient one among them. The winning solver performs elimination on a 26×36 matrix (compared to a 36×46 matrix if using the standard basis and grevlex monomial ordering) and eigenvalue decomposition of a 10×10 matrix. The remaining unknowns are again obtained by the back-substitution to (11).

5   Experiments

We evaluate the performance of both presented solvers on synthetic as well as various real datasets. The main strength of the presented R7Pf and R7Pfr solvers lies in the ability to handle uncalibrated data, which often occurs in the wild.

5.1   Data Setup

Synthetic Data: For the synthetic experiments, we generate random sets of seven points in the cube with side one. We simulate a camera with 60° FOV at random locations facing the center of the cube at a distance between one to four. We generate 1000 samples for each experiment, with 10 increment steps for the parameters that are being varied. We generate the camera motion using constant translational and rotational velocity model and the radial distortion using the one parameter division model. Note that even though our solvers are based on these models, they use approximations to the RS motion and therefore the data is never generated with identical model that is being solved for.

Real Data: We use altogether five datasets. Three outdoor captured by Gopro cameras, two of which are downloaded from Youtube and one was proposed in [4]. Two contain drone footage and one a handheld recording of a rollercoaster ride. We have conducted an offline calibration of the internal parameters and lens distortion of Gopro Hero 3 Black used in dataset Gopro drone 1 from [4] using the Matlab Calibration Toolbox.

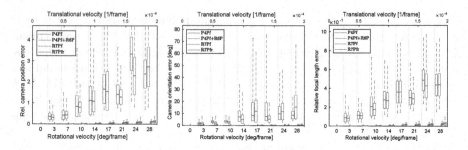

Fig. 2. Increasing camera motion (rotational+translational) on synthetic data with unknown focal length. P4Pf and P4Pf+R6P struggles to estimate the correct pose and focal length in the presence of RS distortions whereas R7Pf (gray-blue) and R7Pfr (magenta) are able to cope with RS effects. (Color figure online)

To create the ground truth we undistorted images grabbed from the entire videos and used them all in an open source SfM pipeline COLMAP [36] to reconstruct a 3D model. Of course, the images containing significant RS distortion were not registered properly or not at all, but the scene was sufficiently reconstructed from the images where RS distortions were insignificant. We then selected the parts of trajectory which have not been reconstructed well and registered the 2D features in those images to the 3D points in the reconstructed scene. This was done for datasets Gopro Drone 2 and Gopro rollercoaster. For Gopro drone 1 we had DSLR images of the scene and we could reconstruct the 3D model using those, which led to much better data overall.

Furthermore, we captured two dataset using the Xiaomi Mi 9 smartphone with both the standard and the wide FOV camera. The standard FOV camera contains virtually no radial distortion whereas the wide FOV camera has moderate radial distortion. We reconstructed the scene using static images from the standard camera and then registered sequences with moving camera to the reconstruction.

5.2 Compared Methods

When neither the focal length nor the radial distortion coefficients are known, the state-of-the-art offers a 4-point solver to absolute pose, focal length (P4Pf) [27] and radial distortion (P4Pfr) [27]. In the presence of RS distortions, one can opt for the R6P algorithm [4] which, however, needs the camera calibration. We solve the problem simultaneously for both the RS parameters, focal length, and radial distortion, which until now could be emulated by running P4Pf or P4Pfr and subsequently R6P on the calibrated and/or undistorted data. That combination is the closest viable alternative to our method, so we consider it as the state-of-the-art and compare with it.

A common practice after robust estimation with a minimal solver and RANSAC is to polish the results with local optimization using all inliers [4]. An important question is whether a simpler model, in our case the baseline

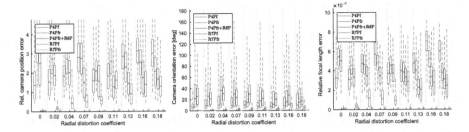

Fig. 3. Increasing radial distortion and unknown focal length. RS motion is kept constant at a value of the middle of Fig. 2. P4Pf and P4Pfr are not able to estimate good pose and focal length under RS distortions, R7Pf slowly deteriorates with increasing radial distortion and R7Pfr performs well across the entire range.

Fig. 4. (left) Original image distorted by radial and RS distortion. (middle) **Image undistorted by our R7Pfr.** (right) Image undistortion by a 3-parametric model estimated by Matlab Calibration Toolbox using a calibration board. We achieve comparable results as a method based on a calibration board.

P4Pf/P4Pfr followed by R6P could be enough to initialize that local optimization and reach the performance of the direct solution with a more complex model, i.e., the proposed R7Pf/R7Pfr. In our experiments we evaluate also the non-linear optimization initialized by RANSAC and see if our solvers outperform the non-linear optimization of the baseline approach.

5.3 Evaluation Metrics

We use various metrics to compare with the state-of-the-art and to show the benefits of our solvers. A common practice [3,5,22] is to use the number of inliers identified by RANSAC as the criterion to demonstrate the performance of minimal solvers on real data. We compare against both the state-of-the-art RANSAC output and the polished result after local optimization.

To highlight the accuracy of the estimated radial distortion and the RS parameters, we use them to remove the radial distortion and the rotational rolling shutter distortion from the images.

Due to the lack of good ground-truth for the camera poses, we evaluate them in two ways. First, we move the camera in place inducing only rotations, which resembles, e.g., an augmented or head-tracking scenario. In this case, the computed camera centers are expected to be almost static and we can show the

Table 1. Average numbers of inliers for different methods.

Dataset	P4Pfr+R6P	P4Pfr+R6P+LO	P4Pfr+R7Pfr	P4Pfr+R7Pfr+LO
Gopro drone 1	45	162	**170**	**203**
Gopro drone 2	124	130	**126**	**131**
Gopro rollerc	130	**137**	**132**	**137**
Xiaomi wide	58	66	**64**	**67**
	P4Pf+R6P	P4Pf+R6P+LO	**P4Pf+R7Pf**	**P4Pfr+R7Pf+LO**
Xiaomi standard	**72**	95	44	**110**

standard deviation from the mean as a measure of the estimated pose error. Second, we evaluate qualitatively the case where the camera moves along a smooth trajectory.

5.4   Results

The experiments on synthetic data verify that the proposed solvers are able to handle unknown focal length, radial distortion and RS distortions. Figure 2 shows results on data with unknown focal length and increasing camera motion. The state-of-the-art P4Pf solver struggles to estimate the camera pose and the focal length accurately as the RS camera rotational and translational velocity increases, resulting in mean orientation errors up to $15°$ and relative focal length error of 40% when the motion is strongest. Given such poor initial focal length estimate from P4Pfr, R6P is not able to recover the pose any better. In contrast, both R7Pf and R7Pfr are able to estimate the pose and focal length accurately, keeping the mean rotation error under $1.0°$ and the relative focal length estimate error under 3% even for the strongest motions.

Next we evaluate the effect of increasing radial distortion and the performance of our R7Pfr solver, see Fig. 3. The magnitude of the RS motion is kept constant through the experiment at the value of about the middle of the previous experiment. First thing to notice is that P4Pfr is less stable under RS distortion, providing worse estimates than P4Pf when radial distortion is close to zero. As the radial distortion increases the performance of P4Pf becomes gradually worse and is outperformed by P4Pfr in the end. R6P initialized by P4Pfr is not able to improve the poor results of P4Pfr. R7Pf slowly deteriorates with increasing radial distortion and R7Pfr provides good results under all conditions.

Additional synthetic experiments, including experiments studying the behaviour of our new solvers for increasing angular errors of the initial orientation, performance on planar scenes, convergence and run-time statistics are in the supplementary material.

The mean number of inliers on real data is summarized in Table 1 and qualitative evaluation of image undistortion is shown in Fig. 5. Camera center precision is evaluated quantitatively in Table 2 and qualitatively in the supplementary material. Our solvers achieve overall better performance in terms of number of

Table 2. Standard deviations from mean position of camera centers. The camera was purely rotating in these datasets; lower deviations mean more precise camera poses.

Dataset	P4Pfr+R6P	P4Pfr+R6P+LO	P4Pfr+R7Pfr	P4Pfr+R7Pfr+LO
Xiaomi wide	25	39	20	20
Xiaomi standard	12	14	14	10

Table 3. Mean relative errors of estimated focal length w.r.t the ground truth focal length $f_{gt} = 800px$ for the Gopro datasets for our method and the baseline and the locally optimized variants (LO).

Dataset	P4Pfr+R6P	P4Pfr+R6P+LO	**P4Pfr+R7Pfr**	**P4Pfr+R7Pfr+LO**
Gopro drone 1	10%	6.5%	**2.37%**	**1.5%**
Gopro rollerc	2.25%	1.75%	**1.75%**	**1.5%**

Fig. 5. Right image shows tentative matches (black), inliers found using P4Pfr followed by R7Pfr (cyan) or R6P (blue) and the inliers after subsequent local optimization (LO) of the R7Pfr result (green) and the R6P result (red). The middle image and right image shows the RS and distortion removal using the R6P parameters after LO and the R7Pfr parameters after LO respectively. (Color figure online)

RANSAC inliers and R7Pfr followed by local optimization provides the best results in all cases. The estimated radial distortion and camera motion is significantly better than that of the baseline methods and can be readily used to remove both radial and RS distortion as shown in Figs. 4 and 5. Relative focal length error compared to ground truth available in dataset Gopro drone 1 in Table 3 shows a significant improvement when using R7Pfr solver.

6 Conclusion

We address the problem of absolute pose estimation of an uncalibrated RS camera, and present the first minimal solutions for the problem. Our two new minimal solvers are developed under the same computational scheme by combining an iterative scheme originally designed for calibrated RS cameras with fast generalized eigenvalue and efficient Gröbner basis solvers for specific polynomial equation systems. The R7Pf solver estimates the absolute pose of a RS camera with unknown focal length from 7 point correspondences. The R7Pfr solver estimates the absolute pose of a RS camera with unknown focal length and unknown radial distortion; also from 7 point correspondences. Our experiments demonstrate the accuracy of our new solvers.

Acknowledgement. This research was supported by OP VVV project Research Center for Informatics No.   CZ.02.1.01/0.0/0.0/16_019/0000765, OP RDE IMPACT No. CZ.02.1.01/0.0/0.0/15 003/0000468, EU H2020 ARtwin No. 856994, and EU H2020 SPRING No. 871245 Projects.

References

1. Abdel-Aziz, Y.I., Karara, H.M.: Direct linear transformation from comparator coordinates into object space coordinates in closerange photogrammetry. In: Symposium on CloseRange Photogrammetry (1971)
2. Ait-Aider, O., Andreff, N., Lavest, J.M., Blaise, U., Ferr, P.C., Cnrs, L.U.: Simultaneous object pose and velocity computation using a single view from a rolling shutter camera. In: European Conference on Computer Vision (ECCV), pp. 56–68 (2006)
3. Albl, C., Kukelova, Z., Pajdla, T.: R6P - rolling shutter absolute pose problem. In: Computer Vision and Pattern Recognition (CVPR), pp. 2292–2300 (2015)
4. Albl, C., Kukelova, Z., Larsson, V., Pajdla, T.: Rolling shutter camera absolute pose. In: IEEE Trans, Pattern Analysis and Machine Intelligence (PAMI) (2019)
5. Albl, C., Kukelova, Z., Pajdla, T.: Rolling shutter absolute pose problem with known vertical direction. In: Computer Vision and Pattern Recognition (CVPR) (2016)
6. Ameller, M.A., Triggs, B., Quan, L.: Camera pose revisited: new linear algorithms. In 14eme Congres Francophone de Reconnaissance des Formes et Intelligence Artificielle. Paper in French (2002)
7. Bouguet, J.Y.: Camera calibration toolbox for matlab, vol. 1080 (2008). http://www.vision.caltech.edu/bouguetj/calib_doc
8. Bujnak, M., Kukelova, Z., Pajdla, T.: A general solution to the p4p problem for camera with unknown focal length. In: Computer Vision and Pattern Recognition (CVPR), pp. 1–8. IEEE (2008)
9. Bujnak, M., Kukelova, Z., Pajdla, T.: New efficient solution to the absolute pose problem for camera with unknown focal length and radial distortion. In: Asian Conference on Computer Vision (ACCV) (2010)
10. Cox, D.A., Little, J., O'shea, D.: Using algebraic geometry, vol. 185. Springer, New York (2006). https://doi.org/10.1007/b138611

11. Demmel, J., Dongarra, J., Ruhe, A., van der Vorst, H., Bai, Z.: Templates for the Solution of Algebraic Eigenvalue Problems: A Practical Guide. SIAM, Philadelphia (2000)
12. Fischler, M.A., Bolles, R.C.: Random sample consensus: a paradigm for model fitting with applications to image analysis and automated cartography. Commun. ACM **24**(6), 381–395 (1981)
13. Fitzgibbon, A.W.: Simultaneous linear estimation of multiple view geometry and lens distortion. In: Computer Vision and Pattern Recognition (CVPR) (2001)
14. Grunert, J.A.: Das Pothenotische Problem in erweiterter Gestalt nebst über seine Anwendungen in der Geodäsie (1841)
15. Guo, Y.: A novel solution to the P4P problem for an uncalibrated camera. J. Math. Imaging Vision **45**(2), 186–198 (2013)
16. Haralick, R., Lee, D., Ottenburg, K., Nolle, M.: Analysis and solutions of the three point perspective pose estimation problem. In: Computer Vision and Pattern Recognition (CVPR), pp. 592–598 (1991)
17. Hartley, R.I., Zisserman, A.: Multiple View Geometry in Computer Vision, second edn.. Cambridge University Press, Cambridge (2004)
18. Hedborg, J., Ringaby, E., Forssen, P.E., Felsberg, M.: Structure and motion estimation from rolling shutter video. In: ICCV Workshops, pp. 17–23 (2011)
19. Hedborg, J., Forssén, P.E., Felsberg, M., Ringaby, E.: Rolling shutter bundle adjustment. In: Computer Vision and Pattern Recognition (CVPR). pp. 1434–1441 (2012)
20. Josephson, K., Byröd, M.: Pose estimation with radial distortion and unknown focal length. In: Computer Vision and Pattern Recognition (CVPR) (2009)
21. Kneip, L., Scaramuzza, D., Siegwart, R.: A novel parametrization of the perspective-three-point problem for a direct computation of absolute camera position and orientation. In: Computer Vision and Pattern Recognition (CVPR) (2011)
22. Kukelova, Z., Albl, C., Sugimoto, A., Pajdla, T.: Linear solution to the minimal absolute pose rolling shutter problem. In: Asian Conference on Computer Vision (ACCV) (2018)
23. Kukelova, Z., Bujnak, M., Pajdla, T.: Automatic generator of minimal problem solvers. In: Forsyth, D., Torr, P., Zisserman, A. (eds.) ECCV 2008. LNCS, vol. 5304, pp. 302–315. Springer, Heidelberg (2008). https://doi.org/10.1007/978-3-540-88690-7_23
24. Kukelova, Z., Bujnak, M., Pajdla, T.: Real-time solution to the absolute pose problem with unknown radial distortion and focal length. In: International Conference on Computer Vision (ICCV) (2013)
25. Lao, Y., Ait-Aider, O., Bartoli, A.: Rolling shutter pose and ego-motion estimation using shape-from-template. In: European Conference on Computer Vision (ECCV), pp. 477–492 (2018)
26. Larsson, V., Åström, K., Oskarsson, M.: Efficient solvers for minimal problems by syzygy-based reduction. In: Computer Vision and Pattern Recognition (CVPR) (2017)
27. Larsson, V., Kukelova, Z., Zheng, Y.: Making minimal solvers for absolute pose estimation compact and robust. In: International Conference on Computer Vision (ICCV) (2017)
28. Larsson, V., Kukelova, Z., Zheng, Y.: Camera pose estimation with unknown principal point. In: Computer Vision and Pattern Recognition (CVPR) (2018)
29. Larsson, V., Oskarsson, M., Astrom, K., Wallis, A., Kukelova, Z., Pajdla, T.: Beyond grobner bases: basis selection for minimal solvers. In: Computer Vision and Pattern Recognition (CVPR) (2018)

30. Magerand, L., Bartoli, A., Ait-Aider, O., Pizarro, D.: Global optimization of object pose and motion from a single rolling shutter image with automatic 2d–3d matching. In: European Conference on Computer Vision (ECCV), pp. 456–469 (2012)
31. Meingast, M., Geyer, C., Sastry, S.: Geometric models of rolling-shutter cameras. arXiv cs/0503076 (2005)
32. Microsoft: Spatial Anchors (2020). https://azure.microsoft.com/en-us/services/spatial-anchors/
33. Sattler, T., Leibe, B., Kobbelt, L.: Efficient & effective prioritized matching for large-scale image-based localization. IEEE Trans. Pattern Anal. Mach. Intell. (PAMI) 39(9), 1744–1756 (2017)
34. Saurer, O., Koser, K., Bouguet, J.Y., Pollefeys, M.: Rolling shutter stereo. In: ICCV, pp. 465–472 (2013)
35. Saurer, O., Pollefeys, M., Lee, G.H.: A minimal solution to the rolling shutter pose estimation problem. In: International Conference on Intelligent Robots and Systems (IROS) (2015)
36. Schönberger, J.L., Frahm, J.M.: Structure-from-motion revisited. In: Computer Vision and Pattern Recognition (CVPR) (2016)
37. Snavely, N., Seitz, S.M., Szeliski, R.: Photo tourism: exploring photo collections in 3d. In: ACM Transactions on Graphics (2006)
38. Wu, C.: P3.5p: Pose estimation with unknown focal length. In: Computer Vision and Pattern Recognition (CVPR) (2015)
39. Zheng, Y., Sugimoto, S., Sato, I., Okutomi, M.: A general and simple method for camera pose and focal length determination. In: Computer Vision and Pattern Recognition (CVPR) (2014)

SRFlow: Learning the Super-Resolution Space with Normalizing Flow

Andreas Lugmayr[✉], Martin Danelljan, Luc Van Gool, and Radu Timofte

Computer Vision Laboratory, ETH Zurich, Zürich, Switzerland
{andreas.lugmayr,martin.danelljan,vangool,radu.timofte}@vision.ee.ethz.ch

Abstract. Super-resolution is an ill-posed problem, since it allows for multiple predictions for a given low-resolution image. This fundamental fact is largely ignored by state-of-the-art deep learning based approaches. These methods instead train a deterministic mapping using combinations of reconstruction and adversarial losses. In this work, we therefore propose **SRFlow**: a normalizing flow based super-resolution method capable of learning the conditional distribution of the output given the low-resolution input. Our model is trained in a principled manner using a single loss, namely the negative log-likelihood. SRFlow therefore directly accounts for the ill-posed nature of the problem, and learns to predict diverse photo-realistic high-resolution images. Moreover, we utilize the strong image posterior learned by SRFlow to design flexible image manipulation techniques, capable of enhancing super-resolved images by, e.g., transferring content from other images. We perform extensive experiments on faces, as well as on super-resolution in general. SRFlow outperforms state-of-the-art GAN-based approaches in terms of both PSNR and perceptual quality metrics, while allowing for diversity through the exploration of the space of super-resolved solutions. Code: git.io/Jfpyu.

1 Introduction

Single image super-resolution (SR) is an active research topic with several important applications. It aims to enhance the resolution of a given image by adding missing high-frequency information. Super-resolution is therefore a fundamentally ill-posed problem. In fact, for a given low-resolution (LR) image, there exist infinitely many compatible high-resolution (HR) predictions. This poses severe challenges when designing deep learning based super-resolution approaches.

Initial deep learning approaches [11,12,19,21,23] employ feed-forward architectures trained using standard L_2 or L_1 reconstruction losses. While these methods achieve impressive PSNR, they tend to generate blurry predictions. This shortcoming stems from discarding the ill-posed nature of the SR problem. The employed L_2 and L_1 reconstruction losses favor the prediction of an *average* over

Electronic supplementary material The online version of this chapter (https://doi.org/10.1007/978-3-030-58558-7_42) contains supplementary material, which is available to authorized users.

A. Vedaldi et al. (Eds.): ECCV 2020, LNCS 12350, pp. 715–732, 2020.
https://doi.org/10.1007/978-3-030-58558-7_42

Fig. 1. While prior work trains a deterministic mapping, SRFlow learns the distribution of photo-realistic HR images for a given LR image. This allows us to explicitly account for the ill-posed nature of the SR problem, and to sample diverse images. (8× upscaling)

the plausible HR solutions, leading to the significant reduction of high-frequency details. To address this problem, more recent approaches [2,15,22,38,46,53] integrate adversarial training and perceptual loss functions. While achieving sharper images with better perceptual quality, such methods only predict a *single* SR output, which does not fully account for the ill-posed nature of the SR problem.

We address the limitations of the aforementioned approaches by learning the conditional *distribution* of plausible HR images given the input LR image. To this end, we design a conditional normalizing flow [10,37] architecture for image super-resolution. Thanks to the exact log-likelihood training enabled by the flow formulation, our approach can model expressive distributions over the HR image space. This allows our network to learn the generation of photo-realistic SR images that are consistent with the input LR image, without any additional constraints or losses. Given an LR image, our approach can sample multiple diverse SR images from the learned distribution. In contrast to conventional methods, our network can thus explore the space of SR images (see Fig. 1).

Compared to standard Generative Adversarial Network (GAN) based SR approaches [22,46], the proposed flow-based solution exhibits a few key advantages. First, our method naturally learns to generate diverse SR samples without suffering from mode-collapse, which is particularly problematic in the conditional GAN setting [17,29]. Second, while GAN-based SR networks require multiple losses with careful parameter tuning, our network is stably trained with a single loss: the negative log-likelihood. Third, the flow network employs a fully invertible encoder, capable of mapping any input HR image to the latent flow-space and ensuring *exact* reconstruction. This allows us to develop powerful image manipulation techniques for editing the predicted SR or any existing HR image.

Contributions: We propose **SRFlow**, a flow-based super-resolution network capable of accurately learning the distribution of realistic HR images corresponding to the input LR image. In particular, the main contributions of this work are as follows: **(i)** We are the first to design a conditional normalizing flow architecture that achieves state-of-the-art super-resolution quality. **(ii)** We harness the strong HR distribution learned by SRFlow to develop novel techniques for controlled image manipulation and editing. (**iii**) Although only trained for super-resolution, we show that SRFlow is capable of image denoising and restoration. (**iv**) Comprehensive experiments for face and general image super-resolution show that our approach outperforms state-of-the-art GAN-based methods for both perceptual and reconstruction-based metrics.

2   Related Work

Single Image SR: Super-resolution has long been a fundamental challenge in computer vision due to its ill-posed nature. Early learning-based methods mainly employed sparse coding based techniques [8,41,51,52] or local linear regression [43,45,49]. The effectiveness of example-based deep learning for super-resolution was first demonstrated by SRCNN [11], which further led to the development of more effective network architectures [12,19,21,23]. However, these methods do not reproduce the sharp details present in natural images due to their reliance on L_2 and L_1 reconstruction losses. This was addressed in URDGN [53], SRGAN [22] and more recent approaches [2,15,38,46] by adopting a conditional GAN based architecture and training strategy. While these works aim to predict *one* example, we undertake the more ambitious goal of learning the distribution of *all* plausible reconstructions from the natural image manifold.

Stochastic SR: The problem of generating diverse super-resolutions has received relatively little attention. This is partly due to the challenging nature of the problem. While GANs provide an method for learning a distribution over data [14], conditional GANs are known to be extremely susceptible to mode collapse since they easily learn to ignore the stochastic input signal [17,29]. Therefore, most conditional GAN based approaches for super-resolution and image-to-image translation resort to purely deterministic mappings [22,35,46]. A few recent works [4,7,30] address GAN-based stochastic SR by exploring techniques to avoid mode collapse and explicitly enforcing low-resolution consistency. In contrast to those works, we design a flow-based architecture trained using the negative log-likelihood loss. This allows us to learn the conditional distribution of HR images, without any additional constraints, losses, or post-processing techniques to enforce low-resolution consistency. A different line of research [6,39,40] exploit the internal patch recurrence by only training the network on the input image itself. Recently [39] employed this strategy to learn a GAN capable of stochastic SR generation. While this is an interesting direction, our goal is to exploit large image datasets to learn a general distribution over the image space.

Normalizing Flow: Generative modelling of natural images poses major challenges due to the high dimensionality and complex structure of the underlying data distribution. While GANs [14] have been explored for several vision tasks, Normalizing Flow based models [9,10,20,37] have received much less attention. These approaches parametrize a complex distribution $p_\mathbf{y}(\mathbf{y}|\boldsymbol{\theta})$ using an invertible neural network f_θ, which maps samples drawn from a simple (e.g. Gaussian) distribution $p_\mathbf{z}(\mathbf{z})$ as $\mathbf{y} = f_\theta^{-1}(\mathbf{z})$. This allows the *exact* negative log-likelihood

$-\log p_{\mathbf{y}}(\mathbf{y}|\boldsymbol{\theta})$ to be computed by applying the change-of-variable formula. The network can thus be trained by directly minimizing the negative log-likelihood using standard SGD-based techniques. Recent works have investigated conditional flow models for point cloud generation [36,50] as well as class [24] and image [3,48] conditional generation of images. The latter works [3,48] adapt the widely successful Glow architecture [20] to conditional image generation by concatenating the encoded conditioning variable in the affine coupling layers [9,10]. The concurrent work [48] consider the SR task as an example application, but only addressing 2× magnification and without comparisons with state-of-the-art GAN-based methods. While we also employ the conditional flow paradigm for its theoretically appealing properties, our work differs from these previous approaches in several aspects. Our work is first to develop a conditional flow architecture for SR that provides favorable or superior results compared to state-of-the-art GAN-based methods. Second, we develop powerful flow-based image manipulation techniques, applicable for guided SR and to editing existing HR images. Third, we introduce new training and architectural considerations. Lastly, we demonstrate the generality and strength of our learned image posterior by applying SRFlow to image restoration tasks, unseen during training.

3   Proposed Method: SRFlow

We formulate super-resolution as the problem of learning a conditional probability distribution over high-resolution images, given an input low-resolution image. This approach explicitly addresses the ill-posed nature of the SR problem by aiming to capture the full diversity of possible SR images from the natural image manifold. To this end, we design a conditional normalizing flow architecture, allowing us to learn rich distributions using exact log-likelihood based training.

3.1   Conditional Normalizing Flows for Super-Resolution

The goal of super-resolution is to predict higher-resolution versions $\mathbf{y}$ of a given low-resolution image $\mathbf{x}$ by generating the absent high-frequency details. While most current approaches learn a deterministic mapping $\mathbf{x} \mapsto \mathbf{y}$, we aim to capture the full conditional distribution $p_{\mathbf{y}|\mathbf{x}}(\mathbf{y}|\mathbf{x}, \boldsymbol{\theta})$ of natural HR images $\mathbf{y}$ corresponding to the LR image $\mathbf{x}$. This constitutes a more challenging task, since the model must span a variety of possible HR images, instead of just predicting a single SR output. Our intention is to train the parameters $\boldsymbol{\theta}$ of the distribution in a purely data-driven manner, given a large set of LR-HR training pairs $\{(\mathbf{x}_i, \mathbf{y}_i)\}_{i=1}^{M}$.

The core idea of normalizing flow [9,37] is to parametrize the distribution $p_{\mathbf{y}|\mathbf{x}}$ using an invertible neural network $f_{\boldsymbol{\theta}}$. In the conditional setting, $f_{\boldsymbol{\theta}}$ maps an HR-LR image pair to a latent variable $\mathbf{z} = f_{\boldsymbol{\theta}}(\mathbf{y}; \mathbf{x})$. We require this function to be invertible w.r.t. the first argument $\mathbf{y}$ for any LR image $\mathbf{x}$. That is, the HR image $\mathbf{y}$ can always be exactly reconstructed from the latent encoding $\mathbf{z}$ as $\mathbf{y} = f_{\boldsymbol{\theta}}^{-1}(\mathbf{z}; \mathbf{x})$. By postulating a simple distribution $p_{\mathbf{z}}(\mathbf{z})$ (e.g. a Gaussian) in the latent space $\mathbf{z}$, the conditional distribution $p_{\mathbf{y}|\mathbf{x}}(\mathbf{y}|\mathbf{x}, \boldsymbol{\theta})$ is implicitly defined by the mapping $\mathbf{y} = f_\theta^{-1}(\mathbf{z}; \mathbf{x})$ of samples $\mathbf{z} \sim p_{\mathbf{z}}$. The key aspect of normalizing flows is that the probability density $p_{\mathbf{y}|\mathbf{x}}$ can be explicitly computed as,

$$p_{\mathbf{y}|\mathbf{x}}(\mathbf{y}|\mathbf{x}, \boldsymbol{\theta}) = p_{\mathbf{z}}\big(f_\theta(\mathbf{y}; \mathbf{x})\big) \left| \det \frac{\partial f_\theta}{\partial \mathbf{y}}(\mathbf{y}; \mathbf{x}) \right|. \tag{1}$$

It is derived by applying the change-of-variables formula for densities, where the second factor is the resulting volume scaling given by the determinant of the Jacobian $\frac{\partial f_\theta}{\partial \mathbf{y}}$. The expression (1) allows us to train the network by minimizing the negative log-likelihood (NLL) for training samples pairs $(\mathbf{x}, \mathbf{y})$,

$$\mathcal{L}(\boldsymbol{\theta}; \mathbf{x}, \mathbf{y}) = -\log p_{\mathbf{y}|\mathbf{x}}(\mathbf{y}|\mathbf{x}, \boldsymbol{\theta}) = -\log p_{\mathbf{z}}\big(f_\theta(\mathbf{y}; \mathbf{x})\big) - \log \left| \det \frac{\partial f_\theta}{\partial \mathbf{y}}(\mathbf{y}; \mathbf{x}) \right|. \tag{2}$$

HR image samples $\mathbf{y}$ from the learned distribution $p_{\mathbf{y}|\mathbf{x}}(\mathbf{y}|\mathbf{x}, \boldsymbol{\theta})$ are generated by applying the inverse network $\mathbf{y} = f_\theta^{-1}(\mathbf{z}; \mathbf{x})$ to random latent variables $\mathbf{z} \sim p_{\mathbf{z}}$.

In order to achieve a tractable expression of the second term in (2), the neural network f_θ is decomposed into a sequence of N invertible layers $\mathbf{h}^{n+1} = f_\theta^n(\mathbf{h}^n; g_\theta(\mathbf{x}))$, where $\mathbf{h}^0 = \mathbf{y}$ and $\mathbf{h}^N = \mathbf{z}$. We let the LR image to first be encoded by a shared deep CNN $g_\theta(\mathbf{x})$ that extracts a rich representation suitable for conditioning in all flow-layers, as detailed in Sect. 3.3. By applying the chain rule along with the multiplicative property of the determinant [10], the NLL objective in (2) can be expressed as

$$\mathcal{L}(\boldsymbol{\theta}; \mathbf{x}, \mathbf{y}) = -\log p_{\mathbf{z}}(\mathbf{z}) - \sum_{n=0}^{N-1} \log \left| \det \frac{\partial f_\theta^n}{\partial \mathbf{h}^n}(\mathbf{h}^n; g_\theta(\mathbf{x})) \right|. \tag{3}$$

We thus only need to compute the log-determinant of the Jacobian $\frac{\partial f_\theta^n}{\partial \mathbf{h}^n}$ for each individual flow-layer f_θ^n. To ensure efficient training and inference, the flow layers f_θ^n thus need to allow efficient inversion and a tractable Jacobian determinant. This is further discussed next, where we detail the employed conditional flow layers f_θ^n in our SR architecture. Our overall network architecture for flow-based super-resolution is depicted in Fig. 2.

3.2 Conditional Flow Layers

The design of flow-layers f_θ^n requires care in order to ensure a well-conditioned inverse and a tractable Jacobian determinant. This challenge was first addressed in [9,10] and has recently spurred significant interest [5,13,20]. We start from the unconditional Glow architecture [20], which is itself based on the RealNVP [10]. The flow layers employed in these architectures can be made conditional in a straight-forward manner [3,48]. We briefly review them here along with our introduced Affine Injector layer.

Conditional Affine Coupling: The affine coupling layer [9,10] provides a simple and powerful strategy for constructing flow-layers that are easily invertible. It is trivially extended to the conditional setting as follows,

$$\mathbf{h}_A^{n+1} = \mathbf{h}_A^n, \qquad \mathbf{h}_B^{n+1} = \exp\big(f_{\theta,\mathrm{s}}^n(\mathbf{h}_A^n; \mathbf{u})\big) \cdot \mathbf{h}_B^n + f_{\theta,\mathrm{b}}^n(\mathbf{h}_A^n; \mathbf{u}). \tag{4}$$

Here, $\mathbf{h}^n = (\mathbf{h}_A^n, \mathbf{h}_B^n)$ is a partition of the activation map in the channel dimension. Moreover, $\mathbf{u}$ is the conditioning variable, set to the encoded LR image $\mathbf{u} = g_\theta(\mathbf{x})$ in our work. Note that $f_{\theta,s}^n$ and $f_{\theta,b}^n$ represent *arbitrary* neural networks that generate the scaling and bias of $\mathbf{h}_B^n$. The Jacobian of (4) is triangular, enabling the efficient computation of its log-determinant as $\sum_{ijk} f_{\theta,s}^n(\mathbf{h}_A^n; \mathbf{u})_{ijk}$.

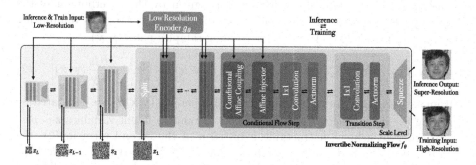

Fig. 2. SRFlow's conditional normalizing flow architecture. Our model consists of an invertible flow network f_θ, conditioned on an encoding (green) of the low-resolution image. The flow network operates at multiple scale levels (gray). The input is processed through a series of flow-steps (blue), each consisting of four different layers. Through exact log-likelihood training, our network learns to transform a Gaussian density $p_\mathbf{z}(\mathbf{z})$ to the conditional HR-image distribution $p_{\mathbf{y}|\mathbf{x}}(\mathbf{y}|\mathbf{x}, \boldsymbol{\theta})$. During training, an LR-HR $(\mathbf{x}, \mathbf{y})$ image pair is input in order to compute the negative log-likelihood loss. During inference, the network operates in the reverse direction by inputting the LR image along with a random variables $\mathbf{z} = (\mathbf{z}_l)_{l=1}^L \sim p_\mathbf{z}$, which generates sample SR images from the learned distribution $p_{\mathbf{y}|\mathbf{x}}$. (Color figure online)

Invertible 1×1 Convolution: General convolutional layers are often intractable to invert or evaluate the determinant of. However, [20] demonstrated that a 1×1 convolution $\mathbf{h}_{ij}^{n+1} = W\mathbf{h}_{ij}^n$ can be efficiently integrated since it acts on each spatial coordinate (i, j) independently, which leads to a block-diagonal structure. We use the non-factorized formulation in [20].

Actnorm: This provides a channel-wise normalization through a learned scaling and bias. We keep this layer in its standard un-conditional form [20].

Squeeze: It is important to process the activations at different scales in order to capture correlations and structures over larger distances. The squeeze layer [20] provides an invertible means to halving the resolution of the activation map $\mathbf{h}^n$ by reshaping each spatial 2×2 neighborhood into the channel dimension.

Affine Injector: To achieve more direct information transfer from the low-resolution image encoding $\mathbf{u} = g_\theta(\mathbf{x})$ to the flow branch, we additionally introduce the affine injector layer. In contrast to the conditional affine coupling layer, our affine injector layer directly affects all channels and spatial locations in the activation map $\mathbf{h}^n$. This is achieved by predicting an element-wise scaling and bias using only the conditional encoding $\mathbf{u}$,

$$\mathbf{h}^{n+1} = \exp\left(f_{\theta,s}^n(\mathbf{u})\right) \cdot \mathbf{h}^n + f_{\theta,b}(\mathbf{u}).\tag{5}$$

Here, $f_{\theta,s}$ and $f_{\theta,s}$ can be any network. The inverse of (5) is trivially obtained as $\mathbf{h}^n = \exp(-f_{\theta,s}^n(\mathbf{u})) \cdot (\mathbf{h}^{n+1} - f_{\theta,b}^n(\mathbf{u}))$ and the log-determinant is given by $\sum_{ijk} f_{\theta,s}^n(\mathbf{u})_{ijk}$. Here, the sum ranges over all spatial i, j and channel indices k.

3.3 Architecture

Our SRFlow architecture, depicted in Fig. 2, consists of the invertible flow network f_θ and the LR encoder g_θ. The flow network is organized into L levels, each operating at a resolution of $\frac{H}{2^l} \times \frac{W}{2^l}$, where $l \in \{1, \ldots, L\}$ is the level number and $H \times W$ is the HR resolution. Each level itself contains K number of flow-steps.

Flow-Step: Each flow-step in our approach consists of four different layers, as visualized in Fig. 2. The Actnorm if applied first, followed by the 1×1 convolution. We then apply the two conditional layers, first the Affine Injector followed by the Conditional Affine Coupling.

Level Transitions: Each level first performs a squeeze operation that effectively halves the spatial resolution. We observed that this layer can lead to checkerboard artifacts in the reconstructed image, since it is only based on pixel re-ordering. To learn a better transition between the levels, we therefore remove the conditional layers first few flow steps after the squeeze (see Fig. 2). This allows the network to learn a linear invertible interpolation between neighboring pixels. Similar to [20], we split off 50% of the channels before the next squeeze layer. Our latent variables $(z_l)_{l=1}^L$ thus model variations in the image at different resolutions, as visualized in Fig. 2.

Low-Resolution Encoding Network g_θ: SRFlow allows for the use of any differentiable architecture for the LR encoding network g_θ, since it does not need to be invertible. Our approach can therefore benefit from the advances in standard feed-forward SR architectures. In particular, we adopt the popular CNN architecture based on Residual-in-Residual Dense Blocks (RRDB) [46], which builds upon [22,23]. It employs multiple residual and dense skip connections, without any batch normalization layers. We first discard the final upsampling layers in the RRDB architecture since we are only interested in the underlying representation and not the SR prediction. In order to capture a richer representation of the LR image at multiple levels, we additionally concatenate the activations after each RRDB block to form the final output of g_θ.

Details: We employ $K = 16$ flow-steps at each level, with two additional unconditional flow-steps after each squeeze layer (discussed above). We use $L = 3$ and $L = 4$ levels for SR factors 4× and 8× respectively. For general image SR, we use the standard 23-block RRDB architecture [46] for the LR encoder g_θ. For faces, we reduce to 8 blocks for efficiency. The networks $f_{\theta,s}^n$ and $f_{\theta,b}^n$ in the conditional affine coupling (4) and the affine injector (5) are constructed using two shared convolutional layers with ReLU, followed by a final convolution.

3.4 Training Details

We train our entire SRFlow network using the negative log-likelihood loss (3). We sample batches of 16 LR-HR image pairs $(\mathbf{x}, \mathbf{y})$. During training, we use an HR patch size of 160×160. As optimizer we use Adam with a starting learning rate of $5 \cdot 10^{-4}$, which is halved at $50\%, 75\%, 90\%$ and 95% of the total training iterations. To increase training efficiency, we first pre-train the LR encoder g_θ using an L_1 loss for 200k iterations. We then train our full SRFlow architecture using only the loss (3) for 200k iterations. Our network takes 5 d to train on a single NVIDIA V100 GPU. Further details are provided in the supplementary.

Source Target Transferred   Source Target Transferred

Fig. 3. Random $8\times$ SR samples generated by SRFlow using a temperature $\tau = 0.8$. LR image is shown in top left.

Fig. 4. Latent space transfer from the region marked by the box to the target image. $(8\times)$

Datasets: For face super-resolution, we use the CelebA [25] dataset. Similar to [18,20], we pre-process the dataset by cropping aligned patches, which are resized to the HR resolution of 160×160. We employ the full train split (160k images). For general SR, we use the same training data as ESRGAN [46], consisting of the train split of 800 DIV2k [1] along with 2650 images from Flickr2K. The LR images are constructed using the standard MATLAB bicubic kernel.

4 Applications and Image Manipulations

In this section, we explore the use of our SRFlow network for a variety of applications and image manipulation tasks. Our techniques exploit two key advantages of our SRFlow network, which are not present in GAN-based super-resolution approaches [46]. First, our network models a distribution $p_{\mathbf{y}|\mathbf{x}}(\mathbf{y}|\mathbf{x}, \boldsymbol{\theta})$ in HR-image space, instead of only predicting a single image. It therefore possesses great flexibility by capturing a variety of possible HR predictions. This allows different predictions to be explored using additional guiding information or random sampling. Second, the flow network $f_\theta(\mathbf{y}; \mathbf{x})$ is a fully invertible encoder-decoder. Hence, *any* HR image $\tilde{\mathbf{y}}$ can be encoded into the latent space as $\tilde{\mathbf{z}} = f_\theta(\tilde{\mathbf{y}}; \mathbf{x})$ and *exactly* reconstructed as $\tilde{\mathbf{y}} = f_\theta^{-1}(\tilde{\mathbf{z}}; \mathbf{x})$. This bijective correspondence allows us to flexibly operate in both the latent and image space.

4.1   Stochastic Super-Resolution

The distribution $p_{\mathbf{y}|\mathbf{x}}(\mathbf{y}|\mathbf{x}, \boldsymbol{\theta})$ learned by our SRFlow can be explored by sampling different SR predictions as $\mathbf{y}^{(i)} = f_{\theta}^{-1}(\mathbf{z}^{(i)}; \mathbf{x})$, $\mathbf{z}^{(i)} \sim p_{\mathbf{z}}$ for a given LR image $\mathbf{x}$. As commonly observed for flow-based models, the best results are achieved when sampling with a slightly lower variance [20]. We therefore use a Gaussian $\mathbf{z}^{(i)} \sim \mathcal{N}(0, \tau)$ with variance τ (also called temperature). Results are visualized in Fig. 3 for $\tau = 0.8$. Our approach generates a large variety of SR images, including differences in e.g. hair and facial attributes, while preserving consistency with the LR image. Since our latent variables $\mathbf{z}_{ijkl}$ are spatially localized, specific parts can be re-sampled, enabling more controlled interactive editing and exploration of the SR image.

4.2   LR-Consistent Style Transfer

Our SRFlow allows transferring the style of an existing HR image $\tilde{\mathbf{y}}$ when super-resolving an LR image $\mathbf{x}$. This is performed by first encoding the source HR image as $\tilde{\mathbf{z}} = f_{\theta}(\tilde{\mathbf{y}}; d_{\downarrow}(\tilde{\mathbf{y}}))$, where $d_{\downarrow}$ is the down-scaling operator. The encoding $\tilde{\mathbf{z}}$ can then be used to as the latent variable for the super-resolution of $\mathbf{x}$ as $\mathbf{y} = f_{\theta}^{-1}(\tilde{\mathbf{z}}; \mathbf{x})$. This operation can also be performed on local regions of the image. Examples in Fig. 4 show the transfer in the style of facial characteristics, hair and eye color. Our SRFlow network automatically aims to ensure consistency with the LR image without any additional constraints.

4.3   Latent Space Normalization

We develop more advanced image manipulation techniques by taking advantage of the invertability of the SRFlow network f_{θ} and the learned super-resolution posterior. The core idea of our approach is to map any HR image containing desired content to the latent space, where the latent statistics can be normalized in order to make it consistent with the low-frequency information in the given LR image. Let $\mathbf{x}$ be a low-resolution image and $\tilde{\mathbf{y}}$ be *any* high-resolution image, not necessarily consistent with the LR image $\mathbf{x}$. For example, $\tilde{\mathbf{y}}$ can be an edited version of a super-resolved image or a guiding image for the super-resolution image. Our goal is to achieve an HR image $\mathbf{y}$, containing image content from $\tilde{\mathbf{y}}$, but that is consistent with the LR image $\mathbf{x}$.

The latent encoding for the given image pair is computed as $\tilde{\mathbf{z}} = f_{\theta}(\tilde{\mathbf{y}}; \mathbf{x})$. Note that our network is trained to predict consistent and natural SR images for latent variables sampled from a standard Gaussian distribution $p_{\mathbf{z}} = \mathcal{N}(0, I)$. Since $\tilde{\mathbf{y}}$ is not necessarily consistent with the LR image $\mathbf{x}$, the latent variables $\tilde{\mathbf{z}}_{ijkl}$ do not have the same statistics as if independently sampled from $\mathbf{z}_{ijkl} \sim \mathcal{N}(0, \tau)$. Here, τ denotes an additional temperature scaling of the desired latent distribution. In order to achieve desired statistics, we normalize the first two moments of collections of latent variables. In particular, if $\{z_i\}_1^N \sim \mathcal{N}(0, \tau)$ are independent, then it is well known [33] that their empirical mean $\hat{\mu}$ and variance $\hat{\sigma}^2$ are distributed according to,

$$\hat{\mu} = \frac{1}{N}\sum_{i=1}^{N} z_i \sim \mathcal{N}\left(0, \frac{\tau}{N}\right), \quad \hat{\sigma}^2 = \frac{1}{N-1}\sum_{i=1}^{N}(z_i - \hat{\mu})^2 \sim \Gamma\left(\frac{N-1}{2}, \frac{2\tau}{N-1}\right). \quad (6)$$

Here, $\Gamma(k,\theta)$ is a gamma distribution with shape and scale parameters k and θ respectively. For a given collection $\tilde{\mathcal{Z}} \subset \{\mathbf{z}_{ijkl}\}$ of latent variables, we normalize their statistics by first sampling a new mean $\hat{\mu}$ and variance $\hat{\sigma}^2$ according to (6), where $N = |\tilde{\mathcal{Z}}|$ is the size of the collection. The latent variables in the collection are then normalized as,

$$\hat{z} = \frac{\hat{\sigma}}{\tilde{\sigma}}(\tilde{z} - \tilde{\mu}) + \hat{\mu}, \quad \forall \tilde{z} \in \tilde{\mathcal{Z}}. \quad (7)$$

Here, $\tilde{\mu}$ and $\tilde{\sigma}^2$ denote the empirical mean and variance of the collection $\tilde{\mathcal{Z}}$.

		Original	Super-Resolved	Restored
DIV2K	PSNR↑	22.48	23.19	27.81
	SSIM↑	0.49	0.51	0.73
	LPIPS↓	0.370	0.364	0.255
CelebA	PSNR↑	22.52	24.25	27.62
	SSIM↑	0.48	0.63	0.78
	LPIPS↓	0.326	0.172	0.143

Source	Target $\mathbf{y}$	Input $\tilde{\mathbf{y}}$	Transferred $\hat{\mathbf{y}}$

Original	Direct SR	Restored

Fig. 5. Image content transfer for an existing HR image (top) and an SR prediction (bottom). Content from the source is applied directly to the target. By applying latent space normalization in our SRFlow, the content is integrated and harmonized.

Fig. 6. Comparision of super-resolving the LR of the original and normalizing the latent space for image restoration.

The normalization in (7) can be performed using different collections $\tilde{\mathcal{Z}}$. We consider three different strategies in this work. **Global normalization** is performed over the entire latent space, using $\tilde{\mathcal{Z}} = \{\mathbf{z}_{ijkl}\}_{ijkl}$. For **local normalization**, each spatial position i,j in each level l is normalized independently as $\tilde{\mathcal{Z}}_{ijl} = \{\mathbf{z}_{ijkl}\}_k$. This better addresses cases where the statistics is spatially varying. **Spatial normalization** is performed independently for each feature channel k and level l, using $\tilde{\mathcal{Z}}_{kl} = \{\mathbf{z}_{ijkl}\}_{ij}$. It addresses global effects in the image that activates certain channels, such as color shift or noise. In all three cases, normalized latent variable $\hat{\mathbf{z}}$ is obtained by applying (7) for all collections, which is an easily parallelized computation. The final HR image is then reconstructed as $\hat{\mathbf{y}} = f_\theta^{-1}(\hat{\mathbf{z}}, \mathbf{x})$. Note that our normalization procedure is stochastic, since a new mean $\hat{\mu}$ and variance $\hat{\sigma}^2$ are sampled independently for every collection of latent variables $\tilde{Z}$. This allows us to sample from the natural diversity of predictions $\hat{y}$, that integrate content from $\tilde{y}$. Next, we explore our latent space normalization technique for different applications.

4.4   Image Content Transfer

Here, we aim to manipulate an HR image by transferring content from other images. Let $\mathbf{x}$ be an LR image and $\mathbf{y}$ a corresponding HR image. If we are manipulating a super-resolved image, then $\mathbf{y} = f_\theta^{-1}(\mathbf{z}, \mathbf{x})$ is an SR sample of $\mathbf{x}$. However, we can also manipulate an existing HR image $\mathbf{y}$ by setting $\mathbf{x} = d_\downarrow(\mathbf{y})$ to the down-scaled version of $\mathbf{y}$. We then manipulate $\mathbf{y}$ directly in the image space by simply inserting content from other images, as visualized in Fig. 5. To harmonize the resulting manipulated image $\tilde{y}$ by ensuring consistency with the LR image $\mathbf{x}$, we compute the latent encoding $\tilde{z} = f_\theta(\tilde{y}; \mathbf{x})$ and perform *local normalization* of the latent variables as described in Sect. 4.3. We only normalize the affected regions of the image in order to preserve the non-manipulated content. Results are shown in Fig. 5. If desired, the emphasis on LR-consistency can be reduced by training SRFlow with randomly misaligned HR-LR pairs, which allows increased manipulation flexibility (see supplement).

4.5   Image Restoration

We demonstrate the strength of our learned image posterior by applying it for image restoration tasks. Note that we here employ the *same* SRFlow network, that is trained only for super-resolution, and not for the explored tasks. In particular, we investigate degradations that mainly affect the high frequencies in the image, such as noise and compression artifacts. Let $\tilde{y}$ be a degraded image. Noise and other high-frequency degradations are largely removed when down-sampled $\mathbf{x} = d_\downarrow(\tilde{y})$. Thus a cleaner image can be obtained by applying any super-resolution method to $\mathbf{x}$. However, this generates poor results since important image information is lost in the down-sampling process (Fig. 6, center).

Our approach can go beyond this result by directly utilizing the original image $\tilde{y}$. The degraded image along with its down-sampled variant are input to our SRFlow network to generate the latent variable $\tilde{z} = f_\theta(\tilde{y}; \mathbf{x})$. We then perform first *spatial* and then *local* normalization of $\tilde{z}$, as described in Sect. 4.3. The restored image is then predicted as $\hat{y} = f_\theta^{-1}(\hat{z}, \mathbf{x})$. By, denoting the normalization operation as $\hat{z} = \phi(\tilde{z})$, the full restoration mapping can be expressed as $\hat{y} = f_\theta^{-1}(\phi(f_\theta(\tilde{y}; d_\downarrow(\tilde{y}))), d_\downarrow(\tilde{y}))$. As shown visually and quantitatively in Fig. 6, this allows us to recover a substantial amount of details from the original image Intuitively, our approach works by mapping the degraded image $\tilde{y}$ to the *closest* image within the learned distribution $p_{\mathbf{y}|\mathbf{x}}(\mathbf{y}|\mathbf{x}, \boldsymbol{\theta})$. Since SRFlow is not trained with such degradations, $p_{\mathbf{y}|\mathbf{x}}(\mathbf{y}|\mathbf{x}, \boldsymbol{\theta})$ mainly models *clean* images. Our normalization therefore automatically restores the image when it is transformed to a more *likely* image according to our SR distribution $p_{\mathbf{y}|\mathbf{x}}(\mathbf{y}|\mathbf{x}, \boldsymbol{\theta})$.

5 Experiments

We perform comprehensive experiments for super-resolution of faces and of generic images in comparisons with current state-of-the-art and an ablative analysis. Applications, such as image manipulation tasks, are presented in Sect. 4, with additional results, analysis and visuals in the supplement.

Evaluation Metrics: To evaluate the perceptual distance to the Ground Truth, we report the default LPIPS [54]. It is a learned distance metric, based on the feature-space of a finetuned AlexNet. We report the standard fidelity oriented metrics, Peak Signal to Noise Ratio (PSNR) and structural similarity index (SSIM) [47], although they are known to not correlate well with the human perception of image quality [16,22,26,28,42,44]. Furthermore, we report the no-reference metrics NIQE [32], BRISQUE [31] and PIQUE [34]. In addition to

Table 1. Results for 8× SR of faces of CelebA. We compare using both the standard bicubic kernel and the progressive linear kernel from [18]. We also report the diversity in the SR output in terms of the pixel standard deviation σ.

LR		↑PSNR	↑SSIM	↓LPIPS	↑LR-PSNR	↓NIQE	↓BRISQUE	↓PIQUE	↑Diversity σ
Bicubic	Bicubic	23.15	0.63	0.517	35.19	7.82	58.6	99.97	0
	RRDB [46]	26.59	0.77	0.230	48.22	6.02	49.7	86.5	0
	ESRGAN [46]	22.88	0.63	0.120	34.04	3.46	23.7	32.4	0
	SRFlow $\tau = 0.8$	25.24	0.71	0.110	50.85	4.20	23.2	24.0	5.21
Prog.	ProgFSR [18]	23.97	0.67	0.129	41.95	3.49	28.6	33.2	0
	SRFlow $\tau = 0.8$	25.20	0.71	0.110	51.05	4.20	22.5	23.1	5.28

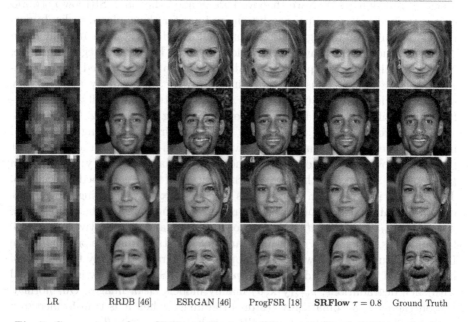

LR   RRDB [46]   ESRGAN [46]   ProgFSR [18]   **SRFlow** $\tau = 0.8$ Ground Truth

Fig. 7. Comparison of our SRFlow with state-of-the-art for 8× face SR on CelebA.

visual quality, consistency with the LR image is an important factor. We therefore evaluate this aspect by reporting the LR-PSNR, computed as the PSNR between the downsampled SR image and the original LR image.

5.1 Face Super-Resolution

We evaluate SRFlow for face SR (8×) using 5000 images from the test split of the CelebA dataset. We compare with bicubic, RRDB [46], ESRGAN [46], and ProgFSR [18]. While the latter two are GAN-based, RRDB is trained using only L_1 loss. ProgFSR is a very recent SR method specifically designed for faces, shown to outperform several prior face SR approaches in [18]. It is trained on the full train split of CelebA, but using a bilinear kernel. For fair comparison, we therefore separately train and evaluate SRFlow on the same kernel.

Results are provided in Table 1 and Fig. 7. Since our aim is perceptual quality, we consider LPIPS the primary metric, as it has been shown to correlate much better with human opinions [27,54]. SRFlow achieves more than twice as good LPIPS distance compared to RRDB, at the cost of lower PSNR and SSIM scores. As seen in the visual comparisons in Fig. 7, RRDB generates extremely blurry

Table 2. General image SR results on the 100 validation images of the DIV2K dataset.

	DIV2K 4×							DIV2K 8×						
	PSNR↑	SSIM↑	LPIPS↓	LR-PSNR↑	NIQE↓	BRISQUE↓	PIQUE↓	PSNR↑	SSIM↑	LPIPS↓	LR-PSNR↑	NIQE↓	BRISQUE↓	PIQUE↓
Bicubic	26.70	0.77	0.409	38.70	5.20	53.8	86.6	23.74	0.63	0.584	37.16	6.65	60.3	97.6
EDSR [23]	28.98	0.83	0.270	54.89	4.46	43.3	77.5	–	–	–	–	–	–	–
RRDB [46]	29.44	0.84	0.253	49.20	5.08	52.4	86.7	25.50	0.70	0.419	45.43	4.35	42.4	79.1
RankSRGAN [55]	26.55	0.75	0.128	42.33	2.45	17.2	20.1	–	–	–	–	–	–	–
ESRGAN [46]	26.22	0.75	0.124	39.03	2.61	22.7	26.2	22.18	0.58	0.277	31.35	2.52	20.6	25.8
SRFlow $\tau = 0.9$	27.09	0.76	0.120	49.96	3.57	17.8	18.6	23.05	0.57	0.272	50.00	3.49	20.9	17.1

| Low Resolution | Bicubic | EDSR [23] | RRDB [46] | ESRGAN [46] | RankSRGAN [55] | **SRFlow** $\tau = 0.9$ | Ground Truth |

Fig. 8. Comparison to state-of-the-art for general SR on the DIV2K validation set.

results, lacking natural high-frequency details. Compared to the GAN-based methods, SRFlow achieves significantly better results in all reference metrics. Interestingly, even the PSNR is superior to ESRGAN and ProgFSR, showing that our approach preserves fidelity to the HR ground-truth, while achieving better perceptual quality. This is partially explained by the hallucination artifacts that often plague GAN-based approaches, as seen in Fig. 7. Our approach generate sharp and natural images, while avoiding such artifacts. Interestingly, our SRFlow achieves an LR-consistency that is even better than the fidelity-trained RRDB, while the GAN-based methods are comparatively in-consistent with the input LR image.

5.2   General Super-Resolution

Next, we evaluate our SRFlow for general SR on the DIV2K validation set. We compare SRFlow to bicubic, EDSR [23], RRDB [46], ESRGAN [46], and RankSRGAN [55]. Except for EDSR, which used DIV2K, all methods including SRFlow are trained on the train splits of DIV2K and Flickr2K (see Sect. 3.3). For the 4× setting, we employ the provided pre-trained models. Due to lacking availability, we trained RRDB and ESRGAN for 8× using the authors' code.

K = 16 Steps   K = 8 Steps   K = 4 Steps   196 Channels   64 Channels

Fig. 9. Analysis of number of flow steps and dimensionality in the conditional layers.

Table 3. Analysis of the impact of the transitional linear flow steps and the affine image injector.

DIV2K 4×	PSNR↑	SSIM↑	LPIPS↓
No Lin. F-Step	26.96	0.759	0.125
No Affine Inj	26.81	0.756	0.126
SRFlow	27.09	0.763	0.125

EDSR and RRDB are trained using only reconstruction losses, thereby achieving inferior results in terms of the perceptual LPIPS metric (Table 2). Compared to the GAN-based methods [46,55], our SRFlow achieves significantly better PSNR, LPIPS and LR-PSNR and favorable results in terms of PIQUE and BRISQUE. Visualizations in Fig. 8 confirm the perceptually inferior results of EDSR and RRDB, which generate little high-frequency details. In contrast, SRFlow generates rich details, achieving favorable perceptual quality compared to ESRGAN. The first row, ESRGAN generates severe discolored artifacts and ringing patterns at several locations in the image. We find SRFlow to generate more stable and consistent results in these circumstances.

5.3   Ablative Study

To ablate the depth and width, we train our network with different number of flow-steps K and hidden layers in two conditional layers (4) and (5) respectively.

Figure 9 shows results on the CelebA dataset. Decreasing the number of flow-steps K leads to more artifacts in complex structures, such as eyes. Similarly, a larger number of channels leads to better consistency in the reconstruction. In Table 3 we analyze architectural choices. The Affine Image Injector increases the fidelity, while preserving the perceptual quality. We also observe the transitional linear flow steps (Sect. 3.3) to be beneficial.

6   Conclusion

We propose a flow-based method for super-resolution, called SRFlow. Contrary to conventional methods, our approach learns the distribution of photo-realistic SR images given the input LR image. This explicitly accounts for the ill-posed nature of the SR problem and allows for the generation of diverse SR samples. Moreover, we develop techniques for image manipulation, exploiting the strong image posterior learned by SRFlow. In comprehensive experiments, our approach achieves improved results compared to state-of-the-art GAN-based approaches.

Acknowledgements. This work was supported by the ETH Zürich Fund (OK), a Huawei Technologies Oy (Finland) project, a Google GCP grant, an Amazon AWS grant, and an Nvidia GPU grant.

References

1. Agustsson, E., Timofte, R.: Ntire 2017 challenge on single image super-resolution: dataset and study. In: CVPR Workshops (2017)
2. Ahn, N., Kang, B., Sohn, K.A.: Image super-resolution via progressive cascading residual network. In: CVPR (2018)
3. Ardizzone, L., Lüth, C., Kruse, J., Rother, C., Köthe, U.: Guided image generation with conditional invertible neural networks. CoRR abs/1907.02392 (2019). http://arxiv.org/abs/1907.02392
4. Bahat, Y., Michaeli, T.: Explorable super resolution. arXiv.vol. abs/1912.01839 (2019)
5. Behrmann, J., Grathwohl, W., Chen, R.T.Q., Duvenaud, D., Jacobsen, J.: Invertible residual networks. In: ICML. Proceedings of Machine Learning Research, vol. 97, pp. 573–582. PMLR (2019)
6. Bell-Kligler, S., Shocher, A., Irani, M.: Blind super-resolution kernel estimation using an internal-gan. In: NeurIPS, pp. 284–293 (2019). http://papers.nips.cc/paper/8321-blind-super-resolution-kernel-estimation-using-an-internal-gan
7. Bühler, M.C., Romero, A., Timofte, R.: Deepsee: deep disentangled semantic explorative extreme super-resolution. arXiv preprint arXiv:2004.04433 (2020)
8. Dai, D., Timofte, R., Gool, L.V.: Jointly optimized regressors for image super-resolution. Comput. Graph. Forum **34**(2), 95–104 (2015). https://doi.org/10.1111/cgf.12544
9. Dinh, L., Krueger, D., Bengio, Y.: NICE: non-linear independent components estimation. In: 3rd International Conference on Learning Representations, ICLR 2015, San Diego, CA, USA, 7–9 May 2015, Workshop Track Proceedings (2015)

10. Dinh, L., Sohl-Dickstein, J., Bengio, S.: Density estimation using real NVP. In: 5th International Conference on Learning Representations, ICLR 2017, Toulon, France, 24–26 April 2017, Conference Track Proceedings (2017)
11. Dong, C., Loy, C.C., He, K., Tang, X.: Learning a deep convolutional network for image super-resolution. In: ECCV, pp. 184–199 (2014). https://doi.org/10.1007/978-3-319-10593-2_13
12. Dong, C., Loy, C.C., He, K., Tang, X.: Image super-resolution using deep convolutional networks. TPAMI 38(2), 295–307 (2016)
13. Durkan, C., Bekasov, A., Murray, I., Papamakarios, G.: Neural spline flows. In: Advances in Neural Information Processing Systems 32: Annual Conference on Neural Information Processing Systems 2019, NeurIPS 2019, 8–14 December 2019, Vancouver, BC, Canada, pp. 7509–7520 (2019)
14. Goodfellow, I.J., et al.: Generative adversarial nets. In: Advances in Neural Information Processing Systems 27: Annual Conference on Neural Information Processing Systems 2014, 8–13 December 2014, Montreal, Quebec, Canada, pp. 2672–2680 (2014)
15. Haris, M., Shakhnarovich, G., Ukita, N.: Deep back-projection networks for super-resolution. In: CVPR (2018)
16. Ignatov, A., et al.: Pirm challenge on perceptual image enhancement on smartphones: report. arXiv preprint arXiv:1810.01641 (2018)
17. Isola, P., Zhu, J., Zhou, T., Efros, A.A.: Image-to-image translation with conditional adversarial networks. In: CVPR, pp. 5967–5976 (2017). https://doi.org/10.1109/CVPR.2017.632
18. Kim, D., Kim, M., Kwon, G., Kim, D.: Progressive face super-resolution via attention to facial landmark. In: arxiv. vol. abs/1908.08239 (2019)
19. Kim, J., Kwon Lee, J., Mu Lee, K.: Accurate image super-resolution using very deep convolutional networks. In: CVPR (2016)
20. Kingma, D.P., Dhariwal, P.: Glow: Generative flow with invertible 1x1 convolutions. In: Advances in Neural Information Processing Systems 31: Annual Conference on Neural Information Processing Systems 2018, NeurIPS 2018, 3–8 December 2018, Montréal, Canada, pp. 10236–10245 (2018)
21. Lai, W.S., Huang, J.B., Ahuja, N., Yang, M.H.: Deep laplacian pyramid networks for fast and accurate super-resolution. In: CVPR (2017)
22. Ledig, C., et al.: Photo-realistic single image super-resolution using a generative adversarial network. In: CVPR (2017)
23. Lim, B., Son, S., Kim, H., Nah, S., Lee, K.M.: Enhanced deep residual networks for single image super-resolution. In: CVPR (2017)
24. Liu, R., Liu, Y., Gong, X., Wang, X., Li, H.: Conditional adversarial generative flow for controllable image synthesis. In: IEEE Conference on Computer Vision and Pattern Recognition, CVPR 2019, Long Beach, CA, USA, 16–20 June 2019, pp. 7992–8001 (2019)
25. Liu, Z., Luo, P., Wang, X., Tang, X.: Deep learning face attributes in the wild. In: Proceedings of International Conference on Computer Vision (ICCV), December 2015
26. Lugmayr, A., Danelljan, M., Timofte, R.: Unsupervised learning for real-world super-resolution. In: ICCVW, pp. 3408–3416. IEEE (2019)
27. Lugmayr, A., Danelljan, M., Timofte, R.: Ntire 2020 challenge on real-world image super-resolution: methods and results. In: Proceedings of the IEEE/CVF Conference on Computer Vision and Pattern Recognition (CVPR) Workshops, June 2020
28. Lugmayr, A., Danelljan, M., Timofte, R., et al.: Aim 2019 challenge on real-world image super-resolution: methods and results. In: ICCV Workshops (2019)

29. Mathieu, M., Couprie, C., LeCun, Y.: Deep multi-scale video prediction beyond mean square error. In: ICLR (2016). http://arxiv.org/abs/1511.05440

30. Menon, S., Damian, A., Hu, S., Ravi, N., Rudin, C.: Pulse: self-supervised photo upsampling via latent space exploration of generative models. In: CVPR (2020)

31. Mittal, A., Moorthy, A., Bovik, A.: Referenceless image spatial quality evaluation engine. In: 45th Asilomar Conference on Signals, Systems and Computers, vol. 38, pp. 53–54 (2011)

32. Mittal, A., Soundararajan, R., Bovik, A.C.: Making a "completely blind" image quality analyzer. IEEE Signal Process. Lett. **20**(3), 209–212 (2013)

33. Murphy, K.P.: Machine Learning: A Probabilistic Perspective. The MIT Press, Cambridge (2012)

34. Venkatanath, N., Praneeth, D., Bh, M.C., Channappayya, S.S., Medasani, S.S: Blind image quality evaluation using perception based features. In: NCC, pp. 1–6. IEEE (2015)

35. Pathak, D., Krähenbühl, P., Donahue, J., Darrell, T., Efros, A.A.: Context encoders: feature learning by inpainting. In: CVPR, pp. 2536–2544. IEEE Computer Society (2016)

36. Pumarola, A., Popov, S., Moreno-Noguer, F., Ferrari, V.: C-flow: conditional generative flow models for images and 3d point clouds. In: CVPR, pp. 7949–7958 (2020)

37. Rezende, D.J., Mohamed, S.: Variational inference with normalizing flows. In: Proceedings of the 32nd International Conference on Machine Learning, ICML 2015, Lille, France, 6–11 July 2015, pp. 1530–1538 (2015)

38. Sajjadi, M.S.M., Schölkopf, B., Hirsch, M.: Enhancenet: single image super-resolution through automated texture synthesis. In: IEEE International Conference on Computer Vision, ICCV 2017, Venice, Italy, 22–29 October 2017, pp. 4501–4510. IEEE Computer Society (2017). https://doi.org/10.1109/ICCV.2017.481

39. Shaham, T.R., Dekel, T., Michaeli, T.: Singan: learning a generative model from a single natural image. In: ICCV, pp. 4570–4580 (2019)

40. Shocher, A., Cohen, N., Irani, M.: Zero-shot super-resolution using deep internal learning. In: CVPR (2018)

41. Sun, L., Hays, J.: Super-resolution from internet-scale scene matching. In: ICCP (2012)

42. Timofte, R., et al.: Ntire 2017 challenge on single image super-resolution: methods and results. In: CVPR Workshops (2017)

43. Timofte, R., De Smet, V., Van Gool, L.: A+: adjusted anchored neighborhood regression for fast super-resolution. In: Cremers, D., Reid, I., Saito, H., Yang, M.-H. (eds.) ACCV 2014. LNCS, vol. 9006, pp. 111–126. Springer, Cham (2015). https://doi.org/10.1007/978-3-319-16817-3_8

44. Timofte, R., Gu, S., Wu, J., Van Gool, L.: Ntire 2018 challenge on single image super-resolution: methods and results. In: CVPR Workshops (2018)

45. Timofte, R., Smet, V.D., Gool, L.V.: Anchored neighborhood regression for fast example-based super-resolution. In: ICCV, pp. 1920–1927 (2013). https://doi.org/10.1109/ICCV.2013.241

46. Wang, X., et al.: Esrgan: Enhanced super-resolution generative adversarial networks. ECCV (2018)

47. Wang, Z., Bovik, A.C., Sheikh, H.R., Simoncelli, E.P.: Image quality assessment: from error visibility to structural similarity. IEEE Trans. Image Process. **13**(4), 600–612 (2004)

48. Winkler, C., Worrall, D.E., Hoogeboom, E., Welling, M.: Learning likelihoods with conditional normalizing flows. arxiv abs/1912.00042 (2019). http://arxiv.org/abs/1912.00042

49. Yang, C., Yang, M.: Fast direct super-resolution by simple functions. In: ICCV, pp. 561–568 (2013). https://doi.org/10.1109/ICCV.2013.75

50. Yang, G., Huang, X., Hao, Z., Liu, M., Belongie, S.J., Hariharan, B.: Pointflow: 3d point cloud generation with continuous normalizing flows. In: ICCV (2019)

51. Yang, J., Wright, J., Huang, T.S., Ma, Y.: Image super-resolution as sparse representation of raw image patches. In: CVPR (2008). https://doi.org/10.1109/CVPR.2008.4587647

52. Yang, J., Wright, J., Huang, T.S., Ma, Y.: Image super-resolution via sparse representation. IEEE Trans. Image Process. **19**(11), 2861–2873 (2010). https://doi.org/10.1109/TIP.2010.2050625

53. Yu, X., Porikli, F.: Ultra-resolving face images by discriminative generative networks. In: ECCV, pp. 318–333 (2016). https://doi.org/10.1007/978-3-319-46454-1_20

54. Zhang, R., Isola, P., Efros, A.A., Shechtman, E., Wang, O.: The unreasonable effectiveness of deep features as a perceptual metric. In: CVPR (2018)

55. Zhang, W., Liu, Y., Dong, C., Qiao, Y.: Ranksrgan: generative adversarial networks with ranker for image super-resolution (2019)

DeepGMR: Learning Latent Gaussian Mixture Models for Registration

Wentao Yuan[1]([✉]), Benjamin Eckart[2], Kihwan Kim[2], Varun Jampani[2],
Dieter Fox[1,2], and Jan Kautz[2]

[1] University of Washington, Seattle, USA
wentaoy@cs.washington.edu
[2] NVIDIA, Santa Clara, USA

Abstract. Point cloud registration is a fundamental problem in 3D computer vision, graphics and robotics. For the last few decades, existing registration algorithms have struggled in situations with large transformations, noise, and time constraints. In this paper, we introduce Deep Gaussian Mixture Registration (DeepGMR), the first learning-based registration method that explicitly leverages a probabilistic registration paradigm by formulating registration as the minimization of KL-divergence between two probability distributions modeled as mixtures of Gaussians. We design a neural network that extracts pose-invariant correspondences between raw point clouds and Gaussian Mixture Model (GMM) parameters and two differentiable compute blocks that recover the optimal transformation from matched GMM parameters. This construction allows the network learn an SE(3)-invariant feature space, producing a global registration method that is real-time, generalizable, and robust to noise. Across synthetic and real-world data, our proposed method shows favorable performance when compared with state-of-the-art geometry-based and learning-based registration methods.

Keywords: Point cloud registration · Gaussian Mixture Model

1 Introduction

The development of 3D range sensors [2] has generated a massive amount of 3D data, which often takes the form of point clouds. The problem of assimilating raw point cloud data into a coherent world model is crucial in a wide range of vision, graphics and robotics applications. A core step in the creation of a world

---

W. Yuan—Work partially done during an internship at NVIDIA.

Electronic supplementary material The online version of this chapter (https:// doi.org/10.1007/978-3-030-58558-7_43) contains supplementary material, which is available to authorized users.

A. Vedaldi et al. (Eds.): ECCV 2020, LNCS 12350, pp. 733–750, 2020.
https://doi.org/10.1007/978-3-030-58558-7_43

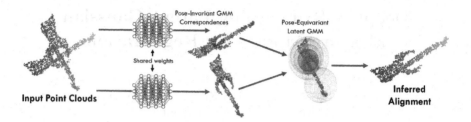

Fig. 1. DeepGMR aligns point clouds with arbitrary poses by predicting pose-invariant point correspondences to a learned latent Gaussian Mixture Model (GMM) distribution.

model is point cloud registration, the task of finding the transformation that aligns input point clouds into a common coordinate frame.

As a longstanding problem in computer vision and graphics, there is a large body of prior works on point cloud registration [30]. However, the majority of registration methods rely solely on matching local geometry and do not leverage learned features capturing large-scale shape information. As a result, such registration methods are often *local*, which means they cannot handle large transformations without good initialization. In contrast, a method is said to be *global* if its output is invariant to initial poses. Several works have investigated global registration. However, they are either too slow for real-time processing [5,47] or require good normal estimation to reach acceptable accuracy [48]. In many applications, such as re-localization and pose estimation, an accurate, fast, and robust global registration method is desirable.

The major difficulty of global registration lies in data association, since ICP-style correspondences based on Euclidean distances are no longer reliable. Existing registration methods provide several strategies for performing data association (see Fig. 2), but each of these prove problematic for global registration on noisy point clouds. Given point clouds of size N, DCP [42] attempts to perform point-to-point level matching like in ICP (Fig. 2a) over all N^2 point pairs, which suffers from $O(N^2)$ complexity. In addition, real-world point clouds don't contain exact point-level correspondences due to sensor noise. FGR [48] performs sparse feature-level correspondences that can be much more efficient (Fig. 2b), but are highly dependent on the quality of features. For example, the FPFH features used in [48] rely on consistent normal estimation, which is difficult to obtain in practice due to varying sparsity or non-rectilinear geometry [41]. Moreover, these sparse correspondences are still point-level and suffer the same problem when exact point-level correspondences don't exist. To solve this problem, one can use probabilistic methods to perform distribution-to-distribution matching (Fig. 2c). However, the distribution parameters are not guaranteed to be consistent across different views due to the well-known problem of Gaussian Mixture *non-indentifiability* [4]. Thus, probabilistic algorithms like HGMR [15] have only local convergence and rely on iterative techniques to update and refine point-to-distribution correspondences.

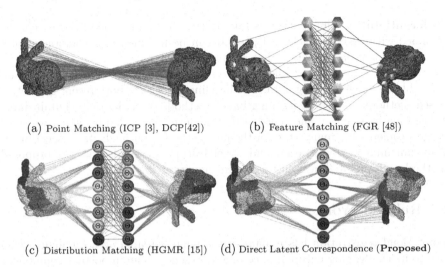

(a) Point Matching (ICP [3], DCP[42])   (b) Feature Matching (FGR [48])

(c) Distribution Matching (HGMR [15])   (d) Direct Latent Correspondence (**Proposed**)

Fig. 2. Comparison of data association techniques in various registration approaches. Unlike previous approaches that attempt to find point-to-point, feature-to-feature, or distribution-to-distribution level correspondences, we learn direct and pose-invariant correspondences to a distribution inside a learned latent space.

In this paper, we introduce a novel registration method that is designed to overcome these limitations by learning pose-invariant point-to-distribution parameter correspondences (Fig. 2d). Rather than depending on point-to-point correspondence [42] and iterative optimization [1], we solve for the optimal transformation in a single step by matching points to a probability distribution whose parameters are estimated by a neural network from the input point clouds. Our formulation is inspired by prior works on Gaussian mixture registration [15, 24], but different from these works in two ways. First, our method does not involve expensive iterative procedures such as Expectation Maximization (EM) [12]. Second, our network is designed to learn a consistent GMM representation across multiple point clouds rather than fit a GMM to a single reference point cloud.

Our proposed method has the following favorable properties:

Global Registration. Point clouds can be aligned with arbitrary displacements and without any initialization. While being accurate on its own, our method can work together with local refinement methods to achieve higher accuracy.

Efficiency. The proposed method runs on the order of 20–50 frames per second with moderate memory usage that grows linearly with the number of points, making it suitable for applications with limited computational resources.

Robustness. Due to its probabilistic formulation, our method is tolerant to noise and different sizes of input point clouds, and recovers the correct transformation even in the absence of exact point-pair correspondences, making it suitable for real-world scenarios.

Differentiability. Our method is fully differentiable and the gradients can be obtained with a single backward pass. It can be included as a component of a larger optimization procedure that requires gradients.

We demonstrate the advantages of our method over the state-of-the-art on several kinds of challenging data. The baselines consist of both recent geometry-based methods as well as learning-based methods. Our datasets contain large transformations, noise, and real-world scene-level point clouds, which we show cause problems for many state-of-the-art methods. Through its connection to Gaussian mixture registration and novel learning-based design, our proposed method performs well even in these challenging settings.

2   Related Work

Point cloud registration has remained a popular research area for many years due to its challenging nature and common presence as an important component of many 3D perception applications. Here we broadly categorize prior work into *local* and *global* techniques, and discuss how emerging *learning-based* methods fit into these categories.

Local Registration. Local approaches are often highly efficient and can be highly effective if limited to regimes where transformations are known *a priori* to be small in magnitude. The most well-known approach is the Iterative Closest Point (ICP) algorithm [3,8] and its many variants [33,34]. ICP iteratively alternates between two phases: point-to-point correspondence and distance minimization. Countless strategies have been proposed for handling outliers and noise [9], creating robust minimizers [18], or devising better distance metrics [26,35].

Another branch of work on local methods concerns probabilistic registration, often via the use of GMMs and the EM algorithm [12]. Traditional examples include EM-ICP [20], GMMReg [24], and methods based on the Normal Distributions Transform (NDT [36]). More recent examples offer features such as batch registration (JRMPC [16]) or robustness to density variance and viewing angle (DARE [23]). Other recent approaches have focused on efficiency, including filter-based methods [19], GPU-accelerated hierarchical methods [15], Monte Carlo methods [13] or efficient distribution-matching techniques [38].

In our experiments, we compare our algorithm against local methods belonging to both paradigms: the Trimmed ICP algorithm with point-to-plane minimization [9,26] and Hierarchical Gaussian Mixture Registration (HGMR) [15], a state-of-the-art probabilistic method.

Global Registration. Unlike local methods, global methods are invariant to initial conditions, but often at the cost of efficiency. Some approaches exhaustively search $SE(3)$ via branch-and-bound techniques (Go-ICP [47], GOGMA [5] and GOSMA [6]). Other approaches use local feature matching with robust optimization techniques such as RANSAC [17] or semidefinite programming [46].

One notable exception to the general rule that global methods must be inefficient is Fast Global Registration (FGR) [48], which achieves invariance to initial pose while remaining as fast or faster than many local methods. We compare against FGR [48], RANSAC [17] and TEASER++ [46] in our experiments as representatives of state-of-the-art geometry-based global methods.

Learning-Based Registration. Deep learning techniques on point clouds such as [31,32,37,44] provide task-specific learned point representations that can be leveraged for robust point cloud registration. PointNetLK [1], DCP [42] and PRNet [43] are the closest-related registration methods to ours. PointNetLK [1] proposes a differentiable Lucas-Kanade algorithm [27] that tries to minimize the feature distance between point clouds. DCP [42] proposes attention-based feature matching coupled with differentiable SVD for point-to-point registration, while PRNet [43] uses neural networks to detect keypoints followed by SVD for final registration. However, as we show in our experiments, due to their iterative nature, PointNetLK and PRNet are local methods that do not converge under large transformations. While DCP is a global method, it performs point-to-point correspondence which suffers on noisy point clouds.

Our proposed approach can be characterized as a *global* method, as well as the first *learning-based, probabilistic* registration method. To further emphasize the difference of our approach in the context of data association and matching, refer to the visual illustrations of various correspondence strategies in Fig. 2.

3   GMM-Based Point Cloud Registration

Before describing our approach, we will briefly review the basics of the Gaussian Mixture Model (GMM) and how it offers a maximum likelihood (MLE) framework for finding optimal alignment between point clouds, which can be solved using the Expectation Maximization (EM) algorithm [15]. We then discuss the strengths and limitations of this framework and motivate the need for learned GMMs.

A GMM establishes a multimodal generative probability distribution over 3D space ($\mathbf{x} \in \mathbb{R}^3$) as a weighted sum of J Gaussian distributions,

$$p(\mathbf{x} \mid \boldsymbol{\Theta}) \overset{\text{def}}{=} \sum_{j=1}^{J} \pi_j \mathcal{N}(\mathbf{x} \mid \boldsymbol{\mu}_j, \boldsymbol{\Sigma}_j), \tag{1}$$

where $\sum_j \pi_j = 1$. GMM parameters $\boldsymbol{\Theta}$ comprise J triplets $(\pi_j, \boldsymbol{\mu}_j, \boldsymbol{\Sigma}_j)$, where π_j is a scalar mixture weight, $\boldsymbol{\mu}_j$ is a 3×1 mean vector and $\boldsymbol{\Sigma}_j$ is a 3×3 covariance matrix of the j-th component.

Given point clouds $\hat{\mathcal{P}}, \mathcal{P}$ and the space of permitted GMM parameterizations $\mathcal{X}$, we can formulate the registration from $\hat{\mathcal{P}}$ to $\mathcal{P}$ as a two-step optimization problem,

$$\textbf{Fitting: } \boldsymbol{\Theta}^* = \underset{\boldsymbol{\Theta} \in \mathcal{X}}{\arg\max}\, p(\mathcal{P} \mid \boldsymbol{\Theta}) \tag{2}$$

$$\textbf{Registration: } T^* = \underset{T \in SE(3)}{\arg\max}\, p(T(\hat{\mathcal{P}}) \mid \boldsymbol{\Theta}^*) \tag{3}$$

where $SE(3)$ is the space of 3D rigid transformations. The fitting step fits a GMM Θ^* to the target point cloud $\mathcal{P}$, while the registration step finds the optimal transformation T^* that aligns the source point cloud $\hat{\mathcal{P}}$ to Θ^*.

Note that both steps maximize the likelihood of a point cloud under a GMM, but with respect to different parameters. In general, directly maximizing the likelihood p is intractable. However, one can use EM to maximize a lower bound on p by introducing a set of latent correspondence variables $\mathcal{C}$. EM iterates between E-step and M-step until convergence. At each iteration k, the E-step updates the lower bound q to the posterior over $\mathcal{C}$ given a guess of the parameters $\Theta^{(k)}$ and the M-step updates the parameters by maximizing the expected joint likelihood under q. As an example, the EM updates for the fitting step (Eq. 2) are as follows.

$$\mathbf{E}_\Theta: q(\mathcal{C}) := p(\mathcal{C} \mid \mathcal{P}, \Theta^{(k)}) \tag{4}$$

$$\mathbf{M}_\Theta: \Theta^{(k+1)} := \underset{\Theta \in \mathcal{X}}{\operatorname{argmax}} \, \mathbb{E}_q[\ln p(\mathcal{P}, \mathcal{C} \mid \Theta)] \tag{5}$$

The key for EM is the introduction of latent correspondences $\mathcal{C}$. Given a point cloud $\mathcal{P} = \{p_i\}_{i=1}^N$ and a GMM $\Theta^{(k)} = \{\pi_j, \boldsymbol{\mu}_j, \boldsymbol{\Sigma}_j\}_{j=1}^J$, $\mathcal{C}$ comprises NJ binary latent variables $\{c_{ij}\}_{i,j=1,1}^{N,J}$ whose posterior factors can be calculated as

$$p(c_{ij} = 1 \mid p_i, \Theta^{(k)}) = \frac{\pi_j \mathcal{N}(p_i \mid \boldsymbol{\mu}_j, \boldsymbol{\Sigma}_j)}{\sum_{j'=1}^J \pi_{j'} \mathcal{N}(p_i \mid \boldsymbol{\mu}_{j'}, \boldsymbol{\Sigma}_{j'})}, \tag{6}$$

which can be seen as a kind of softmax over the squared Mahalanobis distances from p_i to each component center $\boldsymbol{\mu}_j$. Intuitively, if p_i is closer to $\boldsymbol{\mu}_j$ relative to the other components, then c_{ij} is more likely to be 1.

The introduction of $\mathcal{C}$ makes the joint likelihood in the M-step (Eq. 5) factorizable, leading to a closed-form solution for $\Theta^{(k+1)} \stackrel{\text{def}}{=} \{\pi_j^*, \boldsymbol{\mu}_j^*, \boldsymbol{\Sigma}_j^*\}_{j=1}^J$. Let $\gamma_{ij} \stackrel{\text{def}}{=} \mathbb{E}_q[c_{ij}]$. The solution for Eq. 5 can be written as

$$\pi_j^* = \frac{1}{N} \sum_{i=1}^N \gamma_{ij}, \qquad N\pi_j^* \boldsymbol{\mu}_j^* = \sum_{i=1}^N \gamma_{ij} p_i, \tag{7,8}$$

$$N\pi_j^* \boldsymbol{\Sigma}_j^* = \sum_{i=1}^N \gamma_{ij}(p_i - \boldsymbol{\mu}_j^*)(p_i - \boldsymbol{\mu}_j^*)^\top \tag{9}$$

Likewise, the registration step (Eq. 3) can be solved with another EM procedure optimizing over the transformation parameters T.

$$\mathbf{E}_T: q(\mathcal{C}) := p(\mathcal{C} \mid T^{(k)}(\hat{\mathcal{P}}), \Theta^*) \tag{10}$$

$$\mathbf{M}_T: T^{(k+1)} := \underset{T \in SE(3)}{\operatorname{argmax}} \, \mathbb{E}_q[\ln p(T(\hat{\mathcal{P}}), \mathcal{C} \mid \Theta^*)] \tag{11}$$

Many registration algorithms [14,15,19,20,22,23,28] can be viewed as variations of the general formulation described above. Compared to methods using point-to-point matches [8, 18, 35, 42], this formulation provides a systematic way of dealing with noise and outliers using probabilistic models and is fully differentiable. However, the iterative EM procedure makes it much more computationally expensive. Moreover, when the transformation is large, the EM procedure in Eq. 10,11 often gets stuck in local minima. This is because Eq. 6 used in the E Step performs point-to-cluster correspondence based on locality, i.e. a point likely belongs to a component if it is close to the component's center, which leads to spurious data association between $\hat{P}$ and Θ^* when T is large. In the following section, we show how our proposed method solves the data association problem by learning pose-invariant point-to-GMM correspondences via a neural network. By doing so, we also remove the need for an iterative matching procedure.

4 DeepGMR

In this section, we give an overview of Deep Gaussian Mixture Registration (DeepGMR) (Fig. 3), the first registration method that integrates GMM registration with neural networks. DeepGMR features a correspondence network (Sect. 4.1) and two differentiable computing blocks (Sect. 4.2, 4.3) analogous to the two EM procedures described in Sect. 3. The key idea is to replace the E-step with a correspondence network, which leads to consistent data association under large transformations, overcoming the weakness of conventional GMM registration.

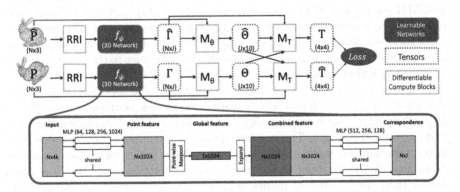

Fig. 3. Given two point clouds $\hat{P}$ and P, DeepGMR estimates the optimal transformation T that registers $\hat{P}$ to P and optionally the inverse transformation $\hat{T}$. DeepGMR has three major components: a learnable permutation-invariant point network f_ψ (Sect. 4.1) that estimates point-to-component correspondences Γ and two differentiable compute blocks $\mathbf{M_\Theta}$ (Sect. 4.2) and $\mathbf{M_T}$ (Sect. 4.3) that compute the GMM parameters Θ and transformation T in closed-form. We backpropagate a mean-squared loss through the differentiable compute blocks to learn the parameters of f_ψ.

4.1   Correspondence Network f_ψ

The correspondence network f_ψ takes in a point cloud with N points, $\mathcal{P} = \{p_i\}_{i=1}^N$, or a set of pre-computed features per point and produces a $N \times J$ matrix $\Gamma = [\gamma_{ij}]$ where $\sum_{j=1}^J \gamma_{ij} = 1$ for all i. Each γ_{ij} represents the latent correspondence probability between point p_i and component j of the latent GMM.

Essentially, the correspondence network replaces the E-step of a traditional EM approach reviewed in Sect. 3. In other words, f_ψ induces a distribution q_ψ over latent correspondences (Eq. 4). Importantly, however, the learned latent correspondence matrix Γ does not rely on Mahalanobis distances as in Eq. 6. This relaxation has several advantages. First, the GMMs generated by our network can deviate from the maximum likelihood estimate if it improves the performance of the downstream task. Second, by offloading the computation of Γ to a deep neural network, higher-level contextual information can be learned in a data-driven fashion in order to produce robust non-local association predictions. Third, since the heavy lifting of finding the appropriate association is left to the neural network, only a single iteration is needed.

Finally, under this framework, the computation of Γ can be viewed as a J-class point classification problem. As a result, we can leverage existing point cloud segmentation networks as our backbone.

4.2 $\mathbf{M_\Theta}$ Compute Block

Given the outputs Γ of f_ψ together with the point coordinates $\mathcal{P}$, we can compute the GMM parameters Θ in closed form according to Eq. 7, 8 and 9. Since these equations are weighted combinations of the point coordinates, the estimated GMM must overlap with the input point cloud spatially, providing an effective inductive bias for learning a stable representation. We refer to the parameter-free compute block that implements the conversion from $(\Gamma, \mathcal{P}) \to \Theta$ as $\mathbf{M_\Theta}$.

In order to have a closed-form solution for the optimal transformation (Sect. 4.3), we choose the covariances to be isotropic, i.e., $\Sigma_j = \mathrm{diag}([\sigma_j^2, \sigma_j^2, \sigma_j^2])$. This requires a slight modification of Eq. 9, replacing the outer product to inner product. The number of Gaussian components J remains fixed during training and testing. We choose $J = 16$ in our experiments based on ablation study results (refer to the supplement for details). Note that J is much smaller than the number of points, which is important for time and memory efficiency (Sect. 5.3).

4.3 $\mathbf{M_T}$ Compute Block

In this section, we show how to obtain the optimal transformation in closed form given the GMM parameters estimated by $\mathbf{M_\Theta}$. We refer to this parameter-free compute block that implements the computation from $(\Gamma, \Theta, \hat{\Theta}) \to T$ as $\mathbf{M_T}$.

We denote the source point cloud as $\hat{\mathcal{P}} = \{\hat{p}_i\}$ and the target point cloud as $\mathcal{P} = \{p_i\}$. The estimated source and target GMMs are denoted as $\hat{\Theta} = (\hat{\pi}, \hat{\mu}, \hat{\Sigma})$ and $\Theta = (\pi, \mu, \Sigma)$. The latent association matrix for source and target are

$\hat{\Gamma} = \{\hat{\gamma}_{ij}\}$ and $\Gamma = \{\gamma_{ij}\}$. We use $T(\cdot)$ to denote the transformation from the source to the target, consisting of rotation R and translation $\mathbf{t}$ $(i.e., T \in SE(3))$.

The maximum likelihood objective of $\mathbf{M_T}$ is to minimize the KL-divergence between the transformed latent source distribution $T(\hat{\boldsymbol{\Theta}})$ and the latent target distribution $\boldsymbol{\Theta}$,

$$T^* = \underset{T}{\operatorname{argmin}} \operatorname{KL}(T(\hat{\boldsymbol{\Theta}}) \mid \boldsymbol{\Theta}) \tag{12}$$

It can be shown that under general assumptions, the minimization in Eq. (12) is equivalent to maximizing the log likelihood of the transformed source point cloud $T(\hat{\mathcal{P}})$ under the target distribution $\boldsymbol{\Theta}$:

$$T^* = \underset{T}{\operatorname{argmax}} \sum_{i=1}^{N} \ln \sum_{j=1}^{J} \pi_j \mathcal{N}(T(\hat{p}_i) \mid \boldsymbol{\mu}_j, \boldsymbol{\Sigma}_j) \tag{13}$$

Thus, Eqs. 12–13 conform to the generalized registration objective outlined in Sect. 3, Eq. 11. We leave the proof to the supplement.

To eliminate the sum within the log function, we employ the same trick as in the EM algorithm. By introducing a set of latent variables $\mathcal{C} = \{c_{ij}\}$ denoting the association of the transformed source point $T(\hat{p}_i)$ with component j of the target GMM $\boldsymbol{\Theta}$, we can form a lower bound over the log likelihood as the expectation of the joint log likelihood $\ln p(T(\hat{\mathcal{P}}), \mathcal{C})$ with respect to some distribution q_ψ over $\mathcal{C}$, where ψ are the parameters of the correspondence network

$$T^* = \underset{T}{\operatorname{argmax}} \mathbb{E}_{q_\psi}[\ln p(T(\hat{\mathcal{P}}), \mathcal{C} \mid \boldsymbol{\Theta})] \tag{14}$$

$$= \underset{T}{\operatorname{argmax}} \sum_{i=1}^{N} \sum_{j=1}^{J} \mathbb{E}_{q_\psi}[c_{ij}] \ln \mathcal{N}(T(\hat{p}_i) \mid \boldsymbol{\mu}_j, \boldsymbol{\Sigma}_j) \tag{15}$$

Note that we parametrize q with the correspondence network f_ψ instead of setting q as the posterior based on Mahalanobis distances (Eq. 6). In other words, we use f_ψ to replace the $\mathbf{E}_T$ step in Eq. 10. Specifically, we set $\mathbb{E}_{q_\psi}[c_{ij}] = \hat{\gamma}_{ij}$ and rewrite Eq. (15) equivalently as the solution to the following minimization,

$$T^* = \underset{T}{\operatorname{argmin}} \sum_{i=1}^{N} \sum_{j=1}^{J} \hat{\gamma}_{ij} \|T(\hat{p}_i) - \boldsymbol{\mu}_j\|_{\boldsymbol{\Sigma}_j}^2 \tag{16}$$

where $\| \cdot \|_{\boldsymbol{\Sigma}_j}$ denote the Mahalanobis distance.

The objective in Eq. (16) contains NJ pairs of distances which can be expensive to optimize when N is large. However, we observe that the output of $\mathbf{M_\Theta}$ $(i.e.,$ using Eqs. 7, 8 and 9), with respect to $\hat{\mathcal{P}}$ and $\hat{\Gamma}$, allows us to simplify this minimization to a single sum consisting only of the latent parameters $\hat{\boldsymbol{\Theta}}$ and $\boldsymbol{\Theta}$. This reduction to a single sum can be seen as a form of barycentric coordinate transformation where only the barycenters' contributions vary with respect to the transformation. Similar types of reasoning and derivations can be found in previous works using GMMs [14, 20] (see the supplement for further details). The final objective ends up with only J pairs of distances, which saves a significant amount of computation. Furthermore, by using isotropic covariances in $\mathbf{M_\Theta}$, the objective simplifies from Mahalanobis distances to weighted Euclidean distances,

$$T^* = \underset{T}{\operatorname{argmin}} \sum_{j=1}^{J} \frac{\hat{\pi}_j}{\sigma_j^2} \|T(\hat{\boldsymbol{\mu}}_j) - \boldsymbol{\mu}_j\|^2 \tag{17}$$

Note that Eq. 17 implies a correspondence between components in the source GMM $\boldsymbol{\Theta}$ and the target GMM $\hat{\boldsymbol{\Theta}}$. Such a correspondence does *not* naturally emerge from two independently estimated GMMs (Fig. 2c). However, by representing the point-to-component correspondence between $\hat{\mathcal{P}}$ and $\hat{\boldsymbol{\Theta}}$ and between $T(\hat{\mathcal{P}})$ and $\boldsymbol{\Theta}$ with the same Γ matrix, an implicit component-to-component correspondence between $\boldsymbol{\Theta}$ and $\hat{\boldsymbol{\Theta}}$ is enforced, which reduces the alignment of point clouds to the alignment of component centroids (weighted by the covariances).

From Eq. 17, we can solve T^* in closed-form using a weighted version of the SVD solution in [40] (refer to the supplement for more detail). Note that our formulation is fundamentally different from previous learning-based approaches such as DCP [42] that use the SVD solver on point-to-point matches. We use a rigorous probabilistic framework to reduce the registration problem to matching latent Gaussian components, which has dimension J that is usually orders of magnitude smaller than the number of points N (in our experiments, $J = 16$ and $N = 1024$). Our formulation is not only much more efficient (SVD has complexity $O(d^3)$ on a $d \times d$ matrix) but also much more robust, since exact point-to-point correspondences rarely exist in real world point clouds with sensor noise.

4.4   Implementation

Our general framework is agnostic to the choice of the particular architecture of f_ψ. In our experiments, we demonstrate state-of-the-art results with a simple PointNet segmentation backbone [31] (see inset of Fig. 3), but in principle other more powerful backbones can be easily incorporated [11,39]. Because it can be challenging for PointNet to learn rotation invariance, we pre-process the input points into a set of rigorously rotation-invariant (RRI) features, proposed by [7], and use these hand-crafted RRI features as the input to the correspondence network. Note that RRI features are only used in the estimation of the association matrix Γ and not in the computation of the GMM parameters inside $\mathbf{M_\Theta}$, which depends on the raw point coordinates. Our ablation studies (see supplement) show that RRI does help, but is not essential to our method.

During training, given a pair of point clouds $\hat{\mathcal{P}}$ and $\mathcal{P}$, we apply the correspondence network f_ψ along with $\mathbf{M_\Theta}$ and $\mathbf{M_T}$ compute blocks to obtain two transformations: T from $\hat{\mathcal{P}}$ to $\mathcal{P}$ and $\hat{T}$ from $\mathcal{P}$ to $\hat{\mathcal{P}}$. Given the ground truth transformation T_{gt} from $\hat{\mathcal{P}}$ to $\mathcal{P}$, our method minimizes the following mean-squared error:

$$L = \|T T_{gt}^{-1} - I\|^2 + \|\hat{T} T_{gt} - I\|^2, \tag{18}$$

where T is the 4×4 transformation matrix containing both rotation and translation and I is the identity matrix. We also experimented with various other loss functions, including the RMSE metric [10] used in our evaluation, but found that the simple MSE loss outperforms others by a small margin (see the supplement).

DeepGMR is fully-differentiable and non-iterative, which makes the gradient calculation more straightforward than previous iterative approaches such as PointNetLK [1]. We implemented the correspondence network and differentiable compute blocks using autograd in PyTorch [29]. For all our experiments, we trained the network for 100 epochs with batch size 32 using the Adam optimizer [25]. The initial learning rate is 0.001 and is halved if the validation loss has not improved for over 10 epochs.

5   Experimental Results

We evaluate DeepGMR on two different datasets: synthetic point clouds generated from the ModelNet40 [45] and real-world point clouds from the Augmented ICL-NUIM dataset [10,21]. Quantitative results are summarized in Table 1 and qualitative comparison can be found in Fig. 4. Additional metrics and full error distribution curves can be found in the supplement.

Baselines. We compare DeepGMR against a set of strong geometry-based registration baselines, including point-to-plane ICP [8], HGMR [15], RANSAC [17], FGR [48][1] and TEASER++ [46]. These methods cover the spectrum of point-based (ICP) and GMM-based (HGMR) registration methods, efficient global methods (FGR) as well as provably "optimal" methods (TEASER++).

Table 1. Average RMSE and recall with threshold 0.2 on various datasets. DeepGMR achieves the best performance across all datasets thanks to its ability to perform robust data association in challenging cases (Sect. 5.2). Local methods are labeled in red; Inefficient global methods are labeled in orange and efficient global methods (average runtime < 1s) are labeled in blue. Best viewed in color.

	ModelNet clean		ModelNet noisy		ModelNet unseen		ICL-NUIM	
	RMSE ↓	Re@0.2 ↑	RMSE ↓	Re@0.2 ↑	RMSE ↓	Re@0.2 ↑	RMSE ↓	Re@0.2 ↑
ICP [8]	0.53	0.41	0.53	0.41	0.59	0.32	1.16	0.27
HGMR [15]	0.52	0.44	0.52	0.45	0.54	0.43	0.72	0.50
PointNetLK [1]	0.51	0.44	0.56	0.38	0.68	0.13	1.29	0.08
PRNet [43]	0.30	0.64	0.34	0.57	0.58	0.30	1.32	0.15
RANSAC10M+ICP	0.01	0.99	0.04	0.96	0.03	0.98	0.08	0.98
TEASER++ [46]	**0.00**	**1.00**	0.01	0.99	**0.01**	**0.99**	0.09	0.95
RANSAC10K+ICP	0.08	0.91	0.42	0.49	0.30	0.67	0.17	0.84
FGR [48]	0.19	0.79	0.2	0.79	0.23	0.75	0.15	0.87
DCP [42]	0.02	0.99	0.08	0.94	0.34	0.54	0.64	0.16
DeepGMR	**0.00**	**1.00**	**0.01**	**0.99**	**0.01**	**0.99**	**0.07**	**0.99**

---

[1] We performed a parameter search over the voxel size v, which is crucial for FGR's performance. We used the best results with $v = 0.08$.

We also compare against state-of-the-art learning based methods [1,42,43]. For RANSAC, we tested two variants with 10K and 10M iterations respectively and refine the results with ICP.

Metrics. Following [10,48], we use the RMSE the metric designed for evaluating global registration algorithms in the context of scene reconstruction. Given a set of ground truth point correspondences $\{p_i, q_i\}_{i=1}^n$, RMSE can be computed as

$$E_{RMSE} = \frac{1}{n}\sqrt{\sum_{i=1}^n \|T(p_i) - q_i\|^2} \tag{19}$$

Since exact point correspondences rarely exist in real data, we approximate the correspondence using $\{p_i, T^*(p_i)\}_{i=1}^n$, where T^* is the ground truth transformation and p_i is a point sampled from the source point cloud, so Eq. 19 becomes

$$E_{RMSE} \approx \frac{1}{n}\sqrt{\sum_{i=1}^n \|T(p_i) - T^*(p_i)\|^2} \tag{20}$$

In our evaluation we use $n = 500$. In addition to the average RMSE, we also calculate the recall across the dataset, i.e. the percentage of instances with RMSE less than a threshold τ. We followed [10] and used a threshold of $\tau = 0.2$.

5.1   Datasets

Synthetic Data. We generated 3 variations of synthetic point cloud data for registration from ModelNet40 [45], a large repository of CAD models. Each dataset follows the train/test split in [45] (9843 for train/validation and 2468 for test across 40 object categories).

ModelNet Clean. This dataset contains synthetic point cloud pairs without noise, so exact point-to-point correspondences exist between the point cloud pairs. Specifically, we sample 1024 points uniformly on each object model in Model-Net40 and apply two arbitrary rigid transformations to the same set of points to obtain the source and target point clouds. Note that similar setting has been used to evaluate learning-based registration methods in prior works [1,42], but the rotation magnitude is restricted, whereas we allow *unrestricted* rotation.

ModelNet Noisy. The setting in *ModelNet clean* assumes exact correspondences, which is unrealistic in the real world. In this dataset, we perturb the sampled point clouds with Gaussian noise sampled from $\mathcal{N}(0, 0.01)$. Note that unlike prior works [1,42], we add noise independently to the source and target, breaking exact correspondences between the input point clouds.

ModelNet Unseen. Following [42,43], we generate a dataset where train and test point clouds come from different object categories to test each method's generalizability. Specifically, we train our method using point clouds from 20 categories and test on point clouds from the held-out 20 categories unseen during training. The point clouds are the same as in *ModelNet noisy.*

Real-World Data. To demonstrate applicability on real-world data, we tested each method using the point clouds derived from RGB-D scans in the Augmented ICL-NUIM dataset [10]. However, the original dataset contains only 739 scan pairs which are not enough to train and evaluate learning-based methods (a network can easily overfit to a small number of fixed transformations). To remedy this problem, we take every scan available and apply the same augmentation as we did for the ModelNet data, i.e. resample the point clouds to 1024 points and apply two arbitrary rigid transformations. In this way, our augmented dataset contains both point clouds with realistic sensor noise and arbitrary transformation between input pairs. We split the dataset into 1278 samples for train/validation and 200 samples for test.

5.2   Analysis

From the quantitative results in Table 1, we can see that DeepGMR consistently achieves good performance across challenging scenarios, including noise, unseen geometry and real world data. TEASER++, the only baseline that is able to match the performance of DeepGMR on synthetic data, is over 1000 times slower (13.3s per instance) than DeepGMR (11ms per instance). In the rest of this section, we analyze common failure modes of the baselines and explain why DeepGMR is able to avoid these pitfalls. Typical examples can be found in the qualitative comparison shown in Fig. 4.

Repetitive Structures and Symmetry. Many point clouds contains repetitive structures (e.g. the bookshelf in the first row of Fig. 4) and near symmetry (e.g. the room scan in the last row of Fig. 4). This causes confusion in the data association for methods relying on distance-based point matching (e.g. ICP) or local feature descriptors (e.g. FGR). Unlike these baselines, DeepGMR performs data association globally. With the learned correspondence, a point can be associated to a distant component based on the context of the entire shape rather than a local neighborhood.

Parts with Sparse Point Samples. Many thin object parts (e.g. handle of the mug in the second row in Fig. 4) are heavily undersampled but are crucial for identifying correct alignments. These parts are easily ignored as outliers in an optimization whose objective is to minimize correspondence distances. Nevertheless, by doing optimization on a probabilistic model learned from many examples, DeepGMR is able to recover the correct transformation utilizing features in sparse regions.

Table 2. Average running time (ms) of efficient registration methods on ModelNet40 test set. DeepGMR is significantly faster than other learning based method [1,42,43] and comparable to geometry-based methods designed for efficiency [15,48]. Baselines not listed (RANSAC10M+ICP, TEASER++) have running time on the order of 10s. OOM means a 16 GB GPU is out of memory with a forward pass on a single instance.

# points	ICP	HGMR	PointNetLK	PRNet	RANSAC10K+ICP	FGR	DCP	DeepGMR
1000	184	33	84	153	95	22	67	**11**
2000	195	35	90	188	101	32	90	**19**
3000	195	37	93	OOM	113	37	115	**26**
4000	198	39	106	OOM	120	40	135	**34**
5000	201	**42**	109	OOM	124	**42**	157	47

Irregular Geometry. Among all the methods, FGR is the only one that performs significantly better on ICL-NUIM data than ModelNet40. This can be attributed to the FPFH features used by FGR to obtain initial correspondences. FPFH relies on good point normal estimates, and since the indoor scans of ICL-NUIM contain mostly flat surfaces with corners (e.g. the fourth row of Fig. 4), the normals can be easily estimated from local neighborhoods. In ModelNet40, however, many point clouds have irregular geometry (e.g. the plant in the third row of Fig. 4) which makes normal estimates very noisy. As a result, FGR's optimization fails to converge due to the lack of initial inlier correspondences.

5.3 Time Efficiency

We compute the average amount of time required to register a single pair of point clouds versus the point cloud size across all test instances in the ModelNet40. The results are shown in Table 2. Two baselines, RANSACK10M+ICP and TEASER++, are not listed because their average running time is over 10s for point clouds with 1000 points. ICP [8], FGR [48] and RANSAC10K+ICP are tested on an AMD RYZEN 7 3800X 3.9 GHz CPU. HGMR [15], PointLK [1] and DeepGMR are tested on a NVIDIA RTX 2080 Ti GPU. DCP [42] and PRNet [43] are tested on a NVIDIA Tesla V100 GPU because they require significant GPU memory. It can be seen that DeepGMR is the fastest among all methods thanks to its non-iterative nature and the reduction of registration to the task of learning correspondences to a low-dimensional latent probability distribution.

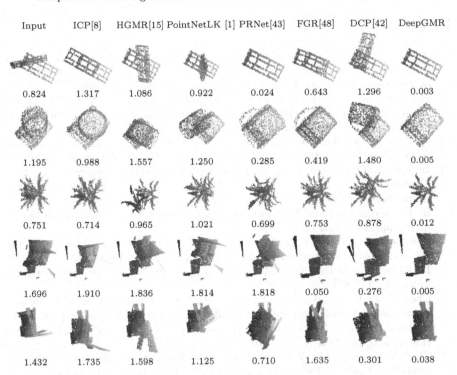

Input	ICP[8]	HGMR[15]	PointNetLK [1]	PRNet[43]	FGR[48]	DCP[42]	DeepGMR
0.824	1.317	1.086	0.922	0.024	0.643	1.296	0.003
1.195	0.988	1.557	1.250	0.285	0.419	1.480	0.005
0.751	0.714	0.965	1.021	0.699	0.753	0.878	0.012
1.696	1.910	1.836	1.814	1.818	0.050	0.276	0.005
1.432	1.735	1.598	1.125	0.710	1.635	0.301	0.038

Fig. 4. Qualitative registration results on noisy ModelNet40 (top 3 rows) and ICL-NUIM point clouds (bottom 2 rows). The RMSE of each example is labeled below the plot. These examples highlight some typical failure modes of existing methods such as 1) ignoring parts with sparse point samples 2) erroneous data association due to repetitive structures and symmetry. DeepGMR avoids these errors by estimating consistent point-to-distribution correspondence and performing robust registration on learned GMMs.

6   Conclusion

We have proposed DeepGMR, a first attempt towards learning-based probabilistic registration. Our experiments show that DeepGMR outperforms state-of-the-art geometry-based and learning-based registration baselines across a variety of data settings. Besides being robust to noise, DeepGMR is also generalizable to new object categories and performs well on real world data. We believe DeepGMR can be useful for applications that require accurate and efficient global alignment of 3D data, and furthermore, its design provides a novel way to integrate 3D neural networks inside a probabilistic registration paradigm.

References

1. Aoki, Y., Goforth, H., Srivatsan, R.A., Lucey, S.: Pointnetlk: robust & efficient point cloud registration using pointnet. In: Proceedings of the IEEE Conference on Computer Vision and Pattern Recognition, pp. 7163–7172 (2019)
2. Beraldin, J.A., Blais, F., Cournoyer, L., Godin, G., Rioux, M.: Active 3d sensing. Modelli e metodi per lo studio e la conservazione dell'architettura storica, University: Scola Normale Superiore, Pisa 10, January 2000
3. Besl, P., McKay, H.: A method for registration of 3-D shapes. IEEE Trans. Pattern Anal. Mach. Intell. **14**(2), 239–256 (1992). https://doi.org/10.1109/34.121791
4. Bishop, C.M.: Pattern recognition and machine learning. Springer, New York (2006)
5. Campbell, D., Petersson, L.: Gogma: globally-optimal gaussian mixture alignment. In: Proceedings of the IEEE Conference on Computer Vision and Pattern Recognition, pp. 5685–5694 (2016)
6. Campbell, D., Petersson, L., Kneip, L., Li, H., Gould, S.: The alignment of the spheres: Globally-optimal spherical mixture alignment for camera pose estimation. In: Proceedings of the IEEE Conference on Computer Vision and Pattern Recognition, pp. 11796–11806 (2019)
7. Chen, C., Li, G., Xu, R., Chen, T., Wang, M., Lin, L.: Clusternet: deep hierarchical cluster network with rigorously rotation-invariant representation for point cloud analysis. In: Proceedings of the IEEE Conference on Computer Vision and Pattern Recognition, pp. 4994–5002 (2019)
8. Chen, Y., Medioni, G.: Object modelling by registration of multiple range images. Image Vis. Comput. **10**(3), 145–155 (1992). Range Image Understanding
9. Chetverikov, D., Stepanov, D., Krsek, P.: Robust euclidean alignment of 3d point sets: the trimmed iterative closest point algorithm. Image Vis. Comput. **23**(3), 299–309 (2005)
10. Choi, S., Zhou, Q.Y., Koltun, V.: Robust reconstruction of indoor scenes. In: Proceedings of the IEEE Conference on Computer Vision and Pattern Recognition, pp. 5556–5565 (2015)
11. Choy, C., Gwak, J., Savarese, S.: 4d spatio-temporal convnets: Minkowski convolutional neural networks. arXiv preprint arXiv:1904.08755 (2019)
12. Dempster, A., Laird, N., Rubin, D.: Maximum likelihood from incomplete data via the EM algorithm. J. R. Statist. Soc. 1–38 (1977)
13. Dhawale, A., Shaurya Shankar, K., Michael, N.: Fast monte-carlo localization on aerial vehicles using approximate continuous belief representations. In: Proceedings of the IEEE Conference on Computer Vision and Pattern Recognition, pp. 5851–5859 (2018)
14. Eckart, B., Kim, K., Troccoli, A., Kelly, A., Kautz, J.: Mlmd: maximum likelihood mixture decoupling for fast and accurate point cloud registration. In: 2015 International Conference on 3D Vision (3DV), pp. 241–249. IEEE (2015)
15. Eckart, B., Kim, K., Kautz, J.: HGMR: Hierarchical gaussian mixtures for adaptive 3d registration. In: Proceedings of the European Conference on Computer Vision (ECCV), pp. 705–721 (2018)
16. Evangelidis, G.D., Horaud, R.: Joint alignment of multiple point sets with batch and incremental expectation-maximization. IEEE Trans. Pattern Anal. Mach. Intell. **40**(6), 1397–1410 (2017)
17. Fischler, M.A., Bolles, R.C.: Random sample consensus: a paradigm for model fitting with applications to image analysis and automated cartography. Commun. ACM **24**(6), 381–395 (1981)

18. Fitzgibbon, A.W.: Robust registration of 2d and 3d point sets. Image Vis. Comput. **21**(13), 1145–1153 (2003)
19. Gao, W., Tedrake, R.: Filterreg: robust and efficient probabilistic point-set registration using gaussian filter and twist parameterization. In: Proceedings of the IEEE Conference on Computer Vision and Pattern Recognition, pp. 11095–11104 (2019)
20. Granger, S., Pennec, X.: Multi-scale EM-ICP: a fast and robust approach for surface registration. ECCV **2002**, 69–73 (2002)
21. Handa, A., Whelan, T., McDonald, J., Davison, A.: A benchmark for RGB-D visual odometry, 3D reconstruction and SLAM. In: IEEE International Conference on Robotics and Automation (ICRA). Hong Kong, China, May 2014
22. Horaud, R., Forbes, F., Yguel, M., Dewaele, G., Zhang, J.: Rigid and articulated point registration with expectation conditional maximization. IEEE Trans. Pattern Anal. Mach. Intell. **33**(3), 587–602 (2011). http://doi.ieeecomputersociety.org/10.1109/TPAMI.2010.94
23. Järemo Lawin, F., Danelljan, M., Shahbaz Khan, F., Forssén, P.E., Felsberg, M.: Density adaptive point set registration. In: Proceedings of the IEEE Conference on Computer Vision and Pattern Recognition, pp. 3829–3837 (2018)
24. Jian, B., Vemuri, B.C.: Robust point set registration using Gaussian mixture models. IEEE Trans. Pattern Anal. Mach. Intell. **33**(8), 1633–1645 (2011). http://gmmreg.googlecode.com
25. Kingma, D.P., Ba, J.: Adam: a method for stochastic optimization. arXiv preprint arXiv:1412.6980 (2014)
26. Low, K.L.: Linear least-squares optimization for point-to-plane ICP surface registration. Chapel Hill, University of North Carolina 4(10) (2004)
27. Lucas, B.D., Kanade, T.: An iterative image registration technique with an application to stereo vision. In: IJCAI, pp. 674–679 (1981)
28. Myronenko, A., Song, X.: Point set registration: coherent point drift. IEEE Trans. Pattern Anal. Mach. Intell. **32**(12), 2262–2275 (2010)
29. Paszke, A., et al.: Automatic differentiation in pytorch (2017)
30. Pomerleau, F., Colas, F., Siegwart, R.: A review of point cloud registration algorithms for mobile robotics. Found. Trends Robot **4**(1), 1–104 (2015)
31. Qi, C.R., Su, H., Mo, K., Guibas, L.J.: Pointnet: Deep learning on point sets for 3d classification and segmentation. In: Proceedings Computer Vision and Pattern Recognition (CVPR), IEEE, vol. 1, no. 2, p. 4 (2017)
32. Qi, C.R., Yi, L., Su, H., Guibas, L.J.: Pointnet++: deep hierarchical feature learning on point sets in a metric space. In: Advances in neural information processing systems, pp. 5099–5108 (2017)
33. Rusinkiewicz, S., Levoy, M.: Efficient variants of the ICP algorithm. In: International Conference on 3-D Digital Imaging and Modeling, pp. 145–152 (2001)
34. Rusinkiewicz, S.: A symmetric objective function for ICP. ACM Trans. Graph. (Proceedings SIGGRAPH) **38**(4) (2019)
35. Segal, A., Haehnel, D., Thrun, S.: Generalized ICP. Robot. Sci. Syst. **2**, 4 (2009)
36. Stoyanov, T.D., Magnusson, M., Andreasson, H., Lilienthal, A.: Fast and accurate scan registration through minimization of the distance between compact 3D NDT representations. Int. J. Robot. Res. (2012)
37. Su, H., et al.: Splatnet: sparse lattice networks for point cloud processing. In: Proceedings of the IEEE Conference on Computer Vision and Pattern Recognition, pp. 2530–2539 (2018)
38. Tabib, W., O'Meadhra, C., Michael, N.: On-manifold GMM registration. IEEE Robot. Automat. Lett. **3**(4), 3805–3812 (2018)

39. Thomas, H., Qi, C.R., Deschaud, J.E., Marcotegui, B., Goulette, F., Guibas, L.J.: Kpconv: flexible and deformable convolution for point clouds. arXiv preprint arXiv:1904.08889 (2019)
40. Umeyama, S.: Least-squares estimation of transformation parameters between two point patterns. IEEE Trans. Pattern Anal. Mach. Intell. **13**(4), 376–380 (1991)
41. Unnikrishnan, R., Lalonde, J., Vandapel, N., Hebert, M.: Scale selection for the analysis of Point-Sampled curves. In: International Symposium on 3D Data Processing Visualization and Transmission, vol. 0, pp. 1026–1033. IEEE Computer Society, Los Alamitos, CA, USA (2006). http://doi.ieeecomputersociety.org/10.1109/3DPVT.2006.123
42. Wang, Y., Solomon, J.M.: Deep closest point: learning representations for point cloud registration. In: The IEEE International Conference on Computer Vision (ICCV), October 2019
43. Wang, Y., Solomon, J.M.: PRNET: Self-supervised learning for partial-to-partial registration. In: Advances in Neural Information Processing Systems, pp. 8812–8824 (2019)
44. Wang, Y., Sun, Y., Liu, Z., Sarma, S.E., Bronstein, M.M., Solomon, J.M.: Dynamic graph CNN for learning on point clouds. ACM Trans. Graph. (TOG) **38**(5), 146 (2019)
45. Wu, Z., et al.: 3d shapenets: a deep representation for volumetric shapes. In: Proceedings of the IEEE Conference on Computer Vision and Pattern Recognition, pp. 1912–1920 (2015)
46. Yang, H., Shi, J., Carlone, L.: Teaser: fast and certifiable point cloud registration. arXiv preprint arXiv:2001.07715 (2020)
47. Yang, J., Li, H., Campbell, D., Jia, Y.: Go-ICP: a globally optimal solution to 3d ICP point-set registration. IEEE Trans. Pattern Anal. Mach. Intell. **38**(11), 2241–2254 (2015)
48. Zhou, Q.-Y., Park, J., Koltun, V.: Fast global registration. In: Leibe, B., Matas, J., Sebe, N., Welling, M. (eds.) ECCV 2016. LNCS, vol. 9906, pp. 766–782. Springer, Cham (2016). https://doi.org/10.1007/978-3-319-46475-6_47

Active Perception Using Light Curtains for Autonomous Driving

Siddharth Ancha$^{(\boxtimes)}$, Yaadhav Raaj, Peiyun Hu, Srinivasa G. Narasimhan,
and David Held

Carnegie Mellon University, Pittsburgh, PA 15213, USA
{sancha,ryaadhav,peiyunh,srinivas,dheld}@andrew.cmu.edu
http://siddancha.github.io/projects/active-perception-light-curtains

Abstract. Most real-world 3D sensors such as LiDARs perform fixed
scans of the entire environment, while being decoupled from the recognition system that processes the sensor data. In this work, we propose a
method for 3D object recognition using light curtains, a resource-efficient
controllable sensor that measures depth at user-specified locations in the
environment. Crucially, we propose using prediction uncertainty of a deep
learning based 3D point cloud detector to guide active perception. Given
a neural network's uncertainty, we develop a novel optimization algorithm to optimally place light curtains to maximize coverage of uncertain regions. Efficient optimization is achieved by encoding the physical
constraints of the device into a constraint graph, which is optimized with
dynamic programming. We show how a 3D detector can be trained to
detect objects in a scene by sequentially placing uncertainty-guided light
curtains to successively improve detection accuracy. Links to code can
be found on the project webpage.

Keywords: Active vision · Robotics · Autonomous driving · 3D vision

1 Introduction

3D sensors, such as LiDAR, have become ubiquitous for perception in
autonomous systems operating in the real world, such as self-driving vehicles
and field robots. Combined with recent advances in deep-learning based visual
recognition systems, they have lead to significant breakthroughs in perception
for autonomous driving, enabling the recent surge of commercial interest in self-
driving technology.

However, most 3D sensors in use today perform *passive perception*, i.e. they
continuously sense the entire environment while being completely decoupled from
the recognition system that will eventually process the sensor data. In such a
case, sensing the entire scene can be potentially inefficient. For example, consider

Electronic supplementary material The online version of this chapter (https://
doi.org/10.1007/978-3-030-58558-7_44) contains supplementary material, which is
available to authorized users.

A. Vedaldi et al. (Eds.): ECCV 2020, LNCS 12350, pp. 751–766, 2020.
https://doi.org/10.1007/978-3-030-58558-7_44

an object detector running on a self-driving car that is trying to recognize objects in its environment. Suppose that it is confident that a tree-like structure on the side of the street is not a vehicle, but it is unsure whether an object turning around the curb is a vehicle or a pedestrian. In such a scenario, it might be beneficial if the 3D sensor focuses on collecting more data from the latter object, rather than distributing its sensing capacity uniformly throughout the scene.

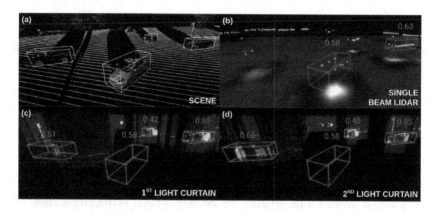

Fig. 1. *Object detection using light curtains.* (a) Scene with 4 cars; ground-truth boxes shown in green. (b) Sparse green points are from a single-beam LiDAR; it can detect only two cars (red boxes). Numbers above detections boxes are confidence scores. Uncertainty map in greyscale is displayed underneath: whiter means higher uncertainty. (c) First light curtain (blue) is placed to optimally cover the most uncertain regions. Dense points (green) from light curtain results in detecting 2 more cars. (d) Second light curtain senses even more points and fixes the misalignment error in the leftmost detection. (Color figure online)

In this work, we propose a method for 3D object detection using *active perception*, i.e. using sensors that can be purposefully controlled to sense specific regions in the environment. Programmable light curtains [2,22] were recently proposed as controllable, light-weight, and resource efficient sensors that measure the presence of objects intersecting any vertical ruled surface whose shape can be specified by the user (see Fig. 2). There are two main advantages of using programmable light curtains over LiDARs. First, they can be cheaply constructed, since light curtains use ordinary CMOS sensors (a current lab-built prototype costs $1000, and the price is expected to go down significantly in production). In contrast, a 64-beam Velodyne LiDAR that is commonly used in 3D self-driving datasets like KITTI [10] costs upwards of $80,000. Second, light curtains generate data with much higher resolution in regions where they actively focus [2] while LiDARs sense the entire environment and have low spatial and angular resolution.

One weakness of light curtains is that they are able to sense only a subset of the environment – a vertical ruled surface (see Fig. 1(c, d), Fig. 2). In contrast, a

LiDAR senses the entire scene. To mitigate this weakness, we can take advantage of the fact that the light curtain is a *controllable* sensor – we can choose where to place the light curtains. Thus, we must intelligently place light curtains in the appropriate locations, so that they sense the most important parts of the scene. In this work, we develop an algorithm for determining how to best place the light curtains for maximal detection performance.

We propose to use a deep neural network's prediction uncertainty as a guide for determining how to actively sense an environment. Our insight is that if an active sensor images the regions which the network is most uncertain about, the data obtained from those regions can help resolve the network's uncertainty and improve recognition. Conveniently, most deep learning based recognition systems output confidence maps, which can be used for this purpose when converted to an appropriate notion of uncertainty.

Given neural network uncertainty estimates, we show how a light curtain can be placed to *optimally* cover the regions of maximum uncertainty. First, we use an information-gain based framework to propose placing light curtains that maximize the sum of uncertainties of the covered region (Sect. 4.3, Appendix A). However, the structure of the light curtain and physical constraints of the device impose restrictions on how the light curtain can be placed. Our novel solution is to precompute a "constraint graph", which describes all possible light curtain placements that respect these physical constraints. We then use an optimization approach based on dynamic programming to efficiently search over all possible feasible paths in the constraint graph and maximize this objective (Sect. 4.4). This is a novel approach to constrained optimization of a controllable sensor's trajectory which takes advantage of the properties of the problem we are solving.

Our proposed active perception pipeline for 3D detection proceeds as follows. We initially record sparse data with an inexpensive single beam LIDAR sensor that performs fixed 3D scans. This data is input to a 3D point cloud object detector, which outputs an initial set of detections and confidence estimates. These confidence estimates are converted into uncertainty estimates, which are used by our dynamic programming algorithm to determine where to place the first light curtain. The output of the light curtain readings are again input to the 3D object detector to obtain refined detections and an updated uncertainty map. This process of estimating detections and placing new light curtains can be repeated multiple times (Fig. 3). Hence, we are able to sense the environment progressively, intelligently, and efficiently.

We evaluate our algorithm using two synthetic datasets of urban driving scenes [9,29]. Our experiments demonstrate that our algorithm leads to a monotonic improvement in performance with successive light curtain placements. We compare our proposed optimal light curtain placement strategy to multiple baseline strategies and find that they are significantly outperformed by our method. To summarize, our contributions are the following:

- We propose a method for using a deep learning based 3D object detector's prediction uncertainty as a guide for active sensing (Sect. 4.2).

- Given a network's uncertainty, we show how to compute a feasible light curtain that maximizes the coverage of uncertainty. Our novel contribution is to encode the physical constraints of the device into a graph and use dynamic-programming based graph optimization to efficiently maximize the objective while satisfying the physical constraints (Sect. 4.3, 4.4).
- We show how to train such an active detector using online light curtain data generation (Sect. 4.5).
- We empirically demonstrate that our approach leads to significantly improved detection performance compared to a number of baseline approaches (Sect. 5).

2   Related Work

2.1   Active Perception and Next-Best View Planning

Active Perception encompasses a variety of problems and techniques that involve actively controlling the sensor for improved perception [1,23]. Examples include actively modifying camera parameters [1], moving a camera to look around occluding objects [4], and next-best view (NBV) planning [5]. NBV refers to a broad set of problems in which the objective is to select the next best sensing action in order to solve a specific task. Typical problems include object instance classification [7,8,18,24] and 3D reconstruction [6,11–13,21]. Many works on next-best view formulate the objective as maximizing information gain (also known as mutual information) [6,7,12,13,21,24], using models such as probabilistic occupancy grids for beliefs over states [6,12,13,21,24]. Our method is similar in spirit to next-best view. One could consider each light curtain placement as obtaining a new "view" of the environment; we try to find the next best light curtain that aids object detection. In Sect. 4.3 and Appendix A, we derive an information-gain based objective to find the next best light curtain placement.

2.2   Object Detection from Point Clouds

There have been many recent advances in deep learning for 3D object detection. Approaches include representing LiDAR data as range images in LaserNet [16], using raw point clouds [19], and using point clouds in the bird's eye view such as AVOD [14], HDNet [26] and Complex-YOLO [20]. Most state-of-the-art approaches use voxelized point clouds, such as VoxelNet [27], PointPillars [15], SECOND [25], and CBGS [28]. These methods process an input point cloud by dividing the space into 3D regions (voxels or pillars) and extracting features from each of region using a PointNet [17] based architecture. Then, the volumetric feature map is converted to 2D features via convolutions, followed by a detection head that produces bounding boxes. We demonstrate that we can use such detectors, along with our novel light curtain placement algorithm, to process data from a single beam LiDAR combined with light curtains.

3   Background on Light Curtains

Programmable *light curtains* [2,22] are a sensor for adaptive depth sensing. "Light curtains" can be thought of as virtual surfaces placed in the environment. They can detect points on objects that intersect this surface. Before explaining how the curtain is created, we briefly describe our coordinate system and the basics of a rolling shutter camera.

Coordinate System: Throughout the paper, we will use the standard camera coordinate system centered at the sensor. We assume that the z axis corresponds to depth from the sensor pointing forward, and that the y vector points vertically

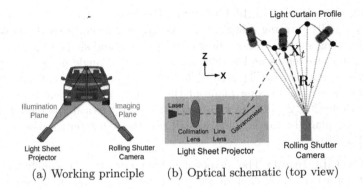

(a) Working principle   (b) Optical schematic (top view)

Fig. 2.   Illustration of programmable light curtains adapted from [2,22]. a) The light curtain is placed at the intersection of the illumination plane (from the projector) and the imaging plane (from the camera). b) A programmable galvanometer and a rolling shutter camera create multiple points of intersection, $\mathbf{X}_t$.

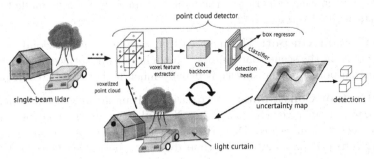

Fig. 3.   *Our method for detecting objects using light curtains.* An inexpensive single-beam lidar input is used by a 3D detection network to obtain rough initial estimates of object locations. The uncertainty of the detector is used to optimally place a light curtain that covers the most uncertain regions. The points detected by the light curtain (shown in green in the bottom figure) are input back into the detector so that it can update its predictions as well as uncertainty. The new uncertainty maps can again be used to place successive light curtains in an iterative manner, closing the loop.

downwards. Hence the xz-plane is parallel to the ground and corresponds to a top-down view, also referred to as the bird's eye view.

Rolling Shutter Camera: A rolling shutter camera contains pixels arranged in T number of vertical columns. Each pixel column corresponds to a vertical imaging plane. Readings from only those visible 3D points that lie on the imaging plane get recorded onto its pixel column. We will denote the xz-projection of the imaging plane corresponding to the t-th pixel column by ray $\mathbf{R}_t$, shown in the top-down view in Fig. 2(b). We will refer to these as "camera rays". The camera has a rolling shutter that successively activates each pixel column and its imaging plane one at a time from left to right. The time interval between the activation of two adjacent pixel columns is determined by the pixel clock.

Working Principle of Light Curtains: The latest version of light curtains [2] works by rapidly rotating a light sheet laser in synchrony with the motion of a camera's rolling shutter. A laser beam is collimated and shaped into a line sheet using appropriate lenses and is reflected at a desired angle using a controllable galvanometer mirror (see Fig. 2(b)). The illumination plane created by the laser intersects the active imaging plane of the camera in a vertical line along the curtain profile (Fig. 2(a)). The xz-projection of this vertical line intersecting the t-th imaging plane lies on $\mathbf{R}_t$, and we call this the t-th "control point", denoted by $\mathbf{X}_t$ (Fig. 2(b)).

Light Curtain Input: The shape of a light curtain is uniquely defined by where it intersects each camera ray in the xz-plane, i.e. the control points $\{\mathbf{X}_1, \ldots, \mathbf{X}_T\}$. These will act as inputs to the light curtain device. In order to produce the light curtain defined by $\{\mathbf{X}_t\}_{t=1}^T$, the galvanometer is programmed to compute and rotate at, for each camera ray $\mathbf{R}_t$, the reflection angle $\theta_t(\mathbf{X}_t)$ of the laser beam such that the laser sheet intersects $\mathbf{R}_t$ at $\mathbf{X}_t$. By selecting a control point on each camera ray, the light curtain device can be made to image any vertical ruled surface [2,22].

Light Curtain Output: The light curtain outputs a point cloud of all 3D visible points in the scene that intersect the light curtain surface. The density of light curtain points on the surface is usually much higher than LiDAR points.

Light Curtain Constraints: The rotating galvanometer can only operate at a maximum angular velocity $\omega_{\max}$. Let $\mathbf{X}_t$ and $\mathbf{X}_{t+1}$ be the control points on two consecutive camera rays $\mathbf{R}_t$ and $\mathbf{R}_{t+1}$. These induce laser angles $\theta(\mathbf{X}_t)$ and $\theta(\mathbf{X}_{t+1})$ respectively. If Δt is the time difference between when the t-th and $(t+1)$-th pixel columns are active, the galvanometer needs to rotate by an angle of $\Delta\theta(\mathbf{X}_t) = \theta(\mathbf{X}_{t+1}) - \theta(\mathbf{X}_t)$ within Δt time. Denote $\Delta\theta_{\max} = \omega_{\max} \cdot \Delta t$. Then the light curtain can only image control points subject to $|\theta(\mathbf{X}_{t+1}) - \theta(\mathbf{X}_t)| \leq \Delta\theta_{\max}, \forall 1 \leq t < T$.

4   Approach

4.1   Overview

Our aim is to use light curtains for detecting objects in a 3D scene. The overall approach is illustrated in Fig. 3. We use a voxel-based point cloud detector [25] and train it to use light curtain data without any architectural changes. The pipeline illustrated in Fig. 3 proceeds as follows.

To obtain an initial set of object detections, we use data from an inexpensive single-beam LiDAR as input to the detector. This produces rough estimates of object locations in the scene. Single-beam LiDAR is inexpensive because it consists of only one laser beam as opposed to 64 or 128 beams that are common in autonomous driving. The downside is that the data from the single beam contains very few points; this results in inaccurate detections and high uncertainty about object locations in the scene (see Fig. 1b).

Alongside bounding box detections, we can also extract from the detector an "uncertainty map" (explained in Sect. 4.2). We then use light curtains, placed in regions guided by the detector's uncertainty, to collect more data and iteratively refine the object detections. In order to get more data from the regions the detector is most uncertain about, we derive an information-gain based objective function that sums the uncertainties along the light curtain control points (Sect. 4.3 and Appendix A), and we develop a constrained optimization algorithm that places the light curtain to maximize this objective (Sect. 4.4).

Once the light curtain is placed, it returns a dense set of points where the curtain intersects with visible objects in the scene. We maintain a *unified point cloud*, which we define as the union of all points observed so far. The unified point cloud is initialized with the points from the single-beam LiDAR. Points from the light curtain are added to the unified point cloud and this data is input back into the detector. Note that the input representation for the detector remains the same (point clouds), enabling the use of existing state-of-the-art point cloud detection methods without any architectural modifications.

As new data from the light curtains are added to the unified point cloud and input to the detector, the detector refines its predictions and improves its accuracy. Furthermore, the additional inputs cause the network to update its uncertainty map; the network may no longer be uncertain about the areas that were sensed by the light curtain. Our algorithm uses the new uncertainty map to generate a new light curtain placement. We can iteratively place light curtains to cover the current uncertain regions and input the sensed points back into the network, closing the loop and iteratively improving detection performance.

4.2   Extracting Uncertainty from the Detector

The standard pipeline for 3D object detection [15, 25, 27] proceeds as follows. First, the ground plane (parallel to the xz-plane) is uniformly tiled with "anchor boxes"; these are reference boxes used by a 3D detector to produce detections. They are located on points in a uniformly discretized grid $G = [x_{min}, x_{max}] \times [z_{min}, z_{max}]$. For example, a $[-40\,\mathrm{m}, 40\,\mathrm{m}] \times [0\,\mathrm{m}, 70.4\,\mathrm{m}]$ grid is used for detecting cars in KITTI [10]. A 3D detector, which is usually a binary detector, takes a point cloud as input, and produces a binary classification score $p \in [0, 1]$ and bounding box regression offsets for every anchor box. The score p is the estimated probability that the anchor box contains an object of a specific class (such as car/pedestrian). The detector produces a detection for that anchor box if p exceeds a certain threshold. If so, the detector combines the fixed dimensions of the anchor box with its predicted regression offsets to output a detection box.

We can convert the confidence score to binary entropy $H(p) \in [0, 1]$ where $H(p) = -p \log_2 p - (1 - p) \log_2(1 - p)$. Entropy is a measure of the detector's uncertainty about the presence of an object at the anchor location. Since we have an uncertainty score at uniformly spaced anchor locations parallel to the xz-plane, they form an "uncertainty map" in the top-down view. We use this uncertainty map to place light curtains.

4.3 Information Gain Objective

Based on the uncertainty estimates given by Sect. 4.2, our method determines how to place the light curtain to sense the most uncertain/ambiguous regions. It seems intuitive that sensing the locations of highest detector uncertainty can provide the largest amount of information from a single light curtain placement, towards improving detector accuracy. As discussed in Sect. 3, a single light curtain placement is defined by a set of T control points $\{\mathbf{X}_t\}_{t=1}^{T}$. The light curtain will be placed to lie vertically on top of these control points. To define an optimization objective, we use the framework of information gain (commonly used in next-best view methods; see Sect. 2.1) along with some simplifying assumptions (see Appendix A). We show that under these assumptions, placing a light curtain to maximize information gain (a mathematically defined information-theoretic quantity) is equivalent to maximizing the objective $J(\mathbf{X}_1, \ldots, \mathbf{X}_T) = \sum_{t=1}^{T} H(\mathbf{X}_t)$, where $H(\mathbf{X})$ is the binary entropy of the detector's confidence at the anchor location of $\mathbf{X}$. When the control point $\mathbf{X}$ does not exactly correspond to an anchor location, we impute $H(\mathbf{X})$ by nearest-neighbor interpolation from the uncertainty map. Please see Appendix A for a detailed derivation.

4.4 Optimal Light Curtain Placement

In this section, we will describe an exact optimization algorithm to maximize the objective function $J(\mathbf{X}_1, \ldots, \mathbf{X}_T) = \sum_{t=1}^{T} H(\mathbf{X}_t)$.

Constrained Optimization: The control points $\{\mathbf{X}_t\}_{t=1}^{T}$, where each $\mathbf{X}_t$ lies on the camera ray $\mathbf{R}_t$, must be chosen to satisfy the physical constraints of the light curtain device: $|\theta(\mathbf{X}_{t+1}) - \theta(\mathbf{X}_t)| \leq \Delta\theta_{\max}$ (see Sect. 3: light curtain constraints). Hence, this is a constrained optimization problem. We discretize the problem by considering a dense set of N discrete, equally spaced points $\mathcal{D}_t = \{\mathbf{X}_t^{(n)}\}_{n=1}^{N}$ on each ray $\mathbf{R}_t$. We will assume that $\mathbf{X}_t \in \mathcal{D}_t$ for all $1 \leq t \leq T$ henceforth unless stated otherwise. We use $N = 80$ in all our experiments which we found to be

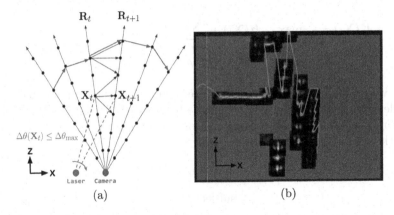

Fig. 4. (a) Light curtain constraint graph. Black dots are nodes and blue arrows are the edges of the graph. The optimized light curtain profile is depicted as red arrows. (b) Example uncertainty map from the detector, and optimized light curtain profile in red. Black is lowest uncertainty and white is highest uncertainty. The optimized light curtain covers the most uncertain regions.

sufficiently large. Overall, the optimization problem can be formulated as:

$$\arg\max_{\{\mathbf{X}_t\}_{t=1}^{T}} \sum_{t=1}^{T} H(\mathbf{X}_t) \tag{1}$$

$$\text{where } \mathbf{X}_t \in \mathcal{D}_t \; \forall 1 \leq t \leq T \tag{2}$$

$$\text{subject to } |\theta(\mathbf{X}_{t+1}) - \theta(\mathbf{X}_t)| \leq \Delta\theta_{\max}, \; \forall 1 \leq t < T \tag{3}$$

Light Curtain Constraint Graph: we encode the light curtain constraints into a graph, as illustrated in Fig. 4. Each black ray corresponds to a camera ray. Each black dot on the ray is a vertex in the constraint graph. It represents a candidate control point and is associated with an uncertainty score. Exactly one control point must be chosen per camera ray. The optimization objective is to choose such points to maximize the total sum of uncertainties. An edge between two control points indicates that the light curtain is able to transition from one control point $\mathbf{X}_t$ to the next, $\mathbf{X}_{t+1}$ without violating the maximum velocity light curtain constraints. Thus, the maximum velocity constraint (Eq. 3) can be specified by restricting the set of edges (depicted using blue arrows). We note that the graph only needs to be constructed once and can be done offline.

Dynamic Programming for Constrained Optimization: The number of possible light curtain placements, $|\mathcal{D}_1 \times \cdots \times \mathcal{D}_T| = N^T$, is exponentially large, which prevents us from searching for the optimal solution by brute force. However, we observe that the problem can be decomposed into simpler subproblems. In particular, let us define $J_t^*(\mathbf{X}_t)$ as the optimal sum of uncertainties of the *tail subproblem* starting from $\mathbf{X}_t$ i.e.

$$J_t^*(\mathbf{X}_t) = \max_{\mathbf{X}_{t+1},\dots,\mathbf{X}_T} H(\mathbf{X}_t) + \sum_{k=t+1}^{T} H(\mathbf{X}_k); \tag{4}$$

$$\text{subject to } |\theta(\mathbf{X}_{k+1}) - \theta(\mathbf{X}_k)| \leq \Delta\theta_{\max}, \ \forall \ t \leq k < T \tag{5}$$

If we were able to compute $J_t^*(\mathbf{X}_t)$, then this would help in solving a more complex subproblem using recursion: we observe that $J_t^*(\mathbf{X}_t)$ has the property of *optimal substructure*, i.e. the optimal solution of $J_{t-1}^*(\mathbf{X}_{t-1})$ can be computed from the optimal solution of $J_t^*(\mathbf{X}_t)$ via

$$J_{t-1}^*(\mathbf{X}_{t-1}) = H(\mathbf{X}_{t-1}) + \max_{\mathbf{X}_t \in \mathcal{D}_t} J_t^*(\mathbf{X}_t)$$

$$\text{subject to } |\theta(\mathbf{X}_t) - \theta(\mathbf{X}_{t-1})| \leq \Delta\theta_{\max} \tag{6}$$

Because of this optimal substructure property, we can solve for $J_{t-1}^*(\mathbf{X}_{t-1})$ via dynamic programming. We also note that the solution to $\max_{\mathbf{X}_1} J_1^*(\mathbf{X}_1)$ is the solution to our original constrained optimization problem (Eq. 1–3).

We thus perform the dynamic programming optimization as follows: the recursion from Eq. 6 can be implemented by first performing a backwards pass, starting from T and computing $J_t^*(\mathbf{X}_t)$ for each $\mathbf{X}_t$. Computing each $J_t^*(\mathbf{X}_t)$ takes only $O(B_{\text{avg}})$ time where B_{avg} is the average degree of a vertex (number of edges starting from a vertex) in the constraint graph, since we iterate once over all edges of $\mathbf{X}_t$ in Eq. 6. Then, we do a forward pass, starting with $\arg\max_{\mathbf{X}_1 \in \mathcal{D}_1} J_1^*(\mathbf{X}_1)$ and for a given $\mathbf{X}_{t-1}^*$, choosing $\mathbf{X}_t^*$ according to Eq. 6. Since there are N vertices per ray and T rays in the graph, the overall algorithm takes $O(NTB_{\text{avg}})$ time; this is a significant reduction from the $O(N^T)$ brute-force solution. We describe a simple extension of this objective that encourages smoothness in Appendix B.

4.5   Training Active Detector with Online Training Data Generation

The same detector is used to process data from the single beam LiDAR and all light curtain placements. Since the light curtains are placed based on the output (uncertainty maps) of the detector, the input point cloud for the next iteration depends on the current weights of the detector. As the weights change during training, so does the input data distribution. We account for non-stationarity of the training data by generating it online during the training process. This prevents the input distribution from diverging from the network weights during training. See Appendix C for algorithmic details and ablation experiments.

5   Experiments

To evaluate our algorithm, we need dense ground truth depth maps to simulate an arbitrary placement of a light curtain. However, standard autonomous driving datasets, such as KITTI [10] and nuScenes [3], contain only sparse LiDAR data, and hence the data is not suitable to accurately simulate a dense light curtain to evaluate our method. To circumvent this problem, we demonstrate our method on two synthetic datasets that provide dense ground truth depth maps, namely the Virtual KITTI [9] and SYNTHIA [29] datasets. Please find more details of the datasets and the evaluation metrics in Appendix D.

Our experiments demonstrate the following: First, we show that our method for successive placement of light curtains improves detection performance; particularly, there is a significant increase between the performance of single-beam LiDAR and the performance after placing the first light curtain. We also compare our method to multiple ablations and alternative placement strategies that demonstrate that each component of our method is crucial to achieve good performance. Finally, we show that our method can generalize to many more light curtain placements at test time than the method was trained on. In the appendix, we perform further experiments that include evaluating the generalization of our method to noise in the light curtain data, an ablation experiment for training with online data generation (Sect. 4.5), and efficiency analysis.

5.1   Comparison with Varying Number of Light Curtains

We train our method using online training data generation simultaneously on data from single-beam LiDAR and one, two, and three light curtain placements. We perform this experiment for both the Virtual KITTI and SYNTHIA datasets. The accuracies on their tests sets are reported in Table 1.

Table 1. Performance of the detector trained with single-beam LiDAR and up to three light curtains. Performance improves with more light curtain placements, with a significant jump at the first light curtain placement.

| | Virtual KITTI | | | | SYNTHIA | | | |
| | 3D mAP | | BEV mAP | | 3D mAP | | BEV mAP | |
	0.5 IoU	0.7 IoU	0.5 IoU	0.7 IoU	0.5 IoU	0.7 IoU	0.5 IoU	0.7 IoU
Single Beam Lidar	39.91	15.49	40.77	36.54	60.49	47.73	60.69	51.22
Single Beam Lidar (separate model)	42.35	23.66	47.77	40.15	60.69	48.23	60.84	57.98
1 Light Curtain	58.01	35.29	58.51	47.05	68.79	55.99	68.97	59.63
2 Light Curtains	60.86	37.91	61.10	49.84	69.02	57.08	69.17	67.14
3 Light Curtains	**68.52**	**38.47**	**68.82**	**50.53**	**69.16**	**57.30**	**69.25**	**67.25**

Note that there is a significant and consistent increase in the accuracy between single-beam LiDAR performance and the first light curtain placement (row 1 and row 3). This shows that actively placing light curtains on the most uncertain regions can improve performance over a single-beam LiDAR that performs fixed scans. Furthermore, placing more light curtains consistently improves detection accuracy.

As an ablation experiment, we train a separate model only on single-beam LiDAR data (row 2), for the same number of training iterations. This is different from row 1 which was trained with both single beam LiDAR and light curtain data but evaluated using only data for a single beam LiDAR. Although training a model with only single-beam LiDAR data (row 2) improves performance over row 1, it is still significantly outperformed by our method which uses data from light curtain placements.

Noise Simulations: In order to simulate noise in the real-world sensor, we perform experiments with added noise in the light curtain input. We demonstrate that the results are comparable to the noiseless case, indicating that our method is robust to noise and is likely to transfer well to the real world. Please see Appendix E for more details.

Table 2. Baselines for alternate light curtain placement strategies, trained and tested on (a) Virtual KITTI and (b) SYNTHIA datasets. Our dynamic programming optimization approach significantly outperforms all other strategies.

	Virtual KITTI				SYNTHIA			
	3D mAP		BEV mAP		3D mAP		BEV mAP	
	.5 IoU	.7 IoU	.5 IoU	.7 IoU	.5 IoU	.7 IoU	.5 IoU	.7 IoU
Random	41.29	17.49	46.65	38.09	60.43	47.09	60.66	58.14
Fixed depth - 15 m	44.99	22.20	46.07	38.05	60.74	48.16	60.89	58.48
Fixed depth - 30 m	39.72	19.05	45.21	35.83	60.02	47.88	60.23	57.89
Fixed depth - 45 m	39.86	20.02	40.61	36.87	60.23	48.12	60.43	57.77
Greedy Optimization (Randomly break ties)	37.40	19.93	42.80	35.33	60.62	47.46	60.83	58.22
Greedy Optimization (Min laser angle change)	39.20	20.19	44.80	36.94	60.61	47.05	60.76	58.07
Frontoparallel + Uncertainty	39.41	21.25	45.10	37.80	60.36	47.20	60.52	58.00
Ours	**58.01**	**35.29**	**58.51**	**47.05**	**68.79**	**55.99**	**68.97**	**59.63**

5.2 Comparison with Alternative Light Curtain Placement Strategies

In our approach, light curtains are placed by maximizing the coverage of uncertain regions using a dynamic programming optimization. How does this compare to other strategies for light curtain placement? We experiment with several baselines:

1. *Random*: we place frontoparallel light curtains at a random z-distance from the sensor, ignoring the detector's uncertainty map.
2. *Fixed depth*: we place a frontoparallel light curtain at a fixed z-distance (15 m, 30 m, 45 m) from the sensor, ignoring the detector's uncertainty map.

3. *Greedy optimization*: this baseline tries to evaluate the benefits of using a dynamic programming optimization. Here, we use the same light curtain constraints described in Sect. 4.4 (Fig. 4(a)). We greedily select the next control point based on local uncertainty instead of optimizing for the future sum of uncertainties. Ties are broken by (a) choosing smaller laser angle changes, and (b) randomly.
4. *Frontoparallel + Uncertainty*: Our optimization process finds light curtains with flexible shapes. What if the shapes were constrained to make the optimization problem easier? If we restrict ourselves to frontoparallel curtains, we can place them at the z-distance of maximum uncertainty by simply summing the uncertainties for every fixed value of z.

The results on the Virtual KITTI and SYNTHIA datasets are shown in Table 2. Our method significantly and consistently outperforms all baselines. This empirically demonstrates the value of using dynamic programming for light curtain placement to improve object detection performance.

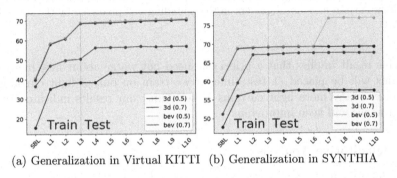

(a) Generalization in Virtual KITTI   (b) Generalization in SYNTHIA

Fig. 5. *Generalization to many more light curtains than what the detector was trained for.* We train using online data generation on single-beam lidar and only 3 light curtains. We then test with placing 10 curtains, on (a) Virtual KITTI, and (b) SYNTHIA. Performance continues to increase monotonically according to multiple metrics. Takeaway: one can safely place more light curtains at test time and expect to see sustained improvement in accuracy.

5.3 Generalization to Successive Light Curtain Placements

If we train a detector using our online light curtain data generation approach for k light curtains, can the performance generalize to more than k light curtains? Specifically, if we continue to place light curtains beyond the number trained for, will the accuracy continue improving? We test this hypothesis by evaluating on 10 light curtains, many more than the model was trained for (3 light curtains). Figure 5 shows the performance as a function of the number of light curtains. We find that in both Virtual KITTI and SYNTHIA, the accuracy monotonically improves with the number of curtains.

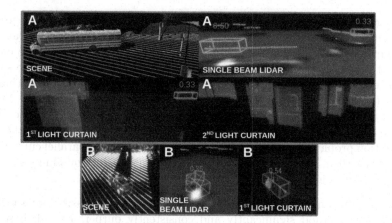

Fig. 6. *Successful cases:* other type of successful cases than Fig. 1. In (A), the single-beam LiDAR incorrectly detects a bus and a piece of lawn as false positives. They get eliminated successively after placing the first and second light curtains. In (B), the first light curtain fixes misalignment in the bounding box predicted by the single beam LiDAR.

This result implies that a priori one need not worry about how many light curtains will be placed at test time. If we train on only 3 light curtains, we can place many more light curtains at test time; our results indicate that the performance will keep improving.

5.4   Qualitative Analysis

We visualized a successful case of our method in Fig. 1. This is an example where our method detects false negatives missed by the single-beam LiDAR. We also show two other types of successful cases where light curtains remove false positive detections and fix misalignment errors in Fig. 6. In Fig. 7, we show the predominant failure case of our method. See captions for more details.

Fig. 7. *Failure cases:* the predominant failure mode is that the single beam LiDAR detects a false positive which is not removed by light curtains because the detector is overly confident in its prediction (so the estimated uncertainty is low). *Middle:* falsely detecting a tree to be a car. *Right:* after three light curtains, the detection persists because light curtains do not get placed on this false positive. False positive gets removed eventually only after six light curtain placements.

6 Conclusions

In this work, we develop a method to use light curtains, an actively controllable resource-efficient sensor, for object recognition in static scenes. We propose to use a 3D object detector's prediction uncertainty as a guide for deciding where to sense. By encoding the constraints of the light curtain into a graph, we show how to optimally and feasibly place a light curtain that maximizes the coverage of uncertain regions. We are able to train an active detector that interacts with light curtains to iteratively and efficiently sense parts of scene in an uncertainty-guided manner, successively improving detection accuracy. We hope this works pushes towards designing perception algorithms that integrate sensing and recognition, towards intelligent and adaptive perception.

Acknowledgements. We thank Matthew O'Toole for feedback on the initial draft of this paper. This material is based upon work supported by the National Science Foundation under Grants No. IIS-1849154, IIS-1900821 and by the United States Air Force and DARPA under Contract No. FA8750-18-C-0092.

References

1. Bajcsy, R.: Active perception. Proc. IEEE **76**(8), 966–1005 (1988)
2. Bartels, J.R., Wang, J., Whittaker, W.R., Narasimhan, S.G.: Agile depth sensing using triangulation light curtains. In: The IEEE International Conference on Computer Vision (ICCV), October 2019
3. Caesar, H., et al.: nuscenes: a multimodal dataset for autonomous driving. arXiv preprint arXiv:1903.11027 (2019)
4. Cheng, R., Agarwal, A., Fragkiadaki, K.: Reinforcement learning of active vision for manipulating objects under occlusions. arXiv preprint arXiv:1811.08067 (2018)
5. Connolly, C.: The determination of next best views. In: Proceedings of 1985 IEEE International Conference on Robotics and Automation, vol. 2, pp. 432–435. IEEE (1985)
6. Daudelin, J., Campbell, M.: An adaptable, probabilistic, next-best view algorithm for reconstruction of unknown 3-D objects. IEEE Robot. Autom. Lett. **2**(3), 1540–1547 (2017)
7. Denzler, J., Brown, C.M.: Information theoretic sensor data selection for active object recognition and state estimation. IEEE Trans. Pattern Anal. Mach. Intell. **24**(2), 145–157 (2002)
8. Doumanoglou, A., Kouskouridas, R., Malassiotis, S., Kim, T.K.: Recovering 6D object pose and predicting next-best-view in the crowd. In: Proceedings of the IEEE Conference on Computer Vision and Pattern Recognition, pp. 3583–3592 (2016)
9. Gaidon, A., Wang, Q., Cabon, Y., Vig, E.: Virtual worlds as proxy for multi-object tracking analysis. In: CVPR (2016)
10. Geiger, A., Lenz, P., Stiller, C., Urtasun, R.: Vision meets robotics: the kitti dataset. Int. J. Robot. Res. **32**(11), 1231–1237 (2013)
11. Haner, S., Heyden, A.: Covariance propagation and next best view planning for 3D reconstruction. In: Fitzgibbon, A., Lazebnik, S., Perona, P., Sato, Y., Schmid, C. (eds.) ECCV 2012. LNCS, vol. 7573, pp. 545–556. Springer, Heidelberg (2012). https://doi.org/10.1007/978-3-642-33709-3_39

12. Isler, S., Sabzevari, R., Delmerico, J., Scaramuzza, D.: An information gain formulation for active volumetric 3D reconstruction. In: 2016 IEEE International Conference on Robotics and Automation (ICRA), pp. 3477–3484. IEEE (2016)
13. Kriegel, S., Rink, C., Bodenmüller, T., Suppa, M.: Efficient next-best-scan planning for autonomous 3D surface reconstruction of unknown objects. J. Real-Time Image Proc. **10**(4), 611–631 (2015)
14. Ku, J., Mozifian, M., Lee, J., Harakeh, A., Waslander, S.L.: Joint 3D proposal generation and object detection from view aggregation. In: 2018 IEEE/RSJ International Conference on Intelligent Robots and Systems (IROS), pp. 1–8. IEEE (2018)
15. Lang, A.H., Vora, S., Caesar, H., Zhou, L., Yang, J., Beijbom, O.: PointPillars: fast encoders for object detection from point clouds. In: Proceedings of the IEEE Conference on Computer Vision and Pattern Recognition, pp. 12697–12705 (2019)
16. Meyer, G.P., Laddha, A., Kee, E., Vallespi-Gonzalez, C., Wellington, C.K.: LaserNet: an efficient probabilistic 3D object detector for autonomous driving. In: Proceedings of the IEEE Conference on Computer Vision and Pattern Recognition, pp. 12677–12686 (2019)
17. Qi, C.R., Su, H., Mo, K., Guibas, L.J.: PointNet: deep learning on point sets for 3D classification and segmentation. In: Proceedings of the IEEE Conference on Computer Vision and Pattern Recognition, pp. 652–660 (2017)
18. Scott, W.R., Roth, G., Rivest, J.F.: View planning for automated three-dimensional object reconstruction and inspection. ACM Comput. Surv. (CSUR) **35**(1), 64–96 (2003)
19. Shi, S., Wang, X., Li, H.: Pointrcnn: 3D object proposal generation and detection from point cloud. In: Proceedings of the IEEE Conference on Computer Vision and Pattern Recognition, pp. 770–779 (2019)
20. Simony, M., Milzy, S., Amendey, K., Gross, H.M.: Complex-YOLO: an euler-region-proposal for real-time 3D object detection on point clouds. In: Proceedings of the European Conference on Computer Vision (ECCV) (2018)
21. Vasquez-Gomez, J.I., Sucar, L.E., Murrieta-Cid, R., Lopez-Damian, E.: Volumetric next-best-view planning for 3D object reconstruction with positioning error. Int. J. Adv. Rob. Syst. **11**(10), 159 (2014)
22. Wang, J., Bartels, J., Whittaker, W., Sankaranarayanan, A.C., Narasimhan, S.G.: Programmable triangulation light curtains. In: Proceedings of the European Conference on Computer Vision (ECCV), pp. 19–34 (2018)
23. Wilkes, D.: Active object recognition (1994)
24. Wu, Z., et al.: 3D shapenets: a deep representation for volumetric shapes. In: Proceedings of the IEEE Conference on Computer Vision and Pattern Recognition (CVPR), June 2015
25. Yan, Y., Mao, Y., Li, B.: SECOND: sparsely embedded convolutional detection. Sensors **18**(10), 3337 (2018)
26. Yang, B., Liang, M., Urtasun, R.: HDNET: exploiting HD maps for 3D object detection. In: Conference on Robot Learning, pp. 146–155 (2018)
27. Zhou, Y., Tuzel, O.: VoxelNet: end-to-end learning for point cloud based 3D object detection. In: Proceedings of the IEEE Conference on Computer Vision and Pattern Recognition, pp. 4490–4499 (2018)
28. Zhu, B., Jiang, Z., Zhou, X., Li, Z., Yu, G.: Class-balanced grouping and sampling for point cloud 3D object detection. arXiv preprint arXiv:1908.09492 (2019)
29. Zolfaghari Bengar, J., et al.: Temporal coherence for active learning in videos. arXiv preprint arXiv:1908.11757 (2019)

Invertible Neural BRDF for Object Inverse Rendering

Zhe Chen[✉], Shohei Nobuhara, and Ko Nishino

Kyoto University, Kyoto, Japan
zchen@vision.ist.i.kyoto-u.ac.jp, {nob,kon}@i.kyoto-u.ac.jp
https://vision.ist.i.kyoto-u.ac.jp

Abstract. We introduce a novel neural network-based BRDF model and a Bayesian framework for object inverse rendering, *i.e.*, joint estimation of reflectance and natural illumination from a single image of an object of known geometry. The BRDF is expressed with an invertible neural network, namely, normalizing flow, which provides the expressive power of a high-dimensional representation, computational simplicity of a compact analytical model, and physical plausibility of a real-world BRDF. We extract the latent space of real-world reflectance by conditioning this model, which directly results in a strong reflectance prior. We refer to this model as the invertible neural BRDF model (iBRDF). We also devise a deep illumination prior by leveraging the structural bias of deep neural networks. By integrating this novel BRDF model and reflectance and illumination priors in a MAP estimation formulation, we show that this joint estimation can be computed efficiently with stochastic gradient descent. We experimentally validate the accuracy of the invertible neural BRDF model on a large number of measured data and demonstrate its use in object inverse rendering on a number of synthetic and real images. The results show new ways in which deep neural networks can help solve challenging radiometric inverse problems.

Keywords: Reflectance · BRDF · Inverse rendering · Illumination estimation

1 Introduction

Disentangling the complex appearance of an object into its physical constituents, namely the reflectance, illumination, and geometry, can reveal rich semantic information about the object and its environment. The reflectance informs the material composition, the illumination reveals the surroundings, and the geometry makes explicit the object shape. Geometry recovery with strong assumptions on the other constituents has enjoyed a long history of research, culminating in

Electronic supplementary material The online version of this chapter (https://doi.org/10.1007/978-3-030-58558-7_45) contains supplementary material, which is available to authorized users.

© Springer Nature Switzerland AG 2020
A. Vedaldi et al. (Eds.): ECCV 2020, LNCS 12350, pp. 767–783, 2020.
https://doi.org/10.1007/978-3-030-58558-7_45

various methods of shape-from-X. Accurate estimation of reflectance and illumination is equally critical for a broad range of applications, including augmented reality, robotics, and graphics where the problem is often referred to as inverse rendering. This paper particularly concerns inverse rendering of object appearance rather than scenes, where the latter would require modeling of global light transport in addition to the complex local surface light interaction.

Even when we assume that the geometry of the object is known or already estimated, the difficulty of joint estimation of the remaining reflectance and illumination, persists. The key challenge lies in the inherent ambiguity between the two, both in color and frequency [35]. Past methods have relied on strongly constrained representations, for instance, by employing low-dimensional parametric models, either physically-based or data-driven (e.g., Lambertian and spherical harmonics, respectively). On top of these compact parametric models, strong analytical but simplistic constraints are often required to better condition the joint estimation, such as a Gaussian-mixture on the variation of real-world reflectance and gradient and entropy priors on nonparametric illumination [25].

While these methods based on low-dimensional parametric BRDF models have shown success in object inverse rendering "in the wild", the accuracy of the estimates are inevitably limited by the expressive power of the models. As also empirically shown by Lombardi and Nishino [25], the estimation accuracy of the two radiometric constituents are bounded by the highest frequency of either of the two. Although we are theoretically limited to this bound, low-dimensional parametric models further restrict the recoverable frequency characteristics to the approximation accuracy of the models themselves. Ideally, we would like to use high-dimensional representations for both the reflectance and illumination, so that the estimation accuracy is not bounded by their parametric forms. The challenge then becomes expressing complex real-world reflectance with a common high-dimensional representation while taming the variability of real-world reflectance and illumination so that they can be estimated from single images. Nonparametric (i.e., tabulated) BRDF representations and regular deep generative models such as generative adversarial networks [16] and variational autoencoders [18] are unsuitable for the task as they do not lend a straightforward means for adequate sampling in the angular domain of the BRDF.

In this paper, we introduce the *invertible neural BRDF* model (*iBRDF*) for joint estimation of reflectance and illumination from a single image of object appearance. We show that this combination of an invertible, differentiable model that has the expressive power better than a nonparametric representation together with a MAP formulation with differentiable rendering enable efficient, accurate real-world object inverse rendering. We exploit the inherent structure of the reflectance by modeling its bidirectional reflectance distribution function (BRDF) as an invertible neural network, namely, a nonlinearly transformed parametric distribution based on normalized flow [29,42,43]. In sharp contrast to past methods that use low-dimensional parametric models, the deep generative neural network makes no assumptions on the underlying distribution and expresses the complex angular distributions of the BRDF with a series of non-linear transformations applied to a simple input distribution. We will show that this provides us with comparable or superior expressiveness to nonparametric representations. Moreover, the invertibility of the model ensures Helmholtz reciprocity and energy conservation, which are essential for physical plausibility. In addition, although we do not pursue in this paper, this invertibility makes iBRDF also suitable for forward rendering applications due to its bidirectional, differentiable bijective mapping. We also show that multiple "lobes" of this nonparametric BRDF model can be combined to express complex color and frequency characteristics of real-world BRDFs. Furthermore, to model the intrinsic structure of the reflectance variation of real-world materials, we condition this generative model to extract a parametric embedding space. This embedding of BRDFs in a simple parametric distribution provides us a strong prior for estimating the reflectance.

For the illumination, we employ a nonparametric representation by modeling it as a collection of point sources in the angular space (*i.e.*, equirectangular environment map). Past methods heavily relied on simplistic assumptions that can be translated into analytical constraints to tame the high-dimensional complexity associated with this nonparametric illumination representation. Instead, we constrain the illumination to represent realistic natural environments by exploiting the structural bias induced by a deep neural network (*i.e.*, deep image prior [45]). We device this deep illumination prior by encoding the illumination as the output of an encoder-decoder deep neural network and by optimizing its parameters on a fixed random image input.

We derive a Bayesian object inverse rendering framework by combining the deep illumination prior together with the invertible neural BRDF and a differentiable renderer to evaluate the likelihood. Due to the full differentiability of the BRDF and illumination models and priors, the estimation can be achieved through backpropagation with stochastic gradient descent. We demonstrate the effectiveness of our novel BRDF model, its embedding, deep illumination prior, and joint estimation on a large amount of synthetic and real data. Both the quantitative evaluation and quantitative validation show that they lead to accurate object inverse rendering of real-world materials taken under complex natural illumination.

To summarize, our technical contributions consist of

- a novel BRDF model with the expressiveness of a nonparametric representation and computational simplicity of an analytical distribution model,
- a reflectance prior based on the embedding of this novel BRDF model,
- an illumination prior leveraging architectural bias of neural networks,
- and a fully differentiable joint estimation framework for reflectance and illumination based on these novel models and priors.

We believe these contributions open new avenues of research towards fully leveraging deep neural networks in solving radiometric inverse problems.

2 Related Work

Reflectance modeling and radiometric quantity estimation from images has a long history of research in computer vision and related areas, studied under the umbrella of physics-based vision, appearance modeling, and inverse rendering. Here we briefly review works most relevant to ours.

Reflectance Models. For describing local light transport at a surface point, Nicodemus [30] introduced the bidirectional reflectance distribution function (BRDF) as a 4D reflectance function of incident and exitant light directions. Since then, many parametric reflectance models that provide an analytical expression to this abstract function have been proposed. Empirical models like Phong [34] and Blinn models [5] offer great simplicity for forward rendering, yet fail to capture complex reflectance properties of real-world materials making them unsuitable for reflectance estimation. Physically-based reflectance models such as Torrance-Sparrow [44], Cook-Torrance [8] and Disney material models [7] rigorously model the light interaction with micro-surface geometry. While these models capture challenging reflectance properties like off-specular reflection, their accuracy is limited to certain types of materials.

Data-driven reflectance models instead directly model the BRDF by fitting basis functions (*e.g.*, Zernike polynomials [19] and spherical harmonics [4,35] or by extracting such bases from measured data [27]. Nishino *et al.* introduce the directional statistics BRDF model [31,32] based on a newly derived hemispherical exponential power distribution to express BRDFs in half-way vector representations and use their embedding as a prior for various inverse-rendering tasks [24,25,33]. Ashikhmin and Premoze [2] use a modified anisotropic Phong fit to measured data. The expressive power of these models are restricted by the underlying analytical distributions. Romeiro *et al.* [37] instead directly use tabulated 2D isotropic reflectance distributions. Although nonparametric and expressive, using such models for estimating the reflectance remains challenging due to their high-dimensionality and lack of differentiability. We show that our invertible neural BRDF model with comparable number of parameters achieves higher accuracy while being differentiable and invertible.

Reflectance Estimation. Joint estimation of reflectance and illumination is challenging due to the inherent ambiguity between the two radiometric quantities. For this reason, many works estimate one of the two assuming the other is known. For reflectance estimation early work such as that by Sato *et al.* [41] assume Torrance-Sparrow reflection and a point light source. Romeiro *et al.* [37] estimate a nonparametric bivariate BRDF of an object of known geometry taken under known natural illumination. Rematas [36] propose an end-to-end neural network to estimate the reflectance map. The geometry is first estimated from the image after which a sparse reflectance map is reconstructed. A convolutional neural network (ConvNet) was learned to fill the holes of this sparse reflectance map. The illumination is, however, baked into and inseparable from this reflectance map. Meka *et al.* [28] models the reflectance as a linear combination of Lambertian, specular albedo and shininess and regress each component with deep convolutional auto-encoders. Lombardi and Nishino [23] use a learned prior for natural materials using the DSBRDF model [31] for multi-material estimation. Kang *et al.* [17] and Gao *et al.* [12] estimate spatially-varying BRDFs of planar surfaces taken under designed lighting patterns with an auto-encoder. These methods are limited to known or prescribed illumination and cannot be applied to images taken under arbitrary natural illumination.

Illumination Estimation. Assuming Lambertian surfaces, Marschner *et al.* [26] recover the illumination using image-based bases. Garon *et al.* [14] encode the illumination with 5th-order spherical harmonics and regress their coefficients with a ConvNet. Gardner *et al.* [13] represent lighting as a set of discrete 3D lights with geometric and photometric parameters and regress their coefficients with a neural network. LeGendre *et al.* [21] trained a ConvNet to directly regress a high-dynamic range illumination map from an low-dynamic range image. While there are many more recent works on illumination estimation, jointly estimating the reflectance adds another level of complexity due to the intricate surface reflection that cannot be disentangled with such methods.

Joint Estimation of Reflectance and Illumination. Romeiro *et al.* [38] use non-negative matrix factorization to extract reflectance bases and Haar wavelets to represent illumination to estimate both. Their method, however, is restricted to monochromatic reflectance estimation and cannot handle the complex interplay of illumination and reflectance across different color channels. Lombardi and Nishino [22,25] introduce a maximum a posterior estimation framework using the DSBRDF model and its embedding space as a reflectance prior and gradient and entropy priors on nonparametric illumination. Our Bayesian framework follows their formulation but overcomes the limitations induced by the rigid parametric reflectance model and artificial priors on the illumination that leads to oversmoothing. More recently, Georgoulis *et al.* [15] extend their prior work [36] to jointly estimate geometry, material and illumination. The method, however, assumes Phong BRDF which significantly restricts its applicability to real-world materials. Wang *et al.* [46] leverage image sets of objects of different materials taken under the same illumination to jointly estimate material and illumination. This requirement is unrealistic for general inverse-rendering. Yu *et al.* [48] introduce a deep neural network-based inverse rendering of outdoor imagery. The method, however, fundamentally relies on Lambertian reflectance and low-frequency illumination expressed in spherical harmonics. Azinović *et al.* [3] introduce a differentiable Monte Carlo renderer for inverse path tracing indoor scenes. Their method employs a parametric reflectance model [7], which restricts the types of materials it can handle. While the overall differentiable estimation framework resembles ours, our work focuses on high fidelity inverse rendering of object appearance by fully leveraging a novel BRDF model and deep priors on both the reflectance and illumination. We believe our method can be integrated with such methods as [3] for scene inverse rendering in the future.

3   Differentiable Bayesian Joint Estimation

We first introduce the overall joint estimation framework. Following [25], we formulate object inverse rendering, namely joint estimation of reflectance and illumination of an object of known geometry, from a single image as maximum a posteriori (MAP) estimation

$$\underset{\mathbf{R},\mathbf{L}}{\operatorname{argmax}}\, p(\mathbf{R},\mathbf{L}|\mathbf{I}) \propto p(\mathbf{I}|\mathbf{R},\mathbf{L})p(\mathbf{R})p(\mathbf{L}), \tag{1}$$

where $\mathbf{I}$ is the RGB image, $\mathbf{R}$ is the reflectance, and $\mathbf{L}$ is the environmental illumination. We assume that the camera is calibrated both geometrically and radiometrically, and the illumination is directional (*i.e.*, it can be represented with an environment map). Our key contributions lie in devising an expressive high-dimensional yet invertible model for the reflectance $\mathbf{R}$, and employing strong realistic priors on the reflectance $p(\mathbf{R})$ and illumination $p(\mathbf{L})$.

We model the likelihood with a Laplacian distribution on the log radiance for robust estimation [25,47]

$$p(\mathbf{I}|\mathbf{R},\mathbf{L}) = \prod_{i,c} \frac{1}{2b_I} \exp\left[-\frac{|\log I_{i,c} - \log E_{i,c}(\mathbf{R},\mathbf{L})|}{b_I}\right], \tag{2}$$

where b_I controls the scale of the distribution, $I_{i,c}$ is the irradiance at pixel i in color channel c, and $E_{i,c}(\mathbf{R},\mathbf{L})$ is the expectation of the rendered radiance.

To evaluate the likelihood, we need access to the forward rendered radiance $E_{i,c}(\mathbf{R},\mathbf{L})$ as well as their derivatives with respect to the reflectance and illumination, $\mathbf{R}$ and $\mathbf{L}$, respectively. For this, we implement the differential path tracing method introduced by Lombardi and Nishino [24].

4   Invertible Neural BRDF

Real-world materials exhibit a large variation in their reflectance that are hard to capture with generic bases such as Zernike polynomials or analytic distributions like that in the Cook-Torrance model. Nonparametric representations, such as simple 3D tabulation of an isotropic BRDF, would better capture the wide variety while ensuring accurate representations for individual BRDFs. On the other hand, we also need to differentiate the error function Eq. 1 with respect to the reflectance and also evaluate the exact likelihood of the BRDF. For this, past methods favored low-dimensional parametric reflectance models, which have limited expressive power. We resolve this fundamental dilemma by introducing a high-dimensional parametric reflectance model based on an invertible deep generative neural network. To be precise, the BRDF is expressed as a 3D reflectance distribution which is transformed from a simple parametric distribution (*e.g.*, uniform distribution). The key property is that this transformation is a cascade of invertible linear transforms that collectively results in a complex nonparametric distribution. The resulting parameterization is high-dimensional having as many parameters as a nonparametric tabulation, thus being extremely expressive, yet differentiable and also can be efficiently sampled and evaluated.

4.1   Preliminaries

The BRDF defines point reflectance as the ratio of reflected radiance to incident irradiance given the 2D incident and exitant directions, ω_i and ω_o, respectively,

$$f_r(\omega_i, \omega_o) = \frac{dL_r(\omega_o)}{L_i(\omega_i)\cos\theta_i d\omega_i}, \qquad (3)$$

where $\omega_i = (\theta_i.\phi_i)$, $\omega_o = (\theta_o.\phi_o)$, $\theta \in [0, \frac{\pi}{2}]$, $\phi \in [0, 2\pi]$.
We use the halfway-vector representation introduced by Rusinkiewicz [40]

$$\omega_h = \frac{\omega_i + \omega_o}{\|\omega_i + \omega_o\|}, \; \omega_d = \mathsf{R}_\mathbf{b}(-\theta_h)\mathsf{R}_\mathbf{n}(-\phi_h)\omega_i, \qquad (4)$$

where $\mathsf{R}_\mathbf{a}$ denotes a rotation matrix around a 3D vector $\mathbf{a}$, and $\mathbf{b}$ and $\mathbf{n}$ are the binormal and normal vectors, respectively. The BRDF is then a function of the difference vector and the halfway vector $f_r(\theta_h, \phi_h, \theta_d, \phi_d)$. We assume isotropic BRDFs, for which ϕ_h can be dropped: $f_r(\theta_h, \theta_d, \phi_d)$. Note that f_r is a 3D function that returns a 3D vector of RGB radiance when given a unit irradiance.

4.2   BRDF as an Invertible Neural Network

Our goal in representing the reflectance of real-world materials is to derive an expressive BRDF model particularly suitable for estimating it from an image. Given that deep neural networks lend us complex yet differentiable functions, a naive approach would be to express the (isotropic) BRDF with a neural network. Such a simplistic approach will, however, break down for a number of reasons.

For one, materials with similar reflectance properties (*i.e.*, kurtosis of angular distribution) but with different magnitudes will have to be represented independently. This limitation can be overcome by leveraging the fact that generally each color channel of the same material shares similar reflectance properties. We encode the BRDF as the product of a normalized base distribution p_r and a color vector $\mathbf{c} = \{r, g, b\}$

$$f_r(\theta_h, \theta_d, \phi_d) = p_r(\theta_h, \theta_d, \phi_d)\mathbf{c}. \qquad (5)$$

The color vector is left as a latent parameter and the base distribution is fit with density estimation during training as we explain later. This way, we can represent materials that only differ in reflectance intensity and base colors but have the same distribution properties by sharing the same base distribution. This separation of base color and distribution also lets us model each BRDF with superpositions of f_r, *i.e.*, employ multiple "lobes". This is particularly important when estimating reflectance, as real-world materials usually require at least two colored distributions due to their neutral interface reflection property (*e.g.*, diffuse and illumination colors) [20].

The bigger problem with simply using an arbitrary deep neural network to represent the BRDF is their lack of an efficient means to sample (e.g., GANs)

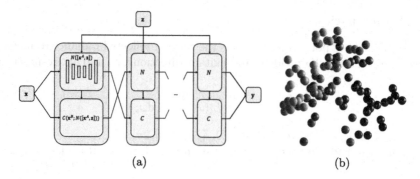

(a)   (b)

Fig. 1. (a) Architecture of the invertible neural BRDF model. The base distribution (p_r in Eq. 5) is represented with a normalizing flow [11,29], which transforms the input 3D uniform distribution $q(\mathbf{x})$ into a BRDF $p_r(\mathbf{y})$ through a cascade of bijective transformations. (b) We condition the parameters with a code $\mathbf{z}$, which let's us learn an embedding space of real-world BRDFs (shown in 2D with PCA). Similar materials are grouped together and arranged in an intuitive manner in this continuous latent space which can directly be used as a reflectance prior.

and restricting latent parametric distribution (e.g., VAEs). For this, we turn to normalizing flow models [42,43] which is one of a family of so-called invertible neural networks [1].

In particular, we represent the normalized base distribution using an extended Non-linear Independent Components Estimation (NICE) model [11,29]. Given a tractable latent (*i.e.*, input) distribution $q(\mathbf{x})$ and a sequence of bijective transformations $f = f_1 \circ f_2 \circ \cdots \circ f_N$, we can get a new distribution $p_r(\mathbf{y})$ by applying transformations $\mathbf{y} = f(\mathbf{x}; \Theta)$. Since f is bijective, under the change of variable formula, $q(\mathbf{x})$ and $p(\mathbf{y})$ are linked with

$$p_r(\mathbf{y}) = q(\mathbf{x}) \left| \det\left(\frac{d\mathbf{x}}{d\mathbf{y}}\right) \right| = q(f^{-1}(\mathbf{y}; \Theta)) \left| \det\left(\frac{df^{-1}(\mathbf{y}; \Theta)}{d\mathbf{y}}\right) \right|, \qquad (6)$$

where $\left| \det\left(\frac{df^{-1}(\mathbf{y};\Theta)}{d\mathbf{y}}\right) \right|$ is the absolute value of the Jacobian determinant. Such a sequence of invertible transformations f is called a normalizing flow. As long as f is complex enough, we can get an arbitrarily complicated $p_r(\mathbf{y})$ in theory. As shown in Fig. 1(a), for each layer f, we use the coupling transformation family C proposed in NICE and each layer is parameterized with a UNet N. The input $\mathbf{x}$ is split into two parts $\mathbf{x}^A$ and $\mathbf{x}^B$, where $\mathbf{x}^A$ is left unchanged and fed into N to produce the parameters of the transformation C that is applied to $\mathbf{x}^B$. Then $\mathbf{x}^A$ and $C(\mathbf{x}^B; N(\mathbf{x}^A))$ are concatenated to give the output. For the input latent distribution $q(\mathbf{x})$, we use a simple 3D uniform distribution.

There are some practical concerns to address before we can train this invertible neural BRDF on measured data. Since the base distribution $p_r(\theta_h, \theta_d, \phi_d)$ of a BRDF inherently has a finite domain ($\theta_h \in [0, \frac{\pi}{2})$, $\theta_d \in [0, \frac{\pi}{2})$, $\phi_d \in [0, \pi)$), it will be easier to learn a transformation mapping from a tractable distribution with the same domain to it. Thus we make $q(\theta_h, \theta_d, \phi_d) \sim \mathcal{U}^3(0, 1)$ and normalize each dimension of $p_r(\theta_h, \theta_d, \phi_d)$ to be in $[0, 1)$ before training. Then we adopt the piecewise-quadratic coupling transformation [29] as C to ensure that the transformed distribution has the same domain. This corresponds to a set of learnable monotonically increasing quadratic curve mappings from $[0, 1)$ to $[0, 1)$ each for one dimension.

4.3   Latent Space of Reflectance

Expressing the rich variability of real-world BRDFs concisely but with a differentiable expression is required to devise a strong constraint (*i.e.*, prior) on the otherwise ill-posed estimation problem (Eq. 1). As a general approach, such a prior can be derived by embedding the BRDFs, for instance, the 100 measured BRDFs in the MERL database [27], in a low-dimensional parametric space. Past related works have achieved this by modeling the latent space in the parameter space of the BRDF model (*e.g.*, linear subspace in the DSBRDF parameter space [25]). More recent works [12,17] train an auto-encoder on the spatial map of analytical BRDF models to model spatially-varying BRDF. The latent space together with the decoder of a trained auto-encoder is then used for material estimation. Our focus is instead on extracting a tight latent space various real-world materials span in the nonparametric space of invertible neural BRDFs.

We achieve this by conditioning the invertible neural BRDF on a latent vector **z** (Fig. 1(a)). We refer to this vector as the embedding code to avoid confusion with the latent distribution (*i.e.*, input) to the invertible BRDF. We jointly learn the parameters Θ of iBRDF and its embedding code **z**

$$\underset{\mathbf{z}, \Theta}{\operatorname{argmax}} \frac{1}{M} \sum_i^M \frac{1}{N} \sum_j^N \log p_r(\theta_h^{ij}, \theta_d^{ij}, \phi_d^{ij} | \mathbf{z}^i; \Theta). \tag{7}$$

Each trained embedding code in the embedding space is associated with a material in the training data (*e.g.*, one measured BRDF of the MERL database). This formulation is similar to the generative latent optimization [6] where the embedding code is directly optimized without the need of an encoder.

We treat each color channel of the training measured data as an independent distribution. Each distribution is assigned a separate embedding code which is initialized with a unit Gaussian distribution. Additionally, after each training step, we project **z** back into the unit hypersphere to encourage learning a compact latent space. This is analogous to the bottleneck structure in an auto-encoder. This conditional invertible neural BRDF is trained by maximizing the likelihood.

During inference, we fix Θ and optimize the embedding code **z** to estimate the BRDF. In practice, we set the dimension of **z** to 16. In other words, the invertible neural BRDFs of real-world materials are embedding in a 16D linear subspace. Figure 1(b) shows the embedding of the 100 measured BRDFs in the MERL database after training. Materials with similar properties such as glossiness lie near each other forming an intuitive and physically-plausible latent space that can directly be used as a strong reflectance prior. Since all materials are constrained within the unit hypersphere during training, we thus define $p(\mathbf{R}) \sim \mathcal{N}(0, \sigma^2 I)$.

5   Deep Illumination Prior

Incident illumination to the object surface can be represented in various forms. With the same reason that motivated the derivation of iBRDF, we should avoid unnecessary approximation errors inherent to models and represent the illumination as a nonparametric distribution. This can be achieved by simply using a latitude-longitude panoramic HDR image $\mathbf{L}(\theta, \phi)$ as an environment map to represent the illumination [10], which we refer to as the illumination map.

The expressive power of a nonparametric illumination representation, however, comes with increased complexity. Even a 1°–sampled RGB illumination map (360×180 in pixels) would have 194400 parameters in total. To make matters worse, gradients computed in the optimization are sparse and noisy and would not sufficiently constrain such a high degree of freedom. Past work have mitigated this problem by imposing strong analytical priors on the illumination $p(\mathbf{L})$, such as sparse gradients and low entropy [25] so that they themselves are differentiable. These analytical priors, however, do not necessarily capture the properties of natural environmental illumination and often lead to overly simplified illumination estimates.

We instead directly constrain the illumination maps to be "natural" by leveraging the structural bias of deep neural networks. Ulyanov *et al.* [45] found that the structure of a deep neural network has the characteristic of impeding noise and, in turn, represent natural images without the need of pre-training. That is, an untrained ConvNet can directly be used as a natural image prior. We adopt this idea of a deep image prior and design a deep neural network for use as a deep illumination prior.

We redefine the illumination map as $\mathbf{L} = g(\theta, \phi; \boldsymbol{\Phi})$, where $\boldsymbol{\Phi}$ denotes the parameters of a deep neural network g. We estimate the illumination map as an "image" generated by the deep neural network g with a fixed random noise image as input by optimizing the network parameters $\boldsymbol{\Phi}$. The posterior (Eq. 1) becomes

$$\underset{\mathbf{R}, \boldsymbol{\Phi}}{\arg\max} \, p(\mathbf{R}, \mathbf{L}|\mathbf{I}) \propto p(\mathbf{I}|\mathbf{R}, g(\theta, \phi; \boldsymbol{\Phi}))p(\mathbf{R}). \tag{8}$$

The advantages of this deep illumination prior is twofold. First, the structure of a deep neural network provides impedance to noise introduced by the differentiable renderer and stochastic gradient descent through the optimization. Second, angular samples of the illumination map are now interdependent through the network, which mitigates the problem of sparse gradients. For the deep neural network g, we use a modified UNet architecture [39]. To avoid the checkerboard artifacts brought by transposed convolutional layer, we use a convolutional layer followed by a bilinear upsampling for each upsampling step. Additionally, to preserve finer details, skip connections are added to earlier layers.

6   Experimental Results

We conducted a number of experiments to validate the effectiveness of 1) the
invertible neural BRDF model for representing reflectance of real-world materials; 2) the conditional invertible neural BRDF model for BRDF estimation; 3)
the deep illumination prior, and 3) the Bayesian estimation framework integrated
with the model and priors for single-image inverse rendering.

6.1   Accuracy of Invertible Neural BRDF

To evaluate the accuracy of invertible neural BRDF, we learn its parameters to
express measured BRDF data in the MERL database and evaluate the representation accuracy using the root mean squared error (RMSE) in log space [25].

As Fig. 2 shows, the invertible neural BRDF achieves higher accuracy than
the nonparametric bivariate BRDF model. The conditional iBRDF, learned on
100%, 80%, and 60% training data all achieve high accuracy superior to other
parametric models namely the DSBRDF and Cook-Torrance models. Note that
all these conditional iBRDFs were trained without the test BRDF data. This
resilience to varying amounts of training data demonstrates the robustness of
the invertible neural BRDF model and its generalization power encoded in the
learnt embedding codes. The results show that the model learns a latent space
that can be used as a reflectance prior without sacrificing its expressive power.

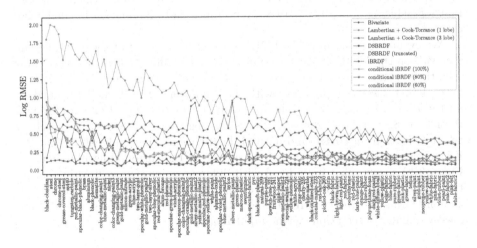

Fig. 2. Log RMSE of iBRDF and conditional iBRDF (*i.e.*, iBRDF with learned latent
space) for 100 MERL materials. The iBRDF has higher accuracy than a nonparametric bivariate model. The conditional iBRDF models using 100%, 80%, and 60% of
the leave-one-out training data achieves higher accuracy than other parametric models (*i.e.*, DSBRDF and Cook-Torrance). These results show the expressive power and
generalizability of the invertible neural BRDF model.

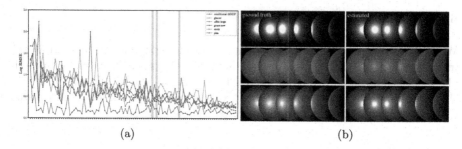

Fig. 3. (a) Log RMSE of iBRDF estimation from single images of a sphere with different materials under different known natural illumination. (b) Renderings of three samples of ground truth and estimated BRDFs. These results clearly demonstrate the high accuracy iBRDF can provide when estimating the full BRDF from partial angular observations under complex illumination.

6.2 Reflectance Estimation with iBRDF

Next, we evaluate the effectiveness of the invertible neural BRDF for single-image BRDF estimation. Unlike directly fitting to measured BRDF data, the BRDF is only partially observed in the input image. Even worse, the differentiable path tracer inevitably adds noise to the estimation process. The results of these experiments tell us how well the invertible neural BRDF can extrapolate unseen slices of the reflectance function given noisy supervision, which is important for reducing ambiguity between reflectance and illumination in their joint estimation.

We evaluate the accuracy of BRDF estimation for each of the 100 different materials in the MERL database rendered under 5 different known natural illuminations each taken from Debevec's HDR environment map set [9]. The BRDF for each material was represented with the conditional iBRDF model trained on all the measured data expect the one to be estimated (*i.e.*, 100% conditional iBRDF in Sect. 6.1). Figure 3(a) shows the log-space RMS error for each of the combinations of BRDF and natural illumination. The results show that the BRDF can be estimated accurately regardless of the surrounding illumination. Figure 3(b) shows spheres rendered with different point source directions using the recovered BRDF. The results match the ground truth measured BRDF well, demonstrating the ability of iBRDF to robustly recover the full reflectance from partial angular observations in the input image.

6.3 Illumination Estimation with Deep Illumination Prior

We examine the effectiveness of the deep illumination prior by evaluating the accuracy of illumination estimation with known reflectance and geometry. We rendered images of spheres with 5 different BRDFs sampled from the MERL database under 10 different natural illuminations. Figure 4 shows samples of the

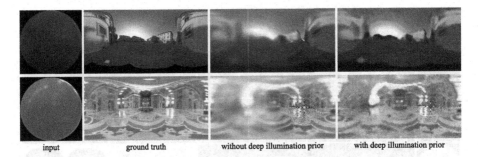

input   ground truth   without deep illumination prior   with deep illumination prior

Fig. 4. Results of illumination estimation without and with the deep illumination prior. The illumination estimates using the prior clearly shows higher details that match those of the ground truth.

ground truth and estimated illumination without and with the deep illumination. These results clearly demonstrate the effectiveness of the proposed deep illumination prior which lends strong constraints for tackling joint estimation.

6.4 Joint Estimation of Reflectance and Illumination

We integrate iBRDF and its latent space as a reflectance prior together with the deep illumination prior into the MAP estimation framework (Eq. 1) for object inverse rendering and systematically evaluate their effectiveness as a whole with synthetic and real images. First, we synthesized a total of 100 images of spheres rendered with 20 different measured BRDFs sampled from the MERL database under 5 different environment maps. Figure 5(a) shows some of the estimation results. Qualitatively, the recovered BRDF and illumination match the ground truth well, demonstrating the effectiveness of iBRDF and priors for object inverse rendering. As evident in the illumination estimates, our method is able to recover high-frequency details that are not attainable in past methods. Please compare these results with, for instance, Fig. 7 of [25].

Finally, we apply our method to images of real objects taken under natural illumination. We use the Objects Under Natural Illumination Database [25]. Figure 5(b) shows the results of jointly estimating the BRDF and illumination. Our reflectance estimates are more faithful to the object appearance than those by Lombardi and Nishino, and the illumination estimates have more details (compare Fig. 5 with Fig. 10 of [25]), which collectively shows that our method more robustly disentangles the two from the object appearance. Note that the color shifts in the BRDF estimates arise from inherent color constancy, and the geometry dictates the recoverable portions of the environment. The estimates are in HDR and exposures are manually set to match as there is an ambiguity in global scaling. Please see the supplemental material for more results.

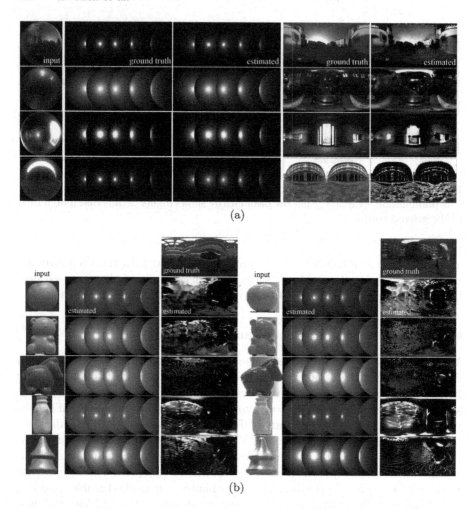

Fig. 5. Results of object inverse rendering using iBRDF and its latent space, and the deep illumination prior from (a) synthetic images and (b) real images. All results are in HDR shown with fixed exposure, which exaggerates subtle differences (*e.g.*, floor tile pattern in Uffizi). The results show that the model successfully disentangles the complex interaction of reflectance and illumination and recovers details unattainable in past methods (*e.g.*, [25]).

7   Conclusion

In this paper, we introduced the invertible neural BRDF model and an object inverse rendering framework that exploits its latent space as a reflectance prior and a novel deep illumination prior. Through extensive experiments on BRDF fitting, recovery, illumination estimation, and inverse rendering, we demonstrated the effectiveness of the model for representing real-world reflectance as well as its use, together with the novel priors, for jointly estimating reflectance and illumination from single images. We believe these results show new ways in which powerful deep neural networks can be leveraged in solving challenging radiometric inverse and forward problems.

Acknowledgement. This work was in part supported by JSPS KAKENHI 17K20143 and a donation by HiSilicon.

References

1. Ardizzone, L., et al.: Analyzing inverse problems with invertible neural networks. In: International Conference on Learning Representations (2019)
2. Ashikhmin, M., Premoze, S.: Distribution-based BRDFs. Unpublished Technical report, University of Utah 2, 6 (2007)
3. Azinović, D., Li, T.M., Kaplanyan, A., Nießner, M.: Inverse path tracing for joint material and lighting estimation. In: Proceedings of the Computer Vision and Pattern Recognition (CVPR). IEEE (2019)
4. Basri, R., Jacobs, D.: Photometric stereo with general, unknown lighting. In: IEEE International Conference on Computer Vision and Pattern Recognition, pp. 374–381 (2001)
5. Blinn, J.F.: Models of light reflection for computer synthesized pictures. In: Proceedings of the 4th Annual Conference on Computer Graphics and Interactive Techniques, pp. 192–198 (1977)
6. Bojanowski, P., Joulin, A., Lopez-Paz, D., Szlam, A.: Optimizing the latent space of generative networks. In: International Conference on Machine Learning (2018)
7. Burley, B., Studios, W.D.A.: Physically-based shading at disney. In: ACM SIGGRAPH 2012, pp. 1–7 (2012)
8. Cook, R.L., Torrance, K.E.: A reflectance model for computer graphics. ACM Trans. Graph. (TOG) 1(1), 7–24 (1982)
9. Debevec, P.: Light probe gallary. http://www.debevec.org/Probes/
10. Debevec, P.: Rendering synthetic objects into real scenes: bridging traditional and image-based graphics with global illumination and high dynamic range photography. In: ACM SIGGRAPH 2008 Classes, SIGGRAPH 2008, pp. 32:1–32:10. ACM, New York (2008). https://doi.org/10.1145/1401132.1401175
11. Dinh, L., Krueger, D., Bengio, Y.: NICE: non-linear independent components estimation. CoRR (2014)
12. Gao, D., Li, X., Dong, Y., Peers, P., Xu, K., Tong, X.: Deep inverse rendering for high-resolution SVBRDF estimation from an arbitrary number of images. ACM Trans. Graph. (TOG) 38(4), 134 (2019)
13. Gardner, M.A., Hold-Geoffroy, Y., Sunkavalli, K., Gagne, C., Lalonde, J.F.: Deep parametric indoor lighting estimation. In: The IEEE International Conference on Computer Vision (ICCV), October 2019
14. Garon, M., Sunkavalli, K., Hadap, S., Carr, N., Lalonde, J.F.: Fast spatially-varying indoor lighting estimation. In: The IEEE Conference on Computer Vision and Pattern Recognition (CVPR), June 2019
15. Georgoulis, S., et al.: Reflectance and natural illumination from single-material specular objects using deep learning. IEEE Trans. Pattern Anal. Mach. Intell. 40(8), 1932–1947 (2017)
16. Goodfellow, I., et al.: Generative adversarial nets. In: Advances in neural information processing systems. pp. 2672–2680 (2014)

17. Kang, K., Chen, Z., Wang, J., Zhou, K., Wu, H.: Efficient reflectance capture using an autoencoder. ACM Trans. Graph. **37**(4), 127-1 (2018)
18. Kingma, D.P., Welling, M.: Auto-encoding variational bayes. In: International Conference on Learning Representations (2014)
19. Koenderink, J.J., van Doorn, A.J., Stavridi, M.: Bidirectional reflection distribution function expressed in terms of surface scattering modes. In: Buxton, B., Cipolla, R. (eds.) ECCV 1996. LNCS, vol. 1065, pp. 28–39. Springer, Heidelberg (1996). https://doi.org/10.1007/3-540-61123-1_125
20. Lee, H.C.: Illuminant color from shading, pp. 340–347. Jones and Bartlett Publishers Inc., USA (1992)
21. LeGendre, C., et al.: DeepLight: learning illumination for unconstrained mobile mixed reality. In: Proceedings of the IEEE Conference on Computer Vision and Pattern Recognition, pp. 5918–5928 (2019)
22. Lombardi, S., Nishino, K.: Reflectance and natural illumination from a single image. In: Fitzgibbon, A., Lazebnik, S., Perona, P., Sato, Y., Schmid, C. (eds.) ECCV 2012. LNCS, vol. 7577, pp. 582–595. Springer, Heidelberg (2012). https://doi.org/10.1007/978-3-642-33783-3_42
23. Lombardi, S., Nishino, K.: Single image multimaterial estimation. In: 2012 IEEE Conference on Computer Vision and Pattern Recognition, pp. 238–245. IEEE (2012)
24. Lombardi, S., Nishino, K.: Radiometric scene decomposition: scene reflectance, illumination, and geometry from RGB-D images. In: International Conference on 3D Vision (2016)
25. Lombardi, S., Nishino, K.: Reflectance and illumination recovery in the wild. IEEE Trans. Pattern Anal. Mach. Intell. **38**(1), 129–141 (2016)
26. Marschner, S.R., Greenberg, D.P.: Inverse lighting for photography. In: Color and Imaging Conference, pp. 262–265. Society for Imaging Science and Technology (1997)
27. Matusik, W.: A data-driven reflectance model. Ph.D. thesis, Massachusetts Institute of Technology (2003)
28. Meka, A., et al.: LIME: live intrinsic material estimation. In: Proceedings of the IEEE Conference on Computer Vision and Pattern Recognition, pp. 6315–6324 (2018)
29. Müller, T., Mcwilliams, B., Rousselle, F., Gross, M., Novák, J.: Neural importance sampling. ACM Trans. Graph. **38**(5), 145:1–145:19 (2019). http://doi.acm.org/10.1145/3341156
30. Nicodemus, F., Richmond, J., Hsia, J., Ginsberg, I., Limperis, T.: Geometric considerations and nomenclature for reflectance. National Bureau of Standards (US) (1977)
31. Nishino, K.: Directional statistics BRDF model. In: 2009 IEEE 12th International Conference on Computer Vision, pp. 476–483. IEEE (2009)
32. Nishino, K., Lombardi, S.: Directional statistics-based reflectance model for isotropic bidirectional reflectance distribution functions. OSA J. Opt. Soc. Am. A **28**(1), 8–18 (2011)
33. Oxholm, G., Nishino, K.: Shape and reflectance estimation in the wild. IEEE Trans. Pattern Anal. Mach. Intell. **38**(2), 376–389 (2016)
34. Phong, B.T.: Illumination for computer generated pictures. Commun. ACM **18**(6), 311–317 (1975)
35. Ramamoorthi, R., Hanrahan, P.: A signal-processing framework for inverse rendering. In: Computer Graphics Proceedings, ACM SIGGRAPH 2001, pp. 117–128 (2001)

36. Rematas, K., Ritschel, T., Fritz, M., Gavves, E., Tuytelaars, T.: Deep reflectance maps. In: Proceedings of the IEEE Conference on Computer Vision and Pattern Recognition, pp. 4508–4516 (2016)
37. Romeiro, F., Vasilyev, Y., Zickler, T.: Passive reflectometry. In: Forsyth, D., Torr, P., Zisserman, A. (eds.) ECCV 2008. LNCS, vol. 5305, pp. 859–872. Springer, Heidelberg (2008). https://doi.org/10.1007/978-3-540-88693-8_63
38. Romeiro, F., Zickler, T.: Blind reflectometry. In: Daniilidis, K., Maragos, P., Paragios, N. (eds.) ECCV 2010. LNCS, vol. 6311, pp. 45–58. Springer, Heidelberg (2010). https://doi.org/10.1007/978-3-642-15549-9_4
39. Ronneberger, O., Fischer, P., Brox, T.: U-net: convolutional networks for biomedical image segmentation. In: Navab, N., Hornegger, J., Wells, W.M., Frangi, A.F. (eds.) MICCAI 2015. LNCS, vol. 9351, pp. 234–241. Springer, Cham (2015). https://doi.org/10.1007/978-3-319-24574-4_28
40. Rusinkiewicz, S.M.: A new change of variables for efficient BRDF representation. In: Drettakis, G., Max, N. (eds.) EGSR 1998. E, pp. 11–22. Springer, Vienna (1998). https://doi.org/10.1007/978-3-7091-6453-2_2
41. Sato, Y., Wheeler, M.D., Ikeuchi, K.: Object shape and reflectance modeling from observation. In: Proceedings of SIGGRAPH 1997, pp. 379–387 (1997)
42. Tabak, E.G., Turner, C.V.: A family of nonparametric density estimation algorithms. Commun. Pure Appl. Math. **66**(2), 145–164 (2013). https://doi.org/10.1002/cpa.21423. https://onlinelibrary.wiley.com/doi/abs/10.1002/cpa.21423
43. Tabak, E.G., Vanden-Eijnden, E.: Density estimation by dual ascent of the log-likelihood. Commun. Math. Sci. **8**(1), 217–233 (2010)
44. Torrance, K.E., Sparrow, E.M.: Theory for off-specular reflection from roughened surfaces. Josa **57**(9), 1105–1114 (1967)
45. Ulyanov, D., Vedaldi, A., Lempitsky, V.: Deep image prior. In: Proceedings of the IEEE Conference on Computer Vision and Pattern Recognition, pp. 9446–9454 (2018)
46. Wang, T., Ritschel, T., Mitra, N.: Joint material and illumination estimation from photo sets in the wild. In: 2018 International Conference on 3D Vision (3DV), pp. 22–31. IEEE (2018)
47. Wilcox, R.R.: Introduction to Robust Estimation and Hypothesis Testing. Academic Press, Cambridge (2011)
48. Yu, Y., Smith, W.A.: InverseRenderNet: learning single image inverse rendering. In: Proceedings of the IEEE/CVF Conference on Computer Vision and Pattern Recognition (CVPR) (2019)

Semi-supervised Semantic Segmentation via Strong-Weak Dual-Branch Network

Wenfeng Luo[1] and Meng Yang[1,2(✉)]

[1] School of Data and Computer Science, Sun Yat-sen University, Guangzhou, China
`luowf5@mail2.sysu.edu.cn, yangm6@mail.sysu.edu.cn`
[2] Key Laboratory of Machine Intelligence and Advanced Computing (SYSU),
Ministry of Education, Beijing, China

Abstract. While existing works have explored a variety of techniques to push the envelop of weakly-supervised semantic segmentation, there is still a significant gap compared to the supervised methods. In real-world application, besides massive amount of weakly-supervised data there are usually a few available pixel-level annotations, based on which semi-supervised track becomes a promising way for semantic segmentation. Current methods simply bundle these two different sets of annotations together to train a segmentation network. However, we discover that such treatment is problematic and achieves even worse results than just using strong labels, which indicates the misuse of the weak ones. To fully explore the potential of the weak labels, we propose to impose separate treatments of strong and weak annotations via a strong-weak dual-branch network, which discriminates the massive inaccurate weak supervisions from those strong ones. We design a shared network component to exploit the joint discrimination of strong and weak annotations; meanwhile, the proposed dual branches separately handle full and weak supervised learning and effectively eliminate their mutual interference. This simple architecture requires only slight additional computational costs during training yet brings significant improvements over the previous methods. Experiments on two standard benchmark datasets show the effectiveness of the proposed method.

Keywords: Semi-supervised · Strong-weak · Semantic segmentation

1 Introduction

Convolutional Neural Networks (CNNs) [11,17,30] have proven soaring successes on the semantic segmentation problem. Despite their superior performance, these CNN-based methods are data-hungry and rely on huge amount of pixel-level annotations, whose collections are labor-intensive and time-consuming. Hence

Electronic supplementary material The online version of this chapter (https://doi.org/10.1007/978-3-030-58558-7_46) contains supplementary material, which is available to authorized users.

researchers have turned to *weakly-supervised learning* that could exploit weaker forms of annotation, thus reducing the labeling costs. Although numerous works [13,15,19,36] have been done on learning segmentation models from weak supervisions, especially per-image labels, they still trail the accuracy of their fully-supervised counterparts and thus are not ready for real-world applications.

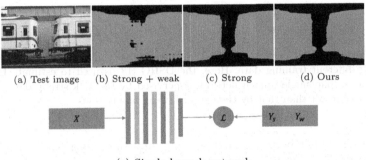

(a) Test image   (b) Strong + weak   (c) Strong   (d) Ours

(e) Single-branch network

Fig. 1. (a) Sample test image; (b) Result using both strong and weak annotations in a single-branch network; (c) Result using only the strong annotations; (d) Result using our strong-weak dual-branch network; (e) Single-branch network adopted by previous methods [19,26,35]; Images (a), (b), (c), (d) in the first row visually demonstrate that using extra weak annotations brings no improvement over only using the strong annotations when a single-branch network is employed. See Fig. 7 for more visual comparisons.

In order to achieve good accuracy while still keeping the labeling budget in control, we focus on tackling a more practical problem under semi-supervised setting, where a combination of strongly-labeled (pixel-level masks) and weakly-labeled (image-level labels) annotations are utilized. However, previous methods (WSSL [26], MDC [35] and FickleNet [19]) simply scratch the surface of semi-supervised segmentation by exploring better weakly-supervised strategies to extract more accurate initial pixel-level supervisions, which are then mixed together with strong annotations to learn a segmentation network, as in Fig. 1(e). However, we discover that the simple combination of strong and weak annotations with equal treatment may weaken the final performance.

To further analyze the roles of strong and weak annotations in conventional single-branch network, we compare the segmentation performance on PASCAl VOC val set using different training data in Fig. 2. When trained on small amount of strong data (1.4k in our experiments), performance of the segmentation network is not as low as people would expect. On the contrary, our implementation achieves a peak mIoU of 68.9% using only 1.4k strong annotations and it is already much better than other methods WSSL [26] (64.6%), MDC [35] (65.7%), FickleNet [19] (65.8%) exploiting extra 9k weak annotations. Moreover, when simply bundling the strong and weak annotations to train a single segmentation network, the

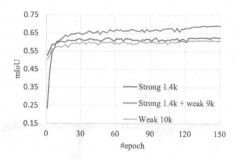

Fig. 2. Segmentation performance of the conventional single-branch network on *val* set using different training data. Here, the DSRG [13] is used to estimate the weak supervisions. The single-branch network supervised by the 1.4k strong data achieved much better result than that by the extra 9k weak annotations.

performance is not better than that using only the strong ones (quantatively shown in Fig. 2 and visually shown in Fig. 1(a), (b), (c)).

Based on the above observations, it can be concluded that such treatment underuses the weakly-supervised data and thus introduces limited improvement, or even worse, downgrading the performance achieved by using only the strong annotations. We further point out two key issues that are previously unnoticed concerning the semi-supervised setting:

1. *sample imbalance*: there are usually much more weak data than the strong ones, which could easily result in overfitting to the weak supervisions.
2. *supervision inconsistency*: the weak annotations Y_w are of relatively poor quality compared to the strong ones Y_s and thus lead to poor performance.

To better jointly use the strong and weak annotations, we propose a novel method of strong-weak dual-branch network, which is a single unified architecture with parallel strong and weak branches to handle one type of annotation data (Fig. 3). To fully exploit the joint discrimination of strong and weak annotations, the parallel branches share a common convolution backbone in exchange for supervision information of different level without competing with each other. The shared backbone enables the free flow of the gradient and the parallel branches can discriminate between the accurate and noisy annotations. Moreover, the dual branches are explicitly learned from strong and weak annotations separately, which can effectively avoid the affect of sample imbalance and supervision inconsistency. This simple architecture boosts the segmentation performance by a large margin while introducing negligible overheads. State-of-art performance has been achieved by the proposed strong-weak network under semi-supervised setting on both PASCAL VOC and COCO segmentation benchmarks. Remarkably, it even boosts the fully-supervised models when both branches are trained with strong annotations on PASCAL VOC.

The main contributions of our paper are three-folds:

1. We for the first time show that segmentation network trained under mixed strong and weak annotations achieves even worse results than using only the strong ones.
2. We reveal that sample imbalance and supervision inconsistency are two key obstacles in improving the performance of semi-supervised semantic segmentation.
3. We propose a simple unified network architecture to address the inconsistency problem of annotation data in the semi-supervised setting.

2   Related Works

In this section, we briefly review weakly-supervised and semi-supervised visual learning, which are most related to our work. Although the idea of multiple branches to capture various context has already been explored in many computer vision tasks, we here highlight the primary difference between previous methods and this work.

Weakly-Supervised Semantic Segmentation. To relieve the labeling burden on manual annotation, many algorithms have been proposed to tackle semantic segmentation under weaker supervisions, including points [2], scribbles [21,32], and bounding boxes [7,31]. Among them, per-image class labels are most frequently explored to perform pixel-labeling task since their collections require the least efforts, only twenty seconds per image [2]. Class Activation Map (CAM) [39], is a common method to extract from classification network a sparse set of object seeds, which are known to concentrate on small discriminative regions. To mine more foreground pixels, a series of methods have been proposed to apply the erasing strategy, either on the original image [36] or high-level class activations [12]. Erasing strategy is a form of strong attention [37] which suppresses selective responsive regions and forces the network to find extra evidence to support the corresponding task. Some other works [4,25,34] also proposed to incorporate saliency prior to ease the localization of foreground objects.

Recently, Huang et al. [13] proposed Deep Seeded Region Growing (DSRG) to dynamically expand the discriminative regions along with the network training, thus mining more integral objects for segmentation networks. And Lee et al. [19] further improved the segmentation accuracy of DSRG by replacing the original CAM with stochastic feature selection for seed generation.

Despite the progress on weakly-supervised methods, there is still a large performance gap (over 10%) from their full-supervised versions [5,6], which indicates that they are unsuitable for the real-world applications.

Semi-supervised Learning. In general, semi-supervised learning [40] addresses the classification problem by incorporating large amount of extra unlabeled data besides the labeled samples to construct better classifiers. Besides earlier methods, like semi-supervised Support Vector Machine [3], many techniques have been proposed to integrate into deep-learning models, such as Temporal Ensembling [18], Virtual Adversarial Training [24] and Mean Teacher [33].

In this paper, we focus on such *semi-supervised learning* setting on semantic segmentation problem, where the training data are composed of a small set of finely-labeled data and large amount of coarse annotations, usually estimated from a weakly-supervised methods. In this configuration, current models [19,20,27,35] usually resort to the sophistication of weakly-supervised models to provide more accurate proxy supervisions and then simply bundle both sets of data altogether to learn a segmentation network. They pay no special attention to coordinating the usage of weak annotations with the strong ones. Such treatment, ignoring the annotation inconsistency, overwhelms the handful yet vital minority and consequently produces even worse results compared to using only the fine data.

Multi-branch Network. Networks with multiple parallel branches have been around for a long time and proven their effectiveness in a variety of vision-related tasks. Object detection models [9,22,28] usually ended with two parallel branches, one for the classification and the other for localization. In addition, segmentation networks, such as Atrous Spatial Pyramid Pooling (ASPP) [5] and Pyramid Scene Parsing (PSP) [38] network, explored multiple parallel branches to capture richer context to localize objects of different sizes. Unlike the above works, we instead utilize parallel branches to handle different types of annotation data.

3   Methods

As aforementioned, the proxy supervisions estimated by weakly-supervised methods are of relatively poor quality in contrast to manual annotations. For finely-labeled and weakly-labeled semantic segmentation task, a natural solution for different supervision is to separately train two different networks, whose outputs are then aggregated by taking the average (or maximum). Although this simple ensemble strategy is likely to boost the performance, it is undesirable to maintain two copies of network weights during both training and inference. Besides, separate training prohibits the exchange for supervision information. To enable information sharing and eliminate the sample imbalance and supervision inconsistency, we propose a dual-branch architecture to handle different types of supervision, eliminating the necessity of keeping two network copies. Figure 3 presents an overview of the proposed architecture.

Notation. Let the training images $X = (x_i)_{i \in [n]}$ be divided into two subsets: the images $X_s = \{(x_1^s, m_1^s), ..., (x_t^s, m_t^s)\}$ with strong annotations provided by the dataset and images $X_w = \{(x_1^w, m_1^w), ..., (x_k^w, m_k^w))\}$, the supervisions of which are estimated from a proxy ground-truth generator G:

$$m_i^w = G(x_i^w) \tag{1}$$

The proxy generator G may need some extra information, such as class labels, to support its decision, but we can leave it for general discussion.

The rest of this section is organized as follows. Section 3.1 discusses in depth why training a single-branch network is problematic. Section 3.2 elaborates on the technical details of the proposed strong-weak dual-branch network.

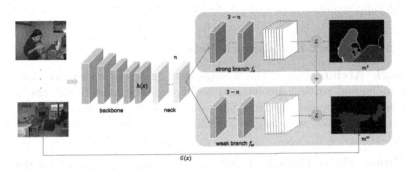

Fig. 3. Overview of the proposed dual-branch network. The proposed architecture consists of three individual parts: backbone, neck module and two parallel branches that share an identical structure but differ in the training annotations. The hyperparameter n controls the number (i.e., 3-n) of individual convolutional layers existing in the parallel branches.

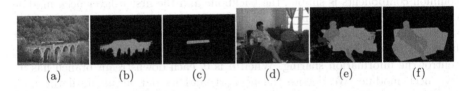

| (a) | (b) | (c) | (d) | (e) | (f) |

Fig. 4. Two inaccurate weak annotations estimated by DSRG [13]. (a), (d): sample training images; (b), (e): masks estimated by DSRG; (c), (f): ground truth. In (b), the *train* mask expands to the background due to color similarity. In (e), large portions of the *human* body are misclassified.

3.1 Oversampling Doesn't Help with Single-Branch Network

Previous works [19,20,27,35] focus on developing algorithm to estimate more accurate initial supervision, but they pay no special attention on how to coordinate the strong and weak annotations. Notably, there are quite a few estimated masks of relatively poor quality (as shown Fig. 4) when image scenes become more complex. Equal treatment biases the gradient signal towards the incorrect weak annotations since they are in majority during the computation of the training loss. Consequently, it offsets the correct concept learned from the strong annotations and therefore leads to performance degradation. In addition, we also conduct experiments via oversampling the strong annotations.

The results shows that oversampling does improve the final segmentation accuracy steadily (62.8%→65.9%) as more strong annotations are duplicated, but it still fails to outperform the result (68.9%) using only the strong annotations. In conclusion, oversampling does not help with the single-branch network either. See the supplementary material for detailed results.

3.2   Strong-Weak Dual-Branch Network

Network Architecture. The strong-weak dual-branch network consists of three individual parts: convolutional backbone, neck module and two parallel branches with identical structure. Since our main experiments centre around the VGG16 network [30], we here give a detailed discussion of the architecture based on VGG16.

Backbone. The backbone is simply the components after removing the fully-connected layers. As in [5], the last two pooling layers are dropped and the dilation rates of the subsequent convolution layers are raised accordingly to obtain features of output stride 8.

Neck Module. The neck module is a series of convolution layers added for better adaptation of the specific task. It could be shared between or added separately into subsequent parallel branches. Let n be the number of convolution layers in the neck module shared by different supervision. Although the design of common components is simple, the backbone and the first n-layer neck module can effectively learn the joint discrimination from the full supervision and weak supervision. The total number of convolution layers in the neck and subsequent branch is fixed, but the hyper-parameter $n \in [0, 3]$ offers greater flexibility to control the information sharing. When n is 0, each downstream branch has its own neck module. We denote the network and its output up until the neck module as $Z = h(X) \in R^{H \times W \times K}$.

Strong-Weak Branches. These two parallel branches have the same structure while differ in the training annotations they receive. The strong branch is supervised by the fine annotation X_s, while the weak branch is trained by the coarse supervisions X_w. The way of separately processing different supervision is quite new because existing semi-supervised semantic segmentation methods adopt a single-branch network and current multi-branch network has never dealt with different types of annotation. The branches $f(Z; \theta_s)$ and $f(Z; \theta_w)$ are governed by independent sets of parameters. For brevity, we will omit the parameters in our notation and simply write $f_s(Z)$ and $f_w(Z)$. The normal cross entropy loss has the following form:

$$\mathcal{L}_{ce}(s, m) = -\frac{1}{|m|} \sum_c \sum_{u \in m_c} \log s_{u,c} \tag{2}$$

where tensor s is the network outputs, m is the annotation mask and m_c denotes the set of pixels assigned to category c. Then the data loss of our method is:

$$s^s = f_s(h(x^s))$$
$$s^w = f_w(h(x^w)) \tag{3}$$
$$\mathcal{L}_{data} = \mathcal{L}_{ce}(s^s, m^s) + \mathcal{L}_{ce}(s^w, m^w)$$

We emphasize that all the loss terms are equally weighted so no hyper-parameter is involved.

3.3   How Does the Dual-Branch Network Help?

During training, we need to construct a training batch with the same amount of strong and weak images. As in semi-supervised semantic segmentation, there are usually much more weak annotations than the strong ones. Consequently, the strong data have been looped through several times before the weak data are exhausted for the first pass, which essentially performs oversampling of the strong data. In this way, the strong data make a difference during training and thus mitigate the effect of sample imbalance.

Our dual-branch network imposes separate treatments on the strong and weak annotations and therefore prevents direct interference of different supervision information, so the supervision inconsistency can be well eliminated. Meanwhile, the coarse ones leave no direct influence on the strong branch, which determines the final prediction. Nevertheless, the extra weak annotations provide approximate location of objects and training them on a separate branch introduces regularization into the underlying backbone to some extent, hence improving the network's generalization capability.

3.4   Implementation Detail

Training. Here we introduce an efficient way to train the strong-weak dual-branch network. A presentation of the processing details can be found in Fig. 5. During training, a batch of $2n$ images $X = [(x_1^s, m_1^s), ..., (x_1^w, m_1^w)...]$ are sampled, with the first half from X_s and the second from X_w. Since the number of weak annotations is usually much bigger than that of strong ones, we are essentially performing an oversampling of the strong annotations X_s. For the image batch $X \in R^{2n \times h \times w}$, we make no distinction of the images and simply obtain the network logits in each branch, namely $S^s, S^w \in R^{2n \times h \times w}$, but half of them (in color gray Fig. 5) have no associated annotations and are thus discarded. The remaining halves are concatenated to yield the final network output $S = [S^s[1:n], S^w[n+1:2n]]$, which are then used to calculate the cross entropy loss irrespective of the annotations employed. We find that this implementation eases the training and inference processes.

Inference. When the network is trained, the weak branch is no longer needed since the information from weak annotations has been embedded into the convolution backbone and the shared neck module. So at inference stage, only the strong branch is utilized to generate final predictions.

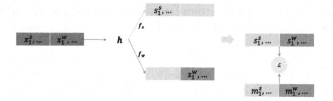

Fig. 5. The images are first forwarded through the network and half of the outputs (in color gray) are dropped before they are concatenated to compute the final loss (only the batch dimension is shown).

4 Experiments

4.1 Experimental Setup

Dataset and Evaluation Metric. The proposed method is evaluated on two segmentation benchmarks, PASCAL VOC [8] and COCO dataset [23]. **PASCAL VOC:** There are 20 foreground classes plus 1 background category in PASCAL VOC dataset. It contains three subsets for semantic segmentation task, *train* set (1464 images), *val* set (1449 images) and *test* set (1456 images). As a common practice, we also include the additional annotations from [10] and end up with a *trainaug* set of 10582 images. For semi-supervised learning, we use the *train* set as the strong annotations and the remaining 9k images as weak annotations. We report segmentation results on both *val* and *test* set. **COCO:** We use the train-val split in the 2016 competition, where 118k images are used for training and the remaining 5k for testing. We report the segmentation performance on the 5k testing images.

The standard interaction-over-union (IoU) averaged across all categories is adopted as evaluation metric for all the experiments.

Proxy Supervision Generator G. To verify the effectiveness of the proposed architecture, we choose the recently popular weakly-supervised method, Deep Seeded Region Growing (DSRG) [13], as the proxy supervision generator G. We use the DSRG model before the retraining stage to generate proxy ground truth for our experiments. Further details could be found in the original paper.

Training and Testing Settings. We use the parameters pretrained on the 1000-way ImageNet classification task to initialize our backbones (either VGG16 or ResNet101). We use Adam optimizer [14] with an initial learning rate of 1e−4 for the newly-added branches and 5e-6 for the backbone. The learning rate is decayed by a factor of 10 after 12 epochs. The network is trained under a batch size of 16 and a weight decay of 1e−4 for 20 epochs. We use random scaling and horizontal flipping as data augmentation and the image batches are cropped into a fixed dimension of 328 × 328.

In test phase, we use the strong branch to generate final segmentation for the testing images. Since fully-connected CRF [16] brings negligible improvements when the network predictions are accurate enough, we do not apply CRF as post refinement in our experiments.

4.2 Ablation Study

To provide more insight into the proposed architecture, we conduct a series of experiments on PASCAL VOC using different experimental settings concerning different network architecture and training data. We use VGG16 as backbone unless stated otherwise.

Using Only 1.4k Strong Data. To obtain decent results with only 1.4k strong annotations, it is important to perform the same number of iterations (instead of epochs) to let the network converge. When trained enough amount of iterations, we could achieve a mIoU of 68.9%, which is much better than the 62.5% reported in FickleNet [19].

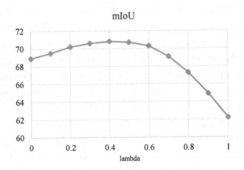

Fig. 6. The segmentation mIoU (%) with respect to different λ's.

Two Separate Networks. As aforementioned, the single-branch network trained under the mixture of strong and weak annotations achieves no better performance than using only the strong ones. Therefore, it is natural to train two different networks on two sets of data since there exists an obvious annotation inconsistency. Specifically, we train two networks, the first supervised by the strong annotations and the second by extra weak annotations. Then their outputs are aggregated through the following equation:

$$F(x) = \lambda * F_w(x) + (1 - \lambda) * F_s(x) \qquad (4)$$

where F_w and F_s denote the weak and strong network respectively. Figure 6 shows the segmentation accuracy under different λ values. Simply training on the strong annotations yields an accuracy of 68.9%, 6.1% higher than the weak one. The result could be improved up to 70.8% with λ equal to 0.4, a 1.9% boost over the strong network. However, separate networks double the computation overhead during both training and inference.

Table 1. Ablation experiments concerning network architectures and training data. Rows marked with "*" are results from the proposed dual-branch network and others are from the single-branch network.

Backbone	Strong branch	Weak branch	mIoU
VGG16	10k weak	–	57.0
VGG16	10k weak (retrain)	–	60.1
VGG16	1.4k strong + 9k weak	–	62.8
VGG16	1.4k strong	–	68.9
VGG16	10k strong	–	71.4
VGG16* (w/o oversampling)	1.4k strong	1.4k strong + 9k weak	66.4
VGG16* (w oversampling)	1.4k strong	1.4k strong + 9k weak	72.2
VGG16* (w oversampling)	1.4k strong	10k strong	73.9

Single Branch vs. Dual Branch. Our VGG16-based implementation of the DSRG method achieves mIoU of 57.0% and 60.1% after retraining, presented in Table 1. With a combination of 1.4k strong annotations and 9k weak annotations estimated from DSRG, the single-branch network only improves the segmentation accuracy by 2.7%. However, it already achieves much higher accuracy of 68.9% under only 1.4k strong annotations, which means the extra 9k weak annotations bring no benefits but actually downgrade the performance dramatically, nearly 6% drop. This phenomenon verifies our hypothesis that equal treatment of strong and weak annotations are problematic as large amount inaccurate weak annotations mislead the network training.

We then train the proposed architecture with the strong branch supervised by the 1.4k strong annotations and the weak one by extra 9k weak annotations. This time the accuracy successfully goes up to 72.2%, a 3.3% improvement over the 1.4k single-branch model. Remarkably, this result is even better than training a single-branch model with 10k strong annotations, which implies that there is an inconsistency between the official 1.4k annotations and the additional 9k annotations provided by [10]. Based on this observation, we conduct another experiment on our dual-branch network with 1.4k strong annotations for the strong branch and 10k strong annotations for the weak branch. As expected, the accuracy is further increased by 1.7%. To see the impact of sample imbalance, we also conduct a experiment with our network without oversampling. And it achieves accuracy of 66.4%, up from mixed training result 62.8% but still worse than that of using only strong annotations. So we conclude that it's more effective to combine dual-branch model with oversampling training strategy.

4.3   Comparison with the State-of-arts

Table 2 compares the proposed method with current state-of-art weakly-and-semi supervised methods: SEC [15], DSRG [13], FickleNet [19], WSSL [26], BoxSup [7], etc. For fair comparison, the result reported in the original paper is listed along with the backbone adopted.

Table 2. Segmentation results of different methods on PASCAL VOC 2012 *val* and *test* set. * - Result copied from the FickleNet paper

Methods	Backbone	Val	Test
Supervision: 10k scribbles			
Scribblesup [21]	VGG16	63.1	–
Normalized cut [32]	ResNet101	74.5	–
Supervision: 10k boxes			
WSSL [26]	VGG16	60.6	62.2
BoxSup [7]	VGG16	62.0	64.2
Supervision: 10k class			
SEC [15]	VGG16	50.7	51.7
AF-SS [36]	VGG16	52.6	52.7
Multi-Cues [29]	VGG16	52.8	53.7
DCSP [4]	VGG16	58.6	59.2
DSRG [13]	VGG16	59.0	60.4
AffinityNet [1]	VGG16	58.4	60.5
MDC [35]	VGG16	60.4	60.8
FickleNet [19]	VGG16	61.2	61.9
Supervision: 1.4k pixel + 9k class			
DSRG [13]*	VGG16	64.3	–
FickleNet [19]	VGG16	65.8	–
WSSL [26]	VGG16	64.6	66.2
MDC [35]	VGG16	65.7	67.6
Ours	VGG16	72.2	72.3
Ours	ResNet101	**76.6**	**77.1**

The weakly-supervised methods are provided in the upper part of Table 2 as reference since many of them used relatively weak supervision, with FickleNet (61.9%) achieving the best performance among other baselines using only class labels. However, the elimination of the demand for pixel-level annotations results in significant performance drop, around 11% compared to their fully-supervised counterparts. There are some recent works exploring other weak supervisions, such as Normalized cut loss [32] and Box-driven method [31]. They improved the segmentation performance significantly with slightly increasing labeling efforts. Our method actually serves as an alternative direction by using a combination of strong and weak annotations to achieve excellent results.

The lower part of Table 2 presents results of the semi-supervised methods. DSRG and FickleNet used the same region growing mechanism to expand the original object seeds. As shown in the table, all previous methods achieved roughly the same and poor performance when learned under 1.4k pixel annotations and 9k class annotations, with the best accuracy 67.6% by MDC approach.

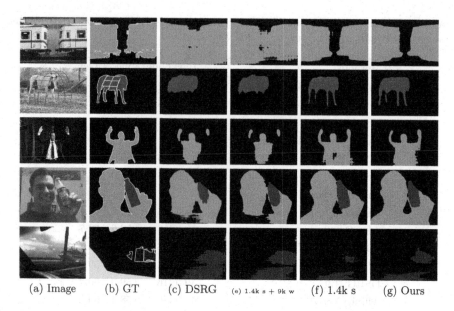

(a) Image   (b) GT   (c) DSRG   (e) 1.4k s + 9k w   (f) 1.4k s   (g) Ours

Fig. 7. Demonstration of sample images. (a) Original images; (b) Ground truth; (c) DSRG; (e) Mixing 1.4k s + 9k w for training; (f) 1.4k strong annotations; (g) Ours under 1.4k s + 9k w.

Our method significantly outperforms all the weakly-and-semi supervised method by a large margin, with state-of-art 77.1% mIoU on the *test* set when ResNet101 backbone is adopted.

4.4   Visualization Result

Figure 7 shows segmentation results of sample images from PASCAL VOC *val* set. As can be seen in the third column, weakly-supervised method (DSRG) generates segmentation maps of relatively poor quality and no improvement is visually significant if combined with 1.4k strong annotations. Our approach manages to remove some of the false positives in the foreground categories, as in the second and third examples. The last line demonstrates a failure case when neither approach is effective to generate correct prediction.

4.5   Results on COCO

To verify the generality of the proposed architecture, we conduct further experiments on the Microsoft COCO dataset, which contains a lot more images (118k) and semantic categories (81 classes), thus posing a challenge even for fully-supervised segmentation approaches. We randomly select 20k images as our strong set and the remaining 98k images as the weak set, whose annotations are estimated from the DSRG method. This splitting ratio is roughly the same compared to PASCAL VOC experiments. We report per-class IoU over all 81

Table 3. Per-class IoU on COCO val set. (a) Single-branch network using 20k strong annotations; (b) Single-branch network using 98k extra weak annotations; (c) Dual-branch network using 98k extra weak annotations.

Cat.	Class	(a)	(b)	(c)	Cat.	Class	(a)	(b)	(c)
BG	background	86.2	78.4	86.7	Kitchenware	wine glass	42.5	36.0	45.2
P	person	74.4	60.7	75.2		cup	38.8	30.9	38.9
Vehicle	bicycle	54.2	48.4	55.3		fork	16.6	0.0	17.2
	car	47.4	38.2	49.5		knife	3.4	0.1	6.9
	motorcycle	70.4	63.7	70.6		spoon	5.9	0.0	5.4
	airplane	63.3	30.5	66.0		bowl	33.0	22.4	34.7
	bus	69.7	64.1	71.5	Food	banana	62.4	53.1	63.3
	train	67.2	46.7	69.8		apple	36.6	29.8	37.3
	truck	43.3	36.4	45.2		sandwich	44.3	35.1	46.0
	boat	42.5	26.1	41.9		orange	55.3	50.3	57.9
Outdoor	traffic light	42.9	27.6	47.1		broccoli	49.9	37.3	53.3
	fire hydrant	74.2	47.3	75.5		carrot	34.4	31.8	37.0
	stop sign	82.3	53.6	87.3		hot dog	38.8	36.0	39.8
	parking meter	48.4	42.7	53.8		pizza	74.8	68.6	76.6
	bench	32.6	25.3	34.9		donut	49.4	48.6	53.9
Animal	bird	56.6	33.9	62.0		cake	45.6	40.6	45.3
	cat	76.7	65.1	77.5	Furniture	chair	24.4	12.3	25.2
	dog	68.7	60.6	69.0		couch	41.0	20.5	42.6
	horse	64.4	50.0	66.2		potted plant	23.4	15.5	24.5
	sheep	70.5	55.5	73.3		bed	46.9	38.2	50.4
	cow	61.7	49.7	65.3		dining table	34.8	9.2	35.0
	elephant	79.9	67.6	81.2		toilet	61.5	45.3	62.7
	bear	79.7	60.4	81.7	Electronics	tv	49.9	22.5	52.5
	zebra	81.7	61.2	82.9		laptop	56.2	40.6	57.4
	giraffe	74.3	47.0	75.0		mouse	38.5	0.7	35.5
Accessory	backpack	11.4	2.5	12.6		remote	37.8	25.7	30.9
	umbrella	57.9	44.3	59.1		keyboard	44.3	35.9	47.1
	handbag	6.8	0.0	8.2		cell phone	44.1	36.8	42.3
	tie	34.5	20.6	35.4	Appliance	microwave	47.2	32.8	44.6
	suitcase	53.1	48.4	57.6		oven	42.9	29.7	47.9
Sport	frisbee	48.2	39.4	50.8		toaster	0.0	0.0	0.0
	skis	14.6	5.3	11.8		sink	40.0	30.5	42.4
	snowboard	37.8	15.6	39.1		refrigerator	55.5	34.8	57.3
	sports ball	27.0	13.3	29.7	Indoor	book	29.9	16.4	29.0
	kite	32.1	23.7	36.2		clock	57.5	16.4	59.5
	baseball bat	10.4	0.0	11.1		vase	45.8	30.1	43.9
	baseball glove	28.4	0.0	37.6		scissors	57.1	34.1	56.4
	skateboard	32.0	20.4	31.6		teddy bear	64.4	57.9	66.1
	surfboard	43.7	32.2	44.5		hair drier	0.0	0.0	0.0
	tennis racket	55.7	47.3	58.1		toothbrush	13.2	8.5	17.6
	bottle	39.7	33.0	39.4		**mean IoU**	**46.1**	**33.4**	**47.6**

semantic categories on the 5k validation images. As shown in Table 3, with 20k strong annotations, the single branch network achieves an accuracy of 46.1%. When we bring in extra 98k weak annotations estimated by DSRG, the performance downgrades by 12.7%, down to only 33.4%, which again verifies our hypothesis. Using our dual-branch network, the performance successfully goes up to 47.6%, which means our approach manages to make use of the weak annotations.

5   Conclusion

We have addressed the problem of semi-supervised semantic segmentation where a combination of finely-labeled masks and coarsely-estimated data are available for training. Weak annotations are cheap to obtain yet not enough to train a segmentation model of high quality. We propose a strong-weak dual-branch network that has fully utilized the limited strong annotations without being overwhelmed by the bulk of weak ones. It manages to eliminate the learning obstacles of sample imbalance and supervision inconsistency. Our method significantly outperforms the weakly-supervised and almost reaches the accuracy of fully-supervised models. We think semi-supervised approaches could serve as an alternative to weakly-supervised methods by retaining the segmentation accuracy while still keeping labeling budget in control.

Acknowledgement. This work is partially supported by National Natural Science Foundation of China (Grants no. 61772568), and the Natural Science Foundation of Guangdong Province, China (Grant no. 2019A1515012029).

References

1. Ahn, J., Kwak, S.: Learning pixel-level semantic affinity with image-level supervision for weakly supervised semantic segmentation. In: CVPR, June 2018
2. Bearman, A., Russakovsky, O., Ferrari, V., Fei-Fei, L.: What's the point: semantic segmentation with point supervision. In: Leibe, B., Matas, J., Sebe, N., Welling, M. (eds.) ECCV 2016. LNCS, vol. 9911, pp. 549–565. Springer, Cham (2016). https://doi.org/10.1007/978-3-319-46478-7_34
3. Bennett, K., Demiriz, A.: Semi-supervised support vector machines. In: NIPs, pp. 368–374. MIT Press, Cambridge (1999). http://dl.acm.org/citation.cfm?id=340534.340671
4. Chaudhry, A., Dokania, P.K., Torr, P., Toor, P.: Discovering class-specific pixels for weakly-supervised semantic segmentation. In: BMVC, vol. abs/1707.05821 (2017)
5. Chen, L., Papandreou, G., Kokkinos, I., Murphy, K., Yuille, A.L.: DeepLab: semantic image segmentation with deep convolutional nets, atrous convolution, and fully connected CRFs. TPAMI **40**, 834–848 (2016)
6. Chen, L., Zhu, Y., Papandreou, G., Schroff, F., Adam, H.: Encoder-decoder with atrous separable convolution for semantic image segmentation. In: ECCV (2018)
7. Dai, J., He, K., Sun, J.: BoxSup: exploiting bounding boxes to supervise convolutional networks for semantic segmentation. In: ICCV, pp. 1635–1643 (2015)

8. Everingham, M., Van Gool, L., Williams, C.K.I., Winn, J., Zisserman, A.: The PASCAL visual object classes challenge 2012 (VOC2012) results (2012). http://www.pascal-network.org/challenges/VOC/voc2012/workshop/index.html
9. Girshick, R.: Fast R-CNN. In: ICCV, pp. 1440–1448 (2015)
10. Hariharan, B., Arbelaez, P., Bourdev, L., Maji, S., Malik, J.: Semantic contours from inverse detectors. In: ICCV (2011)
11. He, K., Zhang, X., Ren, S., Sun, J.: Deep residual learning for image recognition. In: CVPR, pp. 770–778 (2015)
12. Hou, Q., Jiang, P., Wei, Y., Cheng, M.: Self-erasing network for integral object attention. In: NIPS (2018)
13. Huang, Z., Wang, X., Wang, J., Liu, W., Wang, J.: Weakly-supervised semantic segmentation network with deep seeded region growing. In: CVPR, June 2018
14. Kingma, D., Ba, J.: Adam: a method for stochastic optimization. CoRR abs/1412.6980 (2014)
15. Kolesnikov, A., Lampert, C.: Seed, expand and constrain: three principles for weakly-supervised image segmentation. In: ECCV, vol. abs/1603.06098 (2016)
16. Krähenbühl, P., Koltun, V.: Efficient inference in fully connected CRFs with Gaussian edge potentials. In: NIPS (2011)
17. Krizhevsky, A., Sutskever, I., Hinton, G.E.: ImageNet classification with deep convolutional neural networks. In: NIPs, pp. 1097–1105. Curran Associates Inc., USA (2012). http://dl.acm.org/citation.cfm?id=2999134.2999257
18. Laine, S., Aila, T.: Temporal ensembling for semi-supervised learning. In: ICLR, vol. abs/1610.02242 (2016)
19. Lee, J., Kim, E., Lee, S., Lee, J., Yoon, S.: FickleNet: weakly and semi-supervised semantic image segmentation using stochastic inference. In: CVPR, June 2019
20. Li, K., Wu, Z., Peng, K., Ernst, J., Fu, Y.: Tell me where to look: guided attention inference network. In: CVPR, pp. 9215–9223 (2018)
21. Lin, D., Dai, J., Jia, J., He, K., Sun, J.: ScribbleSup: scribble-supervised convolutional networks for semantic segmentation. In: CVPR, pp. 3159–3167 (2016)
22. Lin, T., Goyal, P., Girshick, R., He, K., Dollár, P.: Focal loss for dense object detection. In: ICCV, pp. 2999–3007 (2017)
23. Lin, T.-Y., et al.: Microsoft COCO: common objects in context. In: Fleet, D., Pajdla, T., Schiele, B., Tuytelaars, T. (eds.) ECCV 2014. LNCS, vol. 8693, pp. 740–755. Springer, Cham (2014). https://doi.org/10.1007/978-3-319-10602-1_48
24. Miyato, T., Maeda, S., Koyama, M., Ishii, S.: Virtual adversarial training: a regularization method for supervised and semi-supervised learning. TPAMI 41(8), 1979–1993 (2019). https://doi.org/10.1109/TPAMI.2018.2858821
25. Oh, S., Benenson, R., Khoreva, A., Akata, Z., Fritz, M., Schiele, B.: Exploiting saliency for object segmentation from image level labels. In: CVPR (2017, to appear)
26. Papandreou, G., Chen, L., Murphy, K.P., Yuille, A.L.: Weakly-and semi-supervised learning of a deep convolutional network for semantic image segmentation. In: ICCV, pp. 1742–1750, December 2015. https://doi.org/10.1109/ICCV.2015.203
27. Papandreou, G., Chen, L., Murphy, K.P., Yuille, A.L.: Weakly-and semi-supervised learning of a deep convolutional network for semantic image segmentation. In: ICCV, ICCV 2015, pp. 1742–1750. IEEE Computer Society, Washington, DC (2015). https://doi.org/10.1109/ICCV.2015.203. http://dx.doi.org/10.1109/ICCV.2015.203
28. Ren, S., He, K., Girshick, R., Sun, J.: Faster R-CNN: towards real-time object detection with region proposal networks. TPAMI 39, 1137–1149 (2015)

29. Roy, A., Todorovic, S.: Combining bottom-up, top-down, and smoothness cues for weakly supervised image segmentation. In: CVPR, July 2017
30. Simonyan, K., Zisserman, A.: Very deep convolutional networks for large-scale image recognition. In: ICLR (2015)
31. Song, C., Huang, Y., Ouyang, W., Wang, L.: Box-driven class-wise region masking and filling rate guided loss for weakly supervised semantic segmentation. In: CVPR, June 2019
32. Tang, M., Djelouah, A., Perazzi, F., Boykov, Y., Schroers, C.: Normalized cut loss for weakly-supervised CNN segmentation. In: CVPR, June 2018
33. Tarvainen, A., Valpola, H.: Mean teachers are better role models: weight-averaged consistency targets improve semi-supervised deep learning results. In: ICLR (2017)
34. Wang, X., You, S., Li, X., Ma, H.: Weakly-supervised semantic segmentation by iteratively mining common object features. In: CVPR, June 2018
35. Wei, Y., Xiao, H., Shi, H., Jie, Z., Feng, J., Huang, T.: Revisiting dilated convolution: a simple approach for weakly- and semi-supervised semantic segmentation. In: CVPR, June 2018
36. Wei, Y., Feng, J., Liang, X., Cheng, M.M., Zhao, Y., Yan, S.: Object region mining with adversarial erasing: a simple classification to semantic segmentation approach. In: CVPR, July 2017
37. Zhang, J., Bargal, S.A., Lin, Z., Brandt, J., Shen, X., Sclaroff, S.: Top-down neural attention by excitation backprop. IJCV **126**, 1084–1102 (2016)
38. Zhao, H., Shi, J., Qi, X., Wang, X., Jia, J.: Pyramid scene parsing network. In: CVPR, pp. 6230–6239 (2016)
39. Zhou, B., Khosla, A., A., L., Oliva, A., Torralba, A.: Learning deep features for discriminative localization. In: CVPR (2016)
40. Zhu, X.: Semi-supervised learning literature survey. Technical report 1530, Computer Sciences, University of Wisconsin-Madison (2005)

Author Index

Printed in the United States
by Booksmasters.

Printed in the United States
By Bookmasters